PRINCIPLES OF
NEUROBIOLOGY

PRINCIPLES OF NEUROBIOLOGY

LIQUN LUO

 **Garland Science**
Taylor & Francis Group
NEW YORK AND LONDON

Garland Science
Vice President: Denise Schanck
Development Editor: Monica Toledo
Senior Production Editor: Georgina Lucas
Text Editor: Kathleen Vickers
Illustrator: Nigel Orme
Copyeditor: Sally Huish
Proofreader: Sally Livitt
Indexer: Bill Johncocks
Permissions Coordinator: Sheri Gilbert
Text Design: Xavier Studio
Cover Design: Matthew McClements, Blink Studio, Ltd
Director of Digital Publishing: Michael Morales
Editorial Assistant: Jasmine Ribeaux

© 2016 by Garland Science, Taylor & Francis Group, LLC

About the Author
Liqun Luo earned his bachelor degree from the University of Science & Technology of China and PhD from Brandeis University. Since 1997, Dr. Luo has taught neurobiology to undergraduate and graduate students at Stanford University, where he also directs a lab studying the assembly and function of neural circuits. Dr. Luo is a member of the National Academy of Sciences and the American Academy of Arts and Sciences, and an Investigator of the Howard Hughes Medical Institute.

ISBNs: 978-0-8153-4492-6 (hardcover) 978-0-8153-4494-0 (softcover)

Library of Congress Cataloging-in-Publication Data
Luo, Liqun, 1966-
  Principles of neurobiology / Liqun Luo.
    pages cm
  ISBN 978-0-8153-4492-6 -- ISBN 978-0-8153-4494-0
  1. Neurobiology. I. Title.
  QP355.2.L86 2015
  612.8--dc23
                        2015016826

Published by Garland Science, Taylor & Francis Group, LLC, an informa business, 711 Third Avenue, New York, NY 10017, USA, and 3 Park Square, Milton Park, Abingdon, OX14 4RN, UK.

Printed in the United States of America
15 14 13 12 11 10 9 8 7 6 5 4 3 2

Visit our web site at http://www.garlandscience.com

To my parents, Chongxin Zhong and Kailian Luo,
who have granted me both nature and nurture.

PREFACE

Neurobiology has never seen a more exciting time. As the most complex organ of our body, the brain endows us the ability to sense, think, remember, and act. Thanks to the conceptual and technical advances in recent years, the pace of discovery in neurobiology is continuously accelerating. New and exciting findings are reported every month. Traditional boundaries between molecular, cellular, systems, and behavioral neurobiology have been broken. The integration of developmental and functional studies of the nervous system has never been stronger. Physical scientists and engineers increasingly contribute to fundamental discoveries in neurobiology. Yet we are still far from a satisfying understanding of how the brain works, and from converting this understanding into effective treatment of brain disorders. I hope to convey the excitement of neurobiology to students, to lay the foundation for their appreciation of this discipline, and to inspire them to make exciting new discoveries in the coming decades.

This book is a reflection of my teaching at Stanford during the past 18 years. My students—and the intended audience of this book—include upper division undergraduates and beginning graduate students who wish to acquire an in-depth knowledge and command of neurobiology. While most students reading this book may have a biology background, some may come from physical sciences and engineering. I have discovered that regardless of a student's background, it is much more effective—and much more interesting—to teach students how knowledge has been obtained than the current state of knowledge. That is why I have taken this discovery-based teaching approach from lecture hall to textbook.

Each chapter follows a main storyline or several sequential storylines. These storylines are divided by large section headings usually titled with questions that are then answered by a series of summarizing subheadings with explanatory text and figures. Key terms are highlighted in bold and are further explained in an expanded glossary. The text is organized around a series of key original experiments, from classic to modern, to illustrate how we have arrived at our current state of understanding. The majority of the figures are based on those from original papers, thereby introducing students to the primary literature. Instead of just covering the vast number of facts that make up neurobiology in this day and age, this book concentrates on the in-depth study of a subset of carefully chosen topics that illustrate the discovery process and resulting principles. The selected topics span the entire spectrum of neurobiology, from molecular and cellular to systems and behavioral. Given the relatively small size of the book, students will be able to study much or all of the book in a semester, allowing them to gain a broad grasp of modern neurobiology.

This book intentionally breaks from the traditional division of neuroscience into molecular, cellular, systems, and developmental sections. Instead, most chapters integrate these approaches. For example, the chapter on 'Vision' starts with a human psychophysics experiment demonstrating that our rod photoreceptors can detect a single photon, as well as a physiology experiment showing the electrical response of the rod to a single photon. Subsequent topics include molecular events in photoreceptors, cellular and circuit properties of the retina and the visual cortex, and systems approaches to understanding visual perception. Likewise, 'Memory, Learning, and Synaptic Plasticity' integrates molecular, cellular, circuit, systems, behavioral, and theoretical approaches with the common goal of understanding what memory is and how it relates to synaptic plasticity. The two chapters on development intertwine with three chapters on sensory and motor systems to help students appreciate the rich connections between the development and function of the nervous system. All chapters are further linked by abundant cross-referencing through the text. These links reinforce the notion that topics in neurobiology form highly interconnected networks rather than a linear sequence. Finally and importantly, Chapter 13 ('Ways of Exploring') is dedicated to key methods in neurobiology research and is extensively referenced in all preceding chapters. Students are encouraged to study the relevant methods in Chapter 13 when they first encounter them in Chapters 1–12.

This book would not have been possible without the help of Lubert Stryer, my mentor, colleague, and dear friend. From inception to completion, Lubert has provided invaluable support and advice. He has read every single chapter (often more than once) and has always provided a balanced dose of encouragement and criticism, from strategic planning to word choice. Lubert's classic *Biochemistry* textbook was a highlight in my own undergraduate education and has continued to inspire me throughout this project.

I thank Howard Schulman, Kang Shen, and Tom Clandinin, who, along with Lubert, have been my co-instructors for neurobiology courses at Stanford and from whom I have learned a tremendous amount about science and teaching. Students in my classes have offered valuable feedback that has improved my teaching and has been incorporated into the book. I am highly appreciative of the past and current members of my lab, who have taught me more than I have taught them and whose discoveries have been constant sources of inspiration and joy. I gratefully acknowledge the National Institutes of Health and the Howard Hughes Medical Institute for generously supporting the research of my lab.

Although this book has a single author, it is truly the product of teamwork with Garland Science. Denise Schanck has provided wise leadership throughout the journey. Janet Foltin in the initial phase and Monica Toledo through most of the project have provided much support and guidance, from obtaining highly informative reviews of early drafts to organizing teaching and learning resources. I am indebted to Kathleen Vickers for expert editing; her attention to detail and demand for clarity have greatly improved my original text. I owe the illustrations to Nigel Orme, whose combined artistic talent and scientific understanding brought to life concepts from the text. Georgina Lucas's expert page layout has seamlessly integrated the text and figures. I also thank Michael Morales for producing the enriching videos, and Adam Sendroff and his staff for reaching out to the readers. Working with Garland has been a wonderful experience, and I thank Bruce Alberts for introducing Garland to me.

Finally, I am very grateful for the support and love from my wife, Charlene Liao, and our two daughters, Connie and Jessica. Writing this textbook has consumed a large portion of my time in the past few years; indeed, the textbook has been a significant part of our family life and has been a frequent topic of dinner table conversation. Jessica has been my frequent sounding board for new ideas and storylines, and I am glad that she has not minded an extra dose of neurobiology on top of her demanding high-school courses and extracurricular activities.

I welcome feedback and critiques from students and readers!

Liqun Luo
April 2015

NOTE ON GENE AND PROTEIN NOMENCLATURE

This book mostly follows the unified convention of *Molecular Biology of the Cell* 6th Edition by Alberts et al. (Garland Science, 2015) for naming genes. Regardless of species, gene names and their abbreviations are all in italics, with the first letter in upper case and the rest of the letters in lower case. All protein names are in roman, and their cases follow the consensus in the literature. Proteins identified by biochemical means are usually all in lower case; proteins identified by genetic means or by homology with other genes usually have the first letter in upper case; protein acronyms usually are all in upper case. The space that separates a letter and a number in full names includes a hyphen, and in abbreviated names is omitted entirely.

The table below summarizes the official conventions for individual species and the unified conventions that we shall use in this book.

Organism	Species-Specific Convention		Unified Convention Used in this Book	
	Gene	Protein	Gene	Protein
Mouse	*Syt1*	synaptotagmin I	*Syt1*	Synaptotagmin-1
	Mecp2	MeCP2	*Mecp2*	MeCP2
Human	*MECP2*	MeCP2	*Mecp2*	MeCP2
Caenorhabditis	*unc-6*	UNC-6	*Unc6*	Unc6
Drosophila	*sevenless* (named after recessive phenotype)	Sevenless	*Sevenless*	Sevenless
	Notch (named after dominant mutant phenotype)	Notch	*Notch*	Notch
Other organisms (e.g. jellyfish)		Green fluorescent protein (GFP)	*Gfp*	GFP

RESOURCES FOR INSTRUCTORS AND STUDENTS

The teaching and learning resources for instructors and students are available online. The homework platform is available to everyone, though instructors will need to set up student access in order to use the dashboard to track student progress on assignments. The instructor's resources on the Garland Science website are password-protected and available only to adopting instructors. The student resources on the Garland Science website are available to everyone. We hope these resources will enhance student learning and make it easier for instructors to prepare dynamic lectures and activities for the classroom.

Online Homework Platform with Instructor Dashboard

Instructors can obtain access to the online homework platform from their sales representative or by emailing science@garland.com. Students who wish to use the platform must purchase access and, if required for class, obtain a course link from their instructor.

The online homework platform is designed to improve and track student performance. It allows instructors to select homework assignments on specific topics and review the performance of the entire class, as well as individual students, via the instructor dashboard. The user-friendly system provides a convenient way to gauge student progress, and tailor classroom discussion, activities, and lectures to areas that require specific remediation. The features and assignments include:

- *Instructor Dashboard* displays data on student performance: such as responses to individual questions and length of time required to complete assignments.
- *Tutorials* explain essential or difficult concepts and are integrated with a variety of questions that assess student engagement and mastery of the material.
- *Media Assessments* present movies or explain complex figures from the book and contain a set of questions that assess student understanding of the concepts.
- *Quizzes* test basic reading comprehension and the retention of important terminology and facts. The quizzes are composed of multiple-choice and true-false questions.

The tutorials were created by Andrea Nicholas (University of California, Irvine) and the quizzes were written by Casey Guenthner (Neurosciences Program PhD student in the Luo Lab at Stanford University).

Instructor Resources

Instructor Resources are available on the Garland Science Instructor's Resource Site, located at www.garlandscience.com/instructors. The website provides access not only to the teaching resources for this book but also to all other Garland Science textbooks. Adopting instructors can obtain access to the site from their sales representative or by emailing science@garland.com.

Art of Principles of Neurobiology

The images from the book are available in two convenient formats: PowerPoint® and JPEG. They have been optimized for display on a computer. Figures are searchable by figure number, by figure name, or by keywords used in the figure legend from the book.

Figure-Integrated Lecture Outlines

The section headings, concept headings, and figures from the text have been integrated into PowerPoint presentations. These will be useful for instructors who would like a head start creating lectures for their course. Like all of our PowerPoint presentations, the lecture outlines can be customized. For example, the content of these presentations can be combined with videos and questions from the book or Question Bank, in order to create unique lectures that facilitate interactive learning.

Animations and Videos

The animations and videos that are available to students are also available on the Instructor's Website in two formats. The WMV-formatted movies are created for instructors who wish to use the movies in PowerPoint presentations on Windows® computers; the QuickTime-formatted movies are for use in PowerPoint for Apple computers or Keynote® presentations. The movies can easily be downloaded using the 'download' button on the movie preview page. The movies are related to specific chapters and callouts to the movies are highlighted in color throughout the textbook.

Question Bank

Written by Elizabeth Marin (Bucknell University), and Melissa Coleman (Claremont McKenna, Pitzer, and Scripps Colleges), the Question Bank includes a variety of question formats: multiple choice, fill-in-the-blank, true-false, matching, essay, and challenging 'thought' questions. There are approximately 40–50 questions per chapter, and a large number of the multiple-choice questions will be suitable for use with personal response systems (that is, clickers). The Question Bank provides a comprehensive sampling of questions that require the student to reflect upon and integrate information, and can be used either directly or as inspiration for instructors to write their own test questions.

Diploma® Test Generator Software

The questions from the Question Bank have been loaded into the Diploma Test Generator software. The software is easy to use and can scramble questions to create multiple tests. Questions are organized by chapter and type and can be additionally categorized by the instructor according to difficulty or subject. Existing questions can be edited and new ones added. The Test Generator is compatible with several course management systems, including Blackboard®.

Student Resources

The resources for students are available on the *Principles of Neurobiology* Student Website, located at www.garlandscience.com/neurobio-students

Journal Club

The Journal Club recommends journal articles that complement topics in the textbook to improve students' critical analysis of research and to promote a better understanding of the research process. Each Journal Club document provides background information on the chosen paper as well as questions and discussion points to stimulate in-class discussion. Answers will be provided to instructors only. The Journal Club was developed by Casey Guenthner (Neurosciences Program PhD student in the Luo Lab at Stanford University).

Animations and Videos

There are over 40 narrated movies, covering a range of neurobiology topics, which review key concepts and illuminate the experimental process.

Flashcards

Each chapter contains flashcards, built into the student website, that allow students to review key terms from the text.

Glossary

The comprehensive glossary of key terms from the book is online and can be searched or browsed.

ACKNOWLEDGMENTS

The author and publisher of *Principles of Neurobiology* specially thank Casey Guenthner (Stanford University) for creating the *Journal Club* and *Quizzes*, Melissa Coleman (Claremont McKenna, Pitzer and Scripps Colleges) and Lisa Marin (Bucknell College) for creating the *Question Bank*, and Andrea Nicholas (University of California, Irvine) for creating the *Tutorials*.

The author and publisher of *Principles of Neurobiology* gratefully acknowledge the contributions of the following scientists and instructors for their advice and critique in the development of this book:

Chapter 1: Peter Bergold (SUNY-Downstate Medical Center), Katja Brose (Cell Press), Catherine Dulac (Harvard University), Joachim Hallmayer (Stanford University), Mark Horowitz (Stanford University), Josh Huang (Cold Spring Harbor Laboratory), Eric Knudsen (Stanford University), Eve Marder (Brandeis University), Mike McCloskey (Iowa State University), Kazunari Miyamichi (University of Tokyo), Tim Mosca (Stanford University), Chris Potter (Johns Hopkins University), Annemarie Shibata (Creighton University), Larry Swanson (University of Southern California), Bosiljka Tasic (Allen Institute for Brain Science), Joy Wan (Stanford University), Jian Yang (Columbia University).

Chapter 2: Ben Barres (Stanford University), Peter Bergold (SUNY-Downstate Medical Center), Katja Brose (Cell Press), Laura DeNardo Wilke (Stanford University), Shaul Hestrin (Stanford University), Josh Huang (Cold Spring Harbor Laboratory), Lily Jan (University of California, San Francisco), William Joo (Harvard University), Yulong Li (Peking University), Eve Marder (Brandeis University), Mike McCloskey (Iowa State University), Jing Ren (Stanford University), Tom Schwarz (Harvard University), Kang Shen (Stanford University), Annemarie Shibata (Creighton University), Chuck Stevens (Salk Institute), Tom Südhof (Stanford University), Rachel Wilson (Harvard University), Jian Yang (Columbia University).

Chapter 3: Peter Bergold (SUNY-Downstate Medical Center), Tobias Bonhoeffer (Max Planck Institute of Neurobiology), Katja Brose (Cell Press), Tom Clandinin (Stanford University); Laura DeNardo Wilke (Stanford University), Gord Fishell (New York University), Shaul Hestrin (Stanford University), Josh Huang (Cold Spring Harbor Laboratory), Lily Jan (University of California, San Francisco), William Joo (Harvard University), Yulong Li (Peking University), Eve Marder (Brandeis University), Mike McCloskey (Iowa State University), Jing Ren (Stanford University), Tom Schwarz (Harvard University), Idan Segev (Hebrew University), Kang Shen (Stanford University), Annemarie Shibata (Creighton University), Chuck Stevens (Salk Institute), Tom Südhof (Stanford University), Jian Yang (Columbia University).

Chapter 4: Steve Baccus (Stanford University), Nic Berns (Stanford University), Tobias Bonhoeffer (Max Planck Institute of Neurobiology), Katja Brose (Cell Press), Tom Clandinin (Stanford University), Yang Dan (University of California, Berkeley), Marla Feller (University of California, Berkeley), Andy Huberman (University of California, San Diego), Adi Mizrahi (Hebrew University), Jeremy Nathans (Johns Hopkins University), Bill Newsome (Stanford University), John Pizzey (King's College London), Michael Rosbash (Brandeis University), Botond Roska (Friedrich Miescher Institute), Eric Warrant (University of Lund).

Chapter 5: Nic Berns (Stanford University), Tobias Bonhoeffer (Max Planck Institute of Neurobiology), Tom Clandinin (Stanford University), Claude Desplan (New York University), Dave Feldheim (University of California, Santa Cruz), Josh Huang (Cold Spring Harbor Laboratory), Andy Huberman (University of California, San Diego), Haig Keshishian (Yale University), Alex Kolodkin (Johns Hopkins University), Susan McConnell (Stanford University), Michael Rosbash (Brandeis University), Ed Ruthazer (McGill University), Carla Shatz (Stanford University).

Chapter 6: Katja Brose (Cell Press), Linda Buck (Fred Hutchison Cancer Research Center), John Carlson (Yale University), Xiaoke Chen (Stanford University), Xinzhong Dong (Johns Hopkins University), Catherine Dulac (Harvard University), David Ginty (Harvard University), Casey Guenthner (Stanford University), David Julius (University of California, San Francisco), Eric Knudsen (Stanford University), Kazunari Miyamichi (University of Tokyo), Adi Mizrahi (Hebrew University), Tim Mosca (Stanford University), John Ngai (University of California, Berkeley), Ardem Patapoutian (Scripps Research Institute), John Pizzey (King's College London), Jing Ren (Stanford University), Greg Scherrer (Stanford University), Bosiljka Tasic (Allen Institute for Brain Science), Fan Wang (Duke University), Eric Warrant (University of Lund), Rachel Wilson (Harvard University), Haiqing Zhao (Johns Hopkins University).

Chapter 7: Silvia Arber (University of Basel), Tom Clandinin (Stanford University), Gord Fishell (New York University), Simon Hippenmeyer (Institute of Science & Technology, Austria), Weizhe Hong (Caltech), Josh Huang (Cold Spring Harbor Laboratory), Yuh-Nung Jan (University of California, San Francisco), William Joo (Harvard University), Haig Keshishian (Yale University), Alex Kolodkin (Johns Hopkins University), Jeff Lichtman (Harvard University), Susan McConnell (Stanford University), Ed Ruthazer (McGill University), Kang Shen (Stanford University), Weimin Zhong (Yale University).

Chapter 8: Silvia Arber (University of Basel), Melissa Coleman (Claremont McKenna, Pitzer and Scripps Colleges), Joe Fetcho (Cornell University), Casey Guenthner (Stanford University), Craig Heller (Stanford University), Takaki Komiyama (University of California,

San Diego), Richard Levine (University of Arizona), Eve Marder (Brandeis University), Emmanuel Mignot (Stanford University), Jennifer Raymond (Stanford University), Michael Rosbash (Brandeis University), Krishna Shenoy (Stanford University), Scott Sternson (Howard Hughes Medical Institute Janelia Farm Research Campus), Larry Swanson (University of Southern California), Mark Wagner (Stanford University).

Chapter 9: Bruce Baker (Howard Hughes Medical Institute Janelia Farm Research Campus), Michael Baum (Boston University), Tom Clandinin (Stanford University), Melissa Coleman (Claremont McKenna, Pitzer and Scripps Colleges), Catherine Dulac (Harvard University), Greg Jefferis (Medical Research Council Laboratory of Molecular Biology), William Joo (Harvard), Dev Manoli (University of California, San Francisco), Nirao Shah (University of California, San Francisco), Bosiljka Tasic (Allen Institute for Brain Science), Daisuke Yamamoto (Tohoku University), Larry Young (Emory University).

Chapter 10: Tobias Bonhoeffer (Max Planck Institute of Neurobiology), Tom Clandinin (Stanford University), Laura DeNardo Wilke (Stanford University), Serena Dudek (National Institute of Environmental Health Sciences), Surya Ganguli (Stanford University), Lisa Giocomo (Stanford University), Casey Guenthner (Stanford University), Hadley Wilson Horch (Bowdoin College), Patricia Janak (Johns Hopkins University), Rob Malenka (Stanford University), Karen Parfitt (Pomona College), Mu-ming Poo (University of California, Berkeley), Geert Ramakers (UMC Utrecht), Alcino Silva (University of California, Los Angeles), Malathi Srivatsan (Arkansas State University), Karl Wah Keung Tsim (Hong Kong University of Science and Technology), Charles Yanofsky (Stanford University).

Chapter 11: Sam Gandy (Mt. Sinai Medical School), Aaron Gitler (Stanford University), Casey Guenthner (Stanford University), Wei-Hsiang Huang (Stanford University), Steve Hyman (Harvard University), William Joo (Harvard University), Charlene Liao (Genentech), Rob Malenka (Stanford University), Bill Mobley (University of California, San Diego), Lisa Olson (University of Redlands), Josef Parvizi (Stanford University), David Prince (Stanford University), Martin Raff (University College London), Malathi Srivatsan (Arkansas State University), Karl Wah Keung Tsim (Hong Kong University of Science and Technology), Xinnan Wang (Stanford University), Ryan Watts (Denali Therapeutics), Marius Wernig (Stanford University), Huda Zoghbi (Baylor College of Medicine).

Chapter 12: Richard Benton (University of Lausanne), Nic Berns (Stanford University), Tobias Bonhoeffer (Max Planck Institute of Neurobiology), Sidi Chen (MIT), Tom Clandinin (Stanford University), Russ Fernald (Stanford University), Hunter Fraser (Stanford University), Josh Huang (Cold Spring Harbor Laboratory), Manyuan Long (University of Chicago), Chris Lowe (Stanford University), Jan Lui (Stanford University), Lisa Marin (Bucknell College), Jeremy Nathans (Johns Hopkins University), Dmitri Petrov (Stanford University), Matthew Scott (Carnegie Institution for Science), Brady Weissbourd (Stanford University), Boon-Seng Wong (National University of Singapore).

Chapter 13: Will Allen (Stanford University), Tobias Bonhoeffer (Max Planck Institute of Neurobiology), Tom Clandinin (Stanford University), Karl Deisseroth (Stanford University), Hongwei Dong (University of Southern California), Guoping Feng (MIT), Xiaojing Gao (Caltech), Casey Guenthner (Stanford University), Shaul Hestrin (Stanford University), Josh Huang (Cold Spring Harbor Laboratory), Mark Schnitzer (Stanford University), Mehrdad Shamloo (Stanford University), Krishna Shenoy (Stanford University), Karel Svoboda (Howard Hughes Medical Institute Janelia Farm Research Campus), Larry Swanson (University of Southern California).

SPECIAL FEATURES

CONTENTS

Chapter 7
Wiring of the Nervous System   277

Chapter 8
Motor and Regulatory Systems 325

Chapter 10
Memory, Learning, and Synaptic Plasticity   415

Chapter 11
Brain Disorders   467

Chapter 12
Evolution of the Nervous System   513

CHAPTER 1
An Invitation to Neurobiology

How is behavior controlled by the brain? How is the brain wired up during development? How do we sense the environment? What is changed in the brain when we learn something new? How much of our brain function and behavior is shaped by our genes, and how much reflects the environment in which we grew up? What about the contributions of genes and environment to the brain function and behavior of animals: monkeys, mice, frogs, or fruit flies? How have nervous systems evolved? What goes wrong in brain disorders?

We are about to embark on a journey to explore these questions, which have fascinated mankind for thousands of years. Our ability to investigate and answer these questions experimentally is relatively recent: for example, the concept of the gene is only a century old. What we currently know about the answers to these questions comes mostly from findings made in the past 50 years. In the next 50 years, we will likely learn more about the brain and its control of behavior than in all of prior human history. We are at an exciting time as students of neurobiology, the study of how the nervous system enables our sensation, action, thought, and memory. It is my hope that many readers of this book will be at the forefront of groundbreaking future discoveries.

NATURE AND NURTURE IN BRAIN FUNCTION AND BEHAVIOR

As an entrée to our journey, let's discuss one of the questions raised above regarding the contributions of genes and environment to our brain function and behavior. We know from experience that both genetic inheritance (**nature**) and environmental factors (**nurture**) make important contributions, but how much does each contribute? How do we begin to tackle such a complex question? In scientific research, asking the right questions is often a critical step toward obtaining the right answers. As the evolutionary geneticist Theodosius Dobzhansky put it, "The question about the roles of the genotype and the environment in human development must be posed thus: To what extent are the *differences* observed among people conditioned by the differences of their genotypes and by the differences between the environments in which people were born, grew and were brought up?"

1.1 Human twin studies can reveal the contributions of nature and nurture

Francis Galton first coined the term nature versus nurture in the nineteenth century. He also introduced a powerful method to study the problem: the statistical analysis of human twins. Identical twins (**Figure 1–1**), also called **monozygotic twins**, share 100% of their genes because they are products of the same fertilized egg, or zygote. One can compare specific traits among thousands of pairs of identical twins to see how correlated they are within each pair. For example, if we

Figure 1–1 Identical (monozygotic) twins. Identical twins develop from a single fertilized egg and therefore share 100% of their genes. Most identical twins also share a similar childhood environment. (Courtesy of Christopher J. Potter.)

compare the intelligence quotients (IQs)—an estimate of general intelligence—of any two random people in the population, the correlation is 0. (Correlation is a statistic of resemblance that ranges from 0, indicating no resemblance, to 1, indicating perfect resemblance.) The correlation becomes 0.86 for identical twins (**Figure 1–2**), a striking similarity. However, identical twins also usually grow up in the same environment, so this correlation alone does not help us distinguish between the contributions of genes and the environment.

Fortunately, the human population provides a second group that allows researchers to tease apart the influence of genetic and environmental factors. Non-identical (fraternal) twins occur more often than identical twins in most human populations. These are called **dizygotic twins** because they originate from two independent eggs fertilized by two independent sperm. As full siblings, dizygotic twins are 50% identical in their genes according to Mendel's laws of inheritance. However, like monozygotic twins, dizygotic twins usually share a nearly identical prenatal and postnatal environment. Thus, the differences between traits exhibited by monozygotic and dizygotic twins should result from the differences in 50% of their genes. In our specific example, the correlation of IQ scores between dizygotic twins is 0.60 (Figure 1–2).

Behavioral geneticists use the term **heritability** to describe the contribution of genetic differences to trait differences. Heritability is calculated as the difference between the correlation of monozygotic and dizygotic twins multiplied by two (because the genetic difference is 50% between monozygotic and dizygotic twins). Thus, we have the heritability of IQ = (0.86 − 0.60) × 2 = 0.52. Roughly speaking, then, about half of the IQ score variation among the human population is contributed by genetic differences, or nature. Traditionally, the non-nature component has been presumed to come from environmental factors, or nurture. However, 'environmental factors' as calculated in twin studies include all factors that are not inherited from parents' DNA. These include postnatal environment, which is what we typically think of as nurture, but also prenatal environment, stochasticity

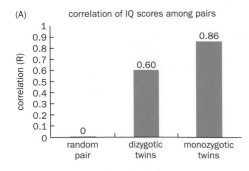

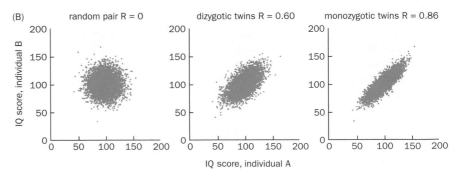

Figure 1–2 Twin studies for determining genetic and environmental contributions to intelligence quotient (IQ). (A) Correlation, or R value, of IQ scores for 4672 pairs of monozygotic twins and 5546 pairs of dizygotic twins. The correlation between the IQ scores of any two randomly selected individuals is zero. The difference in correlation between 100% genetically identical monozygotic twins and 50% genetically identical dizygotic twins can be used to calculate the heritability of traits. The large sample size makes these estimates highly accurate. **(B)** Simulation of IQ score correlation plots for 5000 pairs of unrelated individuals (R = 0), 5000 pairs of dizygotic twins (R = 0.60), and 5000 pairs of monozygotic twins (R = 0.86). The x and y axes of a given dot represent the IQ scores of one individual pair. The simulations are based on a normal distribution of IQ scores (mean = 100, standard deviation = 15). (A, data from Bouchard TJ & McGue M [1981] *Science* 212:1055–1059.)

in developmental processes, somatic mutations (alterations of DNA after fertilization), and gene expression changes due to **epigenetic modifications** (that is, changes made to DNA and chromatin that do not modify the DNA sequence but can alter gene expression; epigenetic modifications typically include DNA methylation and various forms of post-translational modification of histones, the protein component of the chromatin). As we will learn later in the book, all these factors contribute to the wiring of the nervous system that ultimately determines brain function and behavior.

Twin studies have been used to estimate the heritability of many human traits ranging from height (~90%) to the chance of developing schizophrenia (60–80%). An important caveat to these estimates is that most human traits result from complex interactions between genes and the environment, and heritability itself can change with the environment. Still, twin studies offer valuable insights into the relative contributions of genes and nongenetic factors in many aspects of brain function and dysfunction in a given environment. The completion of the Human Genome Project and the development of tools that permit detailed examination of the genome sequence data, combined with the long history of medical and psychological studies of human subjects, has resulted in our own species becoming the subject of a growing body of neurobiological research. However, mechanistic understanding of how genes and the environment influence brain development, function, and behavior requires experimental manipulations that often can only be carried out in animal models. The use of vertebrate and invertebrate model species (see Sections 13.1–13.4) in research has yielded most of what we have learned about brain and behavior to date. Many principles of neurobiology obtained from experiments on specific animal models have turned out to be universally applicable to all organisms, including human beings.

Figure 1–3 Penguin feeding.
The instinctive behaviors of an adult penguin and its offspring photographed in Antarctica, 2009. (Top) The young penguin asks for food by bumping its beak against its parent's beak. (Bottom) The parent releases the food to the young penguin's mouth. (Courtesy of Lubert Stryer.)

1.2 Examples of nature: Animals exhibit instinctive behaviors

Animals exhibit remarkable instinctive behaviors that help them find food, avoid danger, seek mates, and nurture their progeny. For example, directed by its food-seeking instinct, a baby penguin bumps its beak against that of its parent to remind the parent to feed it; in return, the parent instinctively releases the food it has foraged from the sea to feed the baby (**Figure 1–3**).

Instinctive behavior can be elicited by very specific sensory stimuli. For instance, experiments have been conducted to test the response of young chicks to an object that resembles a bird in flight, with the wings placed close to one end of the head–tail axis. When moved in one direction, the object looks like a short-necked, long-tailed hawk; when the direction of movement is reversed, the object appears gooselike, with a long neck and short tail. Upon seeing the object overhead, a young chick produces different responses depending on the direction in which the object moves, running away when the object resembles a hawk, but making no effort to escape when the object resembles a goose (**Figure 1–4**). This escape behavior is **innate**: it is with the chick from birth and is likely genetically

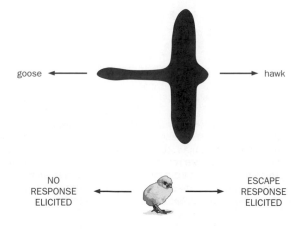

Figure 1–4 Innate escape response of chick to a hawk. A young chick exhibits instinctive escape behavior in response to an object moving overhead that resembles a short-necked, hawk-like bird; this instinctive behavior is triggered by moving the pictured object from left to right. Moving the object from right to left so that it resembles a long-necked goose does not elicit the chick's escape behavior. (Based on Tinbergen N [1951] The Study of Instinct. Oxford University Press.)

Figure 1–5 Barn owls use their auditory system to locate prey in complete darkness. The image was constructed by superimposing a series of photographs taken with an infrared camera. (Courtesy of Masakazu Konishi.)

programmed. The behavior is also **stereotypic**: different chicks exhibit the same escape behavior, with similar stimulus specificity. Once the behavior is triggered, it runs to completion without further sensory feedback. In **neuroethology**, a branch of science that emphasizes the study of animal behavior in the natural environment, such instinctive behavior is referred to as following a **fixed action pattern**. The essential features of the stimulus that activate the fixed action pattern are referred to as **releasers**.

How can genes and developmental programs specify such specific instinctive behaviors? In Chapter 9, we will use sexual behavior as an example to explore this question. For instance, we will learn about how a single gene in the fruit fly named *fruitless* can have a profound control of many aspects of the mating behavior.

1.3   An example of nurture: Barn owls adjust their auditory map to match an altered visual map

Animals also exhibit a remarkable capacity for learning in order to adapt to a changing world. We use the ability of barn owls adjusting their auditory map to changes in their vision to illustrate this capacity.

Barn owls have superb visual and auditory systems that help them catch prey at night when nocturnal rodents are active. In fact, owls can catch prey even in complete darkness (**Figure 1–5**), relying entirely on their auditory system. They can locate accurately the source of sounds made by the prey, based on the small difference in the time it takes for a sound to reach their left and right ears. Using these time differences, the owl's brain creates a map of space, such that the activation of individual nerve cells at specific positions in this brain map informs the owl of the physical position of the prey.

Experiments in which prisms were attached over a barn owl's eyes (**Figure 1–6**A) revealed how the owl responds when its auditory and visual maps provide conflicting information. Normally, the auditory map matches the owl's visual map, so that perceptions of sight and sound direct the owl to the same location (Figure 1–6B). The prisms shift the owl's visual map by 23° to the right. On the first day after the prisms were placed onto a juvenile owl, a mismatch occurred between the owl's visual and auditory maps (Figure 1–6C): sight and sound

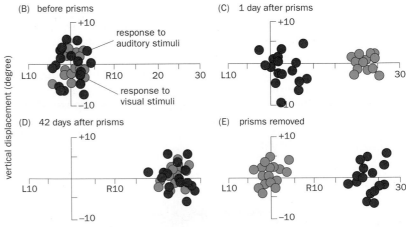

Figure 1–6 Juvenile barn owls adjust their auditory map to match a displaced visual map after wearing prisms. (A) A barn owl fitted with prisms that shift its visual map. **(B)** Before the prisms are attached, the owl's visual map (blue dots) and auditory map (red dots) are matched near 0°. Each dot represents an experimental measurement of owl's final head orientation in response to an auditory or a visual stimulus, presented in the dark. **(C)** One day after the prisms were fitted, the visual map is displaced 23° to

the right of the auditory map. **(D)** After a juvenile owl has worn the prisms for 42 days, its auditory map has adjusted to match its shifted visual map. **(E)** The visual map shifts back immediately after the prisms are removed, causing a temporary mismatch. This mismatch is corrected as the auditory map shifts back soon after (not shown). (A, courtesy of Eric Knudsen; B–E, adapted from Knudsen EI [2002] *Nature* 417:322–328. With permission from Macmillan Publishers Ltd.)

informed the owl of different locations for the prey, resulting in positional confusion. However, a juvenile owl could cope with this situation, perfectly adjusting its auditory map to match its altered visual map by 42 days after starting to wear the prisms (Figure 1–6D). When the prisms were removed, the mismatch happened again (Figure 1–6E), but the owl shifted the auditory map back to its native state shortly afterwards.

The story of the barn owl is an example of how the nervous system learns to cope with the changing world, a phenomenon called **neural plasticity**, that is, changes of the nervous system in response to experience and learning. But the story does not end there. Studies have shown that plasticity declines with age: juvenile owls have the plasticity required to adjust their auditory map to match a visual map displaced by 23°, but adult owls have lost the ability to do so by the time they reach sexual maturity (**Figure 1–7**A). Some human learning capabilities—such as ability to learn a foreign language—likewise decline with age. Thus, experiments on improving plasticity of adult owls may reveal strategies for improving the learning ability of adult humans as well.

Several ways have been found for adult owls to overcome their limited plasticity in shifting their auditory map. If an owl has the experience of adjusting to a 23°-prism shift as a juvenile, it can readjust much more readily to the same prisms as an adult (Figure 1–7B). Alternatively, even adult owls that cannot adjust to a 23° shift all at once can learn to shift their auditory map if the visual field displacement is applied in small increments. Thus, by taking 'baby steps,' adult owls can eventually reach nearly the same magnitude of shift as young owls. Once they have learned to shift in gradual increments, the adult owls can subsequently shift in a single, large step when tested several months after returning to normal conditions (Figure 1–7C).

What are the neurobiological mechanisms that underlie these fascinating plasticity phenomena? In Chapters 4 and 6, we will explore the nature of the visual and auditory maps. In Chapters 5 and 7, we will study how neural maps are formed during development and modified by experience. And in Chapter 10, we will return to this topic when we study memory and learning. Before addressing these concepts, however, we need to learn more basics about the brain and its building blocks. The rest of this chapter will give an overview of the nervous system and will explore how key historical discoveries helped build the conceptual framework of modern neuroscience.

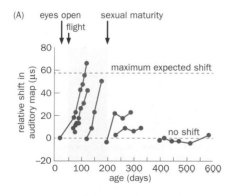
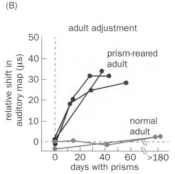
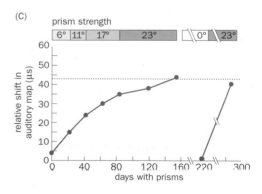

Figure 1–7 Ways to improve the ability of adult barn owls to adjust their auditory map. (A) The ability of owls to adjust their auditory map to match a displaced visual map declines with age. The y axis quantifies the ability to shift the auditory map, measured by the difference in time (μs, or microseconds) it takes for sounds to reach the left and right ears, used by the owl for object location. Each trace represents a single owl, with each dot representing the average of auditory map shift measured at a specific time after the prisms were applied. The shaded zone indicates the sensitive period, during which owls can easily adjust their auditory map in response to visual map displacement. Owls older than 200 days have only limited ability to shift their auditory map. **(B)** Three owls that had successful map adjustment to prism experience when young can also shift their auditory map as adults (red traces). Two owls with no juvenile experience cannot shift their map as adults (blue traces). **(C)** Adult owls can also be trained to shift their auditory map by giving the prism shift in incremental steps, as shown on the left side of the graph. This incremental training enables adult owls to accommodate a sudden shift to the maximum visual displacement of 23° after a period without prism, as shown on the right side of graph. The dotted line at y = 43 μs represents the median shift in juvenile owls in response to a single 23°-prism step. (A & B, adapted from Knudsen EI [2002] *Nature* 417:322–328. With permission from Macmillan Publishers Ltd; C, Linkenhoker BA & Knudsen EI [2002] *Nature* 419:293–296. With permission from Macmillan Publishers Ltd.)

HOW IS THE NERVOUS SYSTEM ORGANIZED?

For all vertebrate and many invertebrate animals, the nervous system can be divided into the **central nervous system** (**CNS**) and **peripheral nervous system** (**PNS**). The vertebrate CNS consists of the **brain** and the **spinal cord** (**Figure 1–8**A, B). Both structures are bilaterally symmetrical; the two sides of the brain are referred as two **hemispheres**. The mammalian brain is composed of morphologically and

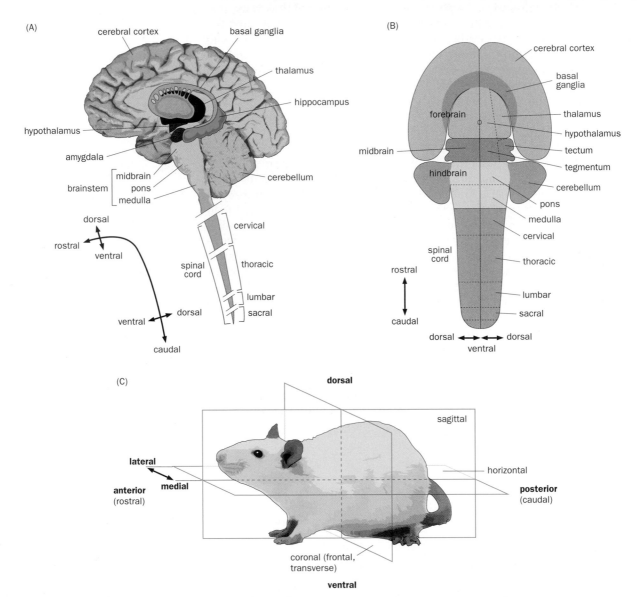

Figure 1–8 The organization of the mammalian central nervous system (CNS). (A) A sagittal (side) view of the human CNS. The basal ganglia (orange), thalamus (purple), hypothalamus (dark blue), hippocampus (light blue), and amygdala (red) from the left hemisphere are superimposed onto a mid-sagittal section of the rest of the CNS (tan background), the left half of which has been cut away to reveal right hemisphere structures (see Panel C for more explanation of the section plane). Major brain structures are indicated, and will be studied in greater detail later in the book. From rostral to caudal, the brainstem is divided into midbrain, pons, and medulla. The spinal cord segments are divided into four groups: cervical, thoracic, lumbar, and sacral. Bottom left, illustration of the rostral–caudal neuraxis (CNS axis). At any given position along the neuraxis in a sagittal plane, the dorsal–ventral axis is perpendicular to the rostral–caudal axis. **(B)** A flatmap of the rat CNS reveals the internal divisions of major brain structures. The flatmap is a two-dimensional representation based on a developmental stage when progenitor cells of the nervous system

are arranged as a two-dimensional sheet. It can be approximated as cutting the CNS along the mid-sagittal plane from the dorsal side and opening the cut surface using the ventral midline as the axis; the ventral-most structures are at the center and the dorsal-most structures are at the sides. (Imagine a book opened to display its pages; the spine of the book—the ventral midline—lays face down.) The left half of the flatmap indicates the major CNS divisions, and the right side indicates major subdivisions. For example, the hypothalamus is ventral to the thalamus, and the midbrain is divided into tectum and tegmentum from dorsal to ventral. **(C)** Schematic illustration of the three principal section planes defined by the body axes. Coronal sections are perpendicular to the rostral–caudal axis; sagittal sections are perpendicular to the medial–lateral axis; and horizontal sections are perpendicular to the dorsal–ventral axis. (B, adapted from Swanson LW [2012] Brain Architecture 2nd ed. Oxford University Press.)

functionally distinct structures that include the **cerebral cortex**, **basal ganglia**, **hippocampus**, **amygdala**, **thalamus**, **hypothalamus**, **cerebellum**, **midbrain**, **pons**, and **medulla**; the last three structures are collectively called the **brainstem**. The brain can also be divided into **forebrain**, **midbrain**, and **hindbrain** according to the developmental origins of each region. The spinal cord consists of repeated structures called segments, which are divided into cervical, thoracic, lumbar, and sacral groups. Each segment gives off a pair of spinal nerves. The PNS comprises **nerves** (that is, discrete bundles of axons) that connect the brainstem and spinal cord of the CNS with the body and internal organs, as well as isolated **ganglia** (clusters of nerve cells) outside the brain and spinal cord. We will study the organization and function of all of these neural structures in subsequent chapters.

The internal structure of the nervous system has traditionally been examined in histological sections. Three types of sections are commonly used and are named following the conventions of histology. In **coronal sections**, also called frontal, transverse, or cross sections, section planes are perpendicular to the **anterior–posterior** axis of the animal (also termed the **rostral–caudal** axis, meaning snout to tail). In **sagittal sections**, section planes are perpendicular to the **medial–lateral** axis (midline to side) of the animal. In **horizontal sections**, section planes are perpendicular to the **dorsal–ventral** (back to belly) axis (Figure 1–8C). In animals with a curved CNS organization, the definition of rostral–caudal axis usually follows the **neuraxis** (axis of the CNS; bottom left of Figure 1–8A) rather than the body axis. The neural axis is defined by the curvature of the embryonic **neural tube**, from which the vertebrate nervous system derives as we will learn in Chapter 7.

1.4 The nervous system consists of neurons and glia

The nervous system is made of two major categories of cells: **neurons** (nerve cells) and **glia**. A typical neuron has two kinds of **neuronal processes** (cytoplasmic extensions). A long, thin process called the **axon** often extends far beyond the cell body (**soma**). In contrast, the thick, bushy processes called **dendrites** are usually close to the soma (**Figure 1–9**A). At the ends of the axons are **presynaptic terminals**, specialized structures that participate in the transfer of information between neurons; the dendrites of many vertebrate neurons are decorated with small protrusions called **dendritic spines**, which likewise function in cell-to-cell information transfer. In the course of this book, we will encounter many neuronal types with distinct morphologies. Most of them have well-differentiated axons and dendrites, which serve distinct functions as will be discussed in Section 1.7.

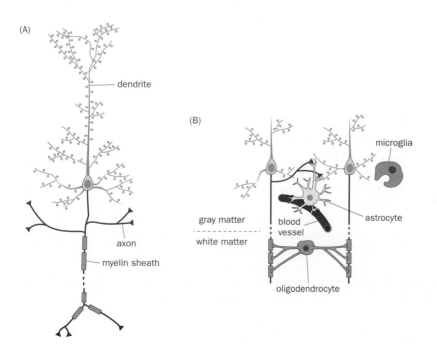

Figure 1–9 Neurons and glia.
(A) Schematic drawing of a typical neuron in the mammalian central nervous system (CNS). Dendrites are in blue, and the axon is in red. The dashed break in the axon indicates that it can travel a long distance from the cell body. The brown structures surrounding the axon are myelin sheath made from glia. The triangles at the ends of the axon branches represent presynaptic terminals, and the protrusions along the dendritic tree are dendritic spines, both of which will be discussed in subsequent sections.
(B) Schematic drawing of glia in the CNS. Oligodendrocytes wrap axons of the CNS neurons. (Schwann cells, not shown here, have a similar function in the peripheral nervous system.) Astrocytes have elaborate processes, whose end-feet wrap around the blood vessels, as well as connections between different neurons. Microglia are immune cells that engulf damaged cells and debris upon activation by injury and during developmental remodeling. (B, see Allen NJ & Barres BA [2009] *Nature* 457:675–677.)

There are four major types of glia in the vertebrate nervous system: **oligodendrocytes**, **Schwann cells**, **astrocytes**, and **microglia** (Figure 1–9B). Oligodendrocytes and Schwann cells play analogous functions in the CNS and PNS, respectively. They wrap axons with their cytoplasmic extensions, called **myelin sheath**, which increase the speed at which information propagates along axons. Oligodendrocytes and myelinated axons constitute the **white matter** in the CNS (because myelin is rich in lipids and appears white). Astrocytes play many roles in the development and regulation of neuronal communication; they are present in the **gray matter** of the CNS, which is enriched in neuronal cell bodies, dendrites, axon terminals, and connections between neurons. Microglia are the resident immune cells of the nervous system, which engulf damaged cells and debris, and help reorganize neuronal connections. Invertebrate nervous systems have a similar division of labor for different glial types.

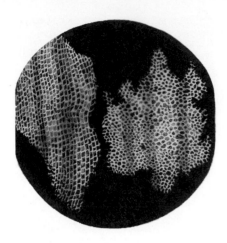

Figure 1–10 The first image of cells. A drawing by Robert Hooke illustrates the repeating units visible in thin sections of cork under a primitive microscope. Hooke thought the units resembled small rooms and coined the term 'cells' to describe them. (From Hooke R [1665] Micrographia. J. Martyn and J. Allestry.)

1.5 Individual neurons were first visualized by Golgi staining in the late nineteenth century

Contemporary students of neurobiology may be surprised to learn that the cellular organization of the nervous system was not uniformly accepted at the beginning of the twentieth century, well after biologists in other fields had embraced the cell as the fundamental unit of life. Robert Hooke first used the term 'cell' in 1665 to describe the repeating units that he observed in thin slices of cork (**Figure 1–10**) using a newly invented piece of equipment—the microscope. Scientists subsequently used microscopes to observe many biological samples and found cells to be ubiquitous structures. In 1839, Matthias Schleiden and Theodor Schwann formally proposed the **cell theory**: all living organisms are composed of cells as basic units. The cell theory was well accepted in almost every discipline of biology by the second half of the nineteenth century except among researchers studying the nervous system. Although cell bodies had been observed in the nervous tissues, many neurobiologists of that era believed that nerve cells were linked together by their elaborate processes to form a giant net, or reticulum, of nerves. Proponents of the **reticular theory** regarded that the reticulum as a whole, rather than its individual cells, constituted the working unit of the nervous system.

Among the neuroscientists who supported the reticular theory of the nervous system was Camillo Golgi, who made many important contributions to science including the discovery of the Golgi apparatus, an intracellular organelle responsible for processing proteins destined for the cell surface, for other intracellular membranous organelles, or for secretion outside the cell. Golgi's greatest contribution, however, was the invention of the **Golgi staining** method. When a piece of neural tissue is soaked in a solution of silver nitrate and potassium dichromate in the dark for several weeks, black precipitates (microcrystals of silver chromate) form stochastically in a small fraction of the nerve cells so that these cells can be visualized against unstained background. Importantly, once black precipitates form within a cell, an autocalatylic reaction occurs such that the entire cell, including most or all of the elaborate extensions, can be visualized in its native tissue (**Figure 1–11**). Golgi staining, for the first time, enabled visualization of the entire morphology of individual neurons. However, despite inventing this key method for neuronal visualization, Golgi remained a believer in the reticular theory (see **Box 1–1**).

It took another great neuroscientist, Santiago Ramón y Cajal, to refute effectively the reticular theory. The work of Ramón y Cajal and several contemporaries instead supported the **neuron doctrine**, which postulated that neuronal processes do not fuse to form a reticulum. Instead, neurons form intimate contact with each other, with communication between distinct neurons occurring at these contact points (see Box 1-1). The term **synapse** was later coined by Charles Sherrington to describe these sites at which signals flow from one neuron to

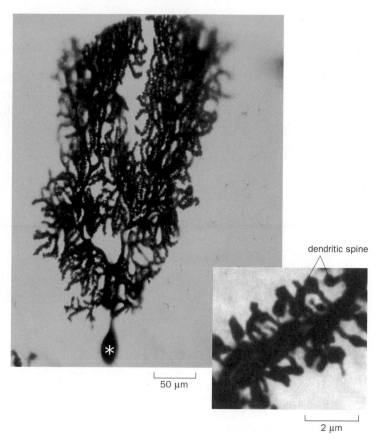

dendritic spine

50 μm

2 μm

Figure 1–11 Golgi staining. An individual Purkinje cell in the mouse cerebellum is stained black by the formation of silver chromate precipitate, allowing visualization of its complex dendritic tree. The axon, which is not included in this image, projects downward from the cell body indicated by an asterisk. The inset shows a higher magnification of a dendritic segment, highlighting the protruding structures called dendritic spines. (Adapted from Luo L, Hensch TK, Ackerman L et al. [1996] *Nature* 379:837–840. With permission from Macmillan Publishers Ltd.)

another. After systematically applying the Golgi staining method to study tissues in many parts of the nervous system, in many organisms ranging from insects to humans, and at many developmental stages, Ramón y Cajal concluded that individual neurons were embryologically, structurally, and functionally independent units of the nervous system.

Box 1–1: The debate between Ramón y Cajal and Golgi: why do scientists make mistakes?

Camillo Golgi and Santiago Ramón y Cajal were the most influential neurobiologists of their time. They shared the 1906 Nobel Prize for Physiology or Medicine, the first to be awarded to neurobiologists. However, their debates on how nerve cells constitute the nervous system—by the reticular network or as individual neurons communicating with each other through synaptic contact—continued during their Nobel lectures (see **Figure 1–12**A, B). We now know that Ramón y Cajal's view was correct and Golgi's view was largely incorrect. For example, utilizing the brainbow labeling method (see Section 13.18 for more details), individual neurons, their dendritic tree, and even their axon terminals can be clearly distinguished by distinct colors (see Figure 1–12C). Interestingly, Ramón y Cajal used Golgi's staining method to refute Golgi's theory. Why didn't Golgi reach the correct conclusion using his own method? Was he not a careful observer? After all, he made many great discoveries including the Golgi apparatus, a protein-processing organelle named after him.

According to Ramón y Cajal's analysis, "Golgi arrived at this conclusion by an unusual blend of accurate observations and preconceived ideas ... Golgi's work actually consists of two separate parts. On the one hand, there is his method, which has generated a prodigious number of observations that have been enthusiastically confirmed. But on the other, there are his interpretations, which have been questioned and rejected."

Before the invention of the Golgi staining method, neurobiologists could not resolve processes from individual nerve cells, and therefore believed that the nerve processes were fused together in a giant net. Golgi was trained in the scientific environment in which this reticular theory was the dominant interpretation of the nervous system organization. Golgi tried to fit his observations with existing theory. For example, even though Golgi was first to discover, using his staining method, that dendritic trees had free endings (Figure 1–12A, top), he thought that dendrites were

(Continued)

Box 1–1: The debate between Ramón y Cajal and Golgi: why do scientists make mistakes?

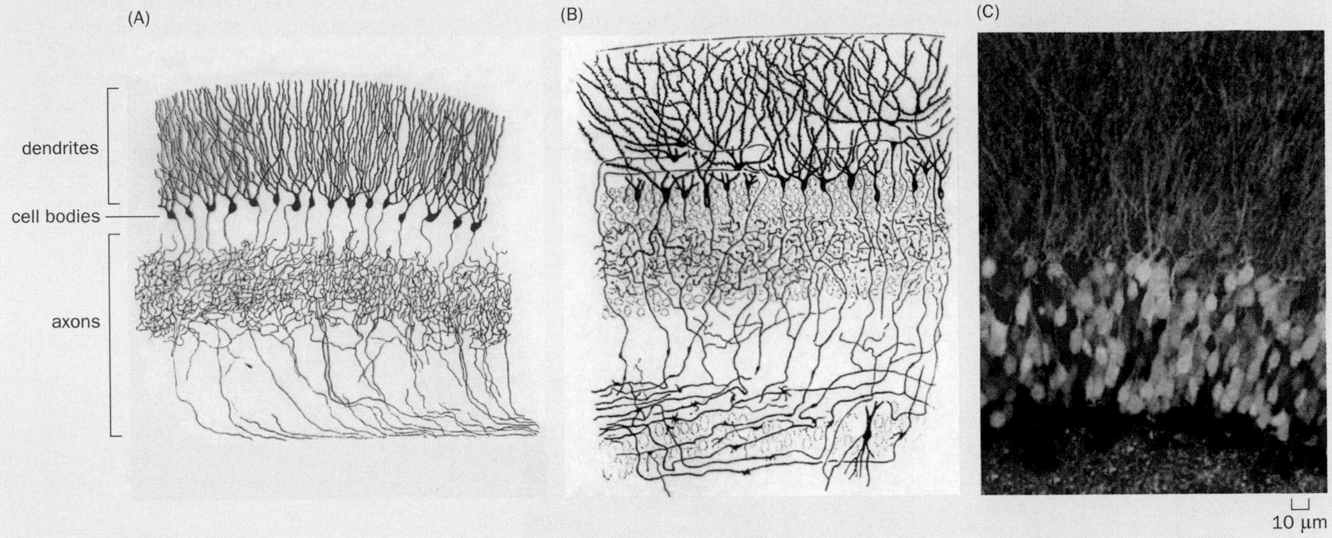

Figure 1–12   Three different views of hippocampal granule cells. (A) Golgi's drawing of granule cells of the hippocampus. The dendritic, cell body, and axonal layers are indicated on the left. In Golgi's drawing, all axons are fused together to form a giant reticulum. **(B)** Ramón y Cajal's depiction of the same hippocampal granule cells. Note that axons below the cell bodies have definitive endings. **(C)** Hippocampal granule cells labeled by the brainbow technique, which allows the spectral separation of individual neurons expressing different mixtures of cyan, yellow, and red fluorescent proteins. Not only cell bodies, but also some of their dendrites above and axon terminals below can be resolved by different colors. (A, adapted from Golgi C [1906] Nobel Lecture; B, adapted from Ramón y Cajal S [1911] Histology of the Nervous System of Man and Vertebrates. Oxford University Press; C, from Livet J, Weissman TA, Kang H et al. [2007] *Nature* 450:56–62. With permission from Macmillan Publishers Ltd.)

used to collect nutrients for nerve cells. It was their axons, which formed an inseparable giant net as he viewed them (Figure 1–12A, bottom), that performed all the special functions of the nervous system.

This story teaches an important lesson: scientists need to be observant, but they also need to be as objective and unbiased as possible when interpreting their own observations.

1.6   Twentieth-century technology confirmed the neuron doctrine

Ramón y Cajal could not convince Golgi to abandon the reticular theory, but many lines of evidence since the Golgi–Ramón y Cajal debate (see Box 1–1) have provided strong support for the neuron doctrine. For example, during embryonic development, neurons start out as having just the cell bodies. Axons then grow out from the cell bodies toward their final destinations. This was demonstrated by observing axon growth *in vitro* in experiments made possible by tissue culture techniques that were invented for the purpose of visualizing neuronal process growth (**Figure 1–13**). Axons are led by a structure called the **growth cone**, which changes its shape dynamically. We will learn more about the function of the growth cone in axon guidance in Chapter 5.

The final proof that neuronal processes do not fuse with each other came from observations made possible through the development of **electron microscopy**, a technique that allows the visualization of structures with nanometer (nm) resolution. (Conventional **light microscopy**, which scientists since Hooke have used to observe biological samples, cannot usually resolve structures less than 200 nm apart because of the physical properties of light). The use of electron microscopy to examine **chemical synapses** (so named because the communication between cells is mediated by the release of chemicals called **neurotransmitters**) revealed that a 20–100 nm gap, the **synaptic cleft**, separates the neuron from its target, which can be another neuron or a muscle cell (**Figure 1–14**A). The synaptic partners are not symmetrical: the presynaptic terminal of the neuron contains small **synaptic vesicles** filled with neurotransmitters, which, upon stimulation, fuse with the plasma membrane and release neurotransmitters into the synaptic cleft. The postsynaptic target cell develops a **postsynaptic specialization** (also called a **postsynaptic density**) that is enriched for the receptors on their plasma membrane surface to receive the neurotransmitters. Chemical synapses are the predominant type of synapses that allows neurons to communicate with each other and with muscle cells. We will study them in greater detail in Chapter 3.

Neurons can also communicate with each other by an alternative means: the **electrical synapse** mediated by the **gap junctions** between neurons (Figure 1–14B). Here, each partner neuron contributes protein subunits to form gap junction channels that directly link the cytoplasm of two adjacent neurons, allowing ions and small molecules to travel from one neuron to the next. These gap junctions probably come closest to what the reticular theory would imagine as a fusion between different neurons. However, macromolecules cannot pass between gap junctions, and the neurons remain two distinct cells with highly regulated communication. The existence of gap junctions therefore does not violate the premise that individual neurons are the building blocks of the nervous system.

1.7 In vertebrate neurons, information generally flows from dendrites to cell bodies to axons

As we introduced in Section 1.4, neurons have two kinds of processes: dendrites and axons. The shapes of dendritic trees and patterns of axonal projections are characteristic for particular types of neurons, and are often used to classify them. For example, the most frequently encountered type of neuron in the mammalian cerebral cortex and hippocampus, the **pyramidal neuron**, has a pyramid-shaped

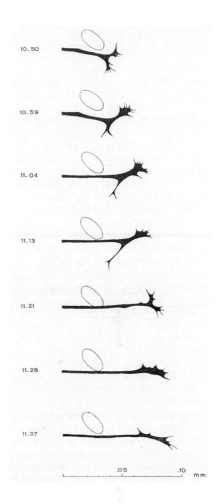

Figure 1–13 The first time-lapse depiction of a growing axon. Frog embryonic spinal cord tissue was cultured *in vitro*. Growth of an individual axon was sketched with an aid of a camera lucida at the time indicated on the left (hour. minute). The stationary blood vessel (oval) provided a landmark for the growing tips of the axon, called growth cones, which undergo dynamic changes in shape including both extensions and retractions. A distance scale (in millimeter) is at the bottom of the figure. (Adapted from Harrison RG [1910] *J Exp Zool* 9:787–846.)

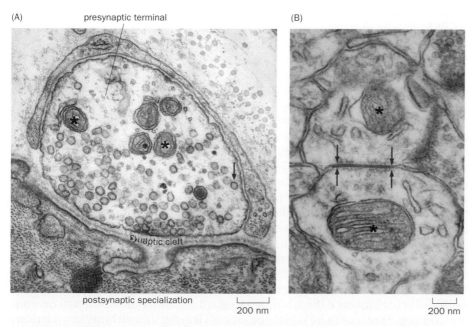

(A)   presynaptic terminal

postsynaptic specialization

200 nm

(B)

200 nm

Figure 1–14 Chemical and electrical synapses. Asterisks indicate mitochondria in both micrographs. **(A)** Electron micrograph of a chemical synapse between the presynaptic terminal of a motor neuron and the postsynaptic specialization of its target muscle cell. A synaptic cleft separates the two cells. The arrow points at a synaptic vesicle. **(B)** Electron micrograph of an electrical synapse (gap junction) between two dendrites of mouse cerebral cortical neurons. Two opposing pairs of arrows mark the border of the electrical synapse. (A, courtesy of Jack McMahan; B, courtesy of Josef Spacek and Kristen M. Harris, Synapse Web.)

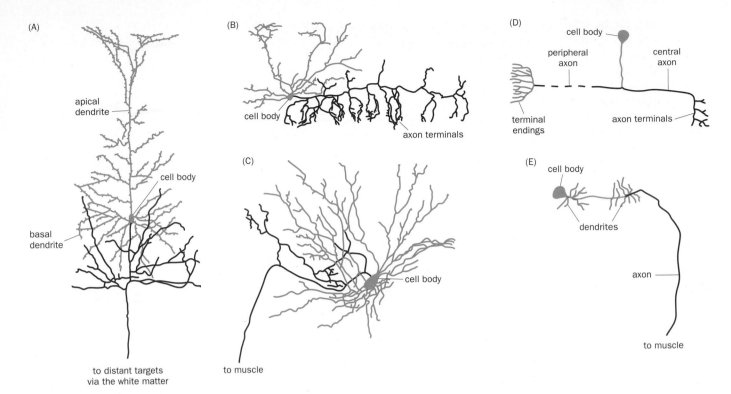

Figure 1–15 Some representative neurons. (A) A pyramidal cell from rabbit cerebral cortex. A typical pyramidal cell has an apical dendrite (blue) that gives off branches as it ascends, several basal dendrites (blue) that emerge from the cell body, and an axon (red) that branches locally and projects to distant targets. **(B)** A basket cell from mouse cerebellum. The basket cell axon (red) forms a series of 'basket' terminals that wrap around Purkinje cell bodies (not drawn). **(C)** A motor neuron from cat spinal cord. Its bushy dendrites (blue) receive input within the spinal cord, and the axon (red) extends outside the spinal cord to the muscle, while leaving behind local branches. **(D)** A mammalian sensory neuron from a dorsal root ganglion. A single process from the cell body bifurcates into a peripheral axon with terminal endings in the skin (equivalent of dendrites that collect sensory information), and a central axon that project into the spinal cord. **(E)** A motor neuron from the fruit fly ventral nerve cord (equivalent to the vertebrate spinal cord). Most invertebrate central neurons are unipolar: a single process comes out of the cell body, which gives rise to dendritic branches (blue) and the axon (red). (A–D, adapted from Ramón y Cajal S [1911] Histology of the Nervous System of Man and Vertebrates. Oxford University Press; E, based on Lee T & Luo L [1999] *Neuron* 22:451–461.)

cell body with an apical dendrite and several basal dendrites that branch extensively (**Figure 1–15**A). A large portion of the dendritic tree is covered with dendritic spines (see Figure 1–11 inset), which contain postsynaptic specializations that are in close contact with presynaptic terminals of partner neurons. Another widely encountered type of neuron, **basket cells**, wrap their axon terminals around the cell bodies of pyramidal cells in the cerebral cortex or **Purkinje cells** (see Figure 1–11) in the cerebellum (Figure 1–15B). The spinal cord **motor neuron** extends bushy dendrites within the spinal cord (Figure 1–15C), and projects its axon out of the spinal cord and into the muscle. Located in a **dorsal root ganglion** just outside the spinal cord, a **sensory neuron** in the **somatosensory system** (which provides bodily sensation) extends a single process that bifurcates, forming a peripheral axon that gives rise to branched terminal endings and a central axon that projects into the spinal cord (Figure 1–15D). Most vertebrate neurons have both dendrites and an axon leaving the cell body, hence they are called **multipolar** (or **bipolar** if there is only a single dendrite); somatosensory neurons are pseudounipolar as there is just one process leaving the cell body but it quickly gives rise to a peripheral and a central branch.

How does information flow within individual neurons? After systematically observing many types of neurons in different parts of the nervous system, Ramón y Cajal proposed a **theory of dynamic polarization**: the transmission of a neuronal signal takes place from dendrites and cell bodies to the axon. Therefore every neuron has (1) a receptive component, the cell body and dendrites; (2) a transmission component, the axon; and (3) an effector component, the axon terminals. With few exceptions (the somatosensory neuron being one), this important principle has been validated by numerous observations and experiments since it was proposed a century ago, and has been used extensively to deduce the direction of information flow in the complex vertebrate central nervous system. We will discuss the cell biological basis of neuronal polarization in Chapter 2.

How did observing the morphology of individual neurons lead to the discovery of this rule? Ramón y Cajal took advantage of the fact that, in sensory systems, information should generally flow from sensory organs to the brain. By examining different neurons along the visual pathway (**Figure 1–16**), for example, one can see that at each connection, dendrites are at the receiving end, facing the external world, whereas axons are oriented so as to deliver such information to more central targets, sometimes at a great distance from the cell body where the axon originates. This applies to neurons in other sensory systems as well. On the other hand, in the motor system, information should generally flow outward from the CNS to the periphery. The morphology of the motor neuron indeed supports the notion that its bushy dendrites receive input within the spinal cord, and its long axon projecting to the muscle provides the output (Figure 1–15C).

Neuronal processes in invertebrates can also be defined as dendrites and axons according to their functions, with dendrites positioned to receive information and axons to send it. However, the morphological differentiation of most invertebrate axons and dendrites, especially in the central nervous system, is not as clear-cut as it is for vertebrate neurons. Most often, invertebrate neurons are **unipolar**, extending a single process that gives rise to both dendritic and axonal branches (Figure 1–15E). Dendritic branches are often, but not always, closer to the cell body. In many cases, the same branches can both receive and send information; this occurs in some vertebrate neurons as well, as we will learn in Chapters 4 and 6. Thus, paradoxically, in the simpler invertebrate nervous systems, it is more difficult to deduce the direction of information flow by examining the morphology of individual neurons.

1.8   Neurons use membrane potential changes and neurotransmitter release to transmit information

What is the physical basis of information flow within neurons? We now know that the nervous system uses electrical signals to propagate information. The first evidence of this came from Luigi Galvani's discovery, in the late eighteenth century, that application of an electric current could produce muscle twitches in frogs. It was known by the beginning of the twentieth century that electrical signals are propagated in neurons by transient changes of the **membrane potential**, the electrical potential difference across the neuronal membrane that reflects the distribution of positive and negative charges on each side of the membrane. As we will learn in more detail in Chapter 2, neurons in the resting state are more negatively charged inside the cells compared with the extracellular environment. When neurons are excited, their membrane potentials change transiently, creating **nerve impulses** that propagate along their axons. But how is information relayed through nerve impulses? Quantitative studies of how sensory stimuli of different magnitudes induce nerve impulses provided important clues.

Studies of muscle contraction in response to electrical stimulation of motor nerves suggested that an elementary nerve impulse underlies different stimulus strengths. An all-or-none conduction principle became clearer when amplifiers for electrical signals were built in the 1920s that made it possible to record nerve impulses from single axon fibers in response to sensory stimulations. Edgar Adrian and co-workers systematically measured nerve impulses from somatosensory neurons (see Figure 1–15D) that convey information about touch, pressure, and pain to the spinal cord. They found that individual nerve impulses were of a uniform size and shape whether they were elicited by weak or strong sensory stimuli; increasing the stimulus strength increased the frequency of such impulses, but not the inherent properties of each impulse (**Figure 1–17**).

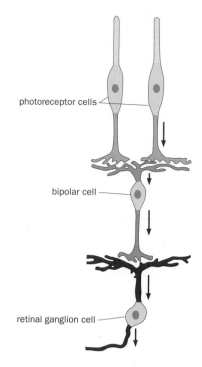

photoreceptor cells

bipolar cell

retinal ganglion cell

Figure 1–16 Neurons and information flow in the vertebrate retina. Visual information is collected by photoreceptor cells in the retina. It is communicated to the bipolar cell and then to the retinal ganglion cell, which projects a long-distance axon into the brain. Note that for both the bipolar cell and retinal ganglion cell information is received by their dendrites (in blue) and sent via their axons (in red). The photoreceptor processes can also be divided into a dendrite-equivalent that detects light (blue) and an axon to send output to the bipolar cell. We will learn more about these cells and connections in Chapter 4. Arrows indicate the direction of information flow. (Adapted from Ramón y Cajal S [1911] Histology of the Nervous System of Man and Vertebrates. Oxford University Press.)

Figure 1–17 Stimulus strengths are encoded by the frequency of uniformly sized nerve impulses. (A) Experimental setup for applying a specified amount of pressure to the toe of a cat, while recording impulses from an associated sensory nerve. **(B)** As the pressure applied to a cat's toe increases, the frequency of impulses measured at the sensory nerve increases, but the size and shape of each impulse remain mostly the same. We now call these nerve impulses action potentials. The x axis shows the timescale in units of seconds (s). (Adapted from Adrian ED & Zotterman Y [1926] *J Physiol* 61:465–483.)

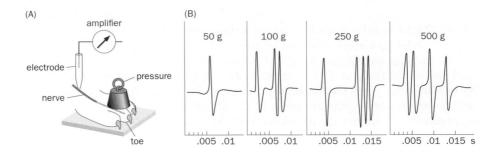

These experiments led to two important concepts of modern neuroscience. The first concept is the presence of an elementary unit of nerve impulses that axons use to convey information across long distances; we now call this elementary unit an **action potential**. In Chapter 2, we will study in greater detail the molecular basis of action potentials, including why they exhibit the all-or-none property. The second concept is that neurons use the frequency of action potentials to convey the intensity of signals being delivered. Whereas the frequency of action potentials is the most widely used means to convey signal intensity throughout the nervous system, the timing of action potentials can also convey important information.

In addition to using action potentials to send signals across long distances within the axons of individual neurons, another important form of communication within neurons are **graded potentials** (also called **local potentials**), referring to membrane potentials that can change in continuous values as opposed to all-or-none. One type of graded potentials, termed **synaptic potentials**, is produced at the postsynaptic sites in response to neurotransmitter release by presynaptic partners. Graded potentials can also be induced at the peripheral endings of sensory neurons by sensory stimuli, such as the pressure on the toe in Adrian's experiment mentioned above; these are termed **receptor potentials**. Unlike action potentials, graded potentials are of different sizes depending on the strength of the input stimuli and the sensitivity of postsynaptic or sensory neurons to those stimuli. Some neurons, including most neurons in the vertebrate retina, do not fire action potentials at all. These **non-spiking neurons** use graded potentials to transmit information even in their axons.

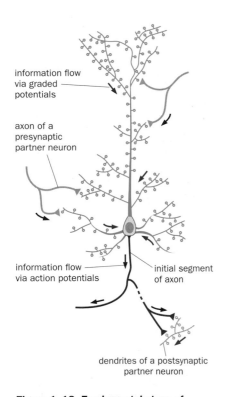

information flow via graded potentials

axon of a presynaptic partner neuron

information flow via action potentials

initial segment of axon

dendrites of a postsynaptic partner neuron

Figure 1–18 Fundamental steps of neuronal communication. A typical neuron in the mammalian CNS receives thousands of inputs distributed along the dendritic tree and dendritic spines (blue), where this neuron is postsynaptic to its presynaptic partners. Inputs are collected in the form of synaptic potentials, which travel toward the cell body (blue arrows) and are integrated at the initial segment of the axon (red) to produce action potentials. Action potentials propagate to the axon terminals (red arrows), and trigger neurotransmitter release that conveys information to many of its postsynaptic partner neurons.

Synaptic potentials are usually produced at the dendritic spines, along the dendrite tree, and at the soma of a neuron. A typical mammalian neuron contains thousands of postsynaptic sites on its dendritic tree, allowing it to collect input from many individual presynaptic partners (**Figure 1–18**). Inputs that are excitatory facilitate action potential production by the postsynaptic neuron, whereas inhibitory inputs dampen action potential production by the postsynaptic neuron. In most neurons, the eventual purpose of all these synaptic potentials is to determine whether, when, and how frequently the neuron will fire action potentials, so that information can propagate along its axon to its own postsynaptic target neurons. The site of action potential initiation is typically at the **initial segment of the axon**, adjacent to the soma. Thus, synaptic potentials generated in dendrites must travel to the cell body in order to have their voices heard.

The rule of action potential initiation at the axon initial segment has exceptions. For example, in the sensory neuron in Figure 1–15D, action potentials are initiated at the junction between terminal endings and the peripheral axon of the sensory neuron such that sensory information can be transmitted by the peripheral and central axon to the spinal cord across a long distance. In invertebrate neurons, which are mostly unipolar, action potential initiation likely occurs at the junction between the dendritic and axonal compartments.

How is information transmitted between neurons? At electrical synapses, membrane potential changes are directly transmitted from one neuron to the next by ion flow across the gap junctions. At chemical synapses, the arrival of action potentials (or graded potentials in non-spiking neurons) at the presynaptic terminals triggers the release of neurotransmitters. Neurotransmitters diffuse across the synaptic cleft and bind to their receptors on the postsynaptic neurons to produce synaptic potentials (Figure 1–18). The process of neurotransmitter release from the presynaptic neuron and neurotransmitter reception by the postsynaptic neuron is collectively referred to as **synaptic transmission**. Thus, while intraneuronal communication is achieved by membrane potential changes in the form of graded potentials and action potentials, interneuronal communication at chemical synapses relies on neurotransmitter release and reception. We will study these fundamental steps of neuronal communication in greater detail in Chapters 2 and 3.

1.9   Neurons function in the context of specific neural circuits

Neurons perform their functions in the context of **neural circuits**, which are ensembles of interconnected neurons that act together to perform specific functions. The simplest circuits in vertebrates, those that mediate the spinal reflexes, can consist of as few as two interconnected neurons: a sensory neuron that receives sensory stimuli from the environment, and a motor neuron that controls muscle contraction. Many fundamental neurobiological principles have been derived from studying these simple circuits.

When a neurologist's hammer hits the knee of a patient during a neurological exam, the lower leg kicks forward involuntarily (**Figure 1–19**). The underlying circuit mechanism for this **knee-jerk reflex** has been identified. Sensory neurons embed their endings in specialized apparatus called **muscle spindles** in an extensor muscle (a muscle whose contraction will extend the knee joint). These sensory neurons detect the stretch of the muscle spindles caused by the physical impact of the hammer and convert the mechanical stimuli into electrical signals, namely receptor potentials at the sensory endings. Next, the peripheral and central axons of the sensory neurons propagate these electrical signals to the spinal cord as action potentials. There, the central axons of the sensory neurons make synaptic connections directly with the dendrites of motor neurons, which extend their own axons outward from the spinal cord and terminate in the same extensor muscle where the sensory neurons embed their endings. (The sensory axons are also called **afferents**, referring to axons projecting from peripheral tissues to the CNS, whereas the motor axons are called **efferents**, referring to axons that project from the CNS to peripheral targets.) Both the sensory and motor neurons in this circuit are **excitatory neurons**. When excitatory neurons are activated, that is, when they fire action potentials and release neurotransmitters, they make their postsynaptic target cells more likely to fire action potentials. Therefore, mechanical stimulation activates sensory neurons. Their excitation in turn activates their postsynaptic motor neurons, causing them to fire action potentials. Action potentials from the motor neurons cause neurotransmitter release at motor axon terminals in the muscle, causing the contraction of the extensor muscle that they innervate.

The knee-jerk reflex involves the coordination of more than one muscle. The flexor muscle, which is antagonistic to the extensor muscle, must *not* contract at the same time in order for the knee-jerk reflex to occur. (As we will learn in Chapter 8, contraction of extensor muscles increases the angle of a joint, whereas contraction of flexor muscles decreases the angle of a joint.) Therefore, in addition to causing the contraction of the extensor muscle, the sensory axons must also inhibit the contraction of the corresponding flexor muscle. This inhibition is mediated by **inhibitory interneurons** in the spinal cord, a second type of postsynaptic neuron targeted by the sensory axons. (Neurobiologists use the term interneuron in two different contexts. In a broad context, all neurons that are not sensory or motor neurons are interneurons. In more a specific context, interneurons refer to neurons that confine their axons within a specific region—as opposed to **projection neurons**, whose axons link two different regions of the nervous system. The spinal inhibitory interneurons fit both criteria.) Activation of sensory neurons excites these inhibitory interneurons, which in turn inhibit the motor neurons that innervate the flexor muscle. The inhibition makes it more difficult for the flexor motor neurons to fire action potentials, causing the flexor

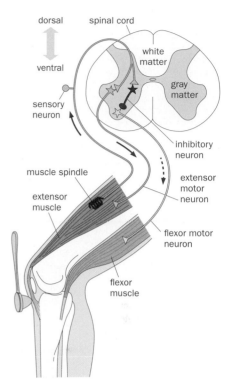

Figure 1–19 The neural circuit underlying the knee-jerk reflex. A simple neural circuit is responsible for the involuntary jerk that results when the front of the human knee is hit with a hammer. In this simplified scheme, a single neuron is used to represent a population of neurons that perform the same task. The sensory neuron extends its peripheral axon to the muscle spindle of the extensor muscle and its central axon to the spinal cord. In the spinal cord, the sensory neuron has two postsynaptic targets: the green motor neuron that innervates the extensor muscle, and the red inhibitory interneuron within the spinal cord, which synapses with the yellow motor neuron that innervates the flexor muscle. When the hammer hits the knee, mechanical force excites the sensory neuron, which results in excitation of the extensor motor neuron to cause contraction of the extensor muscle (following solid arrows) and inhibition of the flexor motor neuron to cause relaxation of the flexor muscle (dashed arrow). The spinal cord is drawn as a cross section. The gray matter at the center contains cell bodies, their dendrites, and the synaptic connections of spinal cord neurons; the white matter at the periphery consists of long-distance axonal projections. The sensory neuron cell body is located in a dorsal root ganglion adjacent to the spinal cord.

muscle to relax. Contraction of the extensor muscle is coordinated with relaxation of the flexor muscle to bring the lower leg forward. First analyzed in the studies of spinal reflexes by Charles Sherrington in the 1890s, the role of inhibition is crucial in coordinating neuronal function throughout the nervous system.

Thus, the knee-jerk reflex involves one of the simplest neural circuits, in which coordinated actions of excitation and inhibition are performed by the monosynaptic connections between sensory neurons and motor neurons and the disynaptic connections between sensory neurons and a different group of motor neurons via inhibitory neuron intermediates. Nervous system functions rely on establishing proper connections between neurons in numerous neural circuits like the knee-jerk reflex circuit; how the nervous system is precisely wired during development is a subject that we will focus on in Chapters 5 and 7.

Most neural circuits are orders of magnitude more complex than the spinal cord reflex circuit. **Box 1–2** discusses commonly used circuit motifs that we will encounter in this book. For example, a subject becomes aware that a hammer has hit her knee because sensory neurons also send branches of axons that ascend along the spinal cord. After passing through relay neurons in the brainstem and thalamus, sensory information eventually reaches the **primary somatosensory cortex**, the part of the cerebral cortex that first receives somatosensory input from the body (**Figure 1–20**). Cortical processing of such sensory input gives her the perception that the knee has been hit. Such information is also propagated to other cortical areas including the **primary motor cortex**. The primary motor cortex sends descending output directly and indirectly to the spinal cord motor neurons to control muscle contraction (Figure 1–20), in case we want to move our leg voluntarily (as compared to the knee-jerk reflex, which is involuntary). We will study these sensory and motor pathways in greater details in Chapters 6 and 8, but in general we know far less about the underlying mechanisms of the ascending, cortical, and descending circuits compared to the reflex circuit in the spinal cord. Deciphering the principles of information processing in complex neural circuits that mediate sensory perception and motor control is one of the most exciting and challenging goals of modern neuroscience.

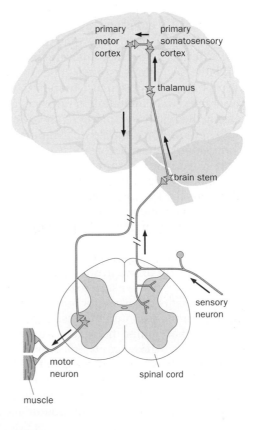

Figure 1–20 Schematic of sensory and motor pathways between the spinal cord and the cerebral cortex. In addition to participating in the spinal cord reflex circuit, sensory neurons also send an ascending branch to connect with relay neurons in the brainstem and thalamus to deliver information to neurons in the primary somatosensory cortex. Through inter-cortical connections, information can be delivered to neurons in the primary motor cortex, which send descending output directly and indirectly to motor neurons for the voluntary control of muscles. Drawn here are the most direct routes for these ascending and descending pathways. The spinal cord is represented as a cross section. The brain in the background is from a sagittal view (not at the same scale as the spinal cord). Arrows indicate the direction of information flow.

Box 1–2: Commonly used neural circuit motifs

The simplest circuit consists of two synaptically connected neurons, such as the sensory neuron–extensor motor neuron connection in the knee-jerk reflex. In circuits that contain more than two neurons, the individual neurons can receive input from and send output to more than one partner. Further complexity arises when some neurons in a circuit are excitatory and others are inhibitory. The nervous system employs many circuit motifs, that is, different ways in which circuits can be configured so that the connection patterns of individual neurons combine to carry out specific functions. We introduce here some of the most commonly used circuit motifs (**Figure 1–21**).

Let's first consider circuits that contain only excitatory neurons. When several neurons synapse onto the same postsynaptic neuron, this constitutes the circuit motif of convergent excitation (Figure 1–21A). Conversely, divergent excitation (Figure 1–21B) refers to the motif where a single neuron synapses onto multiple postsynaptic targets via branched axons (these axon branches are also called **collaterals**). Convergent and divergent connections allow individual neurons to integrate input from multiple presynaptic neurons, and to send output to multiple postsynaptic targets. Serially connected excitatory neurons constitute a **feedforward excitation** motif (Figure 1–21C) to propagate information across multiple regions of the brain, such as the relay of somatosensory stimuli to the primary somatosensory cortex (see Figure 1–20). If a postsynaptic neuron synapses onto its own presynaptic partner, this would produce a feedback excitation motif (Figure 1–21D). Neurons that transmit parallel streams of information can also excite each other, forming a motif of recurrent (lateral) excitation (Figure 1–21E).

When excitatory and inhibitory neurons interact in the same circuit, as is most often the case, many interesting circuit motifs can be constructed and used for different purposes. The names of these motifs that involve inhibitory neurons usually emphasize the nature of the inhibition. In **feedforward inhibition** (Figure 1–21F), an excitatory neuron synapses onto both an excitatory and an inhibitory postsynaptic neuron, and the inhibitory neuron further synapses onto the excitatory postsynaptic neuron. In **feedback inhibition** (Figure 1–21G), the postsynaptic excitatory neuron synapses onto an inhibitory neuron, which synapses back to the postsynaptic excitatory neuron. In both cases, inhibition can control the duration and the magnitude of the excitation of the postsynaptic excitatory neuron. In **recurrent (cross) inhibition** (Figure 1–21H), two parallel excitatory pathways cross-inhibit each other via inhibitory neuron intermediates; the inhibition of the flexor motor neuron in the knee-jerk reflex discussed in Section 1.9 is an example of recurrent inhibition. In **lateral inhibition** (Figure 1–21I), an inhibitory neuron receives excitatory input from one or several parallel streams of excitatory neurons, and sends inhibitory output to many postsynaptic targets of these excitatory neurons. Lateral inhibition is widely used in processing sensory information, as we will study in greater detail in Chapters 4 and 6. Finally, when an inhibitory neuron synapses onto another inhibitory neuron, the excitation of the first inhibitory neuron reduces the inhibitory output of the second inhibitory neuron, causing **disinhibition** of the target of the second inhibitory neuron (Figure 1–21J).

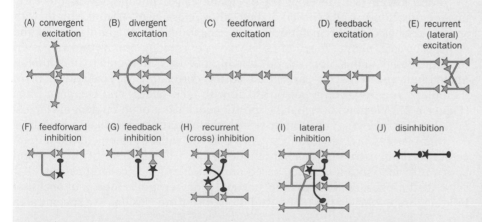

(A) convergent excitation **(B)** divergent excitation **(C)** feedforward excitation **(D)** feedback excitation **(E)** recurrent (lateral) excitation

(F) feedforward inhibition **(G)** feedback inhibition **(H)** recurrent (cross) inhibition **(I)** lateral inhibition **(J)** disinhibition

Figure 1–21 Commonly used circuit motifs. In all panels, the general information flow is from left to right. **(A–E)** Circuit motifs consisting of excitatory neurons. These include convergent excitation (A), in which multiple neurons synapse onto the same neuron; divergent excitation (B), in which a single neuron synapses onto multiple target neurons; feedforward excitation (C), in which neurons are connected in series; feedback excitation (D), in which the postsynaptic neuron synapses onto its presynaptic partner; and recurrent excitation (E), in which two parallel pathways cross excite each other. **(F–J)** Circuit motifs that include inhibitory neurons. In feedforward inhibition (F), the inhibitory neuron receives input from a presynaptic excitatory neuron and sends output to a postsynaptic excitatory neuron, whereas in feedback inhibition (G), the inhibitory neuron receives input from and sends output to the postsynaptic excitatory neuron. For recurrent inhibition (H) and lateral inhibition (I), only feedforward modes are depicted; in the feedback modes of these motifs (not shown), the inhibitory neuron(s) receive input(s) from postsynaptic rather than the presynaptic excitatory neurons in the parallel pathways. In the disinhibition motif (J), the target of the second inhibitory neuron (not shown) can be excitatory or inhibitory. When the first inhibitory neuron is excited, the target of the second inhibitory neuron becomes disinhibited, because excitation of the first inhibitory neuron causes the second inhibitory neuron to be less active.

Circuit motifs containing excitatory and inhibitory neurons are often used in combinations that give rise to many different ways of processing information in complex nervous systems. In Chapter 3, we will encounter another group of neurons, the **modulatory neurons**, which can act on both excitatory and inhibitory neurons to up- or down-regulate their excitability or synaptic transmission, adding further complexity and richness to the information processing functions of neural circuits.

Figure 1–22 Phrenologists' depiction of the brain's organization. According to phrenology, the brain is divided into individual areas specialized for defined mental functions. The size of each area is modified by use. For example, a cautious person would have an enlarged area corresponding to cautiousness.

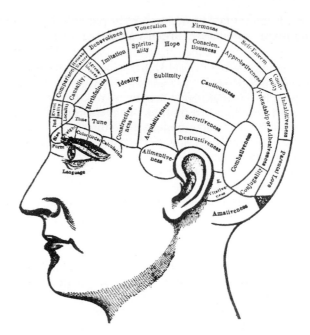

1.10 Specific brain regions perform specialized functions

That specialized functions of the nervous system are performed by specific parts of the brain is well established today. However, throughout prior centuries, philosophers have debated whether brain functions underlie the mind, let alone whether specific regions of the brain are responsible for specific mental activities. Even in the early twentieth century, a prevalent view was that any specific mental function is carried out by neurons across many areas of the cerebral cortex.

Franz Joseph Gall developed a discipline called **phrenology** in the early nineteenth century. Gall believed that all behavior emanates from the brain, with specific regions of the brain controlling specific functions. According to Gall, the centers for each mental function grow with use, creating bumps and ridges on the skull. Based on this proposal, Gall and his followers attempted to map human mental function to specific parts of the cortex, correlating the size and shape of the bumps and ridges on a person's skull with their talents and character traits (**Figure 1–22**). We now know that the specific conclusions of phrenology are largely false, but Gall's thinking about brain specialization was quite advanced for his time.

Brain lesions provided the first clues that specialized regions of the human cerebral cortex carry out specific functions. Each hemisphere of the cerebral cortex is divided into four lobes based on the major folds (called **fissures**) that separate the lobes: the **frontal lobe**, **parietal lobe**, **temporal lobe**, and **occipital lobe** (**Figure 1–23**A). In the 1860s, Paul Broca discovered lesions in a specific area

Figure 1–23 Language centers in the human brain were originally defined by lesions. (A) Major fissures divide each hemisphere of cerebral cortex into frontal, parietal, temporal, and occipital lobes. Broca's area is located in the left frontal lobe adjacent to the part of the primary motor cortex that controls movement of mouth and lips (see Figure 1–25). Wernicke's area is located in the left temporal lobe adjacent to the auditory cortex. **(B)** Photograph of the brain from one of Broca's patients, Leborgne, who could only speak a single syllable, "tan." The lesion site is circled. Observation of similar lesions in language-deficient patients led Broca to propose the area to be essential for language production. (B, from Rorden C & Karnath H [2004] *Nat Rev Neurosci* 5:813–819. With permission from Macmillan Publishers Ltd.)

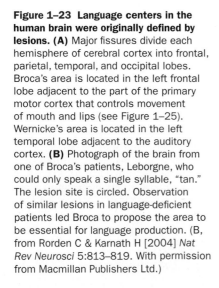
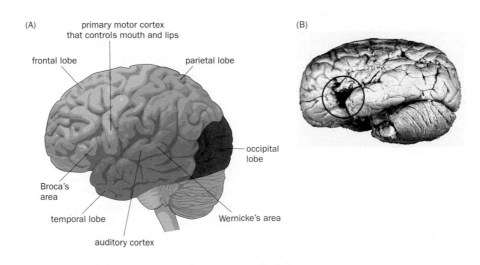

(A)

primary motor cortex that controls mouth and lips

frontal lobe

parietal lobe

occipital lobe

Broca's area

temporal lobe

Wernicke's area

auditory cortex

(B)

of the human left frontal lobe (Figure 1–23B) in patients who could not speak. This area was subsequently named **Broca's area** (Figure 1–23A). Karl Wernicke subsequently found that lesions in a distinct area in the left temporal lobe, now named **Wernicke's area** (Figure 1–23A), were also associated with defects in language. Interestingly, lesions in Broca's area and Wernicke's area give distinct symptoms. Patients with lesions in Broca's area have great difficulty in producing language, whether in speech or in writing, but their understanding of language is largely intact. On the other hand, patients with lesions in Wernicke's area have great difficulty in understanding language, but they can speak fluently, although often unintelligibly and incoherently. These findings led to the proposal that Wernicke's area is responsible for language comprehension whereas Broca's area is responsible for language production. These distinct functions are consistent with the locations of Broca's and Wernicke's areas being close to the motor cortex and the **auditory cortex** (the part of the cortex that first receives auditory sensory input), respectively (Figure 1–23A).

In the twentieth century, two important techniques—brain stimulation and brain imaging—have confirmed and extended findings from lesion studies, revealing in greater detail specific brain regions that carry out distinct functions. Brain stimulation has been used as a standard procedure to map specific brain regions in order to guide brain surgeries, such as severing axonal pathways to treat intractable **epilepsy**. (Epilepsy is a medical condition characterized by recurrent seizures, which are strong surges of abnormal electrical activity that affects part or all of the brain; we will study epilepsy in more detail in Chapter 11.) Such surgeries are often performed without general anesthesia (as brains do not contain pain receptor) so that patients' responses to brain stimulation can be assessed. Stimulation of Broca's area, for instance, causes a transient arrest of speech in patients. These brain stimulation studies have identified additional areas that interfere with language production.

One of the most remarkable advances in the late twentieth century is the non-invasive functional brain imaging of healthy human subjects as they perform specific tasks. The most widely used technique is **functional magnetic resonance imaging (fMRI)**, which monitors signals originating from changes in blood flow that are closely related to local neuronal activity. By allowing researchers to observe whole-brain activity without bias while subjects perform intensive tasks, such as those involving language, fMRI has revolutionized our understanding of specific brain regions that are implicated in specific functions. Such studies have confirmed that Broca's and Wernicke's areas are involved in language production and comprehension, respectively.

fMRI offers higher spatial resolution than do lesion studies, and thus has enabled researchers to ask more specific questions. For example, do bilingual speakers use the same cortical areas for their native and second languages? The answer depends on the cortical area in question and the age at which an individual acquires the second language. For late bilinguals who were first exposed to the second language after 10 years of age, representations of the native and second languages in Broca's area map to adjacent but distinct loci (**Figure 1–24**A). In early bilinguals who learned both languages as infants, the two languages map to the same locus in Broca's area (Figure 1–24B). Thus, the age of language acquisition appears to determine how the language is represented in Broca's area. It is possible that after a critical period during development, native language has already consolidated a space in Broca's area, such that a second language acquired later must utilize other (adjacent) cortical areas. By contrast, the loci in Wernicke's area that represent the two languages are inseparable by fMRI even for late bilinguals (Figure 1–24C).

1.11 The brain uses maps to organize information

Thanks to the combination of anatomical, physiological, functional, and pathological studies on human subjects, we now have a detailed understanding of the gross organization of the human nervous system (see Figure 1–8A). This is complemented by experimental studies in mammalian model organisms, which

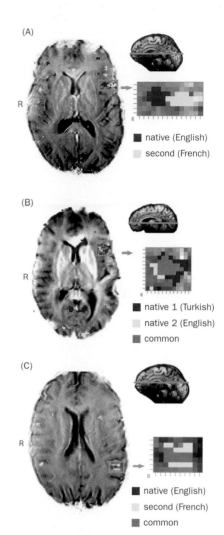

(A)
R
■ native (English)
□ second (French)

(B)
R
■ native 1 (Turkish)
□ native 2 (English)
■ common

(C)
R
■ native (English)
□ second (French)
■ common

Figure 1–24 Representations of native and second languages as revealed by functional magnetic resonance imaging (fMRI). The detection of blood-flow signals associated with brain activity by fMRI provides a means for imaging the brain loci where native and second languages are processed. In the brain scans on the left side of the figure, green rectangles highlight language-processing areas in the left hemisphere; the highlighted areas are magnified on the right. In the miniature brain profiles located at top right, the green lines represent the section plane visualized in the scanned images. R, right hemisphere. **(A)** In a late bilingual, the two languages are represented in separate, adjacent loci within Broca's area. **(B)** In an early bilingual who learned both languages from infancy, the language representations in Broca's area overlap. **(C)** In Wernicke's area, the representations of native and second languages overlap regardless of when the second language was acquired; panels A and C came from the same late bilingual subject. (Adapted from Kim KH, Relkin NR, Lee K-M et al. [1997] *Nature* 388:171–174. With permission from Macmillan Publishers Ltd.)

share this gross organization (see Figure 1–8B). We will study the organization and function of many regions of the nervous system in detail in subsequent chapters.

An important organizational principle worth emphasizing at the very beginning is that the nervous system uses maps to represent information. We have already encountered this phenomenon in our earlier discussion of the auditory and visual maps that barn owls use to target their prey. Two striking examples of maps in the human brain are the **motor homunculus** and **sensory homunculus** (**Figure 1–25**). These homunculi ('little men') were discovered through the use of electrical stimulation during brain surgeries to treat epilepsy, as discussed in Section 1.10. For example, stimulation of cortical neurons in specific parts of the primary motor cortex elicits movement of specific body parts on the contralateral side. (Movement of the left side of the body is controlled by the right side of the brain and vice versa.) Systematic studies revealed a cortical **topographic map** corresponding to movement of specific body parts: nearby neurons in the motor homunculus control the movements of nearby body parts. This map is distorted in

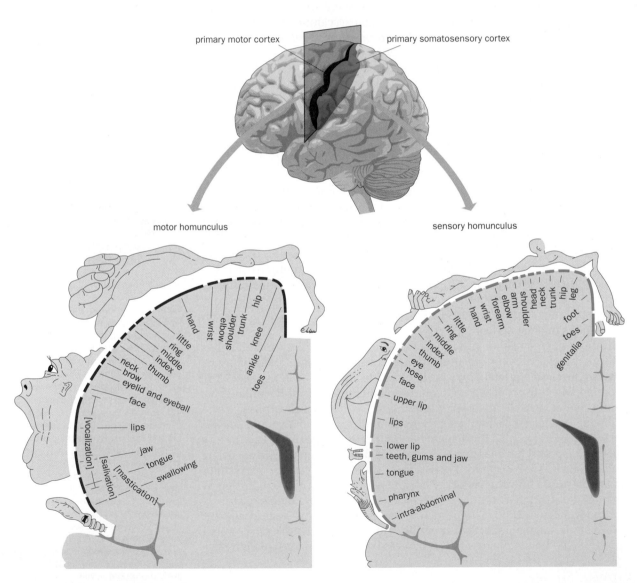

Figure 1–25 Sensory and motor homunculi. Top, the location in the brain of the primary motor cortex and the primary somatosensory cortex. Arrows indicate that sections along the plane with a 90° turn would produce the homunculi in the bottom panels. Bottom left, cortical neurons in the primary motor cortex control movement of specific body parts according to a topographic map. For example, neurons that control movement of lips and jaw are close together, but are distant from neurons that control finger movement. Bottom right, cortical neurons in the primary somatosensory cortex represent a topographic map of the body. For example, neurons that represent touch stimuli on the lips, jaw, and tongue are in adjacent areas, distant from neurons that represent touch stimuli on the fingers. (Adapted from Penfield W & Rasmussen T [1950] The Cerebral Cortex of Man. Macmillan.)

its proportion. The hand, in particular the thumb, is highly overrepresented, as are the muscles surrounding the mouth that enable us to eat and speak (Figure 1–25, bottom left). These distortions reflect disproportional use of different muscles. As we will learn in Chapter 8, the motor homunculus is a simplified representation of a more complex organization of the motor cortex for movement control.

In the adjacent primary somatosensory cortex, there is an equivalent sensory homunculus. Stimulation of specific areas in the primary somatosensory cortex elicits sensations in specific body parts on the contralateral side (Figure 1–25, bottom right). Again, cortical neurons in adjacent areas represent adjacent body parts, forming a topographic map in the primary somatosensory cortex. This preservation of spatial information is all the more striking, considering that the cortical neurons are at least three synaptic connections away from the sensory world that they represent (see Figure 1–20). Obvious distortions are also present, such that some areas of the body (for example, the hand and especially the thumb) are overrepresented compared to other body parts (for example, the trunk). These distortions reflect differential sensitivities of different body parts to sensory stimuli such as touch. Interestingly, cortical neurons in the sensory and motor homunculi that represent the same body part are physically near each other, reflecting a close link between the two cortical areas in coordinating sensation with movement.

Neural maps are widespread throughout the brain. We will learn more about maps in the visual system (Chapter 4); the olfactory, taste, auditory, and somatosensory systems (Chapter 6); the motor system (Chapter 8); and the hippocampus and **entorhinal cortex** (part of the temporal cortex overlying the hippocampus), where maps are used to represent spatial information of the outside world (Chapter 10). We will also study how neural maps are established during development (Chapters 5 and 7).

1.12 The brain is a massively parallel computational device

The brain is often compared to the computer, another complex system that has enormous problem-solving power. Both the brain and the computer contain a large number of elementary units, the neurons and the transistors, respectively, which are wired into complex circuits to process information. The global architectures of the computer and the brain resemble each other, consisting of input, output, central processing, and memory (**Figure 1–26**). Indeed, the comparison between the computer and the brain has been instructive to both computer engineers and neurobiologists.

The computer has huge advantages over the brain in speed and precision of basic operations (**Table 1–1**). Personal computers nowadays can perform elementary arithmetic operations such as addition at a speed of 10^{10} operations per second. However, the speed of elementary operations in the brain, whether measured by the action potential frequency or by the speed of synaptic transmission across a chemical synapse, is at best 10^3 per second. Furthermore, the computer can represent quantities (numbers) with any desired precision according to the bits (binary digits, or 0s and 1s) assigned to each number. For instance, a 32-bit number has a precision of 1 in 2^{32} or 4.2×10^9. Empirical evidence

Figure 1–26 Architectures of the computer and the nervous system. **(A)** Schematic illustration of the five classic components of a computer: input (such as a keyboard or a mouse), output (such a screen or a printer), memory (where data and programs are kept when programs are running), datapath (which performs arithmetic operations), and control (which tells datapath, memory, input, and output devices what to do according to the instructions of the program). Control and datapath together are also called the processor. **(B)** The nervous system can be partitioned in several different ways, one of which is shown here. In this four-systems model, the motor system controls the output of the nervous system (behavior). It is in turn controlled by three other systems: the sensory system that receives input from external environment, the cognitive system that mediates voluntary behavior, and the behavioral state system (such as wake/sleep) that influences the performance of all other systems. Arrows indicate extensive and often bidirectional connections among the four systems. As we will learn in Chapter 10, memory is primarily stored in the form of synaptic connection strengths of neural circuits in these systems. (A, adapted from Patterson DA & Hennessy JL [2012] Computer Organization and Design, 4th ed. Elsevier; B, adapted from Swanson LW [2012] Brain Architecture, 2nd ed. Oxford University Press.)

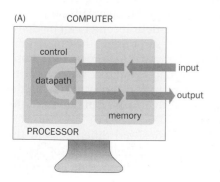

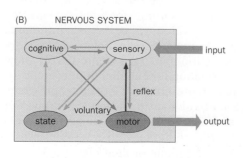

Table 1–1: Comparing the computer and the brain

Properties	Computer[1]	Human brain
Number of basic units	up to 10^9 transistors[2]	~10^{11} neurons; ~10^{14} synapses
Speed of basic operation	10^{10}/s	<10^3/s
Precision	1 in 4×10^9 for a 32-bit number	~1 in 10^2
Power consumption	10^2 watts	~10 watts
Processing method	mostly serial	serial and massively parallel
Input/output for each unit	1–3	~10^3
Signaling mode	digital	digital and analog

[1] Based on personal computers in 2008.

[2] The number of transistors per integrative circuit has doubled every 18–24 months in the past few decades; in recent years the performance gains from this transistor growth has slowed, limited by energy consumption and heat dissipation.

(Data from von Neumann [1958] The Computer and the Brain 1st ed. Yale University Press; Patterson & Hennessy [2012] Computer Organization and Design, 4th ed. Elsevier).

suggests that most quantities in the nervous system have variability of at least a few percent due to biological noise, or a precision of 1 in 10^2 at best. What then enables the brain to outperform the computer in many complex tasks and to do so with a lower power consumption and impressive speed and precision—such as following the trajectory of a tennis ball after it is served at a speed as high as 72 m/s (160 miles per hour), moving to the optimal spot on the court, positioning the arm, and swinging the racket to return the ball in the opponent's court, all within a few hundred milliseconds?

A notable difference between the computer and the brain is the methods by which information is processed within each system. Computer tasks are performed largely in **serial processing** steps; this can be seen by the way engineers program computers by creating a sequential flow of instructions, and by the fact that the operation of each basic unit, the transistor, takes a small number of inputs (1 to 3). For this sequential cascade of operations, high precision is necessary at each step because errors accumulate and amplify in successive steps. The brain also uses serial steps for information processing; in the tennis return example above, information flows from the eye to the brain and then to the spinal cord to control muscle contraction in the legs and arms. However, the nervous system also employs **massively parallel processing**, taking advantage of the large number of neurons and large number of connections each neuron makes. For instance, the moving tennis ball activates many photoreceptors in the eye, which transmit information to different kinds of bipolar and retinal ganglion cells (see Figure 1–16) as well as to inhibitory interneurons in the retina that we will learn about in Chapter 4; information regarding the location, direction, and speed of the ball is extracted by parallel circuits within two to three synaptic connections, and is transmitted in parallel by different kinds of retinal ganglion cells to the brain. Likewise, the motor cortex sends commands in parallel to control motor neuron activity and muscle contraction in the legs, the trunk, and the arms, such that the body and the arms are simultaneously well positioned for the incoming ball.

The massively parallel strategy is possible because each neuron collects inputs from and sends outputs to many other neurons—on average 10^3 for both inputs and outputs for a mammalian neuron. By using the divergent projection motif we discussed in Box 1–2, information from one neural center can be delivered to many parallel downstream pathways. At the same time, the convergent projection motif (see Figure 1–21A) allows many neurons that process the same information to send their inputs to the same postsynaptic neuron. While information represented by individual presynaptic neurons may be noisy, by taking the average of many presynaptic neurons, the postsynaptic neuron can represent the same information with much higher precision.

The computer and the brain also have similarities and differences in the signaling mode of their elementary units. The transistor utilizes **digital** signaling, which uses discrete values (0s and 1s) to represent information. The action potential in neuronal axons is also a digital signal, since it has the all-or-none property; this enables reliable long-distance information propagation. However, neurons also utilize **analog** signaling, which uses continuous values to represent information. In non-spiking neurons, output is transmitted by graded potentials that can transmit more information than can action potentials (we will discuss this in more detail in Chapter 4). Neuronal dendrites also use analog signaling to integrate up to thousands of inputs. Finally, signals for interneuronal communication are mostly analog, as synaptic strengths are continuous variables.

Another important property of the nervous system at play in the tennis return example is that the strengths of synaptic transmission between presynaptic and postsynaptic partners can change in response to activity and experience, as we will study in greater detail in Chapter 10. Repetitive training enables the circuits to become better configured for the tasks being performed, resulting in greatly improved speed and precision.

Over the past decades, engineers have taken inspiration from the brain to improve computer design. For instance, the principles of parallel processing and use-dependent modification of circuits have both been incorporated into modern computers. At the same time, neurobiologists can enhance their understanding of the working of the nervous system and potential strategies it employs to solve complex problems by looking at the brain from an engineering perspective.

GENERAL METHODOLOGY

The development and utilization of proper scientific methodology is essential to advancing our knowledge of neurobiology. We devote the last chapter of this book (Chapter 13) to discussing important methods for exploring the brain. You are highly recommended to study the relevant sections of Chapter 13 when these methods are first introduced and referred to in Chapters 1–12. We conclude this chapter by highlighting a few general methodological principles that will be encountered throughout the book.

1.13 Observations and measurements are the foundations for discovery

At the beginning of this chapter, we noted that asking the right question is often a crucial first step in making important discoveries. A good question is usually specific enough that it can be answered with clarity in the framework of existing knowledge. At the same time, the question's answer should have broad significance.

Careful observation is usually the first step in answering questions. Observations can be made with increasing resolution using ever-improving technology. Our discussion in this chapter about the organization of the nervous system offers good examples. Cells were first discovered because of the invention of the light microscope. The elaborate shapes of neurons were first deciphered because of the invention of the Golgi staining method. The debate between the neuron doctrine and the reticular theory was finally settled with electron microscopy. Inventing new ways of observing can revolutionize our understanding of the nervous system.

Whereas observations can give us a qualitative impression, some questions can be answered only by quantitative measurement. For instance, in order to answer the question of how sensory stimuli are encoded by nerve signals, researchers needed to measure the size, shape, and frequency of the action potentials induced by varying stimulus strengths. This led to the fundamental discovery that stimulus strengths are encoded by the frequency of action potentials, but not by their size or duration. The development of new measurement tools often precedes great discoveries.

Observation and measurement go hand in hand. Observations can be quantitative and often form the basis for measurement. For example, electron microscopy first enabled the visualization of the synaptic cleft. At the same time, it also permitted researchers to measure the approximate distance that a neurotransmitter must travel across a chemical synapse and estimate the physical size of the membrane proteins needed to bridge the two sides of a chemical synapse.

1.14 Perturbation experiments establish causes and mechanisms

While observation and measurement can lead to the discovery of interesting phenomena, they are often not adequate to investigate the underlying mechanisms. Further insight can be obtained by altering key parameters in a biological system and studying the consequences. We refer to these as **perturbation experiments**. Putting prisms on a barn owl is an example of a perturbation experiment. Artificial displacement of the visual map allowed scientists to measure the owl's ability to adjust its auditory map to match an altered visual map. We will encounter numerous perturbation experiments throughout the book.

Most perturbation experiments can be placed into one of the two broad categories referred to as 'loss-of-function' and 'gain-of-function.' In **loss-of-function experiments**, a specific component is taken away from the system. This type of experiment tests whether the missing component is *necessary* for the system to function. As an example, loss of speech caused by specific brain lesions in Broca's patients suggested that Broca's area is necessary for speech production. In **gain-of-function experiments**, a specific component is added to the system. Gain-of-function experiments can test whether a component is *sufficient* for the system to function in a specific context. As an example, electrical stimulations in epileptic patients indicated that activation of specific motor cortical neurons is sufficient to produce twitches of specific muscles. Both loss- and gain-of-function experiments can be used to deduce the causal relationships between components in biological processes.

Originating from the field of genetics, the terms 'loss-of-function' and 'gain-of-function' refer to the deletion or misexpression, respectively, of a specific gene to test its function in a biological process. These experiments are extremely powerful because genes are the basic units for regulating many biological processes, including neurobiological processes. In addition, genetic perturbations can be performed with high precision in many model organisms (see Sections 13.6–13.11). Indeed, we will introduce many examples of gene perturbation experiments that have shed light on the underlying mechanisms of a variety of neurobiological processes.

As the lesion and electrical stimulation examples discussed above illustrate, the concept of loss- and gain-of-function perturbation can be broadened beyond genes. In contemporary neuroscience, a central issue is the analysis of neural circuit function in perception and behavior, and here single neurons or populations of neurons of a particular type are the organizational and operational units. To assess the function of specific neurons or neuronal populations in the operation of a circuit, tools have been developed that can conditionally silence their activity (loss-of-function) or artificially activate them (gain-of-function) with high spatiotemporal precision (see Sections 13.10–13.12 and 13.23–13.25). Given that neurons can participate in neural circuits in a myriad of different ways (see Box 1–2), precise perturbation experiments are crucial in revealing the mechanisms by which neural circuits operate and control neurobiological processes of interest. These experiments also help establish causal relationships between the activities of specific neurons and the neurobiological processes they control.

With these basic concepts and general methodologies in hand, let us begin our journey!

SUMMARY

In this chapter, we have introduced the general organization of the nervous system and some fundamental concepts of neurobiology, framing these topics from a historical perspective. Neurons are the basic building blocks of the nervous system. Within most vertebrate neurons, information—in the form of membrane potential changes—flows from dendrites to cell bodies to axons. Graded potentials in dendrites are summed at the junction between the cell body and the axon to produce all-or-none action potentials that propagate to axon terminals. Neurons communicate with each other through synapses. In chemical synapses, presynaptic neurons release neurotransmitter in response to the arrival of action potentials, and postsynaptic neurons change their membrane potential in response to neurotransmitter binding to their receptors. In electrical synapses, ions directly flow from one neuron to another through gap junctions to propagate membrane potential changes. Neurons act in the context of neural circuits, and form precise connections with their synaptic partners to process and propagate information within circuits. Neural circuits in different parts of the brain perform distinct functions that range from sensory perception to motor control, and the nervous system employs a massively parallel computational strategy to enhance the speed and precision of information processing. In the rest of the book, we will expand our studies of these fundamental concepts in the organization and operation of the nervous system.

FURTHER READING

Books and reviews

Adrian ED (1947) Physical Background of Perception. Clarendon.

Bouchard TJ Jr & McGue M (1981) Familial studies of intelligence: a review. *Science* 212:1055–1059.

Knudsen EI (2002) Instructed learning in the auditory localization pathway of the barn owl. *Nature* 417:322–328.

Penfield W & Rasmussen T (1950) The Cerebral Cortex of Man. Macmillan.

Plomin R, DeFries JC, McClearn GE et al. (2008) Behavioral Genetics, 5th ed. Worth Publishers.

Ramón y Cajal S (1995) Histology of the Nervous System of Man and Vertebrates (1995 translation of the 1911 French version). Oxford University Press.

Swanson LW (2012) Brain Architecture: Understanding the Basic Plan, 2nd ed. Oxford University Press.

Tinbergen N (1951) The Study of Instinct. Oxford University Press.

von Neumann J (1958) The Computer & the Brain, 1st ed. Yale University Press.

Primary research papers

Adrian ED & Zotterman Y (1926) The impulses produced by sensory nerve endings: Part 3. Impulses set up by touch and pressure. *J Physiol* 61:465–483.

Harrison RG (1910) The outgrowth of the nerve fibers as a mode of protoplasmic movement. *J Exp Zool* 9:787–846.

Kim KH, Relkin NR, Lee KM et al. (1997) Distinct cortical areas associated with native and second languages. *Nature* 388:171–174.

Linkenhoker BA & Knudsen EI (2002) Incremental training increases the plasticity of the auditory space map in adult barn owls. *Nature* 419:293–296.

Merolla PA, Arthur JV, Alvarez-Icaza R et al. (2014) A million spiking-neuron integrated circuit with a scalable communication network and interface. *Science* 345:668–673.

CHAPTER 2

Signaling within Neurons

The zoologist is delighted by the differences between animals, whereas the physiologist would like all animals to work in fundamentally the same way.

Alan Hodgkin (1992), *Chance and Design: Reminiscences of Science in Peace and War*

Nervous systems can react rapidly to sensory stimuli. For example, a 5-day-old zebrafish larva shows its first response to a water pulse—a potential threat—within 3 milliseconds, and completely changes its direction of movement within 12 milliseconds, propelling itself away from danger (**Figure 2–1**). This escape behavior relies on the detection of mechanical force by sensory neurons, the transmission of this sensory information through interneurons to motor neurons, and the coordinated contraction or relaxation of appropriate muscles, all within several milliseconds. Because these escape behaviors are crucial for animal survival, such as avoiding a predator (**Movie 2–1**), the speed at which neurons communicate with each other has been subject to strong evolutionary selection pressure.

As we have introduced in Chapter 1, the nervous system uses electrical signals to transmit information within a neuron. Individual neurons are the basic units of the nervous system that receive, integrate, propagate, and transmit signals based on changes in the membrane potential, the difference in electrical potential across the plasma membrane of a neuron (see Figure 1–18). A typical neuron receives inputs at its dendrites and cell body from its presynaptic partners in the form of synaptic potentials. The neuron integrates these synaptic potentials along their path toward the axon initial segment, where action potentials are generated. Action potentials propagate along the axon to the neuron's presynaptic terminals, where they result in the release of neurotransmitters. Neurotransmitters act upon their receptors on the postsynaptic target neurons to produce synaptic potentials, thus completing a full round of neuronal communication.

In this chapter and Chapter 3, we will study the fundamental mechanisms of neuronal communication, focusing on three key steps: (1) the generation and propagation of action potentials, (2) the release of neurotransmitters by presynaptic neurons, and (3) the reception of neurotransmitters by postsynaptic neurons. Before we delve into these key steps of neuronal communication, we will first study the special properties of neurons as large cells with elaborate cytoplasmic extensions and as conductors of electrical signals, because understanding these properties is essential for our study of neuronal communication.

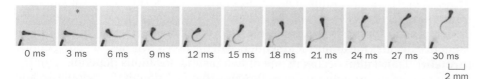

0 ms   3 ms   6 ms   9 ms   12 ms   15 ms   18 ms   21 ms   24 ms   27 ms   30 ms

2 mm

Figure 2–1 Rapid escape response of a zebrafish larva. Time-lapse images of the escape behavior of a zebrafish larva in response to a water pulse from the tube at bottom left. * indicates the first detectable response at 3 milliseconds (ms) after stimulus onset, which is difficult to see in the side-by-side frames shown here but is evident when observed in video clips. (From Liu KS & Fetcho JR [1999] *Neuron* 23:325–335. With permission from Elsevier Inc.)

CELL BIOLOGICAL AND ELECTRICAL PROPERTIES OF NEURONS

Neurons are the largest cells in animals. For instance, the cell body of a sensory neuron that innervates the toe is located in a dorsal root ganglion at about the waist level, but its peripheral branch extends down to the toe, and the ascending branch of the central axon extends up to the brainstem (see Figure 1–20); thus, this sensory neuron spans about 2 m for a tall person and 5 m for a giraffe. Many neurons have complex dendritic trees; for example, the dendritic tree of the cerebellar Purkinje cell (see Figure 1–11) has many hundreds of branches and receives synaptic inputs from tens of thousands of their presynaptic partners. The surface area and volume of axons or dendrites usually exceed those of the cell bodies by several orders of magnitude. These unique architectures enable rapid electrical signaling across long distances, and allow individual neurons to integrate information from many cells.

In order for neurons to initiate, integrate, propagate, and transmit electrical signals, they must continuously synthesize proteins and deliver them to the appropriate subcellular compartments. Thus, there are two ways in which communication happens within the neuron: by the transport of proteins and organelles along their long processes to get those components to the right part of the cell; and by electrical signals moving along those processes. These two might respectively be compared to sending a package through the mail, where a physical object needs to be delivered, and sending a text or email, where only the information is conveyed. Both are critical to the nervous system, but require very different mechanisms. In the following sections, we will study these mechanisms, beginning with the basic molecular and cell biology of the neuron.

2.1 Neurons follow the central dogma of molecular biology and rules of intracellular vesicle trafficking

As with all cells, macromolecule synthesis in neurons follows the **central dogma** of molecular biology, which states that information flows from DNA → RNA → protein (**Figure 2–2**, left). **Genes**, the genetic materials that carry the instructions for how and when to make specific RNAs and proteins, are located in the nucleus on **DNA** molecules, long double-stranded chains of nucleotides that contain the sugar deoxyribose, a phosphate group, and one of four nitrogenous bases: adenine (A), cytosine (C), guanine (G), or thymidine (T). **Transcription** is the process by which RNA polymerase uses DNA as a template to synthesize single-stranded RNAs (chains of ribose-containing nucleotides, in which uracil [U] replaces T); the part of the gene that serves as a template for RNA synthesis is called the gene's **transcription unit**. The pre-messenger RNAs (pre-mRNAs) produced during gene transcription carry information in their specific ribonucleotide sequence, which corresponds to the deoxyribonucleotide sequence of the transcription unit. Pre-mRNAs undergo a series of RNA processing steps. These include **capping** (adding a modified guanosine nucleotide to the 5′ end of the RNA), **RNA splicing** (removing RNA sequences that don't code for protein, called **introns**, and joining together the rest of the sequences, called **exons**), and **polyadenylation** (adding a long sequence of adenosine nucleotides to the RNA's 3′ end). The resulting mature **messenger RNAs** (**mRNAs**) are exported from the nucleus to the cytoplasm, where they are decoded by the ribosomes during protein synthesis (**translation**), such that the information in the mRNA sequence dictates the amino acid sequence of the newly synthesized polypeptide (**protein**). Translation occurs in one of two distinct locations depending on the destination of the protein products. Proteins that function in the cytoplasm and nucleus are synthesized on free ribosomes in the cytoplasm, whereas proteins destined for export from the cell (**secreted proteins**) or that span the lipid bilayer of a membrane (**transmembrane proteins**) are synthesized on ribosomes associated with the **endoplasmic reticulum** (**ER**), a network of membrane-enclosed compartments in eukaryotic cells (Figure 2–2, left).

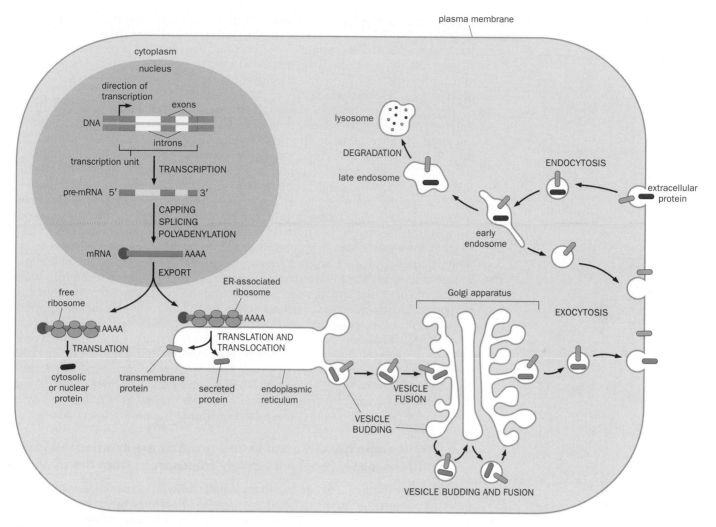

Figure 2–2 Schematic summary of the central dogma of molecular biology and intracellular vesicle trafficking. Left, in the nucleus, double-stranded DNA serves as a template for transcription to produce a pre-mRNA, which grows longer as nucleotides are added to the 3′ end. Pre-mRNA is processed by capping the 5′ end, splicing to remove introns and join exons, and polyadenylation at the 3′ end to produce mature mRNA, which is exported to the cytoplasm. mRNAs encoding cytoplasmic and nuclear proteins (purple) are translated on free ribosomes in the cytosol (left branch). mRNAs encoding secreted (blue) or transmembrane (green) proteins are translated on ribosomes associated with the endoplasmic reticulum (ER). **Bottom right,** after synthesis and translocation across the ER membrane, transmembrane and secreted proteins exit the ER through vesicle budding, pass through the Golgi apparatus via a series of vesicle fusion and budding steps, and are transported to the plasma membrane. Fusion of a vesicle with the plasma membrane (exocytosis) leads to the release of secreted proteins into the extracellular space and delivery of the transmembrane proteins to the plasma membrane. **Top right,** extracellular proteins (red) or transmembrane proteins on the plasma membrane can be internalized through vesicle budding from the plasma membrane (endocytosis) into early endosomes. The content can be recycled back to the plasma membrane through exocytosis, or can be delivered to late endosomes and lysosomes for degradation.

For most secreted proteins or transmembrane proteins destined for the plasma membrane, all or part of their sequence is translocated across the ER membrane as they are translated. Fully translated proteins then undergo a series of trafficking steps via **intracellular vesicles**, which are small, membrane-enclosed organelles in the cytoplasm of eukaryotic cells. Secreted and transmembrane proteins exit the ER via budding of vesicles from the ER membrane, and transit through the Golgi apparatus via a series of vesicle fusion and budding events. Eventually, vesicles that carry these proteins fuse with the plasma membrane in a step called **exocytosis**, so that secreted proteins are released into the extracellular space and transmembrane proteins are retained in the plasma membrane (Figure 2–2, bottom right).

In addition to exocytosis, neurons (like many other cells) can undergo a process called **endocytosis**. Endocytosis allows cells to retrieve fluid and proteins from the extracellular space, or transmembrane proteins from the cell's plasma membrane. Endocytosis products are first delivered to early **endosomes**, which are membrane-enclosed organelles that carry newly ingested materials and newly internalized transmembrane proteins. Proteins from early endosomes can either recycle back to the plasma membrane through exocytosis, or can be transported to late endosomes and **lysosomes**, which contain enzymes for protein degradation (Figure 2–2, top right). In Chapter 3, we will study specific examples of exocytosis and endocytosis in the context of presynaptic neurotransmitter release and postsynaptic neurotransmitter receptor regulation.

While obeying the central dogma and rules of intracellular vesicle trafficking, neurons also have special properties to accommodate their large size and the great distance between the tip of their axonal or dendritic extensions and the cell body (soma). We can ask a simple question: how does a specific protein, such as a neurotransmitter receptor or a protein associated with the presynaptic membrane, get to the dendritic tip or axon terminal? The answer to this 'simple' question is quite complex, and we are far from having complete answers for it. In principle, the corresponding mRNAs can be transported to the final destination before directing protein synthesis there. Alternatively, the protein can be synthesized at the soma and can either diffuse passively or be actively delivered to its final destination. For a transmembrane protein, delivery can take one of the following routes: (1) the intracellular vesicle that carries the protein can fuse with the plasma membrane at the soma, and the transmembrane protein can then diffuse on the plasma membrane to its destination; (2) the vesicle can be transported within dendrites or axons and can then fuse with the plasma membrane at its final destination; or (3) the protein can first be targeted to plasma membrane of one compartment (axon or dendrite), then be endocytosed and trafficked to the final destination, a process called **transcytosis**. All of these protein synthesis and transport mechanisms have been observed. Their relative prevalence depends both on the type of protein and the type of neuronal compartment to which that protein is targeted. We will study some of these mechanisms in the next two sections.

2.2   While some dendritic and axonal proteins are synthesized from mRNAs locally, most are actively transported from the soma

Substantial evidence has demonstrated that mRNAs encoding a subset of proteins are targeted to the dendritic processes, where they direct **local protein synthesis**. Electron microscopic studies revealed the presence of polyribosomes (clusters of ribosomes) in the dendrites, which are suggestive of mRNA being translated (**Figure 2–3A**). *In situ* hybridization studies, which can determine mRNA distribution in native tissues (see Section 13.13), have identified specific mRNAs that are present in dendrites (Figure 2–3B). These dendritically localized

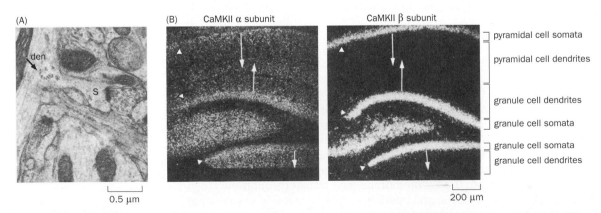

Figure 2–3 Evidence for dendritic protein synthesis. (A) Electron micrograph of part of a rat hippocampal granule cell (see Figure 1–12). At the left of the micrograph is a segment of a dendrite (den), with a dendritic spine (S) branching to the right. A cluster of ribosomes (arrow) is seen at the junction between the dendritic trunk and the spine. **(B)** Localization of mRNA encoding the α and β subunits of the Ca^{2+}/calmodulin-dependent protein kinase II (CaMKII) in a section of the rat hippocampus, detected by *in situ* hybridization using probes specific to each gene. Whereas mRNAs for the CaMKII β subunit are restricted to the layers that contain cell bodies (somata) of both granule cells and pyramidal cells, mRNAs for the CaMKII α subunit are distributed in the somata (arrowheads) as well as on the dendrites of these cells (indicated by the arrows leaving the cell body layers). See Figure 10–6 for a schematic of the hippocampus. (A, from Steward O & Levy WB [1982] *J Neurosci* 2:284–291. With permission from the Society for Neuroscience; B, from Burgin KE, Waxham MN, Rickling S et al. [1990] *J Neurosci* 10:1788–1798. With permission from the Society for Neuroscience.)

mRNAs encode a variety of proteins that are known to function in dendrites and postsynaptic specializations, such as the α subunit of the calcium/calmodulin-dependent protein kinase II (CaMKII; Figure 2–3B, left panel), cytoskeletal elements such as actin and microtubule-associated protein 2 (MAP2), as well as transmembrane receptors that detect neurotransmitters. We will revisit many of these proteins later in this chapter and in Chapter 3. The list of dendritically localized mRNA has greatly expanded in recent years, thanks to more sensitive methods of detecting mRNA such as high-throughput RNA sequencing (see Section 13.13). In addition to polyribosomes and mRNAs, ER and Golgi apparatus-like membrane organelles have also been observed in distal dendrites, enabling locally synthesized transmembrane and secreted proteins to go through the secretory pathway just as in the soma (see Figure 2–2). Finally, local translation from dendrites has been directly demonstrated using a number of *in vitro* preparations.

Local synthesis of proteins in dendrites can solve several problems for the cell: it produces proteins directly in the dendrites without the need of transporting them across long distances, it can cause the protein to be synthesized where it is most needed, and, perhaps most interestingly, it can allow very local regulation of where the protein is made in subcompartments of the dendritic tree. As will be discussed in later chapters, local translation in neuronal dendrites enables rapid synthesis of proteins in response to synaptic signaling; these newly synthesized proteins in turn help modify synaptic signals, which may result in regional remodeling of dendrites and synapses according to synaptic activity. Although much less studied compared to dendritic protein synthesis, local protein synthesis has also been found in developing and even in mature axons. The products of local protein synthesis in developing axons may play important roles in guiding axons toward their targets, a process we will study in greater detail in Chapters 5 and 7.

Even for proteins that are known to be synthesized in dendrites, the mRNA is more enriched in the soma (Figure 2–3B, left panel) suggesting that they are also synthesized in the soma. For many proteins, mRNAs appear exclusively in the soma (for example, Figure 2–3B, right panel). How do these proteins get to their final destination in dendrites and axons? This question has been explored primarily in axons because of the experimental ease of isolating axons at a greater distance from the cell bodies. For example, radioactively labeled amino acids can be injected into regions that house cell bodies of sensory or motor neurons, which extend long axons to distant sites. Newly synthesized proteins that incorporate these radioactively labeled amino acids can be isolated from their axons at different times after injection and at different distances from the cell bodies, and analyzed by biochemical methods such as gel electrophoresis to distinguish their identities (**Figure 2–4**). These studies have identified two major groups of proteins based on the speeds of their appearance in axons. The fast component travels at a speed of 50–400 mm per day (about 0.6–5 μm/s); this

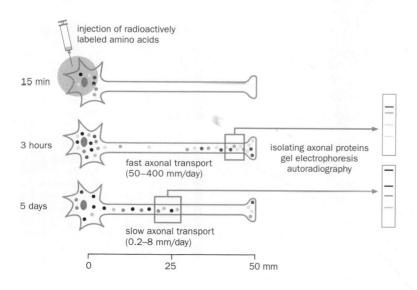

injection of radioactively labeled amino acids

15 min

3 hours

fast axonal transport (50–400 mm/day)

isolating axonal proteins gel electrophoresis autoradiography

5 days

slow axonal transport (0.2–8 mm/day)

0   25   50 mm

Figure 2–4 Studying axonal transport by following radioactively labeled proteins. Top, radioactively labeled amino acids were injected near the neuronal cell body, and were incorporated into newly synthesized proteins shortly after injection or were metabolized. Middle and bottom, at two time periods after the initial injection, proteins were isolated at specific segments of the axons (blue boxes), analyzed by gel electrophoresis, and visualized by autoradiography to distinguish their identities. (Adapted from Roy S [2014] *Neuroscientist* 20:71–81. With permission from SAGE.)

includes mostly transmembrane and secreted proteins. The slow component travels at a speed of 0.2–8 mm per day; this includes mostly cytosolic proteins and cytoskeletal components. These two modes are termed **fast axonal transport** and **slow axonal transport**, respectively. In addition to **anterograde** transport from the cell body to the axon terminal, some proteins, such as those imported via endocytosis, travel in the **retrograde** direction from the axon terminal back to the cell body; the speed of retrograde transport is similar to that of the fast anterograde axonal transport.

What mechanisms account for different modes of axonal transport? Theoretical studies indicate that diffusion within the axons is too slow to account for even slow axonal transport, suggesting that all these transport modes are active processes. Although less well studied, protein and mRNA transport into dendrites likely utilizes similar active processes. In order to understand the mechanisms that underlie active transport, and to appreciate why some proteins are transported to dendrites whereas others to axons, we need to examine the cytoskeletal organization of the neurons.

2.3 The cytoskeleton forms the basis of neuronal polarity and directs intracellular trafficking

Like all eukaryotic cells, neurons rely on two major cytoskeletal elements for structural integrity and motility—**filamentous actin** (**F-actin**, also called **microfilaments**) and **microtubules**. F-actin is composed of two parallel helical strands of actin polymers, whereas microtubules are hollow cylinders consisting of 13 parallel protofilaments made of α- and β-tubulin subunits (**Figure 2–5**). Most cells also have intermediate filaments, referring to cytoskeletal polymers with diameters between those of F-actin (~7 nm) and microtubules (~25 nm). The most prominent intermediate filaments in vertebrate neurons are the **neurofilaments**, which are concentrated in axons and provide stability to axons.

F-actin and microtubules are both polar filaments that have a plus end and a minus end with distinct properties. As in all cells, F-actin is mostly concentrated at the peripheral sites near the plasma membrane of neurons; these include along the axonal or dendritic processes, at the presynaptic terminals and postsynaptic

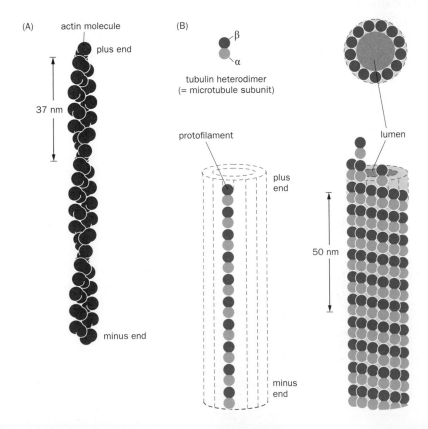

Figure 2–5 Filamentous actin and microtubules are two major cytoskeletal elements. (A) Schematic of filamentous actin (F-actin), which is composed of two helical strands with a repeating unit of 37 nm. **(B)** Schematic diagram of the microtubules. Left, α- and β-tubulin heterodimers assemble into longitudinal protofilaments. Right, the microtubule is made of 13 parallel protofilaments that form a tube with a hollow lumen (cross section is shown above). (Adapted from Alberts B, Johnson A, Lewis J et al. [2015] Molecular Biology of the Cell, 6th ed. Garland Science.)

sites such as dendritic spines in mature neurons, as well as at the growth cones of axons and dendrites in developing neurons. Actin subunits are added to the plus ends of F-actins closer to the plasma membrane, such that actin polymerization can cause membrane protrusions responsible for cell shape changes and cell movement. F-actin is not responsible for long-distance transport along dendritic or axonal processes. Microtubules fill the center of the axonal and dendritic processes, and therefore are the cytoskeletal element along which long-distance transport in neuronal processes occurs. In most nonneuronal cells, the more dynamic plus ends of microtubules point toward the periphery, whereas the more stable minus ends are usually at the center of the cells anchored in microtubule organization centers. The microtubule orientation is more complex in neurons, and indeed contributes to the distinction between dendrites and axons.

As we have introduced in Chapter 1, information generally flows from dendrites to axons. More than one century after Ramón y Cajal's original proposal, there is now a good understanding of the cell biological basis of **neuronal polarity**, which refers to the distinction between axons and dendrites. Generally speaking, subcellular structures and proteins that are related to the reception of information, such as receptors for neurotransmitters, are targeted to dendrites. By contrast, those structures and proteins that are related to the transmission of information, such as synaptic vesicles, are targeted to axons. These are largely achieved by an asymmetric cytoskeletal organization within the neuron; this enables **motor proteins**, which convert energy from ATP hydrolysis to movement along the cytoskeletal polymers, to transport specific cargos to specific destinations.

The organization of microtubules differs in axons and dendrites. Within an axon, parallel arrays of microtubules are oriented with their plus ends pointing away from the cell body and toward the axon terminal, following the 'plus-end-out' rule. However, dendrites have a mixed population of microtubules: the plus–end-out and minus-end-out populations are about equal at the proximal dendrites (**Figure 2–6**). To date, this rule applies to all vertebrate neurons examined in culture or *in vivo*. As discussed in Section 1.7, most invertebrate neurons have a single neurite that exits the cell body and then gives rise to dendritic and axonal branches. However, some sensory and motor neurons in the nematode *C. elegans* and the fruit fly *Drosophila* have distinct axonal and dendritic processes (similar to bipolar neurons in vertebrates). In invertebrate bipolar neurons thus far examined, dendrites appear to have mostly minus-end-out microtubules whereas axons have plus-end-out microtubules. Thus, although details may differ,

Figure 2–6 Cytoskeletal organization and motor proteins in axons and dendrites.
(A) In the axon of a typical vertebrate neuron, microtubules (green) are oriented with plus end (+) pointing toward the axon terminal, or 'plus end out'. In contrast, dendrites contain microtubules with both plus- and minus-end-out orientations. Cargos destined for axon terminals, such as synaptic vesicle precursors (cyan), are preferentially transported toward the axon terminals by plus-end-directed kinesins such as KIF1a (orange). Other cargos are transported by dynein (which moves toward the minus end of microtubules) or other kinesins in the axon and dendrites. For example, mRNA (cyan line) with associated protein complex (light brown) is transported by a kinesin into dendrites. Dynein and most kinesins are dimers with two heads (motor domains), whereas KIF1a acts as a monomer. F-actins (red) are distributed near the plasma membrane along the axons and dendrites (not shown) and are particularly enriched in the dendritic spine and presynaptic terminals. After leaving the microtubules, cargos may be further transported to their local destination by myosin-based movement along F-actins. **(B)** Quick-freeze deep-etch electron microscopy reveals the structure of the axonal cytoskeleton. The arrow points to a structure consistent with kinesin protein moving a cargo along the microtubule. (B, adapted from Hirokawa N, Niwa S & Tanaka Y [2010] *Neuron* 68:610–638. With permission from Elsevier Inc.)

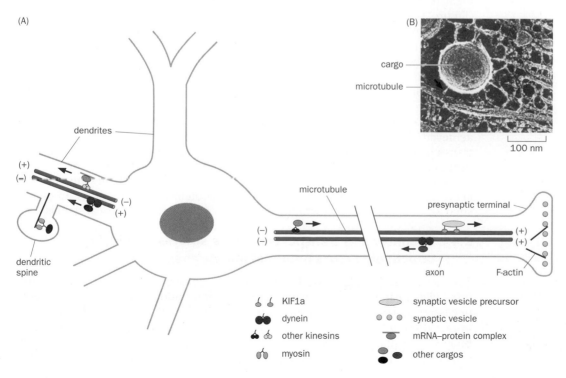

(A)

dendrites

(+)
(−)

dendritic spine

(−)
(+)

microtubule

(−)
(−)

axon

presynaptic terminal

(+)
(+)

F-actin

(B)

cargo

microtubule

100 nm

KIF1a

dynein

other kinesins

myosin

synaptic vesicle precursor

synaptic vesicle

mRNA–protein complex

other cargos

both vertebrate and invertebrate neurons share the principle that dendrites and axons differ in their microtubule orientations.

Two types of motor proteins move cargos along the microtubules: the **kinesin** family of proteins and the cytoplasmic **dynein**. Dynein is a minus-end-directed motor, which means that in the axon it transports cargos from the axon terminal back to the cell body. Most kinesins are plus-end-directed motors, which means that they transport cargos from the cell body to the axon terminal. In dendrites, both dyneins and kinesins can mediate bidirectional transport. Indeed, mRNAs in dendrites for local protein synthesis are transported by dynein and several kinesins on microtubules in the form of mRNA–protein complexes (Figure 2–6).

Dynein and kinesins have specifically associated proteins that link them to specific cargos, and some kinesins may bind directly to cargos. For example, synaptic vesicle precursors are transported from the cell body to the axon terminal by binding directly to a specific kinesin called KIF1a of the kinesin-3 subfamily (Figure 2–6). Certain types of kinesins (such as KIF1a) are highly enriched in axons whereas others are enriched in dendrites, adding to the specificity with which cargo is delivered to defined neuronal compartments. The asymmetric organization of the microtubule cytoskeleton and specific motor–cargo interactions play key roles in the establishment and maintenance of neuronal polarity. Other factors that contribute to neuronal polarity include diffusion barriers at the initial segment of axons for both cytosolic and membrane proteins. In Chapter 7, we will study how polarity is initiated in developing neurons.

In vitro motility studies of kinesins and dyneins indicate that they mediate fast axonal transport. For example, kinesins can move along microtubules at a speed of about 2 μm/s (see **Box 2–1**; **Movie 2–2**), in the same range as fast anterograde axonal transport (see Figure 2–4). Recent studies indicate that slow anterograde axonal transport is also mediated by kinesins. However, slow transport is characterized by much longer pauses between runs (periods when cargos are being transported), whereas fast axonal transport features longer runs and shorter pauses. During its brief runs, slow transport achieves speeds comparable to runs of fast axonal transport.

Microtubules are integral structural components of dendritic trunks and axons, and can be considered the highways that mediate long-distance transport in neurons. However, microtubules are usually absent from dendritic spines and presynaptic terminals. After cargos get off the microtubule highway at their approximate destinations, such as segments of a dendrite, F-actins direct local traffic utilizing a large family of **myosin** proteins as molecular motors (Figure 2–6). We will study the mechanism by which myosin–actin interactions produce motility in the context of muscle contraction in Section 8.1.

In summary, membrane proteins (associated with intracellular vesicles) and cytosolic proteins destined for dendrites or axons are delivered to their destinations by interacting with specific motor proteins, which enable them to be transported along microtubules for long distances and sometimes along actin filaments for local movements. Although we have an outline of the traffic rules, we still do not have complete answers to many questions. How is motor–cargo selection achieved? How is cargo loading and unloading regulated? What regulates the transition between pauses and runs? How are certain motors concentrated in dendrites or axons? While enriching our understanding of the cell biology of neurons, answers to these questions will also elucidate how each neuronal compartment acquires a unique assortment of specialized proteins to carry out its functions, such as receiving input, transmitting output, or propagating electrical signals.

2.4 Channels and transporters move solutes passively or actively across neuronal membranes

The mechanisms we have studied thus far are concerned with how proteins and organelles inside the cell move around, but don't address the question of how necessary molecules from outside the cell get into the cell across the plasma membrane. This requires a different type of transport: across the lipid bilayer. The lipid bilayer of the plasma membrane and membranes of intracellular vesicles is highly impermeable to most charged or polar molecules that are soluble in an aqueous

Box 2–1: How were kinesins discovered?

Breakthroughs in biology often result from utilizing new techniques and appropriate experimental preparations to address important, unsolved questions. The identification of kinesins is a good example to illustrate how the combination of these ingredients drives new discoveries. In the early 1980s, the invention of a new technique called <u>v</u>ideo-<u>e</u>nhanced <u>d</u>ifferential <u>i</u>nterference <u>c</u>ontrast (VE-DIC) microscopy enabled visualization of subcellular organelles in unstained live tissues. When applied to the squid giant axon, a specialized axon whose diameter can reach 1 mm (we will encounter this axon again in the studies of the action potential later in this chapter), organelles were observed to move along filament-like structures running along the

axon's length inside the plasma membrane (**Figure 2–7A**). Furthermore, when the axonal cytoplasm (axoplasm) was extruded from the axon, organelle movement along filaments in the axoplasm, free of axonal membrane, was similarly observed using VE-DIC microscopy.

The ability of the extruded axoplasm to support organelle movement opened the door to many experimental manipulations. For example, researchers could dilute the axoplasm to track organelle movement along a single filament over time and measure its speed (see Movie 2–2). The speed was found to be around 2 μm/s for transporting small organelles, in the same range as the fast axonal transport determined

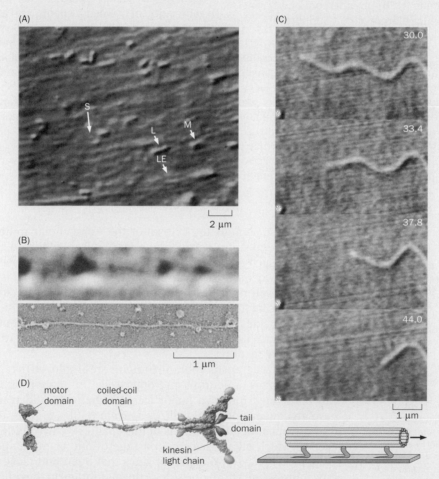

Figure 2–7 Discovery of the first kinesin. (A) An image of a segment from an intact squid giant axon, taken using video-enhanced differential interference contrast (VE-DIC) microscopy, shows horizontal linear elements (LE, which are microtubules) running in parallel with the axon. In video records, many small (S), medium (M), and large (L) organelles can be seen moving along the linear elements. **(B)** Top, VE-DIC image of organelles moving horizontally along what appears to be a single transport filament from extruded squid axoplasm; bottom, the same field of view in electron microscopy taken after VE-DIC study, confirming the presence of a single microtubule. The apparent diameter of the transport filament in the light microscope is inflated by diffraction to about ten times its true diameter. **(C)** Top, time-lapse movie (time at upper right indicates seconds, given to one decimal place) of a single microtubule moving rightward on a glass slide

to which a soluble fraction purified from squid axoplasm had been immobilized, in the presence of ATP. The object marked with an asterisk at the left bottom serves as a stationary marker. Bottom, an interpretive drawing: putative motor proteins from the squid axoplasm attach to the glass and to the microtubule. ATP hydrolysis by multiple motor proteins oriented in the same direction causes microtubule movement relative to the glass. **(D)** Molecular structure of kinesin-1, the kinesin from the squid giant axon. (A, from Allen RD, Metuzals J, Tasaki I et al. [1982] *Science* 218:1127–1129. With permission from AAAS; B, adapted from Schnapp BJ, Vale RD, Sheetz MP et al. [1985] *Cell* 40:455–462. With permission from Elsevier Inc.; C, adapted from Vale RD, Schnapp BJ, Reese TS et al. [1985] *Cell* 40:559–569. With permission from Elsevier Inc.; D, adapted from Vale RD [2003] *Cell* 112:467–480. With permission from Elsevier Inc.)

(Continued)

Box 2–1: How were kinesins discovered?

by tracking radioactively labeled proteins in vertebrate neurons *in vivo* (see Figure 2–4). Chemicals or drugs could be added to the axoplasm to study their effect on the motility assay. It was already known in the early 1980s that the actin-based motor, myosin, utilized ATP hydrolysis to power movement along F-actin. So researchers tested whether ATP hydrolysis is also required for axonal transport. They found that, when ATP was depleted from the axoplasm, or when a non-hydrolyzable ATP analog was added, motility was blocked, indicating that organelle movement along axoplasmic filaments indeed depends on ATP hydrolysis. Finally, following motility studies along a single transport filament using VE-DIC microscopy, electron microscopic analysis of the same filament (Figure 2–7B) provided unequivocal evidence that the individual filaments that support organelle movement are individual microtubules.

The above studies suggested the presence of motor proteins that utilize energy from ATP hydrolysis to move organelles in the squid axoplasm. Indeed, just as microtubules can support organelle movement, a soluble fraction of axoplasm from the squid giant axon containing the putative motor proteins, when immobilized on glass, could also cause individual microtubules to move in the presence of ATP (Figure 2–7C). Using this functional assay, biochemical purification of squid axoplasm led to the identification of a protein complex that could support microtubule movement on glass. Similar protein complexes purified from bovine and chick brains were found to exhibit properties similar to those of the protein complex from squid axoplasm. Members of this protein family were named kinesins (from the Greek *kinein*, to move).

We now know that kinesins are evolutionarily conserved molecular motors found in all eukaryotes. The kinesin complex originally purified from the squid axoplasm belongs to the kinesin-1 subfamily, consisting of two heavy chains and two light chains. Each heavy chain has an N-terminal globular domain that contains the microtubule-binding site and an ATPase, a long coiled-coil domain that mediates the dimerization of two heavy chains, and a C-terminal domain that binds to the light chain and to cargo (Figure 2–7D). Biochemical and biophysical studies have revealed detailed mechanisms of how kinesins move along the microtubules (**Movie 2–3**). Each mammalian genome has about 45 genes that encode different kinesins, many of which are expressed in neurons and are responsible for carrying different cargos to specific subcellular compartments of neurons (see Figure 2–6). As an indication of their importance, mutations in kinesins and proteins associated with kinesins (and dynein) in humans underlie a variety of neurological disorders.

Figure 2–8 Channels and transporters mediate passive and active transport. (A) A channel protein has an aqueous pore that allows solutes to pass through directly when the channel is open, whereas a transporter protein moves the solute across the membrane through sequential opening and closing of at least two gates. In the absence of energy input, solutes move down the electrochemical gradients (see Figure 2–9) through channels and transporters; these are called passive transport. **(B)** Transporter proteins can also mediate active transport by utilizing external energy to move solutes up the electrochemical gradient. (Adapted from Alberts B, Johnson A, Lewis J et al. [2015] Molecular Biology of the Cell, 6th ed. Garland Science.)

environment, such as the cytosol or the extracellular milieu. The lipid bilayers serve as essential compartmental boundaries to delineate cells and intracellular organelles, such as the ER, Golgi apparatus, and synaptic vesicles. Water-soluble molecules—including inorganic ions, nutrients, metabolites, and neurotransmitters, collectively referred to as **solutes**—cannot diffuse across the lipid bilayer and require specific transport mechanisms to move from one side of the lipid bilayer to the other side. Transport across the lipid bilayer is essential for many cellular functions, such as electrical signaling in neurons that we will discuss in detail in the following sections.

Specialized transmembrane proteins are employed to transport solutes across the membranes of neurons and other cells. These membrane transport proteins can be divided into two major classes: **channels** and **transporters**. Channels have an aqueous pore that allows specific solutes to pass directly through when they are open. In later sections of this chapter we will study many **ion channels**, each of which allows selective passage of one or more specific species of ion. Transporters have two separate gates that open and close sequentially, so that they move solutes from one side of the membrane to the other side (**Figure 2–8A**). In general, solutes move through open channels much more rapidly than they do through transporters.

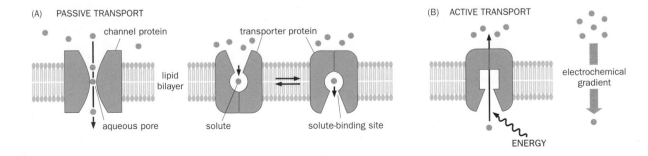

(A) PASSIVE TRANSPORT

channel protein

transporter protein

lipid bilayer

aqueous pore

solute

solute-binding site

(B) ACTIVE TRANSPORT

electrochemical gradient

ENERGY

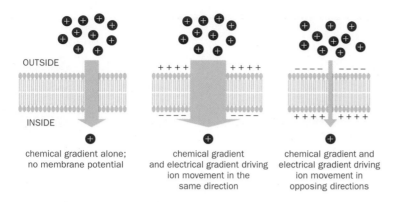

chemical gradient alone;
no membrane potential

chemical gradient
and electrical gradient driving
ion movement in the
same direction

chemical gradient and
electrical gradient driving
ion movement in
opposing directions

Figure 2–9 Electrochemical gradient of charged solutes such as ions. When there is no electrical potential difference across the membrane, the chemical gradient alone determines direction of ion movement, from high to low concentration (left). When there is an electrical potential difference across the membrane, the chemical gradient and the electrical gradient act together to determine the direction and the magnitude of the force governing ion movement. These two components of the electrochemical gradient may work in the same direction (middle) or in opposing directions (right). The size of the arrows symbolizes the magnitude of the force that drives ion movement. (Adapted from Alberts B, Johnson A, Lewis J et al. [2015] Molecular Biology of the Cell, 6th ed. Garland Science.)

All channels and many transporters only allow solutes to move in one direction across the membrane. For solutes that are not charged, they move from the side with higher concentration to the side with lower concentration, or down their **chemical gradient**. If solutes are electrically charged, such as ions, their movement across the membrane creates an **electrical gradient**, that is, a difference in the electrical potential across the membrane. The **electrochemical gradient**, which combines the chemical and the electrical gradients, determines the direction and magnitude of solute movement (**Figure 2–9**). When the electrical gradient and the chemical gradient are in the same direction, they enhance each other in facilitating the movement of the solute (Figure 2–9, middle). When these gradients are in opposing directions, they partially (Figure 2–9, right) or sometimes fully cancel out each other's effect. Transport of solutes down their electrochemical gradients does not require external energy, and is referred to as **passive transport** (see Figure 2–8A).

Some transporters can move a solute across the membrane against its electrochemical gradient (from low to high) using external energy; this process is called **active transport** (Figure 2–8B). The energy for active transport can come from one of the following sources. (1) Chemical reaction: the most frequent form is ATP hydrolysis, in which the transporter is an ATPase and chemical energy from ATP hydrolysis is used to drive conformational change of the protein. (2) Light: in this case energy is derived from photon absorption. Transporters driven by ATP hydrolysis or by light are also called **pumps**. We will discuss light-driven pumps in the context of the evolution of vision in Chapter 12. (3) Coupled transport: in this case the transporter moves two (or more) species of solutes together, and energy gained from transporting one species down the gradient is used to transport another species up the gradient. Coupled transporters (also called **cotransporters**) can be divided into two types: those that move solutes in the same direction are called **symporters**, and those that move solutes in opposite directions are called **antiporters** or **exchangers** (**Figure 2–10; Movie 2–4**).

Figure 2–10 Three types of active transport. An active transporter can be an ATP-driven pump, which utilizes energy from ATP hydrolysis to move solutes against their electrochemical gradient (left). It can be a light-driven pump, which derives its energy from photon absorption (middle). It can also be a coupled transporter, which derives its energy from transporting a second species of solute down its electrochemical gradient. Coupled transporters can be symporters or antiporters (exchangers) depending on whether the two solutes move in the same direction or in opposite directions. The cartoon at far right summarizes the electrochemical gradients of different solutes, with the larger number of a given symbol indicating the high end of an electrochemical gradient. The downward gray arrow shows the electrochemical gradient of the solute represented by yellow circles. (Adapted from Alberts B, Johnson A, Lewis J et al. [2015] Molecular Biology of the Cell, 6th ed. Garland Science.)

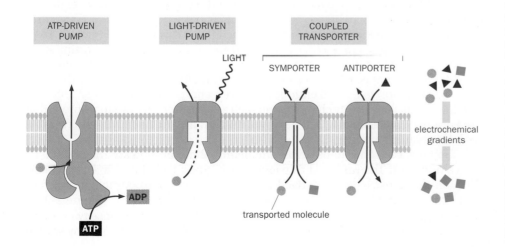

Figure 2–11 Measuring the resting potential of a neuron. (A) Schematic of recording the membrane potential of a neuron at rest. The oscilloscope measures the voltage difference (*V*, in millivolts) between the ground (the electrical potential in the extracellular environment) and the tip of the electrode. **(B)** Prior to the insertion of the electrode tip inside the cell, *V* is zero. After insertion, *V* drops to −75 mV, the resting potential of our model neuron.

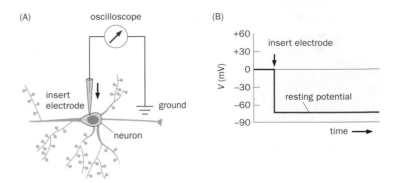

(By analogy to cotransporters, transporters that transport a single species of cargo are also called uniporters.) In the next section, we will encounter specific examples of an ATP-driven pump and cotransporters that play crucial functions in establishing electrochemical gradients of different ions across the plasma membrane, which form the basis of electrical signaling in neurons.

2.5 Neurons are electrically polarized at rest because of ion concentration differences across the plasma membrane and differential ion permeability

Electrical signals in neurons, as well as in other **excitable cells** (cells that produce action potentials, such as muscle cells), rely on a difference in electrical potential across the plasma membrane of the cell. This electrical potential difference between the inside of the cell and the extracellular environment is called the **membrane potential** of the cell. We can measure the membrane potential directly by inserting into the cell a microelectrode, a procedure called **intracellular recording**. A microelectrode for intracellular recording is usually made of glass with a very fine tip filled with a conducting salt solution, so that it makes electrical contact with the inside of a cell; at the other end, the electrode is connected via a wire to an amplifier and oscilloscope (see Figure 13–31 and Sections 13.20–13.21 for different methods of recording neuronal activity.) The membrane potential of a neuron at rest (the **resting potential**) is typically between −50 and −80 millivolts (mV) depending on the cell type (**Figure 2–11**). Thus, neuronal membranes are electrically polarized. A change in the electrical potential inside the cell toward a less negative value is termed **depolarization**. A change in the electrical potential inside the cell toward a more negative value is called **hyperpolarization**.

Neurons are electrically polarized because (1) ion concentrations differ between the two compartments separated by the plasma membrane: the intracellular and the extracellular environments; (2) the **permeability** of the plasma membrane to each of three major ions is different (as we will learn later, permeability is determined by the number of open ion channels that conduct specific ions). For a typical neuron, the concentrations of sodium ions (Na^+) and chloride ions (Cl^-) are ten- to twentyfold higher in the extracellular space relative to the inside of the cell, whereas the intracellular potassium ion (K^+) concentration is thirtyfold higher relative to the extracellular environment (**Figure 2–12A**), although these values vary with types of neurons and animals. At the resting state, the neuronal plasma membrane is less permeable to Na^+ and Cl^-, but is more permeable to K^+. (That is, K^+ crosses the neuronal membrane more readily than Na^+ or Cl^-.) In addition to the two **cations** (positively charged ions such as K^+ and Na^+) and one **anion** (negatively charged ions such as Cl^-), there are organic anions that are enriched intracellularly, but neuronal membranes are usually not permeable to these organic anions.

The ionic concentrations across the neuronal membrane are maintained by active transport. The principal transporter is a pump called an **Na^+-K^+ ATPase**. This ion pump uses energy derived from ATP hydrolysis to pump Na^+ outward and K^+ inward against their respective electrochemical gradients (Figure 2–12B; **Movie 2–5**), thus maintaining the concentration differences of these two important ions across the resting membrane. The Cl^- gradient is maintained by several

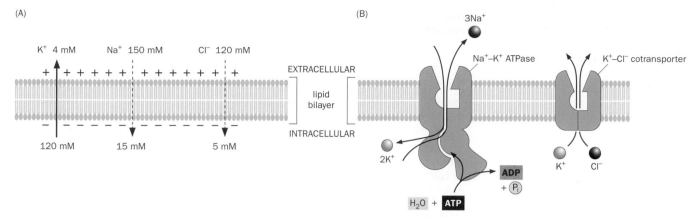

Figure 2–12 Ionic basis of the resting potential. (A) Numbers indicate the concentrations of K+, Na+, and Cl− inside and outside of a typical mammalian neuron, in millimoles per liter (mM). At the resting state, the membrane is permeable to K+ (arrow indicating the direction down the electrochemical gradient) but less permeable to Na+ and Cl− (dashed arrows). The resting membrane potential is largely determined by the balance between the chemical gradient that drives K+ outward, and the electrical gradient that drives K+ inward. + and − signs on opposite sides of the membrane indicate that the intracellular electrical potential is more negative than that of the extracellular environment. **(B)** Left, the Na+–K+ ATPase uses energy from ATP hydrolysis to transport three Na+ ions outward and two K+ inward each cycle, against their electrochemical gradients. The activity of the Na+–K+ ATPase maintains the intracellular concentrations of Na+ and K+ and thus the resting potential by counteracting the leak of Na+ and K+ across the resting membrane. Right, the K+–Cl− cotransporter utilizes energy released as K+ moves down its electrochemical gradient to move Cl− up its electrochemical gradient, helping maintain a concentration gradient of Cl− across the membrane.

cotransporters, such as the **K+–Cl− cotransporter** that couples K+ and Cl− export (Figure 2–12B).

In order to understand how chemical gradients and membrane potentials influence ion movement, let's first consider a hypothetical situation in which the membrane is only permeable to K+. (This situation applies to glia quite well.) The concentration difference across the membrane causes K+ to diffuse outward down the chemical gradient. As K+ flows outward, however, the intracellular compartment becomes more negatively charged, thus increasing the electrical potential difference across the membrane. This deters further outward K+ diffusion, because K+ is a positively charged ion. Eventually, when the chemical and electrical forces reach equilibrium (that is, no net K+ flow because the chemical gradient is balanced by the electrical gradient), the **equilibrium potential** of K+, E_K (the membrane potential when the electrical and chemical forces reach a balance), follows the **Nernst equation**:

$$E_K = \frac{RT}{zF} \ln \frac{[K^+]_o}{[K^+]_i}$$

where $[K^+]_o$ and $[K^+]_i$ are extracellular and intracellular K+ concentrations, R and F are two physical constants, T is the absolute temperature (in kelvin), z is the valence of the ion (+1 for K+), and ln is the natural logarithm. At room temperature, the expression RT/F is about 25 mV. Using the K+ concentration difference of our model neuron (Figure 2–12A), we can calculate the E_K to be about −85 mV, which is close to the resting potential.

The Nernst equation can also be used to determine equilibrium potentials of Cl− and Na+. Using the concentration difference across our model neurons (Figure 2–12A), we can determine that $E_{Cl} = -79$ mV, and $E_{Na} = +58$ mV. Note that despite having a chemical gradient opposite to that of K+, Cl− has a similar equilibrium potential to K+ because it has a negative charge (z = −1). A positive equilibrium potential for Na+ means that if the membrane were only permeable to Na+, the intracellular membrane potential would be positive relative to the extracellular environment. We will see the significance of this when we study the ionic basis of the action potential later in the chapter.

In reality, the resting potential of most neurons is slightly less negative than the K+ equilibrium potential, because the membrane is also somewhat permeable to Na+ and Cl−. When the membrane is simultaneously permeable to multiple ions, the resting membrane potential V_m at equilibrium (that is, when there is no net ion flow) can be calculated by using the **Goldman-Hodgkin-Katz (GHK) equation**:

$$V_m = \frac{RT}{zF} \ln \frac{p_K[K^+]_o + p_{Na}[Na^+]_o + p_{Cl}[Cl^-]_i}{p_K[K^+]_i + p_{Na}[Na^+]_i + p_{Cl}[Cl^-]_o}$$

where p_K, p_{Na}, and p_{Cl} are the permeabilities for K^+, Na^+, and Cl^-, respectively. In essence, the GHK equation states that each ion makes an independent contribution to the resting potential that is weighted according to the permeability of the resting membrane to that ion.

Because the membrane potential is generally not identical to the equilibrium potential for any single ion, there is a force that is tending to push each ion into or out of the cell. This is called the **driving force** for that ion, and it is equal to the difference between the membrane potential and the equilibrium potential for that ion. For example, as the equilibrium potential of Cl^- is generally very close to the resting potential, Cl^- flow is small at rest because of the small driving force, even though the membrane is somewhat permeable to Cl^-. If the membrane is only permeable to K^+ (that is, $p_{Na} = 0$, $p_{Cl} = 0$), then $V_m = E_K$ according to the GHK equation, and there is no driving force for K^+; the net current is zero despite the high permeability for K^+. But since p_{Na} is not negligible, this causes an inward leak of Na^+ down its electrochemical gradient, as both the chemical and electrical gradients favor Na^+ entry into the cell. This raises the membrane potential slightly above E_K and creates a driving force on K^+, causing K^+ to leak outwards. If unopposed, these leak currents would steadily decrease intracellular K^+ concentration and increase intracellular Na^+ concentration; however, they are counterbalanced by the active transport utilizing the Na^+–K^+ ATPase, which pumps Na^+ out and K^+ in (Figure 2–12B). The Na^+–K^+ ATPase resets the intracellular K^+ and Na^+ concentrations, thereby helps maintain the resting potential of neurons.

2.6 Neuronal plasma membrane can be described in terms of electrical circuits

While ions cannot cross the lipid bilayer of the neuronal plasma membrane, the intracellular and extracellular environments are both aqueous solutions that support ion movement. We can model how a neuron functions using terminology developed to describe an **electrical circuit**, that is, an interconnection of electrical elements that contains at least one closed current path. In this section, we introduce basic concepts of electrical circuits; these concepts are instrumental to our discussion of neuronal signaling in subsequent sections.

The simplest electrical circuit consists of two electrical elements: a battery and a resistor. The **battery** maintains a constant voltage (electrical potential difference) across its two terminals and provides the energy source. The **resistor** implements electrical resistance (that is, it opposes the passage of electric current) and produces a voltage across its two terminals when current flows through it (**Figure 2–13**A, left). The electric current (I, the flow of electric charge per unit time) that passes through the resistor follows **Ohm's law**:

$$I = \frac{V}{R}$$

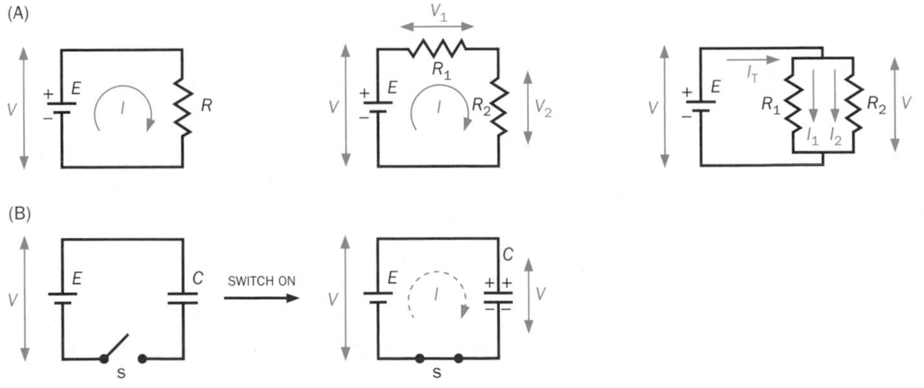

Figure 2–13 Electrical circuits with only resistors or capacitors. (A) Left, this electrical circuit consists of two elements, a battery (E) with a voltage V across its two terminals and a resistor with a resistance R. The current (I, arrow) flows outside the battery from the positive terminal (+, represented by the wider line of the battery) to the negative (−) terminal. Middle, two resistors (with resistance R_1 and R_2) are connected in series. The current that flows through them is the same. The sum of voltages across each resistor ($V_1 + V_2$) equals the voltage across the battery. Right, two resistors are connected in parallel. The voltage across each is the same as the voltage across the battery. The total current (I_T) equals the sum of the currents passing through each path ($I_1 + I_2$). **(B)** The circuit shown here consists of a battery (E), a capacitor (with a capacitance C), and a switch (s). When the switch is turned on, a transient current (dashed arrow) charges the capacitor until the voltage across it equals the voltage across the battery.

in which V is the voltage across the resistor and R is the **resistance** of the resistor. As the electrical wires that connect the battery and the resistor are assumed to have zero resistance, the voltage across the resistor is the same as the voltage across the battery. The units for I, V, and R are ampere (A), volt (V), and ohm (Ω), respectively.

When two resistors are connected in series (Figure 2–13A, middle), the current that passes through each resistor is the same; the voltage across both resistors is the sum of voltages across each resistor, or $V = V_1 + V_2$; and the combined resistance is $R = R_1 + R_2$. An equivalent but more widely used measure of a resistor in electrophysiology is **conductance** (g), which is the inverse of resistance: $g = 1/R$. Thus, when two resistors are connected in series, $1/g = 1/g_1 + 1/g_2$. When the two resistors are connected in parallel (Figure 2–13A, right), the voltages across each resistor are the same; the total current is the sum of the currents that pass through each resistor, or $I = I_1 + I_2$; the combined resistance follows the formula $1/R = 1/R_1 + 1/R_2$, and the combined conductance can be calculated as $g = g_1 + g_2$. The unit for conductance is siemens (S). It follows from the definition of conductance that Ohm's law can also be expressed as:

$$I = gV$$

Note that a resistor is at the same time a **conductor** of electric current, and these two terms are used interchangeably depending on the context; a resistor with high resistance is a poor conductor, and a resistor with low resistance is a good conductor.

We can relate these simple electrical circuits to what we have learned so far about the neuron. The lipid bilayer is an **insulator**, which is a resistor with infinite resistance and thus does not allow electric current to pass through. As noted in Section 2.5, the plasma membrane is a not perfect insulator—even in the resting state, ions can leak through the membrane via specific channels. These ion channels can be modeled as parallel current paths, each consisting of a resistor with a specific resistance and a battery equivalent to the equilibrium potential of the ion. We will discuss this model further in Section 2.7. We will also encounter resistors connected in series when we study propagation of electrical signals along neuronal fibers (dendrites and axons) in Section 2.8.

Another important electrical element is a **capacitor**, consisting of two parallel conductors separated by a layer of insulator in between. A capacitor is a charge-storing device, as it does not allow current to pass through the insulator layer. The lipid bilayer of the plasma membrane, along with the extracellular and intracellular compartments, is an excellent example of a capacitor. In a simple circuit that consists of a battery and a capacitor (Figure 2–13B), when the switch is turned on, current flows from the battery to the capacitor until the capacitor is charged to a voltage that is the same as the battery. Positive charges accumulate on one conductor, while negative charges accumulate on the other conductor; this is how charges are stored. The **capacitance** (C), or the ability of a capacitor to store charge, is defined as $C = Q/V$, where Q is the electric charge stored when the voltage across the capacitor is V. The unit of capacitance is farad (F), and the unit of charge is coulomb (C). When two capacitors are connected in series, the combined capacitor (C) follows the formula $1/C = 1/C_1 + 1/C_2$. When two capacitors are connected in parallel, $C = C_1 + C_2$.

In theory, when a circuit has no resistance (Figure 2–13B), the capacitor is charged instantaneously when the switch is turned on. In reality, circuits always have resistance. In a circuit that contains both resistors and capacitors (an **R-C circuit**), the current that flows through the resistor and the capacitor changes over time after the switch is turned on. The product of resistance and capacitance has the unit of time and is called the **time constant** (designated as τ). The time constant defines how quickly capacitors (such as the plasma membrane) charge or discharge over time in response to external signals, such as a sudden change of current flow (as would result from the opening of channels). The larger the time constant, the longer it takes to charge a capacitor and the more an electrical signal is spread out over time.

Let's examine two examples of R-C circuits to help clarify the important concept of a time constant (see **Box 2–2** for a quantitative treatment of this subject).

Box 2–2: A deeper look at R-C circuits

In Section 2.6 we encountered serial and parallel R-C circuits (Figure 2–14). For students with a background in differential equations, we discuss here how the temporal dynamics of these circuits are derived.

For a serial R-C circuit (Figure 2–14A), the sum of the voltages across the capacitor and the resistor should equal the voltage across the battery, or $V = V_C + V_R$. Since current is a flow of electrical charge (Q) over time (t), at any given time after the switch is turned on, $I(t) = dQ/dt$. Since $Q = CV_C$, we have $I(t) = C\,dV_C/dt = C\,dV/dt - C\,dV_R/dt$. Since V is the constant voltage across the battery and does not change over time, $dV/dt = 0$. According to Ohm's law, $V_R = I(t)R$. Thus, we have the differential equation $I(t) = -RC\,dI(t)/dt$; the solution of this differential equation gives:

$$I(t) = I_0 e^{-\frac{t}{RC}}$$

in which $I(t)$ is the current at time t; $I_0 = V/R$, which is the current at time 0 when the switch is turned on (since $e^0 = 1$); e is the base of the natural logarithm; and RC is the time constant (see Section 2.6). This formula is the basis for the exponential decay of current (Figure 2–14A, middle). The current decays to 37% ($1/e$), 14% ($1/e^2$), or 5% ($1/e^3$), of the initial current I_0 at times equivalent to one, two, or three time constants. We can also determine the time course of capacitor charging (Figure 2–14A, right):

$$V_C = V - V_R = V - RI(t) = V\left(1 - e^{-\frac{t}{RC}}\right)$$

When the parallel R-C circuit is connected to a constant current source (after the switch is turned on in Figure 2–14B, left), there is a redistribution of current over time from the capacitor path (I_C) to the resistor path (I_R), but their sum (I_T) is constant. Also, the voltage (V) across the resistor and the capacitor are the same, which equals $I_R R$. Thus, we have $I_R(t) = I_T - I_C(t) = I_T - dQ/dt = I_T - C\,dV/dt = I_T - RC\,dI_R(t)/dt$. Solving this differential equation gives:

$$I_R(t) = I_T\left(1 - e^{-\frac{t}{RC}}\right)$$

$$V(t) = I_T R\left(1 - e^{-\frac{t}{RC}}\right)$$

$$I_C(t) = I_T e^{-\frac{t}{RC}}$$

The second and third functions above are graphically represented in the middle and right panels of Figure 2–14B. $V(t)$ increases to 63% ($1 - e^{-1}$), 86% ($1 - e^{-2}$), and 95% ($1 - e^{-3}$) of the maximal V at times equivalent to one, two, or three time constants.

Note that we connect the parallel R-C circuit with a constant current source rather than a battery because this is a better model of what happens in neuronal membrane during electrical signaling: the opening of ion channels that provides current flow. It is less interesting to connect a parallel R-C circuit to a battery with a constant voltage, as the capacitor will be charged instantaneously and the circuit is reduced to a circuit with a resistor only, as in Figure 2–13A.

In an R-C circuit where a resistor and a capacitor are connected in series with a battery (**Figure 2–14**A, left), once the switch is turned on, the current decays exponentially over time from the initial value (Figure 2–14A, middle), while the voltage across the capacitor increases exponentially as it approaches the voltage

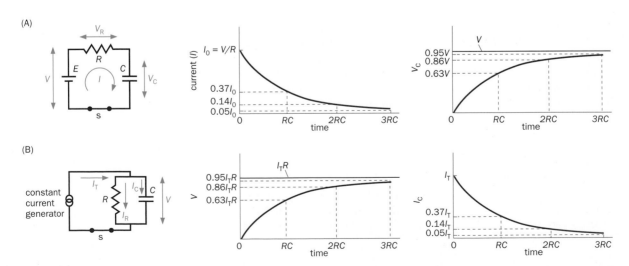

Figure 2–14 Electrical circuits with resistors and capacitors (R-C circuits) exhibit temporal dynamics. (A) Left, a resistor (with a resistance R) and a capacitor (with a capacitance C) are connected in series, supplied with a battery (E). Middle, after the switch (s) is connected, the current (I, following the direction of the arrow) exhibits an exponential decay, with a time constant equal to the product RC. Values of I at times (t) equal to RC, $2RC$, and $3RC$ are shown. Right, the voltage across the capacitor (V_C) gradually approaches the voltage

of the battery (V). Also indicated are values of V_C at $t = RC$, $2RC$, and $3RC$. **(B)** Left, a circuit with a resistor and a capacitor connected in parallel, supplied with a constant current source (with a constant total current I_T). Middle and right, after the switch (s) is connected, V gradually increases to approach its maximal value of $I_T R$ (middle), whereas I_C (right) exhibits exponential decay. V and I_C values are indicated at $t = RC$, $2RC$, and $3RC$.

of the battery (Figure 2–14, right). In an *R-C* circuit where a resistor and a capacitor are connected in parallel and to a constant current source (Figure 2–14B, left), once the switch is turned on, while the sum of the currents (I_T) remains constant throughout, the current that flows through the resistor (I_R) and the voltage ($V = I_R R$) increases over time (Figure 2–14B, middle), and the current that flows through the capacitor (I_C) decreases over time (Figure 2–14B, right).

In both examples, the time constant, *RC*, serves as the unit on the time (x) axis and defines how rapidly electrical signals change over time. For example, in the serial *R-C* circuit, the current decays to 37%, 14%, and 5% of the peak value at times equal to one, two, or three time constants (Figure 2–14A, middle). Likewise, in the parallel *R-C* circuit, the voltage reaches 63%, 86%, and 95% of the peak value at times equal to one, two, or three time constants (Figure 2–14B, middle; see Box 2–2 for how these values are derived).

Whereas the serial *R-C* circuit is more common in electronics, the parallel *R-C* circuit is more widely used in neurobiology, as it is an excellent description of the neuronal plasma membrane. The ion channels function as resistors, and the lipid bilayer together with the extracellular and intracellular environments act as a capacitor, storing electrical charge in the form of ions accumulating near the surface of the membrane. While the membrane capacitance per unit area is mostly constant (~1 μF/cm^2), the membrane resistance can change significantly with membrane potential and time. We will now apply these concepts to examine ion flow across neuronal plasma membrane.

2.7 Electrical circuit models can be used to analyze ion flows across glial and neuronal plasma membrane

Having introduced basic concepts of electrical circuits, we discuss two examples below to illustrate how these electrical circuit models can help us understand current flow across the plasma membrane.

Let's first suppose that the membrane is only permeable to K$^+$, which is a good approximation for glia. This is equivalent to a circuit consisting of two parallel paths, a K$^+$ conducting path and a membrane capacitance path (**Figure 2–15**A). The membrane capacitance path is symbolized by a capacitor (C_m). The K$^+$ conducting path has two electrical elements, a resistor and a battery representing the equilibrium potential of K$^+$. Since resistance is the inverse of conductance, the resistor is symbolized by $1/g_K$, where g_K is the conductance of the K$^+$ path. The battery symbolizes the electrochemical gradient that drives K$^+$ movement along this path, with the voltage across the battery equaling E_K, the equilibrium potential of K$^+$. We can determine the membrane potential, V_m, from the K$^+$ path as the sum of voltage across the resistor (which is I_K/g_K according to Ohm's law) and voltage across the battery, or $V_m = I_K/g_K + E_K$. At the resting state, the influx and efflux of K$^+$ balances out, or $I_K = 0$. Thus, $V_m = E_K$, that is, the resting membrane potential is the same as the equilibrium potential of K$^+$.

Now let's consider a more realistic situation for neurons, in which the membrane is permeable to Cl$^-$ and Na$^+$ in addition to K$^+$. We can add two parallel paths

Figure 2–15 Electrical circuit models of the neuronal plasma membrane. (A) In a simplified model where the membrane is permeable only to K$^+$, the plasma membrane consists of two parallel paths, a membrane capacitance (C_m) path and a K$^+$ path. The resistance of the resistor is indicated by the inverse of the conductance, $1/g_K$. The K$^+$ path also has a battery that corresponds to the equilibrium potential of K$^+$ (E_K). The battery is positive on the extracellular side because the equilibrium potential of K$^+$ is negative intracellularly. V_m is the membrane potential. At rest, since there is no net current flow, $V_m = E_K$. **(B)** In a more realistic model for neurons, the plasma membrane can be considered as four parallel paths: one path for membrane capacitance plus one path each for K$^+$, Cl$^-$, and Na$^+$. Arrows indicate the directions of currents within each path according to the equilibrium potential of each ion and the resting potential of our model neuron in Figure 2–12. Note that while Cl$^-$ ions flow inward (as the resting potential is less negative than the equilibrium potential of Cl$^-$ in our model neuron), the current carried by Cl$^-$ is shown flowing outward; this is because electrical circuit diagrams conventionally indicate the current carried by positive charges. As a result, the current carried by negatively charged ion (like Cl$^-$) is indicated as occurring in the direction opposite the actual direction of ion movement. (B: Adapted from Hodgkin AL & Huxley AF [1952] *J Physiol* 117:500–544.)

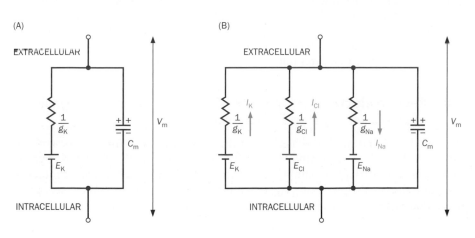

(A)

EXTRACELLULAR

$\frac{1}{g_K}$ E_K C_m V_m

INTRACELLULAR

(B)

EXTRACELLULAR

$\frac{1}{g_K}$ I_K $\frac{1}{g_{Cl}}$ I_{Cl} $\frac{1}{g_{Na}}$ I_{Na} C_m V_m

E_K E_{Cl} E_{Na}

INTRACELLULAR

to Figure 2–15A, one for Cl⁻ and one for Na⁺ (Figure 2–15B). Below, we demonstrate how V_m can be determined based on the conductance and equilibrium potentials for each ion. In this parallel circuit, the voltage across each path is V_m, so we have three equations:

$$V_m = \frac{I_K}{g_K} + E_K \tag{1}$$

$$V_m = \frac{I_{Cl}}{g_{Cl}} + E_{Cl} \tag{2}$$

$$V_m = \frac{I_{Na}}{g_{Na}} + E_{Na} \tag{3}$$

From the above three equations, we have $V_m(g_K + g_{Cl} + g_{Na}) = E_K g_K + E_{Cl} g_{Cl} + E_{Na} g_{Na} + I_K + I_{Cl} + I_{Na}$. At rest, the net current that flows across the membrane should be zero. As the membrane potential is constant, the current flow in the capacitance branch is also zero. Thus,

$$I_K + I_{Cl} + I_{Na} = 0 \tag{4}$$

Accordingly, V_m can be derived as

$$V_m = \frac{E_K g_K + E_{Cl} g_{Cl} + E_{Na} g_{Na}}{g_K + g_{Cl} + g_{Na}}$$

where g and E are the conductance and the equilibrium potential for each ion, respectively. This is in fact the circuit model equivalent of the Goldman-Hodgkin-Katz equation introduced in Section 2.5. This is a more useful formula because conductance and equilibrium potential are easier to determine experimentally than permeability and the absolute ionic concentrations used in the formula in Section 2.5. Note that conductance and permeability are both used to describe how easy it is for an ion to flow cross the plasma membrane, and are often used interchangeably. But there is a subtle difference. Permeability is an intrinsic property of the membrane (reflecting the number of opened channels, as we will learn later), and does not vary whether the ions to be conducted are present or not, whereas conductance depends not only on the permeability but also on the presence of ions.

Once we have determined V_m, we can also determine the currents within each parallel path:

$$I_K = g_K(V_m - E_K)$$

$$I_{Cl} = g_{Cl}(V_m - E_{Cl})$$

$$I_{Na} = g_{Na}(V_m - E_{Na})$$

Note that the values in parentheses represent the driving force for each ion as we defined in Section 2.5. Thus, the current each ion carries is the product of the conductance and the driving force for that ion. As we will learn later in this chapter, g_{Na} and g_K change as a function of the membrane potential, which underlies the production of action potentials. We will also learn in Chapter 3 that synaptic transmission is mediated by a change in the postsynaptic membrane conductance in response to neurotransmitter release from the presynaptic terminal.

2.8   Passive electrical properties of neurons: electrical signals evolve over time and decay across distance

Having introduced the ionic basis of resting potentials and the electrical circuit model of the neuronal plasma membrane, we are now ready to address two key questions in electrical signaling: how neurons respond to electrical stimulation and how electrical signals propagate within neurons. We start with observations from an idealized experiment on a neuronal fiber (a dendrite or an axon), which summarizes the results from real experiments across different preparations and is a good approximation of the properties of neuronal dendrites. In this experiment, an electrode that is connected to a current source is inserted into the neuronal fiber, such that it can 'inject' electric current into the neuronal fiber at the command of the experimenter. We call this electrode a **stimulating electrode**. Inserted into the membrane right next to the stimulating electrode is a **recording electrode** (*a*), which is connected to an amplifier and oscilloscope so it can record the membrane potential change in response to the current injection from the stimulating electrode. Two additional recording electrodes (*b* and *c*) are inserted at different distances along the fiber from the stimulating electrode to record membrane potential spread at distant sites (**Figure 2–16**A; **Movie 2–6**).

We start by injecting a small depolarizing current (that is, injecting positive charges into the neuron) in the form of a rectangular pulse (Figure 2–16B, left). The injected current will flow through the electrode across the membrane and along the inside of the fiber. We observe that the membrane potential at the recording electrode *a* becomes *gradually* depolarized at the beginning of the current pulse, and it returns *gradually* to the resting potential at the end of the current pulse. At the more distant sites, membrane potentials recorded by electrodes *b*

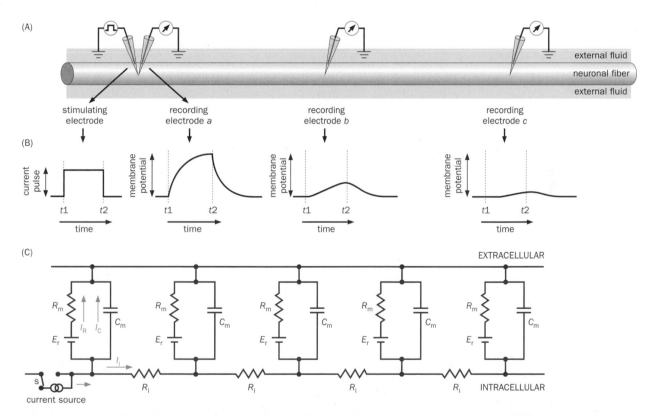

Figure 2–16 Passive electrical properties of neurons observed from an idealized experiment. (A) Illustration of the experimental preparation. A stimulating electrode provides a source of electrical signals in the form of a rectangular current pulse. Three recording electrodes are inserted into the neuronal fiber at different distances from the stimulating electrode. **(B)** When the stimulating electrode delivers a rectangular depolarizing pulse (left), the membrane potential changes at the sites of the three recording electrodes as illustrated (right). Dashed lines represent times (t_1 and t_2) that are aligned with the onset and offset of the current pulse. The *y* axes of the three membrane potential plots have the same scale. Two properties are evident: (1) membrane potential changes are gradual in response to the sharp current pulse, and (2) the magnitudes of the membrane potential changes decay across distance. **(C)** An electrical circuit model of the neuronal fiber. Each membrane segment is approximated as a parallel *R-C* circuit, with a membrane capacitance (C_m), a membrane resistance (R_m) that integrates K$^+$, Cl$^-$, and Na$^+$ conductances, and a battery representing the resting membrane potential (E_r). These segments are joined internally by resistors (with resistance R_i), reflecting internal or axial resistance within the neuronal fiber; the resistance of the external fluid is approximated as 0. Current injection is modeled by transiently connecting the intracellular side with a constant current source through the switch (s). Arrows indicate the direction of current flow caused by injecting positive charges from the current source, including current into the resistor branch (I_R) and capacitance branch (I_C) across the membrane, as well as current that flows within the neuronal fiber (I_i). (Adapted from Katz B [1966] Nerve, Muscle, and Synapse. With permission from McGraw Hill.)

and *c* exhibit similar gradual changes. However, the magnitudes of the membrane potential changes are much diminished (Figure 2–16B, middle and right).

We use an electrical circuit model of the fiber (Figure 2–16C) to explain these results. Each segment of the membrane can be simplified as a parallel *R-C* circuit (see Figure 2–15B), with membrane conductance of different ions in Figure 2–15 combined as R_m, the resting membrane potential as a battery E_r, and the membrane capacitance as C_m. Different segments of the process can be considered as parallel *R-C* circuits, linked by an internal (or axial) resistance (R_i) for ion movement along the longitudinal axis in the fiber interior. Ions can flow more freely in the large extracellular environment; therefore the extracellular resistance is often approximated as zero.

Let's consider the first observation: membrane potentials change gradually in response to a step of injected current. In addition to causing ion flow across the membrane, represented by I_R in the circuit diagram, part of the injected current flows through the capacitance path to charge the membrane capacitor (I_C). As discussed in Section 2.6 and Box 2-2, I_R follows an exponential curve with the product $R_m C_m$ as the time constant, as does the membrane potential change (which is equal to $I_R R_m$ according to Ohm's law; see Figure 2–14B). The smaller the time constant, the faster the membrane potential changes in response to current injection. This experiment reveals a general property of electrical signaling in neurons. Because of the membrane capacitance, electrical signals evolve over time even when current injection is constant, with the product $R_m C_m$ as a key parameter of these temporal dynamics. This property limits the temporal resolution of electrical signals but also provides opportunities for temporal integration: when two individual signals are delivered within a small time interval, they may not be resolved as individual signals, but rather detected as an integrated signal. We will discuss this further in Chapter 3 in the context of dendritic integration of synaptic inputs.

Let's now turn to the second observation: the magnitude of membrane potential change decreases as the distance increases from the site of current injection. Suppose the peak membrane potential change at electrode *a* in response to the current injection is V_0. The spread of this membrane potential change within the fiber is carried by the axial current, which diminishes due to the continual leak of current through membrane conductances along the way. An easy way to visualize this is to simplify the circuit in Figure 2–16C further by considering only the conductance path (**Figure 2–17**A). (This simplification is equivalent to considering the peak magnitude of membrane potential changes after the membrane capacitance is charged.) The axial current (I_i) gradually diminishes in magnitude

Figure 2–17 A simplified circuit model to illustrate electrical signal decay along the distance of a neuronal fiber. (A) In this simplified model of Figure 2–16C, the batteries symbolizing the resting potential are omitted because only the change of the membrane potential from the resting potential is considered here; the capacitors are also omitted because only the peak membrane potential changes are considered. The current injection at position 0 causes a peak membrane potential change of V_0. In addition to passing through the membrane (I_m), part of the injected current also spreads in both directions along the interior of the neuronal fiber as an internal current (I_i) (only the rightward spread is shown here). Along the way, the magnitude of I_i diminishes because of leaky membrane current (I_m), as symbolized by the decreasing sizes of the arrows. **(B)** As a consequence, the membrane potential changes also decay across the distance, following the exponential decay curve. The x axis unit is the length constant of the neuronal fiber. The values for membrane potentials at 1×, 2×, and 3× the length constant are indicated, which equal 0.37 V_0, 0.14 V_0, and 0.05 V_0, respectively. (Adapted from Katz B [1966] Nerve, Muscle, and Synapse. With permission from McGraw Hill.)

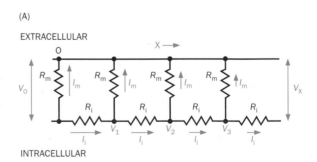
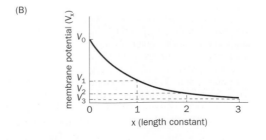

because part of the current leaks away to the outside due to membrane conductance. The membrane potential change $V(x)$ at distance x from electrode a is given by the following formula:

$$V(x) = V_0 e^{-\frac{x}{\sqrt{dR_m/4R_i}}}$$

where R_m is the membrane resistance per unit area of membrane surface, R_i is the internal (axial) resistance per unit volume of the neuronal cytoplasm, and d is the diameter of the fiber. This equation represents an exponential decay of electrical signal across distance (Figure 2–17B), analogous to the exponential decay of current over time in an *R-C* circuit we discussed in Section 2.6 and Box 2–2.

The term $\sqrt{dR_m/4R_i}$ is called the **length constant** or **space constant** (designated as λ). It is expressed in units of length, and one length constant corresponds to the distance across which the peak magnitude of the membrane potential change attenuates to $1/e$, or about 37%, of the original peak magnitude. As specific examples, the positions of the electrodes b and c in our idealized experiment (see Figure 2–16A) were chosen to be 1.5 and 3 length constants away from electrode a. The longer the length constant, the further electrical signals can be transmitted before they decay to a given fraction of their original value. As indicated from the formula, the length constant increases with the diameter of the neuronal fiber. The length constant also increases with the membrane resistance and decreases with the axial resistance. Indeed, animals have evolved various strategies to increase the distance that electrical signals spread, such as enlarging axon diameter as in the squid giant axon, or increasing the unit membrane resistance as in myelination; we will discuss these strategies in more detail in Section 2.13.

The temporal spread of electrical signals and their attenuation across distance are often referred to as **passive electrical properties** of neurons, as opposed to the active properties that we will begin to study in the next section. They are also called the **cable properties** of neuronal fibers, analogous to ocean cables that transmit electrical signals using insulators to separate the interior conductor from the exterior conducting seawater. The time constant, which characterizes the temporal spread of electrical signals, and length constant, which characterizes the attenuation across distance, are the two key passive electrical properties of neurons. **Table 2–1** lists experimentally determined time and length constants of neurons and muscles in various experimental preparations that have played important roles in the history of neurophysiology. We will learn about these specific preparations in later sections and chapters.

2.9 Active electrical properties of neurons: depolarization above a threshold produces action potentials

As is evident from Table 2–1, electrical signals decay considerably across distance if neurons only have passive properties. Even in fibers with very large diameters and consequently a large length constant, one length constant is just over a few

Table 2–1: Time and length constants of axons, dendrites, and muscle cell

Fiber	Diameter (µm)	Length constant (mm)	Time constant (ms)
Squid giant axon[1]	500	5	0.7
Lobster nerve[1]	75	2.5	2
Frog muscle[1]	75	2	24
Apical dendrite of mammalian cortical pyramidal neuron[2]	3	1	~20

[1]Data from Katz B (1966) Nerve, Muscle, and Synapse. McGraw-Hill; length constants were measured in large extracellular volume.

[2]Data from Stuart G, Spruston N & Hausser M (1999) Dendrites. Oxford University Press.

Figure 2–18 Depolarization exceeding a threshold produces action potentials. Using the experimental preparation that was diagrammed in Figure 2–16A, a series of rectangular current pulses were applied through the stimulating electrode (top). The corresponding changes in membrane potential were recorded by electrode *a* and are shown at the bottom. The unit of the *x* axis is 1 ms. For both hyperpolarization pulses and the first two depolarization pulses (current pulses 1–4), the membrane potential changes follow the sign of the current pulses, and their magnitudes are proportional to the magnitudes of the current pulses. In response to the fifth current pulse, the membrane potential change becomes unstable and varies across different trials (as is illustrated by multiple curves). Occasionally the stimulation results in a very large depolarization—the action potential. Action potentials of the same magnitude are always produced in response to the sixth current pulse. (Adapted from Katz B [1966] Nerve, Muscle, and Synapse. With permission from McGraw Hill.)

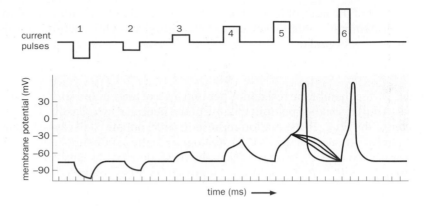

millimeters, meaning that signals decay to 37% of their original magnitude across that short distance. How do electrical signals propagate faithfully over much larger distances? To state the problem concretely, how can signals propagate reliably through a human motor neuron across a distance of approximately a meter in order to control a muscle in the toe?

To answer this question, let's continue our experiment using the setup in Figure 2–16A. Through the stimulating electrode, we inject step current pulses with varying magnitudes and directions into a neuronal fiber, this time an axon (**Figure 2–18**, top). We begin by injecting negative current, which hyperpolarizes the membrane potential. The magnitude of membrane potential changes, as measured by electrode *a* (Figure 2–18, bottom), is proportional to the magnitude of injected negative current. If we reverse the sign and inject positive current into the axon, we see that the membrane potential is depolarized rather than hyperpolarized. The magnitude of depolarization is also proportional to the magnitude of injected positive current, provided that only a small amount of positive current is injected. However, when the injected current exceeds a certain magnitude, a much larger and transient elevation of the membrane potential is produced in some fraction of trials. Above that magnitude of current injection, each current pulse invariably produces a large and transient elevation of the membrane potential. This is called an **action potential** or a **spike**. It is caused by depolarization of the membrane potential above a specific level, the **threshold**, in response to the injection of positive current of a certain magnitude (see Movie 2–6). The stimulus that can cause the neuron to generate an action potential is called a **supra-threshold stimulus**, and the stimulus that cannot is called a **sub-threshold stimulus**.

Note that the size of the action potential does not change with the magnitude of depolarization once the threshold is reached. Furthermore, if an action potential is recorded with electrode *a*, an action potential of similar magnitude and waveform will also be recorded by electrode *b* and electrode *c* (Figure 2–16A). In other words, action potentials propagate with little or no decay. As opposed to the passive spread of electrical signals discussed in Section 2.8, action potentials are an **active electrical property** of neurons. As we will learn later in this chapter, active electrical properties of neurons are a result of voltage-dependent changes in ion conductances.

Not all neurons fire action potentials: some neurons use only graded potentials to transmit electrical signals even in their axons. As you will understand from the earlier discussion of length constants, these must necessarily be neurons with short axons; we will see examples of such neurons in the vertebrate retina in Chapter 4. It should also be noted that active electrical properties are not exclusive to axons: some neurons exhibit active properties in dendrites as well; we will discuss these properties in Chapter 3.

How are action potentials produced? How do they propagate along the axon? We devote the next part of this chapter to addressing these fundamental questions in neuronal signaling.

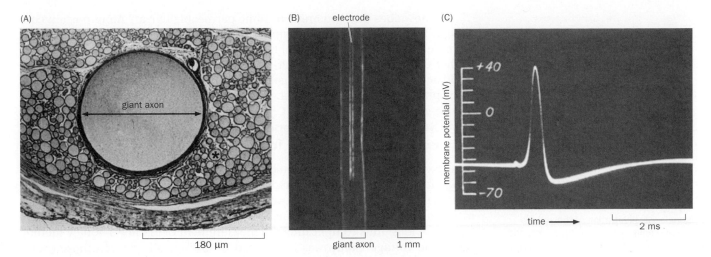

Figure 2–19 Studying action potentials in the squid giant axon. (A) Electron micrograph of a cross section of a squid giant axon showing its large diameter (~180 μm for this sample) as compared to neighboring axons (for example, the axon indicated by *). **(B)** Photograph of an electrode inserted inside a squid giant axon whose diameter is close to 1 mm. **(C)** An action potential recorded from the squid giant axon. (A, courtesy of Kay Cooper and Roger Hanlon; B, from Hodgkin AL & Keyes RD [1956] *J Physiol* 131:592–616; C, from Hodgkin AL & Huxley AF [1939] *Nature* 144:710–711. With permission from Macmillan Publishers Ltd.)

HOW DO ELECTRICAL SIGNALS PROPAGATE FROM THE NEURONAL CELL BODY TO ITS AXON TERMINALS?

In this part of the chapter, we follow the discovery path that has led to our current understanding of mechanisms by which action potentials are produced and propagate. In addition to answering a key question in neuronal communication, that is, how electrical signals propagate from the neuronal cell body to its axon terminals, these studies also established the concept of ion channels and highlighted the mechanisms by which ion channels function.

2.10 Action potentials are initiated by depolarization-induced inward flow of Na⁺

The discovery of the ionic basis of the action potential is a good example of how scientific breakthroughs can result from the introduction of new methods, model organisms, and analytic tools. Squid of the genus *Loligo* have a giant axon whose diameter reaches up to 1 mm, many times larger than nearby axons (**Figure 2–19**A) or the axons of typical mammalian neurons. The giant axon conducts action potentials very rapidly and controls the squid's jet propulsion system, allowing it to quickly avoid danger. The giant axon's large size enabled researchers to insert electrodes and measure action potentials more accurately than before (Figure 2–19B). During such measurements, it was discovered that the membrane potential during the rising phase of the action potential far exceeded zero (Figure 2–19C). Therefore, the action potential is not caused by a transient breakdown of the membrane that allows the resting potential to become zero, a prevalent view held prior to these measurements.

At the peak of the action potential, the membrane potential was observed to approach the Na⁺ equilibrium potential. (Recall that E_{Na} = +58 mV in our model neuron in Figure 2–12.) This finding suggested that the membrane is preferentially permeable to Na⁺ at the peak of the action potential and that the inward Na⁺ flow is responsible for the rising phase of the action potential. To test this hypothesis, the extracellular Na⁺ concentration was systematically reduced. If Na⁺ were responsible for the rising phase of the action potential, one would predict from the Nernst equation that the magnitude of the action potential would decrease with lower concentrations of extracellular Na⁺. This was indeed the case (**Figure 2–20**).

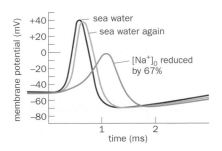

Figure 2–20 Testing the hypothesis that the rising phase of the action potential is caused by Na⁺ influx. The magnitude and speed of the action potential are diminished when the normal extracellular solution (sea water, red trace) was replaced by a solution of 33% sea water and 67% isotonic dextrose (hence the extracellular Na⁺ concentration, or $[Na^+]_o$, was reduced by 67%; blue trace). Reapplication of sea water (green trace) restored the magnitude and speed of the action potential. (Adapted from Hodgkin AL & Katz B [1949] *J Physiol* 108:37–77.)

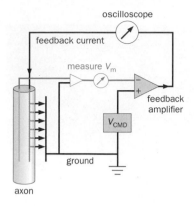

Figure 2–21 Illustration of the voltage clamp technique. The membrane potential (V_m) is measured by the blue wire inserted in the squid giant axon with respect to the ground wire outside the axon. It is then compared to a command voltage set by the experimenter (V_{CMD}) as two different inputs to the voltage clamp feedback amplifier (large triangle). The difference between the two voltages ($V_{CMD} - V_m$) produces a feedback current as the output of the amplifier, which is injected by a second inserted wire (red) into the axon. When V_{CMD} = V_m, there is no feedback current. Upon a step change of V_{CMD}, the feedback current rapidly changes V_m to the new V_{CMD} (within microseconds). Thus, the voltage clamp enables experimenters to control V_m of the axon being studied, and at the same time to measure with an oscilloscope the amount of feedback current needed to hold V_m to the value of V_{CMD}; this quantity of feedback current equals the current that flows across the axon membrane (parallel red arrows). The feedback current can flow in either direction (that is, the red arrows can reverse) depending on the relative values of V_m and V_{CMD}.

But how does the membrane become permeable to Na⁺? An important conceptual breakthrough was the hypothesis that depolarization could induce an increase in membrane permeability to Na⁺, with the influx of Na⁺ resulting in further depolarization. Such a self-reinforcing process (positive feedback loop) could account for the rapid change in membrane potential observed during the rising phase of the action potential.

2.11 Sequential, voltage-dependent changes in Na⁺ and K⁺ conductances account for action potentials

To test whether depolarization could render axonal membranes more permeable to Na⁺, it was important to quantitatively measure ion flows across the membrane under conditions that mimic the action potential. However, ion flow across the membrane changes the membrane potential, which in turn can affect the permeability of ions, thus complicating the measurement of ion flows. A new method called the **voltage clamp** was introduced to distinguish between these events (**Figure 2–21**). The voltage clamp compares the intracellular membrane potential with a command voltage set by the experimenter. A difference between the two voltages automatically produces a feedback current that is injected back into the cell, which rapidly changes the intracellular membrane potential to the value of the command voltage. After the initial stimulation, which is usually in the form of a step change in the command voltage, ion flow across the membrane as a consequence of the membrane potential change can be measured by recording how much current must be injected into the cell in order to maintain the membrane potential at a specified value.

Using the voltage clamp technique, Alan Hodgkin and Andrew Huxley carried out a series of classic experiments around 1950 to determine the ionic basis of the action potential. By subjecting the squid giant axon to depolarizing voltages, they were able to tease apart the ionic flows that underlie an action potential. Importantly, holding the membrane potential at a constant value eliminated the capacitive current (current that charges the membrane in response to voltage change, equivalent to I_C in Figure 2–16C) so that they could measure ionic currents across the membrane (equivalent to I_R in Figure 2–16C) and observe how they changed over time. For example, a 56-mV depolarizing step produced an initial inward current, followed by an outward current (**Figure 2-22**, green trace). The inward current was abolished when extracellular Na⁺ was replaced by choline, an organic ion that carries a +1 charge similar to Na⁺ but is unable to permeate the membrane (Figure 2-22, blue trace). This finding indicated that the initial inward current is indeed caused by Na⁺ influx. Other evidence suggested that in the case of choline replacement, the remaining current is caused by outward K⁺ flux. Thus, the Na⁺ current could be calculated by comparing the difference between the two conditions (Figure 2-22, red trace).

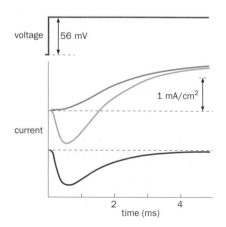

Figure 2-22 Dissociation of Na⁺ and K⁺ currents in voltage clamp experiments. (Top) A voltage step increase of 56 mV was applied to the squid giant axon. (Middle) Ion flow across a unit area of the axonal membrane in response to the depolarizing voltage step was measured by determining how much current was injected into the axon in order to maintain the axon's membrane potential at the command voltage established by the experimenter (the +56 mV step). The green trace shows the current flow under physiological conditions; an initial inward current (the downward portion of the trace) is followed by an outward current (the upward portion). The dotted line demarcates zero net current. The blue trace shows the current flow under conditions in which the external Na⁺ was mostly replaced by choline⁺, which cannot cross the membrane; this trace illustrates the K⁺ current only. (Bottom) The red trace represents the deduced Na⁺ current, which is the difference between the green and blue traces. (Adapted from Hodgkin AL & Huxley AF [1952] *J Physiol* 116:449–472.)

Voltage clamps also allowed a systematic measurement of Na⁺ and K⁺ conductance over a series of different voltages. As we introduced in Section 2.7, conductance is the ratio of the current that passes through the membrane and the driving force, which is the difference between the membrane potential and the equilibrium potential. Thus, the conductances for Na⁺ and K⁺ are given by

$$g_{Na} = \frac{I_{Na}}{V_m - E_{Na}}$$

$$g_K = \frac{I_K}{V_m - E_K}$$

where I_{Na} and I_K are the Na⁺ and K⁺ currents as measured in Figure 2–22, V_m is the membrane potential, and E_{Na} and E_K are the equilibrium potentials of Na⁺ and K⁺. By changing V_m (which equals V_{CMD}) and measuring the currents in the voltage clamp experiments, Hodgkin and Huxley could experimentally determine g_{Na} and g_K at different membrane potentials. They found that both Na⁺ conductance and K⁺ conductance increased when the intracellular membrane potential became more depolarized, and that these conductance changes evolve over time (**Figure 2–23**).

Hodgkin and Huxley made several key discoveries from these experiments. First, they confirmed that the rising phase of the action potential resulted from an influx of Na⁺ and determined that the Na⁺ influx was caused by a rapid increase in Na⁺ conductance as a consequence of membrane depolarization. Second, after the initial depolarization-induced increase, Na⁺ conductance would invariably decrease despite continued depolarization (Figure 2–23A), accounting for the falling phase of the Na⁺ current (see Figure 2–22, red trace). This was termed **inactivation** of the Na⁺ conductance. Third, depolarization also caused an increase of K⁺ conductance, resulting in K⁺ efflux. Importantly, the change of K⁺ conductance lagged behind the change of Na⁺ conductance (Figure 2–23B). Fourth, the Na⁺ and K⁺ conductances appeared to be independent of each other, but both depended on the membrane potential. Well before the molecular mechanisms of membrane transport became known, these findings paved the way for the modern concept of ion channels: transmembrane proteins that are selectively permeable to specific ions (see Section 2.4). Specifically, in the squid giant axon, channels specific to Na⁺ or K⁺ allow these ions to selectively flow through the membrane. The

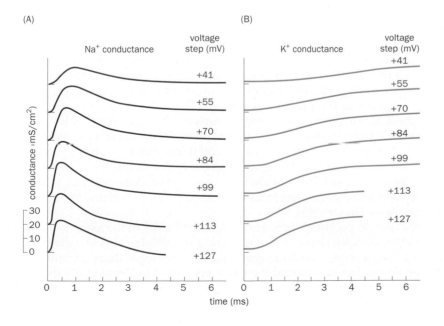

(A) Na⁺ conductance   voltage step (mV)

(B) K⁺ conductance   voltage step (mV)

Figure 2–23 Voltage-dependent change of Na⁺ and K⁺ conductance. Na⁺ (A) and K⁺ (B) conductances (y axis, in millisiemens per square centimeter) change over time following a depolarizing voltage step from the resting potential. A series of measurements were performed, each after a voltage step of a different magnitude (indicated on each trace). Over time, the Na⁺ conductance first increases then decreases, whereas the K⁺ conductance only increases, but more slowly than the initial rise of Na⁺ conductance. Larger voltage steps cause more rapid rise in both conductances. (Adapted from Hodgkin AL & Huxley AF [1952] *J Physiol* 116:449–472.)

(A)

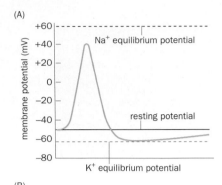

(B)

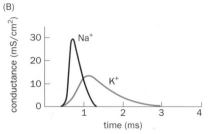

Figure 2–24 A summary of the ionic basis of the action potential.
(A) Schematic of an action potential, with reference to the resting potential and the equilibrium potentials for Na$^+$ and K$^+$. (The K$^+$ equilibrium potential and the resting potential of the squid giant axon are more depolarized than for our model neuron in Figure 2–12.) **(B)** Sequential changes in Na$^+$ and K$^+$ conductance (calculated according to data in Figure 2–23) during the action potential. The rising phase is caused by the increase in Na$^+$ conductance, leading to Na$^+$ influx. The falling phase is accounted for by both the inactivation of the Na$^+$ conductance, which stops the Na$^+$ influx, and the increase in the K$^+$ conductance, leading to K$^+$ efflux. The transition between the rising and falling phase occurs before the rising phase reaches the Na$^+$ equilibrium potential. The falling phase overshoots the resting potential and approaches the K$^+$ equilibrium potential, before the membrane potential gradually returns to the resting potential, which is slightly above the K$^+$ equilibrium potential (see Section 2.5). (Adapted from Hodgkin AL & Huxley AF [1952] *J Physiol* 116:449–472.)

conductance of these ion channels increases when the axon is depolarized. These channels are now called **voltage-gated ion channels** because their conductance changes as a function of the membrane potential.

In summary, the action potential can be accounted for by sequential changes of Na$^+$ and K$^+$ conductance (**Figure 2–24**; **Movie 2–7**), which we now know are caused by the opening and closing of voltage-gated Na$^+$ and K$^+$ channels. At the resting state, the voltage-gated Na$^+$ and K$^+$ channels are both closed. (A different set of K$^+$ channels accounts for K$^+$ permeability at rest.) At the rising phase of the action potential, when the membrane is depolarized, opening of voltage-gated Na$^+$ channels allows Na$^+$ to flow into the cell down its electrochemical gradients. Depolarization also causes an increase in K$^+$ efflux (via the resting K$^+$ channel) because the force produced by the new, smaller electrical gradient is less effective in countering the force produced by the chemical gradient. When the Na$^+$ influx exceeds the K$^+$ efflux, the neuron passes the threshold for firing an action potential. (For most excitable cells the threshold is about 10–20 mV above the resting potential.) More depolarization causes opening of more voltage-gated Na$^+$ channels, which causes further depolarization. This positive feedback loop generates the rapid rising phase of the action potential.

During the falling phase, Na$^+$ channels are inactivated after the initial opening, preventing further Na$^+$ influx. At the same time, voltage-gated K$^+$ channels open, allowing more K$^+$ efflux. These two events together account for the falling phase of the action potential, allowing neurons to repolarize to the resting potential and prepare for the next action potential (Figure 2–24). Importantly, the ionic basis of the action potential, originally discovered in the squid giant axon, applies to neurons and other excitable cells across most of the animal kingdom, including humans.

2.12 Action potentials are all-or-none, are regenerative, and propagate unidirectionally in the axon

The Hodgkin–Huxley model satisfactorily explains several properties of the action potential that ensure faithful transmission of information from the cell body to the axon terminal.

First, action potentials are **all-or-none**. When a stimulus-induced neuronal membrane depolarization is below the threshold, the action potential does not occur. When depolarization exceeds the threshold, the waveform of the action potential is determined by the relative concentrations of Na$^+$ and K$^+$ inside and outside the cell, and these concentrations remain mostly constant for a given neuron. (The Na$^+$ influx and K$^+$ efflux during an action potential causes very small change of intracellular, and even smaller changes of extracellular, Na$^+$ and K$^+$ concentrations. For example, an action potential in the squid giant axon brings in ~10^{-13} mole of Na$^+$ per millimeter of the axon; the same length of the axon contains ~10^{-7} mole of Na$^+$. Thus, intracellular Na$^+$ concentration change as a result of Na$^+$ influx is about one part in a million, which makes a negligible contribution to E_{Na}.) To a first approximation, action potentials assume the same form in response to any supra-threshold stimulus.

Second, action potentials are **regenerative**—they propagate without attenuation in amplitude. Suppose that an action potential occurs at a particular site on the axon. The rising phase creates a substantial membrane depolarization, which spreads down the axon and brings an adjacent region to threshold, which in turn does so for its adjacent downstream region, and so on (**Figure 2–25**). In this way, the action potential propagates in a similar form continuously and faithfully down the axon toward its terminal.

Third, action potentials propagate unidirectionally in the axon, from the cell body to the axon terminals. When an action potential occurs at a given site on the axon (for example, site A in Figure 2–25), in principle depolarization should also spread up the axon toward the cell body (site Z in Figure 2–25) in addition to spreading down the axon toward the axon terminals (site B in Figure 2–25). However, the delayed activation of the K$^+$ channels and the inactivation of the Na$^+$ channels combine to create a **refractory period** after an action potential has

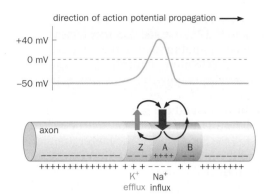

direction of action potential propagation ⟶

+40 mV

0 mV

−50 mV

axon

Z A B

++++

+++++++++++++ +++ − − − − ++ +++++++

K⁺   Na⁺
efflux influx

Figure 2–25 The propagation of an action potential along the axon. A schematic of an action potential as it sweeps across the axon provides a snapshot of electrical signaling events in the axon. The wave front is at site A, where voltage-gated Na⁺ channels open and depolarize the membrane. Positive charges at site A within the axon spread to site B, where they will cause depolarization above the threshold. Thus, at the next moment, the wave front will reach site B. Site Z, where the wave front has just passed, is experiencing a refractory period during which the delayed activation of K⁺ channels and the inactivation of Na⁺ channels prevent the action potential from propagating backward from A to Z. Red and blue arrows indicate Na⁺ influx and K⁺ efflux, respectively. Curved arrows represent current flow that completes the left and right circuit as a result of Na⁺ influx. The charges below represent the membrane potentials at different segments of the axon.

just occurred, during which time another action potential cannot be reinitiated. Because the action potential normally initiates at the axon initial segment and passes through Z before reaching A, another action potential cannot immediately back-propagate from A to Z. This refractory period ensures that the action potential normally only propagates from the cell body down the axon to its terminal, not in the reverse direction.

In most projection neurons, whose axons form synapses on distant target neurons, the action potential first arises at the initial segment of the axon, where voltage-gated Na⁺ channel density per unit membrane area is the highest; this high channel density provides the lowest threshold for action potential initiation. The axon's initial segment is a critical site for the integration of depolarizing and hyperpolarizing synaptic potentials from the dendrites and the cell body; this integrative process is discussed further in Section 3.24. After initiation, action potentials travel unidirectionally within the axon toward the terminals. At the initiation site, however, action potentials can in principle travel in both directions; indeed, in some mammalian neurons, action potentials can back-propagate to dendrites, which like axons contain voltage-dependent Na⁺ and K⁺ channels. In artificial situations where experimenters electrically stimulate the axon or its terminal, action potentials can propagate in a retrograde direction from the axon terminal to the cell body, producing so-called **antidromic spikes** that can be recorded from the cell body. However, these events do not occur under physiological conditions *in vivo*.

Altogether, the above three properties make the action potential an ideal means to transmit information faithfully from neuronal cell bodies across long distances to their axon terminals. But since action potentials are all-or-none, the size of action potentials cannot encode any information about the stimulus. Rather, the information is usually encoded by the rate (number of action potentials/time) or the timing of action potentials in response to a stimulus (see Section 1.8). Action potential frequency is limited by the refractory period. Some neurons, such as fast-spiking inhibitory neurons in the mammalian cortex, can fire up to 1000 Hz, or one action potential per millisecond, which is faster than refractory periods of many neurons. This requires specializing the ion channels so the action potential is repolarized quickly and the refractory period is complete in time for the next action potential. Thus, ion channel properties (such as Na⁺ channel inactivation and the delayed opening of K⁺ channels) have been selected during evolution to ensure unidirectional propagation of action potentials while still permitting high-frequency firing. The broad range of possible spike frequencies expands the information-coding capacity of individual neurons.

2.13 Action potentials propagate more rapidly in axons with larger diameters and in myelinated axons

The speed at which action potentials propagate is not the same for all neurons, but instead depends on the properties of the axon. This is because the rate at which depolarization spreads down the axon (see Figure 2–25) is determined by the cable properties of the axon. Returning to our circuit model of neuronal membranes (see Figures 2–16C and 2–17A), we see that this forward spread is determined by how quickly a change in membrane potential charges the membrane capacitor, as well as the relative distribution of current flowing forward versus leaking out into the extracellular environment. If all other properties were equal, these values would be a function of axon diameter: the larger the axon diameter, the lower the axial resistance of the axon, and the larger the proportion of current would flow forward. You can follow this from the perspective of the length constant (λ), which is equal to $\sqrt{dR_m/4R_i}$ (see Section 2.8). The larger the diameter (d), the larger the length constant, and the further depolarization can spread at a value that is still above threshold to produce the next action potential at a more distant site. This is why the squid has evolved a giant axon that helps it react rapidly to danger.

In principle, increasing R_m (the membrane resistance per unit area) can also increase the length constant. However, increasing R_m also increases the time constant (which is equal to $R_m C_m$ as discussed in Section 2.8) required to charge the membrane along the way, which slows down the propagation of action potentials. One way to compensate for this is to also reduce C_m, the membrane capacitance. Indeed, this is how **axon myelination** works. Many vertebrate axons are wrapped by a **myelin sheath**, which are cytoplasmic extensions of glia—Schwann cells in the PNS and oligodendrocytes in the CNS (see **Box 2–3** for more details). Some axons in large invertebrates are also myelinated, as we will see in Chapter 12. The cytoplasmic extensions of glial cells wrap around myelinated axons many times, with most of the cytoplasm compressed out of the extensions toward the soma to form **compact myelin** consisting of closely packed glial plasma membranes (**Figure 2–26**A). From an electrical circuit perspective, compact myelin is equivalent to having many resistors connected in series, such that the total membrane resistance R_T is equivalent to nR_m, where n is the number of layers of glial membrane. However, in serial connections, the total capacitance C_T is equivalent to C_m/n (recall from Section 2.6 that total capacitance of two capacitors connecting in series follows $1/C_T = 1/C_1 + 1/C_2$; combined capacitance of n identical capacitors connected in series follows $1/C_T = n/C$, or $C_T = C/n$). Thus, myelination greatly increases membrane resistance, and hence length constant, without increasing the time constant (as the product $R_T C_T$ remains unchanged compared to the original $R_m C_m$); this means that once an action potential is produced at a specific site on the axon, depolarization spreads across a large distance to cause far-away axonal membrane to regenerate the action potential.

Despite the increased membrane resistance, small amounts of current still leak out of myelinated axons. Therefore, at **nodes of Ranvier**, which occur at regular intervals usually 200 μm to 2 mm apart, the axon surfaces are exposed to the extracellular ionic environment and contain highly concentrated voltage-gated

Figure 2–26 Axon myelination increases the speed of action potential propagation. (A) An electron micrograph of a cross section of spinal cord axons wrapped by oligodendrocyte membranes. At the center is a single axon wrapped by myelin sheath. **(B)** A fluorescence microscopic image of the rat optic nerve immunostained to visualize three proteins (see Section 13.13 for more details of the immunostaining method). Na$^+$ channels (green) are highly clustered at the center of the node of Ranvier. K$^+$ channels (blue) are distributed peripherally at node. In between are transmembrane proteins named Caspr (red) that help organize channel distribution at the node. **(C)** Schematic of an action potential 'hopping' between nodes of Ranvier. After an action potential occurs at the left node of Ranvier, positive charges rapidly flow to the next node to the right. This is because, as a consequence of myelin wrapping, the internodal membranes have low capacitance, which requires less charging, and high resistance, which allows only a small amount of current to leak through (dashed arrows). The arrival of positive charges at the right node causes rapid depolarization above threshold to regenerate the action potential there. Red arrows indicate the direction of current flow that completes the circuit as a result of Na$^+$ influx at the left node. For simplicity, the circuit resulting from depolarization spread leftward from the left node is omitted. (A, courtesy of Cedric Raine; B, adapted from Rasband MN & Shrager P [2000] *J Physiol* 525:63–73.)

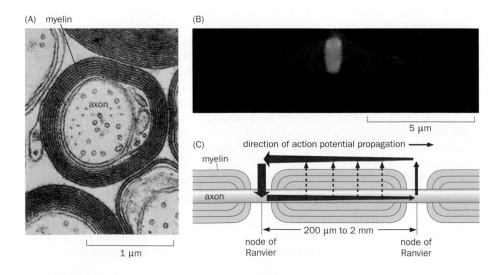

(A)   myelin

axon

1 μm

(B)

5 μm

(C)   direction of action potential propagation ⟶

myelin

axon

200 μm to 2 mm

node of Ranvier   node of Ranvier

Na$^+$ and K$^+$ channels (Figure 2-26B). As a result, depolarizing current spreads rapidly in between the nodes, and action potentials are 'renewed' only at the nodes of Ranvier (Figure 2-26C). Thus, action potentials in a myelinated axon hop from node to node. This is termed **saltatory conduction** (from the Latin *saltare*, to jump), as opposed to continuous propagation in unmyelinated axons (see Figure 2-25).

Myelination greatly increases the conduction speed of action potentials and the capacity for high-frequency firing. Action potentials can travel at a speed of up to 120 m/s in myelinated axons, as compared to <2 m/s in unmyelinated axons. Although unmyelinated axons usually have smaller diameters than myelinated axons (see Box 2-3), the diameter difference alone does not account for the large difference in propagation speeds. Saltatory conduction also saves energy; there is a reduced demand for the Na$^+$-K$^+$ ATPase to pump Na$^+$ outward and K$^+$ inward, as there is little transmembrane current except at the nodes of Ranvier. Because

---

Box 2–3: Axon–glia interactions in health and disease

Axon myelination provides a striking example of the intimate interactions between glia and neurons. In the white matter of the central nervous system (CNS), each oligodendrocyte typically extends several processes that myelinate multiple axons (see Figure 1-9). In the peripheral nervous system (PNS), each Schwann cell is usually dedicated to wrapping a segment of a single axon. During development or remyelination, oligodendrocyte and Schwann cell extensions wrap the axon like a spiral many times and compress the cytoplasm in between layers of the extensions to form the myelin sheath (**Figure 2-27**A).

There is a large diversity of axon–glia interaction in the nervous system. For example, as will be discussed in Chapter 6, the somatosensory system contains distinct types of sensory neurons with characteristic axon diameter, degree of myelination, and action potential conduction speed. The thickness of myelin matches the size of the axon. Sensory neurons that innervate muscle and provide rapid feedback regulation of movement have large-diameter axons, thicker myelin sheaths, and conduct action potentials more rapidly. Sensory neurons that sense touch have intermediate axon diameters and conduction speeds. Many sensory neurons

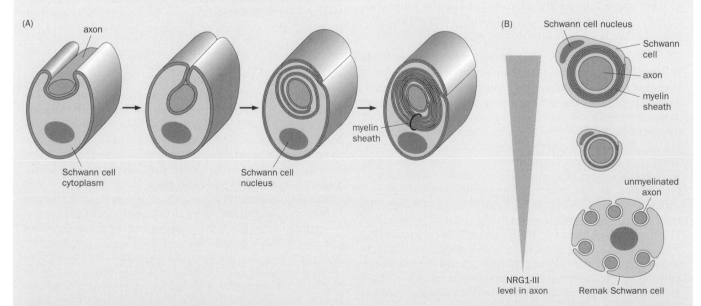

Figure 2–27 Schwann cell wrapping of axons and its regulation by neuregulin signaling. (A) Schematic of sequential steps illustrating Schwann cell wrapping of an axon, forming spiral extensions, and compressing the cytoplasm between the layers of plasma membrane to form myelin sheath. **(B)** Top, large-diameter axons express the highest levels of type III neuregulin-1 (NRG1-III), which directs thicker myelination. Middle, intermediate-diameter axons express intermediate levels of NRG1-III, which directs thinner myelination. Bottom, small-diameter unmyelinated axons express the lowest levels of NRG1-III, which directs their interaction with Remak Schwann cells, forming the Remak bundle. (Adapted from Nave KA & Salzer JL [2006] *Curr Opin Neurobiol* 16:492–500. With permission from Elsevier Inc.)

(Continued)

Box 2–3: Axon–glia interactions in health and disease

that sense temperature and pain are unmyelinated and conduct action potentials more slowly. These unmyelinated axons are nevertheless associated with **Remak Schwann cells**, whose cytoplasm extends in between individual axons to form a Remak bundle (Figure 2–27B). Here the glia's role is simply to segregate individual axons rather than supporting saltatory conduction.

What determines whether an axon should be myelinated or not, and if so, to what degree? These questions have been answered in the PNS. An axonal cell-surface protein called **type III neuregulin-1 (Nrg1-III)** plays a key role: axons that express high levels of Nrg1-III are associated with thick myelin sheaths, axons that express intermediate levels of Nrg1-III are thinly myelinated, and axons that express low levels of Nrg1-III are associated with the Remak bundle (Figure 2–27B). Nrg1-III acts on the erbB2/B3 receptor complex on Schwann cells to direct their differentiation, including the expression of myelin-associated proteins and the spiral wrapping of axons. Nrg1/erbB signaling is not required for myelination by oligodendrocytes, suggesting alternative axon–glia signals in the CNS. Schwann cells and oligodendrocytes also signal back to axons to provide long-term support to their health and integrity.

The importance of myelination in human health is exemplified by **demyelinating diseases**, in which damage to the myelin sheath decreases the resistance between nodes of Ranvier and disrupts the organization of ion channels in the nodal region (see Figure 2–26). This slows down or even stops action potential conduction, causing deficits in sensation, movement, and cognition. Demyelinating diseases can be caused by several factors including autoimmune responses that attack the glial cells and mutations in proteins that are necessary for the function of myelin sheath.

The most common CNS demyelinating disease is **multiple sclerosis (MS)**, an adult-onset inflammation-mediated disease that affects 1 in every 3000 people globally. The hallmarks of MS are the inflammatory plaques in the white matter caused by immune cell attack of myelin. Most MS patients begin with a phase of relapsing–remitting MS, during which patients cycle between inflammatory demyelination with neurological symptoms, and remyelination and recovery. The next phase is characterized by continual and progressive deterioration of the neurological symptoms, which is often irreversible. Although abnormal immune response clearly plays a major role, the cause of MS is mostly unknown. Variants of certain genes such as the major histocompatibility loci confer risks, but environmental factors appear to play a major role. Thanks to the recent development of drugs that inhibit the immune attack on CNS myelin, the life prognosis is far better for a first diagnosis of MS today than it was several decades ago.

Compared to MS, much more is known about the mechanisms of demyelinating diseases in the PNS because many are caused by inherited mutations in specific genes. **Charcot–Marie–Tooth (CMT) disease** (first described by

J.M. Charcot, P. Marie, and H.H. Tooth in 1886) is the most common inherited disorder of the PNS, affecting 1 in 2500 individuals. CMT patients exhibit age-progressive deficits in sensation or movement in a length-dependent manner (that is, distal limbs exhibit the most severe deficits). Genetic alterations in ~30 different genes underlie various forms of the CMT disease that gave similar symptoms; of these, some affect Schwann cells and myelination whereas others affect axons (for example, a specific kinesin that regulates axonal transport; see Section 2.3). We give three examples below to illustrate how different causes can reach the same outcome: the disruption of myelination.

The most common cause of CMT (CMT1A) results from a duplication of a chromosome segment that contains the *Pmp22* gene, causing overexpression of the peripheral myelin protein 22, a membrane protein enriched in compact myelin. Interestingly, spontaneous mutations in the mouse *Pmp22* gene cause neurological phenotypes in the *Trembler* mice, which had been studied for decades as a model of PNS demyelinating disease prior to their association with the *Pmp22* gene. In a revealing experiment, Schwann cells from *Trembler* mice were transplanted into wild-type mice to cover a segment of a regenerating **sciatic nerve** (consisting of sensory and motor axons that innervate the leg); remarkably, only the axon segment associated with the transplanted Schwann cells was hypomyelinated. Conversely, when transplanted into *Trembler* mice, wild-type Schwann cells were able to rescue the hypomyelination defects of the regenerating axons at the transplanted segment (**Figure 2–28**). These experiments demonstrated that the gene product that is defective in *Trembler* mice, later known to be Pmp22, acts in Schwann cells to regulate myelination.

Other CMT mutations have also provided interesting insight into the biology of myelination. For example, CMT1B is caused by mutations in the *Mpz* gene, encoding myelin protein zero. Mpz is an abundant transmembrane protein in Schwann cells that mediates homophilic binding (that is, Mpz proteins expressed on apposing membranes bind to each other and bring membranes together). Mice in which Mpz was disrupted exhibited defective compact myelin, degeneration of myelin, and degeneration of axons. Thus, *Mpz* contributes to the close interactions between adjacent myelin membranes, which are essential for proper myelination and the integrity of the axon. CMT1X is caused by mutations in the *Gjb1* gene, encoding a gap junction channel. Given that myelin is a long spiral that extends from the area around the nucleus to the layer closest to the axon (Figure 2–27A), there is a very long pathway for intracellular transport. Gap junctions between adjacent myelin membranes introduce short cuts across the myelin for intracellular transport but are sufficiently sparse that they do not affect the ability of myelin to act as a superb resistor. CMT1X mutations highlight the importance of gap junction channels in the proper function of Schwann cells in myelination.

Box 2–3: Axon–glia interactions in health and disease

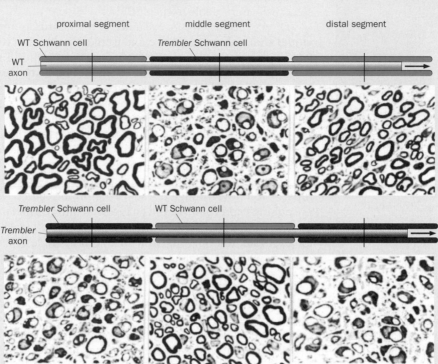

Figure 2–28 The *Trembler* gene product acts in Schwann cells to regulate axon myelination. In these transplantation experiments, the middle segment of sciatic nerve is removed, and replaced with a segment of sciatic nerve from a donor. Axons from the donor degenerate because they are separated from their cell bodies, but the associated Schwann cells survive. Host axons in the original distal segment also degenerate because they are severed from their cell bodies. During regeneration, host axons from the proximal segment (which remained connected to their cell bodies) re-extend into the middle and distal segment (arrows), through the donor and host Schwann cell environment, respectively. **Top**, when wild-type (WT) axons regenerated through Schwann cells from *Trembler* mice, axons were hypomyelinated in the middle segment, as is observed in *Trembler* mice. The micrographs show cross sections of the sciatic nerve (positions indicated by vertical lines in the schematic), where dark rings represent myelin sheaths. Compare the thin myelin sheaths in the middle segment with the thicker sheaths in the proximal and distal segments. **Bottom**, when axons of *Trembler* mice regenerated through the WT Schwann cell environment, they became more myelinated than the proximal and distal segments. Thus, the genotypes of the Schwann cells determine the degree of axon myelination. (Adapted from Aguayo AJ, Attiwell M, Trecarten J et al. [1977] *Nature* 265:73–75. With permission from Macmillan Publishers Ltd.)

These discussions illustrate that fundamental studies in neurobiology and research on disorders of the nervous system are closely related and can greatly benefit from each other. We will encounter this theme throughout the book, culminating in Chapter 11 devoted to brain disorders.

myelin is so important for proper conduction in the axons in vertebrates including humans, improper myelination is responsible for several major neurological disorders, such as multiple sclerosis and Charcot–Marie–Tooth disease (see Box 2–3).

2.14 Patch clamp recording enables the study of current flow across individual ion channels

Studies of the action potential in the squid giant axon suggested the existence of dedicated ion channels for Na⁺ and K⁺. This idea was supported later by the characterizations of toxins that specifically block Na⁺ or K⁺ channels. The most famous is puffer fish **tetrodotoxin (TTX)** (**Figure 2–29**), which potently blocks voltage-gated Na⁺ channels across animal species and is widely used experimentally to silence neuronal firing. In recordings from the squid giant axon, for example, TTX application mimics the replacement of Na⁺ with choline⁺ in the original Hodgkin–Huxley experiment (see Figure 2–22). Other drugs, such as **tetraethylammonium** (**TEA**), selectively block voltage-gated K⁺ channels. Experiments like these helped establish the concept of ion channels that are selectively permeable to specific ions.

Direct support for the existence of ion channels and characterization of individual channel's properties came from an important technical innovation in the late 1970s called **patch clamp recording** (see Section 13.21 and Box 13–2 for more details). In its original form, now called a **cell-attached patch**, a **patch pipette** (also called a patch electrode, which is a specially made glass electrode with a small opening at the tip) forms a high-resistance seal with a small patch of the plasma membrane of an intact cell. This high-resistance seal is called a giga-seal because the resistance exceeds 10^9 ohms, thus preventing ion flow between the pipette and the membrane. Similar to the voltage clamp method we discussed earlier, the experimenter can 'clamp' the voltage in the patch pipette, which corresponds to the extracellular potential for the small patch of the membrane underneath the pipette. Ion flow through the small membrane patch, which sometimes contains only a single ion channel, can be resolved and studied in isolation (**Figure 2–30**A).

When a cell-attached patch clamp was applied to cultured rat muscle cells, for example, the opening and closing of a single channel could be detected as discrete events. When a single channel opened in response to a depolarization of the patch, it produced a unitary inward current of ~1.6 picoampere (circled in Figure 2–30B). When hundreds of these single channel recording traces were summed and averaged over time, a 'macroscopic' Na⁺ current ensemble much like the original Hodgkin–Huxley voltage clamp recording was reconstituted, with characteristic voltage-dependent opening and subsequent inactivation (Figure 2–30B, bottom; compare with the bottom trace of Figure 2–22). This experiment thus revealed the biophysical basis of voltage-dependent changes in Na⁺

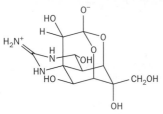

Figure 2–29 Tetrodotoxin (TTX) from the puffer fish. Puffer fish (above), whose resident symbiotic bacteria produce tetrodotoxin (TTX, below), is a delicacy in Japanese cuisine when properly prepared. TTX is a potent blocker of voltage-gated Na⁺ channels across many species. (Image courtesy of Brocken Inaglory/ Wikipedia.)

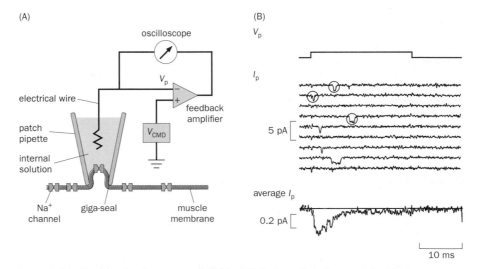

Figure 2–30 Studying ion flow across individual Na⁺ channels using a patch clamp.
(A) A cell-attached patch pipette can record ion flow within a small patch of muscle membrane. The configuration is similar to that of the voltage clamp schematic shown in Figure 2–21. The patch pipette serves two functions: to measure the voltage of the pipette V_p, and to inject current through a feedback circuit such that V_p matches that of the command voltage (V_{CMD}). The oscilloscope measures the current that needs to be injected in order for V_p to match V_{CMD}, which is equivalent to the current that flows through the ion channel(s) under the patch pipette (I_p). **(B)** In response to a depolarization step (V_p) applied to the patch pipette (top), current flows across the patch (I_p). Traces are shown for nine individual current measurements (middle). Na⁺ channel openings can be seen as downward, rectangular steps (for example, circles on the first, second, and fourth traces), as positively charged Na⁺ ions leave the recording pipette and flow into the cell. The sum and average of 300 I_p traces (bottom) resembles the macroscopic Na⁺ current measured by conventional voltage clamp (see red trace in Figure 2–22), including voltage-dependent activation and inactivation. A K⁺ channel blocker was included in the pipette solution to block ion flow through possible K⁺ channel(s) in the membrane patch. (B, adapted from Sigworth FJ & Neher E [1980] *Nature* 287:447–449. With permission from Macmillan Publishers Ltd.)

conductance at the level of single molecules: the proportion of time that an individual voltage-gated Na^+ channel is open and able to conduct current—that is, the channel's **open probability**—is increased by depolarization, and reduced by subsequent inactivation.

Note in the above experiment that each channel opening takes a square-like form, which means that the channel typically transitions between a state that is nonconducting (closed) to one that is conducting (open) without sliding through intermediates. Also note that even though an individual channel makes the closed-to-open transition abruptly, channels do not all open immediately after the membrane potential changes. Indeed, a difference in the delay in depolarization-induced open probability increase in Na^+ and K^+ channels accounts for the temporal difference in Na^+ and K^+ conductance rises during the action potential (see Figure 2–24B).

In general, the current (I) carried by a particular species of ion across a piece of neuronal membrane can be determined from single channel properties by the following formula:

$$I = NP_0\gamma \, (V_m - E)$$

where N is the total number of channels present, P_0 is the open probability of an individual channel, V_m is the membrane potential, E is the equilibrium potential of that ion (hence $V_m - E$ is the driving force), and γ is the **single channel conductance**. Compared to the relationship of current and driving force we learned in Section 2.7, we see that the product $NP_0\gamma$ is equivalent to the macroscopic conductance, g. Thus, the ion conductance across a neuronal membrane is the product of (1) the number of channels present on the membrane, (2) the open probability of each channel, and (3) the single channel conductance. As discussed above, the open probability P_0 is a function of both membrane potential and time, whereas the single channel conductance γ is a physical property of the channel protein but can vary with changes in its ionic milieu.

2.15 Cloning of genes that encode ion channels allows their structure–function relationship to be studied

The molecular structures of ion channels as individual proteins were determined after the cloning of genes that encode specific ion channels, as the revolution in molecular biology spread to neuroscience in the 1980s. Being able to clone a gene requires one or more of the following approaches: (1) purifying the corresponding protein and using the amino acid sequence to deduce nucleotide sequence for designing a probe to screen a **cDNA library** (consists of cloned cDNAs, or complementary DNAs, synthesized from mRNA templates derived from a specific tissue); (2) identifying a mutant defective in the gene product and using molecular genetic techniques to trace the causal gene; or (3) expressing the candidate gene product (by partitioning of a cDNA library) in a host cell and using a functional assay to identify the presence of the gene product. If there is a rich source of the protein and a functional assay (such as a high-affinity ligand) with which to look for the presence of the protein in biochemical fractions, the protein purification route is available and indeed was the one that led to cloning of the first Na^+ channels.

Voltage-gated Na^+ channel proteins were first purified from the electric eel *Electrophorus electricus*, whose electric organ is densely packed with Na^+ channels. (Electric eels use their electric organ to shock their prey with large currents.) Peptide sequences from purified electric eel Na^+ channel proteins were used to identify cDNAs that encode these proteins, and the electric eel cDNAs were then used to identify homologous genes in other organisms, including mammals, leading to the determination of their complete amino acid sequences. These studies revealed a highly conserved primary structure (**Figure 2–31**A): Animals from invertebrates to mammals have voltage-gated Na^+ channels that consist of four repeating modules, each of which contains six transmembrane segments that span the lipid bilayer. This structural conservation explains why toxins such as TTX block Na^+ channels across the animal kingdom. The fourth transmembrane

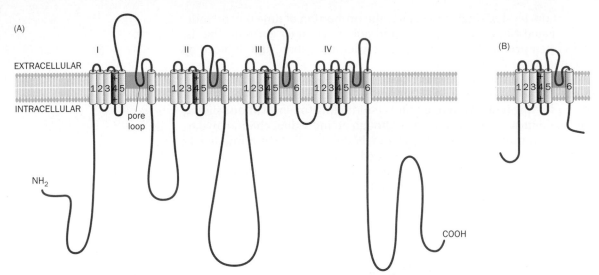

Figure 2–31 Primary structure of voltage-gated Na⁺ and K⁺ channels. (A) A voltage-gated Na⁺ channel is composed of four repeating modules. Each module consists of six helical transmembrane segments (TMs), with the fourth TM (red) containing positively charged amino acids that play a key role in voltage sensing. The pore loop and the adjacent fifth and sixth TMs together constitute the ion conduction pore (green). This structure was originally derived from the voltage-gated Na⁺ channel from electric eel (see Noda M, Shimizu S, Tanabe T et al. [1984] *Nature* 312:121–127). **(B)** A voltage-gated K⁺ channel protein resembles one of the four repeating modules of a voltage-gated Na⁺ channel; note the positively charged amino acids in the fourth TM and the pore loop between the fifth and sixth TMs. Four such subunits constitute a functional channel. The structure was originally derived from a K⁺ channel identified in *Drosophila* after the positional cloning of the *Shaker* gene (see Papazian DM, Schwarz TL, Tempel BL et al. [1987] *Science* 237:749–753 and Tempel BL, Papazian DM, Schwarz TL et al. [1987] *Science* 237:770–775). (Adapted from Yu FH & Catterall WA [2004] *Science STKE* 253:re15. With permission from AAAS; adapted from Sato C, Ueno Y, Asai K et al. [2001] *Nature* 409, 1047–1051.)

segment, termed S4, contains many positively charged amino acids and was hypothesized to be the sensor that detects voltage changes for channel gating. A hydrophobic stretch of amino acids between the fifth and sixth transmembrane segments (S5 and S6) forms an extra pore loop within the membrane. As we will learn in more detail in Section 2.16, the pore loop, S5, and S6 together form the central pore for ion conduction.

While studies in voltage-gated Na⁺ channels benefited from the electric organ, the lack of a similarly enriched source of K⁺ channels and the overall heterogeneity of K⁺ channels made them resistant to similar biochemical approaches. Fortunately, genetic studies in the fruit fly *Drosophila melanogaster* provided an alternative strategy for cloning genes that encode K⁺ channels. A *Drosophila* mutant named **Shaker**, so-called because the mutant flies shake their legs under ether anesthesia, exhibited defects in a fast and transient K⁺ current in muscles and neurons, as well as defects in action potential repolarization. These findings led to the hypothesis that a K⁺ channel was disrupted in the *Shaker* mutant flies. **Positional cloning** of DNA corresponding to the *Shaker* mutant (see Section 13.6 for details) identified the first voltage-gated K⁺ channel (Figure 2–31B). Interestingly, this K⁺ channel protein resembles one of the four repeating modules of the Na⁺ channel; like each of these modules, it has six transmembrane segments, including the positively charged S4 and the pore loop. Subsequent work showed that four such polypeptides (subunits) constitute one functional K⁺ channel.

Cloning of ion channels enabled structure–function studies to investigate the molecular mechanisms underlying different channel properties. One specific example is the inactivation of voltage-gated Na⁺ channel, which contributes to the repolarization phase of the action potential, and allows action potentials to travel unidirectionally with a refractory period (see Sections 2.10 and 2.11). Biophysical studies in the 1970s led to a '**ball-and-chain**' inactivation model in which a cytoplasmic portion of the channel protein (the ball), connected to the rest of the channel by a polypeptide chain, was hypothesized to block the channel pore after the ion channel opens. It was further hypothesized that depolarization not only opens the channel, but also causes movement of charged amino acids to

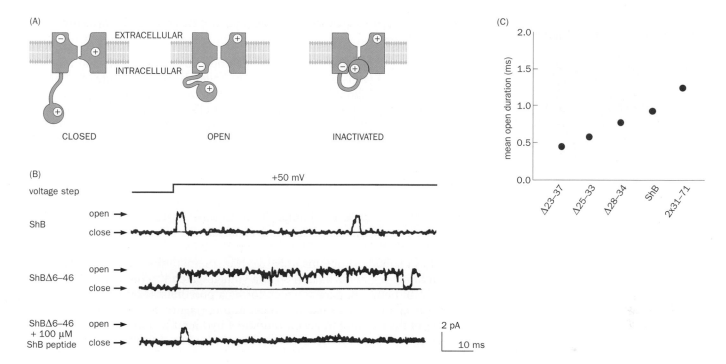

Figure 2–32 **Molecular mechanisms of voltage-gated ion channel inactivation. (A)** The ball-and-chain model of voltage-gated Na$^+$ channel inactivation. Depolarization causes the channel to open, and at the same time moves charged amino acids such that the inner pore of the channel becomes more negatively charged, creating a binding site for a portion of the channel's cytoplasmic domain (the ball), which is positively charged. The binding of the ball to the inner pore inactivates the channel. **(B)** Mutagenesis studies of the Shaker K$^+$ channel (the ShB isoform) support the ball-and-chain model. In patch clamp experiments, the wild-type ShB channel (top) inactivates after initial depolarization induced opening, and remains closed despite continued depolarization at +50 mV. When amino acids 6–46 are deleted from ShB (middle), the channel exhibits defective inactivation, as the channel remains open after depolarization. The inactivation defect of the mutant channel is corrected by supplying a soluble 'ball' peptide comprising the first 20 amino acids of the cytoplasmic domain (bottom). **(C)** Compared to the wild-type channel (fourth column), the mean channel open duration is shortened by deleting 15, 9, and 7 amino acids from the chain that connects the ball to the rest of the channel (first three columns, respectively). The mean channel open duration is lengthened by adding 41 amino acids to the chain (fifth column, duplicating amino acids 31–71). (A, adapted from Armstrong CM & Bezanilla F [1977] *J Gen Physiol* 70:567–590; B & C, adapted from Hoshi T, Zagotta WN & Aldrich RW [1990] *Science* 250:533–538 and Zagotta WN, Hoshi T & Aldrich RW [1990] *Science* 250:568–571.)

create a negatively charged inner channel pore, which facilitates the binding of a positively charged ball (**Figure 2–32**A).

This ball-and-chain model was elegantly validated in the voltage-gated Shaker K$^+$ channel, which undergoes inactivation similar to the voltage-gated Na$^+$ channel. By using a molecular biology technique called *in vitro* **mutagenesis** to alter the DNA sequence of the cloned *Shaker* gene, researchers could express Shaker proteins in which selected stretches of amino acids were deleted, inserted, or replaced. This work revealed that the cytoplasmic N-terminal domain of the Shaker K$^+$ channel was necessary for its inactivation. Deleting a stretch of amino acids in this domain generated a mutant K$^+$ channel that could not undergo inactivation after depolarization-induced channel opening. Application of a peptide containing the first 20 amino acids of the cytoplasmic domain was sufficient to restore the inactivation of this mutant K$^+$ channel (Figure 2–32B). Thus, the first 20 amino acids correspond to the ball. As predicted from the ball-and-chain model, several positively charged amino acids within the first 20 amino acids were found to be crucial for inactivation. Further, decreasing the length of the chain—that is, the intervening polypeptide between the first 20 amino acids and the rest of the channel—resulted in a channel that was open for a shorter period prior to inactivation. Lengthening the chain increased the duration that the channel remained open, suggesting that the length of the chain dictates the 'search' time for the ball to find the open channel (Figure 2–32C). This example illustrates the power of combining molecular biology and electrophysiology to understand the mechanisms of ion channel function.

2.16 Crystal structures reveal the atomic bases of ion channel properties

Ion channels are remarkable molecular machines. Voltage-gated K$^+$ channels can conduct up to 10^8 K$^+$ ions per second, which is near the diffusion rate of K$^+$, while also maintaining a high selectivity for K$^+$: the channel conducts ~10,000 times fewer Na$^+$ ions than K$^+$ ions. Central to the channel's conduction and ion selectivity is the pore loop (see Figure 2–31B). *In vitro* mutagenesis studies have shown that mutations in the pore loop alter ion selectivity. Subsequent structural studies at atomic resolution using X-ray crystallography have provided detailed mechanisms of ion conduction and selectivity.

The amino acid residues at the K$^+$ channel pore loop are highly conserved across a wide range of organisms, from bacteria to humans. Thus, the crystal structure of the bacterial K$^+$ channel KcsA revealed mechanisms of K$^+$ conduction and selectivity that are likely universal. The KcsA channel is not voltage gated, but nevertheless resembles part of the voltage-gated K$^+$ channel: each of the four KcsA subunits contains only two transmembrane helices (equivalent to transmembrane segments S5 and S6 of a voltage-gated K$^+$ channel subunit) and a pore loop in between, part of which forms a pore helix (**Figure 2–33**A). From the side, the channel looks like a conical funnel (Figure 2–33B), with a cavity at the center of the channel to accommodate a hydrated K$^+$ ion. (Ions in solution, including K$^+$ and Na$^+$, are normally present in hydrated form.) Electronegative carboxyl ends of the four pore helices face the cavity and stabilize the K$^+$ ion as it travels through. When the channel is open, intracellular hydrated K$^+$ has free access to the cavity. Between the cavity and the extracellular side is the **selectivity filter**, through which K$^+$ ions pass in dehydrated form. This is accomplished via the interaction of K$^+$ with the electronegative carbonyl groups from the main polypeptide chain corresponding to the most highly conserved amino acids of K$^+$ channels. These close carbonyl interactions, which mimic and replace the water molecules that surround the K$^+$ ion in solution, perfectly match the size of the K$^+$ ion but not the smaller Na$^+$ ion. This accounts for the K$^+$ channel's high degree of ion selectivity and the favorable energetics of K$^+$ conduction through an open

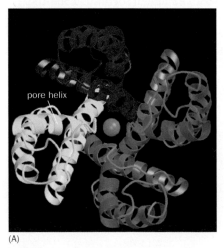
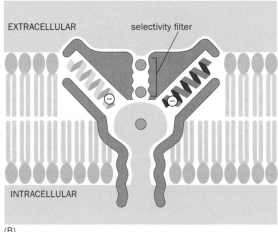
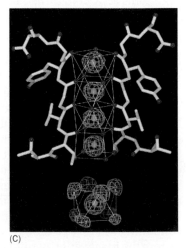

(A)   (B)   (C)

Figure 2–33 Atomic structure of KcsA, a K$^+$ channel from bacteria.
(A) The KcsA channel viewed from the extracellular side. Each of the four subunits (differently colored) consists of two transmembrane helices and a shorter pore helix (indicated for the yellow subunit). A green K$^+$ ion is passing through the central pore. **(B)** From a side view, the KcsA channel resembles a conical funnel. An aqueous passage extends from the intracellular side to a central cavity in the middle of the lipid bilayer. Three K$^+$ ions are shown, one in the cavity and two in the selectivity filter above. The gray shading represents the channel protein viewed from the side. Two of the four pore helices are shown in this side view, with their electronegative carboxyl ends facing the cavity to stabilize the positively charged K$^+$ ion in the

cavity. **(C)** Atomic structure of the selectivity filter. Oxygen atoms from carbonyl groups of the main polypeptide chain (red) create four K$^+$ binding sites within the selectivity filter. At each binding site, the K$^+$ ion is surrounded by eight oxygen atoms (red dots), just as the hydrated K$^+$ ion is surrounded by eight oxygen atoms from water in the cavity shown at the bottom. This mimicry renders conduction of K$^+$ (but not the smaller Na$^+$) more energetically favorable, providing a mechanism for ion selectivity. (A & B, adapted from Doyle DA, Cabral JM, Pfuetzner RA et al. [1998] *Science* 280:69–77. With permission from AAAS; C, adapted from Zhou Y, Morais-Cabral JH, Kaufman A et al. [2001] *Nature* 414:43–48. With permission from Macmillan Publishers Ltd.)

channel (Figure 2–33C; **Movie 2–8**). There are four possible positions for K$^+$ at the selectivity filter, which is usually occupied by two K$^+$ ions at alternate positions. The repulsion between the two K$^+$ ions, coupled with the electrochemical gradient, forces K$^+$ to flow rapidly through the selectivity filter when the channel is open and when there is a driving force.

Following these pioneering studies on the KcsA K$^+$ channel, the structures of many ion channels (see **Box 2–4**) have been solved, including the mammalian homolog of the voltage-gated Shaker K$^+$ channel in its open state. How is the channel gated by voltage? It has been hypothesized since the cloning of voltage-gated Na$^+$ and K$^+$ channels that the fourth transmembrane segment (S4), which contains several positively charged amino acids (see Figure 2–31), plays a key role in voltage sensing. Subsequent structure–function studies indicated the importance of the S4 segment in voltage sensing, and suggested that the segments S1–S4 together constitute a voltage-sensing unit. The X-ray crystal structure revealed detailed arrangements of the S1–S4 helices with regard to the rest of the channel—S5, S6, and the pore loop that constitute the ion conduction pore. Further structural studies on voltage-gated channels in different states promise to provide deeper insights into how the voltage sensor moves in response to depolarization and how that movement triggers channel opening.

Sixty years after the ionic basis of the action potential was revealed, researchers have come a long way in elucidating the mechanisms that underlie this most fundamental form of neuronal communication.

Box 2–4: Diverse ion channels for diverse functions

The voltage-gated Na$^+$ and K$^+$ channels we discussed in the context of the action potential are just two of many kinds of ion channels. In the human genome, more than 200 genes encode ion channels (**Table 2–2**). Ion channels are

Table 2–2: Number of genes encoding ion channels in the human genome

Channel type	Gene number
K$^+$ channels	78
Voltage-gated K$^+$ channels	40
Inward-rectifier K$^+$ channels	15
Two-pore K$^+$ channels	15
Ca^{2+}-activated K$^+$ channels	8
Na$^+$/Ca^{2+} channels	27
Voltage-gated Na^{2+} channels	9
Voltage-gated Ca^{2+} channels	10
Other Ca^{2+} channels	8
Cl$^-$ channels	9
TRP channels	28
Cyclic-nucleotide-gated and HCN channels	10
Neurotransmitter-gated channels	70

Data from the IUPHAR (International Union of Basic and Clinical Pharmacology) database (www.iuphar-db.org).
Abbreviations: TRP, transient receptor potential; HCN, hyperpolarization-activated cyclic-nucleotide-gated.
This table does not include all discovered ion channels.

usually classified based on the ions they conduct and the mechanisms by which they are gated. Many ion channels also share sequence similarities with each other, reflecting their shared evolutionary history (see Chapter 12 for details). **Figure 2–34** depicts a phylogenetic tree for 143 structurally related ion channels, which accounts for the majority of ion channels encoded in the human genome. All 143 channels share a common pore structure with two transmembrane helices (2TMs) and a pore loop. All except two families of K$^+$ channels use a 6TM-unit similar to voltage-gated Na$^+$ and K$^+$ channels (see Figure 2–31).

K$^+$ channels are the most diverse of the channel families and are encoded by at least 78 genes, many of which have multiple alternatively spliced isoforms. They play important roles in many functions of excitable cells, establishing such diverse conditions as the resting potential, the kinetics of repolarization after action potential initiation, and the spontaneous rhythmic firing of pacemaker cells. Many K$^+$ channels are activated by depolarization with differing activation and inactivation kinetics, adapting to neurons with different firing frequencies. Some K$^+$ channels are activated by a rise in intracellular Ca^{2+} or a drop of ATP concentration, allowing cells to alter membrane potentials in response to changes of [Ca^{2+}]$_i$ or energy levels. Members of the **inward-rectifier K$^+$ channel** subfamily preferentially pass inward currents at membrane potentials more hyperpolarized than E_K, and allow minimal outward currents at membrane potentials more positive than E_K. This is because under depolarizing conditions, the inward-rectifier K$^+$ channels are blocked from the intracellular side by positively charged polyamines and Mg^{2+}. Among other

(Continued)

Box 2–4: Diverse ion channels for diverse functions

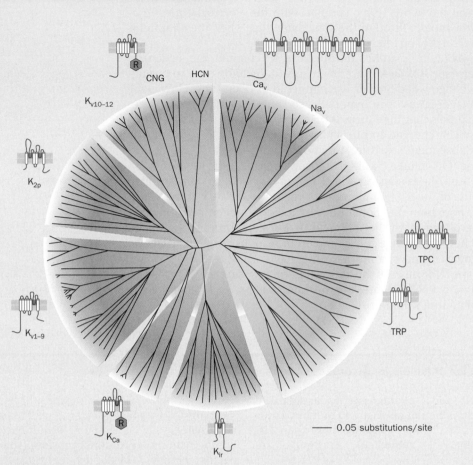

Figure 2–34 Phylogenetic tree of 143 ion channels. This tree is constructed according to similarities in the amino acid sequences of the conserved pore regions of 143 structurally related ion channels, which correspond to the majority of the ion channels listed in Table 2–2; the neurotransmitter-gated channels and chloride channels are not shown here. The scale bar represents a distance on the tree that corresponds to 0.05 amino acid substitutions per site in the sequence. Background shades separate ion channels into related groups: blue, voltage-gated Na$^+$ (Na$_v$) and Ca^{2+} (Ca$_v$) channels; green, transient receptor potential (TRP) and related channels, including two-pore channels (TPC); red, most K$^+$ channels including inward-rectifier K$^+$ channels (K$_{ir}$), Ca^{2+}-dependent K$^+$ channels (K$_{Ca}$), the first nine subfamilies of voltage-gated K$^+$ channels (K$_{V1-9}$), and two-pore K$^+$ channels (K$_{2p}$); orange, cyclic nucleotide-gated (CNG) channels, including structurally related hyperpolarization-activated cyclic nucleotide-gated (HCN) channels and the last three subfamilies of voltage-gated K$^+$ channels (K$_{V10-12}$), which contain a cyclic-nucleotide-binding domain. The schematics surrounding the tree illustrate the membrane topologies of the channel proteins, with the pore loops shaded in dark gray. R within a red hexagon represents cytoplasmic domains that possess cyclic nucleotide- or Ca^{2+}-binding domains. (Adapted from Yu FH & Catterall WA [2004] *Science STKE* 253:re15. With permission from AAAS.)

functions, inward-rectifier K$^+$ channels help maintain the resting potential near E_K. K$^+$ channels are also present in many non-excitable cells. For example, K$^+$ channels are the predominant ion channels in glia.

Like K$^+$ channels, **Cl$^-$ channels** generally stabilize the resting membrane potential, as E_K and E_{Cl} are both near the resting potential. Cl$^-$ channels also play diverse roles in different cell types, often involving intracellular vesicles. Cl$^-$ channel structures are distinct from all other channels (and thus are not represented in Figure 2–34). They consist of two subunits, each of which has 18 membrane-embedded helices and an ion conduction pore. In fact, eukaryotic Cl$^-$ channels are in the same family as bacterial Cl$^-$/H$^+$ exchangers, suggesting that ion channels and transporters can share structural similarities.

Voltage-gated Ca^{2+} channels constitute another important class of proteins in excitable cells. Their primary structure resembles those of voltage-gated Na$^+$ channels, with four repeating modules each containing six transmembrane helices. Different voltage-gated Ca^{2+} channels differ in their activation threshold, single channel conductance, and inactivation speed. Neurons usually maintain a very low intracellular Ca^{2+} concentration of ~0.1 μM, which is >10,000-fold smaller than the extracellular Ca^{2+} concentration (~1.2 mM); this produces a very large E_{Ca} of approximately +120 mV. Opening of voltage-gated Ca^{2+} channels

Box 2–4: Diverse ion channels for diverse functions

thus leads to depolarization due to Ca^{2+} influx driven by both chemical and electrical gradients. Indeed, action potentials in some neurons are mediated by voltage-gated Ca^{2+} channels instead of voltage-gated Na^+ channels. As we will learn in later chapters, voltage-gated Ca^{2+} channels play important roles in regulating neurotransmitter release at the axon terminal. They are also essential for excitation–contraction coupling of muscles, for dendritic integration in some mammalian neurons, and for regulating gene expression and neuronal differentiation in response to neuronal activity. Other Ca^{2+} channels are present on the membrane of internal Ca^{2+} stores and are gated by intracellular messengers.

Whereas K^+, Cl^-, Na^+, and Ca^{2+} channels are so named because of their ion selectivity, some channels are not as selective for the ions they conduct. For example, most **TRP channels** (named after the founding member, transient receptor potential, a *Drosophila* protein essential for visual transduction) and **CNG channels** (for cyclic nucleotide-gated channels) are nonselective cation channels, which means that they are permeable to Na^+, K^+, and sometimes Ca^{2+}. Because the driving force for Na^+ is typically larger than that for K^+, their opening causes more Na^+ influx than K^+ efflux, and therefore produces a net depolarization. As we will learn in Chapters 4 and 6, CNG channels and TRP channels play important roles in sensory neurons to convert environmental stimuli—including light, odorants, pheromones, temperature, and noxious chemicals—into membrane potential changes. TRP channels also contribute to mechanosensation in *Drosophila* and *C. elegans*. The **HCN channels** (for hyperpolarization-activated cyclic nucleotide-gated channel) are structurally related to CNG channels. They are activated by hyperpolarization, usually below –55 mV, in addition to cyclic nucleotides. Because they conduct cations and thus depolarize cells when open

under hyperpolarization conditions, HCN channels are particularly important for rhythmic firing of neurons and rhythmic heart beating.

At least 70 genes in the human genome encode ion channels that are gated by neurotransmitters (Table 2–2); these channels belong to different gene families from those depicted in Figure 2–34. Many of the neurotransmitter-gated channels are nonselective cation channels, and thus their opening depolarizes and causes excitation of postsynaptic neurons. Some neurotransmitter-gated channels are selective for Cl^-; their opening usually mediates inhibition of postsynaptic neurons. We will study the structure and function of these neurotransmitter-gated ion channels in greater detail in Chapter 3.

More than a decade after sequencing the human genome, new ion channels are still being discovered. For instance, as will be discussed in Chapter 6, ion channels that are gated by mechanical forces mediate hearing and touch sensation. In mammals, the molecular nature of mechanosensitive channels is still being intensely investigated and they do not appear to belong to the ion channel families discussed above. For example, a subset of mechanosensitive channels that mediate touch belong to an evolutionarily conserved family of proteins called the **Piezos**, which contain >30 transmembrane segments per subunit with no sequence resemblance to other known ion channels. We expect further additions to the ion channel list (Table 2–2) in the future. Finally, mutations in many human ion channels cause or increase the susceptibility to a variety of nervous system disorders, including epilepsy (see Box 11–4), schizophrenia, autism (see Section 11.26), migraine, and abnormal pain sensitivity, highlighting the importance of ion channels in human health.

SUMMARY

Neurons are extraordinarily large cells. The surface areas and volumes of axons or dendrites often exceed those of cell bodies by several orders of magnitude. To support these structures and their functions, neurons adopt specialized cell biological properties. mRNAs, ribosomes, and secretory pathway components are present in dendrites so that cytosolic and membrane proteins can be synthesized and processed locally; local protein synthesis also occurs at least in developing axons. Organelles and soma-synthesized proteins are actively transported to axons and dendrites by specific microtubule motors. Axonal microtubules are oriented uniformly with their plus ends facing out toward the terminal. A large family of kinesins, which are mostly plus-end-directed microtubule motors, mediate both fast and slow axonal transport to deliver membrane proteins (via intracellular vesicles) and cytosolic proteins from the soma to the axon terminals. Dynein and minus-end-directed kinesins mediate retrograde transport from axon terminals back to the soma. Kinesins and dynein also transport cargos within dendrites. In vertebrate neurons, dendrites possess both plus-end-out and minus-end-out microtubules. The microtubule polarity difference in axons and dendrites is critical for directing specific cargos to appropriate subcellular compartments.

Whereas microtubules run along the centers of dendritic and axonal processes, F-actin is enriched at the periphery, including in dendritic spines and presynaptic terminals. Some cargos reach their final destination via myosin-based transport on F-actin fibers after they leave the microtubule highway.

Electrical signaling in excitable cells is enabled by the properties of the lipid bilayer and activities of special membrane proteins such as transporters and ion channels on the plasma membrane. Active transporters, such as the Na^+–K^+ ATPase, use energy to transport ions across the membrane against their electro-chemical gradients; these transporters maintain concentration differences across the plasma membrane for Na^+, K^+, and Cl^-. In most neurons and muscles, the intracellular compartment is high in K^+ but low in Na^+, Ca^{2+}, and Cl^- compared with the extracellular environment. Because the membrane at rest is more permeable to K^+ than any other ions, the resting membrane potential, which is typically between –50 and –80 mV, is close to the K^+ equilibrium potential.

The neuronal plasma membrane can be effectively described as a parallel *R-C* circuit, with conductance paths for each ion representing ion flow through specific channels, and a capacitance path representing the lipid bilayer. Because of the inherent properties of the *R-C* circuit, electrical signals, such as the membrane potential change in response to current injection, evolve over time, reaching 63% of their maximal magnitude over one time constant. When electrical signals passively propagate along a neuronal fiber, leaky membrane conductance along the way causes the signals to decay across distance; for instance, the membrane potential change attenuates to 37% of the original magnitude across one length constant. In order to propagate electrical signals reliably across a long distance, axons employ active properties such as the action potential.

Action potentials are produced by depolarization above the threshold. Depolarization first opens voltage-gated Na^+ channels, leading to further depolarization and accounting for the rapid rising phase. The falling phase of action potentials is caused by inactivation of Na^+ channels and delayed opening of voltage-gated K^+ channels. This sequence ensures that action potentials are all-or-none, regenerative events that propagate unidirectionally along the axon from the cell body to the axon terminals. Studies utilizing important tools developed in the past decades, such as patch clamp recording, molecular cloning, and membrane protein crystallography, have revealed the molecular and mechanistic basis of how ion channels conduct ions with exquisite selectivity, how channel opening is controlled by voltage, how inactivation occurs, and how the properties of individual ion channels account for macroscopic current in response to membrane potential changes.

In addition to mediating action potentials, ion channels serve diverse functions. We will study these functions in greater detail in subsequent chapters, starting with the central subjects of our next chapter: neurotransmitter release at the presynaptic terminal and neurotransmitter reception at the postsynaptic specialization.

FURTHER READING

Books and reviews

Alberts B, Johnson A, Lewis J et al. (2015) Molecular Biology of the Cell, 6th ed. Garland Science.

Hille B (2001) Ion Channels of Excitable Membranes, 3rd ed. Sinauer.

Hirokawa N, Niwa S & Tanaka Y (2010) Molecular motors in neurons: transport mechanisms and roles in brain function, development, and disease. *Neuron* 68:610–638.

Holt CE & Schuman EM (2013) The central dogma decentralized: new perspectives on RNA function and local translation in neurons. *Neuron* 80:648–657.

Katz B (1966) Nerve, Muscle, and Synapse. McGraw-Hill.

Miller C (2006) ClC chloride channels viewed through a transporter lens. *Nature* 440:484–489.

Yu FH & Catterall WA (2004) The VGL-chanome: a protein superfamily specialized for electrical signaling and ionic homeostasis. *Sci STKE* 2004:re15.

Cell biological properties of neurons

Allen RD, Metuzals J, Tasaki I et al. (1982) Fast axonal transport in squid giant axon. *Science* 218:1127–1129.

Burgin KE, Waxham MN, Rickling S et al. (1990) *In situ* hybridization histochemistry of Ca^{2+}/calmodulin-dependent protein kinase in developing rat brain. *J Neurosci* 10:1788–1798.

Lasek R (1968) Axoplasmic transport in cat dorsal root ganglion cells: as studied with [3-H]-L-leucine. *Brain Res* 7:360–377.

Park HY, Lim H, Yoon YJ et al. (2014) Visualization of dynamics of single endogenous mRNA labeled in live mouse. *Science* 343:422–424.

Steward O & Levy WB (1982) Preferential localization of polyribosomes under the base of dendritic spines in granule cells of the dentate gyrus. *J Neurosci* 2:284–291.

Vale RD, Reese TS & Sheetz MP (1985) Identification of a novel force-generating protein, kinesin, involved in microtubule-based motility. *Cell* 42:39–50.

Vale RD, Schnapp BJ, Reese TS et al. (1985) Organelle, bead, and microtubule translocations promoted by soluble factors from the squid giant axon. *Cell* 40:559–569.

Electrical properties of neurons, action potentials, and ion channels

Aguayo AJ, Attiwell M, Trecarten J et al. (1977) Abnormal myelination in transplanted Trembler mouse Schwann cells. *Nature* 265:73–75.

Armstrong CM & Bezanilla F (1977) Inactivation of the sodium channel. II. Gating current experiments. *J Gen Physiol* 70:567–590.

Doyle DA, Morais Cabral J, Pfuetzner RA et al. (1998) The structure of the potassium channel: molecular basis of K$^+$ conduction and selectivity. *Science* 280:69–77.

Hodgkin AL & Huxley AF (1952) Currents carried by sodium and potassium ions through the membrane of the giant axon of *Loligo*. *J Physiol* 116:449–472.

Hodgkin AL & Huxley AF (1952) A quantitative description of membrane current and its application to conduction and excitation in nerve. *J Physiol* 117:500–544.

Hodgkin AL & Katz B (1949) The effect of sodium ions on the electrical activity of giant axon of the squid. *J Physiol* 108:37–77.

Hoshi T, Zagotta WN & Aldrich RW (1990) Biophysical and molecular mechanisms of Shaker potassium channel inactivation. *Science* 250:533–538.

Noda M, Shimizu S, Tanabe T et al. (1984) Primary structure of *Electrophorus electricus* sodium channel deduced from cDNA sequence. *Nature* 312:121–127.

Papazian DM, Schwarz TL, Tempel BL et al. (1987) Cloning of genomic and complementary DNA from Shaker, a putative potassium channel gene from *Drosophila*. *Science* 237:749–753.

Patel PI, Roa BB, Welcher AA et al. (1992) The gene for the peripheral myelin protein PMP-22 is a candidate for Charcot-Marie-Tooth disease type 1A. *Nat Genet* 1:159–165.

Payandeh J, Scheuer T, Zheng N et al. (2011) The crystal structure of a voltage-gated sodium channel. *Nature* 475:353–358.

Schwarz TL, Tempel BL, Papazian DM et al. (1988) Multiple potassium-channel components are produced by alternative splicing at the Shaker locus in *Drosophila*. *Nature* 331:137–142.

Sigworth FJ & Neher E (1980) Single Na$^+$ channel currents observed in cultured rat muscle cells. *Nature* 287:447–449.

Taveggia C, Zanazzi G, Petrylak A et al. (2005) Neuregulin-1 type III determines the ensheathment fate of axons. *Neuron* 47:681–694.

Zagotta WN, Hoshi T & Aldrich RW (1990) Restoration of inactivation in mutants of Shaker potassium channels by a peptide derived from ShB. *Science* 250:568–571.

Zhou Y, Morais-Cabral JH, Kaufman A et al. (2001) Chemistry of ion coordination and hydration revealed by a K$^+$ channel-Fab complex at 2.0 Å resolution. *Nature* 414:43–48.

CHAPTER 3
Signaling across Synapses

Processes which go through the nervous system may change their character from digital to analog, and back to digital, etc., repeatedly.

John von Neumann (1958),
The Computer & the Brain

In this chapter, we continue the theme of neuronal communication begun in Chapter 2. We discuss first how the arrival of an action potential at the presynaptic terminal triggers neurotransmitter release from synaptic vesicles, and then how neurotransmitters affect the properties of postsynaptic cells. Collectively, these processes are referred to as **synaptic transmission**, through which information is transmitted from the presynaptic cell to the postsynaptic cell across the chemical synapse. In the context of studying postsynaptic reception, we also introduce the fundamentals of signal transduction and describe how synaptic inputs are integrated in postsynaptic neurons. Finally we discuss the electrical synapse, an interneuronal communication form in parallel to the chemical synapse.

HOW IS NEUROTRANSMITTER RELEASE CONTROLLED AT THE PRESYNAPTIC TERMINAL?

In Chapter 2, we addressed the basic cell biological and electrical properties of neurons that are required to understand how molecules, organelles, and action potentials get to the axon terminals. We will now study the major purpose of these movements: to transmit information across synapses to postsynaptic targets, which can be other neurons or muscle cells.

3.1 Action potential arrival at the presynaptic terminal triggers neurotransmitter release

The vertebrate **neuromuscular junction**, the synapse between the motor neuron axon terminals and the skeletal muscle, has been used as a model synapse to explore many basic properties of synaptic transmission that were later found to be widely applicable to other synapses. **Neurotransmitters** are molecules released by presynaptic neurons that act across the synaptic cleft on postsynaptic target cells. The neurotransmitter at the vertebrate neuromuscular junction was identified in the 1930s as **acetylcholine** (**ACh**) (**Figure 3–1**A). An important advantage of studying the neuromuscular synapse is that the postsynaptic muscle cell (also called muscle fiber) is a giant cell that can easily be impaled by a microelectrode for intracellular recording (see Section 13.21); this enables synaptic transmission to be assessed in a sensitive and quantitative manner by recording the resulting current or membrane potential changes in the muscle fiber. The neuromuscular junction is also an unusual synapse in that a single motor axon forms many terminal branches, which harbor hundreds of sites releasing neurotransmitter onto its target muscle, making it a strong and reliable synapse for transmitting action potentials in motor neurons to muscle contraction via depolarization-induced action potentials in the muscle fiber (to be discussed in more detail in Chapter 8). Indeed, in experiments described below, researchers typically adjusted the conditions to prevent muscle action potentials and muscle contraction.

Figure 3–1 Studying synaptic transmission at the vertebrate neuromuscular junction. **(A)** Structure of acetylcholine (ACh), the first identified neurotransmitter. **(B)** Measuring depolarization of a muscle fiber in response to motor axon stimulation or ACh iontophoresis in a neuromuscular junction *in vitro*. The intracellular electrode is inserted in the muscle fiber close to the neuromuscular junction to record the end-plate potential (EPP) in response to motor axon stimulation or focal ACh application at the surface of the muscle close to the neuromuscular junction. The square wave on the motor axon represents an application of current that depolarizes the motor axon, causing it to fire action potentials. The square wave attached to the ACh pipette represents application of positive current that drives positively charged ACh out of the micropipette. **(C)** End-plate potentials (EPPs) in muscle fiber in response to motor axon stimulation (top) or focal ACh application (bottom) are similar in waveform. The first downward dip in the top trace indicates the time of axon stimulation. (C, adapted from Krnjevic K & Miledi R [1958] *Nature* 182:805–806. With permission from Macmillan Publishers Ltd.)

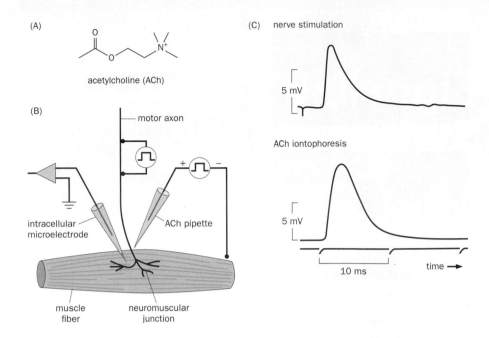

In a typical setup for studying synaptic transmission across the neuromuscular junction, an *in vitro* preparation that contains the muscle and its attached motor nerve was bathed in a solution that mimics physiological conditions. The motor nerve was then stimulated with a stimulating electrode to produce action potentials, and the membrane potential of the muscle fiber was recorded with an intracellular electrode (Figure 3–1B). Motor nerve stimulation was found to induce a transient depolarization in the muscle fiber within a few milliseconds (Figure 3–1C, top panel). This transient depolarization is the synaptic potential produced in the postsynaptic muscle cell, and is called an **end-plate potential**, or **EPP**, as the postsynaptic specialization area of the muscle fiber is also called a motor end plate. We will study the postsynaptic mechanism that produces the EPP in greater detail in the second part of this chapter. For now, we use the EPP as a measure for presynaptic mechanisms that cause neurotransmitter release.

How does motor nerve stimulation produce an EPP? Researchers found that motor nerve stimulation can be mimicked by application of ACh through a micropipette at the contact site between the motor axon terminals and the muscle (Figure 3–1C, bottom panel). (This method is termed **iontophoresis**; here, positively charged ACh is driven out of a micropipette by applying a positive current.) Adding the puffer fish tetrodotoxin (TTX; see Figure 2–29) to the bath, which blocks voltage-gated Na⁺ channels and thus prevents action potential propagation in motor axons, blocked the muscle EPP in response to motor nerve stimulation. However, ACh application could elicit an EPP even when action potentials were blocked by TTX, or when the motor axon was removed altogether. These results indicated that the final effect of action potentials in the motor axon is to trigger ACh release at the axon terminals, and binding of ACh to the muscle membrane triggers depolarization of the muscle fiber in the form of an EPP.

As introduced in Chapter 1, we now know that ACh release is caused by fusion of synaptic vesicles with presynaptic plasma membrane, releasing packets of ACh molecules into the synaptic cleft. The concept that neurotransmitters are released in discrete packets, however, was first deduced prior to the discovery of synaptic vesicles.

3.2 Neurotransmitters are released in discrete packets

Bernard Katz and colleagues applied intracellular recording techniques, then newly invented, to the muscle cells to study the mechanisms of neuromuscular synaptic transmission in the early 1950s. While studying the muscle EPPs evoked by nerve stimulation in the frog neuromuscular junction, they observed that

(A)

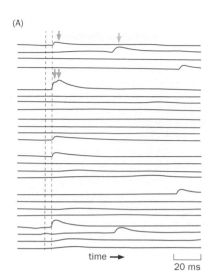

time →

20 ms

(B)

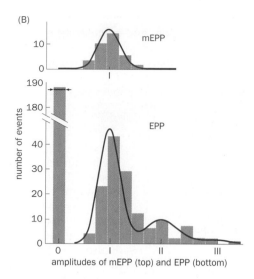

amplitudes of mEPP (top) and EPP (bottom)

Figure 3–2 Miniature end-plate potentials (mEPPs) and a statistical test of quantal neurotransmitter release. (A) At low Ca²⁺ concentration, nerve stimulation (at the time indicated by the first dotted vertical line) infrequently evokes EPPs, each of which follows the nerve stimulus with a specific latency (second dotted vertical line). In the 24 experiments shown here (each represented by a horizontal sweep), five EPPs were evoked (the first is indicated by a yellow arrow; two yellow arrows on the fifth line indicate a doublet, when two quanta were released). Note also the presence of four depolarization events not linked with nerve stimulation; these mEPPs are of a similar size to the evoked EPPs (the mEPP on the second trace is indicated by a cyan arrow). **(B)** Using mEPP (top) as the unitary size, the frequency distribution of evoked EPPs was predicted by the Poisson distribution shown as a continuous curve at the bottom. This fits well with the experimental data plotted as a histogram showing the number of EPPs (y axis) whose amplitudes fit within a certain bin (x axis). Note that the amplitude of mEPP (top) is a continuous variable (the sizes of each neurotransmitter packet are not exactly the same), with the peak around the Roman numeral I representing the average mEPP amplitude (0.875 mV). Likewise, the EPP amplitude (bottom) is also a continuous variable. Roman numerals I, II, and III representing 1×, 2×, and 3× the average mEPP amplitude. The frequency of synaptic failure (0 amplitude) also matches well with the prediction from the Poisson distribution (red line flanked by two arrows). (Adapted from Del Castillo J & Katz B [1954] *J Physiol* 124:560–573.)

muscle fibers also exhibited small EPPs in the absence of any nerve stimulation; these were termed **miniature end-plate potentials**, or **mEPPs**. mEPPs had an intriguing property: for a given neuromuscular preparation, they seemed to have either a defined, unitary size or occasionally a multiple of that defined size. The amplitude of mEPPs, hypothesized to be due to spontaneous release of ACh from motor axon terminals, was usually two orders of magnitude lower than EPPs evoked by nerve stimulation. However, when the extracellular solution contained very low concentrations of Ca²⁺ and high concentrations of Mg²⁺ (both of which inhibit neurotransmitter release, as will be discussed in Section 3.4), a condition could be reached in which most nerve stimulations did not evoke any EPPs. When stimuli did trigger EPPs under these conditions, the amplitude of these evoked EPPs were the same size as mEPPs (**Figure 3–2A**). Further reduction of Ca²⁺ concentrations reduced the frequency of these evoked EPPs, but did not further diminish their amplitude. These observations suggested that mEPPs were the basic unit of synaptic transmission for EPPs induced by nerve stimulation, which under normal conditions were equivalent to the simultaneous occurrence of hundreds of mEPPs. Furthermore, whether spontaneous or triggered by nerve stimulation, most ACh release occurred in the same basic unit with a finite quantal (packet) size, and occasionally two or three times the unit size. These observations led to a **quantal hypothesis of neurotransmitter release**, that is, neurotransmitters are released in discrete quanta of relatively uniform size.

If the quantal hypothesis were true, one could use statistical methods (see **Box 3–1** for details) to predict the frequencies of releasing no quantum, a single quantum, or multiple quanta in response to nerve stimulation. When the release probability is small, which is the case when the neuromuscular junction is in low-Ca²⁺, high-Mg²⁺ medium, the frequency (*f*) that *k* quanta are released per nerve stimulation could be calculated following the **Poisson distribution**:

$$f = \frac{m^k}{k!} e^{-m}$$

where *m* is the mean number of units (quanta) that respond to an individual stimulus. Since mEPPs from each spontaneous release correspond to one unit, *m* can be experimentally determined as the mean EPP amplitude divided by the mean mEPP amplitude. Indeed, the frequency distributions of EPPs calculated above and determined experimentally were an excellent fit: the frequency of the cases when nerve stimulation did not cause any EPP (called **synaptic failure**) matched precisely with the statistical prediction; there was a prominent peak at around the size of the unitary mEPP, and a small peak at twice the mEPP amplitude (Figure 3–2B). Thus, this statistical analysis provided strong support that neurotransmitters are released in discrete packets.

Box 3–1: Binomial distribution, Poisson distribution, and calculating neurotransmitter release probability

The Poisson distribution and the related **binomial distribution** are both probability distributions that describe the frequency of discrete events that occur independently. Let's start our discussion with the binomial distribution. Suppose the probability that an individual event occurs, such as the head faces up after you toss a coin, is p. The binomial distribution describes the frequency (f) in which k events occur (that is, k times heads facing up after coin toss) after n trials:

$$f(k; n, p) = \frac{n!}{k!(n-k)!} p^k (1-p)^{n-k}$$

where $k = 0, 1, 2, \ldots n$, ! is factorial (for example, $4! = 4 \times 3 \times 2 \times 1 = 24$), and $n!/k!(n-k)!$ is the binomial coefficient. Suppose you want to know the likelihood of tossing a coin four times and having only instance of heads. The probability for heads, p, is 0.5 for any given toss of a fair coin. According to the formula above, the binomial coefficient for $k = 0, 1, 2, 3, 4$ is respectively 1, 4, 6, 4, 1 (note that $0! = 1$), and the frequency of occurrence (f) for the five k values can be calculated as 0.0625, 0.25, 0.375, 0.25, 0.0625, respectively. In other words, from the four coin tosses, the probability that the head faces up only once (or three times) is 25%; the probability that the head faces up twice is 37.5%, and the probability that heads faces up four times (or zero time) is 6.25%.

If neurotransmitter release occurs in discrete packets, and if the release of each packet occurs at a probability of p, we can calculate the frequency that k packets out of the total n packets are released using the binomial formula above just as in the example of coin toss. However, researchers did not know the actual values for n (how many quanta are available to be released) or p (how likely is any individual quantum to be released), so it was not possible to apply binomial distribution. Fortunately, according to probability theory, when n is large (>20) and p is small (<0.05), the binomial distribution can be approximated by the Poisson distribution, in which the frequency (f) that k events occur can be determined by a single parameter λ (which equals the product of n and p in the binomial distribution) according to the following formula:

$$f(k; \lambda) = \frac{\lambda^k}{k!} e^{-\lambda}$$

One can experimentally estimate λ (same as m in Section 3.2) because as the product of n and p, it equals the mean number of packets that are released in response to a stimulus, and thus is equivalent to the ratio of evoked EPP and mEPP (assumed to be the quantal unit). Thus, researchers can calculate the probability of release in response to nerve stimulation—estimating the likelihood that no release occurs ($k = 0$), that a single packet is released ($k = 1$), that two packets are released ($k = 2$), and so on—and can then compare these calculations with the actual experimental data, as shown in Figure 3-2B.

Note that in order to apply the Poisson distribution, the release probability (p) must be small and the number of available packets (n) must be large so that p does not change during the measurement of λ. While researchers cannot control n as this is determined by nature (as it turns out, n is very large in the vertebrate neuromuscular junction because there are typically hundreds of neurotransmitter release sites between a motor axon and its muscle target), they can experimentally reduce p by studying neurotransmitter release in low-Ca^{2+} and high-Mg^{2+} media. Synaptic transmission at the neuromuscular junction also follows closely other assumptions required for the Poisson distribution: independent release of each quantum (because of the large number of release sites), the uniformity of the population (p is the same for all quanta), and the relative uniformity of their size (each vesicle contains similar amount of neurotransmitter molecules). In many CNS synapses the assumptions either fail (for example, n is often too small) or cannot be tested adequately. The probability of neurotransmitter release may not follow the Poisson distribution.

3.3 Neurotransmitters are released when synaptic vesicles fuse with the presynaptic plasma membrane

Physiological and anatomical studies often complement each other in driving neuroscience discoveries. The physical basis of the quantal neurotransmitter release became evident when electron microscopy was first applied to the nervous system in the mid 1950s. Thin sections across the nerve terminals revealed that they contain abundant vesicles that are ~40 nm in diameter. At the neuromuscular junction, many such vesicles appear stacked near the presynaptic membrane juxtaposed to the muscle membrane (**Figure 3–3A**). These **synaptic vesicles** were immediately hypothesized to be vesicles that are filled with neurotransmitters. The relatively uniform size of synaptic vesicles explained why neurotransmitters are released in packets with a uniform quantal size. (The quantal size at the frog neuromuscular junction has been estimated to be about 7000 ACh molecules.) The unitary release of neurotransmitters occurs when a single synaptic vesicle fuses with the plasma membrane, dumping its neurotransmitter content into the synaptic cleft and producing an mEPP in the muscle cell. Nerve stimulation under normal conditions (not in low Ca^{2+}) causes hundreds of these vesicle fusion

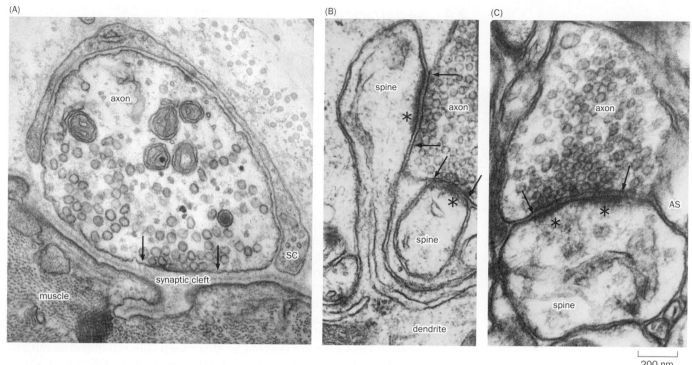

(A)

axon

synaptic cleft

muscle

SC

(B)

spine

axon

*

*

spine

dendrite

(C)

axon

*

*

*

AS

spine

200 nm

Figure 3–3 Structures of synapses revealed by electron microscopy. All images share the scale bar. Red asterisks indicate postsynaptic density. Pairs of arrows define the extent of the active zones in the presynaptic terminals. Note the abundance of ~40 nm diameter synaptic vesicles in each presynaptic terminal; some of these vesicles are 'docked' at the active zone ready for release. **(A)** A frog neuromuscular junction. The synaptic cleft is considerably wider than at the CNS synapses shown in the other two panels. SC indicates a Schwann cell process that wraps around the motor axon terminal. A typical motor axon forms hundreds of such presynaptic terminals onto a muscle fiber. **(B)** Two synapses formed between a single axon and two Purkinje cell dendritic spines in rat cerebellar cortex. **(C)** A synapse from human cerebral cortex. AS indicates an astrocyte process that wraps around many CNS synapses. (A, courtesy of Jack McMahan; B & C, courtesy of Josef Spacek and Kristen M. Harris, Synapse Web.)

events at a given neuromuscular junction, therefore producing EPPs two orders of magnitude higher than when neurotransmitter is released from a single vesicle. Thus, the neuromuscular junction has a high **quantal yield** (that is, a large number of synaptic vesicle exocytosis events per action potential) of several hundreds. By contrast, many synapses in the CNS have much lower quantal yield (a few or just one).

The basic structural elements of chemical synapses are highly similar across the entire nervous system and in different animal species (Figure 3–3B, C). In all cases presynaptic terminals have an electron-dense region called the **active zone**, with clusters of synaptic vesicles 'docked' at the presynaptic membrane ready for release. Across the synaptic cleft from the active zone and underneath the postsynaptic membrane is a structure called **postsynaptic density**, also concentrated with electron-dense structures. We will study the molecular composition of the active zone and postsynaptic density later in the chapter.

While EM studies found plenty of vesicles in the presynaptic terminals, observing a fusion event was necessary to establish the synaptic vesicle hypothesis. Because fusion of synaptic vesicles with the presynaptic plasma membrane, a necessary intermediate step for neurotransmitter release, occurs very transiently, it is difficult to detect such events in an electron microscopic preparation from a steady-state nervous system. To maximize the chance of visualizing such fusion events, experiments were designed to fix the neuromuscular junction samples immediately after the nerve stimulation. This was achieved by stimulating a neuromuscular preparation while the entire sample was falling toward a block that would freeze the tissue immediately upon contact, so that nerve stimulation could be achieved within a few milliseconds prior to fixation. Fusion between synaptic vesicles and the presynaptic plasma membrane were indeed caught in action (**Figure 3–4**). Such studies provide definitive evidence that neurotransmitter release is caused by fusion of synaptic vesicles with the presynaptic plasma membrane.

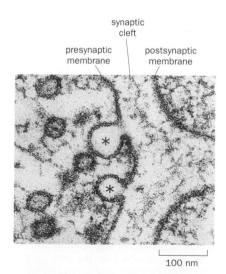

synaptic
cleft

presynaptic
membrane

postsynaptic
membrane

*

*

100 nm

Figure 3–4 Synaptic vesicle fusion caught in action. This electron micrograph was taken from a frog neuromuscular junction preserved 3–5 ms after nerve stimulation, revealing the fusion of two synaptic vesicles (red asterisks) with the presynaptic plasma membrane. (Courtesy of John Heuser. See also Heuser JE & Reese TS [1981] *J Cell Biol* 88:564–580.)

3.4 Neurotransmitter release is controlled by Ca²⁺ entry into the presynaptic terminal

How does action potential arrival cause synaptic vesicle fusion in the presynaptic terminal? As noted in Section 3.2, external Ca^{2+} is essential for action potential-triggered neurotransmitter release: bathing neuromuscular junction preparations in solutions with progressively lower concentrations of Ca^{2+} rendered the stimulation of motor axons increasingly ineffective in generating EPP in muscles. Supplying Ca^{2+} locally at the neuromuscular synapse through iontophoresis provided a means to test when Ca^{2+} was required during action potential-induced synaptic transmission. It was found that a brief application of extracellular Ca^{2+} enabled neurotransmitter release if it occurred immediately before the depolarization pulse, but exogenous Ca^{2+} became ineffective if applied after the depolarization pulse. Thus, extracellular Ca^{2+} is required during a brief period when depolarization occurs, preceding the transmitter release itself.

How does external Ca^{2+} participate in neurotransmitter release? This question was answered with the help of the squid giant synapse, whose presynaptic as well as postsynaptic terminals are large such that researchers can insert electrodes into both compartments for intracellular recordings. (One of the postsynaptic target cells is the neuron that extends the giant axon featured in Chapter 2.) It was found that action potentials could be replaced simply by depolarization, which opens voltage-gated Ca^{2+} channels (see Box 2–4) in the presynaptic plasma membrane, causing an inward flow of Ca^{2+} that triggers neurotransmitter release.

Let's study in more detail one specific experiment (**Figure 3–5**), which showcased the Ca^{2+} dependence of neurotransmitter release and provided information about the timing of different steps of neurotransmitter release. In this experiment, the voltage clamp technique was applied to both the presynaptic terminal and postsynaptic target of the squid giant synapse in the presence of the Na^+ and K^+ channel blockers, such that the only cation that could cross the presynaptic membrane was Ca^{2+}. From a resting potential at –70 mV, a depolarizing voltage step to –25 mV (Figure 3–5A, top) triggered Ca^{2+} influx as measured by presynaptic current (Figure 3–5A, middle), resulting in effective synaptic transmission as measured by an inward postsynaptic current (Figure 3–5A, bottom; we will study the nature of such postsynaptic current in later sections). However, a voltage step to +50 mV did not trigger presynaptic Ca^{2+} influx or postsynaptic current (Figure 3–5B, left portion). This is likely because the +50 mV was close to the equilibrium potential of Ca^{2+} in the presynaptic terminal under the experimental condition, such that even though voltage-gated Ca^{2+} channels were open, there was no driving force for Ca^{2+} influx (see Section 2.5). However, returning the presynaptic membrane potential from +50 mV to –70 mV produced a presynaptic 'tail current' (Figure 3–5B, middle). This is because the membrane potential change was faster than the closure

Figure 3–5 Voltage clamp studies of Ca²⁺ entry into the presynaptic terminal of the squid giant synapse. Voltage steps were applied to the presynaptic terminal (top traces) using the voltage clamp technique (see Figure 2–21). The current injected into the presynaptic terminal to maintain the clamped voltage is equivalent to the Ca^{2+} current across the presynaptic membrane (middle traces), as Na^+ and K^+ channel blockers were applied in these experiments. Postsynaptic current was simultaneously recorded in a voltage clamp setting as a measure of neurotransmitter release (bottom traces). **(A)** A depolarizing step in the presynaptic terminal triggered the opening of the voltage-gated Ca^{2+} channels, which caused Ca^{2+} influx and subsequent postsynaptic response. **(B)** A larger depolarization step of the presynaptic membrane potential close to the equilibrium potential of Ca^{2+} prevented Ca^{2+} entry due to lack of a driving force; no postsynaptic response occurs. A tail current representing Ca^{2+} influx was produced when the presynaptic membrane potential returned to –70 mV, which triggered a postsynaptic response. The pairs of dotted lines represent the presynaptic voltage step (left) and the onset of the postsynaptic response (right). Note a shorter time interval in panel B. (Adapted from Augustine GJ, Charlton MP & Smith SJ [1985] *J Physiol* 367:163–181. See also Llinás RR [1982] *Sci Am* 247:56–65.)

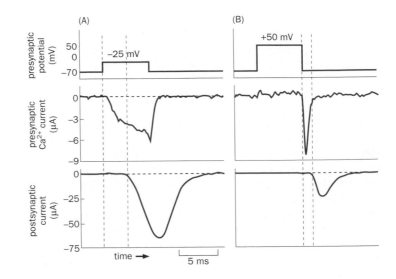

of voltage-gated Ca^{2+} channels; thus there was a transient period when there was a driving force for Ca^{2+} influx while Ca^{2+} channels remained open. The Ca^{2+} influx due to the presynaptic tail current produced a corresponding postsynaptic current response (Figure 3–5B, bottom). Interestingly, the Ca^{2+} tail current triggered a postsynaptic response more rapidly than the presynaptic depolarization did (compare the time interval between the two dotted lines in the two panels). This suggested that the normal synaptic delay between presynaptic depolarization and postsynaptic response consists of two components: a delay between depolarization and opening of voltage-gated Ca^{2+} channels (which was bypassed in the tail current condition as the channels were already open), and a delay between Ca^{2+} entry and the neurotransmitter-triggered postsynaptic response.

The Ca^{2+} hypothesis of neurotransmitter release was further validated by other techniques. In one type of experiment, a chemical dye used as an indicator for changes in Ca^{2+} concentration (see Section 13.22 for more details) was injected into the presynaptic terminal of the squid giant synapse. Nerve stimulation was found to cause a rise of intracellular Ca^{2+} concentration at the presynaptic terminal (**Figure 3–6**). The Ca^{2+} concentration was highest in specific regions of the presynaptic terminal. As will be discussed in Section 3.7, this is because the voltage-gated Ca^{2+} channels are highly concentrated at the active zone, where synaptic vesicles dock and fuse with presynaptic membrane. In another type of experiment, chemical compounds were synthesized that 'cage' Ca^{2+} to prevent the ion's effects; such cages can be triggered by light to release Ca^{2+}. When caged Ca^{2+} was introduced into the presynaptic terminal of the squid giant axon, light could trigger neurotransmitter release in the absence of action potentials or Ca^{2+} entry from the extracellular media.

Together, these experiments firmly established a sequence of events from action potential to neurotransmitter release:

Action potential from the axon → Depolarization of the presynaptic terminal → Opening of voltage-gated Ca^{2+} channels → Ca^{2+} entry into the presynaptic terminal → Fusion of synaptic vesicle with presynaptic plasma membrane → Neurotransmitter release

This sequence of events, which was originally worked out in the frog neuromuscular junction and the squid giant synapse, applies universally to all chemical synapses across the animal kingdom, regardless of the type of synapse and neurotransmitter used.

The short latency between Ca^{2+} entry into the presynaptic terminal and postsynaptic events (~2 ms in Figure 3–5B and often shorter) indicates that there must be a pool of synaptic vesicles that are ready to fuse with the presynaptic plasma membrane immediately upon a rise of intracellular Ca^{2+} concentration. This is consistent with observations in electron microscopy (see Figure 3–3). Furthermore, membrane fusion is energetically unfavorable because breaking two membranes and resealing them necessitates exposing hydrophobic surfaces to water, thus requiring external energy such as ATP hydrolysis. However, the final step of synaptic vesicle fusion is so fast that it is unlikely to involve an ATP hydrolysis-dependent catalytic process. Instead, as we will soon learn, synaptic vesicles are primed for fusion by a specialized protein complex already existing in a high-energy configuration, simply waiting for Ca^{2+} to trigger the sudden change of configuration that permits fusion.

3.5   SNARE and SM proteins mediate synaptic vesicle fusion

We now turn to the molecular mechanisms that mediate the fusion of synaptic vesicles with the plasma membrane (a process also called neurotransmitter **exocytosis**; see Figure 2-2). Our current understanding of these mechanisms came from a convergence of multiple experimental approaches. The first is a biochemical approach to identify presynaptic protein components. Because of the uniform size and buoyancy of synaptic vesicles and their abundance, researchers can purify them to a high degree, which permitted the identification of their

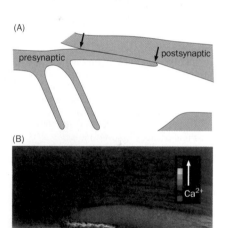

(A)

presynaptic   postsynaptic

(B)

Ca^{2+}

500 µm

Figure 3–6 Nerve stimulation triggers Ca^{2+} entry into the presynaptic terminal of the squid giant synapse. (A) Schematic drawing of the squid giant synapse image in panel B. The presynaptic terminals resemble finger-like extensions in contact with postsynaptic neurons (one of which is shown). The extent of synaptic contact between the two neurons is indicated by the two arrows. **(B)** A brief train of presynaptic action potentials caused the Ca^{2+} concentration to increase in the presynaptic terminal, as reported by fluorescence changes of microinjected fura-2, a Ca^{2+} indicator (see Section 13.22). The Ca^{2+} increase is seen as a shift from cool colors to warm ones. (Adapted from Smith SJ, Buchanan J, Osses LR et al. [1993] *J Physiol* 472:573–593. With permission from the Physiological Society.)

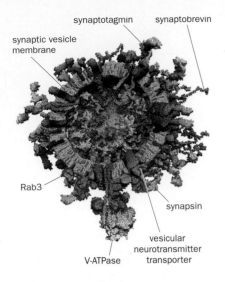

synaptotagmin   synaptobrevin

synaptic vesicle
membrane

Rab3

synapsin

vesicular
neurotransmitter
transporter

V-ATPase

Figure 3–7 The molecular anatomy of a synaptic vesicle. This model is based on quantitative determination of the protein components associated with synaptic vesicles. Each colored structure in this cross section of the synaptic vesicle represents a characterized synaptic vesicle protein. The synaptic vesicle membrane and six synaptic vesicle proteins whose functions are discussed in this and subsequent sections are indicated (see also Table 3–1). (Adapted from Takamori S, Holt M, Stenius K et al. [2006] *Cell* 127:831–846. With permission from Elsevier Inc.)

key components. Indeed, the synaptic vesicle is one of the best-characterized organelles in the cell, with quantitative information about the protein and lipid compositions (**Figure 3–7; Movie 3–1**). We will encounter some of the synaptic vesicle proteins in this and subsequent sections.

The second is a stunning convergence of yeast genetics to identify genes required for secretion and biochemical reconstitution of mammalian vesicle fusion reactions *in vitro*, which led to the conclusion that the fundamentals of membrane fusion were highly conserved and that neurotransmitter exocytosis was a specialized form of membrane fusion that occurs in all cells and in many parts of the cells (see Figure 2–2); we will return to this topic in Chapter 12 in the context of the evolutionary origin of neuronal communication. This convergence led to the identification of many evolutionarily conserved vesicle fusion components and their regulators that will be discussed below. The third is the ability to disrupt genes in *C. elegans, Drosophila*, and mice to test the *in vivo* function of these evolutionarily conserved proteins in synaptic transmission. The fourth is studies of toxins that block specific steps of neurotransmitter release and identification of their protein targets. Together, these approaches have given rise to our current understanding of the neurotransmitter release mechanisms summarized below.

At the core of vesicle fusion are three **SNARE** proteins (SNARE stands for soluble N̲SF-a̲ttachment protein r̲eceptor; NSF is N̲-ethylmaleimide-s̲ensitive fusion protein, named after a chemical inhibitor that blocks vesicle fusion reactions *in vitro*) and **SM proteins** (for S̲ec1/M̲unc18-like proteins; Sec1 was originally identified in a genetic screen in yeast for its requirement in s̲ecretion; Munc18 is the mammalian homology of Unc18, originally identified in a genetic screen in *C. elegans* for mutants that exhibit an u̲ncoordinated phenotype). The first SNARE is a transmembrane protein on the synaptic vesicle called **synaptobrevin** (also named VAMP for v̲esicle-a̲ssociated m̲embrane p̲rotein), which is the most abundant synaptic vesicle protein. As a vesicular protein, synaptobrevin is designated as a **v-SNARE**. The second SNARE is a transmembrane protein on the plasma membrane called **syntaxin**. Owing to its location on the target membrane for vesicle fusion, syntaxin is called a **t-SNARE**. The third SNARE, also a t-SNARE named **SNAP-25** (s̲y̲n̲aptosomal-a̲ssociated p̲rotein with a molecular weight of 25̲ kDa), is anchored onto the cytoplasmic face of the plasma membrane via lipid modification. Once the synaptic vesicle is in the vicinity of the presynaptic plasma membrane, the cytoplasmic domains of synaptobrevin, syntaxin, and SNAP-25 assemble into a very tight complex. How the SNARE complex mediates the fusion is still an active area of research. Current data indicate that the assembly of the SNARE complex proceeds from the membrane-distal to membrane-proximal ends of the SNARE proteins like zipping a zipper. The force generated by the assembly of the SNARE complex drives the synaptic vesicle membrane even closer to the plasma membrane and leads the lipid bilayers to fuse, such that the contents of the synaptic vesicle are exposed to the extracellular space (**Figure 3–8**A; **Movie 3–2**).

The structure of the SNARE complex has been determined at atomic resolution by X-ray crystallography. Three SNARE proteins form a four-helix bundle, with synaptobrevin and syntaxin each contributing one helix and SNAP-25 contributing two helices (Figure 3–8B). Many naturally occurring protease toxins that inhibit neuronal communication target these three SNARE proteins at specific amino acid residues (see **Box 3–2**). Proteolytic cleavage by these proteases is predicted to inhibit the attachment of the four-helix bundle to the membrane, thereby blocking neurotransmitter release.

The SNARE-based mechanism of membrane fusion applies to many fusion reactions in intracellular vesicle trafficking. The v- and t-SNAREs for other specific fusion events (for example, fusion of ER-derived vesicles with the Golgi membrane; see Figure 2–2) resemble the v- and t-SNAREs for synaptic vesicle exocytosis. These findings suggest that the mechanism of synaptic vesicle exocytosis was co-opted from general vesicle trafficking. In all of these reactions, including synaptic vesicle fusion, however, SNARE proteins were found to be insufficient

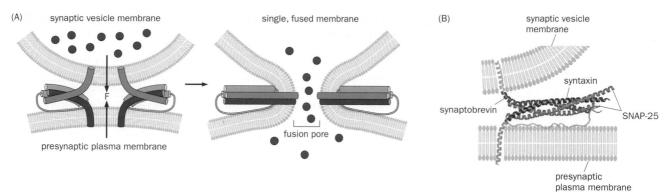

Figure 3–8 Model and structural basis of synaptic vesicle fusion. (A) Schematic models of SNARE complexes before and after membrane fusion. Before fusion (left), the vesicular and target SNAREs are on separate membranes (the synaptic vesicle membrane and the presynaptic plasma membrane, respectively). The strong binding of their cytoplasmic domains, in a zipper-like fashion starting from the two sides and progressing toward the center, produces a force (F) that brings the vesicle and target membranes together, causing them to fuse (right). Colored rods represent helices from the SNARE proteins detailed in panel B. **(B)** Structure of the SNARE complex for synaptic vesicle fusion determined by X-ray crystallography. Blue, red, and green represent α-helices from the cytoplasmic domains of synaptobrevin, syntaxin, and SNAP-25, respectively. Faded blue and red represent transmembrane domains of synaptobrevin and syntaxin, which were not part of the solved crystal structure. The orange strand links the two SNAP-25 helices and is attached to the presynaptic plasma membrane through lipid modification, which was also not part of the crystal structure. (A, adapted from Südhof TC & Rothman JE [2009] *Science* 323:474–477; B, adapted from Sutton RB, Fasshauer D, Jahn R et al. [1998] *Nature* 395:347–353. With permission from Macmillan Publishers Inc.)

to mediate fusion. A partner for SNARE proteins in all fusion reactions is an SM protein, which at the mammalian synapse is a protein called Munc18. This protein binds to SNAREs throughout the fusion reaction and is essential for fusion. The precise function of SM proteins is incompletely understood; a leading hypothesis is that SM proteins act as catalysts for SNARE-mediated fusion.

Box 3–2: From toxins to medicines

Research in neurobiology has benefited tremendously from naturally occurring toxins that have evolved to block specific steps of neuronal communication. These toxins are produced by organisms from a wide range of phylogenetic groups, including bacteria, protists, plants, fungi, and animals. Despite the energy costs of producing them, toxins offer adaptive advantages such as deterring herbivores, fending off predators, or immobilizing prey. Scientists have used these toxins to study the biological functions and mechanisms of action of their target proteins. Some of these toxins have been further developed into medicines.

Virtually all steps of neuronal communication are targets for toxins. Action potentials are potently blocked by tetrodotoxin (TTX, see Figure 2–29), an inhibitor of voltage-gated Na^+ channels produced by symbiotic bacteria in puffer fish, rough-skinned newt, and certain octopi. Synaptic transmission is blocked by a number of proteases produced by the bacteria *Clostridium tetani* and *Clostridium botulinum*. **Tetanus** and **botulinum toxins** specifically cleave SNARE proteins, with each toxin cleaving a specific SNARE at a specific residue, thereby preventing synaptic vesicle fusion with presynaptic membrane (see Figure 3–8). Indeed, identification of the protein targets of tetanus and botulinum toxins was instrumental in establishing that SNARE proteins play a central role in synaptic vesicle fusion. A small peptide from marine snails, **ω-conotoxin**, specifically blocks presynaptic voltage-gated Ca^{2+} channels essential for neurotransmitter release. Other toxins target neurotransmitter receptors that will be discussed later in this chapter. For instance, **curare**, a plant toxin used by native Americans on poisonous arrows, and **α-bungarotoxin** and cobratoxin from snakes, are all potent competitive inhibitors of the acetylcholine receptor at the vertebrate neuromuscular junction, and thereby block motor neuron-triggered muscle contraction. **Picrotoxin**, another plant toxin, is a potent blocker of the GABA$_A$ receptors that mediate fast inhibition in vertebrates and invertebrates alike. **Muscimol**, produced by toxic mushrooms, is a potent activator of the GABA$_A$ receptors. The venoms of predators such as snakes, scorpions, cone snails, and spiders have been a rich source of tools for investigating neuronal communication. The fact that most toxins affect many different animal species also indicates that the molecular machinery of neuronal communication is highly conserved across animals.

Natural toxins and their derivatives have also been used extensively in medicine. Channel blockers have been used to treat epilepsy and intractable pain. Synaptic transmission blockers have been used as muscle relaxants. For example, botulinum toxin A, commonly known as Botox, can be injected into specific eye muscles to treat strabismus (misaligned eyes). Botox injections have also become a popular cosmetic procedure to temporarily remove wrinkles.

3.6 Synaptotagmin serves as a Ca²⁺ sensor to trigger synaptic vesicle fusion

How does Ca^{2+} entry regulate neurotransmitter exocytosis? A prime candidate that links these two events is a class of transmembrane proteins on the synaptic vesicle called **synaptotagmins** (see Figure 3–7), which possess up to five Ca^{2+} binding sites on their cytoplasmic domain. To test the function of synaptotagmin in synaptic transmission, **knockout** mice were created in which synaptotagmin-1, the predominant form of synaptotagmin expressed in forebrain neurons, was disrupted using the gene targeting method in embryonic stem cells (see Section 13.7 for details of the knockout method). To assay for synaptic transmission, embryonic hippocampal neurons from control or knockout mice were dissociated and cultured *in vitro* to allow synapse formation, and pairs of synaptically connected neurons were subjected to a variation of the patch clamp technique called **whole-cell patch recording**. (In whole-cell patch recording, the membrane underneath the patch pipette is ruptured such that the patch pipette is connected to the entire neuron; see Section 13.21 and Box 13–2 for details.) Depolarization of a wild-type presynaptic neuron, which caused it to fire action potentials, resulted in an inward current of the postsynaptic neuron, an indication of successful synaptic transmission. Depolarization of synaptotagmin-1 knockout neurons elicited much smaller postsynaptic responses, indicating that synaptotagmin-1 is required for normal synaptic transmission (**Figure 3–9**A). Earlier studies in *Drosophila* and *C. elegans* indicted that disruption of synaptotagmin homologs in these invertebrates also impaired synaptic transmission.

The knockout experiment did not prove that synaptotagmin acts as a Ca^{2+} sensor, as disrupting other genes encoding proteins essential for synaptic transmission, such as the v-SNARE synaptobrevin, similarly blocked synaptic transmission. Subsequent experiments have provided strong evidence that synaptotagmin is a major Ca^{2+} sensor that regulates neurotransmitter release. For example, a mutant synaptotagmin-1 with a single amino acid change was identified that reduces Ca^{2+} binding by 50% in an *in vitro* biochemical assay. When this mutant synaptotagmin-1 was used to replace the endogenous synaptotagmin-1 in a variation of the knockout procedure called **knock-in** (see Section 13.7), neurons derived from the knock-in mice exhibited a corresponding 50% reduction in the Ca^{2+} sensitivity of neurotransmitter release (Figure 3–9B). Another protein involved in neurotransmitter release is complexin, which has a complex role of both activating the SNARE complex and clamping it at an intermediate step. One

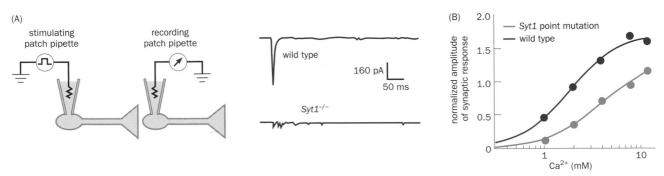

Figure 3–9 Synaptotagmin as a Ca²⁺ sensor in synaptic transmission. **(A)** Left, schematic of experimental preparation to examine the role of synaptotagmin-1 in synaptic transmission. Both hippocampal neurons in culture were subjected to whole-cell patch recording. A depolarizing current was injected into the presynaptic neuron to cause it to fire an action potential, and the postsynaptic response was recorded as an inward current when the membrane potential was clamped at −70 mV. Right, compared with the inward current triggered by a presynaptic action potential between a pair of wild-type neurons (top trace), the synaptic response between a pair of neurons from *Syt1* knockout mice (lacking synaptotagmin-1; meaning both copies of the *Syt1* gene were disrupted) was greatly diminished (bottom trace). **(B)** A point mutation in *Syt1* that reduced Ca^{2+} binding by 50% also reduced the sensitivity of neurotransmitter release of cultured hippocampal neurons to Ca^{2+} by about 50%, as indicated by the downward shift of the mutant curve compared with the wild-type curve, each plotting normalized synaptic transmission amplitude against Ca^{2+} concentration. This finding supports the notion that synaptotagmin-1 acts as a Ca^{2+} sensor for synaptic vesicle fusion in hippocampal neurons. (A, adapted from Geppert M, Goda Y, Hammer RE et al. [1994] *Cell* 79:717–727. With permission from Elsevier Inc.; B, adapted from Fernández-Chacón R, Königstorfer A, Gerber SH et al. [2001] *Nature* 410:41–49. With permission from Macmillan Publishers Ltd.)

current model is that synaptotagmin releases the inhibitory clamp of complexin in a Ca^{2+}-dependent manner, thus allowing SNAREs to complete the vesicle fusion reaction in response to a rise of intracellular Ca^{2+} concentration.

In fast mammalian CNS synapses at physiological temperatures, action potential arrival can cause neurotransmitter release within as little as 150 μs, as measured by postsynaptic depolarization. This interval includes about 90 μs to open voltage-gated Ca^{2+} channels during the action potential upstroke to allow Ca^{2+} influx, and 60 μs in total for Ca^{2+} to trigger vesicle fusion and for neurotransmitter molecules to diffuse across the synaptic cleft and act on post-synaptic cells. To enable this rapid action, synaptic vesicles are docked at the active zone ready for release (see Figure 3–3), with their SNARE proteins already partially preassembled in a high-energy configuration but clamped, waiting for the action of a Ca^{2+} sensor such as synaptotagmin to release the clamp and complete SNARE assembly that drives membrane fusion.

In addition to rapid release of synaptic vesicles after Ca^{2+} entry, it is also important that transmitter release is transient so that the presynaptic terminal can respond to future action potentials with more neurotransmitter release. This requires free Ca^{2+} to be rapidly removed after their entry, and that the Ca^{2+} sensor has a low binding affinity for Ca^{2+}. Indeed, Ca^{2+}-binding proteins and pumps rapidly sequester free Ca^{2+} upon entry. Moreover, synaptotagmin employs multiple low-affinity Ca^{2+}-binding sites that bind Ca^{2+} cooperatively (that is, the binding of one Ca^{2+} facilitates the binding of a second Ca^{2+}); only when multiple sites bind to Ca^{2+} would it be able to trigger neurotransmitter release. Together, these mechanisms ensure that neurotransmitter release is only triggered transiently and locally at the site of Ca^{2+} entry.

3.7 The presynaptic active zone is a highly organized structure

The fast action and transient nature of Ca^{2+}-induced neurotransmitter release relies on the proximity of voltage-gated Ca^{2+} channels and docked synaptic vesicles in the active zone. Indeed, Ca^{2+} imaging of the presynaptic terminals (for example, see Figure 3–6) suggested that the rise of intracellular Ca^{2+} concentration in response to depolarization is highly restricted to microdomains near the active zone. Although the intracellular Ca^{2+} concentration is normally very low (~0.1 μM), it can shoot up transiently to tens or even hundreds of micromolar in the microdomain; this facilitates cooperative binding of Ca^{2+} to multiple Ca^{2+}-binding sites of synaptotagmin to achieve the conformational change necessary for triggering vesicle fusion.

The molecular machinery that organizes the active zone has been extensively characterized in vertebrate neurons (**Figure 3–10**), and many components are conserved in invertebrates. The cytoplasmic domain of the voltage-gated Ca^{2+} channel binds to two active zone core components, RIM (for Rab3-interacting molecule) and RIM-BP (RIM-binding protein). RIM also binds to a synaptic vesicle associated protein Rab3, a small GTPase of the Rab subfamily, thus bringing the synaptic vesicle into proximity with Ca^{2+} channels. In addition, RIM and RIM-BP interact with other active zone proteins, which in turn associate with the actin cytoskeleton that supports the structural integrity of the presynaptic terminal and transports molecules into the presynaptic terminal (see Figure 2–6). The active zone protein complex is also associated with synaptic adhesion molecules. These include the **cadherins** (Ca^{2+}-dependent cell adhesion proteins) present on both pre- and postsynaptic membranes that bind each other (termed **homophilic binding**), and **neurexin** on the presynaptic membrane binding to **neuroligin** on the postsynaptic membrane (termed **heterophilic binding**). These cell-adhesion molecules bring the presynaptic and postsynaptic plasma membranes together, and align the active zone with the postsynaptic membrane rich in neurotransmitter receptors (to be discussed later in the chapter), thus minimizing the distance neurotransmitters need to travel to act on their receptors (Figure 3–10).

Recent studies using super-resolution fluorescent microscopy (see Section 13.17 for more details) have begun to determine where specific molecules are located with respect to each other at the active zone. For example, according to a

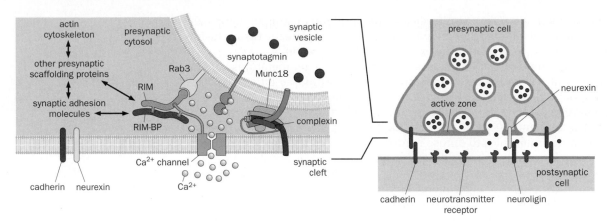

Figure 3–10 Molecular organization of the presynaptic terminal. Left, a magnified model of the presynaptic cell's active zone. The RIM/RIM-BP protein complex binds to the voltage-gated Ca^{2+} channel directly, and to the synaptic vesicle via the Rab3 protein; this allows Ca^{2+} entry to activate synaptotagmin with minimal diffusion, which in turn releases the complexin inhibitory clamp on the SNARE/SM complex and causes neurotransmitter release (the SNARE complex is represented as in Figure 3–8A; Munc18 is the SM protein in mammalian synapses). RIM and RIM-BP are also associated with other presynaptic scaffolding proteins, which are in turn associated with the actin cytoskeleton and with synaptic adhesion molecules. Right, a lower-magnification model of a chemical synapse showing presynaptic and postsynaptic cells. Trans-synaptic adhesion molecules (blue, homophilic binding between cadherins; yellow and red, heterophilic binding between presynaptic neurexin and postsynaptic neuroligin) align the active zone with a postsynaptic density enriched for neurotransmitter receptors, facilitating the rapid action of neurotransmitters. (Adapted from Südhof TC [2012] *Neuron* 75:11–25. With permission from Elsevier Inc.)

model based on super-resolution localization of molecules in the *Drosophila* neuromuscular junction (**Figure 3–11**), the RIM-BP proteins form a ring around a cluster of voltage-gated Ca^{2+} channels at the active zone presynaptic membrane. An active zone scaffolding protein called Bruchpilot (corresponding to a mammalian protein called ELKS) extends from the center of the active zone to the periphery. Glutamate receptors are enriched in the postsynaptic density aligned with the presynaptic active zone (as discussed in Section 3.11, glutamate is used as a neurotransmitter in the *Drosophila* neuromuscular synapse). Future studies on synapses in the central nervous system and in other species will help determine whether all synapses share a similar structural organization, and what variations might exist among different synapses.

3.8 Neurotransmitters are efficiently cleared from the synaptic cleft by enzymatic cleavage or transport into presynaptic and glial cells

In order for the postsynaptic neurons to continually respond to the firing of presynaptic neurons, neurotransmitters released in response to each presynaptic action potential must be cleared from the synaptic cleft efficiently. While diffusion of neurotransmitters away from the synaptic cleft is a major mechanism of clearance, additional mechanisms are employed for neurotransmitter clearance

Figure 3–11 A model of the organization of selected proteins in the *Drosophila* neuromuscular synapse. This model is based on two-color labeling of different pairs of proteins (shown at right) and measurement of their distances apart using a technique called stimulated emission depletion microscopy, which can resolve structures ~50 nm apart (see Section 13.17 for details). For instance, the distance between C-terminal Bruchpilot and the glutamate receptor was estimated by measuring the distance between the fluorescence signal from an antibody against the C-terminus of Bruchpilot and that from an antibody against the glutamate receptor (151 ± 24 nm apart). RIM-BP, Rab3-interacting-molecule binding protein; VGCC, voltage-gated Ca^{2+} channel; GluR, glutamate receptor; N, amino terminus; C, carboxy terminus; SEM, standard error of the mean. (Adapted from Liu KSY, Siebert M, Mertel S et al. [2011] *Science* 334:1565–1569. With permission from AAAS.)

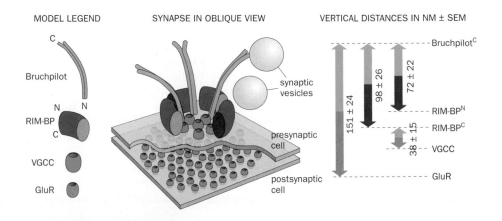

depending on the neurotransmitter system (we will introduce different neurotransmitter systems in more detail in Section 3.11).

ACh at the neuromuscular junction is rapidly degraded by **acetylcholinesterase**, an enzyme enriched in the synaptic cleft. Indeed, this enzyme is so active that most ACh molecules released by motor axon terminals are degraded while diffusing across the short distance of the synaptic cleft. Some of the physiology experiments involving mEPP measurement described in earlier sections actually included acetylcholinesterase inhibitors in the saline to boost the mEPP amplitude.

For most other neurotransmitter systems, excess transmitter molecules in the synaptic cleft are recycled. In a process called **neurotransmitter reuptake**, excess neurotransmitters are first taken back into the presynaptic cytosol using the **plasma membrane neurotransmitter transporters**, which derive energy from co-transporting Na^+ into the presynaptic cell down the Na^+ electrochemical gradient (**Figure 3–12**; see Movie 3-1). Once in the cytosol, neurotransmitters refill new and recycled synaptic vesicles (see Section 3.9) utilizing a second transporter: the **vesicular neurotransmitter transporter** on the synaptic vesicle (see also Figure 3–7). The energy for the vesicular transporters derives from transporting protons in the opposite direction down the proton gradient. The proton gradient (high in the vesicle and low in the cytosol) is created by **V-ATPase**, the largest molecule on the synaptic vesicle membrane (see Figure 3–7), which pumps protons (H^+) into the synaptic vesicle against an electrochemical gradient using energy derived from ATP hydrolysis. In some neurotransmitter systems, excess transmitters are mostly taken up by neurotransmitter transporters on the plasma membrane of glial cells, which wrap around many synapses (see Figure 3–3). In Chapter 11, we will learn more about the neurotransmitter reuptake mechanisms, because drugs altering these mechanisms are widely used to treat psychiatric disorders.

3.9 Synaptic vesicle recycling by endocytosis is essential for continual synaptic transmission

In order to maintain the ability to respond to sustained neuronal firing, presynaptic terminals must be able to replenish the stockpile of synaptic vesicles filled with neurotransmitters. While the synaptic vesicle membrane and proteins are mostly synthesized in the soma (see Sections 2.2–2.3), vesicles are rapidly recycled locally at the synaptic terminals. Considering the distance between the synaptic terminal and the soma, the recycling of synaptic vesicles is critical in order to rapidly recover synaptic vesicles for future rounds of synaptic transmission.

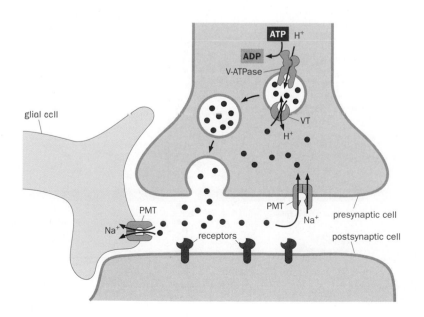

Figure 3–12 Clearance and recycling of neurotransmitters. After being released into the synaptic cleft as a result of synaptic vesicle fusion with the presynaptic plasma membrane, excess transmitters are taken up by plasma membrane transporters (PMTs) on the presynaptic membrane or on the nearby glial plasma membrane; both are symporters that utilize energy from Na^+ entry down its electrochemical gradient. Within the presynaptic cytosol, neurotransmitters are transported into synaptic vesicles by the vesicular neurotransmitter transporters (VTs), which are antiporters that use the energy by transporting protons (H^+) out of the synaptic vesicle down an electrochemical gradient. The V-ATPase on the synaptic vesicle membrane establishes the H^+ gradient in the vesicle using energy from ATP hydrolysis. (See Blakely RD & Edwards RH [2012] *Cold Spring Harb Perspect Biol* 4:a005595.)

After the fusion of synaptic vesicle membrane with presynaptic plasma membrane resulting in the release of neurotransmitter molecules, at least two mechanisms have been proposed to retrieve synaptic vesicles back to the presynaptic cytosol. The first mechanism, called 'kiss and run,' involves a very transient fusion of the synaptic vesicle with the presynaptic plasma membrane to release the neurotransmitters, followed by rapid reformation of the vesicle so that mixing of the vesicle's protein and lipid content with the presynaptic plasma membrane is limited. In the second mechanism, synaptic vesicle membrane becomes part of the presynaptic plasma membrane after full fusion, and is retrieved back to the presynaptic terminal by clathrin-mediated endocytosis. (Clathrin is a protein that assembles into a cage on the cytoplasmic side of a membrane to form a coated pit, which buds off to form a clathrin-coated vesicle.) Whereas full fusion likely applies to most cases of synaptic vesicle recycling, the degree to which the kiss-and-run mechanism is used is still a subject of debate. In both cases, the SNARE complexes are disassembled by NSF in an ATP-dependent manner. (Recall from Section 3.5 that the name SNARE derives from NSF.) Synaptobrevin returns to the synaptic vesicle, while syntaxin and SNAP-25 remain in the presynaptic plasma membrane. The vesicles are then acidified by the proton pump V-ATPase and refilled with neurotransmitters (see Figure 3–12). Filled vesicles join the **reserve pool** of synaptic vesicles. A synaptic vesicle protein called synapsin, a commonly used marker for identifying synapses, is involved in regulating the size of the reserve pool. A small subset of synaptic vesicles constitutes the **readily releasable pool**, which are docked at the active zone, primed by an ATP-dependent process to achieve the high-energy configuration of pre-assembled SNARE complex, and readied for another round of neurotransmitter release in response to depolarization-induced Ca^{2+} entry (**Figure 3–13**).

We use a specific example to illustrate the importance of synaptic vesicle retrieval for continual synaptic transmission and neuronal communication. To identify genes necessary for neuronal communication, forward genetic screens (see Section 13.6) were carried out in the fruit fly *Drosophila* to isolate mutations that caused paralysis when flies were shifted to high temperatures. This led to the discovery of a temperature-sensitive mutation called *Shibire^ts*. *Shibire^ts* flies behave normally at room temperature (~20°C), but are paralyzed shortly after shifting to elevated temperatures (>29°C); their motility returns to normal within a few minutes after the temperature is returned to 20°C. Molecular-genetic analysis

Figure 3–13 The synaptic vesicle cycle. After membrane fusion between the synaptic vesicle and the presynaptic membrane, and release of neurotransmitters into the synaptic cleft, synaptic vesicles can be recycled by two alternative means. In kiss-and-run, synaptic vesicles reform after a very transient fusion with limited exchange of proteins and lipids with presynaptic plasma membrane (1a); in clathrin-mediated endocytosis, the synaptic vesicle membrane fuses fully with the presynaptic plasma membrane and is then retrieved (1b). The interior of vesicles is then acidified by pumping protons (H^+) inside using the V-ATPase on the synaptic vesicle membrane; the synaptic vesicle is then ready to be filled with neurotransmitter using the proton export-coupled vesicular transporter (2a, see also Figure 3–12). Some acidified vesicles go through the early endosome step in this process (2b). Synaptic vesicles filled with neurotransmitters join the reserve pool (3). Some vesicles transit into the readily releasable pool, are docked at the active zone (4), and are primed in an ATP-dependent step (5) ready for exocytosis. Ca^{2+} entry through the voltage-gated Ca^{2+} channel at the active zone then triggers vesicle fusion (6). (Adapted from Südhof TC [2004] *Ann Rev Neurosci* 27:509–547.)

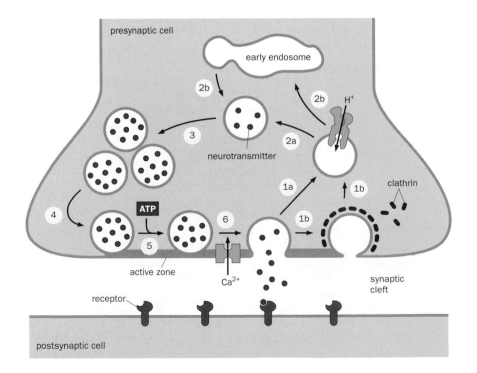

(A)

(B)

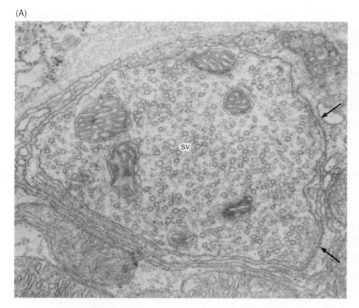

0.1 μm

1 μm

Figure 3–14 Electron micrographs of synapses in temperature-sensitive _Shibire^{ts}_ mutant fruit flies. (A) A _Shibire^{ts}_ mutant fly neuromuscular junction fixed at 19°C. The presynaptic terminal is abundant in synaptic vesicles (sv). Arrows indicate active zones. **(B)** Neuromuscular junction fixed 8 minutes after raising the temperature to 29°C. Note the reduced number of synaptic vesicles in the presynaptic terminal compared with panel A, and the presence of 'collared' vesicles (arrows, see inset for higher magnification) indicating a block of the last step of endocytosis. (From Koenig JH & Ikeda K [1989] _J Neurosci_ 9:3844–3860. With permission from the Society for Neuroscience.)

identified that the _Shibire_ gene encodes a protein called dynamin, which is essential for clathrin-mediated endocytosis of synaptic vesicles. The _Shibire^{ts}_ mutation causes reversible destabilization of dynamin at elevated temperatures. Without vesicle recycling, presynaptic terminals were rapidly deprived of synaptic vesicles after the reserved pool is exhausted (**Figure 3–14**), and became unable to release neurotransmitters in response to further action potentials, thus causing paralysis. The _Shibire^{ts}_ mutation has provided a useful tool for rapidly and reversibly silencing specific neurons _in vivo_ to analyze their function in information processing within neural circuits (see Section 13.23).

As a summary of what we have learned so far, **Table 3–1** provides a list of molecules that play key roles in mediating and regulating the sequence of events required for neurotransmitter release.

3.10 Synapses can be facilitating or depressing

Because synaptic transmission is a key mechanism of interneuronal communication, the **efficacy of synaptic transmission**, measured by the magnitude of the postsynaptic response to a presynaptic stimulus, is subject to many forms of regulation. The ability to change the efficacy of synaptic transmission, or **synaptic plasticity**, is an extremely important property of the nervous system. Depending on the temporal scale, synaptic plasticity is usually divided into **short-term synaptic plasticity**, which occurs within milliseconds to minutes, and **long-term synaptic plasticity**, which can extend from hours to the lifetime of an animal. We discuss below the two simplest forms of short-term plasticity involving changes of neurotransmitter release probability. Long-term synaptic plasticity will be a subject of focus in Chapter 10 in the context of memory and learning.

Although Ca^{2+}-dependent synaptic vesicle fusion provides an essential link between action potential arrival and neurotransmitter release, not every action potential results in the same amount of neurotransmitter release. As discussed earlier, the quantal yield of CNS synapses is much lower than that of the neuromuscular junction because a presynaptic axon may form only a few or a single active zone onto a postsynaptic partner neuron. In some mammalian CNS _in vivo_, the average **release probability**, defined as the probability that an active zone of

Table 3–1: A molecular cast for neurotransmitter release

Molecule	Location	Functions
Synaptic vesicle fusion with presynaptic membrane		
Synaptobrevin/VAMP	synaptic vesicle	mediates vesicle fusion (v-SNARE)
Syntaxin	presynaptic plasma membrane	mediates vesicle fusion (t-SNARE)
SNAP-25	presynaptic plasma membrane	mediates vesicle fusion (t-SNARE)
Sec1/Munc18 (SM)	presynaptic cytosol	likely acts as a catalyst for SNARE-mediated vesicle fusion
Ca^{2+} regulation of synaptic transmission		
Voltage-gated Ca^{2+} channel	active zone of presynaptic membrane	allows Ca^{2+} entry in response to action potential-triggered depolarization
Ca^{2+}	entering from extracellular space to presynaptic cytosol	triggers synaptic vesicle fusion
Synaptotagmin	synaptic vesicle	senses Ca^{2+} to trigger vesicle fusion
Complexin	presynaptic cytosol	binds and regulates SNARE-mediated vesicle fusion
Organization of presynaptic terminal (and alignment with postsynaptic density)		
RIM	active zone	organizes presynaptic scaffold
RIM-BP	active zone	organizes presynaptic scaffold
ELKS/Bruchpilot	active zone	organizes presynaptic scaffold
Rab3	synaptic vesicle	interacts with active zone components
Cadherin	presynaptic and postsynaptic plasma membranes	trans-synaptic adhesion
Neurexin	presynaptic plasma membrane	trans-synaptic adhesion
Neuroligin	postsynaptic plasma membrane	trans-synaptic adhesion
Neurotransmitter and vesicle recycling		
Acetylcholinesterase	synaptic cleft	degrades neurotransmitter acetylcholine
Plasma membrane neurotransmitter transporter (PMT)	presynaptic plasma membrane, glial membrane	transports excess neurotransmitter molecules back to presynaptic cytosol or to nearby glia
Vesicular neurotransmitter transporter (VT)	synaptic vesicle	transports neurotransmitters from presynaptic cytosol to the synaptic vesicle
V-ATPase	synaptic vesicle	establishes proton gradient within the synaptic vesicle
Synapsin	synaptic vesicle	regulates the size of the reserve pool
Clathrin	presynaptic cytosol	retrieves vesicles from presynaptic plasma membrane via endocytosis
Shibire/dynamin	presynaptic cytosol	retrieves vesicles from presynaptic plasma membrane via endocytosis
NSF	presynaptic cytosol	disassembles SNARE complex after fusion

a presynaptic terminal releases the transmitter contents of one or more synaptic vesicles following an action potential, is estimated to be far smaller than 1. If many active zones exist between a presynaptic and a postsynaptic cell, as is the case for the vertebrate neuromuscular junction, the probability that at least one active zone releases a vesicle is close to 1; however, the magnitude of postsynaptic response still depends on the release probability of each active zone.

The release probability can be affected by prior usage of the synapse. In **facilitating synapses**, successive action potentials trigger larger and larger postsynaptic responses. By contrast, in **depressing synapses**, successive action potentials result in smaller and smaller postsynaptic responses (**Figure 3–15**). These changes can be caused by a postsynaptic mechanism, such as altered sensitivity to the release of the same amount of neurotransmitters, as we will discuss later in

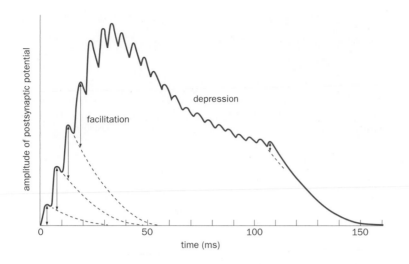

Figure 3–15 Facilitating and depressing synapses. In this schematic, the size of postsynaptic potentials, as indicated by the length of the double arrows parallel to the *y* axis, changes in response to a train of action potentials. The first series exhibit facilitation as each successive action potential produces a larger response; the latter series exhibit depression as responses become smaller and smaller for each successive action potential. The dotted lines represent the natural decay of postsynaptic potentials had there not been a follow-up action potential, and were used as the basis to determine the amplitude of postsynaptic potentials in response to successive action potentials. (Katz B [1966] Nerve, Muscle, and Synapse. With permission from McGraw Hill.)

the chapter and in Chapter 10, but fast facilitation and depression are most often caused by a presynaptic mechanism such as the altered amount of neurotransmitter release. The same synapse can be facilitating or depressing depending on its intrinsic property and its prior history of usage.

In the simplest case, facilitating synapses have a low starting release probability. The amount of release increases during repeated action potentials as active zone Ca^{2+} builds up. Depressing synapses, on the other hand, are usually characterized with a high starting release probability that results in a large amount of release at the beginning of a stimulus train; this exhausts the number of vesicles readily available for release and leads to a decline in the amount of release as the stimulus train proceeds. Because there are typically large numbers of vesicles in the reserve pool that can replenish depleted vesicles in the readily releasable pool, this sort of depression can recover in seconds. In the course of this book, we will encounter many additional mechanisms that adjust synaptic strength using distinct mechanisms and at different temporal scales.

3.11 The nervous system uses many neurotransmitters

To illustrate the basic principles of synaptic transmission, we have focused primarily on the vertebrate neuromuscular junction, which utilizes acetylcholine as the neurotransmitter. The principles we have learned thus far apply to virtually all chemical synapses, regardless of the neurotransmitter they use (**Figure 3–16; Table 3–2**). Two major neurotransmitters used in the vertebrate central nervous system are **glutamate** (glutamic acid), a natural amino acid, and **GABA** (γ-amino butyric acid), derived from glutamate by the enzyme **glutamic acid decarboxylase** (**GAD**). Glutamate is the predominant **excitatory neurotransmitter** in the vertebrate nervous system because its release depolarizes postsynaptic neurons and makes them more likely to fire action potentials. GABA is the predominant **inhibitory neurotransmitter** because its release usually renders postsynaptic neurons less likely to fire action potentials. The amino acid **glycine**, another inhibitory neurotransmitter, is used in a subset of inhibitory neurons in the brainstem and spinal cord of the vertebrate nervous system.

GABA appears to be the major inhibitory neurotransmitter across different species, including many invertebrates such as the nematode *C. elegans*, the fruit fly *Drosophila melanogaster*, and crustaceans (GABA's inhibitory action was first established in the crab). Like vertebrates, *C. elegans* also uses glutamate as the major excitatory neurotransmitter and ACh as the transmitter at the neuromuscular junction. Curiously, *Drosophila* utilizes ACh as the major excitatory neurotransmitter in the CNS and glutamate as the transmitter at the neuromuscular junction (see Figure 3–11). It is important to note that although it is convenient to label a particular neurotransmitter excitatory or inhibitory by its action on postsynaptic cells in most cases, we will see in the next part of the chapter that

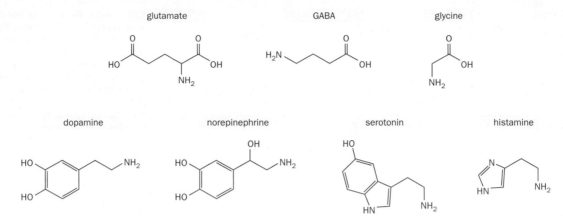

Figure 3–16 Structures of a subset of small-molecule neurotransmitters. Glutamate and glycine are natural amino acids. GABA (γ-amino butyric acid) is produced from glutamate. Dopamine is derived from the amino acid tyrosine. Norepinephrine is produced from dopamine and is a precursor for the hormone epinephrine. Serotonin is derived from the amino acid tryptophan. Histamine is derived from the amino acid histidine. See Figure 3–1A for the structure of acetylcholine.

the same transmitter can be excitatory or inhibitory depending on the properties of its receptor and the ionic composition of the postsynaptic cell.

Another important class of neurotransmitters plays a predominantly modulatory role. Modulatory neurotransmitters (also called **neuromodulators**) can up- or down-regulate the membrane potential, **excitability** (how readily a neuron fires an action potential), or neurotransmitter release by their postsynaptic target neurons, depending on the type of receptors that their postsynaptic neurons express and the subcellular localization of these receptors. Classic neuromodulators include **serotonin** (also called **5-HT** for 5-hydroxytryptamine), **dopamine**, **norepinephrine** (also called noradrenaline), and **histamine** (see Figure 3–16). They are all derived from aromatic amino acids and are collectively called **monoamine neurotransmitters**. In addition to being released into the synaptic cleft, these neurotransmitters can also be secreted into the extracellular space outside the confines of morphologically defined synapses to affect nearby cells; this property is referred to as **volume transmission**. In vertebrates, the cell bodies of neurons that synthesize monoamine neurotransmitters are mostly clustered in discrete nuclei in the brainstem or hypothalamus. They send profuse axons that collectively innervate a large fraction of the nervous system (see Box 8–1). Dopamine and serotonin act as neuromodulators throughout the animal kingdom. In place of norepinephrine, a chemically similar molecule called **octopamine** is used in some invertebrate nervous systems.

Some neurotransmitters have multiple roles in different parts of the nervous system (see Table 3–2). In vertebrates, ACh is used as an excitatory neurotransmitter by motor neurons to control skeletal muscle contraction at the neuromuscular junction. It is also one of the two neurotransmitters employed in the **autonomic nervous system** for neural control of visceral functions such as heart rate, respiration, and digestion. In the brain, ACh can act both as an excitatory neurotransmitter and as a neuromodulator much like the monoamine neurotransmitters. Likewise, norepinephrine functions as the autonomic nervous system's other neurotransmitter, but acts as a neuromodulator in the brain.

The type of neurotransmitter a neuron uses is often used as a major criterion for neuronal classification. Neurons can be broadly classified as excitatory, inhibitory, or modulatory as discussed earlier, and more specifically as glutamatergic, GABAergic, cholinergic, dopaminergic, and so on. Neurons of a given neurotransmitter type express a specific set of genes associated with that type, including enzyme(s) that synthesize the neurotransmitter, a vesicular transporter that pumps the neurotransmitter into synaptic vesicles, and in many cases a plasma-membrane transporter that retrieves the neurotransmitter from the synaptic cleft after release (see Figure 3–12). Some neurons utilize more than one of the neurotransmitters discussed above. For example, some mammalian CNS

Table 3–2: Commonly used neurotransmitters

Neurotransmitter	Major uses in the vertebrate nervous system[1]
Acetylcholine	motor neurons that excite muscle; ANS[2] neurons; CNS excitatory and modulatory neurons
Glutamate	most CNS excitatory neurons; most sensory neurons
GABA	most CNS inhibitory neurons
Glycine	some CNS inhibitory neurons (mostly in the brainstem and spinal cord)
Serotonin (5-HT)	CNS modulatory neurons
Dopamine	CNS modulatory neurons
Norepinephrine	CNS modulatory neurons; ANS[2] neurons
Histamine	CNS modulatory neurons
Neuropeptides	usually co-released from excitatory, inhibitory, or modulatory neurons; neurosecretory cells

[1] See text for variations in invertebrate nervous systems.

[2] ANS, autonomic nervous system; as will be discussed in more detail in Chapter 8, acetylcholine and norepinephrine are used in different subtypes of ANS neurons.

neurons can co-release a modulatory neurotransmitter and the excitatory neurotransmitter glutamate, or a modulatory neurotransmitter and the inhibitory neurotransmitter GABA.

In addition to the small-molecule neurotransmitters we have discussed thus far, some neurons also secrete **neuropeptides** that can act as neurotransmitters to communicate with postsynaptic neurons. The mammalian nervous system utilizes dozens of neuropeptides, with lengths ranging from a few amino acids to several dozen. As we will learn in Chapters 8 and 9, neuropeptides regulate diverse and vital physiological functions such as eating, sleeping, and sexual behaviors. Neuropeptides are usually produced by proteolytic cleavage of precursor proteins in the secretory pathway (see Figure 2–2). They are packaged into **dense-core vesicles** (which are larger than synaptic vesicles and contain electron-dense materials) after vesicles containing neuropeptides bud off from the Golgi apparatus, and are delivered via fast axonal transport to presynaptic terminals. Because they cannot be locally synthesized or recovered after release, but must be transported across long distances from the soma to axon terminals, neuropeptides are used more sparingly. The probability of neuropeptide release seems to be much lower than that of small-molecule neurotransmitters even when they are present in the same terminals. Compared to synaptic vesicles that release small-molecule neurotransmitters, we know far less about the mechanisms that control neuropeptide release from dense-core vesicles. In most cases, neuropeptides play modulatory roles, and are released from neurons that use a small-molecule neurotransmitter. As we will learn in Chapter 8, some neurons secrete neuropeptides into the bloodstream; in these cases, neuropeptides act as hormones to influence the physiology of recipient cells remotely.

The reason why different neurotransmitters have different effects is because the receptors on the postsynaptic membrane have different properties. We now turn to the next step of neuronal communication: how neurotransmitters act on postsynaptic neurons.

HOW DO NEUROTRANSMITTERS ACT ON POSTSYNAPTIC NEURONS?

In the first part of the chapter, we used postsynaptic responses, such as the end-plate potential (see Figure 3–1) or postsynaptic inward current (see Figures 3–5 and 3–9) as assays to investigate the mechanisms of presynaptic neurotransmitter release. In the following sections, we will discuss the mechanisms by which postsynaptic neurons produce these responses. We first discuss rapid responses that occur within milliseconds involving direct change of ion conductance. We then study responses that occur in tens of milliseconds to seconds, involving intracellular signaling pathways. We further highlight responses that occur in hours to days involving new gene expression. Finally, we discuss how postsynaptic neurons integrate different presynaptic input to determine their own firing pattern and neurotransmitter release properties, thus completing a full round of neuronal communication.

3.12 Acetylcholine opens a nonselective cation channel at the neuromuscular junction

We now begin our journey across the synaptic cleft to the postsynaptic side of the synapse, returning first to the vertebrate neuromuscular junction as our model. In Section 3.1, we learned that ACh released from motor axon terminals depolarizes the muscle membrane, and that iontophoretic application of ACh to muscle can substitute for ACh release from presynaptic terminals (see Figure 3–1). How does ACh accomplish this? By locally applying ACh to different regions of muscle fibers, researchers found that exogenous ACh produced the most effective depolarization near the motor axon terminal. These experiments implied that there must be receptors for ACh that are present on the muscle membrane and are concentrated at the neuromuscular junction. Upon ACh binding, ACh receptors trigger a rapid change in the muscle membrane's ion conductance.

To explore the underlying mechanisms, voltage clamp experiments analogous to those carried out on squid giant axons (see Section 2.10) were performed on the muscle fibers to test how ACh release induced by motor axon stimulation changes ion flow across the muscle membrane (**Figure 3–17**A). In these experiments, two electrodes were inserted into the muscle cell, one to measure the membrane potential (V_m) and compare it to a desired command voltage (V_{CMD}), and a second to pass feedback current into the muscle to maintain V_m at the same value as V_{CMD}. The current injected into the muscle, which can be experimentally measured, equals the current that passes through the muscle membrane in response to ACh release, or the **end-plate current**. (Under physiological conditions, that is, when the muscle is not voltage-clamped, the end-plate current would produce a membrane potential change, which is the end-plate potential

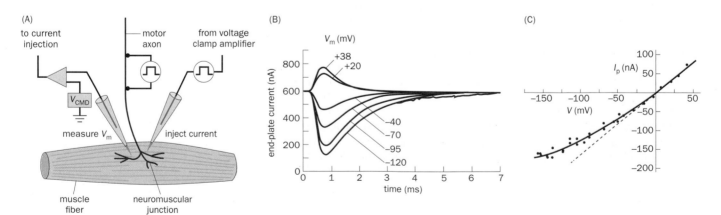

Figure 3–17 Properties of an acetylcholine (ACh)-induced current studied by voltage clamp. (A) Experimental setup. Two intracellular electrodes were inserted into the muscle cell at the frog neuromuscular junction. The first (left) was to record the membrane potential (V_m), which was compared with an experimenter-determined commanding potential (V_{CMD}). The second electrode injected feedback current into the muscle to maintain V_m at V_{CMD}. The end-plate current in response to ACh release caused by motor axon stimulation can be determined from the feedback current that was injected into the muscle cell in order to hold V_m at V_{CMD}. **(B)** The end-plate current elicited by single motor axon stimulation was measured at the six different membrane potentials indicated. At negative potentials

the end-plate current was inward (positive ions flowing into the muscle cell), whereas at positive potentials the end-plate current was outward. **(C)** Peak end-plate current (I_p, y axis) as a function of the muscle membrane potential (V, x axis). Experimental data (represented as dots) fell on a curve (the I–V curve) that was close to linear (dotted line), indicating that the conductance (represented by the slope of the I–V curve) is mostly unaffected by voltage. Note that the current switched sign between negative (inward) and positive (outward) at 0 mV, which is the reversal potential of the channel opened by ACh. (B & C, adapted from Magleby KL & Stevens CF [1972] J Physiol 223:173–197.)

we have used to measure neurotransmitter release in Section 3.1). It was found that ACh release caused an inward current at negative membrane potentials and an outward current at positive membrane potentials (Figure 3–17B). The current–voltage relationship (called an *I–V* **curve**) is nearly linear. The membrane potential at which the current flow reversed direction (called the **reversal potential**) was approximately 0 mV (Figure 3–17C).

If the ACh-induced current were carried by a single ion, the reversal potential should equal the equilibrium potential of that ion, as both reversal potential and equilibrium potential define a state in which the net current is zero. However, the reversal potential of the ACh-induced current is unlike the Na$^+$, K$^+$, or Cl$^-$ currents discussed in Section 2.5, with equilibrium potentials around +58 mV, –85 mV, and –79 mV, respectively. Indeed, experiments that measured the reversal potentials in response to varying extracellular K$^+$, Na$^+$, and Cl$^-$ concentrations suggested that ACh opens a channel that is permeable to both K$^+$ and Na$^+$ and other cations but not to anions such as Cl$^-$. Further evidence indicated that ACh acts on a single channel that is permeable simultaneously to Na$^+$ and K$^+$. At positive membrane potentials, the driving force for K$^+$ efflux is greater than the driving force for Na$^+$ influx (because the V_m is further from E_K than from E_{Na}), and hence K$^+$ efflux exceeds Na$^+$ influx, causing a net outward current. At negative membrane potentials, the driving force for Na$^+$ influx exceeds that for K$^+$ efflux, causing a net inward current. Ca^{2+} influx also makes a small contribution to the inward current. Importantly, since the reversal potential of 0 mV is far above the muscle membrane's resting potential (around –75 mV) or the threshold for action potential production (which is usually 10–20 mV more depolarized than the resting potential), the end-plate current under physiological conditions is always inward, carried by more Na$^+$ influx than K$^+$ efflux (**Figure 3–18A**). This depolarizes the muscle membrane, resulting in the end-plate potential (EPP) we introduced in Section 3.1.

The action of an ACh-induced current can be represented by an electrical circuit model of the muscle membrane, in which the ACh-induced current can be considered as an added branch to the resting muscle membrane (Figure 3–18B). Immediately after the switch is on (representing ACh release), $I_{Na} = g_{Na}(V_m - E_{Na})$, and $I_K = g_K(V_m - E_K)$. Because at rest V_m is around –75 mV, the absolute value $|V_m - E_{Na}|$ far exceeds $|V_m - E_K|$. Assuming that the ACh-activated channel has similar conductance for Na$^+$ and K$^+$ (see below), the inward current from the Na$^+$ branch far exceeds the outward current from the K$^+$ branch. Thus, the ACh release activates a net inward current.

The reversal potential, designated as E_{rev}, is an important property of ion channels that are permeable to more than one ion. It is determined by the relative conductance and equilibrium potential of each ion. Using the electrical circuit model in Figure 3–18B, we can determine their relationship as follows. At the reversal potential ($V_m = E_{rev}$), Na$^+$ influx equals K$^+$ efflux, thus $I_K = I_{Na}$. Since $I_K = g_K(V_m - E_K)$, and $I_{Na} = g_{Na}(V_m - E_{Na})$, we have

$$E_{rev} = \frac{g_{Na}E_{Na} + g_K E_K}{g_{Na} + g_K} = \frac{\frac{g_{Na}}{g_K}E_{Na} + E_K}{\frac{g_{Na}}{g_K} + 1}$$

Figure 3–18 ACh opens a nonselective cation channel on the muscle membrane. **(A)** Schematic of how ACh release causes depolarization of the muscle membrane. At rest (left), the membrane potential of the muscle cell is around –75 mV, similar to the resting membrane potential of many neurons, with higher K$^+$ concentration inside the cell and higher Na$^+$ concentration outside (see Figure 2–12A). ACh binding opens a cation channel on the muscle membrane permeable to both Na$^+$ and K$^+$. This allows more Na$^+$ influx than K$^+$ efflux because of the larger driving force on Na$^+$, thus depolarizing the muscle membrane. **(B)** An electrical circuit model. The left part represents the resting muscle membrane, which includes a membrane capacitance branch (C_m) and a membrane resistance branch (R_m) with a battery representing the resting potential (E_r) (see Sections 2.7 and 2.8). The right part represents the ACh-induced current, with a K$^+$ path and a Na$^+$ path (with a resistance of $1/g_K$ and $1/g_{Na}$, respectively) in parallel. After the switch is turned on (green arrow) by ACh release, the current that passes through the Na$^+$ path is much larger than the current that passes through the K$^+$ path because the driving force for Na$^+$ (= $E_{Na} - V_m$, where E_{Na} is the equilibrium potential for Na$^+$, and V_m is the membrane potential; see Section 2.5) is far greater than the driving force for K$^+$ (= $V_m - E_K$, where E_K is the equilibrium potential for K$^+$). Positive current flows inside the cell, discharges the membrane capacitance, and depolarizes the membrane potential.

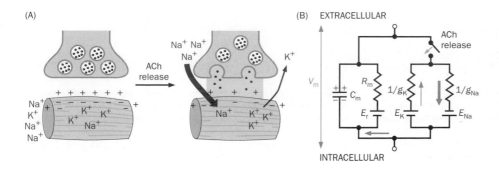

We can see from the above formula that if the conductance for Na^+ and K^+ were equal ($g_{Na}/g_K = 1$), E_{rev} would simply be an average of E_{Na} and E_K. Suppose the ionic concentrations across the muscle membrane are the same as our model neuron in Figure 2–12A, with $E_K = -85$ mV and $E_{Na} = +58$ mV, then E_{rev} should be -13.5 mV. However, since $E_{rev} = 0$ mV as determined in Figure 3–17C, we can calculate that g_{Na}/g_K is approximately 1.5; in other words, the channel that is opened upon ACh binding has higher conductance for Na^+ than for K^+.

3.13 The skeletal muscle acetylcholine receptor is a ligand-gated ion channel

A deeper understanding of the nature of the ACh-induced conductance change requires the identification of the postsynaptic **acetylcholine receptor** (**AChR**) and the ion channel whose conductance is coupled to ACh binding. Further studies indicated that muscle AChR is itself the ion channel. Just as the neuromuscular junction served as a model synapse because of its experimental accessibility, the AChR became a model neurotransmitter receptor because of its abundance, particularly in the electric organ of the *Torpedo* ray, which is highly enriched for an AChR similar to that from the skeletal muscle. Biochemical purification and subsequent cloning of the *Torpedo* AChR revealed that it consists of five subunits: two α, one β, one γ, and one δ (**Figure 3–19**A). Each AChR contains two ACh binding sites, which are respectively located at the α–γ and α–δ subunit interfaces. Both sites need to bind ACh in order for the channel to open. Evidence in support of this heteropentameric receptor as the ACh-activated channel came from a reconstitution experiment in the *Xenopus* oocyte. Co-injection of mRNAs encoding all four AChR subunits into the frog oocyte caused the oocyte, which normally does not respond to ACh, to produce an inward current in response to ACh iontophoresis in voltage clamp experiments. This ACh-induced inward current was reversibly blocked by the AChR **antagonist** (agent that acts to counter the action of an endogenous molecule) curare (see Box 3–2 for more details on curare); washing out the curare restored the inward current (Figure 3–19B). Omitting mRNA for any of the AChR subunits abolished the ACh-induced inward current in the oocyte expression system.

The three-dimensional structure of the *Torpedo* AChR has been determined by high-resolution electron microscopy (**Figure 3–20**). All AChR subunits contain four transmembrane helices, with the M2 helices from all subunits lining the ion conduction pore. The transmembrane helices form a hydrophobic barrier or 'gate' when AChR is closed, preventing ion flow. ACh binding induces the rotation of the α subunits, which causes an alternative conformation of the M2 helices and opens the gate to allow the passage of cations.

To summarize synaptic transmission at the vertebrate neuromuscular junction: action potentials trigger ACh release from motor axon terminals. ACh molecules diffuse across the synaptic cleft and bind to postsynaptic AChRs, which are highly concentrated on the muscle membrane directly apposing the motor axon terminal. Upon ACh binding, muscle AChRs produce a nonselective cation

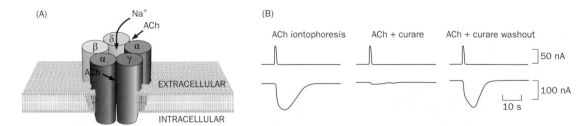

Figure 3–19 Composition of the acetylcholine receptor (AChR). (A) Schematic illustrating the subunit composition of AChR. The two ACh binding sites are at the α–γ and α–δ subunit interfaces. **(B)** Functional expression of AChR was achieved by injecting mRNAs encoding the four AChR subunits into *Xenopus* oocytes. Top traces, current used for iontophoresis of ACh. Bottom traces, inward current measured in a voltage clamp setup in response to ACh application: ACh application led to an inward current (left), which was blocked by curare, an AChR inhibitor (middle), but was reversed after curare was washed out (right). The membrane potential was held at –60 mV. (B, adapted from Mishina M, Kurosaki T, Tobimatsu T et al. [1984] *Nature* 307:604–608. With permission from Macmillan Publishers Ltd.)

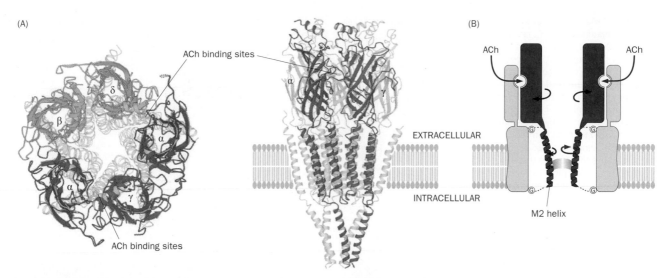

(A)

ACh binding sites

δ

β

α

α

γ

ACh binding sites

EXTRACELLULAR

INTRACELLULAR

(B)

ACh   ACh

M2 helix

Figure 3–20 AChR structure and gating model. (A) Structure of *Torpedo* AChR in a closed state at a resolution of 4 Å as determined by electron microscopy. Left, a surface view from the extracellular side. The tryptophan in the α subunit implicated in ACh binding is highlighted in gold. Only the extracellular portions are colored. Right, a side view showing the transmembrane helices. The front α and γ subunits are highlighted in color. **(B)** A model for AChR activation. ACh binding induces a rotation of part of the extracellular domain of the α-subunit (red). This rotation triggers a conformational change in the transmembrane helix M2 that lines the ion conduction pore, leading to the opening of the ion gate. Dotted lines with circled Gs (for glycine residues) indicate that M2 is connected with the rest of the protein by flexible loops. (A, from Unwin N [2005] *J Mol Biol* 346:967–989. With permission from Elsevier Inc.; B, adapted from Miyazawa A, Fujiyoshi Y & Unwin N [2003] *Nature* 423:949–955. With permission from Macmillan Publishers Ltd.)

conductance that causes more Na^+ influx than K^+ efflux and thus produces depolarization in the form of an EPP. When this depolarization reaches threshold the muscle cell fires action potentials, which results in muscle contraction. We will study the mechanisms of muscle contraction in Section 8.1.

While the open probability of the voltage-gated Na^+ and K^+ channels we studied in Chapter 2 is increased by depolarization, the open probability of muscle AChR channels is increased by ACh binding but not by changes in the membrane potential. The conductance (I/V) is mostly constant across different voltages, as can be seen by the near linear I–V curve in Figure 3–17C. The muscle AChR is therefore called a **ligand-gated ion channel**, and is the prototype of a large family of ligand-gated ion channels (see Table 2–2). Most ligands in the ligand-gated channels are extracellular neurotransmitters such as ACh; however, ligand-gated channels also include channels that are gated by intracellular signaling molecules, some of which will be discussed in later sections.

3.14 Neurotransmitter receptors are ionotropic or metabotropic

Following the pioneering work on vertebrate skeletal muscle AChRs, receptors for several other neurotransmitters were found to be ion channels. All neurotransmitter-gated ion channels in vertebrates belong to one of three subfamilies. GABA-, glycine-, and serotonin-gated ion channels are in the same subfamily as muscle AChRs (**Figure 3–21**, left), with five subunits each possessing four transmembrane segments. Glutamate-gated ion channels constitute a second subfamily, with four subunits each possessing three transmembrane segments (Figure 3–21, middle). Finally, ATP can be used as a neurotransmitter in some neurons, and ATP-gated ion channels are trimers each having just two transmembrane segments (Figure 3–21, right).

Neurotransmitter receptors that function as ion channels, which allow rapid communication across the synapse, are also called **ionotropic receptors** (**Figure 3–22**A). For example, the direct gating of the muscle AChR channel by ACh transmits electrical signals from presynaptic neuron to postsynaptic muscle within a few milliseconds (see Figure 3–1C; Figure 3–17B). Ionotropic receptors are mostly synonymous with the ligand-gated ion channels introduced in the previous section. Both terms encompass receptors that are gated by ligands

Figure 3–21 Three families of ionotropic receptors in vertebrates. Left, like the ionotropic AChR (see Figure 3–20), each subunit of the ionotropic GABA receptor, glycine receptor, and serotonin receptor spans the membrane four times. Five subunits constitute a functional receptor with two neurotransmitter-binding sites (stars). Middle, an ionotropic glutamate receptor has four subunits and four neurotransmitter-binding sites; each subunit spans the membrane three times. Right, an ionotropic P2X receptor consists of three subunits, each of which features an ATP-binding site and spans the membrane twice. (Hille [2001] Ion Channels of Excitable Membranes. With permission from Sinauer.)

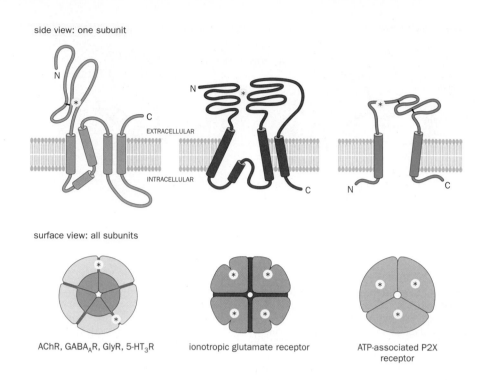

side view: one subunit

EXTRACELLULAR

INTRACELLULAR

surface view: all subunits

AChR, GABA$_A$R, GlyR, 5-HT$_3$R   ionotropic glutamate receptor   ATP-associated P2X receptor

(neurotransmitters) and conduct ion passage across the membrane; the choice of which term to use depends on whether the properties of the receptor or the channel are being emphasized.

In contrast to the fast-acting ionotropic receptors, **metabotropic receptors** (Figure 3–22B), when activated by neurotransmitter binding, trigger intracellular signaling cascades to regulate ion channel conductance, and thus modulate membrane potential indirectly. (The intracellular signaling molecules are often referred to as 'second messengers,' as opposed to the 'first messengers,' the extracellular ligands.) Accordingly, they operate over a longer timescale ranging from tens of milliseconds to seconds. In addition, unlike ionotropic receptors, which are mostly concentrated in the postsynaptic density across the synaptic cleft from the presynaptic active zone, metabotropic receptors are typically not concentrated at the postsynaptic membrane apposing the synaptic presynaptic active zone, and therefore are termed "extrasynaptic."

Many neurotransmitters have both ionotropic and metabotropic receptors (**Table 3–3**). For example, ACh can act on metabotropic receptors in addition to the ionotropic AChR we just studied. To distinguish between the two receptor

Figure 3–22 Ionotropic and metabotropic neurotransmitter receptors. (A) Ionotropic receptors are ion channels that are gated by neurotransmitters. Neurotransmitter binding causes membrane potential change within a few milliseconds. **(B)** Metabotropic receptors act through intracellular second messenger systems to regulate ion channel conductance. Neurotransmitter binding causes membrane potential change in tens of milliseconds to seconds.

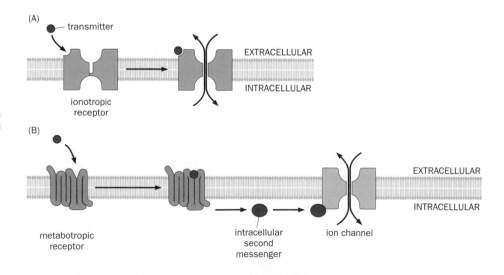

(A)   transmitter

ionotropic receptor

EXTRACELLULAR

INTRACELLULAR

(B)

metabotropic receptor   intracellular second messenger   ion channel

EXTRACELLULAR

INTRACELLULAR

Table 3–3: Ionotropic and metabotropic neurotransmitter receptors encoded by the human genome

Neurotransmitter	Ionotropic		Metabotropic	
	Name	Number of genes	Name	Number of genes
Acetylcholine	nicotinic ACh receptor	16	muscarinic ACh receptor	5
Glutamate	NMDA receptor	7	metabotropic glutamate receptor (mGluR)	8
	AMPA receptor	4		
	others	7		
GABA	GABA$_A$ receptor	19	GABA$_B$ receptor	2
Glycine	glycine receptor	5		
ATP	P2X receptor	7	P2Y receptor	8
Serotonin (5-HT)	5-HT$_3$ receptor	5	5-HT$_{1, 2, 4, 6, 7}$ receptors	13
Dopamine			dopamine receptor	5
Norepinephrine (epinephrine)			α-adrenergic receptor	6
			β-adrenergic receptor	3
Histamine			histamine receptor	4
Adenosine			adenosine receptor	3
Neuropeptides			neuropeptide receptors	dozens

Abbreviations: GABA, γ-aminobutyric acid; P2X receptor, ATP-gated ionotropic receptor; P2Y, ATP-gated metabotropic receptor; 5-HT$_\#$R, serotonin (5-hydroxytryptamine) receptor subtype #; ACh, acetylcholine; NMDA, N-methyl-D-aspartate; AMPA, 2-amino-3-hydroxy-5-methylisoxazol-4-propanoic acid.

Data from the IUPHAR (International Union of Basic and Clinical Pharmacology) database (www.iuphar-db.org).

types, we refer to them according to their specific **agonists** (agents that mimic the action of an endogenous molecule such as a neurotransmitter). Hence, ionotropic AChRs are called **nicotinic AChRs** because they are potently activated by nicotine. Nicotinic AChRs are expressed not only in muscles, but also in many neurons in the brain, where nicotine acts as an addictive stimulant. Metabotropic AChRs are called **muscarinic AChRs** because they are activated by muscarine, a compound enriched in certain mushrooms.

In the following sections, we highlight the actions of key ionotropic and metabotropic receptors for major neurotransmitters in the CNS (Table 3–3).

3.15 AMPA and NMDA glutamate receptors are activated by glutamate under different conditions

Ionotropic glutamate receptors are responsible for the fast action of glutamate, the major excitatory neurotransmitter in the vertebrate CNS. Indeed, glutamatergic excitatory synapses account for the vast majority of synapses in the vertebrate CNS: virtually all neurons—whether they are excitatory, inhibitory, or modulatory—express ionotropic glutamate receptors and are excited by glutamate.

Similar to muscle AChR, ionotropic glutamate receptors are cation channels that do not select between Na$^+$ and K$^+$, with a reversal potential near 0 mV. Under physiological conditions, glutamate binding to ionotropic glutamate receptors produces an inward current called the **excitatory postsynaptic current** (EPSC) (**Figure 3–23**, top), as more positively charged ions flow into the cell than out of it. This is analogous to the end-plate current we saw at the neuromuscular junction (see Figure 3–17). The inward current produces a transient depolarization in the postsynaptic neuron called the **excitatory postsynaptic potential** (EPSP) (Figure 3–23, bottom), analogous to the EPP at the neuromuscular junction. The recordings shown in Figure 3–23 were made in acutely prepared **brain slices** (fresh sections of brain tissue about a few hundred micrometers thick) that preserve local three-dimensional architecture and neuronal connections while allowing experimental access, such as whole-cell patch recording of individual neurons and control of extracellular media.

Historically, ionotropic glutamate receptors have been divided into three subtypes that are named for their selective responses to three agonists: **AMPA**

excitatory postsynaptic current (EPSC)

excitatory postsynaptic potential (EPSP)

Figure 3–23 Excitatory postsynaptic current (EPSC) and excitatory postsynaptic potential (EPSP) at a glutamatergic synapse. Representative EPSC (top) and EPSP (bottom) recorded using whole-cell patch clamping from hippocampal pyramidal neurons in an *in vitro* slice preparation, in response to electrical stimulation of glutamatergic input axons. The EPSC was recorded in a voltage clamp mode when the membrane potential was held at –90 mV, and the EPSP was recorded in a current clamp mode (see Box 13–2 for details). The vertical ticks before the EPSC and EPSP are artifacts of electrical stimulation. (Adapted from Hestrin S, Nicoll RA, Perkel DJ et al. [1990] *J Physiol* 422:203–225.)

Figure 3–24 Properties of AMPA and NMDA subtypes of ionotropic glutamate receptors. (A) When the postsynaptic neuron (represented by a dendritic spine) is near the resting potential, glutamate (GLU) released from the presynaptic neuron opens only the AMPA receptor channel (AMPAR), causing Na⁺ entry and producing excitatory postsynaptic potentials (EPSPs). The NMDA receptor channel (NMDAR) is blocked by external Mg²⁺ and therefore cannot be opened by glutamate binding alone. **(B)** When the postsynaptic neuron is depolarized, the Mg²⁺ block is relieved. Both NMDAR and AMPAR can now be opened by glutamate binding. The NMDAR is highly permeable to Ca²⁺. For simplicity, the smaller K⁺ efflux through open AMPAR and NMDAR channels is omitted. (Adapted from Cowan WM, Südhof TC & Stevens CF [2001] Synapses. Johns Hopkins University Press.)

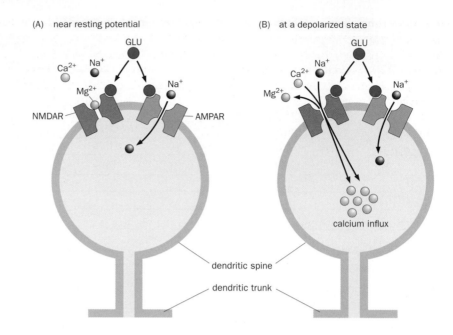

(A) near resting potential

(B) at a depolarized state

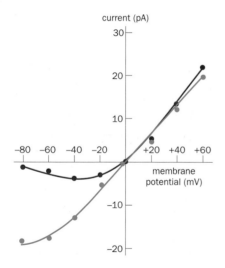

Figure 3–25 Current–voltage relationship of the NMDA receptor in the presence or absence of external Mg²⁺. Blue curve, in Mg²⁺-free media, the conductance of the NMDA receptor is nearly constant between −60 mV and +60 mV, as indicated by the near linear slope of the I–V curve. This indicates that the NMDA receptor per se is not gated by voltage, similar to the nicotinic acetylcholine receptor (see Figure 3–17C). Red curve, the presence of physiological concentrations of extracellular Mg²⁺ markedly diminishes the inward current at negative membrane potentials, because Mg²⁺ blocks cation influx. Data were obtained using whole-cell patch recording of cultured mouse embryonic neurons. (Adapted from Nowak L, Bregestovski P & Ascher P [1984] *Nature* 307:462–465. With permission from Macmillan Publishers Ltd.)

(2-amino-3-hydroxy-5-methylisoxazol-4-propanoic acid), **kainate** (kainic acid), and **NMDA** (N-methyl-D-aspartate). Molecular cloning of these receptors revealed that they are encoded by distinct gene subfamilies of ionotropic glutamate receptors (see Table 3–3). Because the properties of AMPA and kainate receptors are more similar to each other, they are collectively called non-NMDA receptors. In contrast, NMDA receptors have distinctive properties. Below we use the AMPA and NMDA receptors to illustrate these differences.

AMPA receptors are the fast glutamate-gated ion channels that conduct Na⁺ and K⁺. Depending on the subunit composition, some AMPA receptors are also permeable to Ca²⁺ in addition to Na⁺ and K⁺ (see below). They mediate synaptic transmission at the glutamatergic synapses when the postsynaptic neuron is near the resting potential. Because the driving force of Na⁺ is much greater than that of K⁺ near the resting potential, AMPA receptor opening causes a net influx of positively charged ions, resulting in depolarization of postsynaptic neurons (**Figure 3–24**A).

NMDA receptors have the unusual property of not only being gated by glutamate but also being influenced by the membrane potential; they also require glycine as a co-agonist. However, the mechanism by which membrane potential affects NMDA receptor conductance is different from the voltage-gated Na⁺ and K⁺ channels discussed in Chapter 2. At the extracellular face of the membrane, the mouth of the NMDA receptor is blocked by Mg²⁺ at negative potentials, such that the channel remains closed despite glutamate binding (Figure 3–24A). However, depolarization of the postsynaptic membrane relieves the Mg²⁺ block (Figure 3–24B). In the absence of external Mg²⁺, the NMDA receptor conductance is not affected by the membrane potential, as can be seen by the near linear I–V curve (**Figure 3–25**, blue line), similar to the I–V curve for AChR (see Figure 3–17C). By contrast, under physiological external Mg²⁺ condition, the conductance is greatly reduced when the membrane potential is negative (Figure 3–25, red line). Thus, the NMDA receptor acts as a **coincidence detector**, and opens only in response to concurrent presynaptic glutamate release *and* postsynaptic depolarization. This property is very important in synaptic plasticity and learning, as well as activity-dependent wiring of the nervous system, subjects we will return to in Chapters 5 and 10. Once opened, the NMDA receptors have high Ca²⁺ conductance. While AMPA receptors provide initial depolarization to release the Mg²⁺ block of nearby NMDA receptors—these two glutamate receptors are often co-expressed in the same postsynaptic site—NMDA receptors contribute additional depolarization alongside AMPA receptors. Importantly, Ca²⁺ influx via NMDA receptors contributes to many biochemical changes in postsynaptic cells, as will be discussed later in the chapter.

Structural studies have revealed how glutamate receptor subunits are arranged and how ligand binding might trigger channel opening. All ionotropic glutamate receptors are composed of four subunits (see Figure 3–21). Each subunit consists of several modular domains (**Figure 3–26**): an amino terminal domain, a ligand-binding domain, a transmembrane domain that comprises three membrane-spanning helices (M1, M3, and M4) plus an additional pore loop (M2), and a carboxy-terminal intracellular domain. AMPA receptors can form functional homo-tetramers (composed of four identical subunits) although they are usually found *in vivo* as hetero-tetramers of two or more of the four variants, GluA1, GluA2, GluA3, and GluA4. Crystal structures of tetramers composed of GluA2 suggest that glutamate binding results in a large conformational change in the ligand-binding domain, which causes a corresponding conformational change in the adjacent transmembrane domain to open the ion conductance pore. NMDA receptors are obligatory hetero-tetramers composed of two **GluN1** (also called NR1) subunits, each with a binding site for the co-agonist glycine, and two **GluN2** (also called NR2) subunits, each with a glutamate-binding site. GluN1 is encoded by a single gene, whereas GluN2 has four variants, GluN2A, GluN2B, GluN2C, and GluN2D, encoded by four separate genes.

The subunit composition of both AMPA and NMDA receptors has important functional consequences. For example, most AMPA receptors contain the GluA2 subunit; most GluA2-containing AMPA receptors are impermeable to Ca^{2+} due to a post-transcriptional modification called **RNA editing**, which changes the mRNA sequence encoding a key residue in GluA2's channel pore. AMPA receptors that lack GluA2 or contain unedited GluA2 subunits are permeable to Ca^{2+} (though not as permeable as are the NMDA receptors). AMPA receptors that lack GluA2 are also susceptible to a voltage-dependent block by intracellular polyamines, preventing Na^+ influx when the neuron becomes more depolarized. These AMPA receptors are thus inward-rectifiers analogous to the inward-rectifier K^+ channels we discussed in Box 2–4. NMDA receptors containing different GluN2 variants also have distinct channel conductances and cytoplasmic signaling properties, and bind differentially to postsynaptic scaffolding proteins (see the next section). Combinations of different subunits thus offer both AMPA and NMDA receptors a rich repertoire of functional and regulatory properties. Indeed, the subunit compositions of AMPA and NMDA receptors differ in different types of neurons, undergo developmental changes in the same types of neurons, and can be regulated by synaptic activity.

3.16 The postsynaptic density is organized by scaffolding proteins

Just as the presynaptic terminal is highly organized by active-zone scaffold proteins (see Section 3.7), the postsynaptic density is highly organized by postsynaptic proteins. At the glutamatergic synapses, for example, the postsynaptic density consists of not only glutamate receptors but also a large number of associated proteins (**Figure 3–27**). These include (1) trans-synaptic adhesion proteins that align active zones with postsynaptic densities (see also Section 3.7), (2) proteins that participate in signal transduction cascades, and (3) a diverse array of scaffolding proteins that connect the glutamate receptors and trans-synaptic adhesion molecules to signaling molecules and cytoskeletal elements. The resulting protein network controls glutamate receptor localization, density, trafficking, and signaling, all of which affect synaptic transmission and synaptic plasticity. Synaptic scaffolds are also present in GABAergic postsynaptic terminals, utilizing scaffolding proteins that only partially overlap with those found in glutamatergic synapses. We will learn more about the postsynaptic density protein network in the context of development and synaptic plasticity in Chapters 7 and 10, and how their dysfunction contributes to brain disorders in Chapter 11.

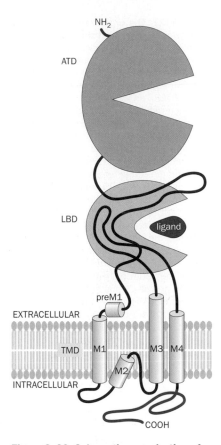

Figure 3–26 Schematic organization of an ionotropic glutamate receptor. Each of the four subunits of the ionotropic glutamate receptor (only one is shown here) is composed of an amino-terminal domain (ATD), a ligand-binding domain (LBD), a transmembrane domain (TMD), and a carboxy-terminal intracellular domain (red). The line represents the polypeptide chain from the extracellular amino terminus (NH_2) to the intracellular carboxy-terminus (COOH). The M1–M4 cylinders represent helices that span across (M1, M3, M4) or loop into (M2) the plasma membrane. The cartoon is based on the crystal structure of a homotetramer of GluA2. (Adapted from Sobolevsky AI, Rosconi MP & Gouaux E [2009] *Nature* 462:745–756. With permission from Macmillan Publishers Ltd.)

Figure 3–27 The organization of postsynaptic density at the glutamatergic synapse. At the cell surface, the postsynaptic density of a mature glutamatergic synapse is enriched in AMPA and NMDA glutamate receptors (AMPAR and NMDAR), as well as trans-synaptic cell adhesion molecules such as cadherins and neuroligins (which respectively bind to presynaptic cadherins and neurexins; see Figure 3–10). Named for their localization to the postsynaptic density and their molecular weight, the scaffolding proteins of the PSD-95 family bind to many proteins, including the GluN2 subunit of the NMDAR, the AMPAR-associated TARPs (transmembrane AMPAR regulatory proteins), the synaptic adhesion molecule neuroligins, the signal-transducing enzyme CaMKII, and other scaffolding proteins that bind to metabotropic glutamate receptors (mGluR) and other postsynaptic density proteins (not shown). The diagram only depicts a subset of known components and interactions in the postsynaptic density. (Adapted from Sheng M & Kim E [2011] *Cold Spring Harb Perspect Biol* 3:a005678.)

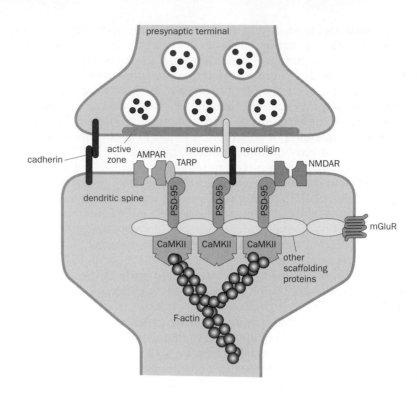

We use one of the most abundant scaffolding proteins at the glutamatergic synapse, **PSD-95** (<u>p</u>ostsynaptic <u>d</u>ensity protein-<u>95</u> kDa), to illustrate the organizational role of scaffolding proteins in the dendritic spine where glutamatergic synapses are usually located (Figure 3–27). PSD-95 contains multiple protein–protein interaction domains, including three **PDZ domains**, which bind to C-terminal peptides with a specific sequence motif that occurs in many transmembrane receptors. (PDZ is an acronym for three proteins that share this domain: <u>P</u>SD-95 originally identified from biochemical analysis of the postsynaptic density; <u>D</u>iscs-large in *Drosophila* that regulates cell proliferation and is also associated with the postsynaptic density; and <u>Z</u>O-1, an epithelial tight junction protein.) These protein–protein interaction domains enable PSD-95 to bind directly to the GluN2 subunit of the NMDA receptor, a family of AMPA receptor-associated proteins called TARPs (<u>t</u>ransmembrane <u>A</u>MPA receptor <u>r</u>egulatory <u>p</u>roteins), the trans-synaptic adhesion molecule neuroligin, and Ca2+/calmodulin-dependent protein kinase II (CaMKII, an enzyme highly enriched in postsynaptic densities, whose role in signal transduction will be introduced in Section 3.20). PSD-95 also binds other PDZ-domain-containing scaffolding proteins that in turn associate with other postsynaptic components such as metabotropic glutamate receptors and the actin cytoskeleton. Thus, the scaffold protein network stabilizes neurotransmitter receptors at the synaptic cleft by placing them close to the trans-synaptic adhesion complex apposing the active zone (see Figure 3–10), brings enzymes (for example, CaMKII) close to their upstream activators (for example, Ca^{2+} entry through the NMDA receptor) and downstream substrates, and organizes the structure of the dendritic spine by bridging the trans-synaptic adhesion complex and the underlying actin cytoskeleton.

3.17 Ionotropic GABA and glycine receptors are Cl⁻ channels that mediate inhibition

The role of inhibition in nervous system function was first established in the study of spinal cord reflex over a century ago (see Section 1.9). In the 1950s, when intracellular recording techniques were applied to the study of spinal motor neurons, it was found that stimulating their inhibitory input axons resulted in a rapid membrane potential change due to current flow across the motor neuron membrane. These current and membrane potential changes are called the **inhibitory postsynaptic current** (**IPSC**) and **inhibitory postsynaptic potential** (**IPSP**), respectively. In a revealing experiment (**Figure 3–28**), the membrane potential of the motor neuron was set by the experimenter at different initial values by injecting currents through an electrode, while the membrane potential was measured by a second electrode in response to stimulation of its inhibitory input axons. It was found that when the initial membrane potential was equal to or more depolarized than the resting potential of –70 mV or so, stimulation of the inhibitory input caused hyperpolarization, whereas when the initial membrane potential was set more hyperpolarized than –80 mV, stimulation of the inhibitory input produced depolarization. The reversal potential, around –80 mV, is close to the equilibrium potential for Cl⁻ (E_{Cl}), suggesting that the IPSC is carried by Cl⁻ flow. Indeed, by increasing intracellular Cl⁻ concentration, the reversal potential became less negative following the change of E_{Cl} as predicted from the Nernst equation. This experiment suggested that inhibition of the spinal motor neuron is mediated by an increase of Cl⁻ conductance across the motor neuron membrane.

Subsequent studies have shown that the fast inhibitory action is mediated by the neurotransmitters glycine (used by a subset of inhibitory neurons in the spinal cord and brainstem) and GABA (used by most inhibitory neurons), which act on ionotropic **glycine receptors** or **GABA$_A$ receptors**, respectively. The structure of GABA$_A$ receptors is similar to that of the nicotinic AChRs (see Figure 3–20), consisting of a pentamer with two α subunits, two β subunits, and one γ subunit. Each subunit has multiple isoforms encoded by several genes (see Table 3–3), and other subunits such as δ and ε can be used in lieu of γ. Many pharmaceutical drugs act on GABA$_A$ receptors to modulate inhibition in the brain. As we will learn in Chapter 11, the most widely used anti-epilepsy, anti-anxiety, and sleep-promoting drugs bind to and enhance the functions of GABA$_A$ receptors. Glycine receptors are also composed of a pentamer with two α subunits. Both GABA$_A$ and glycine receptors are ligand-gated ion channels that are selective for anions, primarily Cl⁻.

How does an increase of Cl⁻ conductance that results from the opening of GABA$_A$ (or glycine) receptor channels on postsynaptic neurons cause inhibition? In most neurons, E_{Cl} is slightly more hyperpolarized than the resting potential as in the case of the spinal motor neuron we just studied. Thus, an increase of Cl⁻ conductance causes Cl⁻ influx (which is equivalent to an outward current because Cl⁻ carries a negative charge), resulting in a small hyperpolarization (**Figure 3–29**A, left panel). Importantly, if the neuron also receives simultaneously an excitatory input (for example, opening of glutamate receptor channels), which produces an EPSP, the relatively depolarized potential increases the driving force for Cl⁻ influx. This increases the outward current triggered by GABA, which counters the EPSP-producing inward current, making it more difficult for the cell's membrane potential to reach the threshold for firing action potentials (Figure 3–29A, middle and right panels).

The interaction of excitatory and inhibitory input can also be seen in an electrical circuit model, where each is represented by a branch consisting of a switch (representing neurotransmitter release), a conductance (g_e or g_i representing EPSC or IPSC conductance), and a battery (representing the reversal potential for the excitatory glutamate receptors, E_{e-rev}, or the GABA$_A$ receptor, which equals E_{Cl}). When only the inhibitory input is switched on, because E_{Cl} is more hyperpolarized than the resting potential (E_r), a small outward current is produced from the g_i branch, resulting in a small hyperpolarizing IPSP (Figure 3–29B, left). When only the excitatory input is switched on, a large inward current is produced from the g_e branch because E_{e-rev} is much more depolarized than E_r, resulting in a large depolarizing EPSP (Figure 3–29B, middle). When both the excitatory and the inhibitory inputs are switched on, part of the inward current in the g_e branch flows outward through the g_i branch (Figure 3–29B, right), leading to a smaller depolarization effect than when the g_e branch is active alone. Indeed, as can be seen from the circuit model, even when E_{Cl} equals E_r, which means that there is no net influx or efflux of Cl⁻ at rest, GABA$_A$ receptor opening creates an

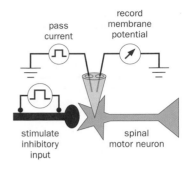

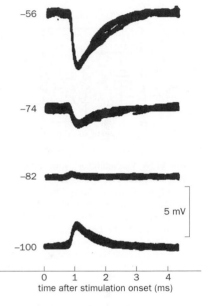

Figure 3–28 Inhibitory postsynaptic potentials (IPSPs). Top, experimental setup. Two electrodes were inserted into a spinal motor neuron, one for passing current to change the holding membrane potential, and the other to measure the membrane potential in response to electrical stimulation of the inhibitory input. Bottom, IPSPs recorded at four different holding membrane potentials. Each record represents the superposition of about 40 traces. At the membrane potentials of –74 mV or above, stimulation of inhibitory input resulted in hyperpolarizing IPSPs, with increasing amplitudes as the membrane potentials became less negative. At the membrane potentials of –82 mV or below, stimulation of inhibitory input resulted in depolarizing IPSPs, with increasing amplitudes as the membrane potentials became more negative. (Graphs adapted from Coombs JS, Eccles JC & Fatt P [1955] *J Physiol* 130:326–373.)

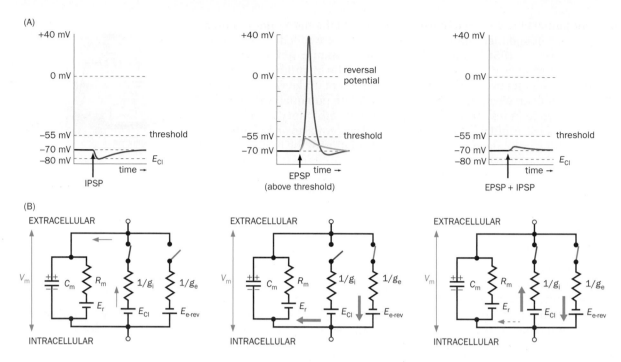

Figure 3–29 The inhibitory effect of Cl⁻ conductance mediated by GABA_A receptor. (A) In this neuron, the Cl⁻ equilibrium potential, E_{Cl}, is slightly more hyperpolarized than the resting potential. Left, IPSP from GABA_A receptors causes hyperpolarization of this postsynaptic neuron toward E_{Cl}. Middle, EPSP from glutamate receptors causes depolarization of the postsynaptic neuron, as the reversal potential at ~0 mV is far above the resting potential. If the amplitude of EPSP exceeds the threshold, it produces an action potential. Right, IPSP can cancel the effect of EPSP when both excitatory and inhibitory inputs are present at the same time, thus preventing the postsynaptic neuron from firing. **(B)** Circuit models for the three situations in panel A. To the resting neuronal model represented by the membrane capacitance (C_m), resistance (R_m) and resting potential (E_r), two additional branches are added, which represent the inhibitory and excitatory neurotransmitter receptors with conductance of g_i and g_e when neurotransmitter binding opens the receptor channels. Left, when only the inhibitory branch

is switched on (GABA release activating GABA_A receptors), a small outward current results (vertical upward arrow) because E_{Cl} is more hyperpolarized than E_r. This causes more charges to build up at C_m, thus hyperpolarizing the membrane potential (V_m). Middle, when only the excitatory branch is switched on (glutamate release activating glutamate receptors), a large inward current results (vertical downward arrow) because the reversal potential for the excitatory ionotropic glutamate receptors ($E_{e\text{-rev}}$) is far more depolarized than E_r. This causes discharge of C_m, thus depolarizing the membrane potential. Right, when both the inhibitory and excitatory branches are switched on (GABA and glutamate are released at the same time), a large fraction of the inward current in the excitatory branch is diverted by the outward current in the inhibitory branch. As a result, the current to discharge C_m (dashed arrow) is smaller. (Note that the more depolarized V_m is, the larger the outward current is due to the larger driving force for Cl⁻).

extra path that tends to hold the membrane potential near E_{Cl} so as to counteract the inward current created by the excitatory input, and therefore diminishes the voltage change across the membrane. This so-called 'shunting' contributes to GABA's potent inhibitory effect.

A noteworthy exception to GABA's inhibitory effects can occur in developing neurons. The intracellular Cl⁻ concentration is high in many developing neurons because their Cl⁻ exchangers (see Figure 2–12B) are not yet fully expressed. When the intracellular Cl⁻ concentration is sufficiently elevated, E_{Cl} is substantially more depolarized than the resting potential so that an increase of Cl⁻ conductance results in Cl⁻ efflux, causing depolarization that can exceed the threshold for action potential generation. Under these circumstances, GABA can act as an excitatory neurotransmitter.

As we will learn soon, another inhibitory action of GABA is mediated by metabotropic **GABA_B receptors**, which usually act through intracellular signaling pathways to cause the opening of K⁺ channels. Because E_K is always more negative than the resting potential, opening of K⁺ channels always causes hyperpolarization, making the neurons less likely to reach the threshold for an action potential in response to excitatory input. GABA_B receptors are not only distinct from GABA_A receptors in the channels that they open but as metabotropic receptors they are also distinct in their mode of action.

3.18 All metabotropic neurotransmitter receptors trigger G protein cascades

We now turn to metabotropic receptors, which act through intracellular signaling pathways rather than mediating ion conduction directly (see Figure 3–22B). These receptors, all of which belong to the **G-protein-coupled receptor** (**GPCR**) **superfamily**, participate in signaling cascades that involve a heterotrimeric guanine nucleotide-binding protein (**trimeric GTP-binding protein,** or simply **G protein**). ACh, glutamate, and GABA all bind to their own metabotropic receptors: muscarinic AChRs, metabotropic GluRs (mGluRs), and GABA$_B$ receptors, respectively, each with several variants. Additional GPCRs include the receptors for dopamine, norepinephrine, serotonin (most subtypes), ATP (P2Y subtypes), adenosine, and all neuropeptides (see Table 3–3), as well as the sensory receptors for vision, taste, and olfaction that we will study in Chapters 4 and 6. Indeed, GPCRs constitute the largest gene family in mammals that encompass receptors of diverse functions (**Figure 3–30**). GPCRs are crucial for neuronal communication, for responding to external stimuli, and for regulating many other physiological processes. Many pharmaceutical drugs currently in use target GPCRs, demonstrating their importance to human physiology and health.

All GPCRs share a common structure with seven transmembrane helices (**Figure 3–31**A). Almost all GPCRs are activated by binding of specific extracellular ligands. (The notable exception is rhodopsin in photoreceptors, which is activated by light absorption, as will be discussed in greater detail in Chapter 4.) Ligand binding triggers conformational changes in the transmembrane helices and allows the cytoplasmic domain to associate with a trimeric G protein complex consisting of three different subunits: **Gα**, **Gβ**, and **Gγ** (Figure 3–31B).

Prior to GPCR activation, the G protein heterotrimer preassembles and binds GDP via the Gα nucleotide-binding site (Figure 3–31C, Resting state). Because Gα and Gγ are both lipid-modified, this ternary complex associates with the plasma membrane. Ligand activation of the GPCR triggers the binding of its cytoplasmic domain to Gα. This stabilizes a nucleotide-free conformation of Gα and thereby catalyzes the replacement of GDP with GTP (Figure 3–31C, Steps 1 and 2). Next, GTP binding causes Gα to dissociate from Gβγ. Depending on the cellular context, Gα-GTP, Gβγ, or both can trigger downstream signaling cascades (Figure 3–31C, Step 3). Gα not only binds to GDP and GTP but also carries an intrinsic **GTPase** activity that hydrolyzes GTP to GDP. This GTPase activity provides a built-in termination mechanism for G protein signaling (Figure 3–31C, Step 4), and is often facilitated by additional proteins (see **Box 3–3**). GDP-bound Gα has a strong affinity for Gβγ that promotes the reassembly of the ternary complex; this return to the resting state (Figure 3–31C, Step 5) readies the trimeric G protein for the next

Figure 3–30 The superfamily of G-protein-coupled receptors (GPCRs) in the human genome. The human genome contains more than 700 GPCRs that are separated into five major branches according to sequence similarities of their transmembrane domains. The dot at the center represents the root of the branches. Numbers in parentheses indicate the number of genes within a specific branch. Names of some representative GPCRs discussed in this book are given for each branch. The GLUTAMATE branch includes mGluRs and GABA$_B$ receptors, as well as sweet and umami taste receptors. The FRIZZLED/TAS2 branch includes bitter taste receptors. The SECRETIN branch includes neuropeptide corticotropin-releasing factor (CRF) receptors involved in the stress response. The ADHESION branch includes several receptors that signal across the synaptic cleft. The largest branch, RHODOPSIN, is further divided into four clusters. These include many GPCRs important in neurobiology: opsins and melanopsins for vision, and receptors for serotonin, dopamine, acetylcholine (muscarinic), and epinephrine/norepinephrine (α cluster); many neuropeptide receptors (β and γ clusters); receptors for thyroid-stimulating hormone (TSH), follicle-stimulating hormones (FSH), and a large number of rapidly evolving odorant receptors (δ cluster). (Based on Fredriksson R, Lagerström MC, Lundin LG et al. [2003] *Mol Pharmacol* 63:1256–1272.)

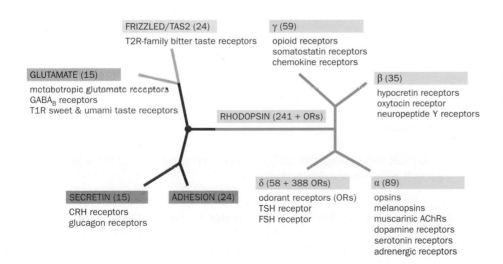

FRIZZLED/TAS2 (24)
T2R-family bitter taste receptors

γ (59)
opioid receptors
somatostatin receptors
chemokine receptors

GLUTAMATE (15)
metabotropic glutamate receptors
GABA$_B$ receptors
T1R sweet & umami taste receptors

β (35)
hypocretin receptors
oxytocin receptor
neuropeptide Y receptors

RHODOPSIN (241 + ORs)

SECRETIN (15)
CRH receptors
glucagon receptors

ADHESION (24)

δ (58 + 388 ORs)
odorant receptors (ORs)
TSH receptor
FSH receptor

α (89)
opsins
melanopsins
muscarinic AChRs
dopamine receptors
serotonin receptors
adrenergic receptors

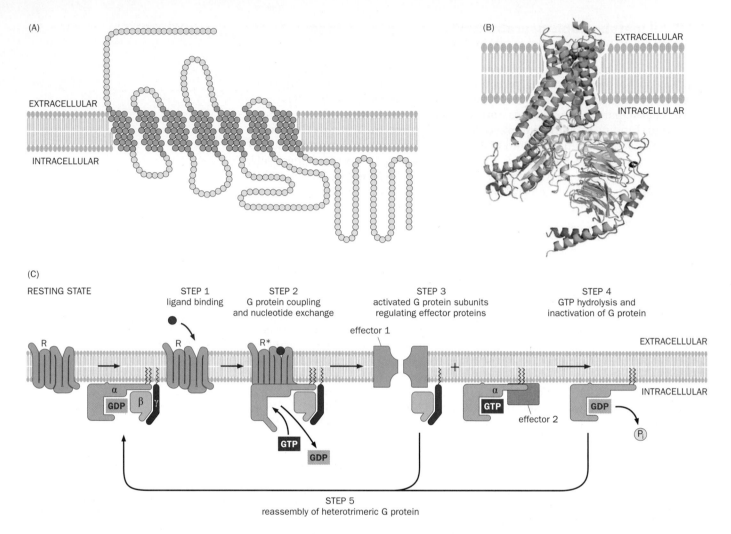

Figure 3–31 Structure and signaling cascade of a G-protein-coupled receptor (GPCR). (A) Primary structure of the β2-adrenergic receptor, the first cloned ligand-gated GPCR. Each circle represents an amino acid. All GPCRs span the lipid membrane seven times, with the N-terminus on the extracellular side and C-terminus on the intracellular side. **(B)** Crystal structure of the β2-adrenergic receptor in complex with trimeric G protein. Green, β2-adrenergic receptor, with seven transmembrane helices spanning the lipid bilayer; yellow, agonist in its binding pocket; orange, Gα; cyan, Gβ; blue, Gγ. The part of Gα in the foreground contains binding sites for GDP/GTP, Gβ, and the β2-adrenergic receptor. The part of the Gα in the background can swing relative to the part in the foreground, allowing exchange of GDP and GTP (see panel C). **(C)** Schematic of the GPCR signaling cascade. Resting state: The preassembled trimeric G protein in the GDP-bound state associates with the plasma membrane because Gα and Gγ are covalently attached to lipids (zigzag lines). Step 1: Ligand binding. Step 2: A conformational change of the GPCR induced by ligand binding (R → R*) creates a binding pocket for Gα. R* (activated GPCR) catalyzes the exchange of GDP for GTP on Gα. Step 3: GTP-bound Gα dissociates from R*, releasing Gα-GTP and Gβγ to trigger their respective effector proteins that transduce and amplify signals; here, effector 1 is an ion channel that binds Gβγ, and effector 2 is an enzyme that binds Gα-GTP. Step 4: The intrinsic GTPase activity of Gα converts Gα-GTP to Gα-GDP. Step 5: Gα-GDP re-associates with Gβγ, returning to the resting state. (A, adapted from Dohlman HG, Caron MG & Lefkowitz RJ [1987] *Biochemistry* 26:2657–2668; B & C, adapted from Rasmussen SG, DeVree BT, Zou Y et al. [2011] *Nature* 477:549–555. With permission from Macmillan Publishers Ltd.)

round of GPCR activation cycle (**Movie 3–3**). GPCRs are the largest subfamily of a superfamily of G proteins that alternate between GDP-bound and GTP-bound forms to transmit cellular signals (Box 3–3).

3.19 A GPCR signaling paradigm: β-adrenergic receptors activate cAMP as a second messenger

β-Adrenergic receptors (see Figure 3–31A, B) are the most extensively studied ligand-activated GPCRs. They are activated by epinephrine and norepinephrine (also known as adrenaline and noradrenaline, from which the name of the receptors originates). Whereas norepinephrine is produced by neurons and acts as a neurotransmitter in both the CNS and the autonomic nervous system

Box 3–3: G proteins are molecular switches

The G protein cycle outlined in Figure 3–31C is a universal signaling mechanism of GPCRs, which are widely used in many biological contexts. Indeed, the switch between a GDP-bound form and a GTP-bound form defines the G protein superfamily, which includes not only trimeric G proteins but also small monomeric GTPases such as the **Rab**, **Ras**, and **Rho** families. These small GTPases resemble part of the Gα subunit of the trimeric G protein. Rab GTPases regulate different steps of intracellular vesicular trafficking (see Section 2.1); we have encountered a family member, Rab3, in the context of bridging the synaptic vesicle with the presynaptic active zone scaffolding proteins (see Figure 3–10). The Ras family of GTPases contains key signaling molecules involved in cell growth and differentiation. As will be discussed in Box 3–4, Ras GTPases play crucial roles in transducing signals from the cell surface to the nucleus. Rho GTPases are pivotal regulators of the cytoskeleton; we will study them in the context of nervous system wiring in Chapter 5.

All members of the G protein superfamily are molecular switches. For the trimeric G proteins as well as the Ras and Rho families of GTPases, the GDP-bound form is inactive and the GTP-bound form is active in downstream signaling. The transitions between the GTP-bound and GDP-bound forms are usually facilitated by two types of proteins: the **guanine nucleotide exchange factors** (**GEFs**), which switch GTPases on by catalyzing the exchange of GDP for GTP, and **GTPase activating proteins** (**GAPs**), which switch GTPases off by speeding up the endogenous GTPase activity, converting GTP to GDP (**Figure 3–32**; **Movie 3–4**). As will be discussed in Section 3.22, proper signal termination is an important aspect of signaling.

In the context of trimeric G protein signaling discussed in Section 3.18, ligand-activated GPCRs act as GEFs for the

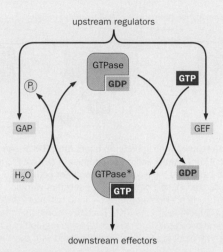

Figure 3–32 The GTPase cycle. GTPases cycle between a GDP-bound form and a GTP-bound form. For signaling GTPases such as trimeric G proteins as well as Ras and Rho subfamilies of small GTPases, the GTP-bound form usually binds effectors and activates downstream signaling. The guanine nucleotide exchange factor (GEF) catalyzes the exchange of GDP for GTP, thus activating the GTPases. The GTPase activating protein (GAP) speeds up the rate at which the G protein's endogenous GTPase activity hydrolyzes the bound GTP, thus inactivating the GTPases. GEFs and GAPs are regulated by upstream signals.

trimeric G proteins. By stabilizing the transition state of the nucleotide-free conformation of Gα (Figure 3–31B), GPCRs catalyze the exchange of GDP for GTP on Gα (Step 2 of Figure 3–31C). The reaction is driven in the direction of Gα-GTP production by the dissociation of Gα-GTP from the GPCR and from Gβγ. We will learn more about GAPs in GPCR signaling in the context of visual transduction in Chapter 4.

(see Section 3.11), **epinephrine** is produced primarily by chromaffin cells in the adrenal gland, circulates through the blood, and acts as a hormone to mediate systemic responses to extreme conditions, the so-called 'fright, fight, and flight' responses. (A small number of CNS neurons also produce epinephrine as a modulatory neurotransmitter.) Classic biochemical studies demonstrated that epinephrine activates β-adrenergic receptors to produce an intracellular second messenger called **cyclic AMP** (**cAMP**). cAMP is synthesized from ATP by the action of a membrane-associated enzyme called **adenylate cyclase** (**Figure 3–33**). In fact, studies of mechanisms by which β-adrenergic receptors activate adenylate cyclase, together with parallel investigations of the signal transduction pathways downstream from rhodopsin activation (to be discussed in Section 4.4), first led to the discovery that trimeric G proteins act as essential intermediates in GPCR signaling.

Originally identified as a second messenger in the context of epinephrine action, cAMP is a common downstream signal for many GPCRs. In Chapters 4 and 6, we will learn that cAMP and its cousin **cyclic GMP** (**cGMP**) can directly gate ion channels in visual and olfactory systems. However, the most widely used cAMP effector is the **cAMP-dependent protein kinase** (also called **A-kinase**, **protein kinase A**, or **PKA**). PKA is a **serine/threonine kinase**, which means that it adds phosphate onto specific serine or threonine residues of target proteins and

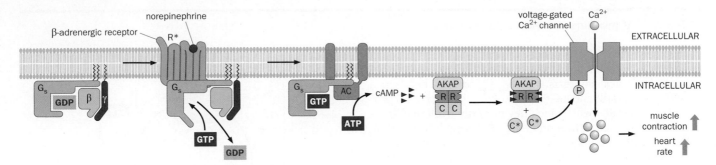

Figure 3–33 Norepinephrine speeds up heart rate: GPCR signaling through cyclic AMP (cAMP) and protein kinase A (PKA). From left, norepinephrine binding to the β-adrenergic receptor activates G_s, a Gα variant, in the cardiac muscle cell. G_s-GTP associates with and activates the membrane-bound adenylate cyclase (AC). AC catalyzes the production of cAMP from ATP. cAMP activates PKA. Each PKA consists of two regulatory (R) and two catalytic (C) subunits. Each regulatory subunit contains two cAMP binding sites, and is associated with the A-kinase anchoring protein (AKAP). When all four cAMP binding sites on the regulatory subunits are occupied, the catalytic subunits are released from the complex, become active (C*), and phosphorylate their substrates. In the cardiac muscle cell, a key PKA substrate is a voltage-gated Ca^{2+} channel. PKA phosphorylation of the Ca^{2+} channel increases its open probability, facilitating Ca^{2+} influx. A rise in intracellular Ca^{2+} increases cardiac muscle contraction and heart rate. The pathway from G_s to PKA activation is used in many other cellular contexts.

thereby changes their properties. PKA is composed of two regulatory and two catalytic subunits; in the absence of cAMP, these subunits form an inactive tetramer that is usually associated with various **AKAPs** (for A-kinase anchoring proteins) located in specific parts of the cell. cAMP binding to the regulatory subunits triggers the dissociation of the catalytic subunits from the regulatory subunits; the catalytic subunits become free to phosphorylate their substrates (Figure 3–33; **Movie 3–5**). PKA phosphorylates many substrates with short- or long-lasting effects on neuronal excitability.

As a specific example, we discuss the mechanism by which norepinephrine released from the axon terminals of neurons in the **sympathetic nervous system** (a branch of the autonomic nervous system) induces cardiac muscle contraction and increases heart beat rate. Norepinephrine binds to and activates the cardiac muscle β-adrenergic receptor, which associates with a Gα variant called **G_s** (for **stimulatory G protein**). G_s-GTP triggers cAMP production by binding to and activating an adenylate cyclase. Elevated cAMP levels lead to PKA activation. PKA phosphorylates a voltage-gated Ca^{2+} channel on the plasma membrane of cardiac muscle, which increases its open probability. The increased Ca^{2+} entry through the Ca^{2+} channel facilitates the cardiac muscle cells to contract and speeds up the heart rate (Figure 3–33). Thus, this signaling cascade provides a paradigm of how a neurotransmitter (norepinephrine) elicits a physiological response (an increase in heart rate) through a second messenger (cAMP) and its downstream effectors (PKA, voltage-gated Ca^{2+} channel).

3.20 α and βγ G protein subunits trigger diverse signaling pathways that alter membrane conductance

The human genome encodes twenty Gα, six Gβ, and three Gγ variants. Their different combinations give rise to a myriad of trimeric G proteins that are coupled to different GPCRs and can trigger diverse signaling pathways. For example, in addition to the G_s we just discussed, a variant of Gα called **G_i** (for **inhibitory G protein**) also binds to adenylate cyclase but inhibits its activity, resulting in a decrease of intracellular cAMP concentration. Different Gα variants are associated with different receptors and regulate distinct downstream signaling pathways. With regards to regulating postsynaptic neuronal function, the ultimate effectors are usually ion channels that regulate membrane potential or neurotransmitter release, most notably K^+ and Ca^{2+} channels (see Box 2–4). In the previous section, we discussed a classic example of how norepinephrine activates a Ca^{2+} channel via cAMP and PKA to increase cardiac muscle contraction. In the next two sections, we examine more examples to highlight the diverse outcomes of GPCR signaling.

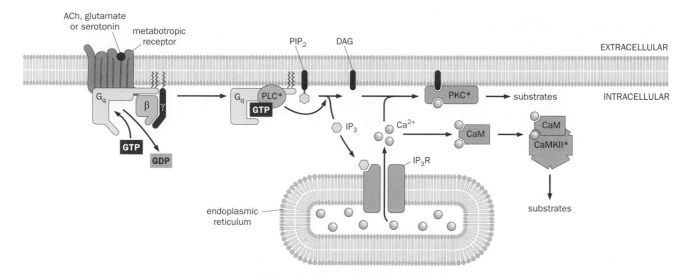

Figure 3–34 GPCR signaling through phospholipase C (PLC) and Ca^{2+}. From left, activation of metabotropic receptors in response to a variety of neurotransmitters (for example, ACh, glutamate, serotonin) activates G$_q$, a variant of Gα. G$_q$-GTP in turn activates PLC, which catalyzes the conversion from PIP$_2$ to DAG and IP$_3$. IP$_3$ activates the IP$_3$ receptor (IP$_3$R, an IP$_3$-gated Ca^{2+} channel) on the ER membrane, allowing Ca^{2+} to be released from ER to the cytosol. DAG and Ca^{2+} co-activate PKC. Ca^{2+} also binds to calmodulin (CaM), and the resulting complex activates CaMKII and other CaM kinases. Asterisks represent activated components.

An important G protein effector of many metabotropic receptors (for example, receptors for ACh, glutamate, serotonin) is a membrane-associated enzyme called **phospholipase C** (**PLC**) (**Figure 3–34**; **Movie 3–6**). PLC is activated by **G$_q$**, a Gα variant. Activated PLC cleaves a membrane-bound phospholipid called PIP$_2$ (phosphatidyl 4,5-bisphosphate) to produce two important second messengers: **diacylglycerol** (**DAG**) and **inositol 1,4,5-triphosphate** (**IP$_3$**). DAG binds to and activates **protein kinase C** (**PKC**), a serine/threonine kinase. PKC activation also requires a rise of intracellular Ca^{2+} concentration. This is achieved by IP$_3$, which binds to an IP$_3$-gated Ca^{2+} channel (the **IP$_3$ receptor**) on the membrane of the endoplasmic reticulum (ER) and triggers the release of ER-stored Ca^{2+} into the cytosol. In addition to activating PKC, Ca^{2+} interacts with many additional effectors. A key effector is a protein called **calmodulin**. The Ca^{2+}/calmodulin complex can regulate diverse signaling pathways, including the activation of Ca2+/calmodulin-dependent protein kinases (CaM kinases), another important group of serine/threonine kinases. Like PKA, both PKC and CaM kinases phosphorylate many downstream target proteins, including ion channels and receptors, to modulate their activity. A specific subtype of CaM kinases, **CaM kinase II** (**CaMKII**), is one of the most abundant proteins in the postsynaptic density (see Section 3.16). Ca^{2+} can also directly increase the open probability of Ca^{2+}-dependent K$^+$ channels (see Box 2–4). Thus, activation of PLC activates PKC and at the same time causes a rise of intracellular Ca^{2+} concentration, both of which can alter neuronal excitability (Figure 3–34).

Historically, Gα was identified first as the signaling intermediate between the GPCR and the effector. However, Gβγ can also mediate signaling. This was first shown in the case of ACh regulation of heartbeat. In fact, the concept of a chemical neurotransmitter was first established in this context in a classic experiment conducted in 1921 by Otto Loewi. It was known that stimulating the **vagus nerve**, a cranial nerve that connects the brainstem with internal organs, would slow the heartbeat. Loewi collected the fluid from a frog heart that had been stimulated by the vagus nerve, added it to an unstimulated heart, and found that beating of the second heart also was slowed. This experiment showed that vagus nerve stimulation released a chemical transmitter, identified afterwards as ACh, to slow the heartbeat. (According to Loewi, the initial idea of this experiment came from a dream in the middle of the night. He wrote it down on a piece of paper and went back to sleep. The next morning he remembered dreaming about something

Figure 3–35 ACh slows down heart rate: direct action of Gβγ on a cardiac muscle K⁺ channel. From left, ACh activation of a muscarinic ACh receptor (mAChR) on a cardiac muscle cell causes the dissociated Gβγ to bind directly and activate a G-protein-coupled inward-rectifier K⁺ (GIRK) channel, leading to K⁺ efflux and hyperpolarization of the muscle cell. Hyperpolarization decreases muscle contraction and reduces the heart rate.

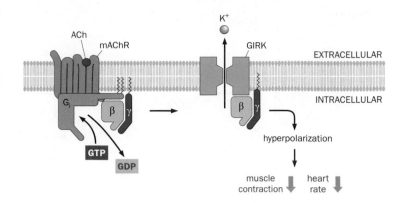

important but could not remember what it was or decipher what he had written. After spending a desperate day, he remembered what the dream was when he woke up in the middle of the next night. This time he got up immediately and went to the lab to perform the experiment.)

Subsequent work has shown that ACh binds to a specific muscarinic AChR and triggers the dissociation of the trimeric G protein complex. βγ subunits then bind to and activate a class of K⁺ channels called GIRKs (<u>G</u>-protein-coupled <u>i</u>nward-<u>r</u>ectifier <u>K</u>⁺) channels, resulting in K⁺ efflux, hyperpolarization of cardiac muscle cells, and a slowing of the heartbeat (**Figure 3–35**).

We have seen that two different neurotransmitters, norepinephrine and ACh, act on different receptors, G proteins, and effectors to speed up or slow down the heart rate, respectively (compare Figures 3–33 and 3–35). In fact, the Gα that is activated by the muscarinic AChR is a G_i variant, which inhibits adenylate cyclase and counteracts the effect of β-adrenergic receptors. These neurotransmitters are used in the two opposing branches of the autonomic nervous system. The sympathetic branch, which uses norepinephrine as a neurotransmitter, and the **parasympathetic** branch, which uses ACh as a neurotransmitter, usually have antagonistic functions (see Section 8.12 for more details).

3.21 Metabotropic receptors can act on the presynaptic terminal to modulate neurotransmitter release

In addition to acting on the dendrites and cell bodies, metabotropic receptors can also act on the presynaptic terminals of their postsynaptic target neurons to modulate neurotransmitter release. In the simplest form, neurons can use metabotropic receptors to modulate their own neurotransmitter release, as in the case of sympathetic neurons that release norepinephrine (**Figure 3–36**A). The presynaptic terminals of these neurons express α-adrenergic receptors that can bind norepinephrine released into the synaptic cleft. Activation of these presynaptic α-adrenergic receptors rapidly inhibits the voltage-gated Ca^{2+} channel at the active zone, which reduces the depolarization-induced Ca^{2+} entry that is essential for triggering neurotransmitter release. This negative feedback loop results in diminishing levels of neurotransmitter release, leading to presynaptic depression. As noted in Section 3.10, many such mechanisms of short-term plasticity act on the presynaptic cell by altering the probability of neurotransmitter release.

To provide further insight into the mechanism by which norepinephrine inhibits presynaptic Ca^{2+} channels, cell-attached patch clamp recordings were performed under different conditions. In the control condition where no norepinephrine was applied, a depolarizing voltage step induced an inward current through the presumed voltage-gated Ca^{2+} channels within the membrane patch (Figure 3–36B, left). When norepinephrine was included only in the patch pipette, depolarization-induced inward current was greatly reduced (Figure 3–36B, middle). When norepinephrine was applied in the media but not in the pipette, depolarization-induced inward current was similar to the control (Figure 3–36B, right). Because activation of α-adrenergic receptors outside the patch did not inhibit

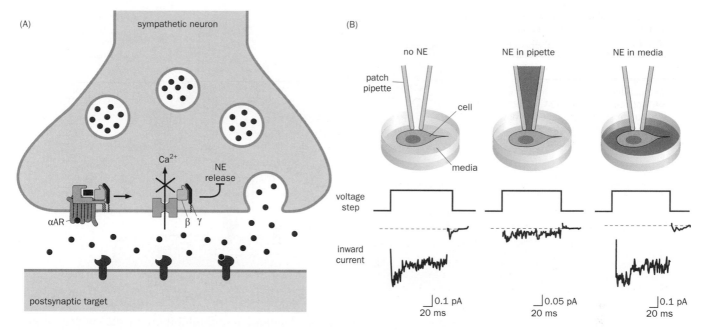

Figure 3–36 Local action of norepinephrine on a presynaptic Ca²⁺ channel. (A) From left, released norepinephrine (NE, blue) binds and activates a presynaptic α-adrenergic receptor (αAR). Activated Gβγ binds and inhibits the presynaptic voltage-gated Ca²⁺ channel, which reduces Ca²⁺ influx in response to depolarization and thereby inhibits neurotransmitter release. **(B)** A patch pipette was used to record the current flow across a patch of plasma membrane from a frog sympathetic neuron in culture in the cell-attached mode (see Figure 2–30A). A depolarizing voltage step (from –80 mV to –10 mV) was applied to activate voltage-gated Ca²⁺ channels in the membrane patch underneath the electrode (Na⁺ and K⁺ channels were inhibited by blockers included in the pipette solution). Left, in

the absence of norepinephrine (NE), the depolarization step induced the opening of the Ca²⁺ channel within the patch, as seen by the inward current. Middle, when the patch pipette was filled with NE, the depolarization-induced Ca²⁺ current was largely eliminated. Right, if NE was excluded from the patch pipette but was applied everywhere else, depolarization still induced the inward Ca²⁺ current. Together, these experiments indicated that the action of norepinephrine on the Ca²⁺ channel was confined to the patch of membrane underneath the pipette. (B, adapted from Lipscombe D, Kongsamut S & Tsien RW [1989] *Nature* 340:639–642. With permission from Macmillan Publishers Ltd.)

voltage-gated Ca²⁺ channels within the patch, these data argued against involvement of an intracellular second messenger that can diffuse in the cytoplasm, and for a mechanism by which α-adrenergic receptors act locally on the voltage-gated Ca²⁺ channels within the membrane patch. Indeed, subsequent experiments determined that this inhibition is mediated by direct binding of Gβγ to the Ca²⁺ channel (Figure 3–36A), as in the case of ACh activation of GIRK.

A presynaptic terminal of a given neuron can also contain metabotropic receptors for neurotransmitters produced by other neurons. In this case, the presynaptic terminal of a neuron acts as the postsynaptic site for these other neurons (**Figure 3–37**). Depending on the nature of the neurotransmitter, the type of the receptor, the signaling pathway, and final effector, the net effect could either be facilitation or inhibition of neurotransmitter release. Accordingly, these effects are termed **presynaptic facilitation** or **presynaptic inhibition**. Presynaptic facilitation can be achieved by closing K⁺ channels, which depolarizes the presynaptic membrane potential and makes it easier to activate voltage-gated Ca²⁺ channels so that Ca²⁺ entry can trigger neurotransmitter release; we will see an example of this in Chapter 10 where serotonin mediates presynaptic facilitation in the sea slug *Aplysia* to enhance the magnitude of a reflex to a noxious stimulus. Presynaptic inhibition can be achieved by opening K⁺ channels or by closing voltage-gated Ca²⁺ channels, both of which inhibit neurotransmitter release. For example, in Chapter 6, we will learn that *Drosophila* olfactory receptor neurons (ORNs) activate GABAergic local interneurons, which synapse onto ORN axon terminals to provide negative feedback control of ORN neurotransmitter release through the GABA_B receptors. (Presynaptic inhibition can also be achieved through GABA acting on ionotropic GABA_A receptors that are present on the presynaptic terminals

Figure 3–37 Presynaptic facilitation and inhibition. Left, an example of presynaptic facilitation by activation of a metabotropic receptor, such as a serotonin receptor; facilitation can be achieved by decreasing K^+ conductance (red inhibitory sign). Right, examples of presynaptic inhibition by GABA. Activation of the $GABA_A$ receptor increases Cl^- conductance and therefore counters depolarization. Activation of the $GABA_B$ receptor can act by increasing K^+ conductance (green arrow) or by inhibiting Ca^{2+} conductance (red inhibitory sign).

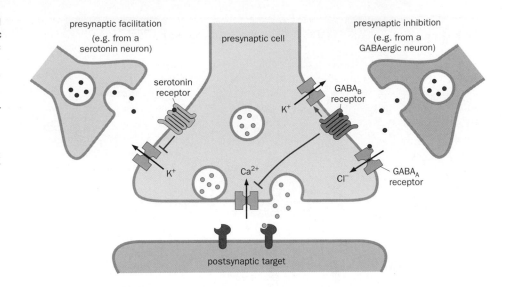

of some neurons.) Presynaptic facilitation and inhibition are also widely used in the vertebrate nervous systems.

3.22 GPCR signaling features multiple mechanisms of signal amplification and termination

As we have seen in previous sections, metabotropic neurotransmitter receptors have diverse functions that depend on their locations and their coupling to different G proteins, signaling pathways, and effectors. Their effects unfold more slowly than the rapid ion conduction of ionotropic receptors. However, second messenger systems contribute an important property: signal amplification. For example, activation of a single adrenergic receptor can trigger multiple rounds of G protein activation; each activated adenylate cyclase can produce many cAMP molecules; and each activated PKA can phosphorylate many substrate molecules.

Signals need to be properly terminated in order for cells to respond to future stimuli. Indeed, all signaling events we have discussed so far are associated with built-in termination mechanisms. For example, the GPCR is deactivated when its ligand dissociates; Gα-GTP is deactivated by its intrinsic GTPase activity, often facilitated by GTPase-activating proteins (GAPs); Gβγ is deactivated by re-association with Gα-GDP; adenylate cyclase is deactivated in the absence of Gα-GTP; the cAMP produced by adenylate cyclase is metabolized into AMP by an enzyme called **phosphodiesterase**; the catalytic subunits of PKA re-associate with regulatory subunits and become inactive when cAMP concentration declines; and **protein phosphatases** remove phosphates from phosphorylated proteins, thus counteracting the actions of kinases. While some of these termination mechanisms are constitutive, others are regulated by signals.

Signal amplification and termination apply generally to signal transduction pathways (**Box 3–4**). In Chapter 4, we will see a salient example of signal amplification and termination in the context of studying how photons are converted to electrical signals in vision.

3.23 Postsynaptic depolarization can induce new gene expression

In addition to changing the membrane potentials and excitability of postsynaptic neurons on timescales of milliseconds (through ionotropic receptors) or tens of milliseconds to seconds (through metabotropic receptors), neurotransmitters can also trigger long-term (hours to days) changes in the physiological state of

Box 3–4: Signal transduction and receptor tyrosine kinase signaling

In response to extracellular signals, cells utilize many pathways to relay such signals to varied effectors and produce specific biological effects; this process is generally referred to as **signal transduction**. In the context of synaptic transmission, we have focused on the actions of ionotropic and metabotropic receptors that change the membrane potential of the postsynaptic cell. In this box, we expand the scope by placing neurotransmitter receptor signaling in the general framework of signal transduction, and by discussing receptor tyrosine kinase signaling pathways, which are crucial for both nervous system development and function.

In a typical signal transduction pathway (**Figure 3–38**A), an extracellular signal (a **ligand**) is detected by a **cell-surface receptor** in the recipient cell. (We will learn an exception to this in Chapter 9: steroid hormones diffuse across the cell membrane to bind receptors within the cell.) The signal is then relayed through one or a series of intracellular signaling proteins to reach their effector(s), producing cellular responses to the extracellular signal. The final effectors are diverse, but usually fall into one of the following categories: (1) enzymes that alter the metabolism of the cell;

(2) regulators of gene expression that change chromatin structure, gene transcription, mRNA metabolism, or protein translation and degradation; (3) cytoskeletal proteins that regulate cell shape, cell movement, and intracellular transport; (4) ion channels that alter the cell's membrane potential and excitability. Indeed, we can map what we have learned about metabotropic and ionotropic receptor signaling onto this general scheme of signal transduction (Figure 3–38B).

The extracellular signal can come from different sources. If the signal is produced by the recipient cell itself (as is the case of presynaptic norepinephrine receptor signaling; see Figure 3–36), it is termed an **autocrine** signal. If the signal comes from nearby cells, it is termed a **paracrine** signal; neurotransmitters can be considered as specialized paracrine signals where the target cells are restricted to postsynaptic partners. If the signal comes from a remote cell through circulating blood, it is called an **endocrine** signal or a hormone (as is the case of epinephrine). When the signal comes from a neighboring cell, it can either be a diffusible molecule such as a neurotransmitter or a secreted protein,

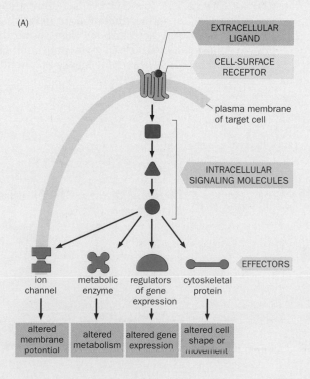

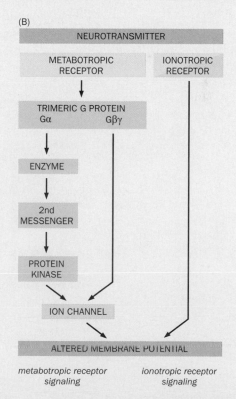

Figure 3–38 Signal transduction pathways. (A) A cartoon of a generic signaling pathway. Binding of an extracellular ligand to its cell-surface receptor elicits a signal that is transduced by intracellular signaling proteins to various effectors, such as an enzyme that modifies metabolism, a gene regulatory protein that alters gene expression, a cytoskeletal protein that affects cell shape or motility, or an ion channel that influences membrane potential. **(B)** Metabotropic and ionotropic receptor signaling

pathways are mapped on the generic signaling pathway in panel A, with colors matching the components of the signaling pathway. Note that in ionotropic receptor signaling, the ionotropic receptor is both a receptor and an effector for changing membrane potential, thus representing the shortest and fastest (within milliseconds) signaling pathway. (A, adapted from Alberts B, Johnson A, Lewis J et al. [2015] Molecular Biology of the Cell, 6th ed. Garland Science.)

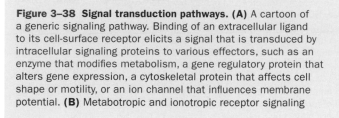

(Continued)

Box 3–4: Signal transduction and receptor tyrosine kinase signaling

or a membrane-bound protein that requires cell–cell contact in order to send the signal. Secreted and membrane-bound protein ligands are widely used in cell–cell communication during development, which will be discussed in detail in Chapters 5 and 7.

In addition to the ionotropic receptors and metabotropic receptors (GPCRs), a major class of cell-surface receptors used in the nervous system is enzyme-coupled receptors, where the receptor has an enzymatic activity in its intracellular domain. As an example, we discuss here a widely used class of enzyme-coupled receptors called **receptor tyrosine kinases** (**RTKs**), which are transmembrane proteins with an N-terminal extracellular ligand-binding portion and a C-terminal intracellular portion possessing a tyrosine kinase domain as well as tyrosine phosphorylation sites (**Figure 3–39**A). About 60 genes in the mammalian genome encode RTKs. We focus below on RTK signaling involving the neurotrophin receptors, but the general principles apply to other RTK signaling. **Neurotrophins** are a family of secreted proteins that regulate the survival, morphology, and physiology of target neurons (we will discuss the biological effects of these proteins in Section 7.15). They bind to and activate a family of RTKs called the **Trk receptors**.

How does neurotrophin binding to Trk activate signaling? Neurotrophins naturally form dimers. When each neurotrophin binds a Trk receptor, the neurotrophin dimer brings

two Trk receptors in close proximity, such that the tyrosine kinase of one Trk can phosphorylate the tyrosine residues on the other Trk. Phosphorylation of key tyrosine residues creates binding sites for specific adaptor proteins. These adaptor proteins contain either an **SH2** (**s**rc **h**omology **2**) domain or a PTB (**p**hospho**t**yrosine **b**inding) domain, which enables the adaptors to bind phosphorylated tyrosine in the context of specific amino acid sequences and thereby initiate downstream signaling. In the Trk receptors, for instance, two key tyrosine residues recruit the binding of several specific adaptor proteins, eliciting separate transduction pathways that also cross-talk with each other (Figure 3–39A).

One such signaling pathway is initiated by binding of the adaptor Shc (Figure 3–39B), which binds tyrosine-phosphorylated Trk via its PTB domain and becomes tyrosine phosphorylated by Trk. This recruits the binding of Grb2, an SH2-domain-containing adaptor protein. Grb2 is associated with Sos, a guanine nucleotide exchange factor for the small GTPase Ras (see Box 3–3). Ras is normally associated with the membrane because of lipid modification analogous to Gα. Thus, Trk activation recruits Sos to the plasma membrane to catalyze the exchange of GDP for GTP on Ras. Ras-GTP binds a downstream effector called Raf, a serine/threonine protein kinase. Raf phosphorylates and activates another serine/threonine protein kinase Mek, which in turn phosphorylates and activates a third serine/threonine kinase Erk. Activated Erk

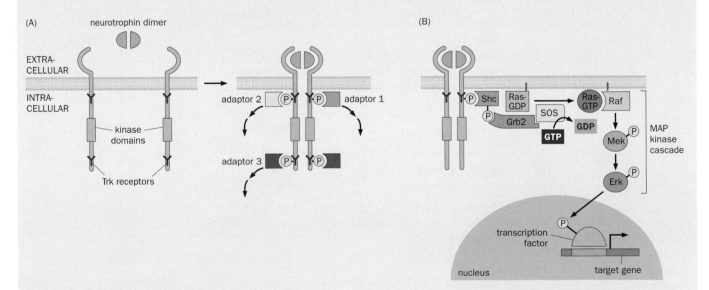

Figure 3–39 Neurotrophin receptor as an example of receptor tyrosine kinase (RTK) signaling. (A) In the absence of neurotrophin, Trk receptors are present as monomers and tyrosine residues (Y) are not phosphorylated. Binding of the neurotrophin dimer brings two Trk receptors in close proximity, allowing the kinase domain of each Trk to phosphorylate the tyrosine residues on the other Trk. Tyrosine phosphorylation recruits binding of specific adaptor proteins, each eliciting a downstream signaling event. Different adaptors can bind the same phosphorylated tyrosine (as in the case of adaptor 1 and adaptor 2). **(B)** Details of one adaptor pathway. Shc binds

to a membrane-proximal phosphorylated tyrosine on Trk, leading to tyrosine phosphorylation of Shc. This helps recruit the binding of the Grb2-Sos complex. Sos acts as a guanine nucleotide exchange factor that catalyzes the conversion of Ras-GDP to Ras-GTP (red zigzag lines indicate lipid modification of Ras). Ras-GTP binds to Raf, which phosphorylates and activates Mek, which in turn phosphorylates and activates Erk. Activated Erk can directly and indirectly phosphorylate a number of transcription factors, which activate or repress transcription of target genes. Raf, Mek, and Erk (also called MAP kinase) constitute the MAP kinase cascade.

Box 3–4: Signal transduction and receptor tyrosine kinase signaling

phosphorylates and activates a number of **transcription factors** (DNA-binding proteins that activate or repress transcription of target genes), which leads to transcription of specific genes that promote neuronal survival and differentiation, two major biological effects of neurotrophin signaling during development.

Erk is also called MAP kinase (for <u>m</u>itogen-<u>a</u>ctivated <u>p</u>rotein <u>k</u>inase), and therefore Mek is a MAP kinase kinase (since

it phosphorylates MAP kinase), and Raf is a MAP kinase kinase kinase. The Raf–Mek–Erk kinase cascade is often referred to as a **MAP kinase cascade**, which acts downstream of Ras and a number of other signaling molecules. The Ras-MAP kinase cascade is a widely used signaling pathway that serves many functions, including cell survival and differentiation discussed above, cell fate determination (see Section 5.17), as well as cell proliferation. It is also used in activity-dependent transcription (see Section 3.23).

postsynaptic neurons by inducing expression of new genes. As a specific example, transcription of *Fos* was induced by ionotropic AChR activation within 5 minutes of nicotine application (**Figure 3–40**). *Fos* encodes a transcription factor, and its transient activation can change the expression of many downstream target genes.

Fos is the prototype of a class of genes called **immediate early genes** (**IEGs**), whose transcription is rapidly induced by external stimuli in the presence of protein synthesis inhibitors; this means that no new protein synthesis is required to turn on IEGs. In neurons, IEGs can be rapidly induced by neuronal activity in postsynaptic neurons in response to presynaptic neurotransmitter release. Some IEGs, such as *Fos* or *Egr1* (<u>e</u>arly <u>g</u>rowth <u>r</u>esponse-<u>1</u>), encode transcription factors that regulate the expression of other genes. Other IEGs encode regulators of neuronal communication more directly. Among them, **brain-<u>d</u>erived <u>n</u>eurotrophic <u>f</u>actor** (**BDNF**) is a secreted neurotrophin that regulates the morphology and physiology of target neurons (see Box 3–4). **Arc** (<u>a</u>ctivity-<u>r</u>egulated <u>c</u>ytoskeleton-associated protein) is a cytoskeletal protein present at the postsynaptic density that regulates trafficking of glutamate receptors, thus contributing to synaptic plasticity. As will be discussed in later chapters of this book, **activity-dependent transcription** (that is, regulation of gene expression by neuronal activity) plays a prominent role in the maturation of synapses and neural circuits during development and in their modulation by experience in adulthood. Because of their rapid induction by neuronal activity, expression of IEGs has also been widely used as a means to identify which neurons in the brain are activated by specific experiences or behavioral episodes (see Box 13–3).

Many signaling pathways have been identified that link neurotransmitter reception to transcription. A rise in intracellular Ca^{2+} concentration ($[Ca^{2+}]_i$) is often a key step. $[Ca^{2+}]_i$ rise can be accomplished by several means: through the NMDA receptor at the postsynaptic density in the dendritic spines (see Figure 3–24), through voltage-gated Ca^{2+} channels enriched on the dendritic trunk and the cell body, and through the IP_3 receptors (see Figure 3–34) or the related **ryanodine receptors** on the ER membrane. (Instead of being activated by IP_3, ryanodine receptors are activated by a rise in $[Ca^{2+}]_i$, and thus amplify the Ca^{2+} signal; ryanodine is a plant-derived agonist of this ER-resident Ca^{2+} channel.) Although free Ca^{2+} ions usually do not diffuse far from the source of entry into the cytosol, they can associate with various Ca^{2+}-binding proteins, most notably calmodulin (CaM) (see Figure 3–34), and initiate signals that can be transduced to the nucleus (**Figure 3–41**). For example, Ca^{2+}/CaM can activate a number of CaM kinases, including CaMKII enriched in postsynaptic density and CaMKIV enriched in the nucleus. A specific isoform of CaMKII, γCaMKII, has recently been shown to act as a shuttle that transports Ca^{2+}/CaM from the plasma membrane near the voltage-gated Ca^{2+} channel to the nucleus so that Ca^{2+}/CaM can activate nuclear effectors such as CaMKIV. In addition, Ca^{2+}/CaM can activate several subtypes of adenylate cyclase, leading to the production of cAMP and activation of PKA. The Ras-MAP kinase cascade (see Box 3–4) is yet another signaling pathway that can be activated by Ca^{2+}/CaM.

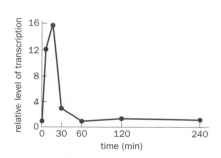

Figure 3–40 Nicotinic AChR activation induces transcription of *Fos*, an immediate early gene. Nicotine application to a cultured neuronal cell line at time 0 induces a rapid and transient transcription of *Fos*, as quantified by newly synthesized *Fos* RNA. (Adapted from Greenberg ME, Ziff EB & Greene LA [1986] *Science* 234:80–83. With permission from AAAS.)

Figure 3–41 Signaling pathways from the synapse to the nucleus. Shown here are pathways from the postsynaptic terminals and somatodendritic plasma membrane to the nucleus that involve Ca^{2+} and lead to the phosphorylation and activation of a transcription factor, CREB. An increase of intracellular Ca^{2+} concentration can result from an influx of extracellular Ca^{2+} through the NMDA receptors concentrated in dendritic spines or the voltage-gated Ca^{2+} channels enriched on the somatodendritic plasma membrane, or can be mediated by the release of Ca^{2+} from internal stores in the ER through the ryanodine receptors. Ca^{2+} together with its binding partner calmodulin (CaM) activates CaM kinases, Rsk (via the Ras-MAP kinase cascade), and PKA (via Ca^{2+}-activated adenylate cyclase and cAMP production). CaM kinases, Rsk, and PKA can all phosphorylate CREB, promoting its activity to induce transcription of target genes with cAMP-response elements (CREs) in their promoters. (Based on Cohen S & Greenberg ME [2008] *Annu Rev Cell Dev Biol* 24:183–209 and Deisseroth K, Mermelstein PG, Xia H et al. [2003] *Curr Opin Neurobiol* 13:354–365.)

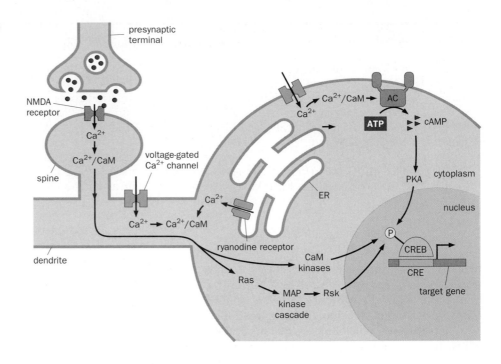

As a specific example, we discuss how these pathways lead to activation of a transcription factor called **CREB**. CREB was originally identified because it binds to a DNA element (**CRE**) in the promoter of the gene that produces a neuropeptide **somatostatin**, rendering somatostatin's transcription responsive to cAMP regulation. (CRE stands for <u>c</u>AMP <u>r</u>esponse <u>e</u>lement; CREB for <u>CRE</u> <u>b</u>inding protein.) CRE was subsequently found in the promoter of many IEGs including *Fos*. Phosphorylation at amino acid serine-133 is crucial for the activity of CREB as a transcriptional activator. Biochemical experiments have shown that serine-133 can be phosphorylated by a number of kinases, including PKA, CaMKIV, and a protein kinase called Rsk (<u>r</u>ibosomal protein <u>S6 k</u>inase); Rsk is in turn activated by MAP kinase phosphorylation. Although all of these kinases can be activated by Ca^{2+} (Figure 3–41), each pathway has unique properties. For example, the CaM kinase-mediated pathway is more rapid, resulting in CREB phosphorylation that peaks within minutes after a transient neuronal depolarization, whereas the MAP kinase pathway mediates a gradual increase of CREB phosphorylation over 60 minutes following a transient neuronal depolarization.

In addition to CREB, other Ca^{2+}-responsive transcription factors are known to bind to different IEG promoters. Thus, neuronal activity has many routes to access the nucleus and change the transcriptional program of postsynaptic cells. Furthermore, neuronal activity and Ca^{2+} can also affect chromatin structures through enzymes that control the methylation of DNA and the post-translational modification (for example, methylation, demethylation, acetylation, and deacetylation) of histones, the protein component of the chromatin. These **epigenetic modifications** also alter gene expression patterns through regulation of chromatin structures and accessibility of promoters to specific transcription factors. As will be discussed in Chapter 11, mutations in many components of the synapse-to-nucleus signaling pathways have been found to cause human brain disorders, highlighting the important role of activity-dependent transcription in human mental health.

3.24 Dendrites are sophisticated integrative devices

Regulating gene expression aside, the primary function of synaptic transmission is to influence the firing patterns of postsynaptic neurons. This is the means by which information is propagated from one layer of neurons to the next within neural circuits. As a way of integrating what we've learned about neuronal communication in Chapter 2 and this chapter, in the final two sections we discuss how synaptic inputs are integrated in the postsynaptic neuron to produce its firing pattern, thus completing a full round of neuronal communication (see Figure 1-18). We start our discussion with excitatory inputs in this section.

Most excitatory inputs to a neuron are provided on the dendrites via transient changes of conductance (for example, opening of ionotropic glutamate receptor channels), producing EPSCs and consequently EPSPs (see Figure 3-23). In order to influence the firing pattern, these electrical signals need to travel to the axon initial segment to contribute to the depolarization there. As we learned in Section 2.8, electrical signals evolve over time and decay across distance, specified by the passive (cable) properties of neuronal fibers such as the time constant (τ) and length constant (λ). Theoreticians have used model neurons to calculate EPSPs at the soma produced by synaptic input at different locations in dendrites. In the model neuron shown in **Figure 3-42**A, for example, the complex dendritic tree is simplified to 10 compartments with varying distances from the soma in order to calculate somatic EPSPs in response to dendritic input. A transient increase of synaptic conductance, equivalent to a transient opening of excitatory neurotransmitter receptor channels, produces somatic EPSPs with different shapes and amplitudes when applied to different locations in the dendrites (Figure 3-42B). The further distant the synaptic input is, the slower is the rise of the somatic EPSP and the broader the EPSP spreads temporally. Furthermore, the further away the synaptic input, the smaller is the amplitude of somatic EPSP. This is because EPSPs produced from more distant synapses decay more substantially as they need to travel longer distances to reach the soma. In this model neuron, a synaptic input given at compartment 4 or 8 produces a peak somatic EPSP amplitude that is only 29% or 10% of the peak somatic EPSP amplitude when the same input is given at the soma (Figure 3-42B).

A mammalian CNS neuron receives on average thousands of excitatory synaptic inputs along its dendritic tree. A single EPSP at one synapse is usually insufficient to depolarize the postsynaptic neuron above the threshold for firing action potentials, due to the small size of an individual EPSP when it arrives at the axon initial segment. Indeed, at any given time, the postsynaptic neuron integrates many excitatory inputs in order to reach the firing threshold. Such integration takes two forms. In **spatial integration**, nearly simultaneously activated synapses at different spatial locations sum their excitatory postsynaptic currents when they converge along the path to the soma, producing a larger EPSP (**Figure 3-43**A). In **temporal integration**, synapses activated within a specific window (including successive activation at the same synapse) sum their postsynaptic currents to produce a larger EPSP (Figure 3-43B).

Figure 3–42 Somatic EPSPs from dendritic inputs in a model neuron.
(A) The soma and dendritic tree of this neuron are simplified to 10 compartments for the purpose of mathematical modeling. Compartment 1 represents the soma, and compartments 2–10 represent dendritic segments with increasing distance from the soma, with the length constant (λ) as the unit. Dotted lines illustrate divisions between every two compartments.
(B) When a transient excitatory input of the same size and shape (dotted curve, with y axis to the right) is provided at compartment 1, 4, or 8, the shapes of EPSPs at the soma show distinct profiles. The somatic EPSP produced by the input given at the soma (compartment 1) has the largest amplitude and fastest rising and decay time, the somatic EPSP produced by the input given at compartment 8 has the smallest amplitude and slowest rising and decay time, and the somatic EPSP produced by the input given at compartment 4 has the intermediate amplitude and temporal spread. Time is represented in the unit of the time constant τ. (Adapted from Rall W [1967] *J Neurophysiol* 30:1138–1168.)

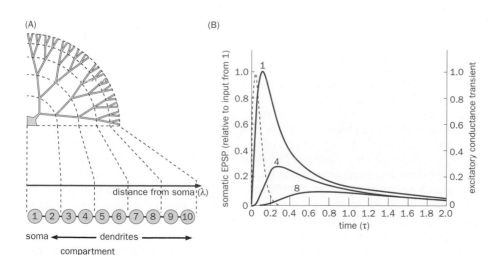

(A)

distance from soma (λ)

① ② ③ ④ ⑤ ⑥ ⑦ ⑧ ⑨ ⑩

soma ◄──── dendrites ────►
compartment

(B)

somatic EPSP (relative to input from 1)

excitatory conductance transient

time (τ)

Figure 3–43 Schematic representation of spatial and temporal integration of synaptic inputs. (A) Spatial integration. Two excitatory inputs from two branches of the dendritic tree that arrive shortly after one another (vertical dotted lines in the EPSP traces indicate the same time) summate their signals when they converge, producing depolarization of the membrane potential at the axon initial segment that exceeds the amplitude produced by each alone (compare the heights of the second and the first peak). **(B)** Temporal integration. Two discrete EPSPs produced at the same synapse (top left) become gradually integrated as they travel from distal dendrites toward the soma due to the temporal spread of electrical signals (see Figure 3–42B). At the axon initial segment, the integrated EPSP produces a peak potential (the second peak) that is greater than that produced by a single EPSP (the first peak).

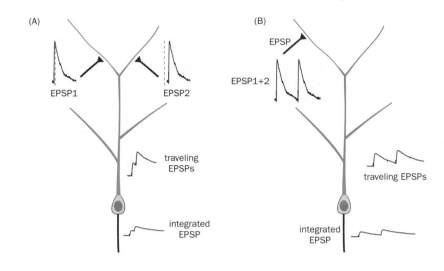

As we see from the model neuron in Figure 3–42, inputs from proximal synapses contribute more to the firing of the neuron because they are less attenuated. In some mammalian neurons, distal synapses are stronger in order to compensate for such distance-dependent attenuation. Importantly, inputs from distal synapses also have a longer window during which to contribute to temporal integration (Figure 3–43B). On the basis of excitatory inputs alone, we can already see that individual dendrites act as sophisticated integrative devices, given the complex morphology and abundant excitatory synapses of typical mammalian CNS neurons. On a moment-by-moment basis, a spiking neuron converts analog signals from the many inputs it receives into a digital signal of whether or not to fire an action potential.

While the passive properties of neuronal membranes we have discussed thus far provide a foundation for understanding how synaptic inputs contribute to firing of postsynaptic neurons, recent studies indicate that dendritic integration is more complex and nuanced. Voltage-gated Na^+, Ca^{2+}, and K^+ channels are present on the dendrites of many mammalian CNS neurons, which endow active properties to the dendritic membrane (that is, voltage-dependent conductance changes). For example, opening of voltage-gated Na^+ or Ca^{2+} channels by EPSPs causes further depolarization in dendrites, and thereby amplifies the EPSP signal. Co-activation of nearby excitatory synapses in dendritic branches can produce dendritic spikes that actively propagate across dendritic segments, conceptually similar to the action potentials we studied in Chapter 2. Although these dendritic spikes may not propagate all the way to the soma (because of the lower density of voltage-gated channels in dendrites compared to axons), they nevertheless amplify synaptic input and propagate membrane potential changes across a large distance with smaller attenuation compared to passive decay. Finally, action potentials produced in the axon initial segment can back-propagate in neurons with voltage-gated channels in dendrites, and these back-propagated action potentials can interact with EPSPs in interesting ways.

As a specific example, we study an experiment in which a cortical pyramidal neuron in an *in vitro* brain slice was subjected to dual patch clamp recording at the soma and at the apical dendrites (**Figure 3–44**A). Electrical stimulation of its input axons produced a synaptic potential, but it was significantly attenuated in the soma, below the threshold of firing an action potential (Figure 3–44B). However, if the recorded neuron was induced to fire an action potential by current injection 5 ms prior to input stimulation, the back-propagated action potential synergized with the dendritic synaptic potential to reach the threshold of firing a dendritic spike, which greatly amplified the synaptic potential, allowing it to trigger two additional action potentials that were propagated to distant target neurons (Figure 3–44C). Assuming that under physiological conditions, the

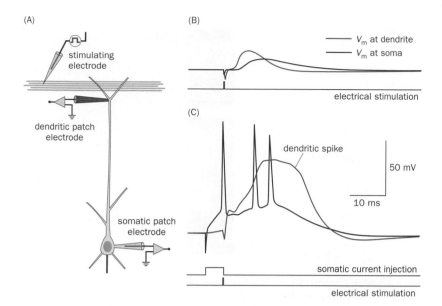

Figure 3–44 Interactions between synaptic input and a back-propagating action potential. (A) Experimental setup. A cortical pyramidal neuron in a brain slice is being recorded by a patch electrode at the apical dendrite (red) and a patch electrode at the soma (gray), both in whole-cell mode. A stimulating electrode delivers electrical stimulation to the input axons. **(B)** Electrical stimulation (bottom trace) produces a dendritic EPSP recorded by the dendritic patch electrode (red), and an attenuated somatic EPSP recorded by the somatic patch electrode. The somatic EPSP is below the threshold for firing an action potential. **(C)** A 5-ms depolarizing current pulse injected into the soma prior to electrical stimulation produces an action potential (the first black spike), which propagates back to the dendrites and integrates with the dendritic EPSP to reach the threshold for producing the dendritic spike (red trace). The propagation of the dendritic spike to the soma produces two additional action potentials (the second and third black spikes). Thus, the back-propagating action potential synergizes with the dendritic EPSP to produce additional output spikes. (Adapted from Larkum ME, Zhu JJ & Sakmann B [1999] *Nature* 398:338–341. With permission from Macmillan Publishers Ltd.)

pyramidal neuron fires action potentials in response to proximal dendritic inputs, this integration mechanism can enable the neuron to amplify near-synchronous input at the proximal and distal dendrites by producing a burst of action potentials that could not be generated by either the distal or the proximal input alone.

It appears that many mammalian CNS neurons have active properties, which often differ in different neurons or even in different compartments of the same neuron because of differential distribution and density of voltage-gated ion channels. We are far from a complete understanding of how synaptic potentials are integrated in light of these active properties; indeed this is a highly active area of research.

3.25 Synapses are strategically placed at specific locations in postsynaptic neurons

In addition to excitatory inputs discussed in the previous section, each neuron also receives inputs from inhibitory neurons and modulatory neurons. How these inputs shape the output of the postsynaptic neuron depends on which subcellular compartments of the postsynaptic neuron these inputs synapse onto.

In general, most excitatory synapses are located on dendritic spines distributed throughout the dendritic tree (**Figure 3–45**). The various presynaptic terminals that target a given postsynaptic neuron may originate from many different presynaptic partner neurons, but each dendritic spine typically receives synaptic input from a single excitatory presynaptic terminal. The thin spine neck creates chemical and electrical compartments for each synapse such that it can be modulated independently from neighboring synapses. These largely independent compartments enable neurons to encode information in the strengths of individual synapses with different input neurons. A neuron can thus modulate its connection strengths with different input neurons independently according to prior experience; this property is crucial for memory, as will be discussed in more detail in Chapter 10.

In contrast to excitatory synapses that are most enriched on spines, inhibitory synapses form broadly across the postsynaptic membrane at dendritic shafts, dendritic spines, the cell body, and the axon initial segment. These distributions allow inhibitory synapses to generate IPSPs at strategic places to oppose the action of EPSPs as they pass by (see Figure 3–29). Let's use a typical pyramidal neuron in the cerebral cortex to illustrate inhibitory inputs it receives from three

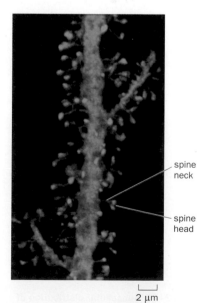

spine neck

spine head

2 μm

Figure 3–45 Dendritic spines. A dendritic segment of a human cortical pyramidal neuron ~100 μm from the cell body, showing dendritic spines with long necks. Imaged after intracellular injection of a fluorescent dye. (Yuste, Rafael. Dendritic Spines, Cover Image, © 2010 Massachusetts Institute of Technology, by permission of The MIT Press.)

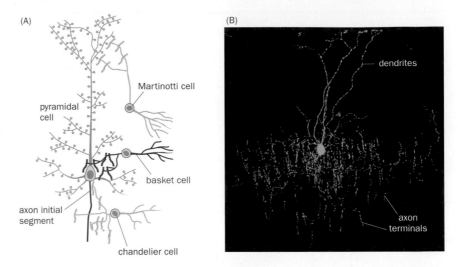

Figure 3–46 Inhibitory inputs to a cortical pyramidal neuron. (A) Schematic drawing of presynaptic terminals (shown as short strings of beads) from three different types of GABAergic inhibitory neuron onto a cortical pyramidal neuron, whose dendrites are in cyan and axons in red. The Martinotti cell (yellow), basket cell (blue), and chandelier cell (green) form synapses respectively onto the distal dendrites, the cell body and proximal dendrites, and the axon initial segment of the pyramidal neuron and other pyramidal neurons not shown. **(B)** A chandelier cell in the mouse cerebral cortex. Each group of presynaptic terminals, which appears as a candle on an old-fashioned chandelier, wraps around the initial segment of a pyramidal neuron. Each chandelier cell thus controls the firing of many pyramidal neurons. (B, courtesy of Z. Josh Huang. See also Taniguchi H, Lu J & Huang ZJ [2013] *Science* 339:70–74.)

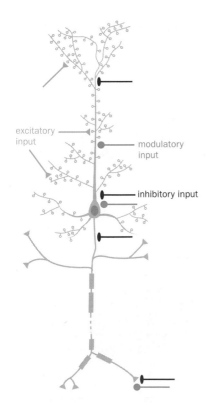

Figure 3–47 Subcellular distribution of synaptic input. For a typical mammalian neuron, excitatory inputs are received mostly at the dendritic spines (and along the dendrites for neurons that lack dendritic spines). Inhibitory inputs are received at the dendritic spines and shaft, cell body, axon initial segment, and presynaptic terminals. Modulatory inputs are received at dendrites, cell bodies, and presynaptic terminals.

types of GABAergic neurons (**Figure 3–46**A). The **Martinotti cell** targets its presynaptic terminals on distal dendrites of the pyramidal cell, and thereby affects the integration of synaptic potentials along specific dendritic segments. For instance, activation of the Martinotti cells can inhibit the production or propagation of the dendritic spikes discussed in previous section. The **basket cell** targets its presynaptic terminals around the cell body of the pyramidal neuron, and thereby influences the overall integration of synaptic input from all dendritic branches. The **chandelier cell** targets its presynaptic terminals specifically to the initial axon segment of many pyramidal cells (Figure 3–46B), such that its synaptic inputs to the pyramidal cell have the most direct impact on the production of action potentials.

A postsynaptic neuron can receive synaptic input at its own presynaptic terminals, as discussed in Section 3.21. Here, inputs do not control the action potential firing rate, but rather the efficacy with which action potentials in the postsynaptic neuron lead it to release neurotransmitters. The presynaptic partners in these cases are mostly modulatory neurons that use transmitters such as acetylcholine, dopamine, serotonin, and norepinephrine. Some GABAergic neurons also exert their action on the presynaptic terminals of their target neurons (see Figure 3–37).

In summary, individual neurons are complex and highly organized integrators. Each neuron receives inputs from its numerous presynaptic partners at its complex dendritic tree, its cell body, its axon initial segment, and its presynaptic terminals (**Figure 3–47**). The interactions of excitatory, inhibitory, and modulatory inputs together shape the neuron's output patterns, which are communicated to its own postsynaptic target neurons by the frequency and timing of action potentials and the probability of neurotransmitter release induced by each action potential. Some neurons also receive input (and send output) through electrical synapses (**Box 3–5**). At a higher level, individual neurons are parts of complex neural circuits that perform diverse information-processing functions, from sensory perception to behavioral control. Having studied the basic concepts and principles of neuronal communication, we are now ready to apply them to fascinating neurobiological problems in the following chapters.

Box 3–5: Electrical synapses

Although chemical synapses are the predominant form of interneuronal communication, electrical synapses are also prevalent in both vertebrate and invertebrate nervous systems. The morphological correlate of the electrical synapses is the **gap junction**, which usually contains hundreds of closely clustered channels that bring the plasma membranes of two neighboring cells together (see Figure 1–14B) and allow passage of ions and small molecules between the two cells. In mammalian neurons, electrical synapses usually occur at the somatodendritic compartments of two partner neurons.

In vertebrates, gap junctions are made predominantly by a family of **connexin** proteins, encoded by about 20 genes in the mammalian genome. Each gap junction channel is composed of 12 connexin subunits, with 6 subunits on each apposing plasma membrane forming a hemi-channel. Each connexin subunit has four transmembrane domains with an additional N-terminal domain embedded in the membrane. As revealed by the crystal structure for connexin-26 (**Figure 3–48**), extensive interactions between the extracellular loops of the hemi-channels bring the two apposing membranes from neighboring cells within 4 nm of each other, and align the two hemi-channels to form a pore with an innermost diameter of 1.4 nm. Invertebrate gap junctions are made by a different family of proteins called **innexins** (<u>innex</u>tebrate con<u>nex</u>in). A third family of proteins called pannexins may contribute to gap junctions in both vertebrates and invertebrates.

Electrical synapses differ from chemical synapses in a number of important ways. First, whereas chemical synapses transmit signals with a delay on the order of 1 ms between depolarization in the presynaptic terminal and synaptic potential generation in the postsynaptic cell, electrical synapses transmit electrical signals with virtually no delay. Second, whereas chemical synapses transmit only depolarizing signals (and, in spiking neurons, only supra-threshold signals that produce action potentials), electrical synapses transmit both depolarizing and hyperpolarizing signals. Third, whereas chemical synapses are asymmetrical—membrane potential changes in the presynaptic neuron produce membrane potential changes in the postsynaptic neuron, but not vice versa—electrical signals can flow in either direction across electrical synapses. Exceptions exist to this rule, however; some electrical synapses prefer one direction over the opposite direction, and are called rectifying electrical synapses. Finally, many electrical synapses allow small molecules such as peptides and second messengers to pass through; indeed, the diffusion of small-molecule dye from one cell to another, called **dye-coupling**, is often used as a criterion to identify the presence of gap junctions between two cells. The conductance of electrical synapses can be modulated by a number of factors such as the membrane potential, the transjunctional voltage (the difference between membrane potentials across the electrical synapse), and chemical factors such as phosphorylation, pH, and Ca^{2+} concentration.

The special properties of electrical synapses discussed above are utilized in many circuits in invertebrates and vertebrates. For instance, electrical synapses are found in circuits where rapid transmission is essential, such as in the vertebrate retina for processing motion signals (where they transmit analog signals between non-spiking neurons), and in escape circuits to avoid predators. Indeed, electrical synapses were first characterized between the giant axon and motor neuron of the crayfish escape circuit in the 1950s. Another utility of electrical synapses is to facilitate synchronized firing between electrically coupled neurons (another term for neurons that form electrical synapses with each other). As a specific example, we study below electrical synapses in the mammalian cerebral cortex.

Using whole-cell patch recording techniques (see Box 13–2 and Section 13.21 for details) in a cortical slice preparation, researchers found that when two fast-spiking (FS) inhibitory neurons (corresponding mostly to basket cells in Figure 3–46A) were recorded simultaneously with patch electrodes, current injection into one

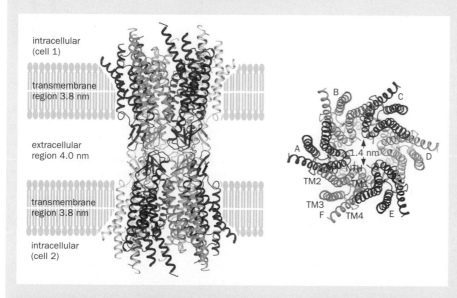

Figure 3–48 Structure of a gap junction channel. Summary of crystal structures of connexin-26 obtained at a resolution of 3.5 Å. Left, view from the side. Each hemi-channel consists of six subunits that are differentially colored. Each subunit has four transmembrane helices and an N-terminal helix (NTH) embedded in the membrane. Right, surface view, with the transmembrane helices and NTH labeled for subunit F. The central passage allows molecules with a linear dimension smaller than 1.4 nm to pass freely between two cells. (Adapted from Maeda S, Nakagawa S, Suga M et al. [2009] *Nature* 458:597–602. With permission from Macmillan Publishers Ltd.)

(Continued)

Box 3–5: Electrical synapses

cell caused nearly synchronous membrane potential changes in both cells; both depolarization and hyperpolarization could result depending on the sign of the injected current (**Figure 3–49**A), indicating that these two cells formed electrical synapses. Paired recording of many cell types indicated that electrical synapses form with exquisite cell-type specificity. For example, low-threshold-spiking (LTS) inhibitory neurons (corresponding mostly to Martinotti cells in Figure 3–46A) also formed electrical synapses with each other at a high probability, but they rarely formed electrical synapses with FS neurons. Excitatory pyramidal neurons, which could be identified as regular-spiking (RS) cells by their electrophysiological properties, did not form any electrical synapses with themselves or with FS or LTS

neurons (Figure 3–49B). Furthermore, whereas injecting sub-threshold depolarizing currents into one of the two electrically coupled FS cells did not elicit action potentials, injecting the same sub-threshold depolarizing currents into both FS cells elicited synchronous action potentials (Figure 3–49C). This suggested that a network of FS cells can act as detectors for synchronous activities, and their synchronous firing can further strengthen synchronous cortical activities. Subsequent work indicated that in addition to FS and LTS cells, other specific types of inhibitory neurons also form type-specific electrical synapse networks, thus providing rich substrates for coordinating electrical activities in the cerebral cortex.

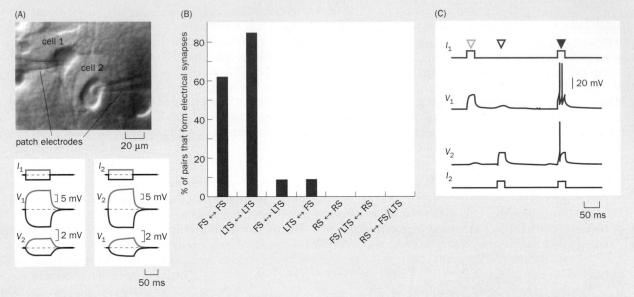

Figure 3–49 Electrical synapses between inhibitory neurons in the rat cerebral cortex. (A) Top, image of a rat cortical slice with two cells and two patch electrodes, taken with differential interference contrast microscopy. Bottom left, when positive (blue) or negative (red) current (I_1) was injected into cell 1, it depolarized or hyperpolarized the membrane potential of cell 1 (top or bottom traces of V_1). In addition, cell 2 was also depolarized or hyperpolarized at the same time (top or bottom traces of V_2). Note the reduced amplitude and slower rising time of V_2 compared to V_1 due to attenuation across the gap junction and the time taken to charge the membrane capacitance of cell 2. Bottom right, positive or negative current injected into cell 2 (I_2) also caused depolarization or hyperpolarization of both cells. Thus, these two cells form electrical synapses. **(B)** Quantification of electrical

synapses between specific types of cells based on paired recording in panel A. Cells were classified based on their firing patterns into fast spiking (FS), low-threshold spiking (LTS), or regular spiking (RS), corresponding roughly to basket cells, Martinotti cells, and pyramidal cells, respectively. Arrows indicate the directionality of electrical synapse tested in paired recording. **(C)** The top and bottom traces show the injection of small depolarizing currents into cell 1 (top) or cell 2 (bottom). Injection into a single cell (open arrowheads) did not cause firing of either cell. Injecting the same size current into both cells simultaneously (filled arrowhead) caused both cells to fire action potentials. (Adapted from Galarreta M & Hestrin S [1999] *Nature* 402:72–75. With permission from Macmillan Publishers Ltd. See also Gibson JR, Beierlein M & Connors BW [1999] *Nature* 402:75–79.)

SUMMARY

Neurons communicate with each other using electrical and chemical synapses. Electrical synapses allow rapid and bidirectional transmission of electrical signals between neurons via the gap junction channels. Although less prevalent than chemical synapses, electrical synapses are widely used in both invertebrates and vertebrates—for example in neural circuits that require rapid information propagation or synchronization. Chemical synapses are unidirectional: electrical signal in the presynaptic neuron is transmitted to the postsynaptic neuron or muscle via the release of a chemical intermediate, the neurotransmitter.

At the presynaptic terminal, neurotransmitter release is mediated by fusion of the synaptic vesicle with the presynaptic plasma membrane. Action potential arrival depolarizes the presynaptic terminal, causing the opening of voltage-gated Ca^{2+} channels at the active zone. Ca^{2+} influx, acting through a synaptic vesicle-associated Ca^{2+} sensor synaptotagmin, releases the molecular clamp on a partially assembled SNARE complex. The full assembly of the SNARE complex, consisting of one v-SNARE on the synaptic vesicle and two t-SNAREs on presynaptic plasma membrane, provides the force that drives membrane fusion and release of transmitter from within the synaptic vesicle to the synaptic cleft. Excess neurotransmitter molecules are rapidly degraded or recycled through reuptake mechanisms. After neurotransmitter release, synaptic vesicles are rapidly recycled and refilled with neurotransmitters, enabling continual synaptic transmission in response to future action potentials.

Nervous systems across the animal kingdom utilize a common set of neurotransmitters. In the vertebrate CNS, glutamate is the main excitatory neurotransmitter, whereas GABA and glycine are the main inhibitory neurotransmitters. Acetylcholine is the excitatory neurotransmitter in the vertebrate neuromuscular junction (and some CNS neurons), but can also act as a modulatory neurotransmitter in the CNS. Other neuromodulators include dopamine, serotonin, norepinephrine, and neuropeptides. The specific actions of neurotransmitters are determined by the properties of their receptors on the postsynaptic neurons.

Neurotransmitter receptors are either ionotropic or metabotropic. Ionotropic receptors are ion channels that are gated by neurotransmitter binding, and act rapidly to produce synaptic potentials within a few milliseconds of presynaptic action potential arrival. The ionotropic acetylcholine and glutamate receptors are nonselective cation channels with reversal potentials around 0 mV; upon neurotransmitter binding these receptors produce depolarization in the form of excitatory postsynaptic potentials. The NMDA receptor acts as a coincidence detector as its channel opening depends both on presynaptic glutamate release and a depolarized state of the postsynaptic neuron. Ionotropic GABA and glycine receptors are Cl^- channels, with reversal potentials near or below the resting potential. Their opening usually produces Cl^- influx, which counters excitatory postsynaptic potentials and inhibits postsynaptic neurons from reaching the threshold at which they fire action potentials.

Metabotropic receptors for acetylcholine, glutamate, GABA, monoamines, and neuropeptides are all G-protein-coupled receptors. Neurotransmitter binding leads to the association of the metabotropic receptor with the trimeric G protein, exchange of GDP for GTP, and dissociation of Gα-GTP and Gβγ subunits from each other and from the receptor. Gα-GTP and Gβγ each can activate different effectors depending on specific G protein variants and cellular context. Gβγ can act on K^+ and Ca^{2+} channels directly, whereas Gα usually acts via second messengers such as cAMP and Ca^{2+} to activate protein kinases that phosphorylate ion channels to change membrane potentials and excitability of the postsynaptic neurons. Metabotropic receptor activation usually causes membrane potential changes within tens of milliseconds to seconds. Longer-term changes of postsynaptic neurons in response to neurotransmitter release and neuronal activity involve synapse-to-nucleus signaling and alterations of gene expression.

Chemical synapses are highly organized structurally. At the presynaptic terminal, the active zone protein complexes bring synaptic vesicles to the immediate vicinity of voltage-gated Ca^{2+} channels such that Ca^{2+} influx rapidly triggers neurotransmitter release. Trans-synaptic cell adhesion proteins, by interacting with scaffolding proteins in both the presynaptic active zone and the postsynaptic density, help align active zone and high-density neurotransmitter receptors across the synaptic cleft. Postsynaptic density scaffolding proteins further link neurotransmitter receptors to their regulators and effectors for efficient synaptic transmission and for regulating synaptic plasticity.

Integration of excitatory, inhibitory, and modulatory inputs at the dendrites, cell bodies, and axon initial segments of postsynaptic neurons collectively determine their own action potential firing patterns. Synaptic input to the axon terminals of postsynaptic neurons further modulates the efficacy with which postsynaptic action potentials lead to neurotransmitter release. All of these mechanisms are amply used in the nervous system to produce functions from sensation to action that we will study in the following chapters.

FURTHER READING

Books and reviews

Cohen S & Greenberg ME (2008) Communication between the synapse and the nucleus in neuronal development, plasticity, and disease. *Annu Rev Cell Dev Biol* 24:183–209.

Hille B (2001) Ion Channels of Excitable Membranes, 3rd ed. Sinauer Associates Inc.

Huang EJ & Reichardt LF (2003) Trk receptors: roles in neuronal signal transduction. *Annu Rev Biochem* 72:609–642.

Isaac JT, Ashby MC & McBain CJ (2007) The role of the GluR2 subunit in AMPA receptor function and synaptic plasticity. *Neuron* 54:859–871.

Katz B (1966) Nerve, Muscle, and Synapse. McGraw-Hill.

Llinas RR (1982) Calcium in synaptic transmission. *Sci Am* 247:56–65.

Sheng M, Sabatini BL & Südhof TC (2012) Cold Spring Harbor Perspectives in Biology: The Synapse. Cold Spring Harbor Laboratory Press.

Südhof TC & Rothman JE (2009) Membrane fusion: grappling with SNARE and SM proteins. *Science* 323:474–477.

Unwin N (2013) Nicotinic acetylcholine receptor and the structural basis of neuromuscular transmission: insights from *Torpedo* postsynaptic membranes. *Q Rev Biophys* 46:283–322.

Presynaptic mechanisms

Augustine GJ, Charlton MP & Smith SJ (1985) Calcium entry and transmitter release at voltage-clamped nerve terminals of squid. *J Physiol* 367:163–181.

Bennett MK, Calakos N & Scheller RH (1992) Syntaxin: a synaptic protein implicated in docking of synaptic vesicles at presynaptic active zones. *Science* 257:255–259.

Del Castillo J & Katz B (1954) Quantal components of the end-plate potential. *J Physiol* 124:560–573.

Fernandez-Chacon R, Konigstorfer A, Gerber SH et al. (2001) Synaptotagmin I functions as a calcium regulator of release probability. *Nature* 410:41–49.

Geppert M, Goda Y, Hammer RE et al. (1994) Synaptotagmin I: a major Ca^{2+} sensor for transmitter release at a central synapse. *Cell* 79:717–727.

Heuser JE & Reese TS (1981) Structural changes after transmitter release at the frog neuromuscular junction. *J Cell Biol* 88:564–580.

Ichtchenko K, Hata Y, Nguyen T et al. (1995) Neuroligin 1: a splice site-specific ligand for beta-neurexins. *Cell* 81:435–443.

Katz B & Miledi R (1967) The timing of calcium action during neuromuscular transmission. *J Physiol* 189:535–544.

Koenig JH & Ikeda K (1989) Disappearance and reformation of synaptic vesicle membrane upon transmitter release observed under reversible blockage of membrane retrieval. *J Neurosci* 9:3844–3860.

Krnjevic K & Miledi R (1958) Acetylcholine in mammalian neuromuscular transmission. *Nature* 182:805–806.

Kuffler SW & Yoshikami D (1975) The number of transmitter molecules in a quantum: an estimate from iontophoretic application of acetylcholine at the neuromuscular synapse. *J Physiol* 251:465–482.

Liu KS, Siebert M, Mertel S et al. (2011) RIM-binding protein, a central part of the active zone, is essential for neurotransmitter release. *Science* 334:1565–1569.

Llinas R, Sugimori M, Silber RB (1992) Microdomains of high calcium concentration in a presynaptic terminal. *Science* 256:677–679.

Sabatini BL & Regehr WG (1996) Timing of neurotransmission at fast synapses in the mammalian brain. *Nature* 384:170–172.

Schiavo G, Benefenati F, Poulain B et al. (1992) Tetanus and botulinum-B neurotoxin block neurotransmitter release by proteolytic cleavage of synaptobrevin. *Nature* 359:832–835.

Schneggenburger R & Neher E (2000) Intracellular calcium dependence of transmitter release rates at a fast central synapse. *Nature* 406:889–893.

Söllner T, Whiteheart SW, Brunner M et al. (1993) SNAP receptors implicated in vesicle targeting and fusion. *Nature* 362:318–324.

Sutton RB, Fasshauer D, Jahn R et al. (1998) Crystal structure of a SNARE complex involved in synaptic exocytosis at 2.4 Å resolution. *Nature* 395:347–353.

Takamori S, Holt M, Stenius K et al. (2006) Molecular anatomy of a trafficking organelle. *Cell* 127:831–846.

Watanabe S, Trimbuch T, Camacho-Perez et al. (2014) Clathrin regenerates synaptic vesicles from endosomes. *Nature* 515:228–233.

Postsynaptic mechanisms

Coombs JS, Eccles JC & Fatt P (1955) The specific ionic conductances and the ionic movements across the motoneuronal membrane that produce the inhibitory post-synaptic potential. *J Physiol* 130:326–374.

Fredriksson R, Lagerstrom MC, Lundin LG et al. (2003) The G-protein-coupled receptors in the human genome form five main families. Phylogenetic analysis, paralogon groups, and fingerprints. *Mol Pharmacol* 63:1256–1272.

Galarreta M & Hestrin S (1999) A network of fast-spiking cells in the neocortex connected by electrical synapses. *Nature* 402:72–75.

Greenberg ME, Ziff EB & Greene LA (1986) Stimulation of neuronal acetylcholine receptors induces rapid gene transcription. *Science* 234:80–83.

Hestrin S, Nicoll RA, Perkel DJ et al. (1990) Analysis of excitatory synaptic action in pyramidal cells using whole-cell recording from rat hippocampal slices. *J Physiol* 422:203–225.

Larkum ME, Zhu JJ & Sakmann B (1999) A new cellular mechanism for coupling inputs arriving at different cortical layers. *Nature* 398:338–341.

Lipscombe D, Kongsamut S & Tsien RW (1989) Alpha-adrenergic inhibition of sympathetic neurotransmitter release mediated by modulation of N-type calcium-channel gating. *Nature* 340:639–642.

Ma H, Groth RD, Cohen SM et al. (2014) γCaMKII shuttles Ca^{2+}/CaM to the nucleus to trigger CREB phosphorylation and gene expression. *Cell* 159:281–294.

Magee JC & Cook EP (2000) Somatic EPSP amplitude is independent of synapse location in hippocampal pyramidal neurons. *Nat Neurosci* 3:895–903.

Magleby KL & Stevens CF (1972) A quantitative description of end-plate currents. *J Physiol* 223:173–197.

Mishina M, Kurosaki T, Tobimatsu T et al. (1984) Expression of functional acetylcholine receptor from cloned cDNAs. *Nature* 307:604–608.

Miyazawa A, Fujiyoshi Y & Unwin N (2003) Structure and gating mechanism of the acetylcholine receptor pore. *Nature* 423:949–955.

Montminy MR, Sevarino KA, Wagner JA et al. (1986) Identification of a cyclic-AMP-responsive element within the rat somatostatin gene. *Proc Natl Acad Sci USA* 83:6682–6686.

Nowak L, Bregestovski P, Ascher P et al. (1984) Magnesium gates glutamate-activated channels in mouse central neurones. *Nature* 307:462–465.

Rall W (1967) Distinguishing theoretical synaptic potentials computed for different soma-dendritic distributions of synaptic input. *J Neurophysiol* 30:1138–1168.

Rasmussen SG, DeVree BT, Zou Y et al. (2011) Crystal structure of the β2 adrenergic receptor-Gs protein complex. *Nature* 477:549–555.

Sheng M, Thompson MA & Greenberg ME (1991) CREB: A Ca^{2+}-regulated transcription factor phosphorylated by calmodulin-dependent kinases. *Science* 252:1427–1430.

Sobolevsky AI, Rosconi MP & Gouaux E (2009) X-ray structure, symmetry and mechanism of an AMPA-subtype glutamate receptor. *Nature* 462:745–756.

Takeuchi A & Takeuchi N (1960) On the permeability of end-plate membrane during the action of transmitter. *J Physiol* 154:52–67.

Taniguchi H, Lu J & Huang ZJ (2013) The spatial and temporal origin of chandelier cells in mouse neocortex. *Science* 339:70–74.

CHAPTER 4
Vision

The primary purpose of the nervous system is for animals to sense the world and respond accordingly. Thus, sensory systems are of prime importance for animals to survive as individuals and species. Studies of sensory systems have also revealed many of the principles of nervous system operation. We therefore devote this chapter and Chapter 6 to studying how sensory systems work.

For most humans, vision is the sense on which we are most reliant, as suggested by the epigraph. Indeed, vision is utilized by most multicellular organisms, from jellyfish to mammals (see Chapter 12 for more details of its evolution). The ultimate purpose of vision for most animals is to identify food sources, seek mates, and avoid dangers such as predators. All visual systems share a common source of input, namely photons hitting photoreceptors in spatial and temporal patterns. The job of the visual system is to extract useful features from these light signals in forms that help guide animal behavior, in order to optimize their opportunities for survival and reproduction. Some of the key tasks the visual system performs include differentiating objects from background, locating objects of interest, detecting motion, and navigating in the environment.

This chapter's organization mirrors the sequence of events that make vision possible. We will trace the visual pathway from photoreceptors to retinal circuits to the visual cortex. Given the complexity of the topic and the knowledge we have obtained, the details can sometimes seem overwhelming, but at each level of analysis it is useful to keep in mind the ultimate purpose of vision: to navigate the world and optimize survival.

HOW DO RODS AND CONES DETECT LIGHT SIGNALS?

Human and most vertebrate animals use the eyes (**Figure 4–1**) as the only means to sense light. Light enters the eye through the pupil—the opening or aperture at the center of the iris—and is focused through the lens. The pupil size adjusts automatically according to the light level, and the curvature of the lens also

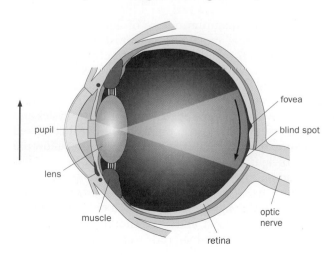

Figure 4–1 Cross section of a human eye. Light travels past the pupil and is focused by the lens onto the retina. An external object (arrow) is projected as an inverted image on the retina. Visual signals are transmitted to the brain via the optic nerve. A blind spot is formed at the head of the optic nerve. At the center of the retina, the fovea has the densest population of cones, which are the photoreceptors responsible for high-acuity vision and color vision (see Section 4.8 for more details).

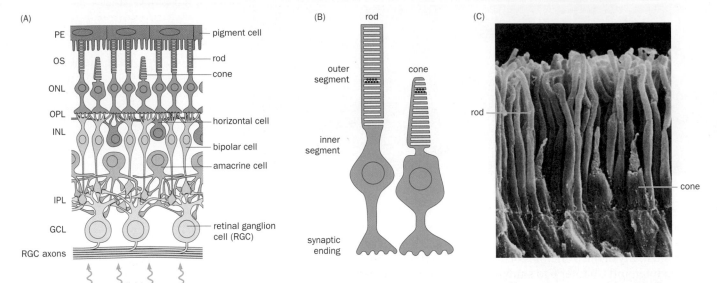

Figure 4–2 Organization of the retina and photoreceptor cells. (A) The layered structure of the retina. In this schematic, light enters the retina from the bottom. PE, pigment epithelium; OS, outer segments of photoreceptors; ONL, outer nuclear layer that contains rod and cone cell bodies; OPL, outer plexiform layer where rods and cones synapse with bipolar and horizontal cells; INL, inner nuclear layer that contains the cell bodies of bipolar, horizontal, and most amacrine cells; IPL, inner plexiform layer where bipolar, amacrine, and retinal ganglion cells (RGCs) form synapses; GCL, ganglion cell layer that contains cell bodies of RGCs and some amacrine cells. The major cell types are listed on the right. Whereas photoreceptors (rods and cones), bipolar cells, and RGCs are all glutamatergic excitatory neurons, horizontal cells and amacrine cells are mostly inhibitory neurons. We will encounter many subtypes later in the chapter. **(B)** Schematic drawing of a rod and a cone. Within the outer segment, photosensitive molecules (dots) are highly concentrated in an orderly stack of membranes. In rods these membranes are intracellular discs, whereas in cones they are continuous with plasma membrane. (Dots reflect the distribution of only a small fraction of photosensitive molecules to highlight the type of membrane on which photoreception occurs.) **(C)** A scanning electron micrograph showing three cones in the foreground amongst many rods. (B, adapted from Baylor DA [1987] *Inv Ophthal Vis Sci* 28:34–49; C, courtesy of William Miller.)

automatically adjusts such that the eye can focus on a range of object distances. The outside world is projected as images onto a thin layer of cells at the back of the eye, the **retina**, which for neurobiologists is the most important part of the eye. The retina is where visual input is received as spatiotemporal patterns of photons, converted to electrical signals, processed by exquisitely precise retinal circuits, and then sent to the brain through the **optic nerve**.

The vertebrate retina is a layered structure made of five major types of neurons (**Figure 4–2**A). The input layer, which is actually at the back of the retina, consists of **photoreceptors** that detect photons and convert them to electrical signals, a universal form of information that is understood by the rest of the nervous system as we have learned in Chapter 2. The output layer is composed of **retinal ganglion cells** (**RGCs**) that transmit information from the eye to the brain through their axons, which collectively make up the optic nerve. In between are **bipolar cells** that transmit information from the photoreceptors to the RGCs, as well as **horizontal cells** and **amacrine cells**, whose actions influence the signals that are transmitted from photoreceptors to bipolar cells and then to the RGCs.

We start our journey through the visual system with the photoreceptors. Two types of photoreceptors are present in vertebrates, the **rods** and **cones**, named after their shapes (Figure 4–2B, C). As will be discussed in detail later, cones are responsible for high acuity and color vision; in primates, cones are concentrated in the central part of the retina called the **fovea** (Figure 4–1). Rods are more numerous and are the more sensitive photon detectors specialized for night vision, and this is where our story begins.

4.1 Psychophysical studies revealed that human rods can detect single photons

How sensitive are our rods? Given the quantum nature of light, we can ask a more precise version of this question: how many photons must be absorbed by a rod in order to elicit a biological effect? Using human **psychophysical studies**, which investigate the relationship between physical stimuli and the sensations or behaviors they elicit, Selig Hecht and co-workers obtained a decisive answer to this question.

Taking into consideration the available knowledge from both the physics and biology of vision, these investigators utilized experimental conditions that offered the highest sensitivity for the human retina, that is, the conditions that allow the detection of the lowest amount of light input. These included having the subjects (the investigators themselves) adapt to the dark by sitting in a completely dark room for at least 30 minutes before the onset of the experiments (dark adaptation is discussed later in the chapter), shining lights on a small area of the peripheral retina where rods were known to be the densest, and using short (1 ms) exposures of light with a wavelength (510 nm) to which human rods were known to be most

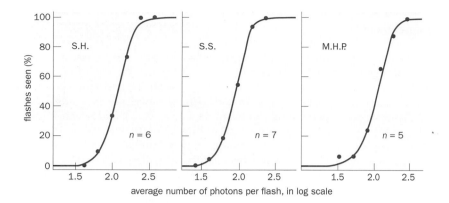

Figure 4–3 Psychometric plots for three investigators. The percentage of light flashes seen by the observers (*y* axis) is plotted against the average number of photons per flash (*x* axis, in log scale—log scales are all 10-based in this book unless otherwise noted). Each circle represents the average of 35–50 independent measurements. These data were fitted with a series of Poisson distributions, where *n* is the number of quanta needed to elicit a flash sensation. The best fits (curves) for each of the investigators (designated by their initials) were *n* = 6, 7, and 5, respectively. (Adapted from Hecht S, Shaler S & Pirenne MH [1942] *J Gen Physiol* 25:819–840. With permission from Rockefeller University Press.)

sensitive. They then varied the number of photons in these flashes, and scored whether the subjects said they saw the light flash or not.

Physical measurement and statistical analysis of frequency-of-seeing plots (**Figure 4–3**)—**psychometric functions** that quantify the frequency of light flashes seen against the relative light intensity—reached similar conclusions. To reliably perceive a light flash, an average of five to seven independent events (photon absorptions) must occur in a retinal field of about 500 rods. Since the probability of two photons being absorbed by the same rod under these circumstances is very small, each rod must be able to report the absorption of a single photon.

4.2 Electrophysiological studies identified the single-photon response of rods: light hyperpolarizes vertebrate photoreceptors

Rods have cytoplasmic extensions at both ends (see Figure 4–2B). As is true for most other neurons, a rod's output end is the presynaptic terminal, where information is transmitted to the rod's postsynaptic partners—the bipolar cells and horizontal cells. At the input end is a highly specialized photon detection apparatus, the **outer segment**, which is made of tightly stacked membrane disks enriched in photosensitive molecules.

Given the prediction from human psychophysical studies that individual rods can detect single photons, experiments were designed to measure the light sensitivity of single rods. The outer segment of a single rod from dissected frog retina (see **Box 4–1** for the use of different animal models to study vision) was sucked by negative pressure into an electrode (a suction electrode) that formed a tight seal with the rod at the bottom. In this configuration, electric current passing through ion channels in the plasma membrane of the rod outer segment could be measured in response to a light beam directed at the outer segment (**Figure 4–4**A). This design enabled the measurement of response of a single rod to the light stimulation applied to only a small section of the rod outer segment.

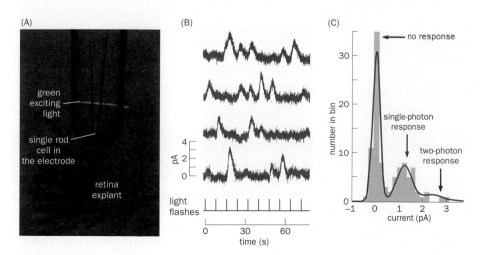

Figure 4–4 Single-photon responses of the rod. (A) The outer segment of a single rod from a piece of frog retina was collected in a suction electrode to measure its electrical response to stimulation by a narrow beam of green light. **(B)** Responses of a single rod (top) to light flashes (bottom, marked as spikes), measured as currents (in picoamperes, pA). The currents constitute outward flow as indicated by the positive peaks. Each row represents responses to 10 light flashes. A total of 40 responses are shown. At this light intensity, most flashes produced no response. The remaining stimuli produced unitary responses. The timescale of light flashes is given at the bottom. **(C)** A histogram of light-induced current was fitted to a Poisson distribution (the smooth curve), revealing the magnitude of single-photon responses. (A, courtesy of Denis Baylor; B & C, adapted from Baylor DA, Lamb TD & Yau KW [1979] *J Physiol* 288:613–634. With permission from the Physiological Society.)

Box 4–1: Vision research uses diverse animal models

Researchers use different animal models to study vision because each model offers unique experimental advantages or special properties (see Sections 13.1–13.5 for a general discussion). For example, the retinal explants of amphibians (see Figure 4–4) and reptiles (see Figure 4–12) are relatively easy to maintain *in vitro*, enabling researchers to characterize electrophysiological properties of photoreceptors and other retinal cells with exquisite experimental control. Bovine retinas provide abundant outer segments for protein purification, facilitating structural studies of rhodopsin (see Figure 4–6B) and biochemical analysis of phototransduction (see Figure 4–8). The mouse is increasingly being used for vision research because it offers precise deletion of individual genes (see Figure 4–13), labeling of specific neuronal populations and their axonal projections (see Figure 4–36), and the ability to manipulate the activities of specific cell types (see Figure 4–47). Cats (for example, Figure 4–24) and monkeys (for example, Figure 4–43) are widely used in studying retinal, thalamic, and cortical

processing of visual signals because they possess superior visual acuity. Rhesus monkeys are also used because they can be trained to perform sophisticated behavioral tasks that allow researchers to probe neural mechanisms of visual perception (for example, Figure 4–52), and because among all model organisms their visual system most resembles the human visual system. Finally, humans can be subjects of vision research because they can directly report what is seen in psychophysical studies (see Figure 4–1), and because human genetic variants can inform how the visual system functions (see Figure 4–23). In Chapters 5 and 12, we will further introduce the use of invertebrate visual systems to study their development and function.

In addition to the experimental advantages each model organism offers, comparative studies of diverse visual systems also provide insight into how visual systems came about and how diverse solutions may be used to tackle a common problem. We will expand on this evolutionary perspective in Chapter 12.

By systematically reducing the light intensity, a condition was reached in which most light flashes did not produce any response. The occasional responses that did occur had a rather uniform size (Figure 4–4B). The probability distribution of the response amplitude fits nicely with the Poisson distribution, which we have encountered in the context of quantal neurotransmitter release (see Box 3–1), indicating that the responses were caused mostly by absorption of single photons and occasionally by absorption of two photons (Figure 4–4C).

These single-rod measurements confirmed earlier findings that photon absorption results in *hyperpolarization* of rods; that is, current flows out of the rod in response to light (Figure 4–4B). This is an unusual strategy for reporting sensory stimulations; one would think that sensory stimuli should depolarize and activate the neurons, rather than hyperpolarize and inhibit them. This is indeed the case for invertebrate photoreceptors and for most other sensory systems (see Chapter 12), but vertebrate rods and cones use hyperpolarization to report light stimulation.

The single-photon responses not only offered a satisfying physiological explanation for the human psychophysical observations, but also provided a quantitative measurement of exactly what a single photon does to a rod photoreceptor. Each photon results in ~1 picoampere (pA) of net outward current across the membrane of the rod (Figure 4–4C), which is equivalent to a blockade of ~10^7 positive ions (mostly Na$^+$) that would otherwise flow into the cell (**Figure 4–5**). How is this feat achieved? To answer this question, we now turn to the biochemistry of the rods.

4.3 Light activates rhodopsin, a prototypical G-protein-coupled receptor

The photosensitive molecules that are highly enriched in the stacked disk membranes of the rod outer segment (see Figure 4–2B) are the **rhodopsins**. Each rhodopsin consists of a protein called **opsin** (from *opsis*, meaning vision in Greek) and a small molecule called **retinal** that is closely related to, and derived from, vitamin A. Retinal is covalently attached to a lysine residue of the opsin.

Retinal is the **chromophore** (light-absorbing portion of a molecule) that gives rhodopsin its rosy color (in Greek, *rhodo* means rose). Retinal exists in two isomers in photoreceptors: all-*trans* and 11-*cis*. Photon absorption by rhodopsin

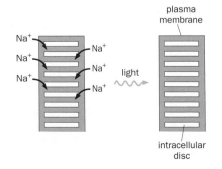

Figure 4–5 Light hyperpolarizes rods.
Schematic drawing of a portion of the rod outer segment. Absorption of each photon blocks the inward flow of about 10^7 cations in total.

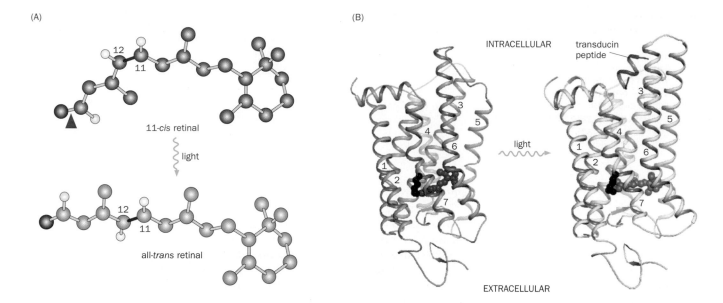

(A)

11-*cis* retinal

light

all-*trans* retinal

(B)

INTRACELLULAR   transducin peptide

light

EXTRACELLULAR

Figure 4–6 The structure of rhodopsin.
(A) Light triggers the chromophore retinal to switch from an 11-*cis* to an all-*trans* configuration. Note the difference to the left of the dark bond between 11 and 12. The red arrowhead indicates the bond between retinal and opsin (the black circle is from a lysine residue of the opsin). **(B)** Crystal structures of inactive rhodopsin with 11-*cis* retinal (red, left panel) and active rhodopsin with all-*trans* retinal (blue, right panel) in complex with a fragment of the G protein transducin at the intracellular face. The lysine residue of the opsin (black) to which the retinal attaches is rendered in a space-filling format. Light-induced isomerization of retinal creates a binding site for transducin by causing conformational changes in the transmembrane helices, which are numbered 1–7. (A, structures based on Wald G [1968] *Science* 162:230–239; B, adapted from Choe HW, Kim YJ, Park JH et al. [2011] *Nature* 471:651–656. With permission from Macmillan Publishers Ltd.)

causes a switch of 11-*cis* retinal to all-*trans* retinal (**Figure 4–6**A). This large structural change of retinal then triggers a corresponding conformational change of its protein partner, the opsin.

Opsin is a prototypical G-protein-coupled receptor (GPCR). It has seven transmembrane helices, a design commonly used in receptors for sensory stimuli, hormones (such as epinephrine), and some neurotransmitters (see Section 3.18). Among all GPCRs, rhodopsin was the first to be purified to homogeneity and to have its primary structure determined. It was also the first GPCR whose structure was solved with atomic resolution by X-ray crystallography. Retinal isomerization from 11-*cis* to all-*trans* induces a structural change in the transmembrane helices (Figure 4–6B), which creates a binding site on the cytosolic side for the α subunit of a heterotrimeric G protein called transducin, which we will discuss in detail below.

4.4 Photon-induced signals are greatly amplified by a transduction cascade

Studies of rhodopsin have been greatly facilitated by the abundance of rhodopsin proteins in the rod outer segments and the ease of obtaining large amounts of rod outer segments from bovine retinas. In fact, rod outer segments can respond to light activation *in vitro*, which enabled detailed studies of the **phototransduction** process: the biochemical reactions triggered by photon absorption.

Biochemical investigations first revealed that light induced a decline in the level of **cyclic GMP (cGMP)**, a cyclic nucleotide similar to the second messenger cAMP we studied in the context of adrenergic receptor signaling (see Section 3.19). cGMP is produced from GTP by the enzyme **guanylate cyclase** and is hydrolyzed by a second enzyme, **phosphodiesterase (PDE)** (**Figure 4–7**A). Importantly, direct injection of cGMP into rod outer segments caused rod depolarization that mimics darkness. So a decline of the cGMP level appeared to be responsible for light-triggered rod hyperpolarization, suggesting that light activates PDE (Figure 4–7B). But how is this achieved?

Biochemical experiments revealed that light-activated rhodopsin in rod outer segments catalyzes the exchange of bound GDP for GTP. A GTP-binding protein complex named **transducin** was identified as the intermediate that links light-induced rhodopsin activation to PDE activation and subsequent decrease of cGMP levels. Specifically, purified rhodopsin and purified transducin (composed of Tα, Tβ, and Tγ subunits), when placed together, could reconstitute light-induced GDP–GTP exchange without the involvement of PDE (**Figure 4–8**, Steps 1 and 2). Furthermore, purified Tα, when GTP bound, was sufficient to trigger PDE activation in the absence of light stimulation (Figure 4–8, Step 3). The intrinsic GTPase

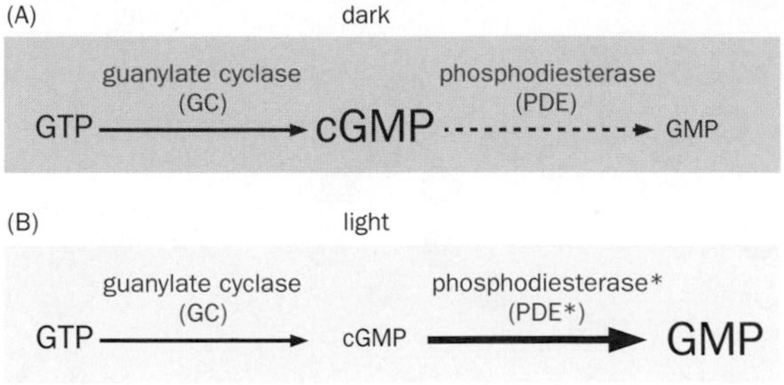

Figure 4–7 Light activates phosphodiesterase, which hydrolyzes cGMP. Guanylate cyclase (GC) catalyzes the synthesis of cGMP from GTP. Phosphodiesterase (PDE) catalyzes cGMP hydrolysis. Both enzymes are membrane-bound. **(A)** In the dark, PDE is inactive (dashed arrow), and cGMP is abundant (larger font). **(B)** Light triggers activation of PDE (* designates active), which hydrolyzes cGMP (smaller font).

(A)   dark

guanylate cyclase   phosphodiesterase
(GC)   (PDE)

GTP ──────────────→ cGMP ·············→ GMP

(B)   light

guanylate cyclase   phosphodiesterase*
(GC)   (PDE*)

GTP ──────────→ cGMP ──────────→ GMP

activity of Tα results in its return to the GDP-bound state and its re-association with Tβγ, forming the trimeric complex needed for the next round of activation (Figure 4–8, Step 4). This mechanism bears a striking resemblance to the activation of adenylate cyclase by the β-adrenergic receptor via trimeric G proteins (see Figures 3–31 and 3–33; Movie 3-3). Indeed, transducin is a trimeric G protein complex, and Tα in phototransduction is equivalent to G_s in β-adrenergic receptor signaling. These studies in parallel established the paradigm for how GPCRs, a widely distributed class of receptors, transduce signals.

In the phototransduction cascade (Figure 4–8), a single light-activated rhodopsin (R*) catalyzes the activation of >20 transducin molecules, each of which activates a PDE, together resulting in the hydrolysis of tens of thousands of cGMP per second. Signal amplification in this biochemical cascade begins to explain how a single photon triggers a large decrease in the cGMP level.

Figure 4–8 Schematic diagram of the phototransduction mechanism. Step 1: Light triggers activation of rhodopsin (R to R*). Step 2: Activated rhodopsin (R*) catalyzes the exchange of the transducin (T)-bound nucleotide, replacing GDP with GTP; Tα-GTP subsequently dissociates from Tβγ and R*. Step 3: Tα-GTP binds to and activates phosphodiesterase (PDE*) by sequestering the inhibitory PDE γ subunits. (Each Tα-GTP molecule binds to one PDE γ subunit; for simplicity only one Tα-GTP/PDE γ subunit complex is drawn here). This leads to the hydrolysis of cGMP to GMP. Step 4: The intrinsic GTPase activity of Tα hydrolyzes Tα-GTP to Tα-GDP, which re-associates with Tβγ and is ready for another round of phototransduction. The deactivation of R* and regeneration of R is discussed in Section 4.6. (Adapted from Stryer L [1983] *Cold Spring Harbor Symp Quant Biol* 48:841–852.)

4.5 Light-triggered decline of cyclic-GMP level directly leads to the closure of cation channels

How does cGMP hydrolysis cause a change in current flow across the rod membrane? It had been assumed that cGMP, like its famous cousin cAMP, acts as a second messenger by activating protein kinases (see Figure 3–33). However, in the case of phototransduction, cGMP binds directly to the cytoplasmic side of a cation channel and increases its conductance. This mechanism of action was revealed using the patch clamp technique (**Figure 4–9**A; see Box 13-2). When a piece of rod outer segment plasma membrane containing the channel was exposed from the cytoplasmic side to solutions containing increasing concentrations of cGMP, the channel conductance increased correspondingly (Figure 4–9B). This effect did not depend on the presence of ATP, ruling out the involvement of protein kinases. The effect of cGMP was specific: the channel could not be opened by cAMP with concentrations orders of magnitude higher than that of cGMP. This direct

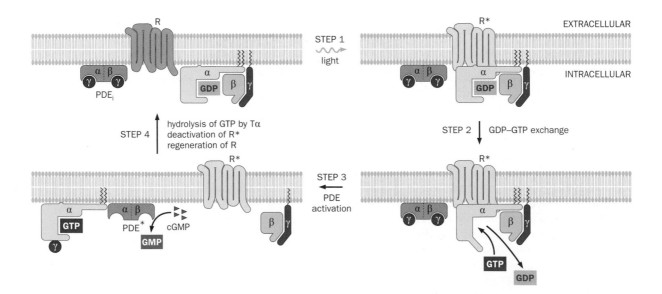

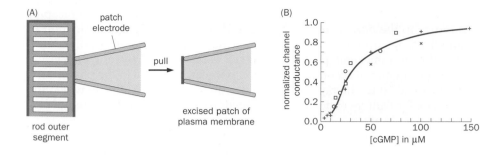

activation of ion channels by cGMP allows the change of cGMP concentration to rapidly affect the membrane potential of rods in response to light.

Subsequent cloning of the gene that encodes the rod cGMP-gated channel—which is a member of the cyclic nucleotide-gated channel (**CNG channel**) family—revealed that its primary structure resembles that of voltage-gated K^+ channels we discussed in Chapter 2, with six transmembrane domains and a pore loop between S5 and S6 (see Figure 2–34 and Box 2–4). The C-terminal cytoplasmic domain of all CNG channels contains an additional cyclic nucleotide-binding domain. Like K^+ channels, a functional CNG channel is made of four subunits with four cyclic nucleotide-binding sites. Channel opening is gated by direct binding of cyclic nucleotide, rather than by changes of membrane potential. Like the nicotinic acetylcholine receptor (nAChR) discussed in Section 3.13, the rod CNG channel is a nonselective cation channel: its opening in the dark causes influx of both Na^+ and Ca^{2+} and efflux of K^+. Because the rod CNG channel has a reversal potential near 0 mV (also like the nAChR), the resulting Na^+ influx far exceeds K^+ efflux, so the net effect of channel opening is depolarization (see Figure 3–18). As will be discussed shortly, Ca^{2+} entry in the dark is crucial for recovery and adaptation.

The identification of the cGMP-gated cation channel provides the final piece in our story of how light is converted to electrical signals in the rod photoreceptor (**Figure 4–10**). The light-triggered conformational change of rhodopsin activates transducin, which then activates PDE, resulting in cGMP breakdown. A decline in cGMP concentration causes the closure of cGMP-gated cation channels, decreasing a steady influx of cations. This leads to the hyperpolarization of the rod, which causes a decline in the release of glutamate (the neurotransmitter used by rods and cones) at rod → bipolar cell and rod → horizontal cell synapses.

4.6 Recovery enables the visual system to respond to light continually

We have so far emphasized the sensitivity of rods, but an optimized sensory system incorporates many additional features such as speed, dynamic range, selectivity, and reliability. The combination of superb single-photoreceptor electrophysiology and biochemistry in rods has enabled precise measurement and mechanistic understanding of many of these features. In the following two sections, we study two related processes: recovery and adaptation.

Figure 4–9 cGMP directly activates a cation channel. (A) A small piece of the plasma membrane of the rod outer segment can be excised by a patch clamp electrode to study its properties *in vitro*. **(B)** The excised membrane from the rod outer segment, when placed in a defined solution, increases its conductance in response to increasing concentrations of cGMP in the solution (equivalent to the intracellular side of the membrane patch). Different symbols represent measurements performed at different Ca^{2+} concentrations, which did not affect the conductance. (B, adapted from Fesenko EE, Kolesnikov SS & Lyubarsky AL [1985] *Nature* 313:310–313. With permission from Macmillan Publishers Ltd.)

Figure 4–10 A summary of the phototransduction cascade. * indicates an activated component. Rightward arrows indicate the successive steps of signal transduction.

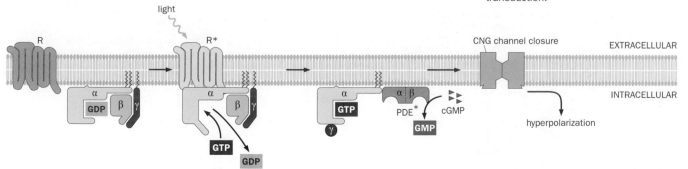

Recovery refers to the process by which light-activated photoreceptor cells return to the dark state (see a general discussion of signal termination in Section 3.22). Recovery ensures that each visual stimulus produces only a transient signal, and the same photoreceptor cell is ready to receive the next visual stimulus in a matter of seconds. All rod components activated in the phototransduction cascade must be rapidly returned to the dark state, a state in which the rod is ready for light reception and transduction. This requires the action of several parallel molecular pathways.

The cGMP level must increase during recovery to reopen the CNG channels that depolarize the rod. This is achieved by the activation of guanylate cyclase (GC), the enzyme that catalyzes the synthesis of cGMP from GTP (see Figure 4–7). GC activity is highly sensitive to intracellular Ca^{2+} concentration ($[Ca^{2+}]_i$). It remains largely inactive when $[Ca^{2+}]_i$ is greater than 100 nM, but becomes highly active when $[Ca^{2+}]_i$ drops below 60 nM (**Figure 4–11**A). In the dark, $[Ca^{2+}]_i$ is ~400 nM, balanced by an influx via CNG channels and an efflux via Na^+/Ca^{2+} exchange proteins on the plasma membrane. When light causes cGMP levels to decline, so that the CNG channels close, $[Ca^{2+}]_i$ drops below 50 nM because the exchange proteins remain active. This decrease in $[Ca^{2+}]_i$ activates GC through an intermediate protein called **GCAP** (**guanylate cyclase activating protein**), a calcium-binding protein that in its calcium-free form binds to and activates GC. Thus, closure of cGMP-gated channels by light triggers a negative-feedback loop leading to elevated cGMP, which opens the cGMP-gated channels and restores the dark state (Figure 4–11B, pathway 1).

Tα-GTP must be deactivated. As already discussed, Tα-GTP has an intrinsic GTPase activity and can convert itself to a GDP-bound state. Tα-GDP then dissociates from PDE, inactivating that enzyme. The intrinsic GTPase of Tα-GTP is greatly facilitated by RGS9 (**RGS** stands for **regulator of G protein signaling**) (Figure 4–11B, pathway 2). RGS9 acts as a GTPase activating protein (GAP) for Tα-GTP (see Figure 3–32).

Rhodopsin must be deactivated to prevent further activation of the transduction cascade. To shut off rhodopsin, rhodopsin kinase specifically binds to and phosphorylates the cytoplasmic tail of R* but not R; this phosphorylation recruits a protein called arrestin to R*; binding of arrestin to phosphorylated R* competes with Tα binding and therefore prevents R* from activating more transducin molecules (Figure 4–11B, pathway 3; Movie 3–3).

Finally, for rhodopsin to return to the dark state, all-*trans* retinal must be converted back to 11-*cis* retinal. This process is slow (measured in minutes) because it takes place in **pigment cells** in the pigment epithelium (see Figure 4–2) and involves the exchange of retinal molecules between rods and pigment cells. (The conversion of all-*trans* retinal to 11-*cis* retinal for cones occurs in a specialized cell type in the retina called the **Müller glia**.) However, given that only a small fraction of a rod's rhodopsin molecules are typically activated by low-light stimulation, the activities of rhodopsin kinase and arrestin ensure that within seconds the rod is ready for visual excitation again using the remaining rhodopsin molecules in the same rod.

Figure 4–11 Mechanisms of recovery.
(A) Guanylate cyclase activity is highly sensitive to intracellular Ca^{2+} concentration, with GC activity increasing as $[Ca^{2+}]_i$ decreases. **(B)** Summary of three major mechanisms of recovery (red) superimposed on the phototransduction pathway (black) reproduced from Figure 4–10. (1) The closure of CNG channels reduces $[Ca^{2+}]_i$, which causes GCAP to activate guanylate cyclase (GC*), leading to increased cGMP production. cGMP binding to CNG channels leads to channel opening and membrane depolarization. (2) RGS9 facilitates the hydrolysis of Tα-bound GTP, thereby deactivating PDE. (3) Rhodopsin kinase (R* kinase) specifically phosphorylates R*. Phosphorylated R* recruits binding of arrestin. R* phosphorylation and arrestin binding deactivate R*. (A, adapted from Koch KW & Stryer L [1988] *Nature* 334:64–66. With permission from Macmillan Publishers Ltd.)

(A)

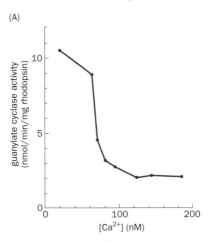

(B)

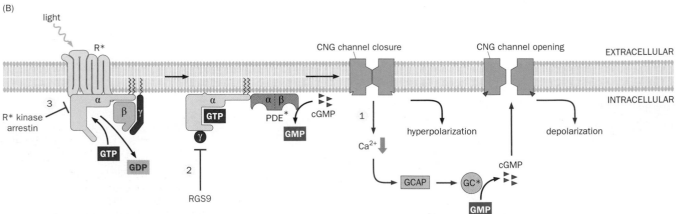

4.7   Adaptation enables the visual system to detect contrast over a wide range of light levels

Object detection requires separating the object from its background. Luminance contrast is a primary means for object detection (we will discuss the other primary means, color contrast, in later sections). Our visual system has a very large **dynamic range** (which is defined as the ratio between the largest and smallest value of light intensity), enabling us to detect single photons in near darkness and to discriminate objects from background over a 10^{11}-fold range of ambient light level. Rods are responsible for the lower part of this dynamic range, discerning luminance contrast over a 10^4 range of ambient light level. How can rods detect single photons, and at the same time detect luminance contrast and not be saturated by light intensity four orders of magnitude greater? This is achieved by **adaptation**, the adjustment of a photoreceptor's sensitivity according to the background light level.

Adaptation is a universal phenomenon in sensory biology, originally described by Ernst Weber in the 1820s in an observation that is now called **Weber's Law** (or the Weber–Fechner relation). Although it doesn't hold under all stimulus conditions, it applies when the just-noticeable difference between two sensory stimuli is proportional to the magnitude of the stimulus. For example, a human subject can distinguish the weight of 100 g from 105 g (a just-noticeable difference of 5 g, or 5% of the total weight) but cannot distinguish 1000 g from 1005 g. It typically takes a difference of 50 g for the distinction to be clear at the 1000-g range (a just-noticeable difference of 50 g, or 5% of the total weight again).

In the visual system, light adaptation means that photoreceptors become less sensitive to the same intensity of stimulation when the background illumination is higher. In other words, to achieve the same amount of hyperpolarization at higher background illumination, a stronger stimulus is required. This is illustrated quantitatively in the experiment shown in **Figure 4–12**. To reach a 1 mV hyperpolarization (indicated in the figure by the dashed line), a light flash with a relative intensity of ~$10^{-6.5}$ was adequate for dark-adapted rods. As the background light intensity increased, the flash intensity to reach the same 1 mV hyperpolarization also increased. Note also that at low background light, the light response saturated at low light intensity. That is, the photocurrent did not increase in response to further increases in light intensity; this saturation appears as a flattening of the response curve. However, as background light increased, saturation shifted rightward on the curve. Thus, adaptation expands the dynamic range of light intensity within which rods can distinguish the intensity differences between background and objects.

What is the basis of visual adaptation? Interestingly, Ca^{2+} is a key signal for adaptation as it is for recovery (see Figure 4–11B). High levels of background illumination cause some cGMP-gated channels to be closed, resulting in a decline

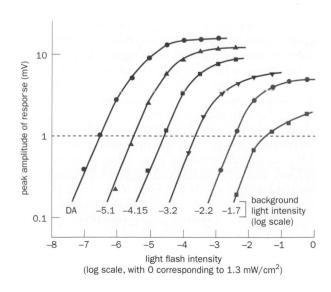

Figure 4–12 Light adaptation of gecko rods. Each dot corresponds to peak amplitude of hyperpolarization measured by intracellular recording (*y* axis) in response to a 0.1 s light flash of a given intensity (*x* axis). Light intensity log of 0 corresponds to 1.3 mW/cm² for both light flash and background. Experiments were grouped in a series of curves according to background light intensities given at the bottom of curves. DA, dark adapted. Four groups on the left (red) and two groups on the right (brown) were recorded from two separate photoreceptors. As background light increases, producing the same magnitude of response (for example, a 1 mV hyperpolarization indicated by the dashed line) requires increasing intensity of light stimulation. (Adapted from Kleinschmidt J & Dowling JE [1975] *J Gen Physiol* 66:617–648. With permission from Rockefeller University Press.)

Figure 4–13 Defects in recovery and adaptation in guanylate cyclase activating protein (GCAP) knockout mice. (A) Dark-adapted flash response in wild-type and GCAP knockout mice. Photocurrents (in picoamperes, pA) were measured from single rods using the suction electrode (see Figure 4–4A). Loss of GCAP increases the response magnitude and slows the time course of recovery. **(B)** Light adaptation of wild type (blue) and GCAP knockout (red) mice. Values on the y axis represent the rod response to a light flash at a given background light intensity relative to the same flash when given after dark adaption (x axis; darkness is equivalent to 1, or 10^0, on the x axis). As background light intensity increases, the same flash gives increasingly smaller responses, but the mutant curve drops faster and saturates at a light intensity two orders of magnitude lower than wild type. The gray curve is a theoretical prediction of the flash response in the absence of adaptation. (Adapted from Mendez A, Burns ME, Sokal I, et al. [2001] *Proc Natl Acad Sci USA* 98:9948–9953.)

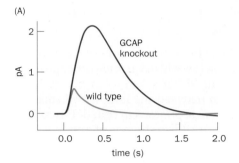

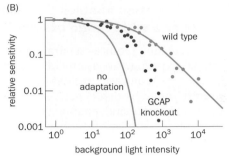

of $[Ca^{2+}]_i$. This increases the basal level of GC activity because of the action of GCAP. To counteract this heightened GC activity, stronger activation of PDE by more light is required to reduce the cGMP level and thereby promote effective hyperpolarization.

Analysis of GCAP knockout mice demonstrated that protein's essential function—and the role of Ca^{2+} regulation of GC activity—in both recovery and adaptation. Compared with wild type, GCAP knockout mice exhibited much higher sensitivity in response to light flashes (**Figure 4–13**A). This is because without GCAP activation of GC, a small amount of PDE activation can cause cGMP levels to decline and the CNG channels to close, leading to hyperpolarization. However, the knockout mice required much longer periods of time to return to baseline, as GC won't be activated by declining Ca^{2+} for the synthesis of new cGMP (see Figure 4–11B). Thus, Ca^{2+} regulation of GC enabled by GCAP is essential to shorten light responses and ensure a speedy recovery.

GCAP mutant mice also exhibited severe defects in light adaptation. In wild-type mice, the sensitivity to light flash relative to the dark-adapted condition decreases as background light intensity increases, following Weber's law. In the GCAP mutant, this decrease in sensitivity was much more rapid and the response curve saturated at a light intensity two orders of magnitude lower than the wild type (Figure 4–13B). The GCAP mutants still exhibit some degree of adaptation, suggesting the contribution of additional mechanisms. Two of these additional mechanisms also involve Ca^{2+}: (1) with its calmodulin partner, Ca^{2+} reduces the affinity of CNG channels for cGMP; and (2) Ca^{2+} elevation also inhibits R* phosphorylation. Thus, Ca^{2+} serves as a central coordinator of adaptation.

In summary, elaborate biochemical mechanisms have evolved not only to provide rods with exquisite sensitivity to convert light stimuli to electrical signals, but also to negatively regulate the phototransduction cascade once it is triggered. These mechanisms ensure that light-induced signals are transient and have a wide dynamic range.

4.8   Cones are concentrated in the fovea for high-acuity vision

We now turn to the second type of photoreceptors, the cones (see Figure 4–2B). The signaling pathway from light to electrical signals in cones follows the same theme as in rods. However, variations on the theme allow the cones to serve complementary functions to rods: vision in bright light, vision with high spatial acuity, a wider range of motion vision, and color vision. Indeed, cones are much more important in the daily life of humans than rods.

Here is an experiment you can do to test your high-acuity and color vision. You fix your eye to the front, extend your arm on your side at the eye level, place on your palm small objects with different colors and shapes, and move your extended arm slowly to the front. You will find that you cannot tell the exact color and shape of the objects until they are quite close to the front. The reason for this observation is that the cones, responsible for high-acuity and color vision, are highly concentrated in the fovea at the central part of your retina (**Figure 4–14**A). The retinal eccentricity is used to define the location of the retina that corresponds to a particular space in the **visual field** (which is the entire external world that can be seen at a given time). Retinal eccentricity is measured by the angle (in degrees) formed

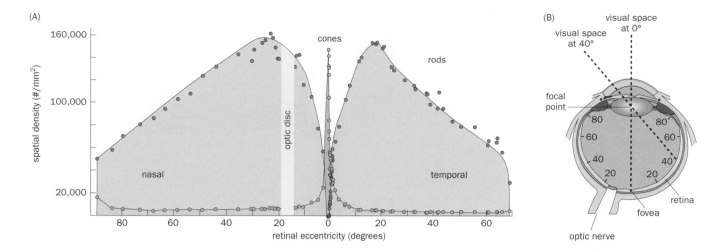

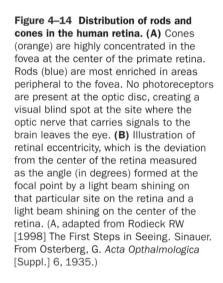

at the focal point by a light beam shining on a particular site on the retina and a light beam shining on the center of the retina (0 degree in Figure 4–14B). Cones are highly concentrated within a few degrees of retinal eccentricity. By contrast, rods are densest in areas of the retina adjacent to the fovea (Figure 4–14B). That is why a dim star in the night sky can be seen more easily if you look slightly to the side of the star.

High-acuity vision in primates is supported by the high density of cones packed in the fovea. Neighboring cones at the fovea can thus resolve much smaller spots of light compared with the peripheral retina where the density of photoreceptors is sparser. The primate retina devotes a high proportion of its processing power to the fovea: although the fovea only contains 1% of the total area of the retina, information from the fovea to the brain is transmitted by about half of the retinal ganglion cell axons. In addition, unlike the rest of the retina, where light must pass through layers of cells and synapses before reaching the photoreceptors (see Figure 4–2A), at the center of the fovea, other cell and synaptic layers are displaced peripherally (**Figure 4–15**). The fovea is also devoid of blood vessels. These properties enable light to reach cones with minimal obstruction, further enhancing high-acuity vision. In patients with **macular degeneration**, a disease that causes photoreceptors in the fovea to die, high-acuity vision is preferentially impaired.

4.9   Cones are less sensitive but faster than rods

Comparing responses to light flashes in single rods or cones (**Figure 4–16**) reveals important differences in their sensitivity and speed. Cones are much less sensitive than rods. Individual cones require flash intensities about two orders of magnitude greater in order to produce currents similar to those produced by individual rods. However, cones are much faster than rods in their light responses. Cones reach their peak current 50 ms after a light flash, compared to about 150 ms in rods. Moreover, cone responses return to baseline much more rapidly than rod responses, so cones can react to repeated light stimulation faster than rods and therefore have better temporal resolution, which is crucial for motion detection. These differences reflect variations in phototransduction cascades and recovery mechanisms (see Figure 4–11B), as well as morphological differences between rods and cones. For example, the opsin kinase expressed by cones has a much

Figure 4–14 Distribution of rods and cones in the human retina. (A) Cones (orange) are highly concentrated in the fovea at the center of the primate retina. Rods (blue) are most enriched in areas peripheral to the fovea. No photoreceptors are present at the optic disc, creating a visual blind spot at the site where the optic nerve that carries signals to the brain leaves the eye. **(B)** Illustration of retinal eccentricity, which is the deviation from the center of the retina measured as the angle (in degrees) formed at the focal point by a light beam shining on that particular site on the retina and a light beam shining on the center of the retina. (A, adapted from Rodieck RW [1998] The First Steps in Seeing. Sinauer. From Osterberg, G. *Acta Opthalmologica* [Suppl.] 6, 1935.)

Figure 4–15 Inner retinal layers are displaced at the human fovea. Retina near the fovea of a live human eye imaged by optical coherence tomography. At the foveal pit, the inner retinal cell layers as well as the plexiform layers are displaced such that light can access the cones directly (compare with Figure 4–2A). ONL, outer nuclear layer; OPL, outer plexiform layer; INL, inner nuclear layer; IPL, inner plexiform layer; GCL, ganglion cell layer. (Adapted from Cucu RG, Podoleanu AG, Rogers JA, et al. [2006] *Optics Letters* 31:1684–1686.)

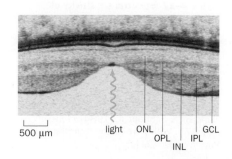

Figure 4–16 Flash responses of single rod and cone in the rhesus monkey. Currents from a rod and a cone in response to increasing magnitude of light flashes are plotted in a series of curves. Each curve represents response to 2× intensity of flash compared to the curve immediately below it. The bottom curve for the cone starts at a light intensity level about 65× higher than the bottom curve for the rod, indicating that cones are less sensitive than rods. Cones recover faster than rods. Cones also produce an 'undershoot' during the recovering phase. (Adapted from Baylor DA [1987] *Inv Ophthal Vis Sci* 28:34–49. With permission from the Association for Research in Vision and Ophthalmology.)

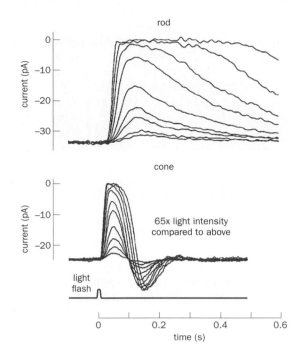

higher specific activity toward cone opsins than rhodopsin kinase has toward rhodopsin. RGS9, which facilitates the hydrolysis of Tα-bound GTP and is a rate-limiting factor for recovery in rods, is expressed at a much higher level in cones. Finally, the larger plasma membrane surface-to-volume ratio of cone cells compared with rods (see Figure 4–2B) makes the $[Ca^{2+}]_i$ decline and feedback more rapidly in response to light stimulation. All these factors contribute to a lower sensitivity but speedier recovery in cones compared with rods.

Cones can also adapt to a broader range of light intensity than rods. As we discussed in Section 4.7, rods are responsible for discerning luminance contrast over a 10^4 range of ambient light level from near darkness. Cones extend this range by an additional factor of 10^7, enabling us to see over a 10^{11}-fold range of ambient light level. The above differences in rods and cones contribute to this disparity. In addition, bright light causes retinal isomerization in significant fractions of cone opsins, a process called pigment bleaching. Pigment bleaching reduces the effective concentration of cone opsins that can still respond to light, which also contributes to adaptation as more light is needed to produce the same photocurrents from the cone population.

4.10 Photoreceptors with different spectral sensitivities are needed to sense color

In principle, the cone functions we have discussed thus far could be accomplished by a new type of photoreceptor that differs from rods in density, distribution, morphology of the outer segment, and phototransduction pathway parameters. However, if this hypothetical photoreceptor responded to light as a function of wavelength in the same way that rods do—that is, if it shared the rods' **spectral sensitivity**—it would fail to serve one important cone function: color vision.

Our sense of color depends on the detection and *comparison* of light with different wavelengths. Visible light corresponds to wavelengths ranging from ~370 nm (violet) to ~700 nm (red). When the white light from the sun that contains a mixture of different wavelengths shines on an object, some wavelengths (that is, a subset of spectra) are absorbed more than others; the unabsorbed subset gives a specific color to the object. Color vision enables the distinction between objects based on the wavelengths of light they reflect, refract, transmit, or fluoresce.

The spectral sensitivity of rhodopsin covers a large part of the visible light range, with the peak sensitivity around 500 nm (**Figure 4–17**); for this reason, Hecht and colleagues used 510-nm light for their psychophysics experiment (see

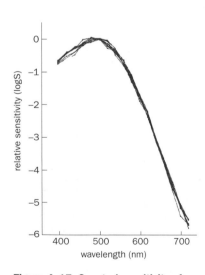

Figure 4–17 Spectral sensitivity of monkey rods. This was determined by sensitivity of single rods in response to light flashes of different wavelengths (λ), in log of sensitivity (S), expressed as a ratio to sensitivity to a 501 nm wavelength light flash [logS = $\log_{10}(S_\lambda/S_{501})$]. Spectral sensitivity curves of 10 different rods are superimposed. (Adapted from Baylor DA, Nunn BJ & Schnapf JL [1984] *J Physiol* 357:575. With permission from the Physiological Society.)

Section 4.1). Using rhodopsin alone, however, it is not possible to differentiate between a certain amount of light with a wavelength of 500 nm, or 10 times that amount of light with a wavelength of 570 nm, because both conditions would excite rods to the same extent (Figure 4–17). The presence of multiple photoreceptors with distinct spectral sensitivities makes it possible for an animal to extract color information by comparing the relative excitation of two different kinds of photoreceptors. According to the color-opponency theory proposed by Ewald Hering in the 1870s, human color vision is achieved by three sets of antagonistic color receptors interacting with each other: black–white, blue–yellow, and green–red. Although seemingly at odds with what we will study in the next few sections, the color-opponency theory successfully explains many phenomena, such as the influence that visual background has on our color perception (for example, **Figure 4–18**). We will return to the neural circuit basis of color opponency in Section 4.19.

4.11 Humans have three types of cones

In addition to rods, most mammals have two types of cones with different spectral sensitivities: a short-wavelength S-cone, and a longer-wavelength cone. Catarrhines, a group of primates that comprises humans, apes, and Old World monkeys, are fortunate to have three different cones—the S-cone, M-cone, and L-cone (in colloquial terms, the blue, green, and red cones)—and therefore are called **trichromats**. Trichromacy enables primates to more easily identify objects of similar luminance but different colors, such as red fruits among green leaves. The idea of trichromacy was first suggested in 1802 by noted polymath Thomas Young, who proposed that "as it is almost impossible to conceive of each sensitive point of the retina to contain an infinite number of particles, each capable of vibrating in perfect unison with every undulation, it becomes necessary to suppose the number limited... each sensitive filament of the nerve may consist of three portions, one for each principal color."

Studies since the 1960s have confirmed Young's hypothesis and identified three types of human cones with different spectral sensitivities. We summarize these findings by returning to a familiar preparation: electrophysiological recording of single photoreceptors using the suction electrode (see Figure 4-4). In response to stimulation by monochromatic light flashes with systematically varied wavelengths, it was found that individual cones from rhesus monkey (an Old World monkey) belonged to one of three types with distinct spectral sensitivities: S, M, or L (**Figure 4–19**A). The variance of spectral sensitivity was so small among the cones of the same type that each cone could be unambiguously assigned to one of these three types. Experiments using retina from humans with normal color vision displayed spectral sensitivity similar to that of the monkey (Figure 4-19B).

How are the three different cones distributed in the retina? This was determined in living human eyes using a technique called retinal densitometry. By comparing reflections from the retina before and after providing a saturating amount of light at a specific wavelength, images of 'bleached' retina revealed distribution

Figure 4–18 Color perception is influenced by the background. The two crosses in this image are of the same color (see their junction at the top). However, the left cross appears gray on a yellow background, and the right cross appears yellow on a gray background. (Adapted from Albers J [1975] Interaction of Color. Yale University Press.)

Figure 4–19 Spectral sensitivity of cones in the monkey and human. **(A)** Spectral sensitivity of three types of cones from rhesus monkeys as determined by recording photocurrent from individual cones in response to light flashes of different wavelengths. *y* axis is the relative sensitivity to peak in logarithm, as in Figure 4–17. **(B)** Spectral sensitivity of human cones superimposed on the spectrum of visible light. A micro-spectrophotometer was used to measure the absorption of different wavelengths by individual cones obtained during eye surgery. (A, Adapted from Baylor DA, Nunn BJ & Schnapf JL [1987] *J Physiol* 390:145–160. With permission from The Physiological Society; B, Adapted from Nathans J [1989] *Sci Am* 260:42–49. With permission from Macmillan Publishers Ltd.)

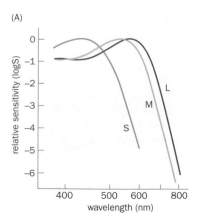

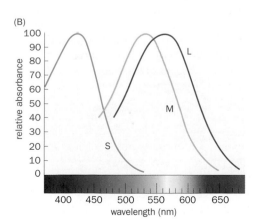

Figure 4–20 Spatial distribution of three types of cones in the human retina. Pseudocolor images of trichromatic cone mosaics in the fovea of two individuals determined by comparing retinal images before and after treatment with a saturating amount of light with a specific short, medium, or long wavelength. Blue, green, and red dots represent S, M, and L cones, respectively. (From Roorda A & Williams DR [1999] *Nature* 397:520–522. With permission from Macmillan Publishers Ltd.)

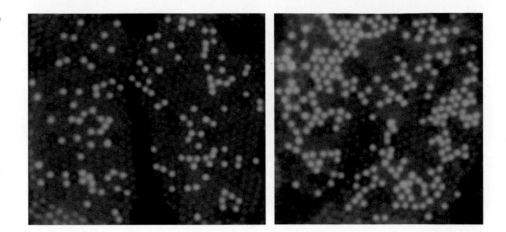

of each of the pigments at single-cone resolution (**Figure 4–20**). S-cones are much sparser than M- and L-cones, making up less than 5% of total cones at the fovea. This explains why we don't have high spatial resolution for blue color. Indeed, because lights with different wavelength refract differentially through the single lens we have in front of our eye, they cannot be focused with equal sharpness on the retina; this property is called **chromatic aberration**. We have evolved sharper focus for long-wavelength light, so we would not gain better resolution of blue light even if the blue cone density were higher. The ratio of M- and L-cones differs among individuals, and the spatial distributions of M- and L-cones appear random. In Chapter 12, we will learn how evolution has shaped these properties.

4.12 Cloning of the cone opsin genes revealed the molecular basis of color detection

By the early 1980s, experimental evidence had provided strong support for the existence of three types of cones in some primate retinas, each with distinct spectral sensitivity (see Figure 4–19). But what is the molecular basis for this? A simple hypothesis was that there exist three separate cone opsin genes, with each cone cell expressing one and only one opsin with a specific spectral sensitivity. A further hypothesis was that the genes encoding the cone opsins are similar to the rod opsin gene, as they likely derive from a common ancestor. The isolation of cone opsin genes validated these hypotheses.

The abundance of rhodopsin in the outer segments of rods (see Section 4.3) enabled the purification and determination of partial amino acid sequences for bovine rod opsin protein, which led to the identification of the corresponding gene. The DNA encoding the bovine rod opsin was then used as a probe to identify human opsin genes using a strategy called low-stringency hybridization, which enables the isolation of DNA based on similar but not identical sequences. Four separate genes were discovered that encode human opsins: one for the rod opsin and three for the S-, M-, and L-cone opsins (**Figure 4–21**), as predicted by the original hypothesis.

Figure 4–21 Schematic and pair-wise comparisons of human opsin proteins. Amino acid sequences of rod opsin (the protein component of rhodopsin) and three cone opsins (S, M, L) were optimally aligned in a generic G-protein-coupled receptor configuration with seven transmembrane helices (see Figure 4–6B). The N-termini face the disc lumen or extracellular environment, and the C-termini face the cytoplasm. In these pair-wise comparisons, open circles indicate identical amino acids, while filled circles indicate different amino acids. (Adapted from Nathans J, Thomas D, & Hogness DS [1986] *Science* 232:193–202. With permission from AAAS.)

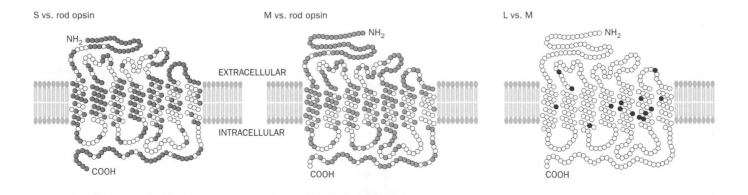

S vs. rod opsin

M vs. rod opsin

L vs. M

The three cone opsins each share about ~40% amino acid sequence identity with the rod opsin. The S opsin also shares ~40% sequence identity with the M and L opsins. The M and L opsins, however, are 96% identical (Figure 4–21). Among the sequence variations are charged amino acids located in the transmembrane domains that interact with retinal; these variations account for the different spectral sensitivities of different opsins when coupled to retinal.

The M and L opsin genes map closely on the X chromosomes. This property together with their very high sequence identity suggest that the M and L opsins evolved from a common ancestor fairly recently. This is in keeping with the notion that trichromacy developed about 35 million years ago during primate evolution. It is likely a result of a gene duplication event that occurred in a common ancestor of Old World monkeys, apes, and humans, after Old World monkeys diverged from New World monkeys (we will discuss this subject in greater detail in Section 12.16).

4.13 Defects in cone opsin genes cause human color blindness

It is well known that red–green color blindness (that is, individuals cannot distinguish red from green; **Figure 4–22**) is much more common than blue color blindness and occurs much more frequently in men than women (~7% of men and 0.4% women in the United States are red–green color-blind). Prior to opsin gene cloning, genetic mapping had already pinpointed green and red color blindness to nearby regions on the X chromosome, hence explaining the predominance of color blindness in males, who have only one copy of the X chromosome compared to two copies in females. After the M and L opsin genes were cloned, molecular alterations in these genes were found in a majority of color-blind males tested (**Figure 4–23**A). Given the high degree of sequence similarity and short distance between the M and L opsin genes and their flanking DNA, unequal crossing over during meiotic recombination could occur at a high frequency, resulting in the loss or alteration of the M or L opsin genes (Figure 4–23B).

Thus, the identification and analysis of human cone opsin genes provided a satisfying explanation for the molecular-genetic basis of color detection as well as variant color vision including color blindness in humans.

HOW ARE SIGNALS FROM RODS AND CONES ANALYZED IN THE RETINA?

During image detection, photoreceptors report light intensity as two-dimensional arrays that change over time. But the visual system does not simply provide a faithful representation of the external world. Its purpose is to extract useful

trichromat   dichromat

Figure 4–22 Test for color blindness. The Ishihara color plates can test for green–red color blindness. The numbers 5 and 29, which are obvious to trichromats (left), are difficult to distinguish for patients missing the M cone receptor (right).

Figure 4–23 L and M opsin gene rearrangements responsible for the most common forms of variant human color vision. (A) Top, individuals who have normal L-cone sensitivity and are missing M-cone sensitivity (M⁻L⁺) have an intact L opsin gene (red) in the head-to-tail array and an altered or missing M opsin gene (green). Bottom, by contrast, individuals who have normal M-cone sensitivity and are missing L-cone sensitivity (M⁺L⁻) have an intact M opsin gene and an altered or missing L opsin gene. (Note that amino acids that determine the spectral sensitivity are clustered at the C-terminal half of the opsin protein, such that a hybrid with N-terminal L opsin and C-terminal M opsin usually confers M-cone sensitivity.) **(B)** Recombination between L and M opsin genes at meiosis leads to gene arrangements that produce variant color vision. Because of high sequence similarity between the L and M opsin genes and their flanking DNA, unequal crossing over (dashed line) can occur occasionally between two copies of the X chromosome in maternal germ-line cells. This results in gene duplication in one gamete and gene deletion in the other gamete. In some cases—as shown here—the recombination event occurs within the genes and generates L/M and M/L hybrid genes, which code for cone opsins with sensitivities that differ from normal M and L cone opsins (for example, M′ refers to an altered M-cone sensitivity). Since males have only one X chromosome, a male child who inherits a variant X chromosome from his mother would exhibit variant color vision that depends on the number and nature of remaining opsin genes. (Adapted from Nathans J, Thomas D, & Hogness DS. [1986] *Science* 232:203–210. With permission from AAAS.)

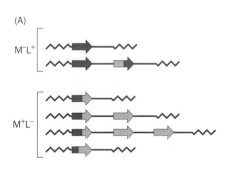

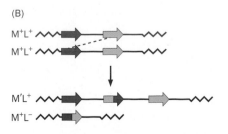

features, such as motion, depth, color, and size of visual objects, from the light signal in order to help animals to survive and reproduce. Furthermore, animals can devote only finite neural resources to vision due to various biological constraints. For example, humans have ~100 million photoreceptors, but only ~1 million retinal ganglion cells (RGCs) to transmit all visual information from the retina to the brain. So the visual system discards as much irrelevant information as it can and encodes the remaining, behaviorally relevant information as efficiently as possible.

The rest of the chapter deals with a central question: how do the retina and the brain extract behaviorally relevant information from the visual stimuli? Specifically, how do the two-dimensional arrays of electrical signals from the photoreceptors collectively inform the brain as to what is seen? Where and how does feature extraction occur?

4.14 Retinal ganglion cells use center–surround receptive fields to analyze contrast

As we highlighted in Chapter 1, asking the right question is often an important step toward obtaining the right answer. The broad questions we have just raised are obviously significant to our understanding of vision, but they do not necessarily help us to get started. Instead, we need questions that can be clearly answered with decisive experiments. One key question that has facilitated the understanding of visual information processing is: what visual signals best excite a given cell at a specific stage of the visual processing pathways?

Before answering this question, we must introduce an important concept in sensory physiology, the **receptive field**. This concept originated from somatosensation, referring to the area of the body where stimuli could influence the firing of a neuron in the somatosensory pathway. The receptive field of a neuron in the visual system refers to the area of visual field (or area of the retina as they correspond, see Figure 4–14B) from which activity of a neuron can be influenced by light. To measure the receptive field of a neuron, one can place an electrode in the extracellular space next to the neuron and measure its firing pattern (action potentials), a technique called **single-unit extracellular recording** (see Section 13.20 for details). The experimenter can then systematically alter the visual stimuli that the retina is receiving, to identify the patterns of stimuli that can influence the firing pattern of the neuron being recorded. Thus, identifying the receptive field of a neuron means to identify the visual signals that best excite (or inhibit) that neuron. As we move along the neural pathways that process visual signals, we will find that the receptive field changes in interesting and illuminating ways. We start with the RGCs, which carry visual information from the retina to the brain.

Steve Kuffler performed a classic experiment to identify the RGC receptive field in the mammalian retina in the early 1950s, shortly after electrodes for single-unit extracellular recording were invented. An electrode was inserted from the side of the eye into the retina of an anesthetized cat, and placed next to an RGC. Then a spot of light was moved across the retina to identify the area that corresponded to the RGC being recorded, as indicated by changes in its firing rate. The size and pattern of the spot were then systematically varied to determine how such variations changed the firing pattern of the RGC.

Figure 4–24 shows two types of cells that Kuffler identified. For each of these cells, the receptive field can be described in terms of two concentric circles: a small circular center and a broader ring located around the center, called the surround. Both cells had basal firing in the absence of visual stimuli (top traces). Cell 1 was best excited by a small spot of bright light at the center, as measured by the number of action potentials fired during the stimulation period (second trace). As the diameter of the spot increased, the firing rate *decreased* (third trace); after the spot reached a certain size, further increase of the diameter no longer changed the firing rate. This is because light stimuli that fall in the small circle at the center of Cell 1's receptive field activate the cell, whereas stimuli that fall in the surround inhibit the cell. The inhibitory effect was best illustrated by the fourth stimulus: a central dark spot surrounded by light failed to elicit action potentials and instead

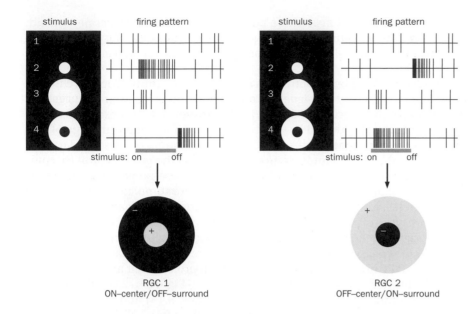

Figure 4–24 Receptive fields of two major types of retinal ganglion cells. A series of stimuli (1–4) were applied to the retina. The firing patterns for two different types of RGCs were recorded as action potentials (vertical bars) in response to these stimuli. The stimulus duration is indicated by the horizontal bars below the fourth action potential trace. The receptive fields of the two cells are characterized by two concentric circles: the center and the surround. +, excited by light; −, inhibited by light. (Adapted from Hubel DH [1995] Eye, Brain, and Vision [Scientific American Library, No 22], W.H. Freeman. See also Kuffler SW [1953] *J Neurophysiol* 16:37–68.)

decreased the RGC's basal firing rate. By contrast, Cell 2 displayed the opposite properties: it was best excited by the fourth stimulus that best inhibited Cell 1, and best inhibited by the second stimulus that best excited Cell 1. In addition to responses to light onset, both cells also responded to light offset—for example, inhibitory responses to the fourth light stimulus for Cell 1 and to the second stimulus for Cell 2 were both followed by increased firing shortly after light offset, before settling into the baseline firing patterns.

Kuffler found that it was not possible to convert one cell type to the other by changing stimulus conditions. Instead, Cell 1 and Cell 2 in Figure 4–24 represent two distinct types of RGCs. Using their response to light onset as the criterion, the Cell 1 type is termed ON-center/OFF-surround, while the Cell 2 type is termed OFF-center/ON-surround. We now know that there are many subtypes of RGCs within these two major types (see Figure 4–28).

That RGCs have **center–surround** receptive fields has important implications for visual information processing. RGCs do not simply respond to light, but start to analyze the spatial pattern, contrasting light and dark over a small area of the retina. Similar principles, namely that visual neurons respond to changes in light, were also found in earlier studies of invertebrates and lower vertebrates. But how do RGCs acquire their center–surround receptive field?

4.15 Bipolar cells are either depolarized or hyperpolarized by light based on the glutamate receptors they express

To understand how RGCs acquire their receptive field, we need a better understanding of information flow from cell to cell within the retinal circuit (see Figure 4–2). We use pathways downstream of cones as examples, as most of the retinal circuit is devoted to analyzing information delivered from the cones.

Before getting started, one important property of information processing in the retina is worth emphasizing. As discussed in Section 1.8, neurons use two forms of electrical signaling: all-or-none action potentials and graded local potentials. Most electrical signaling within the vertebrate retina utilizes graded potentials. Only RGCs and a subset of amacrine cells fire action potentials. Two disadvantages of graded potentials compared to action potentials are (1) reduced speed of conduction and (2) attenuation during conduction. However, with the exception of RGCs, retinal neurons synapse on partners that are located within the retina, so the distances that these graded potentials travel are short and the losses associated with them are not significant. On the other hand, graded signals can transmit more information than action potentials. This is because the unit of information transfer for graded potentials is individual synaptic vesicles as opposed to individual spikes for action potentials. Graded potentials can transmit more vesicles per second than action potentials can transmit spikes per second. So neurons within the retina use graded potentials to transmit information before the most behaviorally relevant information is selected for transmission to the brain by RGC axons.

Another useful concept to introduce when we trace information flow through neural circuits is the **sign** of signals. We can give any given neuron a sign (for example, positive for becoming depolarized in response to a sensory stimulus). If a second neuron is activated by the given neuron, it maintains the sign of that neuron (also becoming depolarized by the sensory stimulus); if a second neuron is inhibited by the given neuron, it inverts the sign of the given neuron (becoming hyperpolarized by the sensory stimulus). With these preparations, let's now examine what happens to bipolar cells when cones are stimulated by light.

Cones release the excitatory neurotransmitter glutamate that directly activates two types of postsynaptic target cells: the bipolar cells and the horizontal cells (see Figure 4–2A). Bipolar cells are themselves glutamatergic excitatory neurons and synapse directly onto RGCs. There are two major subtypes of bipolar cells, **OFF bipolar** and **ON bipolar**, which project their axons to different laminae (sublayers) of the inner plexiform layer (**Figure 4–25**). OFF bipolars express ionotropic glutamate receptors (see Section 3.15) and are depolarized by glutamate release from the photoreceptors. Therefore, OFF bipolars follow the sign of photoreceptors and are hyperpolarized by light. In ON bipolars, glutamate activates a metabotropic glutamate receptor and the G-protein cascade (see Section 3.18) that result in the closure of cation channels. Hence a sign inversion occurs between the photoreceptors and ON bipolars. ON bipolars are inhibited by glutamate release from the photoreceptors and are therefore depolarized by light.

These two distinct bipolar cell types enable the visual system to analyze dark-to-light and light-to-dark transitions in parallel. Parallel processing of information is a common strategy in sensory biology that we will expand later in this chapter and in Chapter 6 (see also Section 1.12).

4.16 Lateral inhibition from horizontal cells constructs the center–surround receptive fields

The second type of cell that receives direct synaptic input from the cone is the horizontal cell. Horizontal cells are inhibitory neurons that extend their processes laterally to connect with many photoreceptors (see Figure 4–2A). Individual

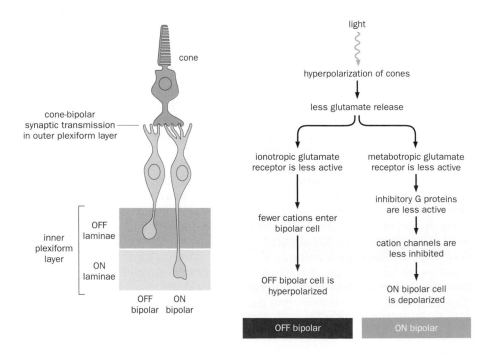

Figure 4–25 OFF and ON bipolar cells respond to light in opposite manners. Left, schematic of the pathway from cones to OFF and ON bipolar cells, which project their axons to OFF and ON laminae of the inner plexiform layer, respectively. Right, flow charts of responses to light stimulation from the cone to the OFF and ON bipolar cells. OFF bipolar cells maintain the sign of photoreceptors by using an ionotropic glutamate receptor. ON bipolar cells invert the sign of photoreceptors by using metabotropic glutamate receptors coupled to inhibitory G proteins, activation of which leads to closure of cation channels.

horizontal dendrites both receive excitatory input from the photoreceptors and send inhibitory output back to the presynaptic terminals of photoreceptors to reduce their glutamate release. (Note that a major exception to the general rule discussed in Section 1.7 regarding information being received by dendrites and sent by axons applies to the horizontal cells, as well as the amacrine cells we will soon discuss: dendrites can both receive and send information.) Furthermore, the inhibitory output also spreads to presynaptic terminals of neighboring photoreceptors (**Figure 4–26**). This inhibitory spread, or **lateral inhibition** (see Figure 1–21I), enhances the difference between the light-stimulated photoreceptor and the surrounding unstimulated photoreceptors, and provides a mechanism by which their downstream bipolar cells can acquire a center–surround receptive field. Indeed, lateral inhibition is used throughout the nervous system to sharpen the differences among parallel inputs and to increase signal-to-noise ratio as information is propagated through neural circuits.

Let's study how lateral inhibition produces a center–surround receptive field by examining three conditions of light input. First, suppose that a small spot of light is directed on a center cone that synapses onto a downstream bipolar cell

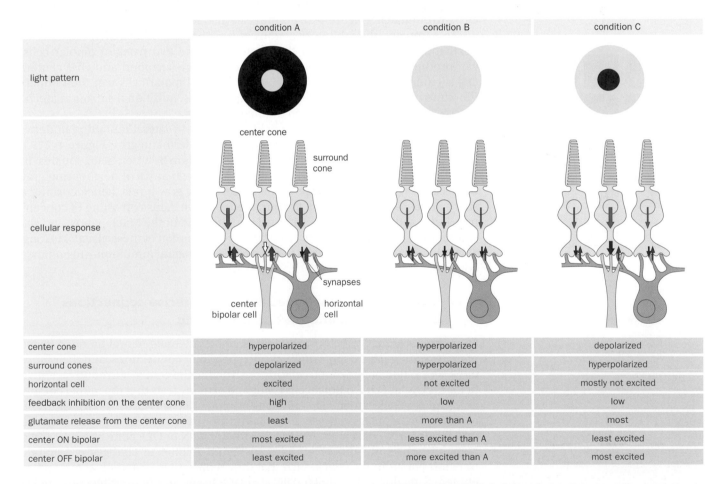

	condition A	condition B	condition C
center cone	hyperpolarized	hyperpolarized	depolarized
surround cones	depolarized	hyperpolarized	hyperpolarized
horizontal cell	excited	not excited	mostly not excited
feedback inhibition on the center cone	high	low	low
glutamate release from the center cone	least	more than A	most
center ON bipolar	most excited	less excited than A	least excited
center OFF bipolar	least excited	more excited than A	most excited

Figure 4–26 Bipolar cells acquire their center–surround receptive field through lateral inhibition of horizontal cells. Condition A: a small spot of light is directed on the center cone. The center cone is hyperpolarized whereas many surround cones (only two are drawn here for simplicity) are depolarized. The horizontal cell, which receives excitatory input from and sends inhibitory feedback output to many cones, is nearly maximally excited. The combined action of light and the horizontal cell causes the center cone to release the least amount of glutamate. Condition B: a larger spot of light covers both center and surround cones. All cones become hyperpolarized, which causes the horizontal cell to hyperpolarize and thereby reduces its inhibitory output. The center cone releases more glutamate than that

in condition A. Condition C: light activates the surround cones but not the center cone. The center cone releases most glutamate because of its own depolarized state and minimal inhibition from the horizontal cell. In all panels, sizes of green arrows represent the depolarization status of the cones as a consequence of light response (the larger, the more depolarized). Sizes of red arrows represent different levels of inhibition from the horizontal cell, which is a reflection of the amounts of glutamate the horizontal cell receives. Sizes of the black arrows represent the amount of glutamate release from the cones, which is affected positively by green arrows and negatively by the red arrows. The open arrow in condition A represents the least amount of glutamate release from the center cone.

(Figure 4–26, condition A). Light hyperpolarizes the center cone, causing it to release less glutamate. At the same time, the surround cones do not receive light input, are depolarized, and release lots of glutamate. Because each horizontal cell receives inputs from many cones, it is nearly maximally excited, and consequently inhibits all cones it connects with, including the center cone. Thus, due to intrinsic hyperpolarization by light, and inhibition by horizontal cells, the center cone releases the least amount of glutamate. Second, when the spot of light becomes larger (Figure 4–26, condition B), nearby cones also become hyperpolarized. The horizontal cell is less excited and its inhibitory action is diminished. This relief of inhibition, or **disinhibition**, of the center cone causes it to release more glutamate compared to condition A. Third, when the surrounding cones but not the center cone receive the light input (Figure 4–26, condition C), the horizontal cell is still mostly hyperpolarized because it receives many more inputs from the surround cones than the center cone. The center cone is still mostly disinhibited, while it is intrinsically depolarized due to not receiving light input. Therefore the center cone releases the highest level of glutamate. If the bipolar cell postsynaptic to the center cone were an ON bipolar, it would have a receptive field similar to the ON-center/OFF-surround RGC that Kuffler described. If the center bipolar cell were an OFF bipolar, it would have a receptive field similar to the OFF-center/ ON-surround RGC (Figure 4–26 bottom; see also Figure 4–24).

In summary, the inhibitory actions of horizontal cells and their lateral connectivity, in combination with the existence of the two types of bipolar cells, can produce bipolar cells that have ON-center/OFF-surround and OFF-center/ ON-surround receptive fields (**Movie 4–1**). These bipolar cells can then impart the same receptive fields to their postsynaptic RGCs. Additional surround mechanisms for RGCs can be mediated by amacrine cells.

The center–surround receptive fields mediated by lateral inhibition underlie some striking visual illusions. For instance, in the Hermann grid (**Figure 4–27**A), gray blobs appear at the intersections of white bands in between black squares. In the Mach bands (Figure 4–27B), the borders between any two neighboring gray bands appear to have a stronger difference in their gray levels, hence creating a sharp edge. As an exercise, try to compare receptive fields consisting of concentric circles at different locations in the Hermann grid or the Mach bands, and see whether you can come up with a satisfactory explanation of these visual illusions. The Mach bands also illustrate a useful function of lateral inhibition—to enhance edge detection.

4.17 Diverse retinal cell types and their precise connections enable parallel information processing

The retina is arguably the best studied piece of vertebrate neural tissue in terms of understanding the composition of cell types, their connections, and their functions. The action of horizontal cells in constructing the center–surround receptive fields for ON and OFF bipolar cells is but one of many feats that the retinal circuits accomplish. There are in fact ~10 types of bipolar cells (**Figure 4–28**). Of these, one—an ON bipolar—receives information from the rods, and the rest receive information from the cones. Among the different ON or OFF cone bipolars, some respond transiently and others in a sustained fashion to changes of synaptic transmission from the cones, such that they represent information across different temporal domains. Bipolar cells also vary in the size of their receptive fields and therefore the spatial coverage of photoreceptors.

The parallel streams of information carried by different bipolar cells are delivered to distinct types of RGCs through precise connections in the retinal circuits. For example, the ON and OFF bipolar cells form synapses respectively with ON and OFF RGCs at distinct laminae in the inner plexiform layer (see Figure 4–25, left). Some RGCs (bistratified or ON–OFF RGCs) elaborate their dendrites in two separate laminae and therefore receive input from both ON and OFF bipolars (for example, the right RGC in the penultimate row in Figure 4–28). The bipolar → RGC synapses are further modulated by the actions of ~30 different

(A)

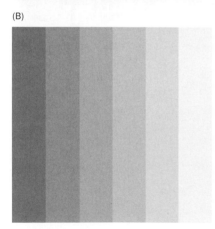

(B)

Figure 4–27 Visual illusions due to lateral inhibition. (A) The Hermann grid consists of black squares separated by vertical and horizontal white bands. Ghost gray blobs appear at the intersections of the white bands. **(B)** The Mach bands consist of a series of juxtaposed grayscale bands. At each border, the edge on the dark side appears darker and the edge on the light side appears lighter, thus exaggerating the contrast.

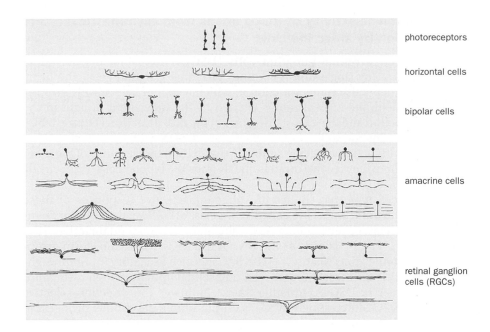

photoreceptors

horizontal cells

bipolar cells

amacrine cells

retinal ganglion cells (RGCs)

Figure 4–28 Cell types in the mammalian retina. Retinal cell types have been defined by a combination of morphological and physiological properties. Photoreceptors include rods and two or three types of cones with different spectral sensitivities. Horizontal cells provide lateral inhibition to construct the center–surround receptive fields of bipolar and ganglion cells. One type of horizontal cell (right) has separate axonal and dendritic processes based on their distances from the soma, which form reciprocal synapses with rods and cones, respectively. Different types of bipolar cells have receptive fields of different sizes, and connect with distinct types of RGCs via terminations of their axons in specific laminae of the inner plexiform layer. Amacrine and ganglion cells have the most types. Many amacrine and ganglion cells of the same type cover the retina once and only once, with varying lamina specificities and receptive field sizes. (From Masland RH [2001] *Nat Neurosci* 4:877–886. With permission from Macmillan Publishers Ltd.)

types of amacrine cells, most of which are GABAergic or glycinergic inhibitory neurons that elaborate their dendrites within the inner plexiform layer. (Like the horizontal cells introduced earlier, dendrites of amacrine cells both receive input and send output, and many amacrine cells do not have axons; in fact, the name amacrine means a cell without a long process, or an axon). As a consequence, different types of RGCs carry to the brain distinct visual information regarding contrast, size, motion, and color.

Interestingly, dendrites of many types of RGCs and amacrine cells cover the entire retina once and only once, in the same way that tiles cover a kitchen floor (**Figure 4–29**). This property is aptly named **dendritic tiling**; it allows a particular type of retinal neuron to sample the entire visual world without redundancy. With the introduction of new tools in molecular genetics in recent years (see Sections 13.10 and 13.12), an expanding number of RGC and amacrine types have been identified and their functions in retinal information processing determined. As specific examples, we discuss in the next two sections how retinal circuits extract motion and color information through the actions of specific retinal cells and their connection patterns.

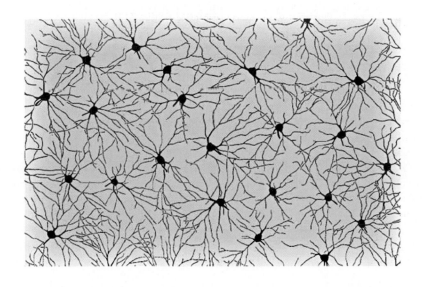

Figure 4–29 Dendritic tiling of a specific RGC subtype in the cat. RGC cell bodies are regularly spaced. Dendrites from neighboring cells have minimal overlap, and together they cover the retina uniformly. (From Wässle H [2004] *Nat Rev Neurosci* 5:1–11. With permission from Macmillan Publishers Ltd.)

4.18 Direction-selectivity of RGCs arises from asymmetric inhibition by amacrine cells

Direction-selective retinal ganglion cells (**DSGCs**) were first discovered in 1960s in rabbit using single-unit recordings of retinal ganglion cells. The firing rate of DSGCs depends on the direction in which a spot of light travels through the receptive field: the highest rate is elicited by the movement of light in the **preferred direction** and the lowest firing rate in the **null direction** (**Figure 4–30**). The most abundant and well-studied DSGCs are the ON–OFF DSGCs: they send one dendritic tuft each to a specific lamina in the ON and OFF laminae of the inner plexus layer, where they receive input from ON and OFF bipolar cells, respectively (**Figure 4–31**A).

A crucial cell type that shapes the direction-selective responses of these DSGCs is the **starburst amacrine cell** (**SAC**), a class of GABAergic inhibitory neurons that also releases acetylcholine. Two subclasses of SACs send their dendrites to the same ON- or OFF-lamina as the ON–OFF DSGCs, where they form extensive synaptic connections (Figure 4–31B). SACs receive excitatory input throughout their dendritic trees from bipolar cells, and send output from their distal third of dendritic trees to DSGCs. Ablating SACs eliminates the direction-selective response of DSGCs. Interestingly, a SAC's output from a dendrite is strongest in response to light stimulation in a centrifugal direction (excitation spreading from soma to the tip of the dendrite). So how do SACs regulate the direction selectivity of DSGCs?

We use an experiment that employed paired **whole-cell patch recording** of SAC and DSGC to illustrate their relationship. Whole-cell patch clamp recording not only can measure the DSGC output in the form of action potentials, but also can separately measure the excitatory and inhibitory synaptic inputs, and thereby can be used to investigate synaptic mechanisms that produce neuronal output (see Section 13.21 and Box 13–2 for details). Stimulating SACs and recording from DSGCs revealed that SACs did not produce detectable excitatory input to DSGCs, as depolarization of SACs (Figure 4–31C, bottom) did not produce appreciable excitatory postsynaptic synaptic currents in DSGCs (Figure 4–31C, middle). SACs produce inhibitory input to DSGCs that depended on the position of the cells in the retina (Figure 4–31C, top). Strong inhibition only occurred if the SAC was located on the null side of the DSGC, but not on the preferred side (see Figure 4–32B for illustrations of the null and preferred side).

Given the symmetrical shape of the dendritic field for both SACs and DSGCs (Figure 4–31B), how is this asymmetric inhibition achieved? This question was investigated using **laser-scanning two-photon imaging** of Ca^{2+} signals followed by **serial electron microscopic (EM) reconstruction** (see Sections 13.22 and 13.19 for details of these techniques, respectively). Two-photon imaging of Ca^{2+} signals enabled the determination of direction selectivity of many individual DSGCs in a retinal explant. Their synaptic connections with SACs were subsequently reconstructed in serial EM sections. It was found that SACs preferentially formed

Figure 4–30 Direction-selective ganglion cells (DSGCs). Single-unit recording of a ganglion cell in the rabbit reveals that it is most active when a light spot moves upward across its receptive field, and is mostly silent when the light spot moves downward. Arrows on the left indicate the direction in which the light moves. Numbers correspond to the spikes fired during the period of motion stimulus (bar at the bottom). (Adapted from Barlow HB, Hill RM, & Levick WR [1964] *J Physiol* 173:377–407.)

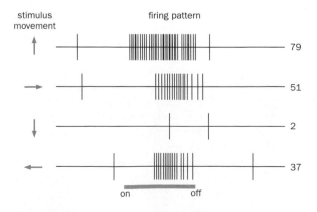

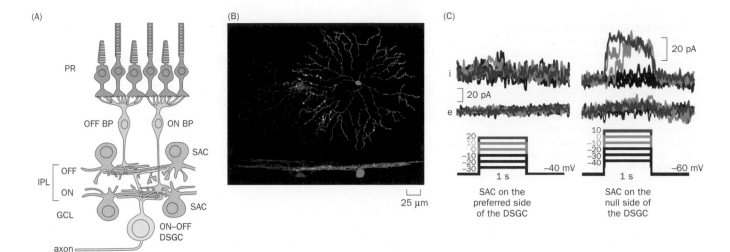

Figure 4–31 Asymmetric inhibition of direction-selective ON–OFF ganglion cells by starburst amacrine cells. (A) Schematic diagram of input to the ON–OFF direction-selective ganglion cell (DSGC). SAC, starburst amacrine cells. BP, bipolar cells. PR, photoreceptors. IPL, inner plexiform layer. GCL, ganglion cell layer. **(B)** Visualization of an ON–OFF DSGC (red) and a neighboring SAC (green) after fluorescent dye fill. The top view is taken from the retinal surface, while the bottom view shows a retinal cross section as in panel A. White dots indicate potential synaptic contacts between the two cells. **(C)** In the patch clamp experiment on an SAC/DSGC pair, excitatory (e) and inhibitory (i) inputs were separately measured in the DSGC (top two series of traces) in response to the depolarizing steps applied to the SAC (below). Strong inhibition occurred when SAC was located on the null side of the DSGC (right), but did not occur when SAC was located on the preferred side of the DSGC (left). No excitatory input was detected in either direction. (B, from Wei W & Feller MB [2011] *Trends Neurosci* 34:638–645. With permission from Elsevier Inc.; C, from Fried SI, Münch TA & Werblin FS [2002] *Nature* 420:411–414. With permission from Macmillan Publishers Ltd.)

synaptic connections onto the null side of the DSGC dendrites, despite being in physical proximity with DSGC dendrites in other directions (**Figure 4–32**A). This asymmetric connectivity provides a potential anatomical basis for asymmetric inhibition of DSGCs by SACs. The mechanism by which this asymmetric inhibition is established remains to be explored.

With these different lines of information, we can now piece together a model to account for direction selectivity of DSGCs. Let's use the ON response of a DSGC as an example to illustrate (Figure 4–32B). When light moves in the null direction, SACs located on the null side of the DSGC are excited first by ON bipolar cells. The centrifugal excitation of SACs' dendrites cause them to release GABA, and the preferential connections with the DSGC provide strong inhibitory input right before, and at the same time, as the DSGC receives excitatory input from bipolar cells. This effectively cancels out light-induced excitation. When light moves in the preferred direction, the DSGC receives excitatory input from bipolar cells before it receives GABAergic input from SACs because SACs located on the preferred side do not deliver strong inhibitory input. This sequence causes the maximal excitation of the DSGC.

4.19 Color is sensed by comparing signals from cones with different spectral sensitivities

As discussed in Section 4.11, humans and some primates have three types of cones: S, M, and L (see Figures 4–19 and 4–20). Light with different wavelengths differentially activates the three types of cones. How do these differential activations give rise to the perception of different colors?

Mechanisms similar to those that support the center–surround receptive field for detecting luminance contrast are used to compare differential excitations of cones of distinct spectral sensitivities, the output of which is delivered to the brain by **color-opponent RGCs**. Two major types of color-opponent RGCs have been identified, the blue–yellow and green–red, as presciently predicted by Hering's color opponent theory over a century ago (see Section 4.10). (The luminance contrast-detecting RGCs we discussed in Section 4.14 have mixed color input to

Figure 4–32 Asymmetric synaptic connections between SAC and DSGC and a model of direction selectivity. (A) The cumulative distribution of output synapses of 24 SACs is displayed relative to their respective somata (center). Synapses are color-coded according to the direction-selectivity of DSGCs (bottom left). SAC output synapses are preferentially found at the distal branches and align anti-parallel with the preferred direction of DSGCs. For instance, purple synapses (onto east-preferred DSGCs) are mostly to the west of the SAC soma. Thus, from the perspective of a DSGC, SAC input synapses are preferentially formed on DSGC's branches in the null direction. **(B)** A model that accounts for direction selectivity of a DSGC, whose directional preference (pref) is indicated by the arrow underneath. Although each DSGC receives input from many SACs, only two are shown for simplicity—one synapses on the dendrites on the preferred side of the DSGC (SAC1) and the other synapse on the null side (SAC2). Only the ON response is modeled. Both the DSGC and SACs receive excitatory synaptic input from ON bipolars (green dots). The DSGC receives preferential inhibitory input from the SAC (red dots) on the null side. The centrifugal preference of SACs' synaptic output is also indicated by the arrows. When the light stimulus moves from left to right, SAC1 is excited first, followed by the DSGC, and then SAC2. Although the dendrites of SAC1 are maximally excited, its inhibitory input to the DSGC is minimal because of the weak SAC1→DSGC connection. The inhibitory input from SAC2 to the DSGC comes after DSGC's own excitation, and is also diminished because the direction is opposite of the preferred direction for SAC2's dendritic output. Thus the DSGC is maximally activated. When the light stimulus comes from right to left, SAC2 is activated first, and its dendrites release most GABA because its preferred direction matches that of the light stimulus. Because of the strong SAC2 → DSGC connection, the DSGC receives strong inhibitory input from SAC2 right before, and at the same time, as it receives excitatory bipolar input. Inhibition cancels excitation (see Figure 3–29) and thus the DSGC is minimally activated. (A, from Briggman KL, Helmstaedter M & Denk W [2011] *Nature* 471:183–188. With permission from Macmillan Publishers Ltd; B, adapted from Wei W & Feller MB [2011] *Trends Neurosci* 34:638–645. With permission from Elsevier Inc.)

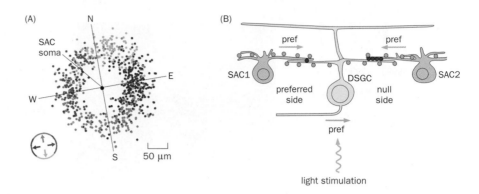

both center and surround, and hence are equivalent to the black–white opponency system in Hering's theory.) The blue–yellow opponent system is used by all mammals to compare short- and long-wavelength light, whereas the green–red opponent system is used by trichromatic primates to further distinguish the two long-wavelength colors. Below we first illustrate how the blue–yellow opponent system is established, followed by the red–green opponent system.

Glutamate release from each S-cone is already 'color opponent.' This results from a combination of its intrinsic phototransduction property and its inhibition by horizontal cells, using mechanisms analogous to the center–surround receptive field construction discussed earlier (see Figure 4–26). Specifically, when a retinal field that contains mosaic of S-, M-, and L-cones (**Figure 4–33**A) receives pure short-wavelength (blue) light input, S-cones are hyperpolarized, which reduces their glutamate release. Furthermore, horizontal cells are activated because M- and L-cones, which constitute the vast majority of all cones (see Section 4.11), are depolarized for receiving minimal light input; activation of horizontal cells further inhibits glutamate release from S-cones. As long-wavelength light (green, red, or a mixture which produces yellow) is added to the short-wavelength light input, horizontal cells become less excited; S-cones are disinhibited and release more glutamate even though their intrinsic phototransduction status is not changed. When the retinal field receives only long-wavelength light, S-cones release most glutamate because of their depolarization status and the disinhibition from horizontal cells. Thus, the output of S-cones is influenced by short-wavelength (blue) and long-wavelength (yellow) light in an antagonistic manner, or is blue–yellow color opponent (Figure 4–33A).

The S-cone output is received by a specialized type of **blue-ON bipolar cells**, which selectively connect with the S-cones (Figure 4–33B). These blue-ON bipolars inherit the color opponency of the S-cones, except that they are excited by blue light and inhibited by yellow light because of the sign inversion at the cone → ON bipolar synapse (see Figure 4–25). A major target of the blue–yellow color opponent system is the blue-ON/yellow-OFF **small bistratified RGCs**, which use their ON-layer dendrites to receive input from blue-ON bipolars, and their OFF-layer dendrites to receive input from OFF bipolars. Thus, the small bistratified RGCs are excited by blue light and inhibited by yellow light based on their input from blue-ON bipolars alone. This color opponency is further enhanced by their collection of input from OFF bipolars, which carry mostly additional yellow signals (Figure 4–33B).

In addition to the blue-ON/yellow-OFF RGCs, physiological studies have also identified blue-OFF/yellow-ON RGCs that are excited by yellow light and inhibited by blue light. Recent studies have suggested that they too derive their color opponency from the blue-ON bipolars, except that the connection between the bipolar and RGC involves an intermediate, a glycinergic inhibitory amacrine cell, which inverts the sign.

The green–red color-opponent signal in trichromatic primate retina is read out by **midget ganglion cells**. In the fovea, which contains the highest cone density, each midget RGC is excited by a single midget bipolar cell, which in turn

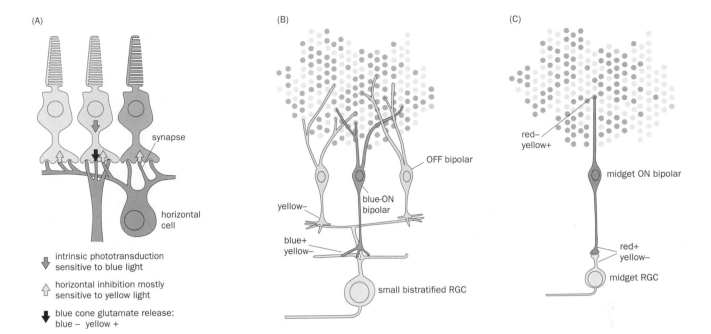

(A)

synapse

horizontal
cell

⇩ intrinsic phototransduction
sensitive to blue light

⇧ horizontal inhibition mostly
sensitive to yellow light

⬇ blue cone glutamate release:
blue − yellow +

(B)

OFF bipolar

yellow−

blue-ON
bipolar

blue+
yellow−

small bistratified RGC

(C)

red−
yellow+

midget ON bipolar

red+
yellow−

midget RGC

receives input from a single M- or L-cone (Figure 4–33C). Just like the S-cones (Figure 4–33A), the M- and L-cones themselves already produce color-opponent output as a consequence of their own phototransduction cascade that preferentially reports green or red light stimulus, and inhibition from horizontal cells reporting information from a mixture of green and red (and a small contribution from blue) stimuli. Thus, the green–red color opponent system, at least in the fovea, does not require further specialized retinal circuitry or cell types: the bipolars and RGCs simply inherit the color-opponency of the cone output. Perhaps because of this property, as will be discussed in Section 12.17, expression of a second long-wavelength cone opsin in mice, a species that normally has only one long-wavelength cone opsin, could enhance spectral discrimination of transgenic mice in the long-wavelength range.

4.20 The same retinal cells and circuits can be used for different purposes

None of the color-opponent RGCs we just discussed is dedicated exclusively to color vision. The green–red opponent system is also used for high-acuity vision in the center of the retina: the midget RGCs in the fovea receive information from a single cone, which offers these RGCs the ultimate spatial resolution (see Figure 4–33C). Trichromatic primates took advantage of this existing system for another use—green–red opponency—after the introduction of a new type of cone during their evolution.

Even the more ancient blue-ON/yellow-OFF opponent RGC system has other functions. In the dark, blue-ON/yellow-OFF RGCs collect information from rods. Indeed, even though rods are more numerous than cones, they do not have dedicated ganglion cells. They send signals through the rod bipolar cells, which are ON bipolars. Rod bipolars excite the **AII amacrine cells**, whose depolarization spreads to blue-ON bipolar cells via gap junction. Light signals from rods thus excite the blue-ON/yellow-OFF RGCs we just studied (**Figure 4–34**). This pathway provides the neural basis for a psychophysical observation: in dim light, visual scenes have a bluish hue.

Remarkably, at high light levels, when rods are saturated, the rod pathway can still be activated when the stimulation area is sufficiently large, but with an inverted sign. This is because light hyperpolarizes cones, reduces lateral inhibition of horizontal cells, and in turn depolarizes rods. The inverted rod pathway carries surround inhibition for the cone signals.

Figure 4–33 The blue–yellow and green–red color-opponent systems. **(A)** S-cone output is color opponent due to a combined effect of its intrinsic phototransduction and horizontal-cell-mediated lateral inhibition. It intrinsically responds to short-wavelength (blue) light. Because the surround consists mostly of M- and L-cones, the inhibitory signal from horizontal cells carry long-wavelength (yellow) signal, and hence the S-cone output can be designated as blue− yellow+ with reference to whether light increases (+) or decreases (−) its glutamate release. **(B)** The retinal circuit that transmits information from S-cones to blue-ON/yellow-OFF small bistratified RGC. Blue-ON bipolars collect information only from S-cones. They send excitatory input to the small bistratified retinal ganglion cells, and their output is blue+ yellow− as they invert the sign of the S-cones. The small bistratified RGC also receives excitatory input from OFF bipolars, which carry additional yellow− signal. **(C)** An example of the green–red color-opponent system in the trichromat fovea. The L-cone output is red− yellow+. The midget ON bipolar inverts the sign, such that the midget RGC is excited by long-wavelength (red) light and inhibited by medium-wavelength (green) light. (Adapted from Mollon JD [1999] *Proc Acad Natl Sci USA* 96:4743–4745. See also Dacey DM and Lee BB [1994] *Nature* 367:731–735; Verweij J, Hornstein EP & Schnapf JL [2003] *J Neurosci* 23:10249–10257; Packer OS, Verweij J, Li PH et al. [2010] *J Neurosci* 30:568–572.)

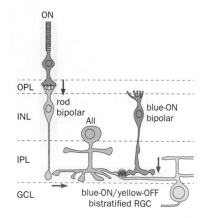

Figure 4–34 Rods piggyback on the cone circuit. Rod signals are transmitted by a single type of rod ON bipolar cell. When light is directed onto the rod, the rod becomes hyperpolarized. The sign-inverting rod → rod bipolar synapses (red arrow) depolarize the rod bipolar, which excites the AII amacrine cell through excitatory synapses (horizontal green arrow). The AII amacrine cell spreads depolarization to the blue-ON bipolar via gap junctions (symbolized by the zigzagged line). Depolarization of the blue-ON bipolar excites ON-center RGCs including the bistratified RGC through excitatory synapses (vertical green arrow). Thus, the blue-ON/yellow-OFF bistratified RGC is excited by the light signal through the rod. (Adapted from H. Kolb H & Famiglietti EV [1974] *Science* 186:47–49. See also Field GD, Greschner M, Gauthier JL et al [2009] *Nat Neurosci* 12:1159–1166.)

In summary, the retinal circuits and their output RGCs, while separating the visual information into parallel streams, are also multifunctional in processing and delivering to the brain different forms of information under different light conditions. **Box 4–2** provides another fascinating example of multifunctionality of retinal cells.

HOW IS INFORMATION PROCESSED IN THE VISUAL CORTEX?

We will now travel along the visual pathway from the eye to the brain. Axons of RGCs, which collect all output of the visual information from the retina, form the optic nerves (**Figure 4–35**A). At the **optic chiasm**, where the optic nerves from the left and right eyes converge, axons of RGCs on the **nasal** side (close to the nose) project **contralaterally** (refers to an axonal projection that crosses the midline and therefore terminates on the opposite side as the soma), whereas axons on the **temporal** side (close to the temple) project **ipsilaterally** (refers to an axonal projection that stays on the same side as the soma). (As will be discussed in Section 5.6, the fractions of RGC axons that cross—or do not cross—the midline differ between species.) After passing the optic chiasm, RGC axons continue to travel along the **optic tract**, while sending branches that terminate at several distinct central targets that serve different functions. These include the suprachiasmatic nucleus in the hypothalamus for the regulation of circadian rhythms (see Box 4–2), the **pretectum** in the brainstem for the regulation of pupil and lens reflexes and for reflexive eye movement, and the **superior colliculus** for the regulation of head orientation and eye movement. We will focus our discussion on one important central target, the **lateral geniculate nucleus** (**LGN**) of the thalamus, which acts as a processing station for information that is transmitted to the **visual cortex** for conscious vision.

4.21 Retinal information is topographically represented in the lateral geniculate nucleus and visual cortex

RGC axons form synapses with LGN neurons, which send their own axons into the visual cortex (Figure 4–35B). These axonal projections are organized

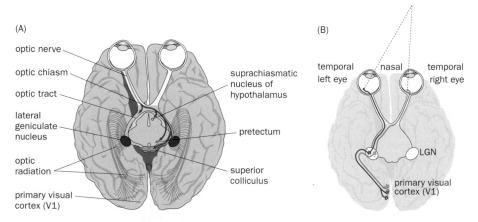

Figure 4–35 Organization of the visual pathways from the eye to the brain. (A) Visual system as seen from the bottom side of the brain. Axons of retinal ganglion cells (red line) together form the optic nerve, which travels past the optic chiasm and continues within the optic tract to reach distinct targets in the brain. These include the suprachiasmatic nucleus (SCN) of the hypothalamus that specializes in regulating the circadian rhythm in response to light, the pretectum that provides reflex control of pupil and lens, the superior colliculus that orients the movements of the head and eyes, and the lateral geniculate nucleus (LGN) in the thalamus that relays information to the visual cortex. For clarity, only RGC axons from the eye on the left side are drawn; a subset of those that stay on the left side of the brain after optic chiasm are incompletely drawn. Gray lines indicate the axons of LGN neurons projecting to V1, forming the optic radiation. **(B)** Illustrations of how visual information from the two eyes is organized. RGCs from the left and right eyes looking at the same point in space project their axons to similar locations but different layers in the LGN (illustrated are two layers separated by a dashed line). Visual information from the two eyes first converges in cortical neurons, either through directly convergent LGN axonal input from both eyes to the same cortical neuron (purple), or through convergent input from cortical neurons (red and blue) that receive input from only one eye.

Box 4–2: Intrinsically photosensitive retinal ganglion cells have multiple functions

The **intrinsically photosensitive retinal ganglion cells**, or **ipRGCs**, are an extreme example of the multifunctionality of retinal cells and circuits discussed in Section 4.20. Like other RGCs, ipRGCs extend dendrites into the inner plexiform layer and receive input from rods and cones via bipolar cells. They also send axons to common targets of many RGC types, such as the lateral geniculate nucleus that relays information to the visual cortex (see Section 4.21). Thus, ipRGCs participate in aspects of the retina's image analysis function just like other RGCs. But these cells also possess a novel trait.

What distinguishes ipRGCs from all other RGCs is that they can be directly activated by light! In other words, the ipRGCs are also photoreceptors. This remarkable discovery was made in 2002 from several convergent lines of evidence. In addition to forming images, light also plays an important role in entraining the circadian clock via the RGC axon projection to the **suprachiasmatic nucleus (SCN)** in the hypothalamus, a subject that will be discussed in more detail in Chapter 8. While the eye is essential for this function, blind mice in which rods and cones are degenerated can still sense light and entrain their clock. It was discovered first in the frog and then in the mouse that some RGCs express a special G-protein-coupled receptor called **melanopsin** (**Figure 4–36**A), which resembles opsins used in invertebrate vision (see Section 12.14 for details). Using mice that were genetically engineered to express an axon label specifically in melanopsin-expressing RGCs, researchers found that their axon terminals were highly enriched in the SCN (Figure 4–36B). Finally, both melanopsin-expressing RGCs and RGCs that project axons to the SCN were strongly depolarized by light (Figure 4–36C) even when the transmission from rods and cones was blocked completely, indicating that melanopsin-expressing RGCs are intrinsically photosensitive. Subsequent analyses of melanopsin knockout mice confirmed the function of ipRGCs in circadian clock entrainment. In addition, ipRGCs also play an essential role in the pupillary light reflex that enables the automatic adjustment of pupil size in response to different light intensity. Researchers are still discovering new ways in which ipRGCs impact the visual system.

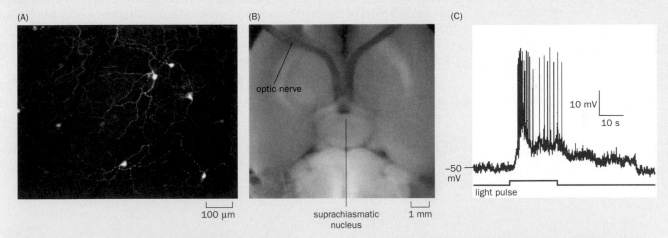

(A) 100 μm

(B) optic nerve   suprachiasmatic nucleus   1 mm

(C) 10 mV   10 s   –50 mV   light pulse

Figure 4–36 Intrinsically photosensitive retinal ganglion cells (ipRGCs). (A) Mouse retina stained with an antibody against melanopsin. Melanopsins are distributed throughout the dendritic trees of ipRGCs. **(B)** The optic nerve is visualized by the activity of β-galactosidase (blue), expressed from a transgene knocked in at the locus that expresses melanopsin, which labels the axonal projections of ipRGCs (see Section 13.7 for the knock-in method). Melanopsin-expressing ipRGCs terminate their axons at the suprachiasmatic nucleus. **(C)** Whole-cell patch clamp recording from an ipRGC retrogradely labeled by injecting fluorescent beads into the SCN. A light pulse (bottom trace) depolarizes the ipRGC, causing it to fire action potentials (top trace). This response persisted after synaptic transmission was blocked (not shown), indicating that ipRGCs are directly depolarized by light. (A, from Provencio I, Rollag MD & Castrucci A [2002] *Nature* 415:493–494. With permission from Macmillan Publishers Ltd; B, from S. Hattar S, Kumar M, Park A, et al [2002] *Science* 295:1065–1070. With permission from AAAS; C, adapted from Berson DM, Dunn FA & and Takao M [2002] *Science* 295:1070–1073. With permission from AAAS.)

topographically, such that RGCs that neighbor each other in the retina project to LGN neurons that neighbor each other, which in turn connect with neurons that neighbor each other in the **primary visual cortex** (also called **V1**, or striate cortex). **Retinotopy**—the topographical arrangement of cells in the visual pathway according to the position of the retinal ganglion cells that transmit signals to them—enables the spatial information in the retina to be faithfully represented in the brain, and is a fundamental organizational principle in the visual system. We will study how this is achieved in great detail in Chapter 5.

Since nasal RGC axons project contralaterally and temporal RGC axons project ipsilaterally the left LGN receives input from the temporal side of the left retina and nasal side of the right retina, both of which correspond to the right half of the visual field (Figure 4–35B). Thus, the LGN and visual cortex in the left half of the brain receive input from the right visual field, whereas the LGN and visual cortex in the right half of the brain receive input from the left visual field.

In addition to retinotopy, the primate LGN is organized into six layers that can be visualized by density of cell staining (**Figure 4–37**). Layers 1–2 have larger cell bodies (termed magnocellular layers) and receive input from RGCs with large receptive fields. Layers 3–6 have smaller cell bodies (termed parvocellular layers) and receive input from RGCs with small receptive fields. In addition, each layer receives input from one specific eye: layers 1, 4, and 6 receive contralateral input, whereas layers 2, 3, and 5 receive ipsilateral input. Thus, the LGN layers help segregate inputs of RGC axons according to their cell types and eye of origin. In other mammals with fewer LGN layers (such as cats and ferrets) or no anatomically apparent layers (such as rodents), inputs from RGCs representing different cell types and eye of origin are nevertheless segregated in LGNs, suggesting that this is a universal organizational principle. Importantly, individual LGN neurons represent information received from only one eye. Information from the left and right eyes is first integrated in the primary visual cortex (Figure 4–35B).

In primates, the fovea is highly expanded in V1; as the eccentricity increases (equivalent to the visual field moving toward the periphery, see Figure 4–14), the corresponding area in V1 for the unit area of the retina decreases (**Figure 4–38**). Thus, as in the sensory and motor homunculi discussed in Section 1.11, the visual cortex does not represent the retina proportionally. Visual stimuli from the central part of the visual field are processed in larger cortical space.

4.22 Receptive fields of LGN neurons are similar to those of RGCs

What visual stimuli best excite LGN neurons? David Hubel and Torsten Wiesel investigated this question using the same single-unit recording technique to monitor neurons from the LGN that their mentor Steve Kuffler had applied to the study of RGCs in the retina. In order to precisely control visual stimuli, they

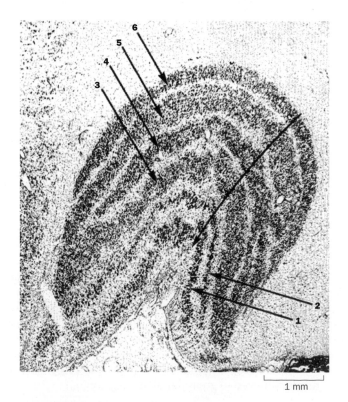

Figure 4–37 Cell layers in the monkey lateral geniculate nucleus. Nissl staining (see Section 13.15 for details), which labels cell bodies, shows that the monkey LGN is organized into six layers. Layers 1 and 2 have larger cell bodies, and are termed magnocellular layers; layers 3–6 have smaller cell bodies, and are termed parvocellular layers. Layers 1, 4, and 6 receive contralateral input, whereas layers 2, 3, and 5 receive ipsilateral input. The six maps of the visual field are in precise register; for example, the arrow in the center bisects LGN neurons in different layers that correspond to the same space in the visual field. (From Hubel DH & Wiesel TN [1977] *Proc R Soc Lond B* 198:1–59. With permission from the Royal Society.)

1 mm

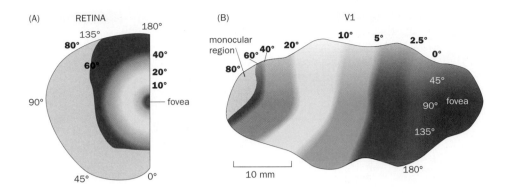

Figure 4–38 Topographic representation of the retina in the primary visual cortex (V1) of rhesus monkey. (A) The retina is diagrammed in a polar coordinate system, with the radius representing eccentricity (in bold) and circumference representing the dorsoventral position (0° being most ventral). **(B)** V1 is diagrammed in a Cartesian coordinate system, with the x axis representing eccentricity (bold) and the y axis representing the dorsoventral position. Colors in the V1 map correspond to the same colors in the retina map. For instance, representation of the fovea (purple), which occupies a small area in the retina, is highly enlarged in V1. As eccentricity increases, the corresponding area in V1 for a unit area of the retina decreases. Gray areas represent monocular region where only the left eye (shown) receives visual stimuli. The retinotopy of V1 was determined by extracellular single-unit recording of V1 neurons in anesthetized monkey in response to stimuli targeted to different parts of the retina. The recording sites were determined by histology, and V1 is represented as a two-dimensional map that maintains the spacing in the three-dimensional cortex. (Adapted from Van Essen DC, Newsome WT & Maunsell JHR [1984] *Vis Res* 24:429–448. With permission from Elsevier Inc.)

anesthetized their subject (cat) to minimize eye movement such that they could systematically map the receptive field by altering the locations of visual stimuli on a screen facing the anesthetized cat. LGN neurons were found to have similar response properties as RGCs. They have center–surround concentric receptive fields, with the ON-center/OFF-surround receptive field or the reverse (see Figure 4-24). Thus, to a first degree of approximation, LGN neurons represent visual information in a similar manner as RGCs.

LGN neurons likely function as much more than a simple relay station from the eye to the visual cortex. Indeed, only about 10% of synapses onto LGN neurons come from RGCs; the remaining synapses are from cortical or brainstem neurons, or from inhibitory neurons in the thalamus. The activity of LGN neurons can be modulated by cortical feedback and by the brain state, which enables the LGN to control information flow into the cortex. The thalamus is also well suited for multisensory integration, as information for all senses except olfaction passes through specific nuclei in the thalamus before entering the cortex. Much work is still needed to investigate these functions of the thalamus.

4.23 Primary visual cortical neurons respond to lines and edges

Although the receptive fields of LGN neurons resemble those of RGCs, Hubel and Wiesel encountered something very different when they applied the single-unit recording techniques to the primary visual cortex (V1), where LGN axons terminate. **Figure 4–39** and **Movie 4–2** illustrate the mapping procedure of the receptive field of a V1 neuron. After the microelectrode was inserted into the cortex, visual stimuli were projected onto a screen that represents the entire visual field of the cat to identify an area that could excite a cortical neuron next to the electrode tip. Then, the shape, size, and position of the stimulus were systematically varied with the aim of finding the stimuli that most effectively changed the firing pattern of the neuron being recorded. Circular spots of light were no longer the best stimuli no matter where they were placed and what sizes they were (Figure 4–39A). The best stimulus was a bar of light with a specific orientation; for example, the neuron being tested in Figure 4–39 responds best to a vertical bar (Figure 4–39B). Once the bar reached a certain height, further lengthening had no additional effect, but widening the bar in either direction reduced the firing rate. Thus, the receptive field of this cell can be drawn as shown in Figure 4–39C, with an ON-center region, flanked by two OFF-surround regions. Cells with such a receptive field in V1 are called **simple cells**.

Different simple cells can have distinct orientation preferences and sizes of the ON/OFF regions. The OFF regions on either side of the ON region are not necessarily symmetrical; in fact some cells have only one OFF region surround. Despite these variations, the receptive fields of simple cells share common properties: (1) they are spatially restricted to a narrow area of the visual field; (2) they are best stimulated by bars of light; (3) they are highly sensitive to orientations; and (4) they have ON–OFF antagonism. Diffuse light that covers both the ON and OFF regions does not excite these cells. Simple cells can therefore be considered as line or edge detectors in specific regions of the visual field.

(A)

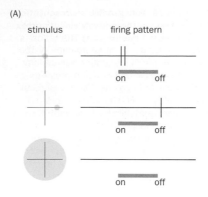

Figure 4–39 Receptive field of a simple cell in the cat primary visual cortex (V1). (A) In response to visual stimuli at left, the vertical bars in the plots at right represent action potentials recorded by extracellular electrodes; stimulation durations (1 s) are represented as horizontal bars below the action potential plots. Circular spots of various sizes elicit minimal response in this V1 simple cell. **(B)** Simple cells are most responsive to a bar of light with a specific orientation. In this example, a bar of light oriented vertically elicits the most intense burst of action potential firing during the stimulus period. Response plots are formatted as in (A). **(C)** The receptive field of the cortical neuron in (B). Triangles represent excitation, and circles represent inhibition. (Adapted from Hubel DH & Wiesel TN [1959] *J Physiol* 148:574–591.)

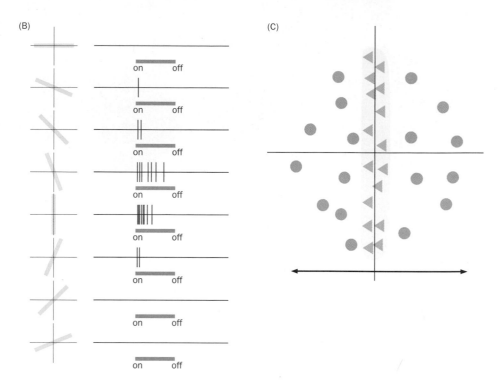

A different class of cells, also encountered in V1, is termed **complex cells**. These have more complex and more variable receptive fields than simple cells (see **Figure 4–40** for a specific example). In general, complex cells do not have mutually antagonistic ON and OFF regions as do simple cells. But like simple cells, complex cells are also highly orientation-selective, responding to light bars on a dark background or dark bars on an illuminated background. Complex cells on average have larger receptive fields than simple cells, and a bar with a specific orientation falling on any part of the receptive fields would excite the cell. Thus, complex cells can be considered more abstract line/edge detectors than simple cells, as they generalize over space, responding to the line/edge ON or OFF anywhere in the receptive field.

4.24 How do visual cortical neurons acquire their receptive fields?

Hubel and Wiesel proposed that the receptive field of a simple cell can form by combining the receptive fields of a series of LGN neurons arranged in a line as a consequence of their specific connection patterns (**Figure 4–41**A). Likewise, the receptive field of a complex cell can be constructed by combining a series of simple cells located in adjacent areas with the same orientation (Figure 4–41B). These proposals have been called a feed-forward model, as the receptive field of a neuron is shaped primarily by neurons at an earlier visual processing stage. Accordingly, receptive fields of neurons along the visual pathway are organized in a hierarchy, from light intensity detection (photoreceptors), to contrast detection (RGC, LGN), to line and edge detection (simple and complex cells in V1). Through each transformation, light information comes closer to the ultimate purpose of vision we emphasized at the beginning of the chapter: extraction of features that are useful to guide animal behaviors essential for survival. By encoding lines and edges in individual V1 neurons, it is more efficient to send this information to multiple higher visual cortical areas for specialized functions such as face recognition and motion detection, as will be discussed later.

More than half a century since these classic studies, the neural circuit basis of these feed-forward models in the visual cortex is still not fully elucidated, although experimental data are consistent with simple cells deriving their receptive field properties mostly from feed-forward projections of LGN neurons. It is also unclear how these receptive fields form during development. Alternative

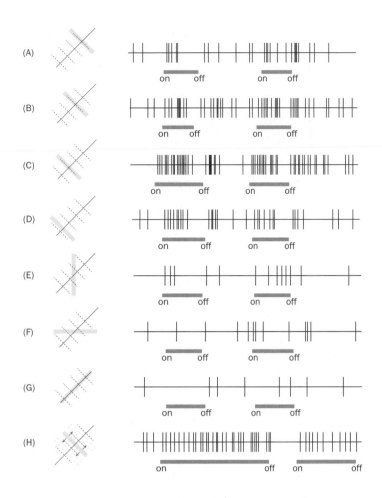

Figure 4–40 Receptive field of a complex cell. This complex cell is best excited by light bar oriented at a 135° angle. It is not sensitive to the exact position of the bar within a relatively broad area (**A–D**), is highly orientation selective (**E–G**), and responds to motion of the 135°-bar within the receptive field (**H**). Horizontal bars below each of the firing pattern plots indicate the period of stimulation. (Adapted from Hubel DH & Wiesel TN [1962] *J Physiol* 160:106–154.)

models have been proposed, which emphasize the contribution of either the role of cortical inhibition, or inter-connections among V1 cortical neurons. The resolution of these models will require a combination of physiology and circuit tracing exemplified by the study of direction-selective RGCs in the retina discussed in Section 4.18. In the cortex, these approaches remain highly challenging from a technical standpoint. Regardless of the final answers, it is clear that the responses of cortical neurons must reflect circuit properties, which include the physiological characteristics of individual neurons and their precise wiring pattern in the visual cortex.

4.25 Cells with similar properties are vertically organized in the visual cortex

As we introduced in Chapter 1, the cerebral cortex is organized into different areas that serve distinct functions. Within V1, the first organization is represented by the retinotopy: specific parts of the visual cortex receive information from specific parts of the retina and therefore represent specific parts of the visual

Figure 4–41 Models of receptive field construction for simple and complex cells. (A) The receptive field of a simple cell can arise from the convergent input of a series of LGN neurons, so that several circular receptive fields in combination form a line. **(B)** The receptive field of a complex cell can arise from convergent input of a series of spatially aligned simple cells with the same orientation selectivity. These hypotheses have not been proven. (Adapted from Hubel DH & Wiesel TN [1962] *J Physiol* 160:106–154. With permission from the Physiological Society.)

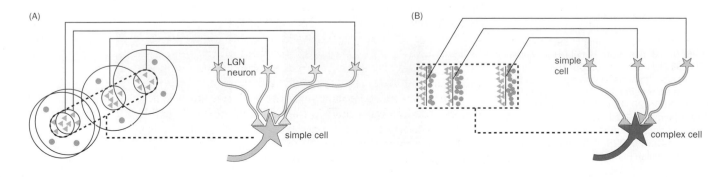

field (see Figure 4–38). The second level of organization is so-called **functional architecture**: within a given retinotopic area, neurons that share similar properties are further organized along the axis perpendicular to the cortical surface.

When Hubel and Wiesel performed their single-unit recording studies, they not only recorded the visual responses of individual cells, but also their positions in the cortex. In a typical experiment, they inserted an electrode from the cortical surface until they found a responding cell, measured its receptive field and orientation selectivity, then advanced the electrode further until they found and recorded the next cell, and so on. At the end of the recording, a large amount of current was passed to create a lesion marking the end of the electrode, and brain was examined in histological sections afterwards (**Figure 4–42**A). Plotting the orientation selectivity of individual cells along the electrode path revealed a remarkable organization. If the electrode was inserted perpendicular to the cortical surface, then cells exhibited similar orientation selectivity at different depths of the cortex. However, if the electrode was inserted at an oblique angle to the surface of the cortex, then the orientation selectivity of recorded cells changed gradually (Figure 4–42A). Together with earlier findings in the somatosensory cortex, these studies suggest that the cerebral cortex is organized as vertical columns from the surface to the white matter underneath, and cells within the same column share similar properties.

Modern imaging experiments have greatly improved the resolution at which orientation columns can be observed. Whereas single-unit recordings can only record one cell a time, optical imaging can record activities of many neurons simultaneously (see Section 13.22 for details). For example, **intrinsic signal imaging** utilizes metabolic activities, including blood flow and oxygenation levels, as markers of neuronal activity. When applied to the visual cortex in response to stimuli of different orientations, intrinsic signal imaging revealed that orientation-selective columns are organized in a pinwheel-like structure in cats and monkeys. The transitions between different orientations are continuous (rather than discrete as Hubel and Wiesel originally imagined), except at the center of pinwheels, where cells with different orientations converge at singular points

Figure 4–42 Orientation columns in the cat visual cortex revealed by three different methods. (A) Two examples of single-unit recordings followed by histological tracing of the electrode paths. Lines intersecting the electrode paths represent the preferred orientation, with longer lines representing individual cortical cells and shorter lines representing regions with unresolved background activity. The left electrode penetration path is more perpendicular to the cortical surface, and the orientation preferences change little for different cells along the path. The right electrode penetration path is more oblique, and orientation preferences change markedly along the path. **(B)** Surface view of orientation columns visualized by intrinsic signal imaging. Visual stimuli with different orientations (color coded at right) resulted in activation of different cell patches superimposed onto the same image. The white square highlights a pinwheel, with the center of the square representing the center of the pinwheel. **(C)** Surface view of orientation columns near the pinwheel center visualized by two-photon Ca^{2+} imaging. Visual stimuli using bars with different orientations (right) increased Ca^{2+} dyes fluorescence, which enabled individual cells with specific orientation selectivity to be mapped. This image is an overlay of cells recorded 130 to 290 μm from the surface corresponding to layers 2 and 3 of V1; the significance of these layers is discussed further in Section 4.26. (A, adapted from Hubel DH & Wiesel TN [1962] *J Physiol* 160:106–154. With permission from the Physiological Society; B, from Bonhoeffer T & Grinvald A [1991] *Nature* 353:429–431. With permission from Macmillan Publishers Ltd; C, from Ohki K, Chung S, Kara P et al. [2006] *Nature* 442:925–928. With permission from Macmillan Publishers Ltd.)

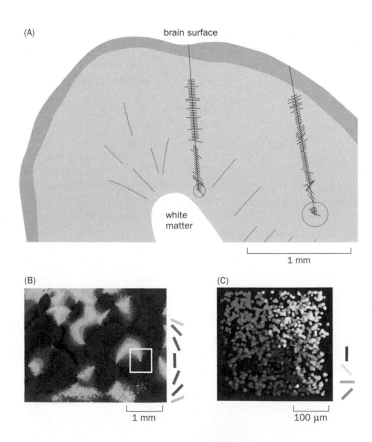

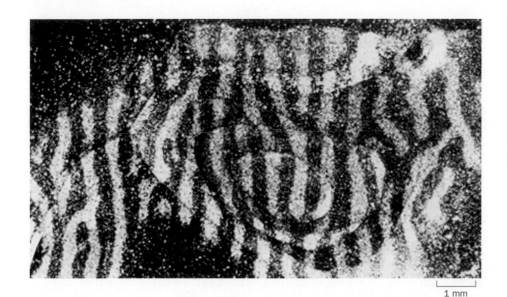

1 mm

Figure 4–43 Ocular dominance columns in the primary visual cortex. Autoradiograph of V1 layer 4c of a monkey whose one eye was injected with radioactively labeled amino acids. Populations of neurons that received input from the injected eye (white stripes) alternate with populations of neurons that received input from the un-injected eye (dark stripes). (From Hubel DH & Wiesel TN [1977] *Proc R Soc Lond B* 198:1–59. With permission from the Royal Society.)

(Figure 4–42B). Two-photon fluorescence imaging of intracellular Ca^{2+} concentration changes, which permits neuronal activity to be recorded with single-cell resolution, further demonstrated that the orientation columns and pinwheel centers are highly ordered at the level of individual neurons (Figure 4–42C). The significance of this pinwheel organization and how such a precise organization is constructed are not yet understood.

Vertical organization was also found when cortical neuron responses to the left and right eyes were examined in the primary visual cortex. Whereas individual LGN neurons respond to stimuli from one eye, individual visual cortical neurons can respond to input from both eyes that corresponds to the same spatial position (see Figure 4–35B), creating **binocular vision**, which is important for depth perception. However, many cortical neurons in the input layer 4 (see next section) still have a strong preference for input either from the left or the right eye, a property termed **ocular dominance**. When Hubel and Wiesel examined their electrode tracks, they imagined cells with the same ocular dominance to form column-like structures and referred to them as ocular dominance columns, a name that is still used today. Further studies using a technique called trans-neuronal tracing revealed that regions of similar responses are not narrow like columns but have a more extensive lateral structure. In this technique, radioactively labeled amino acids were injected into one eye. They were taken up and incorporated into proteins in retinal neurons, transported by RGC axons to the LGN, and then further transferred to V1 through the RGC → LGN synapses and LGN axons. Eventually, LGN axons and cortical neurons connected to the injected eye were preferentially labeled by radioactivity. V1 sections were then exposed on a film to reveal the spatial distribution of radioactive labeling (**Figure 4–43**). This yielded a striking image of strongly labeled stripes (representing input from the injected eye) alternating with largely unlabeled stripes (representing input from the un-injected eye).

Not all mammals exhibit anatomically recognizable orientation or ocular dominance columns. Ocular dominance columns are not evident even in some primate species. In rodents, many individual cells are orientation-selective, but no apparent columnar organization is discernible: neurons with the same orientation selectivity do not cluster, and neighboring neurons often have distinct orientation selectivity (for example, compare **Figure 4–44** with Figure 4–42C). At present, researchers can only speculate on the reason for these species differences. One hypothesis is that grouping of neurons with similar properties increase the sharpness of orientation discrimination, which cats are better at than rats. The second hypothesis is that only when the visual cortex is large enough does it become necessary to group together neurons of similar properties, so as to limit the length of the axons needed to connect them. A third hypothesis is that columns do not

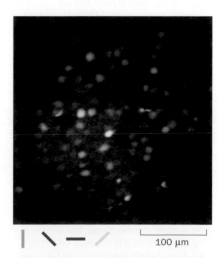

100 μm

Figure 4–44 Orientation-selective cells in the rat primary visual cortex. Cells are color coded according to their orientation selectivity (below). In the rat, cells with the same orientation do not form organized columns but instead appear scattered. (From Ohki K, Chung S, Ch'ng YH et al. [2005] *Nature* 433:597–603. With permission from Macmillan Publishers Ltd.)

serve specific physiological functions, and are simply developmental by-products of large cortices. Whatever the outcomes of these debates may be, the vertical organization of neurons with similar properties—whether clustered into columns or not—appears to be a cortical organization motif. We now turn to how signals flow among these vertically organized cortical neurons.

4.26 Information generally flows from layer 4 to layers 2/3 and then to layers 5/6 in the neocortex

The mammalian **neocortex** (the largest part of the cerebral cortex, including the visual cortex) is typically divided into six layers based on cell density differences that can be visualized using histological stains (**Figure 4–45**A). Studies of the visual cortex using *in vivo* intracellular recording (see Section 13.21 for details) have provided general insights into how information flows between different layers of the neocortex. Intracellular recordings can not only map the receptive field as single-unit extracellular recordings do but also fill the recorded neurons with histochemical tracers such that their dendritic and axonal projection patterns can be reconstructed. These experiments revealed that neurons located at different layers of the primary visual cortex have distinct dendritic and axonal projection patterns (Figure 4–45B), which can be used to deduce a model of information flow (Figure 4–45C).

Let's start with the LGN input to the visual cortex. The LGN axons from the thalamus predominantly terminate in layer 4 of the primary visual cortex. Depending on the types of LGN neurons, their axons terminate in different sub-layers within layer 4 (axons of cells *a* and *b* in Figure 4–45B). Layer 4 cortical neurons (for example, cells *c* and *d*) restrict their dendrites mostly within layer 4 so as to receive the LGN input; they send their axons predominately within their own layer and vertically to layers 2/3. Layer 2/3 neurons (for example, cell *e*) send their axons within layers 2/3, to layer 5, and through the white matter to other visual cortical areas. Layer 5 neurons (for example, cell *f*) project their axons to layer 6, to other cortical and subcortical areas such as the basal ganglia, and to superior colliculus for cortical control of eye movement. Layer 6 neurons (for example, cell *g*) send axons back to layer 4, and to the LGN, providing feedback control of LGN input. Thus, information flows generally from layer 4 to layer 2/3 and then to layer 5/6 within the cortex. Physiological data support this model of general information flow. For instance, simple cells are enriched in layer 4, consistent with them being direct recipients of LGN input. Complex cells are not usually found in layer 4, but are present in layers 2/3 and layers 5/6, consistent with them receiving input from simple cells.

Figure 4–45 Information flow in the primary visual cortex. (A) Nissl staining of human V1 revealing that cortical neurons are organized in distinct layers based on cell size, shape, and density. **(B)** Representative axons of the cat LGN neurons that provide input to V1 (*a–b*), as well as axonal (red) and dendritic (blue) morphology of V1 neurons from layers 4ab (*c*), 4c (*d*), 2/3 (*e*), 5 (*f*), and 6 (*g*), respectively, revealed by tracer injected during intracellular recordings. Axons and dendrites were distinguished from each other according to their morphological characteristics. **(C)** Schematic representation of information flow, with the assumption that overlapping axons and dendrites in the same cortical layer form synaptic connections. The cells illustrated in panel B correspond to the cells immediately beneath them that are diagrammed in panel C. The feed-forward information flow within V1 is highlighted in red. (A, adapted from Brodmann K [1909] *Vergleichende Localisationslehre der Grosshirnrinde in ihren Prinzipien dargestellt auf Grund des Zellenbaues* [Barth: Leipzig]; B & C, adapted from Gilbert CD [1983] *Ann Rev Neurosci* 6:217–247. With permission from Annual Reviews.)

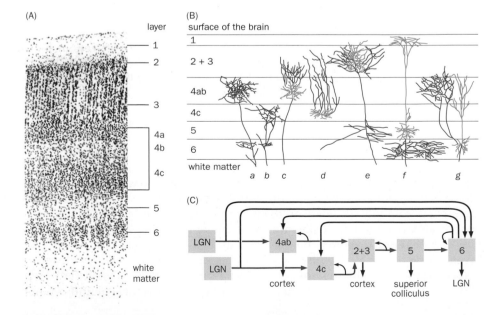

It is useful to keep in mind that reality is more complex than the LGN → L4 → L2/3 → L5 → L6 feed-forward model, given the extensive interconnections of neurons within the same layer (recurrent connections; see Box 1-2), the complex dendritic trees for some neurons that span multiple layers, and the participation of many kinds of inhibitory neurons. Nevertheless, information flow within an area of the neocortex, first outlined in studies of V1, appears to apply to other sensory cortical areas as well (**Box 4–3**). Identifying the principle of information processing in the cerebral cortex remains one of the most important challenges in modern neuroscience.

Box 4–3: Cracking neocortical microcircuits

We've learned that visual signals at V1 follow a general direction from thalamus → L4 → L2/3 → L5/6. Since the mammalian neocortex has the six-layered organization, is this pattern of information flow a general property of neocortex? Studies in other cortical areas have lent some support for this proposal. For example, in the mouse somatosensory cortex that represents whiskers (see Box 5-3 for more details), excitatory connections between neurons in different layers have been determined by paired whole-cell patch clamp recording in brain slices. As summarized in **Figure 4–46**, the signal flow, as determined by the direction and strength of synaptic connections between pair of neurons, follows the general direction of L4 → L2/3 → L5/6. Other prominent connections exist within specific cortical areas, including direct connections from L4 to L5/L6 in the somatosensory cortex (Figure 4–46) and L6 → L4 in the visual cortex (see Figure 4–45C), but the comparison of V1 and somatosensory cortex supports the notion of a 'canonical microcircuit' that applies to information flow within a neocortical area. Depending on the specific input to and output from a given cortical area, there could be significant variations on this canonical microcircuit in different parts of the neocortex. For example, primary sensory cortices tend to have a well-elaborated layer 4, reflecting its roles in receiving and processing thalamic sensory input. The primary motor cortex, on the other hand, has a less prominent layer 4 but a more prominent layer 5, reflecting the importance of its subcortical projections for motor control (see Chapter 8).

In addition to excitatory glutamatergic neurons, which constitute about 80% of cortical neurons and have been the focus of our discussion so far, GABAergic neurons, which make up most of the remaining 20% of cortical neurons, also play many essential functions in shaping cortical activity. Most cortical GABAergic neurons act locally: they receive local and long-range excitatory input plus local inhibitory input from other GABAergic neurons, and they send inhibitory output to nearby glutamatergic or GABAergic neurons. Cortical GABAergic neurons can be subdivided into many subtypes based on their morphology, electrophysiological properties, layer and synapse distributions, and expression of molecular markers (see Figure 3–46).

We use a specific example of investigating visual information processing in mouse V1 (**Figure 4–47**A) to integrate several concepts we have learned, and to showcase modern techniques for mapping circuit function. A powerful approach to investigate the function of specific neuronal types in circuits is to selectively enhance or silence their activity and quantitatively measure the consequences. Researchers have established many transgenic mouse lines that express the Cre recombinase in specific neuronal populations. These Cre lines can be used in combination with viral transduction of Cre-dependent expression of effector proteins (see Sections 13.10–13.12 and 13.23–13.25). Some of the most powerful effectors are light-activated ion channels and pumps from single-cell eukaryotes and prokaryotes (see Section 12.13), which enable researchers to control neuronal activity by shining light directly on neurons that express these effectors. For example, blue light directly activates neurons expressing the effector **channelrhodopsin-2 (ChR2)**, a light-activated cation channel originating from single-cell green alga. Orange light directly silences neurons expressing different effectors, **archaerhodopsin** and **halorhodopsin**, which are proton and chloride pumps from archaea, respectively. Genetic targeting of these effectors

(Continued)

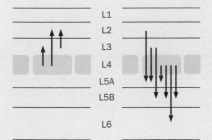

Figure 4–46 Connectivity between excitatory neurons of mouse primary somatosensory cortex. Arrows indicate the origin and destination of direct connections as determined by paired whole-cell patch clamp recording (see Section 13.26). Rectangles in layer 4 represent anatomically distinct areas within the somatosensory cortex that represent individual whiskers; a detailed discussion about the organization of the somatosensory cortex representing whiskers can be found in Box 5–3. Layer 5 has been subdivided to L5A and L5B. (Adapted from Lefort S, Tomm C, Floyd Sarria JC et al. [2009] *Neuron* 61:301–316. With permission from Elsevier Inc.)

Box 4–3: Cracking neocortical microcircuits

allows researchers to control the activity of specific neuronal populations by light stimulation, a procedure referred to as **optogenetics** (see Section 13.25).

When V1 layer 6 pyramidal neurons engineered to express ChR2 were artificially activated by blue light, visual responses of V1 neurons from other layers were attenuated during the blue light stimulation period (Figure 4–47B, top). When layer 6 neurons engineered to express archaerhodopsin and halorhodopsin were silenced by orange light, visual responses of V1 neurons from other layers were enhanced (Figure 4–47C, top). Interestingly, the orientation selectivity of other cortical neurons was unaffected

by manipulating the activity of layer 6 neurons, as seen by the similarities in their tuning curves despite the magnitude differences (Figure 4–47, bottom panels). Thus, both loss-of-function (neuronal silencing) and gain-of-function (neuronal activation) experiments suggested that layer 6 modulates the rate at which the firing frequency of other cortical neurons increases in response to increasing excitatory input. This function is called **gain control** and is discussed at greater length in Section 6.9. It remains to be determined whether layer 6 neurons in other cortical areas serve a similar function.

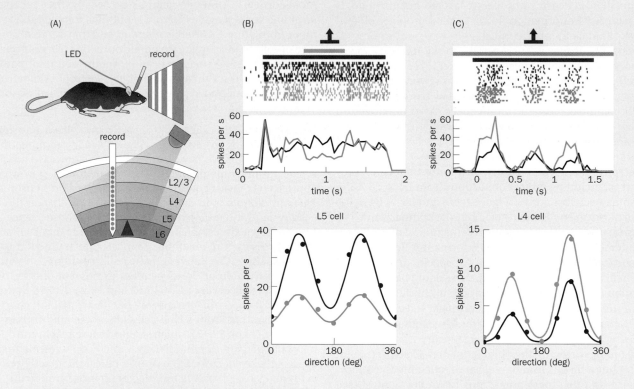

Figure 4–47 Effects of activating or silencing layer 6 pyramidal neurons on visual response of V1 neurons. (A) Experimental setup. An anesthetized transgenic mouse that expresses the Cre recombinase in a large subset of layer 6 pyramidal neurons was given visual stimuli of drifting gratings on the screen. An extracellular multi-electrode array was inserted into V1 to record visual responses of neurons in various layers (L2/3 to L6; each dot corresponds to a recording site). A light-emitting diode (LED) was used to photostimulate channelrhodopsin-2 (ChR2) with blue light, or a mixture of archaerhodopsin and halorhodopsin with orange light in separate experiments (not shown). These optogenetic effectors were expressed from viral vectors in a Cre-dependent manner, such that these effectors were only expressed in a subset of layer 6 pyramidal neurons (red triangle). **(B)** A layer 5 (L5) cell in mouse V1 is activated by an upward drifting grating (indicated by the arrow on the top). The black horizontal bar represents the visual stimulation period. Individual spikes are shown below the horizontal bar as vertical ticks; each row is a repetition. Rows in blue indicate trials in which a blue-light stimulation was given (during the time indicated

by the blue horizontal bar at the top); blue light depolarizes layer 6 neurons that express ChR2. The plot below quantifies the spike frequency. As is evident, the spike frequency of this L5 cell is decreased by blue light activation of L6 neurons. The bottom plot is a tuning curve of the L5 cell, obtained by stimulating the neuron with drifting gratings of different orientations (90° represents motion in the upward direction, 270° represents motion in the downward direction). Thus, optogenetic activation of layer 6 neurons does not change the orientation selectivity of the L5 neurons, but reduces its response magnitude. **(C)** Similar experiments as in (B) except that visual response was recorded from a layer 4 (L4) cell, and that orange light was applied to suppress the activity of layer 6 neurons that express a mixture of archaerhodopsin and halorhodopsin throughout the visual stimulation period. Optogenetic silencing of layer 6 neurons does not change the orientation selectivity of the L4 neurons, but enhances its response magnitude. (Adapted from Olsen SR, Bortone DS, Adesnik H et al [2012] *Nature* 483:47–54. With permission from Macmillan Publishers Ltd.)

4.27 Visual information is processed in parallel streams

Vision consists of perception of different aspects of the visual world. As discussed in Section 4.17, visual information is processed in parallel in the retina. At the level of the retinal output, at least 20 different RGC subtypes, each having different morphology and connectivity within the retina, carry different kinds of processed information to the brain. Although some RGCs project their axons to the hypothalamus or superior colliculus for non-conscious functions (see Figure 4–35A), most RGC subtypes also send information via the LGN to the cortex, eventually giving rise to conscious perception of object color, shape, depth, and motion. How are these different streams of information organized?

Traditionally, based on major RGC subtypes and their respective projections to different LGN layers, visual pathways have been divided into two main components: the P and M pathways (**Figure 4–48**; see Section 4.21). The **P pathway** originates from RGCs that have small receptive fields (such as the midget cells) and carries information about high-acuity and color vision. The **M pathway** originates from RGCs that have large receptive fields and carries information about

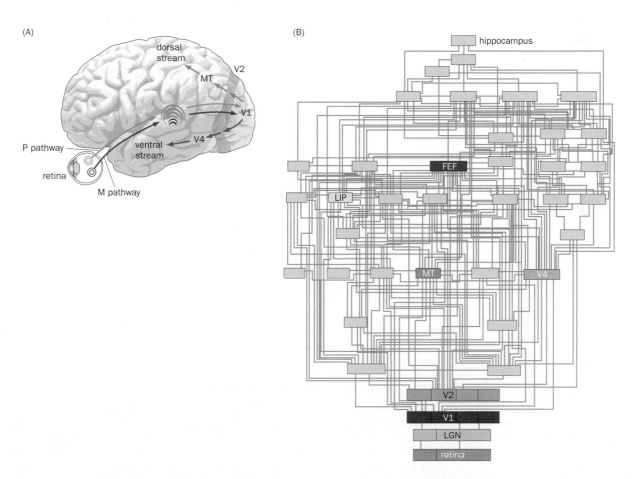

Figure 4–48 Parallel organization of the visual pathways.
(A) Schematic of two major parallel pathways from the eye to the cortex in primates. The P and M pathways start with retinal ganglion cells with small (P) and large (M) receptive fields. The P and M pathways engage different sublayers in the LGN and in layer 4 of the primary cortex. After V1, information diverges into the dorsal and ventral streams. The dorsal stream derives information mainly from the M pathway, and is responsible for analyzing motion and depth. The ventral stream derives information equally from the P and M pathways, and analyzes shape and color. V2 (gray band), an area immediately adjacent to V1 that relays information to both ventral and dorsal streams. V4, a major analysis station in the ventral

stream. MT, the middle temporal visual area specializes for analyzing motion signals (see Section 4.29 for more details). **(B)** A hierarchical processing model based on the connection diagram of different visual cortical areas derived from anatomical tracing experiments in rhesus monkey. The >30 areas form >300 connections, mostly reciprocal, and establish highly intertwined cortical processing streams. The dorsal stream going to the parietal lobe is to the left, and the ventral stream going to the temporal lobe is to the right. FEF, frontal eye field; LIP, lateral intraparietal area (see Section 4.29 for more details). (B, adapted from Felleman DJ & Van Essen DC [1991] *Cereb Cortex* 1:1–47. With permission from OUP.)

luminance and has excellent contrast and temporal sensitivity; thus M signals are particularly appropriate for computing motion. Recent findings of new RGC subtypes and their projections indicate that this two-pathway system is a gross simplification—each pathway likely contains many sub-pathways carrying distinct information. Nevertheless, the major division of labor between the M and P pathways illustrates the principle that visual information is processed along parallel paths from the retina to the cortex.

When these parallel information streams arrive at V1, they terminate in different sublayers of layer 4 (see Figure 4–45). Information exiting V1 is again divided roughly into two major streams (Figure 4–48A). The **ventral stream** (often called the 'what' stream) leads to the temporal cortex, which is responsible for analyzing form and color, whereas the **dorsal stream** (the 'where' stream) leads to the parietal cortex, which is responsible for analyzing motion and depth. The M and P pathways each contribute to both the dorsal and ventral streams—with the dorsal stream receiving input mostly from the M pathway and ventral stream receiving about equal amount of input from the M and P pathways.

The general direction of information flow among cortical layers within an area we discussed in Section 4.26 and Box 4–3 can be extended to infer information flow between different cortical areas. For instance, when the connections of V1 and the adjacent V2 (secondary visual cortex) were examined by **anterograde** (from cell bodies to axon terminals) and **retrograde** (from axon terminals to cell bodies) tracing methods (see Section 13.18), it was found that V1 → V2 axonal projections terminated mostly in layer 4 of V2, whereas V2 → V1 axonal projections terminate in superficial and deep layers but avoid layer 4. Since visual information in the cortex originates from V1 and spreads to higher cortical areas, these experiments suggest a general connection rule between intercortical areas: feed-forward projections (for example, V1 → V2) terminate in layer 4, whereas feedback projections (for example, V2 → V1) avoid layer 4.

Extensive anatomical tracing experiments based on the above rule have coarsely subdivided the visual cortex of the rhesus monkey into more than 30 areas (Figure 4–48B). These cortical areas are extensively interconnected, often via reciprocal connections (such as V1 → V2 and V2 → V1). They nevertheless appear to follow a hierarchical organization, where visual information from V1 undergoes ~10 levels of processing through cortical areas with multiple distinct processing streams that are highly intertwined (Figure 4–48B).

The bewildering complexity of these connections raises important questions: at what level and in what form does visual perception arise? One can ask a more specific question about the parallel processing streams. Suppose that one sees a white cat chasing after a black mouse, which is itself going after a piece of yellow cheese. According to what we have learned, the motion, color, and form of these three objects are processed in separate streams going to different cortical areas. How does one correctly associate 'white' and 'fast moving' with 'cat', or 'yellow' and 'stationary' with 'cheese'? This poses a binding problem: how does the visual system correctly link up (or bind) different features of the same objects?

At present we do not have a satisfying answer to this question. One property that can contribute to the solution to the binding problem is visual attention. **Attention** refers to the cognitive function in which a subset of behaviorally relevant sensory information is selected for further processing at the expense of irrelevant information. Electrophysiological recording experiments in awake, behaving monkeys (Section 4.29 below provides more discussion of this approach) have indicated that visual responses to an object being attended are characterized by an increased firing rate and stimulus selectivity and decreased trial-to-trial variability, compared to objects not being attended. One area implicated in this attentional control is the **frontal eye field** (FEF; Figure 4–48B) in the **prefrontal cortex**, a frontal neocortical area anterior to the motor cortex implicated in high-level cognitive functions. The frontal eye field receives extensive feed-forward connections from visual areas in both the dorsal and ventral streams, and sends feedback projections to many visual cortical areas. Evidence has suggested that the frontal eye field is causally related to attentional modulation of processing of signals, such as firing rate and stimulus selectivity, in specific areas of the visual cortex.

How the cerebral cortex processes information is still mostly unknown. This is a major unsolved question in science, and the study of vision has brought us closest to its frontier. In the final two sections of this chapter, we give two specific examples of research to elucidate the function of higher visual cortical areas.

4.28 Face recognition cells form a specialized network in the primate temporal cortex

Starting in the 1970s, single-unit recordings in monkeys have shown that some neurons in the ventral visual processing stream of the temporal lobe recognize specific objects including faces (**Figure 4–49**). The advance of **functional magnetic resonance imaging** (**fMRI**) (see Section 1.10) led to the discovery that a specific area of human temporal cortex, the **fusiform face area**, is preferentially activated by images of human faces. Subsequently, face recognition areas were also discovered in rhesus monkeys using fMRI. Because fMRI samples the average activity of many neurons, these findings implied that the face recognition areas may represent concentrated populations of face-recognition cells. This hypothesis was tested by combining fMRI with single-unit recording in the rhesus monkey: fMRI permitted accurate targeting of the microelectrode at the face recognition area. Strikingly, a large majority of recorded neurons in certain areas of the monkey temporal cortex appeared to be preferentially tuned to faces (**Figure 4–50**).

Six patches of face recognition cells—known as PL, ML, MF, AF, AL, and AM (named after their relative anatomical locations)—have been identified in the inferior temporal lobe of rhesus monkeys (**Figure 4–51**A). Although the exact locations of these patches differ from individual to individual, their relative positions are more stereotyped such that each patch can be individually recognized. This stereotyped positioning permitted further investigation of the relationship between these face recognition patches using electrical **microstimulation**—the application of currents to activate neurons nearby the recording electrode (see Section 4.29 below for further details about this method). Microstimulation of a given face recognition patch was followed by fMRI to determine which other brain areas were activated in response. Interestingly, stimulation of one face recognition patch preferentially activated other face recognition patches (Figure 4–51A), whereas stimulating an area adjacent to a face recognition patch activated patches of other, non-face recognition areas (Figure 4–51B). These experiments suggested that the face recognition patches are preferentially connected with each other.

Further studies suggested a hierarchical processing model for face recognition. Neurons in the middle patches (ML and MF) recognize faces from specific angles. Neurons in the anterior-lateral (AL) patch not only recognize faces from specific angles, but also recognize the mirror images of those faces. Neurons in

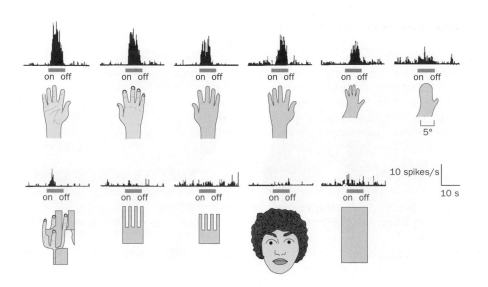

Figure 4–49 Object-selective neurons in the temporal cortex. Single-unit recording in anesthetized monkey revealed that inferior temporal cortex contains neurons that selectively respond to specific objects. In this example shown, the neuron is activated (fires more spikes) in response to human hands (top row, first four images from left) and a monkey hand (top row, fifth image). A monkey hand with digits fused (top row, sixth image) and the images shown in the bottom row were less effective in stimulating the neuron. Horizontal bars underneath action potential plots indicate the stimulation period. Other neurons (not shown) responded selectively to faces. (Adapted from Desimone R, Albright TD, Gross CG, et al [1984] *J Neurosci* 8:2051–2062. With permission from The Society for Neuroscience.)

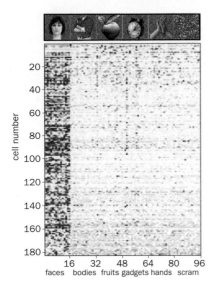

Figure 4–50 Face recognition areas are highly enriched for face recognition cells. Targeted single-unit recording after functional magnetic resonance imaging revealed that face recognition cells are highly concentrated in the middle face patch of the monkey temporal cortex. Firing patterns of 182 cells in response to 16 images each of faces, bodies, fruits, gadgets, hands, and scrambled images. Orange pixels represent action potentials. (From Tsao DY, Freiwald WA, Tootell RBH, et al [2006] *Science* 311:670–674. With permission from AAAS.)

Figure 4–51 Face recognition patches are preferentially interconnected. Face recognition patches were first identified by functional magnetic resonance imaging (fMRI) and shown as a 'flat map' (by flattening the cortical surface as a two-dimensional representation) outlined in green: PL, ML, MF, AF, AL, AM. **(A)** A microelectrode was inserted at the ML (marked by x) and first used to confirm that the neuron being recorded was face-selective. The electrode was then used to stimulate that area, and activation of other brain areas as indicated by false color (see bottom right for a key in arbitrary units) was superimposed on the face recognition patches. Note the coincidence between the face recognition areas identified by fMRI with those activated by microstimulation. **(B)** As a control, a stimulation electrode was placed outside the face recognition patches. Microstimulation also activated two patches outside the stimulation area, but these did not overlap with the face recognition areas. (From Moeller S, Freiwald WA & Tsao DY [2008] *Science* 320:1355–1359. With permission from AAAS.)

the anterior-medial (AM) patch respond to faces from all views. Thus, information flow appears to follow the general direction of ML/MF → AL → AM, achieving more abstract face recognition (face recognition independent of viewing angles) at the AM area. The latency (time delay) of neuronal activation in response to face stimuli supports this model, with ML/MF neurons activated prior to AL neurons, and AL neurons activated prior to AM neurons.

Studies of face recognition cells and patches have raised many interesting questions about visual perception for future investigations. How are the receptive fields of neurons along the ventral processing stream (see Figure 4–48) sequentially transformed from line and edge detection in V1 to face recognition in inferior temporal cortex? Why are face recognition areas clustered into distinct patches? What are the functions of the cortical areas located between these face recognition patches? Do different face recognition cells in one area recognize different faces? The fact that primates devote cortical areas specifically to face recognition may reflect the abundance and importance of social interactions in these species. Whatever the evolutionary significance may be, face recognition offers an outstanding system to further explore the general question of visual perception.

4.29 Linking perception to decision and action: microstimulation of MT neurons biased motion choice

The ultimate purpose of visual perception is to guide animal behavior. So where and how is sensation converted to action? How can we study this? Thus far we have discussed insights on visual information processing obtained by studying what stimuli can best excite visual pathway neurons. But how do we know that the activities of specific neurons contribute to the animal's perceptions? One approach is to design behavioral assays that are relevant to the activity of these neurons and then use those behavioral assays to test the functional consequences of perturbing the neuronal activity.

Along the dorsal stream in the higher-order visual cortex of primates is an area called the **middle temporal visual area** (**MT**; see Figure 4–48). MT neurons are especially sensitive to the direction of motion: individual neurons fire in response to motion in a particular direction, and neurons tuned to the same direction of motion are clustered in vertical columns. In order to investigate the function of these neurons in motion perception, researchers trained monkeys to report the direction of motion in a random dot display on a TV screen (**Figure 4–52**A). The monkey would first fix its eyes on a screen where the moving dots were displayed. After the display was over, it was given two choices represented by two indicator lights on the screen, one located in the direction of motion, the other in the opposite direction. The monkey would make a **saccade** (that is, move the fixation point of its eyes rapidly) toward what it judged was the direction of motion (Figure 4–52B, C) and would receive a juice reward for

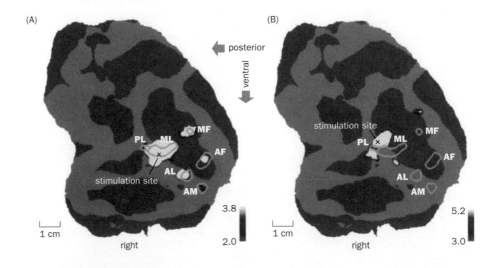

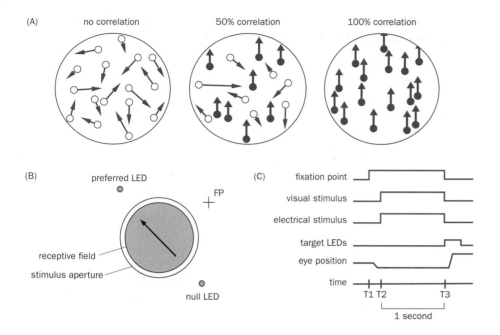

Figure 4–52 Experimental paradigm to test monkey's motion perception. (A) Schematic illustration of the random dot display. The direction (indicated by arrows) and correlation of moving dots are adjusted by experimenters. Shown here are three examples in which 0, 50, or 100% of the dots exhibited correlated upward motion. (B) In an actual experiment, the monkey was trained to fix his eyes on the fixation point (FP). A neuron in MT was first recorded to determine its receptive field and motion preference. Then the moving dot stimuli were applied in the receptive field of the recorded neuron; the direction of coherent motion was either in the cell's preferred direction, or in the opposite or null direction. (C) The temporal sequence of the microstimulation experiment. Monkeys first fixed their eyes on the fixation point (T1). Then the random dot display was applied to the receptive field, along with electrical stimulation (microstimulation) for 1 second (T2 to T3). At the end of the visual and electrical stimulation, the monkeys chose between one of two light-emitting diode (LED) indicator lights, moving their eyes to one of the LEDs to signal its perception of the direction of motion stimulus. Microstimulations were applied in half of the trials and withheld in the other half. (A, adapted from Salzman CD, Murasugi CM, Britten KH, et al. [1992] *J Neurosci* 12:2331–2355. With permission from The Society for Neuroscience; B & C, adapted from Salzman CD, Britten KH & Newsome WT [1990] *Nature* 346:174–177. With permission from Macmillan Publishers Ltd.)

each correct choice. Dots with correlated motion (indicated by filled circles in Figure 4–52A) were mixed with dots that move in random directions (indicated by open circles in Figure 4–52A). The task is easy when 100% or 50% of moving dots are correlated, is difficult if only 10% of moving dots are correlated, and becomes impossible when 0% of moving dots are correlated. After extensive training, monkeys could make correct choices above chance if about 10% of dots were correlated. Using this behavioral assay, researchers found that MT lesions markedly elevated correlation needed for monkeys to make correct choices, indicating that MT is required for motion perception.

To further probe the underlying neural mechanism, an electrode was placed in MT to identify the receptive field and preferred direction of the neuron recorded by the electrode. The correlated motion of moving dots in the behavioral experiment was then applied either in the preferred direction of the recorded neuron, or the opposite (null) direction. A psychometric function could be established by plotting the percentage eye movement in the preferred direction by the monkey against the percentage of dots with correlated motion (**Figure 4–53**, purple lines). The same electrode was then used for microstimulation to activate nearby neurons

Figure 4–53 Microstimulation of MT neurons bias motion perception.
Psychometric curves plotting the proportion of decisions in which two different monkeys (top and bottom graphs) moved their eyes toward the recorded neuron's preferred direction against the percentage correlation of the moving dots, following the experimental set-up in Figure 4–52. Positive values on the *x* axis represent correlated motion applied in the same direction as the preferred direction of the neuron; negative *x* axis values correspond to correlated motion in the null direction. The monkeys had different biases when no electrical microstimulation was applied. The first monkey displayed a small bias for the preferred direction (~56% at 0% correlation), while the second monkey displayed a large bias for the null direction (~33% at 0% correlation); an unbiased result would be ~50% at 0% correlation. In both cases, microstimulation shifted the curve leftwards. For example, after viewing 5%-correlated dots moving in the preferred direction (green dashed line), the first monkey chose the preferred direction ~70% of the time in the absence of microstimulation; these correct choices increased to ~80% with microstimulation. On the other hand, after viewing 5%-correlated dots moving in the null direction (red line), the monkey's erroneous preferred-direction choices increased from 40% without microstimulation to 65% with microstimulation. Microstimulation therefore consistently increased the number of decisions that favored the neuron's preferred direction. (Adapted from Salzman CD, Britten KH & Newsome WT [1990] *Nature* 346:174–177. With permission from Macmillan Publishers Ltd.)

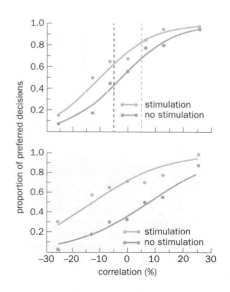

while the monkey was viewing the moving dots in a random half of the trials (Figure 4–52C). Microstimulation shifted the psychometric function leftwards (Figure 4–53, orange lines). In other words, microstimulation helped the monkey to make correct choices when correlated motion was in the preferred direction of stimulated neurons, but caused the monkey to make more errors when correlated motion was in the null direction. Both examples supported the notion that the net effect of microstimulation is to bias the monkey's choice toward the preferred direction of the stimulated neurons.

This is an important conceptual advance, as we have moved from correlation to causation. The activity of MT neurons not only reports the direction of motion; activation of MT neurons is also sufficient to cause a change in the animal's behavior. Given that there are many neurons in the MT and elsewhere that are influenced by direction of motion, it is remarkable that the effect of microstimulation of a small area could cause a significant behavioral bias. One possibility is that since MT neurons with similar properties (that is, similar direction-of-motion selectivity) are arranged in columns, as we have seen for V1 neurons, microstimulation may have activated many nearby neurons that encode the same information and can thereby collectively affect the animal's perceptual choice.

Subsequent studies have shown that motion information from MT is analyzed in the **lateral intraparietal area** (**LIP**) of the parietal cortex before being sent to areas in the frontal cortex (such as the frontal eye field) and superior colliculus that control eye movement. LIP neurons appear to be particularly important in forming the decision to make a saccade in a particular direction after analyzing the sensory information. Their firing rates were predictive of which direction the eyes would move, signaling the monkey's decision about its perception of the moving dots' direction. LIP neurons that corresponded to eye movement toward the direction of the monkey's eventual choice (response direction) ramped up their firing rate as the monkey was viewing the moving dots and during the delay period before the saccade, as if LIP neurons were accumulating evidence. The power of LIP neurons to predict the direction of the saccade improved with time and stimulus strength (**Figure 4–54**). Microstimulation of LIP neurons during the decision process could speed up the reaction time of eye movement toward the response direction, demonstrating a causal relationship between LIP neuron activity and the decision process.

Thus, by focusing on a specific and quantitative behavioral task and identifying relevant neurons through physiological recording and activity manipulation, researchers have gained important insight into the neural basis of motion perception and decision making. Microstimulation during behavioral tasks has become a powerful method that complements physiological recordings of neuronal

Figure 4–54 Neurons in the lateral intraparietal area (LIP) and perceptual decisions. (A) Behavioral task. Monkeys fix their eye on the fixation point (+). Two targets appear, one of which is within the receptive field (RF) of a LIP neuron being recorded. After a period of moving-dot stimulus and a variable delay, the monkey signals its perception of the direction of motion by making a saccade toward one of the two targets. **(B)** Average firing rates of 104 LIP neurons during the motion-viewing period (left panel), during the delay period before the eye movement (before saccade on the right panel), and afterward (after saccade on the right panel). When the monkey judged the direction as being toward the receptive field (solid lines), the firing rate of LIP neurons increased over time during the motion-viewing and delay periods. In the color-coded inset key, the percentages refer to the percent correlation of the moving dots. The strength of the stimulus (percentage of correlated moving dots) affected the speed and magnitude of LIP neurons' firing rate increase during the motion-viewing period. The firing rate of LIP neurons decreased when the monkey judged the direction as being away from the receptive field (dashed lines). Not shown here, LIP neurons in the opposite half of the brain (with receptive fields corresponding to target 2) demonstrated the opposite response—ramping up for target 2 choices and ramping down for target 1 choices. (Adapted from Shadlen MN & Newsome WT [2001] *J Neurophysiol* 86:1916–1936. With permission from the American Physiological Society.)

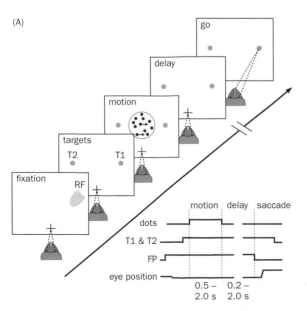

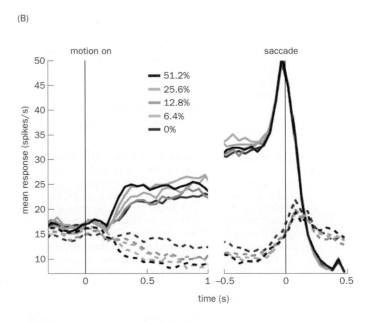

activity, as it tests the causal relationship between the activity of the neurons and their function. Importantly, these kinds of causal link between neuronal activity and animal behavior require the experiments to be performed in awake, behaving animals. Indeed, neurons in the **association cortex** (cortical areas that associate information from multiple sensory areas and link to motor output), such as LIP, tend to be silent in anesthetized animals but become highly active in awake, behaving animals.

In recent years, exciting new technological developments have made possible manipulations of neuronal activities that are more precise than microstimulation by recording electrodes. For example, as illustrated in Box 4–3, channelrhodopsin-2 (ChR2) can be induced to depolarize neurons that express it by shining light of appropriate wavelength to produce action potentials with millisecond precision (see Section 13.25). ChR2 can be genetically targeted to specific populations of neurons, enabling researchers to establish causal relationships between the activity of the targeted neurons and circuit or behavioral output.

SUMMARY

All sensory systems transform environmental stimuli into electrical signals. These signals are transmitted to the brain to form an internal representation, or perception, of the sensory stimuli, with the ultimate function of helping animals to survive and reproduce. In vision, the sensory stimuli are the spatiotemporal patterns of light projected onto a two-dimensional retina, from which animals extract information about the identity, location, and motion of objects in the external world that are of behavioral significance.

The first step in vision is to detect light stimuli and convert them into electrical signals. This is achieved by photoreceptors—rods and cones—which together cover a very wide range of light intensities. In rods and cones, photon absorption triggers isomerization of the retinal, which leads to conformational changes in the covalently linked opsin, a G-protein-coupled receptor. Photoexcited opsin then activates the trimeric G protein transducin, which in turn activates phosphodiesterase, leading to cGMP hydrolysis. A decline of cGMP concentration directly closes cGMP-gated cation channels and hyperpolarizes rods and cones. cGMP-gated channel closure also causes a decline of intracellular Ca^{2+} concentration, which triggers a series of biochemical events that lead to the recovery of phototransduction components to their dark state, and the adjustment of the phototransduction efficiency for adaptation to different background light levels.

Variations of the phototransduction components and their properties enable rods and cones to serve different functions. Rods are more sensitive to light due to greater amplification of the phototransduction cascade, and are used mostly for night vision. Cones recover more rapidly and have a larger adaptation range, and serve daylight vision and motion vision. The high density of cones in the primate fovea enables cones to serve high-acuity vision. Finally, the presence of different kinds of cones with opsins that confer different spectral sensitivities—short-, medium-, and long-wavelength for trichromatic primates including humans—enables color vision.

Signals from rods and cones are analyzed by exquisitely precise retinal circuits before information is delivered by retinal ganglion cell (RGC) axons to the brain. Two types of bipolar cells receive synaptic input from photoreceptors and deliver excitatory output to RGCs. The OFF bipolars maintain the sign of photoreceptors and are thus hyperpolarized by light, whereas the ON bipolars invert the sign and are thus depolarized by light. Horizontal cells are excited by photoreceptors, and send inhibitory feedback laterally to many photoreceptors. This lateral inhibition causes the output of the photoreceptor—glutamate release—to reflect not only their intrinsic phototransduction, but also the activities of surrounding photoreceptors. This center–surround antagonism creates the concentric receptive fields for bipolar cells and RGCs to detect luminance contrast. Lateral inhibition also contributes to color detection: a comparison of light signals from cones with different spectral sensitivities.

In addition to detecting luminance and color contrast, retinal circuits extract many other kinds of signals through parallel actions of different types of bipolar cells, amacrine cells, and RGCs. For example, asymmetric inhibition by starburst amacrine cells makes certain RGCs sensitive to the direction of motion. At the same time, many retinal cells and circuits are multifunctional under different conditions. For example, intrinsically photosensitive RGCs not only receive light input from rods and cones, but also can be directly depolarized by light through melanopsin. One of many functions of ipRGCs is to directly entrain the circadian clock.

A key brain target for RGC axons is the lateral geniculate nucleus (LGN) of the thalamus, which relays information to the visual cortex for further analysis of form, color, and motion. A principal organization in the LGN and visual cortical areas is retinotopy—neighboring neurons represent neighboring points in visual space in an orderly manner. The visual receptive fields of individual neurons, however, are transformed along the visual pathway. The receptive fields of LGN neurons are similar to those of RGCs, relaying contrast information to the cortex. The receptive fields of simple and complex cells in V1 become bars with specific orientations, suggesting that V1 cells detect lines and edges. Cells in the MT are highly tuned to motion in specific directions, whereas cells in patches of the inferior temporal cortex are highly tuned to faces, suggesting further specialized functions in these higher cortical visual areas.

Studies of the visual cortex have led the exploration of mammalian neocortex function. In V1, information generally flows from LGN axons → layer 4 → layers 2/3 → layers 5/6 vertically. Between cortical areas, feed-forward or feedback inputs usually terminate in or avoid layer 4 of the recipient areas, respectively. These rules have enabled the construction of a hierarchical model for visual information streams beyond V1. Our understanding of the general principles of information processing in neocortex, and specifically how visual response properties in one visual area are transformed to those of another area, are still rudimentary. Electrophysiological recordings, circuit tracing, activity manipulation, computational modeling, and quantitative studies of behavior must be combined to tackle the complexity of neocortical function.

FURTHER READING

Books and reviews

Baylor DA (1987) Photoreceptor signals and vision. Proctor lecture. *Invest Ophthalmol Vis Sci* 28:34–49.

Hubel DH & Wiesel TN (2004) Brain and Visual Perception: The Story of a 25-Year Collaboration. Oxford University Press.

Luo DG, Xue T & Yau KW (2008) How vision begins: an odyssey. *Proc Natl Acad Sci USA* 105:9855–9862.

Nathans J (1989) The genes for color vision. *Sci Am* 260:42–49.

Reynolds & Desimone (1999) The role of neural mechanisms of attention in solving the binding problem. *Neuron* 24:19–29.

Rodieck RW (1998) The First Steps in Seeing. Sinauer.

Stryer L (1988) Molecular basis of visual excitation. *Cold Spring Harb Symp Quant Biol* 53 Pt 1:283–294.

Wässle H (2004) Parallel processing in the mammalian retina. *Nat Rev Neurosci* 5:747–757.

Wei W & Feller MB (2011) Organization and development of direction-selective circuits in the retina. *Trends Neurosci* 34:638–645.

Light detection in rods and cones

Baylor DA, Lamb TD & Yau KW (1979) Responses of retinal rods to single photons. *J Physiol* 288:613–634.

Baylor DA, Nunn BJ & Schnapf JL (1987) Spectral sensitivity of cones of the monkey *Macaca fascicularis*. *J Physiol* 390:145–160.

Choe HW, Kim YJ, Park JH et al. (2011) Crystal structure of metarhodopsin II. *Nature* 471:651–655.

Fesenko EE, Kolesnikov SS & Lyubarsky AL (1985) Induction by cyclic GMP of cationic conductance in plasma membrane of retinal rod outer segment. *Nature* 313:310–313.

Fung BK, Hurley JB & Stryer L (1981) Flow of information in the light-triggered cyclic nucleotide cascade of vision. *Proc Natl Acad Sci USA* 78:152–156.

Hecht S, Shlaer S & Pirenne MH (1942) Energy, quanta, and vision. *J Gen Physiol* 25:819–840.

Kleinschmidt J & Dowling JE (1975) Intracellular recordings from gecko photoreceptors during light and dark adaptation. *J Gen Physiol* 66:617–648.

Koch KW & Stryer L (1988) Highly cooperative feedback control of retinal rod guanylate cyclase by calcium ions. *Nature* 334:64–66.

Mendez A, Burns ME, Sokal I et al. (2001) Role of guanylate cyclase-activating proteins (GCAPs) in setting the flash sensitivity of rod photoreceptors. *Proc Natl Acad Sci USA* 98:9948–9953.

Nathans J, Piantanida TP, Eddy RL et al. (1986) Molecular genetics of inherited variation in human color vision. *Science* 232:203–210.

Nathans J, Thomas D & Hogness DS (1986) Molecular genetics of human color vision: the genes encoding blue, green, and red pigments. *Science* 232:193–202.

Palczewski K, Kumasaka T, Hori T et al. (2000) Crystal structure of rhodopsin: A G protein-coupled receptor. *Science* 289:739–745.

Roorda A & Williams DR (1999) The arrangement of the three cone classes in the living human eye. *Nature* 397:520–522.

Signal analysis in the retina

Barlow HB, Hill RM & Levick WR (1964) Retinal ganglion cells responding selectively to direction and speed of image motion in the rabbit. *J Physiol* 173:377–407.

Berson DM, Dunn FA & Takao M (2002) Phototransduction by retinal ganglion cells that set the circadian clock. *Science* 295:1070–1073.

Briggman KL, Helmstaedter M & Denk W (2011) Wiring specificity in the direction-selectivity circuit of the retina. *Nature* 471:183–188.

Dacey DM & Lee BB (1994) The 'blue-on' opponent pathway in primate retina originates from a distinct bistratified ganglion cell type. *Nature* 367:731–735.

Euler T, Detwiler PB & Denk W (2002) Directionally selective calcium signals in dendrites of starburst amacrine cells. *Nature* 418:845–852.

Field GD, Greschner M, Gauthier JL et al. (2009) High-sensitivity rod photoreceptor input to the blue-yellow color opponent pathway in macaque retina. *Nat Neurosci* 12:1159–1164.

Fried SI, Munch TA & Werblin FS (2002) Mechanisms and circuitry underlying directional selectivity in the retina. *Nature* 420:411–414.

Hattar S, Liao HW, Takao M et al. (2002) Melanopsin-containing retinal ganglion cells: architecture, projections, and intrinsic photosensitivity. *Science* 295:1065–1070.

Kolb H & Famiglietti EV (1974) Rod and cone pathways in the inner plexiform layer of cat retina. *Science* 186:47–49.

Kuffler SW (1953) Discharge patterns and functional organization of mammalian retina. *J Neurophysiol* 16:37–68.

Packer OS, Verweij J, Li PH et al. (2010) Blue-yellow opponency in primate S cone photoreceptors. *J Neurosci* 30:568–572.

Szikra T, Trenholm S, Drinnerberg A et al. (2014) Rods in daylight act as relay cells for cone-driven horizontal cell-mediated surround inhibition. *Nat Neurosci* 17:1728-1735.

Verweij J, Hornstein EP & Schnapf JL (2003) Surround antagonism in macaque cone photoreceptors. *J Neurosci* 23:10249–10257.

Information processing in the visual cortex

Bonhoeffer T & Grinvald A (1991) Iso-orientation domains in cat visual cortex are arranged in pinwheel-like patterns. *Nature* 353:429–431.

Desimone R, Albright TD, Gross CG et al. (1984) Stimulus-selective properties of inferior temporal neurons in the macaque. *J Neurosci* 4:2051–2062.

Felleman DJ & Van Essen DC (1991) Distributed hierarchical processing in the primate cerebral cortex. *Cereb Cortex* 1:1–47.

Ferrera VP, Nealey TA & Maunsell JH (1992) Mixed parvocellular and magnocellular geniculate signals in visual area V4. *Nature* 358:756–761.

Gilbert CD & Wiesel TN (1979) Morphology and intracortical projections of functionally characterised neurones in the cat visual cortex. *Nature* 280:120–125.

Hanks TD, Ditterich J & Shadlen MN (2006) Microstimulation of macaque area LIP affects decision-making in a motion discrimination task. *Nat Neurosci* 9:682–689.

Hubel DH & Wiesel TN (1959) Receptive fields of single neurones in the cat's striate cortex. *J Physiol* 148:574–591.

Hubel DH & Wiesel TN (1962) Receptive fields, binocular interaction and functional architecture in the cat's visual cortex. *J Physiol* 160:106–154.

Moeller S, Freiwald WA & Tsao DY (2008) Patches with links: a unified system for processing faces in the macaque temporal lobe. *Science* 320:1355–1359.

Newsome WT, Britten KH & Movshon JA (1989) Neuronal correlates of a perceptual decision. *Nature* 341:52–54.

Ohki K, Chung S, Ch'ng YH et al. (2005) Functional imaging with cellular resolution reveals precise micro-architecture in visual cortex. *Nature* 433:597–603.

Olsen SR, Bortone DS, Adesnik H et al. (2012) Gain control by layer six in cortical circuits of vision. *Nature* 483:47–52.

Rockland KS & Pandya DN (1979) Laminar origins and terminations of cortical connections of the occipital lobe in the rhesus monkey. *Brain Res* 179:3–20.

Salzman CD, Britten KH & Newsome WT (1990) Cortical microstimulation influences perceptual judgements of motion direction. *Nature* 346:174–177.

Shadlen MN & Newsome WT (2001) Neural basis of a perceptual decision in the parietal cortex (area LIP) of the rhesus monkey. *J Neurophysiol* 86:1916–1936.

CHAPTER 5
Wiring of the Visual System

There are billions of neurons in our brains, but what are neurons? Just cells. The brain has no knowledge until connections are made between neurons. All that we know, all that we are, comes from the way our neurons are connected.

Tim Berners-Lee (2000),
Weaving the Web: The Original Design and Ultimate Destiny of the World Wide Web

The human brain contains about 10^{11} neurons, and each makes on average 10^3 synapses with other neurons. Thus, there are about 10^{14} synaptic connections in the brain that give rise to our ability to sense, think, remember, and act. How are such vast numbers of connections correctly established during development?

The brain is often compared with another complex system that has impressive problem-solving power, the computer (see Section 1.12). Their functions are both specified by wiring diagrams constructed from much simpler units: the neuron and the transistor. A key difference is the wiring process. Each connection on a computer chip is specified by engineers. By contrast, the brain is self-assembled during development, starting from a single cell. It is its own engineer. The instructions for the entire developmental process, including brain wiring, are embedded in an organism's DNA sequences. Indeed, the information contained in the human genome sequence (3×10^9 nucleotides of DNA with four possible bases for each nucleotide, equivalent to 750 megabytes) can easily fit within a DVD. How does information in the DNA direct the wiring of the brain? Is that information sufficient?

An enduring debate among neuroscientists, psychologists, and even philosophers is whether brain wiring is shaped more by nature (genes) or nurture (experience). Given that both factors contribute, as we discussed in Chapter 1, more relevant questions are: How do genes and environment specify brain wiring? Do their relative contributions differ in different parts of the nervous system, at different stages of neural development, and in different animals?

In this chapter and Chapter 7, we will explore the mechanisms of nervous system development with a particular emphasis on brain wiring. Many fundamental insights about brain wiring have come from research on the visual system, which will be the focus of this chapter. In Chapter 7, we will expand our scope to study the wiring of other parts of the nervous system and summarize the general principles involved.

HOW DO RETINAL GANGLION CELL AXONS FIND THEIR TARGETS?

Imagine that you are one of about a million retinal ganglion cells (RGCs) near the center of a human retina, for example a midget RGC with a red-ON center receptive field (see Figure 4–33C) located in the fovea of the left eye (**Figure 5–1**). Your function is to transmit information to the brain about the presence of long-wavelength light within your visual field. Proper wirings of your dendrites and axon are essential for fulfilling this function. Your dendrites must connect with the correct types of bipolar and amacrine cells to construct a center–surround receptive field, while your long axon must exit the eye, travel along the left optic nerve, decide whether to stay in the left hemisphere or cross the midline at the optic chiasm, travel along the optic tract, and send branches to the superior colliculus and the lateral geniculate nucleus (LGN). As we learned in Chapter 4, visual information in the retina is topographically represented in the brain as retinotopic maps. In order to preserve the spatial relationships of the visual image on the retina, your axon must terminate at appropriate retinotopic positions within the superior

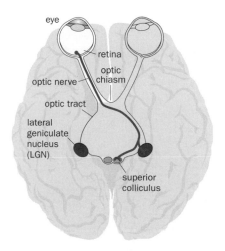

Figure 5–1 The journey of an axon of a retinal ganglion cell. The axon of an RGC must make a series of choices during development: to exit the eye and join the optic nerve, to cross (or not cross) the midline at the optic chiasm, to form terminal branches at specific brain targets (for example, the lateral geniculate nucleus and superior colliculus), and to terminate at specific layers and positions within these targets, according to the RGC's subtype and retinal location.

Figure 5–2 The retinotopic map.
The retina forms a two-dimensional visual image of an object in the external world. This image is reconstructed in a brain target, the tectum, by point-to-point topographic projections of retinal ganglion cell axons. The tectum in non-mammalian vertebrates is equivalent to the superior colliculus in mammals.

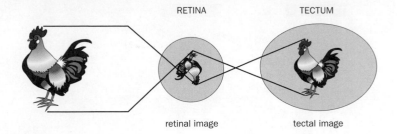

colliculus and LGN according to your location in the eye. Finally, you must choose the correct synaptic partners from specific layers within these targets according to your identity as a red-ON center midget cell.

As this example illustrates, a single neuron makes many decisions during the wiring process. It is astonishing to consider that in the developing nervous system, millions of neurons are making so many decisions simultaneously. In the first part of this chapter, we will focus largely on one specific aspect of visual system wiring: how do RGC axons terminate at appropriate spatial locations in their target areas, such that the two-dimensional image of the visual scene on the retina is properly represented in the brain (**Figure 5–2**)? Studies of this specific problem have yielded important insights about general principles of brain wiring.

5.1 Optic nerve regeneration experiments suggested that RGC axons are predetermined for wiring

Historically, two broad mechanisms by which neurons become connected to their targets were proposed. In one mechanism, axons at first connect to many different targets; through trial and error, a subset of these initially **exuberant connections** (that is, excess connections made during development that are not retained into adulthood) are selected by their functions to establish the final connection pattern (**Figure 5–3**, left). In the other mechanism, axons are predetermined to choose their targets directly without functional selection (Figure 5–3, right). The dominant view in the early twentieth century was that functional selection plays a predominant role in establishing neuronal connections. A decisive experiment on optic nerve regeneration performed by Roger Sperry strongly suggested that neurons are predetermined to choose their targets.

Figure 5–3 Two contrasting mechanisms for axon targeting. In the first mechanism (left), axons initially overproduce branches and make exuberant connections. The correct connections are later selected by function, and inappropriate connections are pruned. In the second mechanism (right), axons are predetermined to find their targets.

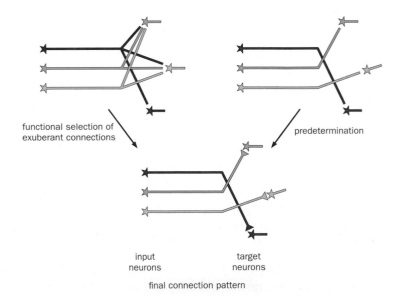

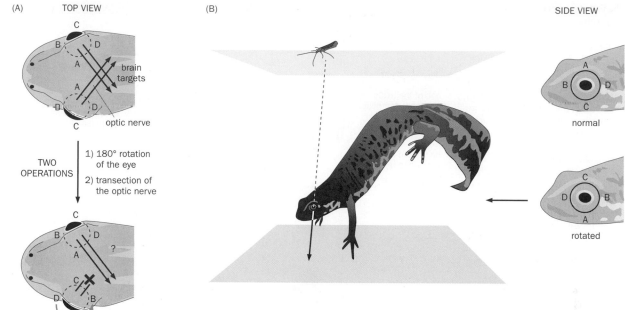

Figure 5–4 Sperry's optic nerve regeneration experiment. (A) Schematic of experimental design. The left eye was rotated 180° followed by axon transection. Letters surrounding the eyes are with respect to the original position. Red lines and arrows represent RGC axons and the direction of their brain target. '?' at the target of regenerating axons represents different outcomes to be tested by the experiment, which are listed at the bottom. Note that RGC axons project mostly to contralateral brain targets; we will discuss this topic in Section 5.6. **(B)** Illustration of newt's behavior based on its inverted vision from the manipulated eye. When given food above, the newt swam downward to fetch it. (Note that before the behavioral test, the other eye was removed so the newt could only see with the rotated eye.) Side views of the normal and rotated eyes are drawn on the right. (Adapted from Sperry RW [1943] *J Comp Neurol* 79:33–55.)

Amphibians have the amazing ability to **regenerate** their nerve connections after damage by re-extending truncated axons back to their targets. The optic nerve, which consists of the RGC axons that connect the eye to the brain, can be transected and then allowed to regenerate so that vision is completely restored. Taking advantage of this property, Sperry surgically rotated one of a newt's eyes 180° such that the visual world was upside down and front-side back; each RGC now looked at a point in space that was 180° different from the point it looked at before surgery. He then severed the optic nerve, and allowed the RGC axons from the rotated eye to re-grow into the brain to form connections with their targets (**Figure 5–4**A). After appropriate time for regeneration, he used behavioral experiments to test what the newt saw with its manipulated eye.

Three possible outcomes can be envisioned from this experiment. First, vision could be restored completely. This would provide strong evidence supporting the functional selection hypothesis. Second, vision could be blurred. This might mean that the regenerated axons are only partially successful in finding their targets in the brain due to the eye's rotation (recall that vision could be restored completely if no eye rotation was performed). Third, vision could be restored, but in an inverted manner, suggesting that RGC axons connected to the same brain targets before and after the rotation. The third outcome was what Sperry observed from behavioral experiments after regeneration. When food was provided at the surface of the aquarium, above the newt's head, the newt would swim downward to fetch it, bumping itself against the bottom of the tank (Figure 5–4B).

This experiment strongly suggested that RGC axons carry specific information corresponding to their *original* positions in the eye, and use such information to find their targets in the brain. Despite the rotation of the eye, such positional information still enabled these RGC axons to connect with the original target neurons in the brain, restoring vision but in an inverted manner.

5.2 Point-to-point connections between retina and tectum arise by chemoaffinity

Over the next 20 years, Sperry and colleagues collected more information on how regenerated RGC axons grew into the **tectum**, the major target of RGCs in the brain of amphibians and lower vertebrates that is equivalent to the mammalian

Figure 5–5 Orderly projections of regenerating retinal axons in the tectum. Each of the four pairs of drawings illustrates the tectum targeting of RGC axons from a portion of a retina, following the transection of the retina's optic nerve, the ablation of half the retina's RGCs, and the regeneration of axons by the remaining RGCs. For example, the top pair illustrates that RGCs from the ventral retina project their axons to the medial tectum, while the second pair shows that axons of the dorsal RGCs project to the lateral tectum. RGCs from the posterior (temporal) retina project their axons to the anterior tectum (third pair), whereas RGCs from the anterior (nasal) retina their project to the posterior tectum (fourth pair). Gray areas represent ablated RGCs in the retina and empty target field in the tectum. (Adapted from Sperry RW [1963] *Proc Natl Acad Sci USA* 50:703–710.)

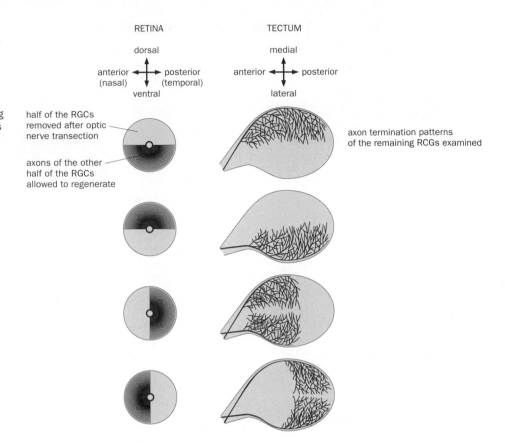

RETINA
dorsal
anterior (nasal) ◄────► posterior (temporal)
ventral

TECTUM
medial
anterior ◄────► posterior
lateral

half of the RGCs removed after optic nerve transection

axons of the other half of the RGCs allowed to regenerate

axon termination patterns of the remaining RCGs examined

superior colliculus. They transected the optic nerve, ablated half of the RGCs, and then examined the axon terminations of the remaining RGCs in the tectum. Their key findings are summarized in **Figure 5–5**. Ventral RGCs project to the medial half of the tectum, dorsal RGCs to the lateral half. Anterior RGCs (also called **nasal** RGCs because they are close to the nose) project to the posterior tectum, whereas posterior RGCs (also called **temporal** RGCs because they are close to the temple) project to the anterior tectum. Thus, the point-to-point map between the retina and the tectum is enabled by the orderly projections of RGC axons. The axonal projection of anterior (nasal) RGCs to the posterior tectum is particularly illuminating: here the axons pass through an empty field of tectum—because the posterior (temporal) RGCs that normally terminate here are ablated—and home in on their original targets, providing conclusive evidence that RGC axons are predetermined to connect with specific targets in the brain, at least in the context of regeneration.

This evidence led Sperry to propose the **chemoaffinity hypothesis** in 1963: "It seems a necessary conclusion from these results that cells and fibers of the brain and (spinal) cord must carry some kind of individual identification tags, presumably cytochemical in nature, by which they are distinguished one from another almost, in many regions, to the level of the single neuron; and further, that the growing fibers are extremely particular when it comes to establishing synaptic connections, each axon linking only with certain neurons to which it becomes selectively attached by specific chemical affinity."

An apparent problem with the chemoaffinity hypothesis is that our genome does not have enough information to encode as many tags, or cell-surface recognition proteins used to guide axons, as there are neurons and connections, especially if every single neuron must carry an individual identification tag to be distinguished from every other neuron. Sperry realized this difficulty and proposed, specifically for the retinotectal mapping, that protein gradients can be used to provide positional information in the retinal and tectal fields. This means that different levels of the same protein can be used to specify the precise targeting of many different neurons. (We will see below that these predictions are prescient, and will discuss additional mechanisms that can be used to solve this problem in Chapter 7.)

5.3 The posterior tectum repels temporal retinal axons

The chemoaffinity hypothesis inspired scientists to search for the cytochemical tags that guide growing axons toward their targets. Retinotectal mapping has been one of the leading model systems in this search. In this particular context, the chemoaffinity hypothesis predicts that there are molecular differences among cells from different parts of the tectum, in order for retinal axons to differentially select their targets. Likewise, there must be molecular differences among retinal axons originating from distinct parts of the retina so that they react differently to the molecular cues in the tectum.

Indeed, biochemical studies showed that membrane proteins extracted from the posterior tectum of chicks differed from those extracted from the anterior tectum. (The chick has a large tectum, making it an excellent source of starting material for biochemical analysis.) This was elegantly illustrated using the stripe assay (**Figure 5–6**). Membranes from anterior or posterior tectum, which presumably include cell-surface recognition proteins, were laid down in 50-μm alternating stripes to serve as substrates for retinal axons to grow onto *in vitro*. Temporal retinal axons grew preferentially on the membranes from the anterior tectum, whereas nasal retinal axons exhibited no selectivity. (Without alternating stripes, even temporal retinal axons can grow on membranes from posterior tectum; this demonstrates the importance of alternative choices to reveal selectivity in axon growth.)

The above observation could be caused by two alternative possibilities: either temporal retinal axons are attracted by membranes derived from the anterior tectum, where they normally terminate *in vivo*, or they are repelled by membranes derived from posterior tectum. A simple experiment was carried out to distinguish between these two alternatives. Since these putative attractants or repellents are most likely proteins, and protein activities are usually abolished by heating, anterior or posterior tectal membrane proteins were selectively heat-inactivated before placing on the stripes. It was found that temporal RGC axons lost selectivity when posterior (but not anterior) tectal membrane proteins were heat-inactivated. Thus, it appeared that temporal axons respond to an activity

Figure 5–6 Temporal RGC axons are repelled by the posterior tectal membrane. (A) Experimental setup. Membranes from anterior (A) and posterior (P) tectum were laid down in alternating 50-μm stripes. Explants of temporal (T) or nasal (N) retina were placed above the stripes, allowing RGC axons to grow down on the tectal membrane stripes. **(B)** Temporal axons (bottom image) grow on stripes that contain anterior tectal membrane proteins (labeled A in the top image) and avoid posterior tectal membrane (P). **(C)** Nasal axons grow indiscriminately. The top and bottom images in panels B and C are from the same samples double-labeled by fluorescent markers that recognize posterior tectal membrane preparation (top) and retinal axons (bottom). (B & C, from Walter J, Kern-Veits B, Huf J et al. [1987] *Development* 101:685–696. With permission from The Company of Biologists Ltd.)

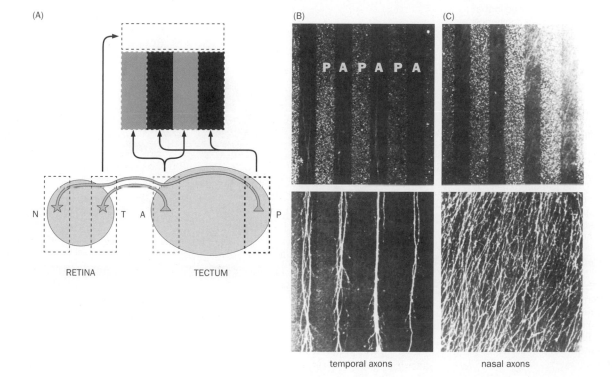

(A)   RETINA   TECTUM

(B)   temporal axons   (C)   nasal axons

on the posterior tectal membranes: they normally target to the anterior tectum because they are repelled by protein component(s) present on the posterior tectal membrane.

A further experiment showed that after treatment of the posterior tectal membrane with an enzyme called phosphatidylinositol-specific phospholipase C (PI-PLC), which releases membrane proteins that are attached to the membrane by a **glycosylphosphatidylinositol** (**GPI**) lipid anchor, resulted in a loss of the repellent activity by the posterior tectal membrane. This experiment suggested that the repellent activity comes from a protein that is anchored to the tectal membrane by GPI.

5.4 Gradients of ephrins and Eph receptors instruct retinotectal mapping

The repellent activity in the posterior tectal membrane was next used in a bioassay to biochemically purify the specific protein(s) that are responsible for repulsion. This led to the identification of a protein now called an **ephrin** (in this case, ephrin-A5). The protein sequence indicates that ephrin-A5 is an extracellular protein attached to the plasma membrane via a GPI lipid anchor, as predicted from the PI-PLC experiment discussed above. Sequence analysis of ephrin-A5 further suggested that it may act as a ligand for the **Eph receptors**, which are transmembrane proteins with a cytoplasmic tyrosine kinase domain (**receptor tyrosine kinases**; see Box 3–4) that are highly expressed in the nervous system. Importantly, purified ephrin-A5 mimicked the posterior tectal membrane in repelling temporal retinal axons in the stripe assay.

Consistent with previous findings suggesting that a repellent is present in the posterior tectal membrane, mRNA of ephrin-A5 is expressed in the tectum as a posterior > anterior gradient, that is, more abundant in the posterior and less abundant in the anterior tectum. Remarkably, EphA3, which is a receptor for ephrin-A5, is expressed in the retina in a temporal > nasal gradient (**Figure 5–7**A). Thus, a scenario emerged that could account for the selectivity in targeting: RGC axons that originate from the most temporal part of the retina express the highest amount of the EphA3 receptor, and therefore are most sensitive to its repellent ligand ephrin-A5 distributed as a posterior > anterior gradient in the tectal target. Thus, the temporal-most RGCs target the anterior-most region of the tectum. RGCs from progressively more nasal positions in the retina express progressively

Figure 5–7 Expression gradients for an ephrin in the tectum and an Eph receptor in the retina. (A) Left, in the retina, *in situ* hybridization (see Section 13.13) shows that EphA3 receptor mRNA is distributed in a temporal > nasal gradient (top panel). The hybridization signal, which is schematically illustrated in the dashed box in the middle panel, is quantified in the bottom panel. The cup-shaped spherical retina was cut along several radial lines so as to be mounted flat (like a clover leaf) for imaging and quantification. N, nasal; T, temporal; D, dorsal; V, ventral. Right, in the tectum, *in situ* hybridization shows that mRNAs for ephrin-A5 are distributed in a posterior > anterior gradient (P > A, top panel). The hybridization signal, which is schematically illustrated in the dashed box in the middle panel, is quantified in the bottom panel. **(B)** A model for retinotectal mapping. Temporal RGCs express high levels of EphA3, and can only target to the anterior tectum, which expresses low levels of the repellent ephrin-A5. Nasal axons express low levels of EphA3 and can target to the posterior tectum because they are less sensitive to repulsion by high levels of ephrin-A5. (A, adapted from Cheng H, Nakamoto M, Bergemann AD et al. [1995] *Cell* 82:371–381. With permission from Elsevier Inc. See also Drescher U, Kremoser C, Handwerker C et al. [1995] *Cell* 82:359–370; B, adapted from Tessier-Lavigne M [1995] *Cell* 82:345–348. With permission from Elsevier Inc.)

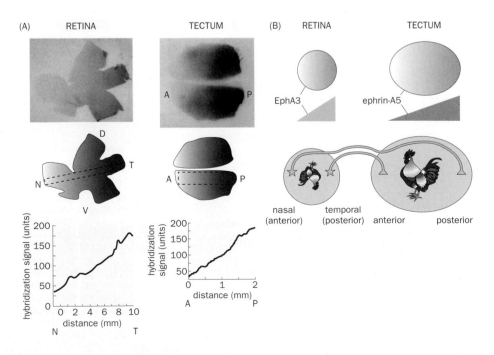

less EphA3 receptor, and therefore are less sensitive to repulsion by ephrin-A5. As a consequence, they target to progressively more posterior regions of the tectum (Figure 5–7B; **Movie 5–1**).

The function of ephrins in retinal axon targeting was confirmed by viral misexpression in chick and knockout experiments in mouse. In the mouse, both ephrin-A5 and a related ephrin-A2 are expressed in a posterior > anterior gradient in the superior colliculus (the mammalian equivalent of the tectum). In mice that were double knockout for *EphrinA5* and *EphrinA2*, temporal axons no longer exhibited selective targeting to the posterior superior colliculus; instead, their axons were scattered along the entire anterior–posterior axis (**Figure 5–8**). Curiously, nasal axons were also affected, and the mistargeted nasal or temporal axons formed clusters; we will return to these observations later in the chapter. The double knockout experiments demonstrated that ephrin-A2 and ephrin-A5 indeed play essential roles *in vivo* to instruct RGC axon targeting along the anterior–posterior axis of the superior colliculus.

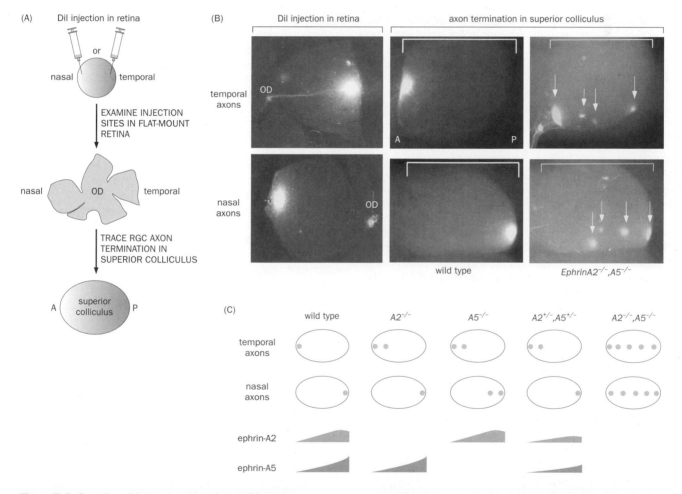

Figure 5–8 Genetic validation for ephrin-A function *in vivo*.
(A) Experimental procedure. Top, fluorescently labeled DiI, a lipophilic dye that diffuses along the lipid bilayer and serves as an axon tracer, was injected into either the temporal or nasal retina. Middle, the retina was flat-mounted by cutting along several radial lines for validating sites of DiI injection. OD, optic disc where RGC axons exit the retina to project to the brain. Bottom, DiI-labeled axon terminations were examined in the superior colliculus. A, anterior; P, posterior.
(B) Analysis of retinotectal projection in wild-type and *EphrinA2/A5* double mutant mice. Left panels show DiI injection sites at the temporal and nasal retina. Right panels show RGC axon termination patterns in the superior colliculus to reveal their distribution. In wild type, temporal or nasal axons project to anterior or posterior superior colliculus, respectively. In *EphrinA2/A5* double knockout mice, both temporal and nasal axons are scattered along the anterior–posterior axis of the superior colliculus (white arrows). **(C)** Schematic summary of axon targeting for five different genotypes: wild type, single *EphrinA* knockouts (*A2^{-/-}* and *A5^{-/-}*), double heterozygotes (*A2^{+/-};A5^{+/-}*), and double knockouts (*A2^{-/-};A5^{-/-}*). Dots represent the terminations of temporal or nasal axons. Bottom panels summarize the protein gradients of ephrin-A2 and -A5. As the number of disrupted ephrin gene copies increases, the targeting defects become more severe. (Adapted from Feldheim DA, Kim Y, Bergemann AD et al. [2000] *Neuron* 25:563–574. With permission from Elsevier Inc.)

The identification of ephrins and Eph receptors as ligand–receptor pairs that are expressed in complementary gradients in the tectum/superior colliculus and retinal axons provided a satisfying proof of a key aspect of Sperry's chemoaffinity hypothesis, proposed 30 years earlier. More broadly, these studies also suggested that similar mechanisms could operate in the wiring of the nervous system in general. These discoveries, along with several other axon guidance molecules identified in the 1990s, launched a new era in the molecular biology of axon guidance (**Box 5–1**).

5.5   A single gradient is insufficient to specify an axis

The posterior > anterior ephrin-A gradient in the tectum and EphA gradient in the retina satisfactorily account for the targeting of temporal axons to the anterior tectum, because temporal axons express the highest amounts of the EphA receptor

---

Box 5–1: Molecular biology of axon guidance

In the mid-1990s, biochemical approaches in vertebrates, cellular studies in grasshopper embryos, and genetic approaches in the roundworm *C. elegans* and the fruit fly *Drosophila* led to the identification of a set of classic **axon guidance molecules**. Studies of these guidance molecules resulted in a general framework regarding the mechanisms by which axons are guided toward their targets (**Figure 5–9**A).

Axons can be guided away from **repellents** (repulsive molecular cues) or toward **attractants** (attractive molecular cues). Each of these two categories includes both **long-range cues**, which are secreted proteins and can act at a distance from their cells of origin, and **short-range cues**, which are cell-surface-bound proteins and require contact between the cells that produce them and the axons they guide in order to exert their effects. These guidance cues act as **ligands** to activate **receptors** expressed on the surface of axonal growth cones (Figure 5–9B; see Box 5–2 for more discussion of growth cones). For example, the ephrin-As are contact-mediated repulsive ligands, since these proteins are bound to membranes by a GPI anchor and repel those RGC axons that express EphA receptors. Some families of

axon guidance cues contain both secreted and membrane-bound proteins; this is the case for the **semaphorins**, which were independently identified by monoclonal antibodies against antigens expressed in specific subsets of axon fascicles in grasshopper embryos and by biochemical purification of a repellent activity in vertebrate neurons. Some guidance cues can act as an attractant in one context and as a repellant in a different context. An important class of contact-mediated attractive molecules is the **cell adhesion molecules**, which were traditionally thought to cause cells to adhere to other cells or to the extracellular matrix. These include <u>i</u>mmunoglobulin superfamily <u>c</u>ell <u>a</u>dhesion <u>m</u>olecules (**Ig CAMs**) and Ca^{2+}-dependent cell adhesion proteins (**cadherins**). These cell adhesion molecules also play important roles in axon guidance. Remarkably, many of these axon guidance cues are evolutionarily conserved across worms, flies, and mammals, not only in their mechanisms of action, but also in certain cases in the biological processes they regulate.

We use **netrin/Unc6** as examples to illustrate evolutionarily conserved mechanisms of axon guidance. In the vertebrate spinal cord, **commissural neurons** are located in the dorsal

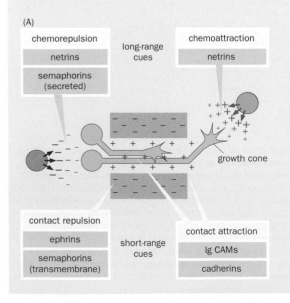

(A)

chemorepulsion
netrins
semaphorins (secreted)

long-range cues

chemoattraction
netrins

growth cone

contact repulsion
ephrins
semaphorins (transmembrane)

short-range cues

contact attraction
Ig CAMs
cadherins

(B)

ligand	receptor
ephrin-A	EphA
ephrin-B	EphB
EphA	ephrin-A
EphB	ephrin-B
netrin/UNC-6	DCC/Unc40; Unc5
semaphorin	plexin; neuropilin; integrin
plexin; semaphorin	semaphorin
cadherin	cadherin
Ig CAM	Ig CAM
Hedgehog	Ptc; Ihog/Boc
Wnt	Frizzled; Derailed/Ryk

Figure 5–9 Molecular mechanisms of axon guidance. (A) Axons can be guided by repulsion or attraction. These repellents or attractants can either be secreted (from the red or green cell, respectively), and therefore able to act at long range, or bound to the cell surface so that they require contact with axons (blue) to exert their effect. **(B)** A list of axon guidance cues (ligands) and their receptors. Note that some molecules can serve as both ligands and receptors. We will encounter many of these proteins in this chapter and Chapter 7. (A, adapted from Tessier-Lavigne M & Goodman CS [1996] *Science* 274:1123–1133. With permission from AAAS.)

spinal cord, and send their axons ventrally toward the **floor plate**, a structure at the ventral midline of the spinal cord, as an intermediate target before crossing the midline to relay sensory information to the brain (we will discuss the function of these neurons in Chapter 6). Biochemical purification of the activity that attracts commissural axons toward the ventral midline identified two related proteins called netrins. Netrins are normally expressed from the floor plate, and when expressed from heterologous cells could induce the turning of commissural axons from spinal cord explants toward the netrin source (**Figure 5–10**A). These experiments demonstrated that netrins act as diffusible attractants to guide commissural axons toward the floor plate (Figure 5–10B, left panel).

Sequence analysis indicated that netrins are secreted proteins that share strong similarity with the *C. elegans* protein Unc6. The *Unc6* gene was originally identified by its mutant phenotypes of uncoordinated movement (hence the name '*Unc*'). Unc6 was shown to be required for circumferential axon guidance along the dorsal–ventral axis in *C. elegans*. In wild-type worms, dorsally located sensory neurons target their axons ventrally, whereas ventrally located motor neurons target their axons dorsally (Figure 5–10B, right panel). In *Unc6* mutant worms, sensory axons exhibited defects in their ventral guidance whereas motor axons exhibited defects in their dorsal guidance. The bi-functionality of Unc6 inspired researchers to test whether vertebrate netrins can also repel axons that normally project dorsally, in addition to attracting commissural axons ventrally. Indeed, floor-plate-derived netrin-1 was shown to be responsible

for the projection of trochlear motor axons away from the ventral midline of the brainstem (Figure 9-10B, left).

Thus, Unc6 and netrins are expressed in the ventral midline, can be chemoattractants or chemorepellents for different axons, and regulate analogous axon guidance events along the dorsal–ventral axis in both worms and vertebrates, including mammals (Figure 5–10B). Further studies indicate that utilization of different receptors in the recipient axons (Figure 5–9B) accounts for the differing responses, that is, attraction versus repulsion; this feature is also conserved across worms and mammals. Specifically, *C. elegans* **Unc40** and its mammalian homolog **DCC** (deleted in colon cancer), when acting alone, mediate attraction to an Unc6/netrin source. In the presence of **Unc5**, a transmembrane protein that acts as a co-receptor of Unc40/DCC for Unc6/netrin from *C. elegans* to mammals, axons turn away from the Unc6/netrin source.

Since axon guidance cues were first identified in the mid-1990s, the list of cues has expanded considerably. Receptors for most of these axon guidance cues have also been identified (Figure 5–9B). Interestingly, some of these ligand–receptor pairs also function earlier in development to pattern cells, tissues, and body axes, as well as to guide migration of cells including neurons. These axon guidance molecules are widely used to wire many different parts of the nervous system and also to regulate later steps of target selection. Some of these molecules are also used to regulate synapse development. We will return to many of these molecules later in this chapter and in Chapter 7.

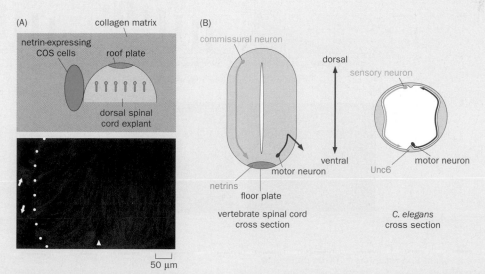

Figure 5–10 Netrin and evolutionarily conserved axon guidance mechanisms from nematodes to mammals. (A) Top, schematic of experimental preparation to test the activity of netrin on commissural axons. The dorsal spinal cord explant that contains the dorsal-most roof plate and commissural neurons (circles) were cultured *in vitro*. Commissural axons would normally grow ventrally, as illustrated. Bottom, when netrin-expressing COS cells were placed along the dotted line to the left of the dorsal spinal cord explant, the commissural axons (stained in red) that were located close to the lateral edge turned left toward the COS cells, and some exited the explant (arrows). Thus, netrin expressed from COS cells attracts commissural axons at a distance. The arrowhead marks the border where netrin has a turning effect on commissural axons. **(B)** Schematic summary of the evolutionarily conserved functions of netrin. Produced at the ventral midline, netrin and Unc6 act as attractive cues for the ventral guidance of commissural neurons in vertebrates and sensory neurons in *C. elegans*. They also act as repellents for the dorsal guidance of motor neuron axons in both vertebrates and *C. elegans*. (A, image from Kennedy TE, Serafini T, de la Torre JR et al. [1994] *Cell* 78:425–435. With permission from Elsevier Inc.; B, based on Serafini T, Kennedy TE, Galko MJ et al. [1994] *Cell* 78:409–424; Colamarino SA & Tessier-Lavigne M [1995] *Cell* 81:621–629; and Hedgecock EM, Culotti JG & Hall DH [1990] *Neuron* 2:61–85.)

for the ephrin-A repellent. But nasal axons also express the EphA receptors, albeit at a low level. Why are they not repelled by the ephrin-A gradient, but instead target to the posterior tectum where there is a high amount of the repellent activity? Additional mechanisms must be employed to explain this puzzle.

One possibility is competition among RGC axons, resulting in the filling of target space. As a consequence of axon–axon competition, nasal axons are 'pushed' to more posterior tectum because the anterior tectum is already occupied by temporal axons that express the highest concentration of a receptor for a posterior repellent. Studies in genetically engineered mice that overexpress EphA in a subset of RGCs provided strong evidence for this model. About 40% of RGCs across the retina express a transcription factor called Islet2. Using a knock-in strategy (see Section 13.7), a constant level of additional EphA receptor was expressed in RGCs under the control of the *Islet2* promoter. This created a gain-of-function condition: intercalated among the RGCs that expressed EphA receptors in a natural gradient were a population of about 40% transgenic RGCs that expressed a constant level of EphA3 driven by the *Islet2* promoter (**Figure 5–11**A, top), thus generating two intercalated gradients with different levels of EphA receptors. So what happened to the retinotopic map in these knock-in mice?

Unlike wild-type mice, which form one-to-one retina–target connections, the knock-in mice had a duplicated map in the superior colliculus: dye injections at a single small spot in the retina labeled two separate target areas along the anterior–posterior axis of the superior colliculus (Figure 5–11A, bottom). Both target areas continued to follow the retinotopy (Figure 5–11B). The simplest interpretation of these data is that, in these knock-in mice, Isl2-negative RGCs formed a map in the superior colliculus according to their endogenous EphA levels, while Isl2-positive RGCs formed a separate map according to EphA levels that were the sum of the

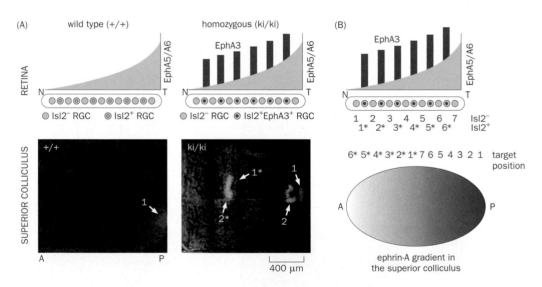

Figure 5–11 Relative levels of Eph receptors in RGCs determine their axon targeting positions. (A) Top, schematic of Eph levels in RGCs from wild-type or homozygous knock-in (ki/ki) mice. Roughly 40% of RGCs (symbolized by cells with open circles) express the transcription factor Islet2 (Isl2), and Isl2⁺ cells are intercalated among Isl2⁻ RGCs across the retina. In the ki/ki mice, these Isl2⁺ cells (symbolized by cells with dark blue circles) express additional EphA3 under the control of the *Isl2* promoter. This EphA3 overexpression (dark blue) is superimposed on the endogenous EphA gradients (light blue), mostly composed of EphA5 and EphA6. N, nasal; T, temporal. Bottom, targeting positions of RGC axons in the superior colliculus (A, anterior; P, posterior) in mice with corresponding genotypes after dye injection in the nasal retina at the positions indicated by numbers in panel B. In the wild-type mouse, a red dye was injected into the retinal area 1. In the ki/ki mouse, a red dye was injected into retinal

area 1+1*, and a green dye was injected into the retinal area of 2+2*. Knock-in mice have a duplicated map in the superior colliculus, which corresponds to the respective terminations of Isl2⁻ (1, 2) and Isl2⁺ (1*, 2*) RGC axons. **(B)** Schematic interpretation of the data from the ki/ki mice. Isl2⁺ RGC axons (1* to 6*) push the Isl2⁻ RGC axons toward the posterior superior colliculus. For example, temporal-most RGC#7 (Isl2⁻) normally targets its axon to the anterior-most superior colliculus because RGC#7 expresses the highest amount of EphA receptors. In the ki/ki mice, RGC#7 is pushed toward the middle of the superior colliculus because its Eph receptor level is exceeded by all of the Isl2⁺ RGCs. Despite the fact that neither its Eph receptor nor the target ephrin levels are altered, RGC#7 targeting is altered because of axon competition. (Adapted from Brown A, Yates PA, Burrola P et al. [2000] *Cell* 102:77–88. With permission from Elsevier Inc.)

graded endogenous EphAs and the constant additional EphA3 expressed from the *Islet2* promoter. This experiment illustrated an important property regarding EphA–ephrin-A interaction: the target positions of RGC axons in the superior colliculus are determined by the relative rather than absolute levels of EphA. For example, an Isl2-negative temporal RGC (number 7 in Figure 5–11B) normally targets to the anterior-most of the superior colliculus because of its high level of EphA expression. However, in the knock-in mice its axon mistargeted to the middle of the superior colliculus despite the fact that its EphA expression level had not changed at all. The mistargeting occurred because its EphA level was surpassed by those of Isl2-positive RGCs overexpressing EphA3. Thus, RGC axons must communicate with each other (directly or indirectly) to make their target selection. Exactly how this communication is achieved is not known.

Besides axon–axon competition, other mechanisms also contribute to targeting along the anterior–posterior axis. One mechanism is bidirectional signaling of ephrin-A–EphA. As well as expressing a temporal > nasal EphA gradient, RGC axons also express a nasal > temporal ephrin-A counter-gradient. Likewise, the tectum (superior colliculus) also expresses an anterior > posterior EphA counter-gradient in addition to the posterior > anterior ephrin-A gradient (**Figure 5–12**A). Besides the ephrin-A → EphA signaling, which was discovered first and called **forward signaling**, EphA can also serve as a repulsive ligand for growth cones expressing ephrin-A, a process called **reverse signaling** (Figure 5–12B). Reverse signaling can explain why nasal axons prefer posterior superior colliculus, as these axons express the highest concentration of ephrin-A, which forces them to choose target regions that express the lowest concentration of EphA. When ephrin-As were knocked out everywhere, no receptors in the nasal RGCs were present to detect the EphA gradient in the superior colliculus, thus abolishing the selectivity.

How are graded expression patterns of axon guidance molecules established during development? This question has been investigated in the context of ephrin-A gradients in the chick tectum. It was found that a transcription factor called Engrailed2 is also expressed in a posterior > anterior gradient, and misexpression of Engrailed2 can cause misexpression of ephrin-A, suggesting that expression of ephrin-A is up-regulated by Engrailed2. Graded Engrailed2 expression is in turn regulated by members of the **fibroblast growth factor** (FGF) family of secreted proteins. During early development, FGF mRNA is produced at the midbrain–hindbrain junction coinciding with the posterior edge of the tectum, such that secreted FGF proteins form posterior > anterior gradients in the tectum. In tectum explant cultures, FGFs can up-regulate Engrailed2 and ephrin-A

Figure 5–12 Counter-gradients and bidirectional signaling of ephrin-A and EphA. (A) Schematic illustration that both ephrin-A and EphA are expressed as counter-gradients in the retina and the superior colliculus (SC). Forward signaling from ephrin-A expressed in SC to EphA-expressing RGCs causes temporal axons to avoid posterior SC, whereas reverse signaling from EphA expressed in SC to ephrin-A-expressing RGCs can cause nasal axons to avoid anterior SC. The arrows indicate RGC axon targeting, not the direction of signaling. **(B)** Illustration of forward and reverse ephrin-A/EphA signaling. In forward signaling, ephrin-A acts as a ligand to send a signal to EphA-expressing growth cones, where the tyrosine kinase domain in the intracellular portion of EphA plays an important role. In reverse signaling, EphA acts as a ligand to send a signal to ephrin-A-expressing growth cones; because ephrin-A is anchored to the membrane by glycosylphosphatidylinositol (red zigzag line) and lacks an intracellular domain, it requires a co-receptor to transduce signals to the growth cone. (A, adapted from Rashid T, Upton L, Blentic A et al. [2005] *Neuron* 47:57–69. With permission from Elsevier Inc.; B, adapted from Egea J & Klein R [2007] *Trends Cell Biol* 17:230–238. With permission from Elsevier Inc.)

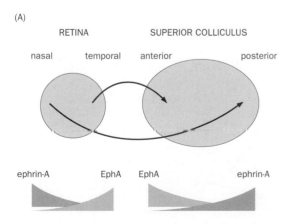

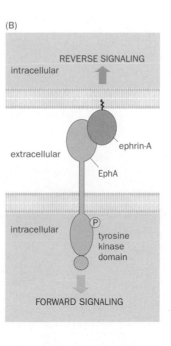

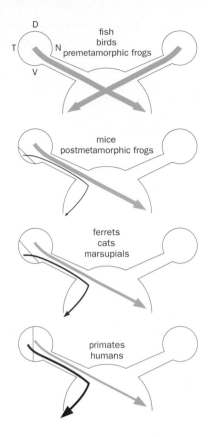

Figure 5–13 Midline crossing of RGC axons. In different animals, different fractions of RGCs project ipsilaterally (that is, to the hemisphere on the same side as the eye). As the fraction of ipsilateral projection increases (represented by the thickness of the red arrow), the animal's degree of binocular vision increases correspondingly. Thus, binocular vision is absent in fish, birds, and tadpoles (top panel), but is highly developed in primates (bottom panel). (Adapted from Petros TJ, Rebsam A & Mason CA [2008] *Annu Rev Neurosci* 31:295–315. With permission from Annual Reviews.)

expression, and down-regulate EphA expression. Thus, developmental patterning molecules such as FGFs help set up graded expression of axon guidance molecules via transcriptional regulation, a theme we will revisit in Chapter 7.

5.6 To cross, or not to cross: that is the question

In all vertebrates, an **optic chiasm** forms as a consequence of many RGC axons crossing the midline to the other side of the brain (**Figure 5–13**; see also Figure 5–1). This is likely an ancestral state because in fish, tadpoles, and birds, all RGC axons cross the chiasm. In these animals, the eyes are typically lateral (located on the sides of the head) and sample mutually exclusive regions of visual space. Information is therefore sent from the left eye to the right brain and from the right eye to the left brain. However, for animals whose eyes have overlapping receptive fields (as in humans), the left and right eyes view partially overlapping regions of visual space to achieve binocular stereovision. Hence, information from each eye is sent to both the left and right brain, and converges in the primary visual cortex (see Figure 4–35). This binocular vision allows us to compute the depth of objects in our visual field.

The degrees of binocular vision in different animals result from different percentages of RGC axons crossing the midline at the optic chiasm (Figure 5–13) as well as different eye positions. In humans, axons that originate from the nasal retina (~60% of total RGCs) cross the midline, constituting the **contralateral** projection, whereas axons from the temporal retina (~40% of total RGCs) remain on their own side, constituting the **ipsilateral** projection. This enables the temporal RGCs from the ipsilateral eye and nasal RGCs from the contralateral eye that look at the same visual space to converge their projections in the vicinity of the same brain hemisphere for binocular vision (see Figure 4–35). In carnivores, such as cats and ferrets, 15–30% of axons are ipsilateral. In mice, only 3–5% of the RGCs located in the ventrotemporal retina project ipsilaterally, so mice have only a small degree of binocular vision. How do RGC axons decide whether or not to cross when they reach the optic chiasm?

Visualizing RGC axon growth cones traveling near the optic chiasm in mouse embryos was revealing (**Figure 5–14**). Growth cones from both ipsilateral and contralateral retinal axons slow down considerably at the chiasm. There, actin-rich structures at the leading edge of the growth cones—thin projections (filopodia) that are surrounded by sheet-like webbing (lamellipodia)—undergo cycles of extension and retraction that are driven by the polymerization and depolymerization of actin and by the motor protein myosin (see more discussion of this subject in **Box 5–2**). The contralateral axons then quickly move forward, whereas the ipsilateral axons continue to extend and retract until ipsilaterally directed filopodia are consolidated and give rise to the new growth cone.

Figure 5–14 Growth cone behavior at the optic chiasm. Time-lapse images of a growth cone from an ipsilaterally projecting RGC in an embryonic mouse retina-chiasm explant, taken as the RGC commits to not crossing the midline (dotted line to the right of the last image). min, minutes. (Adapted from Godement P, Wang L & Mason CA [1994] *J Neurosci* 14:7024–7039. With permission from the Society for Neuroscience.)

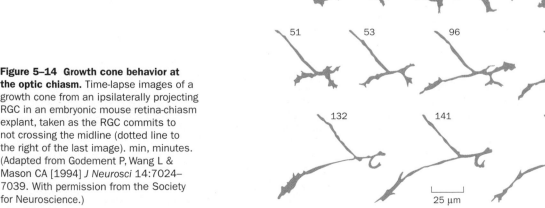

Box 5–2: Cell biology and signaling at the growth cone

How do guidance cues exert their effect on axons to direct their trajectory? The secret lies at the growing tip of the axon, the **growth cone**. Originally discovered by Santiago Ramón y Cajal in his Golgi staining of developing nerve tissues, and subsequently studied by live imaging in tissue culture (see Figure 1–13; **Movie 5–2**), the growth cone contains two prominent structures that consist primarily of filamentous actin (F-actin; see Figure 2–5A). The thin, protruding **filopodia** are made of bundled F-actin, whereas the veil-like **lamellipodia** are a meshwork of branched F-actin (**Figure 5–15**A). Rapid polymerization and depolymerization of F-actin, in combination with myosin-induced flow of F-actin from the periphery to the center of the growth cone (**retrograde flow**), cause the growth cone to exhibit dynamic shape changes, including extension, retraction, and turning. The central part of the growth cone is rich in microtubules (see Figure 2–5B), which provide structural support and a stabilizing force. Like F-actins, microtubules are also highly dynamic; a subset of microtubules protrudes to the leading edge of the growth cone and interacts intimately with the actin cytoskeleton.

The essence of growth cone guidance is to transform the detection of extracellular cues to changes in the underlying cytoskeleton. Suppose that the axon in Figure 5–15B turns downward in response to attractive cues below, repulsive cues above, or both. Turning can be achieved by preferentially stabilizing the filopodia that extend toward the bottom, destabilizing the filopodia that extend toward the top, or a combination of these events. This bias can then be reinforced by the microtubule cytoskeleton via stabilizing dynamic microtubules that extend to the stabilized filopodia and by transporting membranous organelles and vesicles to fuel the growing filopodia via microtubule motors (see Section 2.3). A growing axon extends long filopodia in different directions to better detect the differences of attractive and repulsive cues in its environment, allowing the growth cone to make the most informed comparison to direct its next move.

Axon guidance cues eventually exert their function by modulating the actin cytoskeleton through their receptors and downstream signaling pathways. Many signaling pathways have been documented that link diverse axon guidance receptors to the regulation of the cytoskeleton. For example, small GTPases of the Rho family play crucial roles. Rho GTPases, including Rho, Rac, and Cdc42, regulate the polymerization and depolymerization of actin and

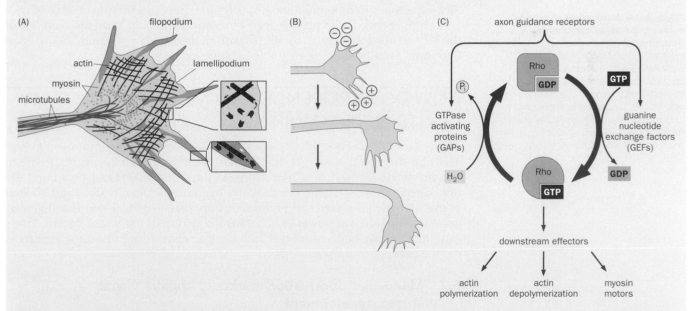

Figure 5–15 Cell biology and signaling at the growth cone.

(A) A typical growth cone contains microtubules at the center and F-actin at the periphery. The F-actin fibers exist in two major forms: branched networks that underlie the lamellipodia, and bundles present in the filopodia. Insets highlight that actin polymerization occurs at the leading edge in both filopodia and lamellipodia, with plus ends (the concave or hollowed-out sides of each arrowhead-shaped monomer) facing the leading edge of the cell close to the membrane. Myosins are distributed throughout the growth cone but are most concentrated at the interface between actin- and microtubule-rich domains. At the periphery, the myosin motors are responsible for retrograde flow of F-actin fibers (white arrow in the lower inset). A subset of microtubules also protrudes to the leading edge, where they interact intimately with the actin cytoskeleton. **(B)** Attractive guidance cues (plus signs) act to stabilize the filopodia,

while repulsive guidance cues (minus signs) act to destabilize the filopodia. The combination of these effects causes the growth cone to turn. **(C)** One of many growth cone signaling mechanisms involves the Rho family small GTPases. Receptors that detect axon guidance cues transmit signals via Rho GTPases to downstream effectors that modulate cytoskeletal activity. Guanine nucleotide exchange factors (GEFs) convert Rho to the active (GTP-bound) form, whereas GTPase activating proteins (GAPs) inactivate Rho by hydrolyzing the bound GTP to GDP. By regulating GEFs and GAPs, axon guidance receptors control switching between the active and inactive forms of Rho. (Adapted from Luo L [2002] *Ann Rev Cell Dev Biol* 18:601–635. See also Dent EW, Gupton SL & Gertler FB [2011] *Cold Spring Harb Perspect Biol* 3:a001800.)

(Continued)

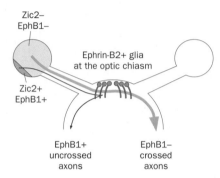

Figure 5–16 Mechanism of midline crossing of RGC axons in mice. Ventromedial retina specifically expresses the transcription factor Zic2, which activates EphB1 expression. At the chiasm, EphB1-expressing RGCs choose the ipsilateral trajectory because they are repelled by glial cells that express ephrin-B2. (Adapted from Petros TJ, Rebsam A & Mason CA [2008] *Ann Rev Neurosci* 31:295–315.)

The molecular basis for the differing behaviors of ipsilaterally and contralaterally projecting axons has been identified in mice (**Figure 5–16**). During the developmental window when RGC axons project to the brain, EphB1 (another member of the Eph receptor family we encountered in previous discussions), is specifically expressed in the ventrotemporal areas of the retina precisely coincident with the position where ipsilateral RGC axons originate. Glial cells at the optic chiasm express ephrin-B2, which repels EphB1-expressing RGCs and prevents them from crossing the midline. The restricted expression of EphB1 is enabled by the specific expression of the transcription factor Zic2 in the ventrotemporal retina. Misexpression of Zic2 in other parts of the retina resulted in misexpression of EphB1, forcing otherwise contralateral axons to project ipsilaterally. Interestingly, Zic2 expression in different animal species precisely predicts the degree of ipsilateral projection, and hence the degree of binocular vision. Thus, the evolution of the Zic2 expression pattern, combined with changes in eye position, was likely an important force behind the evolution of binocular vision.

HOW DO EXPERIENCE AND NEURONAL ACTIVITY CONTRIBUTE TO WIRING?

So far, our studies of how RGC axons select their targets in the brain seem to suggest that the visual system is 'hard-wired,' with molecular recognition able to explain nearly all aspects of how axons find their targets. However, around the time Sperry proposed the chemoaffinity hypothesis, David Hubel and Torsten Wiesel performed a series of experiments on visual cortical neurons that suggested a different, yet powerful, force for the wiring of the visual system. These experiments have had just as much influence on generations of neuroscientists as did the chemoaffinity hypothesis.

5.7 Monocular deprivation markedly impairs visual cortex development

To refresh what we learned in Chapter 4, recall that some neurons in the primary visual cortex manifest binocular responses by combining inputs from neurons of the lateral geniculate nucleus (LGN). Each LGN neuron receives input from only one eye, and neurons representing different eyes are segregated into eye-specific layers at the LGN. Due to RGC axons crossing at the optic chiasm (see Figure 5–13), some LGN neurons receive input from the temporal part of the ipsilateral eye, whereas other LGN neurons receive input from the nasal part of the contralateral eye. Thus, in each LGN, different neurons receive information from each eye viewing the same visual space. LGN signals from the left or right eye eventually converge onto individual cortical neurons to produce binocular vision (see Figure 4–35B).

When Hubel and Wiesel mapped receptive fields of single neurons in the primary visual cortex (V1) of the cat (see Section 4.25), they found that individual

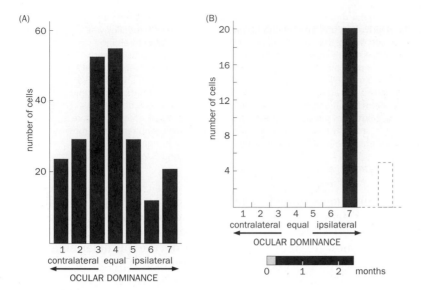

Figure 5–17 Visual response of cortical neurons under normal and monocular deprivation conditions. (A) Neurons in the cat primary visual cortex (V1) are divided into seven categories based on their response to visual stimuli from the ipsilateral and contralateral eyes. Those that respond exclusively to contralateral or ipsilateral stimuli are given a value of 1 or 7, respectively, while cells that respond equally well to stimulus from either eye receive the value 4. Values 2, 3, 5, and 6 represent cortical neurons that exhibit different degrees of contralateral or ipsilateral bias. The plot shows the distribution of scores given to 233 cells recorded from V1 of a normally reared cat. **(B)** When the contralateral eye was deprived during the period indicated by the black bar underneath by eyelid suture, virtually all neurons responded only to visual stimulation from the ipsilateral eye after the suture was reopened. Five cells (represented in the distribution by a dashed bar) could not be driven by either eye. (Adapted from Wiesel TN & Hubel DH [1963] *J Neurophysiol* 26:1003–1017. With permission from the American Physiological Society.)

cells responded preferentially to stimuli from the ipsilateral eye, the contralateral eye, or both eyes that see the same visual space. They plotted these responses according to a 1 to 7 **ocular dominance** scale (**Figure 5–17**A). Cells that received a score of 1 or 7 responded exclusively to visual input from the contralateral or ipsilateral eye, respectively. A score of 4 indicated that the cell responded equally well to both eyes, whereas scores of 2, 3, 5, and 6 represented differing degrees of contralateral or ipsilateral bias. Cells that possessed the same ocular dominance properties were found to cluster spatially along the vertical axis of the cortex, which were named **ocular dominance columns**.

To probe whether visual experience has anything to do with the development of the receptive field of cortical neurons, Hubel and Wiesel sutured one eyelid of the cat from birth, thus depriving much of its visual input. They then opened the sutured eye when the cat was several months old, and recorded the visual responses along the visual pathway. Little difference was found between responses recorded from the retinal neurons of the sutured and open eyes or from the neurons of the two LGNs, but a profound difference was found in V1 neurons. The great majority of cortical cells were driven by the normal eye; very few cortical cells responded properly to stimulation from the deprived eye (Figure 5–17B). Essentially the same results were obtained in later experiments performed on monkeys. Behaviorally, these animals lost visual functions in their deprived eye, consistent with physiological recordings in the cortex.

Hubel and Wiesel further performed a series of time-course experiments to determine when during development monocular deprivation had the most significant effect. They found a window that they termed the **critical period**, which in cats begins abruptly at about the fourth week after birth (coincident with the kittens opening their eyes) and ends gradually around 12 weeks of age. Monocular deprivation after the critical period had little effect compared to the large effect during the critical period. For example, closing one eye for just a few days during the period of the fourth to fifth week had devastating effects both on vision and on cortical responses to the deprived eye, while much longer closure later in life had no significant consequences. For the most part, visual system defects arising from monocular deprivation during the critical period were irreversible.

The drastic effect of visual deprivation was demonstrated when radioactive amino acid was injected into a single eye to visualize the ocular dominance columns in layer 4 of V1 by trans-neuronal tracing (see Section 4.25). In normal animals, radioactively labeled and unlabeled stripes had similar widths, representing the equal contribution of inputs from left and right eyes to layer 4 of V1 (**Figure 5–18**, bottom). In monocularly deprived monkeys and cats, most layer 4 of V1 came to be dominated by input from the non-deprived eye (Figure 5–18, top).

monocular deprivation

normal

└─┘
1 mm

Figure 5–18 Ocular dominance columns visualized by autoradiography under monocular deprivation and normal conditions. Top, one eye of a monkey was surgically closed during the critical period, and radioactively labeled amino acids were subsequently injected into the open eye. These amino acids were incorporated into proteins, transported from RGC axons to postsynaptic targets in the LGN and ultimately to the axon terminals of LGN neurons in layer 4 of V1, and visualized by autoradiography. Note that the bright bands (representing input to the cortex from the injected, open eye) are much wider than the dark bands (representing input from the deprived eye). Bottom, the same labeling procedure performed on a normally reared monkey revealed that ocular dominance columns receiving input from the injected and non-injected eyes are of equal width. (From Hubel DH, Wiesel TN & LeVay S [1977] *Phil Trans R Soc Lond B* 278:377–409. With permission from the Royal Society of London.)

These experiments have had profound implications on disciplines ranging from neuroscience and developmental psychology to ophthalmology. They revealed the powerful force of nurture, or experience, in sculpting the developing brain. They also indicated that chemoaffinity is not the sole determinant for brain wiring. But how is brain wiring shaped by experience?

5.8   Competing inputs are sufficient to produce spatial segregation at the target

An important clue came from a control experiment by Hubel and Wiesel. When all vision was deprived during the critical period by suturing the eyelids of both eyes, the cortical cells demonstrated visual responses that, although still highly abnormal, were not as drastically affected as those resulting from monocular deprivation. Thus, the effect on deprivation is not simply atrophy due to disuse. One possible explanation of these data is that input from the left and right eyes, via their thalamocortical axons originating from eye-specific layers of the lateral geniculate nuclei (see Figure 4–35B), compete for space in the visual cortex. Visual experience, in the form of correlated neuronal activity (see Sections 5.11–5.12 below), may provide a competitive advantage to the non-deprived eye. Under binocular deprivation, inputs from both eyes are weakened and the resulting defect is not as severe as that caused by monocular deprivation.

How could one experimentally strengthen the causality between competitive inputs from the two eyes and the formation of ocular dominance columns? As we noted in Section 1.14, two kinds of perturbation experiments, loss-of-function and gain-of-function, can often be used to test causality between two phenomena. The experiments we have described so far belong to the loss-of-function category, testing whether experience from—or activity in—the eyes is necessary for the development of ocular dominance columns in V1. Can one design a gain-of-function sufficiency experiment to strengthen the causality between competing inputs and the generation of ocular dominance columns?

The key is to identify a normal condition in which there is minimal competition between inputs from two eyes. This condition can readily be found in nature. Recall that binocular vision exists to varying degrees in different vertebrates (see Figure 5–13). The frog has few RGCs that project ipsilaterally, so its tecta receive input mostly from the contralateral eye, and there is little opportunity for inputs from both eyes to compete. Remarkably, researchers created novel circumstances for such competition by utilizing the superb embryological techniques available in frogs. A third eye was created by transplanting an eye primordium to a normal frog embryo at the appropriate location and developmental stage (**Figure 5–19A**). The retinal axons from the third eye projected to one of the tecta, targeting appropriate

Figure 5–19 Competing retinal inputs from the third eye produce segregated RGC axons in the tectum. **(A)** Transplantation of an eye primordium to a host embryo at an appropriate developmental stage can produce a frog with three eyes. **(B)** When axonal input from the third eye competes for tectal targets with axonal input from one of the normal eyes, segregation of axons appears, as indicated by trans-neuronal tracing of radioactively labeled amino acids injected into a normal eye. Inset, enlargement showing axonal input segregation in dark field, with the labeled axon termini appearing white. (From Constantine-Paton M & Law MI [1978] *Science* 202:639–641. With permission from AAAS.)

(A)

(B)

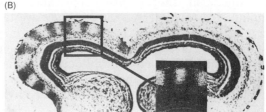

retinotopic positions in response to the molecular cues discussed earlier in the chapter. The third-eye axons then had to compete for space with axons from a normal eye. When radioactively labeled amino acids were injected into the normal eye that projected to the same tectum as the third eye, inputs from two eyes that projected to the same tectum were segregated into bands of a characteristic width (Figure 5–19B), resembling the ocular dominance columns in the primary visual cortex of normal cats and monkeys.

This experiment provided an elegant demonstration that competing inputs are sufficient to generate segregated bands at their postsynaptic targets. Follow-up experiments showed that blocking neuronal firing (by injecting tetrodotoxin, or TTX, into the eye to block voltage-gated Na⁺ channels and thus action potential propagation) resulted in the desegregation of eye-specific inputs into the tecta of three-eyed frogs. These findings supported a role for neuronal activity in the generation and maintenance of segregated tectal inputs from each eye, similar to the role of neuronal activity in the development of ocular dominance columns in the visual cortex, to which we now return.

5.9 Ocular dominance columns in V1 and eye-specific layers in LGN develop by gradual segregation of eye-specific inputs

The striking appearance of ocular dominance columns in layer 4 of V1 in cats and monkeys, and their perturbation during the critical period by monocular visual deprivation (see Figures 5–17 and 5–18), led researchers to ask how the columns first form during development. The autoradiography method of visualizing ocular dominance columns was used to probe this question. It was found that during early development, monocular injection of radioactive amino acids into either eye labeled a continuous band in layer 4 of V1. This band gradually became intermittent as development proceeds, until it achieved the degree of segregation observed in adults. Likewise, physiological recordings found that during early development, most cortical neurons respond, albeit weakly, to inputs from both eyes. However, subsequent experiments revealed that radioactive amino acids show much greater lateral diffusion (or spillover) in young brains, leading to an overestimate of initial exuberance. Experiments utilizing intrinsic signal imaging and anterograde tracing from individual LGNs suggest that ocular dominance columns form earlier than revealed by the autoradiography method. The exact time of ocular dominance formation is still not completely resolved due to technical limitations, but researchers agree that they occur early during development (see below) and the boundaries of these columns become sharper as development proceeds.

Inputs from the eyes are conveyed to the cortex via LGN neurons (see Figure 4–35), where the process of eye-specific input segregation has been more precisely defined. The autoradiography method suggested that RGC axonal projections from both eyes were intermixed in the LGN during early development. This input gradually became refined and eventually segregated into eye-specific LGN layers. These observations were confirmed by visualizing the terminal arborizations of individual RGC axons. For example, an RGC axon that terminates in the LGN initially forms branches along its entire path across the LGN. As layer-specific segregation occurs, the RGC axon branches more densely within what will become the layer that is specific for its eye; at the same time, its branches within the layer destined to be occupied by the other eye are pruned (**Figure 5–20**).

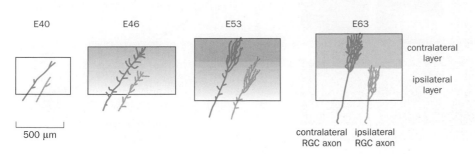

E40   E46   E53   E63

contralateral layer

ipsilateral layer

500 μm

contralateral   ipsilateral
RGC axon   RGC axon

Figure 5–20 Formation of eye-specific layers of RGC axons in the LGN in the cat. Gradual appearance of eye-specific layers (segregation of gray and white background, as determined by labeling of radioactive amino acids injected into one eye) in the cat LGN occurs as individual axons prune their branches in the incorrect layer and elaborate their branches in the correct layer. Embryonic days after fertilization are noted above. (Adapted from Sretavan DW and Shatz CJ [1986] *J Neurosci* 6:234–251.)

Thus, it appears that eye-specific layers in the LGN and ocular dominance columns in V1 form by a process of exuberant connections followed by subsequent pruning. Comparing the onset of the ocular dominance columns with the critical period defined by monocular deprivation revealed that the ocular dominance columns form well before the onset of the critical period. This suggested that the ocular dominance columns are consolidated by experience and neuronal activity during the critical period, but that some earlier processes contribute to initial column formation. What might these earlier processes be?

5.10 Retinal neurons exhibit spontaneous waves of activity before the onset of vision

In monkeys, ocular dominance columns form prenatally, during a period when the eyes are naturally occluded. Eye-specific segregation in the LGN occurs even earlier, well before the onset of vision. If competing inputs from the two eyes are responsible for eye-specific segregation of RGC axons in the LGN and for initial development of the ocular dominance columns in V1, what is the nature of such competition?

Experiments utilizing TTX injection to block activity suggested the importance of **spontaneous neuronal activity** in ocular dominance column development. Binocular TTX injection blocked the formation of ocular dominance columns in V1 much more completely than dark rearing or binocular eyelid suturing, suggesting that in addition to patterned visual stimulation, which is degraded by dark rearing or suturing, spontaneous retinal activity plays an important role. Likewise, blocking prenatal spontaneous activity in the retina prevented eye-specific segregation of RGC axons at the LGN. But how does spontaneous activity create competition?

It turns out that before the onset of vision, the retina exhibits spontaneous neuronal activation, with nearby RGCs exhibiting correlated activities. These findings were first described in prenatal retina of rats, and subsequently investigated in more detail in ferrets. This is because ferrets have more advanced binocular visions than rodents (see Figure 5–13), and are born in a much more immature state compared with cats and monkeys, such that experiments investigating their visual system development can be carried out in postnatal animals. The correlated activities of RGCs could be seen by multi-electrode recording of isolated retina from young ferrets: extracellular electrodes 50–500 μm apart detected sequential action potentials of nearby retinal neurons (**Figure 5–21**A). This spread of spontaneous excitation of retinal neurons (which includes RGCs and amacrine cells) across the retina has since been called a **retinal wave**.

These retinal waves were more dramatically demonstrated by Ca²⁺ imaging experiments (Figure 5–21B; **Movie 5–3**), which enabled researchers to visualize

Figure 5–21 Retinal neurons exhibit waves of spontaneous action potentials. **(A)** A retinal wave is shown sweeping through the retina from bottom right to top left in 3 seconds (s). Each dot represents one electrode in a multi-electrode array on which a postnatal day 5 ferret retinal explant was placed. Dots with larger size represent higher frequencies of RGC action potential firing. **(B)** Retinal waves can also be detected by Ca²⁺ imaging of retinal explants from young ferrets. Dark signals represent increased intracellular Ca²⁺ concentration, which reports higher neuronal activity. Two separate waves are shown (top and bottom panels), with different origins and directions of spread. **(C)** Cross correlation of retinal neuron activity as a function of distance from Ca²⁺ imaging experiments. When a neuron in a defined central position is active, cross correlation measures how often another neuron at a given distance from the central neuron is also active; a cross correlation of 1 means that two neurons are always co-active. As is evident, neurons are more likely to be co-active if they are nearby. (A, adapted from Meister M, Wong ROL, Baylor DA et al. [1991] *Science* 252:939–943. With permission from AAAS; B, from Feller MB, Wellis DP, Stellwagen D et al. [1996] *Science* 272:1182–1187. With permission from AAAS; C, adapted from Feller MB, Butts DA, Aaron HL et al. [1997] *Neuron* 19:293–306. With permission from Elsevier Inc.)

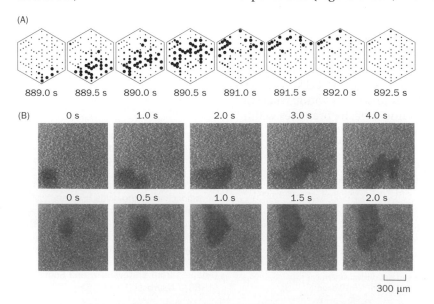

(A)

889.0 s   889.5 s   890.0 s   890.5 s   891.0 s   891.5 s   892.0 s   892.5 s

(B)   0 s   1.0 s   2.0 s   3.0 s   4.0 s

0 s   0.5 s   1.0 s   1.5 s   2.0 s

300 μm

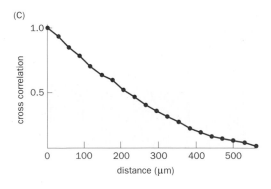

(C)

activity of many neurons simultaneously by imaging fluorescence changes in response to a rise of intracellular Ca^{2+} concentration (see Section 13.22). As waves of neuronal activation sweep across the retina, nearby RGCs in the same retina are activated at about the same time (Figure 5–21C). By contrast, activities from RGCs in different retinas are uncorrelated. As we will learn in Section 5.12, correlated activity patterns from the same eye strengthen synaptic connections, whereas uncorrelated activity patterns weaken synaptic connections. Thus, correlation of activity patterns in the same eye by retinal waves could account for the segregation of RGC axons into eye-specific layers in the LGN.

Further studies of the retinal waves identified several distinct phases of wave propagation during development, before eye opening. The most prominent phase, which produces cholinergic retinal waves involving connections between cholinergic amacrine cells and RGCs, coincides with RGC axons making their target choices in the LGN and superior colliculus. During this period, rods and cones have not been born, and bipolar connections to the RGCs have not been established, consistent with the idea that these retinal waves are spontaneous in RGCs rather than evoked by visual experience. Interfering with cholinergic transmission from cholinergic amacrine cells to RGCs can disrupt cholinergic retinal waves; this provided an opportunity to test the role of retinal waves in the wiring of the visual system.

5.11 Retinal waves and correlated activity drive segregation of eye-specific inputs

Epibatidine, an alkaloid found on the skin of an endangered species of poison dart frog, is a long-lasting cholinergic agonist that blocks retinal waves by binding tightly to the nicotinic acetylcholine receptor (nAChR) expressed by RGCs and amacrine cells and causing receptor desensitization. Thus, epibatidine blocks synaptic transmission and hence cholinergic retinal waves. Injection of epibatidine into the eye therefore provided a loss-of-function condition to test the physiological role of retinal waves in the wiring of the visual system. Unlike the cat where eye-specific segregation of RGC axons occur prenatally (see Figure 5–20), eye-specific segregation of RGC axons in the ferret LGN occurs during the first 10 postnatal days, allowing these perturbation experiments. Eye-specific segregation was visualized by injecting the left and right eyes with different axonal tracers to differentially label their RGC axons (**Figure 5–22**A). At postnatal day 1, RGC axons from the left and the right eyes were completely intermixed in the binocular region of the LGN (Figure 5–22B). Nine days later, axons from the ipsilateral and

Figure 5–22 Retinal waves regulate eye-specific segregation of RGC axons at the LGN. (A) Schematic of experimental procedure. Two different axonal tracers (pseudocolored in red and green) were injected into the left or right retina of the ferret to label RGCs from each eye, and RGC axon termination patterns at lateral geniculate nuclei (LGNs) were examined a day later. Shown in the schematic is a control experiment examined at postnatal day 10 (P10). The binocular zone, where eye-specific inputs are segregated into distinct layers, is bracketed. Panels B–F show RGC axon terminals at LGNs under different experimental conditions. **(B)** In a control (saline injected) ferret at postnatal day 1 (P1), RGC axons are not segregated, as seen by the yellow patches in the future binocular zone. **(C)** By P10, the binocular zone is well segregated in a control ferret, with ipsilateral axons occupying the A1 layer, flanked by the contralateral A and C layers. Between the arrowheads lies a monocular zone that represents only the contralateral eye. **(D)** When both eyes were injected with epibatidine to block retinal waves, eye-specific segregation did not occur; the binocular zone is largely yellow as in P1. **(E)** When epibatidine was injected monocularly into the right eye, axons from the left eye (green) take over the binocular zone in both LGNs. **(F)** The cAMP agonist CPT-cAMP augments retinal waves. Injecting CPT-cAMP into the left eye expands the area occupied by RGC axon terminals from the left eye (green) in the binocular zones of both LGNs. (B–E, from Penn AA, Riquelme PA, Feller MB et al. [1998] *Science* 279:2108–2112. With permission from AAAS; F, from Stellwagen D & Shatz CJ [2002] *Neuron* 33:357–367. With permission from Elsevier Inc.)

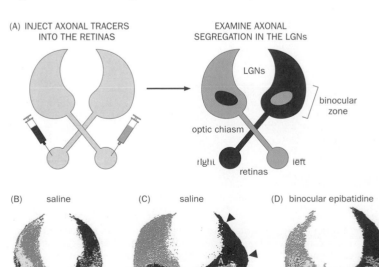

(A) INJECT AXONAL TRACERS INTO THE RETINAS

EXAMINE AXONAL SEGREGATION IN THE LGNs

LGNs

binocular zone

optic chiasm

right   left

retinas

(B)   saline   (C)   saline   (D)   binocular epibatidine   (E)   monocular epibatidine   (F)   monocular CPT-cAMP

A1   A

500 µm

P1   P10

contralateral eyes were normally well segregated (Figure 5–22C). Binocular epi-batidine injection prevented the segregation of eye-specific input (Figure 5–22D), suggesting that cholinergic retinal waves are essential for the eye-specific segrega-tion of RGC axons in the LGN.

Interestingly, monocular injection of epibatidine caused RGC axons from the injected eye to lose territory to axons from the uninjected eye (Figure 5–22E). On the other hand, a gain-of-function condition in which one eye was injected with a cAMP agonist known to increase the frequency of retinal waves caused RGCs from the injected eye to outcompete axons from the normal eye for territory in the LGN (Figure 5–22F). Together, these experiments demonstrated that competitive interactions between inputs from two eyes, dependent upon spontaneous activity generated by retinal waves, plays a critical role in the proper segregation of eye-specific axons in the LGN.

As discussed in Section 5.10, retinal waves likely act in the context of eye-specific segregation by providing correlated activity patterns within the same eye. In the mouse, a direct test for the role of correlated activity in eye-specific layer segregation of RGC axons in the superior colliculus has been performed using optogenetic stimulation. In the small binocular zone of the mouse superior col-liculus, ipsilateral and contralateral RGCs normally segregate their axons in dif-ferent layers between postnatal days 5 and 9, before the onset of natural vision. Combining tools in viral delivery and mouse genetics, light-activated channelrho-dopsin (ChR2; see Box 4–3 and Section 13.25) was expressed in the mouse RGCs, and blue light that activates ChR2 was delivered to both eyes in either synchro-nous or asynchronous patterns from postnatal day 5 on. Synchronous activation of RGCs from the two eyes disrupted eye-specific segregation of their inputs in the superior colliculus, whereas asynchronous activation facilitated the segrega-tion (**Figure 5–23**). The simplest interpretation for these results is that correlated activity patterns, whether provided naturally by retinal waves within the same eye or artificially by synchronous optogenetic stimulations in different eyes, cause axons to stabilize projections in the same layer. By contrast, uncorrelated activity patterns, provided naturally by uncorrelated retinal waves from different eyes or artificially by asynchronous optogenetic stimulations, facilitate axon segregation.

Recent *in vivo* imaging experiments in neonatal mice (**Figure 5–24**A), before eye opening, revealed that waves of spontaneous activity occur across many areas

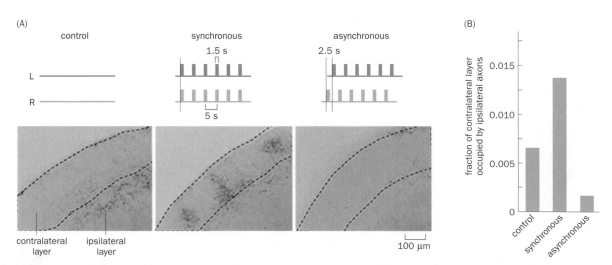

Figure 5–23 Synchrony of firing patterns and eye-specific segregation of axons. (A) Top panels, schematic of the experimental procedure. Channelrhodopsin (ChR2) was expressed in RGCs from both eyes of transgenic mice. In the two experimental treatments, light was delivered to the two eyes either in synchrony (middle panel: vertical bars indicate 1.5 s synchronous optogenetic stimulations to both eyes every 5 s) or in asynchrony (right panel: 1.5 s optogenetic stimulations every 5 s, with the onset of stimuli to each eye 2.5 s apart) from postnatal days 5 through 7. In the control treatment at left, no light was delivered. Bottom panels, ipsilateral axons labeled in black were visualized in the contralateral layer between the two dashed lines; images were taken at P9. In control animals, the contralateral layer contained few residual ipsilateral axons. The abundance of ipsilateral axons in the contralateral layer was enhanced by synchronous optogenetic activation of both eyes but was diminished by asynchronous activation of the left and right eyes. **(B)** Quantification of these effects. (Adapted from Zhang J, Ackman JB, Xu H et al. [2012] *Nat Neurosci* 15:298–309. With permission from Macmillan Publishers Ltd.)

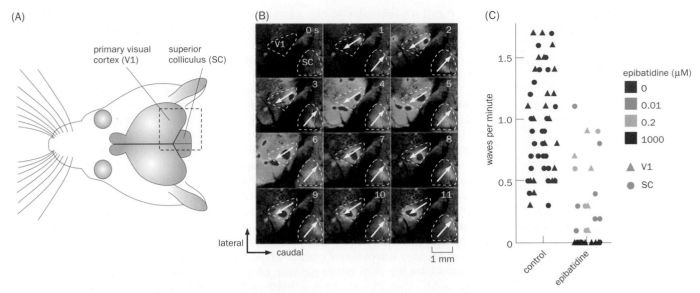

Figure 5–24 Retinal waves propagate to the superior colliculus and V1. (A) Experimental condition. A genetically encoded Ca^{2+} indicator, GCaMP3 (see Section 13.22), is expressed by Cre/loxP mediated binary system (see Section 13.10) in the RGCs as well as in cortical neurons in transgenic mice (green). Ca^{2+} imaging was performed in the boxed region, which includes the superior colliculus (SC, where RGC axons terminate) and V1 (which receives RGC input via intermediate neurons in the LGN). **(B)** Time-lapse imaging (with time indicated at the top right corner of each image) shows a wave of activity from rostromedial to caudolateral superior colliculus, reflecting the retinal waves detected at RGC axon terminals. At the same time, a wave of activity sweeps from caudolateral to rostromedial V1. Arrows in SC and V1 indicate the direction of wave propagation; they correspond to the same retinotopy (see Figure 5–31). This suggests that the V1 wave is a result of the retinal wave propagating to the cortex. **(C)** The frequency of waves detected in the SC and V1 is reduced or abolished by application of low or high concentration of epibatine in the contralateral eye. Each circle or triangle represents waves per minute in SC or V1 measured from one movie. Whereas the loss of collicular waves in this experiment reflects that fact that Ca^{2+} signals in the collicular imaging came from RGC axons, the loss of V1 waves suggests that spontaneous waves in V1 originate mostly from cholinergic retinal waves propagating to the cortex. (Adapted from Ackman JB, Burbridge TJ & Crair MC [2012] *Nature* 490:219–227. With permission from Macmillan Publishers Ltd.)

of the visual pathways, including the superior colliculus, the primary visual cortex (V1), and the higher-order visual cortical areas. Waves of spontaneous activity in the superior colliculus and V1 are often coordinated not only in timing but also with respect to their retinotopy (Figure 5–24B), suggesting that they originate from the retina. Indeed, epibatidine injection into the contralateral eye (recall that >95% RGCs in the mouse project contralaterally) severely diminished spontaneous waves in V1 (Figure 5–24C), supporting the notion that they are triggered by cholinergic retinal waves propagating along the ascending visual pathway.

5.12 Hebb's rule: correlated activity strengthens synapses

How do correlated or uncorrelated activity patterns affect neuronal wiring? To answer this question, we begin with a psychologist named Donald Hebb. In his 1949 book entitled *The Organization of Behavior*, Hebb made a 'neurophysiological postulate' to explain how learning can be transformed into lasting memory: "When an axon of cell *A* is near enough to excite a cell *B* and repeatedly or persistently takes part in firing it, some growth process or metabolic change takes place in one or both cells such that *A*'s efficiency, as one of the cells firing *B*, is increased."

Hebb had no biological data to support his postulate—he simply thought that such a rule, if it existed, would provide a structural basis to explain memory. He could not have imagined that his postulate would have a long-lasting influence not only on the study of memory and learning (which will be discussed in detail in Chapter 10), but also on activity-dependent wiring of the brain. His postulate, now called **Hebb's rule** (**Figure 5–25**A), has repeatedly received experimental support. So has an important extension to the converse (Figure 5–25B): when the presynaptic axon of cell *A* repeatedly and persistently fails to excite the postsynaptic cell *B* while cell *B* is firing under the influence of other presynaptic axons, metabolic change takes place in one or both cells such

(A)

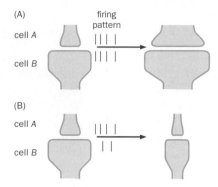

(B)

Figure 5–25 Schematic depiction of Hebb's rule. The presynaptic terminal of cell A and postsynaptic site of cell B are shown. **(A)** According to Hebb's rule, if cell A repeatedly or persistently takes part in exciting cell B, so that their activity patterns are correlated as shown, the synaptic connection between them strengthens. **(B)** Hebb's rule has been extended for the converse: if cell A repeatedly and persistently fails to excite cell B, so that their activity patterns are uncorrelated, then the synaptic connection between them weakens. Changes in connection strength are depicted here as changes in the size of the synapse, but connection strength changes in the developing nervous system can also be realized by turning transient contacts into new synapses or by breaking existing synapses between two cells. (See Hebb DO [1949] The Organization of Behavior. LEA Inc.; Stent GS [1973] *Proc Natl Acad Sci USA* 70:997–1001.)

that cell A's efficiency at firing cell B is decreased. A further extension of Hebb's rule that applies particularly to activity-dependent wiring during development is that strengthening a weak synapse can convert a transient connection to a more stable one, whereas progressive weakening of an existing synapse can eventually eliminate the connection altogether.

As a specific example, let's examine how Hebb's rule and its extensions can explain the segregation of RGC axons into eye-specific layers in the binocular zone of the LGN (**Figure 5–26**). Early during development, RGC axons from both the left and right retinas form connections with postsynaptic target neurons residing in areas that will eventually develop into left or right eye-specific layers (see Figure 5–20). Consider as an example an LGN neuron in the left-eye-specific layer. Although it receives input from RGCs from the left and right eyes, it receives more synaptic inputs from left-eye RGCs (Figure 5–26, top), perhaps in response to molecular cues. The retinotopic mapping we studied in the superior colliculus also applies to the LGN (see Section 5.15), such that inputs originating from RGCs that are physically close to each other in the retina are activated together by the same retinal waves (see Figure 5–21C). These coordinated inputs are sufficient to depolarize the LGN target neuron, causing it to fire action potentials. According to Hebb's rule, the synapses between left-eye RGCs and the LGN neuron will be strengthened as a result. RGCs originating from the right eye fire action potentials with a different temporal pattern, since their retinal waves are not correlated with the retinal waves from the left eye. Given the relative weakness of their initial connections, the right-eye RGCs frequently fail to depolarize the target neuron sufficiently to cause it to fire, further weakening these connections according to the extension of Hebb's rule. Repeated rounds of this process, enabled by persistent retinal waves, ensure that the left-eye RGCs win the battle for connecting to this neuron; the right-eye RGCs withdraw their connections (Figure 5–26, bottom). In another part of the LGN, right-eye RGCs have the initial advantage and out-compete the left-eye RGCs. This explains why competition between the inputs from two eyes to drive the firing of the postsynaptic neurons is essential in the segregation of eye-specific layers. Colloquially, these phenomena are often expressed as "Firing together, wiring together. Out of sync, lose your link."

The same principles apply to the formation of ocular dominance columns in layer 4 of V1, although the particular neurons involved would differ, with the presynaptic LGN neurons targeting postsynaptic V1 layer 4 neurons. It is significant that, during development, eye-specific segregation of RGC axons at the LGN

Figure 5–26 Application of Hebb's rule to eye-specific connections between RGCs and an LGN neuron. At an early developmental stage, the LGN neuron (green) receives RGC input from both eyes (top). More RGCs from the left eye (blue) are connected to the LGN neuron, making the LGNs more likely to fire according to left-eye RGC firing patterns (as indicated by the similarity of vertical bars in blue and green). Coincident firings of pre- and postsynaptic neurons strengthen the synapses between left-eye RGCs and the LGN neuron. The non-coincident firings of right-eye RGCs (orange) and the LGN neuron weaken their synapses. As development proceeds, right-eye RGC synapses are lost while left-eye RGC axons and LGN dendrites form new branches and connections; eventually, the LGN neuron connects only to the left eye (bottom). Molecular cues likely create the initial connection bias between LGN neurons and left- or right-eye RGC axons, because eye-specific layers are stereotyped across animals.

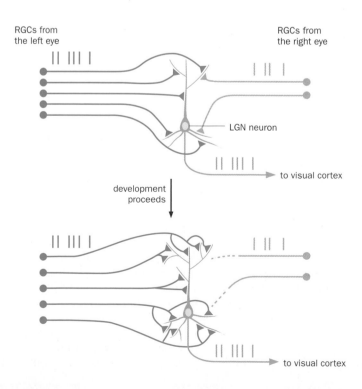

precedes eye-specific segregation of LGN axons in V1. This ensures that LGN axons have already solidified their ocular preference by the time they are making their target choices in V1, and nearby LGN axons may have already acquired similar firing patterns. The LGN axons can then impose these patterns on individual V1 layer 4 neurons, just as RGC axons impose their patterns on LGN neurons.

We can further extend the same logic to explain the sensitivity of these eye-specific connections during the critical period in V1. The ocular dominance columns take a long time to solidify, such that the vision-driven activity of the RGCs and LGNs can participate in this process after retinal waves recede and the retina is dominated by light- and photoreceptor-initiated activity. When deprived of this activity, application of Hebb's rule and its extensions would result in the strengthening of connections with the open eye and the withering of connections with the closed eye.

5.13 A Hebbian molecule: the NMDA receptor acts as a coincidence detector

What are the mechanisms that underlie Hebb's rule? More specifically, how is the correlation of pre- and postsynaptic activity detected? How does correlated firing of pre- and postsynaptic neurons lead to growth? What leads to the withering of synaptic connections when coincident firing does not occur? How widely applicable is Hebb's rule in the developing nervous system?

Recall we learned in Chapter 3 that most excitatory transmission in the vertebrate CNS is carried out by the neurotransmitter glutamate. This includes synaptic transmission by RGCs and LGN neurons to their postsynaptic targets. One of the ionotropic glutamate receptors we studied in Section 3.15, the **NMDA receptor**, has an interesting property. The pore of the NMDA receptor channel is normally blocked by Mg^{2+}, such that extracellular glutamate is not sufficient to open the channel (see Figure 3–24). The Mg^{2+} block can be relieved by depolarization of the postsynaptic cells. Therefore the NMDA receptor is essentially a molecular coincidence detector: it is activated only when (1) the presynaptic neuron is releasing the neurotransmitter glutamate and (2) the postsynaptic neuron is simultaneously depolarized to an extent that relieves the Mg^{2+} block. Importantly, the NMDA receptor is not the only glutamate receptor in postsynaptic neurons; glutamate can also activate the AMPA receptor, which results in depolarization.

We can use Figure 5–26 to explain how the NMDA receptor can execute Hebb's rule. Suppose that a series of action potentials from RGCs arrive at their axon terminals on the LGN neuron, reflecting retinal waves or visual stimuli. Glutamate release induced by the first batch of action potentials causes the LGN neuron to depolarize to the extent that the Mg^{2+} block of the NMDA receptor is relieved. The next set of action potentials now has the opportunity to open the NMDA receptor channel. Unlike most AMPA receptors, which depolarize the postsynaptic cell but do not conduct Ca^{2+}, the NMDA receptor allows Ca^{2+} entry that can trigger a series of biochemical reactions in the postsynaptic cells. These include local changes at the synapse as well as altered gene expression in the nucleus (see Section 3.23). These changes can lead to the stabilization and growth of synapses between the RGC axons and LGN dendrites.

Pharmacological experiments have supported a role of the NMDA receptor in activity-dependent wiring of the visual system, from eye-specific segregation in three-eyed frogs to shifts in ocular dominance columns induced by monocular deprivation. Recent live imaging experiments further suggest a role of the NMDA receptor in stabilizing RGC axonal branches in the developing frog tectum in response to synchronous visual stimuli. The molecular mechanisms by which NMDA receptor activation strengthens the synapses and lack of correlated firing weakens the synapses during development are still not well understood. We will revisit this topic in Chapter 10, where the role of NMDA receptor in synaptic plasticity and learning will be discussed in more detail.

A powerful means of establishing the necessity of a molecule in a biological process is a gene knockout experiment. The role of the NMDA receptor in the normal development of eye-specific layers and ocular dominance columns has not been examined by such an experiment due to technical limitations. Hebb's rule plays instrumental roles in activity-dependent wiring in many other parts of the nervous system. One salient example is the development of the whisker-barrel system in the mouse, where the role of the NMDA receptor has been particularly well demonstrated (**Box 5–3**).

HOW DO MOLECULAR DETERMINANTS AND NEURONAL ACTIVITY WORK TOGETHER?

In the first part of this chapter, we used retinotopic projections of RGCs as an example to illustrate how hard-wired molecular determinants enable developing axons to find their targets. In the second part, we used ocular dominance columns and layer-specific segregation of RGC axons to illustrate how experience and spontaneous neuronal activity shape the final connectivity. If activity-independent and

---

Box 5–3: Activity-dependent wiring of the rodent whisker-barrel system depends on the NMDA receptor

Rodents use their whiskers to perform many tasks, such as assessing the quality and distance of objects via the sense of touch. Whiskers are represented as discrete units, termed **barrels**, in layer 4 of the mouse somatosensory cortex. (The part of the rodent somatosensory cortex that represents whiskers is called the **barrel cortex**.) Each barrel represents input primarily from a single whisker (**Figure 5–27**). Discrete units corresponding to individual whiskers are also found along the ascending pathways from whisker to barrel cortex. Sensory neurons that innervate whiskers project their axons to the brainstem in discrete units called barrelettes, with each barrelette composed of terminals of axons originating from the sensory neurons that mostly innervate a single whisker. Brainstem neurons then carry whisker-specific information to the thalamus, where their endings form discrete units called barreloids, again with each barreloid representing sensory information mostly from a single whisker. Lastly, thalamic neurons relay such information via their axonal projections to specific barrels in the somatosensory cortex (Figure 5–27).

Barrel formation in layer 4 of the somatosensory cortex occurs during the first postnatal week in mice. At birth

(postnatal day zero, or P0), when **thalamocortical axons** (**TCAs**, that is, axons of thalamic neurons that project to the cortex) first arrive at the somatosensory cortex, the TCA terminals and their major postsynaptic targets, the stellate cells, are evenly distributed across layer 4 (**Figure 5–28**, left). By postnatal day 7 (P7), the barrels are clearly evident (Figure 5–28, right), and the TCAs representing a single whisker are confined to a single barrel center. The cell bodies of layer 4 stellate cells are mostly clustered at the periphery of the barrels, with each stellate cell sending dendrites into a single barrel. Thus, the segregation of TCAs into discrete barrels is somewhat analogous to segregation of eye-specific layers and formation of ocular dominance columns in the visual system.

How is the barrel cortex patterned within the first postnatal week? As in the visual system, sensory experience plays an essential role in the segregation of TCAs into discrete barrels. Newborn mice already have well-developed whiskers. When whisker follicles were ablated in newly born mice to deprive sensory inputs, the segregations of TCAs and stellate cells did not occur. Ablating a specific row of whisker follicles was especially revealing. TCAs corresponding

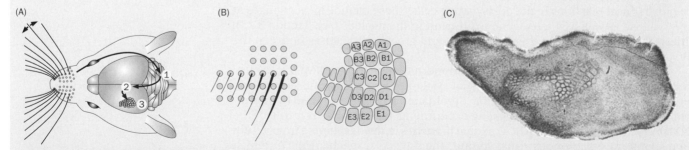

(A)   (B)   (C)

Figure 5–27 The whisker-barrel system in the mouse. (A) Information flow from the whiskers to the barrel cortex, from top view of a mouse. Sensory neurons that innervate individual whiskers project their axons to the brainstem barrelettes (1). Brainstem neurons project axons to the contralateral somatosensory thalamus to form the barreloids (2). Thalamocortical axons then project to the primary somatosensory cortex to form the barrels (3). **(B)** Correspondence of individual whiskers and barrels. The C2 whisker and its corresponding barrel are highlighted in yellow. **(C)** The organization of barrels in the somatosensory cortex can be seen in a mouse brain section with Nissl staining (see Section 13.15), which labels the cell bodies. (A & B, Petersen CCH [2007] *Neuron* 56:339–355. With permission from Elsevier Inc.; C, Woolsey TA & van der Loos H [1970] *Brain Res* 17:205–242. With permission from Elsevier Inc.)

activity-dependent mechanisms both make essential contributions to nervous system development, when and where is each mechanism used, and how do they work together?

Generally speaking, activity-independent molecular determinants are responsible for the coarse wiring of the nervous system, directing axons to specific regions, although in some species and circuits they can direct very precise connections, down to single cells and subcellular compartments. Neuronal activity and experience play an important role in refining the coarse wiring and determining the quantitative features, such as the number and strengths of synapses between two connecting neurons, perhaps to optimize neuronal circuits for their proper functions. Activity-independent mechanisms usually precede activity-dependent mechanisms, though they can overlap considerably. The ways in which these two mechanisms work together have not been examined systematically. Wiring of the visual system has offered one of the few cases in which both mechanisms have been investigated in sufficient detail to offer a glimpse of how they work together.

---

to the ablated whiskers did not segregate into individual barrels. They also occupied much smaller cortical areas compared to those that represent adjacent, intact whiskers (**Figure 5–29**A). In fact, it was this experiment that initially revealed which whisker corresponds to which barrel.

The NMDA receptor plays an essential role during multiple steps in the formation of the whisker-barrel system. When the obligatory GluN1 subunit of the NMDA receptor was knocked out in mice, the animals died at birth, so barrel cortex development could not be examined. The barrelettes in the brainstem normally develop prenatally, but were absent in GluN1-knockout mice, indicating an essential role for the function of the NMDA receptor in barrelette formation. Using a conditional knockout strategy (see Section 13.7), GluN1 was knocked out only in cortical neurons, allowing the knockout mice to survive past the first week. Barrels in layer 4 somatosensory cortex were diffusely organized: TCAs no longer formed tight clusters, and stellate cells were no longer displaced to the sides of the barrels (Figure 5–29B). In these mice, barrelettes and barreloids in the brainstem and thalamus developed normally, since the NMDA receptor was nonfunctional only in the cortex. Thus, this experiment indicated that the NMDA receptor in cortical neurons is essential for segregating the TCA terminals and stellate cells.

Since almost every neuron in the cortex expresses the NMDA receptor, it is still difficult to interpret the exact function of the NMDA receptor in patterning the barrel cortex when it is knocked out in all cortical neurons at once. To investigate NMDA receptor function in individual cells, a subunit of the NMDA receptor that plays a prominent role during development, GluN2B, was knocked out in isolated single cells using a genetic mosaic strategy that also specifically labels these knockout cells (see Section 13.16 for details); this allows the assessment of the function of the NMDA receptor made by individual neurons. Since only a very small fraction of isolated neurons lost GluN2B, barrel cortex pattern formation was normal. However, whereas control stellate neurons confined their dendrites to a specific barrel, the stellate neurons that lost GluN2B extended dendrites to multiple adjacent barrels (Figure 5–29C), indicating that each stellate neuron requires its own NMDA receptor to pattern its dendrites and ensure that the TCA inputs it receives represent sensory information from one whisker. These results support the notion that the NMDA receptor function is required for patterning dendritic arborizations of individual neurons according to the input they receive (Figure 5–29D), likely through a Hebbian mechanism analogous to the sorting of axons into eye-specific layers or ocular dominance columns (see Figure 5–27).

(*Continued*)

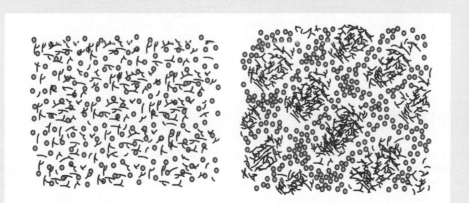

Figure 5–28 Patterning of the barrel cortex during the first postnatal week. At postnatal day 0 (left), thalamocortical axons (TCAs, red) and stellate cells (blue) are evenly distributed in layer 4 of the somatosensory cortex. One week later (right), TCAs are segregated into individual barrels according to the whiskers they represent, and stellate cells are displaced to the sides of the barrels. (From Molnár Z & Molnár E [2006] *Neuron* 49:639–651. With permission from Elsevier Inc.)

Box 5–3: Activity-dependent wiring of the rodent whisker-barrel system depends on the NMDA receptor

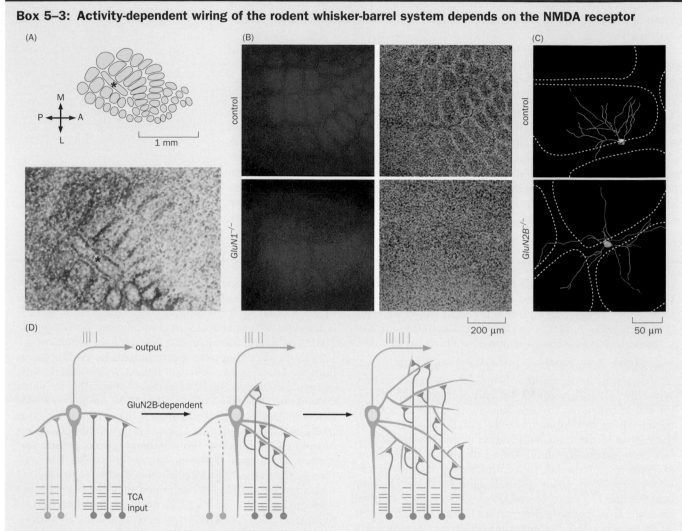

Figure 5–29 Sensory experience and the NMDA receptor regulate the patterning of both the barrel cortex and individual neuron's dendrites. (A) When a particular row of whiskers follicles was ablated in a newly born mouse, the barrels representing those whiskers were fused. In both the schematic (top) and the Nissl-stained brain section (bottom), the star marks the fused barrels. A, anterior; P, posterior; M, medial; L, lateral. **(B)** Compared to control mice (top panels), knockout mice whose cortical cells lack the NMDA receptor subunit GluN1 (bottom panels) no longer properly segregate their thalamocortical axons (TCAs) and stellate cells. In the left panels, TCAs are labeled by bulk injection of the axon tracer DiI in the thalamus; in the right panels, nuclear staining labels the stellate cells green. **(C)** A control stellate cell (top) extends its dendrites to a single barrel; dotted lines outline the barrels. In contrast, when the NMDA receptor subunit GluN2B is knocked out in an individual stellate cell (bottom), the cell distributes its dendrites to multiple barrels. **(D)** Interpretation of the results in panel C according to Hebb's rule. At an early stage of development, dendrites of a stellate cell (green) are contacted by TCAs representing multiple whiskers. If TCAs representing one whisker (blue) provide more input to the stellate cell compared to TCAs representing another whisker (orange), the stellate cell is more likely to fire action potentials (represented as bars perpendicular to the axon) according to the blue TCA firing pattern. Over time, correlated firing leads to strengthening of the synapses and growth of dendritic branches representing the blue whisker; uncorrelated firing in other dendritic branches leads to destabilization of synapses and pruning of the dendrites representing other whiskers. (A, adapted from van der Loos H & Woolsey TA [1973] *Science* 179:395–398. With permission from AAAS; B, from Datwani A, Iwasato T, Itohara S et al. [2002] *Mol Cell Neurosci* 21:477–492. With permission from Elsevier Inc.; C & D, adapted from Espinosa JS, Wheeler DG, Tsien RW et al. [2009] *Neuron* 62:205–217. With permission from Elsevier Inc.)

5.14 Ephrins and retinal waves act in parallel to establish the precise retinocollicular map

From what we've learned in the first part of the chapter, gradients of ephrin-As and EphA receptors would seem to satisfactorily account for the precise targeting of RGC axons in the superior colliculus along the anterior–posterior axis (see Figures 5–7 and 5–8). Researchers were therefore surprised by the discovery that knockout mice for **β2 nAChR**, a subunit for nicotinic acetylcholine receptor

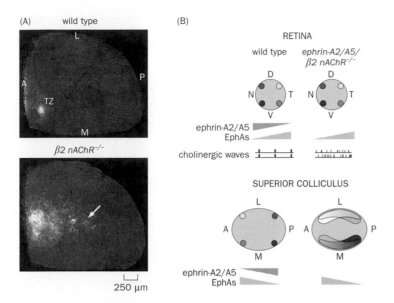

Figure 5–30 Retinal waves and ephrins work together to establish the retinocollicular map. (A) In a wild-type mouse (top panel), a focal injection of the fluorescent axon tracer DiI in the temporal retina resulted in a focal axon termination zone (TZ) in the anterior superior colliculus. A, anterior; P, posterior; L, lateral; M, medial. Focal injection of DiI into the temporal retina of a mouse lacking the β2 nicotinic acetylcholine receptor (*β2 nAChR*−/−, in which cholinergic retinal waves during the first postnatal week are disrupted and RGCs fire asynchronously) resulted in more diffuse labeling of RGC axon terminals in the superior colliculus beyond the normal termination zone (arrow in the bottom panel), indicating a defect in the refinement of RGC axons. **(B)** Summary of experiments examining the combined roles of ephrin/Eph and retinal waves in RGC axon projection. Relative concentrations of ephrin-As (blue) and EphA receptors (orange) across tissues are represented by colored triangles (indicating gradients). Groups of RGCs from four positions in the retina are shown as colored dots. In wild-type animals (left), these four groups of RGC axons project in an orderly fashion to four different areas of the superior colliculus. When ephrin-A2, ephrin-A5, and cholinergic retinal waves are disrupted in knockout mutants (right), RGC axons from a specific retinal position along the nasal–temporal (N–T) axis project diffusely along the entire anterior–posterior (A–P) axis of the superior colliculus. Remarkably, RGC axon projections along the dorsoventral (D–V) axis are unperturbed, indicating that different mechanisms must be employed. (A, from McLaughlin T, Torborg CL, Feller MB et al. [2003] *Neuron* 40:1174–1160. With permission from Elsevier Inc.; B, adapted from Pfeiffenberger C, Yamada J & Feldheim DA [2006] *J Neurosci* 26:12873–12884.)

essential for cholinergic retinal wave propagation, exhibited profound RGC axon targeting errors at the superior colliculus. In β2 nAChR knockout mice, although the coarse topography was maintained, RGC axons originating from a small part of the retina produced a more diffuse termination zone at the superior colliculus compared to controls (**Figure 5–30**A).

A likely explanation for the β2 nAChR knockout phenotypes is that cholinergic retinal waves confer correlated activity on neighboring RGC neurons to strengthen their connections with neighboring target neurons, ensuring that adjacent RGCs project to target neurons that are likewise adjacent. This explanation fits well with a curious finding from ephrin-A2/A5 double knockout mice: axon terminals from nearby RGCs still exhibited clustering at incorrect positions rather than completely diffuse projections along the anterior–posterior axis (see Figure 5–8); this most likely occurred because intact retinal waves in these mutant mice still ensured that nearby RGC axons target to nearby targets. To test the relationship between ephrins and retinal waves, triple knockout mice were generated that lacked ephrin-A2, ephrin-A5, and β2 nAChR. In these mice, a diffuse pattern of RGC axons along the anterior–posterior axis was observed, whether from RGCs originating in nasal or temporal retina (Figure 5–30B). This phenotype was much more severe than phenotypes exhibited in ephrin-A2/A5 double knockout (see Figure 5–8) or β2 nAChR knockout, indicating that ephrins and retinal waves work synergistically: while ephrins direct global topographic RGC axon projections in the superior colliculus, retinal waves refine and preserve local neighboring relationships of RGC axons.

5.15 Ephrins and retinal waves also work together to establish the retinotopic map in the visual cortex

As we learned in Chapter 4, the visual system is organized as a series of retinotopic maps such that spatial information in the retina is prominently represented in the LGN, the primary visual cortex (V1), and higher areas of the visual cortex. We've also learned that spontaneous retinal waves can propagate to V1 (see Figure 5–24). Do the mechanisms of retinotopic projections to the superior colliculus/tectum we have focused on thus far also apply to constructing retinotopic maps in other areas? Indeed, gradients of ephrin-As and EphA receptors have been found in both the LGN and V1 (**Figure 5–31**). Through both forward and reverse signaling (see Figure 5–12B), these gradients help establish topographic maps in multiple visual areas. They also help align different visual maps. For instance, information in the superior colliculus is used for reflexive control of eye movement, whereas information to the LGN is transmitted to V1 for detailed analysis of visual features during visual perception. V1 also sends descending input to the superior colliculus.

Figure 5–31 Ephrin-A and EphA gradients in multiple regions of the visual system. Ephrin-A and EphA counter-gradients are found in each area of the developing mammalian visual system, including the retina, the superior colliculus (SC), the lateral geniculate nucleus (LGN), and the primary visual cortex (V1). These gradients provide positional cues that allow input from nasal (red) and temporal (blue) RGCs to be represented correctly at the appropriate sites within their central targets. Arrows represent the direction of information flow. N, nasal; T, temporal; A, anterior; P, posterior; D, dorsal; V, ventral; M, medial; L, lateral. (Adapted from Cang J & D.A. Feldheim DA [2013] *Ann Rev Neurosci* 36:51–77.)

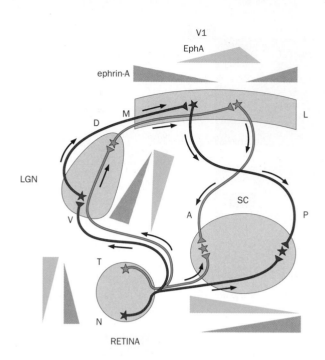

The topographic maps of the descending input from V1 and the ascending RGC projections to the superior colliculus align with each other (Figure 5–31).

The roles of ephrins and retinal waves have been examined in the construction of retinotopic maps in V1 by using intrinsic signal imaging (see Figure 4–42 and Box 13–3) to map receptive fields of cortical neurons in response to grating bar stimuli in knockout mice (**Figure 5–32**). In wild-type animals, V1 has an orderly and continuous retinotopic map. The map was less organized in ephrin-A2/A5 or β2 nAChR knockout mice, and was not detectable in triple knockout mice, suggesting that ephrin-As and retinal waves collaborate in establishing the V1 map, just as they do for the superior colliculus map. The abnormal retinotopic map in V1 can result from a combination of an abnormal RGC → LGN map and additional defects in the LGN → V1 projections.

A general conclusion can be made based on these studies. Ephrin-A–EphA interactions specify the global topographic target position for RGCs, and retinal waves impose an additional rule that neurons with correlated activity project to nearby targets. Indeed, if molecular gradients such as Ephrin-A and EphA were to act alone to specify precise positions, axons from neighboring RGCs would have to use a slight difference in their receptor levels to detect very small molecular differences in the target area. Such a mechanism might be challenging to establish. Acting together, molecular determinants and neuronal activity ensure that the retinal positions along the anterior–posterior axis are precisely represented at multiple visual areas. Other molecules and mechanisms must be employed to establish retinotopic map along the orthogonal dorsal–ventral axis; we know far less about the nature of these mechanisms compared to our knowledge about retinotopic projections along the anterior–posterior axis.

Figure 5–32 Ephrins and retinal waves are required for the retinotopic map in the visual cortex. A grating bar that moves along the nasal–temporal axis was used as a visual stimulus (left) in mice. Cortical responses were measured by intrinsic signal imaging (see Section 4.25). Colors correspond to positions in the 77° of visual space shown on the left. Compared to the orderly map in wild type, the maps in *ephrin-A2/A5* double knockout or *β2 nAChR* knockout mice are more scattered; the map is completely disrupted in the triple knockout. (Adapted from Cang J, Niell CM, Liu X et al. [2008] *Neuron* 57:511–523. With permission from Elsevier Inc.)

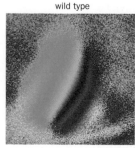

wild type

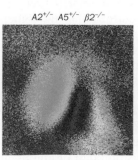

A2⁻/⁻ A5⁻/⁻ β2⁺/⁻

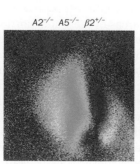

A2⁺/⁻ A5⁺/⁻ β2⁻/⁻

A2⁻/⁻ A5⁻/⁻ β2⁻/⁻

77°

1 mm

5.16 Different aspects of visual system wiring rely differentially on molecular cues and neuronal activity

Are molecular cues and activity-dependent mechanisms employed equally in wiring different parts of the nervous system? In this final section on the vertebrate visual system, we will use a few examples to explore the relative contributions of these two wiring mechanisms.

As we learned in Chapter 4, specific types of RGCs connect with specific types of bipolar and amacrine cells in the inner plexiform layer of the retina, which consists of distinct laminae. Dendrites from most RGCs and amacrine cells are limited to one or two specific lamina(e), as are axons from all bipolar cell types (**Figure 5–33**A). To test whether neuronal activity and competition are involved in lamina-specific targeting, neurotransmission was blocked specifically from the ON bipolar cells in genetically modified mice. This produced a competitive disadvantage for the ON pathway, a manipulation conceptually similar to suturing one eye shut in the context of competition between two eyes. Surprisingly, the connection specificity was not altered: ON bipolar cells still connected with ON RGCs at the correct lamina, albeit with reduced synapse numbers. Even the ON–OFF bistratified RGCs, which normally elaborate two dendritic trees to two distinct laminae and synapse with both ON and OFF bipolar cells, still did so despite the selective silencing of ON bipolar cells (Figure 5–33A). This experiment suggested that the lamina-specific targeting of dendrites and axons in the retina is more hard-wired than the formation of eye-specific layers or ocular dominance columns.

Indeed, molecular determinants have been partially identified in specifying the lamina-specific targeting of retinal neurons. In the chick, two pairs of immunoglobulin superfamily homophilic cell adhesion proteins, Sidekicks and **Dscams**, are expressed in RGCs and in bipolar and amacrine cells that target to the same lamina (Figure 5–33B). (Sidekick was named after a *Drosophila* homolog that functions in photoreceptor differentiation; Dscam was named after a human homolog <u>D</u>own syndrome <u>c</u>ell <u>a</u>dhesion <u>m</u>olecule, the corresponding gene of which is located in a region of human chromosome 21 that has an extra copy in Down syndrome, although it is unclear whether an extra copy of the *Dscam* gene contributes to symptoms of Down syndrome.) Homophilic cell adhesion proteins expressed from different cells can bind each other through their extracellular domains and facilitate cell adhesion. Misexpression and RNAi knockdown of Sidekicks and Dscams suggest that these proteins regulate lamina-specific process targeting of retinal neurons in the chick.

Evolutionarily conserved repulsive axon guidance molecules, the semaphorins, and their receptors, the **plexins** (see Box 5–1), also play key roles in lamina-specific targeting. For example, in the mouse, PlexA4 is expressed in retinal neurons that target processes to the OFF layer, whereas its repulsive ligand is expressed in retinal neurons that target processes to the ON layer (Figure 5–33C). Deleting either PlexA4 or Sema6A caused OFF layer-targeting amacrine and RGC dendrites to mistarget to ON layer (Figure 5–33D). Additional semaphorins and plexins are involved in constraining inner plexiform layer-targeting retinal neurons not to target their processes to the outer plexiform layer. Thus, both attractive (adhesive) and repulsive molecular interactions jointly ensure the layer- and lamina-specific targeting of retinal neurons.

RGCs of a given subtype not only extend their dendrites to specific laminae of the inner plexiform layer in the retina to receive information, but also project their axons to specific laminae of their central targets, the superior colliculus and LGN, to send information. Tracing axonal projections of a specific RGC subtype labeled in a transgenic mouse line revealed both lamina-specific targeting as well as column-like lateral organization of the RGC axon terminations in the superior colliculus. Remarkably, blocking cholinergic retinal waves affected the refinement of the columnar organization, but not the lamina-specific targeting of these axon terminations (**Figure 5–34**). Thus, neuronal activity can selectively affect a subset of wiring properties of the same population of neurons.

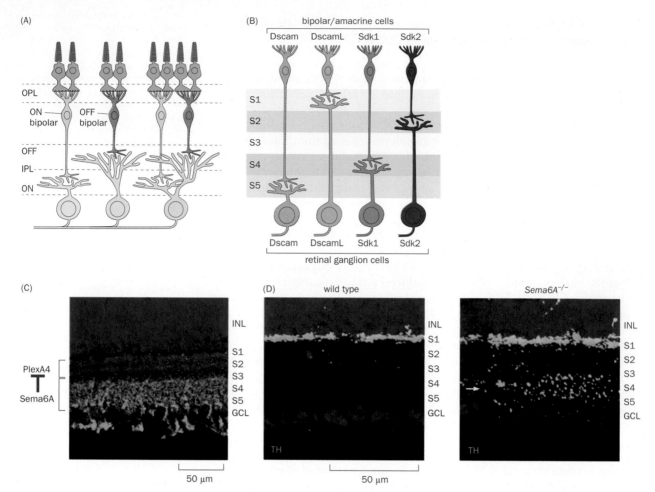

Figure 5–33 Lamina-specific targeting in the retina is determined by attractive and repulsive molecular interactions. (A) In wild-type mice, ON bipolar cells connect with ON ganglion cells (left), and a subset of the dendrites of ON-OFF ganglion cells (right) in the ON laminae of the inner plexiform layer (IPL). OFF bipolar cells connect with OFF ganglion cells (middle) and a subset of ON-OFF ganglion cell dendrites in the OFF laminae (far right). When neurotransmission is specifically blocked from the ON bipolar cells by expression of a transgene encoding tetanus toxin, which produces disparate input to the ON and OFF laminae, connection specificity is not altered. **(B)** The IPL is divided into five laminae (S1–S5), shown here in different colors. Each of the four types of chick immunoglobulin family cell adhesion molecules—Dscam, Dscam-like (DscamL), and Sidekicks 1 and 2 (Sdk1, Sdk2)—can bind to itself but not to the other three. Each is expressed in corresponding bipolar/amacrine cells and retinal ganglion cells that arborize processes in the same lamina. These molecules likely regulate lamina-specific targeting by selective adhesion with their synaptic partners. **(C)** Sema6A (green) is expressed in mouse retinal neurons that target processes to the ON laminae, whereas its repulsive receptor PlexA4 (red) is expressed in retinal neurons that target processes to the OFF laminae S1 and S2. The repulsive interaction ensures that PlexA4-expressing neurons restrict their processes to the OFF lamina. INL, inner nuclear layer; GCL, ganglion cell layer. **(D)** A subclass of amacrine cells that express an enzyme tyrosine hydroxylase (TH) target their processes to the S1 lamina in wild type, as visualized by antibody staining against TH (green). In *Sema6A* mutants, TH-positive processes mistarget to S3–S5 laminae (yellow arrow). Blue stains nuclei in both panels C and D, which label the INL and GCL. (A, adapted from Kerschensteiner D, Morgan JL, Parker ED et al. [2009] *Nature* 460:1016–1022. With permission from Macmillan Publishers Ltd; B, adapted from Yamagata M & Sanes JR [2008] *Nature* 451:465–471. With permission from Macmillan Publishers Ltd; C & D, from Matsuoka RL, Nguyen-Ba-Charvet KT, Parray A et al. [2011] *Nature* 470:259–264. With permission from Macmillan Publishers Ltd.)

Studies thus far have suggested that wiring in the visual cortex relies more on spontaneous activity and visual experience. For example, orientation selectivity, a salient feature in the primary visual cortex (see Section 4.23), is also influenced by spontaneous neuronal activity before eye opening and by visual experience afterwards, although less plastic than ocular dominance. The direction selectivity that certain cortical neurons possess appears late during development after eye opening. Visual experience is required for the proper development of direction-selective columns. For example, exposing naive ferrets to motion in a specific direction can selectively strengthen the direction selectivity of cortical neurons and their organization into columns according to the direction of training motion.

These examples suggest that the relative contributions of activity-independent molecular determinants, spontaneous activity, and visual experience differ

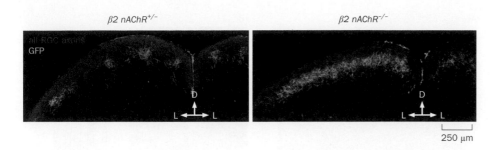

Figure 5–34 Retinal waves are required for refinement of RGC axons into columns but not specific layers. Left, in control mice heterozygous for the *β2 nAChR* gene, a genetically defined subpopulation of RGCs targets their axons (labeled in green) to a specific layer of the superior colliculus. These axons are additionally clustered in column-like organization. All RGC axons are labeled red using an anterograde tracer. Right, when cholinergic retinal waves are blocked in *β2 nAChR* knockout mice, columnar organization is disrupted, but layer-specific targeting remains normal. D, dorsal, L, lateral. (From Huberman AD, Manu M, Koch SM et al. [2008] *Neuron* 59:425–438. With permission from Elsevier Inc.)

for the wiring of different visual circuits. An important parameter that determines whether activity-independent or activity-dependent mechanisms are employed may be whether the connection specificity reflects qualitative difference between discrete cell types or quantitative difference among the same cell types. Connection specificity that distinguishes discrete cell types, such as ON- or OFF-types of bipolar cells, amacrine cells, and RGCs in the retina, is more likely specified by molecular determinants. On the other hand, connection specificity that involves quantitative difference among similar cell types, such as the connection of a specific LGN neuron to the same types of RGCs from the left or right eye or a V1 neuron to the same type of projection neurons from the left or right LGN, may more likely utilize activity-dependent mechanisms. Because cell types are better defined in the retina than in the cortex at present, we see more hard-wiring examples in the retina. As more effort is being devoted to identifying cell types in the cortex at present, it will be interesting to determine in the future whether hard-wiring examples are similarly prevalent in the cortex, and how they interact with activity-dependent mechanisms.

Given the importance of cell type in our discussion of wiring specificity, a key question is how cell types are determined during development. This question has been particularly well studied in the *Drosophila* visual system, to which we turn as the final topic of the chapter.

VISUAL SYSTEM DEVELOPMENT IN *DROSOPHILA*: LINKING CELL FATE TO WIRING SPECIFICITY

We have so far focused on the visual systems of vertebrates. Many invertebrates such as insects also have superb vision: consider the difficulty of swatting an annoying housefly! Powerful molecular genetic approaches in the fruit fly *Drosophila* have made its visual system a model to address general questions regarding cell fate determination and wiring specificity.

The compound eye of *Drosophila* (**Figure 5–35**A) is made of ~800 repeating units called **ommatidia**. Each ommatidium contains eight photoreceptors: R1–R6 at the periphery, and R7 and R8 at different depths in the center of the ommatidia (Figure 5–35B). R1–R6 photoreceptors are responsible for motion detection; they send their axons to the **lamina**, the first **neuropil** layer underneath the retina, where they synapse with lamina neurons (Figure 5–35C–D). (Neuropils are structures composed mostly of synapses between axons and dendrites in invertebrate nervous systems.) R7 and R8 are responsible for color vision; they send their respective axons to the M6 and M3 layers of the **medulla**, a neuropil that lies beneath the lamina. Lamina neurons also send their axons to specific medulla layers (Figure 5–35D). Different transmedullary neurons send dendrites to specific medulla layers to receive input from R7, R8, and lamina neurons; the axons of these transmedullary neurons project to higher visual centers in the **lobula complex**. Thus, layer-specific organization in neuronal processes in the fly **optic lobe** (consisting of retina, lamina, medulla, and lobula complex) resembles that of the vertebrate retina (see Figure 5–33A). We can infer that the *Drosophila* visual circuit is mostly hard-wired, because mutations that disrupt phototransduction, action potential propagation, and synaptic transmission do not lead to detectable wiring defects.

Figure 5–35 The *Drosophila* eye and visual circuit. (A) Scanning electron micrograph of a *Drosophila* compound eye, which consists of ~800 ommatidia. **(B)** Electron micrographs of ommatidial cross sections at a superficial (top) and deep (bottom) level as indicated by the horizontal lines on the right. The electron-dense structures are the rhabdomeres, which are enriched in rhodopsins, analogous to the outer segments of the vertebrate photoreceptors (see Section 12.14). Rhabdomeres from each numbered photoreceptor (R1–R8) occupy stereotyped positions in the ommatidium; note that the central photoreceptor in the superficial cross section is R7, whereas R8 is central in the deep cross section. **(C)** Cross section of the *Drosophila* optic lobe showing its neuronal organization. The photoreceptors are stained in green. Their cell bodies occupy the outermost layer, the retina. Depending on the cell types (see Panel D), their axons project to two neuropils, the lamina and the medulla (stained in red with a synapse marker; because the extensive overlap between the photoreceptor axons and the synapse marker staining, the lamina appears yellow). The lobula complex receives input from medulla neurons. Blue, nuclear staining indicating the location of cell bodies. **(D)** Schematic diagram of the fly visual circuit. Photoreceptors R1–R6 are responsible for analyzing motion signals. They target their axons to the lamina, where they synapse with L1 and L2 neurons that send axons to specific layers of the medulla. R7 and R8 send axons directly to layer 6 and layer 3 of the medulla, respectively, for analyzing color signals. This layer-specific targeting resembles that of amacrine cells, bipolar cells, and RGCs in the vertebrate retina (see Figure 5–33A). TM, transmedullary neurons. The medulla contains additional layers (7–10) that are not drawn. Note that all neurons except the photoreceptors shown here are unipolar, with dendrites and axons originating from a single process. This is a property of most invertebrate neurons (see Section 1.7). (A, adapted from Ready DF, Hanson TE & Benzer S [1976] *Dev Biol* 53:217–240. With permission from Elsevier Inc.; B, adapted from Reinke R & Zipursky SL [1988] *Cell* 55:321–330. With permission from Elsevier Inc.; C, from Williamson WR, Wang D, Haberman AS et al. [2010] *J Cell Biol* 189:885–899. With permission from Rockefeller University Press; D, adapted from Sanes JR & Zipursky SL [2010] *Neuron* 66:15–36. With permission from Elsevier Inc.)

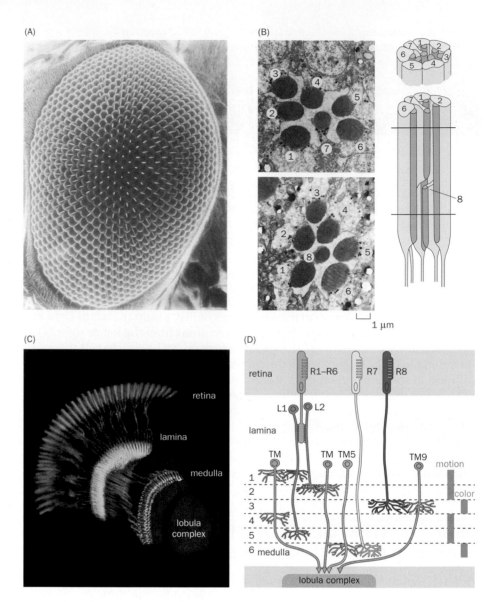

5.17 Cell–cell interactions determine photoreceptor cell fates: R7 as an example

The nervous system contains many different types of neurons and glia. For example, the vertebrate retina has more than 60 types of neurons (see Figure 4–28), each with its unique morphology, connection pattern, and neurotransmitter profile. How a neuron or a glia acquires its **cell fate** (that is, the type of cell it is to become) is a fundamental question in developmental neurobiology, and studies of photoreceptor fate determination in *Drosophila* have provided rich mechanistic insights.

Cell fates can be determined by two distinct mechanisms. The first mechanism is **asymmetric cell division**; a cell is born with a fate different from its sibling because an intrinsic determinant segregates asymmetrically during cell division. In this case, cell fates are correlated with **cell lineage**, or the relationship of cells by birth. The second mechanism is **induction**, or cell–cell interactions; a cell is born with the same potential as its sibling or cousins, and its eventual fate is acquired by receiving external inductive signals. The stereotyped positions of R1–R8 within each ommatidium (see Figure 5–35B), as well as a wealth of molecular markers, enabled researchers to examine when specific R cells become neurons, and whether R cells are related by birth.

Using a pan-neuronal marker to stain developing ommatidia, R cells were found to undergo neuronal differentiation in a stereotyped sequence. R8 is the first to become a neuron, followed by the R2/R5 pair, the R3/R4 pair, and the R1/R6 pair. R7 is the last cell to adopt a neuronal fate (**Figure 5–36**A). To probe their lineage relationships, one can use a method called **clonal analysis**, which randomly labels an early progenitor in such a way that all its progeny are also labeled (see Section 13.16 for details). If specific R cells within an ommatidium are always co-labeled, we can deduce that those cells share an immediate common progenitor. Strikingly, no such relations were found in any R cell pairs, suggesting that their fates are acquired by cell–cell interactions, rather than through cell lineage. Indeed, the inductive mechanisms for R cell fate determinations are among the best-studied in developmental biology. Here we examine the determination of R7 cell fate as an example.

Analyses of two *Drosophila* mutations, *Sevenless* and ***Bride of sevenless*** (**Boss**), have been particularly instructive. Flies lacking either gene have a very specific phenotype: they lack the R7 photoreceptor in each ommatidium. The presumptive R7 cell turns into a nonneuronal support cell in the absence of induction. **Mosaic analysis** was used to determine which cells require *Sevenless* or *Boss* in

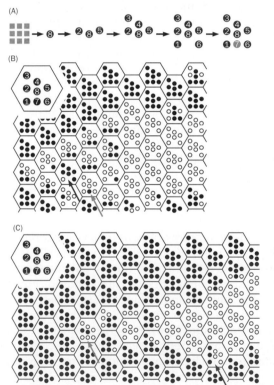

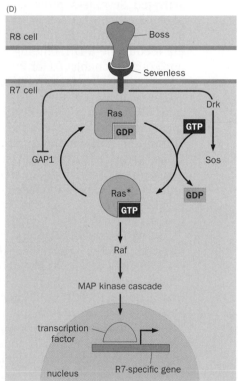

Figure 5–36 Determination of the R7 fate. (A) From a field of undifferentiated cells (gray squares), R8 is the first cell to become a neuron (red) in an ommatidium. R8 is followed by R2/R5, R3/R4, R1/R6, and finally R7 (blue) **(B)** Mosaic analysis of *Sevenless*. The scheme represents a cross section through many ommatidia, each enclosed by a hexagon with R1–R8 arranged according to the key at top left. Open dots represent *Sevenless* mutant cells marked by their lack of a cell marker, *White*; these were derived from a single precursor cell (a clone). Filled dots represent wild-type cells. These mosaic clones can be used to determine which cell must be wild type in order for an ommatidium to have an R7 (the middle cell in the bottom row within each ommatidium). The red arrow points at an ommatidium in which R1–R5 and R8 are all wild type, yet R7 is missing. The blue arrow points to an ommatidium in which all cells except R7 are mutant, yet R7 is present. Collectively, these data indicate that *Sevenless* is needed only in the presumptive R7 for R7 determination. **(C)** Mosaic analysis of *Boss* follows the same scheme introduced in panel B except that open dots represent *Boss* mutant cells marked by

lacking *White*. The blue arrow points to an ommatidium in which only R2 and R8 are wild type, and R7 is present. The red arrow points to an ommatidium in which R2 is wild type, R8 is mutant for *Boss*, and R7 is absent. Collectively, these data indicate that *Boss* is required in R8 for R7 determination. **(D)** A summary of the signal transduction pathway for R7 determination. Boss binding to Sevenless causes tyrosine phosphorylation of Sevenless, which recruits binding of Drk and Son of sevenless (Sos) to the membrane. Sos acts as a guanine nucleotide exchange factor for the small GTPase Ras. GTP-bound Ras activates Raf and the MAP kinase cascade, which in turn activates transcription factors to initiate R7-specific transcription. Activated Sevenless also inhibits GAP1, a GTPase activating protein that inactivates Ras. (B, adapted from Tomlinson A & Ready DF [1987] *Dev Biol* 123:264–275. With permission from Elsevier Inc.; C, adapted from Reinke R & Zipursky SL [1988] *Cell* 55:321–330. With permission from Elsevier Inc.; D, based on Simon MA, Bowtell DD, Dodson GS, et al. [1991] *Cell* 67:701; Zipursky SL & Rubin GM [1994] *Ann Rev Neurosci* 17:373–397.)

order for an ommatidium to develop an R7 cell. Genetic mosaic flies were created in which a specific population of marked cells carried mutations whereas the rest of the cells were wild type (see Section 13.9 for details). Although *Sevenless* is normally expressed in several R cells, if the wild-type gene is present in the presumptive R7 cell, the ommatidium develops an R7 cell; the *Sevenless* gene can be absent in any other R cell without affecting R7 cell development in a given ommatidium (Figure 5–36B). Thus, *Sevenless* acts **cell autonomously**. It is needed only in the presumptive R7 cell to make an R7 photoreceptor. By contrast, analysis of ommatidia that are mosaic for *Boss* indicated that as long as the R8 in an ommatidium has wild-type *Boss*, an R7 develops (Figure 5–36C). Thus *Boss* acts **cell nonautonomously**: it is required in R8 for R7 formation. Indeed, molecular cloning and biochemical studies indicated that *Sevenless* encodes a receptor tyrosine kinase that acts in R7 to receive inductive signals, and *Boss* encodes a ligand with multiple transmembrane domains that is expressed only in R8 cells; the Boss protein, which is localized to the R8 cell membrane, binds to and activates the Sevenless protein in the neighboring presumptive R7 cell (Figure 5–36D).

Genetic studies have also elucidated an intracellular signaling pathway by which Sevenless transduces signals to regulate R7 fate (Figure 5–36D). As with many receptor tyrosine kinases such as neurotrophin receptors (see Figure 3–39), activated Sevenless proteins dimerize, and the homodimer subunits cross-phosphorylate each other at tyrosine residues. Phosphorylated tyrosine on Sevenless recruits the binding of an adaptor protein called Drk, which brings to the membrane a Drk-binding partner called Son of sevenless (Sos), a guanine nucleotide exchange factor for the small GTPase Ras. This switches Ras into the GTP-bound, active form. At the same time, phosphorylated Sevenless inhibits the GTPase activating protein GAP1 and thereby helps maintain Ras in its GTP-bound state. Ras activates Raf and the MAP kinase cascade, leading to activation of specific transcription factors to execute an R7-specific gene expression program. Remarkably, this signaling pathway from receptor tyrosine kinase to Ras and MAP kinase cascade is highly conserved in many cell fate determination processes across the animal kingdom (see Box 3–4). It is also used in synaptic activity-induced gene expression (see Figure 3–41). Lastly, it is a major cell proliferation pathway; many components, such as Ras and Raf, when abnormally activated, lead to cancer.

As we learned in Chapter 4, color vision requires animals to have photoreceptors with different spectral sensitivities. *Drosophila* R7 and R8 express different rhodopsin (Rh) proteins that allow them to detect light in the UV and blue/green ranges, respectively. The onset of Rh expression occurs well after photoreceptors have acquired their R7 or R8 fates. In fact, both R7 and R8 have two subtypes. About 30% of R7 cells express Rh3; the remaining 70% express Rh4. The distribution of Rh3- and Rh4-expressing R7s appears stochastic (**Figure 5–37**A), analogous to the distribution of the L- and M-cones in the human retina (see Figure 4–20). The distribution of R7 rhodopsins originates from stochastic expression of a transcription

Figure 5–37 Determination of R7 and R8 subtypes that express rhodopsins with different spectral sensitivities. **(A)** About 30% of R7 cells express Rh3, and 70% express Rh4. The image shows that the retinal distribution of Rh3- and Rh4-expressing R7s appears random. (Note that color labels here do not correspond to the spectral sensitivities of R7 photoreceptors.) **(B)** Stochastic expression of Spineless in about 70% of R7 cells activates Rh4 and inhibits Rh3 expression. Spineless also inhibits an R7 → R8 signal that activates Rh5 and inhibits Rh6 expression in R8. Thus, expression of Spineless in R7 gives rise to Rh4 expression in R7 coupled with Rh6 expression in R8 (left). If Spineless protein is not expressed in R7, the resulting expression of Rh3 in R7 is coupled with Rh5 expression in R8 within the same ommatidium (right). Black, active components; gray, inactive components. (B, after Johnston RJ & Desplan C [2010] *Ann Rev Neurosci* 26:689–719.)

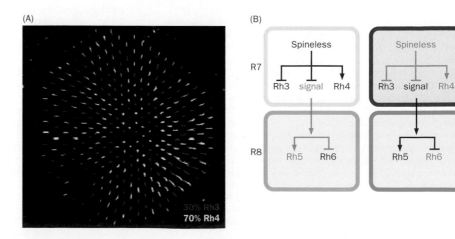

factor called Spineless in R7, which activates Rh4 expression and represses Rh3 expression (Figure 5–37B). There are also two populations of R8: 30% of R8 cells express Rh5, and 70% express Rh6. Remarkably, rhodopsin expression in R7 and R8 is coordinated within each ommatidium: an Rh3-expressing R7 always couples with an Rh5-expressing R8, whereas an Rh4-expressing R7 always couples with an Rh6-expressing R8. In *Sevenless* or *Boss* mutant flies where R7 is absent, all R8 cells express Rh6, indicating that signaling from Rh3-expressing R7 is essential to induce Rh5 expression in R8. In *Spineless* mutants, all R8 cells express Rh5. Thus, in addition to regulating Rh expression in R7, Spineless also inhibits the R7 → R8 signal that orchestrates the coupling of R7 and R8 subtypes within each ommatidium (Figure 5–37B).

In summary, cell fates of *Drosophila* photoreceptors are determined by a series of cell–cell interaction events. Some fate decisions are made soon after a cell is born, such as whether to become an R8 or an R7. Other decisions are made well after a cell is born, such as whether an R7 expresses Rh3 or Rh4. R7 receives an inductive signal from R8 to acquire the R7 fate; at a later time, an R7 subtype sends an inductive signal back to R8 to instruct it to become a specific subtype. Similar cell–cell interaction events occur throughout the developing nervous system to determine numerous cell types and subtypes.

5.18 Multiple parallel pathways participate in layer-specific targeting of R8 and R7 axons

Connecting with proper synaptic partners is one of the most important facets of a neuron's fate. Thus, a major outcome of a neuron's fate decision is the expression of a unique repertoire of guidance receptors, such that its growth cones react to environmental cues differently from the growth cones of a different neuronal type. *Drosophila* photoreceptors have been used to identify molecules and mechanisms that regulate targeting specificity. We use layer-specific targeting of R7 and R8 to highlight some of these findings.

In wild-type flies, R7 axons target to the M6 layer of the medulla, whereas R8 axons target to the M3 layer (see Figure 5–35D). Genes necessary for the correct targeting of R7 or R8 axons have been identified using forward genetic screens (see Section 13.6 for detail). For example, N-cadherin, a Ca^{2+}-dependent homophilic cell adhesion protein (see Box 5–1) enriched in the nervous system, is essential for R7 targeting to M6. Genetic mosaic analysis indicated that when N-cadherin was removed from isolated single R7 neurons, their axons mistargeted to the M3 layer (**Figure 5–38**). This demonstrates that N-cadherin acts cell autonomously in R7 for its layer selection. However, N-cadherin is expressed in all R cells and in most target neurons as well. Thus, it likely works with additional molecules to specify R7 targeting. Indeed, disruption of genes encoding several other proteins including Lar, a receptor tyrosine phosphatase, also caused R7 axons to mistarget to the M3 layer (Figure 5–38). These additional factors, as well as the timing of expression of these cell-surface proteins, can contribute to the specificity of mutant phenotypes due to loss of widely expressed proteins such as N-cadherin.

While the ability of R8 axons to target M3 is not affected by the loss of N-cadherin or Lar, other mechanisms have been identified that specifically regulate R8 targeting. For example, netrin, a well-known axon guidance attractant

Figure 5–38 Selected cell-surface proteins regulate layer-specific targeting of R7 axons. All photoreceptor axons are labeled red. Green label (which appears yellow as it overlaps with red) highlights isolated wild-type R7 axons (left) or R7 axons mutant for *N-cadherin* (middle) or *Lar* (right) in a genetic mosaic animal (see Section 13.16 for details of the labeling method). Wild-type R7 axons (arrowheads) extend their growth cones to the M6 layer (lower dashed line), whereas the individual R7 axons mutant for *N-cadherin* or *Lar* expression (arrows) in an otherwise normal environment terminate their axons near the M3 layer (upper dashed line), where R8 axons normally terminate. Thus, N-cadherin and Lar are cell autonomously required for layer-specific targeting of R7 axons. (From Clandinin TR, Lee C-H, Herman T et al. [2001] *Neuron* 32:237–248. With permission from Elsevier Inc.)

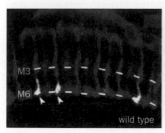
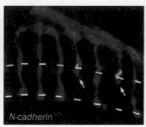
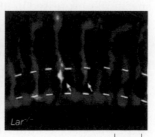

M3
M6
wild type   *N-cadherin*⁻/⁻   *Lar*⁻/⁻

20 μm

Figure 5–39 Selected cell-surface proteins that regulate layer-specific targeting of R8 axons. (A) In wild type, R8 axons target to the M3 layer. Deletion of netrin, which is normally expressed in lamina neurons that project to the M3 layer, or deletion of the netrin receptor Frazzled in photoreceptor cells in mosaic animals, causes R8 axons to stall at the border of the medulla, or mistarget to the M1 or M2 layers. Misexpressing netrin in the M1 and M2 layers causes R8 axons to mistarget to the M1 and M2 layer. Loss of Capricious, a transmembrane protein with leucine-rich repeats, causes R8 axons to mistarget (mostly to the M2 layer). Note that R7 targeting in all above conditions is unaltered. Misexpressing Capricious in R7 cells misdirects R7 axons to terminate at the M3 layer. These experiments indicate that both netrin and Capricious play instructive roles in R8 axon targeting. **(B)** In a genetic mosaic fly where most cells are wild type, green cells are specifically devoid of Frazzled. All R cells and their axons are labeled in red, revealing terminations in the lamina by R1–R6 axons or in the medulla by R7 and R8 axons. A subset of R8 axons that express the Rh6 rhodopsin are labeled in blue, showing their termination in the M3 layer. The box highlights a single *frazzled* mutant axon (arrow), which terminates at more superficial layer. Right panels are magnified boxed region with green, red, and blue labeling in separate panels. **(C)** Left, in wild type, all Rh6-expressing R8 axons terminate at M3. Right, when netrin is ectopically expressed in target neurons that terminate in M1 and M2 layers, many Rh6-expressing R8 axons (blue) terminate at the border of M1 (arrowhead) or the M2 layer (arrow). Red, all R cell axons. Bottom panels are high magnification showing only Rh6-expressing R8 axons in the medulla. These data are schematically summarized in panel A. (A, adapted from Sanes & Zipursky [2010] *Neuron* 66:15–36. With permission from Elsevier Inc.; see also Shinza-Kameda M, Takasu E, Sakurai K et al. [2006] *Neuron* 49:205–213; B & C, from Timofeev K, Holy W, Hadjieconomou D et al. [2012] *Neuron* 75:80–93. With permission from Elsevier Inc.)

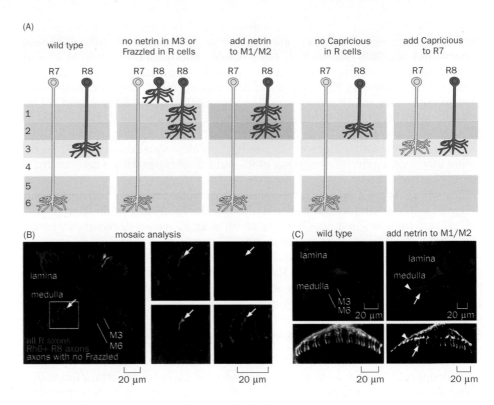

(see Box 5–1), is normally expressed in the M3 layer by specific lamina neurons that target that layer. R8 cells express the *Drosophila* netrin receptor **Frazzled** (homologs of DCC/Unc40), and the attraction mediated by the netrin–Frazzled interaction contributes to the layer specificity of R8 axon targeting (**Figure 5–39**A). In mosaic animals in which Frazzled was removed from singly labeled R8 cells, these mutant R8 axons terminated at more superficial layers (Figure 5–39B). When netrin was misexpressed in target layers M1 and M2, R8 axons were mistargeted to these layers (Figure 5–39C). In addition to netrin/Frazzled signaling, R8 but not R7 cells also specifically express **Capricious**, a transmembrane protein that contains extracellular leucine-rich repeats. Removal of Capricious affected axon targeting of R8, whereas misexpression of Capricious in R7 caused mistargeting of R7 axons to the M3 layer, where R8 axons normally target (Figure 5–39A). These loss- and gain-of-function experiments, along with their cell-type-specific expression, indicate that netrin/Frazzled and Capricious play an instructive role in layer-specific targeting, that is, the specific expression of these guidance molecules instructs the axons to make specific targeting decisions. Since disruption of each of these molecules leads to targeting errors of a subset of R8 axons, these studies also suggest that multiple ligand–receptor systems act in parallel to ensure the fidelity of axon targeting in wild-type animals, a common theme in axon guidance and target selection.

How does Capricious acquire its R8-specific expression pattern, which appears to be a key determinant in targeting specificity? Researchers found that Capricious expression is activated by the transcription factor Senseless, which is specifically expressed in R8 and acts as a key R8 fate determinant. Senseless also promotes the expression of R8-specific rhodopsins and inhibits the expression of R7-specific rhodopsins. Co-regulation of axon targeting receptors and rhodopsins ensures that wiring specificity is closely linked with spectral sensitivity (**Figure 5–40**). R7 cells express their own specific transcription factor, Prospero, which represses R8-specific rhodopsin expression. Senseless expression is further inhibited in R7 by the activity of a transcriptional repressor called NFYC, the fly homolog of a vertebrate nuclear factor Y complex subunit (Figure 5–40). When NFYC was deleted from R7, misexpression of Senseless in mutant R7 turned on Capricious and led to mistargeting of R7 axons to the M3 layer.

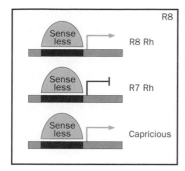 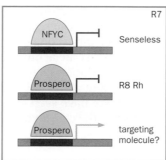

Figure 5–40 Linking cell fate and targeting specificity. In R8 (left), the R8-specific transcription factor Senseless activates the expression of Capricious, and at the same time ensures R8-specific rhodopsin (Rh) expression by activating expression of R8 rhodopsins and inhibiting expression of R7 rhodopsins. In R7 (right), NFYC represses Senseless expression, and R7-specific transcription factor Prospero represses R8 rhodopsin expression. The link between Prospero and targeting molecules remains to be established. (Based on Morey M, Yee SK, Herman T et al. [2008] *Nature* 456:795–799. With permission from Macmillan Publishers Ltd.)

The studies of layer-specific targeting of R7 and R8 axon targeting illustrate a general principle: cell fate determination ensures that each cell type expresses a unique set of guidance molecules, which in turn instruct their axons to make specific wiring choices. We will revisit this topic in Chapter 7, where we extend the lessons learned from studies of the visual system to brain wiring in general.

SUMMARY

Two major mechanisms are used for wiring the nervous system. The first involves specific recognitions of extracellular molecular cues by receptors on neuronal growth cones, originally proposed as the chemoaffinity hypothesis. The extracellular cues can be attractive or repulsive. They can be secreted and act at a distance or be cell-surface-bound and require contact between growth cones and cue-producing cells. Guidance receptors act by modulating cytoskeletal changes in growth cones to mediate attraction and repulsion. Many guidance mechanisms, such as those involving netrin and semaphorin, are highly conserved in evolution.

The formation of the retinotopic map in the visual system offers an excellent experimental paradigm to dissect the mechanisms by which molecular cues instruct connection specificity. Retinal ganglion cell (RGC) axons express a temporal > nasal gradient of EphA receptors. The tectum/superior colliculus, a major target of RGC axons, expresses a posterior > anterior gradient of the repellent ephrin-A. Ephrin-A–EphA interactions play a major role in defining the areal positions at which RGC axons terminate along the anterior–posterior axis in the target according to their cell body positions in the retina. Molecular cues play essential roles in many other aspects of visual system wiring, as in determining whether RGC axons cross or do not cross the optic chiasm and in defining the lamina(e) within the retina to which RGCs, amacrine cells, and bipolar cells target their dendritic and axonal processes.

Activity-dependent wiring constitutes the second major mechanism for establishing neuronal connections. This has been particularly well-studied in ocular dominance columns in the primary visual cortex (V1) and eye-specific segregation of RGC axons in the lateral geniculate nucleus (LGN). Spontaneous activities in the form of retinal waves play a crucial role in driving eye-specific segregation in the LGN and V1. Visual experience during the critical period consolidates ocular dominance columns. Hebb's rule of 'firing together, wiring together' provides a cellular mechanism by which neuronal activity can influence wiring specificity. The NMDA receptor acts as a molecular coincidence detector for correlated firing. Studies in the whisker-barrel system have provided strong support for the function of NMDA receptor in executing Hebb's rule during the activity-dependent wiring process.

Molecular determinants alone can specify connectivity with great precision, as exemplified by the wiring of the fly visual system. However, in wiring much of the vertebrate visual system, molecular determinants and activity-dependent wiring act in concert, as exemplified by the collaboration of ephrins/Eph receptors and retinal waves in establishing global positions and refining local near-neighbor relations in the retinotopic maps. Different circuits rely differentially on the contributions of molecular cues and activity-dependent mechanisms. Defining when, how much, and why each mechanism is used in wiring different parts of the nervous system will be interesting topics for future research.

Neuronal wiring is a crucial output of a cell fate decision. Fate determinants, which often are transcription factors, control the expression of guidance receptors such that different types of neurons chart different paths through the same environment. We have seen examples in the layer-specific targeting of fly photoreceptors and in the chiasm crossing of vertebrate RGCs. We will revisit and expand on many of these principles in our study of general neural development in Chapter 7.

FURTHER READING

Books and reviews

Cang J & Feldheim DA (2013) Developmental mechanisms of topographic map formation and alignment. *Annu Rev Neurosci* 36:51–77.

Debb DO (1949) The Organization of Behavior: A Neuropsychological Theory. John Wiley & Sons Inc.

Dent EW, Gupton SL & Gertler FB (2011) The growth cone cytoskeleton in axon outgrowth and guidance. *Cold Spring Harb Perspect Biol* 3:a001800.

Huberman AD, Feller MB & Chapman B (2008) Mechanisms underlying development of visual maps and receptive fields. *Annu Rev Neurosci* 31:479–509.

Petros TJ, Rebsam A & Mason CA (2008) Retinal axon growth at the optic chiasm: to cross or not to cross. *Annu Rev Neurosci* 31:295–315.

Sanes JR & Zipursky SL (2010) Design principles of insect and vertebrate visual systems. *Neuron* 66:15–36.

Sperry RW (1963) Chemoaffinity in the orderly growth of nerve fiber patterns and connections. *Proc Natl Acad Sci USA* 50:703–710.

Tessier-Lavigne M & Goodman CS (1996) The molecular biology of axon guidance. *Science* 274:1123–1133.

Axon guidance of retinal ganglion cells

Brown A, Yates PA, Burrola P et al. (2000) Topographic mapping from the retina to the midbrain is controlled by relative but not absolute levels of EphA receptor signaling. *Cell* 102:77–88.

Cheng HJ, Nakamoto M, Bergemann AD et al. (1995) Complementary gradients in expression and binding of ELF-1 and Mek4 in development of the topographic retinotectal projection map. *Cell* 82:371–381.

Drescher U, Kremoser C, Handwerker C et al. (1995) *In vitro* guidance of retinal ganglion cell axons by RAGS, a 25 kDa tectal protein related to ligands for Eph receptor tyrosine kinases. *Cell* 82:359–370.

Feldheim DA, Kim YI, Bergemann AD et al. (2000) Genetic analysis of ephrin-A2 and ephrin-A5 shows their requirement in multiple aspects of retinocollicular mapping. *Neuron* 25:563–574.

Godement P, Wang LC & Mason CA (1994) Retinal axon divergence in the optic chiasm: dynamics of growth cone behavior at the midline. *J Neurosci* 14:7024–7039.

Sperry RW (1943) Visuomotor coordination in the newt (*Triturus viridescens*) after regeneration of the optic nerve. *J Comp Neurol* 79:33–55.

Walter J, Kern-Veits B, Huf J et al. (1987) Recognition of position-specific properties of tectal cell membranes by retinal axons *in vitro*. *Development* 101:685–696.

Activity-dependent wiring in the visual system

Constantine-Paton M & Law MI (1978) Eye-specific termination bands in tecta of three-eyed frogs. *Science* 202:639–641.

Feller MB, Wellis DP, Stellwagen D et al. (1996) Requirement for cholinergic synaptic transmission in the propagation of spontaneous retinal waves. *Science* 272:1182–1187.

Hubel DH, Wiesel TN & LeVay S (1977) Plasticity of ocular dominance columns in monkey striate cortex. *Philos Trans R Soc Lond B Biol Sci* 278:377–409.

Meister M, Wong RO, Baylor DA et al. (1991) Synchronous bursts of action potentials in ganglion cells of the developing mammalian retina. *Science* 252:939–943.

Munz M, Gobert D, Schohl A et al. (2014) Rapid Hebbian axonal remodeling mediated by visual stimulation. *Science* 344:904–909.

Penn AA, Riquelme PA, Feller MB et al. (1998) Competition in retinogeniculate patterning driven by spontaneous activity. *Science* 279:2108–2112.

Sretavan DW & Shatz CJ (1986) Prenatal development of retinal ganglion cell axons: segregation into eye-specific layers within the cat's lateral geniculate nucleus. *J Neurosci* 6:234–251.

Stryker MP & Harris WA (1986) Binocular impulse blockade prevents the formation of ocular dominance columns in cat visual cortex. *J Neurosci* 6:2117–2133.

Wiesel TN & Hubel DH (1963) Single-cell responses in striate cortex of kittens deprived of vision in one eye. *J Neurophysiol* 26:1003–1017.

Zhang J, Ackman JB, Xu HP et al. (2012) Visual map development depends on the temporal pattern of binocular activity in mice. *Nat Neurosci* 15:298–307.

Collaboration of activity-independent and -dependent wiring in the visual system

Cang J, Niell CM, Liu X et al. (2008) Selective disruption of one Cartesian axis of cortical maps and receptive fields by deficiency in ephrin-As and structured activity. *Neuron* 57:511–523.

Huberman AD, Manu M, Koch SM et al. (2008) Architecture and activity-mediated refinement of axonal projections from a mosaic of genetically identified retinal ganglion cells. *Neuron* 59:425–438.

Kerschensteiner D, Morgan JL, Parker ED et al. (2009) Neurotransmission selectively regulates synapse formation in parallel circuits *in vivo*. *Nature* 460:1016–1020.

Li Y, Van Hooser SD, Mazurek M et al. (2008) Experience with moving visual stimuli drives the early development of cortical direction selectivity. *Nature* 456:952–956.

Matsuoka RL, Nguyen-Ba-Charvet KT, Parray A et al. (2011) Transmembrane semaphorin signalling controls laminar stratification in the mammalian retina. *Nature* 470:259–263.

McLaughlin T, Torborg CL, Feller MB et al. (2003) Retinotopic map refinement requires spontaneous retinal waves during a brief critical period of development. *Neuron* 40:1147–1160.

Pfeiffenberger C, Yamada J & Feldheim DA (2006) Ephrin-As and patterned retinal activity act together in the development of topographic maps in the primary visual system. *J Neurosci* 26:12873–12884.

Yamagata M & Sanes JR (2008) Dscam and Sidekick proteins direct lamina-specific synaptic connections in vertebrate retina. *Nature* 451:465–469.

Drosophila visual system development

Clandinin TR, Lee CH, Herman T et al. (2001) *Drosophila* LAR regulates R1-R6 and R7 target specificity in the visual system. *Neuron* 32:237–248.

Hiesinger PR, Zhai RG, Zhou Y et al. (2006) Activity-independent prespecification of synaptic partners in the visual map of *Drosophila*. *Curr Biol* 16:1835–1843.

Morey M, Yee SK, Herman T et al. (2008) Coordinate control of synaptic-layer specificity and rhodopsins in photoreceptor neurons. *Nature* 456:795–799.

Ready DF, Hanson TE & Benzer S (1976) Development of the *Drosophila* retina, a neurocrystalline lattice. *Dev Biol* 53:217–240.

Reinke R & Zipursky SL (1988) Cell-cell interaction in the *Drosophila* retina: the *bride of sevenless* gene is required in photoreceptor cell R8 for R7 cell development. *Cell* 55:321–330.

Simon MA, Bowtell DD, Dodson GS et al. (1991) Ras1 and a putative guanine nucleotide exchange factor perform crucial steps in signaling by the sevenless protein tyrosine kinase. *Cell* 67:701–716.

Timofeev K, Joly W, Hadjieconomou D et al. (2012) Localized netrins act as positional cues to control layer-specific targeting of photoreceptor axons in *Drosophila*. *Neuron* 75:80–93.

Tomlinson A & Ready DF (1987) Cell fate in the *Drosophila* ommatidium. *Dev Biol* 123:264–275.

Wernet MF, Mazzoni EO, Celik A et al. (2006) Stochastic spineless expression creates the retinal mosaic for colour vision. *Nature* 440:174–180.

Axon guidance and activity-dependent wiring in other systems

Colamarino SA & Tessier-Lavigne M (1995) The axonal chemoattractant netrin-1 is also a chemorepellent for trochlear motor axons. *Cell* 81:621–629.

Datwani A, Iwasato T, Itohara S et al. (2002) NMDA receptor-dependent pattern transfer from afferents to postsynaptic cells and dendritic differentiation in the barrel cortex. *Mol Cell Neurosci* 21:477–492.

Espinosa JS, Wheeler DG, Tsien RW et al. (2009) Uncoupling dendrite growth and patterning: single-cell knockout analysis of NMDA receptor 2B. *Neuron* 62:205–217.

Hedgecock EM, Culotti JG & Hall DH (1990) The unc-5, unc-6, and unc-40 genes guide circumferential migrations of pioneer axons and mesodermal cells on the epidermis in *C. elegans*. *Neuron* 4:61-85.

Serafini T, Kennedy TE, Galko MJ et al. (1994) The netrins define a family of axon outgrowth-promoting proteins homologous to *C. elegans* UNC-6. *Cell* 78:409–424.

Van der Loos H & Woolsey TA (1973) Somatosensory cortex: structural alterations following early injury to sense organs. *Science* 179:395–398.

CHAPTER 6

Olfaction, Taste, Audition, and Somatosensation

All sensory systems share common tasks, including the transformation of sensory stimuli into electrical signals, the optimization of detection sensitivity, selectivity, speed, and reliability, and the extraction of salient features with the ultimate purpose of helping animals survive and reproduce. At the same time, each sensory system has unique properties related to the physical nature of the sensory stimuli and how the sense serves the organism. Our detailed study of vision in Chapter 4 provides a framework that we will expand here as we examine the remaining major senses: olfaction, taste, audition, and somatosensation. We will study olfaction first and in greatest detail, as it differs from the other senses in some important ways, such as the large number of odorant receptors and the direct path by which olfactory signals are transmitted to the cortex. We begin our story with salmon homing.

HOW DO WE SENSE ODORS?

Salmon have a fascinating life cycle. They are born and live their first 1–2 years in freshwater streams. Then, they swim out into the ocean to feed on rich food sources, grow, and accumulate fat. When they are reproductively mature, they swim back from the ocean, often for thousands of miles, to the very stream in which they were born, passing many obstacles and other similar streams along the way (**Figure 6–1**). They spawn in their native stream, completing the cycle, and die shortly afterwards.

This remarkable homing behavior of salmon relies primarily on their sense of smell. Local differences in soil and vegetation give each stream a unique chemical composition and a distinctive odor. The memory of these odors becomes imprinted in young salmon, particularly during the period when they leave their native streams and swim toward the ocean. Adult salmon then use these odor memories to find their native streams during the journey home.

In a field study, young Coho salmon that were hatched and raised at a Wisconsin State Fish Hatchery were exposed to one of two non-natural chemicals (either morpholine or phenethyl alcohol [PA]) in trace amounts early in their life. They were then released into Lake Michigan at an age when salmon normally leave their native streams. Eighteen months later, when the salmon returned to spawn, one stream was scented with morpholine and another stream with PA at concentrations similar to those the salmon were exposed to when young. More than 90% of captured salmon previously exposed to morpholine were caught in the stream scented with morpholine, and more than 90% of captured salmon previously exposed to PA were caught in the stream scented with PA (**Figure 6–2**). This striking specificity provided strong support for the odor-imprinting hypothesis.

The salmon homing behavior provides just one example of how the sense of smell shapes animal life. Because the olfactory system has somewhat regressed during recent primate evolution, we humans may be less aware of its importance.

Figure 6–1 Salmon homing. Adult salmon swimming back from the ocean toward their native streams to reproduce. (Courtesy of Marvina Munch/USFWS.)

Figure 6–2 Salmon homing relies on olfaction. (A) Map of a field study site where salmon homing was monitored. Hatchery-raised salmon that had been exposed to a trace amount of either morpholine (M) or phenethyl alcohol (PA) when young were monitored after they matured and swam upstream to spawn. The map shows the release site for the three groups of tagged young salmon (M-exposed, PA-exposed, and untreated controls). The two labeled streams were scented with M or PA, respectively, at the time when mature salmon returned to spawn. More than 20 rivers along the coast of Lake Michigan (beyond the map shown) were also monitored for salmon capture. **(B)** Quantification of where M-treated, P-treated, and control salmon were captured, as a percentage of total salmon captured from each treatment group. The vast majority of M-treated and PA-treated salmon were caught in the M-treated and PA-treated streams, respectively. By contrast, most control salmon were captured in other streams. (Adapted from Scholz AT, Horrall RM, Cooper JC, et al. [1976] *Science* 192:1247–1249.)

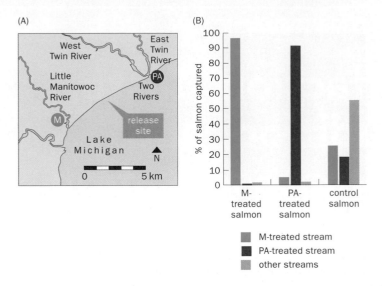

For many animals, however, the sense of smell is the primary means of finding food, avoiding predators, and seeking mates, all of which are essential for the survival of individuals and the propagation of species.

6.1 Odorant binding leads to opening of a cyclic nucleotide-gated channel in olfactory receptor neurons

Odorants, that is, molecules that elicit olfactory perception, are usually airborne and volatile (except for aquatic animals such as salmon). Diffusing through the air and into the nose, odorants pass through a layer of mucus and land on the surfaces of **olfactory cilia**. Each cilium is a dendritic branch of an **olfactory receptor neuron** (ORN, also called an olfactory sensory neuron) that resides within the **olfactory epithelium**. Olfaction begins with the binding of odorants to **odorant receptors** on the olfactory cilia. This event triggers the depolarization of ORNs and produces action potentials that are propagated by ORN axons to the **olfactory bulb**, the first olfactory processing center in the brain. Within the olfactory bulb, ORN axons terminate in discrete, ball-like structures called **glomeruli**, where they form synapses with the dendrites of their postsynaptic target neurons (**Figure 6–3**).

A key question to investigate in any sensory system is how sensory stimuli are converted to electrical signals, the universal form of neuronal communication. For the olfactory system, the question becomes: how does the binding of an odorant to its receptor(s) initiate an electrical signal? Before their identification, evidence already suggested that odorant receptors were **G-protein-coupled receptors** (**GPCRs**), similar to rhodopsins in photoreceptor cells. One key difference between these two systems is the trigger: odorant receptors are activated by binding of odorants, whereas rhodopsins are activated by absorption of photons.

Figure 6–3 Organization of the peripheral olfactory system. (A) Odorants pass through the nostrils and bind to the odorant receptors on the olfactory cilia, which are dendrites of the olfactory receptor neurons (ORNs). ORN axons terminate in discrete glomeruli in the olfactory bulb, the first olfactory processing center in the brain. **(B)** Scanning electron micrograph of rat olfactory cilia. (B, from Menco BPM & Farbman AI [1985] *J Cell Sci* 78:283–310. With permission from the Company of Biologists, Ltd.)

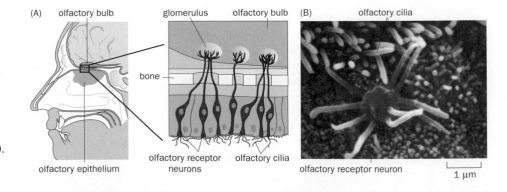

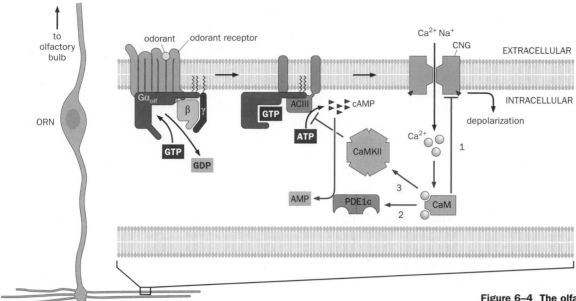

The signal transduction pathway through which odorant receptor activation generates an electrical signal has been delineated in detail (**Figure 6–4**). Odorant binding to odorant receptors triggers the activation of $G\alpha_{olf}$, a specific kind of $G\alpha$. $G\alpha_{olf}$ then activates a specific type of membrane-bound adenylate cyclase, ACIII, which leads to an increase in the concentration of cyclic AMP (cAMP). cAMP binds to a cyclic nucleotide-gated (CNG) channel and causes it to open, thereby allowing the influx of cations (Na^+, Ca^{2+}) down their electrochemical gradients into ORNs, causing depolarization. An additional step of amplification via opening of Ca^{2+}-activated Cl^- channels may also contribute to the depolarization (Cl^- is more concentrated in the ORN than in the extracellular environment, thus opening of a Cl^- channel leads to Cl^- efflux). Hence, ORNs are depolarized as a result of odorant binding and receptor activation.

The importance of the signal transduction pathway for olfaction has been demonstrated by analysis of mutant mice. For example, CNG channel knockout mice are **anosmic**, that is, unable to perceive odors, because they cannot transform odorant binding to electrical signals (**Figure 6–5**). Mice lacking $G\alpha_{olf}$ or ACIII are similarly anosmic.

Figure 6–4 The olfactory transduction pathway and its regulation. Left, an olfactory receptor neuron (ORN). Right, a magnified image of the cilium, a dendritic extension of the ORN, where the odorant binds to the odorant receptor. This binding activates a G protein cascade involving a special $G\alpha$ ($G\alpha_{olf}$) and an adenylate cyclase (ACIII) that catalyzes the production of cAMP. cAMP binds to and activates a cyclic nucleotide-gated (CNG) channel, leading to the influx of Na^+ and Ca^{2+} and depolarization of the ORN. Negative regulatory pathways (see Section 6.2) are colored red, including (1) inhibition of CNG channel by Ca^{2+}/ calmodulin complex (CaM), (2) activation of a phosphodiesterase (PDE1c) by CaM, and (3) inhibition of ACIII by CaM kinase II (CaMKII). (Courtesy of Haiqing Zhao. See also Firestein S [2001] *Nature* 413:211–218.)

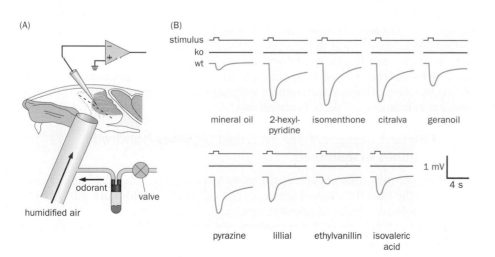

Figure 6–5 A cyclic nucleotide-gated (CNG) channel is essential for olfaction. (A) Illustration of the electroolfactogram (EOG) in a side view of the mouse olfactory system. An extracellular electrode was inserted in the olfactory epithelium (pink) to measure the collective activity of many ORNs in response to odorant delivered to the nostrils. **(B)** Mice with a knockout (ko) of CNG channel expressed in the olfactory epithelium show no EOG response to any of the odorants (red traces). By comparison, EOG traces from wild-type (wt) animals (blue) illustrate a reduction of extracellular potential (a consequence of ORN depolarization), followed by a recovery phase. Odorant stimuli are shown in black. (B, adapted from Brunet LJ, Gold GH & Ngai J [1996] *Neuron* 17:681–693. With permission from Elsevier Inc.)

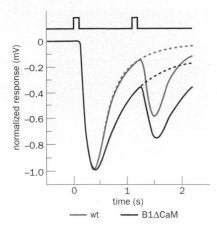

Figure 6–6 Calmodulin binding to the cyclic nucleotide-gated (CNG) channel regulates recovery. Electroolfactograms illustrate that, compared with the wild-type (wt), B1ΔCaM mice in which the CNG channel's calmodulin binding site is mutated show slower decay of the olfactory response, indicating defective recovery. As measured by the difference between the solid and dashed lines, the response of these mutant mice to the second odorant pulse is relatively normal, suggesting that adaptation is not as affected. The two odorant pulses (100 µM amyl acetate for 100 ms each) are in black. (Adapted from Song Y, Cygnar KD, Sagdullaev B et al. [2008] *Neuron* 58:374–386. With permission from Elsevier Inc.)

6.2   Ca²⁺ coordinates olfactory recovery and adaptation

As in the visual system, olfactory responses must be terminated after the odorant is withdrawn so that ORNs can respond to future olfactory stimulation. This process is termed olfactory recovery. The response to olfactory stimulation is also modified by prior experience of the same odorant, a phenomenon called olfactory adaptation. Although olfactory recovery and adaptation refer to different biological phenomena, they share molecular pathways, as do visual recovery and adaptation (see Sections 4.6 and 4.7). Indeed, the entry of Ca^{2+} via the CNG channel plays a central role in both systems, but a comparison of olfaction and vision reveals that the recovery and adaptation pathways for each process act on different effectors (compare Figures 6–4 and 4–11).

Intracellular Ca^{2+}, in a complex with the Ca^{2+}-binding protein calmodulin, has at least three independent functions in regulating olfactory recovery and adaptation. First, Ca^{2+}/calmodulin binds to and directly inhibits the CNG channel. Second, Ca^{2+}/calmodulin activates the phosphodiesterase PDE1c, facilitating the hydrolysis of cAMP. Third, Ca^{2+}/calmodulin activates Ca^{2+}/calmodulin-dependent protein kinase II, which phosphorylates ACIII and thereby down-regulates the production of new cAMP. Altogether, these negative feedback loops return cAMP to the baseline level, close the CNG channels, and prepare the ORNs for future stimuli (Figure 6–4).

The relative contribution of each negative feedback mechanism to olfactory recovery and adaptation has been examined by recording odorant responses in genetically modified mice. For example, the knock-in procedure (see Section 13.7) has been used to produce mice in which the calmodulin-binding site on the CNG channel is mutated, disrupting regulation of the channel by calmodulin. Such mice have been used to study the decay of response to odorant stimulation (a measure for recovery) and the magnitude of response to a second odorant pulse (a measure for adaptation). These studies suggest that Ca^{2+}/calmodulin regulation of the CNG channel plays a more prominent role in recovery than in adaptation (**Figure 6–6**). Knockout studies have revealed that lack of PDE1c causes a more severe defect in adaptation than in recovery.

6.3   Odorants are represented by combinatorial activation of olfactory receptor neurons

We now turn to a central question in olfaction: how are odorants recognized by the olfactory system? Odorants are mixes of volatile chemicals with diverse structures and properties (**Figure 6–7**). The mammalian olfactory system can detect and discriminate a vast number of odorants. How is this feat accomplished?

Physiological studies of the response of individual ORNs to odorants suggest that odorant recognition is a complex process. Comparing the responses of different ORNs to a panel of chemically diverse odorants (**Figure 6–8**), it was found that each odorant activates many ORNs, and each ORN is activated by multiple odorants. Therefore, the identity of an odorant is not encoded by a single ORN, but rather by the outputs of many different ORNs. In other words, odorants are represented combinatorially at the level of ORNs. This is reminiscent of the color coding we studied in Chapter 4, where color information is extracted by comparing the activities of cones expressing opsins that confer different spectral sensitivities. However, whereas color vision utilizes only two or three input channels, olfaction uses many hundreds of parallel channels.

6.4   Odorant receptors are encoded by many hundreds of genes in mammals

Just as the identification of cone opsin genes revealed the molecular basis of color detection, the identification of genes for the odorant receptors provided insight into the molecular basis of odorant detection. Furthermore, it was also instrumental in the discovery of the organizational principles of the olfactory system (see Sections 6.6–6.8).

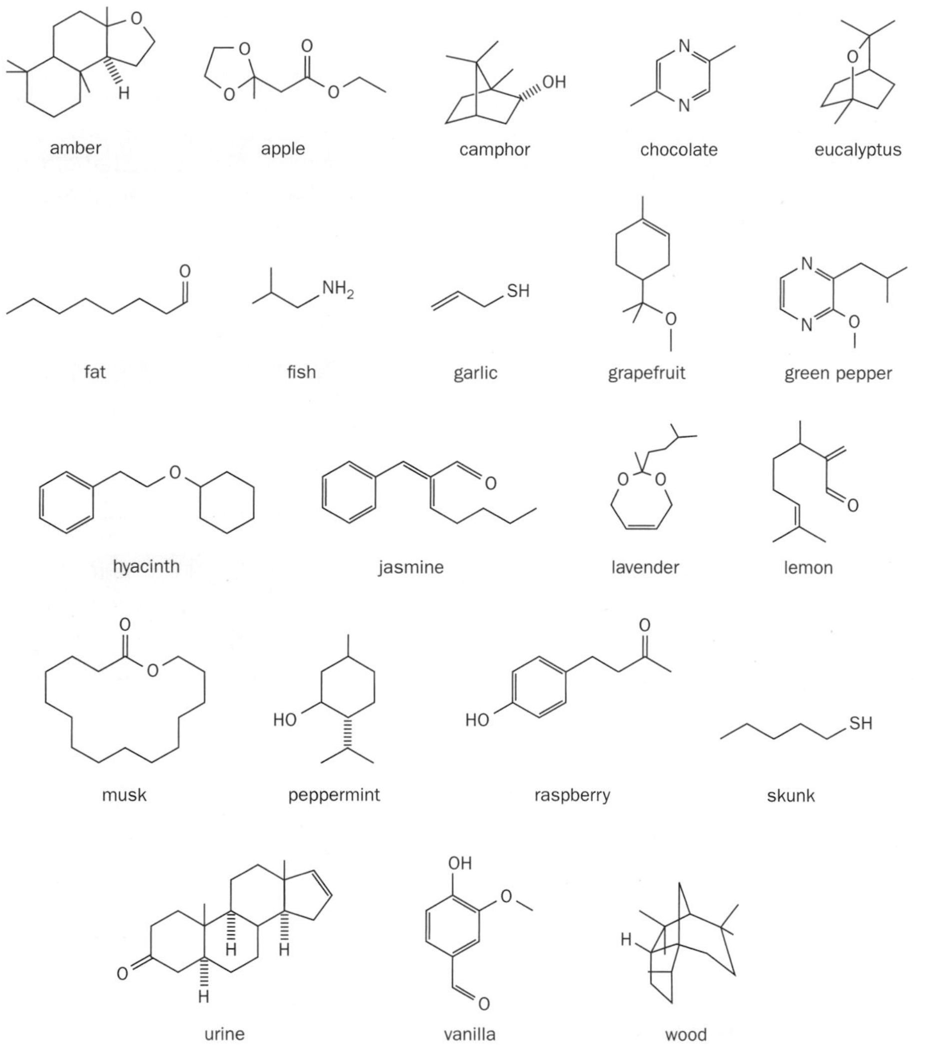

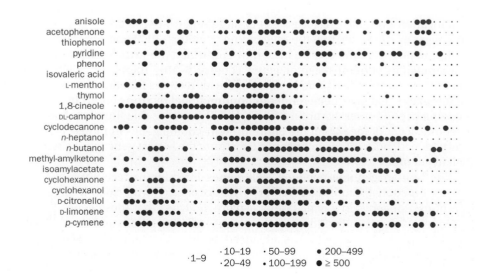

amber   apple   camphor   chocolate   eucalyptus

fat   fish   garlic   grapefruit   green pepper

hyacinth   jasmine   lavender   lemon

musk   peppermint   raspberry   skunk

urine   vanilla   wood

Figure 6–7 Structural formulae of odorants that give selected smells. Odorants have diverse chemical structures. Shown here are characteristic compounds that contribute to specific smells. (Courtesy of Linda Buck.)

As noted in Section 4.12, known amino acid sequences derived from purified bovine rhodopsin were used in the cloning of opsin genes. However, no purified odorant receptor was available to provide any amino acid sequence information. To isolate odorant receptor cDNAs, an assumption was made that odorant

anisole
acetophenone
thiophenol
pyridine
phenol
isovaleric acid
L-menthol
thymol
1,8-cineole
DL-camphor
cyclodecanone
n-heptanol
n-butanol
methyl-amylketone
isoamylacetate
cyclohexanone
cyclohexanol
D-citronellol
D-limonene
p-cymene

·1–9   ·10–19   ·50–99   ● 200–499
       ·20–49   ·100–199 ● ≥ 500

Figure 6–8 Combinatorial coding of odorants by ORNs. Responses of 60 individual frog ORNs (columns) to a panel of 20 odorants (rows). The different odorants, which exhibit diverse structures, were delivered at high concentrations. Spot sizes relate to action potentials per minute, with the key located below. Each odorant stimulates a different set of ORNs, and the intensity of the response depends on both the ORN and the odorant. (Adapted from Sicard G & Holley A [1984] *Brain Res* 292:283–296. With permission from Elsevier Inc.)

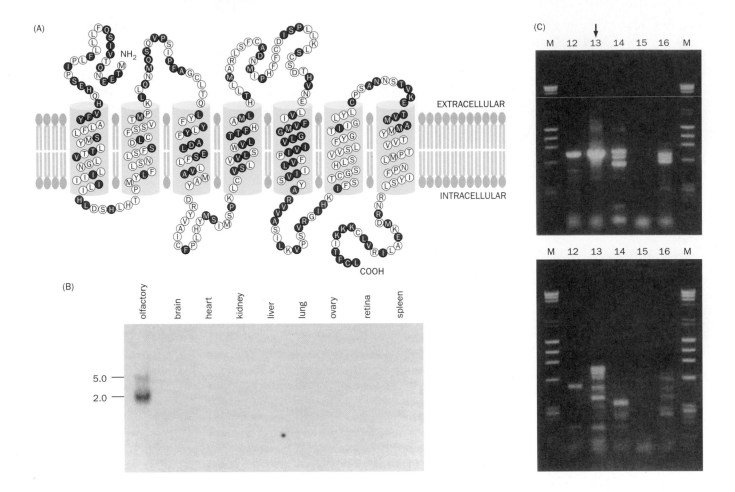

Figure 6–9 Identification of genes encoding odorant receptors. (A) Primary structure of an odorant receptor. The seven cylinders represent seven transmembrane helices. Yellow circles represent conserved amino acids. Blue circles represent amino acids that are highly variable among different odorant receptors, many of which are located in the transmembrane helices where odorants bind. **(B)** A northern blot (see Section 13.13) showing that mRNAs of odorant receptors are expressed specifically in the olfactory epithelium. Molecular weight markers (in kilobases) are shown on the left. **(C)** PCR products (top) and their restriction enzyme digestion patterns (bottom). In lane 13 (arrow) the sum of the molecular weights of the PCR digestion products greatly exceeds the molecular weight of the PCR product itself, suggesting that the original PCR product is a mixture of different DNA species. M, molecular weight marker. (Adapted from Buck L & Axel R [1991] *Cell* 65:175–187. With permission from Elsevier.)

receptors were G-protein-coupled receptors (**Figure 6–9**A) based on earlier studies implicating G protein, adenylate cyclase, and cyclic nucleotide-gated channels in olfactory transduction (see Section 6.2). In 1991, researchers took advantage of what was then a recent innovation in molecular biology, a highly sensitive DNA amplification technique called **polymerase chain reaction** (**PCR**). Oligonucleotide primer pairs were designed to correspond to highly conserved amino acids within the transmembrane domains of known GPCRs. These primers were used to amplify cDNAs that had been generated from mRNAs isolated from rat nasal epithelia. Probes from such amplified cDNAs detected mRNAs that were expressed exclusively in the olfactory epithelium (Figure 6–9B), supporting the idea that these cDNAs corresponded to odorant receptors. Analysis of the PCR amplification products suggested a very large repertoire (Figure 6–9C): the first cloning study estimated that the rat genome contains ~1000 odorant receptors.

In the post-genome era, we have now confirmed the extraordinary diversity of mammalian odorant receptor genes (**Figure 6–10**). There are ~1400 genes that encode odorant receptors in mice; of these, 1063 are functional and 328 are **pseudogenes** that have been rendered nonfunctional by stop codons in the coding sequences or other disrupting mutations. This means that functional odorant receptors account for >4% of the entire protein-coding genes in mouse (~23,000 total). Indeed, odorant receptor genes constitute the largest gene family in mammalian genomes.

The human genome contains ~800 genes for odorant receptors, but only 388 are functional; more than half are pseudogenes. Having such a large fraction of pseudogenes in a gene family usually indicates that these genes are not under selection pressure and are in the process of being eliminated. Accordingly, our sense of smell may not be as advanced as that of mice, rats, or pigs, whose genomes each encode >1000 functional odorant receptors (Figure 6–10).

6.5 Polymorphisms in odorant receptor genes contribute to individual differences in odor perception

It has been well documented that significant individual differences exist in people's olfactory perception. For instance, some people can detect a strong smell in their urine after eating asparagus while others cannot. As another example, androstenone, a mammalian social odorant commonly used in animal husbandry to induce female pigs to mate, smells pleasant (vanilla-like) or hardly at all to some people and sickening to others. What accounts for individual differences in smell? Twin studies suggested a strong genetic component, as identical twins have a much more similar threshold for androstenone detection than fraternal twins (**Figure 6–11**A). Differences in androstenone perception are now known to be associated with **polymorphisms**—DNA sequence variations among individuals—in the human odorant receptor OR7D4 (Figure 6–11B), which is strongly activated by androstenone. In the human population, 16% of alleles encode substitutions at amino acid positions 88 and 133 that change the more common arginine/threonine (RT) variant to the tryptophan/methionine (WM) variant. Biochemical studies indicated that the WM variant of OR7D4 has a much lower affinity for androstenone compared to the RT variant. Psychophysical tests showed that humans who carry two RT alleles are more sensitive to androstenone and are more likely to consider the smell sickening, whereas humans heterozygous or homozygous for the WM allele are more likely to consider androstenone vanilla-like (Figure 6–11B). Thus, amino acid changes in a single odorant receptor contribute to different odor perceptions.

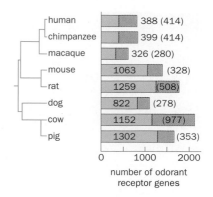

Figure 6–10 Odorant receptor genes are the largest gene family in the mammalian genome. Hundreds of odorant receptor genes (blue) and pseudogenes (red, in parentheses) have been identified in the sequenced genomes of mammalian species. At left is a phylogenetic tree for these mammalian species (see Chapter 12 for more details). (Adapted from Nei M, Niimura Y & Nozawa M [2008] *Nat Rev Genet* 9:951–963. With permission from Macmillan Publishers Ltd. The data for pig is from Groenen MA, Archibald AL, Uenishi H et al. [2012] *Nature* 491:393–398.)

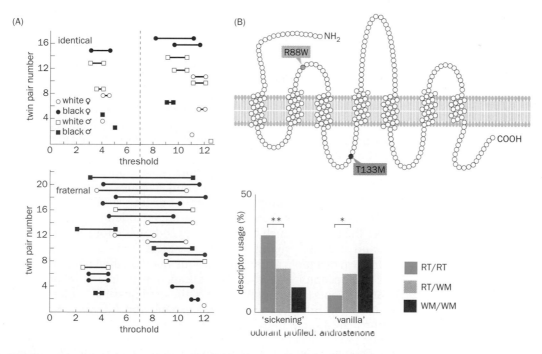

Figure 6–11 Polymorphisms in an odorant receptor contribute to individual differences in androstenone smell. (A) Twin studies revealed a strong genetic component in the detection threshold of androstenone. Results for the two members of each twin pair are linked by a horizontal line. A single entry for a twin pair indicates identical threshold for each. The *x* axis shows the response thresholds to androstenone in a series of binary steps, with the most diluted (1) concentration being 1.79 µM and the most concentrated (12) being 3.67 mM. The concentration indicated by the dashed vertical line divides subjects into two groups: androstenone-sensitive (to the left of the line) and androstenone-insensitive (to the right of the line). As is evident, the threshold varies much less for identical twins than for fraternal twins, regardless of race and sex. **(B)** Top, the structure of human OR7D4, with the location of the polymorphic amino acids indicated. Bottom, humans with different OR7D4 alleles differ in their use of descriptors for androstenone. Human subjects with one or two copies of the WM allele have an increased probability of considering androstenone vanilla-like. The differences are statistically significant (*, P<0.05; **, P<0.01; χ^2 test with Bonferroni correction). (A, adapted from Wysocki CJ & Beauchamp GK [1984] *Proc Natl Acad Sci USA* 81:4899–4902. With permission from the authors; B, adapted from Keller A, Zhuang H, Chi Q et al. [2007] *Nature* 449:468–472. With permission from Macmillan Publishers Ltd.)

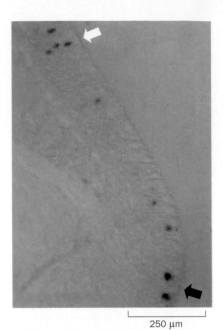

250 μm

Figure 6–12 Each odorant receptor gene is expressed in a small fraction of ORNs. *In situ* hybridization of mouse olfactory epithelium with odorant receptor-specific probes. The light (white arrow) and dark (black arrow) signals represent two different probes recognizing two distinct odorant receptors. Each spot represents a single ORN cell body. A small fraction of the olfactory epithelium is shown here. The vast majority of ORNs are not labeled by either of the probes, indicating that these two odorant receptors are expressed in a small fraction of ORNs. (From Vassar R, Ngai J & Axel R [1993] *Cell* 74:309–318. With permission from Elsevier Inc.)

6.6   Each olfactory receptor neuron (ORN) expresses a single odorant receptor

So far, we have examined how odorant binding leads to the activation of individual ORNs, and how odorant molecules are detected by up to 1000 different receptors. But how does the brain recognize odors as specifically perceived objects (**percepts**)? How do animals distinguish different odors so that they respond differently to the smells of food, mates, or predators? The organization of the olfactory system offers insights that help us begin to answer these questions.

There are about 5 million ORNs in the nasal epithelium of the mouse. How is the expression of 1000 different odorant receptors partitioned over the 5 million ORNs? Early studies of ORN physiology had shown that each ORN responds to many odorants (see Figure 6–8). These data suggested two possible modes of organization: (1) each ORN expresses multiple odorant receptors, or (2) each ORN expresses a single odorant receptor that is capable of binding multiple odorants.

The identification of odorant receptors and characterization of their expression led to the unequivocal resolution of this issue. *In situ* **hybridization**, in which nucleic acid probes are used to visualize sites of gene expression in native tissues (see Section 13.13), revealed how the cells that express mRNA of a particular odorant receptor are distributed in the nasal epithelia (**Figure 6–12**). On average, each odorant receptor is expressed in about 0.1% of ORNs. This frequency is consistent with the hypothesis that each ORN expresses only a single odorant receptor. Indeed, within a given ORN, the functional odorant receptor mRNA is transcribed exclusively from one chromosome of a homologous pair; this property is called **allelic exclusion**. Recent studies indicate that genes that encode odorant receptors are in chromosomal regions that are subjected to complex histone modifications, such that they are inaccessible to transcriptional machinery. Transient activity of a histone modification enzyme allows just one odorant receptor gene to be expressed. This is then followed by a feedback mechanism: once an ORN expresses a functional odorant receptor, a negative feedback pathway involving an activated adenylate cyclase is triggered so no other odorant receptor can be expressed in that same ORN.

Given that each ORN expresses a single odorant receptor, mouse ORNs can be divided into ~1000 types based on the specific odorant receptor that they express. Since individual ORNs respond to many odorants, each odorant receptor must be activated by multiple odorants. Indeed, when dissociated ORNs expressing identified odorant receptors were stimulated by a panel of odorants to assay for activity based on Ca^{2+} imaging, this was verified: each odorant receptor is activated by a multitude of odorants, and each odorant activates a multitude of odorant receptors. This revealed a property similar to previous studies of individual ORNs (see Figure 6–8): the identities of odorants are encoded by a combinatorial receptor code. The power of such combinatorial coding is enormous. For example, if each odorant activates two independent receptors, and each combinatorial activation of any two receptors generates a unique percept, then a repertoire of 1000 receptors (and receptor-expressing ORNs) can, in principle, generate half a million different percepts ($1000 \times 999/2$; that is, the total possible choices for selecting 2 out of 1000). This number can be further expanded if different percepts are generated by the activation of more than two receptors, by the differential extent of activation of individual receptors, or by different time courses of activation. This is how the olfactory system can in principle recognize, encode, and distinguish a vast number of odorant molecules.

6.7   ORNs expressing a given odorant receptor are broadly distributed in the nose

How does the brain decipher the mixture of odorants in the nose? ORN axons terminate in spherical structures called glomeruli in the olfactory bulb (see Figure 6–3A). Within each glomerulus, ORN axons form synaptic connections with the dendrites of **mitral cells** and **tufted cells**, which are distinct subtypes of output neurons of the olfactory bulb. Mitral cells and tufted cells share many

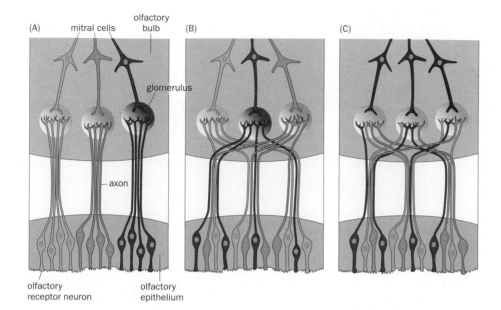

Figure 6–13 Three possible models of olfactory system organization. ORNs of the same color express the same odorant receptor, and therefore represent the same olfactory information. **(A)** ORNs that express the same odorant receptor have spatially clustered cell bodies and send axons to the same glomerulus. **(B)** ORNs that express the same receptor have spatially dispersed cell bodies, but their axons converge to the same glomerulus. **(C)** ORNs that express the same receptor have spatially dispersed cell bodies, and their axons project to different glomeruli; each glomerulus receives axons from ORNs that express different odorant receptors. (Adapted from Axel R [1995] *Sci Am* 273:154–159. With permission from Macmillan Publishers Ltd.)

similarities—both are glutamatergic excitatory neurons that send apical dendrites to a single glomerulus and project long axons to relay information to multiple areas of the **olfactory cortex**, where olfactory information is further analyzed. However, these cell types differ in their cell body locations, response properties, and detailed axonal projection patterns (see Figure 6-19). For simplicity, we use mitral cells to represent olfactory bulb output neurons in subsequent discussions.

What kind of organizational scheme could allow the brain to determine which odorants are present within an animal's surroundings? At least three different scenarios can be envisioned (**Figure 6–13**). First, ORNs expressing the same odorant receptors could be spatially clustered in the olfactory epithelium and project their axons to the same glomerulus in the olfactory bulb. Second, ORNs expressing different odorant receptors could be spatially intermingled in the olfactory epithelia, with their axons being sorted out during their targeting process, such that ORNs expressing the same odorant receptor project to the same glomerulus. In both of the above cases, each mitral cell receives direct input from one type of ORN. Third, information received by different odorant receptors could be intermingled at the level of individual glomeruli; in this case, each mitral cell would receive direct input from many types of ORNs, and deciphering the ORN activation pattern would be left for the olfactory cortex.

In situ hybridization studies in mouse and rat using specific probes to odorant receptor mRNAs revealed that the cell bodies of a given ORN type are broadly distributed along the long axis of the nasal epithelium without apparent spatial organization. The distribution along the orthogonal axis exhibits some patterns, with each ORN type distributed within a zone about a quarter the length of the axis (**Figure 6–14**). Thus, different ORN types expressing different receptors are completely intermingled along the long axis and considerably intermingled even along the short axis. These findings ruled out the scenario modeled in Figure 6-13A: ORN cell bodies expressing the same odorant receptor are not spatially clustered. The broad distribution of a given ORN type enables the sampling of odorant molecules across a large surface area of nasal epithelium, and therefore may increase the sensitivity of odorant detection.

6.8 ORNs expressing the same odorant receptor project their axons to the same glomerulus

Interestingly, *in situ* hybridization also detected mRNA of a given odorant receptor in specific glomeruli located in stereotyped positions in the olfactory bulb. The simplest interpretation is that while the cell bodies of ORNs expressing a given

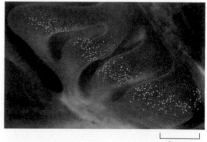

⊢——⊣ 1 mm

Figure 6–14 Distribution of ORNs expressing a given receptor in the olfactory epithelia. Three different types of ORNs, each expressing a different odorant receptor, are shown here as blue, yellow, and green, respectively. Note that the three ORN types are broadly distributed along the long axis of the olfactory epithelia. Along the short axis, they are expressed in different quarters and therefore are mostly segregated. Summarized from *in situ* hybridization data. (From Vassar R, Ngai J & Axel R [1993] *Cell* 74:309–318. With permission from Elsevier Inc.)

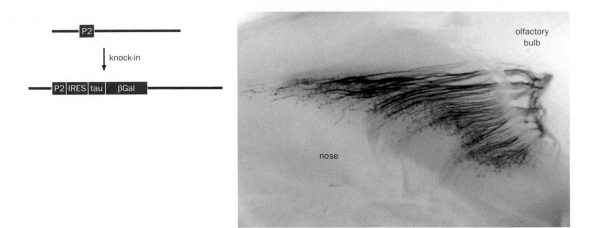

Figure 6–15 Visualizing ORN axon convergence. Left, the gene encoding the odorant receptor P2 (the box indicates its coding region) was engineered in embryonic stem cells by inserting a piece of DNA that enables the tracing of P2-expressing ORNs. IRES, internal ribosome entry sequence that allows the translation of the axon tracer protein tau-βgal from the modified P2 mRNAs; tau, a microtubule-binding protein for efficient labeling of axons; βgal, β-galactosidase enzyme from bacteria. Right, β-galactosidase activity, visualized with a colored substrate, reveals ORN cell bodies in the nose (dots) and the convergence of their axonal projections onto one glomerulus in the olfactory bulb. (Adapted from Mombaerts P, Wang F, Dulac C et al. [1996] *Cell* 87:675–686. With permission from Elsevier Inc.)

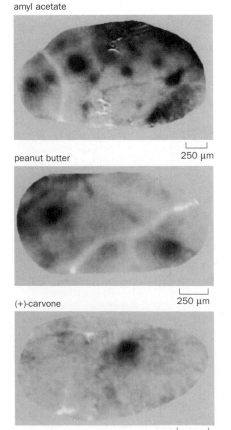

odorant receptor are distributed broadly in the olfactory epithelium, their axons are targeted to the same glomeruli (Figure 6–13B). As a result, their mRNAs can be visualized as concentrated spots in the olfactory bulb. This convergent projection of ORN axons was directly visualized with an elegant genetic labeling method in the mouse. Using the knock-in procedure (see Section 13.7), a specific odorant receptor gene was engineered to co-express an axonal tracer, the β-galactosidase enzyme with a color substrate fused with a microtubule binding protein **tau** that is highly enriched in axons. In this manner, ORNs expressing the engineered odorant receptor gene can be traced from their cell bodies all the way to their axon terminals in the olfactory bulb by visualizing the color substrate (**Figure 6–15**).

Many ORN types have been tagged with this method thus far. These data indicate that the axons of ORNs that express the same odorant receptor converge on specific glomeruli, usually one each in the medial and lateral surface of the olfactory bulb. Moreover, in different animals, the target glomeruli for the same ORN type (expressing the same odorant receptor) are positioned with a large degree of **stereotypy**, that is, little variation across different individuals.

Thus, olfactory information is represented as a spatial map of glomerular activation in the olfactory bulb. The problem of identifying which odorants are present in the environment is equivalent to deciphering which glomeruli are activated. Indeed, using neuronal activity monitoring techniques such as the intrinsic signal imaging we encountered in Chapter 4 (see Figure 4–42B), specific odorants have been shown to activate specific ensembles of glomeruli (**Figure 6–16**).

These remarkable discoveries illustrate a general principle in the organization of the nervous system: the use of maps to represent information. In studying the visual system, it was intuitive to assume that the brain uses a spatial map to represent information gathered in the retina, as vision is inherently a representation of the spatial property of the world. In contrast, it was not apparent, *a priori*, how odorants should be represented in the brain. Despite the similarity of using neural maps to represent sensory information, however, the glomerular map of odorants is qualitatively different from the visual map: it does not correspond to space in the external world, but rather to the nature of the chemicals present.

Figure 6–16 Intrinsic signal imaging of odorant response in the mouse olfactory bulb. Each dark patch represents a glomerulus activated in response to specific odorants as indicated. Each odorant elicits a specific spatial pattern of glomerulus activation. (From Rubin BD & Katz LC [1999] *Neuron* 23:499–511. With permission from Elsevier Inc.)

6.9 Olfactory bulb circuits transform odor representation through lateral inhibition

We now have a good understanding of the first steps of olfaction. Olfactory information is represented by patterns of glomerular activation in the olfactory bulb, made
possible by the convergent axonal projections of ORNs expressing the same odorant receptors. How do glomerular activation patterns inform higher olfactory centers, eventually leading to odor perception and odor-driven behavior? To date, there
are no satisfactory answers to these important questions. The next two sections will
address progress and challenges in this area of research, focusing on the olfactory
bulb and the olfactory cortex, respectively.

The neural circuitry of the olfactory bulb resembles that of the retina
(**Figure 6–17**). Within each glomerulus, ORN axons form glutamatergic synapses
with the apical dendrites of mitral cells. Each mitral cell sends apical dendrites
to a single glomerulus. ORN → mitral cell thus constitutes an excitatory pathway analogous to the photoreceptor → bipolar cell → ganglion cell pathway in
the retina. Because each glomerulus receives input from ORNs that express the
same odorant receptor, and because each mitral cell sends apical dendrites to a
single glomerulus, each mitral cell can be identified with a single odorant receptor. Given this orderly anatomy, each glomerulus can be thought of as a discrete
olfactory processing channel. On average, each glomerulus receives input from
many thousands of ORNs, and the output is conveyed by several dozen mitral
cells. There is therefore an information convergence—each mitral cell receives
pooled input from many hundreds of ORNs, which increases the sensitivity

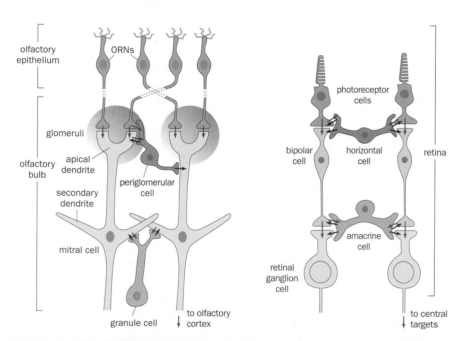

**Figure 6–17 Schematic diagrams comparing the olfactory bulb and
retinal circuits.** Simplified diagram of neurons and connections in
the olfactory bulb (left) and analogous connection patterns of cells
in the retina (right, see Sections 4.16–4.20 for more details). In
both diagrams, thin lines represent axons and thick lines represent
dendrites. Axon and dendrite terminals are enlarged to show the
connections in greater detail. Green represent excitatory neurons,
and orange inhibitory neurons. In the olfactory bulb circuit, olfactory
receptor neurons (ORNs) and mitral cells constitute the excitatory
pathway that delivers olfactory signals from the nose to the brain
through discrete olfactory processing channels, two of which are shown
here. This is made possible because each glomerulus receives axonal
input from ORNs that express the same odorant receptor, and mitral
cells send their apical dendrites to a single glomerulus. Granule cells
and most periglomerular cells are GABAergic. They spread inhibition
laterally across different glomeruli. The green and red arrows indicate
excitatory and inhibitory synaptic connections between these cells,
respectively. Dashed ORN axons indicate that the distance from cell
bodies to glomeruli is much longer than schematized. Not shown are
many additional types of olfactory bulb local interneurons, some of
which extend processes across larger areas of the olfactory bulb than
periglomerular and granule cells, and some of which use dopamine
and glutamate as neurotransmitters. Note that the dendrites of
mitral, granule, periglomerular, horizontal, and amacrine cells not only
receive but also send information. Indeed, many GABAergic neurons
lack axons and are exceptions to the theory of dynamic polarization
discussed in Section 1.7. Despite the apparent similarities, there
are notable differences. The olfactory circuit utilizes one fewer
excitatory neuron compared with the retinal circuit. Horizontal cells
mainly inhibit presynaptic terminals of photoreceptors (although they
can also directly inhibit bipolar cells), whereas periglomerular cells
primarily inhibit mitral cell dendrites. (Adapted from Shepherd GM
[ed] [2004] The Synaptic Organization of the Brain, 6th ed. Oxford
University Press.)

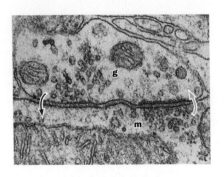

Figure 6–18 Dendrodendritic synapses between mitral cells and granule cells. Electron micrograph of adjacent reciprocal synapses between a granule cell process (g) and a mitral cell process (m). Arrows indicate the direction of information flow. (From Rall W, Shepherd GM, Reese TS et al. [1966] *Exp Neurol* 14:44–56. With permission from Elsevier Inc.)

and decreases the noise for mitral cell representation of olfactory information (see Box 1–2; Section 1.12).

The olfactory bulb performs more functions than simply information pooling. Just as in the retina, the olfactory bulb contains many local interneurons, whose projections are confined to the olfactory bulb; these are mostly GABAergic and therefore inhibitory, although some local interneurons use glutamate or dopamine as neurotransmitters. Indeed, these local interneurons outnumber mitral (and tufted) cells by ~100-fold in the mammalian olfactory bulb. The **periglomerular cells** are a large subgroup of interneurons that receive direct input from ORN axons within the glomeruli or from apical dendrites of mitral cells, and spread inhibition to nearby glomeruli (Figure 6–17A). The **granule cells** are another large subgroup that receive input not from ORNs, but rather from mitral cells. In addition to having apical dendrites that innervate single glomeruli, mitral cells also send out **secondary dendrites** that cover a sizeable fraction of the olfactory bulb. Dendrites of granule cells and secondary dendrites of mitral cells form reciprocal **dendrodendritic synapses**: mitral cells excite granule cells, which in turn inhibit both the same mitral cells that excited them and other mitral cells with secondary dendrites nearby (Figure 6–17A; **Figure 6–18**). Thus, the periglomerular and granule cells in the olfactory bulb circuit resemble the horizontal and amacrine cells in the retinal circuits (Figure 6–17).

The similarity of olfactory bulb circuitry to that of the retina suggests that olfactory bulb GABAergic neurons play a role in lateral inhibition (see Section 4.16). Specifically, each mitral cell collects at its apical dendrites excitatory input from the axons of presynaptic ORNs in its cognate glomerulus, and receives at both its apical and its secondary dendrites inhibitory input from nearby glomeruli representing other olfactory processing channels, analogous to the center–surround receptive fields of retinal ganglion cells. Thus, for each olfactory processing channel, olfactory bulb interneurons transform the 'receptive field' of input ORNs to that of the output mitral cells. (Receptive field in olfaction can be defined as the odorant repertoire that can alter the firing pattern of a cell in the olfactory system.) However, unlike the retina—where the center–surround receptive field has clear functions in analyzing contrast, color, or motion by comparing activation levels of nearby photoreceptors—the exact role of lateral inhibition in the olfactory bulb is less clear. A simple model has been proposed that lateral inhibition sharpens odorant representation against similar odorants. This model is unlikely to be generally applicable, in part because neighboring glomeruli do not necessarily represent chemically similar odorants. Furthermore, the 'odor space' (that is, the entire repertoire of odorant molecules) has many dimensions: there are numerous chemically distinct odorants, each of which has multiple functional groups that activate multiple ORNs and glomeruli (see Figures 6–8 and 6–16). This multidimensionality makes it difficult to project odorants onto the two-dimensional olfactory bulb surface in an orderly fashion.

An alternative view is that lateral inhibition can still serve a useful function in **gain control** of the mitral cells, even if surround inhibition is nonselective. (Gain control refers to a process that modulates the relationship between the input to a system and its output, thereby keeping output within a limited dynamic range; see also Box 4–3). For example, when the animal encounters a strong odorant environment where many ORNs are activated simultaneously, many GABAergic inhibitory neurons would be activated so that the gain of all olfactory processing channels is reduced, thereby preventing saturation in neuronal firing rates. If gain control preferentially suppresses weak responses, then the olfactory processing channel that is activated most strongly will tend to stand out from the rest. Another possibility is that lateral inhibition serves mainly to regulate the timing of mitral cell activity. Because the timing of mitral cell spiking is known to vary with odorant stimuli, this timing might carry information about odorant identity.

6.10 Olfactory inputs are differentially organized in distinct cortical areas

Among all sensory systems, the olfactory pathway is the only one that does not have to pass through the thalamus: mitral cells and tufted cells receive input from

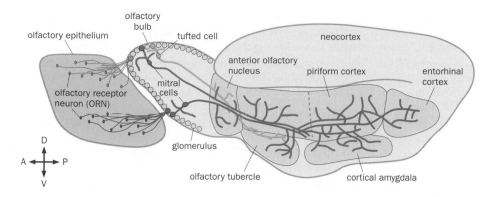

Figure 6–19 Schematic of the central olfactory system in the mouse. In this side view of the mouse brain, the organization of the olfactory system from the olfactory epithelium to the olfactory bulb is highlighted by the convergent axonal projections of two types of olfactory receptor neurons to two discrete glomeruli. In the olfactory bulb, each mitral (blue, red) and tufted (green) cell sends apical dendrites into a single glomerulus, and secondary dendrites laterally. Mitral cells project long-distance axons to multiple olfactory cortical areas including the anterior olfactory nucleus, piriform cortex, olfactory tubercle, cortical amygdala, and entorhinal cortex. Despite the fact that two mitral cells originate from different parts of the olfactory bulb, their axonal projection patterns are not apparently distinguishable in the piriform cortex. Tufted cell axons innervate anterior olfactory cortices and terminate in a special region of the olfactory tubercle. The dotted line in the piriform cortex divides this largest region of the olfactory cortex into anterior (A) and posterior (P) parts. D, dorsal; V, ventral. The cortical amygdala receives input preferentially from the dorsal olfactory bulb. (Courtesy of Kazunari Miyamichi. The mitral and tufted cell drawings follow original data from Igarashi KM, Ieki N, An M et al. [2012] *J Neurosci* 32:7970–7985.)

ORN axons and project their axons directly to various olfactory cortical areas (**Figure 6–19**). Thus, the sense of smell has more direct access to the cortex than any other sense, as cortical neurons are only two synapses away from the sensory world. Convergent axonal projections to the same glomerulus from ORNs of the same type enable the organization of olfactory information into a spatial map in the olfactory bulb. How is this glomerular map represented in the olfactory cortex?

Electrophysiological and optical imaging studies have suggested that the largest olfactory cortical region, the **piriform cortex**, displays a surprising lack of spatial organization. For example, when two-photon Ca^{2+} imaging (see Section 13.22 and Figure 4–42C) was applied to the piriform cortex, it was found that individual cortical neurons activated by specific odorants were distributed broadly across the piriform cortex with no discernible spatial patterns (**Figure 6–20**). These physiological studies are consistent with anatomical studies suggesting that the axons of individual mitral cells project broadly throughout the piriform cortex (Figure 6–19). This is in sharp contrast to the primary visual cortex discussed in Chapter 4, where cortical neurons are spatially organized according to retinotopy, and in some species further according to ocular dominance and orientation selectivity.

Why does the piriform cortex discard the spatial organization from the olfactory bulb? One possibility is that the unorganized connectivity between mitral

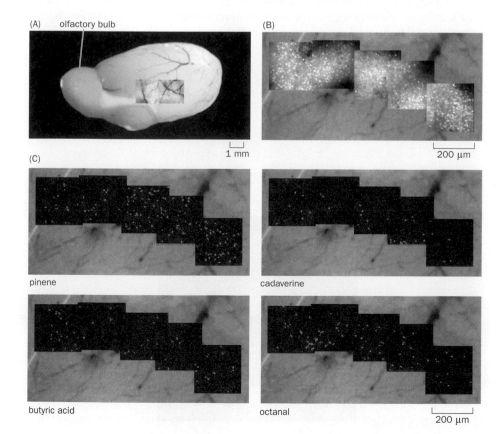

Figure 6–20 Piriform cortical neurons activated by specific odorant stimuli are broadly distributed with no apparent order. (A) Window of *in vivo* imaging in the mouse piriform cortex; the olfactory bulb is to the left. **(B)** More than 1000 cells can be imaged in the same anesthetized animal using a fluorescent indicator that reports changes in intracellular Ca^{2+} concentration. **(C)** Piriform cortical cells responsive to four different odorants— pinene, cadaverine, butyric acid, and octanal—are intermingled and scattered throughout the imaging window, indicating that the spatial map of discrete glomeruli in the olfactory bulb is not retained in the piriform cortex. (From Stettler DD & Axel R [2009] *Neuron* 63:854–864. With permission from Elsevier, Inc.)

cells and piriform cortical neurons creates a blank slate for piriform cortical neurons to collectively receive a large number of combinations of different olfactory processing channels from the olfactory bulb. This stochastic connection matrix can be a substrate on which olfactory experiences act by consolidating or eliminating connections, or by adjusting the strengths of connections, to create odor representation in the piriform cortex that reflects an individual's experience. We will discuss these concepts in greater detail in Chapter 10.

Not all cortical areas (see Figure 6–19) resemble the piriform cortex in their odor representations. Tracing the axons of mitral cells revealed projection patterns in the piriform cortex that were similar across mitral cells originating from different glomeruli, consistent with the physiological data discussed above; however, projections to the **cortical amygdala**, part of the olfactory amygdala complex that receives direct mitral cell input, displayed more orderly patterns. These findings were reinforced by a study utilizing a technique called **retrograde trans-synaptic tracing**, which allows for the labeling of direct presynaptic partners of a given neuron (see Section 13.19). Individual neurons in different areas of the olfactory cortex were found to receive direct input from multiple mitral cells representing broadly distributed glomeruli from the olfactory bulb, allowing cortical neurons to potentially decode the combinatorial glomerular code. Cortical neurons in the piriform cortex receive direct input from glomeruli throughout the olfactory bulb with no obvious bias, consistent with the stochastic connectivity hypothesis. However, many neurons from the cortical amygdala receive strongly biased input that favors glomeruli in the dorsal olfactory bulb (**Figure 6–21**).

What is special about the dorsal olfactory bulb? In an experiment where ORNs that project to the dorsal olfactory bulb were selectively ablated, mice no longer avoided odorants that would elicit innate avoidance in normal mice, such as chemicals from spoiled food or fox urine. These observations suggest that some olfactory processing channels in the dorsal bulb may be particularly important in transforming detection of specialized odorants to innate avoidance behavior.

In summary, outputs of the olfactory bulb are differentially represented in different cortical areas, which could serve distinct purposes. Analogous findings were made in the fruit fly olfactory system, which will be discussed in the next part of this chapter. In mammals, an entirely parallel system to what we have discussed thus far, the accessory olfactory system, is used for detecting nonvolatile chemicals of special biological significance (**Box 6–1**).

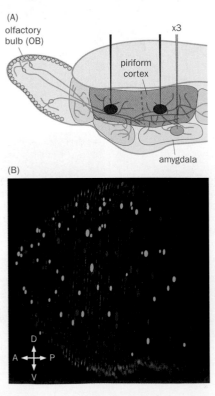

Figure 6–21 Differentially organized olfactory bulb input to the piriform cortex and cortical amygdala. (A) Viruses that enable retrograde trans-synaptic tracing were injected into small volumes of different olfactory cortical areas. Transduction of cortical neurons leads to the labeling of synaptically connected mitral cells and their corresponding glomeruli in the olfactory bulb. **(B)** Two separate injections in the piriform cortex (magenta, red) resulted in labeled glomeruli that are distributed throughout the olfactory bulb; injections in the cortical amygdala (green; three injections combined; ×3) resulted in labeled glomeruli that are located preferentially in the dorsal olfactory bulb. Labeled glomeruli for each experiment were reconstructed and superimposed onto a standard olfactory bulb model. A, anterior; P, posterior; D, dorsal; V, ventral. (Adapted from Miyamichi K, Amat F, Moussavi F et al. [2011] *Nature* 472:191–196. With permission from Macmillan Publishers Ltd.)

Box 6–1: The mammalian accessory olfactory system is specialized for detecting pheromones and predator cues

Most mammalian species possess an **accessory olfactory system** (also called the **vomeronasal system**), which is anatomically and biochemically distinct from the main olfactory system. Sensory neurons of the accessory olfactory system are housed in a special structure called the **vomeronasal organ (VNO)** located at the front of the nose (**Figure 6–22**A). Whereas ORNs in the main olfactory epithelia detect airborne odorants, VNO neurons detect nonvolatile stimuli through a narrow aqueous duct. VNO neurons project their axons to the glomeruli of the **accessory olfactory bulb**, which is adjacent to the main olfactory bulb (Figure 6–22A). Mitral cells of the accessory olfactory bulb deliver information to the olfactory cortex, but their axons innervate different areas from those innervated by the mitral cells of the main olfactory bulb.

Each VNO neuron expresses a single vomeronasal receptor that belongs to either the V1R or the V2R family. These two GPCR families are distinct from each other and from the odorant receptors in the main olfactory system. The signal transduction pathways downstream of V1R and V2R are also distinct from that used by the main olfactory system: whereas activation of ORNs in the main olfactory system is mediated by the opening of a cyclic nucleotide-gated (CNG) channel (see Figure 6–4), VNO neuron activation is mediated by the opening of a specific TRP channel (see Box 2–4) called TRPC2. *Trpc2* knockout mice lose the function of the entire accessory olfactory system.

The number of genes encoding intact V1Rs and V2Rs varies considerably among different mammalian species, with ~300 in mice but only a few in primates. Thus, the function of the accessory olfactory system varies significantly across species. Using immediate early gene expression as a measure of neuronal activity (see Section 3.23), stimuli that activate ~90 mouse vomeronasal receptors have been identified (Figure 6–22B). Some of these vomeronasal receptors are activated by **pheromones** present in the urine, tear, and skin secretions of mice. (Pheromones are substances produced by an individual to elicit a specific reaction from other individuals of the same species. The name pheromone is derived from the Greek *pherein*, to transfer, and *hormon*, to excite.) A large number of mouse vomeronasal receptors are devoted to the detection of chemical cues from other species such as mammalian and avian predators (Figure 6–22B).

We use a specific example to illustrate functions of the accessory olfactory system. Urine is an important chemical source for social communication of certain mammals including mice. For example, male mice are known to mark their territories by spreading urine. A resident male exhibits aggression toward a sexually mature male intruder, but not toward a castrated male. However, a castrated intruder swabbed with urine from a sexually mature male can elicit aggression from the resident male (**Figure 6–23**A). This behavioral response allowed biochemical purification of the active components in mouse urine that elicited aggression when swabbed onto a castrated male. **Major urinary proteins (MUPs)**, which are highly stable proteins that can mark territory for a long duration, were identified as one active component. Swabbing castrated male intruders with purified MUPs that had been expressed in *E. coli* was sufficient to elicit aggression from wild-type resident males (Figure 6–23A). This response requires the function of the accessory olfactory system; *Trpc2*-knockout resident mice did not exhibit aggression toward intruder males.

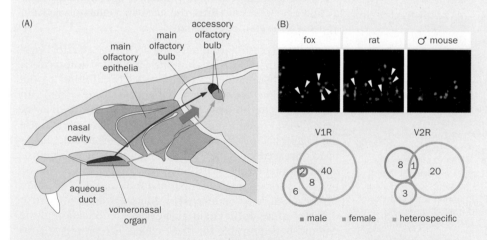

Figure 6–22 The accessory olfactory system and the vomeronasal receptors. (A) The main and accessory olfactory systems in the mouse viewed from the side. The vomeronasal organ (VNO) is located at the front of the nose, connected with the nasal cavity by an aqueous duct. Apically (red) and basally (green) located VNO neurons express the V1R and V2R classes of receptors, and project their axons (red and green arrows) to the anterior and posterior accessory olfactory bulb, respectively. The blue arrow represents ORN axon projection from the main olfactory epithelium (pink) to the main olfactory bulb. **(B)** Top, double *in situ* hybridization to identify receptors for specific chemical cues. VNO neuronal expression of immediate early gene *Egr1* (see Section 3.23) in response to beddings of fox, rat, or male mice for the same female recipient appears in green; expression of a specific subclass of V2R receptors is shown in red. The yellow signals (arrowheads) represent overlap between the V2R probe and neurons that are activated by fox or rat bedding (but no yellow cells with male mouse stimuli). Thus, these specific V2R receptors are activated by predator odors but not by conspecifics. Bottom, Venn diagrams for 56 V1R and 32 V2R receptors with respect to their activation by stimuli from male mice, female mice, or heterospecifics (including predators and other species of mice). V2Rs are more selectively tuned to one of the three types of cues, as only one V2R responds to more than one type of cues. (A, adapted from Brennan PA & Zufall F [2006] *Nature* 444:308–315. With permission from Macmillan Publishers Ltd; B, adapted from Isogai Y, Si S, Pont-Lezica L et al. [2011] *Nature* 478:241–245. With permission from Macmillan Publishers Ltd.)

(Continued)

Box 6–1: The mammalian accessory olfactory system is specialized for detecting pheromones and predator cues

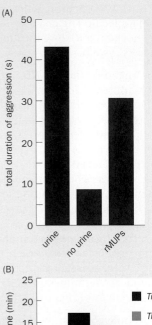

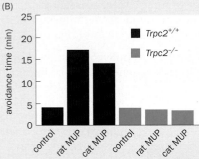

Figure 6–23 Major urinary proteins (MUPs) elicit aggressive or defensive responses. (A) Duration of aggression (in a 10-minute interval) exhibited toward a castrated male intruder by a resident male. Castrated male intruders that had been swabbed with urine from a sexually mature male elicit more aggression from the resident than those not swabbed with urine. Swabbing the castrated male intruders with recombinant MUPs expressed in bacteria (rMUPs) also increases the aggression of resident males. **(B)** Mice avoid an area in a cage that contains rat or cat MUP. This behavior is abolished in mice lacking TRPC2, and is thus dependent on the function of VNO neurons. (A, adapted from Chamero P, Marton TF, Logan DW et al. [2007] *Nature* 450:899–902. With permission from Macmillan Publishers Ltd; B, adapted from Papes F, Logan DW & Stowers L [2010] *Cell* 141:692–703. With permission from Elsevier Inc.)

MUPs represent a means not only for intraspecies social communication but also for communication between species. Mice are innately fearful of cats and rats, which are their predators. Rat urine elicited robust defensive responses in mice, including avoidance and the release of stress hormones. Biochemical purification of rat urine identified the active component that elicits defensive response in mice to be a specific rat MUP. Again, *Trpc2* mutant mice lost their responses toward rat and cat MUPs (Figure 6–23B), indicating that these MUPs act through the accessory olfactory system. Future identification of specific vomeronasal receptors that mediate the effect of these MUPs, and tracing the neural circuits in the brain, will be fruitful in deciphering the neural basis by which MUPs from different species elicit aggressive or defensive behaviors.

As discussed in Section 6.4, a large fraction of the genes encoding the main olfactory system's odorant receptors have become pseudogenes in humans. The situation is more drastic in the accessory olfactory system: humans have five intact V1R and zero intact V2R genes; the remainder (>100) are all pseudogenes. The gene encoding the TRPC2 channel has also become a pseudogene in humans. The VNO itself is not present in adult humans, though it appears transiently during development. Thus, the accessory olfactory system appears to be extinct in humans.

Given the evolutionary loss of the accessory olfactory system, do humans use pheromones for social interaction? Recent studies in mice indicate that, in addition to the accessory olfactory system, the main olfactory system can also detect pheromones (see Chapter 9 for more details), which leaves open the possibility that humans can still use the main olfactory system for pheromone signaling. It has been observed that women living together tend to synchronize their menstrual cycles. Such synchrony can be induced by compounds from the armpits of donor women. Ovulation can be accelerated or delayed by donor extracts obtained at different times in the menstrual cycle. Identification of the chemicals responsible for these effects is required to consolidate our understanding of possible human pheromones and to explore their biological functions.

HOW DO WORMS AND FLIES SENSE ODORS?

An understanding of the mammalian brain can come from studies of simpler organisms. We have seen an excellent example of this in Chapter 2: the use of the squid giant axon to determine the ionic basis of the action potential. The principles we learned from squid apply nearly universally to all animals with a nervous system. Can an understanding of complex neurobiological problems, such as olfactory perception, also benefit from studies of simpler organisms?

The use of model organisms (see Sections 13.1–13.5) in neurobiological research has considerable merit. Model organisms usually offer biological simplicity, technical ease, or both. We can borrow insights learned from simpler organisms to investigate analogous problems in mammals, perhaps also using similar approaches that have proven successful in simpler organisms. If the mechanisms are similar, simpler systems can inform us about common solutions to neurobiological problems and speed up discoveries in mammals. Even if the solutions in simpler organisms are different, we still learn about alternative solutions to a common problem and, in the process, enrich our understanding about the diversity of life.

6.11 *C. elegans* encodes olfactory behavioral choices at the sensory neuron level

The nematode *Caenorhabditis elegans* has emerged as an important model organism for the study of the nervous system (see Section 13.2). *C. elegans* has 302 neurons, and the wiring diagram (**connectome**) for its nervous system at the synaptic level—the approximately 6393 chemical synapses between neurons, 1410 neuromuscular junctions, and 890 gap junctions—has been mapped by serial electron microscopic reconstructions. Neurons essential for *C. elegans* olfaction were identified by behavioral studies of odorant-induced attraction and repulsion coupled with laser ablation and genetic manipulation. Three pairs of sensory neurons—two AWAs, two AWBs, and two AWCs—are responsible for detection of most volatile chemicals. These neurons send their dendrites to the anterior opening of the worm for odorant detection (**Figure 6–24**). Interestingly, there is a clear functional separation for these three pairs of neurons: AWAs and AWCs are required for odor-mediated attraction, while AWBs are required for odor-mediated repulsion. These three pairs of neurons collectively express many of the ~100 G-protein-coupled receptors, which are the presumed odorant receptors. This simple organization poses an interesting question about odor perception and odor-mediated behavior: is it the activation of a particular neuron, or is it rather the activation of a particular odorant receptor (independent of the neuron that expresses it), that determines the behavioral output?

Odr10 has been identified as a receptor responsible for detecting an attractant, diacetyl, and is expressed in AWAs. Worms without Odr10 (*Odr10* mutants) are no longer attracted to diacetyl. Using a transgenic approach (see Section 13.10), Odr10 was introduced back into specific neuronal types. If an *Odr10* transgene was introduced back into both AWAs of *Odr10* mutant worms, the transgene rescued attraction to diacetyl, indicating that Odr10 normally acts in AWAs to mediate attraction. But if the *Odr10* transgene was introduced into the AWBs of *Odr10* mutant worms, the transgenic worms were now repelled by diacetyl (**Figure 6–25**). Thus, it is neither the odorant nor its receptor, but rather the responding neuron, which determines the behavioral output.

This elegant experiment suggested a simple logic for *C. elegans* odor perception: receptors for attractants are expressed in the AWAs or AWCs, and receptors for repellents are expressed in the AWBs. The response of a particular type of

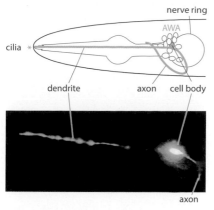

Figure 6–24 An olfactory neuron in *C. elegans*. Schematic (top) and fluorescence image (bottom) of the one of the two AWA sensory neurons. The cell body of the AWA is in the nerve ring; its dendrite extends as a cilium, which ends in the anterior opening of the worm (*). (Adapted from Sengupta P, Chou JH & Bargmann CI [1996] *Cell* 84:875–887. With permission from Elsevier Inc.)

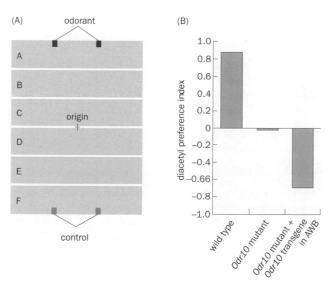

Figure 6–25 *C. elegans* **olfactory neurons determine the behavioral output.** **(A)** Odorant preference assay. Worms start at the center of a plate (origin), and are allowed to move freely for a defined period of time. A preference index is calculated according to the number of worms distributed in different sectors: (A + B − E − F)/total number. **(B)** Wild-type worms are attracted to diacetyl (left). Worms that lack the *Odr10* odorant receptor gene are insensitive to diacetyl (middle). When an *Odr10* transgene is introduced into the AWB neurons of *Odr10* mutant worms, these worms are repelled by diacetyl (right). (B, adapted from Troemel ER, Kimmel BE & Bargmann CI [1997] *Cell* 91:161–169. With permission from Elsevier Inc.)

sensory neuron then determines the behavioral output, presumably through its specific connections with the motor programs mediating attraction or repulsion. This solution may not be easily extended to the more complex mammalian olfactory system, which has many more sensory neurons, the capacity to distinguish between many different odorants, and a different organizational principle (that is, one receptor per neuron). It nevertheless illustrates how a simple organism with severe constraints on neuronal number manages to solve the problem of olfactory perception to guide its behavior: by encoding the **hedonic values** (whether potentially beneficial or harmful to the animal) of odorants at the starting point of the sensory system—the sensory neurons. Indeed, a similar logic applies to the mammalian taste system (see Section 6.21).

6.12 *C. elegans* sensory neurons are activated by odorant withdrawal and engage ON- and OFF-pathways

Given that we know all the synaptic connections of the 302 *C. elegans* neurons, one might expect that the circuit mechanisms that transform the activity of AWA/AWC or AWB to attractive or repulsive behavior should be readily deciphered. However, the following studies illustrate that having a wiring diagram is just a first step in understanding the operation of a neural circuit.

The first surprise came when it was found that the AWC sensory neuron is activated by odorant withdrawal, rather than by odorant application. *C. elegans* neurons are very small and notoriously difficult for electrophysiological recordings. The discovery was made by using Ca^{2+} imaging of a restrained worm in response to odorants delivered in a microfluidic device (**Figure 6–26**A, B, top). A genetically encoded Ca^{2+} indicator, GCaMP (see Section 13.22), was selectively expressed in an AWC neuron such that an increase in fluorescence intensity reflected its activation. Whereas odorant application caused a slight decrease of the Ca^{2+} signal (Figure 6–26A, blue trace), odorant withdrawal caused a large increase of the Ca^{2+}

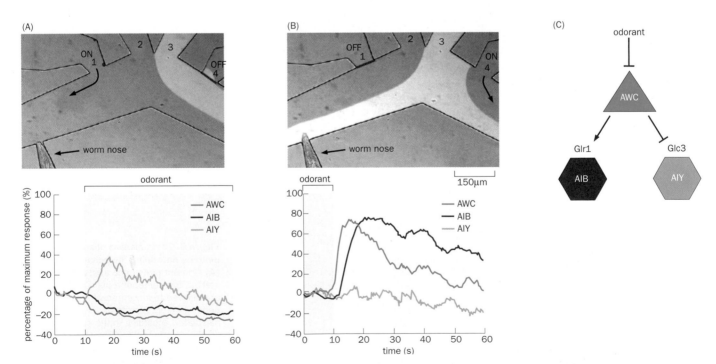

Figure 6–26 Olfactory responses of AWC and downstream interneurons. (A, B) Top, schematic of the microfluidic device used to control odorant delivery to a restrained worm that is being imaged for Ca^{2+} signal in specific neurons. When streams 1–3 are open (A), odorant (gray) reaches the worm nose. When streams 2–4 are open (B), buffer from stream 3 (white) reaches the worm nose. Bottom, AWC and AIB are inhibited by the application of an attractive odorant (isoamyl alcohol), and activated by odorant withdrawal. AIY is activated by odorant application and unresponsive to odorant withdrawal. The *y* axis measures the percentage of maximum response in the fluorescence intensity of a genetically encoded Ca^{2+} indicator expressed in the respective neurons. These curves represent averages from separate experiments. **(C)** Summary of the circuit diagram. AWC releases glutamate as a neurotransmitter and is inhibited in the presence of odorants. AWC activates AIB through GLR-1, an AMPA-like ionotropic glutamate receptor, but inhibits AIY through GLC-3, a glutamate-gated chloride channel. (Adapted from Chalasani SH, Chronis N, Tsunozaki M et al. [2007] *Nature* 450:63–70. With permission from Macmillan Publishers Ltd.)

signal (Figure 6–26B, blue trace), indicating that AWC is inhibited by odorants, and activated by odorant withdrawal.

According to the wiring diagram, AWC synapses with a small number of interneurons, including AIB and AIY. When Ca^{2+} imaging was performed in AIB or AIY neurons in response to odorant stimulation (by expressing GCaMP selectively in AIB or AIY, respectively), it was found that AIB was also activated by odorant withdrawal, with a longer latency than AWC (Figure 6–26B, red trace). By contrast, AIY was activated by odorant application (Figure 6–26A, green trace). Genetic perturbation experiments indicated that AWC uses glutamate as a neurotransmitter; when glutamate release by AWC was inhibited, odorant responses in AIB and AIY were abolished, indicating that AIB and AIY receive odorant information via AWC. How could AWC signal to two downstream neurons with opposite effects? Further analysis indicated that AIB expresses Glr1, an ionotropic glutamate receptor similar to the AMPA receptor, which causes depolarization of AIB in response to glutamate. By contrast, AIY expresses Glc3, a glutamate-gated chloride channel, which causes hyperpolarization of AIY in response to glutamate (Figure 6–26C).

Thus, the AWC circuit resembles the vertebrate rod and cone circuits we studied in Chapter 4 in two ways. (1) Sensory neurons are hyperpolarized by olfactory stimuli. (2) Parallel ON- and OFF-pathways are engaged downstream from sensory neurons, utilizing the same neurotransmitter (glutamate) but different receptors (compare Figure 6–26C with Figure 4–25). The second property reflects a convergent strategy (discussed in detail in Chapter 12) to solve diverse sensory processing problems. The parallel ON- and OFF-pathways increase the contrast for odorant detection and the sensitivity of *C. elegans* to odorant onset and offset. Both are beneficial for navigation toward food sources and away from harmful substances.

6.13 The olfactory systems in insects and mammals share many similarities

Unlike *C. elegans*, which has a limited number of neurons to address its olfactory needs, insect brains, with hundreds of thousands to many millions of neurons, possess olfactory systems that are remarkably similar to those of mammals (**Figure 6–27**). The insect **antennal lobes**, equivalent to the mammalian olfactory bulbs as sites of ORN axon termination, have a similar glomerular organization. Recent studies, particularly in the fruit fly *Drosophila melanogaster*, a genetic model organism, have provided insight into the function of the glomerular organization in the antennal lobe for olfactory information processing.

Two pairs of external sensory organs, the antennae and the maxillary palps, are the 'noses' of the fly and house the ORN cell bodies (Figure 6–27 bottom left). Like mammals, most fruit fly ORNs express a single odorant receptor. Also mirroring the mammalian system discussed earlier, fruit fly ORNs that express the same odorant receptor also project their axons to the same glomerulus in the antennal lobe (Figure 6–27 bottom right). Olfactory information is then relayed to higher olfactory centers by antennal lobe **projection neurons** (**PNs**). Like the mammalian mitral/tufted cells, most PNs send their dendrites to a single glomerulus and thus receive direct input from a single ORN type. The antennal lobe also has many **local interneurons** (**LNs**), most of which are GABAergic. Thus, the organizations of the fly and mammalian olfactory systems share remarkable similarities.

There are only ~50 ORN types and ~50 glomeruli in each fly antennal lobe, compared with ~1000 ORN types and ~2000 glomeruli in each mouse olfactory bulb; thus the fly olfactory system is considerably simpler numerically. All glomeruli are recognizable by their stereotyped size, shape, and relative position. The correspondence of ORN type and glomerular identity has been completely mapped. Furthermore, the odorant response properties of a large fraction of the ORN repertoire have been determined *in vivo* by an elegant strategy that combines mutant and transgene expression in a similar way to that discussed in the *C. elegans* Odr10 experiment. Specifically, in a mutant strain that lacks the endogenous *Or22a* gene (which encodes a specific odorant receptor), the

Figure 6–27 Schematic of olfactory systems in the fly and mouse. The olfactory systems of the mouse (top, a side view) and fly (bottom, a frontal view) share similar glomerular organizations, as can be seen by comparing the mouse olfactory bulb and fly antennal lobe, both of which are targeted by olfactory receptor neuron (ORN) axons. Additional similarities include: (1) each ORN expresses one specific odorant receptor; (2) ORNs expressing the same odorant receptor project their axons to the same glomeruli (on both antennal lobes in the fly); and (3) each second-order mitral cell or projection neuron (PN) sends dendrites to a single glomerulus, and thus receives direct input from a single ORN type. (The fly olfactory system also contains atypical PNs not shown here that innervate multiple glomeruli.) Antennae (AT) and maxillary palps (MP) house olfactory receptor neurons (ORNs). Local interneurons (LN) in the antennal lobe extend elaborate dendrites that cover many glomeruli (see Figure 13–23 for a singly labeled LN). PNs send axons mainly to two higher olfactory centers, the mushroom body (MB) and the lateral horn (LH). Numbers in parentheses indicate the numbers of cells (ORN, PN, LN, mitral cells), types (ORN, PN, mitral cells), or structures (glomeruli). (Adapted from Komiyama T & Luo L [1996] *Curr Opin Neurobiol* 16:67–73. With permission from Elsevier Inc. The scanning electron microscopic image of the fly head is courtesy of John R. Carlson.)

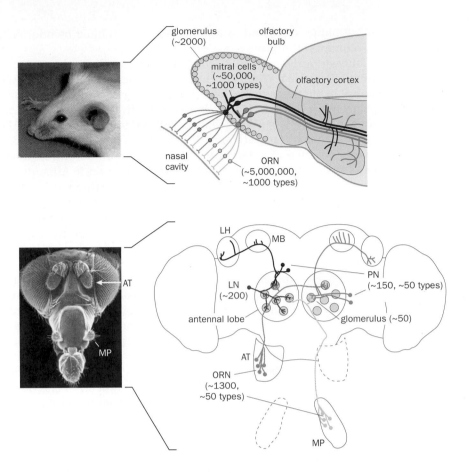

Or22a promoter is used to drive the expression of a different odorant receptor gene, such that the odor response of Or22a-expressing ORNs now reflects the properties of the odorant receptor expressed from the transgene (**Figure 6–28**A). The firing properties of a majority of odorant receptors encoded in the *Drosophila* genome in response to a large panel of odorants have been determined in this way (Figure 6–28B).

Similar to mammalian studies discussed earlier (see Figure 6–8), each odorant activates multiple odorant receptors, and each odorant receptor is activated by multiple odorants. Some odorant receptors are broadly tuned, meaning that they are activated by many odorants, whereas others are narrowly tuned, meaning that they are more selectively activated by a small number of odorants. The tuning curve is also shaped by the concentration of odorants—lower concentrations result in more selective tuning (Figure 6–28C). Whereas most odorant receptors are activated by odorants, some odorant receptors are *inhibited* by specific odorants (Figure 6–28B)—the basal firing rates of ORNs expressing these odorant receptors are decreased when the fly is exposed to specific odorants. (Because of the high sensitivity of *in vivo* recording in flies, both activation and inhibition were detected here in contrast to the studies exemplified in Figure 6–8, where only activations by odorants were reported.) These studies have made the fruit fly olfactory system the most completely characterized with regard to sensory stimuli and to olfactory input channels (ORN types).

6.14 The antennal lobe transforms ORN input for more efficient representation by projection neurons

How is olfactory information processed in the antennal lobe following delivery by ORN axons? This question can be answered by systematically comparing the

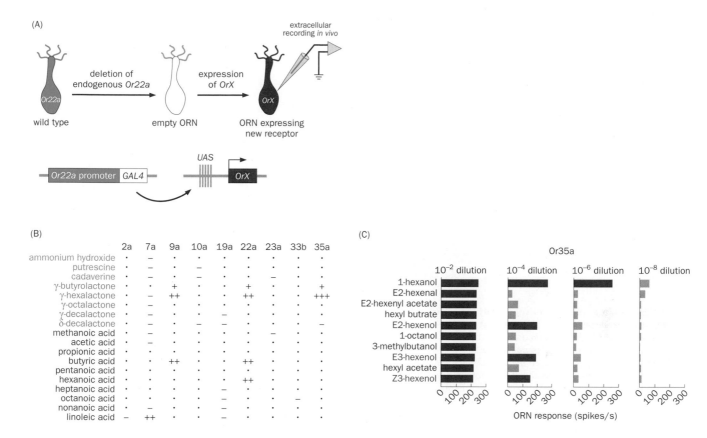

Figure 6–28 Odorant coding by the *Drosophila* odorant receptor repertoire. (A) Experimental strategy. Left, in wild-type flies, Or22a ORNs express the odorant receptor Or22a. Middle, in mutant flies that lack the *Or22a* gene, Or22a ORNs become 'empty ORNs' that do not express a functional odorant receptor. Right, a transgene encoding a different odorant receptor (OrX) driven by the *Or22a* promoter can be introduced into the *Or22a* mutant background, such that the Or22a ORNs now express OrX; *in vivo* extracellular recordings can determine the property of OrX in response to odorant stimuli. Bottom, the transgene expression actually utilizes the GAL4/UAS binary system; from the first transgene, the *Or22a* promoter drives the expression of the yeast GAL4 transcription factor, which binds to UAS and activates the OrX expression from a second transgene (see Section 13.10 for more details of this strategy). **(B)** Olfactory responses for specific odorant receptors as assayed by measuring ORN firing rates *in vivo*. •, +, ++, and +++ represent, respectively, <50, 50–100, 100–150, and 150–200 spikes per second elicited by 10^{-2} dilution of a specific odorant (rows) for 22a ORNs lacking an endogenous *Or22a* gene and expressing a specific odorant receptor (columns). – represents a >50% decrease of the basal firing rate in response to odorant application. This figure illustrates only a fraction of the results from this experiment, which assayed 110 odorants and 24 odorant receptors, that is, about half of the receptor repertoire in fruit fly. Subsequent studies have characterized the response profiles of the remainder of the receptor repertoire. **(C)** Concentration-dependent response of ORNs expressing the Or35a odorant receptor. As the concentration of odorant decreases, the tuning of Or35a ORNs becomes narrower. (Adapted from Hallem EA & Carlson JR [2006] *Cell* 125:143–160. With permission from Elsevier Inc.)

firing patterns of ORNs and PNs that are direct synaptic partners, to assess whether and how odorant coding is transformed between the ORNs and PNs (**Figure 6–29**A). These experiments are very difficult in mammals because of technical limitations but can be readily performed in flies because of the identifiable glomeruli and tools that allow genetic access of many ORN and PN types. The following properties have been identified from such comparisons.

First, firing rates from individual PNs are less variable than those of their presynaptic ORNs when comparing responses to the same stimulus (Figure 6–29B, C). This is because, on average, 60 ORNs (30 each from ipsi- and contralateral side) synapse onto three PNs within each glomerulus (see Figure 6–27 bottom), and the axon from each ORN forms strong synapses with the dendrites of all PNs within the glomerulus. In addition, PNs projecting to the same glomerulus form electrical synapses with each other to further synchronize their activity. These properties allow individual PNs to pool information from many ORNs and thus encode sensory information as rates of action potential firing (see Section 1.8) more reliably than their presynaptic partner ORNs.

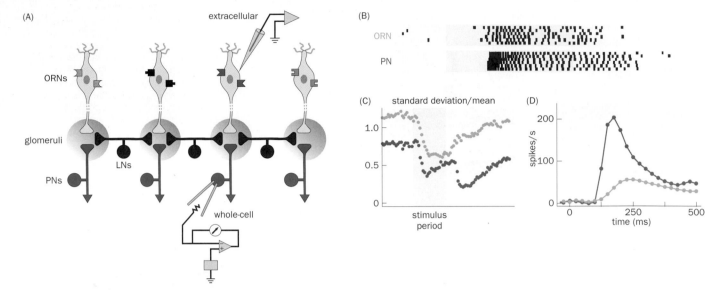

Figure 6–29 PN firing rates selectively report the onset of ORN firing and are less variable than ORN firing rates. (A) Schematic of experimental setup. Extracellular recording in the antenna or maxillary palp can measure the firing rates of specific ORN types in response to olfactory stimuli. Whole-cell patch clamp recording can measure the firing rate of their partner PNs that are genetically engineered to express a fluorescent marker. The basic circuit is shown where ORNs and partner PNs form type-specific excitatory connections in specific glomeruli. Local interneurons (LNs), mostly GABAergic, laterally link multiple glomeruli. **(B)** ORN and PN firing in response to 500 ms odorant stimulation (yellow shading). Each row represents a separate trial with identical stimuli. Each vertical tick is a spike. **(C)** PN responses are less variable than ORN responses, as indicated by the lower coefficient of variation, a ratio of standard deviation and mean of spike counts. **(D)** Peri-stimulus time histogram for ORNs and PNs, which plots firing rates of ORN (light green) and PN (dark green) as a function of time after odorant onset (at 0 ms). The PN firing rate peaks and decays faster than the ORN firing rate. (B, adapted from Bhandawat V, Olsen SR, Gouwens NW et al. [2007] *Nat Neurosci* 10:1474–1482. With permission from Macmillan Publishers Ltd.)

Second, PNs preferentially report the rising phase of ORN activity. This can be seen from the raw data on spikes (Figure 6–29B) or in the **peri-stimulus time histogram** (**PSTH**), which plots firing rates of neurons as a function of time after stimulus onset (Figure 6–29D). At least two separate factors account for this property: (1) ORN → PN synapses exhibit short-term depression such that, during a train of action potentials, ORN spikes that arrive later cause smaller postsynaptic potentials than earlier spikes (see Section 3.10); and (2) in addition to activating PNs, ORNs also activate local interneurons (LNs; see Figures 6–26 and 6–27), which are mostly GABAergic inhibitory neurons. LNs synapse onto ORN axon terminals and hence reduce the efficiency of transforming ORN action potentials to neurotransmitter release, a process termed **presynaptic inhibition** (see Section 3.21). Therefore, LN activation (resulting from ORN activation) provides a negative feedback control of the synaptic output of ORNs and, hence, of the input to PNs. Together, these factors allow PNs to be highly selective to odorant onset rather than to the constant presence of the same odorant, thereby informing the organism about changes in the environment. This is olfactory adaptation at the circuit level, distinct from adaptation at the level of ORN signal transduction discussed in Section 6.2.

Third, by spreading their representations of different odorants across a larger range of firing rates than ORNs, PNs use the available coding space more efficiently than ORNs. In the **coding space** of a neuronal population, the activity state is represented as a point in a multi-dimensional space; each axis typically represents the firing rate of one constituent neuron (**Figure 6–30**A; see the next paragraph for a specific example). When the firing rates of individual ORN types are plotted against those of partner PNs for different odorants, most ORN/PN pairs produce nonlinear curves. Weak ORN signals are selectively amplified in PNs, whereas strong ORN signals produce saturating levels of PN activation (Figure 6–30B). The two factors already discussed, short-term depression and presynaptic inhibition by GABAergic LNs, contribute to these properties. In addition, at low odorant concentrations, a small number of excitatory LNs boost the weak ORN signal.

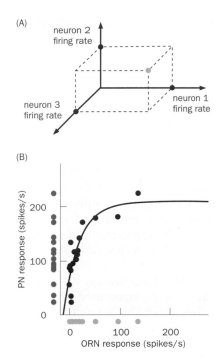

(A) neuron 2 firing rate

neuron 3 firing rate

neuron 1 firing rate

(B)

Figure 6–30 Coding space and relationship between ORN and PN firing rates. (A) Illustration of coding space for a neuronal population consisting of just three neurons. The activity state of this neuronal population at any given time is represented by the green dot in the three-dimensional space; its x, y, or z value (black dots) corresponds to the firing rate of each component neuron. **(B)** Nonlinear transformation of projection neuron responses from olfactory receptor neuron responses. Each black dot represents one of the 18 odorants whose evoked responses were measured independently for a specific ORN–PN pair. In this nonlinear transformation, weak ORN inputs are selectively amplified, while strong ORN inputs produce saturated PN responses. Thus, the firing rate is more distributed in the PN space than the ORN space for this ORN–PN pair. Most ORN–PN pairs across different glomerular channels exhibit a similar curve. (B, adapted from Bhandawat V, Olsen SR, Gouwens NW et al. [2007] *Nat Neurosci* 10:1474–1482. With permission from Macmillan Publishers Ltd.)

Together, these properties allow PNs to use the coding space more efficiently than ORNs, which can be beneficial for odor discrimination.

Let's use a specific example to illustrate these points. The responses to 18 diverse odorants were measured for seven pairs of ORNs and PNs. The odorant identity could thus be represented by the firing rates of seven ORN types in a seven-dimensional ORN space, or alternatively, by the firing rates for seven PN types in a seven-dimensional PN space. While it is difficult to visualize a seven-dimensional space, a statistical method called **principal component analysis** can be used for data analysis and representation. Principal component analysis transforms the representation of a complex dataset in high dimensional space using a set of orthogonal axes called principal components according to data spread; data are most spread along the axis of the first principal component, followed by the axis of the second principal component, and so forth. High-dimensional datasets can then be represented in lower-dimensional space along the first few principal components. When the 18 different odorants were plotted in the space of the first two principal components, their representations were more clustered in the ORN space than in the PN space (**Figure 6–31**). This means that it is easier for a downstream decoder to discriminate these odorants based on the PN population activity than the ORN population activity.

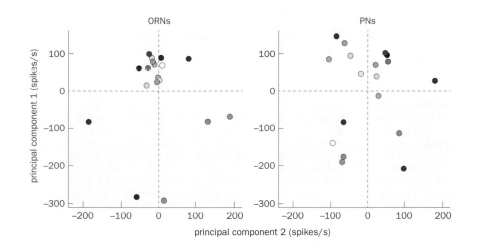

Figure 6–31 Odorant coding is more widely distributed among populations of PNs than ORNs. Eighteen distinct odorants (here assigned different colors) are represented by two principal components in the seven-dimensional coding space (from odorant response data of seven ORN–PN pairs) of firing rates of ORNs (left) or PNs (right). Different odorants are more widely distributed in the PN coding space compared to the ORN coding space, which means that they are more easily distinguishable by downstream neurons. (Adapted from Bhandawat V, Olsen SR, Gouwens NW et al. [2007] *Nat Neurosci* 10:1474–1482. With permission from Macmillan Publishers Ltd.)

6.15 Odors with innate behavioral significance use dedicated olfactory processing channels

As we discussed earlier in the chapter, combinatorial coding, in which the odor identity is distributed among several olfactory processing channels, expands the coding capacity beyond the number of parallel olfactory processing channels that are defined by the number of odorant receptors. This strategy allows the 50 processing channels in flies to recognize and discriminate many more than 50 odorants. However, an important challenge lies in having 'decoder' neurons in higher olfactory centers that are capable of distinguishing the activation of different combinations of PN populations. So far, the principle of how this is accomplished has been limited to theoretical analysis because the decoder neurons in higher olfactory centers have not been investigated systematically. However, certain odorants are of such special importance to an organism that dedicated olfactory processing channels may be devoted to the representation of those odorants. This strategy simplifies the process of decoding, as specific higher-order neurons receive information predominantly from a single processing channel. We provide two examples below that suggest the use of this strategy within the fly olfactory system.

When stressed (such as being vigorously shaken in a test tube), fruit flies release odors that naive flies avoid. This behavior can be recapitulated in the laboratory by placing naive flies in a T-maze (**Figure 6–32**A), where one arm contains air released from shaken flies and the other arm contains fresh air. Ninety percent of naive flies choose the side that contains fresh air (Figure 6–32B, see also **Movie 6–1**). A major component of this stress odor was identified as carbon dioxide (CO_2), which flies avoid in a concentration-dependent manner (Figure 6–32B).

An ORN type that expresses the odorant receptor Gr21a appears to be dedicated to CO_2 sensation. Two different kinds of experiments have been used to test the function of Gr21a-expressing ORNs in mediating the CO_2 avoidance behavior. In the first experiment, Gr21a ORNs were acutely and selectively silenced by expression of a temperature-sensitive mutant protein, $Shibire^{ts}$, which blocks synaptic transmission at high temperatures (see Sections 3.9 and 13.23). Flies no longer avoided CO_2 under this condition. This loss-of-function experiment proved the necessity of Gr21a ORNs for CO_2 avoidance. In the second experiment, a

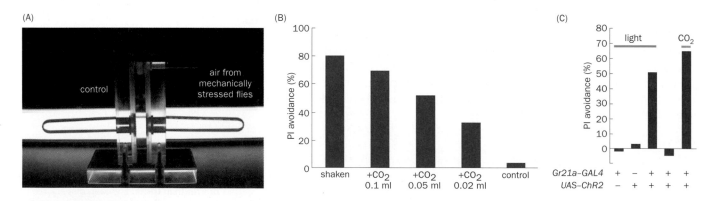

Figure 6–32 Activation of CO_2-sensitive ORNs produces avoidance behavior. (A) A T-maze can be used to test flies' preference to odorants. Flies were introduced at the center of the choice point from an 'elevator' in the middle, and allowed to choose between two arms for 1 minute (see Movie 6–1). Performance index (PI) = [(number of flies in the control arm − number of flies in the experimental arm) / total number] × 100 (%). Thus, if the distribution of 100 flies is 90 versus 10, PI = 80%. **(B)** Flies avoid the tube that contains air from mechanically stressed flies (shaken). They also avoid the CO_2-containing tubes in a concentration-dependent manner (columns 2–4). In the control, both tubes contain air. **(C)** In this experiment, a light-activated channelrhodopsin-2 (ChR2) transgene is expressed from Gr21a neurons using the GAL4/UAS binary system (see Section 13.10). Five

experimental conditions are assessed; in the first three conditions, blue light is delivered to one of the two tubes. Control flies that have only the GAL4 or the UAS transgene do not express ChR2 and do not respond to light. Flies that have both transgenes such that ChR2 is expressed in Gr21a ORNs avoid the tube that is exposed to the blue light (column 3), just as they avoid the tube that contains CO_2 (column 5); column 4 is a no-light control. Thus, Gr21a ORNs encode a repulsive behavioral response, whether activated naturally by CO_2 or artificially by light-induced depolarization. (A, courtesy of David J. Anderson; B, adapted from Suh GSB, Wong AM, Hergarden AC et al. [2004] *Nature* 431:854–859. With permission from Macmillan Publishers Ltd; C, adapted from Suh GSB, Ben-Tabou de Leon S, Tanimoto H et al. [2007] *Curr Biol* 17:905–908. With permission from Elsevier Inc.)

light-activated channelrhodopsin, ChR2, was expressed only in Gr21a ORNs such that they would be selectively activated by blue light (see Section 13.25). When flies were placed in a T-maze in which one of the two arms was exposed to blue light that depolarized the CO_2-sensitive ORNs, the flies avoided the arm exposed to the blue light, much in the same way they avoided the arm that contained CO_2 (Figure 6–32C). This gain-of-function experiment indicated that activation of Gr21a ORNs is sufficient for the CO_2 avoidance behavior. Together, these experiments suggest that repulsion by CO_2 is already encoded in this special ORN class, much like repulsion is encoded in AWB neurons in *C. elegans* (see Figure 6–25).

Flies use mating pheromones to communicate with each other about their sex and mating status (see Chapter 9 for more details). A pheromone produced by male *Drosophila*, 11-*cis*-vaccenyl acetate (cVA), an ester of a long-chain alcohol, inhibits the courtship of males toward other males. cVA also inhibits courtship of males toward mated females, which contain cVA transferred from their previous male partner during mating (a cunning way for males to minimize sperm competition). ORNs that express the Or67d odorant receptor play a major role in this behavior. Silencing the activity of Or67d ORNs reduced the male–male courtship inhibition, whereas expression of a moth pheromone receptor in Or67d ORNs reduced courtship of male flies to virgin females that had been treated with the corresponding moth pheromone. Thus, Or67d ORNs are specialized for cVA sensing, just as Gr21 ORNs are specialized for CO_2 sensing. Indeed, electrophysiological experiments indicated that the postsynaptic partner PNs of Or67d ORNs are also narrowly tuned to cVA and are minimally activated by other odorants tested.

Thus, while combinatorial coding allows fruit flies to distinguish thousands of different odorants using only ~50 odorant receptors and processing channels, some processing channels may be devoted principally to detection of single, behaviorally important stimuli.

6.16 Odor representation in higher centers is stereotyped or stochastic depending on whether the center directs innate or learned behavior

How is the olfactory information carried by PNs represented in higher olfactory centers for odor detection, discrimination, and odor-induced behavior? Much future work is needed to satisfactorily answer these fundamental questions, but for now the numerical simplicity of the fly olfactory system and the abundance of genetic tools have yielded more advanced understanding of higher olfactory centers in *Drosophila* compared with the mammalian olfactory cortex.

The output neurons of the antennal lobe, PNs, send their axons to two major structures: the **mushroom body** and the **lateral horn** (see Figure 6–27 bottom). The mushroom body is a center for olfactory learning and memory, whereas the lateral horn is involved in odor-mediated innate behaviors. To investigate how olfactory input to these higher-order olfactory centers is organized, PNs that send dendrites to different glomeruli were individually labeled using genetic techniques (see Section 13.16). Systematic analyses revealed striking stereotypy of axonal branching patterns and the locations of terminal axonal arbors in the lateral horn for any given PN type (**Figure 6–33**). Individual PN axons send branches to invade different parts of the lateral horn, and different types of PN axons project to overlapping parts of the lateral horn. These features enable the integration of information conveyed by different PNs to the same postsynaptic third order olfactory neurons (convergence), while at the same time allowing individual PNs to deliver information to multiple third order neurons (divergence).

The availability of data on odorant response profiles for most ORN types (see Figure 6–28) and their glomerular targets enables the assignment of odorant specificity to most PN types. One can then model odorant response maps based on PN axon termination patterns. Results from such studies indicate that the lateral horn can be divided into at least two parts (**Figure 6–34**), a large part for sensing fruit odors, and a smaller part for sensing mating pheromones such as cVA. This spatial segregation enables higher-order neurons to more easily distinguish

Figure 6–33 Stereotyped PN axon termination patterns in the lateral horn but not the mushroom body. Left panels: Labeling of singly labeled PNs that send dendrites to the DL1, DM5, and VA1v glomeruli of the antennal lobe, respectively. Right panels: axonal projection patterns in the mushroom body (MB) and the lateral horn (LH) are shown for three DL1 PNs (top), three DM5 PNs (middle), and three VA1v PNs (bottom) from nine different flies. Note that LH arborization patterns are similar among PNs of the same glomerular type and distinct among different PN types. In contrast, MB arborization patterns are much more variable among PNs of the same type. (From Marin EC, Jefferis GSXE, Komiyama T et al. [2002] *Cell* 109:243–255. With permission from Elsevier Inc.)

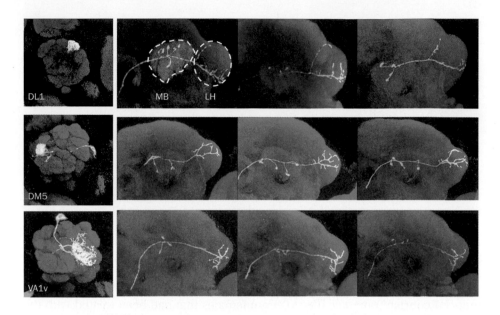

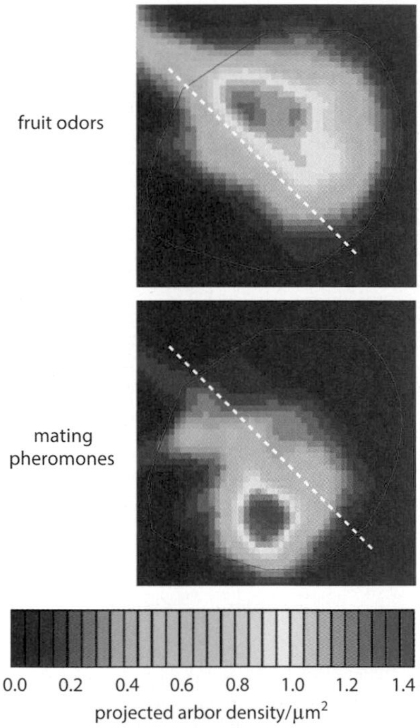

Figure 6–34 Spatial segregation of fruit odorant and mating pheromone representations in the lateral horn. Presynaptic terminal density maps for PNs that represent fruit odors (upper panel) and mating pheromones (lower panel). Density is mapped within the lateral horn (outlined in red) in false color according to the key at the bottom. Fruit odors and mating pheromones are largely separable by the dotted white line, indicating spatial segregation for the representations of these olfactory stimuli that have distinct behavioral significance. (From Jefferis GSXE, Potter CJ, Chan AI et al. [2007] *Cell* 128:1187–1203. With permission from Elsevier Inc.)

these two types of olfactory information and direct different behavioral outputs (for example, foraging versus mating).

Interestingly, PN axon arborization in the mushroom body is much less stereotyped (Figure 6–33). Further detailed mapping of the connectivity between PN and mushroom body neurons did not identify any apparent structure: individual mushroom body neurons appear to stochastically sample the glomerular repertoire. Physiological studies also indicate that odorant responses of the mushroom body neurons lack stereotypy: the odorant response of a genetically defined small subpopulation is as variable as the entire mushroom body neuronal population. Thus, each fly may generate a unique odor representation in the mushroom body, the meaning of which is likely acquired by experience.

In summary, the two higher olfactory centers in the fly, the lateral horn and mushroom body, represent odors in distinct manners. The highly organized representation in the lateral horn suits its role in regulating innate behavior; the connectivity is selected in the course of evolution and develops according to a predetermined genetic plan. By contrast, the stochastic representation in the mushroom body suits its role in associative learning; the meaning of the representation is likely acquired during an individual's life experience. A similar principle may apply to the vertebrate olfactory system. For example, the smell of native stream may be imprinted in salmon's equivalent of mushroom bodies in their juvenile period, enabling them to find the same stream to spawn years later. Indeed, the mammalian piriform cortex appears to resemble the fly mushroom body in its representation of odorants and mitral cell input (see Section 6.10). It will be interesting to determine in future studies whether other olfactory cortical regions in mammals have more organized olfactory bulb input that resembles the lateral horn for mediating innate odor-induced behavior.

TASTE: TO EAT, OR NOT TO EAT?

Chemicals can also be detected by the taste and trigeminal chemosensory systems in vertebrates. The trigeminal chemosensory system (a part of the somatosensory system, which will be discussed in detail later) warns the animal of noxious stimuli. The taste system detects **tastants**—nonvolatile and hydrophilic molecules in saliva—using taste receptors in the tongue and oral cavity. Whereas olfaction can detect chemicals at a distance, taste does so at closer range. In mammals, taste is dedicated to the regulation of feeding by revealing the content and safety of potential food. (In other species such as the fruit fly, taste is also used to regulate mating behavior; see Chapter 9.) The identification of the taste receptors has enabled a clear understanding of how different tastes are sensed on the tongue.

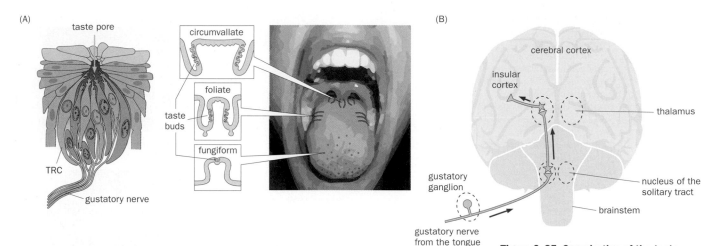

6.17 Mammals have five classic taste modalities: sweet, bitter, umami, salty, and sour

Taste begins with the binding of chemicals to the tip of the **taste receptor cells** on the surface of the tongue and oral cavity. Clusters of tens of taste receptor cells form a **taste bud**, with the apical extensions of taste receptor cells forming the **taste pores** at the surface of the tongue (**Figure 6–35**A). Groups of taste buds form several kinds of papilla, which are distributed at different parts of the tongue. Tastant binding to taste receptors induces depolarization of taste receptor cells. Unlike olfactory receptor neurons, taste receptor cells do not have their own axons. They release neurotransmitter at their base, which activates the terminal branches of the **gustatory nerves** that innervate the taste buds. Information is relayed to the **nucleus of the solitary tract** (**NTS**) in the brainstem, then to the thalamus, and eventually reaches part of the **insular cortex** specialized for taste sensation (Figure 6–35B).

Taste has been divided into five classic modalities based largely on human perception: **bitter**, **sweet**, **umami**, **salty**, and **sour**. (Note the distinction between taste and flavor; **flavor** is a synthesis of taste and olfaction, and therefore has many more than five kinds.) These modalities are likely to be universal. Bitter and sour tastes usually warn animals of potential toxic chemicals or spoiled food, and are generally aversive. Sweet and umami (savory, meaty) tastes provide information about the nutritional content of the food, such as sugar and amino acids, and are generally appetitive. The salty taste allows animals to regulate their sodium ion level, and can be appetitive or aversive depending on the salt concentration and the animal's current physiological state.

Receptors for all five taste modalities are distributed throughout the tongue, and each taste bud can contain cells responsive to different modalities. This raises questions as to how the taste system is organized at the periphery. Is each taste receptor tuned to a specific modality, or is it more broadly tuned? What kind of taste information does each afferent axon carry to the brain?

Figure 6–35 Organization of the taste system. (A) Left, each taste bud consists of up to ~100 taste receptor cells (TRCs) that extend apical processes toward the taste pore, which opens to the oral cavity. Terminals of the gustatory nerve that innervate the taste bud contact the basal side of the taste receptor cells. Middle and right, taste buds are located in several kinds of papillae—circumvallate, foliate, and fungiform—which are located at different parts of the tongue. **(B)** Schematic of the central taste system. Gustatory nerves from the front of the tongue, back of the tongue, and pharynx (not shown) originate from neurons in separate ganglia, all of which project their central axons to the nucleus of the solitary tract (NTS) in the brainstem. Taste information is relayed by NTS neurons to the thalamic neurons and then to the insular cortex. Arrows indicate the direction of information flow. (A, adapted from Chandrashekar J, Hoon MA, Ryba NJP et al. [2006] *Nature* 444:288–294. With permission from Macmillan Publishers Ltd.)

6.18 Sweet and umami are sensed by heterodimers of the T1R family of G-protein-coupled receptors

The quest for taste receptors began in the late 1990s, following the spectacular success of the identification of odorant receptors and the insights they brought to the study of olfaction. Using molecular biology techniques that allow the identification of genes differentially expressed in taste receptor cells, two G-protein-coupled receptors with large N-terminal extracellular domains, **T1R1** and **T1R2**, were identified as specifically expressed in taste receptor cells. The mRNAs of T1R1 and T1R2 were detected in taste buds, and T1R1 and T1R2 proteins were concentrated in the taste pore (**Figure 6–36**). Thus, these GPCRs were strong candidates for taste receptors. But what modalities do they detect?

In mice, the gene for a third member of the T1R family, **T1R3**, was subsequently identified that corresponds to the *Sac* locus, which is the genomic location of a

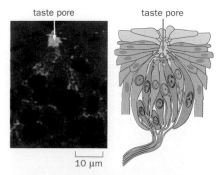

Figure 6–36 T1R1 protein is concentrated at the taste pore. T1R1 protein (green in the image on the left, revealed by immunostaining using an antibody against T1R1) is concentrated at the taste pore; red staining is for filamentous actin that labels the cytoplasm. The schematic on the right shows a taste bud from a similar view. (Adapted from Hoon MA, Adler E, Lindemeier J et al. [1999] *Cell* 96:541–551. With permission from Elsevier Inc.)

Figure 6–37 Identification of the sweet receptor. (A) Mice were given two bottles from which to drink, one containing water and the other containing sucrose solutions of specified concentration (*x* axis). Higher *y*-axis scores indicate stronger preference; a *y*-axis score of 50 means no preference. Wild-type (orange circles) or *Sac* mutant mice expressing a T1R3 transgene (red circles) prefer sucrose solution at much lower concentrations (red trace) than *Sac* mutant mice (purple trace). **(B)** HEK293 cells transfected with T1R2 and T1R3, as well as a promiscuous G protein, respond to sucrose application with a rise of intracellular Ca^{2+} concentration as measured by a fluorescence indicator (scale to the left of the top panel, in nM). (Adapted from Nelson G, Hoon MA, Chandrashekar J et al. [2001] *Cell* 106:381–390. With permission from Elsevier Inc.)

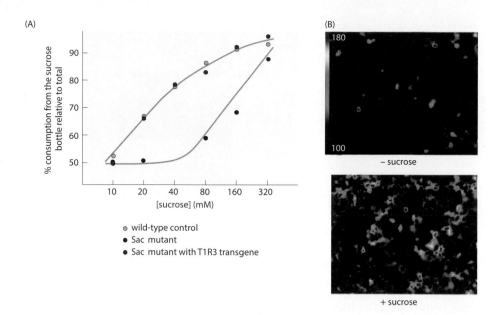

6.19 Bitter is sensed by a family of ~30 T2R G-protein-coupled receptors

In 1931, a synthetic chemist at DuPont discovered that the dust of phenylthio-carbamide (PTC; **Figure 6–39**) he synthesized tasted bitter to the occupant of the

Figure–38 Guess what taste receptor is missing in each of these animals? Cats and tigers have lost T1R2 function, and hence the ability to taste sweetness. Giant pandas do not have functional T1R1, and hence cannot taste umami. (Left, courtesy of Sumeet Moghe/Wikipedia; Right, courtesy of Chen Wu/Wikipedia.)

spontaneous mutation that made mice insensitive to sweet tastes. Introducing a normal T1R3 transgene into the *Sac* mutant mouse restored sugar sensitivity, as assayed by preference in a two-bottle choice test (**Figure 6–37**A). The function of T1Rs was further tested by introducing these receptors along with a promiscuous G protein into heterologous cells that normally do not respond to sugars. In this assay, cells expressing T1R3 alone did not respond to sugar application, but cells expressing T1R2 and T1R3 together responded robustly (Figure 6–37B). Cells expressing T1R2 + T1R3 responded not only to natural sweets such as sucrose, but also to artificial sweeteners such as saccharin. Thus, T1R2 and T1R3 together constitute the mammalian sweet taste receptors.

Similar heterologous expression studies identified T1R1 + T1R3 as the mammalian umami receptor. These results were confirmed by physiological and behavioral responses to specific tastants in mutant mice. Mice lacking T1R1 were insensitive to umami stimuli such as the meaty tastant monosodium glutamate (MSG), but maintained the response to sweet stimuli. Mice lacking T1R2 still responded to umami but not to sweet. Mice lacking T1R3 responded to neither sweet nor umami, but had intact bitter, sour, and salty responses. Thus, T1R3 acts as a co-receptor for both sweet and umami tastants, conferring sweet taste when acting together with T1R2 and umami taste when acting together with T1R1.

Not all mammals have intact sweet or umami taste (**Figure 6–38**). Cats are indifferent to sweet food. The genes encoding T1R2 in domestic cats, as well as their wild feline relatives, were all found to carry the same mutations, rendering them pseudogenes. These mutations occurred in a common ancestor of these feline species. Similarly, when the giant panda genome was sequenced, T1R1 was found to have become a pseudogene. This means that giant pandas do not have umami taste. It is possible that these mutations were the evolutionary drivers that resulted in cats being obligate carnivores or that led pandas to a vegetarian diet even though they are closely related to bears and dogs, which are omnivores. Alternatively, the special diets of cats and panda may have evolved first and removed the selection pressure on their sweet and umami receptors, respectively, so that the unutilized receptors mutated and became pseudogenes.

next bench, but was tasteless to himself even though he was much closer to the source of the compound. Subsequent genetic studies have shown that PTC tasters and non-tasters exist among diverse human populations. These two groups exhibit a difference of at least 16-fold in their sensitivity thresholds for PTC detection, and these traits transmit from parents to progeny in a Mendelian fashion. Genetic mapping of taste sensitivity to PTC and other bitter tastants provided clues that led to the identification of bitter taste receptors as a family of G-protein-coupled receptors called **T2Rs**. Mutations in specific human T2R genes account for different sensitivities to specific bitter tastants.

Although sweet, umami, and bitter tastes are all conferred by G-protein-coupled receptors, there are significant differences between bitter taste receptors and sweet and umami receptors. These differences are instructive about the behavioral and evolutionary relevance of these different taste modalities to animals. First, the T2Rs are encoded by a large family of proteins—about 30 in humans and 40 in mice—compared with the three T1R receptors that account for all sweet and umami tastes. This diversity of bitter receptors may reflect animals' needs to detect a diverse repertoire of potentially toxic compounds through the bitter taste system.

Second, T2Rs usually have much higher affinity for their tastants than do T1Rs. For example, the T2R5 receptor from a normal mouse strain is activated by its ligand cycloheximide (a potent protein synthesis inhibitor produced by bacteria) in the hundreds of nanomolar range. By contrast, T1R2 + T1R3 is activated by sucrose at the tens of millimolar range (see Figure 6–37A), or $>10^5$ higher concentration. This difference illustrates the different functions of these taste modalities. For bitter compounds that are potentially toxic and to be avoided (such as a protein synthesis inhibitor), the higher the affinity, the less the animal has to taste before rejecting it. For sweet and umami tastants that bring nutrients to animals, low affinity is beneficial to ensure that the amount of nutrient is sufficiently high to be worth the effort of eating.

Third, different bitter taste receptors are co-expressed in the same taste receptor cells (**Figure 6–40**A). This may make it difficult for animals to distinguish between different bitter tastants, but it may hardly matter. It is far more important for animals to avoid toxic compounds than to distinguish among different bitter toxins. However, T2Rs are not co-expressed in the same taste receptor cells as T1Rs (Figure 6–40B); indeed, T1R1 and T1R2, conferring umami and sweet tastes, respectively, are not co-expressed in the same cells. This segregation makes the distinction between umami, sweet, and bitter modalities unequivocal at the level of individual taste receptor cells.

Figure 6–39 Structure of phenylthiocarbamide (PTC), a compound famous in taste research. PTC tastes extremely bitter to some people but is tasteless to others. Mutations in a single gene encoding a T2R receptor account for the difference between tasters and non-tasters. (See Fox AL [1932] *Proc Natl Acad Sci USA* 18:115–120; Blakeslee AF [1932] *Proc Natl Acad Sci USA* 18:120–130; Kim et al. [2003] *Science* 299:1221–1225.)

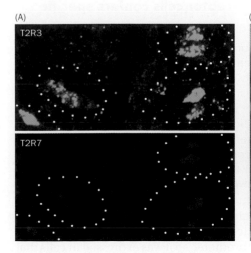
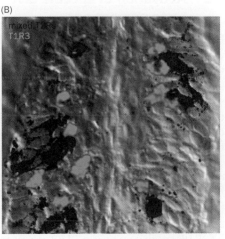

Figure 6–40 Bitter receptors are co-expressed in the same cells, which do not co-express the sweet or umami receptor. (A) Two different T2R receptors are co-expressed in the same taste cells. The top and bottom panels are the same tongue section showing *in situ* hybridization signals using T2R3 and T2R7 as probes. **(B)** Bitter receptors (red, a mix of 20 T2R probes) and the T1R3 co-receptor for sweet or umami (green) are expressed in different cells. (A, from Adler E, Hoon MA, Mueller KL et al. [2000] *Cell* 100:693–702. With permission from Elsevier Inc.; B, from Nelson G, Hoon MA, Chandrashekar J et al. [2001] *Cell* 106:381–390. With permission from Elsevier Inc.)

Figure 6–41 Summary of mammalian taste receptors and cells. Umami and sweet taste are each sensed by two GPCRs, with T1R3 as a shared co-receptor. Bitter taste is sensed by the T2R family of GPCRs. Sour taste depends on the function of cells expressing TRP channel PKD2L1 and a related PKD1L3. Attractive salty taste (low sodium) requires the epithelial Na⁺ channel ENaC, which consists of the α, β, and γ subunits. (Adapted from Yarmolinsky DA, Zuker CS & Ryba NJP [2009] *Cell* 139:234–244. With permission from Elsevier Inc. Not shown here, aversive high-salt taste requires both bitter and sour taste receptor cells; see Oka Y, Butnaru M, von Buchholtz L et al. [2013] *Nature* 494:472–475.)

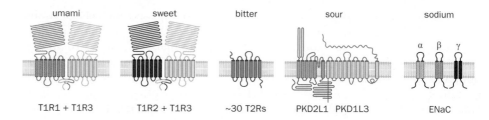

6.20 Sour and salty tastes involve specific ion channels

Whereas sweet, umami, and bitter tastes are mediated by GPCRs, the sour and salty (sodium) tastes involve specific ion channels (**Figure 6–41**). Sour taste involves a TRP channel PKD2L1, and a TRP-channel-like protein PKD1L3, named after their similarity to ion channels associated with polycystic kidney disease. PKD2L1 and PKD1L3 are co-expressed in a subset of taste receptor cells, and when co-expressed in heterologous cells, they caused the intracellular Ca^{2+} concentration to rise in response to acid application. In mice, genetic ablation of PKD2L1-expressing taste receptor cells abolished physiological responses to sour tastants but did not affect responses to sweet, umami, bitter, or low-salt tastants, indicating that these cells are tuned to sour taste.

The salty taste has two subsystems. The first salty taste system responds only to Na^+, a physiologically important ion, and usually elicits appetitive response at least when the salt concentration is relatively low (<100 mM NaCl). This low-salt or sodium taste is inhibited by amiloride, an inhibitor for the **epithelial Na⁺ channel (ENaC)** (Figure 6–41). Indeed, ENaC is essential for sodium taste in mice. Salt-deprived ENaC knockout mice were no longer attracted to NaCl solution. These mutant mice were still repelled by high concentration salt solutions, revealing a second salty taste system, which responds to a high concentration of NaCl (>300 mM) and other salts. This high-salt system is not inhibited by amiloride and usually elicits an aversive response. Recent studies suggest that the high-salt system activates both the bitter- and sour-taste receptor cells to elicit the aversive response. The mechanisms by which high concentrations of salt activate bitter- and sour-taste receptor cells remain to be determined.

The human tongue does not appear to contain ENaC channels, and salty taste in humans is not affected by amiloride. These observations suggest additional mechanisms at work for human salty taste. Indeed, even in the best-characterized mouse taste system a significant fraction of taste receptor cells do not express molecules characteristic of sweet, umami, bitter, sour, or salty tastes (Figure 6–41), raising the possibility that taste modalities beyond the classic five exist and remain to be explored.

6.21 Activation of specific taste receptor cells confers specific taste perceptions

Having studied the receptors and cells for five different modalities (Figure 6–41), we now turn to the question raised in Section 6.17 regarding the cellular organization for taste perception. We have already discussed evidence that sweet, umami, and bitter receptors were expressed in distinct taste receptor cells, and that killing sour-sensing cells did not affect the other modalities. These findings indicate that each modality is represented by its specific taste receptor cells, with the exception of high-salt taste. The following experiment provided a functional demonstration that the activation of specific taste cells confers specific taste perceptions.

Phenyl-β-D-glucopyranoside (PDG) tastes bitter to humans, but does not elicit responses in wild-type mice. The human receptor for PDG was identified as hT2R16. This receptor was used to produce two kinds of transgenic mice. In the first kind, hT2R16 was driven by a promoter from a mouse T2R receptor, so it was expressed in cells that normally responded to bitter tastants. In the second kind, hT2R16 was driven by the T1R2 receptor promoter, so it was expressed in cells that normally responded to sweet tastants. In the two-bottle choice test, transgenic mice expressing hT2R16 in bitter cells appeared to have acquired a new bitter taste for PDG because they avoided PDG-containing water. However, transgenic mice expressing hT2R16 in sweet cells were *attracted* to PDG-containing water, as if PDG were a sweet tastant (**Figure 6–42**). Thus, introducing a bitter receptor into a sweet cell can reprogram the behavioral response to the bitter tastant. This experiment shows that the sweet or bitter taste is a reflection of the selective activation of sweet or bitter cells, rather than a property of the receptors or the tastant. This logic is remarkably similar to olfactory perception in *C. elegans* discussed earlier (see Figure 6–25).

How activations of specific taste receptor cells are represented in the brain is still largely an open question. Given the separation of taste modalities at the level of taste receptor cells, it seems likely that individual taste afferents would represent specific modalities. Indeed, Ca^{2+} imaging experiments suggested that different parts of the insular cortex were enriched for cells that were preferentially activated by sweet, bitter, umami, and NaCl tastants, forming a gustatory map (**Figure 6–43**). However, individual cells that respond to different taste modalities have also been found along the ascending pathway, and at least the high-salt taste engages two different kinds of taste receptor cells at the periphery. To what extent, where, and how information from different taste modalities is integrated remain to be explored.

Indeed, perceptions of different tastes, such as sweet, umami, and bitter, need to be integrated for the animals to make the ultimate decision: to eat or not to eat. Taste perception also interacts extensively with other sensory modalities—such as olfaction (for flavor) and trigeminal somatosensory system (for texture, spice, and temperature)—to shape our food choices and enjoyment of fine dining (for example, **Figure 6–44**). Finally, past experience and current physiological states also play significant roles in our food preference. Thus, studies of taste and eating provide opportunities not only to delineate the neural mechanisms of taste

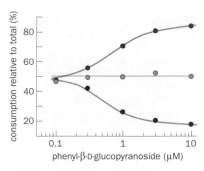

● hT2R16 expressed from the T1R2 promoter
● control (no hT2R16)
● hT2R16 expressed from a T2R promoter

Figure 6–42 Bitter or sweet taste is determined by activation of specific taste receptor cells. In the two-bottle choice assay, one bottle contains water and the other contains a phenyl-β-D-glucopyranoside (PDG) solution at concentrations specified on the x axis. A value of 50 on the y axis indicates no preference; >50 indicates appetitive; <50 indicates aversive. PDG is a natural ligand for the human bitter receptor T2R16, but does not elicit a response in control mice (blue trace). Transgenic mice expressing the T2R16 under the control of a bitter receptor (T2R) promoter avoid PDG (purple trace). Transgenic mice expressing the T2R16 under the control of a sweet receptor (T1R2) promoter prefer PDG (red trace). (Adapted from Mueller K, Hoon MA, Erlenbach I et al. [2005] *Nature* 434:225–229. With permission from Macmillan Publishers Ltd.)

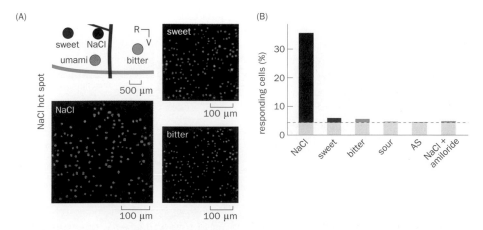

Figure 6–43 A gustatory map in the insular cortex. (A) Top left, schematic representation of a gustatory map in the insular cortex, where sweet, umami, low-salt (NaCl), and bitter tastants activate spatially clustered cells at different regions of the insular cortex. R, rostral, V, ventral. Red and blue lines indicate the middle cerebral artery and the rhinal vein, respectively, as landmarks. In the NaCl hot spot, many cells are activated by NaCl (bottom left) but few by sweet or bitter tastants (right), as measured by two-photon Ca^{2+} imaging. Gray spots are cells visualized by a fluorescent Ca^{2+} indicator injected into the insular cortex. Colored cells represent those that are activated by specific tastants above a certain threshold. **(B)** Quantification of the percentage of cells responding to different tastants at the NaCl hot spot. The gray portions of the bars below the dashed line represent the percentage of cells responding to artificial saliva (AS), which serves as a baseline. Note that in the NaCl hot spot, most responsive cells are activated by NaCl; these responses are abolished by amiloride, which inhibits the ENaC channel that is required for behavioral attraction to salt. Hot spots for other taste modalities were similarly identified. (Adapted from Chen X, Gabitto M, Peng Y et al. [2011] *Science* 333:1262–1266. With permission from AAAS.)

Figure 6–44 Multisensory integration in our choice of food. Ma-Po Tofu, a famous dish in Sichuan cuisine, engages multiple sensory modalities including olfaction, taste, vision (colorful display), and trigeminal somatosensation (for texture, temperature, hot chili pepper, and Sichuan peppercorn).

perception, but also to reveal how multiple senses and physiological states are integrated. We will return to the subject of eating in Chapter 8.

AUDITION: HOW DO WE HEAR AND LOCALIZE SOUNDS?

Audition is a particularly important sense in humans, as we communicate extensively using spoken languages and music. In the animal world, audition is also most widely used to communicate with conspecifics—to identify and locate mating partners, competitors, parents, and progeny. Other major functions of audition include to alert animals to the presence of predators or to help them identify prey. Thus, these functions are similar to the functions of vision and olfaction; indeed these senses complement each other. Vision depends on light, whereas audition and olfaction operate in light or dark, but with vastly different temporal precision. Odors, usually carried by wind, travel at a speed on the order of 1 m/s, whereas sounds travel at a speed of 340 m/s in the air, two orders of magnitude faster. Indeed, a prominent feature of audition is the representation and extraction of information, such as object location, in the temporal domain.

Sounds are transmitted as airborne pressure waves at specific frequencies (**Figure 6–45**A). In mammals, sounds are collected by the outer ear and cause vibrations of the tympanic membrane (**eardrum**) at the intersection of the outer and middle ear. The vibrations are transmitted by three tiny bones in the air-filled middle ear, which move like pistons tapping on the elastic membrane of the fluid-filled chambers of the **cochlea** in the inner ear (Figure 6–45B, C). Each cycle of a sound stimulus evokes a cycle of up-and-down movement of a small volume of fluid in each of the three chambers in the inner ear.

The cochlea (from the Greek *cochlos*, meaning snail) is a coiled structure that resembles a snail's shell anchored in the temporal bone. In a cross section of the cochlea (Figure 6–45C), the **hair cells**, the sensory cells for audition, are embedded in a sheath of epithelial cells sitting on top of an elastic membrane, the **basilar membrane**. Hair cells, the surrounding support cells, and the basilar membrane constitute the **organ of Corti**, which sits between the three fluid-filled chambers.

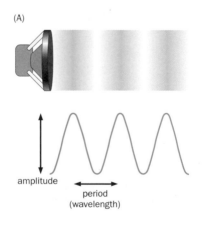

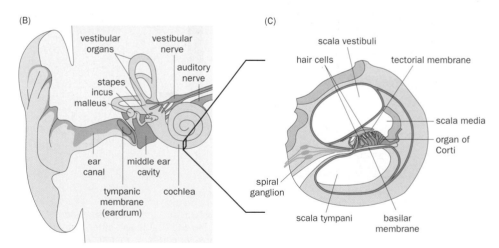

Figure 6–45 Sounds and their reception by the mammalian ear.
(A) Sounds are transmitted as fluctuations of air particles. For 1000 Hz (1 kHz, or 1000 cycles per second) sounds, the period between two adjacent wave peaks is 1 ms. The wavelength is 0.34 m given the speed of sound at 340 m/s. The amplitude of the wave reflects the loudness of the sound. **(B)** After passing through the ear canal, air particle fluctuations are transmitted through the tympanic membrane to cause vibrations of three bones in the air-filled cavity of the middle ear, the malleus, incus, and stapes. These vibrations are transmitted in the form of up-and-down movement of fluid in the fluid-filled chambers of the cochlea in the inner ear. Above the cochlea are the vestibular organs and their associated nerve, which will be discussed in detail in Box 6–2.

(C) Magnified cross section of the cochlea. In between three fluid-filled chambers (scala vestibuli, scala media, and scala tympani) is the organ of Corti, which consists of hair cells (light green), support cells (brown), and the basilar membrane (yellow); see Figure 6–50 for a further magnified schematic of the organ of Corti. Displacement of the basilar membrane due to the fluid movement produces the relative movement of the hair cells against the overlying tectorial membrane (blue). This changes the membrane potential of the hair cells (see Figure 6–47 for more details), and the resulting electrical signals are transmitted to the brain by the axons of the spiral ganglion neurons, which make up the majority of the auditory nerve.

The cyclic up-and-down fluid movement induced by sound stimuli displaces the basilar membrane in a cyclical fashion, which produces shear force between the hair cells and the overlying **tectorial membrane**. These mechanical stimuli are then converted to electrical signals via membrane potential changes in the hair cells, as we discuss below (see **Movie 6–2**).

6.22 Sounds are converted to electrical signals by mechanically gated ion channels in the stereocilia of hair cells

Humans can sense air pressure waves up to 20,000 Hz (20 kHz), or 1 cycle every 50 μs; the sensitivity of some animals (for example, bats) extends to frequencies well above 100 kHz. How quickly are sounds converted to electrical signals? In the late 1970s, it was demonstrated in an *in vitro* preparation that frog hair cells depolarized in response to mechanical stimuli with a latency of 40 μs (**Figure 6–46**); subsequent studies indicated that in mammals the latency could be <10 μs. As second messenger systems operate on the time scale of tens to hundreds of milliseconds, these measurements ruled out the possibility that **mechanotransduction**, the process by which mechanical stimuli are converted to electrical signals, is mediated by a second messenger system as is the case for vision and olfaction. Instead, they suggested that mechanical stimuli directly opened ion channels to depolarize the membrane potential of hair cells. How is this achieved?

The hair cells in all vertebrates share a similar structure and transduce mechanical stimuli in the same way. Each hair cell extends a bundle of hairs, **stereocilia**, from its apical surface. Each stereocilium is a rigid cylinder, rich in bundled F-actin fibers, with a narrow constriction at the base that enables it to pivot. Stereocilia from the same hair cell bundle are arranged in rows of increasing heights, much like a staircase (**Figure 6–47A**). In the 1980s, electron microscopy studies revealed that at the apical end, each stereocilium is connected to its

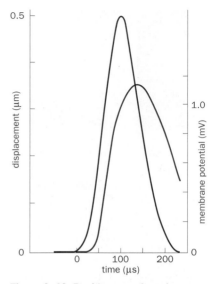

Figure 6–46 Rapid conversion of mechanical stimuli to electrical signals. In this experiment, hair cells from a dissected bullfrog vestibular saccule (see Box 6–2) were stimulated with a glass probe attached to an electrically driven motor. The hair cell response was measured by recording electrical potentials across the two chambers between which the dissected saccule was placed. The latency of the response is about 40 μs. (Adapted from Corey DP & Hudspeth JA [1979] *Biophys J* 26:499–506. With permission from Elsevier Inc.)

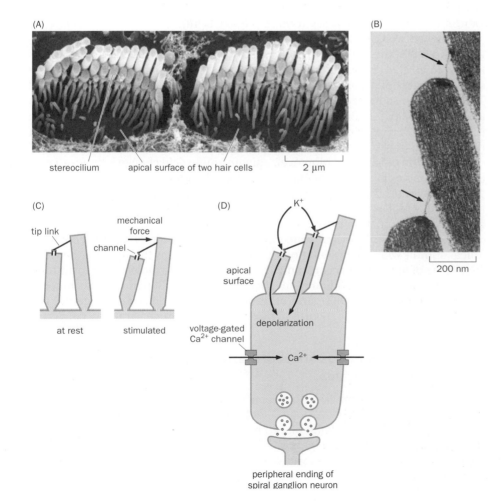

stereocilium   apical surface of two hair cells   2 μm

(C)

tip link   mechanical force   channel   K⁺   apical surface   voltage-gated Ca²⁺ channel   depolarization   Ca²⁺   peripheral ending of spiral ganglion neuron

at rest   stimulated

Figure 6–47 Mechanisms by which the hair cell converts mechanical stimuli to electrical signals. (A) A scanning electron micrograph shows the stereocilia of two adjacent inner hair cells of the mouse cochlea. These stereocilia are made of a series of actin-based bundles of successive heights that are arranged like staircases. **(B)** Two tip links (arrows) between three adjacent stereocilia visualized by electron microscopy. **(C)** A model of mechanotransduction channel gating by the relative movement of two adjacent stereocilia. Mechanical force in the direction of the ascending staircase of stereocilia opens the channel. **(D)** Schematic of a vertebrate hair cell in the inner ear. The stereocilia protrude from the apical surface and serve as detectors of mechanical stimuli. A mechanical deflection facing the ascending staircase causes the opening of mechanotransduction channels, entry of K⁺ ions that are highly enriched in the apical extracellular fluid, and depolarization of the hair cell. This triggers the opening of voltage-gated Ca²⁺ channels, Ca²⁺ entry, and neurotransmitter (glutamate) release at the base of the hair cell. Glutamate depolarizes the peripheral endings of spiral ganglion neurons, which transmit signals as action potentials to the brain. (A, from Kazmierczak P & Müller U [2011] *Trends Neurosci* 36:220–229. With permission from Elsevier Inc.; B, from Hudspeth AJ [2013] *Neuron* 80:536–537. With permission from Elsevier Inc.)

200 nm

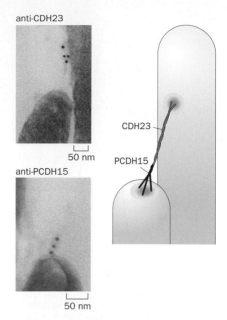

anti-CDH23

50 nm

anti-PCDH15

CDH23

PCDH15

50 nm

Figure 6–48 Molecular constituents of the tip link. Left, gold particles associated with antibodies against CDH23 (top) and PCDH15 (bottom) show that cadherin CDH23 and protocadherin PCDH15 preferentially localize to the top and bottom parts of the tip link, respectively. Right, a model: the tip link is composed of cadhesin CDH23 from the taller stereocilium and protocadherin PCDH15 from the shorter stereocilium. These two proteins bind to each other via the N-termini in their extracellular domains. (Adapted from Kazmierczak P, Sakaguchi H, Tokita J et al. [2007] *Nature* 449:87–91. With permission from Macmillan Publishers Ltd.)

taller neighbor by a structure called the **tip link** (Figure 6–47B). The staircase-like arrangement of stereocilia and the tip links explain how mechanical force causes changes in the membrane potentials of hair cells.

According to a well-accepted model, sound-wave-induced relative movements between neighboring stereocilia causes the tip links to open stretch-sensitive **mechanotransduction channels** (Figure 6–47C). The stereocilia are exposed to a fluid that has a high K^+ concentration. Opening of the mechanotransduction channel causes K^+ influx and thus depolarization of the hair cell. This triggers Ca^{2+} entry through voltage-gated Ca^{2+} channels and glutamate release at the base of the hair cell (Figure 6–47D), leading to depolarization of the peripheral endings of the **spiral ganglion neurons** (see Figure 6–45C). Spiral ganglion neurons (whose cell bodies reside in the spiral ganglion next to the cochlea) are bipolar: their peripheral axons collect auditory information from the hair cells, and their central axons form the **auditory nerve**, which transmits auditory information to the brain. Sound intensity regulates the magnitude of hair cell depolarization and transmitter release, which triggers firing of spiral ganglion neurons at specific rates.

The recent convergence of human genetic studies and physiological and cell biological studies in animal models has shed light on the molecular nature of the mechanotransduction apparatus. Deafness is the most common sensory deficit of genetic origin, affecting one in every 500 people. Mutations in about 100 genes have been associated with syndromic and non-syndromic hearing loss. Many deafness genes affect the development, structure, and function of hair cells in the organ of Corti, such as those that encode actin, actin-associated proteins, and myosin motors. Two deafness genes encode Ca^{2+}-dependent cell adhesion molecules, called cadherin-23 (CDH23) and protocadherin-15 (PCDH15), which constitute the structural components of the tip link. Biochemical and immuno-electron microscopy studies (see Section 13.17 for details) identified that the tip link is formed by CDH23 from the taller stereocilium binding to PCDH15 from the shorter stereocilium via their N-termini (**Figure 6–48**).

Although well characterized by electrophysiological and biophysical studies since the 1980s, the molecular identity of the mechanotransduction channel has remained elusive long after the human genome was sequenced in 2001. Recent studies suggest that transmembrane channel-like 1 and 2 proteins (Tmc1 and Tmc2), whose sequences do not resemble known ion channels (see Box 2–4), are likely to be part of the mechanotransduction channel. Different mutations in the human *Tmc1* gene cause recessive and dominant forms of deafness. Mutant mice lacking both *Tmc1* and *Tmc2* genes are deficient in mechanically induced depolarization of hair cells. Moreover, a single amino acid change in mouse Tmc1 alters the single channel conductance of the mechanotransduction channel, suggesting that Tmc1 contributes to the pore of the channel. Definitive proof requires a sufficiency test, as has been performed for a mammalian mechanotransduction channel in the somatosensory system (see Section 6.29).

In summary, auditory transduction is mediated by a direct coupling of mechanical stimuli with the opening of ion channels through the tip links between stereocilia of hair cells. Recent studies have identified the molecular constituents of the tip link and candidate mechanotransduction channel components. Further studies of how these and other associated proteins work together will uncover a more complete picture of how mechanical force is rapidly converted into electrical signals.

6.23 Sound frequencies are represented as a tonotopic map in the cochlea

The hearing range of humans is between 20 Hz and 20,000 Hz, with the highest sensitivity around 4000 Hz. A piano's 88 keys range from 27.5 to 4186 Hz, with the middle C at 261.6 Hz. When played in sequence, most people can easily distinguish middle C from the neighboring B or C♯ keys, which are 15.7 Hz below or 15.5 Hz above the middle C, respectively. Individuals with perfect pitch can identify the exact key on listening to a single note. How are these feats accomplished?

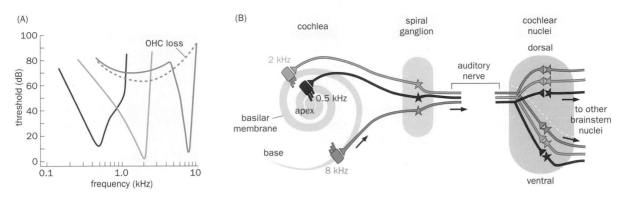

Figure 6–49 Sound frequencies are represented as tonotopic maps. (A) Schematic of frequency tuning curves of three auditory nerve fibers. The x axis represents sound frequency, while the y axis represents the sound level in dB (decibel, sound pressure level in a logarithmic unit). Zero dB approximates the lowest sound audible by humans, and every 10 dB increment represents a tenfold increase in power. These curves are generated by presenting sound at different frequencies and identifying the minimal sound level that causes the auditory nerve to fire action potentials above baseline. Thus, the troughs represent the most sensitive frequencies for the auditory nerve fibers (0.5, 2, and 8 kHz for the three fibers, respectively). The dotted blue line represents the tuning curve of the 8-kHz auditory nerve after loss of outer hair cells (OHCs), which is discussed in detail in Section 6.24. **(B)** In the cochlea, hair cells are tuned to progressively lower frequencies from the base to the apex because of the property of the underlying basilar membrane. From the base to the apex, even though the diameter of the cochlea becomes smaller, the basilar membrane (in blue) becomes progressively wider and less stiff, and therefore resonates with progressively lower vibration frequencies. Three representative hair cells tuned to 8 kHz, 2 kHz, and 0.5 kHz are shown. Spiral ganglion neurons collect input from individual hair cells via orderly axonal projections, and send output through the auditory nerve to the dorsal and ventral cochlear nuclei (separated by the dashed line) in the brainstem. (A, adapted from Fettiplace R & Hackney CM [2006] *Nat Rev Neurosci* 7:19–29. With permission from Macmillan Publishers Ltd.)

Electrophysiological recordings revealed that individual auditory nerve fibers (each fiber corresponds to the axon of a spiral ganglion neuron) are most sensitive to sound at a **characteristic frequency**. This property is called **frequency tuning**, and can be displayed on a frequency–intensity plot as a V-shaped curve (**Figure 6–49**A). Different auditory fibers exhibit different characteristic frequencies. Indeed, individual hair cells are tuned to specific frequencies based on their locations along the cochlea (Figure 6-49B), which causes frequency tuning in their postsynaptic spiral ganglion neurons. In mammals, the thickness and stiffness of the basilar membrane vary systematically along the curved structure of the cochlea. As a result of these mechanical variations, different frequencies of sound resonate with vibrations at different locations along the basilar membrane. The narrow, rigid base of the basilar membrane (near the eardrum) is most sensitive to high-frequency sounds, whereas the wide, flexible apex (furthest from the eardrum) is most sensitive to low-frequency sounds. Hair cells along the cochlea are thus tuned to different frequencies along the length of the basilar membrane, from its high-frequency base to its low-frequency apex, forming a **tonotopic map** (Figure 6-49B).

The mammalian organ of Corti contains two types of hair cells—one row of inner hair cells and three rows of outer hair cells (**Figure 6–50**). More than 95% of peripheral axons from spiral ganglion neurons contact inner hair cells. Each individual spiral ganglion neuron sends a peripheral axon (afferent fiber) that terminates on a single hair cell, and is thereby tuned to the frequency of just one hair cell. The tonotopic map in the cochlea is relayed by the orderly, central projections of spiral ganglion neurons to the **cochlear nuclei** in the brainstem (Figure 6-49B). Each inner hair cell is innervated by ~10 afferents from spiral ganglion neurons, and depending on the position of the termination on the hair cell, each afferent is differentially sensitive to the magnitude of hair cell depolarization. Sound intensity can thus be encoded by which of the ~10 spiral ganglion neurons that innervate a particular inner hair cell is activated, in addition to the firing rate of these spiral ganglion neurons. In summary, the inner hair cells and the spiral ganglion neurons are respectively responsible for detecting auditory signals and for transmitting information about sound frequency and intensity to the brain.

(A)

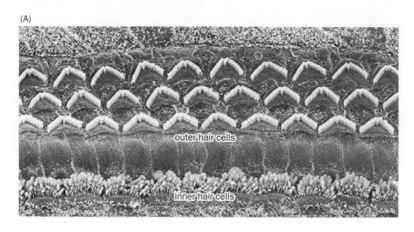

(B)

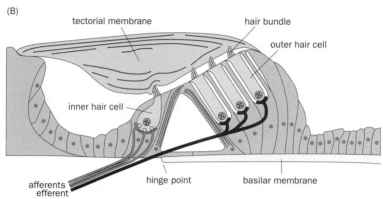

Figure 6–50 The inner and outer hair cells in the mammalian organ of Corti. (A) Scanning electron micrograph of the apical surface of a cochlear segment. Stereocilia from three rows of outer hair cells and one row of inner hair cells are evident. (A magnified surface view of two inner hair cells can be seen in Figure 6–47A). **(B)** Schematic of a cochlea cross section (magnified view of Figure 6–45C). Each inner hair cell receives ~10 afferent axons (three are shown in dark green), each originating from one neuron in the nearby spiral ganglion; these afferents deliver the auditory information to the brainstem. A small fraction of afferent axons (not shown) innervate outer hair cells, but most of the axons that project to outer hair cells are efferents (one is shown in red) originating from brainstem neurons, which also contribute to the auditory nerve. The hinge point anchors the basilar membrane, which moves up and down in response to sound stimulation, causing relative motion of hair bundles and the tectorial membrane. (A, from Hudspeth AJ [2013] *Neuron* 80:536–537. With permission from Elsevier Inc.; B, adapted from Fettiplace R & Hackney CM [2006] *Nat Rev Neurosci* 7:19–29. With permission from Macmillan Publishers Ltd.)

An important property of spiral ganglion neurons representing low-frequency sounds is that their firing pattern is periodic, reflecting the cyclic nature of sound stimuli. This can be seen by the periodic distribution of inter-spike intervals recorded from individual auditory fibers, with the period equaling the sound period (**Figure 6–51**A). This property is called **phase locking**, that is, spikes of spiral ganglion neurons occur at a specific phase of each cycle of the sound stimulus. Phase locking originates from the fact that the membrane potential of inner hair cells exhibits cyclic fluctuations following each cycle of low-frequency sounds (Figure 6–51B). [Phase locking does not work at frequencies higher than 2–3 kHz, constrained by the time constant (see Section 2.6) of hair cell membrane.] As neurotransmitters are released during the depolarized phase of each cycle, this produces action potentials in the postsynaptic spiral ganglion neurons in a cyclic manner (Figure 6–51C). Phase locking enables the timing of action potential onset to have higher temporal precision than the duration of the action potential itself. For instance, for a 1000-Hz sound whose period is 1 ms, the onset of action potentials occurs at a specific phase of the 1-ms cycle, which can be much less than 1 ms. Phase locking is also observed in central auditory neurons, endowing the temporal precision of neuronal firing in the auditory system essential for sound localization (see Sections 6.26 and 6.27).

The outer hair cells are innervated by <5% of afferent fibers, each of which innervates multiple outer hair cells. The function of these afferent fibers is not well

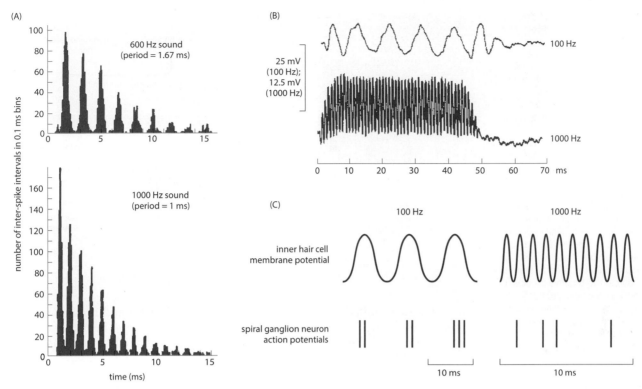

Figure 6–51 Phase-locking properties of spiral ganglion neurons and inner hair cells. (A) Histogram of inter-spike intervals recorded from a single auditory fiber of a squirrel monkey in response to 600-Hz (top) or 1000-Hz (bottom) sound stimuli. Note the periodic nature of the distribution; the period of the histogram in each case matches the sound period noted in the top right of the graph. This means that firing of the auditory nerve occurs at a specific phase of the sound wave, even though not every cycle of sound produces a spike (for instance, for the bottom graph, the events corresponding to the peak at 2 ms represent instances in which the intervening cycle did not produce a spike). **(B)** Membrane potential of a guinea pig inner hair cell in response to 100-Hz (top) or 1000-Hz (bottom) sound stimuli, measured by intracellular recording. Note that the period of the membrane potential fluctuation matches the period of sound stimuli. **(C)** Schematic drawing illustrating that membrane potential fluctuations of an inner hair cell (top) lead to periodic action potentials (vertical bars) in a postsynaptic spiral ganglion neuron (bottom). In the case of the 100-Hz sound, several action potentials are produced following each cycle of membrane potential elevation. In the case of the 1000-Hz sound, because of the refractory period of action potentials (100s of μs to 1 ms; see Section 2.12), each cycle of membrane potential fluctuation produces at most one action potential. Not every spiral ganglion neuron fires a spike at every cycle (thus producing the various inter-spike intervals illustrated in panel A), but firing of a population of spiral ganglion neurons can represent all cycles. The timing of each action potential corresponds to a specific phase of hair cell membrane potential, thus is more precise (within 10s of μs) than the duration of action potential itself (~1 ms); this property enhances precision of sound localization discussed in Section 6.26. (A, adapted from Rose JE, Brugge JF, Anderson DJ et al. [1967] *J Neurophysiol* 30:769–793; B, adapted from Palmer AR & Russell IJ [1986] *Hearing Res* 24:1–15. With permission from Elsevier Inc.)

understood. On the other hand, outer hair cells receive the bulk of efferent fibers from the brainstem (Figure 6–50). Thus, outer hair cells do not appear to play a prominent role in transmitting information to the brain. Instead, their major function is to amplify auditory signals through their interesting motor properties, as we discuss in the next section.

6.24 Motor properties of outer hair cells amplify auditory signals and sharpen frequency tuning

As discussed in Section 4.4, signals derived from photon absorption in photoreceptors are greatly amplified by a second messenger system in the visual transduction process. Given that the mechanotransduction channels responsible for changing membrane potentials of the hair cells are directly gated by sound stimuli, are auditory signals also amplified? If so, how?

Ample evidence indicates that auditory signals are indeed greatly amplified in the cochlea. In mammals, this is due to the remarkable function of the outer hair cells. Experiments carried out in the 1970s indicated that when outer hair cells representing certain frequencies of sounds were ablated by focal injection of high-dose antibiotics, the remaining inner hair cells representing the same

Figure 6–52 Motor functions of the outer hair cells amplify the auditory signal. (A) Schematic illustration of two motor functions of the outer hair cells. Hair bundle motility refers to the phenomenon that when force is applied to open the mechanotransduction channel, channel opening produces force in the direction of the stimulus, causing hair bundles to move (dotted red). Somatic motility (electromotility) refers to the phenomenon that when the outer hair cell is depolarized, conformational changes of the highly enriched membrane protein prestin shorten the hair cell along its long axis. **(B, C)** Schematic illustration of positive feedback loops (on yellow background) involving hair bundle motility (B) and electromotility (C). Note that the positive feedback loop in panel B is part of one step (red arrow) in the positive feedback loop in panel C. The amplification of auditory signals in the outer hair cells from these positive feedback loops is transmitted to nearby inner hair cells through the enhanced basilar membrane displacement. (A, adapted from Fettiplace R & Hackney CM [2006] *Nat Rev Neurosci* 7:19–29. With permission from Macmillan Publishers Ltd.)

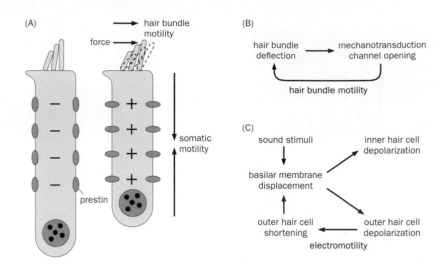

frequencies exhibited greatly increased detection threshold and broadened frequency selectivity (see Figure 6–49A). Two mechanisms have been discovered since, both involving motility of the outer hair cells triggered by sound stimuli (**Figure 6–52**A).

The first form of outer hair cell motility derives from the hair bundle motor. When a positive deflection of the hair bundle (that is, along the direction of the ascending staircase) opens the mechanotransduction channel, the opening of the channel itself generates force in the direction of the stimulus, which leads to the opening of more mechanotransduction channels and thus creates a positive feedback loop (Figure 6–52B). Ca^{2+} entry into the stereocilia through mechanotransduction channels plays an important role in hair bundle motility, as do myosin motors at the tip of the stereocilia.

The second form of outer hair cell motility comes from the shape change of the outer hair cell. Hyperpolarization causes the outer hair cell to lengthen along its long axis, whereas depolarization causes it to shorten; this remarkable property, discovered in the 1980s, is called **electromotility**. The outer hair cell movement is in turn transmitted back to the basilar membrane to enhance its displacement, creating another positive feedback loop (Figure 6–52C). These positive feedback loops amplify the sound signals that are transmitted to the nearby inner hair cells via enhanced displacement of the basilar membrane.

The electromotility of outer hair cells has not been seen in any other parts of the nervous system, including the neighboring inner hair cells. Indeed, this difference of outer and inner hair cells enabled a differential gene expression screen, leading to the discovery of a gene encoding the protein **prestin** (after the music notation *presto* for fast tempo) that mediates electromotility. Ectopic expression of prestin in cultured cells derived from human kidney is sufficient to confer these cells high-frequency motility in response to membrane potential changes (**Figure 6–53**A). Prestin has a primary structure resembling that of Cl– transporters, with multiple transmembrane domains, and is highly enriched on the outer hair cell plasma membrane. Membrane potential changes are thought to alter the conformation of prestin (Figure 6–52A), which underlies its function in electromotility.

To examine the function of electromotility *in vivo*, knock-in mice were created in which two amino acids of prestin necessary for electromotility were replaced. While mechanotransduction was unaffected, these knock-in mice had drastically elevated thresholds and severely degraded frequency tuning (Figure 6–53B), similar to the case of outer hair cell loss (see Figure 6–49A). These experiments provide compelling evidence that prestin-mediated electromotility is essential for auditory signal amplification and sharpened frequency tuning.

(A)

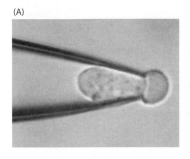

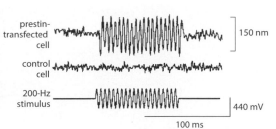

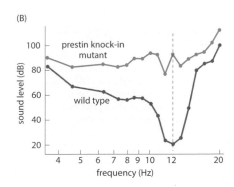

Figure 6–53 Prestin is a transducer of electromotility essential for auditory signal amplification. (A) Left, micrograph of the experimental set up to test prestin-mediated motility. Control or prestin-transfected cultured cells were partially drawn into a micro-chamber that mimics the elongated outer hair cells. Voltage pulses were applied between the solutions in the micro-chamber and the surroundings, and movement of cell edge was recorded by a video camera. Right, 200-Hz voltage pulses produce 200-Hz motility of a prestin-transfected cell but not a control cell. **(B)** Average tuning curves of wild-type (brown) and *prestin* knock-in mutant mice in which two amino acids essential for electromotility were replaced (blue). The knock-in mutants exhibit a greatly increased threshold and degraded frequency selectivity. In this experiment, a technique different from single fiber recording shown

in Figure 6-49A was used to produce tuning curves. The compound action potentials were measured from a small group of auditory fibers in response to a 12-kHz tone (vertical dashed line). The frequency (*x* axis) and level (*y* axis) of a co-applied second tone that could mask the 12 kHz-induced compound action potentials are plotted. In wild type, the closer the frequency of the second tone is to 12 kHz, the more effective is the masking, as can be seen by the low sound level required near 12 kHz. (A, adapted from Zheng J, Shen W, He DZZ et al. [2000] *Nature* 405:149–155. With permission from Macmillan Publishers Ltd; B, adapted from Dallos P, Wu X, Cheatham MA et al. [2008] *Neuron* 58:333–339. With permission from Elsevier Inc.)

6.25 Auditory signals are processed by multiple brainstem nuclei before reaching the cortex

So far, we have learned how the cochlea detects and processes auditory signals. Hair cells convert mechanical stimuli into electrical signals. Positions of activated hair cells along the cochlea represent sounds of different frequencies in a tonotopic map. Whereas the outer hair cells amplify auditory signals and sharpen the frequency tuning, the inner hair cells are responsible for producing electrical signals that are sent to the brain via ordered projections of the spiral ganglion neurons. In the following sections, we discuss how the brain analyzes these auditory signals and extracts behaviorally relevant information.

Axons of the spiral ganglion neurons terminate in the dorsal and ventral cochlear nuclei in the brainstem. Projection neurons in the dorsal cochlear nucleus send information directly to the contralateral **inferior colliculus** in the midbrain, or via intermediate neurons in the lateral lemniscus. Projection neurons in the ventral cochlear nucleus send information to **superior olivary nuclei** on both the ipsi- and contralateral sides of the brainstem, where auditory signals from the left and right ears are first integrated. Projection neurons from the superior olivary nuclei also terminate in the inferior colliculus, an important center that integrates all auditory information. Auditory information is relayed from the inferior colliculus to the **medial geniculate nucleus** of the thalamus, and then to the auditory cortex in the temporal lobe of the cerebral cortex (**Figure 6–54**).

A prominent organizational feature along the ascending auditory processing pathway is the preservation of the tonotopic map originated in the cochlea. The orderly projections of spiral ganglion neuron axons impose the tonotopic maps onto their postsynaptic targets in both the dorsal and ventral cochlear nuclei (see Figure 6–49B). Orderly tonotopic maps are also found in the superior olivary nucleus, inferior colliculus, medial geniculate nucleus, and primary auditory cortex, enabled by serial topographic projections that probably result from mechanisms similar to the retinotopic projections discussed in Chapter 5. The maintenance of tonotopy at multiple levels indicates that sound frequency is an important feature of the auditory system, just like spatial position in the visual system.

Figure 6–54 Central auditory pathways.
Axons of spiral ganglion neurons terminate in the dorsal and ventral cochlear nuclei. Projection neurons of the dorsal cochlear nucleus send axons directly to the contralateral inferior colliculus, or indirectly via intermediate neurons in the lateral lemniscus. Projection neurons of the ventral cochlear nucleus terminate in ipsilateral and contralateral superior olivary nuclei. Neurons from superior olivary nuclei integrate ipsilateral and contralateral auditory signals and project to the inferior colliculus. Inferior colliculus projection neurons send axons to the superior colliculus (not shown), and to the medial geniculate nucleus of the thalamus, which in turn projects to the auditory cortex. For simplicity, only auditory signals from the left cochlea are depicted, and only excitatory projections are represented (each by a single neuron). See Figure 6–57 for more detailed circuits of the superior olivary nuclei.

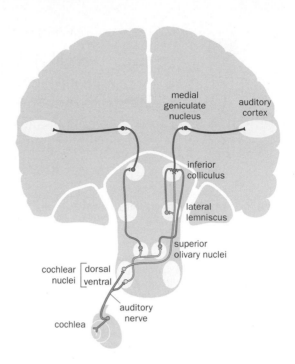

In contrast to the visual system, where information leaving the retina reaches the primary visual cortex through just one intermediate synapse in the thalamus (see Figure 4–35B), the auditory system employs a number of additional brainstem nuclei for auditory information processing (Figure 6–54). These brainstem nuclei have many different functions. For instance, the superior olivary nuclei send axons back to the cochlea, particularly to the outer hair cells (see Figure 6–50) to regulate auditory amplification. The dorsal cochlear nucleus, as well as receiving auditory signals from the cochlea, also receives input from the somatosensory and vestibular systems. Thus, output from the dorsal cochlear nucleus already carries integrated signals from multiple senses; one proposed function is to help localize sounds in the vertical plane, as will be discussed in Section 6.27. In addition to the excitatory projections outlined in Figure 6–54, inhibitory neurons at multiple stages of the auditory pathways play important roles in shaping auditory signals. For instance, inhibitory circuits acting on the inferior colliculus underlie the **precedence effect**, which refers to the ability of the first-arriving sound to suppress the perception of later-arriving sounds. The precedence effect helps distinguish a sound coming from its original source (which arrives first) from its echoes (which arrive later). Indeed, the best studied function of the brainstem auditory nuclei is to extract information about sound location, which is discussed in the next two sections.

6.26 In the owl, sound location is determined by comparing the timing and levels of sounds reaching two ears

As well as sensing the frequency and intensity of sound, audition provides animals with information about sound location. Recall the superb ability of the barn owl to catch prey in complete darkness based on auditory information alone (see Figure 1–5). A number of different mechanisms have been identified for sound localization. We start with sound localization in the horizontal plane.

When a sound comes from the left, it reaches the left ear earlier than it reaches the right ear, generating an **interaural time difference** (**ITD**). (For humans, the distance between the two ears is about 20 cm; given the speed of sound is 340 m/s, this gives the maximal ITD of about 600 μs. The maximal ITD for barn owls is about 200 μs.) As the direction of sound moves either toward the front or back of the head, the ITD decreases systematically, and when it is located straight ahead (or straight behind), ITD is zero. This strict correspondence between ITD and the horizontal positions of sounds enables the auditory system to construct a map of space based on ITDs.

A theory was proposed as to how the brain can construct such a map. Suppose that some central neurons in the brain act as **coincidence detectors**, that is, their activity peaks when signals from the left and the right ear activate these neurons simultaneously. These coincidence detector neurons can be arranged in a manner such that different neurons receive input from left and right ears with slightly different time delays. Peak activity of specific neurons can then report specific sound locations in the horizontal plane (**Figure 6–55A; Movie 6-3**).

This theory, known as the Jeffress model after its original proposer, was beautifully validated decades later in the barn owl, where a brainstem structure called the **nucleus laminaris** (**NL**; analogous to the mammalian medial superior olivary nucleus) receives input from the cochlear nuclei on both sides. Ipsilateral

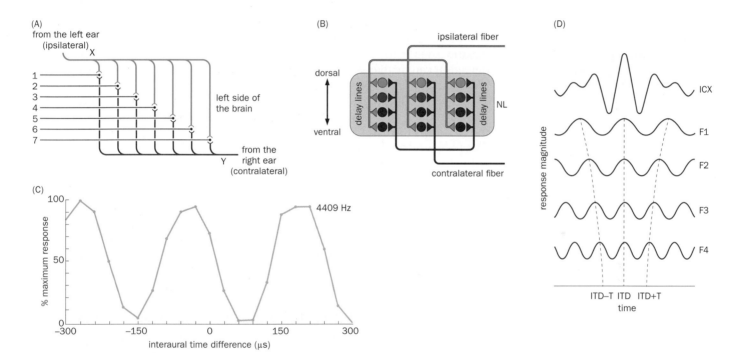

Figure 6–55 Interaural time difference can be used to determine sound localization. (A) The Jeffress model for the neural basis of sound localization. Suppose that sound signals from the left (ipsilateral) and right (contralateral) ears reach point X and Y simultaneously. Neurons 1–7 are postsynaptic targets of both ipsilateral and contralateral fibers. Given its location, neuron 1 receives ipsilateral input prior to contralateral input; neuron 4 receives both inputs at the same time; neuron 7 receives contralateral input prior to ipsilateral input. Further suppose that neurons 1–7 are best excited when the two inputs are precisely coincident. Thus, neuron 1 is best excited (inputs to neuron 1 are coincident) when the sound is closer to the contralateral ear than the ipsilateral ear. Likewise, neuron 4 is best excited by sounds at the mid-plane between the two ears, and neuron 7 is best excited by sounds closer to the ipsilateral ear than to the contralateral ear. The output signals from neurons 1–7 therefore carry information about specific interaural time differences (ITDs) that reflect sound locations in the horizontal plane. **(B)** Schematic of barn owl nucleus laminaris (NL). Fibers from the ipsilateral and contralateral cochlear nucleus enter NL from the dorsal and ventral side, respectively. Sound signals from the ipsilateral or contralateral ear reach the dorsal or ventral NL surface at the same time, and then travel within the NL (delay lines). Both fibers form synapses with NL neurons along their paths. NL neurons located along the dorsoventral axis are therefore tuned to different ITDs. (NL neurons along the orthogonal axis, left-to-right in this schematic, are tuned to different frequencies.) **(C)** An ITD curve of an NL neuron obtained from *in vivo* intracellular recording of action potentials (*y* axis, firing rate relative to maximum) in response to systematically varying ITDs (*x* axis). This neuron is best activated when the contralateral ear leads the ipsilateral ear by 30 μs (contralateral lead is represented by negative values). Note that the neuron also fires maximally at −270 μs or +180/210 μs. This is because sound waves are periodic, and the peaks at −270 μs and +210 μs correspond to coincidence with the two neighboring wave peaks (the 240 μs time difference reflects the frequency of the sound stimuli at 4409 Hz). **(D)** Model for resolving phase ambiguity by ICX neurons. An ICX neuron (top) integrates inputs from NL neurons with the same ITD but representing multiple frequencies (F1–F4). Dotted lines show how peaks at the ITD line up across frequency, whereas the peaks at ITD ± T (for tonal period) diverge with frequency. Therefore, the ICX neuron is maximally active only at the ITD in response to broadband stimuli. Experimental data (not shown) support this model. (A, adapted from Jeffress LA [1948] *J Comp Physiol Psychol* 41:35–39. With permission from the American Psychological Association; B, adapted from Ashida G & Carr CE [2011] *Curr Opin Neurobiol* 21:745–751. With permission from Elsevier Inc.; C, adapted from Carr CE & Konishi M [1990] *J Neurosci* 10: 3227–3245. With permission from the Society for Neuroscience; D, after Peña JL & Konishi M [2000] *Proc Natl Acad Sci USA* 97:11787–11792. Copyright National Academy of Sciences, USA.)

input enters the NL from the dorsal side and contralateral input enters the NL from the ventral side. Both pathways traverse the NL, forming synapses onto the same neurons along each path (Figure 6–55B). *In vivo* extracellular recordings showed that sound signals reach the dorsal surface of the NL from the ipsilateral ear and the ventral surface of the NL from the contralateral ear at about the same time, ~3 ms after onset of sound stimuli. (The contralateral fibers are larger, with greater action potential conduction speed to compensate for the longer distance traveled.) It then takes ~200 μs for signals to travel through thin axon fibers across the NL; this duration matches the barn owl's maximal ITD. These thin axon fibers form **delay lines**, because each axon fiber carries auditory signals to NL neurons located at different dorsoventral positions with different time delays. Specifically, NL neurons located closer to the dorsal surface receive ipsilateral input prior to contralateral input, whereas NL neurons located closer to the ventral surface receive contralateral input prior to ipsilateral input (Figure 6–55B). *In vivo* intracellular recordings confirmed that individual NL neurons were most strongly activated by sounds with specific ITDs (Figure 6–55C). Because the inputs to NL neurons from two ears are phase-locked to sound stimuli (see Section 6.23; phase locking extends to ~9 kHz in barn owls), the temporal precision for coincidence detection can be tens of microseconds, far shorter than the duration of an action potential (see Figure 6–51C).

Thus, the properties of NL neurons in the barn owl match very closely the Jeffress model, and the distribution of ITD signals along the dorsoventral axis of the NL forms the first spatial map of the auditory environment in the owl's brain. This map is relayed to the inferior colliculus in the midbrain. Because of the cyclic nature of sound waves, the coincidence detector neurons in the NL reach maximal response either when signals from the left and right ears arrive at the same time or when they arrive one (or more) tonal period apart from each other (Figure 6–55C). This creates a phase ambiguity for ITD-based sound localization, because individual NL neurons are maximally activated by more than one time difference. Interestingly, while individual NL neurons are also frequency-tuned (frequency tuning is along an axis orthogonal to the dorsoventral axis in Figure 6–55B), space-specific neurons in the external nucleus of the inferior colliculus (ICX) integrate ITD information across multiple frequencies. This integration helps resolve the phase ambiguity: when the owl hears a broadband signal (that is, a signal containing a wide range of frequencies), a space-specific neuron in the ICX is most excited when the time difference between the two ears equal ITD but not one or more tonal period apart. This is because only at ITD do the peak responses of all NL neurons with the same ITD but representing different frequencies align (Figure 6–55D).

As aerial predators, barn owls must locate sounds in both horizontal and vertical planes in order to catch prey in complete darkness. Indeed, single-unit recordings indicated that individual ICX neurons in the owl are tuned to sounds that originate from a specific vertical and a specific horizontal position (**Figure 6–56**A). Neurons tuned to different horizontal and vertical positions collectively form an auditory spatial map in the ICX. How do owls localize sounds in the vertical plane? The barn owl's left and right ears are vertically displaced from each other, such that they are differentially sensitive to sounds from above and below. The **interaural level differences (ILDs)**, arising from the differences in the amount of attenuation and amplification of signals between the ears, provide information about the vertical position of sounds. ILD signals are analyzed along an auditory pathway in parallel with the ITD pathway we just discussed. A brainstem nucleus called posterior dorsal lateral lemniscus (LLDp) (analogous to the lateral lemniscus in the mammals; see Figure 6–54) receives excitatory input from the contralateral cochlear nucleus and inhibitory input from the ipsilateral cochlear nucleus. Thus, LLDp neurons are best excited by contralateral input, and their spontaneous activity is diminished by ipsilateral input. Different neurons along the dorsoventral axis of the LLDp are sensitive to different ILDs, creating a map for sounds from different vertical positions. LLDp neurons also project to the inferior colliculus, sending level difference signals to the same ICX neurons that

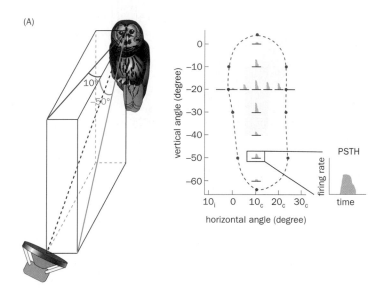

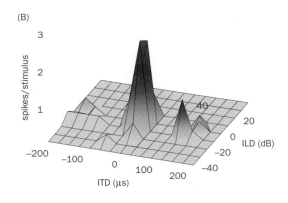

receive time difference signals from the NL. This way, individual space-specific ICX neurons are tuned to both ITD and ILD (Figure 6–56B). ICX neurons project to the nearby tectum (equivalent to the mammalian superior colliculus), so the auditory and visual maps can be aligned (see Section 1.3; the mechanisms of map plasticity will be discussed further in Section 10.25).

6.27 Mechanisms of sound location in mammals differ from those in the owl

Although the general principles of sound location revealed in the barn owl may apply to other animals, detailed circuits and mechanisms differ considerably in different species. In mammals, including humans, information about sound location in the vertical plane is mostly derived from monaural cues (signals from a single ear) rather than binaural comparisons. Because of the complex shape of the outer ear, sounds with different frequencies from above or below are differentially attenuated or amplified, changing the amplitude spectrum of the sound in a characteristic manner (**Figure 6–57A**). These changes in sound spectrum provide cues for determining the vertical position of sounds. Sound location in the horizontal plane is achieved by a combination of ITDs and ILDs in mammals. ITDs are preferentially used to localize low-frequency sounds (Figure 6–57B), because the precise timing provided by phase locking of auditory neurons only works well for sounds with frequency <2 kHz (see Section 6.23). ILDs are preferentially used to localize high-frequency sounds (>2 kHz) because high-frequency (short-wavelength) sounds are more readily deflected, bouncing off the head to create larger intensity differences between the two ears (Figure 6–57C).

In the mammalian auditory system, ITDs are analyzed in the medial superior olivary nucleus (MSO) in the brainstem (Figure 6–56D). The MSO receives excitatory inputs from both ipsilateral and contralateral ventral cochlear nuclei, and plays an analogous role to that of the owl's NL in measuring ITDs, although the mechanisms are more complex and less well understood. ILDs are analyzed in the lateral superior olivary nucleus (LSO), where individual neurons receive excitatory input from the ipsilateral ventral cochlear nucleus, and inhibitory input from the ipsilateral medial nucleus of the trapezoid body (MNTB), which in turn receives excitatory input from the contralateral ventral cochlear nucleus (Figure 6–57D). Thus, LSO neurons are excited by ipsilateral auditory signals and inhibited by contralateral auditory signals, opposite to the LLDp neurons in the owl. Whereas auditory signals regarding the horizontal position are encoded by a combination of ITDs and ILDs, signals regarding the vertical position are encoded in frequency shifts and analyzed mostly by pathways originating from the dorsal cochlear nucleus. These two sets of signals converge in the inferior colliculus to create a map of auditory space (see Figure 6–54).

Figure 6–56 Auditory space map and integration of ITD and ILD signals in the owl's exterior nucleus of the inferior colliculus (ICX). (A) Left, schematic illustration of two angles that define the direction of a sound in the owl's auditory space. Draw a line that connects the owl with the source of the sound (dark dashed line) and project this line onto the horizontal plane (magenta line) and vertical plane (blue line). The angles between these projected lines and a line straight forward are used to define the sound direction on the horizontal (magenta angle) and vertical (blue angle) planes. Right, receptive field of an ICX neuron measured by single-unit recording in a lightly anesthetized owl in response to sounds from a moving speaker that was systematically placed at different locations. Dashed lines mark the borders of the neuron's receptive field projected from the actual measurement sites (red dots). The neuron is most excited by sounds from the dark yellow area (best area). Each small diagram is a peri-stimulus time histogram (PSTH, magnified in the inset on the right) for sound stimulus given at that position. The best area for this particular neuron is around 10° contralateral (10_c) on the horizontal plane (x axis) and −20° on the vertical plane (y axis; negative values represent positions below the straight forward horizontal plane). **(B)** *In vivo* intracellular recording of an ICX neuron from an anesthetized owl in response to auditory stimuli given from microphones placed on the two ears, such that ITDs and ILDs could be precisely controlled and systematically varied. The neuron is maximally active at a specific combination of ITD and ILD levels. Minor peaks on the ITD axis represent ITD ± tonal period (see Figure 6–55). (A, adapted from Knudsen El & Konishi M [1978] *Science* 200:795–797; B, adapted from Peña JL & Konishi M [2001] *Science* 292:249–252. With permission from AAAS.)

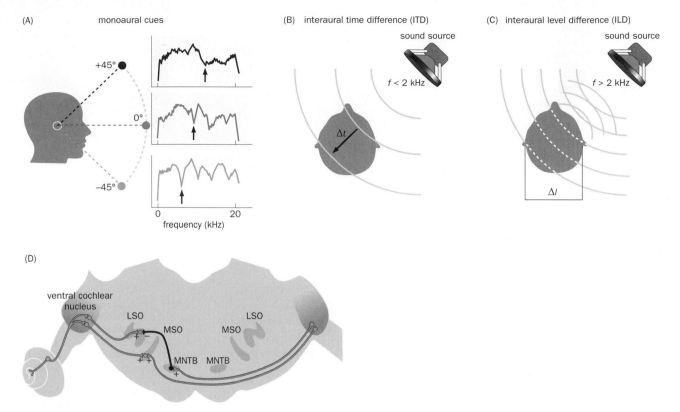

(A) monoaural cues

+45°

0°

−45°

frequency (kHz)

0   20

(B) interaural time difference (ITD)

sound source

f < 2 kHz

Δt

(C) interaural level difference (ILD)

sound source

f > 2 kHz

Δl

(D)

ventral cochlear nucleus

LSO   LSO

MSO   MSO

MNTB MNTB

Figure 6–57 Sound localization systems in mammals. (A) Spectral analysis for sound localization in the vertical plane in mammals. Illustrated in colors are interactions of a broadband sound with the vertically asymmetric outer ear; the same trough (black arrow) in the effective spectrum of the sound shifts to higher frequencies (that is, rightward on the *x* axis) when the sound source is shifted from below to above the horizon. **(B)** Interaural time difference (ITD) is preferentially used to localize sounds of low frequency (*f* < 2 kHz) because low-frequency sounds create larger time difference (Δ*t*). **(C)** Interaural level difference (ILD) is preferentially used to localize sounds of high frequency (*f* > 2 kHz) because high-frequency sounds are more effectively blocked by the head, creating larger level difference (Δ*l*). **(D)** Schematic brainstem circuits that analyze ITD and ILD in mammals. ITDs are analyzed in the medial superior olivary nuclei (MSO), which receive excitatory input from both ipsilateral and contralateral cochlear nuclei. ILDs are analyzed in the lateral olivary nuclei (LSO), which receive excitatory input from the ipsilateral cochlear nucleus and inhibitory input from the ipsilateral medial nucleus of trapezoid body (MNTB). MNTB receives excitatory input from the contralateral cochlear nucleus. For simplicity, only inputs to the left LSO and MSO are shown. (A–C, adapted from Grothe B, Pecka M & McAlpine D [2010] *Physiol Rev* 90:983–1012. With permission from the American Physiological Society.)

6.28 The auditory cortex analyzes complex and biologically important sounds

By the time auditory signals reach the inferior colliculus, information regarding sound frequency, intensity, and location has been extracted and represented by specific neurons. Such information is channeled to the superior colliculus to regulate head and eye orientation, and to the auditory cortex via the medial geniculate nucleus (see Figure 6–54). What is the function of the mammalian auditory cortex? Compared to the visual cortex discussed in Chapter 4, much less is known about the function of the auditory cortex. One prominent organization in the **primary auditory cortex (A1)** is a coarse tonotopy along the rostral–caudal axis. A1 neurons also represent many other auditory properties. Some are excited by signals from one ear whereas others are excited by signals from both ears. Some A1 neurons respond to auditory signals whose frequencies fall within a specific bandwidth, within a certain level of loudness, or with a specific latency to sounds. Some A1 neurons respond to the rate and direction of frequency modulation (for example, sounds with increasing or decreasing frequencies). Syntheses of these properties can, in principle, allow neurons in higher-order auditory cortices to analyze complex sounds such as those used in human speech. Much work is required to understand the neural basis of how these different properties of A1 neurons arise and how they are integrated by higher-order auditory cortical neurons for analysis of complex sounds that enable us to comprehend speech and enjoy music.

The best understood examples of auditory cortical organization came from studies of **echolocation** in bats, which can detect flying insects, their nightly meal, in complete darkness. Echolocating bats emit ultrasonic pulses (biosonar), and use echoes of these pulses to derive information about the size, distance, flying velocity, and wing beating of their targets. The auditory cortex of echolocating bats has specialized areas dedicated to analyzing different aspects of their ultrasonic echoes. Some bats emit constant-frequency (CF) pulses. Others emit frequency-modulated (FM) pulses. Yet others, such as the mustached bat, which hunts insects in vegetation, emit pulses that have both CF and FM components

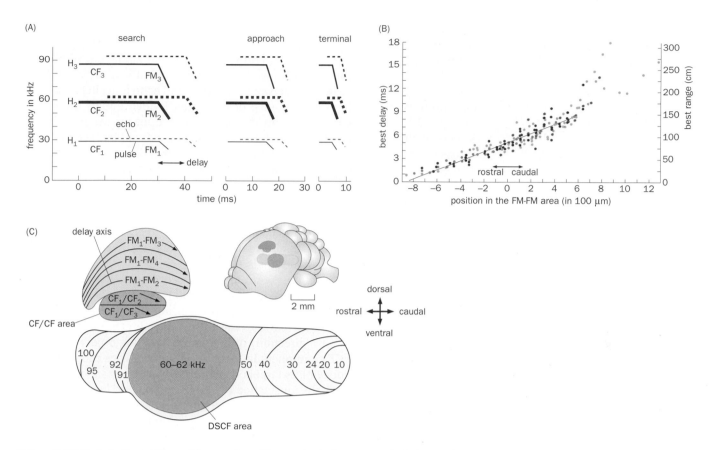

Figure 6–58 Organization of the auditory cortex of the mustached bat for analyzing ultrasonic echoes. **(A)** Ultrasonic pulses emitted by the mustached bat. In these time–frequency graphs, each pulse consists of a constant frequency (CF) component that appears as a horizontal line, and a frequency-modulated (FM) component that is the diagonal line at the end of each pulse. The pulses are emitted at a fundamental frequency of ~30 kHz (H_1), and second (H_2), third (H_3), and fourth (not shown) harmonics, with the second harmonic being the strongest (represented by the heaviest lines). Their respective echoes are shown in red, with a time delay. The duration of the pulses (and their echoes) progressively shortens from the search phase to the approach phase and the terminal phase of hunting. **(B)** Relationship of neuronal positions along the rostral–caudal axis in the FM-FM area (x axis) and their best delay (the left y axis) or best target range (the right y axis). 0 corresponds to 5-ms best delay. The regression line is based on 152 neurons with 0 to 10 ms best delay recorded from six different animals (neurons from each animal are designated by a specific color). Thus, neurons in the rostral part of the FM-FM area represent shorter best delay, and likely function in the

terminal phase, whereas neurons in the caudal part represent longer best delay, and likely function in the search phase. **(C)** Schematic of mustached bat's auditory cortex (top right) and a magnified view (left and bottom). The FM-FM area (green) is organized along the rostral–caudal axis according to the best delay (delay axis, with arrows pointing to direction of increasing best delay), as illustrated in panel B. The CF/CF area (orange) is organized along the rostral–caudal axis according to target velocity (with arrows pointing to the direction of increasing velocity). In both FM-FM and CF/CF areas, bands along the dorsal–ventral axis correspond to echoes in the frequency range of different harmonics. Representation of 60–62 kHz, corresponding to the echoes of the second (and dominant) harmonic, is drastically expanded compared to other frequency ranges in the tonotopic map, constituting the Doppler-shifted constant frequency area (DSCF) area. (A, adapted from O'Neill WE & Suga N [1982] *J Neurosci* 2:17–31; B, adapted from Suga N & O'Neill WE [1979] *Science* 206:351–353; C, adapted from Suga N [1990] *Sci Am* 262(6):60–68. With permission from Macmillan Publishers Ltd.)

(Figure 6–58A). The mustached bat emits ultrasonic pulses at a **fundamental frequency** of ~30 kHz and its second, third, and fourth **harmonics** (sounds with frequencies that are 2, 3, and 4 times that of the fundamental frequency), with the second harmonic being the most dominant harmonic. During the search phase, these pulses are emitted at low rates (~10/s) with long durations (~35 ms). The rate increases and duration shortens as the bat approaches its target. At the terminal phase, the rate increases to ~100/s and duration shortens to just a few milliseconds (Figure 6–58A).

In order to determine the neural basis of echolocation, single-unit recordings were performed from the auditory cortex of lightly anesthetized bats in response to experimenter-applied ultrasonic stimuli (much like mapping receptive fields of

V1 neurons discussed in Section 4.23). In a cortical area called the FM-FM area, specialized in analyzing the FM part of the echo, it was found that many neurons did not respond to the pulse or the echo when presented alone, but responded selectively when the pulse was followed by an echo with a specific time delay. The 'best delay,' the time interval between the pulse and echo that elicits the most excitation, corresponds to the distance of the target (which equals time delay multiplied by the speed of sound divided by 2, to account for the round trip). Moreover, the value of the best delay for individual neurons varied systematically across the cortex, creating a map of target distances along the rostral–caudal axis of the brain (Figure 6–58B). Further, each FM-FM neuron is tuned to analyze echoes of a specific frequency: either the second, third, or fourth harmonic. Neurons tuned to different echo harmonics are arranged in bands orthogonal to the axis of best delays (the green structure in Figure 6–58C). The relative strength of the different echo harmonics indicates the size of the object. Thus, the auditory cortex of the mustached bat is highly organized for analyzing object distance and size based on echoes of the sonar signals.

What is the function of the CF component? According to a wave property called the **Doppler effect**, sound frequency detected by an observer increases if the sound-emitting object moves toward the observer, and decreases if the sound-emitting object moves away from the observer. The magnitude of frequency change reflects the velocity (speed + direction) of the sound-emitting object relative to the observer. Thus, the CF component in a bat's ultrasonic pulses can be used to analyze the relative velocity of its targets based on frequency change caused by the Doppler effect. Indeed, the mustached bat devotes a second specialized area in the auditory cortex, the CF/CF area, for velocity analysis. Like the FM-FM area, one axis of the CF/CF area (the orange structure in Figure 6–58C) is organized according to the harmonics: the CF_1/CF_2 or CF_1/CF_3 area is tuned to a CF_1 pulse followed by Doppler-shifted second or third harmonic echo, respectively. The other axis is organized according to the magnitude of the Doppler shift, reflecting the relative velocity of the object.

A third specialized area in the bat's auditory cortex is a highly exaggerated representation of sound frequencies between 60 and 62 kHz, a structure called the Doppler-shifted constant frequency area (DSCF area; the pink structure in Figure 6–58C). Sounds at 60–62 kHz cover the frequency range of the second harmonic echoes, the dominant frequency from which bats collect information about object details, such as wing beats of insects. The overrepresentation of the second harmonic frequency in the DSCF area is analogous to the overrepresentation of the foveal portion of the visual field in the primate visual cortex for analyzing high-acuity and color signals (see Figure 4–38). Collectively, auditory cortical neurons in the FM-FM, CF/CF, and DSCF areas allow the mustached bat to extract information regarding size, distance, velocity, and other details about objects, enabling the bat to identify and capture flying insects in vegetated areas at night.

Lessons learned from studying auditory specialists such as the mustached bat can be instructive in illustrating auditory cortex function in general. For instance, the auditory cortex of other mammals is similarly tuned to analyzing sounds of particular biological significance, such as the social calls of mates, parents, and progeny. Likewise, the human auditory cortex has specialized areas that allow us to analyze the phonetic elements of our language. Indeed, as introduced in Section 1.10, patients with damage to the Wernicke area, a high-order auditory cortical area, exhibit selective defects in language comprehension. It remains a future challenge to understand the organizational principles of the auditory cortex in other mammals, including humans, to the level of detail that we understand the auditory cortex of echolocating bats.

Whereas signals in the auditory system originate from the cochlea, its sister organs in the inner ear are the sensory organs for the vestibular system, which serves entirely different functions (**Box 6–2**).

Box 6–2: The vestibular system senses movement and orientation of the head

In addition to the cochlea, the inner ear also contains sensory organs of the **vestibular system**, which in many ways parallels the auditory system. For example, both systems use stereocilia in hair cells to convert mechanical stimuli into electrical signals. In fact, the vestibular system is the more ancient of the two systems and is present in all chordates; the auditory end organ was derived from the vestibular sensory organs in bony fish, and became more elaborate in terrestrial animals, especially mammals. The vestibular system senses the movement and orientation of the head, and uses these signals to regulate a variety of functions including balance, spatial orientation, coordination of head and eye movement, and perception of self-motion. Because we are unaware of most functions it performs, the vestibular system is sometimes called a silent sense.

The vestibular system has five sensory organs: two **otolith organs** and three **semicircular canals** (**Figure 6–59**A). The otolith organs, called the utricle and the saccule, are responsible for sensing linear acceleration and stationary head tilts.

The stereocilia of hair cells in both the utricle and the saccule are embedded in a membrane that is densely packed with calcium carbonate crystals known as otoliths (meaning 'ear stone' in Greek). While the hair cells are anchored in the bony structure of the inner ear, and therefore move in sync with the head, the inertia of otoliths created by linear acceleration in a specific direction creates a relative movement between the hair cells and the overlying otoliths, which bend stereocilia whose staircases ascend in the same direction as the direction of motion (Figure 6–59B). This opens the mechanotransduction channels in the hair bundles and causes depolarization of the hair cells. Stationary head tilts also activate hair cells whose ascending stereocilia are aligned with the direction of the tilt due to gravity of the otoliths. Because of the different orientations of the hair cell stereocilia with respect to the three axes of space, hair cells in the utricle and saccule are collectively sensitive to linear accelerations and head tilts in all directions (Figure 6–59A, middle). Left–right accelerations and head tilts are sensed by the utricular hair cells, up–down accelerations (for

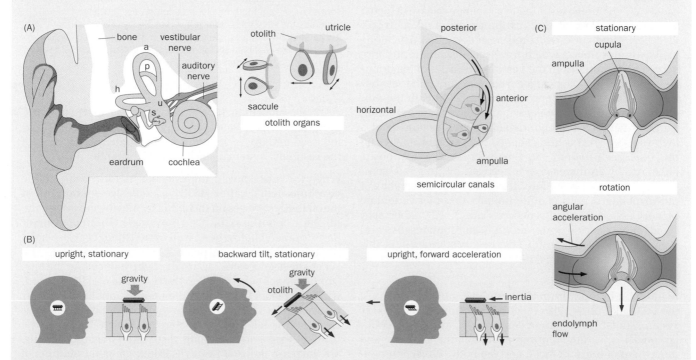

Figure 6–59 The vestibular sensory organs. (A) Left, schematic drawing of the structure of the vestibular organs and the cochlea in the inner ear. u, utricle; s, saccule; h, p, a, horizontal, posterior, anterior semicircular canals. Middle, magnified utricle and saccule. The stereocilia are anchored in the membranes rich in calcium carbonate crystals (otoliths). As a result of the specific orientations of hair cell stereocilia, the utricle senses linear accelerations or head tilts along the left–right and front–back axes, whereas the saccule senses linear accelerations along the top–bottom and front–back axes. Right, magnified semicircular canals. The horizontal semicircular canal senses angular accelerations (rotations) in the horizontal plane, whereas the anterior and posterior semicircular canals sense angular accelerations in the vertical planes. The hair cells are located in the bulges at the bases of the semicircular canals called ampulla. **(B)** Compared to the upright stationary position, both stationary backward tilt and upright forward acceleration bend the stereocilia because of gravity or the inertia of the otoliths; this opens mechanotransduction channels, depolarizes the hair cells, and increases synaptic transmission to vestibular ganglion neurons (red arrows). **(C)** Compared to the stationary position, angular acceleration toward the left causes the stereocilia, which are enriched in a structure called cupula, to bend rightwards because of the flow created by the inertia of the fluid (endolymph) within the semicircular canal. This relative movement opens the mechanotransduction channels, depolarizes the hair cells, and increases synaptic transmission to vestibular ganglion neurons (red arrow). (A, adapted from Day BL & Fitzpatrick RC [2005] *Curr Biol* 15:R583–R586. With permission from Elsevier Inc.)

(Continued)

Box 6–2: The vestibular system senses movement and orientation of the head

example, the beginning and end of an elevator ride) are sensed by saccular hair cells, and front–back accelerations and head tilts are sensed by hair cells of both organs.

The three semicircular canals sense angular accelerations. The horizontal semicircular canal senses head rotation in the horizontal plane; the anterior and posterior semicircular canals, which are orthogonal to each other, collectively sense head rotations in various vertical planes (Figure 6–59A, right). The semicircular canals are filled with fluid, which contacts the stereocilia of hair cells that are anchored in the bulges at the bases of the canals (Figure 6–59C). Angular acceleration in the same plane as a semicircular canal creates relative movement of fluid that bends the stereocilia of hair cells. This causes the opening or closing (depending on the direction) of the mechanotransduction channels in hair cells to initiate electrical signaling. Signals from three semicircular canals collectively provide information about head rotations in all axes in the three-dimensional space.

As in the auditory system (see Figure 6–47), depolarization of hair cells in the vestibular system activates voltage-gated Ca^{2+} channels, resulting in glutamate release detected by afferent nerves of the **vestibular ganglion neurons**. Electrical signals are delivered as action potentials along the **vestibular nerve**, which travels alongside the auditory nerve and terminates in the **vestibular nuclei** of the brainstem. Electrophysiological recordings of individual vestibular nerve fibers connecting with hair cells in both the otolith organs and semicircular canals indicate that vestibular ganglion neurons fire spontaneous action potentials at a high rate (~40 Hz) in the absence of head movement and tilt. Thus, each neuron can report movement in opposite directions by an increase or a decrease in firing rate compared to the basal firing rate. This bidirectional signaling of the vestibular system is quite unique—in other vertebrate sensory systems, sensory neurons are mostly either activated (as in olfaction, taste, audition, and somatosensation) or are inhibited (as in vision) by sensory stimuli. Thus, this silent sense is in fact constantly at work even though we are unaware of it.

In addition to input from the vestibular ganglion neurons, the vestibular nuclei also receive input from other sensory systems, notably the somatosensory system. Thus, vestibular nuclei integrate information about the position and movement of both the head and the body. Projection neurons in the vestibular nuclei send their axons to many parts of the CNS to carry out distinct functions (**Figure 6–60**). For example, descending axons to the spinal cord regulate the activity of motor neurons, which in turn control muscle contraction in the neck, limb, and body to regulate balance and posture (we will discuss the motor system in more detail in Chapter 8). Ascending axons deliver information to the ventroposterior thalamic nucleus, and to a number of cortical areas for multi-sensory integration. For example, vestibular signals reach specific areas of the parietal cortex, where they are integrated with signals from the visual system to help distinguish motion in the external world from self-motion. Other important targets of projection neurons from the

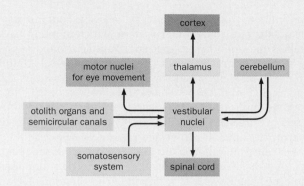

Figure 6–60 Flow of vestibular signals in the CNS. Axons of the vestibular ganglion neurons deliver vestibular signals from the sensory organs to the vestibular nuclei in the brainstem, which have several divisions that receive differential inputs from the otolith organs and semicircular canals, as well as input from the somatosensory system. Descending vestibular signals to the spinal cord regulate posture, balance, and head position through a number of reflexes. Ascending signals to the thalamus and cortex contribute to the integration of vestibular signals with signals from other senses, such as visual signals. Vestibular signals are also sent to brainstem motor nuclei that control eye movement, which underlies the vestibulo-ocular reflex (see Figure 6–61), and to the cerebellum for movement modulation, such as providing gain control for the vestibulo-ocular reflex via feedback signal from the cerebellum to the vestibular nuclei.

vestibular nucleus include brainstem nuclei that control eye movement (Figure 6–60), which are essential for coordinating head and eye movement. We illustrate this function of the vestibular system using a specific example.

When the head moves, the eye sockets move with it. The **vestibulo-ocular reflex** (**VOR**) is a reflexive eye movement that stabilizes images on the retina by moving the eyes in the direction opposite to the head movement. The neural circuit basis for the VOR has been well established. Let's use head rotation in the horizontal plane as an example (**Figure 6–61**). When the head rotates toward the right, hair cells in the horizontal semicircular canal on the right side are depolarized, which increases the firing of their corresponding vestibular ganglion neurons. Hair cells in the horizontal semicircular canal on the left are hyperpolarized, which decreases the firing of their corresponding vestibular ganglion neurons. The increase in firing from vestibular ganglion neurons of the right horizontal semicircular canal is transmitted by excitatory neurons in the medial vestibular nucleus to the contralateral (left) abducens nucleus, which increases the firing of motor neurons that cause contraction of the lateral rectus muscle, leading to leftward movement of the left eye. The left abducens nucleus neurons also increase the firing rate of the motor neurons at the contralateral (right) oculomotor nucleus, which causes contraction of the medial rectus muscles of the right eye, also leading to its leftward movement. At the same time, the decrease in firing from vestibular ganglion neurons of the left horizontal semicircular canal is transmitted by a parallel pathway, which causes the relaxation of the medial rectus muscle of the left eye and

Box 6–2: The vestibular system senses movement and orientation of the head

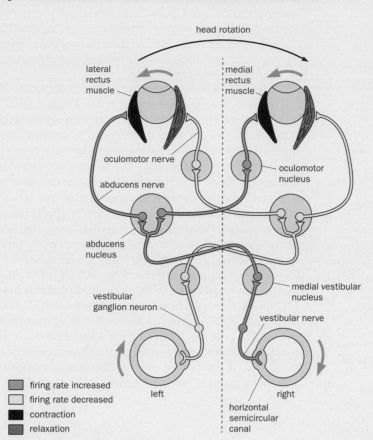

Figure 6–61 Circuit basis of the vestibulo-ocular reflex. When the head rotates to the right, it increases the firing rate of the vestibular ganglion neurons associated with the right horizontal semicircular canal, and decreases the firing rate of the vestibular ganglion neurons associated with the left horizontal semicircular canal. The signals from the right vestibular nuclei increase the firing rate of neurons in the left abducens nucleus, which causes contraction of the lateral rectus muscle of the left eye, and contraction of the medial rectus muscle on the right eye (via motor neurons in the oculomotor nucleus). In parallel, a decrease in firing rate from the left vestibular nuclei causes relaxation of the medial rectus muscle of the left eye and the lateral rectus muscle of the right eye. These muscle actions cause both eyes to rotate toward the left, so that images in the retinas remain stable while the head rotates to the right.

the lateral rectus muscle of the right eye (Figure 6–61). Thus, the VOR enables image stabilization on the retina during head movement by producing compensatory eye movements. The **VOR gain**, or the degree to which the eye moves in response to head movement, can be adjusted precisely by experience; for instance, wearing a pair of near-sighted eyeglasses makes the images look smaller, which requires a smaller VOR gain compared to the situation without eyeglasses. As will be discussed in Section 8.8, the cerebellum, which also receives vestibular input and sends output back to the vestibular nuclei (see Figure 6–60), adjusts VOR gain.

SOMATOSENSATION: HOW DO WE SENSE BODY MOVEMENT, TOUCH, TEMPERATURE, AND PAIN?

Of all the senses, the somatosensory system has the largest sensory organs—the entire skin and musculature. It also responds to the most diverse set of sensory stimuli, including mechanical, thermal, and chemical stimuli. From these stimuli, the somatosensory system derives information about body position and movement (**proprioception**), temperature (**thermosensation**), various forms of touch perceptions, as well as pain (**nociception**) and itch (**pruriception**). (The sense of internal organ function, or **interoception**, will be discussed in Chapter 8 when we study the autonomic nervous system.) As was introduced in Chapter 1, some of the fundamental discoveries in neuroscience, such as the use of action potential firing rates to encode information (see Figure 1–17), the circuit basis of reflex behavior (see Figure 1–19), and the topographic organization of cortex (see Figure 1–25), were made in the context of studying the somatosensory system. In the following sections, we first discuss the general organization of the somatosensory system, and then focus on touch and pain research that elucidates general principles of sensory systems.

The cell bodies of all sensory neurons of the vertebrate somatosensory system are located in either the **dorsal root ganglia** (**DRG**; 31 pairs in humans) parallel to the spinal cord or a pair of **trigeminal ganglia** adjacent to the brainstem, for sensation of the body or the face, respectively (**Figure 6–62**). Each DRG contains 10,000–14,000 sensory neurons in the mouse; this number is much larger in humans to accommodate the larger body size. Each sensory neuron extends a single process, which bifurcates to give rise to a peripheral axon that collects information from the skin or musculature, and a central axon that sends information to the spinal cord and/or the brainstem (**Figure 6–63**). The receptor potentials produced by mechanical, thermal, and chemical stimuli are converted into action potentials near a sensory neuron's peripheral endings; action potentials transmit sensory information along the peripheral and central axons to the CNS. The spatiotemporal patterns of the action potentials of somatosensory neurons inform the brain about which kind of stimulus is being sensed, where the stimulus comes from, and how intense the stimulus is. As discussed in Sections 1.8 and 1.11, the stimulus intensity is encoded in the sensory neuron by the rate of action potentials, and the brain utilizes a somatotopic map to represent the stimulus location (see also Box 5–3 for a specific example of the whisker-barrel system in rodents). We focus next on how the somatosensory system distinguishes which kind of stimulus is being sensed.

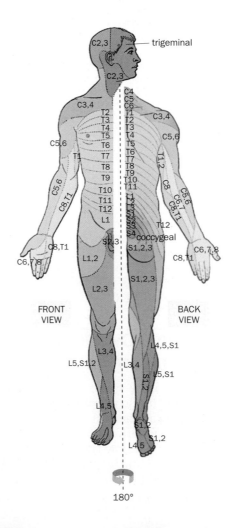

Figure 6–62 Human dermatome map. Each dermatome is an area of skin that is mainly innervated by a single nerve. The face dermatome is innervated by the trigeminal nerve originating from the trigeminal ganglion. Dermatomes in the rest of the body are numbered according to the dorsal root of the spinal nerves (axons of the sensory neurons in the dorsal root ganglia) that innervate them (C2–C8, cervical; T1–T12, thoracic; L1–L5, lumbar; S1–S5, sacral; coccygeal). The first cervical nerve (C1) does not innervate the skin.

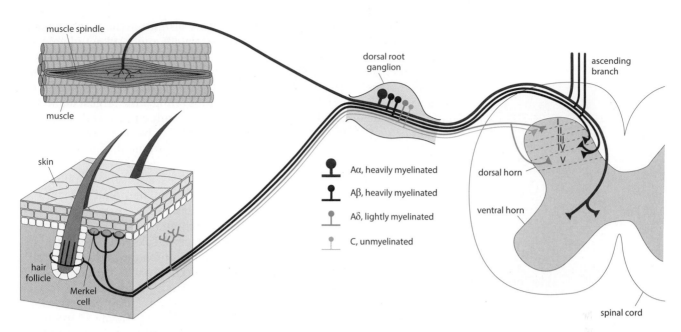

Figure 6–63 Parallel organization of the somatosensory system. All sensory neurons in the somatosensory system representing the body are located in the dorsal root ganglia (DRG). Each sensory neuron exhibits unique termination patterns of its peripheral axon in the skin or musculature, and projects its central axon to specific lamina(e) of the spinal cord. (The central projections of sensory neurons will be discussed in more detail in Section 6.33.) Five kinds of sensory neurons are shown in this diagram. The red proprioceptive neuron has the thickest axon with heavy myelination (Aα fiber), innervates the muscle spindle, and terminates in the central and ventral spinal cord. Two purple low-threshold mechanoreceptors (LTMRs) for sensing touch also have thickly myelinated axons (Aβ fibers), innervate a hair follicle or Merkel cells, respectively, and terminate in laminae III–V of the dorsal horn. The blue temperature-sensing neuron has a thinly myelinated axon (Aδ fiber) and free peripheral endings, and terminates in laminae I and V of the dorsal horn. The green nociceptive neuron has an unmyelinated axon (C fiber) and free peripheral endings, and terminates in dorsal horn laminae I–II. The proprioceptive neuron and LTMRs also form branches that ascend into the brainstem.

6.29 Many types of sensory neurons are used to encode diverse somatosensory stimuli

The somatosensory system is one of the most complex and least well understood sensory systems. This complexity arises because there are many types of sensory neurons; each type can act alone or in combination with other types to encode information about a specific type of sensory stimulus. New types of sensory neurons are still being discovered. Historically, sensory neurons are divided by their peripheral endings, which determines what kind of sensory stimulus they respond to, by the size of their axon fibers and degree of myelination, which determines the speed of action potential conduction (see Box 2–3), and by the adaptation property of sensory neurons in response to sustained stimuli (**Table 6–1**).

Table 6–1: Different types of sensory neurons in the somatosensory system

Sensation	Stimulus type	Axon fiber[1]	Peripheral endings
Proprioception	mechanical	Aα, Aβ	muscle spindles, tendons, joints
Touch	mechanical	Aβ, rapidly or slowly adapting	Merkel cells, Meissner corpuscles, Ruffini endings, Pacinian corpuscles, hair follicles
		Aβ, Aδ, C	hair follicles
Temperature, pain, itch[2]	mechanical	Aδ, C (Aβ)	free endings in skin
	heat, cold, chemical	Aδ, C (Aβ)	free endings in skin and internal organs

All categories above include multiple cell types.

[1] The diameter of the axon fiber correlates with action potential conduction speeds. The conduction speed follows the order of Aα (70–120 m/s), Aβ (30–70 m/s), Aδ (5–30 m/s), and C (0.2–2 m/s), reflecting their descending degrees of myelination (see Figure 6–63). The values of conduction speed above are from humans.

[2] These categories are grouped together because they often share the same sensory neurons.

Most sensory neurons in the somatosensory system are the **mechanosensory neurons**, which account for proprioception, touch, and a subset of pain sensations (Table 6–1). **Proprioceptive neurons** embed their peripheral endings in the muscle spindles, tendons, and joints, which allow these neurons to sense muscle stretch and tension (Figure 6–61, top left; see also Figure 1–19 for their function in the knee-jerk reflex). Proprioceptive neurons employ the largest diameter and most heavily myelinated axons with the fastest conduction speed. The fast conduction of proprioceptive axons provides rapid feedback to the motor system regarding muscle strength and tension during movement, the importance of which will be discussed in detail in Chapter 8.

Touch sensory neurons innervate hair follicles, specialized epithelial cells, and encapsulated corpuscles in the skin. Different neuronal subtypes sense innocuous touch including vibration, indentation, pressure, and stretch of the skin, as well as movement or deflection of hairs (Figure 6–63, bottom left). Compared with neurons that sense noxious mechanical stimuli discussed below, touch sensory neurons are more sensitive to low-force mechanical stimuli, and are therefore called **low-threshold mechanoreceptors** (**LTMRs**). LTMRs that innervate the glabrous (hairless smooth) skin, such as the primate's hand, are categorized based on the peripheral ending and adaptation property during a sustained stimulus (**Figure 6–64**). For example, the SAI (s̲lowly a̲dapting type I̲) LTMRs innervate **Merkel cells** at the junction between the epidermis and dermis and fire more persistently during a sustained stimulus (hence slowly adapting); SAI touch neurons sense skin indentation and are responsible for fine discrimination of textures because each SAI LTMR has a relatively small receptive field (the area of skin that influences the firing of a sensory neuron) compared with other types of LTMRs. The r̲apidly a̲dapting type I̲ (RAI) LTMRs terminate their sensory endings in a special structure called **Meissner corpuscle**, fire only at the onset and offset of a stimulus (thus rapidly adapting), and sense movement of the skin such as low-frequency vibration. The type II LTMRs terminate in encapsulated structures deep in the dermis: SAII LTMRs sense pressure and have **Ruffini endings**, whereas RAII LTMRs sense high-frequency vibration and terminate in **Pacinian corpuscles** (Figure 6–64). All these LTMRs employ the heavily myelinated **Aβ fibers** to transmit signals rapidly to the CNS.

As well as innervating the glabrous skin, several distinct types of LTMRs innervate hair follicles and Merkel cells in the hairy skin; while some subtypes use Aβ fibers, others use lightly myelinated **Aδ fibers** or unmyelinated **C fibers** (Table 6–1). Whereas LTMRs with A fibers (A-LTMRs) predominantly play a role in discriminating objects, C-LTMRs respond to gentle touch with slow conduction speeds, and are thought to mediate affective aspects of touch. Indeed, a specific

Figure 6–64 Different types of mechanosensory neurons and their sensory endings in the glabrous skin. Left, schematic drawing of nerve termination patterns for five types of neurons; right, names of the neuronal type, their nerve ending, and their adaption property. SAI and SAII low-threshold mechanoreceptors (LTMRs) innervate Merkel cells and Ruffini endings, respectively, and their firing pattern (each vertical bar represents an action potential) is more persistent in response to a sustained stimulus (bottom right). RAI and RAII neurons innervate Meissner and Pacinian corpuscles, respectively, and their firing pattern is more transient in response to a sustained stimulus. Note that SAI and RAI fibers terminate more superficially than SAII and RAII fibers, all of which employ thickly myelinated Aβ fibers (myelin sheaths are represented in gray). Mechanosensory nociceptive neurons (high-threshold mechanoreceptors, or HTMRs) have free nerve endings. (Adapted from Abraira VE & Ginty DD [2013] *Neuron* 79:618–639. With permission from Elsevier Inc.)

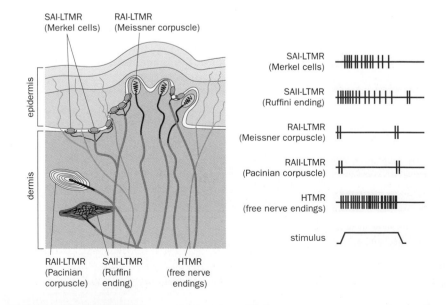

class of C-LTMRs has been identified in mice that mediates pleasurable touch such as gentle stroking.

Some mechanosensory neurons are **nociceptive neurons** that sense noxious mechanical insult. They usually require a high threshold of mechanical stimuli to elicit action potentials, and are thus also called **high-threshold mechanoreceptors (HTMRs)**. Most HTMRs have free nerve endings in the skin and fire persistently during a sustained stimulus (Figure 6-64). Activation of HTMRs elicits a pain sensation.

In addition to the mechanosensory neurons, many somatosensory neurons sense temperature and chemicals. A large fraction of **thermosensory neurons** are also nociceptive neurons; their activation by hot or cold temperatures produces a pain sensation as a protective mechanism for animals to avoid noxious temperatures. Most chemosensory neurons in the somatosensory system are also nociceptive; they respond to environmental irritants and/or endogenously released chemicals due to injury or inflammation. For example, the 'taste' of spices such as hot chili pepper (see Figure 6–44) is caused by activation of trigeminal chemosensory neurons that innervate the oral cavity. Activation of some chemosensory neurons produces an itch sensation. Notably, as we will discuss in more detail in Section 6.31, some neurons can sense stimuli from more than one sensory modality, and are thus called **polymodal neurons**.

Thermosensory and nociceptive neurons have free endings in the periphery and are heterogeneous in their axon fibers and stimulus specificity (Table 6-1). Nociceptive neurons can be broadly classified into two groups based on their axon fiber size: those that have myelinated Aδ fibers are activated by heat, noxious mechanical stimuli, or both, and mediate acute and well-localized pain (also called first or fast pain); those that have smaller, unmyelinated C fibers are activated by hot and cold temperatures as well as endogenous chemicals released by injury and tissue inflammation, and mediate poorly localized slow pain and chronic inflammatory pain.

Although pain is predominantly mediated by C and Aδ fibers, and touch by Aβ fibers, all three fibers contribute to both touch and pain (Table 6-1). The categorization of somatosensory neuron types is very much still a work in progress. For example, a recent study in mice has grouped DRG neurons into 11 types based on gene expression patterns in single cells using statistical analysis of RNA-seq data (see Section 13.13). The identifications of sensory receptors have begun to provide mechanisms by which somatosensory neurons convert sensory stimuli into electrical signals, and shed light onto the organization of the somatosensory system. We discuss these advances in the next three sections.

6.30 Merkel cells and some touch sensory neurons employ Piezo2 as a mechanotransduction channel

Since a large fraction of somatosensory neurons respond to mechanical stimuli, a key question is how mechanical forces are converted to electrical signals. Just as in the case for audition discussed in Section 6.22, studies have shown that ion channels directly gated by mechanical forces are responsible for mechanotransduction. It was not until recently that the first mammalian mechanotransduction channel was identified (see **Box 6–3** for candidate mechanotransduction channels in *C. elegans* and *Drosophila*).

None of the classic ion channels (see Box 2–4), including homologs of invertebrate candidate mechanotransduction channels, were known to mediate mechanotransduction in mammals. Researchers therefore hypothesized that mechanotransduction in mammals might be mediated by new types of ion channels, which should be transmembrane proteins. Using a neuronal cell line that produces inward currents in response to mechanical stimuli, a systematic **RNA interference (RNAi)** screen (see Section 13.8 for details) was carried out to knock down the expression of predicted transmembrane proteins in the mouse genome. Knockdown of one protein named **Piezo1** (after *piesi* in Greek, meaning pressure) led to a marked reduction of mechanically induced inward current

Box 6–3 Mechanotransduction channels in worms and flies

Mechanotransduction is essential for many physiological processes in diverse organisms. Researchers have therefore carried out forward genetic screens (see Section 13.6 for details) in *C. elegans* and *Drosophila* for mutants that are defective in sensing mechanical stimuli, with the expectation that some of the corresponding genes would encode mechanotransduction channels. A large number of *Mec* genes have been identified in *C. elegans* based on the abnormal <u>mec</u>hanosensory phenotypes in mutant worms. A protein complex consisting of four membrane-associated proteins—Mec2, Mec4, Mec6, and Mec10—is essential for mechanotransduction. Mec2 and Mec6 are accessory proteins, while Mec4 and Mec10 both belong to the epithelial Na⁺ channel (ENaC) family and appear to be the pore-forming subunits of a mechanotransduction channel. Loss of Mec4 abolishes mechanical stimulus-induced inward current in touch sensory neurons *in vivo* (**Figure 6–65**A), while single amino acid changes in MEC-4 alter conductance and ion selectivity.

In *Drosophila*, genetic screens for touch insensitive mutant larvae identified *Nompc* (<u>no</u> <u>m</u>echanoreceptor <u>p</u>otential <u>C</u>), which encodes a TRP channel (see Box 2–4) and is necessary for mechanical stimulus-induced current flow into sensory neurons. Recent work demonstrated that NompC acts in a specific type of larval body wall sensory neurons, called

class III dendritic arborization neurons, to sense gentle touch, and when ectopically expressed, confers touch sensitivity to another type of sensory neuron or mechanically induced inward current to cultured cells (Figure 6-67B).

Mechanotransduction in *C. elegans* and *Drosophila* appears to utilize similar proteins, albeit in different contexts. For instance, TRP4, the *C. elegans* homolog of *Drosophila* NompC, acts as a stretch-activated mechanotransduction channel. The *Drosophila* homolog of Piezo and one of the ENaC channels work together in class IV dendritic arborization neurons of the larval body wall to sense nociceptive mechanical stimuli; thus, gentle touch and noxious mechanical stimuli are sensed independently by two different types of sensory neurons using different mechanotransduction channels. With the exception of Piezo, none of the mechanotransduction channels in worms and flies have been shown to transduce mechanical signals in mammals. (The ENaC channels are involved in salty taste in mammals, see Figure 6–41.) At the same time, Piezos are clearly not the only mechanotransduction channels in mammals (see Section 6.30). Thus, fundamental discoveries still lie ahead as we advance toward a more complete characterization of mechanotransduction channels, their mechanisms of action, and the cellular logic and evolutionary processes underlying their differential utilization in diverse organisms.

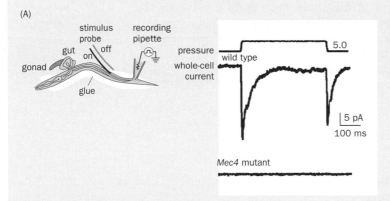

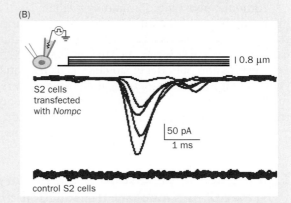

Figure 6–65 Mechanotransduction channels in worms and flies. **(A)** Left, schematic of experimental preparation for measuring mechanotransduction in *C. elegans*. A larva was immobilized on glue. Internal hydrostatic pressure was released (some internal organs such as the gut and gonad were externalized) such that a touch neuron (green) could be subjected to whole-cell patch recording in response to mechanical stimulation from a stimulus probe. Right, a touch sensory neuron from a wild-type larva responded to the onset and offset of pressure application (top trace, in nanoNewton/μm²) by passing inward current (middle trace), whereas a touch sensory neuron from a *Mec4* mutant larva did not respond to pressure

(bottom trace). **(B)** Cultured *Drosophila* S2 cells were subjected to whole-cell patch recording while mechanical displacement of the membrane was applied with an electrically driven mechanical probe. In response to an increasing series of displacements, S2 cells transfected with *Nompc* cDNA exhibited inward currents with increasing magnitudes (superimposed traces), whereas control S2 cells did not respond to mechanical stimuli. (A, adapted from O'Hagan R, Chalfie M & Goodman MB [2005] *Nat Neurosci* 8:43–50. With permission from Macmillan Publishers Ltd; B, adapted from Yan Z, Zhang W, He Y et al. [2013] *Nature* 493:221–225. With permission from Macmillan Publishers Ltd.)

(**Figure 6–66**A). Moreover, expression of Piezo1 or its close homolog Piezo2 in cells that normally did not respond to mechanical stimuli could confer mechanically induced inward current to these cells, indicating that either of the Piezos is sufficient for mechanotransduction. Mechanotransduction channels that were active in cells expressing Piezo were permeable nonselectively to cations, with a reversal potential around 0 mV (Figure 6–65B), much like nicotinic acetylcholine

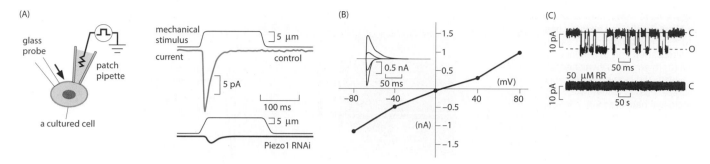

Figure 6–66 Piezos are mechanotransduction channels. (A) Left, experimental setup: a cell (from the mechanosensitive Neuro2A cell line) is stimulated with an electrically driven glass probe (stimulation indicated by a red arrow) while whole-cell patch recording is performed with a patch pipette. Top right, in a control cell, mechanical stimulus (top trace) induced an inward current (bottom trace) when the cell was clamped at −80 mV. Bottom right, in a Piezo1 knockdown cell, inward current is greatly diminished despite larger mechanical force (note the different scales). Mechanical stimulus is measured as distance of the stimulating probe from the skin; positive values indicate indentation. **(B)** The current–voltage relationship measured from whole-cell recordings of mechanical stimulus-induced current flow (y axis) at different membrane potentials (x axis). The current traces in response to mechanical stimuli are shown in the top left quadrant; each trace corresponds to a current measurement at a different membrane potential. These experiments were conducted by transfecting DNA that encoded Piezo2 into cells that were not otherwise mechanosensitive; thus, the current–voltage relations reflect the property of Piezo2. **(C)** Single-channel current recorded from a patch of membrane consisting of artificial lipid bilayer and purified Piezo1 protein. Single-channel openings (O) and closings (C) can be easily seen (top trace). Channel openings were abolished after the application of ruthenium red (RR), a blocker of Piezo channels (bottom trace). (A & B, adapted from Coste B, Mathur J, Scmidt M et al. [2010] *Science* 330:55–60; C, adapted from Coste B, Xiao B, Santos JS et al. [2012] *Nature* 483:176–181. With permission from Macmillan Publishers Ltd.)

receptors in the skeletal muscle and AMPA-type glutamate receptors in neurons (see Figure 3–17); their opening causes depolarization of cells with a negative resting potential. Both Piezo1 and Piezo2 are predicted to have >30 transmembrane segments, though they do not exhibit sequence similarity to known ion channels. To test if the Piezos are ion channels themselves, rather than activators of mechanotransduction channels that are made from a different protein but are somehow inactive in the absence of the Piezos, researchers purified Piezo1 proteins and placed them in artificial lipid bilayer. Patch recordings from such reconstituted lipid bilayer validated that Piezo1 by itself forms an ion channel (Figure 6–65C).

As we discussed in Section 6.29, the peripheral endings of LTMRs are associated with specialized structures. For instance, endings of SAI fibers in both glabrous and hairy skin form close contact with Merkel cells, which are specialized epidermal cells. Whether the Merkel cell or the SAI terminal is the site where mechanical stimuli are converted to electrical signals has long been a subject of debate. Recent studies demonstrated that isolated Merkel cells can respond to mechanical stimuli by conducting inward current, and this response was completely eliminated in mice in which *Piezo2* was conditionally knocked out in Merkel cells (**Figure 6–67**A). Thus, Merkel cells are intrinsically mechanosensitive, and depend on Piezo2 for mechanotransduction. In a skin-nerve explant preparation in which mechanical stimulation of Merkel cells induced robust action potentials recorded from SAI fibers in wild type, SAI fibers from mice in which *Piezo2* was knocked out in Merkel cells still exhibited an initial response to a mechanical stimulus, but were defective in a more sustained response (Figure 6–66B). This experiment indicated that both Merkel cells and SAI nerve terminals independently transduce mechanical signals: SAI nerve terminals are responsible for rapid and dynamic response to mechanical stimulus, whereas Merkel cells are necessary for the sustained firing activity of SAI axons, presumably via synaptic transmission from Merkel cells to SAI terminals.

What molecule is responsible for mechanotransduction in the SAI terminal? Expression studies indicated that mouse Piezo2 (but not Piezo1) is expressed in a subset of DRG neurons, including LTMRs that innervate Merkel cells and form Meissner corpuscles, as well as Aβ-, Aδ-, and C-LTMRs that innervate hair follicles. Indeed, mice in which the *Piezo2* gene was conditionally knocked out in DRG neurons exhibited profound defects in their behavioral responses to low-force mechanical stimuli. However, these mice still exhibited normal response to

Figure 6–67 Mechanotransduction in Merkel cells and sensory neurons.
(A) Left, illustration of the experiment. A dissociated and GFP-labeled Merkel cell (from genetically engineered mice in which GFP is specifically expressed in Merkel cells) in culture was subjected to whole-cell patch recording in response to a mechanical stimulus applied through a glass probe. Right, mechanical stimuli applied on the Merkel cell induced robust inward current when the membrane potential was held at −80 mV (black trace), but this inward current was abolished when *Piezo2* was conditionally knocked out in Merkel cells (blue trace). Thus, Merkel cells transduce mechanical stimulus into electrical signals in a Piezo2-dependent manner. **(B)** Left, illustration of the skin–nerve preparation. Fluorescently labeled Merkel cells (yellow) were stimulated with a glass probe while action potentials from the associated SAI fibers (peripheral axons of SAI LTMRs) were being recorded. Right, the mechanical stimulus is represented by distance of the stimulating probe from the skin (top). At 0 (dotted line), the probe touches the skin; above 0, the probe creates an indentation within the skin. An SAI fiber from a wild-type mouse responded to the stimulus with sustained firing, whereas in a mutant mouse in which *Piezo2* was conditionally knocked out (CKO) from Merkel cells, sustained firing of SAI fiber was disrupted but the initial response persisted. (A, adapted from Maksimovic S, Nakatani M, Baba Y et al. [2014] *Nature* 509:617–621. With permission from Macmillan Publishers Ltd; B, adapted from Woo S, Ranade S, Weyer AD et al. [2014] *Nature* 509:622 –626. With permission from Macmillan Publishers Ltd.)

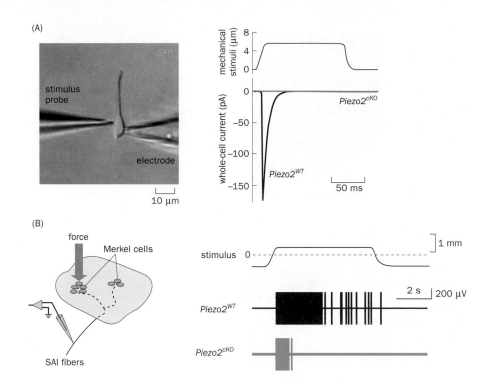

high-force mechanical stimuli. Thus, Piezo2 is required for mechanotransduction in neurons that sense gentle touch. These studies also indicated the presence of other, yet-to-be-identified, mechanotransduction channels that work in parallel with Piezo2 to convert mechanical stimuli to electrical signals in sensory neurons.

6.31 TRP channels are major contributors to temperature, chemical, and pain sensation

In addition to mechanosensation, the somatosensory system also detects temperature at the innocuous range, noxious heat or cold that causes temperature-induced nociception, and chemical irritants. Remarkably, these diverse stimuli can be sensed by the same receptors.

To identify receptors that mediate nociception, researchers took advantage of the fact that capsaicin, the main pungent ingredient in 'hot' chili peppers, activates nociceptive neurons. cDNA pools made from dorsal root ganglia were transfected into cultured cells, and Ca^{2+} imaging was used to identify the presence of a cDNA encoding an active capsaicin receptor from the pool (**Figure 6–68**A). This **expression cloning** strategy identified the capsaicin receptor as a TRP family channel (see Box 2–4), **TRPV1**, which is permeable to Ca^{2+}, Na^+, and K^+. TRPV1 is activated not only by capsaicin, but also by temperatures >43°C (Figure 6–68B, C), in the noxious range that produces heat pain. A similar expression cloning strategy led to the identification of another TRP channel, **TRPM8**, which is activated by menthol (known to evoke a cool sensation) and a wide range of cool to cold temperatures (<26°C). When expressed in the *Xenopus* oocyte, TRPV1 and TRPM8 were activated by hot and cool temperature, respectively (Figure 6–68C; **Movie 6–4**). These TRP channels are expressed in largely non-overlapping populations of DRG neurons, suggesting that TRPV1- and TRPM8-expressing DRG neurons constitute two independent pathways to deliver information regarding noxious heat and cool/cold temperature sensation to the CNS.

This hypothesis has been supported by physiological and behavioral defects in knockout mice. For example, compared with wild type, the fraction of DRG neurons that can be activated with a threshold of ~45°C (a temperature that produces heat pain) is markedly reduced in *Trpv1* mutant mice but was unaffected in the *Trpm8* mutant mice. On the other hand, the percentage of DRG neurons

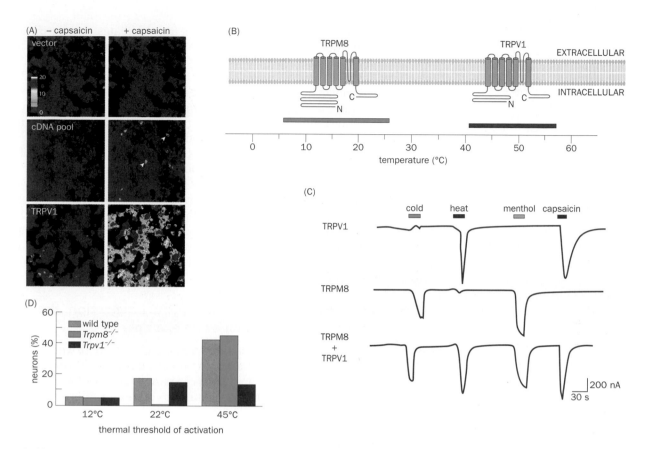

Figure 6–68 TRP channels sense heat, cold, and chemicals.
(A) Discovery of TRPV1 channel by expression cloning. Top row, control HEK293 cells did not respond to 3 μM capsaicin application; the relative intracellular calcium concentration ($[Ca^{2+}]_i$) is indicated in the heat map on the left. Middle row, a small fraction of cells (arrowheads) transfected with a pool of cDNAs derived from dorsal root ganglia responded to capsaicin with a rise of $[Ca^{2+}]_i$, as measured by the Fura-2 Ca^{2+} indicator (see Section 13.22). Bottom row, after repeatedly dividing the cDNA pool into smaller and smaller subdivisions, a single cDNA from the pool, encoding TRPV1, was identified that produced capsaicin-induced $[Ca^{2+}]_i$ rise. **(B)** Schematic of TRPV1 and TRPM8 cation channels and the temperature ranges that activate these channels (blue for TRPM8, red for TRPV1). Other temperature ranges are covered by other TRP or Cl⁻ channels.
(C) Response of *Xenopus* oocytes injected with mRNA encoding TRPV1, TRPM8, or both to cold, heat, menthol, and capsaicin.

As measured by the production of inward current in a voltage-clamp setting, TRPV1 is activated by heat and capsaicin, whereas TRPM8 is activated by cold temperature and menthol. **(D)** Activation of trigeminal ganglion neurons in wild-type, *Trpv1* mutant (*Trpv1⁻/⁻*), or *Trpm8* mutant (*Trpm8⁻/⁻*) mice was measured by Ca^{2+} imaging in response to temperature gradients. *Trpv1* mutant mice exhibited a marked reduction in the fraction of neurons that were activated with a threshold of 45°C, whereas neurons with an activation threshold of 22°C were absent in *Trpm8* mutant mice. (A, From Caterina MJ, Schumacher MA, Tominaga M et al. [1997] *Nature* 389:816–824. With permission from Macmillan Publishers Ltd; B & C, adapted from McKemy DD, Neuhausser WM & Julius D [2002] *Nature* 416:52–58. With permission from Macmillan Publishers Ltd; D, adapted from Bautista DM, Siemens J, Glazer JM et al. [2007] *Nature* 448:204–208. With permission from Macmillan Publishers Ltd.)

that can be activated with a threshold of ~22°C (an innocuous cool temperature) was eliminated in *Trpm8* mutant but was unaffected in *Trpv1* mutant mice. However, a sizable fraction of DRG neurons were still activated by noxious heat in the *Trpv1* mutant, and a small population of neurons with an activation threshold of ~12°C (a temperature that produces cold pain) was unaffected in *Trpm8* mutant (Figure 6–68D). These data indicate the existence of additional sensors for noxious heat and cold temperatures. Several additional hot or cold temperature-activated TRP channels have been reported. In addition, a Ca^{2+}-activated Cl⁻ channel is activated by high temperature and contributes to noxious heat sensation *in vivo*. Thus, temperature sensing may engage a variety of receptors that act in parallel. Some heat-activated channels are co-expressed with TRPV1 in the same DRG neurons; this explains why behavior in response to sensing noxious heat is more strongly affected by depleting TRPV1-expressing neurons than by deleting the *Trpv1* gene.

Additional TRP channels have been identified that sense chemicals. For instance, TRPA1 is a receptor for pungent components of mustard oil and garlic, and for environmental irritants such as acrolein. TPRA1 can also act as an intracellular-signal-activated ion channel that mediates depolarization in response to activation of other receptors by sensory stimuli (analogous to the TRPC2 channel in the accessory olfactory system, see Box 6–1). Expression of TRPA1 overlaps with the expression of TRPV1 in DRG neurons. Thus, DRG neurons can be polymodal by at least two distinct mechanisms: first, they can express a receptor (such as TRPV1) that senses stimuli in more than one modality; second, they can express more than one kind of receptor, each of which responds to sensory stimuli belonging to one or more modalities.

In summary, studies of the TRP channels have unified the sensation of temperature, pain, and chemicals. Heat and cool/cold temperatures appear to be sensed mainly by two separate types of DRG neurons that express TRPV1 and TRPM8, respectively. However, we still have an incomplete understanding of temperature sensation across all ranges and for the mechanistic basis for the distinction of innocuous and noxious temperatures. For example, there are to date no identified sensory neurons dedicated to sensing warmth; it is possible that warm sensation is synthesized from the integrated activity of heat- and cool-sensing neurons in the central nervous system, a topic we turn to next.

6.32 Sensation can be a product of central integration: the distinction of itch and pain as an example

Given the numerous types of distinct sensory stimuli, one simple way of organizing the somatosensory system is to assign a dedicated type of sensory neuron to each type of sensory stimulus; the activation of a particular type of sensory neuron informs the brain that a particular sensory stimulus is present. This type of organization resembles how tastes are sensed in mammals (see Section 6.21), odors are sensed in *C. elegans* (see Section 6.11), or some odors that have innate behavioral significance are sensed in *Drosophila* (see Section 6.15). Our discussion so far has suggested that the somatosensory system employs this organization. An alternative strategy of sensory processing is to compare signals from multiple types of sensory neurons in order to extract specific information, as exemplified by color perception (see Sections 4.10 and 4.19). The somatosensory system also employs this integrative strategy, as is exemplified by the studies of itch.

Itch is defined as an unpleasant sensation that elicits the desire or reflex to scratch. It is usually induced by **pruritogens**, which include endogenously released chemicals such as histamine, environmentally produced chemicals such as spicules from the tropical legume *Mucuna pruiens*, or drugs such as chloroquine, which is widely used to treat malaria. Itch has long been considered a sub-modality of pain, and indeed the histamine receptors responsible for histamine-induced itch are expressed mostly in a subset of TRPV1-expressing neurons. However, the following experiments suggest that itch is distinct from pain.

Members of the Mrgpr (Mas-related G protein-coupled receptors) family of G protein-coupled receptors (GPCRs) are expressed in a subset of DRG neurons. A particular member, **MrgprA3**, has been found to be a receptor for chloroquine and mediates chloroquine-induced itch. MrgprA3 is expressed in about 4% of DRG neurons, which co-express both histamine receptor and TRPV1. Indeed, *in vivo* recording indicated that MrgprA3-expressing neurons are highly polymodal—they are activated not only by application of chloroquine or histamine, but also by capsaicin and even by noxious mechanical stimuli. However, ablation of MrgprA3 neurons selectively reduced itch behavior induced by chloroquine and histamine without affecting behavior elicited by noxious heat or mechanical stimuli. One interpretation could be that MrgprA3-expressing neurons normally contribute to both itch and pain sensation; in their absence, other neurons can compensate for the pain sensation but not as well for the itch sensation. To further investigate this question, genetically engineered mice were produced in which *TRPV1* was deleted in the entire animal, but was added back only in MrgprA3 neurons (**Figure 6–69**A). These mice were then given capsaicin through cheek injection and assayed for behavior. In this paradigm, mice elicit distinct, stereotypical responses to painful or itchy stimuli: they wipe the cheek with forepaw in response to pain but scratch the cheek with hindpaw in response to itch. In accord with the

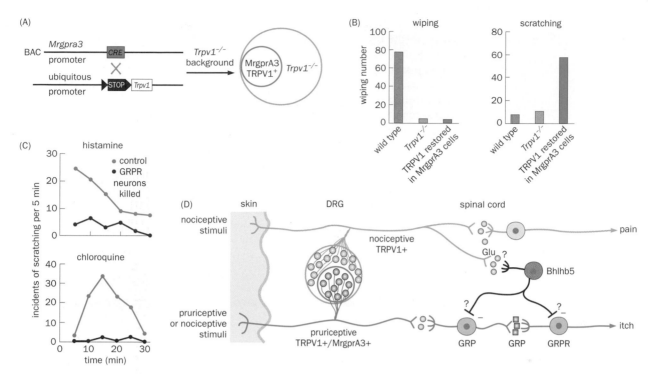

Figure 6–69 The itch-sensing pathway. (A) Schematic of expressing TRPV1 only in MrgprA3-expressing neurons. Transgenic mice expressing Cre recombinase under the control of *Mrgpra3* regulatory region on a bacterial artificial chromosome (BAC) vector (which is integrated into the mouse genome) were crossed to mice that carry a second transgene in which TRPV1 expression is conditional upon Cre/*loxP* (triangles)-mediated excision of the stop sequence between the ubiquitous promoter and the *Trpv1* coding sequence (see Section 13.10 for details of the expression strategy). In addition, these mice were homozygous mutant for the endogenous *Trpv1* gene. Thus, only MrgprA3 neurons express TRPV1 as indicated by the Venn diagram on the right. In the Venn diagram, the green circle represents neurons that normally express TRPV1 but were devoid of TRPV1 expression because of the *Trpv1* mutant background. The blue circle represents neurons that normally express MrgprA3 (a subset of TRPV1-expressing neurons), but expressed TRPV1 because of action of the two transgenes on the left. **(B)** In response to cheek injection of capsaicin, wild-type controls exhibited a robust wiping response but minimal scratching, consistent with a pain response; *Trpv1* mutant mice displayed minimal response; mice in which TRPV1 was only expressed in MrgprA3 neurons displayed minimal wiping but a robust scratching response, suggesting that activation of MrgprA3 neurons produced

the itch sensation. **(C)** Compared with controls, mice in which gastrin-releasing peptide receptor (GRPR)-expressing neurons in the spinal cord were ablated did not respond to any tested pruritogens, including histamine (top) and chloroquine (bottom). These experiments suggested that GRPR neurons are necessary for pruriception. **(D)** A model for the relationship between pain and itch sensation. Pruriceptive stimuli such as chloroquine activate MrgprA3-expressing neurons, which activates GRPR neurons either directly (not shown) or indirectly through spinal cord gastrin-releasing peptide (GRP)-releasing neurons. Activation of GRPR neurons produces itch sensation. Nociceptive stimuli activate both MrgprA3-expressing neurons that co-express TRPV1 (blue) and other TRPV1-expression neurons that are nociceptive (green). Nociceptive TRPV1 neurons may activate spinal cord inhibitory neurons that suppress the pruriceptive pathway. Bhlhb5-expressing neurons are candidate inhibitory neurons because their loss causes excessive scratching in mutant mice (? indicates that these connections remain hypothetical). (A & B, adapted from Han L, Ma C, Liu Q et al. [2013] *Nat Neurosci* 16:174–182. With permission from Macmillan Publishers Ltd; C, after Sun Y, Zhao Z, Meng X et al. [2009] *Science* 325:1531–1534; D, adapted from Han L & Dong X [2014] *Annu Rev Biophys* 43:331–355. With permission from Annual Reviews.)

notion that capsaicin activates nociceptive neurons, cheek injection of capsaicin elicited a strong wiping response but minimal scratching response in wild-type mice. *Trpv1* mutant mice produced a minimal wiping or scratching response. Mice that expressed TRPV1 only in MrgprA3 neurons, however, produced a robust scratching response and minimal wiping response (Figure 6–69B), suggesting that activation of MrgprA3 neurons produces itch but not pain sensation.

Further insights into itch sensation came from studies of a group of spinal cord neurons that express the **gastrin-releasing peptide receptor** (**GRPR**), a G-protein-coupled receptor that is activated by gastrin-releasing peptide. Pharmacological ablation of GRPR neurons selectively eliminated scratch responses to all tested pruritic agents including histamine and chloroquine (Figure 6–69C) without affecting pain behavior. MrgprA3 neurons can send signals either directly or indirectly to GRPR neurons in the spinal cord. Thus, a model that can account for the experimental data discussed thus far is that pruriceptive neurons, such as those that express MrgprA3, are activated by both itchy and painful stimuli. Painful stimuli also activate other nociceptive neurons, which suppress the activity of GRPR neurons through inhibition within the spinal cord, therefore producing pain but not itch. Itchy stimuli only activate pruriceptive neurons, which in turn activate GRPR neurons dedicated for itch sensation (Figure 6–69D). In support of this model, selective loss of a subset of inhibitory interneurons that express a transcription factor Bhlhb5 exhibited profoundly elevated spontaneous scratching behavior, suggesting a role of these neurons in inhibiting itch. Further investigations of the synaptic interactions of these neuronal types are required to validate this model. Activity of mechanosensory neurons may also send an inhibitory signal to the itch pathway, which can explain why scratching can temporarily relieve itch.

It is likely that the strategy employed in distinguishing pain and itch is widely used for other sensory modalities in the somatosensory system, such as thermosensation and touch. Integration of the activity of different neuron types can produce the many types of sensation we derive from our somatosensory system. These integrations can occur in the spinal cord and along the ascending pathway, which we now turn to.

6.33 Touch and pain signals are transmitted by parallel pathways to the brain

How does the central nervous system make sense of the activity of different types of sensory neurons? The answer partly comes from the fact that different types of sensory neurons in the dorsal root ganglia terminate in specific locations within the spinal cord (see Figure 6–63, right), and synapse with distinct types of post-synaptic neurons. The gray matter of the spinal cord is traditionally divided into 10 different laminae along the dorsal–ventral axis. Laminae I–V are located in the **dorsal horn** (the dorsal part of the spinal gray matter), which is devoted to processing somatosensory information. Laminae I–II are predominantly targeted by axons of nociceptive and thermosensory neurons, which mostly use unmyelinated C fibers and some lightly myelinated Aδ fibers. Some LTMRs with C and Aδ fibers also terminate in layer II at sites ventral to the terminations of nociceptive fibers. Laminae III–V are targeted mostly by LTMRs with Aδ and Aβ fibers. Proprioceptive sensory neurons send their axons further into the middle of the spinal cord and the **ventral horn** (the ventral part of the spinal gray matter where motor neurons reside); some proprioceptive neurons form synapses directly with motor neurons to form reflex circuits (see Figure 1–19).

Within each lamina, further connection specificity must exist between the input axons from different sensory neuron subtypes and their spinal cord target neurons. Such connection specificity enables sensory information to be processed and integrated in the dorsal horn, and transmitted by specific projection neurons to the brain. Although a general outline is understood (see Figure 6–63), little is known about specific cell types and connection patterns. Recent efforts to identify specific cell types of both sensory and spinal cord neurons using molecular-genetic approaches can significantly advance our understanding. For instance, genetic labeling of different types of mouse sensory neurons in the same animal has revealed clear separation of the spinal cord terminations of those neurons (**Figure 6–70**A). Genetic manipulations of these specific types of neurons, such as selective ablation or altering receptor expression (for example, see Figure 6–69), can help dissect the function of these neurons in sensory perception.

As well as terminating in specific spinal cord laminae, Aβ-LTMRs (and proprioceptive neurons; see Figure 6–63) also send an ascending branch through the **dorsal column pathway** of the spinal cord to the **medulla** (the most caudal part of the brainstem); this constitutes the direct dorsal column pathway. In parallel, LTMR terminals in the spinal cord also synapse with **dorsal horn projection neurons** (that is, neurons located in the dorsal horn of the spinal cord that project axons to the brain), which send ascending axons along the dorsal column pathway that terminate in the same medulla area as axons in the direct pathway; this constitutes the indirect dorsal column pathway. Since C and Aδ fibers do not have ascending branches, the indirect pathway is a major carrier of

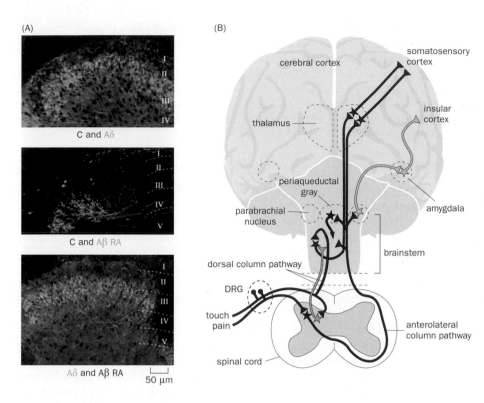

Figure 6–70 Spinal cord endings of sensory neurons and central pathways for touch and pain. (A) In this example, central projections of LTMRs belonging to a subset of C, Aδ, or Aβ classes are labeled with specific genetic markers in transgenic mice. Each image is a dorsal horn spinal cord section, where axon terminations from the two LTMR classes named beneath the panel are visualized in the same animal, with one class in green and the other in red. (The Aδ channel has relatively high background; only intense green color represents signal.) Roman letters indicate spinal cord laminae. C fibers terminate in lamina II, Aδ fibers terminate mostly in lower part of lamina II and lamina III, whereas the rapidly adapting (RA) Aβ fibers terminate in layers III, IV, and V. **(B)** Simplified schematic drawing of major central pathways for touch (red) and pain (blue), where each neuron represents one or a few subtypes of neurons. The touch sensory neuron (Aβ-LTMR) sends a branch that ascends to the brainstem within the dorsal column pathway (the direct pathway). It also synapses with a dorsal horn neuron (orange), which sends an ascending branch in the dorsal column pathway (the indirect pathway). Neurons in both the direct and indirect pathways synapse with a medulla neuron that projects its axon across the midline to terminate at the contralateral thalamus. (Not shown is the spinocervical tract pathway that transmits touch information to the contralateral thalamus in parallel with the dorsal column pathway.) The nociceptive sensory neuron synapses with a second-order dorsal horn spinal neuron, which sends its axon across the midline of the spinal cord and ascends within the anterolateral column pathway to the contralateral thalamus. Distinct thalamic neurons relay the touch and pain information to the primary somatosensory cortex. Some second-order pain pathway neurons terminate at different brainstem nuclei (blue branches). At the parabrachial nucleus, pain information is relayed by a projection neuron to the amygdala, and subsequently to the insular cortex for affective aspects of pain sensation (the cyan pathway). At the periaqueductal gray, target neurons of ascending anterolateral column pathway initiate descending modulation of pain through the autonomic and neuromodulatory systems (purple). (A, from Li L, Rutlin M, Abraira VE et al. [2011] *Cell* 147:1615. With permission from Elsevier Inc.; B, based on Abraira VE & Ginty DD [2013] *Neuron* 79:618–639; Basbaum AI, Bautista DM, Scherrer G et al. [2009] *Cell* 139:267–284.)

ascending touch information originating from C-LTMRs and Aδ-LTMRs. Axons of both direct and indirect dorsal column pathways synapse with medulla neurons that project to the contralateral thalamus. Thalamic neurons then relay these touch signals to the primary somatosensory cortex in a topographic manner to form the sensory homunculus (Figure 6–70B, red pathway; see also Figures 1–20 and 1–25). In addition to the dorsal column pathway, which transmits information from both glabrous and hairy skin, touch information from hairy skin is also relayed by distinct dorsal horn projection neurons with ascending axons that form the **spinocervical tract pathway** and terminate in the lateral cervical nucleus. Projection neurons from the lateral cervical nucleus then relay information to the contralateral thalamus.

Pain, itch, and temperature sensations are mostly conveyed through a distinct ascending path: lamina I dorsal horn projection neurons receive input from sensory axons, send their own axons across the ventral spinal cord, and ascend in the contralateral **anterolateral column pathway**. Some axons in the anterolateral column terminate in thalamic nuclei, through which information is relayed to the primary somatosensory cortex (Figure 6–70B, blue pathway). This pathway provides information about the location of pain, itch, and temperature sensation. Other axons in the anterolateral column pathway innervate brainstem target neurons. One target is the **parabrachial nucleus**, which relays pain information to the amygdala and hypothalamus to control the autonomic nervous system response to pain stimuli (see Chapter 8 for more details on the autonomic nervous system). From the amygdala, pain information is also sent to the insular cortex (Figure 6–70B, cyan). This pathway is responsible for the affective aspects of pain perception. Another major brainstem target for the anterolateral pathway is the **periaqueductal gray** in the midbrain, which initiates the descending modulation of pain (Figure 6–70B, purple) that will be discussed in the next section.

Thus, the organization of central pathways that deliver touch and pain information to the brain is somewhat analogous to the M and P pathways in the visual system that respectively transmit motion and high-acuity/color information from the retina to the visual cortex (see Section 4.27). Just as the two-pathway system in vision is a gross simplification given the presence of many types of retinal ganglion cells, so too is this simple division of the touch and pain pathways a gross simplification in the somatosensory system. We've already seen that touch information is transmitted by several parallel pathways (the direct and indirect dorsal column pathways, and the spinocervical tract pathway). In fact, in addition to the major contribution by lamina I neurons, the anterolateral column pathway also contains the axons of projection neurons in deep laminae that might also transmit touch information. Ongoing efforts to identify specific projection neuron types will enable precise manipulation of their activity patterns using modern circuit analysis tools (see Sections 13.12 and 13.23–13.25), thereby helping to dissect the ascending pathways and their functions in transmitting rich somatosensory information to the brain.

6.34 Pain is subjected to peripheral and central modulation

In our studies of sensory systems, we have placed considerable emphasis on information flow from sensory organs in the periphery to higher-order neurons in the brain. However, information can also flow from higher-order neurons back to lower-order neurons for feedback modulation of the sensory system, or from sensory neurons to their peripheral endings. Sensory systems are also powerfully modulated by brain states (see Figure 1–26B); for example, the sensitivity of our sensory systems drops markedly when we are drowsy or asleep. The study of pain has offered insight about these modulations, which also have important clinical relevance.

Pain modulation begins as early as the peripheral endings of sensory neurons. Inflammation is responsible for much of the pain after tissue damage or pathogen infection. The inflammatory response, while important for repairing damaged tissues, also produces molecules such as **bradykinin** (a peptide) and **prostaglandin** (a lipid), which bind to specific GPCRs on the sensory endings of nociceptive neurons (**Figure 6–71**A). These compounds can activate nociceptive neurons through second messenger systems to open TRP channels; in this example, TRP channels serve as effectors for other sensory receptors. Inflammatory molecules can also sensitize nociceptive neurons by decreasing the threshold of TRP channel opening in response to sensory stimuli; in this instance, the TRP channels themselves serve as sensory receptors. In fact, nociceptive neurons can secrete neuropeptides such as **substance P** and calcitonin gene-related peptide (**CGRP**) from the peripheral endings, which act on local cells to produce bradykinin and prostaglandin that in turn activate or sensitize the nociceptive neurons (Figure 6–71A). This phenomenon, called **neurogenic inflammation**, underscores the bidirectional signaling of nociceptive neurons: in addition to transmitting pain signals from the periphery to the CNS, their peripheral ends can also release neuropeptides in response to their own activity, which contribute to inflammation. Under inflammatory conditions, exposure of injured tissue to gentle touch or innocuous temperature can cause pain, a phenomenon called **allodynia**. Some of the most widely used analgesic (pain-relieving) drugs, such as aspirin and ibuprofen, target Cox-1 and Cox-2, the cyclooxygenase enzymes that

Figure 6–71 Peripheral and central modulation of pain. (A) Inflammation induces the release of molecules such as bradykinin and prostaglandin that bind to their respective G-protein-coupled receptors (blue) at the peripheral endings of nociceptive neurons. This triggers various second messenger systems that activate TRPV1 and TRPA1 channels (green), or sensitize these channels to reduce their activation threshold, causing hypersensitivity to pain. Second messenger systems can regulate other local effectors and gene expression, resulting in long-term change of pain sensitivity. Sensory nerve endings can also release neuropeptides such as substance P and calcitonin gene-related peptide (CGRP), which can act on local cells to trigger additional release of bradykinin and prostaglandin. **(B)** At the dorsal horn of the spinal cord, synaptic transmission between the nociceptive sensory neuron and the dorsal horn projection neuron is regulated by local neurons that release endogenous opioid peptides, which act through opioid receptors (cyan dots) to inhibit both the release of glutamate from sensory neuron terminals and the depolarization of postsynaptic dorsal horn projection neurons in response to glutamate release. Opioid-releasing neurons in turn are modulated by descending axons from brainstem serotonin and norepinephrine neurons. Serotonin and norepinephrine neurons can also directly modulate the activity of dorsal horn projection neurons. + indicates excitation; − indicates inhibition. Differential responses to the same neurotransmitters are caused by expression of different receptors that couple to different second messenger systems in the recipient neurons.

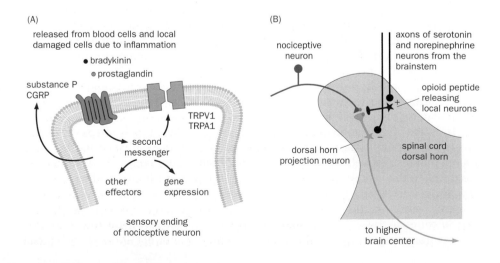

synthesize prostaglandin, thereby reducing inflammation-induced activation and sensitization of nociceptive neurons at the periphery.

Our understanding of pain modulation in the central pathway came from studies of mechanisms of another class of powerful analgesic drugs, the **opioids**. The analgesic effect of opiates, natural opioids harvested from the opium poppy, *Papaver somniferum*, has been known for thousands of years. Although the active ingredient, **morphine**, was isolated in 1804, the mechanism of opiate action was not known until the 1970s, when radioactively labeled opiates were found to bind to specific receptors in the nervous system. We now know that **opioid receptors** constitute a subfamily of GPCRs that are distributed widely across the nervous system and are highly concentrated in laminae I-II of the dorsal horn, both in the presynaptic terminals of sensory neurons and in the postsynaptic dorsal horn neurons that relay pain signals to the brain. Activation of opioid receptors reduces neurotransmitter release from sensory neurons, decreases the responsiveness of dorsal horn neurons, or both (Figure 6–71B), explaining the opioids' analgesic effects. The discovery of opioid receptors also triggered the search for their endogenous ligands. Several peptides including enkephalin, endorphin, and dynorphin have been identified that modulate the pain pathway just as exogenous opioids do (Figure 6–71B).

Application of opioids to the brain can also relieve pain—indeed the most effective regions subject to opioid-mediated analgesic effect are specific brainstem nuclei where opioid receptors are highly expressed. It is thought that activation of opioid receptors activates neuromodulatory systems including norepinephrine and serotonin neurons, which send descending projections into the spinal cord. Serotonin and norepinephrine neurons regulate the release of local opioid peptides and modulate sensitivity of dorsal horn neurons to input from nociceptive sensory neurons (Figure 6–71B). These modulations by endogenous opioid peptides, serotonin, and norepinephrine may underlie the **placebo effect**, in which the suggestion of pain medication without actual medication can effectively reduce the pain experienced by some patients. The detailed circuit and cellular mechanisms of pain modulation still require future investigations, which should have large payoffs both in understanding how sensory systems are modulated in general and in developing more effective and specific medicines to alleviate pain.

6.35 Linking neuronal activity with touch perception: from sensory fiber to cortex

Sensory systems transform external stimuli into internal representations in the brain, which gives rise to perception. We have assumed that the activity of sensory neurons and neurons in the ascending pathways contribute to perception, but how do we actually know this? Experiments in the somatosensory system have contributed important insights toward answering this important question.

A powerful approach to establishing a link between the activity of specific neurons in the sensory system and perception is to compare the detection threshold of neurons to specific stimuli and the perceptual threshold of the organism to the very same stimuli. As was discussed in Chapter 4, the perceptual threshold can be determined by testing the psychometric functions of human subjects (see Figure 4–3) or trained monkeys (see Figure 4–52). Vernon Mountcastle and co-workers applied sinusoidal mechanical stimuli (flutter) to the fingertips of human and monkey subjects, and varied the stimulation frequency to determine the threshold amplitude at which these stimulations could be detected. Interestingly, the thresholds at which specific frequencies of flutter produced perception— plotted as the frequency–amplitude curves—were very similar for monkey and human subjects (**Figure 6–72**A). In parallel, electrophysiological recordings of single sensory axons (see Figure 1–17) were carried out to determine the detection threshold of individual sensory neurons in response to the same set of sensory stimuli applied to the fingertips of anesthetized monkeys. Two types of sensory neurons were activated depending on the frequency of mechanical stimuli; low- and high-frequency stimuli activate sensory neurons that terminate at Meissner and Pacinian corpuscles, respectively (see Figure 6–64). Remarkably, the lower

Figure 6–72 Comparing the threshold of perception with the thresholds of sensory neurons to the same stimuli. **(A)** Frequency–amplitude curves for perceptual threshold based on psychophysical experiments on six monkey (red) and five human (blue) subjects. Each point of the curve represents the lowest amplitude of sinusoidal mechanical stimuli of a given frequency applied to the fingertip that was perceived by human and monkey subjects. The human subjects would answer whether they felt the stimulus or not, whereas the monkeys had been trained to indicate the presence of a stimulus applied to the fingertip of a restrained hand by using the other hand to press a button. The remarkable similarities indicate that similar physiological mechanisms underlie the flutter perception in both monkeys and humans. **(B)** Superimposed on the perceptual threshold curve (red, average from data in panel A, error bars indicate standard error of the mean) are thresholds of two types of sensory neurons recorded from individual Aβ fibers innervating the fingers in response to the same sinusoidal mechanical stimuli as in the psychophysical experiments in panel A. The blue and green curves represent the lowest bound, or the most sensitive of each type of sensory neurons, at a given frequency. Comparing the curves for both neuron types and the perceptual threshold curve across a range of frequencies, we can see that at each frequency the perceptual threshold matches well with the lower of the two neuronal thresholds, suggesting that the limits of the psychophysical performance are set by the most sensitive individual neurons. (Adapted from Mountcastle VB, LaMotte RH & Carli G [1972] *J Neurophysiol* 35:122–136.)

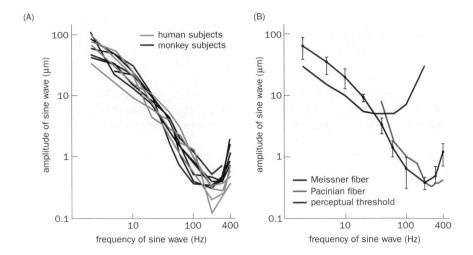

bound of the detection threshold of each type of sensory neurons matched very well with the perceptual threshold of the monkeys in the respective frequency ranges (Figure 6–72B). These experiments suggested that these two types of sensory neurons are responsible for the flutter perception across different frequency ranges, and provided a striking example for the **lower envelope principle**: the limits of psychophysical performance are set by the most sensitive individual neurons. Similar observations were made in subsequent **microneurography** experiments in waking human subjects, where individual mechanosensory axons were recorded while the subject attempted to detect brief indentation of the skin at the most sensitive point of the fiber's receptive field. The threshold of neuronal response matches with the perceptual threshold, suggesting that the recorded neural activity in the sensory fiber carries the information used for the perception of the skin indentation.

Similar approaches that compare the threshold of stimulus perception with the threshold of neuronal firing that can be influenced by the same stimulus have also been applied in the somatosensory cortex. As introduced in Chapter 1, the somatosensory cortex has a gross topographic map of the body surface (see Figure 1–25). Within a particular area of the somatosensory cortex, neurons that represent the same kinds of information—for instance rapidly adapting neurons that detect vibratory stimuli in a fingertip—are organized in vertical columns, a discovery made by Mountcastle that inspired similar findings in the primary visual cortex (see Section 4.25). To explore the relationship between neuronal activity and perception, thirsty monkeys were trained to perform a behavioral task in which they must indicate correctly the presence or absence of a stimulus in order to receive a drop of liquid as a reward (**Figure 6–73**A, B). At the same time, single-unit recordings were performed in specific areas of the somatosensory, premotor, and motor cortices (Figure 6–73C). It was found that rapidly adapting neurons in the primary somatosensory cortical areas (corresponding to area 1/3b of Brodmann's map; see Section 13.15 for details) that represent the fingertip indeed had a detection threshold that matched the perceptual threshold, suggesting that these neurons carry the information used for perception as do the sensory fibers in the periphery.

By definition, neuronal response to near-threshold stimulus is variable (on average, a neuron responds 50% of the time to a stimulus at the threshold). Likewise, perceptual response is also variable (on average, the monkey detects a stimulus at the perceptual threshold 50% of the time). Thus, co-variability between different trials provides valuable information about the relationship between neuronal activity and perception: how well can a neuron's response predict the monkey's perceptual response? It was found that neuronal activity in the primary somatosensory cortex could only predict behavioral outcome at chance level (Figure 6–73D), suggesting that while contributing essential information to perception, the primary somatosensory cortex is not the site where perceptual

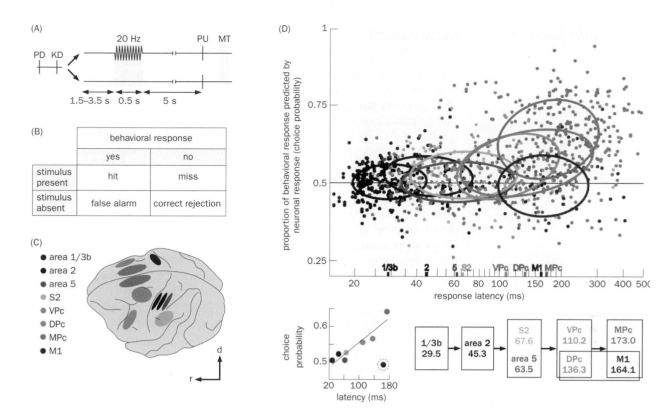

Figure 6–73 Neuronal activity and touch perception in the cortex.
(A) Behavioral task. The trial began when a stimulator probe indented the skin of one fingertip of the restrained right hand (probe down, PD). The monkey then placed its left hand on a key (key down, KD). After a variable interval, on half of the randomly selected trials, a 20-Hz vibratory stimulus near the detection threshold was applied. After a fixed interval, the probe was moved up (PU), signaling to the monkey that it could move its left hand to target (MT), which is one of the two push buttons to indicate whether it felt the stimulus (yes) or not (no). **(B)** The trial outcome is classified into one of the four categories depending on whether the stimulus is present or not and whether the behavioral response is "yes" or "no." During training, the monkey would receive a reward for trials that resulted in a hit or correct rejection, but not for trials that resulted in a miss or false alarm. **(C)** While the monkey was performing the behavioral task, single-unit recordings were performed in different areas of the primary somatosensory cortex (areas 1/3b, 2, 5), secondary somatosensory cortex (S2), ventral, dorsal, and medial premotor cortex (VPc, DPc, MPc), as well as the primary motor cortex (M1).

These areas are indicated by different colors in a schematic monkey brain (r, rostral; d, dorsal). **(D)** Top, summary of response latency (x axis) and perceptual signals (y axis) across cortical areas. Each dot represents a single unit recorded in a particular cortical area color matched with the schematic in panel C. The y-axis value represents a choice-probability index, which is an indicator of the accuracy with which neuronal response predicts behavioral response, with 0.5 being at chance level and 1 being perfect prediction. Neurons in each area were fitted with two-dimensional normal distributions with ovals as the standard deviations. Color marking at the x axis represents the mean response latency of neurons in each cortical area. Bottom left, response latency positively correlates with proportion of predicted behavioral response in all cortical areas except M1 (dotted circle). Bottom right, proposed information flow across cortical areas based on mean latencies (numbers underneath the cortical area, in milliseconds). Each rectangle groups the areas with latencies that are statistically indistinguishable from each other. (Adapted from de Lafuente V & Romo R [2006] *Proc Natl Acad Sci USA* 39:14266–14271. Copyright National Academy of Sciences, USA.)

decision is made. By contrast, activity of neurons in the medial premotor cortex (MPc) can predict behavioral response well above chance. Indeed, there is a gradual increase in the ability for neuronal activity along a pathway from somatosensory cortex in the parietal lobe to premotor cortex in the frontal lobe to predict behavioral response, correlating with increasing latency of their firing to stimulus onset (Figure 6–73D). Control experiments suggested that activity in premotor cortical neurons did not simply represent motor preparation; indeed, neuronal activity in the primary motor cortex (M1) did not predict behavioral outcome.

In summary, these experiments suggest that perception develops over time and across cortical areas, from representing sensory stimulus in the primary somatosensory cortex to representing a perceptual report in the premotor cortex. A fascinating future direction is to investigate the mechanisms by which such representation is achieved through the activity and connections of cortical neurons. These investigations will shed light on a central question in neurobiology: how do sensory stimuli produce perceptions, which in turn influence actions?

SUMMARY

All sensory systems share common tasks. They transform environmental (and bodily) stimuli into electrical signals, and transmit these signals to the brain to form internal representations of the sensory stimuli. However, the physical nature of the sensory stimuli ranges widely from chemical, mechanical, thermal, to light. This necessitates the evolution of multiple sensory systems with distinct properties best suited for detecting and extracting features from different types of stimuli.

The olfactory system detects numerous volatile chemicals in the environment to provide animals with information about food, mates, and danger from a distance. The mammalian olfactory system employs many hundreds of G-protein-coupled receptors (GPCRs) for odorant detection. The binding of odorants to their receptors activates a cyclic nucleotide-gated cation channel, leading to depolarization of olfactory receptor neurons (ORNs) in the nose and propagation of action potentials by ORN axons to the olfactory bulb. Each odorant activates multiple receptors, and each receptor is activated by multiple odorants. This evolutionarily conserved combinatorial coding strategy enables the olfactory systems to distinguish many more odorants than the odorant receptors they possess.

From insects to mammals, most individual ORNs express a single odorant receptor. ORNs expressing the same receptor project their axons to the same glomerulus, and synapse with second-order projection neurons that send dendrites to single glomeruli. This creates parallel and discrete information processing channels. Through convergent ORN axon projections, synaptic properties, and lateral inhibition by local interneurons in the insect antennal lobe (and very likely in the vertebrate olfactory bulb as well), ORN input is transformed for more reliable and efficient representation by second-order projection neurons. In parallel with the use of a combinatorial coding strategy, odors with innate behavioral significance, such as danger signals or mating pheromones, utilize dedicated information processing channels in flies and a specialized accessory olfactory system in most mammals. Studies in both mice and flies suggest that different higher olfactory centers have different input organization: an organized and stereotyped set regulating innate behavior, and a variable set representing individual experience. Due to its limited number of neurons, the *C. elegans* olfactory system uses a different coding strategy: many odorant receptors are expressed in the same sensory neurons, and these neurons encode the hedonic values of odorants to guide behavior.

The primary job of the taste system, a chemical sense in parallel to olfaction, is to detect nutrient levels in food and avoid harmful substances. It operates at a closer range than does olfaction. Each of the five different kinds of tastants in mammals—sweet, umami, bitter, sour, and salty—is mostly detected by a distinct kind of taste receptor cell. Sweet and umami tastants, which inform animals about nutrient levels, are each detected by a heterodimeric partner of two GPCRs of the T1R family with low-binding affinity. About 30 GPCRs of the T2R family are used to detect bitter compounds, often with high affinity, enabling animals to avoid trace amount of toxins. Separable areas in the taste cortex are differentially activated by different kinds of tastants. Taste and olfaction work together to give rise to perceived flavor.

Two parallel sensory systems in the inner ear convert mechanical stimuli into electrical signals, but differ in their functions. The vestibular system utilizes otolith organs and semicircular canals to sense head orientation and movement; these sensory signals are used, often in combination with signals from other sensory systems, to regulate balance and spatial orientation, to coordinate head and eye movement, and to perceive self-motion. The auditory system utilizes the cochlea to detect sounds for communication with conspecifics and for detecting predators and prey.

Temporal precision is a key feature of the auditory system. Sounds are rapidly converted to electrical signals by directly opening and closing mechanotransduction channels at the tip of stereocilia in the hair cells. Cyclic hair cell depolarization causes postsynaptic spiral ganglion neurons to fire in sync with specific phases of sound cycles up to several kilohertz. This temporal precision is instrumental for sound localization, which relies on sub-millisecond interaural time differences extracted by coincidence-detecting neurons in the brainstem. Along with interaural level differences (and other cues in mammals), animals construct spatial map in the inferior colliculus for sound locations. Different frequency sounds activate hair cells at different locations in the cochlea. Through ordered projections, this tonotopic map is maintained through the ascending auditory pathway, highlighting another important feature of the auditory system. Besides representing sound frequency, level, and location, the auditory cortex analyzes biologically important sounds. This is exemplified by specialized areas in the auditory cortex of echolocating bats dedicated to analyzing the time delay and frequency shift of echoes, allowing bats to identify the size, distance, and relative speed of their insect prey.

The somatosensory system employs many types of sensory neurons to detect diverse mechanical, thermal, and chemical stimuli, giving rise to proprioception, thermosensation, pruriception, nociception, and touch sensation. Touch is sensed by different types of low-threshold mechanoreceptors (LTMRs) that differ in their specialized terminal endings, adaptation properties, and degrees of myelination. Mechanotransduction channels identified to date include ENaC- and TRP-family channels in invertebrates and the multi-pass transmembrane proteins Piezos in mammals; additional mechanotransduction channels remain to be discovered. Temperature, chemicals, and pain are sensed mainly by unmyelinated C fibers and thinly myelinated Aδ fibers with free sensory endings, with TRP-family channels being the key receptors. TRPV1 and TRPM8 are the major sensors for heat and cool/cold temperatures, respectively. TRPV1 is also expressed in many nociceptive and pruriceptive multimodal neurons.

Somatosensory signals are transmitted to the CNS via ordered projections of specific types of somatosensory neurons to specific laminae of the spinal cord. Signal integration in the spinal cord contributes to sensory processing, as exemplified by the distinction between pain and itch. Touch and temperature/pain signals are transmitted mostly through parallel pathways to the brainstem, thalamus, and somatosensory cortex. The perception of pain is modulated at the sensory endings by molecules released during the inflammatory response, and at the spinal cord connections by endogenous opioids and central feedback pathways. Parallel physiological and psychophysical studies of touch have enriched our understanding of the relationship between neuronal activity and perception: the perceptual threshold is determined by the most sensitive sensory neurons at the periphery, and perceptual decision making develops over time and across cortical areas from the somatosensory to the premotor cortex.

FURTHER READING

Books and reviews

Abraira VE & Ginty DD (2013) The sensory neurons of touch. *Neuron* 79:618-639.

Axel R (1995) The molecular logic of smell. *Sci Am* 273:154-159.

Bargmann CI (2006) Comparative chemosensation from receptors to ecology. *Nature* 444:295-301.

Basbaum AI, Bautista DM, Scherrer G et al. (2009) Cellular and molecular mechanisms of pain. *Cell* 139:267-284.

Firestein S (2001) How the olfactory system makes sense of scents. *Nature* 413:211-218.

Konishi M (2003) Coding of auditory space. *Annu Rev Neurosci* 26:31-55.

Parker AJ & Newsome WT (1998) Sense and the single neuron: probing the physiology of perception. *Annu Rev Neurosci* 21:227-277.

Shepherd GM (2004) The Synaptic Organization of the Brain, 5th ed. Oxford University Press.

Su CY, Menuz K & Carlson JR (2009) Olfactory perception: receptors, cells, and circuits. *Cell* 139:45-59.

Suga N (1990) Biosonar and neural computation in bats. *Sci Am* 262:60-68.

Vollrath MA, Kwan KY & Corey DP (2007) The micromachinery of mechanotransduction in hair cells. *Annu Rev Neurosci* 30:339-365.

Yarmolinsky DA, Zuker CS & Ryba NJ (2009) Common sense about taste: from mammals to insects. *Cell* 139:234-244.

Vertebrate olfaction

Brunet LJ, Gold GH & Ngai J (1996) General anosmia caused by a targeted disruption of the mouse olfactory cyclic nucleotide-gated cation channel. *Neuron* 17:681-693.

Buck L & Axel R (1991) A novel multigene family may encode odorant receptors: a molecular basis for odor recognition. *Cell* 65:175-187.

Igarashi KM, Ieki N, An M et al. (2012) Parallel mitral and tufted cell pathways route distinct odor information to different targets in the olfactory cortex. *J Neurosci* 32:7970-7985.

Isogai Y, Si S, Pont-Lezica L et al. (2011) Molecular organization of vomeronasal chemoreception. *Nature* 478:241-245.

Keller A, Zhuang H, Chi Q et al. (2007) Genetic variation in a human odorant receptor alters odour perception. *Nature* 449:468-472.

Lyons DB, Allen WE, Goh T et al. (2013) An epigenetic trap stabilizes singular olfactory receptor expression. *Cell* 154:325-336.

Malnic B, Hirono J, Sato T et al. (1999) Combinatorial receptor codes for odors. *Cell* 96:713-723.

Miyamichi K, Amat F, Moussavi F et al. (2011) Cortical representations of olfactory input by trans-synaptic tracing. *Nature* 472:191-196.

Mombaerts P, Wang F, Dulac C et al. (1996) Visualizing an olfactory sensory map. *Cell* 87:675-686.

Papes F, Logan DW & Stowers L (2010) The vomeronasal organ mediates interspecies defensive behaviors through detection of protein pheromone homologs. *Cell* 141:692-703.

Rall W, Shepherd GM, Reese TS et al. (1966) Dendrodendritic synaptic pathway for inhibition in the olfactory bulb. *Exp Neurol* 14:44-56.

Rubin BD & Katz LC (1999) Optical imaging of odorant representations in the mammalian olfactory bulb. *Neuron* 23:499-511.

Scholz AT, Horrall RM, Cooper JC et al. (1976) Imprinting to chemical cues: the basis for home stream selection in salmon. *Science* 192:1247-1249.

Stettler DD & Axel R (2009) Representations of odor in the piriform cortex. *Neuron* 63:854-864.

Invertebrate olfaction

Bhandawat V, Olsen SR, Gouwens NW et al. (2007) Sensory processing in the *Drosophila* antennal lobe increases reliability and separability of ensemble odor representations. *Nat Neurosci* 10:1474-1482.

Caron SJ, Ruta V, Abbott LF et al. (2013) Random convergence of olfactory inputs in the *Drosophila* mushroom body. *Nature* 497:113-117.

Chalasani SH, Chronis N, Tsunozaki M et al. (2007) Dissecting a circuit for olfactory behaviour in *Caenorhabditis elegans*. *Nature* 450:63-70.

Hallem EA & Carlson JR (2006) Coding of odors by a receptor repertoire. *Cell* 125:143-160.

Jefferis GS, Potter CJ, Chan AM et al. (2007) Comprehensive maps of *Drosophila* higher olfactory centers: spatially segregated fruit and pheromone representation. *Cell* 128:1187-1203.

Kurtovic A, Widmer A & Dickson BJ (2007) A single class of olfactory neurons mediates behavioural responses to a *Drosophila* sex pheromone. *Nature* 446:542-546.

Olsen SR, Bhandawat V & Wilson RI (2010) Divisive normalization in olfactory population codes. *Neuron* 66:287-299.

Suh GS, Ben-Tabou de Leon S, Tanimoto H et al. (2007) Light activation of an innate olfactory avoidance response in *Drosophila*. *Curr Biol* 17:905-908.

Troemel ER, Kimmel BE & Bargmann CI (1997) Reprogramming chemotaxis responses: sensory neurons define olfactory preferences in *C. elegans*. *Cell* 91:161-169.

Taste

Blakeslee AF (1932) Genetics of sensory thresholds: taste for phenyl thio carbamide. *Proc Natl Acad Sci USA* 18:120-130.

Chandrashekar J, Mueller KL, Hoon MA et al. (2000) T2Rs function as bitter taste receptors. *Cell* 100:703-711.

Chen X, Gabitto M, Peng Y et al. (2011) A gustotopic map of taste qualities in the mammalian brain. *Science* 333:1262-1266.

Fox AL (1932) The relationship between chemical constitution and taste. *Proc Natl Acad Sci USA* 18:115-120.

Hoon MA, Adler E, Lindemeier J et al. (1999) Putative mammalian taste receptors: a class of taste-specific GPCRs with distinct topographic selectivity. *Cell* 96:541-551.

Mueller KL, Hoon MA, Erlenbach I et al. (2005) The receptors and coding logic for bitter taste. *Nature* 434:225-229.

Nelson G, Hoon MA, Chandrashekar J et al. (2001) Mammalian sweet taste receptors. *Cell* 106:381-390.

Audition

Carr CE & Konishi M (1990) A circuit for detection of interaural time differences in the brain stem of the barn owl. *J Neurosci* 10:3227-3246.

Chan DK & Hudspeth AJ (2005) Ca^{2+} current-driven nonlinear amplification by the mammalian cochlea *in vitro*. *Nat Neurosci* 8:149-155.

Corey DP & Hudspeth AJ (1979) Response latency of vertebrate hair cells. *Biophys J* 26:499-506.

Jeffress LA (1948) A place theory of sound localization. *J Comp Physiol Psychol* 41:35-39.

Kazmierczak P, Sakaguchi H, Tokita J et al. (2007) Cadherin 23 and protocadherin 15 interact to form tip-link filaments in sensory hair cells. *Nature* 449:87-91.

Knudsen EI & Konishi M (1978) A neural map of auditory space in the owl. *Science* 200:795-797.

O'Neill WE & Suga N (1979) Target range-sensitive neurons in the auditory cortex of the mustache bat. *Science* 203:69-73.

Pan B, Geleoc GS, Asai Y et al. (2013) TMC1 and TMC2 are components of the mechanotransduction channel in hair cells of the mammalian inner ear. *Neuron* 79:504-515.

Pena JL & Konishi M (2001) Auditory spatial receptive fields created by multiplication. *Science* 292:249-252.

Pickles JO, Comis SD & Osborne MP (1984) Cross-links between stereocilia in the guinea pig organ of Corti, and their possible relation to sensory transduction. *Hear Res* 15:103-112.

Rose JE, Brugge JF, Anderson DJ et al. (1967) Phase-locked response to low-frequency tones in single auditory nerve fibers of the squirrel monkey. *J Neurophysiol* 30:769-793.

Zheng J, Shen W, He DZ et al. (2000) Prestin is the motor protein of cochlear outer hair cells. *Nature* 405:149-155.

Somatosensation

Bautista DM, Siemens J, Glazer JM et al. (2007) The menthol receptor TRPM8 is the principal detector of environmental cold. *Nature* 448:204-208.

Caterina MJ, Schumacher MA, Tominaga M et al. (1997) The capsaicin receptor: a heat-activated ion channel in the pain pathway. *Nature* 389:816-824.

Coste B, Mathur J, Schmidt M et al. (2010) Piezo1 and Piezo2 are essential components of distinct mechanically activated cation channels. *Science* 330:55-60.

de Lafuente V & Romo R (2006) Neural correlate of subjective sensory experience gradually builds up across cortical areas. *Proc Natl Acad Sci USA* 103:14266-14271.

Han L, Ma C, Liu Q et al. (2013) A subpopulation of nociceptors specifically linked to itch. *Nat Neurosci* 16:174–182.

Li L, Rutlin M, Abraira VE et al. (2011) The functional organization of cutaneous low-threshold mechanosensory neurons. *Cell* 147:1615–1627.

Maksimovic S, Nakatani M, Baba Y et al. (2014) Epidermal Merkel cells are mechanosensory cells that tune mammalian touch receptors. *Nature* 509:617–621.

McKemy DD, Neuhausser WM & Julius D (2002) Identification of a cold receptor reveals a general role for TRP channels in thermosensation. *Nature* 416:52–58.

Mountcastle VB, LaMotte RH & Carli G (1972) Detection thresholds for stimuli in humans and monkeys: comparison with threshold events in mechanoreceptive afferent nerve fibers innervating the monkey hand. *J Neurophysiol* 35:122–136.

O'Hagan R, Chalfie M & Goodman MB (2005) The MEC-4 DEG/ENaC channel of *Caenorhabditis elegans* touch receptor neurons transduces mechanical signals. *Nat Neurosci* 8:43–50.

Ranade SS, Woo SH, Dubin AE et al. (2014) Piezo2 is the major transducer of mechanical forces for touch sensation in mice. *Nature* 516:121–125.

Sun YG, Zhao ZQ, Meng XL et al. (2009) Cellular basis of itch sensation. *Science* 325:1531–1534.

Usoskin D, Furlan A, Islam S et al. (2015) Unbiased classification of sensory neuron types by large-scale single-cell RNA sequencing. *Nat Neurosci* 18:145–153.

Vrontou S, Wong AM, Rau KK et al. (2013) Genetic identification of C fibres that detect massage-like stroking of hairy skin *in vivo*. *Nature* 493:669–673.

Walker RG, Willingham AT & Zuker CS (2000) A *Drosophila* mechanosensory transduction channel. *Science* 287:2229–2234.

CHAPTER 7

Wiring of the Nervous System

In this chapter, we return to the subject of nervous system wiring, a problem of central importance in neurobiology. We will apply what we learned in Chapter 5's discussion of visual system wiring to the rest of the nervous system and expand on the principles that govern these wiring processes. We will examine the daunting task of establishing trillions of synaptic connections among billions of neurons by asking two deceptively simple questions: (1) How do individual neurons differentiate and connect with their partners, and (2) how do groups of neurons wire in a coordinated manner to form a functional circuit, such as a neural map?

From the perspective of an individual neuron, the wiring problem can be viewed as a sequential differentiation process (**Figure 7–1**). We will first follow this differentiation process in the context of general nervous system development. Given the evolutionary conservation of many processes in neural development (for example, axon guidance; see Box 5–1), we will study neuronal differentiation using particularly well-understood examples from different parts of the nervous systems in diverse animals. Next, we will use the wiring of the olfactory system as a model to explore how neural circuits are assembled, comparing and contrasting with the visual system wiring discussed in Chapter 5. We will end the chapter by summarizing the strategies encountered in this chapter and Chapter 5, discussing how animals can use a small number of genes to establish a much larger number of connections.

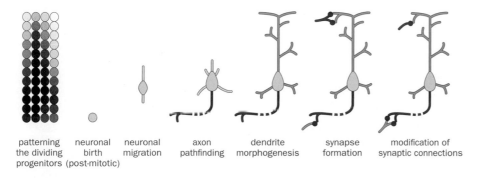

patterning the dividing progenitors | neuronal birth (post-mitotic) | neuronal migration | axon pathfinding | dendrite morphogenesis | synapse formation | modification of synaptic connections

embryonic

early postnatal

throughout life

Figure 7–1 Schematic summary of the neuronal differentiation process. The nervous system is patterned by early developmental events, such that neural progenitor cells at different positions along the body axes produce different types of neurons. Neurons are born when they exit the final cell cycle (never to divide again). A post-mitotic neuron undergoes a series of differentiation processes. These include migrating to its final destination, extending an axon toward its postsynaptic partners, establishing dendrite branching patterns, forming synaptic connections with its pre- and postsynaptic partners, and modifying those connections. Although the general sequence of events follows the left-to-right progression shown in the figure, some steps can overlap considerably. The bottom timeline shows that for most mammalian neurons, the first six steps are accomplished during embryonic and early postnatal development, with the last step stretching from late embryogenesis to the end of life.

HOW DOES WIRING SPECIFICITY ARISE IN THE DEVELOPING NERVOUS SYSTEM?

7.1 The nervous system is highly patterned as a consequence of early developmental events

Development (**Figure 7–2**A) begins with **fertilization**, the fusion of sperm and egg to create a genetically new organism. The resulting fertilized egg, or **zygote**, undergoes a process called **cleavage**, in which rapid cell division (typically without cell growth) generates a **blastula**, a hollow ball containing thousands of cells. Cleavage is followed by **gastrulation**, during which extensive cell rearrangement converts a blastula into a **gastrula** with three distinct germ layers: outer (**ectoderm**), middle (**mesoderm**), and inner (**endoderm**). In addition to creating three separate germ layers, gastrulation also establishes or consolidates the anterior–posterior and dorsal–ventral axes in vertebrates. All vertebrates and most invertebrates follow these developmental steps, often using conserved molecular mechanisms (see Chapter 12), although with large variations in details and timing. For example, a fruit fly embryo progresses from fertilization to the completion of gastrulation in less than three hours, whereas the same developmental events in a human embryo require nearly three weeks.

The construction of the nervous system in vertebrates begins during the next stage of development, called **neurulation** (Figure 7–2A, B). At the dorsal side of the embryo, the ectodermal cells overlying a mesoderm structure called the **notochord** thicken and form the **neural plate**. The center of the neural plate then drops toward the interior of the embryo while the two edges move toward each other; this process of invagination creates the neural fold. Two edges of the neural fold eventually fuse at the center, creating the **neural tube**, a hollow tube surrounded by layers of neuroectodermal cells. The neural tube is subsequently covered by additional layers of ectoderm-derived epidermal cells (Figure 7–2B). The lumen of the hollow tube develops into the **ventricles** that house the cerebrospinal fluid. The neuroepithelial cells that enclose the ventricles are **neural progenitors**; they divide next to the ventricles to produce intermediate precursors and post-mitotic neurons and glia. Neurons then migrate away from their birthplace near the ventricles to occupy different layers of neural tissues (see Section 7.2). A special group of cells at the junction between the dorsal neural tube and the overlying epidermal cells, the **neural crest cells** (Figure 7–2B), migrate away from the neural tube to produce diverse cell types that include cells of the peripheral nervous system, such as sensory neurons in the dorsal root ganglia and sympathetic neurons of the autonomic nervous system.

Figure 7–2 An overview of early steps of development. (A) Early development of the frog embryo is schematized to illustrate key developmental stages. The egg and blastula (product of cleavage) are shown as external side views. The gastrula (product of gastrulation) is shown as a sagittal section to reveal the internal tissues that constitute the three germ layers: ectoderm, mesoderm, and endoderm. Also labeled is the future notochord, a mesoderm-derived structure that plays an important role in patterning the nervous system. The neurula (product of neurulation) is shown from a dorsal view. The red dotted line corresponds to the cross-sectional views shown in panel B. **(B)** Cell movement during neurulation. Cells of the neuroectoderm (cyan) overlying the notochord invaginate to form the neural fold and neural tube. Surrounding epidermal cells (red) migrate to the center to cover the neural tube. Arrows indicate directions of cell movement. Neural crest cells, which are located at the dorsal neural tube at the final stage of this sequence, migrate out at a later stage (not shown) to produce neurons in the peripheral nervous system, among other cell types. Somites are mesoderm-derived and give rise to skeletal muscles, cartilages, tendons, and connective tissues. (A, adapted from Wolpert L, Jessell TM, Lawrence P et al. [2006] Principles of Development 3rd ed. Oxford University Press; B, adapted from Schroeder TE [1970] *J Embryol Exp Morphol* 23:427–462.)

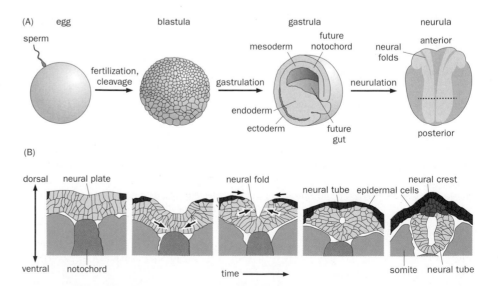

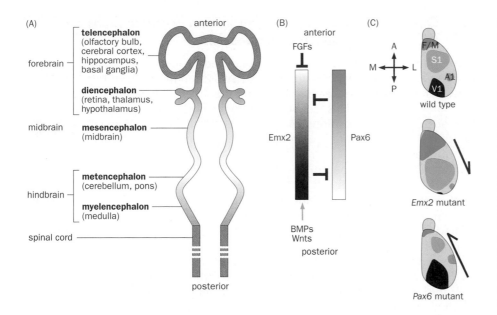

Figure 7–3 Patterning of the neural tube along the anterior–posterior axis. (A) The anterior neural tube is initially subdivided into the forebrain, midbrain, and hindbrain. It subsequently develops into the five-division stage shown here. Parentheses include notable adult structures that each of these divisions gives rise to (see also Figure 1–8). **(B)** The telencephalon is patterned by secreted morphogens—fibroblast growth factors (FGFs) from the anterior neural tube and bone morphogenetic proteins and Wnt family proteins (BMPs/Wnts) from the posterior—that specify expression gradients of the transcription factors Emx2 and Pax6 in neural progenitors along the anterior–posterior axis. Emx2 and Pax6 mutually inhibit each other. **(C)** Compared to wild-type mice, *Emx2* mutant mice expand the anterior cortical areas at the expense of posterior cortical areas, whereas in *Pax6* mutant mice the anterior cortical areas are diminished relative to wild type and the posterior cortical areas are enlarged. These data indicate that Pax6 and Emx2 are essential for the development of the anterior and posterior cortical areas, respectively. F/M, frontal/motor cortex (blue); S1, primary somatosensory cortex (green); A1, primary auditory cortex (orange); V1, primary visual cortex (red). A, anterior; P, posterior; M, medial; L, lateral. In schematics for the knockout mutants, arrows indicate the direction of cortical area expansion due to loss of Emx2 or Pax6. (B & C, adapted from O'Leary DDM, Chou SJ & Sahara S [2007] *Neuron* 56:252–269. With permission from Elsevier.)

The neural tube, which is located on the dorsal side of vertebrate embryos, is highly patterned along both its long anterior–posterior (rostral–caudal) axis and the orthogonal dorsal–ventral axis. For example, along the anterior–posterior axis, the neural tube is subdivided into the **forebrain**, **midbrain**, **hindbrain**, and spinal cord, each giving rise to different neural structures (**Figure 7–3**A; also see Figure 1–8). The cell types produced by the neural tube differ according to their position of origin along the dorsal–ventral and anterior–posterior axes. In the **telencephalon** (the anterior part of the forebrain), for instance, the dorsal (or palial) neural tube produces glutamatergic excitatory neurons of the cerebral cortex, whereas the ventral (or subpalial) neural tube produces GABAergic neurons that migrate into the cerebral cortex and the basal ganglia (see Section 7.2). In the spinal cord, the dorsal neural tube produces dorsal horn projection neurons that relay sensory information to the brain (see Section 6.33), the ventral neural tube produces motor neurons that innervate muscles, and both the dorsal and ventral neural tubes produce many distinct types of interneurons (see Section 7.4).

Neural tubes are patterned by developmental **morphogens**. These secreted proteins are synthesized by signaling centers (cells at specific locations such as junctions between two brain compartments) and diffuse across tissues to regulate the expression of target genes in a concentration-dependent manner. Key morphogen targets are often transcription factors in progenitors, which control gene expression programs according to where the progenitors are located along the body axes. For example, the developing telencephalon is patterned into different functional areas by **fibroblast growth factors** (**FGFs**) from the anterior neural tube and a combination of **bone morphogenetic proteins** (**BMPs**) and **Wnt** family proteins (**Wnts**) from the posterior neural tube. Through activation, repression, and mutual inhibition of target genes, these morphogens establish gradients of transcription factor expression in progenitors across the telencephalon. For example, the transcription factor Emx2 exhibits a high posterior–low anterior expression gradient in cortical progenitors, whereas the transcription factor **Pax6** has a high anterior–low posterior gradient (Figure 7–3B). These expression patterns suggest that Emx2 and Pax6 play important roles in specifying posterior and anterior cortical areas, respectively. This hypothesis was supported by gene knockout experiments in mice. *Pax6* mutant mice exhibited an expansion of posterior cortical areas such as the primary visual cortex, with a concomitant shrinkage of anterior frontal and motor cortices. *Emx2* mutant mice exhibited the opposite phenotype, that is, an expansion of the frontal and motor cortices and primary somatosensory cortex at the expense of the visual cortex (Figure 7–3C). Thus, these transcription factors specify cortical areas according to their expression patterns.

7.2 Orderly neurogenesis and migration produce many neuronal types that occupy specific positions

The adult nervous system has a highly organized structure. For example, neurons in different layers of the cerebral cortex have different connection patterns and serve distinct functions (see Box 4–3). How are different types of cortical neurons generated from an initial sheet of progenitor cells along the neural tube? How do they occupy specific positions? We use studies of the mammalian cerebral cortex to explore these questions. As we will see, excitatory and inhibitory neurons follow different rules.

Cortical excitatory neurons, which use glutamate as the neurotransmitter and are mostly pyramidal in cell body shape, originate in the dorsal telencephalon from progenitor cells that reside in the **ventricular zone**, a layer of cells next to the ventricles. The final layer in which a given cortical excitatory neuron resides is predicted by the neuron's birthday, that is, the time when a neuron exits the last cell cycle and becomes post-mitotic. This was originally discovered by injecting pregnant mice with radioactively labeled thymidine at specific developmental stages and using autoradiography to track the positions of strongly labeled cells in the postnatal cortex. (Radioactively labeled thymidine incorporates into newly synthesized DNA during cell division cycles; the cells that are most strongly labeled are those that became post-mitotic shortly after injection and never divided again.) Cells that were born early reside in deep cortical layers; cells born later reside in progressively more superficial layers (**Figure 7–4**A). This inside-out pattern ruled out the possibility that early-born cells might be passively displaced away from the ventricular zone by late-born cells.

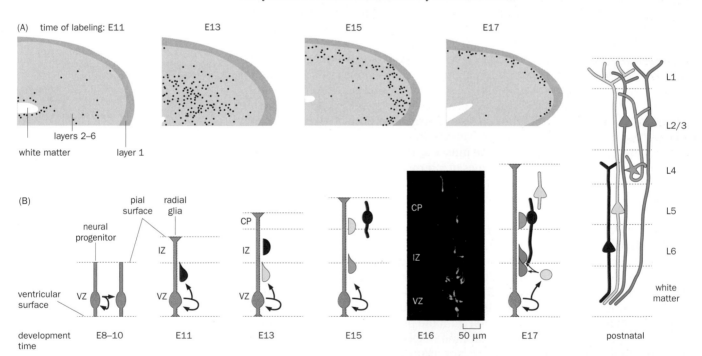

Figure 7–4 Neurogenesis and migration of cortical excitatory neurons. (A) Autoradiography of ^{3}H-thymidine-labeled neurons in the visual cortex of mice examined at postnatal day 10. Sections are labeled according to the embryonic day on which ^{3}H-thymidine was injected into the pregnant mouse (E11, E13, E15, or E17). Each dot represents a heavily labeled neuron. Thus, neurons born on later days reside in progressively more superficial cortical layers. **(B)** Schematic summary of cortical neurogenesis and migration using the mouse as a model. At early embryonic stages (E8–10), neural progenitors at the ventricular zone (VZ) divide symmetrically to expand the pool of progenitors; at later stages they become radial glia and extend processes that span from the ventricular surface to the pial surface. The neurons produced by asymmetric division from radial glia at E11 (red) are the first to initiate and complete migration along the radial glia; as a result, these E11-born neurons settle in the deepest layer (layer 6, or L6). As neurons born at E13 (yellow), E15 (green), and E17 (blue) follow this radial migration pattern, they migrate past the early-born neurons to settle in increasingly superficial layers (L5, L4, and L2/3, respectively). Note that asymmetric division can produce intermediate progenitors (for example, the cyan cell at E17) that divide further to produce post-mitotic neurons. IZ, intermediate zone (which disappears after neurogenesis is complete); CP, cortical plate (which develops into cortical layers). Inset, section of developing mouse cerebral cortex at E16. Blue staining labels all nuclei, outlining the three layers of cells (VZ, IZ, and CP). Green stains developing neurons from a single radial glia clone produced by a mosaic labeling method (see Section 13.16). (A, adapted from Angevine JB & Sidman RL [1961] *Nature* 192:766–768. With permission from Macmillan Publishers Inc.; B, image courtesy of Simon Hippenmeyer. See also Gao P, Postiglione MP, Krieger TG et al. [2014] *Cell* 159:775-788.)

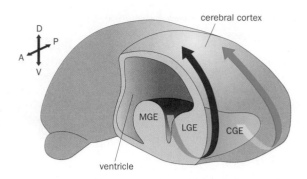

Figure 7–5 Origin and migration of cortical GABAergic neurons. Cortical GABAergic neurons are generated within the ventral telencephalon in the medial and caudal ganglionic eminences (MGE, CGE) and migrate dorsally and tangentially to the developing cerebral cortex around the cerebral ventricle hollows in the right hemisphere, which has been exposed by removing the anterior structure for illustration. D, dorsal; V, ventral; A, anterior; P, posterior. The lateral ganglionic eminence (LGE) gives rise to interneurons in the olfactory bulb and GABAergic projection neurons in the striatum. (Courtesy of Edmund Au & Gord Fishell. See also Batista-Brito R & Fishell G [2009] *Curr Top Dev Biol* 87:81–118 for more details.)

Subsequent studies have provided a detailed understanding of orderly cortical neurogenesis and migration. At an early stage, neural progenitors undergo symmetric division to expand the neural progenitor pools. As development proceeds, the division patterns become asymmetric (see Section 7.3 for more discussion about asymmetric cell division). At this stage, the progenitors are called **radial glia** because each extends two radial processes—one to the ventricle and the other to the pial surface of the developing cortex (Figure 7–4B). Each radial glia progenitor divides to produce another radial glia progenitor and a post-mitotic neuron, or an **intermediate progenitor** that further divides to produce post-mitotic neurons. Post-mitotic neurons migrate radially away from the ventricle along the radial glia processes toward the pial surface. Neurons that are born later actively migrate past their early-born cousins and settle progressively farther from their birthplace (Figure 7–4B). As will be discussed in Chapter 12, variations of this basic plan in different mammalian species account for the remarkable differences in brain size among mammals.

Whereas cortical glutamatergic neurons originate in the dorsal telencephalon at the ventricular zone and migrate radially, cortical GABAergic neurons are produced in the ventral telencephalon from structures called **ganglionic eminences** and migrate dorsally and tangentially across the developing cortex (**Figure 7–5**). Accumulating evidence has suggested that different types of GABAergic neurons are produced in different regions of the ganglionic eminences and are born at different times. They then migrate in an orderly fashion to their final destination. For instance, neurons derived from the medial ganglionic eminence (MGE) account for ~70% of GABAergic interneurons in the cerebral cortex, including fast-spiking basket and chandelier cells that express the marker parvalbumin, as well as Martinotti cells that express somatostatin (see Section 3.25 and Box 3–5). Neurons derived from the caudal ganglionic eminence (CGE) account for the remaining 30% of cortical GABAergic neurons, which express different markers and play distinct functions from MGE-derived neurons. MGE and CGE also collectively produce all GABAergic interneurons in the basal ganglia and amygdala. The lateral ganglionic eminence (LGE) gives rise to many different types of interneurons that occupy the olfactory bulb and most GABAergic projection neurons in the striatum, the function of which we will discuss in Chapter 8.

Many interesting questions are currently being investigated: What controls the total number of cortical neurons being born in a given species? How are their migrations directed and terminated? What ensures that each piece of cerebral cortex has the correct number and relative abundance of specific types of glutamatergic and GABAergic neurons? How do GABAergic neurons from a common origin obtain the necessary information to acquire appropriate connectivity within a wide variety of target areas?

7.3 Cell fates are diversified by asymmetric cell division and cell–cell interactions

By the time a neuron extends its axon and dendrites, it has usually acquired a **fate**, the outcome of the developmental decision as to what kind of cell it is. Some of the important (and related) fate choices include: Is a neuron going to be excitatory,

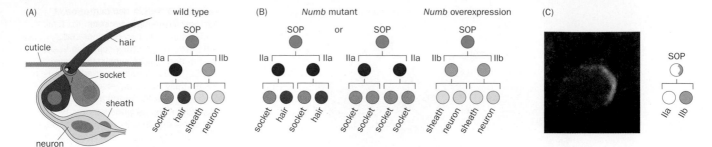

Figure 7–6 Asymmetric segregation of Numb during cell division diversifies sensory organ cell fates. (A) Schematic drawing of an external sensory organ in *Drosophila* comprising four cells—socket cell, hair cell, sheath cell, and neuron—that are produced from a sensory organ precursor (SOP) via intermediate precursors IIa and IIb. More recent study has identified an extra division after IIb that produces a glial cell (not depicted here) and the precursor that gives rise to the sheath and the neuron. **(B)** In *Numb* mutants, neuron and sheath cell are transformed to socket and hair cells (left), or all four cells become socket cells (middle). When *Numb* is transiently overexpressed in the SOP, the opposite transformation occurs: socket and hair cells are replaced by sheath cells and neurons (right). **(C)** The Numb protein (green) is asymmetrically distributed to one side of the SOP shortly before it divides. The schematic on the right shows that this asymmetric segregation produces two daughter cells with different levels of the Numb protein. (Adapted from Rhyu MS, Jan LY & Jan YN [1994] *Cell* 76:477–491. With permission from Elsevier. See also Uemura et al. [1989] *Cell* 58:349–360.)

inhibitory, or modulatory? What neurotransmitter will it use? What presynaptic and postsynaptic connections will it make? As was discussed in Section 5.17, cell fate can be determined through asymmetric segregation of fate determinants or cell–cell interactions. Investigations of these questions in vertebrates have greatly benefited from pioneering studies in worms and flies because of the relative simplicity and abundant genetic tools available in these invertebrate model organisms. For example, the study of cell fate determination in the *Drosophila* sensory organ provides an excellent example of the interplay between asymmetric segregation of fate determinants and cell–cell interactions (**Movie 7–1**).

Each external sensory organ in *Drosophila* consists of a socket cell, a hair cell, a sheath cell, and a sensory neuron. These cells derive from a common **sensory organ precursor** (**SOP**) through a series of asymmetric cell divisions (**Figure 7–6**A). In a *Drosophila* mutant called *Numb*, sensory neurons and sheath cells were missing (hence the name), and were replaced by extra socket and hair cells (Figure 7–6B, left); in extreme cases all cells became sockets (Figure 7–6B, middle). Overexpression of a *Numb* transgene during a specific time window caused an opposite transformation of cell fates—extra neurons and sheath cells were produced at the expense of socket and hair cells (Figure 7–6B, right). These data suggested that the **Numb** protein product normally acts to diversify cell fate within the SOP lineage.

The mechanisms by which Numb does so became clear when the distribution of the Numb protein was examined. Shortly before the SOP divides, Numb becomes localized to one side of the SOP such that it is preferentially inherited by one of the two daughter cells (Figure 7–6C). Asymmetric segregation of Numb occurs in subsequent divisions as well, giving rise to cells with distinct fates within the SOP lineage.

How does asymmetric segregation of Numb confer distinct fates on different daughter cells? Biochemical and genetic experiments indicated that a principal mechanism of action of Numb is to inhibit the function of **Notch**, a transmembrane receptor widely used for diversifying cell fate during development. Notch is activated by a transmembrane ligand **Delta** from a neighboring cell. This induces an intramembrane proteolytic cleavage, which releases the intracellular domain of Notch to the nucleus, where it binds to its partners, called Mastermind and Suppressor of Hairless in *Drosophila*, and activates target gene expression (**Figure 7–7**A). Among these target genes are *Notch* and *Delta* themselves—*Notch* transcription is up-regulated and *Delta* transcription is down-regulated by Notch signaling. These feedback loops can account for how Notch–Delta signaling diversifies cell fates through a process called **lateral inhibition** (which is distinct from the lateral inhibition in information processing by neural circuits that we discussed in Sections 4.16 and 6.9).

Suppose that two neighboring cells, *A* and *B*, initially express the same amount of Notch and Delta. If Notch activity is inhibited in *B* by intrinsic factors (such as Numb) or extrinsic signals, inhibition of Delta expression is relieved. As Delta expression in *B* increases, Notch signaling in *A* increases, while Delta expression in *A* decreases, so that *B* receives less ligand for its own Notch (Figure 7–7B). Thus, the Notch–Delta signaling amplifies an initial small difference between neighboring cells, leading to activation of Notch signaling in one cell and suppression of it in the other. This results in differential expression of Notch target genes and eventually different fates among neighboring cells.

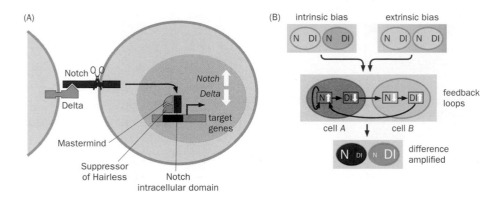

Figure 7–7 Notch–Delta mediated lateral inhibition diversifies cell fates. (A) In response to binding of ligand Delta from a neighbor cell, Notch activation leads to the cleavage and translocation of its intracellular domain to the nucleus (red), where it interacts with its nuclear partners, Mastermind and Suppressor of Hairless, resulting in target gene expression. Among these gene expression changes, *Notch* is up-regulated and *Delta* is down-regulated. **(B)** A model for Notch–Delta mediated lateral inhibition. Two cells initially have the same Notch (N) and Delta (Dl) levels. Due to an intrinsic (left) or extrinsic (right) biasing signal, Notch is somewhat inhibited in cell *B* on the right; this leads to further decrease of Notch expression and increase of Delta expression. The neighboring cell *A* is thus exposed to more Notch ligands and has higher Notch activity, which further enhances cell *A*'s expression of Notch and decreases its expression of Delta. These feedback loops amplify the initial differences in Notch and Delta levels. (Based on Artavanis-Tsakonas S, Rand MD & Lake RJ [1999] *Science* 284:770–776.)

While lateral signaling utilizes feedback loops to amplify small initial differences, Numb biases the Notch–Delta signaling to create stereotyped decisions. Mechanisms similar to those used for sensory organ cell fate specification are widely used throughout development. Indeed, at an earlier developmental stage, the fate determination that differentiates the SOP from neighboring epidermal cells also utilizes Notch–Delta mediated lateral inhibition. Vertebrate homologs of Notch, Delta, and Numb, as well as nuclear partners of Notch and some of the Notch target genes, are implicated in many fate determination events, including the process of cortical neurogenesis discussed in Section 7.2.

7.4 Transcriptional regulation of guidance molecules links cell fate to wiring decision

From a molecular perspective, the fate decision is usually realized by the expression of a unique set of fate determinants that distinguish a cell from other cells expressing different sets of fate determinants. Important among these determinants are transcription factors. Cell-type-specific transcription factors can dictate the expression and relative abundance of guidance molecules. As a result, axons from neurons with different fates may contain different populations of guidance molecules and respond differently to a common environment during their guidance and targeting.

We have already encountered such examples in the wiring of the visual system. For instance, in the mammalian retina, retinal ganglion cells (RGCs) that project ipsilaterally express a transcription factor Zic2, which specifies the expression of EphB1 so these axons can respond to a repellent ephrin-B2 in the chiasm and turn toward the ipsilateral side (see Figure 5–15). In the chick tectum, graded expression of ephrin-As along the anterior–posterior axis is regulated by graded expression of the transcription factor Engrailed2, which is in turn regulated by the FGF morphogens produced from the midbrain–hindbrain junction at the posterior edge of the developing tectum (see Section 5.5).

The mechanism by which neuronal fates are specified in vertebrates is best understood in the spinal cord, thanks to a combination of studies using chick embryology and mouse genetics. Morphogens secreted from the dorsal roof plate or ventral floor plate reach different parts of the spinal cord at different concentrations and specify distinct cell fates along the dorsal–ventral axis in a concentration-dependent manner. As a particularly well-studied example, the floor-plate-derived morphogen **Sonic Hedgehog (Shh)** is responsible for determining different fates of neuronal progenitors at different positions in the ventral spinal cord by regulating the expression of specific transcription factors (**Figure 7–8**A, B). Shh acts by repressing a group of class I transcription factors, Pax7, Dbx1, Dbx2, Irx3, and Pax6, at different concentrations. Shh also activates class II transcription factors Nkx6.1 and Nkx2.2, again at different concentrations (Figure 7–8C). (All of these transcription factors contain a DNA-binding domain called **homeodomain**, the origin of which will be discussed further in Chapter 12.) Furthermore, class I and class II proteins mutually repress each other's expression (Figure 7–8C), which sharpens the borders between different progenitor pools created initially

Figure 7–8 Morphogens and transcription factors specify spinal cord neuronal fates. (A) Micrograph of Sonic hedgehog (Shh) protein distribution. Shh (in white) is produced in the notochord and floor plate. It diffuses dorsally within the spinal cord, forming a concentration gradient. D, dorsal; V, ventral. **(B)** Shh acts by turning off (⊣) and turning on (→) different transcription factors listed on the top (class I) and bottom (class II) in a concentration-dependent manner, thereby specifying the expression of these transcription factors in progenitor cells at different D–V positions. For example, a low level of Shh can turn off Pax7 expression, whereas a high level is needed to turn off Pax6 expression. Thereby Pax7 expression is restricted to the dorsal cells, whereas Pax6 expression extends more ventrally. **(C)** Two pairs of class I/class II transcription factors mutually repress each other's expression, thereby sharpen the borders of their expression domains along the D–V axis. **(D)** Five progenitor domains created by differential expression of these transcription factors give rise to five types of spinal neurons, the V0, V1, V2, and V3 interneurons, and motor neurons (MN). **(E)** Motor neurons are produced in the D–V domain that expresses Pax6 and Nkx6.1, but not Irx3 and Nkx2.2. (Adapted from Jessell TM [2000] *Nat Rev Genet* 1:20–29. With permission from Macmillan Publishers Ltd. See also Briscoe J, Pierani A, Jessell TM et al. [2000] *Cell* 101:435–445.)

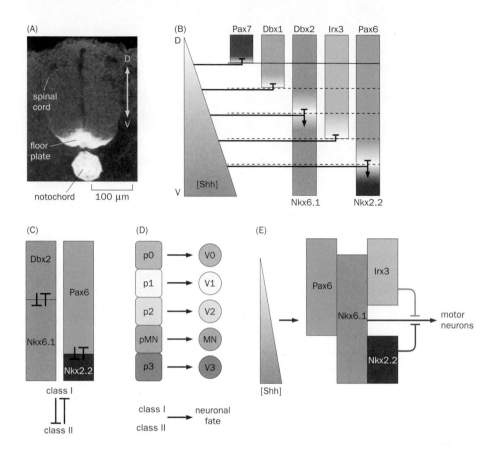

by differential responses to the Shh morphogen gradient. Collectively, differential expression of these transcription factors specifies the fates of five progenitor types, four of which produce ventral spinal cord interneurons and one produces motor neurons (Figure 7–8B, D).

As a specific example of the transcription factor code, progenitors that express Pax6 and Nkx6.1 but do not express Irx3 or Nkx2.2 become motor neuron progenitors (Figure 7–8E). They produce motor neurons by turning on motor-neuron-specific transcriptional programs as their progeny exit the cell cycle. These include the expression of a series of motor-neuron-specific transcription factors, which in turn regulate the expression of specific guidance receptors that direct their axons out of the spinal cord and into the muscle field. Motor neurons that innervate specific muscles are further diversified. For example, the fate of a motor neuron that projects into the dorsal or ventral limb is determined by its expression of one of the two specific transcription factors that contain the LIM domain: Lim1 and Isl1 (**Figure 7–9**A). Lim1-expressing motor neurons are located laterally and project axons to the dorsal limb, whereas Isl1-expressing motor neurons are located medially and project axons to the ventral limb. Lim1 and Isl1 mutually repress each other's expression, such that each motor neuron expresses only one of these two transcription factors (Figure 7–9B). At the target field, the dorsal limb expresses a third LIM transcription factor called Lmx1b. Genetic perturbations of Lim1 in motor neurons or Lmx1b in targets disrupt the correct choice of motor axon innervation of muscles, indicating that these transcription factors play essential roles in the wiring decisions of these motor neurons.

How do these transcription factors regulate connectivity between motor axons and their targets? It turns out that they regulate the expression of ephrins and Eph receptor guidance molecules, which we first encountered in our discussions of the visual system. In the dorsal limb, Lmx1b promotes expression of ephrin-B2 and represses expression of ephrin-As. The target field is thus patterned by the binary expression of two ephrins: ephrin-As in the ventral limb and ephrin-B2 in the dorsal limb. In the motor neurons, Lim1 promotes the expression of EphA4 receptor whereas Isl1 promotes the expression of EphB receptors (Figure 7–9B). When axons arrive at the target field, Lim1/EphA4-expressing axons are repelled

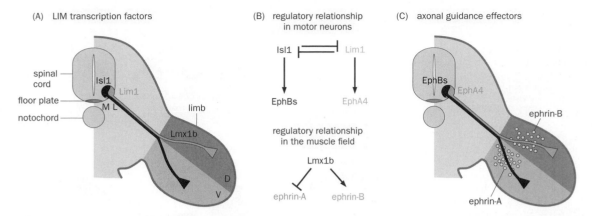

Figure 7–9 LIM transcription factors control the expression of ephrin/Eph to specify motor axon targeting. (A) Motor neurons located in the more medial position (M) of the spinal cord express the LIM transcription factor Isl1 (red) and target their axons selectively to the ventral limb (V), whereas motor neurons located more laterally (L) express a different LIM transcription factor, Lim1 (green), and target their axons to the dorsal limb (D). A third LIM transcription factor, Lmx1b, is expressed in the dorsal limb. **(B)** Mutual repression of Isl1 and Lim1 ensures that they are expressed in different motor neurons.

Furthermore, Isl1 promotes the expression of EphBs, whereas Lim1 promotes the expression of EphA4. In the limb, Lmx1b activates ephrin-B expression and represses ephrin-A expression. **(C)** EphB-expressing motor axons are repelled by dorsal limb ephrin-B and selectively target the ventral limb, whereas EphA-expressing motor axons are repelled by ventral limb ephrin-A and target the dorsal limb. (Adapted from Kania A & Jessell TM [2003] *Neuron* 38:581–596. With permission from Elsevier Inc. See also Luria V, Krawchuk D, Jessell TM et al. [2008] *Neuron* 60:1039–1053.)

by ephrin-As in the ventral limb and therefore target dorsally, whereas Isl1/EphBs-expressing axons are repelled by ephrin-B2 in the dorsal limb and therefore target ventrally (Figure 7–9C). Thus, the coordinated expression and function of these dual ephrin/Eph receptor systems provides a robust mechanism to ensure that two subpopulations of motor neurons connect correctly with two distinct groups of muscle targets.

Following the lead of spinal cord research, transcription factors that specify neuronal fates and axon projection patterns in more complex tissues such as the mammalian cerebral cortex have also been discovered. The projection patterns of cortical pyramidal neurons can be divided into three major types (**Figure 7–10**A). The first type, including all layer 2/3 and a fraction of layers 5 and 6 pyramidal neurons, projects their axons to other cortical areas. Of these corticocortical projection neurons, a subset project axons across the **corpus callosum** (which is composed of axon bundles that link the two cerebral hemispheres) to the contralateral cortex; these neurons are called **callosal projection neurons** (**CPNs**). The second type of neurons project their axons to subcortical targets, such as the pons, superior colliculus, and spinal cord; these are called **subcerebral projection neurons** (**SCPNs**). SCPNs constitute a large subset of layer 5 neurons. The third type of neurons project their axons to the thalamus, and are therefore called **corticothalamic projection neurons** (**CTPNs**). CTPNs constitute a large subset of layer 6 neurons (Figure 7–10A).

Like the spinal cord examples we just discussed, fates and projection patterns of cortical projection neurons are specified by a set of transcription factors that mutually inhibit each other. For example, the transcription factor **Satb2** is expressed in CPNs and is required for their callosal projection patterns. Satb2-expressing CPNs normally project axons across the corpus callosum but not to subcerebral structures. However, axons of these same neurons in *Satb2* mutant mice failed to project across the corpus callosum, and instead projected to subcortical structures (Figure 7–10B). Likewise, the transcription factor **Fezf2** is expressed in SCPNs and specifies subcerebral projections whereas the transcription factor **Tbr1** is expressed in CTPNs and regulates their thalamic projections. Fezf2 and Tbr1 mutually repress each other; Fezf2 also represses Satb2, which in turn represses Ctip2, a downstream transcription factor that regulates SCPN identity and subcerebral projections (Figure 7–10C). Mutual repression of transcription factors ensures that each type of projection neuron expresses a unique set of guidance molecules that specify their distinct projections.

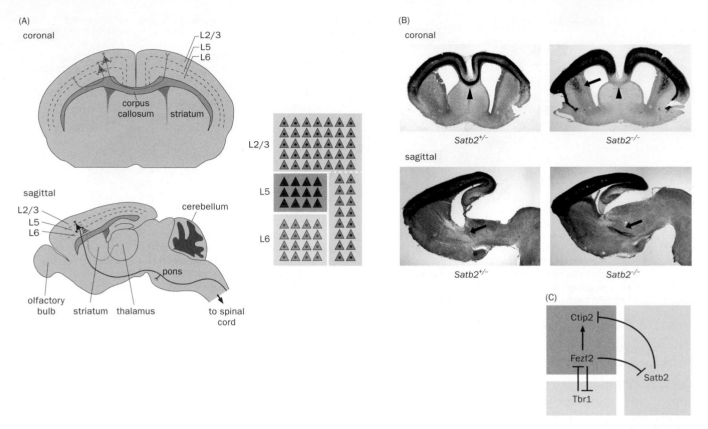

Figure 7–10 Transcription factors that specify axonal projections of cortical neurons. (A) Left, schematic coronal (top) and sagittal (bottom) views of the adult mouse brain, showing axonal projection patterns of three major types of cortical projection neurons. Corticocortical projection neurons (blue) either project axons across the corpus callosum to the contralateral hemisphere (CPNs), or to other cortical areas within the same hemisphere. Subcerebral projection neurons (SCPNs, red) project their axons to subcortical areas such as the pons and spinal cord. Corticothalamic projection neurons (CTPNs, green) project their axons to the thalamus. Right, schematic layer distribution of three projection neuron types. Note that in reality, layer 5 (L5) or L6 SCPNs or CTPNs are intermingled with corticocortical projection neurons. **(B)** Satb2 is required for callosal projections. In heterozygous controls, axons of the Satb2-expressing neurons (visualized in blue, as a result of a βGal insertion in the *Satb2* locus) cross the corpus callosum (arrowhead) but do not extend to subcerebral structures (arrow). In *Satb2* homozygous mutants, axons no longer project across the corpus callosum (arrowhead), but instead extend to subcerebral structures (arrows). These samples were examined at embryonic day 18.5, thus the sections look different from the adult schematic in panel A. **(C)** An oversimplified summary of transcription factors that regulate the projection patterns of CPNs (blue), SCPNs (red), and CTPNs (green) in cortical layers 5/6, as well as their mutually repressive relationships. (A & C, adapted from Greig LC, Woodworth MB, Galazo MJ et al. [2013] *Nat Rev Neurosci* 14:755–769. With permission from Macmillan Publishers Ltd; B, from Alcamo EA, Chirivella L, Dautzenberg M et al. [2008] *Neuron* 57:364–377. With permission from Elsevier Inc.)

7.5 Crossing the midline: Combinatorial actions of guidance receptors specify axon trajectory choice

As we learned in Chapter 5, axons are guided by attractive or repulsive guidance cues. These cues act on guidance receptors at the surface of the growth cones to regulate the underlying cytoskeletal dynamics (see Boxes 5-1 and 5-2). In the next two sections, we use the midline crossing and pathway choices in insects and vertebrates as specific examples to explore the mechanisms of axon guidance that expand on these principles.

Axons often follow specific trajectories in a highly complex environment. This has been well documented in the embryonic **ventral nerve cord** of insects, which is analogous to the spinal cord of vertebrates. The grasshopper embryo has been used as a model system because of its large neurons, allowing dyes to be injected with a microelectrode into single, **identified neurons** (meaning that the same neurons can be recognized from animal to animal because of their stereotyped location, size, and shape). The ventral nerve cord explant with dye-injected neurons can be placed on a microscope slide for live imaging as development unfolds. It was found that axons from individually identified neurons follow specific pathways—for instance, the vast majority of neurons first extend their

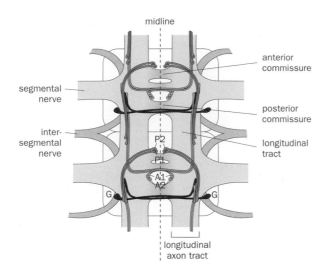

Figure 7–11 Axons follow specific pathways in the grasshopper embryonic ventral nerve cord. Schematic of two adjacent segments of a grasshopper ventral nerve cord, showing axon bundles that form the anterior and posterior commissures, the longitudinal tracts on each side of the midline (dashed line), and the segmental and inter-segmental nerves that connect to peripheral tissues. Five pairs of neurons are drawn within each segment. Axons of P1/P2 neurons (blue) do not cross midline and project posteriorly. Axons of A1/A2 neurons (green) and G neurons (purple) cross the midline and project anteriorly. (Adapted from Goodman CS & Bastiani MJ [1984] *Sci Am* 251:58–66. With permission from Macmillan Publishers Ltd.)

axons across the midline; their growth cones then turn either anteriorly or posteriorly at a specific distance away from the midline (**Figure 7–11**).

Many of the identified neurons in the grasshopper ventral nerve cord have subsequently been found in *Drosophila*. Although neurons are much smaller in size (and more difficult to inject dye into), *Drosophila* offers the power of forward genetic screens to identify new molecules required for axon guidance from their loss-of-function phenotypes (see Section 13.6 for more details). This approach complements well the biochemical approaches to identify axon guidance molecules discussed in Section 5.4 and Box 5–1. When stained with an antibody that recognizes all axons of the embryonic ventral nerve cord, wild-type axons can be seen as segmentally repeated anterior and posterior commissures across the midline, as well as longitudinal axonal tracts that run along the anterior–posterior axis (**Figure 7–12**A). This is because, as in grasshopper embryos, axons of most embryonic ventral nerve cord neurons cross the midline once, forming the commissures, and then turn either anteriorly or posteriorly, forming the longitudinal tracts. Mutants have been identified that exhibit interesting phenotypes with regard to midline crossing. For example, in *Slit* mutants, all axons collapse at the midline (Figure 7–12B). In *Roundabout* (*Robo*) mutants, individual axons cross the midline multiple times instead of turning anterior or posterior after midline crossing, thus forming thick commissures and thin longitudinal tracts (Figure 7–12C). In *Commissureless* (*Comm*) mutants, no axons ever cross the midline (Figure 7–12D).

Subsequent molecular genetic, biochemical, and cell biological analyses have identified the mechanisms by which the protein products of *Slit*, *Robo*, and *Comm* regulate midline crossing. **Slit** is a secreted protein produced from midline glia (stained in blue in Figure 7–12A) and acts as a repulsive axon guidance ligand. **Robo** is a receptor for Slit. **Comm** acts in the secretory pathway (see Figure 2–2)

Figure 7–12 Ventral nerve cord of wild-type and mutant *Drosophila* embryos. **(A)** The wild-type *Drosophila* embryonic ventral nerve cord is organized as repeated segments of anterior and posterior commissures (AC and PC, composed of axon bundles that cross the midline), segmental nerves (S), and inter-segmental nerves (IS), as well as longitudinal tracts (L). Note the striking resemblance to the grasshopper embryo in Figure 7–11. Brown shows antibody staining that recognizes all axons. Blue stains midline glia. **(B)** In homozygous *Slit* mutant embryos, all axons collapse at the midline. **(C)** In homozygous *Roundabout* (*Robo*) mutant embryos, longitudinal tracts are thinner and commissures are thicker. **(D)** In homozygous *Commissureless* (*Comm*) mutant embryos, axons do not cross the midline and therefore commissures are missing. (From Seeger M, Tear G, Ferres-Marco D et al. [1993] *Neuron* 10:409–426. With permission from Elsevier Inc.)

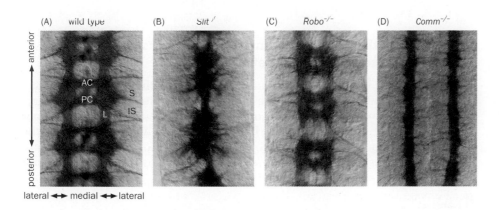

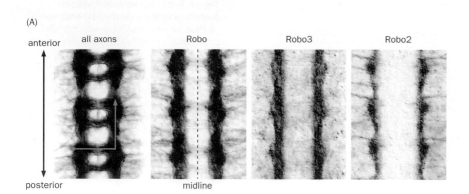

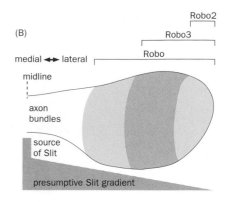

Figure 7–13 The combined activity of Robo proteins specifies axon trajectories in *Drosophila* embryos. (A) The first micrograph shows the same antibody staining as in Figure 7-12A, which labels all axons. The axon path of a typical neuron is shown in cyan. The 2nd to 4th micrographs show that three Robo proteins are expressed in different subsets of longitudinal axon tracts. Note that Robo proteins are absent from the commissures because they are down-regulated by Comm. **(B)** Model for how expression of Robo family proteins specifies axon position along the medial–lateral axis, seen from a cross section of the fly ventral nerve cord. In response to the presumptive Slit gradient, axons are repelled to different distances from the midline according to the total amount of Robo proteins they express. (Adapted from Simpson JH, Bland KS, Fetter RD et al. [2000] *Cell* 103:1019–1032. With permission from Elsevier Inc. See also Rajagopalan S, Vivancos V, Nicholas E. et al. [2000] *Cell* 103:1033–1045.)

to down-regulate cell-surface expression of Robo. When axons initiate midline crossing, expression of Comm prevents Robo from being on the cell surface, thus enabling axons to cross the midline. After midline crossing, down-regulation of Comm results in up-regulation of Robo on the cell surface, thus preventing axons from re-crossing the midline because of the repulsive Slit/Robo interaction. These mechanisms can explain the mutant phenotypes as follows. In the absence of Comm, Robo is always expressed on the cell surface and axons cannot cross the midline. (In fact, a small fraction of neurons whose axons do not cross the midline naturally do not express Comm.) In the absence of Robo, a large number of axons re-cross the midline after initial crossing because they have lost the receptor for a midline repellent. In the absence of Slit, no midline repellent is present so all axons stay at the midline because of midline attractants, which are the fly homologs of *C. elegans* Unc6 and vertebrate netrins introduced in Box 5–1. Indeed, as will be discussed in Section 7.6, the function of the Slit/Robo system in regulating midline axon guidance is also conserved in the vertebrate spinal cord. The reason why *Slit* and *Robo* mutants do not exhibit identical phenotypes is because additional Robo proteins act together with Robo, as discussed below.

After midline crossing, axons join the longitudinal tract to project anteriorly or posteriorly at specific distances away from the midline. What determines the distance? This is achieved by a second function of the Slit/Robo system. Flies have three *Robo* genes that encode for receptors for Slit: Robo, Robo2, and Robo3. Antibody staining indicated that these three proteins are expressed in different axons as a function of their distance from the midline. Robo is expressed in all axons; Robo3 is expressed in the lateral two-thirds of axons; Robo2 is expressed in the lateral-most one-third of axons (**Figure 7–13**A). Given that the repellent Slit is concentrated at the midline, a model was proposed that the net concentration of Robo proteins on the axons determines the lateral position of their trajectory: the more total Robo proteins an axon expresses, the more lateral the position that axon takes (Figure 7–13B). This model was validated by a series of loss-of-function and gain-of-function experiments: removing and adding Robo proteins caused medial and lateral shifts of axon positions, respectively. Thus, the Robo proteins expressed by an axon specify the medial–lateral position of the trajectory it takes.

7.6 Crossing the midline: Axons switch responses to guidance cues at intermediate targets

Many axons travel long distances to reach their final destinations. Intermediate targets are frequently used to divide these long journeys into segments. For example, commissural axons in the spinal cord are first attracted ventrally to the floor plate (see Box 5–1); they then cross the floor plate and extend anteriorly along the spinal cord. The floor plate is thus an intermediate target for commissural axons. An interesting question arises: if commissural axons are so attracted to the floor plate, why do they ever leave after reaching it?

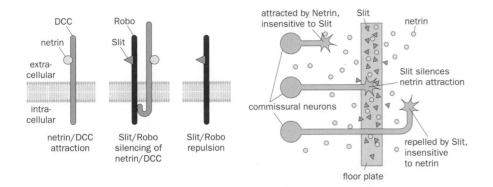

Figure 7–14 Commissural axons switch responses to guidance cues at the floor plate, an intermediate target. Left, netrin binds to the DCC receptor to elicit attraction. The presence of Slit induces binding of Robo to DCC via their cytoplasmic domains, which inactivates the DCC response to netrin. In addition, binding of Slit to the Robo receptor elicits repulsion. Right, a schematic summary of how the growth cones of commissural axons respond to guidance cues expressed in the floor plate—the attractant netrin and the repellent Slit—before, during, and after midline crossing. (Adapted from Stein E & Tessier-Lavigne M [2001] *Science* 291:1928–1938.)

Commissural axons are attracted to the floor plate by the chemoattractant netrin (see Box 5–1). As another striking example of evolutionary conservation of axon guidance molecules and mechanisms, the floor plate also makes the vertebrate homolog of *Drosophila* **Slit** that acts as a chemorepellent. Slit is presumed to be less diffusible than netrin and therefore acts within a shorter range. Growth cones of commissural axons use **DCC** (deleted in colon cancer) and Robo as receptors to respond to netrin and Slit, respectively (**Figure 7–14**). Before an axon crosses the midline, the attractive signaling of netrin/DCC is dominant over the repulsive signaling of Slit/Robo. Once it reaches the midline, the axon comes under the influence of Slit. Activation of the axon's Robo receptor by Slit not only results in Slit/Robo-mediated repulsion, but also silences the netrin/DCC-mediated attraction through direct binding of Robo to DCC, which inactivates DCC's cytoplasmic signaling domain. Thus, Robo-mediated repulsion vetoes DCC-mediated attraction by switching off the attractive response to netrin, and thereby ensures that commissural axons leave the floor plate soon after their arrival.

As discussed in Section 7.4, the floor plate produces the morphogen Shh to pattern the spinal cord for cell fate determination (see Figure 7–8). Later in neural development, Shh is used in addition to netrin as a floor-plate-derived axon guidance cue that attracts commissural axons toward the midline. Shh signaling can also induce repulsion by yet another floor-plate-produced guidance cue, the secreted semaphorin repellent **Sema3A** (see Box 5–1). Binding of Shh initiates an intracellular signaling cascade that potentiates signaling by Sema3A in commissural neurons. Thus, as axons approach the midline, Sema3A-induced repulsion, potentiated by Shh signaling, contributes to the extension of midline-crossing axons away from the floor plate.

After crossing the midline, commissural axons turn anteriorly to project toward the brain (commissural neurons include dorsal horn projection neurons that transmit temperature and pain signals to the brain; see Figure 6–70B). What causes them to turn anteriorly rather than posteriorly? At the developing ventral spinal cord, a secreted morphogen, Wnt4 (see Section 7.1), is distributed in an anterior > posterior gradient, and acts as an attractant for commissural axons through the Frizzled3 receptor (see Box 5–1). In an 'open book' preparation where the behavior of commissural axons after midline crossing can be examined *in vitro*, wild-type axons all turned anteriorly, whereas axons from *Frizzled3* mutant mice did not exhibit selectivity (**Figure 7–15**). This and other experiments confirmed that the Wnt–Frizzled interaction plays an instructive role for directing commissural axons anteriorly after midline crossing.

Several general lessons can be learned from the examples discussed in this and previous sections. First, molecules and mechanisms are highly conserved from invertebrates to mammals, as exemplified by the netrin/DCC and Slit/Robo systems. Second, the same molecules can be used at multiple stages of development, as exemplified by the fact that morphogens such as Shh and Wnts used to pattern early embryonic development are reused as axon guidance molecules, and by the dual function of Slit/Robo in midline crossing and lateral positioning of axonal trajectories. Third, attractants and repellents often act in parallel to

Figure 7–15 Wnt–Frizzled signaling directs commissural axons to turn anteriorly after midline crossing.
(A) Schematic of the axonal path of commissural neurons, which first grows ventrally toward the midline, and, after midline crossing, turns anteriorly.
(B) Illustration of the open book preparation, where the spinal cord is bisected from the dorsal side and laid open. Turning of fluorescently labeled commissural axons can be examined *in vitro*. **(C)** Whereas all wild-type commissural axons turn anteriorly in the open book preparation (left), commissural axons in *Frizzled3* mutants (*Fz3*$^{-/-}$) randomly turn anteriorly and posteriorly (middle), or stall (quantified in the right panel). A, anterior; P, posterior; fp, floor plate. (Adapted from Lyuksyutova AI, Lu CC, Milanesio N et al. [2003] *Science* 302:1984–1988. With permission from AAAS.)

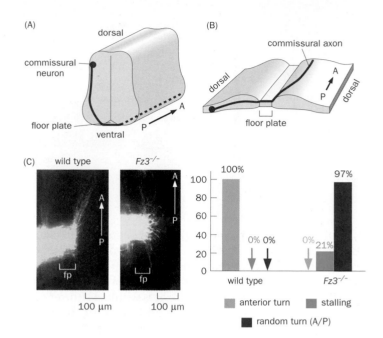

ensure the accuracy of decisions made by the growth cone, as exemplified by action of netrin and Shh in attracting axons to the midline, and by the action of Slit and Sema3A in repelling axons away from the midline. Fourth, guidance molecules can also act in a hierarchical manner to guide axons sequentially, as exemplified by the initial attraction of commissural axons to the floor plate and their subsequent repulsion during midline guidance. Together, these actions guide axons with remarkable precision along their complex journey. In the case commissural neurons, for example, these actions ensure the axons first grow ventrally toward the midline, then cross the midline, and finally turn anteriorly toward their brain targets.

7.7 The cell polarity pathway participates in determining whether a neuronal process becomes an axon or a dendrite

Thus far we have focused on the role of axons in determining wiring specificity. In general, axons travel longer distances than dendrites, so they perform a greater share of the work involved in wiring the nervous system. Dendrites can also play an active role in determining wiring specificity, as will be discussed later in the chapter. However, axons and dendrites serve distinct functions in sending and receiving signals, respectively, as discussed in Sections 1.7, 3.24, and 3.25. In addition, axons and dendrites have different morphologies: a typical neuron has a single axon but has multiple dendrites with elaborate branches near the cell body. Before we study mechanisms of dendrite morphogenesis in the following sections, we first take a step back to an earlier stage of development and address in this section the following question: how are axons and dendrites specified during development, such that they acquire different morphologies and functions?

Studies of hippocampal neurons in culture have significantly contributed to our understanding of these events. When dissociated embryonic hippocampal neurons are plated on a culture dish under appropriate conditions, they first extend numerous lamellipodia (see Box 5–2) that consolidate into a few short processes. One of these processes extends rapidly and becomes the axon. Subsequently, the remaining processes start to extend at a slower rate and branch; these become dendrites (**Figure 7–16**A). Hippocampal cultures can be maintained for weeks, during which time axons and dendrites develop characteristic pre- and postsynaptic properties and form synaptic connections.

The ease with which this culture system can be observed and manipulated has enabled researchers to probe the mechanisms that establish neuronal

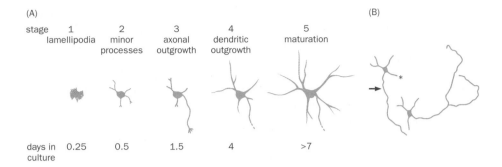

(A)

stage	1	2	3	4	5
	lamellipodia	minor processes	axonal outgrowth	dendritic outgrowth	maturation

(B)

days in culture: 0.25   0.5   1.5   4   >7

Figure 7–16 Differentiation of the axon and dendrites in dissociated hippocampal neurons. (A) After embryonic hippocampal neurons are dissociated and plated in a culture dish with an appropriate medium, they undergo the characteristic sequence of developmental events that is schematized here in five stages. Time after plating is indicated below. By day 1.5, one process becomes an axon. The rest of the processes develop into dendrites during subsequent days. **(B)** When an axon is severed (marked by the arrow in the left drawing), another process (*) develops into an axon; the drawing to the right shows the same neuron 24 h later. (A, adapted from Dotti CG, Sullivan CA & Banker GA [1988] *J Neurosci* 8:1454–1468. With permission from the Society for Neuroscience; B, adapted from Dotti CG & Banker GA [1987] *Nature* 330:254–256. With permission from Macmillan Publishers Ltd.)

polarity, most notably the distinction between axons and dendrites (see Sections 2.2 and 2.3). For example, when a growing axon is severed close to the cell body, a new axon develops from one (and only one) of the remaining short processes (Figure 7–16B). This experiment suggested that axon determination is plastic, and that once one process becomes the axon, other processes are inhibited from taking the same fate. Subsequent time-lapse imaging studies have revealed that shortly before one of the processes extends rapidly and becomes an axon, its growth cone is particularly dynamic in extending and retracting filopodia and lamellipodia, accompanied by rapid turnover of actin polymerization and depolymerization. Indeed, researchers can turn any process of a stage-2 neuron into an axon by locally applying at its growth cone a low dose of cytochalasin D, which causes depolymerization of the actin cytoskeleton. This experiment suggested that destabilization of the actin cytoskeleton in the growth cone is sufficient to induce axon formation. Likewise, locally stabilizing microtubules at the growth cone of a stage-2 process can induce axon formation.

Many molecular manipulations can perturb neuronal polarity in the hippocampal culture system. The function of a small subset of these molecules has been validated *in vivo* using gene knockout in mice. A central pathway involves *Par* genes, which were originally identified in a genetic screen in *C. elegans* for their role in asymmetric partitioning of cytoplasmic components in the early embryo. These *Par* genes have subsequently been shown to regulate cell polarity in many different tissues and cell types. In cultured hippocampal neurons, the mammalian **LKB1** kinase, a homolog of the *C. elegans* protein Par4, is concentrated in the short process that is destined to become the axon, but not in the other short processes (**Figure 7–17**A). Conditional knockout of the mouse *Lkb1* gene drastically reduces the abundance of axons *in vivo* (Figure 7–17B). LKB1 is a component of a protein kinase cascade. Protein kinase A phosphorylates and activates LKB1, which in turn phosphorylates a set of SAD kinases related to Par1, another cell

Figure 7–17 A cell polarity kinase cascade regulates axon specification.
(A) Prior to overt axon differentiation in cultured hippocampal neuron, the LKB1 protein (top) is highly concentrated in only one process (arrows) out of several processes that are equally enriched with a different marker, Tuj-1 (bottom). **(B)** Conditional knockout of *Lkb1* in the embryonic neocortex leads to a severe reduction of axon abundance (green, indicated by arrows in control). **(C)** A working model for the signaling pathway from extracellular polarizing cues to the regulation of cytoskeleton for axon specification. (A, from Shelly M, Cancedda L, Heilshorn S et al. [2007] *Cell* 129:565–577. With permission from Elsevier Inc.; B, from Barnes AP, Lilley BN, Pan YA et al. [2007] *Cell* 129:549–563. With permission from Elsevier Inc.)

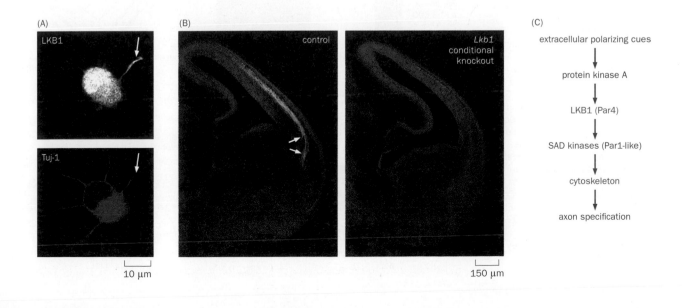

(A) LKB1 — Tuj-1 — 10 μm

(B) control — *Lkb1* conditional knockout — 150 μm

(C)
extracellular polarizing cues
↓
protein kinase A
↓
LKB1 (Par4)
↓
SAD kinases (Par1-like)
↓
cytoskeleton
↓
axon specification

polarity regulator identified in *C. elegans* (Figure 7–17C). Many of these components are conserved from worm to mammal, suggesting that the establishment of neuronal polarity uses similar mechanisms across the animal kingdom.

A conceptual framework for establishing neuronal polarity has emerged from these and other studies. Asymmetry in signaling, due to asymmetric exposure to an external signal, asymmetric localization of an intrinsic factor, or both, singles out one process to become the axon. This initial determination is likely reinforced by a positive feedback loop to augment the growth of the axon, and by a negative (inhibitory) signal that spreads to neighboring processes to prevent them from becoming axons. An important consequence of these signaling events is the modulation of the cytoskeletal elements—actin filaments and microtubules—that enable the rapid extension of axons. The polarity signal also leads to the differential organization of microtubule polarity in axons and dendrites; this helps maintain neuronal polarity by facilitating the transport of different cargos to specific compartments appropriate for their functions, as will be discussed in the next section.

7.8 Local secretory machinery is essential for dendrite morphogenesis and microtubule organization

What makes dendrites morphologically different from axons (see Figure 1–15)? Like the search for axon guidance molecules discussed in Section 7.5, researchers can tease out answers to this question by conducting unbiased forward genetic screens (see Section 13.6) to determine which genes, when mutated, preferentially affect dendrite, but not axon, growth. One such screen in *Drosophila* embryonic sensory neurons revealed that mutations in components of a well-studied cell biological process, vesicle trafficking from the endoplasmic reticulum (ER) to the Golgi apparatus (see Section 2.1), preferentially disrupted dendrite growth (**Figure 7–18**A). The Golgi apparatus is traditionally thought to be near the nucleus. However, recent studies have revealed that fragments of the Golgi apparatus (termed **Golgi outposts**) are distributed in the dendrites of cultured mammalian hippocampal neurons (Figure 7–18B) and *Drosophila* sensory neurons, but are absent from axons. Disruption of Golgi outposts impaired dendrite growth and branch extension in mammalian as well as *Drosophila* neurons.

As discussed in Section 2.3, a key distinction between dendrites and axons is the orientation of their microtubules. In an axon, the growing (plus) ends of the microtubules face toward the axon terminus, that is, in axons microtubules are

Figure 7–18 Golgi outposts, dendrite elaboration, and microtubule nucleation. (A) Mutations in *Dar3* (<u>d</u>endritic <u>ar</u>bor <u>r</u>eduction), which encodes a *Drosophila* Sar1 GTPase essential for ER → Golgi vesicle trafficking, preferentially affect dendrite elaboration with minimal effect on axon growth. WT, wild type. **(B)** Golgi outposts (arrowheads), visualized by a fluorescently tagged Golgi marker, are distributed along the dendrites (outlined) of a cultured rat hippocampal neuron. **(C)** Schematic summary of the function of the secretory pathway in dendrites. Vesicle trafficking (black arrows) from ER to plasma membrane via Golgi outposts fuels the growing dendrites with new membrane lipids and proteins. Some of these proteins may be synthesized locally on the rough ER in dendrites. Golgi outposts may also serve as nucleation centers for microtubules, producing both plus-end-out (orange) and minus-end-out (blue) microtubules, with their minus ends interacting with the Golgi outposts and accounting for their bidirectional organization in the dendrites. (A, data from Ye B, Zhang Y, Song W et al. [2007] *Cell* 130:717–729; B, from Horton AC & Ehlers MD [2003] *J Neurosci* 23:6188–6199. With permission from the Society for Neuroscience. See also Ori-McKenney KM, Jan LY & Jan YN [2012] *Neuron* 76:921–930.)

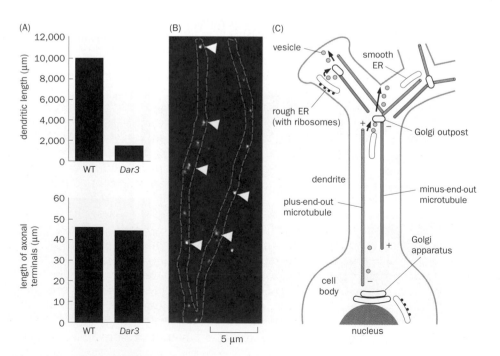

plus-end out, whereas dendrites contain mixed populations of plus-end-out and minus-end-out microtubules. In nonneuronal cells, the minus ends are usually nucleated in the centrosome near the center of the cells. Interestingly, Golgi outposts can also serve as microtubule nucleation centers. Thus, Golgi outposts likely serve two separate functions: they fuel the growth of dendrites by facilitating the delivery of membrane from the ER to the plasma membrane, and they organize microtubules to facilitate bidirectional transport of cargos between cell bodies and dendritic terminals (Figure 7–18C).

ER → Golgi trafficking is an essential step in the secretory pathway that delivers transmembrane proteins and secreted proteins to the plasma membrane. In principle, neuronal proteins destined for the secretory pathway can be delivered to axonal and dendritic compartments after being synthesized and processed by the ER and Golgi apparatus in the cell body. However, the presence of mRNA, ribosomes, ER, and Golgi outposts in the dendrites enables neurons to perform local synthesis of transmembrane proteins and secreted proteins in dendrites far from the cell bodies (see Section 2.2). Local protein synthesis confers flexibility in response to synapse-specific signaling and plays an important role in synaptic plasticity, as will be further discussed in Chapter 10.

7.9 Homophilic repulsion enables self-avoidance of axonal and dendritic branches

Dendrites, as their name (from *dendron* in Greek, meaning tree) indicates, are usually highly branched, enabling them to efficiently sample the input space. Axons also often branch along their long journey so that neurons can send information to postsynaptic targets at distinct locations. Dendritic and axonal branching utilizes one of two mechanisms: (1) growth cone splitting or (2) **interstitial branching**, that is, extending a collateral from the side of a process (**Figure 7–19**). In either case, the two branches must be segregated in order for them to extend in different directions. Studies of *Drosophila* **Dscam** proteins suggest that, in many neurons, an active process is employed to prevent axonal or dendritic branches of the same neuron from sticking to each other.

Dscam proteins are evolutionarily conserved homophilic cell adhesion proteins, with multiple extracellular immunoglobulin (Ig) domains and fibronectin repeats. Remarkably, 38,016 different isoforms are encoded by a single *Dscam* gene and are generated via alternative splicing; the extracellular domain alone has 19,008 different variants (**Figure 7–20**A). Important insights into the function of Dscam diversity came from the convergence of analysis of the *Dscam* mutant phenotypes, the expression pattern of Dscam isoforms, and biochemical and structural studies of Dscam proteins. When Dscam was deleted from whole flies or from individual neurons within genetically mosaic flies, sister dendritic and axonal branches from the same neuron stuck together after branching, as exemplified by the dendrites of sensory neurons that innervate the *Drosophila* larval body wall (Figure 7–20B). Furthermore, the Ig domains encoded by variable exons enable Dscam proteins to bind one another *in trans* (from opposing membranes) in an isoform-specific manner, such that only two identical Dscam isoforms bind strongly to each other (**homophilic binding**; Figure 7–20C). Interestingly, homophilic binding of Dscam extracellular domains leads to cytoplasmic domain signaling that results in repulsion between binding partners. Another clue to the puzzle is that individual neurons appear to express a stochastic set of alternatively spliced Dscam isoforms.

Taken together, a model emerges by which molecular diversity is used to regulate dendritic and axonal branching. Each neuron expresses multiple Dscam isoforms. When a nascent branch forms, the sister branches have the same Dscam proteins and hence bind to and repel each other. This homophilic repulsion ensures **self-avoidance**: different axonal branches from the same neuron go their own ways to innervate distinct targets, and different dendritic branches from the same neuron maximize the territory they cover and minimize overlap. Different neurons express distinct sets of Dscam isoforms, so no strong repulsion exists between their axons or dendrites. This enables different neurons to fasciculate

Figure 7–19 Two distinct branching mechanisms. A dendrite or an axon can generate two branches via growth cone splitting (left), or interstitial branching, where a branch extends from the trunk of a process (right).

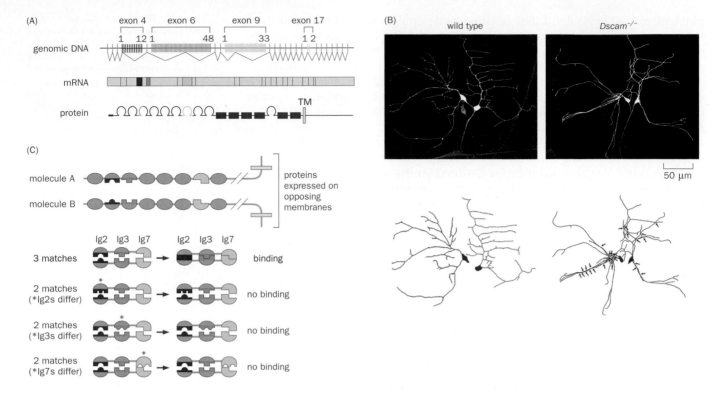

Figure 7–20 Dscam diversity and self-avoidance of axonal and dendritic branches. (A) Top, genomic structure of the *Drosophila Dscam* gene. Each vertical bar represents an exon. Middle, alternative mRNA splicing ensures that in the mature mRNA, exon 4 (red) is chosen from one of 12 variants, exon 6 (blue) from one of 48 variants, exon 9 (green) from one of 33 variants, and exon 17 (yellow) from one of 2 variants. Because these choices are made independently, the *Dscam* gene can produce 38,016 isoforms (= 12 × 48 × 33 × 2). Bottom, protein structure of Dscam. The extracellular portion of Dscam protein is composed of 10 immunoglobulin (Ig) domains (arches) and six fibronectin type III domains (rectangles). The second, third, and seventh Ig domains contain variable exons 4, 6, and 9, respectively. TM, transmembrane domain. **(B)** Dendritic branches of wild-type sensory neurons extend away from each other without overlap (left). In a *Dscam* mutant (right), dendritic branches from the same neuron stick together (arrows). Original images (top) and interpretive drawings (bottom) are shown for two neighboring neurons. **(C)** Summary of Dscam binding. Top, structural studies reveal that the three variable Ig domains contribute to the binding interface of two Dscam molecules located on opposing membranes. Bottom, biochemical studies show that Dscam can mediate strong homophilic binding only when all three variable Ig domains are identical. (A, adapted from Schmucker D, Clemens JC, Shu H et al. [2000] *Cell* 101:671–684. With permission from Elsevier Inc.; B, from Matthews BJ, Kim ME, Flanagan JJ et al. [2007] *Cell* 129:593–604. With permission from Elsevier Inc.; C, adapted from Sawaya MR, Wojtowicz WM, Andre I et al. [2008] *Cell* 134:1007–1018. With permission from Elsevier Inc.)

(bundle) their axons along a common pathway, or to have overlapping dendritic territories.

Are similar mechanisms used in mammals? There are only two Dscam isoforms in mammals, not enough to distinguish dendrites of the same neuron from many other neurons nearby. In a remarkable example of convergent evolution, analogous cell-surface molecules that exhibit molecular diversity in mammals, the **protocadherins**, function in a manner similar to *Drosophila* Dscam to regulate dendrite morphogenesis. Cadherins are ancient calcium-dependent cell adhesion proteins that play numerous important roles in cell–cell adhesion and morphogenesis during development (see Box 5–1 and Sections 5.18 and 6.22. Cadherin proteins form a sizeable family that can be used for cell–cell recognition: a typical vertebrate genome contains about 20 different cadherin genes and additional genes encoding protocadherins that share sequence homologies with cadherins. In particular, three genetic loci, *Pcdha*, *Pcdhb*, *Pcdhg*, produce 14 α-, 22 β-, and 22 γ-protocadherin isoforms, respectively, with a total of 58 extracellular domain variants in mice (**Figure 7–21**A). Biochemical studies suggest that different protocadherins from the same cell can assemble into hetero-multimers, and that these hetero-multimers mediate isoform-specific interactions *in trans*. These combinations in principle can produce many more than 58 distinct protein complexes for specific binding.

Knockout of the *Pcdhg* cluster (which encodes 22 γ-protocadherins) in the retina caused the dendrites of individual amacrine cells to form clumps

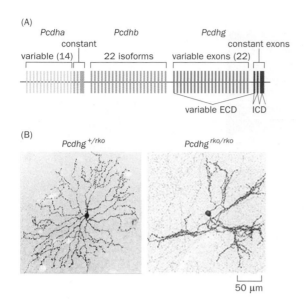

Figure 7–21 Clustered protocadherins in dendritic self-avoidance of mouse retinal neurons. (A) The mouse genome encodes three protocadherin clusters. In the *Pcdha* and *Pcdhg* clusters, an individual variable exon encodes the extracellular domain (ECD), the transmembrane domain, and a portion of the intracellular domain (yellow or cyan), and is joined with exons that encode the constant intracellular domain (ICD) via splicing. In the *Pcdhb* cluster, each exon (green) encodes the entire protein. **(B)** While the dendrites of a control starburst amacrine (heterozygous for a retinal knockout, or *rko*) spread broadly outward (left), homozygous retinal knockout of the entire *Pcdhg* cluster causes the dendrites of a starburst amacrine cell to clump together (right). (Adapted from Lefebvre JL, Kostadinov D, Chen WV et al. [2012] *Nature* 488:517–521. With permission from Macmillan Publishers Ltd.)

(Figure 7–21B), suggesting that protocadherins mediate homophilic repulsion just as *Drosophila* Dscam does. It will be of great interest to examine whether the homophilic repulsion enabled by Dscam, protocadherins, and possibly other molecules plays a general role in the elaboration of complex dendritic trees and branching of complex axonal arbors throughout the nervous system.

A phenomenon related to dendritic self-avoidance is dendritic tiling, which was introduced in Section 4.17 in the context of retinal neuronal types. Here, dendrites from different individual retinal neurons of the same type avoid each other, such that collectively, any retinal neuronal type samples the entire visual world without redundancy (see Figure 4–29). Dendritic tiling also occurs in sensory neurons that innervate the *Drosophila* larval body wall and belong to the same type (see Figure 7–20B). Ablating one neuron can cause dendrites of neighboring neurons of the same type to invade the empty space that was occupied by the ablated neuron, suggesting that mutual repulsion of dendrites from neighboring neurons of the same type accounts for dendritic tiling. Despite playing a key role in dendritic self-avoidance, neither Dscam in flies nor clustered protocadherins in the mammalian retina have been shown to regulate dendritic tiling. Thus, dendritic tiling might involve yet-to-be-discovered cell-surface recognition molecules and mechanisms.

7.10 Subcellular site selection of synaptogenesis uses both attractive and repulsive mechanisms

As axons reach their targets, and dendrites elaborate their branches, the next phase of neural development begins: the formation of synapses that allow neurons to communicate with each other or with their muscle targets (see Chapter 3). In this section, we discuss site selection for synapse formation, and subsequent sections will address how synaptogenesis occurs.

Many axons in the mammalian central nervous system (CNS) not only must find their postsynaptic partners, but also must target synapses to specific subcellular sites on the postsynaptic neurons. For instance, as we introduced in Section 3.25, three classes of neocortical GABAergic interneurons form synapses on different subcellular compartments of the pyramidal neurons: basket cells, chandelier cells, and Martinotti cells form synapses at the somata, axon initial segments, and distal dendrites of the target pyramidal neurons, respectively (see Figure 3–46). How do GABAergic neurons selectively form synapses on specific subcellular compartments of their target neurons? Insights have come from studies of this problem in the cerebellar basket cells, which synapse selectively onto the somata as well as axon initial segments of their postsynaptic target Purkinje cells (there are no chandelier cells in the cerebellum). **Neurofascin**, an Ig superfamily

Figure 7–22 Subcellular distribution of neurofascin directs basket cell presynaptic terminal formation. (A) In a wild-type Purkinje cell labeled in green by the Purkinje marker calbindin (Calb), neurofascin (NF, in red) is concentrated at the axon initial segment (AIS, between the two arrows). Neurofascin is also distributed along the Purkinje cell body (arrowheads) in a gradient, with the highest concentrations being close to the axon. **(B)** In an *Ankg* mutant, neurofascin is distributed evenly along the Purkinje cell body and is present in the axon at distances farther from the soma as indicated by the distance between the two arrows. The bracket indicates the length of AIS in wild-type Purkinje cells. **(C)** Schematic showing basket cell axon terminals (green) and neurofascin (red) distribution in a wild type and in an *Ankg* mutant. In the mutant, the neurofascin gradient in the Purkinje cell body and its restriction to the AIS segment of the axon are disrupted, causing basket cell axon terminals to overshoot the AIS and to branch abnormally. (Adapted from Ango F, di Cristo G, Higashiyama H et al. [2004] *Cell* 119:257–272. With permission from Elsevier Inc.)

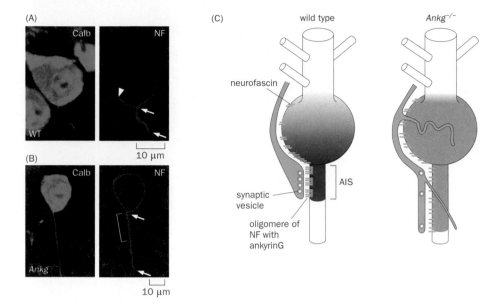

molecule, serves as a cue to target the basket cell axon. Neurofascin forms a subcellular gradient along the surface of the Purkinje cell, with the highest concentration at the axon initial segment. This is because the cytoplasmic domain of neurofascin is associated with an intracellular scaffolding protein called **ankyrinG**, which is highly concentrated at the axon initial segment. In *Ankg* mutant mice, which do not produce ankyrinG, neurofascin distribution became diffuse, as did the distribution of basket cell synapses (**Figure 7–22**). Thus, the precise subcellular targeting of basket cell axons is directed by selective subcellular distribution of an attractive cue on the target neuron.

A different solution to subcellular synaptic targeting has been uncovered in studies of the DA9 motor neuron of *C. elegans*. The DA9 neuron forms presynaptic terminals only along a specific segment of its axon's contact with its muscle targets (**Figure 7–23**). Synapse distribution in DA9 is directed by a posteriorly secreted Wnt protein. (In *C. elegans* as in mammals, Wnt proteins are used as morphogens to specify global patterning in early developmental stages.) Wnt signaling mediated by a receptor expressed in the DA9 neuron inhibits synapse formation in the posterior segment of the axon. Synaptogenesis occurs only in an anterior zone further away from the influence of posterior Wnt. Thus, an extracellular protein that acts to pattern the global body plan is also used to confine synapse formation to a specific segment along the axon.

Figure 7–23 Presynaptic terminal distribution in *C. elegans* DA9 neuron is regulated by an external Wnt gradient. (A) Schematic drawing of the DA9 motor neuron in the posterior worm, depicting the location of its cell body, dendrite, axon, and presynaptic terminals. Lin44 is one of the *C. elegans* Wnt homologs and is produced by posterior cells. A, anterior; P, posterior; D, dorsal; V, ventral. **(B)** In wild-type second instar larvae (L2), Lin44 expression at the posterior (green) prevents presynaptic terminals (marked in red by the synaptic vesicle marker Rab3) from forming in the bracketed axon segment. **(C, D)** Compared with wild-type worm (C), the bracketed synapse exclusion zone is shorter in a *Lin44* mutant worm (D), as presynaptic terminal distribution (marked by Rab3) extends more posteriorly. (From Klassen MP & Shen K [2007] *Cell* 130:704–716. With permission from Elsevier Inc.)

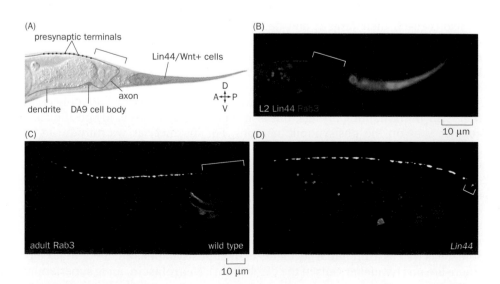

These two examples illustrate that, as is the case for axon guidance (see Box 5–1), the targeting of synapses to specific subcellular locations can be achieved through attractive or repulsive mechanisms and by using secreted or cell-surface-bound cues.

7.11 Bidirectional trans-synaptic communication directs the assembly of synapses

Once axons reach their final target, they are ready to form synapses with their postsynaptic partners. What are the mechanisms that convert a dynamic growth cone to a stable presynaptic terminal? What triggers the differentiation of the postsynaptic specialization? Extensive communication between pre- and post-synaptic partners has been identified that triggers the differentiation of both compartments, converting an initial contact into what can be a life-long union as synaptic partners.

The vertebrate neuromuscular junction has been successfully used as a model to study both synaptic transmission (see Chapter 3) and synapse development because of its experimental accessibility. A key step in postsynaptic development is the clustering of acetylcholine receptors (AChRs) apposing the presynaptic motor axon terminal, which allows efficient synaptic transmission (see Section 3.1). Motor axon terminals produce a protein called **agrin**, named after its ability to cause AChR aggregation in cultured muscle cells (**Figure 7–24**A). Agrin acts through a receptor complex in muscle composed of **LRP4** (low-density lipoprotein receptor-related protein 4) and **MuSK** (muscle-specific receptor tyrosine kinase), which activates a signaling cascade that facilitates AChR clustering. Indeed, mice mutant for *Agrin*, *Lrp4*, or *Musk* all exhibit severe defects in AChR clustering (Figure 7–24B) and die at the neonatal stage, providing *in vivo* support for the function of this ligand–receptor complex in neuromuscular junction development.

Interestingly, muscle can be pre-patterned in the mouse as AChR clusters form at the right location in the absence of innervating motor axons, and motor axons appear to seek pre-patterned AChR clusters to form synapses. Indeed, the lack of AChR clustering in *Agrin* mutant mice can be suppressed by a lack of motor axon innervation altogether. These experiments led to a current model: motor axon terminals send at least two signals to the muscle, one (likely the neurotransmitter ACh itself) to disperse AChR, and the other (agrin) to counteract the dispersing signal and thereby reinforce AChR clustering near a functional (that is, ACh-producing) motor nerve terminal (Figure 7–24C). These findings highlight the extensive trans-synaptic communication necessary to establish just one aspect of synapse maturation, the clustering of AChR. Indeed, in addition to serving as a co-receptor with MuSK for agrin, muscle-derived LRP4 also induces presynaptic

Figure 7–24 Communication between motor axons and muscles in establishing the neuromuscular junction. (A) When chicken myotubes (control, top) are exposed to the agrin-containing fraction of *Torpedo* electric organ, AChR clusters form (bottom). AChRs are visualized by fluorescently tagged α-bungarotoxin, a snake-derived toxin that binds to and inhibits nicotinic AChR (see Box 3–2). **(B)** Drawings depict the results of staining with α-bungarotoxin (red) in heterozygous control mice (top) and *Agrin* mutant mice (bottom). *Agrin* knockout results in severe disruption of AChR clustering (spread of red signals) and abnormal branching of motor axons (gray). *Musk* or *Lrp4* mutant mice exhibit similar phenotypes. **(C)** Schematic summary of communication between the motor axon terminal and the muscle cell. Muscle AChRs spontaneously form clusters (1). The motor axon produces ACh (2) to disperse AChR clusters by endocytosis (3). A second product from the motor axon, agrin (4), acts through the MuSK/LRP4 receptor complex to inhibit AChR dispersion (5). Muscle LRP4 also serves as a ligand that signals back to the motor axon (6) to induce presynaptic differentiation. (A, From Godfrey EW, Nitkin RM, Wallace BG et al. [1984] *J Cell Biol* 99:615–627; B, from Gautam M, Noakes PG, Moscoso L et al. [1996] *Cell* 85:525–535. With permission from Elsevier Inc.; C, adapted from Kummer TT, Misgeld T & Sanes JR [2006] *Curr Opin Neurobiol* 16:74–82. With permission from Elsevier Inc. See also Zhang B, Luo S, Wang Q et al. [2008] *Neuron* 60:285–297; Kim N, Stiegler AL, Cameron TO et al. [2008] *Cell* 135:334–342; Yumoto N, Kim N & Burden SJ [2012] *Nature* 489:438–442.)

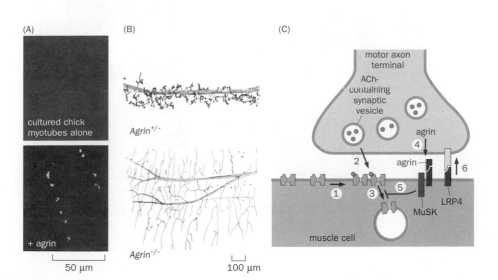

(A)

cultured chick myotubes alone

+ agrin

50 μm

(B)

Agrin$^{+/-}$

Agrin$^{-/-}$

100 μm

(C)

motor axon terminal

ACh-containing synaptic vesicle

agrin

4

2

agrin

6

1

3

5

LRP4

MuSK

muscle cell

terminal differentiation independent of MuSK (Figure 7–24C). Additional motor axon- and muscle-derived signals likely act to stabilize and maintain the synaptic connections.

Of the trans-synaptic signaling molecules identified in the CNS, the transmembrane proteins **neurexin** and **neuroligin** have been studied most extensively. Neurexins are enriched on the presynaptic membrane, and their binding partner neuroligins are located on the postsynaptic membrane (**Figure 7–25**A; see Figures 3–10 and 3–27 for the roles of neurexin and neuroligin in organizing presynaptic terminal and postsynaptic density). Analogously to agrin at the neuromuscular junction, neurexins expressed from cultured nonneuronal cells can induce opposing neurons to develop postsynaptic specializations, including the clustering of postsynaptic scaffolding proteins and receptors for GABA (Figure 7–25B) or glutamate. Conversely, neuroligins expressed in cultured nonneuronal cells can induce apposing neurons to develop presynaptic specializations, including active zones and synaptic vesicle clusters (Figure 7–25C). Neuroligins can also determine the type of synapses. For example, neuroligin2 is enriched at inhibitory postsynaptic sites and in *in vitro* assays can induce the differentiation of GABAergic presynaptic assemblies, whereas neuroligin-1 is enriched at excitatory postsynaptic sites and can induce glutamatergic presynaptic assemblies. These data suggest that neuroligin and neurexin act as bidirectional trans-synaptic signals for the induction of both pre- and postsynaptic assemblies.

Despite the robust synapse-inducing activity demonstrated by neuroligins *in vitro*, mice lacking all three neuroligin genes still have the normal number of synapses, albeit with severely reduced synaptic transmission. One possible explanation is that neuroligins act in parallel with other trans-synaptic signaling pathways to induce synapse development; other pathways may compensate for synapse development in the absence of neuroligins. These knockout experiments nevertheless revealed an essential role of neuroligin in synapse maturation

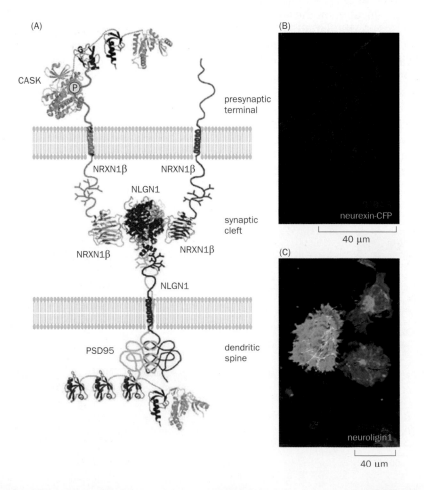

Figure 7–25 Neurexin/neuroligin-mediated trans-synaptic interactions can induce synapse assembly. (A) A structural model of trans-synaptic binding between a dimer of neuroligin1 (NLGN1) and two molecules of NRXN1β, a neurexin isoform. The cytoplasmic domain of NLGN1 binds one of the three PDZ domains (red) of PSD95, a postsynaptic scaffolding protein (see Figure 3–27); the cytoplasmic domain of NRXN1β binds the PDZ domain of CASK, a presynaptic scaffolding protein. **(B)** Expression of neurexin1β (tagged with cyan fluorescent protein, or CFP) on the surface of a fibroblast (blue) induces clustering of GABA receptors (red) in co-cultured neurons. **(C)** Expression of neuroligin1 (green) in HEK293 cells (the cell on the left express the highest amount) induces clustering of synapsin, a synaptic-vesicle-associated protein (red), in passing axons of co-cultured neurons. Yellow indicates co-localization of synapsin and neuroligin. (A, from Südhof TC [2008] *Nature* 455:903–911. With permission from Macmillan Publishers Ltd.; B, from Graf E, Zhang X, Jin SX et al. [2004] *Cell* 119:1013–1026.With permission from Elsevier Inc.; C, from Scheiffele P, Fan J, Choih J et al. [2000] *Cell* 101:657–669. With permission from Elsevier Inc.)

and synaptic transmission. Finally, as an indication of the importance of trans-synaptic signaling molecules in human health, mutations in neurexins and neuroligins as well as their associated scaffolding proteins, have been linked with neurodevelopmental disorders including autism and schizophrenia, which will be discussed further in Chapter 11.

7.12 Astrocytes stimulate synapse formation and maturation

Synapses are often wrapped by glia, such as the astrocytes that surround CNS synapses and the Schwann cells that surround neuromuscular junctions (see Figure 3–3). In addition to supporting many aspects of synaptic functions, such as neurotransmitter recycling (see Figure 3–12), glia also regulate synapse development.

The roles of astrocytes in synapse development have been systematically examined in cultured retinal ganglion cells (RGCs). Purified RGCs cultured *in vitro* can extend axons and dendrites, but exhibit little synaptic activity as measured by spontaneous postsynaptic currents. (In this artificial culture system, RGCs constitute both the presynaptic and postsynaptic cells even though *in vivo* RGCs do not form synapses with each other.) When RGCs were co-cultured with astrocytes, postsynaptic currents increased dramatically in both magnitude and frequency (**Figure 7–26**A). This was due both to increased synapse number and enhanced synaptic strength. Conditioned media from cultured astrocytes could mimic the effect of astrocytes, which provided a means for biochemical identification of astrocyte-derived factors responsible for promoting synapse number and function. A class of astrocyte-derived factor—the **thrombospondins** (**TSPs**)—was identified that enhances synapse number: adding TSP1 to the RGC culture mimicked the synaptogenic effects of astrocytes or astrocyte-conditioned media, as measured by RGC synapse numbers (Figure 7–26B). However, although TSP-induced synapses appeared structurally normal, they were functionally silent as measured by spontaneous postsynaptic currents. Another class of astrocyte-derived factor—the glypicans, which are glycosylated extracellular proteins anchored to the cell surface by GPI (the anchor can be cleaved by endogenous phospholipases)—was necessary to produce functionally mature synapses (Figure 7–26C). Further studies suggested that TSPs use a postsynaptic calcium channel subunit called α2δ-1 as a receptor to promote synapse formation. (The role of α2δ-1 as a TSP receptor is unrelated to its calcium channel function.) Glypican was found to promote postsynaptic surface expression of GluA1, an AMPA glutamate receptor subunit, thereby enhancing synaptic efficacy. Importantly, mouse knockout mutants for TSPs, glypicans, or α2δ-1 exhibit reduced synapse numbers or reduced efficacy in synaptic transmission, supporting the function of these proteins in synapse formation and maturation *in vivo*.

In summary, synapse development relies on extensive communication between pre- and postsynaptic partners as well as with glia. These communications can be achieved by secreted factors (for example, agrin and thrombospondin) or transmembrane proteins that interact across the synaptic cleft (for example, neurexin and neuroligin). Such signaling events ensure that presynaptic terminals develop active zones for efficient neurotransmitter release and that the apposing postsynaptic densities are enriched in neurotransmitter receptors. As was discussed in Chapter 5, neuronal activity plays a major role in refining synaptic connections during development. A prime example of how neuronal activity

Figure 7–26 Secreted factors from astrocytes promote synapse formation and maturation. (A) Whole-cell recording of spontaneous postsynaptic currents from a retinal ganglion cell (RGC) that has been cultured with other purified RGCs (left) or from an RGC that has been co-cultured with purified RGCs and astrocytes (right). The presence of astrocytes increases the frequency and magnitude of spontaneous postsynaptic currents. **(B)** Co-culture with astrocytes (astros) causes a 6-fold increase in RGC synapse number as quantified from electron micrographs. This effect is mimicked by astrocyte-conditioned media (ACM) or purified thrombospondin1 (TSP1). **(C)** TSP is insufficient to enhance synaptic strength as measured by postsynaptic currents (top). Addition of media from COS-7 cells expressing glypican4 enhances synaptic strength (bottom). (A, data from Ullian EM, Sapperstein SK, Christopherson KS et al. [2001] *Science* 291:657–661; B, data from Christopherson KS, Ullian EM, Stokes CCA et al. [2005] *Cell* 120:421–433; C, data from Allen NJ, Bennett ML, Foo LC et al. [2012] *Nature* 486:410–414.)

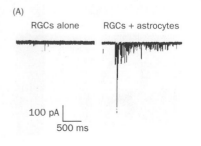

(A)
RGCs alone   RGCs + astrocytes

100 pA
500 ms

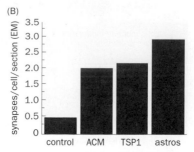

(B)

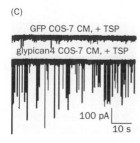

(C)
GFP COS-7 CM, + TSP

glypican4 COS-7 CM, + TSP

100 pA
10 s

influences synapse maturation can be observed in neuromuscular junction connectivity, our next topic under discussion.

7.13 Activity and competition refine neuromuscular connectivity

In mammals, not all synapses that form during development are long lasting. We've already seen salient examples in our study of the visual system. For example, retinal ganglion cells (RGCs) from both eyes initially connect with the same target neuron in the lateral geniculate nuclei (LGN); through activity-dependent mechanisms following Hebb's rule, the connections are refined such that each LGN neuron is only innervated by RGCs from a single eye in an adult (see Figure 5–26). A similar phenomenon has been observed in the maturation of neuromuscular connectivity. In mature mice, each muscle fiber (an individual muscle cell) is innervated by a single motor axon. But in newborn mice, each muscle fiber is innervated by as many as 10 motor axons. During the first two weeks of postnatal life, the poly-innervation pattern is refined to mono-innervation through a process called **synapse elimination** (**Figure 7–27**A). How does this refinement occur? Easier access to the neuromuscular junction compared with synapses in the central nervous system has allowed the synapse elimination process to be characterized with greater detail.

Time-lapse imaging studies have charted a detailed time course of synapse elimination at the neuromuscular junction. In newborn mice, multiple axons form synapses at the same neuromuscular junction, intermingling their terminals. Next, terminals that originate from different motor axons gradually segregate: one terminal expands while the others shrink. Eventually, all of the motor axons except one lose their territory and withdraw (Figure 7–27B). These observations suggest a competitive process among different motor axons to sculpt the final connectivity pattern. Synapse elimination is an activity-dependent process, because blocking neuromuscular synaptic transmission slows down the process that leads from poly-innervation to mono-innervation. If synaptic transmission is eliminated in a subset of axons by conditionally deleting choline acetyltransferase, an enzyme essential for ACh synthesis, the mutant axons cannot release ACh and are usually unable to compete effectively against a wild-type axon (Figure 7–27C).

As we will learn in Chapter 8, each muscle (composed of many muscle fibers) is innervated by multiple motor neurons that belong to the same 'motor pool.' We already alluded to the specificity of motor neuron–muscle connections in Section 7.4 in the context of transcription factors specifying trajectory choice of motor axons innervating dorsal and ventral limbs (see Figure 7–9). Indeed, classic embryological experiments in the chick have revealed that motor neurons innervating different muscles already exhibit high specificity when they first invade the muscle field, making minimal errors. This is likely mediated by molecular determinants that match different motor neurons and muscles. However, within each muscle, an individual motor neuron only innervates a subset of muscle fibers, and collectively motor neurons of the same motor pool innervate all muscle fibers through exuberant connections followed by synapse elimination (Figure 7–27A). What does the outcome look like after this process?

Figure 7–27 Competitive synapse elimination shapes the motor neuron–muscle connectivity. (A) Schematic of neuromuscular junction maturation. In newly born mice, each muscle fiber is innervated by multiple motor neuron axons. During the first two postnatal weeks, synapses are eliminated through a local competitive process until each muscle fiber is innervated by a single motor neuron. **(B)** Summary of synapse elimination steps. Inputs from different motor neurons are initially intermingled. Their terminals are then segregated. As one terminal expands, others shrink and withdraw. **(C)** Motor axons that lack choline acetyltransferase (ChAT), and thus cannot release ACh, do not compete effectively with co-innervating axons that express ChAT. In this micrograph, two axons innervating the same neuromuscular junction are stained in red. Only one axon contains the ChAT protein (labeled cyan by an anti-ChAT antibody). The ChAT containing axon is thicker and occupies more territories, and is predicted to win the competition. (A, adapted from Tapia JC, Wylie JD, Kasthuri N et al. [2012] *Neuron* 74:816–829. With permission from Elsevier Inc.; B, adapted from Sanes JR & Lichtman JW [1999] *Annu Rev Neurosci* 22:389–442. With permission from Annual Reviews; C, adapted from Buffelli M, Burgess RW, Feng G et al. [2003] *Nature* 424:430–434. With permission from Macmillan Publishers Ltd.)

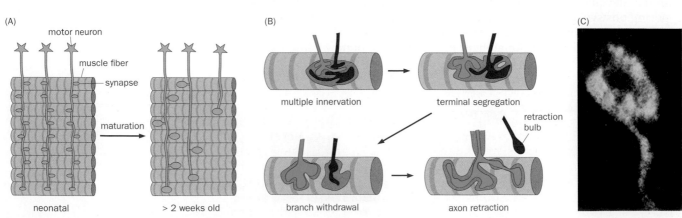

(A)

motor neuron

muscle fiber

synapse

maturation

neonatal   > 2 weeks old

(B)

multiple innervation   terminal segregation

retraction bulb

branch withdrawal   axon retraction

(C)

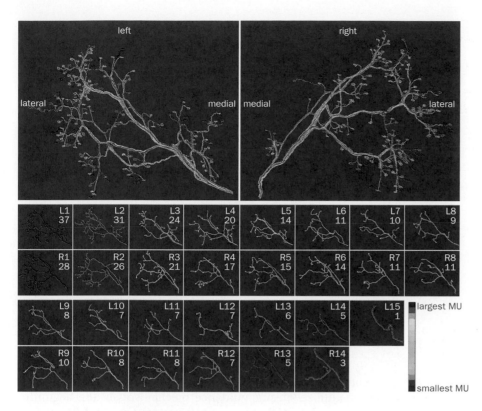

Figure 7–28 A connectome consisting of a muscle pair and complete set of innervating motor axons. (Top) Complete reconstruction of the connectivity of all motor neurons (15 on the left, 14 on the right) with a pair of small muscles (~200 muscle fibers per side) in the ear of a one-month-old mouse. Each motor axon is differentially colored, with branches ending on individual muscle fibers (not shown). (Bottom) Branching patterns of each motor neuron from the left (L1–L15) and right (R1–R14) are arranged according to the number of muscle fibers the neuron innervates (the size of motor unit, or MU), shown below the motor neuron number on the top right. The right-side neurons are flipped for ease of comparison with the left-side neurons. At the level of individual motor neurons, no similarities in the left- and right-side branching patterns are apparent. (From Lu J, Tapia JC, White OL et al. [2009] *PLoS Biol* 7:e1000032.)

The **connectome** of a small ear muscle—the pattern of synaptic connections formed between the entire set of motor neurons (about 15) and the entire set of muscle fibers they innervate (about 200)—has been reconstructed for each side of a mouse (**Figure 7–28**). Although the motor neurons that target the ear muscles on the left or right side share an identical genome, the branching patterns of the left-side and right-side motor axons differ considerably. These observations demonstrate that the detailed connection pattern is not genetically specified and is likely a result of synapse elimination based on local, activity-dependent competition. Patterns nevertheless emerge. For example, each muscle is innervated by a collection of motor neurons that synapse with a few to several dozen muscle fibers. The number of muscle fibers a motor neuron innervates correlates with its axon diameter. These properties provide the anatomical basis for the 'size principle' that we will examine further in Chapter 8: motor neurons are usually progressively activated by increasing activity, with the smaller motor neurons (that innervate fewer muscle fibers) becoming activated before larger motor neurons (that innervate more muscle fibers). Future analysis of how this relatively simple connectome emerges during development will shed light on the rules by which neuronal activity sculpts synaptic connectivity.

7.14 Developmental axon pruning refines wiring specificity

The developing nervous system also employs **stereotyped axon pruning** to sculpt connectivity. Stereotyped axon pruning differs from neuromuscular refinement in that the outcomes are the same from animal to animal and the process usually involves elimination of long-distance axonal projections. One of the most striking examples was discovered while studying mammalian cortical layer 5 subcerebral projection neurons (SCPNs, see Section 7.4) by injecting retrograde tracers into the target areas during different stages of development. SCPNs from both the primary motor cortex and the visual cortex first project axons to the spinal cord, and then extend interstitial branches (see Figure 7–19) to innervate the superior colliculus and several brainstem targets. Subsequently, SCPNs from the motor cortex prune their branches that innervate the superior colliculus, whereas visual cortical SCPNs prune their branches that innervate the spinal cord, consistent with their respective functions in controlling body and eye movement (**Figure 7–29**). It is unclear why certain neurons employ stereotyped axon pruning after initial

Figure 7–29 Sculpting connection specificity of subcerebral projection neurons through stereotyped axon pruning. Cortical layer 5 subcerebral projection neurons (SCPNs) in the primary motor cortex and the visual cortex initially develop similar projection patterns. Both extend axons to the spinal cord prior to sending interstitial branches to the superior colliculus (SC) and several brainstem structures (mes, mesencephalon; p, pons; IO, inferior olive; DCN, dorsal column nuclei). The wiring specificity is shaped by subsequent axon pruning. Motor cortical SCPNs selectively prune their SC branch, whereas visual cortical SCPNs selectively prune their branches to the spinal cord and most brainstem targets. (Adapted from Luo L & O'Leary DM [2005] *Annu Rev Neurosci* 28:127–156. With permission from Annual Reviews.)

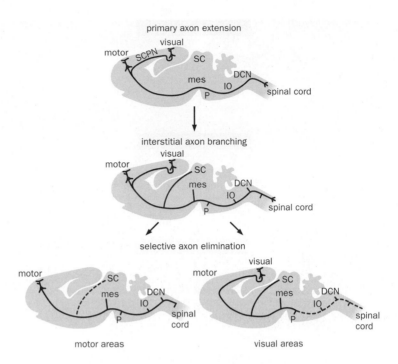

exuberant connections to achieve wiring specificity in long-distance projections, whereas most neurons utilize specific guidance mechanisms at the beginning (see Figure 5–3 for a comparison of these two mechanisms). One speculation for a stereotyped axon pruning strategy employed by SCPNs is evolutionary constraint: ancestral SCPNs might have innervated all subcortical targets; after SCPNs diversified in their functions in controlling body or eye movement, it was easier to evolve selective axon pruning than to alter initial guidance for achieving the final connection specificity of different subtypes of SCPNs.

Stereotyped axon pruning also occurs extensively in the insect nervous system during metamorphosis, when the larval connectivity is transformed to the adult connectivity to suit their different life styles (for example, crawling versus flying). Studies of axon pruning in *Drosophila* mushroom body neurons (whose function will be encountered in Chapters 9 and 10) revealed that the larva-specific axonal branches are eliminated by a process called **developmental axon degeneration**, during which axons are fragmented into pieces that are subsequently engulfed by surrounding glia (**Figure 7–30**A). Developmental axon degeneration is morphologically similar to **Wallerian degeneration**, a process that eliminates the severed axonal segments originally described by Augustus Waller in 1850. Interestingly, developmental axon degeneration and Wallerian degeneration also share molecular mechanisms. For example, microtubule disassembly is an early step in both processes. Further, axon pruning was inhibited in mushroom body neurons in which the **ubiquitin-proteasome system**, a universally used protein degradation system in eukaryotes, was disrupted (Figure 7–30B). Proteasome inhibitors also slowed the degeneration of distal axons of rat optic nerve following transection (Figure 7–30C). These studies raised the possibility that an axon self-destruction program may be employed during developmental axon degeneration, and is reactivated in adults in response to injury. Indeed, the axon degeneration program may also be abnormally activated in certain neurodegenerative disorders, causing axon degeneration that contributes to clinical symptoms.

7.15 Neurotrophins from target cells support the survival of sensory, motor, and sympathetic neurons

Our discussion so far indicates that nervous system development employs both progressive events (for example, neurogenesis, axon extension, dendrite elaboration, and synapse formation) and regressive events (for example, synapse elimination and axon pruning). Another widely occurring regressive event is

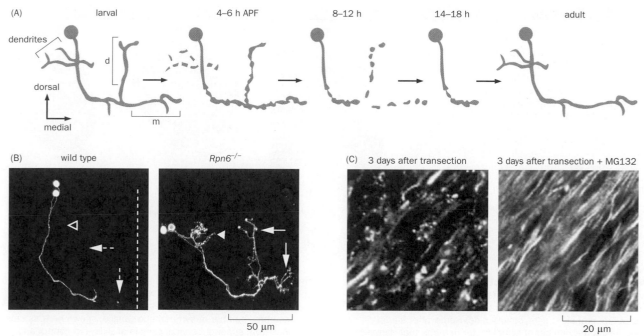

Figure 7–30 Stereotyped axon pruning during *Drosophila* metamorphosis via developmental degeneration, which shares similarities with Wallerian degeneration. (A) Schematic illustration of dendrite and axon pruning of a single mushroom body (MB) neuron during metamorphosis, transforming larval-specific to adult-specific projection patterns. In the larva (left panel), each MB neuron extends a single process that gives rise to several dendritic branches near the cell body. The axon bifurcates at a specific location to produce a dorsal branch (d) and a medial branch (m). During the first 18 hours after puparium formation (APF, the middle three panels), dendrites as well as the medial and dorsal axonal branches degenerate by breaking into pieces, retaining the axon before the original branching point. This axon then extends only medially to form the adult-specific pattern (right). **(B)** Two wild-type (left) or *Rpn6* mutant (right) MB neurons from *Drosophila* at 18 h

APF visualized by a mosaic labeling method (see Section 13.16). Whereas the larval-specific axonal branches (dashed arrows) and dendrites (open triangles) have been pruned in the wild type, they are inappropriately retained in *Rpn6* mutant neurons (arrows and arrowheads). Rpn6 is a key proteasome subunit conserved from yeast to human. Dotted lines indicate the midline. **(C)** Rat optic nerves are stained with an antibody against tubulin. Three days after transection, control optic nerves distal to the transection show considerable degeneration (left), whereas degeneration is delayed by treatment with MG132, a proteasome inhibitor (right). (A, adapted from Luo L & O'Leary DM [2005] *Annu Rev Neurosci* 28:127–156. With permission from Annual Reviews; B, from Watts RJ, Hoopfer ED & Luo L [2003] *Neuron* 38:871–885. With permission from Elsevier Inc.; C, from Zhai Q, Wang J, Kim A et al. [2003] *Neuron* 39:217–225. With permission from Elsevier Inc.)

programmed cell death (apoptosis). Indeed, the phenomenon and core mechanisms of apoptosis are conserved from *C. elegans* to mammals and are employed in the development of most tissues and organs, including the nervous system. For example, about 40% of the spinal motor neurons that innervate a chick's limbs die during the course of normal development.

Classic embryological manipulation experiments in chicks suggested that peripheral targets affect motor neuron numbers. Removal of a limb bud caused a reduction in the number of motor neurons that innervated that limb, whereas transplantation of an extra limb bud increased the number of limb-innervating motor neurons (**Figure 7–31**A). The number of sensory neurons in the dorsal root ganglia is similarly affected by peripheral targets. In principle, changes of neuronal number in these experiments could result from peripheral targets influencing the proliferation, differentiation, or survival of motor and sensory neurons. Systematic examinations of how the number and morphology of neurons change during the time course of development suggested that regulation of neuronal survival plays a predominant role. Searching for the peripheral agent responsible for neuronal survival led to the discovery of a secreted protein named **nerve growth factor** (**NGF**), which has a dramatic stimulating effect on the outgrowth of axons of sensory and sympathetic ganglia cultured *in vitro* (Figure 7–31B). The effect of NGF on neuronal survival *in vivo* was validated by the observation that daily infusion of purified NGF into newly born mice greatly increased the size of sympathetic ganglia, whereas daily infusion of NGF antiserum, which inhibits the activity of NGF, drastically decreased their size due to neuronal death (Figure 7–31C). Together, these experiments led to the influential **neurotrophic**

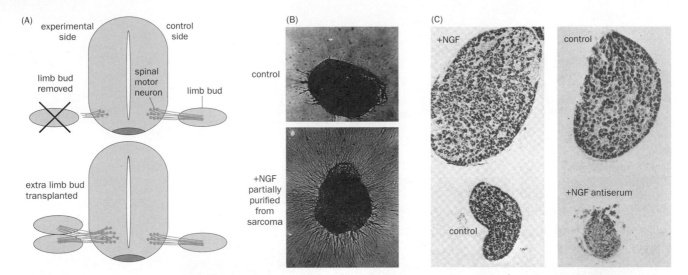

Figure 7–31 Neuron–target interactions and the effect of nerve growth factor. (A) Schematic summary of interactions between spinal motor neurons and their limb targets in chick embryos. Left, experimental side; right, control side. Removal of a limb bud decreases the number of spinal motor neurons that normally innervate the limb bud, whereas transplantation of an extra limb bud increases the number of corresponding spinal motor neurons. **(B)** Compared with an untreated control (top), a cultured sympathetic ganglion incubated for 18 h *in vitro* with nerve growth factor (NGF) partially purified from sarcoma exhibits abundant axonal outgrowth (bottom). **(C)** Left, injecting mice daily with purified NGF, starting at birth, causes dramatic increases in the size and neuronal number of sympathetic ganglia. Shown here are cross sections from thoracic ganglia of 19-day-old control and NGF-injected mice. Right, daily injection of NGF antiserum causes a drastic reduction in the size and neuronal number of sympathetic ganglia, shown here as cross sections from superior cervical ganglia of 9-day-old control and NGF antiserum-injected mice. Detailed analyses indicate that these changes are caused by NGF's effect on the survival of sympathetic neurons. (A, see Hamburger V [1934] *J Exp Zool* 68:449–494; Hamburger V [1939] *Physiol Zool* 12:268–284; B, from Cohen S, Levi-Montalcini R & Hamburger V [1954] *Proc Natl Acad Sci USA* 40:1014–1018; C, from Levi-Montalcini R & Booker B [1960] *Proc Natl Acad Sci USA* 46:373–384 & 384–391.)

hypothesis: the survival of developing neurons depends on neurotrophic factors produced by target cells. This ensures number matching between properly connected neurons and their targets.

NGF is the founding member of the **neurotrophin** family, which also includes **brain-derived neurotrophic factor** (**BDNF**), **neurotrophin-3** (**NT3**), and **neurotrophin-4** (**NT4**). All neurotrophins are the cleavage products of proneurotrophins. The proneurotrophins, as well as all mature neurotrophins, bind to the **p75NTR** neurotrophin receptor, whereas each neurotrophin binds preferentially to one of the three high-affinity **Trk** receptor tyrosine kinases (**Figure 7–32**A). Elegant *in vitro* culture studies have demonstrated that NGF must be supplied to the axonal compartment, rather than the cell-body compartment, in order to maintain the survival of cultured sympathetic neurons and to stimulate axon growth (Figure 7–32B). As introduced in Box 3–4 in the context of receptor tyrosine signaling, neurotrophins act as dimers to bring two Trk receptor molecules into close proximity so that their cytoplasmic tyrosine kinases can cross-phosphorylate each other (see Figure 3–39). This leads to the activation of a number of downstream signaling pathways. One of these, which involves the Sos, Ras, and MAP kinase cascade we encountered in Section 5.17, activates transcription of genes that promote survival and differentiation. Trk and Ras also activate phosphatidylinositol 3-kinase (PI3K) and subsequently a protein kinase named Akt (also known as protein kinase B), which inhibits apoptotic signaling and thus promotes survival. Interestingly, much of the signaling occurs while neurotrophins and their receptors are being transported from the axon terminal to the cell body (Figure 7–32C).

The role of neurotrophins in supporting the differentiation and survival of sensory, sympathetic, and motor neurons is widely supported by experimental evidence. Neurotrophins and Trks are also widely expressed in the brain during development and are continually expressed in the adult. However, it is unclear whether most CNS neurons require target-derived trophic support for their survival. On the other hand, extensive evidence has suggested that neurotrophins in

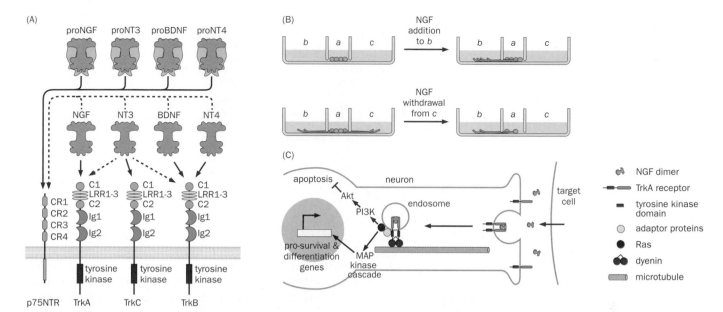

Figure 7–32 Neurotrophin signaling mechanisms. (A) Neurotrophins and their receptors. NGF, BDNF, NT3, and NT4 are proteolytic products of secreted proneurotrophins (all shown as dimers), all of which bind to p75NTR (neurotrophin receptor). Mature neurotrophins bind to p75NTR with low affinity (dashed arrows), and bind with high affinity to one of the three Trk receptors. Ig, immunoglobulin domain; LRR, leucine-rich repeat; CR, C1, C2, cysteine-rich domains. **(B)** Testing where NGF is required for axon growth and neuronal survival. In a three-compartment culture dish, sympathetic neurons are placed in the center compartment *a*, and axons are allowed to grow into compartments *b* and *c*. However, media exchange between compartments is prevented by grease at the bottom of the compartment dividers. Top, neurons in chamber *a* extend axons only toward the chamber where NGF is supplied. Bottom, at the start, sympathetic neurons have already extended their axons to NGF-supplied chambers *b* and *c*. If NGF is subsequently withdrawn from chamber *c*, the axons in chamber *c* degenerate, and neurons that sent axons only to compartment *c* die. These experiments indicate that NGF is required at the axon terminal to support neuronal survival and axon growth. **(C)** Schematic of NGF signaling pathway. Target-derived NGF binds TrkA receptors at the axon terminal. The NGF/TrkA complex is endocytosed and retrogradely transported in endosomes from the axon terminal to the cell body. Activated TrkA receptors recruit adaptors and signaling molecules to the endosome. These signaling events alter the gene expression program and promote survival of the neuron. (A, adapted from Reichardt LF [2006] *Phil Trans R Soc B* 361:1545–1564. With permission from the Royal Society; B, See Campenot RB [1977] *Proc Natl Acad Sci USA* 74:4516–4519; C, see Zweifel LS, Kuruvilla R & Ginty DD [2005] *Nat Rev Neurosci* 6:615–625.)

the CNS regulate many other important functions, including dendrite morphogenesis, synapse development, and synaptic plasticity. Thus, this important class of proteins has evolved to serve diverse roles in vertebrate neural development and function.

ASSEMBLY OF OLFACTORY CIRCUITS: HOW DO NEURAL MAPS FORM?

So far in the chapter, we have followed the developmental sequence of individual neurons from patterning the progenitors and regulating neurogenesis to establishing and refining their synaptic connections. We have utilized examples from *C. elegans* motor neurons and *Drosophila* sensory neurons to mammalian spinal cord and cortical neurons to illustrate general principles of neural development. In the next part of the chapter, we use the specific example of how olfactory maps are established to address the second question raised at the beginning of the chapter: How do groups of neurons wire in a coordinated manner to form a functional circuit? This focus on the olfactory system provides an informative comparison to the wiring of the visual system we studied in detail in Chapter 5.

7.16 Neural maps can be continuous, discrete, or a combination of the two

As was introduced in Chapter 1, one of the most important organizational principles of the brain is that neurons are organized as maps to represent information. In the visual system, physical space in the world is first represented as a

Figure 7–33 Continuous and discrete neural maps. (Top) Schematic of a continuous neural map, exemplified here by retinotopic projection in the visual system. Neighboring neurons in the input field connect with neurons that neighbor each other in the target field, thereby preserving the spatial organization of the visual image. (Bottom) Schematic of a discrete neural map, exemplified here by the glomerular map in the olfactory system. Spatial organization in the target field does not reflect the position of input neurons, but instead is determined by input neuron type. Each of the three colors represents a set of olfactory receptor neurons that express the same odorant receptor, and target their axonal projections to the same glomerulus. (Adapted from Luo L & Flanagan JG [2007] *Neuron* 56:284–300. With permission from Elsevier Inc.)

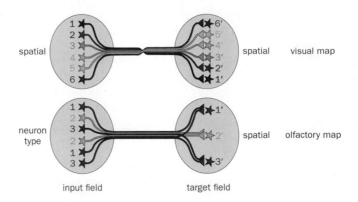

Figure 7–34 Three developmental mechanisms for establishing wiring specificity. Three wiring strategies **(A–C)** can give rise to the identical outcome **(D)**. In the first strategy (A), input neurons are prespecified, and target neurons acquire their identity in the process of connecting with specific input neurons. In the second strategy (B), target neurons are prespecified, and input neurons acquire their identity from the target neurons with which they connect. In the third strategy (C), input and target neurons are independently prespecified and find each other during the connection process. (Adapted from Jefferis G, Marin EC, Stockers RF et al. [2001] *Nature* 414:204–208. With permission from Macmillan Publishers Ltd.)

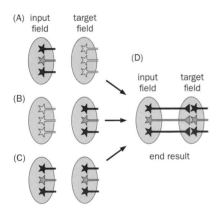

two-dimensional map in the retina. Through ordered retinotopic projections, this retinal map is relayed to and transformed in multiple higher-order visual areas (see Chapters 4 and 5). Similarly, the sensory and motor homunculi of the somatosensory and motor cortices represent maps of bodily sensation and motor output, respectively (see Figure 1–25). The auditory system uses maps to represent the location and frequency of sounds (see Sections 6.26 and 6.23). Even in the olfactory system, where the quality of stimuli rather than their spatial location is usually the primary concern, the brain utilizes a glomerular map to represent different odorants (see Section 6.8). Neural maps are also used elsewhere in the brain. For example, as will be discussed in Box 10-2, place cells in the hippocampus and grid cells in the entorhinal cortex represent the spatial location of an animal in its environment. Thus, if we can understand how neural maps are established, we will have made considerable progress in solving the problem of brain wiring.

Neural maps are of two kinds. The first kind, a **continuous map**, is typified by the retinotopic projections in which spatial information in the input field, the retina, is topographically represented in the target field, the tectum (**Figure 7–33**, top), and throughout visual centers of the brain. The key property of continuous maps is that both the input and target fields use spatial relationships among neurons to represent information: input field neurons that neighbor each other connect with target field neurons that neighbor each other. The second kind of map is a **discrete map**, in which spatially dispersed neurons of a specific type project to a discrete target, as exemplified by the relationship between olfactory receptor neurons (ORNs) and glomeruli (Figure 7–33, bottom). No matter where they are spatially located, ORNs that express the same odorant receptor project their axons to the same glomerulus.

Most neural maps resemble one of these two extremes, or combine some aspects of both. The continuous tonotopic map that represents sound frequency in the auditory system is similar to the retinotopic map. Taste systems use discrete information-processing channels for different taste modalities, comparable to the olfactory map. The somatosensory cortex is coarsely a continuous representation of the bodily surface. However, representations of discrete body surface components can be superimposed on this continuous somatosensory map, as we have seen in our discussion of the whiskers of rodents being represented by discrete barrels in the somatosensory cortex (see Box 5-3). Even in the visual system, which is dominated by continuous retinotopic maps, discrete retinal ganglion cell types representing different types of information (for example, ON/OFF responses, motion, color) connect with discrete types of bipolar and amacrine cells in the retina, and project their axons to discrete layers in their central targets.

We can envision three distinct developmental strategies for constructing neural maps (**Figure 7–34**). In the first strategy, input neurons are specified prior to connection whereas target neurons are naive before connections form and acquire their identity via their connections with specific input neurons. The second strategy reverses this relationship, so that naive input neurons acquire their identity by connecting with specified target neurons. In the third strategy, both input and target neurons are specified independently prior to forming the connections.

The example of the retinotectal map we studied in Chapter 5 uses the third strategy above, as neurons in both the input field (the eye) and target field (the tectum) of the retinal ganglion cells are pre-patterned by expression of graded EphA and ephrin-A according to their spatial position (see Figure 5–7). What about the formation of discrete maps in the olfactory system? We will now examine how these general concepts apply to the wiring of the olfactory systems in the mouse and *Drosophila*. Each system has fascinating properties that have taught us general lessons about the establishment of wiring specificity in the assembly of neural circuits.

7.17 In mice, odorant receptors instruct ORN axon targeting by regulating expression of guidance molecules

As we learned in Chapter 6, mouse ORNs that express a given odorant receptor converge their axons to only two of the 2000 specific glomeruli in the olfactory bulb. The positions of these glomerular targets are largely stereotyped from animal to animal. This feat is repeated for 1000 ORN types. Thus, ORN axon targeting represents one of the most striking examples of wiring precision in the developing nervous system. Along the long axis of the nasal epithelium, ORNs expressing the same odorant receptor are randomly distributed, yet their axons target to a precise location along the anterior–posterior (A–P) axis of the bulb. We will focus on how ORN axons achieve this feat, highlighting differences with the position-based topographic targeting that characterizes the visual system.

What molecular determinants could be responsible for the precise targeting of 1000 distinct ORN types, each with an average population of ~5000 individual ORNs scattered randomly along the long axis of the nasal epithelium? The most obvious hypothesis is that the odorant receptors themselves are involved, since these distinctive molecules are shared by all of the ORNs that project axons to a given glomerulus. This hypothesis was tested by changing the odorant receptor of a given ORN type to that of a different odorant receptor. The receptor swap was achieved by modifying the knock-in strategy that enabled tagging of all ORNs expressing a given receptor (**Figure 7–35**A; see also Figure 6–15). For example, the genetic locus that normally expresses the odorant receptor P2 was engineered to express the coding sequence of another odorant receptor (M12 or M71) tagged with an axonal marker. Axons from ORNs in which M12 had replaced P2 (M12 → P2 ORNs) targeted to a new glomerulus that did not correspond to the normal targets for either P2 or M12 ORNs; the M71 → P2 ORNs yielded similar results

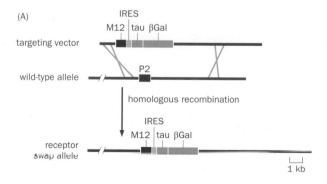
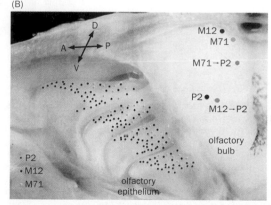

Figure 7–35 Alteration of ORN axon targeting after swapping the odorant receptors. (A) Genetic scheme of receptor swapping. The coding sequence of the P2 odorant receptor is replaced by that of the M12 odorant receptor at the P2 locus. In addition, an axonal marker, tau-βGal, followed by an internal ribosome entry sequence (IRES), was inserted to label ORN axons expressing the transgenic M12 odorant receptor (see legend to Figure 6–15 for more details.) The pair of crosses symbolizes that homologous recombination between the targeting vector and the wild-type allele produces the receptor swap allele. **(B)** Replacing P2 with M12 or M71 caused retargeting of ORNs to new sites designated as M12 → P2 and M71 → P2, respectively, which are distinct from the P2, M12, and M71 target glomeruli. ORN cell bodies in the olfactory epithelium and their target glomeruli in the olfactory bulb are color coded. A, anterior; P, posterior; D, dorsal; V, ventral. (Adapted from Wang F, Nemes A, Mendelsohn M et al. [1998] *Cell* 93:47–60. With permission from Elsevier Inc.)

(Figure 7–35B). Since altering the odorant receptor changes the target choice of an ORN type, these experiments suggest that odorant receptors play an instructive role in ORN axon targeting. However, these instructions are not sufficient to direct the engineered ORNs to the target glomeruli corresponding to the donor receptors.

How do odorant receptors instruct ORN axon targeting? As discussed in Section 6.1, odorant receptors are G-protein-coupled receptors that regulate cAMP production in response to odorant binding. To test whether changes in G protein coupling and cAMP signaling affect ORN axon targeting, researchers generated a series of transgenic mice that express a specific odorant receptor (I7) tagged with an axon marker. When a mutation was introduced in the I7 receptor that blocked its ability to bind G proteins, then axon convergence was disrupted. However, at the time of initial ORN targeting in embryos, mice have no sense of smell and do not yet express the olfactory-specific G_{olf} proteins necessary for olfactory transduction. How then do odorant receptors transduce signals? An alternative signaling pathway was suggested by the finding that the mutant I7 targeting defect could be corrected by co-expressing constitutively active forms of G_s protein, cAMP-dependent protein kinase A (PKA), or the CREB transcription factor (a substrate of PKA known to mediate cAMP-regulated gene expression; see Section 3.23). Thus, the role of odorant receptors in regulating ORN axon targeting is distinct from their role in odorant detection, in which cAMP is used directly to regulate a cyclic nucleotide-gated channel (**Figure 7–36**A; see also Figure 6–4). For ORN axon targeting, it appears that odorant receptors regulate gene expression via G_s, cAMP, PKA, and CREB, all of which are present in immature ORNs (Figure 7–36B).

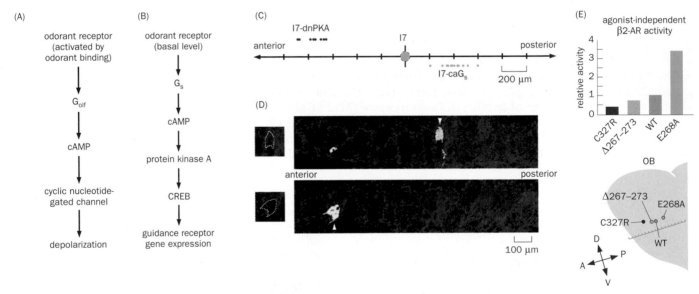

Figure 7–36 Signaling mechanisms by which odorant receptors regulate ORN targeting. (A) Signaling mechanism by which odorant receptors mediate olfaction (simplified from Figure 6–4). **(B)** Proposed signaling mechanism by which odorant receptors regulate ORN axon targeting (see also Figures 3–33 and 3–41). **(C)** Summary of the effects of diminished or enhanced cAMP signaling on ORN axon targeting. The target glomerulus (green) of ORNs that express wild-type I7 serves as a landmark. Co-expressing I7 with a dominant-negative protein kinase A (dnPKA) causes ORNs to shift their glomerular targeting anteriorly (blue dots), whereas co-expressing I7 with a constitutively active G_s (caG_s) causes ORNs to shift their glomerular targeting posteriorly (yellow dots). Each dot represents data from an individual mouse. **(D)** These two panels are different olfactory bulb sections from the same transgenic mouse. Glomeruli (outlined by blue nuclear staining) in the posterior olfactory bulb have higher axonal neuropilin-1 expression (red), indicating that they are targeted by ORNs expressing higher level of neuronpilin-1. The cyan-stained glomerulus (top panel, arrowhead) is targeted by axons that express the I7 olfactory receptor with normal cAMP signaling; these axons express neuropilin-1 (top left inset). The yellow-stained glomerulus (bottom panel, arrowhead) is targeted by cAMP-signaling-deficient axons that co-express the I7 receptor and dnPKA; these axons express negligible amount of neuropilin-1 (bottom left inset), and their targeting is shifted to a more anterior glomerulus. **(E)** Target positions along the anterior–posterior axis of the olfactory bulb from ORNs that transgenically express one of four β2 adrenergic receptor (β2-AR) variants (bottom) correlate with the relative basal activity of these β2-AR variants measured in vitro (top). These variants are either single amino acid changes (C327R, E236A) or a deletion of a few amino acids (Δ267–273). (C and D, adapted from Imai T, Suzuki M & Sakano H [2006] *Science* 314:657–661. With permission from AAAS; E, adapted from Nakashima A, Takeuchi H, Imai T et al. [2013] *Cell* 154:1314–1325. With permission from Elsevier Inc.)

When the glomerular target positions of these genetically engineered ORNs were examined, a correlation was found between the strength of G_s/cAMP/PKA signaling in the engineered ORNs and the location of their glomerular target along the anterior–posterior axis of the olfactory bulb: the stronger the signaling, the more posterior their glomerular target (Figure 7–36C). Gene expression profiling identified target genes that are differentially expressed in ORNs whose axons target different positions along the anterior–posterior axis. Among these genes was one for the axon guidance receptor **neuropilin-1 (Nrp1)** (see Box 5–1), which was expressed in a posterior > anterior gradient in ORN axon terminals (Figure 7–36D). These data supported a model in which odorant receptors regulate ORN axon targeting by regulating the expression of axon guidance molecules via the G protein/cAMP/CREB pathway.

How do different odorant receptors activate different levels of cAMP/PKA/CREB signaling? Because ORN axon targeting initiates prenatally prior to odor exposure, it has been proposed that ligand-independent basal activity of odorant receptors as GPCRs specifies how strong the signal is. This idea was put to the test by using β2 adrenergic receptor (β2-AR), the best characterized GPCR (see Sections 3.18 and 3.19), as a substitute for an odorant receptor. Transgenic mice in which β2-AR was used to replace an endogenous odorant receptor resulted in ORNs targeting to a specific glomerulus at the middle, along the A–P axis of the olfactory bulb. When different β2-AR mutants that exhibited different basal GPCR activity were used instead of the wild-type β2-AR, a strong correlation was found between the basal activity and targeting along the A–P axis of the olfactory bulb (Figure 7–36E). This experiment provided strong support for the hypothesis that basal activity of odorant receptors determines the GPCR signaling strength, which in turn regulates the expression of specific levels of guidance molecules.

To summarize the current model (Figure 7–36B), each odorant receptor is presumed to have a specific basal (that is, ligand-independent) GPCR activity that can be coupled to G_s, a Gα protein present in developing ORNs. Therefore, ORNs expressing a given odorant receptor should have similar basal level signaling, resulting in expression of similar levels of guidance molecules. No matter where the cell bodies are scattered along the A–P axis for a given type of ORN, the same levels of guidance molecules they express can guide their axons to their glomerular target in the olfactory bulb.

7.18 ORN axons sort themselves by repulsive interactions before reaching their target

Given the widespread cell body distribution of ORNs expressing the same odorant receptor, it still seems a daunting task for them to target their axons to a single glomerulus, even if they express similar levels of guidance receptors. Detailed investigation of Nrp1 and its repulsive ligand semaphorin 3A (Sema3A; see Section 7.6 and Box 5–1) offer an interesting mechanism: axons sort themselves out before they reach the target.

Both Nrp1 and Sema3A are regulated by cAMP signaling in ORNs, but in opposite directions. Strong cAMP signaling leads to high Nrp1 and low Sema3A levels, whereas weak cAMP signaling promotes high Sema3A and low Nrp1 levels (**Figure 7–37**A). Thus, each ORN axon has a specific level of Nrp1 and Sema3A at the outset of its journey. ORN axon bundles leaving the olfactory epithelium were examined at different intermediate stages of their journey toward the olfactory bulb. In the early stages, axons with high levels of Sema3A were intermingled with high-Nrp1 axons, as would be expected given the random distribution of their cell bodies in the olfactory epithelium. As axons progressed toward the olfactory bulb, high-Sema3A and high-Nrp1 axons gradually segregated. Before they entered the olfactory bulb, high-Nrp1 and high-Sema3A axons had already formed separate bundles and were poised to target to posterior and anterior parts of the olfactory bulb, respectively (Figure 7–37B). Repulsive interactions between Sema3A- and Nrp1-expressing ORN axons appear to be essential for this sorting: in conditional knockout mice in which either Sema3A (Figure 7–37C) or Nrp1 were deleted from

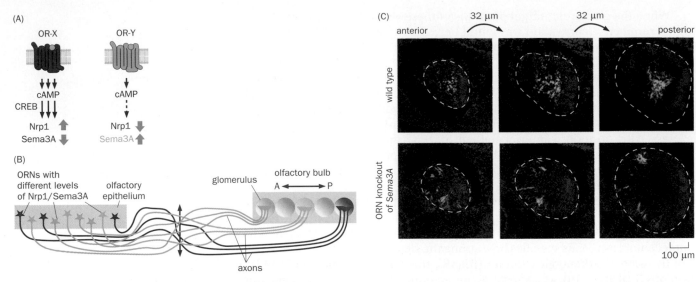

Figure 7–37 Sorting of ORN axons along the path through semaphorin/neuropilin mediated axon–axon repulsion. (A) Axon guidance cue Sema3A and its receptor Nrp1 are both regulated by cAMP signaling, but in opposite directions. ORNs with high basal levels of cAMP signaling activity express low levels of Sema3A and high levels of Nrp1, while ORNs with low levels of cAMP signaling are Nrp1-poor and Sema3A-rich. **(B)** At the start of the journey from the olfactory epithelium to the olfactory bulb, ORN axons with different levels of Sema3A/Nrp1 are intermingled. They sort themselves out along the journey by Sema3A/Nrp1-mediated repulsive axon–axon interactions. **(C)** Cross sections of ORN axon bundles taken at 32-μm intervals along the anterior (toward the nose) to posterior (toward the olfactory bulb) path. Nrp1-expressing axons (red) and Sema3A-expressing axons (green) gradually sort themselves out en route to the bulb in the wild type (top). However, when Sema3A is conditionally knocked out in ORNs (bottom), sorting is disrupted. (A & B, courtesy of Kazunari Miyamichi; C, from Imai T, Yamazaki T, Kobayakawa R et al. [2009] *Science* 325:585–590. With permission from AAAS.)

ORNs, both the order of axons traveling to the olfactory bulb and the final targeting positions of these axons were disrupted.

The use of signaling among developing axons to constrain their targeting choices has proved to be an important mechanism in brain wiring. We already introduced this idea in the context of retinotopic mapping (see Section 5.5) and will see more examples in the wiring of the fly olfactory circuit in Section 7.22. The pre-target sorting of ORNs provided a striking example of the power of repulsive axon–axon interactions in establishing order among axons projected from random starting points.

7.19 Activity-dependent regulation of adhesion and repulsion refines glomerular targeting

Axon sorting provides patterned entry of ORN axons into the olfactory bulb along the anterior–posterior axis. Separate studies have indicated that the patterning of ORN axons along the orthogonal dorsal–ventral axis also involves axon–axon interactions that utilize a different semaphorin/neuropilin pair. But what is responsible for the eventual convergence of ORN axons expressing the same odorant receptors to *discrete* glomeruli located at various positions along the anterior–posterior and dorsal–ventral axes? Are molecules based on genetic hardwiring sufficient to achieve the precise wiring, or do experience and spontaneous neuronal activity play a role in refining these targeting events, as is the case for visual system wiring?

The role of odor-evoked activity in ORN axon targeting was examined by combining the cyclic nucleotide gated channel mutant that causes anosmia (see Figure 6–4) with genetically labeled individual ORN types. In most cases, axon convergence of individual ORN types appeared largely normal, suggesting that olfactory experience does not play a major role in ORN axon convergence. However, subtle defects were observed in a subset of ORN types. Indeed, even in normal mice, for some ORN types the convergence of axons to single glomeruli is incomplete at birth and requires refinement during postnatal development. This refinement was blocked by naris closure, a surgical procedure that prevents odorants from entering one of the two nostrils, suggesting that olfactory experience

Figure 7–38 Refinement of glomerular targeting requires ORN activity. (A)
ORNs refine their axonal targets to a single glomerulus with normal experience
(bulb°). When one of the nostrils is closed so that ORNs projecting to the
corresponding bulb (bulb^x) are no longer exposed to odorants, ORN axons retain
multiple glomerular targets (arrows). See Figure 6–15 for the strategy to label
axons of an ORN type. **(B)** Quantification of axon refinement defect, as measured
by percentage of bulbs that exhibit multiple glomerular targets, in normal and
experience-deprived conditions for two ORN types. (Adapted from Zou DJ,
Feinstein P, Rivers AL et al. [2004] *Science* 304:1976–1979. With permission
from AAAS.)

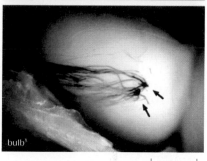

is required for uniglomerular convergence of ORN axons (**Figure 7–38**). Interfering
with spontaneous activity of specific ORN types by expressing an inward rectifier
K⁺ channel (see Box 2–4), which hyperpolarizes ORNs and renders them less able
to reach the threshold to fire action potentials, had a more drastic effect on ORN
axon convergence even though the general positioning of axon targeting was less
perturbed.

The above experiments indicate that spontaneous and odor-evoked ORN
neuronal activities play a role in refining ORN axon targeting to individual glomeruli. What is the molecular basis? Searching for molecules whose expression was
affected by naris closure identified cell-surface proteins whose expression levels
are regulated in an activity-dependent manner. These include ephrin-A and the
EphA receptor as well as immunoglobulin superfamily cell adhesion molecules
Kirrel2 and Kirrel3, which exhibit homophilic binding activity. Because they
express the same odorant receptor, ORNs of the same type should have similar
odor-evoked activity. They therefore are endowed with an 'identity code' through
the expression of specific levels of ephrin-A, EphA, and homophilic cell adhesion
molecules. Adhesion among ORNs of the same types and repulsion between different types can thus help refine the targeting of individual ORN classes to discrete glomeruli (**Figure 7–39**).

Thus, in this context of activity-dependent ORN axon targeting, postnatal
olfactory experience sharpens the expression of cell–cell interaction proteins in
ORNs according to the odorant receptors they express. It would be interesting to
test whether activity-dependent regulation of the expression of cell–cell interaction molecules in the mammalian olfactory system applies generally to activity-
dependent wiring processes, thereby contributing to the Hebbian mechanism of
'fire together, wire together' (see Figure 5–26).

So far, we have discussed ORN axon targeting to the olfactory bulb without
considering the kinds of patterning that might exist in the bulb before ORN axons

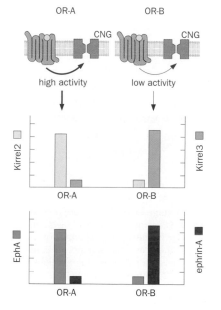

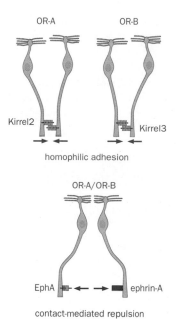

**Figure 7–39 Activity-dependent
expression of cell-surface proteins refines
glomerular targeting.** (Left) High levels
of odorant-induced activity for a given
olfactory receptor (for example, OR-A)
induce expression of Kirrel2 and EphA,
whereas low levels of odorant-induced
activity (for example, OR-B) cause Kirrel3
and ephrin-A to be highly expressed. Thus,
each ORN type has an identity code based
on its levels of these surface protein
expression. (Right) Kirrel2 and Kirrel3
can cause axons of the same ORN type
to adhere, whereas ephrin-A and EphA
expressed on axons of different ORN types
cause axon segregation. Both processes
can help refine glomerular targeting.
(Adapted from Serizawa S, Miyamichi K,
Takeuchi H et al. [2006] *Cell* 127:1057–
1069. With permission from Elsevier Inc.)

invade. Is the target area pre-patterned or a blank slate (see Figure 7–34)? For the mouse olfactory bulb, little is known about the answers to these questions. We now turn to the fly antennal lobe, the counterpart of the vertebrate olfactory bulb, where this topic has been studied in detail.

7.20 *Drosophila* projection neurons' lineage and birth order specify the glomeruli that their dendrites target

As we learned in Chapter 6, the glomerular organization of the insect olfactory system is strikingly similar to that of mammals (see Figure 6–27). ORNs that express the same odorant receptor project axons to the same glomerulus in the antennal lobe. Information is then relayed to higher olfactory centers by olfactory **projection neurons** (**PNs**), most of which send dendrites to a single glomerulus. The fly olfactory system offers many advantages for studying the wiring specificity of neural circuits. Glomeruli are individually identifiable based on their stereotyped shape, size, and relative positions. Genetic tools such as **MARCM** (**m**osaic **a**nalysis with a **r**epressible **c**ell **m**arker, see Section 13.16) allow investigators to visualize individual neurons or neurons in the same lineage, and to delete or mis-express any gene specifically in these labeled neurons in genetically mosaic animals. Given that each ORN type connects to one PN type at a specific glomerulus and vice versa, developmental mechanisms like those proposed in Figure 7–34 can be tested with precise, single-cell resolution.

Neurons in the insect brain derive from progenitor cells termed **neuroblasts**. MARCM-based lineage analysis has revealed that most PNs derive from two neuroblasts, one located dorsal to, the other lateral to, the antennal lobe. PNs from the dorsal and lateral lineages project dendrites to a stereotyped subset of glomeruli that are mutually exclusive (**Figure 7–40**). This observation suggested that a PN's lineage limits its glomerular choice. Furthermore, PNs within a lineage project dendrites to different glomeruli following a stereotyped birth order. Thus, lineage and birth order predict to which glomerulus a PN sends dendrites, with which ORN type it makes synaptic connections, and, ultimately, which odorants it represents.

Further examination of olfactory circuit wiring during *Drosophila* development revealed that dendrites from individual PNs extend to specific regions of the developing antennal lobe before the arrival of ORN axons. These developing dendritic projections occupy specific areas of the developing antennal lobe that

(A)

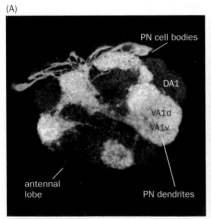
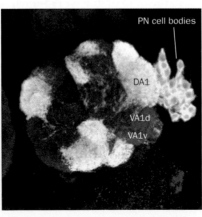

(B)

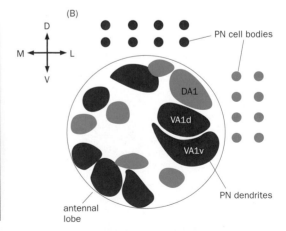

Figure 7–40 Lineage constrains glomerular choice of *Drosophila* projection neuron dendrites. (A) PNs derived from dorsal and lateral neuroblast lineages target different glomeruli. Left, dorsal PNs and their dendrites express a membrane-targeted GFP and are labeled green, revealing the location of their target glomeruli; magenta staining with a synapse marker labels all glomeruli. Right, green-labeled lateral PNs and their dendrites highlight a complementary set of glomerular targets. These images are single optical sections of two different antennal lobes at similar depth. Cell bodies of most dorsal PNs are in a different focal plane and thus not visible. **(B)** Schematic summary of a subset of intercalated glomeruli that are targeted by dorsal (red) and lateral (blue) PNs, respectively. D, dorsal; V, ventral; M, medial; L, lateral. Three glomeruli, DA1, VA1d, VA1v (named after their positions such as dorsal-anterior 1), are labeled in the schematic and the corresponding images in panel A. (Adapted from Jefferis G, Marin EC, Stockers RF et al. [2001] *Nature* 414:204–208. With permission from Macmillan Publishers Ltd.)

correspond to their future target glomeruli. Thus, not only do PNs have their own identity independent of ORNs, they also establish an initial spatial map within the developing antennal lobe before the arrival of their presynaptic partners. Establishment of wiring specificity of the olfactory circuit at the antennal lobe can therefore be divided into two phases. First, PN dendrites use ORN-independent cues to coarsely target their dendrites in a type-specific manner. Second, ORN axons enter the pre-patterned antennal lobe to find their partners. We will study both processes in the following sections.

7.21 Graded determinants and discrete molecular labels control the targeting of projection neuron dendrites

Semaphorins were found to play important roles in PN dendrite targeting in flies, akin to their function in ORN axon targeting in mammals. Specifically, a transmembrane form of semaphorin, **Sema1A**, is distributed in a dorsolateral > ventromedial gradient in the developing antennal lobe, which prior to ORN axon arrival consists mostly of PN dendrites (**Figure 7–41**A). Removing Sema1A from a single PN shifted its dendrite targeting ventromedially down the Sema1A gradient (Figure 7–41B), indicating that Sema1A plays a cell-autonomous role. Thus, while semaphorins serve as guidance ligands in most cases, Sema1A acts as a receptor to regulate PN dendrite targeting. Furthermore, different PNs express different levels of Sema1A, and are differentially repelled by two secreted semaphorins, **Sema2A** and **Sema2B**, which are distributed as ventromedial > dorsolateral counter-gradients. Opposing gradients of different semaphorins thus cooperate to specify where PN dendrites target along the dorsolateral–ventromedial axis.

What mechanisms ensure that each PN targets its dendrites to a single, discrete glomerulus? Graded determinants are useful for setting up a coarse map, but

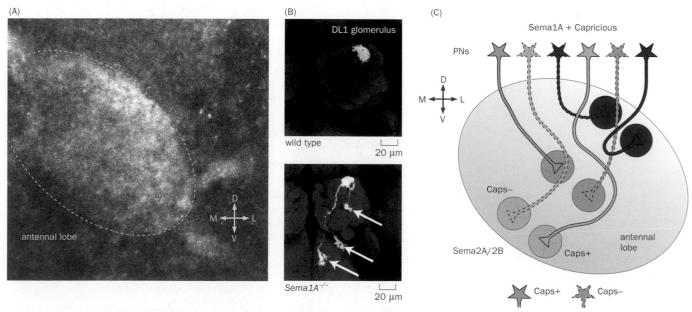

Figure 7–41 PN dendrite targeting is instructed by global gradients and local binary determinants. (A) Sema1A protein is distributed in a dorsolateral > ventromedial gradient in the developing antennal lobe (encircled with dashed line), which prior to the arrival of ORN axons consists mostly of PN dendrites. **(B)** Top, a single DL1 PN targets its dendrites exclusively to the DL1 glomerulus (located at the dorsolateral antennal lobe). Bottom, when Sema1A is removed from a single DL1 PN, while some dendrites still target to the DL1 glomerulus, a large subset of dendrites mistarget ventromedially (bottom, arrows). Thus, Sema1A is cell autonomously required in DL1 PNs to restrict their dendrites to the DL1 glomerulus. Both wild-type and mutant DL1 PNs are labeled green by the MARCM method; a synapse marker for all glomerular structures is stained magenta. **(C)** Summary of molecules and mechanisms that control PN dendrite targeting. Different PNs express different amounts of Sema1A, shown here as different intensities of red. The blue-to-yellow background shading represents the ventromedial > dorsolateral gradient of Sema2A and Sema2B, two secreted semaphorins that repel Sema1A-expressing PN dendrites. Thus, PNs that express progressively higher levels of Sema1A target their dendrites to progressively more dorsolateral glomeruli, accounting for the Sema1A gradient seen in panel A. In addition, some PNs express Capricious (Caps, solid outline), while others do not (dotted outline). Caps-positive and Caps-negative PNs segregate into discrete glomeruli at different regions of the antennal lobe. D, dorsal; V, ventral; M, medial; L, lateral. (Adapted from Komiyama T, Sweeney LB, Schuldiner O et al. [2007] *Cell* 128:399–410. With permission from Elsevier Inc. See also Sweeney LB, Chou YH, Wu Z et al. [2011] *Neuron* 72:734–747; Hong W, Zhu H, Potter CJ et al. [2009] *Nat Neurosci* 12:1542–1550.)

are likely insufficient to specify discrete outcomes with high precision, as neighboring glomeruli may exhibit only a small differences in level. A second kind of PN molecular label was found. About half of PNs express the leucine-rich-repeat containing cell-surface protein Capricious (Caps), which we encountered in our earlier discussion of visual system wiring in *Drosophila* (see Figure 5–39). Glomerular targets of Caps-positive and Caps-negative PNs form a salt-and-pepper pattern. Loss of Caps in Caps-positive PNs caused them to mistarget selectively to Caps-negative glomeruli. Conversely, misexpression of Caps in Caps-negative PNs caused them to mistarget selectively to Caps-positive glomeruli. Genetic analysis suggested that Caps mediate dendrite–dendrite interactions. Thus, Caps acts as a binary determinant to segregate PN dendrites into two distinct subgroups. This function is analogous to that of mouse Kirrel2/3- and ephrin-A/EphA-mediated axon–axon interactions in ORN axon refinement (see Figure 7–39).

In summary, global graded determinants, such as semaphorins, and local binary determinants, such as Capricious, work together to specify the dendrite targeting of different PN types to discrete areas of the antennal lobe (Figure 7–41C).

7.22 Sequential interactions among ORN axons limit their target choice

Odorant receptors are not used for ORN axon targeting in *Drosophila*, in contrast to their use in mammals. The use of odorant receptors to instruct ORN targeting likely occurred during vertebrate or mammalian evolution that accompanied the large expansion of ORN types. Indeed, *Drosophila* ORNs express odorant receptors only after wiring specificity is achieved. Thus, the wiring mechanisms for fly ORNs differ from the odorant-receptor-dependent ORN axon targeting in mammals discussed in Section 7.19.

However, like their mammalian counterparts, *Drosophila* ORNs do rely on extensive axon–axon interactions and pre-target sorting, and even utilize the same family of molecules, the semaphorins. When pioneering ORN axons from the antenna first arrive at the antennal lobe, individual axons circumnavigate the antennal lobe by taking either a dorsolateral or a ventromedial trajectory (**Figure 7–42**). The secreted semaphorin Sema2B is enriched in axons that take the ventromedial trajectory but is absent from axons that take the dorsolateral trajectory. In flies lacking Sema2B, ORN axons that normally take the ventromedial trajectory took the dorsolateral trajectory instead and eventually targeted ectopic glomeruli in the dorsolateral antennal lobe, far from their normal ventromedial targets. Genetic analyses suggest that Sema2B-expressing ORNs are first attracted to a ventromedial trajectory by the antennal lobe gradients of Sema2A and Sema2B that pattern the targeting of PN dendrites in an earlier stage of development (see Figure 7–41). Sema2B from these initial ventromedial ORN axons then attracts other Sema2B-expressing ORN axons to the same trajectory (Figure 7–42, left panels). These axon–axon interactions, analogous to those observed in vertebrates, ensure correct trajectory choice, which is essential for eventual target

Figure 7–42 Schematic model illustrating that sequential axon–axon interactions limit ORN target choice. When antennal ORN axons first reach the antennal lobe, a ventromedial > dorsolateral gradient of antennal lobe Sema2A and Sema2B (red shading) attracts axons that express Sema2B (red) to a ventromedial trajectory (arrow), whereas axons that do not express Sema2B (orange) take a dorsolateral trajectory. Sema2B from early-arriving ORN axons then attracts other Sema2B-expressing ORN axons to the same trajectory (curved arrow). When maxillary palp (MP) axons (cyan) arrive later, the antennal lobe has been largely occupied by antennal ORNs. Sema1A from antennal ORN axons repels MP ORN axons (⊣), such that MP axons innervate glomeruli complementary to those innervated by antennal ORNs. D, dorsal; V, ventral; M, medial; L, lateral. (Adapted from Sweeney LB, Couto A, Chou YH et al. [2007] *Neuron* 53:185–200; Joo WJ, Sweeney LB, Liang L et al. [2013] *Neuron* 78:673–686. With permission from Elsevier Inc.)

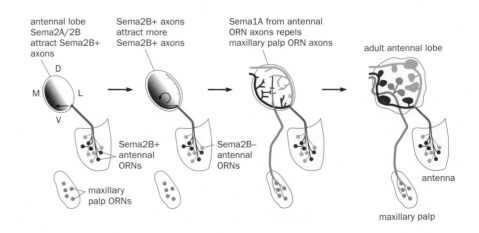

selection. Interestingly, the specification of Sema2B expression in the half of the ORNs that take the ventromedial trajectory, but not in the other half that take the dorsolateral trajectory, is achieved by the Notch–Delta-mediated lateral inhibition discussed in Section 7.3.

Axon–axon interactions also occur at the final target regions, as exemplified by targeting of ORN axons from the maxillary palp (MP), which reach the antennal lobe well after antennal ORN axons. (Flies have two spatially separated 'noses' that house different ORN types: the antenna and the maxillary palp; see Figure 6–27). When antennal ORNs were ablated during development, MP axons mistargeted to glomeruli normally occupied by antennal ORNs, suggesting that antennal ORN axons normally repel later-arriving MP ORN axons to constrain their target choice (Figure 7–42, right panels). Sema1A plays an essential role in ORNs by serving as a repulsive ligand, in contrast to its function in PNs as a receptor that instructs dendrite targeting. This conclusion was derived from the observation that deleting Sema1A from antennal ORN axons, but not MP axons, affected the targeting choice of MP axons. Thus, axon–axon interactions at multiple stages limit the target choices of axons from a given ORN type.

7.23 Homophilic matching molecules instruct connection specificity between synaptic partners

What mechanisms are used to establish the final, one-to-one connection between 50 types of ORNs and PNs? Given the extensive ORN axon–axon interactions we just discussed, it is conceivable that ORN axons could be highly patterned to create an ORN axon map, which is then superimposed onto the PN dendrite map established earlier. Alternatively, or in addition, ORN axons may recognize cues on their partner PN dendrites. Indeed, experimentally induced PN dendrite shift causes corresponding shift of partner ORN axons, suggesting that ORN axons can recognize cues on their postsynaptic partner PNs and match those cues appropriately even when PN dendrites mistarget.

This phenomenon prompted genetic screens for 'matching' molecules that, when misexpressed, would force nonsynaptic partners to match, or prevent normal synaptic partners from matching. Such screens identified two evolutionarily conserved Teneurins as synaptic-partner-matching molecules. Each of the two Teneurins in the fly, Ten-m and Ten-a, is highly expressed in a distinct subset of matching ORN and PN types (**Figure 7–43**A, B). Both molecules exhibit homophilic adhesion properties. These data suggest that matching expression of Ten-m

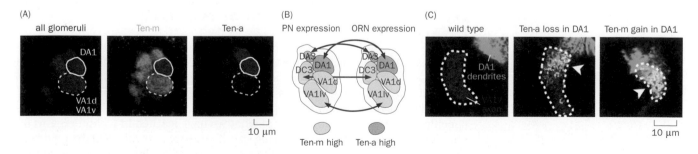

Figure 7–43 Teneurin-mediated homophilic attractions direct PN–ORN synaptic partner matching. (A) Expression patterns of Ten-m (green) and Ten-a (red) in the developing antennal lobe, with the magenta neuropil marker labeling all glomeruli evenly. The solid circles highlight the DA1 glomerulus, which has high levels of Ten-a and low levels of Ten-m. The dashed circles highlight the VA1d and VA1v glomeruli, which have high levels of Ten-m and low levels of Ten-a. **(B)** Schematic summary of matching Teneurin expression in five glomeruli. PNs and ORNs that target the same glomerulus exhibit matching levels of Ten-m (blue) and Ten-a (orange) expression. Those that target DA1 are Ten-a high and Ten-m low; those that target VA1d and VA1v are Ten-m high and Ten-a low; those that target DA3 are high for both; those that target DC3 are low for both. Double-ended arrows symbolize the hypothesis that the matching levels of Teneurins instruct synaptic partner matching between cognate ORNs and PNs. **(C)** Genetic evidence that supports the matching hypothesis. In wild type, DA1 dendrites (green) and VA1v axons (red) do not match (left). When Ten-a is knocked down in DA1 PNs, which normally express high levels of Ten-a, DA1 dendrites mismatch with VA1v axons, which normally express low levels of Ten-a (middle). When Ten-m is overexpressed in DA1 PNs, which normally express low Ten-m, their dendrites also mismatch with VA1v axons, which normally express high levels of Ten-m (right). (Adapted from Hong W, Mosca TJ & Luo L [2012] *Nature* 484:201–207. With permission from Macmillan Publishers Ltd.)

and Ten-a can be used to instruct synaptic partner matching (Figure 7–43B). This hypothesis has been supported by loss- and gain-of-function experiments. For example, loss of Ten-a in DA1 PNs, which normally express high level of Ten-a, caused their dendrites to mismatch with VA1v ORN axons, which normally express low level of Ten-a (Figure 7–43C, middle). Likewise, overexpression of Ten-m in DA1 PNs, which normally express low level of Ten-m, also caused their dendrites to mismatch with VA1v ORN axons, which normally express high level of Ten-m (Figure 7–43C, right). Further, simultaneous expression of Ten-m in non-partner ORNs and PNs caused their axons and dendrites to form ectopic connections. These synaptic-partner-matching molecules thus ensure final connection specificity between cognate ORNs and PNs.

Teneurins act in a different manner than Caps discussed earlier; whereas Caps mediate PN dendrite–dendrite interaction to sort PNs into discrete glomeruli, Teneurins mediate ORN axon–PN dendrite interaction to match synaptic partners. Teneurin's mode of action may be analogous to that described in the chick retina, where homophilic cell adhesion molecules Sidekicks and Dscams are expressed in matching retinal neurons targeting to the same laminae and regulate lamina-specific process targeting (see Figure 5–33B).

In summary, genetic analyses of fly olfactory circuit assembly have identified a wealth of cellular and molecular mechanisms by which wiring specificity of a moderately complex neural circuit is achieved during development (**Movie 7–2**). The fly olfactory circuit clearly employs an independent specification mechanism (see Figure 7–34C) that requires more complex molecular recognition strategies than the alternative mechanisms (see Figure 7–34A, B). Why? As we learned in Chapter 6, PNs that project dendrites to specific glomeruli also exhibit stereotyped axon terminal arborizations in the lateral horn, a higher olfactory center (see Figure 6–33). Indeed, many molecules involved in PN dendrite targeting also regulate PN axon terminal arborizations. This hard-wiring strategy can ensure that specific olfactory information received by specific ORN types is always delivered to specific regions of the lateral horn via independently specified PNs. This may be essential for innate odor-mediated behaviors such as feeding or courtship, which are robust and do not depend on learning.

By the same token, mammalian mitral/tufted cells may also be genetically prespecified to some degree, such that information from specific odorant receptors can be faithfully transmitted to specific targets in the olfactory cortex where it can drive innate olfaction-mediated behavior, such as avoidance of a predator odor. These predictions can be experimentally tested.

HOW DO ~20,000 GENES SPECIFY 10^{14} CONNECTIONS?

As we have seen, developing nervous systems employ many different strategies to precisely wire themselves. Some strategies are highly conserved through evolution, using the same molecules and mechanisms; others are more varied and adapted to the developmental programs of specific parts of the nervous system or specific organisms. As a summary of what we have learned in Chapter 5 and this chapter so far, we come back to 'wiring specificity problem' raised at the beginning of Chapter 5. The human brain contains ~10^{11} neurons that make ~10^{14} synaptic connections. If each neuron carries a specific 'identification tag' that allows it to be wired differently from any other neuron, as Sperry proposed in his chemoaffinity hypothesis, how is wiring specificity achieved given that there are only ~20,000 protein-coding genes in the human genome? The following sections discuss various solutions to this problem (see **Table 7–1** for a summary of the examples discussed in this chapter and Chapter 5).

7.24 Some genes can produce many protein variants

We encountered an extraordinary example of molecular diversity in the *Drosophila* Dscam. This single gene can encode cell-surface proteins with 19,008 variants of

Table 7–1: Strategies for maximizing the number of wiring specification signals from a limited genome

Strategy	Example (Section)	Known function	Animal model
Many protein products from a single gene	Dscam (7.9)	dendrite and axon self-avoidance	fly
	protocadherins (7.9)	dendrite self-avoidance	mammal
	neurexins (7.11, 7.24)	synapse formation	vertebrate
Many levels from a single protein	ephrin-A/EphA (5.4)	retinotopic map	vertebrate
	semaphorins (7.18, 7.21)	olfactory maps	mouse, fly
Multiple functions from a single protein	ephrin-A and EphA as both ligands and receptors (5.5, 5.15)	retinotopic map formation	vertebrate
	Sema1A as a receptor or a ligand (7.21, 7.22)	olfactory map formation	fly
	Unc40 without or with Unc5 (Box 5–1)	attractive or repulsive midline guidance	*C. elegans* to vertebrate
Same protein used in multiple places and times	ephrin-A/EphA (5.4, 7.4, 7.19)	retinotopic map; motor axon guidance; glomerular refinement	vertebrate
	Sema1A, Sema2B (7.21, 7.22)	dendrite targeting; axon targeting in olfactory map formation	fly
	Sema3A (7.6, 7.18)	midline repulsion; ORN axon sorting	mouse
	Sonic hedgehog (7.4, 7.6)	dorsoventral patterning in the spinal cord; midline guidance	vertebrate
	Wnt (7.1, 7.6, 7.10)	developmental patterning along the A–P axis; direction of turning along the A–P axis; subcellular presynaptic site selection	*C. elegans*, vertebrate
	Slit (7.5, 7.6)	midline crossing; pathway choice	fly to vertebrate; fly
	netrin (Box 5–1, 5.18)	midline guidance; photoreceptor axon targeting	*C. elegans* to vertebrate; fly
	Capricious (5.18, 7.21)	visual system; olfactory system	fly
Combinatorial use of proteins	semaphorins and Capricious (7.21)	projection neuron dendrite targeting	fly
	semaphorins and ephrin-A/EphA (7.18, 7.19)	ORN axon targeting	mouse
Use of experience and spontaneous neuronal activity	ocular dominance column in the visual cortex (5.7, 5.9)		cat, ferret, monkey
	eye-specific layer segregation in the LGN (5.9–5.11)		mammal
	neuromuscular synapse refinement (7.13)		mouse
	wiring of the whisker-barrel system (Box 5–3)		mouse
	refinement of glomerular targeting (7.19)		mouse

the extracellular domains (see Figure 7–20). In addition to their established role in self-avoidance, these variants can in principle be used for matching synaptic partners, given their striking isoform-specific binding specificity. Indeed, chick Dscams likely serve as homophilic recognition molecules for layer-specific targeting in the retina (see Figure 5–33B).

Although vertebrate Dscams do not possess the extraordinary alternative splicing potential of the insect Dscams, other classes of vertebrate proteins have been shown to exhibit molecular diversity. For example, the clustered protocadherins play a role analogous to Dscam in regulating self-avoidance of dendritic branches. Indeed, protocadherin α and γ clusters also utilize the same cytoplasmic domain but alternative extracellular domains to achieve molecular diversity (see Figure 7–21). Another example of extraordinary diversity was found in the trans-synaptic signaling molecule neurexin (see Figure 7–25). Six major neurexin isoforms are encoded by three separate genes, each of which has two alternative promoters. Altogether, ~3000 neurexin variants can be produced through independent selection of several alternative exons encoding the extracellular domain. Recent studies have shown that molecular diversity in neurexins is used to specify different synaptic properties.

7.25 Protein gradients can specify different connections

The use of protein gradients reduces the number of molecular species required to wire the nervous system: different quantities of the same protein can specify different connections. This was first demonstrated for the ephrin/Eph receptor gradients in retinotopic mapping (see Figure 5–7). The use of molecular gradients to specify continuous maps seems intuitive. Two neurons exhibiting small differences in receptor concentration might project to targets a short distance apart; as the receptor levels of two neurons increase, the distance between their projection targets also increases.

A surprising finding is that molecular gradients are also used to specify discrete maps, such as the olfactory map. In the case of mammalian ORN axon targeting, levels of Sema3A and its repulsive receptor neuropilin are used to sort different ORN axons along the path toward their final destination (see Figure 7–37). In the fly antennal lobe, a projection neuron's level of Sema1A, a transmembrane semaphorin, specifies the position of the PN's dendritic target along a particular axis according to extracellular gradients of two secreted semaphorins (see Figure 7–41). However, in each case, these gradients specify coarse targeting, which is further refined by cell-type-specific interactions to create discrete targets.

7.26 The same molecules can serve multiple functions

The strategy of using the same molecule for multiple functions is exemplified by ephrin-A and EphA in the retinotopic mapping. Each molecule can serve as either a ligand or a receptor in forward or reverse signaling (see Figure 5–12). One function for this bidirectional signaling is to sharpen retinotopic map with more than one gradient (see Section 5.5). Bidirectional signaling also enables the graded expressions of ephrin-A and EphA to serve as ligands to instruct the targeting of their presynaptic partners, and as receptors to instruct their own axon targeting to match with their postsynaptic partners across multiple stages of visual pathways (see Figure 5–31). Likewise, Sema1A in *Drosophila* can serve as either a ligand or a receptor in different olfactory neurons (see Figures 7–41 and 7–42). As a third example, expression of Unc40/DCC alone produces an attractive response to midline guidance cue Unc6/netrin, whereas co-expression of Unc40/DCC with Unc5 produces a repulsive response to the same guidance cue (see Box 5–1).

Although we have not discussed specific examples, activity and specificity of axon guidance receptors can also be modified extracellularly by glycosylation and proteolysis, and intracellularly by changes in intracellular signaling molecules (such as Ca^{2+} or cAMP) or downstream effectors. These regulations can allow axon guidance receptors to produce distinct responses to guidance cues.

7.27 The same molecules can be used at multiple times and places

We have seen many examples of this strategy. In vertebrates, for instance, ephrin/Eph ligand/receptor pairs are used in the visual system for retinotopic mapping (see Figure 5–7), in the olfactory system for refining ORN axon targeting in an odorant receptor- and activity-dependent manner (see Figure 7–38), and again in spinal cord motor neurons to specify their limb innervation patterns (see Figure 7–9). The spatial separation of these circuits allows the same molecules to be reused without causing ambiguity. In other cases, temporal differences between targeting events make it possible for the same molecular signals to perform multiple, unambiguous functions. One such example can be found in the *Drosophila* olfactory circuit, where the same semaphorins (Sema1A, Sema2B) participate in the wiring decisions of PN dendrites and ORN axons (see Figures 7–41 and 7–42). Furthermore, molecules active in early developmental patterning can find additional uses in neuronal wiring. For example, Sonic Hedgehog is used for patterning the vertebrate spinal cord along the dorsal–ventral axis earlier in development (see Figure 7–8), and is later used as a midline guidance cue for commissural neurons. In mice and *C. elegans*, Wnt is used for global patterning of the anterior–posterior axis, and is reused later for specifying the turning direction of commissural axons after midline crossing in mice (see Figure 7-15) and site selection of presynaptic terminals of *C. elegans* motor neurons (see Figure 7-23).

In addition to conserving molecules, this pattern of multiple uses suggests that it may be simpler to evolve new spatial and temporal expression patterns for existing molecules than to invent new molecules for the purpose of wiring different parts of the nervous system. This idea will be discussed further in Chapter 12.

7.28 Combinatorial use of wiring molecules can reduce the number of wiring molecules needed

Another strategy for maximizing the uses of a finite number of signaling molecules takes advantage of the combinatorial actions of different wiring molecules. For the sake of simplicity, let's assume that input and target neurons use homophilic adhesion molecules. If each input neuron–target neuron pair uses only one homophilic adhesion molecule to specify its connection, then 25 molecules are needed to specify 25 connections. If molecules 1 through 5 and A through E can be independently recognized by the synaptic partners, and if the partners preferentially form synapses only when both molecules match, then the combinatorial use of these 10 molecules can specify 25 connections (**Figure 7–44**). By the same token, to specify 10^{12} connections, 2,000,000 molecules are sufficient if there is a mechanism for the combinatorial recognition of two molecules, 30,000 are sufficient for combinatorial recognition of three molecules, and a mere 600 are sufficient if combinatorial recognition of six molecules is used. This strategy can rapidly reduce the number of molecules needed to specify connections, but it also imposes new difficulties. Cells must impose thresholds for connections so that only when multiple labels simultaneously match will connections be allowed or highly preferred, and must also devise mechanisms to interpret multiple signals independently and unambiguously. Indeed, distinct signaling mechanisms have been found to act downstream of guidance receptors (see Box 5-2).

One can envision many instances in which this strategy might be used. For instance, ephrin-A/EphA specifies anterior–posterior positions for the retinotopic map. Molecules that specify dorsal–ventral positions likely work with the ephrin-A/EphA system combinatorially, so that each retinal axon has a specific value on each of the two axes. In this way, a two-dimensional retinal map can project in an orderly fashion to higher-order visual centers in the brain. A number of strategies, including temporal separation (see Section 7.27) and step-wise assembly (see Section 7.29), can be used to overcome the difficulties that the combinatorial strategy imposes on signal interpretation or thresholding.

7.29 Dividing wiring decisions into multiple steps can conserve molecules and increase fidelity

We have seen how combining multiple molecular recognition events can decrease the number of molecules needed to specify a wiring decision. The same molecule-conserving result can also be achieved by combining sequential steps. Imagine that 25 input neurons are to be connected to 25 target neurons. If all connections are made in a single step simultaneously, 25 specificity molecules are needed. If the decisions are divided into two steps, with each step having five possible outcomes, then the first step can use five molecules to divide 25 input neurons into five subgroups. In the second step, the same five new molecules can be used to specify the connection specificity of each subgroup (**Figure 7–45**). In addition

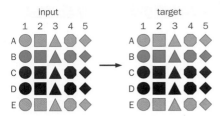

Figure 7–44 Combinatorial strategy. If each input or target neuron expresses two molecules (represented here as a number and a letter), and if both molecules must match for a connection to be specified or highly preferred, then 10 molecules can specify 25 connections. As illustrated here, this means that five possible colors (letters) and five possible shapes (numbers) can generate a matrix of 25 unique colored shapes (letter–number combinations). A blue triangle input neuron (that is, a neuron that expresses B+3) would preferentially connect with a blue triangle (B+3) target neuron, but would fail to connect with target neurons that express only one of the two identifying molecules, for example, red triangles (which express C+3), or blue squares (which express B+2).

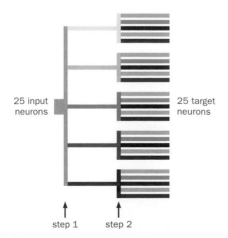

Figure 7–45 Sequential decisions can conserve molecules and increase fidelity. At step 1, input neurons are divided into five subgroups specified by five molecules symbolized by different gray levels. At step 2, five molecules symbolized by different colors are utilized to divide each subgroup into five sub-subgroups.

to conserving molecules, such step-wise decisions may increase robustness and reduce errors at each decision point, as there are fewer simultaneous choices.

These steps can be temporal or spatial, and we have encountered examples of both. For example, in the dendrite targeting of the fly olfactory circuit, an initial coarse map is established by global gradients of Sema1A and Sema2A/2B. Next, local binary choices are specified by cell-type-specific expression of Capricious. Capricious-positive glomeruli are distributed throughout the antennal lobe in a salt-and-pepper fashion, suggesting that the presence or absence of Capricious couples with graded semaphorin levels to locally sort out dendrite targeting (see Figure 7–41). Although the strict temporal order has not been proven, it is likely that semaphorin-mediated coarse targeting occurs prior to Capricious-mediated glomerular sorting.

The pre-target sorting of mammalian ORN axons provides an example of spatial steps. Instead of delaying the decision to the final destination at the target, the growing axons sort themselves out along their journey from the olfactory epithelium to the olfactory bulb. As they travel toward their targets, axons expressing identical odorant receptors—and thus similar levels of neuropilin-1 and Sema3A—are mutually repelled by axons expressing different levels of these molecules, so that ORNs of the same receptor type selectively associate with each other. These repulsive interactions can be thought of as a continuous series of steps. When the axons finally reach the olfactory bulb, each axon is constrained by its pre-sorting and can only choose from a limited set of targets (see Figure 7–37).

7.30 Many connections do not need to be specified at the level of individual synapses or neurons

The combined use of molecular and developmental strategies described above allows a limited set of molecules to wire many complex neural circuits with precision. However, neuronal wiring only needs to be specified as precisely as is useful. In the neuromuscular system, the connections between specific motor neuron pools and muscles are precisely specified. However, within a motor pool–muscle pair, the details of connections between individual motor neurons and muscle fibers vary considerably, as exemplified by the connectome of the mouse ear muscle (see Figure 7–28). In the olfactory system, great precision is required during ORN–PN matching. The wiring of each ORN or PN type must not be confused with any other type, as the different types represent distinct sensory information. However, the wiring of individual ORNs that express the same odorant receptor does not need to be defined further, as these ORNs project to the same glomerulus and represent the same sensory information. Thus, a nearly identical set of wiring instructions can be applied to ~30 or ~5000 ORNs of the same type in the fly or the mouse, respectively.

In certain neural circuits, it is not necessary or even beneficial for wiring to be specified precisely. The connection matrix between *Drosophila* olfactory PNs and mushroom body neurons (see Section 6.16) provides a good example. Here, the circuit function can be served as long as individual PNs connect stochastically with some mushroom body neurons. The meaning of these connections is acquired by individual experience. It is possible that the role of stochastic wiring increases with increasing brain complexity; this interesting hypothesis remains to be explored.

7.31 Wiring can be instructed by neuronal activity and experience

An important mechanism for neuronal wiring is spontaneous and experience-driven neuronal activity. This allows the strength and pattern of synaptic transmission to directly shape neuronal wiring. We have studied this most extensively in the wiring of the visual system (see Sections 5.7–5.11), but have also encountered it in wiring the somatosensory system (see Box 5–3) and the olfactory system (see Section 7.19), and in synapse refinement at the vertebrate neuromuscular junction (see Section 7.13). In complex circuits, molecular determinants guide the wiring of the basic circuit scaffold, with cellular and sometimes subcellular specificity. The number and strength of connections, which may contribute to the optimization of circuit function, can be refined by the powerful Hebbian rule—'fire together, wire together.' Indeed, as we learn new things throughout life, the brain constantly changes its wiring, mostly in the form of synaptic strengths between connecting neurons, as will be discussed further in Chapter 10.

A future challenge is to determine the relative contributions of the different wiring strategies described in this chapter and Chapter 5, in particular the relative contributions of activity-independent molecular specification and activity-dependent processes in brain wiring. Are there general rules that determine their differential use? How do these two mechanisms act together? One possible generalization might be that the proportion of activity-dependent wiring increases as the nervous system gets more complex, as might be observed by comparing the wiring of worm or fly with that of the mammalian brain. Yet another generalization might be that neuronal circuits that are used for innate behavior, such as avoiding a predator odor, are more hard-wired by molecular determinants selected by evolution, whereas those that underlie acquired ability, such as mastering a language, are sculpted more by an individual's life experience and neuronal activity.

SUMMARY

Wiring of the nervous system occurs in the context of a highly coordinated developmental program. The vertebrate nervous system derives from the neural tube, an ectodermal structure. Extrinsic morphogens and intrinsic transcription factors pattern the anterior–posterior and dorsal–ventral axes of the neural tube, from which diverse types of neural progenitors and neurons arise. Cell fate diversification can be achieved by both asymmetric cell division and cell–cell interactions. Coordinated expression and mutual repression of cell-type-specific transcription factors are frequently used to specify fates of distinct cell types. Post-mitotic neurons often migrate long distances from their birthplace to their final destination. By the time individual neurons begin to extend their axons and dendrites and start the wiring process, the nervous system already has a rough blueprint that specifies the numbers and types of neurons at specific locations.

The first step of a neuron's morphological differentiation is to send out processes, one of which becomes the axon. Axons often travel long distances toward their postsynaptic targets, guided on the way by intermediate cues. A critical output of the neuronal fate decision is the expression of a specific repertoire of guidance receptors. Guidance receptors in the growth cones of axons respond to attractive and repulsive cues along their path, and can change responses at intermediate targets. Multiple cue–receptor systems often act simultaneously and sequentially to guide axons to their final destinations. Dendrites differ from axons in their length, branching pattern, microtubule polarity, and dependence on local secretory machinery. Dendrites and axons are similar in the need for branching to effectively cover the receptive space or to send output to distinct targets. Homophilic repulsion by molecules with extraordinary diversity enables self-avoidance of axonal and dendritic branches.

A critical step of brain wiring is synapse formation. Extensive bidirectional communication between pre- and postsynaptic partners occurs both in neuron–muscle and in neuron–neuron synapses, utilizing trans-synaptic molecular interactions. Signaling from surrounding glia is also crucial for synapse formation and maturation. Subcellular site selection of synapses uses molecules and mechanisms similar to those involved in axon guidance and developmental patterning. Brain wiring also utilizes a series of regressive events to achieve its final connectivity. These include elimination of extra synapses, pruning of exuberant axons and dendrites, and death of excess neurons.

The precise one-to-one pairing of axons of a given type of olfactory receptor neuron (ORN) with dendrites of a given type of postsynaptic projection neuron (PN) in the mouse and fly olfactory systems provides an excellent model to study how wiring specificity arises in the context of assembling a functional neural circuit. In the mouse, the basal level signaling of odorant receptors plays an instructive role in ORN axon targeting by regulating the expression levels of guidance molecules. Repulsive axon–axon interactions sort ORN axons along their path. Activity-dependent expression of adhesive and repulsive molecules further refines uniglomerular targeting of ORN axons. In the fly, ORNs and PNs are independently specified, and both neuronal types play active roles in regulating wiring specificity. PN dendrites coarsely pre-pattern the antennal lobe, utilizing a combination of global gradients and local binary determinants. ORN axons sort themselves through axon–axon interactions and recognize cues on partner PNs to establish the final one-to-one matching.

The molecular and cellular mechanisms we have studied in this chapter and in Chapter 5 begin to explain how a limited number of recognition molecules can specify an astronomical number of synaptic connections. A small set of cell-surface receptors have been identified that exhibit extraordinary molecular diversity. Different concentrations of the same protein can be used to specify different connections. The same molecule can have multiple functions, and can be used in different parts of the nervous system and at different developmental stages to specify distinct connections. The combinatorial strategy can conserve molecules, as can the strategy of dividing wiring decision into multiple steps. Many connections can be instructed by neuronal activity and experience. Some connections do not need to be specified at the level of individual neurons or synapses, as neuronal connections need to be specified only as precisely as are useful.

For ease of experimental observation and manipulation, most studies of brain wiring have thus far been conducted with relatively simple model neurons and organisms. The lessons learned and the tools acquired can now be applied to more complex neural circuits in the mammalian brain. Deciphering how neural circuits are organized in the adult is often the prelude to determining how wiring specificity arises. Elucidating the organizational principles of neural circuits and the mechanisms of their assembly will in turn help us understand the function of these neural circuits in perception and behavior.

FURTHER READING

Reviews and books

Jan YN & Jan LY (2010) Branching out: mechanisms of dendritic arborization. *Nat Rev Neurosci* 11:316–328.

Jessell TM (2000) Neuronal specification in the spinal cord: inductive signals and transcriptional codes. *Nat Rev Genet* 1:20–29.

Levi-Montalcini R (1987) The nerve growth factor 35 years later. *Science* 237:1154–1162.

Luo L & Flanagan JG (2007) Development of continuous and discrete neural maps. *Neuron* 56:284–300.

O'Leary DD, Chou SJ & Sahara S (2007) Area patterning of the mammalian cortex. *Neuron* 56:252–269.

Rakic P (1988) Specification of cerebral cortical areas. *Science* 241:170–176.

Sanes JR & Lichtman JW (1999) Development of the vertebrate neuromuscular junction. *Annu Rev Neurosci* 22:389–442.

Zipursky SL & Grueber WB (2013) The molecular basis of self-avoidance. *Annu Rev Neurosci* 36:547–568.

General neural development

Alcamo EA, Chirivella L, Dautzenberg M et al. (2008) Satb2 regulates callosal projection neuron identity in the developing cerebral cortex. *Neuron* 57:364–377.

Anderson SA, Eisenstat DD, Shi L et al. (1997) Interneuron migration from basal forebrain to neocortex: dependence on *Dlx* genes. *Science* 278:474–476.

Angevine JB, Jr & Sidman RL (1961) Autoradiographic study of cell migration during histogenesis of cerebral cortex in the mouse. *Nature* 192:766–768.

Ango F, di Cristo G, Higashiyama H et al. (2004) Ankyrin-based subcellular gradient of neurofascin, an immunoglobulin family protein, directs GABAergic innervation at Purkinje axon initial segment. *Cell* 119:257–272.

Barnes AP, Lilley BN, Pan YA et al. (2007) LKB1 and SAD kinases define a pathway required for the polarization of cortical neurons. *Cell* 129:549–563.

Bradke F & Dotti CG (1999) The role of local actin instability in axon formation. *Science* 283:1931–1934.

Briscoe J, Pierani A, Jessell TM et al. (2000) A homeodomain protein code specifies progenitor cell identity and neuronal fate in the ventral neural tube. *Cell* 101:435–445.

Buffelli M, Burgess RW, Feng G et al. (2003) Genetic evidence that relative synaptic efficacy biases the outcome of synaptic competition. *Nature* 424:430–434.

Campenot RB (1977) Local control of neurite development by nerve growth factor. *Proc Natl Acad Sci USA* 74:4516–4519.

Charron F, Stein E, Jeong J et al. (2003) The morphogen sonic hedgehog is an axonal chemoattractant that collaborates with netrin-1 in midline axon guidance. *Cell* 113:11–23.

Christopherson KS, Ullian EM, Stokes CC et al. (2005) Thrombospondins are astrocyte-secreted proteins that promote CNS synaptogenesis. *Cell* 120:421–433.

Cohen S, Levi-Montalcini R & Hamburger V (1954) A nerve growth-stimulating factor isolated from sarcomas 37 and 180. *Proc Natl Acad Sci USA* 40:1014–1018.

Dotti CG, Sullivan CA & Banker GA (1988) The establishment of polarity by hippocampal neurons in culture. *J Neurosci* 8:1454–1468.

Gautam M, Noakes PG, Moscoso L et al. (1996) Defective neuromuscular synaptogenesis in agrin-deficient mutant mice. *Cell* 85:525–535.

Godfrey EW, Nitkin RM, Wallace BG et al. (1984) Components of *Torpedo* electric organ and muscle that cause aggregation of acetylcholine receptors on cultured muscle cells. *J Cell Biol* 99:615–627.

Graf ER, Zhang X, Jin SX et al. (2004) Neurexins induce differentiation of GABA and glutamate postsynaptic specializations via neuroligins. *Cell* 119:1013–1026.

Horton AC & Ehlers MD (2003) Dual modes of endoplasmic reticulum-to-Golgi transport in dendrites revealed by live-cell imaging. *J Neurosci* 23:6188–6199.

Kania A & Jessell TM (2003) Topographic motor projections in the limb imposed by LIM homeodomain protein regulation of ephrin-A:EphA interactions. *Neuron* 38:581–596.

Klassen MP & Shen K (2007) Wnt signaling positions neuromuscular connectivity by inhibiting synapse formation in *C. elegans*. *Cell* 130:704–716.

Lefebvre JL, Kostadinov D, Chen WV et al. (2012) Protocadherins mediate dendritic self-avoidance in the mammalian nervous system. *Nature* 488:517–521.

Levi-Montalcini R & Booker B (1960) Destruction of the sympathetic ganglia in mammals by an antiserum to a nerve-growth protein. *Proc Natl Acad Sci USA* 46:384–391.

Levi-Montalcini R & Booker B (1960) Excessive growth of the sympathetic ganglia evoked by a protein isolated from mouse salivary glands. *Proc Natl Acad Sci USA* 46:373–384.

Lu J, Tapia JC, White OL et al. (2009) The interscutularis muscle connectome. *PLoS Biol* 7:e32.

Lyuksyutova AI, Lu CC, Milanesio N et al. (2003) Anterior-posterior guidance of commissural axons by Wnt-frizzled signaling. *Science* 302:1984–1988.

Ori-McKenney KM, Jan LY & Jan YN (2012) Golgi outposts shape dendrite morphology by functioning as sites of acentrosomal microtubule nucleation in neurons. *Neuron* 76:921–930.

Rhyu MS, Jan LY & Jan YN (1994) Asymmetric distribution of numb protein during division of the sensory organ precursor cell confers distinct fates to daughter cells. *Cell* 76:477–491.

Scheiffele P, Fan J, Choih J et al. (2000) Neuroligin expressed in nonneuronal cells triggers presynaptic development in contacting axons. *Cell* 101:657–669.

Schmucker D, Clemens JC, Shu H et al. (2000) *Drosophila* Dscam is an axon guidance receptor exhibiting extraordinary molecular diversity. *Cell* 101:671–684.

Spitzweck B, Brankatschk M & Dickson BJ (2010) Distinct protein domains and expression patterns confer divergent axon guidance functions for *Drosophila* Robo receptors. *Cell* 140:409–420.

Tapia JC, Wylie JD, Kasthuri N et al. (2012) Pervasive synaptic branch removal in the mammalian neuromuscular system at birth. *Neuron* 74:816–829.

Tosney KW & Landmesser LT (1985) Specificity of early motoneuron growth cone outgrowth in the chick embryo. *J Neurosci* 5:2336–2344.

Watts RJ, Hoopfer ED & Luo L (2003) Axon pruning during *Drosophila* metamorphosis: evidence for local degeneration and requirement of the ubiquitin-proteasome system. *Neuron* 38:871–885.

Wojtowicz WM, Wu W, Andre I et al. (2007) A vast repertoire of Dscam binding specificities arises from modular interactions of variable Ig domains. *Cell* 130:1134–1145.

Assembly of the olfactory circuits

Hong W, Mosca TJ & Luo L (2012) Teneurins instruct synaptic partner matching in an olfactory map. *Nature* 484:201–207.

Imai T, Suzuki M & Sakano H (2006) Odorant receptor-derived cAMP signals direct axonal targeting. *Science* 314:657–661.

Imai T, Yamazaki T, Kobayakawa R et al. (2009) Pre-target axon sorting establishes the neural map topography. *Science* 325:585–590.

Jefferis GS, Marin EC, Stocker RF et al. (2001) Target neuron prespecification in the olfactory map of *Drosophila*. *Nature* 414:204–208.

Joo WJ, Sweeney LB, Liang L et al. (2013) Linking cell fate, trajectory choice, and target selection: genetic analysis of Sema-2b in olfactory axon targeting. *Neuron* 78:673–686.

Komiyama T, Sweeney LB, Schuldiner O et al. (2007) Graded expression of semaphorin-1a cell-autonomously directs dendritic targeting of olfactory projection neurons. *Cell* 128:399–410.

Nakashima A, Takeuchi H, Imai T et al. (2013) Agonist-independent GPCR activity regulates anterior-posterior targeting of olfactory sensory neurons. *Cell* 154:1314–1325.

Serizawa S, Miyamichi K, Takeuchi H et al. (2006) A neuronal identity code for the odorant receptor-specific and activity-dependent axon sorting. *Cell* 127:1057–1069.

Wang F, Nemes A, Mendelsohn M et al. (1998) Odorant receptors govern the formation of a precise topographic map. *Cell* 93:47–60.

CHAPTER 8

Motor and Regulatory Systems

To move things is all mankind can do, for such the sole executant is muscle, whether in whispering a syllable or in felling a forest.

Charles Sherrington, 1924

The sensory systems that we studied in Chapters 4 and 6 enable animals to know about the world they live in. But sensation is only the first step: to adapt to the world, animals must act upon sensory knowledge. For example, animals have evolved the ability to sense the presence of food—through a combination of olfaction, vision, and taste—so that they can eat the food and obtain energy essential for life. Likewise, animals have evolved the ability to detect danger so that they can fight, hide, or run away from it. In this chapter, we study the output of the nervous system that makes these active responses to sensation possible.

A principal output is the **motor system**, which controls the contraction of skeletal muscles and thereby enables movement, such as reaching for and grasping an object, walking, talking, or maintaining the body's posture. **Figure 8–1** illustrates just one example of how the nervous system can exquisitely control movement, in this case enabling a horse to trot gracefully. A second output system is the **autonomic nervous system**, which controls the contraction of smooth and cardiac muscles and which regulates the function of internal organs. The

Figure 8–1 The movement of a trotting horse. A complete stride of a trotting horse in 12 frames, engraved after photos by Eadweard Muybridge. These photos (exposure time 1/500 s) showed clearly for the first time that the trotting horse is entirely in the air for part of the stride (frames 4, 5, 9, and 10). (From Muybridge E [1878] *Sci Am* 39:241. With permission from Macmillan Publishers Ltd.)

third output system is the **neuroendocrine system**, which secretes hormones in response to sensory stimuli and brain states and which regulates an animal's physiology and behavior, including the response to food and daily cycle of light and darkness. A regulatory center for both autonomic and neuroendocrine systems is the hypothalamus. The autonomic and neuroendocrine systems are key components of the body's regulatory systems.

We start the chapter with the motor system, following a theme of locomotion control in vertebrates. We then discuss the autonomic nervous system and neuroendocrine system. In the final parts of the chapter, we use the examples of eating, circadian rhythm, and sleep to illustrate how these basic functions are regulated.

HOW IS MOVEMENT CONTROLLED?

Movement is produced by the coordinated activation of motor neurons, which causes the coordinated contraction of skeletal muscles. In vertebrates, motor neuron cell bodies are located in the spinal cord and the brainstem; their axons exit the CNS and innervate specific muscles in the body and head, respectively. Motor neurons are themselves under elaborate control: they receive direct input from proprioceptive somatosensory neurons, from local circuit interneurons, from brainstem nuclei specialized in initiating and modulating movement, and from the motor cortex. These descending motor control centers are organized in a hierarchical manner, with the somatosensory system providing feedback signals at multiple levels within this hierarchy. In addition, there are two important loops involving the basal ganglia and cerebellum that add to the sophistication of motor controls (**Figure 8–2**). In Chapters 4 and 6, we traced the sensory systems by following the direction of information flow, starting with sensory neurons and moving into the brain. We will take the same peripheral-to-central approach to our studies of the motor system, but this time we will address topics in the direction opposite the information flow, starting with muscles and motor neurons and then describing how layers of upstream neurons and circuits contribute to movement control.

8.1 Muscle contraction is mediated by sliding of actin and myosin filaments and is regulated by intracellular Ca²⁺

Muscle contraction underlies all bodily movement, as reflected by the epigraph. The mechanism of muscle contraction is understood in molecular detail. The muscle cell contains **myofibrils**, which are thread-like longitudinal structures (**Figure 8–3**A). Under the electron microscope, myofibrils can be seen to consist of repeating units called **sarcomeres**. Each sarcomere is made of overlapping thick and thin filaments arranged in an orderly fashion. The thin filaments are composed of **filamentous actin** (**F-actin**) fibers with several associated proteins (Figure 8–3B). The thick filaments are composed of the **myosin** protein (Figure 8–3C). The sliding of thick and thin filaments over each other, mediated by the physical interactions of actin and myosin, forms the basis of muscle contraction (Figure 8–3D).

As was introduced in Section 2.3, myosin is an F-actin-binding **motor protein**. Myosin has a long tail embedded in the thick filament and a globular head that forms the cross bridge between the thick and thin filaments (Figure 8–3C, D). The head of myosin has an ATPase domain that can hydrolyze ATP. The hydrolysis cycle requires interaction of the myosin head with actin and converts chemical energy from ATP hydrolysis into mechanical force in the form of a **power stroke**, which underlies how myosin and the actin filaments move relative to each other (**Figure 8–4**A; **Movie 8–1**). This motility can be visualized in reduced preparations; for example, fluorescently labeled actin filaments were observed to move on a glass slide coated with pure myosin proteins (Figure 8–4B; see also **Movie 8–2**). Indeed, the movement produced by the interaction of myosin and actin, which was first investigated in the context of muscle contraction, underlies

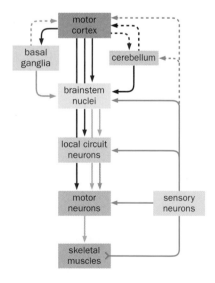

Figure 8–2 Hierarchical organization of movement control. Arrows indicate the direction of information flow. Solid arrows between two areas indicate that at least some connections are direct. Dashed arrows indicate connections through intermediate neurons. The Y-shaped terminal of sensory neurons in the skeletal muscle symbolizes the peripheral ending of proprioceptive sensory neurons.

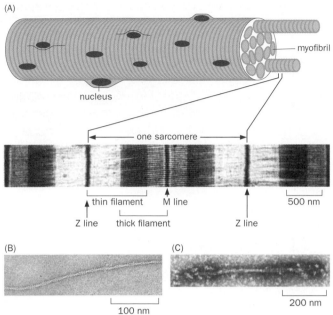

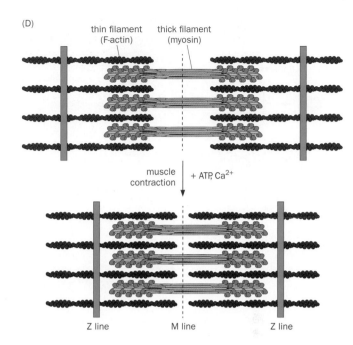

Figure 8–3 The molecular organization of the muscle cell and the sliding filament model of muscle contraction. (A) Top, schematic of a muscle cell, which has multiple nuclei and consists of parallel myofibrils. Bottom, as seen under the electron microscope, myofibrils are composed of repeating units called sarcomeres. Each sarcomere is formed by the intersection of thin filaments originating from the Z line with thick filaments originating from the M line. **(B)** A negatively stained electron micrograph shows a microfilament consisting of two F-actin polymer strands; in muscle, F-actin constitutes the major component of the thin filament. **(C)** A negatively stained electron micrograph shows an aggregation of purified myosin molecules, which forms a bare region in the middle and thick protrusions at each end, resembling the organization of the thick filaments. **(D)** Illustration of the sliding filament model. The relative movement of thick and thin filaments, caused by myosin motors moving on the actin filaments, underlies muscle contraction. Ca^{2+} is required for the myosin–actin interaction. ATP hydrolysis powers the movement. See Figure 8–4A for more details. (A–C, micrographs from Huxley HE [1965] *Sci Am* 213:18–27. With permission from Macmillan Publishers Ltd. See also Huxley AF & Niedergerke R [1954] *Nature* 173:971–973 and Huxley H & Hanson J [1954] *Nature* 173:973–976.)

many aspects of cell motility including cytokinesis, cell migration, and growth cone guidance (see Box 5–2).

The actin/myosin-mediated contraction also requires Ca^{2+} in addition to ATP. This is because the F-actin in thin filaments is coated by two proteins called tropomyosin and troponin, which prevent the actin from binding with the myosin head under normal (that is, low) intracellular Ca^{2+} concentrations. A rise in $[Ca^{2+}]_i$ causes a conformational change in the actin–tropomyosin complex, exposing the actin surface that interacts with myosin.

Ca^{2+} regulation of actin/myosin-mediated contraction forms the link between motor neuron activity and muscle contraction, a process termed **excitation–contraction coupling** (**Figure 8–5**). As discussed in Sections 3.1 and 3.12, the arrival of an action potential at the motor axon terminal in the vertebrate neuromuscular junction causes its depolarization and release of the neurotransmitter acetylcholine (ACh). ACh binds to and opens the nicotinic ACh receptor/cation channel at the neuromuscular junction, which causes depolarization of the muscle cell and production of action potentials within the muscle cell itself. Muscle cell depolarization triggers the release of Ca^{2+} from the **sarcoplasmic reticulum**, a special endoplasmic reticulum derivative that extends throughout the muscle cells. **Transverse tubules** (**T tubules**), which are invaginations of the plasma membrane that extend into the muscle cell interior, bring plasma membrane close to the sarcoplasmic reticulum, such that depolarization effectively triggers Ca^{2+} release throughout the entire large muscle cell. This causes nearly synchronous contraction of all sarcomeres within the same muscle cell, enabling muscles to respond rapidly to commands from motor neurons. Efficient reuptake of Ca^{2+} by the sarcoplasmic reticulum enables muscle cells to respond to repeated commands from motor neurons (see **Movie 8–3**).

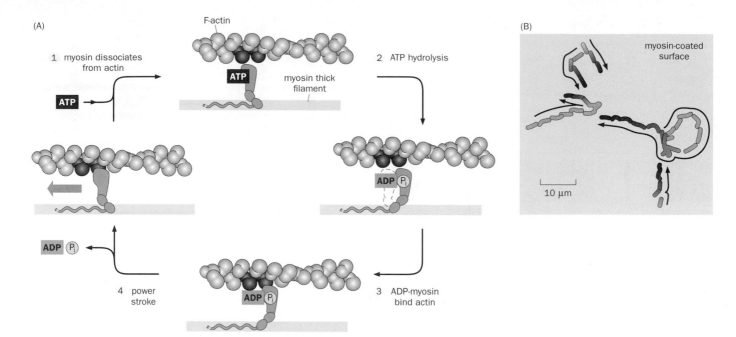

(A)

F-actin

1   myosin dissociates
    from actin

ATP

ATP

myosin thick
filament

2   ATP hydrolysis

ADP Pᵢ

4   power
    stroke

ADP Pᵢ

3   ADP-myosin
    bind actin

ADP Pᵢ

(B)

myosin-coated
surface

10 µm

Figure 8–4 ATP hydrolysis powers the relative movement of myosin and actin filaments. (A) The coupled cycles of actin–myosin interaction and ATP hydrolysis. (1) ATP binding to myosin triggers the dissociation of myosin and actin; (2) ATP hydrolysis causes conformational change of the myosin head such that it aligns with the binding surface of the next actin subunit (circles); (3) Myosin-ADP binds to actin again; (4) ADP and Pᵢ release produces a power stroke—sliding of F-actin against the myosin (pink arrow). Note that the myosin head's orientation relative to the actin changes with the power stroke. The highlighted actin subunits are shifted one subunit to the right with respect to the actin filament before and after Step 1

to accommodate a new cycle. **(B)** Movement paths of fluorescently labeled actin filaments (rectangles) on a myosin-coated glass slide. Five actin filaments were tracked over the course of 38 s, and their positions are indicated at successive short intervals as they appeared on the video monitor. Arrows indicate the movement direction (the later in time, the darker the actin filaments appear). See also Movie 8–2. (A, based on Lymn RW & Taylor EW [1971] *Biochemistry* 10:4617. With permission from the American Chemical Society; B, adapted from Kron J & Spudich JA [1986] *Proc Natl Acad Sci USA* 83:6272.)

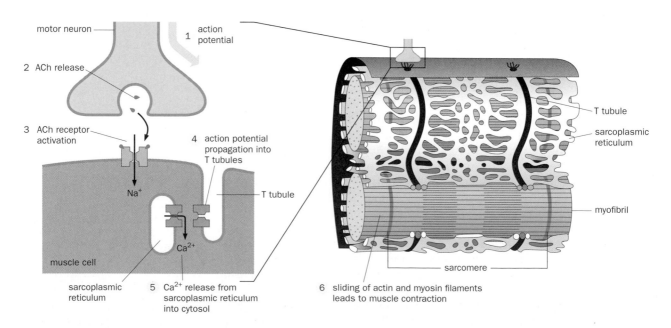

motor neuron

1   action
    potential

2   ACh release

3   ACh receptor
    activation

Na⁺

4   action potential
    propagation into
    T tubules

T tubule

muscle cell

sarcoplasmic
reticulum

5   Ca²⁺ release from
    sarcoplasmic reticulum
    into cytosol

Ca²⁺

T tubule

sarcoplasmic
reticulum

myofibril

sarcomere

6   sliding of actin and myosin filaments
    leads to muscle contraction

Figure 8–5 Sequence of events from motor neuron excitation to skeletal muscle contraction. The arrival of an action potential at the motor axon terminal (1) triggers acetylcholine (ACh) release (2). ACh binds to the nicotinic ACh receptor on the postsynaptic muscle surface, which opens the nicotinic ACh receptor channel and triggers depolarization and action potential production in the muscle cell (3).

Action potentials propagate within the muscle cell to the T tubules (4), and trigger Ca²⁺ release from nearby sarcoplasmic reticulum into the cytosol (5). Elevated [Ca²⁺]ᵢ causes muscle contraction (6). (The magnified drawing on the right is adapted from Alberts B, Johnson A, Lewis J et al. [2015] Molecular Biology of the Cell, 6th ed. Garland Science.)

8.2 Motor units within a motor pool are recruited sequentially from small to large

Each muscle consists of a few hundred to over a million individual muscle cells (also called **muscle fibers**). As discussed in Section 7.13, each muscle fiber is innervated by a single motor neuron in adults. However, each motor neuron innervates multiple muscle fibers, ranging from a few in an eye muscle to a few thousand in a leg muscle. Individual muscle fibers that are innervated by a single motor neuron are dispersed within a given muscle, such that activation of that motor neuron produces force evenly across the muscle (**Figure 8–6**; see also Figures 7–27 and 7–28). A motor neuron and the set of muscle fibers it innervates are collectively called a **motor unit**. As the neuromuscular junctions are powerful synapses that almost always convert presynaptic action potentials to neurotransmitter release and muscle contraction, all muscle fibers within a motor unit are nearly always activated together. Thus, the motor unit is the elementary unit of force that can be activated by the motor system.

Motor neurons that innervate the same muscle are physically clustered together in the ventral spinal cord or brainstem and constitute a **motor pool** (Figure 8–6); the number of neurons in a motor pool ranges from dozens to thousands. The **motor unit size** (that is, the number of muscle fibers a motor neuron innervates) varies considerably for neurons within the same motor pool (see Figure 7–28). Collectively, motor units within a motor pool follow a **size principle**: neurons that have smaller motor unit sizes, which usually have smaller axon diameters and cell bodies, fire before neurons with larger motor unit sizes.

This size principle can be illustrated by recording motor axon bundles in response to sensory or electrical stimulation. For example, pressing on the Achilles tendon elicits a stretch reflex that activates the triceps surae muscle in the lower leg (**Figure 8–7A**). The magnitude of triceps surae activation increases as the pressure on the Achilles tendon increases. By recording motor axon bundles that innervate the triceps surae muscle in response to different magnitudes of pressure, researchers found that the motor axons of the smaller units were excited by a small amount of pressure and that motor axons corresponding to increasingly larger motor units were recruited sequentially as the pressure increased. In response to decremental pressure, motor axons with the largest units ceased firing first, followed in an orderly manner by axons of decreasing motor unit size (Figure 8–7B). Importantly, this order of motor neuron recruitment does not usually vary whether the muscles are stimulated by the natural stretch reflex or by direct electrical stimulation, suggesting that it is an intrinsic property of the motor pool.

The size principle of motor units enables incremental control of the magnitude of an individual muscle's contraction in response to excitatory and inhibitory inputs received by its motor pool. In this way, the size principle resembles sensory adaptation and Weber's law that was discussed in the context of sensory perception (see Section 4.7). When the magnitude of muscle contraction is small,

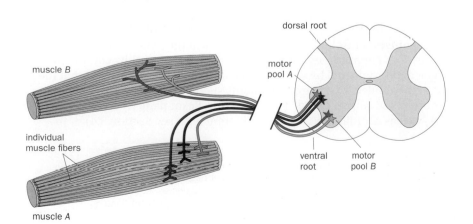

Figure 8–6 Motor pools and motor units. Each muscle is innervated by a cluster of motor neurons in the ventral spinal cord that constitute a motor pool; the axons of these motor neurons exit the spinal cord via the ventral root. Two motor pools, A and B, are shown here. Within each motor pool, individual motor neurons innervate multiple muscle fibers, whereas each muscle fiber is innervated by a single motor neuron. A motor neuron and the set of muscle fibers it innervates together constitute a motor unit. Motor units vary in size, as is illustrated for the two motor units from motor pool B.

(A)

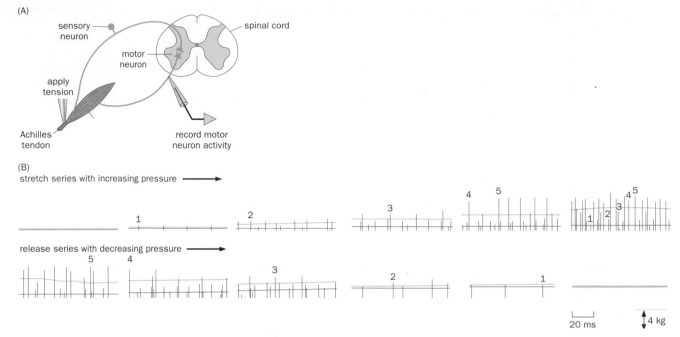

(B)

stretch series with increasing pressure ⟶

release series with decreasing pressure ⟶

20 ms   ⬍ 4 kg

Figure 8–7 Motor units of different sizes are recruited in an orderly manner. (A) Schematic illustration of the experimental setup. Nerves that innervate all hindlimb muscles except the triceps surae muscle were surgically severed. Pressure was applied to the Achilles tendon via a mechanoelectric transducer to elicit the stretch reflex. Responses of motor neurons that innervate the triceps surae muscle were recorded at the ventral root of the spinal cord, where groups of motor axons can be recorded simultaneously (only one is illustrated). The size of an action potential (vertical bar in panel B) correlates with axon diameter and hence with motor unit size. **(B)** Top, as the pressure increases (indicated by the distance between the blue and black horizontal lines, scale at bottom right), the smallest motor neuron (1) fires first, followed by motor neurons of increasing sizes (2 through 5). Bottom, as the pressure decreases, the largest motor neuron (5) ceases firing first, followed by motor neurons of descending size (4 through 1). (B, adapted from Henneman E, Somjen G, & Carpernter DO [1965] *J Neurophysiol* 28:560.)

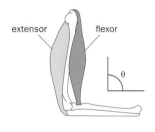

extensor contraction

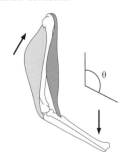

flexor contraction

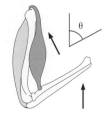

adding or subtracting a small motor unit makes a notable difference; these differences can be used for fine motor control. As the magnitude of muscle contraction increases, the size of motor unit required to make a notable difference in the total contraction strength also becomes larger. The size principle is also important from an energetics perspective. Most movements are small and only use small motor units that consume less energy, whereas large motor units, which exert greater forces and consume more energy, are used more rarely.

8.3 Motor neurons receive diverse and complex input

We now turn to a key question in movement control: how are activations of different muscles coordinated? One principal output of muscle contraction is to change the angle of a joint, as in the case of the knee-jerk reflex (see Figure 1–19). The angle change is achieved by coordinated action of the **extensor** muscle, whose contraction increases the angle (and thus extends the joint), and the **flexor** muscle, whose contraction decreases the angle (**Figure 8–8**). Extensors and flexors are **antagonistic muscles** because they perform opposite actions and often fire in succession. For example, extension of a joint is initiated by contraction of the extensor and terminated by subsequent contraction of the flexor, so that the joint is not overextended. Complex movements such as the trotting of a horse involve the coordinated contraction of many extensor–flexor pairs within each of the four legs, as well as coordination between different legs (see Section 8.4).

Figure 8–8 Extensor and flexor muscle pairs control the joint angle. Contraction of the extensor muscle increases the joint angle (θ), leading to the extension of the joint. Contraction of the flexor muscle decreases the joint angle. The extent and timing of extensor and flexor contraction are coordinated to control the change of joint angles precisely.

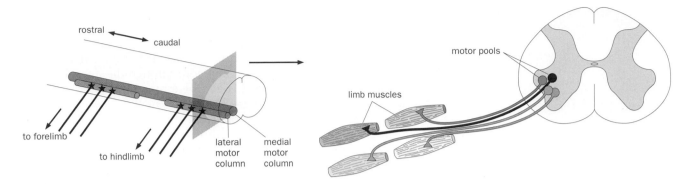

Figure 8–9 Organization of the motor columns and motor pools in the spinal cord. Left, motor neurons are organized in motor columns in the ventral spinal cord along the rostral–caudal axis. The medial motor columns regulate trunk muscles, whereas the lateral motor columns are present at the levels of the spinal cord corresponding to the location of the limbs, and innervate limb muscles. Right, a cross section of the spinal cord at the level of the hindlimb. Motor pools that innervate specific muscles (shown as four color-matched pairs) are located in stereotyped positions within the ventral spinal cord. This organization is bilaterally symmetrical and only the left side of the motor columns, motor pools, and muscle targets are illustrated.

Our discussions so far have indicated that coordination of muscle contraction must be due to coordinated firing of specific motor pools. Therefore, we need to learn more about how motor pools are organized and how motor neuron firing is controlled by their inputs. Motor pools that control trunk muscles are located in two bilaterally symmetrical medial motor columns that run the length of the rostral–caudal axis of the spinal cord, while motor pools that regulate muscles within each limb form lateral motor columns at the rostral–caudal positions in the spinal cord that innervate the limbs (**Figure 8–9**). Within the lateral motor columns, the motor pools that innervate specific limb muscles are located in stereotyped positions. These were originally determined by retrograde tracing (see Section 13.18): dyes injected into a specific muscle can be taken up by the axon terminals of motor neurons that innervate that muscle and transported back to the neuronal cell bodies, causing the corresponding motor pool to be labeled.

Motor neurons are one of the most complex neuronal types in the nervous system with regard to their input sources. Each motor neuron elaborates a dendritic tree that covers a large area of the ventral spinal cord (see Figure 1–15C), enabling it to receive direct input from diverse sources (see Figure 8–2). One major source of input is from 'local' excitatory and inhibitory interneurons, whose cell bodies are located in the spinal cord; these are collectively termed spinal cord **premotor neurons**. The distribution of spinal cord premotor neurons is complex, and has recently been analyzed comprehensively by applying the retrograde trans-synaptic tracing method discussed in Section 6.10 (see Section 13.19 for details). **Figure 8–10** shows the distribution of premotor neurons for the motor pool that innervates the quadriceps muscle in the dorsal thigh of the mouse hindlimb. Premotor neurons can be ipsilateral or contralateral to the motor pools and are spread across many spinal cord segments. Promoter neurons consist of multiple subtypes that use different neurotransmitters including glutamate, GABA, glycine, and ACh. Motor pools that innervate different muscles receive input from distinct populations of premotor neurons.

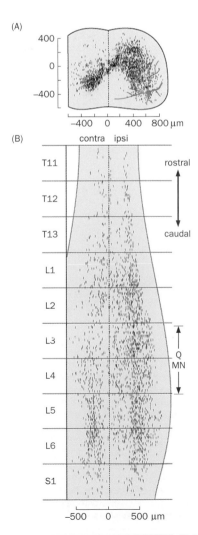

Figure 8–10 Distribution of spinal neurons presynaptic to the motor pool innervating the quadriceps muscle in the mouse right hindlimb. Even though the cell bodies of quadriceps motor pools (Q MN) are restricted to the two segments as indicated, their presynaptic premotor neurons (dots) are widely distributed as seen from **(A)** a transverse projection of the spinal cord and **(B)** a longitudinal projection covering thoracic (T), lumbar (L), and sacral (S) segments. A drawing of the dendritic arbor of a typical Q motor neuron (cyan) is superimposed in panel A. The dotted red line represents the midline of the spinal cord. See Figure 8–18A for more details of the method used to obtain these data. (Adapted from Stepien AE, Tripodi M & Arber S [2010] *Neuron* 68:456–472. With permission from Elsevier Inc. Motor neuron drawing courtesy of Silvia Arber.)

In addition to receiving input from spinal cord premotor neurons, motor neurons also receive monosynaptic input from a small subset of proprioceptive somatosensory neurons, which innervate muscle spindles and form simple reflex arcs (see Figure 1–19). These sensory neurons are located in the dorsal root ganglia, and their central axons enter the spinal cord through the **dorsal root** (see Figure 6–63), distinct from the **ventral root** where motor axons exit the spinal cord to innervate the muscles (see Figure 8–6). Motor neurons also receive direct descending input from neurons in the brainstem and motor cortex, whose axons travel down the spinal cord along distinct pathways in the spinal cord white matter that are composed of myelinated axons. And, like the motor neurons they regulate, spinal cord premotor neurons themselves receive input from sensory neurons and descending neurons from the brainstem and motor cortex (see Figure 8–2).

The bewildering complexity of spinal cord premotor neurons, along with descending input and sensory feedback, offers exquisite control of motor neuron firing and muscle contraction. It also poses great challenges for researchers to discover the underlying principles of motor coordination. In the next three sections, we will study an important principle by which rhythmic activation of motor neurons and muscle contraction is achieved.

8.4 Central pattern generators coordinate rhythmic contraction of muscles during locomotion

Locomotion involves coordinated and rhythmic contraction of many distinct muscles. For example, in a trotting horse, different leg muscles are activated in a specific sequence to enable each leg to step on the ground, leave the ground, extend forward, and step on the ground again. The four legs are highly coordinated. When the horse slows down to a walk, or speeds up to a gallop, the cycle speed for each leg and the synchrony among the legs differ from when the horse is trotting (**Figure 8–11**). How are these motor programs controlled?

Recall the knee-jerk reflex discussed in Section 1.9. Proprioceptive sensory neurons transduce the mechanical stimulation in the spindle of the extensor muscle to activate the extensor motor neurons through monosynaptic excitation,

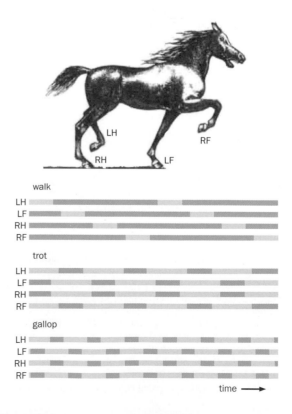

Figure 8–11 Stepping patterns of a horse during walking, trotting, and galloping. From left to right, each horizontal bar indicates for a single leg the time off the ground (gray segments) and on the ground (green segments). During trotting, the left hindleg (LH) and right foreleg (RF) move in sync, as do the right hindleg (RH) and left foreleg (LF). During galloping, the two forelegs are in sync, as are the two hindlegs. (Adapted from Pearson K [1976] *Sci Am* 235:72–86. With permission from Macmillan Publishers Ltd.)

and at the same time inhibit the flexor motor neurons through an intermediate spinal cord inhibitory neuron (see Figure 1–19). An early hypothesis of 'chained reflexes' proposed that rhythmic movement such as walking was due to sequential activation of spinal reflexes. Specifically, movement of a leg caused by contraction of a muscle would produce feedback from proprioceptive sensory neurons, which would activate a second motor pool and its corresponding muscle through the spinal cord reflex circuits. This would trigger the activation of a second set of sensory neurons that in turn would activate a third motor pool and muscle, and so on, until the original muscle would be activated again, completing the chained reflex cycle.

The chained reflex hypothesis predicts that (1) the spinal cord with intact sensory feedback should be sufficient for rhythmic activation of muscles, and (2) when sensory feedback is prevented, rhythmic activation of muscles should cease. The organization of the spinal cord made testing the sensory feedback possible, since transecting the dorsal root would block sensory feedback to motor neurons without affecting the connections between motor neurons and muscles, which leave the spinal cord through the ventral root (see Figure 8–6). An experiment to test these hypotheses was performed a century ago. In this experiment, all the muscles of the hindlimbs of an anesthetized cat were surgically inactivated except for one extensor–flexor pair in the lower leg, so that contraction of the extensor (gastrocnemius) and flexor (tibialis anterior) could be measured accurately. A spinal transection rostral to the segments that control the hindlimb was found to induce a transient rhythmic and alternate contraction of the remaining extensor and flexor that mimicked normal walking. Importantly, when the dorsal roots had been transected to eliminate sensory feedbacks, the pattern and frequency of alternate contraction of the extensor and flexor induced by spinal transection persisted (**Figure 8–12**A). This experiment supported prediction (1), as rhythmic contractions of the extensor and flexor could indeed be supported by isolated spinal cord disconnected from descending control. (As we will learn later, the rhythmic activity must be *initiated* by the brainstem; in this case the act of spinal cord lesion initiated the rhythmic activity.) In contrast to prediction (2), however, this experiment showed that rhythmic contractions persisted in the absence of sensory feedback.

The idea that an autonomous spinal cord mechanism could produce rhythmic output was further supported when technical advances in the 1960s and 1970s made it possible to measure the rhythmic contraction of many muscles

Figure 8–12 Rhythmic and coordinated muscle contraction in the absence of sensory feedback. (A) Contractions of a flexor muscle and an extensor muscle induced by spinal cord transection were recorded. The dorsal roots had been transected prior to the experiment to remove sensory feedback. The same numbers below the two contraction records indicate the same times (linked to the trace by dotted lines). Within each cycle, the flexor contracted prior to the extensor, followed by a period when neither muscle contracted. Thus, the rhythmic activation of these muscles persisted in the absence of sensory feedback. **(B)** A mesencephalic cat, in which the brainstem/spinal cord and the cerebral cortex/thalamus had been surgically disconnected, was induced to walk on a treadmill by brainstem stimulation. Action potentials of four muscles during the step cycle were measured by electromyogram (EMG). The four muscles showed similar activity patterns when sensory feedback was intact (left) or removed by dorsal root transection (right), as seen in the EMG record (top), or by their timing of activation during the step cycles (bottom; 15 EMG recordings per muscle were compiled, with step cycle as the x-axis unit). The four muscles from top to bottom are an angle extensor, a hip flexor, a knee flexor, and a second hip flexor. (A, adapted from Brown TG [1911] *Proc Roy Soc London [B]* 84:308; B, adapted from Grillner S & Zangger P [1975] *Brain Res* 88:367–371. With permission from Elsevier Inc.)

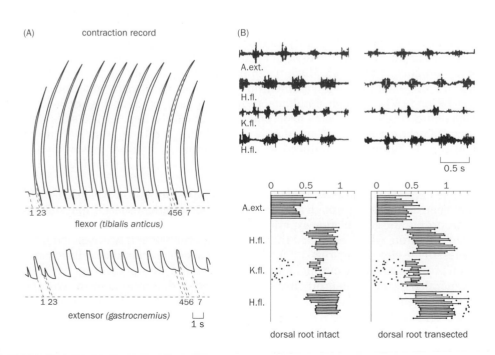

(A)   contraction record

flexor (tibialis anticus)

extensor (gastrocnemius)

(B)

A.ext.

H.fl.

K.fl.

H.fl.

0.5 s

A.ext.

H.fl.

K.fl.

H.fl.

dorsal root intact   dorsal root transected

during actual locomotion. For example, in a widely used experimental preparation, an incision is made at the level of the midbrain (also called mesencephalon, see Figure 1–8) of a cat such that the cerebral cortex/thalamus and the brainstem/spinal cord are disconnected. Although the resulting 'mesencephalic cat' could no longer voluntarily control its movement, it was still able to walk on a treadmill after brainstem stimulation (see Section 8.7). The contractions of many muscles during the walk can be recorded simultaneously by their action potential patterns in electromyograms. The coordinated contractions of different leg muscles during stepping were found to be similar before and after dorsal root transection, as quantitatively measured by the timing and duration of contraction for each muscle during the stepping cycle (Figure 8–12B).

Studies in invertebrate systems of rhythmic movements including locomotion likewise found that rhythmicity originates from specific segments or ganglia of the central nervous system (see Section 8.5). Collectively, these experiments led to the concept of the **central pattern generator** (**CPG**), which refers to a central nervous system circuit that is capable of producing rhythmic output for coordinated contraction of different muscles without sensory feedback. The existence of CPGs does not mean that sensory feedback is unimportant. On the contrary, sensory feedback modulates and can override the CPG rhythm. For example, in the mesencephalic cat, sensory feedback produced by increasing the speed of the treadmill could modulate the speed of the stepping cycle and even trigger a transition of the motor patterns from walking to trotting or galloping. Nevertheless, experiments such as those described in Figure 8–12 indicate that rhythmic output can originate from neural circuits in the spinal cord.

The concept of CPGs has been extended beyond control of locomotion; CPGs have been proposed that control breathing, swallowing, and many other rhythmic movements. Indeed, the phenomenon of neural network oscillation goes beyond motor control. For example, different frequencies of rhythmic activity observed in the thalamus, cerebral cortex, and hippocampus have been proposed to play important roles in perception, cognition, and memory. How do neural circuits produce rhythmic output?

8.5 Intrinsic properties of neurons and their connection patterns produce rhythmic output in a model central pattern generator

The best mechanistic understanding of rhythmic output production by CPGs has come from studies of several invertebrate model circuits. These circuits usually consist of a small number of individually identifiable neurons that are large in size and easily accessible for electrophysiological recordings (see Section 13.1). For example, the **stomatogastric ganglion** (**STG**) of crustaceans (for example, lobster and crab) produces a pyloric rhythm to control the cyclic movement of a portion of the stomach. The pyloric rhythm can be seen in the triphasic firing patterns of four types of neurons—an interneuron (AB) and three types of motor neurons (PD, LP, and PY)—through simultaneous intracellular recording (**Figure 8–13**A). Each neuron cycles between a hyperpolarized state and a depolarized state with bursts of action potentials. Importantly, the pattern seen in an intact lobster or crab can be faithfully reproduced when the stomatogastric nervous system is studied *in vitro* in the complete absence of sensory feedback, indicating that the rhythmic firing pattern is intrinsic to the STG.

The connection patterns among these four types of neurons (Figure 8–13B) have been established through a combination of simultaneous electrophysiological recording and cell ablation experiments. At the core of the pyloric rhythm is the AB interneuron, which exhibited rhythmic firing even when it was isolated from the rest of the circuit (Figure 8–13C). Thus, AB is called a **pacemaker** cell, because it can produce rhythmic output in the absence of input. This pacemaker property is generated by the AB neuron's **intrinsic properties**, determined by the composition, concentration, and biophysical properties of the ion channels it expresses. Based on studies of other STG neurons, the transition from the depolarized to the hyperpolarized state in AB is presumed to be caused by the inactivation of

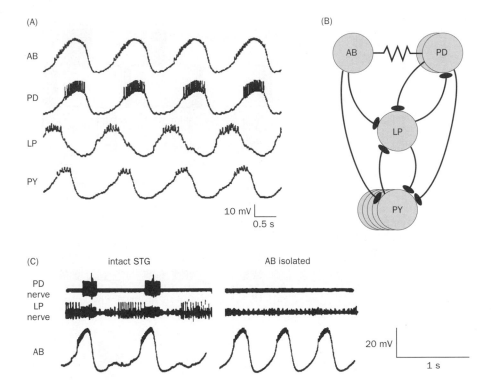

Figure 8–13 The pyloric circuit in the crustacean stomatogastric nervous system. (A) Simultaneous recordings of AB, PD, LP, and PY neurons in the stomatogastric ganglion (STG) show that each neuron cycles between depolarized and hyperpolarized states, with the AB/PD, LP, and PY neurons having offset activation phases. Action potentials (vertical spikes) are associated with the depolarized states. **(B)** The connection diagrams between 1 AB, 2 PD, 1 LP, and 5 PY neurons. AB and PD are electrically coupled through gap junctions (zigzag line). All chemical synapses are inhibitory. **(C)** In this experiment, the STG was first dissected out but remained connected with its central input nerve. The output spikes of PD and LP neurons were measured from extracellular recording of their motor nerves, while the AB neuron was recorded by an intracellular electrode. All exhibited rhythmic output patterns (left). The conduction of the input nerve was then blocked, and the PD and LP neurons were killed, as seen by the lack of spike output. The AB neuron continued to oscillate (albeit faster) in the absence of all functional connections (right). (A & B, adapted from Marder E & Bucher D [2007] *Annu Rev Physiol* 69:291–316. With permission from Annual Reviews; C, Adapted from Miller JP & Selverston A [1982] *J Neurophysiol* 48:1378.)

voltage-gated cation channels and delayed opening of K$^+$ channels, analogous to the voltage-gated Na$^+$ and K$^+$ channels discussed in the context of action potential production (see Section 2.10). The rebound from the hyperpolarized state to the depolarized state is likely caused by the opening of hyperpolarization-activated cation channels such as the HCN channels (see Box 2–4), which causes depolarization and leads to activation of voltage-gated cation channels.

As diagrammed in Figure 8–13B, AB is electrically coupled with the PD motor neurons through gap junctions; this causes the PDs to fire in sync with AB. AB and PD also form inhibitory chemical synapses onto LP and PY, and thereby inhibit their firing. The LP and PY neurons mutually inhibit each other; LP also inhibits PD. Thus, when AB and the PDs are in their depolarized state, their firing inhibits LP and PY neurons, forcing them to fire out of sync with AB/PD. When AB/PD stop firing, LP rebounds from inhibition before PY (this is because LP receives less inhibition from PD and expresses fewer K$^+$ channels and more hyperpolarization-activated cation channels than does PY); at this point, LP inhibits the firing of PD and PY. When the PY neurons eventually rebound from inhibition, they inhibit the firing of LP, and as a consequence disinhibit PD and facilitate the beginning of the next cycle. This chain of mutual inhibition among the three types of motor neurons produces the triphasic rhythm used to control coordinated contraction of stomach muscles innervated by PD, LP, and PY. In summary, the rhythmic firing of STG neurons is determined by the intrinsic biophysical properties of the constituent neurons as well as their connection patterns and connection strengths.

The simplicity of the crustacean STG system has enabled researchers to generate quantitative models based on the intrinsic properties of individual neurons and the connection strengths of the gap junction and inhibitory synapses. Such models have been used to simulate rhythmic output that fits experimental data. In one such study, the pyloric rhythm was simplified by treating AB and PD as equivalent such that there are three neuronal types in the model: AB/PD, LP, and PY. From 20 million combinations of ion channel compositions (which determine membrane conductance) and connection strengths (which determine synaptic conductance) among the three types of neurons, nearly half a million distinct 'solutions' were found that closely resembled the pyloric rhythm observed in animals. For example, two such solutions produced nearly identical triphasic

Figure 8–14 Similar network activity can be produced by distinct circuit parameters. Membrane potential traces from two model pyloric networks closely resemble each other (top), despite being produced by very different combinations of ion channels and synaptic strengths (bottom). Only a small subset of ion channel properties (represented as membrane conductance) and synaptic connection parameters (represented as synaptic conductance) are listed here. CaS, a voltage-gated slow and transient Ca^{2+} current; A, a voltage-gated transient K^+ current; KCa, a Ca^{2+}-dependent K^+ current; Kd, a delayed rectifier K^+ current; Na, a voltage-gated Na^+ current; H, a hyperpolarization-activated inward current. See Box 2–4 for more details about the ion channels that produce some of these currents. (Adapted from Prinz AA, Bucher D & Marder E [2004] *Nat Neurosci* 7:1345–1352. With permission from Macmillan Publishers Ltd.)

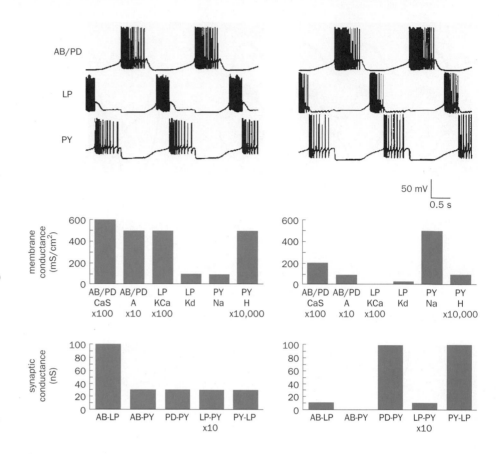

rhythms (**Figure 8–14**, top), using notably different combinations of membrane and synapse conductance (Figure 8-14, bottom). For example, the Na^+ conductance in the PY neuron is low for the case on the left and high for the case on the right, whereas the KCa conductance (reflecting the property of K^+-activated Ca^{2+} channels, see Box 2–4) in the LP neuron is high on the left and low on the right. Variable parameters have indeed been observed experimentally in natural populations of animals. Thus, distinct combinations of neuronal intrinsic properties and connection strengths can give similar network activities, reflecting the flexibility and robustness of the network that produces the pyloric rhythm.

The crustacean STG system is also known to be regulated by multiple modulatory neurotransmitters, which act by changing membrane and synaptic conductance in specific neurons that express their receptors (see Section 3.11; neuromodulation will be further discussed in Box 8–1). Since many solutions can produce the same rhythmic output pattern, this implies that each neuromodulator can regulate a different set of ion channels or synaptic transmission components in different cells to achieve the same outcome.

8.6 The spinal cord uses multiple central pattern generators to control locomotion

Compared with the crustacean STG, the mechanism by which CPGs in the vertebrate spinal cord control locomotion is much more complex and much less well understood. Nevertheless, insights obtained from less complex invertebrate circuits can be applied in testing specific hypotheses. As discussed earlier, an elementary step in motor coordination is the alternate contraction of extensors and flexors that control the same joint (see Figure 8–8). One model to explain this is that the extensor and flexor motor neurons are each activated in a mutually exclusive manner through mutually inhibitory premotor circuits (**Figure 8–15**A). At a higher level, different extensor–flexor pairs that control different joints of the same limb might be coordinated by analogous interactions among excitatory and inhibitory neurons to bring about the coordinated movement of a limb; these

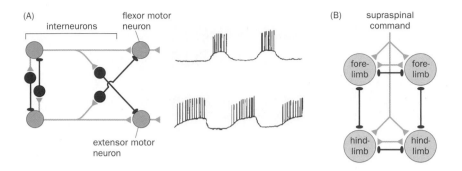

Figure 8–15 Conceptual framework for mammalian central pattern generators that control locomotion. (A) In this model (left), flexor and extensor motor pools (represented by a single motor neuron each) are excited by their corresponding excitatory premotor neurons (green). These excitatory premotor neurons inhibit each other and their antagonistic motor neurons through inhibitory interneurons (red), thus creating alternating patterns of excitation (right). **(B)** At a higher level, CPG networks for different limbs are proposed to be connected via inhibitory (red) or excitatory (green) interactions. The excitatory and inhibitory interactions between the limbs may be switched on or off depending on modes of locomotion. All CPG networks also receive descending excitatory input from the brainstem and motor cortex. While serving as good working hypotheses, these models are hypothetical in the sense that specific neuronal elements have not been identified. (A, adapted from Pearson K [1976] *Sci Am* 235:72–86. With permission from Macmillan Publishers Ltd; B, adapted from Grillner S [2006] *Neuron* 52:751–766. With permission from Elsevier Inc.)

constitute a CPG network for an individual limb. At an even higher level, the four limb CPG networks might be further coordinated to control different kinds of locomotion. For instance, during walking of most mammals, the left and right limbs are out of sync, as are forelimbs and hindlimbs on the same side. This may involve a mutual inhibition of CPGs that control these different limbs. In these models, mutual inhibition of left versus right CPGs is proposed to switch to mutual excitation during hopping or galloping such that left and right limbs are in sync (Figure 8–15B; see also Figure 8–11).

Physiological recording and perturbation experiments in preparations such as the mesencephalic cat have provided support for this conceptual framework of CPG organization. However, a deeper understanding of locomotion control requires identification of the constituent neurons and elucidation of the connection patterns of the CPG network, which is currently a highly active area of investigation. Since these CPG elements are likely composed of many individual neurons of multiple cell types, the methods that allowed researchers to crack the simpler STG circuit—such as simultaneous recording of multiple neurons and systematic ablation of identified neurons—is not adequate. The circuit functions analogous to the ones of a single neuron in the STG are likely carried out by a population of neurons of a specific type in the vertebrate spinal cord. One promising approach is to use findings from gene expression and developmental studies (see Figure 7–8) to identify and gain genetic access to individual spinal neuron types. Genetic access could then allow connectivity tracing and activity manipulation of entire populations of specific neuronal cell types using modern neural circuit analysis tools (see Chapter 13).

We give one specific example to illustrate this approach. As discussed in Section 7.4, the transcription factor Dbx1 is expressed in developing spinal cord progenitors that eventually become the V0 interneurons. Transgenic mice were produced in which Dbx1-expressing cells conditionally express diphtheria toxin A following Cre-induced recombination. These mice were crossed to a second transgenic mouse line in which the Cre recombinase is expressed from Dbx1-expressing progenitors. The resulting double transgenic mice, in which V0 neurons were specifically ablated, exhibited a striking phenotype. Wild-type mice alternate their left and right limbs during walking, whereas transgenic mice synchronized their limbs such that they hopped rather than walked at all speeds tested (see **Movie 8–4**). Recording the hindlimb flexor and extensor activities in an *in vitro* spinal cord explant preparation indicated that the left–right alternating firing pattern was switched to a synchronous firing pattern in the absence of V0 interneurons (**Figure 8–16**). Thus, the V0 interneurons play an essential role in alternating activities of left and right limbs during normal walking. The flexor–extensor cycles within each limb remained intact despite the disruption of left–right alternation in the absence of V0 interneurons, suggesting that they are regulated by independent CPGs.

V0 interneurons contain both an excitatory and an inhibitory subpopulation. Interestingly, ablating inhibitory V0 interneurons preferentially affected left–right alternation at low speeds of locomotion, whereas ablating excitatory V0 interneurons preferentially affected alternation at high speeds of locomotion. (These specific ablations utilize intersectional genetic methods discussed in Section 13.12.) These observations suggest that different subpopulations are recruited to the

Figure 8–16 Ablation of the V0 spinal cord interneurons disrupts left–right alternation. (A) Schematic of experimental setup. Simultaneous recordings of four ventral roots, which connect to extensors and flexors of hindlimb muscles as indicated, were performed in an *in vitro* explant. The locomotor-like pattern in this preparation is triggered by application of NMDA and serotonin, mimicking brainstem activation (see Figure 8–17). **(B)** Top, in wild-type animals, the activities of LL2 and RL2 alternate with each other, as do the activities of LL5 and RL5; these alternating activities account for the alternating movement of the left and right limbs during walking (see Movie 8–4). Bottom, when V0 neurons were ablated, the LL2 and LL5 activities are flipped compared to wild type (compare up and down arrows). Thus, the activities of LL2 and RL2 are synchronized, as are those of LL5 and RL5; this produces a hopping movement. Note that the activities of LL2 and LL5, and those of RL2 and RL5, also alternate due to their flexor–extensor relationship, and remain unchanged when V0 is ablated. Dark and light alternating stripes represent the alternating phases of the locomotor cycles. (Adapted from Talpalar AE, Bouvier J, Borgius L et al. [2013] *Nature* 500:85–88. With permission from Macmillan Publishers Ltd.)

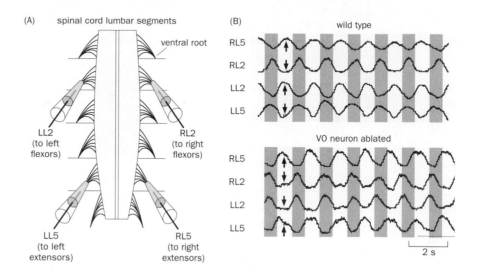

same regulatory network under different locomotion speeds; in other words, the composition of CPGs for regulating the same motor pattern is dynamic. Similar observations have been made in physiological studies of zebrafish swimming. As the swimming speed increases, while new interneuron populations become active, interneurons that were active during low speeds become inactive. This is distinct from motor neuron recruitment discussed in Section 8.2, where small motor units remain consistently active even after large motor units are recruited for more powerful muscle contractions. Furthermore, studies in the zebrafish larva suggested an ordered differentiation process, where neurons that drive the fastest, most powerful swimming mature prior to neurons that drive increasingly weaker and slower movement. This property likely enables zebrafish larvae to escape predators effectively at a young age (see Figure 2–1 and Movie 2–1).

8.7   The brainstem contains specific motor control nuclei

As discussed in Section 8.4, although isolated spinal cord preparations can produce rhythmic output for locomotion, the initiation of rhythmic output in these preparations requires excitatory stimuli such as application of the excitatory neurotransmitter glutamate. Where does the endogenous excitation come from *in vivo*? Electrical stimulation of an area called the **mesencephalic locomotor region (MLR)** in the brainstem was found to initiate locomotion in the mesencephalic cat. Indeed, increasing the intensity of stimulation to this area can speed up the walking and trigger transitions to trotting or galloping (**Figure 8–17**). Thus,

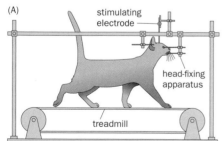

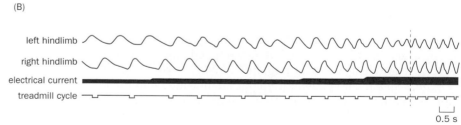

Figure 8–17 Electrical stimulation in the mesencephalic locomotor region of the brainstem initiates stepping. (A) Experimental setup. An incision is made in the midbrain such that the cerebral cortex/thalamus is disconnected from the brainstem/spinal cord. The resulting mesencephalic cat cannot voluntarily control its locomotion. When placed on the treadmill, whose speed is driven by the step cycle of the cat, electrical stimulation can be applied to specific parts of the brain to initiate locomotion, which can be measured by the limb movement and treadmill speed. **(B)** Electrically stimulating the mesencephalic locomotor region (MLR) initiated movement, as seen by the cycles of the hindlimbs. Furthermore, increasing the current of electrical stimulation (thick trace below the cycles) sped up the step cycle, as seen by the increasing frequency of the step cycles as well as the treadmill cycle (bottom). Note that toward the end (after the vertical dashed line), trotting became galloping, with both hindlimbs moving in sync. (Adapted from Shik ML, Severin FV & Orlovski GN [1966] *Biophysics* 11:756–765. With permission from Springer Science & Business Media.)

the MLR appears to be an important center that controls locomotion. MLR stimulation activates reticulospinal neurons in the brainstem; these neurons send descending axons through the reticulospinal tract to innervate spinal interneurons and motor neurons. Lesion and anatomical tracing using anterograde and retrograde dyes (see Section 13.18) have implicated multiple brainstem regions in motor control.

The classic electrical stimulation, lesion, and anatomical tracing studies are limited by a lack of cellular resolution (with respect to the types of cells in the implicated regions). As in the case of spinal cord CPG research, recent advances in neural circuit dissection begin to change that picture. We use a specific example in mice to showcase how modern viral-genetic tools can be used to dissect brainstem nuclei in motor control. To identify brainstem nuclei that are directly presynaptic to forelimb or hindlimb motor neurons, modified rabies virus was injected to forelimb or hindlimb muscles, and was taken up by motor neurons and trans-synaptically spread to premotor neurons (see Figure 8–10 and Section 13.19), including those in the brainstem (**Figure 8–18**A). Among the many nuclei that send direct input to forelimb and hindlimb motor neurons, a brainstem nucleus called MdV (which stands for medullary reticular formation ventral part) preferentially innervates forelimb over hindlimb motor neurons (Figure 8–18B). Further

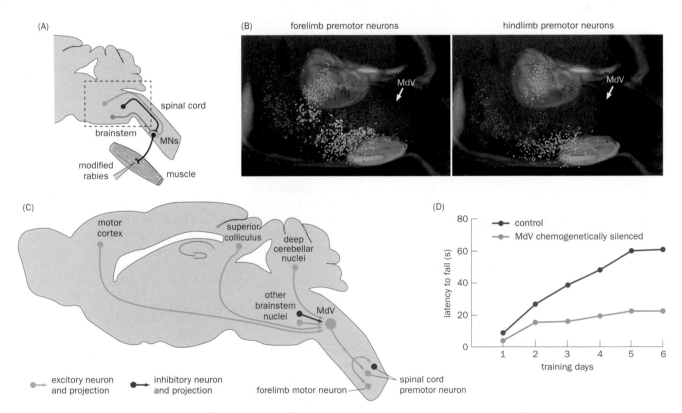

Figure 8–18 Anatomical and functional dissection of brainstem motor control nuclei. (A) Experimental strategy to identify brainstem premotor neurons. Modified rabies virus (see Figure 13–30 and Section 13.19 for more details) was injected into a specific muscle, was taken up by motor neuron (MN) axons that innervate that muscle, and trans-synaptically spread to premotor neurons that are presynaptic to those MNs. The dashed box indicates the approximate location of the brainstem model in panel B. **(B)** Distribution of brainstem premotor neurons that innervate the forelimb (left) or hindlimb (right) muscles revealed by injecting modified rabies virus to forelimb or hindlimb muscles. Dots represent premotor neurons, which are differentially colored according to their locations in specific brainstem nuclei. Yellow area, vestibular nuclei; green area, inferior olive. While most brainstem nuclei contain premotor neurons that innervate both the forelimb and hindlimb (some with differing density),

MdV has almost exclusively forelimb premotor neurons. **(C)** Summary of major input to and output from excitatory MdV neurons in a sagittal schematic of the mouse brain and rostral spinal cord. Arrows indicate monosynaptic input. **(D)** Silencing excitatory MdV neurons disrupts skilled motor tasks. Mice were placed on a rotating rod whose speed accelerated from 5 to 50 rotations per min in 5 min. The y axis quantifies the time mice stayed on the rod. The x axis represents training days (four trials per day). The chemical PSEM was injected daily prior to training to both groups of mice. Only the experimental group expressed PSEM-responsive Cl⁻ channels in excitatory MdV neurons, which were transiently silenced during task performance. The performance of these experimental mice was markedly impaired compared to controls. (Adapted from Esposito MS, Capelli P & Arber S [2014] *Nature* 508:351–356. With permission from Macmillan Publishers Ltd.)

analysis indicated that MdV premotor neurons are primarily glutamatergic excitatory neurons. They are presynaptic to only a specific subset of forelimb motor neurons, and they send direct input to both excitatory and inhibitory spinal cord interneurons in addition to motor neurons (Figure 8–18C). The trans-synaptic tracing method was also applied to mapping inputs to excitatory MdV neurons themselves, finding that they receive direct input from subcerebral projection neurons in the motor cortex, deep cerebellar nuclei neurons (to be discussed in more detail in the next section), output neurons from the superior colliculus, and many other brainstem nuclei. Thus, MdV premotor neurons integrate descending motor control from the motor cortex and cerebellum, and in turn preferentially regulate specific forelimb muscles (Figure 8–18C).

To explore the function of MdV in motor control, a chemogenetic method was employed to determine their loss-of-function phenotypes in forelimb-dependent motor tasks. Adeno-associated virus that expresses a Cre-dependent, chemically gated Cl^- channel that can be activated by application of a chemical called PSEM (see Section 13.23 and Figure 13–42) was injected into the MdV region of transgenic mice that express the Cre recombinase in glutamatergic MdV neurons. PSEM application leads to hyperpolarization of these MdV neurons, which transiently silences their output (from ~20 min after drug application until the chemical is metabolized in a few hours). A rotarod assay was used to test skilled motor behavior. In this assay, a mouse is placed onto a rotating rod; the mouse must hold onto the rod and move in coordination with the rotating rod in order not to fall (see Section 13.29 and Figure 13–51). Control mice learned the task after daily practice, as seen by the gradual increase of latency to fall over one week. Mice in which excitatory MdV neurons were silenced failed to do so (Figure 8–18D), while performing normally in simpler motor behavior such as running around their home cage. Further experiments showed that transient silencing of MdV neurons after mice had already acquired proficient motor skill execution also degraded their performance, indicating that glutamatergic MdV neurons are required for the execution of skilled motor tasks. Thus, while the control of innate behaviors such as locomotion appears to be contained within the spinal cord, skilled behaviors require higher brain centers.

We are only at the beginning of functionally dissecting specific neurons in specific brainstem nuclei. Much remains to be learned about their circuit organization, input, and output. The example above illustrates the approaches scientists are currently taking to reach these goals. In the next three sections, we will turn to three higher centers that control the motor system we have discussed so far: the motor cortex, cerebellum, and basal ganglia. Whereas the motor cortex is considered an ultimate motor command center, the cerebellum and basal ganglia constitute regulatory circuits essential for movement control (see Figure 8–2). We start with the cerebellum.

8.8 The cerebellum is required for fine control of movement

The cerebellum (Latin for 'little brain') is evolutionarily ancient in vertebrates, and occupies a sizable chunk of the mammalian brain (see Figure 1–8). Although the cerebellum has been implicated in diverse functions, including cognitive functions, its best-characterized functions are fine control of movement and motor learning. Cerebellar defects in human patients and experimental animals cause various kinds of motor system problems, such as **ataxia**, an abnormality in coordinated muscle contraction and movement. For example, transgenic mice with defective cerebellar Purkinje cells cannot walk a straight path; instead, they wobble from side to side (**Figure 8–19**). How does the cerebellum control movement? Before answering this question, we need to first examine the circuit organization of the cerebellum, which is in fact one of the best understood in the mammalian brain due to the small number of participating cell types (**Figure 8–20A**).

The most morphologically complex neuron in the cerebellum is the **Purkinje cell** (see Figure 1–11). Each Purkinje cell extends an elaborate planar dendritic tree that receives 10^4–10^5 excitatory synapses from **parallel fibers** that intersect the Purkinje cell dendrites at right angles (Figure 8–20A). Parallel fibers originate

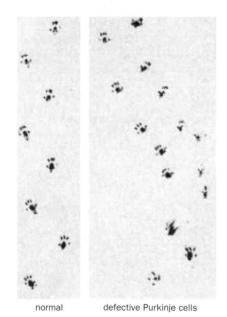

normal   defective Purkinje cells

Figure 8–19 Ataxia caused by a cerebellar defect. The hindfeet of the mice were dipped in paint and their footprints were recorded on paper. Compared with normal mice that walk straight, mice with cerebellar defects—in this case due to connection abnormalities of Purkinje cells—typically wobble from side to side, with inconsistent step sizes and the hindfeet more widely spaced. (From Luo L, Hensch TK, Ackerman L et al. [1996] *Nature* 379:837–840. With permission from Macmillan Publishers Ltd.)

(A)

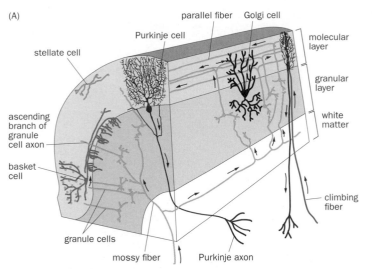

(B)

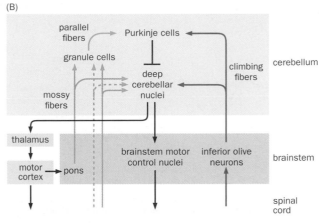

Figure 8–20 Organization of the cerebellar circuit. (A) Organization of the cerebellar cortex. Purkinje cell bodies form a single layer between the molecular and granular layers, and send output to the deep cerebellar nuclei via the white matter. Their planar dendrites extend across the entire molecular layer. Granule cells are located in the granular layer. Their axons first ascend into the molecular layer, and then bifurcate to form parallel fibers that intersect the Purkinje dendrites (they appear as dots in the cross section). Two external inputs, mossy fibers and climbing fibers, synapse onto granule cells and Purkinje cells, respectively. Also drawn are three major types of local GABAergic neurons, the basket, stellate, and Golgi cells. Arrows indicate the flow of information. **(B)** Schematic summary of major connections within the cerebellum and with other neural elements involved in spinal cord motor control. Cerebellar granule cells receive input from mossy fibers originating directly (solid arrow) or indirectly (dotted arrow) from the spinal cord, or originating from the pons that receives descending input from the motor cortex. The inferior olive neurons receive input from the spinal cord and send climbing fibers to the Purkinje cells. The deep cerebellar nuclei, which receive input from the Purkinje cells and the collaterals of mossy fibers and climbing fibers, send output to brainstem nuclei involved in motor control and to the motor cortex via the thalamus.

from **granule cells**, the most numerous type of neuron in the brain, which are located in the granular layer and receive excitatory input from **mossy fibers** originating from neurons outside the cerebellum. Each Purkinje cell is also innervated by a single **climbing fiber**, an axon branch originating from a neuron in the brainstem's **inferior olive** nucleus (see the green structure in Figure 8–18B) that 'climbs' the major branch of the Purkinje dendritic tree, forming numerous excitatory synapses along the way. Purkinje cells are GABAergic and send inhibitory signals to neurons in the output nuclei of the cerebellum, which are called the **deep cerebellar nuclei**. Major projection targets of the deep cerebellar nuclei include the brainstem motor nuclei for descending motor control and the thalamus for communication with the cortex. The two input pathways, mossy fibers and climbing fibers, also send collateral branches to the deep cerebellar nuclei (Figure 8–20B). In addition to these projection neurons, the cerebellar cortex also contains three major types of local interneurons—the basket cell (see Figure 1-15B and Figure 7-22), the **stellate cell**, and the **Golgi cell** (Figure 8–20A). All three cells receive input from parallel fibers in the molecular layer. The basket and stellate cells send inhibitory output to Purkinje cells at the somata and distal dendrites, respectively, thus serving feed-forward inhibition. The Golgi cell sends inhibitory output back to granule cells, thus serving feedback inhibition (see Box 1–2). This circuit organization is repeated across the entire cerebellum, with complex somatotopic maps. For instance, the medial and lateral parts of the cerebellum regulate trunk and limb movement, respectively. Purkinje cells from the most posterior division directly innervate the vestibular nuclei (see Box 6–2) instead of the deep cerebellar nuclei; we will discuss their function shortly.

Around 1970, the circuit architecture of the cerebellum led theoreticians to propose, on the basis of Hebb's rule (see Section 5.12), that the coincident firing of climbing fibers and parallel fibers could modify the parallel fiber–Purkinje cell synapses and, in so doing, allow climbing fiber input to modulate how Purkinje cells respond to activation of granule cell populations. Indeed, electrophysiological experiments in the 1980s validated synaptic modification both *in vivo* and in

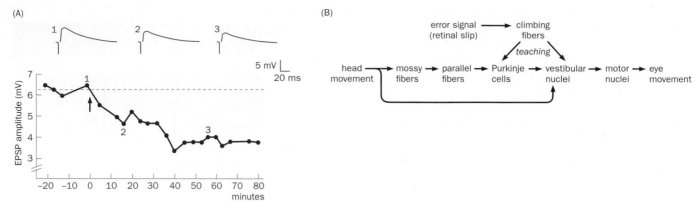

Figure 8–21 Synaptic plasticity in the cerebellum and its role in motor calibration. (A) Top, three excitatory postsynaptic potential (EPSP) traces from a Purkinje cell in response to parallel fiber stimulation in a rabbit cerebellar slice preparation. Trace 1 was taken before the parallel fiber and climbing fiber co-stimulation (upward arrow in the bottom graph), and traces 2 and 3 were taken after the co-stimulation. The EPSP amplitudes of traces 2 and 3 are smaller than that of trace 1. The bottom graph quantifies these effects. The dotted line shows the average EPSP magnitude before pairing. **(B)** Cerebellar circuit for regulating the vestibulo-ocular reflex (VOR). In the VOR, the head movement signal is relayed to the vestibular nuclei (bottom branch) to control eye movement (see Figure 6–61 for more details), but the signal strength can be adjusted by the cerebellar loop detailed here. The head movement signal reaches the cerebellum via the mossy fibers, and an error signal caused by image movement on the retina (retinal slip) reaches the cerebellum via the climbing fibers. Pairing of these signals modifies the parallel fiber → Purkinje cell synaptic strength, which alters the signals sent to the vestibular nuclei, thereby adjusting the VOR strength. Thus, the error signal 'teaches' the circuit through modification of synaptic strength. (A, adapted from Sakurai M [1987] *J Physiol* 394:463; B, based on Boyden ES, Katoh A, & Raymond JL [2004] *Ann Rev Neurosci* 27:581.)

cerebellar slices. Specifically, coincident firing of the climbing fiber and parallel fibers that innervate the same Purkinje cell was found to decrease the synaptic strength between parallel fibers and the Purkinje cell (**Figure 8–21**A) in a process called **long-term depression**. Other forms of synaptic plasticity have since been identified in the cerebellar cortex and deep cerebellar nuclei. (We will study synaptic plasticity and learning in more detail in Chapter 10.)

We use the example of the vestibulo-ocular reflex (VOR, see Figure 6–61) to illustrate how synaptic plasticity in the cerebellar circuit can be used to calibrate the strength of the VOR (gain control), a form of motor learning. Recall that the VOR causes a compensatory eye movement in the opposite direction to the head turn, thus stabilizing visual images on the retina during head turns. However, when a change in circumstances (for example, wearing a pair of glasses that shrink the size of visual images) causes a mismatch between the magnitude of eye rotation and head turn, the VOR can be adjusted by cerebellum-based motor learning. Specifically, head movement produces vestibular signals in the semicircular canals that are sent to the cerebellum via the mossy fibers. An error signal produced from retinal slip (that is, imperfect VOR that fails to stabilize images during a head turn) is sent to the cerebellum via the climbing fibers. Repeated pairings of the error signal with the vestibular signal modify the synaptic connection strength between parallel fibers and Purkinje cells, which alters the signals sent to the vestibular nuclei, thereby adjusting the strength of VOR signals sent to motor nuclei that control eye movement (Figure 8–21B). Evidence suggests that plasticity in the vestibular nuclei also contributes to VOR regulation.

Compared to our understanding of motor learning, such as gain control of the VOR, much less is known about how the cerebellum coordinates ongoing movement such as locomotion. Nevertheless, an analogy can be made between VOR gain control and movement coordination (Figure 8–20B, compared with Figure 8–21B). In the part of the cerebellum involved in spinal cord-mediated movement, mossy fibers carry signals from spinal cord about motor intent (a copy of output signals from premotor and motor neurons) and motor performance (feedback from proprioceptive sensory neurons). Mossy fibers also carry command signals from the motor cortex relayed through the pons. Thus, the cerebellum is an excellent position to integrate these different kinds of information. One hypothesis is that climbing fibers deliver an error signal to adjust parallel fiber → Purkinje cell synaptic strengths, and thereby Purkinje cell output, in response to input from granule cell populations as in the case of VOR gain control. This modifies moment-by-moment cerebellar output to the brainstem nuclei to adjust descending motor control in response to changing circumstances such as a perturbed step cycle. Much future work is needed to test this hypothesis and to decipher circuit mechanisms underlying fine control of movement.

8.9 The basal ganglia participate in initiation and selection of motor programs

The basal ganglia, also called cerebral nuclei, are a collection of nuclei underneath the cerebral cortex (see Figure 1–8). Two neurological disorders, Parkinson's disease and Huntington's disease, which primarily affect the basal ganglia, highlight the importance of these structures in motor control. Patients with Parkinson's disease have difficulty initiating movement, whereas patients with Huntington's disease cannot stop excessive movement. (We will discuss these diseases in more detail in Chapter 11.) Like the cerebellum, the basal ganglia use a generic circuit design (**Figure 8–22**) for a wide range of functions that vary according to the

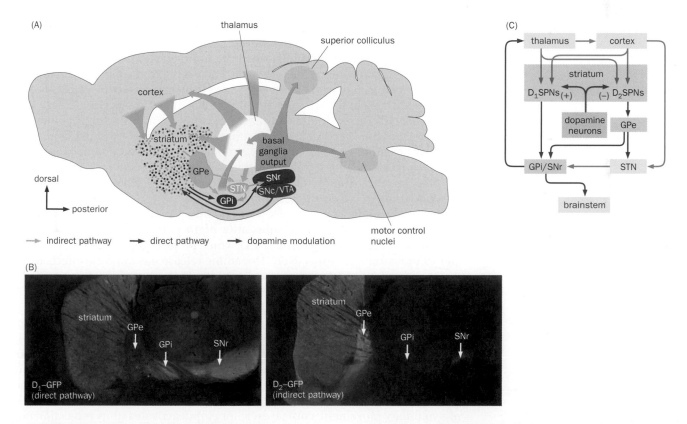

Figure 8–22 Organization of the basal ganglia circuits.
(A) A simplified model of basal ganglia circuits from a sagittal perspective of a mouse brain. The striatum receives excitatory inputs from the cortex and thalamus, and sends output via two types of GABAergic spiny projection neurons (SPNs) that express two different dopamine receptors. SPNs that express the D_1 receptor (D_1SPN, red) constitute the direct pathway, projecting mainly to the globus pallidus internal segment (GPi) and substantia nigra pars reticulata (SNr). SPNs that express the D_2 receptor (D_2SPN, orange) constitute mainly the indirect pathway, projecting to the globus pallidus external segment (GPe). The GPe sends GABAergic projections directly to the GPi, and via the subthalamic nucleus (STN, which also receives direct cortical input) to the GPi and SNr. The GPi and SNr send basal ganglia output to the thalamus, superior colliculus, and brainstem motor control nuclei. Dopamine neurons in

the substantia nigra pars compacta (SNc) and the ventral tegmental area (VTA) also receive input from the striatum (and other inputs not shown here), and send modulatory output back to the striatum.
(B) The projection patterns of D_1SPN (left) or D_2SPN (right) are visualized in sagittal sections of transgenic mice in which green fluorescent protein (GFP) expression is driven by the regulatory elements for the D_1 or D_2 receptor, respectively. Cell bodies of both neuronal types are within the striatum; D_1SPNs project mainly to the GPi and SNr whereas D_2SPNs project to the GPe, as seen by GFP fluorescence intensity. **(C)** A simplified basal ganglia circuit diagram. Green arrows represent excitatory projections. Red arrows represent inhibitory projections. Blue arrows represent dopamine neuron projections, which promote D_1SPN (+) and inhibit D_2SPN (−) firing. (A & B, adapted from Gerfen CR & Surmeier DJ [2011] *Annu Rev Neurosci* 34:441–466. With permission from Annual Reviews.)

specific sources of input and output, with motor control being one of these functions. Before discussing their function in motor control, we first outline the basal ganglia circuit.

The input nucleus of the basal ganglia is the **striatum**, also called the caudate-putamen because in primates the striatum consists of two separable structures, the caudate and the putamen. The striatum receives convergent excitatory input from the cerebral cortex and the thalamus. The dorsolateral striatum preferentially receives input from sensory and motor cortices and thus is most directly involved in motor control. The dorsomedial striatum preferentially receives input from association cortices and is involved in cognitive processes. The ventral striatum (also called the **nucleus accumbens**) preferentially receives input from the prefrontal cortex, hippocampus, and amygdala, and is implicated in regulating motivational behavior.

The striatum consists mostly of two types of intermingled GABAergic **spiny projection neurons** that are distinguished by their expression of different G-protein-coupled dopamine receptors. Those that express the D_1 receptor constitute the **direct pathway**, which projects predominantly to the output nuclei of the basal ganglia—the **globus pallidus internal segment** (**GPi**) and the **substantia nigra pars reticulata** (**SNr**). Those that express the D_2 receptor constitute the **indirect pathway**. They predominantly project to the **globus pallidus external segment** (**GPe**), which in turn sends GABAergic input to the GPi either directly or through the **subthalamic nucleus** (**STN**) (Figure 8–22A). This distinction between two different pathways can be visualized in transgenic mice that express green fluorescence protein (GFP) under the control of regulatory elements from the D_1 or D_2 dopamine receptor genes (Figure 8–22B). A major recipient of GPi and SNr GABAergic projection is the thalamus, which itself projects to the cerebral cortex and directly back to the striatum, forming two feedback loops (Figure 8–22A, C). SNr also sends direct output to brainstem motor control nuclei (including the MLR discussed in Section 8.7) and superior colliculus. The actual connection patterns are more complex than the simplified model outlined above, but the model captures the major features of the basal ganglia circuit.

The neurotransmitter dopamine modulates the excitatory synaptic connections through which the cerebral cortex and thalamus provide input to the striatal spiny projection neurons. The dopamine neurons responsible for this modulation are located in the **substantia nigra pars compacta** (**SNc**) and the adjacent **ventral tegmental area** (**VTA**), which project preferentially to the dorsal and ventral striatum, respectively. Dopamine release has opposite effects on the direct and indirect pathways. Activation of the D_1 receptor, which is coupled to a stimulatory G protein, depolarizes the D_1-expressing spiny projection neurons and hence activates the direct pathway. Activation of the D_2 receptor, which is coupled to an inhibitory G protein, causes hyperpolarization of the D_2-expressing spiny projection neurons and inhibition of the indirect pathway (Figure 8–22C).

Given these circuit properties, how do the basal ganglia regulate movement? Striatal GABAergic projection neurons are mostly silent at rest. By contrast, output neurons in the GPi and SNr are active, sending **tonic** (that is, regularly timed and repetitive) inhibitory output to their targets. Immediately prior to the onset of a voluntary movement, cortical excitatory input activates spiny projection neurons (for example, see **Figure 8–23**). This inhibits the firing of GPi and SNr output neurons through the direct pathway, causes a disinhibition of the motor control centers in the superior colliculus and brainstem (Figure 8–22C), and thus facilitates movement initiation. Likewise, the tonic inhibitory output to the thalamus is also relieved, further promoting movement initiation. The role of the indirect pathway is less clear. One hypothesis is that its activation sharpens the specificity of the direct pathway, akin to the lateral inhibition discussed in sensory systems (see Sections 4.16 and 6.9).

The causal relationship between movement and activation of the direct and indirect pathways has been tested using optogenetics (see Section 13.25). D_1- or D_2-expressing spiny projection neurons in mice were engineered to express channelrhodopsin ChR2 so that they could be activated by light. As predicted from

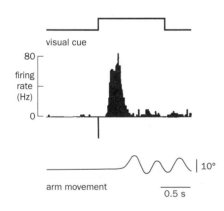

Figure 8–23 Firing of some striatal neurons anticipates movement onset. In this experiment, a monkey was trained to move its arm three times (bottom trace) in response to the onset of a visual cue (top trace). The firing rate of a neuron in the arm control region of the striatum was recorded by an extracellular electrode in the behaving monkey. This striatal neuron did not track the movement per se, but instead fired most vigorously prior to movement onset. Other striatal neurons in the same study did track movement (not shown). (Adapted from Kimura M [1990] *J Neurophysiol* 63:1277.)

the above circuit model, activation of the direct pathway decreased firing of SNr output neurons, while activation of the indirect pathway increased firing of SNr output neurons (**Figure 8–24**A). Behaviorally, direct pathway activation enhanced locomotion, whereas indirect pathway activation suppressed locomotion (Figure 8–24B). These experiments support the model that global activation of the basal ganglia direct or indirect pathway facilitates or inhibits movement, respectively. More recent Ca^{2+} imaging experiments suggest that when mice initiate a specific motor task, both direct and indirect pathways are in fact activated at the same time. One hypothesis is that activating the direct pathway initiates a motor program specific for the task, whereas activating the indirect pathway suppresses competing or unwanted motor programs.

In addition to regulating the initiation and execution of movement, evidence has suggested that the feedback sent by the basal ganglia to the thalamus and cortex plays important roles in motor skill learning and habit formation. These roles are facilitated by modulation of cortical synaptic input to spiny projection neurons by the dopamine neurons. Projection of VTA dopamine neurons to the ventral striatum is particularly important in carrying signals related to reward. We will return to the function of the basal ganglia in the context of reward-based learning in Chapter 10, and in movement disorders and addiction in Chapter 11.

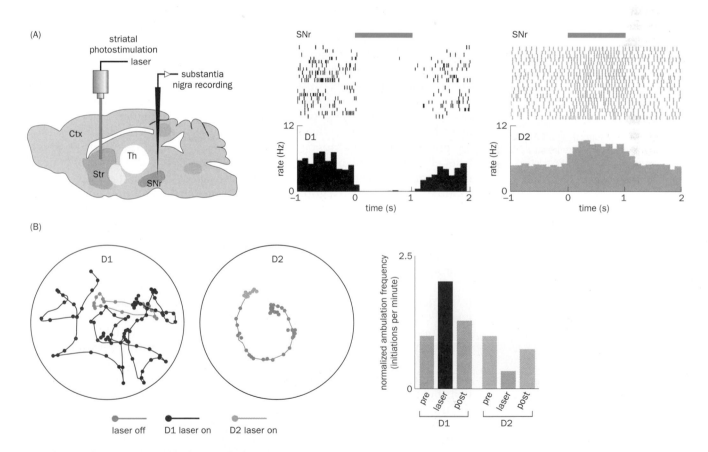

Figure 8–24 Function of the direct and indirect pathways investigated by optogenetic stimulation. (A) Left, schematic of the experiment depicted in a sagittal perspective of the mouse brain. Channelrhodopsin (ChR2) was specifically expressed in D$_1$ or D$_2$ dopamine receptor-expressing spiny projection neurons (SPNs) in the striatum (Str). Middle and right, the top graphs are action potential plots in which each row represents a trial; the bottom graphs illustrate the firing rate of the SNr neurons prior to, during, or after photostimulation. Photostimulation (blue bar period) of D$_1$SPNs suppressed the firing of SNr neurons (middle panels), whereas stimulating D$_2$SPNs enhanced SNr neuron firing (right panels).

Ctx, cortex; Th, thalamus. **(B)** Photostimulation of D$_1$SPNs or D$_2$SPNs enhances or suppresses movement, respectively. The location of a mouse in a circular arena is measured at 300-ms intervals. The gray paths represent 20 s of activity before photostimulation; the colored paths (red or orange) represent 20 s of activity during photostimulation. When D$_1$SPNs were activated, the mouse ran more, as indicated by the longer distances between the red dots. When D$_2$SPNs were activated, the mouse moved less, as indicated by the clustered orange dots. These effects are quantified in the right panel. (Adapted from Kravitz AV, Freeze BS, Parker PR et al. [2010] *Nature* 466:622–626. With permission from Macmillan Publishers Ltd.)

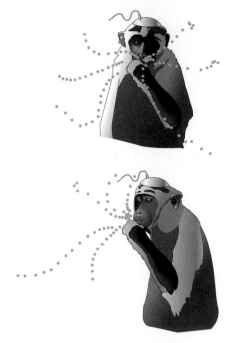

Figure 8–25 Microstimulation of the motor cortex can cause complex movements. In this example, 500-ms microstimulations at a specific site in the arm area of the motor cortex caused the contralateral arm to move toward the mouth regardless of its initial position. At the same time, the hand was changed to a grip posture toward the mouth, and the mouth opened, mimicking a feeding behavior. The dotted lines represent 11 different trajectories traced from video recordings at 30 frames per second. (Adapted from Graziano M, Taylor C & Moore T [2002] *Neuron* 34:841–851. With permission from Elsevier Inc.)

8.10 Voluntary movement is controlled by the population activity of motor cortical neurons in a dynamical system

In mammals, the motor cortex, which includes the **primary motor cortex (M1)** and the more anterior **premotor cortices**, provides the ultimate command to initiate voluntary movement and to control complex movement. Returning to our example of locomotion, if a cat sees an obstacle in its walking path, it must modify its stepping cycle to bypass the obstacle. Likewise, 'simple' acts in our daily life, such as catching a ball or reaching for a glass of water and grasping it, require integration of the motor system with the visual system (to see the ball or glass) and the somatosensory system (to know where the arm and hand are with respect to the glass or the ball). The motor cortex, often with the help of the parietal cortex, integrates information from multiple sensory systems and sends descending axons to the motor control regions of the brainstem, to spinal cord interneurons, and to motor neurons themselves (see Figure 8-2). The relative contributions of these pathways for motor control are largely unclear at present. Evidence suggests that the motor cortex tends to control fine movement of distal limbs more directly and trunk muscles more indirectly via the brainstem and spinal cord interneurons. The motor cortex also communicates extensively with the basal ganglia and cerebellar circuits (see Sections 8.8 and 8.9) to modulate motor output. How these different circuits interact together to orchestrate movement control is a challenging topic for future investigation.

How is the motor cortex organized? As discussed in Section 1.11, the primary motor cortex contains a somatotopic map, the motor homunculus (see Figure 1–25); this was originally discovered in primates and in human patients undergoing neurosurgery: application of brief electrical stimulations to specific areas of the motor cortex caused specific muscles to twitch. The depiction of the motor homunculus gives the impression that there is a point-to-point representation of body muscles in the primary motor cortex. While regions that control legs, arms, trunk, and head are clearly segregated in the motor cortex, more recent studies indicate that a precise topography does not exist at a finer scale. Indeed, whereas brief electrical stimulation causes muscle twitching, longer stimulation can produce complex and coordinated movements, such as moving the arm toward the mouth with the hand in a grip position and opening the mouth (**Figure 8–25**). As discussed in Section 4.29, an important caveat of microstimulation experiments is that the number and types of neuronal cell bodies and axons being stimulated are poorly defined. However, the finding that specific and ethologically relevant movement can be elicited by stimulating a single site suggests that motor cortical neurons and circuits can encode motor programs for specific behavior.

In vivo electrophysiological recordings in behaving monkeys have provided important insights into the organization of the motor cortex at the single-neuron level. For example, individual cortical neurons in the finger representation region of the motor cortex were recorded during an experiment in which monkeys were trained to move one finger at a time. It was found that most individual neurons were broadly tuned to the movement of multiple fingers (**Figure 8–26**A). At the same time, neurons maximally tuned to the movement of a specific finger were distributed at multiple locations, intermingled with neurons maximally tuned to the movement of other fingers (Figure 8–26B).

Given the broad tuning of individual neurons, how does the motor cortex control specific movement parameters? We use the control of arm reaching to illustrate. In a revealing experiment, monkeys were trained to move an arm from a start position in the center of a cube to one of the eight corners of the cube, representing eight directions with equal distances in a three-dimensional space. The spiking activities of hundreds of individual cortical neurons in the arm area of the motor cortex were recorded using a **multi-electrode array** (see Figure 13–33) during each of the eight kinds of movement. Activity of most individual neurons varied during movement in more than one direction (**Figure 8–27**A), just as neurons in the finger area are active during movement of more than one finger. The 'preferred direction' for each neuron, as a vector in a three-dimensional space, could nevertheless be determined based on its firing rates during the eight

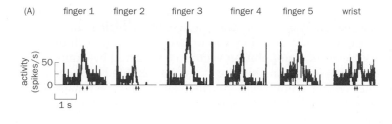

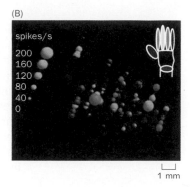

Figure 8–26 Representation of finger movement in the primary motor cortex of the monkey. (A) Activity of an individual motor cortical neuron during instructed movement of five fingers and the wrist, as determined by *in vivo* extracellular recording. The first and second arrows beneath the plot signify the beginning and end of the movement, respectively. The cell being recorded was maximally tuned to movement of finger 3, but it was active during other movements as well. **(B)** Spatial distribution of active neurons in the hand area of the motor cortex.

Each neuron is color-coded according to its maximal tuning to the movement of one of the five fingers or the wrist, following the color scheme at top right. The size of the sphere represents the firing rate according to the key shown on the left. Recorded neurons form parallel lines in the same direction as that of electrode penetrations, from top right to bottom left. Neurons tuned to the movement of the same finger(s) are not clustered at this scale. (Adapted from Schieber MH & Hibbard LS [1993] *Science* 261:489.)

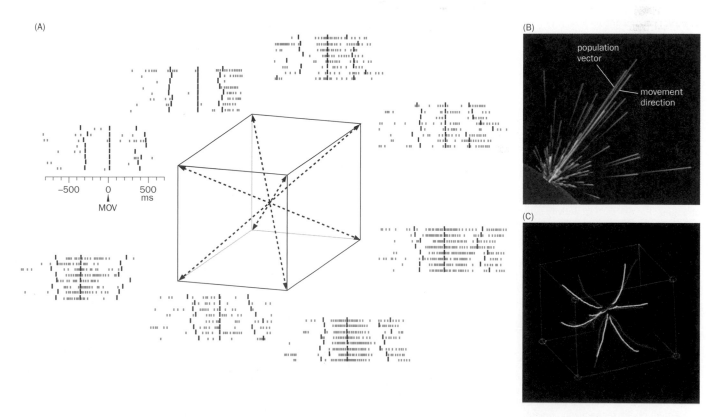

Figure 8–27 Movement directions are determined by population activity of motor cortical neurons. (A) The monkey was trained to move its arm from the start position at the center of the cube to one of the eight corners upon the onset of a visual cue. The eight plots shown here illustrate the action potentials fired by a single neuron during the arm movement in eight directions. Within a plot, each horizontal line represents the neuronal activity recorded during one trial. Each blue vertical bar represents an action potential. The red vertical bars in the middle, which are used to align the plots, indicate the onset of the movement (MOV) and are labeled as *t* = 0 ms (milliseconds) in the timescale beneath the upper left plot. The red vertical bars to the left and right represent the onset of the visual cue and the end of arm movement, respectively. The firing rate of

this neuron changed most prior to and during arm movement toward the two directions at bottom right, but the firing rate also changed when the arm moved in other directions. **(B)** The firing rates of individual neurons are represented as individual vectors (blue). The direction of the population vector of 224 individual neurons (orange) approximates the direction of movement (green). **(C)** The neural representations of trajectories (yellow; constructed on the basis of the population vector over time) resemble the actual trajectories (red). (A & B, adapted from Georgopoulos AP, Schwartz AB & Kettner RE [1986] *Science* 233:1416–1419. With permission from AAAS; C, from Georgopoulos AP, Kettner RE, & Schwartz AB [1988] *J Neurosci* 8:2928. With permission from The Society for Neuroscience.)

different kinds of movement. The direction of an arm movement on any given trial could not be reliably predicted from the activity of single neurons, as each neuron had its own preferred direction, was broadly tuned, and exhibited variable spiking rates. However, a **population vector**, constructed by summing the preferred direction vectors of several hundred neurons weighted by the magnitude of their firing rates during a movement, provided an excellent estimate of the actual arm movement (Figure 8–27B). In other words, it was possible to predict the direction of arm movement based on the population activity of motor cortical neurons. Remarkably, even the trajectory of arm movement from the reception of the start cue to the end of the movement could be approximated from time-varying firing rates of the motor cortical neuronal population (Figure 8–27C). These experiments provide strong evidence that movement direction is determined by the population activity of motor cortical neurons.

In vivo recordings indicate that activity of most motor cortical neurons correlates with contraction of specific muscles as well as specific movement parameters such as direction, magnitude, and velocity. These observations raised a question: do individual motor cortical neurons 'encode' muscle action (for example, contraction of the bicep) or movement parameters (for example, movement of the forearm toward the body at a certain speed)? Instead of asking this question from the perspective of individual neurons, an alternative perspective of motor control is to treat the motor cortex as a **dynamical system**, composed of different **dynamical states** that evolve over time according to specific rules. Recall that in our discussion of olfactory coding, we have introduced the concept that the activity state of a neuronal population can be represented as a point in a multidimensional space, with each axis representing the firing rate of one constituent neuron (see Figure 6–30A). Likewise, the dynamical state of the motor cortex at any given time can be described as the firing rates of a population of motor cortical neurons at that specific time. Motor cortical activity during a motor task over time can be represented as changes of dynamical states following a specific trajectory in the state space (**Figure 8–28A**).

We use a specific example to illustrate how the dynamical systems perspective provides insights into cortical control of movement. In a reaching task, a monkey was trained to move its arm toward a target appearing on the screen only after a subsequent 'go cue' appeared (Figure 8–28B, top). Trajectories of neural states, computed from the activities of hundreds of simultaneously recorded motor cortical neurons, were followed during the entire period for many individual trials (Figure 8–28B, bottom). Target appearance caused the trajectories to move to a smaller region in the state space during movement preparation (from the blue ellipse to the green ellipse). The onset of go cues caused the neural states to move into yet another small region following similar trajectories across different trials (from the green ellipse to the gray ellipse). These observations suggest that stimulus onset reduces the variability of neuronal population activity. Further analysis of population dynamics during arm reaching, a non-periodic motor task, uncovered a strong oscillatory component indicative of rhythmic neural activity that is normally associated with periodic motor tasks such as locomotion. This finding suggests that rhythmic neural activity is a commonly employed strategy in motor control.

Figure 8–28 A dynamical systems perspective of motor control. (A) The dynamical state of the motor cortex at a given time is a vector in a high-dimensional state space, with values on each axis representing the firing rate of an individual neuron. The first three dimensions are shown here as solid arrows, and other dimensions are symbolized by the dashed gray arrows in the background. During a motor task, changes in motor cortical activity over time can be plotted as trajectories in the state space. As an example, two trajectories representing two trials of a behavioral task in panel B are schematized. **(B)** Top, behavioral task. The monkey is trained to move its arm to reach a target shown on a screen only after a subsequent go cue appears. Bottom, motor cortical activity for 18 different trials, represented by 18 trajectories in the neural state space. Blue, green, and black dots represent neural states (computed from the activities of hundreds of recorded cortical neurons) just prior to target appearance, at the appearance of the go cue, and at movement onset, respectively. Trajectories between blue and green dots represent movement planning, whereas trajectories between green and black dots represent movement initiation. The high-dimensional state space in panel A is simplified by projecting it onto two dimensions while preserving the relative size of the ellipsoids, which represent variance across 47 trials. (Adapted from Shenoy KV, Sahani M, & Churchland MM [2013] *Ann Rev Neurosci* 36:337. See also Churchland MM, Yu BM, Cunningham JP, et al. [2010] *Nat Neurosci* 13:369.)

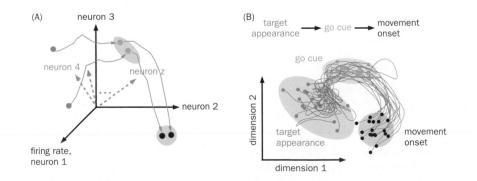

8.11 Population activity of motor cortical neurons can be used to control neural prosthetic devices

The ability to predict movement directions using the firing patterns of cortical neurons has inspired intense research in **neural prosthetic devices**, or brain–machine interfaces. A major goal of these devices is to help patients who suffer from paralysis due to spinal cord injury or motor neuron disease to regain voluntary motor control. These patients' motor cortices are still active and can presumably send commands to control body movement. Unfortunately, either the axons that deliver the commands to the spinal cord are damaged, or the motor neurons have degenerated, such that the brain and muscles are disconnected. The general strategy for creating neural prosthetic devices starts with extracting the activity of motor cortical neurons. The most effective approach thus far is an implanted multi-electrode array that can directly record the action potentials of hundreds of neurons. The activities of these neurons are then fed into a computer to extract movement intent, which is used to control an output device such as a robotic arm or a computer cursor (**Figure 8–29**A). Visual feedback to the patients can help them learn to change the firing patterns of the recorded neurons for better performance, in a **closed-loop** configuration. Remarkable progress has been made in recent years in the accuracy, complexity, and speed of such neural prosthetic devices. We give two examples to illustrate these advances.

In the first example, spike patterns of 116 single neurons were recorded from an implanted multi-electrode array in a monkey's motor cortex (Figure 8–29B, top).

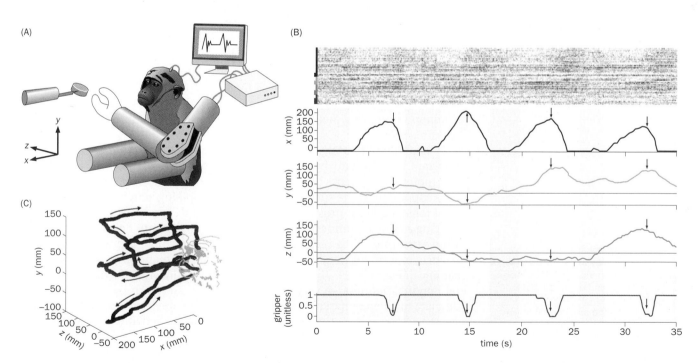

Figure 8–29 Cortical control of a prosthetic arm for self-feeding.
(A) Schematic of a brain–machine interface for prosthetic control. A multi-electrode array was implanted in the motor cortex of the monkey. The spiking patterns of many individual motor cortical neurons were used for real-time control of a robotic arm using a population vector algorithm (see Figure 8–27). The monkey had first been trained to control the robotic arm with a joystick using its real arm with visual feedback, before graduating to cortical control in this experiment, during which the real arms were restrained.
(B) Top, spiking activity of 116 single neurons used to control the robotic arm during four trials of self-feeding, grouped by their tuning in different dimensions of movement (red, x; green, y; blue, z; purple, gripper) in negative (thin bar) and positive (thick bar) directions. The bottom four traces correspond to robotic arm movement in the x, y, and z directions, and the position of the gripper (1 = open, 0 = closed) during the four trials shown in panel C. Arrows indicate the gripper closing on the target. **(C)** Spatial trajectories of the same four trials as in panel B bottom; the positions of the marshmallow varied. Red and blue portions of the trajectory represent open and closed position of the gripper, respectively. Note that the monkey opened the gripper before the food reached its mouth because it learned that marshmallow tended to stick to the gripper so it did not need to close the gripper for the full duration of the return trip. (Adapted from Velliste M, Perel S, Spalding MC et al. [2008] Nature 453:1098–1101. With permission from Macmillan Publishers Ltd.)

A computer decoded these patterns in real time by extracting movement intent using a method similar to the weighted sum population vector discussed in Section 8.10. The movement intent was then executed in the form of a robotic arm that had five degrees of freedom: three at the shoulder, one at the elbow, and one at the hand for gripping. After training with fetched marshmallows as a reward, the monkey was able to control the extension of the robotic arm by 'thinking' about the movement (Figure 8–29B, bottom). Such thinking caused the monkey to extend the robotic arm toward a marshmallow placed at different locations in the three-dimensional space, grab the marshmallow, and bring it to its mouth (Figure 8–29C; see also **Movie 8–5**). Remarkably, these actions were performed within a few seconds, the same order of magnitude as using the real arm, with a 60% success rate. This and other similar examples provide encouragement that this approach can help human patients with motor deficits to regain movement control.

The second example is the first human clinical trial that used neural prosthetics to help a patient with tetraplegia to move a computer cursor via cortical control. The patient suffered a spinal cord injury three years prior to having a 100-electrode array surgically implanted into his motor cortex (**Figure 8–30**A). During the training period following surgery, the patient was asked to imagine moving his hand along the trajectory of a cursor on a computer screen controlled by a technician (Figure 8–30B). Population activity of motor cortical neurons recorded during the imagined movement was then used to construct an algorithm that best matched the motor intent. During the next phase, the patient imagined moving the cursor, and the algorithm transformed his motor cortical activity, recorded from the multi-electrode array, into real-time control of the computer cursor. After extensive training, the patient could perform a variety of tasks, such as moving the cursor to designated targets while avoiding obstacles (Figure 8–30C), opening files in an email inbox, and drawing circles (see Movie 8–5), all using imagined motion. Despite these remarkable achievements, human clinical trials have also revealed important limitations of this approach. For example, the speed, accuracy, and level of control are considerably less than the control of a computer cursor by hand using a standard mouse. The long-term stability of electrode arrays is also an important limiting factor. Researchers are making notable progress tackling these limitations.

In addition to helping patients, neural prosthesis research has also provided important insights into how the motor cortex controls voluntary movements. Each motor task appears to be controlled by an ensemble of cortical neurons—this is supported by the successful extraction of spiking activity from populations of neurons to control prosthetic devices and the dependence of this success on engaging a critical number of neurons (which ranges from dozens to hundreds). At the same time, each individual neuron can contribute to multiple tasks—for instance, movement in all three dimensions of space and grip, as shown in the example in Figure 8–29. These properties may explain why information from several hundred neurons—a very small fraction of total motor cortical neurons—is adequate to allow remarkable control of prosthetic devices. From the dynamical systems perspective, neural states determined by the activities of a few hundred

Figure 8–30 Movement control with neural prosthetics by a human tetraplegia patient. (A) A magnetic resonance image of the patient's brain prior to surgical implantation of a 4 mm × 4 mm multi-electrode array in the arm control region of the right motor cortex. The square pointed at by the red arrow denotes the position of the future implant. **(B)** The computer cursor was controlled based on motor cortical activity of the patient recorded by the multi-electrode array, with closed-loop visual feedback during training. **(C)** Example of a cursor control task performed by the patient involving the acquisition of targets (green circles) and the avoidance of obstacles (red squares). Blue lines denote the cursor trajectory in four separate performances. (Adapted from Hochberg LR, Serruya MD, Friehs GM et al. [2006] *Nature* 442:164–171. With permission from Macmillan Publishers Ltd.)

(A)   (B)   (C)

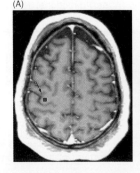

neurons can be useful approximations of neural states of an entire subregion in the relevant motor homunculus. The improvement of performance by learning through closed-loop feedback suggests that cortical neuron ensembles that control movement are plastic—indeed this was observed by monitoring changes in the tuning properties of individual neurons and ensembles of neurons as animals learn to improve their performance during training. We will return to the subject of learning and plasticity in Chapter 10.

HOW DOES THE BRAIN REGULATE THE FUNCTIONS OF INTERNAL ORGANS?

We have so far discussed the motor system as an output of the nervous system that controls body movement. A second output of the nervous system is the autonomic nervous system, also called the **visceral motor system**, which controls the internal organs that perform an animal's basic physiological functions, such as the digestion of food and the circulation of blood. The name 'autonomic' reflects that many of these functions are involuntary and cannot be consciously controlled. Perhaps because they are so vital to an animal's well-being, such functions may have been selected by evolution to be beyond voluntary control, which can be capricious or unreliable. A high center of the autonomic nervous system is the hypothalamus, which also controls the third output of the nervous system, the neuroendocrine system.

In the following sections, we begin with the two major divisions of the autonomic nervous system: the **sympathetic system** and the **parasympathetic system**. (A third division of the autonomic nervous system, the **enteric nervous system**, is associated with the gastrointestinal tract and acts rather independently to regulate digestion, with some input from the sympathetic and parasympathetic systems.) We then provide an overview of the central control of the autonomic nervous system. Lastly we focus on the function of the hypothalamus, including its role in controlling the neuroendocrine system.

8.12 The sympathetic and parasympathetic systems play complementary roles in regulating body physiology

The final effectors of the autonomic nervous system are (1) the **smooth muscle** that controls movement within the digestive, respiratory, vascular, excretory, and reproductive systems; (2) the **cardiac muscle** that controls heartbeat; and (3) various glands that excrete fluids locally through specific ducts (the **exocrine system**) or release hormones into the bloodstream that circulate throughout the body (the **endocrine system**). Whereas the effector of the motor system—the skeletal muscle—is controlled by motor neurons, the effector of the autonomic nervous system is controlled by the sympathetic and parasympathetic systems in parallel, each with two layers of output neurons (**Figure 8–31**).

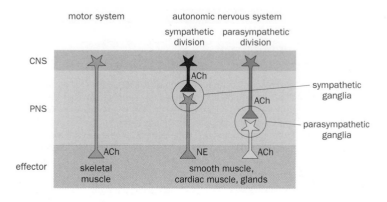

Figure 8–31 Comparison of motor and autonomic output systems. The motor system consists of motor neurons (green) projecting to their effectors, the skeletal muscle. The sympathetic and parasympathetic divisions of the autonomic nervous system both use two layers of output neurons: (1) preganglionic neurons in the CNS that project to the ganglia (darker red and blue), and (2) postganglionic neurons in the ganglia that project to their effectors (lighter red and blue). The parasympathetic ganglia are closer to the final effector organs than are the sympathetic ganglia. While postganglionic sympathetic neurons use norepinephrine (NE) as a neurotransmitter, all other types of output neurons use the neurotransmitter acetylcholine (ACh).

Imagine a mouse that, while searching for food, suddenly smells cat urine. Its heart rate, breathing rate, and blood pressure all go up, while most of its house-keeping functions such as salivation and digestion are down-regulated. The mouse is preparing for emergency responses, such as running away to escape or freezing to avoid a predator's attention. These responses result from the activation of the sympathetic system. When the danger is over, the mouse's heart rate, breathing rate, and blood pressure return to baseline, and it resumes its normal activities. These recovery processes are enabled by the parasympathetic system.

The parasympathetic and sympathetic divisions are the yin and yang of the autonomic nervous system, as their actions on most effector organs are complementary and often in opposition (**Figure 8–32**). For example, the parasympathetic system stimulates salivation and digestion, while the sympathetic system inhibits

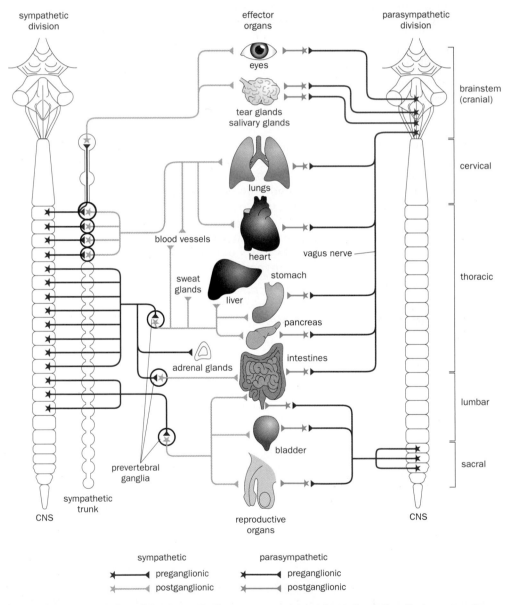

Figure 8–32 Organization of the sympathetic and parasympathetic systems. In most cases, the sympathetic system (red) and parasympathetic system (blue) regulate the actions of the same effector organs, often with opposing effects. In the sympathetic system, preganglionic neurons (dark red) are located in the thoracic and lumbar segments of the lateral spinal cord; postganglionic neurons (light red) are in the nearby sympathetic trunk or prevertebral ganglia. In the parasympathetic system, preganglionic neurons are located in the brainstem and sacral spinal cord, whereas postganglionic neurons are located near the effector organs they control. Many of the parasympathetic preganglionic axons, such as those that exit the brainstem and constitute part of the vagus nerve, project long distances to their target neurons. Some effectors, such as blood vessels, sweat glands, and the adrenal glands, receive only sympathetic input. The adrenal glands are innervated directly by preganglionic neurons. Only a subset of effectors of the autonomic system is depicted here. (Adapted from Jänig W [2006] The Integrative Action of the Autonomic Nervous System.)

these processes. The parasympathetic system slows down the heartbeat, while the sympathetic system speeds it up. The parasympathetic system constricts airways in the lungs, while the sympathetic system causes airways to relax. In general, the sympathetic system facilitates energy expenditure to support enhanced activity in the face of extreme situations, such as the 'fright, fight, or flight' responses of the mouse we just discussed. The sympathetic system inhibits physiological functions not immediately required for survival and activates emergency responses. Indeed, one output of the sympathetic system is to produce **epinephrine** (adrenaline) through innervation of the adrenal glands; epinephrine is a hormone that circulates in the bloodstream to increase energy expenditure so that the body is ready for rapid action. On the other hand, the parasympathetic system returns the body's physiology to its non-emergency state and facilitates energy conservation ('rest and digest').

In both sympathetic and parasympathetic systems, the neurons that innervate the effectors are called **postganglionic neurons**. Their cell bodies are located in the sympathetic or parasympathetic ganglia in the PNS. Postganglionic neurons are controlled by **preganglionic neurons**, whose cell bodies are located within the CNS and which project axons that synapse onto the postganglionic neurons in the ganglia (Figures 8–31 and 8–32). (Pre- and postganglionic neurons are also collectively called **visceral motor neurons**, as opposed to somatic motor neurons that control skeletal muscle contraction.) In the sympathetic system, the preganglionic neurons are located in the thoracic and lumbar segments of the lateral spinal cord gray matter, and the postganglionic neurons are in the nearby sympathetic trunk or prevertebral ganglia. In contrast, the parasympathetic preganglionic neurons are located in the brainstem and the caudal spinal cord, and their axons, such as those that constitute the **vagus nerve**, travel long distances to innervate parasympathetic ganglia, which are usually closely associated with their effector organs (Figure 8–32).

As in somatic motor neurons in the motor system, preganglionic neurons of both the sympathetic and parasympathetic systems use ACh as their neurotransmitter. Postganglionic parasympathetic neurons also use ACh, while postganglionic sympathetic neurons use the neurotransmitter norepinephrine (Figure 8–31; also see Sections 3.19 and 3.20 for the opposing actions of norepinephrine and ACh on cardiac muscle contraction). Neuropeptides are also widely used to modulate the actions of the sympathetic and parasympathetic systems.

8.13 The autonomic nervous system is a multilayered regulatory system

Like the motor system (see Figure 8–2), the sympathetic and parasympathetic systems are also controlled at multiple levels (**Figure 8–33**). One level of control originates from **visceral sensory neurons**. One population of visceral sensory neurons is located in the dorsal root ganglia along with the somatosensory neurons (see Figure 6–63); their peripheral axons terminate in internal organs, and their central axons terminate in the spinal cord. These axons can synapse onto preganglionic neurons directly or via interneurons for local feedback control of the autonomic output. They also transmit information via second-order neurons to the **nucleus of the solitary tract** (**NTS**), a nucleus in the caudal brainstem that specializes in collecting sensory information from internal organs (as well as information about taste; see Figure 6–35). A second population of visceral sensory neurons is located in ganglia associated with the vagus nerve; their peripheral axons constitute part of the vagus nerve and terminate in various organs, and their central axons terminate at the NTS. NTS neurons provide feedback to autonomic output at multiple levels and at the same time send ascending information through the **parabrachial nucleus** in the brainstem to the thalamus and the **insular cortex**, a part of the cerebral cortex that represents taste (see Figure 6–35), pain (see Figure 6–70), and sensation from internal organs, among other functions it serves. The insular cortex interacts extensively with the prefrontal cortex, and together they constitute the cerebral control center of the autonomic nervous system (Figure 8–33).

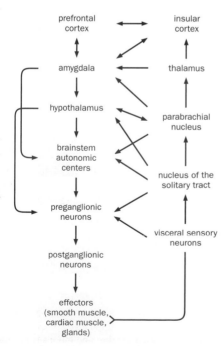

Figure 8–33 Multi-layered control of the autonomic nervous system. Arrows indicate the direction of information flow through direct synaptic connections. Additional multi-synaptic connections are not shown.

The central control of the autonomic nervous system also involves two important structures, the **amygdala** and the **hypothalamus**. The amygdala—an almond-shaped structure located in the ventral forebrain—receives extensive input from the thalamus, the insular cortex, and the prefrontal cortex. It provides feedback to the cortex and sends descending input to brainstem autonomic control centers and to the hypothalamus. Through its control of the autonomic nervous system, the amygdala regulates physiological responses to fear and emotional states, as in the example of the mouse that smells cat urine. We will study the function and circuit organization of the amygdala in Chapter 10 in the context of emotion-related signal processing and memory. For now, we turn to the hypothalamus, a key regulator of the sympathetic and parasympathetic systems, as well as a key coordinator of hormone secretion.

8.14 The hypothalamus regulates diverse basic body functions via homeostasis and hormone secretion

Named for its location beneath the thalamus, the hypothalamus is composed of many discrete nuclei, each with specialized functions (**Figure 8–34**A). These functions include the control of (1) energy expenditure—through regulation of eating, digestion, and metabolic rates; (2) blood pressure and electrolyte composition—through regulation of drinking and salt intake; (3) reproduction—through hormonal regulation of sexual maturation, mating, pregnancy, and lactation; (4) body temperature—through various thermoregulatory processes; (5) emergency responses—through stress hormone secretion and sympathetic activation; and (6) circadian rhythms and sleep.

The hypothalamus is a collection of parallel circuits that regulate each of these basic bodily functions. Some hypothalamic nuclei receive input from sensory systems that provide relevant information about the environment, such as light from the visual system to control circadian rhythms, or odors and pheromones from the main and accessory olfactory systems to indicate the presence of a potential mate or predator. A significant fraction of the input received by many hypothalamic nuclei is **interoceptive** (that is, reflecting the state of the body, including all the internal organs). Many hypothalamic circuits regulate **homeostasis**, which refers to coordinated physiological processes that maintain the steady state of the organism. Specifically, input to hypothalamic nuclei provides quantitative indicators of physiological states, such as blood pressure, body temperature, or

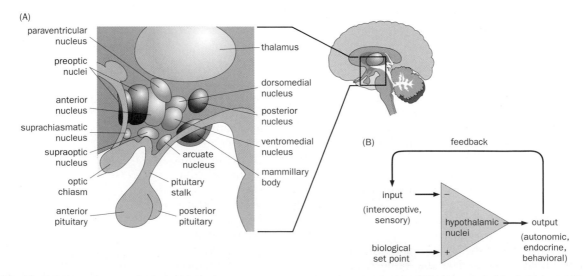

Figure 8–34 Organization and homeostatic function of the hypothalamus. (A) Schematic of the hypothalamus and pituitary from a sagittal perspective, depicting only a subset of nuclei close to the mid-sagittal plane. The functions of some of these nuclei will be discussed in the remainder of this chapter and in Chapter 9. For example, the arcuate nucleus, paraventricular nucleus, and lateral hypothalamus (not shown) regulate food intake and energy

expenditure. The suprachiasmatic nucleus regulates circadian rhythms, while the preoptic nuclei and the ventromedial nucleus regulate sexual behavior. **(B)** In this homeostatic model, the magnitude of the hypothalamic output is proportional to the difference between the input and the set point (as symbolized by the amplifier). The net effect of the output is to reduce this difference, as indicated by the feedback arrow.

nutritional levels. Differences between these values and biological set points lead to activation of specific hypothalamic nuclei that control autonomic, behavioral, and endocrine output, causing a reduction of such differences (Figure 8–34B). This homeostatic regulation maintains physiological states within a narrow range around the biological set points. For example, if our body temperature rises above the biological set point of 37°C due to intense exercise, the hypothalamus sends signals to the sympathetic system to enhance sweating (which cools the body via evaporation) and vasodilation (which increases heat loss via enhanced blood flow to the skin). The hypothalamus also regulates behavior—its action causes us to feel hot so that we seek a cool place or remove extra layers of clothing, facilitating a return to normal body temperature.

In parallel with its output to downstream neural circuits (see Figure 8–33), another important output of the hypothalamus is the control of hormone secretion from the nearby **pituitary**, the endocrine center of the brain. Two distinct mechanisms are employed (**Figure 8–35**). Some hypothalamic neurons produce the peptide hormones **oxytocin** and **vasopressin**, project axons to the **posterior pituitary**, and release these hormones from their axon terminals in the posterior pituitary directly into the bloodstream for systemic circulation. (Oxytocin and vasopressin can also be released from synaptic terminals as peptide neurotransmitters that signal to postsynaptic neurons.) Other hypothalamic neurons release pre-hormones into local blood circulation. Carried by specialized portal vessels through the pituitary stalk, these pre-hormones travel to the **anterior pituitary**, where they act on endocrine cells and induce them to release their hormones that then circulate throughout the body. Some hypothalamic neurons also release signaling molecules, such as somatostatin and dopamine, that inhibit hormone release by endocrine cells in the anterior pituitary. Hormones from the anterior and posterior pituitary regulate a variety of physiological functions, including growth, metabolism, reproduction, and stress response (**Table 8–1**).

The issue of how specific hypothalamic circuits maintain homeostasis can be divided into several questions: How is input sensed by hypothalamic nuclei? How is the biological set point determined? How is the difference between the set point and the actual physiological state compared and translated into output? In the remaining sections of this chapter, we will study how these regulatory mechanisms control eating, sleep, and circadian rhythms; we will revisit these questions in Chapter 9 in the context of sexual behavior.

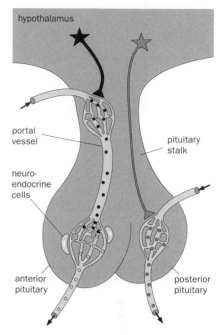

Figure 8–35 Hypothalamic control of hormone release through the pituitary. Oxytocin- and vasopressin-producing neurons (blue) release hormones (blue dots) directly to the bloodstream in the posterior pituitary, whereas other hypothalamic neurons (red) release pre-hormones (red dots) that circulate to the anterior pituitary through the portal vessel, where they activate neuroendocrine cells (yellow) to release hormones (yellow dots) that circulate throughout the body. Arrows indicate the direction of blood flow.

Table 8–1: Hormones released from hypothalamic neurons and pituitary endocrine cells

Hormone		Function
Direct hormone release by hypothalamic neurons at the posterior pituitary		
Oxytocin		regulates maternal and social behavior
Vasopressin		regulates water balance; regulates social behavior
Stimulatory action of hypothalamic neurons on anterior pituitary endocrine cells		
Hypothalamus	*Anterior pituitary*	
Corticotropin-releasing hormone (CRH)	adrenocorticotropin (ACTH)	stimulates glucocorticoid release from the adrenal gland and regulates stress response
Gonadotropin-releasing hormone (GnRH)	luteinizing hormone (LH); follicle-stimulating hormone (FSH)	stimulates production of sex hormones, sexual maturation, and sexual behavior
Growth hormone-releasing hormone (GHRH)	growth hormone (GH)	stimulates growth
Thyrotropin-releasing hormone (TRH)	thyroid-stimulating hormone (TSH, also called thyrotropin); prolactin	stimulates metabolism; stimulates milk production
Inhibitory action of hypothalamic neurons on anterior pituitary endocrine cells		
Hypothalamus	*Anterior pituitary*	
Somatostatin	growth hormone	
Dopamine	prolactin	

HOW IS EATING REGULATED?

We all have this experience daily: when we are hungry, food becomes attractive and we crave it; after eating, we lose interest in even the most delicious food. One possible explanation of the above phenomena is that a feedback signal from the body to the brain regulates our desire to eat. Indeed, as we will learn from the following sections, there are two types of feedback signals. The first type acts immediately after our stomach is full. The second type acts over a longer time-scale, signaling nutritional status. Experiments in a wide variety of animal species have shown that, while starvation caused body weight to decrease and forced overfeeding causes body weight to increase, if food were provided *ad libitum* afterwards (that is, if the animals have free access to food and can eat as much or as little as they like), their body weight would return to a stable, pre-experimental value. When normal rats were fed food that was diluted with non-nutritious substances, the amount they ate corresponded not to the volume of the food but to its caloric content, such that their body weight was maintained. These striking examples of homeostasis suggest the existence of feedback signals relating to nutritional status that control body weight. Body weight is maintained through the regulation of two processes: food intake and energy expenditure. Although these processes are closely linked, they can also be controlled separately. In the following sections, we focus on the neural mechanisms that control eating.

8.15 Hypothalamic lesion and parabiosis experiments suggested that eating is inhibited by a negative feedback signal from the body

Unlike normal rats, rats with lesions in the ventromedial hypothalamus significantly increased their eating when food was provided *ad libitum*, leading them to become obese. These experiments suggested that the ventromedial hypothalamus regulates eating. But how is eating regulated such that normal rats can maintain their body weight?

The initial insight to this question came from a **parabiosis** experiment, in which two rats were surgically joined at the abdomen when young, such that they had limited exchange of substances that circulated in their blood during the rest of their lives together. After successful surgeries, parabiotically linked rats grew well and were healthy (**Figure 8–36**A). The ventromedial hypothalamus was then lesioned in one rat of the pair. After this neurosurgery, the lesioned rat started to

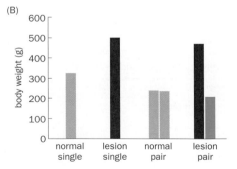

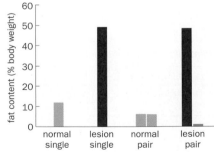

Figure 8–36 Parabiosis experiments suggest that a feedback signal from the body inhibits feeding. (A) A pair of parabiotic rats. The abdomens of these rats were surgically joined when they were young so that they had limited exchange of signaling molecules in their blood afterwards. **(B)** Postmortem analysis of average body weight and fat content for four experimental conditions: normal single rats (green), single rats with lesions in the ventromedial hypothalamus (red), parabiotic pairs in which neither partner is lesioned (green/green), and parabiotic pairs in which one of the partners has a lesion in the ventromedial hypothalamus (red) and the other does not (blue). Both single and partnered rats with lesions overeat, leading to dramatic gains of weight and fat content. In parabiotic pairs, a hypothalamic lesion in one partner causes the non-lesioned partner to refuse food and starve, as indicated by the drastic loss of fat content. This experiment suggests that a circulating signal, which the lesioned rat can't sense but constantly produces as a result of overeating, is sensed by and inhibits feeding of the non-lesioned partner. (Adapted from Hervey GR [1959] *J Physiol* 145:336–352. With permission from the Physiological Society.)

eat more food and became obese. Remarkably, its parabiotic partner lost interest in food even when the food was provided under its nose; as a consequence, the non-lesioned partner lost weight and became emaciated over time. Postmortem analysis of weight and fat content revealed that the partners with hypothalamic lesions had dramatically increased body weight and fat content, as did single rats that had similar lesions. By contrast, the non-lesioned partners had lost weight and had dramatically reduced fat content (Figure 8–36B). The simplest interpretation of these results is that the lesioned partner lost the ability to respond to a negative feedback signal from the body and therefore kept eating, causing obesity. The excessive food intake led to continual production of the negative feedback signal that, circulating through both partners, instructed the non-lesioned partner with a normal hypothalamus to stop eating, leading to its starvation.

8.16 Studies of mouse mutants led to the discovery of the leptin feedback signal from adipose tissues

Clues to the nature of the feedback signal came from studies of two spontaneously occurring mutant strains of mice, *Ob* (for *Obese*) and *Db* (for *Diabetic*). Mice homozygous for each of these mutations displayed almost identical obese phenotypes (**Figure 8–37**A), but their corresponding genes were mapped to different genetic loci. Important insight regarding the nature of these mutations came again from a series of parabiosis experiments (Figure 8–37B). When *Ob/Ob* (homozygous for the *Ob* mutation) and wild-type (+/+) mice were joined parabiotically, the *Ob/Ob* mouse ate less and became less obese than non-parabiotic *Ob/Ob* mice, and the +/+ mouse gained more weight than non-parabiotic +/+ mice. When *Db/Db* and +/+ mice were joined parabiotically, the *Db/Db* mouse's weight gain was similar to that of non-parabiotic *Db/Db* mice, whereas the +/+ mouse stopped eating and became emaciated, similar to the partner of the ventromedial hypothalamus-lesioned rat. Finally, when parabiosis was carried out between *Db/Db* and *Ob/Ob* mice, the *Db/Db* mouse kept gaining weight whereas the *Ob/Ob* mouse stopped eating and lost weight (Figure 8–37B).

These remarkable results led researchers to propose that the normal *Ob* gene encodes a circulating negative feedback signal to the brain to inhibit eating, whereas the normal *Db* gene encodes a receptor in the hypothalamus that responds to this signal. Thus, when *Db/Db* mice were parabiotically linked with +/+ or *Ob/Ob* mice, the *Db/Db* partner could not respond to the negative feedback signal and therefore kept eating. The continued production of the negative feedback signal inhibited the +/+ or *Ob/Ob* partners from eating, leading to starvation. When *Ob/Ob* mice were parabiotically linked with +/+ mice, the signal supplied by the +/+ partner helped the *Ob/Ob* partner to control its eating and body weight; the +/+ mice gained weight because the signal was diluted.

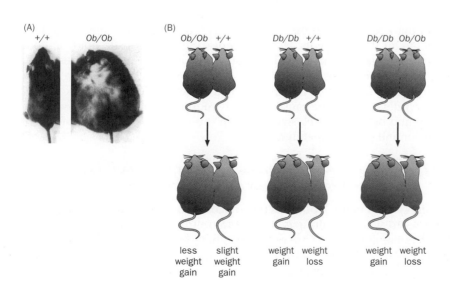

(A) +/+   *Ob/Ob*

(B) *Ob/Ob* +/+   *Db/Db* +/+   *Db/Db* *Ob/Ob*

less weight gain   slight weight gain   weight gain   weight loss   weight gain   weight loss

Figure 8–37 Parabiosis experiments with mutant mice that exhibit obesity. (A) At age 10 months, a normal mouse (left) weighed 29 grams. A mouse homozygous for the *Ob* mutation (right), which arose spontaneously in the Jackson Laboratory, a mouse repository center, weighed 90 grams. **(B)** Summary of three parabiotic experiments. The consequences to each individual after parabiotic pairing are shown below. Collectively, these experiments suggest that *Ob* might produce a feedback signal, and *Db* might be necessary for the interpretation of the signal. (A, from Ingalls AM, Dickie MM & Snell GD [1950] *J Hered* 41:317–318. With permission from OUP. B, based on Coleman DL & Hummel KP [1969] *Am J Physiol* 217:1298–1304 and Coleman DL [1973] *Diabetologia* 9:294–298.)

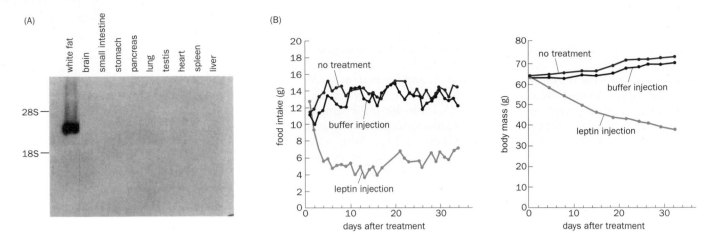

Figure 8–38 Leptin as a feedback signal to control food intake.
(A) A northern blot shows that the *Leptin* mRNA is specifically
produced in white fat but not in any other tissue. 28S and 18S
ribosome RNAs are used as molecular weight markers. **(B)** Compared
to no treatment (red traces) or buffer injection (purple traces), daily
injection of 5 µg leptin per gram of body weight (blue traces) into
Ob/Ob mice results in drastic reduction of food intake (left) and body
mass (right). (A, from Zhang Y, Proenca R, Maffei M et al. [1994]
Nature 372:425–432. With permission from Macmillan Publishers
Ltd; B, adapted from Halaas JL, Gajiwala KS, Maffei M, et al. [1995]
Science 269:543.)

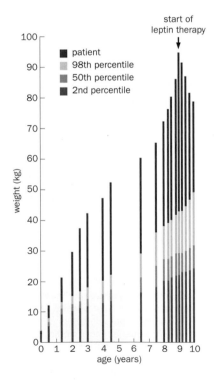

**Figure 8–39 An example of leptin
therapy in humans.** A patient
homozygous for a frameshift mutation in
the *Leptin* gene developed early-onset
obesity. She constantly felt hungry,
demanded food, and was disruptive when
denied food. At age 9, her body weight
was about twice the 98th percentile
level. Daily injection of recombinant
leptin steadily reduced her body weight.
(Adapted from Farooqi IS, Jebb SA,
Langmack G, et al. [1999] *New Engl J
Med* 341:879.)

The proof of the above hypothesis came when a molecular-genetic method
called **positional cloning** (see Section 13.6) was used to identify the gene that
corresponds to the *Ob* mutation. The *Ob* gene was found to encode a secreted protein, named **leptin** (from the Greek *leptos*, for thin), which is produced specifically
by fat tissue (**Figure 8–38**A) and which circulates throughout the body. When
Ob/Ob mice were injected with leptin protein produced *in vitro* from the cloned
cDNA, their food intake and body mass were markedly reduced (Figure 8–38B).
Injection of leptin into normal mice caused them to eat less and became thin also.
Shortly following the cloning of the *Ob* gene, its receptor was identified and found
to correspond to the *Db* gene. The *Db* gene is highly expressed in the ventromedial
hypothalamus, in particular the **arcuate nucleus** (see Figure 8–34A), which lesion
studies had implicated in the control of eating. The discoveries of leptin and its
receptor provided a satisfying proof that the adipose tissue produces a negative
feedback signal for hypothalamic control of food intake and body weight. They
also enabled further studies of the neural mechanism by which this is achieved
(see below).

Leptin's amino acid sequence is highly conserved among mammals, suggesting that the protein's function has been conserved. Indeed, following the cloning
of mouse *Leptin*, disruption of the human *Leptin* gene was found to be the cause
of a rare population of early-onset obesity cases. Daily injection of recombinant
leptin markedly reduced the weight gain (**Figure 8–39**) and improved the quality of life for these patients. However, most obese patients do not react to leptin
treatment; in fact serum leptin level usually correlates with the degree of obesity.
This suggests that most obese patients do not have deficits in leptin production
but rather are deficient in their response to the leptin signal, much like the *Db/Db*
mice. Understanding the cause of obesity requires deeper understanding of how
leptin, and other feedback signals, act in the nervous system to control feeding
behavior and energy expenditure.

8.17 POMC and AgRP neurons in the arcuate nucleus
are central regulators of eating

The expression of the leptin receptor, encoded by the *Db* gene, provides the first
clues as to where leptin acts in the brain. The best-studied leptin target cells
that regulate food intake are two groups of intermingled neurons in the arcuate
nucleus of the ventromedial hypothalamus. Leptin activates the first group of neurons, called **POMC neurons** because they express pro-opiomelanocortin. POMC

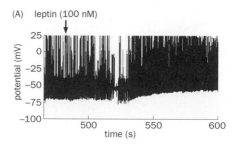

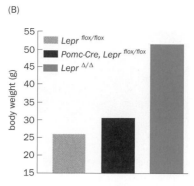

Figure 8–40 Leptin activates POMC neurons to inhibit eating. (A) Whole-cell patch clamp recording of a POMC (pro-opiomelanocortin) neuron in a hypothalamic slice from a transgenic mouse in which POMC neurons were labeled with GFP to aid their identification. Leptin application increases action potential frequency within a minute. **(B)** Removal of the gene encoding the leptin receptor (*Lepr*, same as *Db*) in POMC neurons using *Pomc-Cre*-mediated conditional knockout (red) leads to a significant increase in body weight compared with control mice without the *Pomc-Cre* transgene (orange). However, the weight gain associated with loss of POMC leptin reception is only a small fraction of the weight gain observed in mice in which the leptin receptor is deleted in all cells (blue). *Lepr^flox/flox* refers to the conditional allele of *Lepr*, which is only inactivated in Cre-expressing cells. (A, adapted from Cowley MA, Smart JL, Rubinstein M et al. [2001] *Nature* 411:480–484. With permission from Macmillan Publishers Ltd; B, adapted from Balthasar N, Coppari R, McMinn J et al. [2004] *Neuron* 42:983–991. With permission from Elsevier Inc.)

is a precursor protein that is cleaved to produce multiple neuropeptides, including the anorexigenic (appetite-reducing) α-melanocyte-stimulating hormone (**α-MSH**). α-MSH reduces food intake by activating the melanocortin-4 receptor (**MC4R**), a G-protein-coupled receptor expressed by target neurons. Disruption of the genes encoding POMC or MC4R leads to obesity in both mice and humans; indeed, losing one copy of the MC4R gene in humans is sufficient to cause obesity, and MC4R disruption is the largest single genetic contribution to human obesity, accounting for up to 5% of severe obesity cases.

It is instructive to study how researchers have established a direct link between leptin and POMC neurons in the arcuate nucleus in the context of body weight control. First, the leptin receptor is highly expressed in POMC neurons, suggesting that leptin can act directly on POMC neurons. Second, electrophysiological recordings of POMC neurons in hypothalamic slices *in vitro* showed an increased firing of POMC neurons within minutes of leptin application (**Figure 8–40**A). Recent work suggests that binding of leptin to the leptin receptor activates a signaling pathway that opens a cation-selective TRP channel. Leptin application also reduces GABAergic inhibition onto POMC neurons (see below). Both promote depolarization of the POMC neurons. Third, while leptin application reduces the body weight of *Ob/Ob* mice (see Figure 8–38B), it fails to do so in arcuate-lesioned *Ob/Ob* mice. This experiment demonstrates that leptin reception in the arcuate nucleus is essential for body weight control. Fourth, conditional knockout of the leptin receptor in POMC neurons is sufficient to cause weight gain (Figure 8–40B). This last line of evidence directly establishes a causal relationship between the action of leptin on POMC neurons and body weight control. However, the body weight gain produced by deletion of the leptin receptor in POMC neurons is much milder than that produced by leptin receptor deletion in the entire animal (Figure 8–40B). Together with the third line of evidence, this indicates that leptin must also act directly in other neurons in the arcuate nucleus to regulate body weight.

The second group of arcuate nucleus neurons that express the leptin receptor are the **AgRP neurons**, whose activity is inhibited by leptin binding; the mechanism of this inhibition remains unclear. AgRP neurons release two orexigenic (appetite-stimulating) neuropeptides, the agouti-related protein (AgRP) and neuropeptide Y (NPY). In addition, AgRP neurons also release the inhibitory neurotransmitter GABA. The importance of AgRP neurons in regulating eating has been demonstrated by both loss- and gain-of-function experiments. In a loss-of-function approach, knock-in mice were created in which AgRP neurons selectively express a human receptor for the bacterial diphtheria toxin. (The human receptor is much more sensitive to the toxin than the endogenous mouse receptor. Recall that diphtheria toxin can also be directly expressed to kill cells, as in the experiment described in Figure 8–16. The strategy of expressing a diphtheria toxin receptor allows temporal control of cell killing via toxin application.) Application of diphtheria toxin in adult mice selectively killed AgRP neurons within days; this caused the mice to stop eating and rapidly lose body weight (**Figure 8–41**A). Thus, AgRP neurons are necessary for food intake. In a gain-of-function experiment, optogenetic activation of channelrhodopsin-expressing AgRP neurons

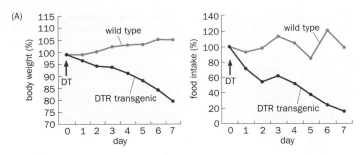

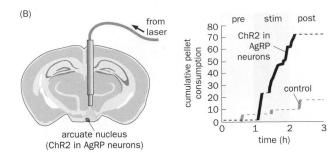

Figure 8–41 AgRP neurons promote eating. (A) Application of diphtheria toxin (DT) to wild-type mice (blue) does not affect body weight (left) or food intake (right). However, in transgenic mice (red) that express human diphtheria toxin receptor (DTR) in AgRP neurons, acute DT application selectively kills these neurons and results in a drastic reduction of body weight and food intake. **(B)** Left, an adeno-associated virus expressing Cre-dependent channelrhodopsin (ChR2) was injected into the arcuate nucleus of transgenic mice that express Cre only in AgRP neurons, allowing selective optogenetic activation of AgRP neurons by blue light from a laser. Right, photostimulation (stim) leads to robust food pellet consumption during the stimulation period, compared to before pre) and after (post) the stimulation (dotted lines). The light trace shows a control mouse without ChR2 expression. (A, adapted from Luquet S, Perex FA, Hnasko TS, et al. [2005] *Science* 310:683; B, adapted from Aponte Y, Atasoy D & Sternson SM [2011] *Nat Neurosci* 14:351–355. With permission from Macmillan Publishers Ltd.)

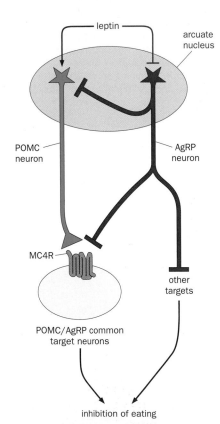

Figure 8–42 Leptin acts on two antagonistic populations of arcuate neurons to suppress eating. POMC neurons, which inhibit eating by activating the melanocortin-4 receptor (MC4R) in target neurons, are activated by leptin. At the same time, leptin inhibits AgRP neurons, which promote eating via three different mechanisms: (1) by inhibiting POMC neurons directly; (2) by antagonizing POMC's effect on their common target neurons through AgRP's competition with α-MSH produced by POMC neurons for binding the MC4R receptor; and (3) through GABA-mediated inhibition of additional target neurons that inhibit eating (see Figure 8–43 for more details).

induced voracious eating within minutes (Figure 8–41B), regardless of whether the mice had just eaten or not. Thus, acute activation of AgRP neurons is sufficient to induce eating and can override inhibitory feedback signals.

In summary, leptin acts directly on two intermingled neuronal populations within the arcuate nucleus: it activates POMC neurons that normally inhibit eating, and it inhibits AgRP neurons that normally promote eating. Both of these mechanisms contribute to leptin's appetite-reducing effect (**Figure 8–42**). AgRP and POMC neurons have extensive interactions. AgRP neurons directly inhibit POMC neurons through GABA and NPY release. In addition, the AgRP neuropeptide competes with α-MSH for MC4R binding, such that AgRP neurons antagonize the action of POMC neurons at their common targets. However, the function of AgRP neurons cannot be fully accounted for by their role in antagonizing POMC neurons. For example, neither the anorexigenic effect of ablating AgRP neurons nor the orexigenic effect of stimulating AgRP neurons was affected by blocking MC4R signaling. These observations suggest that AgRP neurons act on additional targets to exert their orexigenic effect, likely through their release of GABA to inhibit target neurons (Figure 8–42). We discuss these targets in the next section.

8.18 Multiple feedback signals and neural pathways act in concert to regulate eating

The search for targets that mediate the orexigenic effects of AgRP neurons has begun to define additional circuits and pathways for eating regulation (**Figure 8–43**). The arcuate AgRP neurons project to multiple areas within and outside the hypothalamus and inhibit their target neurons. Glutamatergic neurons in the parabrachial nucleus (PBN) of the brainstem have been identified as a critical target of AgRP neurons. Starvation due to ablation of AgRP neurons (see Figure 8–41A) could be rescued by (1) providing GABA agonist to the PBN, (2) blocking the excitatory input to the PBN from the nucleus of the solitary tract (NTS), or (3) inhibiting the excitatory output of the PBN. These data suggest that the PBN integrates visceral and taste sensory information from the NTS (see Section 8.13) and nutritional information from the hypothalamus. On the other hand, the **paraventricular hypothalamic nucleus (PVH)** is a critical target involved in the voracious eating triggered by optogenetic stimulation of AgRP neurons, since the optogenetic stimulation of axon terminals of channelrhodopsin-expressing AgRP neurons in the PVH was sufficient to induce eating, and silencing the PVH neurons mimicked the effect of activating presynaptic inhibitory AgRP neurons. According to a current model, overall activation of the NTS → PBN pathway inhibits appetite; this pathway is normally antagonized by GABAergic input from AgRP neurons. At the same time, inhibition of PVH neurons by AgRP neurons activates a hunger circuit, likely mediated by PVH projections to the caudal brainstem that regulates

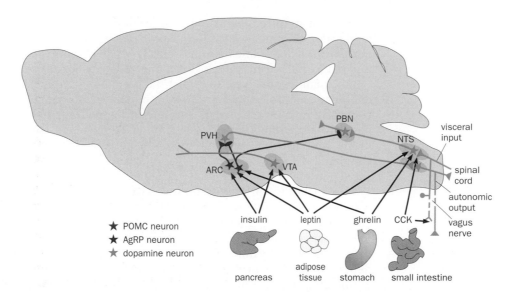

Figure 8–43 A working model of neural pathways involved in eating and their regulation by feedback signals from the body. The body produces multiple feedback signals to the brain to control feeding. Leptin and insulin reflect nutritional status (relating to fat and sugar, respectively), whereas ghrelin and cholecystokinin (CCK) reflect the gastrointestinal status (signaling hunger and satiety, respectively). These signals act at multiple sites in the brain (arrows). A prominent site is the arcuate nucleus (ARC), which contains the POMC and AgRP neurons (see Figure 8–42 for details). Leptin and insulin also act on dopamine neurons in the ventral tegmental area (VTA) that may regulate the reward value of food. Another major target site is the nucleus of the solitary tract (NTS). CCK also acts as a neurotransmitter to activate NTS through the vagus nerve. The AgRP → PVH (paraventricular hypothalamus nucleus) circuit regulates acute eating, whereas the AgRP → PBN (parabrachial nucleus) circuit counteracts an NTS → PBN excitation-induced appetite reduction. Other, less understood targets of POMC and AgRP neurons that are implicated in eating are not depicted here. (The AgRP circuit dissection is based on Wu Q, Clark MS & Palmiter RD [2012] *Nature* 483:594–597 and Atasoy D, Betley N, Su HH et al. [2012] *Nature* 488:172–177.)

parasympathetic preganglionic neurons (Figure 8–43). Other target neurons of AgRP neurons, such as the POMC neurons we discussed in the previous section, participate in long-term homeostatic regulation of food intake and energy expenditure. Eating is also regulated by lateral hypothalamus neurons that express the neuropeptide orexin, which also functions in sleep regulation (see Section 8.23).

In addition to leptin, the body produces a number of other feedback signals to the brain to regulate eating; these act on some of the same neurons and circuits we have studied so far (Figure 8–43). For example, the pancreas produces **insulin** in response to a rise of blood glucose level after meals, and insulin receptors are widely distributed in the brain, including in AgRP neurons of the arcuate nucleus. Leptin and insulin levels reflect the nutritional status of the body, and their actions in controlling food intake involve relatively long timescales. Short-term signals from the gastrointestinal system stimulate or inhibit food intake before or after each meal; of these, the two best-studied are the neuropeptides **ghrelin** and **cholecystokinin (CCK)** (Figure 8–43). Ghrelin is produced by stomach-associated glands in response to a reduced glucose level, and acts as a hunger signal to stimulate eating. CCK is produced in small intestine in response to a rise in fatty acid concentration, and acts as a satiety signal to inhibit eating. A major target for these short-term signals is the nucleus of the solitary tract (NTS) in the brainstem. CCK can act as a neurotransmitter on the sensory endings of the vagus nerve to transmit signal directly to NTS. Both CCK and ghrelin also act as hormones that circulate in the blood to reach the NTS and other brain targets. The effects of these short-term signals can be influenced by long-term signals. For instance, leptin can act directly through its receptor on NTS neurons, or indirectly through its action on hypothalamic neurons, to increase the sensitivity of the NTS to the CCK satiety signal, such that when the nutritional level is high, animals are more likely to terminate their meals in response to a satiety signal.

Other sites of action for both leptin and insulin are dopamine neurons in the ventral tegmental area (VTA) in the midbrain (Figure 8–43). As we will learn in Chapters 10 and 11, VTA dopamine neurons regulate reward-based action selection. Consistent with an important role of dopamine in the regulation of food intake, mutant mice that cannot produce dopamine do not voluntarily consume food even if they are hungry. Both leptin and insulin have been shown to reduce the activity of dopamine neurons. A likely action of leptin and insulin in the VTA is to reduce the reward value of food, thus reducing the desire to eat.

In summary, we have seen that the seemingly simple acts that we perform effortlessly multiple times a day—the initiation and conclusion of eating—are controlled by multiple signals and interacting neural pathways. Some of these pathways can compensate for one another. For instance, despite the importance of the AgRP neurons, ablating them during early development has little effect compared to the severe effect of ablation in adult mice (see Figure 8–41A). Even in starvation induced by AgRP neuronal ablation in adults, transient rescue by GABA agonist application in the PBN for several days can lead to permanent rescue after withdrawal of the GABA agonist, likely by providing enough time to activate an alternative pathway. These studies highlight the flexibility of the nervous system to compensate for lost neurons with alternative pathways. They also underscore the importance of eating in the life of an animal—multiple and partially redundant pathways have evolved to ensure that it is properly controlled.

Investigating the molecular and neural mechanisms of eating control has shed light on obesity, which is a pressing health problem in our modern society because it is a predisposing factor for many diseases and has become increasingly prevalent in recent years. As the extreme example of human mutations in leptin illustrates, genetic makeup contributes significantly to obesity. After all, eating is a basic drive essential for life. For our hunter-gatherer ancestors in the not-so-distant past, genotypes that generated less effective feedback to suppress eating might have promoted greater nutrient storage and been favored by natural selection during times of food scarcity. In an age of food abundance and sedentary lifestyle, genetic compositions that had been advantageous may now be prone to obesity. As the story of leptin shows, understanding the basic mechanisms of eating regulation is key to bringing obesity under control.

HOW ARE CIRCADIAN RHYTHMS AND SLEEP REGULATED?

We've just used eating as an example to illustrate the molecular and neural basis of hypothalamus-mediated homeostatic regulation of body weight. We use below the examples of circadian rhythms and sleep to expand further on how regulatory systems operate. While eating has a clear purpose of bringing nutrients to the body to maintain energy balance, the exact physiological function of sleep remains enigmatic despite the fact that we spend about a third of our lives doing it. Sleep is timed by **circadian rhythms**, self-sustained oscillations in an organism's behavior, physiology, and biochemistry, with a period close to 24 hours. In the following sections, we first study the mechanisms that regulate circadian rhythms, which are well understood thanks to inspiring research in the past decades. We then turn to sleep regulation and discuss its possible functions.

8.19 Circadian rhythms are driven by an auto-inhibitory transcriptional feedback loop that is conserved from flies to mammals

Circadian rhythms are found in many branches of life from bacteria to humans. These rhythms reflect the interaction of organisms living on Earth with a salient environmental signal—the daily cycle of light and dark caused by Earth's rotation around its own axis once every 24 hours. For example, most adult fruit flies eclose (emerge from their pupal case) in the morning, so that young flies have a whole day of light to search for their first meal. Mice are active at night because they are safer from their predators. However, daily rhythms are not just passive reactions to environmental stimuli: circadian rhythms are self-sustained even in the absence of light. For instance, mice kept in constant darkness continue to be active only during their 'subjective nights.' The period of the circadian rhythm, as best seen by the onset of the running wheel activity, is slightly shorter than 24 hours but with minimal variation (**Figure 8–44**). (During the light–dark cycle, the period returns to 24 hours due to a process called light entrainment that will be discussed later.) These phenomena indicate that an internal clock is running in the absence of environmental cues. How can we find out what this biological clock is?

The first insight into the molecular nature of this biological clock came from genetic studies in fruit flies around 1970. Hypothesizing that circadian rhythms are regulated by specific genes and would be disrupted by mutations in those genes, Seymour Benzer and colleagues conducted a forward genetic screen (see Section 13.6) in fruit flies, utilizing their eclosion as an assay to identify which genes, when mutated, would disrupt circadian rhythms. Three mutations that mapped to the same gene (named *Period*) were found that sped up, slowed down, or abolished circadian rhythms (**Figure 8–45**A). This finding immediately suggested not only the existence of genes that control circadian rhythm, but also that *Period* must be related to the central control mechanism, since different disruptions of this single gene could adjust the period of the clock in opposite directions. A similar genetic screen later carried out in mice identified a mutation named ***Clock*** (for <u>c</u>ircadian <u>l</u>ocomotor <u>o</u>utput <u>c</u>ycles <u>k</u>aput), which slowed down the circadian rhythm in heterozygous mice, and disrupted the rhythm in homozygotes (Figure 8–45B).

Subsequent studies have shown that the protein products of the *Period* and *Clock* genes are key components of an auto-inhibitory transcriptional network conserved from flies to mammals. CLOCK is a DNA-binding transcription factor that together with its partner (CYC in fly and BMAL1 in mouse, which are orthologs) binds to the promoters of *Period* and many other rhythmically expressed genes to activate their transcription. Among the resulting products are the fly protein PER (encoded by the *Period* gene) and its binding partner TIM (encoded by the ***Timeless*** gene), and the two mouse PER proteins (encoded by two genes, *Per1* and *Per2*) and their binding partners, the CRY proteins (for **cryptochrome**;

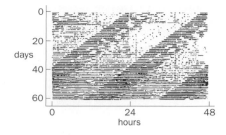

Figure 8–44 Circadian rhythm of running wheel utilization by a mouse. Each horizontal line is a 2-day record of running wheel utilization (vertical ticks) by a mouse in constant darkness. These records are double plotted for easier visualization; the record for day *n* + 1 is both to the right of day *n* on the same horizontal line and immediately below day *n* on the next horizontal line. Despite being kept in constant darkness for 60 days, the mouse's wheel utilization exhibits a remarkably consistent circadian rhythm with a period slightly shorter than 24 h, as seen by the leftward shift of the daily onset of running wheel utilization. (Adapted from Pittendrigh CS [1993] *Annu Rev Physiol* 55:17–54. With permission from Annual Reviews.)

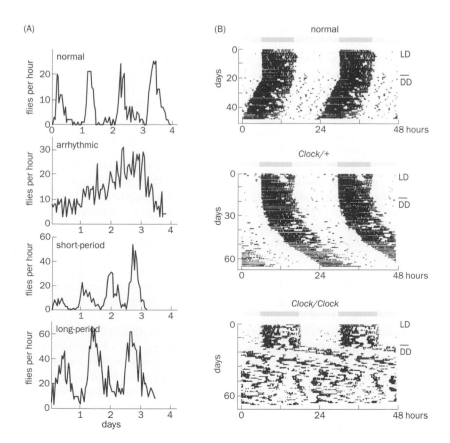

Figure 8–45 Circadian rhythm mutants in *Drosophila* and mice. (A) Eclosion rhythms of populations of normal and mutant flies. Normal adult flies eclose (emerge from their pupal case) in the morning. After being exposed to a 12 h light–12 h dark cycle during development and then kept in constant darkness, the eclosion of normal adult flies still occurred during the subjective morning, with a ~24 h rhythm. Arrhythmic mutant flies lost the circadian clock altogether, whereas the short- and long-period mutant flies had rhythms of ~19 h and ~28 h, respectively. All three mutants disrupt the same gene, *Period*. **(B)** Rhythms of running wheel utilization by mice as described in Figure 8–44. During the first 20 days, mice were kept in a 14 h light–10 h dark cycle (LD, indicated by the alternating light and dark bars above). They were then switched to constant darkness (DD). During LD, the rhythms were entrained by light, and the mice utilized the running wheel during the night. Upon shifting to DD, the period for the normal mouse was 23.7 h (the onset of running wheel activity shifted leftward). By contrast, the *Clock/+* heterozygous mutant had a period of 24.8 h (activity onset shifted rightward), and the *Clock/Clock* homozygous mutant had a period of 27.1 h during the first 10 days in DD, and lost the circadian rhythm afterwards. (A, adapted from Konopka RJ & Benzer S [1971] *Proc Natl Acad Sci USA* 68:2112; B, adapted from Vitaterna MH, King DP, Chang A, et al. [1994] *Science* 264:719.)

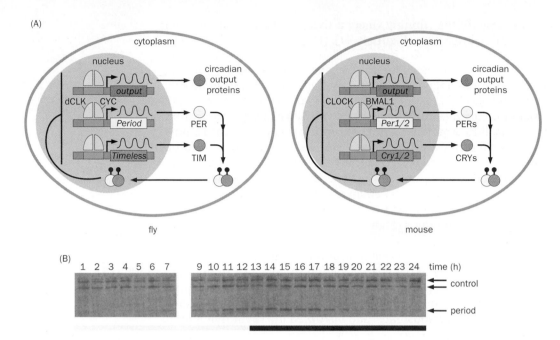

Figure 8–46 Auto-inhibitory transcriptional feedback in the central clock of *Drosophila* and mice. (A) The central clock mechanisms are strikingly similar in the fly and mouse. A pair of transcription activators—dCLK (*Drosophila* CLOCK) and CYC in flies, CLOCK and BMAL1 in mice—activates transcription of *Period* and *Timeless* in flies or two *Per* and two *Cry* genes in mice. The resulting protein products form a complex, become phosphorylated (black dots attached to the proteins), and enter into the nucleus to negatively regulate their own transcription. Many other genes are similarly regulated in a circadian fashion, which accounts for circadian output. **(B)** *Per* mRNA levels cycle in a circadian fashion, with a peak at dusk and a trough at dawn. The levels of two control mRNA species remain constant. Populations of flies were maintained in a 12 h light (1–12) and 12 h dark (13–24) cycle as indicated at the bottom. Heads of flies were harvested hourly for mRNA extraction. (A, adapted from Mohawk JA, Green CB, & Takahashi JS [2012] *Ann Rev Neurosci* 35:445; B, adapted from Hardin PE, Hall JC & Rosbash M [1990] *Nature* 343:536–540. With permission from Macmillan Publishers Ltd.)

encoded by two *Cry* genes). Heterodimeric complexes of PER/TIM (fly) and PER/CRY (mouse) enter the nucleus and negatively regulate the activity of CLOCK, thereby forming negative feedback loops that down-regulate their own production (**Figure 8–46**A). An initial insight leading to this model came from the finding that *Period* mRNA levels cycled in a circadian fashion (Figure 8–46B), suggesting that PER may negatively regulate its own transcription. Specifically, when the level of PER protein is high, PER inhibits its own transcription, so that the *Period* mRNA level declines. This leads to less PER protein being made, which then relieves the negative regulation of *Period* transcription. This in turn causes more *Period* mRNA to be produced, leading again to a high level of PER protein, and the cycle continues (see **Movie 8–6**).

Research in recent decades has revealed details of the circadian clock and molecular basis of its regulation. In addition to the negative transcriptional feedback loop for regulating the production of PER and its partners (TIM or CRY proteins), other regulatory mechanisms control the phosphorylation, nuclear entry, and degradation of these circadian rhythm regulators. Another transcriptional feedback loop produces a circadian rhythm in the expression of the CLOCK partner BMAL1 in mice, adding robustness to the network. One of the most striking findings is that the circadian regulation of gene expression is not restricted to neurons in the brain but instead is found in nearly all cells in peripheral tissues, with a large fraction of all transcripts in flies and mammals being regulated in a circadian cycle. For example, under the right conditions, mammalian fibroblasts in culture can maintain circadian gene expression that utilizes the same molecules and mechanisms found in the neurons that control the animal's circadian behavior. Indeed, circadian cycles of gene expression and auto-inhibitory transcriptional networks are also essential to circadian rhythm regulation in plants and unicellular fungi. Although the proteins utilized in these organisms differ from PER and CLOCK, the auto-inhibitory transcriptional feedback loop appears to be a universal strategy for diverse eukaryotic organisms to regulate circadian rhythms.

8.20 Entrainment in flies is accomplished by light-induced degradation of circadian rhythm regulators

A universal feature of circadian rhythm is that its phase can be reset by light, a process called **entrainment**. (Phase refers to specific time during the 24-hour period a given physiological, behavioral, or biochemical activity reaches a particular stage of the cycle, such as the peak or trough.) This is how we can recover from jetlag after being in a new time zone for a few days. For a typical 12-hour day–12-hour night cycle, light exposure during the first half of the night shifts the phase of the circadian rhythm to an earlier time (phase delay), whereas light exposure during the second half of the night shifts the phase to a later time (phase advance). How does light interact with the circadian clock to reset the phase? And given that many cells throughout the body have their own clocks, how are different clocks synchronized?

Unlike the conserved central clockwork mechanisms, light entrainment mechanisms differ considerably in flies and mice. Because of the thin insect cuticle, light can penetrate directly into most cells including dedicated clock neurons in the brain. The photoreceptor protein is cryptochrome (CRY), an ancient blue-light sensor. (The mammalian homologs of fly CRY do not function as photoreceptors, but curiously function as an integral part of the core negative feedback loop as seen in Figure 8–46A.) In response to light, fly CRY forms a complex with and induces the rapid degradation of the PER binding partner, TIM. Thus, light removes the negative regulatory PER/TIM complex. In the first half of the night, before a critical amount of the complex has enter the nucleus to down-regulate its own transcription, degradation of the complex means that more time is needed to build up PER/TIM proteins from the existing mRNAs. This resets the clock back to the beginning of the night, causing a phase delay. By the second half of the night, the PER/TIM complex has accumulated in the nucleus and down-regulated transcription of *Period* and *Timeless*, such that little mRNA is left. A light pulse given at this time also degrades PER/TIM proteins, but in this case mimics the natural light degradation of the PER/TIM complex that would normally occur at the beginning of the day, which initiates a new transcription and circadian cycle. Thus, a light pulse given during the second half of the night resets the clock forward to dawn, causing a phase advance (**Figure 8–47**).

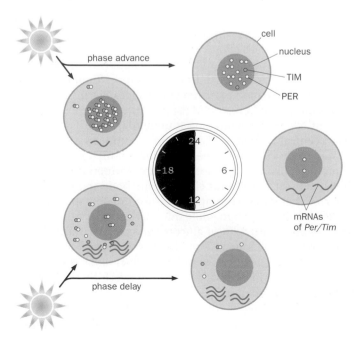

Figure 8–47 A model for light entrainment in flies. Light triggers TIM and PER degradation and thereby relieves the inhibition of *Per*/*Tim* mRNA production. Thus, following a full day of accumulation, levels of *Per* and *Tim* mRNA peak in the early evening. In the middle of the night, PER/TIM proteins enter the nucleus and inhibit *Per*/*Tim* transcription. A light pulse given before PER/TIM nuclear entry degrades PER/TIM proteins while *Per*/*Tim* mRNA levels remain high, thus mimicking the state of dusk. A light pulse given after PER/TIM nuclear entry degrades proteins while mRNA levels are already low, mimicking the state of dawn. (Adapted from Hunter-Ensor M, Ousley A & Sehgal A [1996] *Cell* 84:677–685. With permission from Elsevier Inc.)

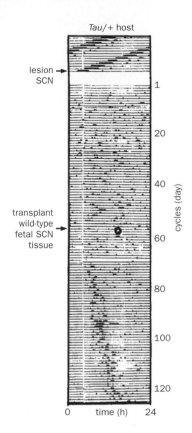

Tau/+ host

lesion → SCN

transplant wild-type fetal SCN tissue →

cycles (day)

time (h)

Figure 8–48 The genotype of suprachiasmatic nucleus (SCN) determines the behavioral phenotype of mosaic animals. In this example, a golden hamster heterozygous for the *Tau* mutation served as a host for SCN transplantation after lesion. All experiments were recorded in constant darkness after the animal had established circadian rhythm assayed by running wheel utilization. Before the SCN lesion, it had a period of 21.7 h (top rows). After the lesion, the behavioral rhythm was abolished. After transplantation of fetal SCN tissue from wild-type golden hamster (circle indicates time of transplantation), not only was the rhythm restored, it also had a wild-type period of ~24 h. (From Ralph MR, Foster RG, Davis FC et al. [1990] *Science* 247:975–978. With permission from AAAS.)

8.21 Pacemaker neurons in the mammalian suprachiasmatic nucleus integrate input and coordinate output

In mammals, even though many cells have their own clocks, most are not accessible to light. The **suprachiasmatic nucleus (SCN)** of the hypothalamus (see Figure 8–34A) is a master regulator of both circadian rhythms and light entrainment. SCN lesions disrupt behavioral rhythmicity, which can be restored by transplanted fetal SCN tissues. The following experiment established that the SCN occupies a position at the top of circadian hierarchy in mammals. A spontaneous mutation (named *Tau*) was found in golden hamsters that shortened the circadian period to ~22 hours when heterozygous and to ~20 hours when homozygous. When fetal SCN tissues from *Tau* mutant were transplanted to SCN-lesioned wild-type hosts, the restored rhythm also had a shortened period. By contrast, a normal circadian period was produced in mutant hosts that had wild-type fetal SCN tissues transplanted after lesion (**Figure 8–48**). Thus, the genotype of the SCN determines the circadian phenotype of the mosaic animals, highlighting a central role of the SCN in the control of rhythmic behavior.

The SCN receives light input from intrinsically photosensitive retinal ganglion cells (ipRGCs) in the retina (see Box 4–2). Light directly activates ipRGCs, which release glutamate at their axon terminals on SCN neurons. Glutamate receptor activation triggers Ca^{2+} entry, kinase activation, and phosphorylation of the transcription factor CREB (see Section 3.23) in the SCN neurons. Phosphorylated CREB binds to the promoters of *Per1* and *Per2* and activates their transcription. Thus, transcriptional regulation of clock genes appears to play a major role in mammalian light-induced phase resetting, whereas the connection between light and clock proteins is predominantly post-transcriptional in the fly.

While a direct connection to the retina affords the SCN a privileged position to interact with the environment for phase resetting, the following properties contribute to the SCN being the master regulator of circadian rhythms. (1) The electrical activity of SCN neurons oscillates in a circadian fashion even in isolation, qualifying them as **circadian pacemaker neurons** (analogous to the pacemaker neurons in the central pattern generators discussed in Section 8.5). This is in part due to circadian regulation of K^+ channel activity that determines the resting membrane potential of SCN neurons. (2) SCN neurons form a close network with other SCN neurons through gap junctions and neuropeptide signaling, such that SCN neurons are highly synchronized in their circadian electrical activity and gene expression. This network property has been observed in electrophysiological recordings and by imaging the expression of bioluminescent circadian reporter generated in transgenic mice. Whereas individually cultured SCN neurons exhibited circadian cycles with different phases and periods, neurons in SCN slices that had preserved local connections displayed robust synchronization of circadian period and phase between cells (**Figure 8–49**A). (3) SCN neurons secrete signaling molecules rhythmically and make extensive synaptic connections with many hypothalamic nuclei, such that SCN neurons can impose their coherent rhythmic activity onto these targets, which in turn regulate clocks in the peripheral tissues as well as the physiology and behavior of the organism (Figure 8–49B).

For example, SCN neurons send output to the periventricular hypothalamus nucleus (PVH), a central regulator of autonomic output (see Figure 8–43). SCN neurons also send output to the temperature control nucleus of the hypothalamus and regulate organismal physiology through temperature cycles, a powerful entrainment agent for peripheral clocks. (Our body temperature fluctuates about 0.5°C daily, with the lowest temperature around 4:30 a.m. and the highest around 7 p.m.) SCN neurons also regulate the activity of hypothalamic CRH neurons, which control the production of glucocorticoids (see Table 8–1) in a circadian fashion. Peripheral clocks, in addition to receiving signals from the SCN, can also be entrained by other cues. For example, the liver clock is entrained by food intake; in SCN-lesioned animals, food can be a powerful entrainment signal to regulate many aspects of physiology.

In summary, circadian regulation of physiology, biochemistry, and behavior of multicellular organisms—including mammals—reflects a combination of

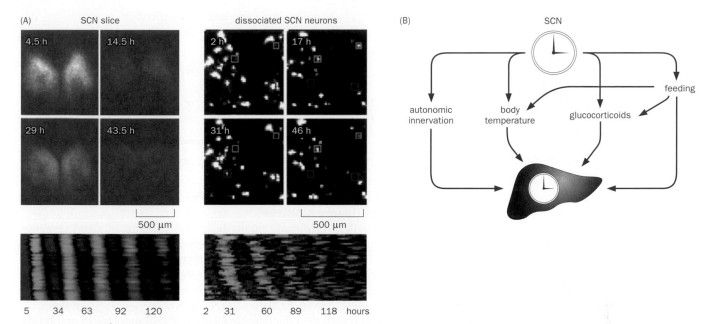

Figure 8–49 Local connections and global actions of suprachiasmatic nucleus (SCN) neurons. (A) Top, bioluminescence images of *in vitro* cultured SCN slices or dissociated SCN neurons from mice in which the firefly luciferase was knocked-in at the *Per2* locus; the production of light by luciferase in the presence of its substrates reports the level of endogenous cyclic expression of *Per2*. Images were taken at different times after culture preparation, as indicated in the upper-left corner of each panel. SCN neurons from a slice (left) exhibit synchronized activity; most cells have high *Per2* transcriptional activity at 4.5 h and 29 h after the onset of slice culture and low activity at 14.5 h and 43.5 h. In cultures of dissociated SCN neurons (right), local connections between cells are severed. Individual cells in these cultures still exhibit cycles of high and low expression but do not oscillate in the same phase

(for example, compare the cells within red versus green squares). Bottom, raster plots tracking the bioluminescence of 40 SCN cells over 5 days (each cell is represented by one horizontal line, with values above or below the average in red and green, respectively), showing synchronized rhythms in the SCN slice but much less so in dissociated neurons. **(B)** Summary of pathways from the SCN to peripheral clocks (using the liver clock as an example). Through secreted signals and neuronal projections, SCN neurons send rhythmic output to many hypothalamic nuclei that control autonomic innervation, body temperature, feeding, and hormone secretion, all of which influence peripheral clocks. (A, from Liu AC, Welsh DK, Ko CH et al. [2007] *Cell* 129:605–616. With permission from Elsevier Inc.; B, adapted from Mohawk JA, Green CB, & Takahashi JS [2012] *Ann Rev Neurosci* 35:445.)

cell-autonomous auto-inhibitory transcriptional regulation and extensive communication that links SCN pacemaker neurons, other hypothalamic nuclei, and peripheral tissues. A major contribution to our current understanding originated from the identification of genes that control circadian rhythms in fruit flies and mice and follow-up studies of their mechanisms of action. Indeed, mutations in the human homologs of the fly *Period* gene and its regulators are responsible for **advanced sleep phase syndrome**—a disorder characterized by very early morning waking and an early evening sleep onset. These findings highlight how studying basic neurobiological problems in invertebrate model organisms can have profound relevance to human health.

8.22 Sleep is widespread in the animal kingdom and exhibits characteristic electroencephalogram patterns in mammals

We all know what sleep is in humans and are familiar with sleep in other mammals. But sleeplike phenomena are widespread in the animal kingdom. Researchers have agreed upon a set of behavioral criteria that define the sleep state: sleep is rapidly reversible (as opposed to coma); it is associated with decreased responsiveness to sensory stimuli; it is subjected to homeostatic regulation—sleep deprivation results in more sleep afterwards; it is also timed by the circadian system, which governs a more rigid pattern of daily alertness and sleepiness, including nocturnal and diurnal patterns. Using these behavioral criteria, the concept of sleep has been extended to organisms such as zebrafish and fruit flies. For example, fruit flies exhibit bouts of inactivity and characteristic postures that occur mostly at night. During this sleeplike state, flies are less easily aroused by mechanical stimulation. This sleeplike state is also homeostatically regulated; application of strong stimuli that prevented flies from sleeping at night

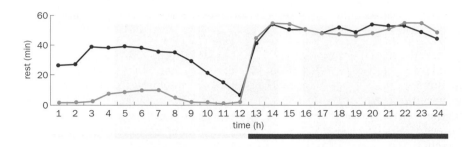

Figure 8–50 Rest in *Drosophila* is timed by circadian rhythm and regulated by homeostasis. During a typical circadian cycle, normal fruit flies rest mostly at night (blue trace). Minutes of rest (*y* axis) are plotted for each assayed hour (*x* axis; 1–12 h = lights on, 13–24 h = lights off). When rest was disrupted during the previous night (not shown), marked rebound of rest occurred during the next day (red trace). (Adapted from Shaw PJ, Cirelli C, Green RJ, et al. [2000] *Science* 287:1834.)

can induce a robust sleep rebound (that is, an increase of sleep after deprivation) during the next day (**Figure 8–50**).

In mammals, which have traditionally been the subjects of sleep research, sleep and sleep stages can be defined using **electroencephalography** (**EEG**). EEG was first discovered in the 1920s. It reports the collective activities of many neurons recorded by electrodes placed on the surface of the scalp (see Figure 11-47). During the waking period, EEG patterns are characterized by oscillations with high frequency and low amplitude. As we descend into deeper stages of sleep, the oscillations measured with EEG decrease in frequency and increase in amplitude, until we reach the deepest stage—slow wave sleep, when low-frequency, large-amplitude oscillations are most evident (**Figure 8–51**A). In the 1950s, it was discovered that a certain stage of sleep, termed **REM sleep**, is characterized by rapid eye movements and complete muscle relaxation. The rest of sleep is accordingly termed non-REM sleep, or **NREM sleep**. REM sleep has been used to define sleep cycles (Figure 8–51B). When we fall asleep, we first enter light NREM sleep and descend to deep NREM sleep. We then ascend to lighter stages of NREM sleep before transitioning to REM sleep. We subsequently descend back to deep NREM sleep, and repeat this cycle several times before waking up. On average each cycle in humans lasts for about 90 minutes (Figure 8–51B). REM sleep is also known to be associated with having vivid dreams. Adult humans typically spend 25% of sleep time in REM sleep, but REM sleep occupies up to 80% of infants' total sleep time.

The synchronized oscillatory EEG patterns during NREM sleep are contributed both by activities produced by local recurrent connections among cortical neurons, and by long-range input from the thalamus. Indeed, the thalamic nuclei contain the best-characterized pacemaker cells in mammals, which can produce rhythmic firing patterns in the absence of input because of their intrinsic ion channel properties, similar to the pacemaker cells in the crustacean stomatogastric ganglion (see Section 8.5). During the waking period and during REM sleep, inputs from sensory and neuromodulatory systems alter the firing pattern from the synchronous oscillatory to high-frequency asynchronous firing. In the next section, we discuss how neuromodulatory systems regulate the sleep–wake cycle.

Figure 8–51 EEG patterns and sleep cycles in human. (A) Representative EEG patterns for different phases of sleep. From wakefulness to stage 4 sleep, there is a progressive increase in amplitude and reduction in frequency in the EEG patterns. **(B)** Recording of a human subject during a 7-hour period of normal sleep. Sleep stages are shown on the left. Blue bars represent periods of REM sleep. Arrows point to the end of a sleep cycle (the end of the REM sleep) and the beginning of the new sleep cycle. Vertical bars below represent body movements, with tall bars representing major body movements. Note that sleep researchers have recently combined stages 3 and 4 as a new stage 3 (slow wave sleep). (Adapted from Dement W & Kleitman N [1957] *Electroenceph Clin Neurophysiol* 9:673–690. With permission from Elsevier Inc.)

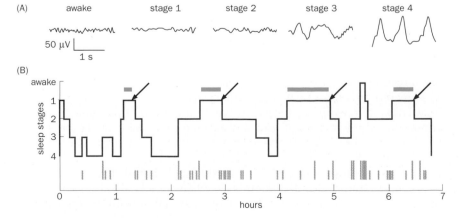

8.23 The mammalian sleep–wake cycle is regulated by multiple neurotransmitter and neuropeptide systems

What is the neural basis of sleep regulation? This question has so far been investigated primarily in mammals, in which both behavior and EEG patterns can be used to monitor the sleep state. Lesion and electrical stimulation studies have identified an **ascending arousal system** from the brainstem to the forebrain that is essential for maintaining wakefulness (**Figure 8–52**). The ascending arousal system consists of several parallel streams, each using a different neurotransmitter. One stream is the cholinergic neurons in the tegmental nuclei that project to the thalamus and the basal forebrain cholinergic neurons. These tegmental and basal forebrain cholinergic neurons are most active during wakefulness and REM sleep. The activity of these neurons promotes the firing of thalamic and cortical neurons. Additional streams in the ascending arousal system originate from groups of neurons that use monoamines as neurotransmitters and that project directly to the cerebral cortex, hippocampus, and basal forebrain (**Box 8–1**). These streams include the norepinephrine neurons in the **locus coeruleus**, the serotonin neurons in the **raphe nuclei**, the dopamine neurons near the dorsal raphe nucleus, and the histamine neurons in the **tuberomammillary nucleus** of the hypothalamus. These neurons are all active while animals are awake and decrease their activity during both non-REM and REM sleep. Yet another stream involves neurons that express the neuropeptide hypocretin, which will be discussed in greater detail below.

A second system important for regulating the sleep–wake cycle consists of the sleep-active neurons. These neurons are by definition most active while animals are asleep. The best studied are the GABAergic neurons located in the preoptic area of the hypothalamus (Figure 8–52; see also Figure 8–34A). Lesions of these neurons, in particular those located in the ventrolateral preoptic area (VLPO), caused a more than 50% reduction in NREM and REM sleep. Hence they are not only sleep-active neurons but also sleep-promoting neurons.

Interestingly, neurons in the VLPO and nearby medial preoptic area (MPOA) project axons to the tegmental cholinergic neurons and all classes of the monoamine neurons in the ascending arousal system, and inhibit their activity by releasing GABA. At the same time, VLPO/MPOA neurons receive input from many

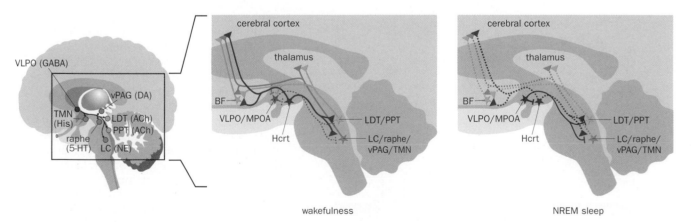

wakefulness   NREM sleep

Figure 8–52 Neural systems that maintain states of wakefulness and sleep. Schematic of neurons that are active during wakefulness (middle) and sleep (right) in a mid-sagittal view of the human brain, magnified from the box at left. Middle, during wakefulness, cholinergic (ACh) neurons in the laterodorsal and pedunculopontine tegmental nuclei (LDT and PPT), norepinephrine neurons (NE) in the locus coeruleus (LC), dopamine neurons (DA) in the ventral periaqueductal gray (vPAG), serotonin neurons (5-HT) in the raphe nuclei, and histamine (His) neurons in the tuberomammillary nucleus (TMN) are all active. These neurons relay their activity to the cerebral cortex—either directly or indirectly via basal forebrain (BF) cholinergic neurons—and to the thalamus; they also inhibit GABAergic neurons in the ventrolateral and medial preoptic areas (VLPO and MPOA) in the hypothalamus. Hypothalamic hypocretin (Hcrt) neurons are likewise active and send excitatory output to many neurons including the wake-promoting cholinergic and monoamine neurons. Right, during NREM sleep, VLPO/MPOA GABAergic neurons are active, and inhibit tegmental cholinergic neurons, LC/DR/TMN monoamine neurons, and Hcrt neurons. In REM sleep (not shown), the activity states of these neurons are similar to NREM except that cholinergic neurons are active. In these schemes, active neurons are represented by solid projections and inactive neurons are represented by dashed projections. These modulatory neurons can excite or inhibit their target neurons depending on the receptors; the symbol ⊣ is used to denote known inhibition. For simplicity, TMN neurons are grouped here with LC/raphe neurons in the brainstem but are actually located in the hypothalamus. (Adapted from Saper CB, Fuller PM, Pederson NP et al. [2010] *Neuron* 68:1023–1042. With permission from Elsevier Inc.)

Box 8–1: Neuromodulatory systems

Whereas neurotransmitters such as glutamate or GABA can elicit rapid excitation or inhibition of postsynaptic neurons by activating ionotropic receptors, modulatory neurotransmitters, also termed **neuromodulators**, act on a slower timescale. The classic neuromodulators include monoamines (dopamine, serotonin, norepinephrine, and histamine), acetylcholine (when it acts on muscarinic receptors), and numerous neuropeptides (see Section 3.11). In this chapter, we have seen many examples of these neuromodulators, such as the regulation of striatal output by dopamine; regulation of eating by α-MSH, NPY, AgRP, and CCK; and the regulation of sleep by hypocretin, acetylcholine, and monoamines. With few exceptions, these modulators act on G-protein-coupled receptors expressed by their target neurons and therefore have slower and longer-lasting effects compared to the more rapid actions of ionotropic glutamate and GABA receptors (see Table 3–3).

Neuromodulators can have diverse effects on their targets depending on the location and downstream signaling events of their receptors; **Figure 8–53** uses the example of dopamine's action as an illustration. Acting on dopamine receptors at the presynaptic terminal of a pair of synaptically connected neurons, dopamine can alter neurotransmitter release probability by regulating presynaptic membrane

potential, Ca^{2+} entry, or neurotransmitter release machinery. Acting on dopamine receptors at the postsynaptic membrane, dopamine can alter transmitter detection by regulating post-translational modification and trafficking of neurotransmitter receptors. Dopamine can also alter how synaptic potentials are integrated in postsynaptic neurons, as well as the threshold and waveform of action potentials, by modulating ion channels that control resting potential and membrane properties. All of these effects can be bidirectional depending on the specific dopamine receptors these target neurons express; for example, D_1 receptors are coupled to G_s and therefore their activation increases cAMP concentration, whereas D_2 receptors are coupled to G_i and their activation decreases cAMP concentration. Finally, both the presynaptic and postsynaptic partners of the synapse that dopamine modulates (Figure 8–54) can be excitatory, inhibitory, or other modulatory neurons. Thus, dopamine and neuromodulators in general can have diverse effects on the functions of their target neurons and circuits that these neurons serve.

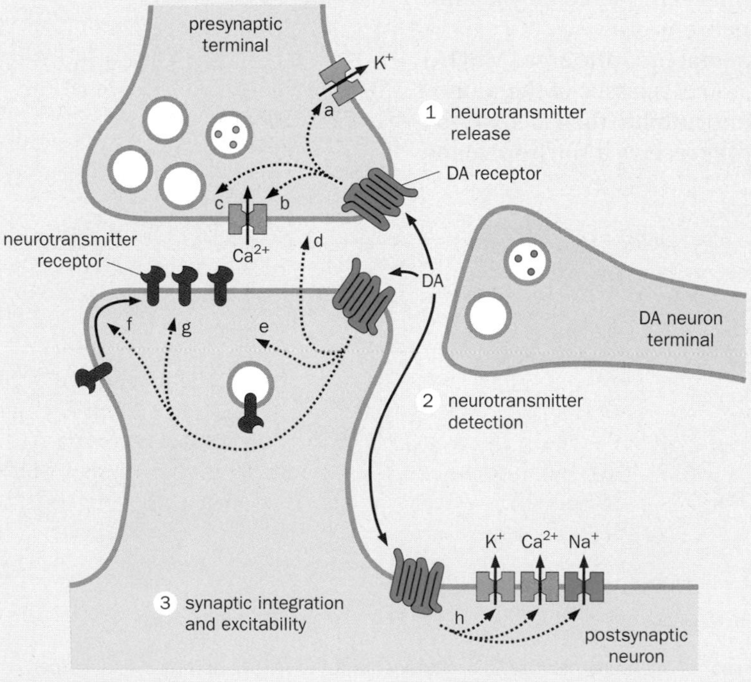

A striking feature of neuromodulatory systems is their broad reach. We highlight here four neuromodulatory systems that regulate forebrain function: the norepinephrine, serotonin, dopamine, and acetylcholine systems. Although the cell bodies of these neurons are clustered in discrete nuclei in the brainstem and basal forebrain, their axons elaborate through much of the forebrain (**Figure 8–54**). For example, an estimated 3000 norepinephrine neurons clustered in the locus coeruleus in the rat brain give rise to axons that innervate throughout the neocortex, olfactory bulb, olfactory cortex, hippocampus, amygdala, and thalamus (Figure 8–55A). Likewise, serotonin neurons from the raphe nuclei project diffusely throughout the forebrain (Figure 8–55B), acting on up to 18 different receptors (see Table 3–3) that are differentially expressed in different target neurons. (In addition to forebrain targets, both norepinephrine neurons from the locus coeruleus and serotonin neurons from the raphe nuclei also send extensive projections that innervate broadly the brainstem, cerebellum, and spinal cord.) Midbrain dopamine neurons have relatively more focused projections compared to the norepinephrine and serotonin systems, but still project widely across the striatum, olfactory cortex, and prefrontal cortex (Figure 8–55C). Acetylcholine-producing neurons in the basal forebrain also project broadly across the neocortex, hippocampus, amygdala, and olfactory bulb (Figure 8–55D); while acetylcholine also signals through nicotinic receptors which are ionotropic, at least part of its action is mediated by muscarinic receptors and is therefore modulatory in nature.

Figure 8–53 Diverse effects of dopamine on its target neurons. Dopamine (DA) can exert its modulatory effect at multiple sites of its target neurons. It can affect (1) neurotransmitter release by modulating presynaptic K^+ channels (a), Ca^{2+} channels (b), or release machinery (c), or regulating the production of retrograde messengers (d; will be discussed in Chapter 10). It can affect (2) postsynaptic neurotransmitter detection by regulating insertion (e), recruitment (f), or properties (g) of neurotransmitter receptors. It can also affect (3) synaptic integration and excitability of postsynaptic neurons by modulating voltage-gated K^+, Na^+, and Ca^{2+} channels (h). (Adapted from Tritsch NX & Sabatini BL [2012] *Neuron* 76:33–50. With permission from Elsevier Inc.)

We are only at the beginning of understanding the diverse functions of these neuromodulatory systems and their mechanisms of action. We have

Box 8–1: Neuromodulatory systems

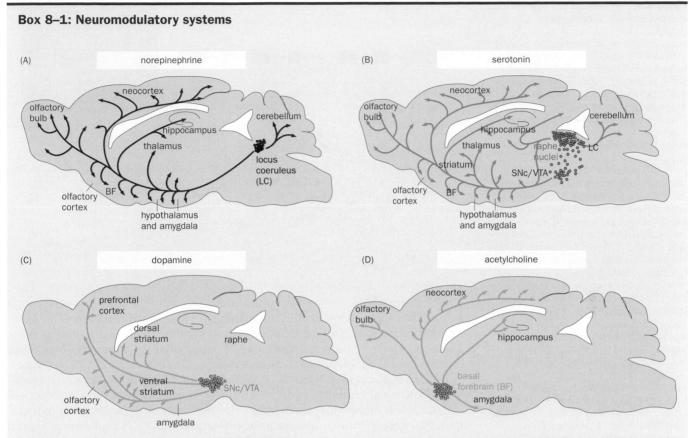

Figure 8–54 Neuromodulatory neurons in the brainstem and basal forebrain project to sites throughout the forebrain.
(A) Norepinephrine neurons in the locus coeruleus (LC) project their axons diffusely across most areas of the forebrain (except the striatum) and the cerebellum. **(B)** Serotonin neurons from the raphe nuclei likewise project extensively to almost the entire forebrain and cerebellum. **(C)** Dopamine neurons from the ventral tegmental area (VTA) and substantia nigra pars compacta (SNc) project widely to the striatum, olfactory cortex, and prefrontal cortex. **(D)** Acetylcholine-producing neurons in the basal forebrain (BF) project broadly to the neocortex, hippocampus, amygdala, and olfactory bulb. Note that only the major target sites of the forebrain and cerebellum projections are depicted. Although LC, raphe, SNc/VTA, and BF respectively contain the largest clusters of norepinephrine, serotonin, dopamine, and acetylcholine neurons in the brain, neurons that utilize these transmitters are also present in smaller clusters in other parts of the brain (not shown). (Courtesy of Brandon Weissbourd, Lindsay Schwarz, and Kevin Beier. The drawings are sagittal views of the mouse brain based on axon projection data from the Allen Brain Atlas (http://www.brain-map.org/)

discussed the function of the dopamine system in the context of modulating movement (see Figure 8–22) and eating (see Figure 8–43). We've also discussed the function of norepinephrine, serotonin, dopamine, and acetylcholine neurons in regulating the sleep–wake cycle (see Figure 8–52).

These are just the tip of the iceberg. Their importance in human health is highlighted by the fact that most drugs that are used currently to treat psychiatric disorders interfere with the function of these modulatory systems, as will be discussed in more detail in Chapter 11.

neurons in the arousal system and are inhibited by acetylcholine, norepinephrine, dopamine, and serotonin. Thus, the arousal and sleep-promoting systems form a mutually inhibitory circuit to maintain the stability of each state, and to facilitate rapid and complete transitions between states. This mutually inhibitory circuit motif resembles that of the central pattern generators controlling locomotion rhythms discussed earlier in this chapter (see Figures 8–13 and 8–15). The logic of these systems is similar: animals activate flexor or extensor muscles in alternation but not at the same time; likewise, animals are either awake or asleep but not both at once.

An important insight into sleep regulation came from studying a sleep disorder called **narcolepsy**. Narcoleptic patients have trouble staying awake during the day, especially after being happy or excited. Narcoleptic patients can also switch from being awake directly into REM sleep without going through

Figure 8–55 Hypocretin and hypocretin neurons promote wakefulness.
(A) Recessive mutations identified in narcoleptic dogs (pictured on the right) map to the hypocretin receptor gene. The narcoleptic mutation in the Doberman strain is caused by an insertion in the 3rd intron (long red bar) that results in the skipping of exon 4 during splicing. The narcoleptic mutation in the Labrador strain occurs in intron 6 near the 5′ splice site (short red bar), causing exon 6 to be skipped. Both mutations produce nonfunctional receptor proteins. **(B)** Hypocretin-expressing (Hcrt) neurons (purple dots) are clustered in a small area of the lateral hypothalamus, but their axons (arrows) project to many parts of the brain. **(C)** Channelrhodopsin (ChR2, blue trace) or a control fluorescent protein (mCherry, red trace) were expressed in hypocretin-expressing neurons in the lateral hypothalamus via viral targeting of Cre-dependent expression vector into *Hcrt-Cre* transgenic mice. Photostimulation of ChR2-expressing-hypocretin neurons increases the probability of transition to wakefulness from REM sleep (as seen by the leftward shift of the probability distribution curve compared to control) and from non-REM sleep (not shown). (A, adapted from Lin L, Faraco J, Li R et al. [1999] *Cell* 98:365–376. With permission from Elsevier Inc.; B, adapted from Peyron C, Tighe DK, van den Pol AN, et al. [1998] *J Neurosci* 18:9996; C, adapted from Adamantidis AR, Zhang F, Aravanis AM et al. [2007] *Nature* 450:420–424. With permission from Macmillan Publishers Ltd.)

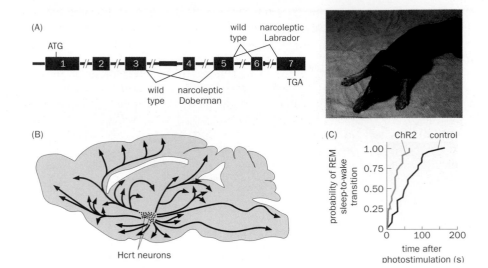

NREM sleep, which does not occur normally (see Figure 8–51). A breakthrough in understanding the narcolepsy mechanism came from the identification of mutations that give rise to narcolepsy-like symptoms in certain breeds of dogs. Positional cloning, the same method used to identify the *Obese* and *Clock* genes (see Section 13.6), identified in narcoleptic dogs mutations that disrupt the normal splicing of a gene (**Figure 8–55**A) that encodes a G-protein-coupled receptor (GPCR) for neuropeptides derived from **hypocretin**. Hypocretin is also called **orexin**; it was independently identified from biochemical purification of hypothalamic neuropeptides that activate orphan GPCRs (GPCRs without known ligands), and found to stimulate rats' food intake when administered through the ventricle to the brain. Indeed, about the same time narcoleptic mutations in dogs were mapped to the hypocretin receptor, orexin-knockout mice were found to exhibit phenotypes resembling narcolepsy. Thus, hypocretin/orexin has dual functions in regulating sleep and food intake. Loss of hypocretin neurons and, in rare cases, mutations in the hypocretin gene have since been found to be causes of human narcolepsy.

Hypocretin-producing neurons are located exclusively in the lateral hypothalamus, but their axons project widely in the brain, including to acetylcholine-, norepinephrine-, serotonin-, and histamine-producing neurons in the ascending arousal pathway (Figure 8–53B). *In vivo* recording indicated that hypocretin neurons are most active when animals explore the environment during the waking period and stop firing during sleep. Furthermore, optogenetic activation of hypocretin-producing neurons in mice increased the probability that animals wake up from NREM or REM sleep (Figure 8–53C). Thus, the hypocretin system promotes wakefulness, likely through its activation of the ascending arousal system (Figure 8–52) as well as its direct action on target neurons throughout the brain (Figure 8–53B).

We only have a coarse outline of the neural systems that regulate sleep, and many questions remain to be answered by future investigations. How do different sleep-regulating neurons act? How are their activities modulated by the circadian rhythm? What are the molecular and neural bases of homeostatic regulation (sleep rebound)? And how are REM and non-REM sleep respectively regulated?

8.24 Why do we sleep?

Perhaps the most enigmatic question regarding sleep is why we need it. As introduced earlier, sleep is ubiquitous in mammals, birds, and reptiles, including many prey animals that risk their lives to sleep. Indeed, some marine mammals and birds have developed uni-hemispheric sleep, a form of sleep in which half of the brain undergoes slow wave sleep while the other half remains awake for the purpose of evading predators or maintaining flying or swimming during migration.

Sleeplike states also extend to invertebrates (see Figure 8–50). "If sleep does not serve an absolutely vital function, then it is the biggest mistake evolutionary process has ever made," remarked Allan Rechtschaffen, a sleep researcher who showed the importance of sleep in the following deprivation study.

To separate the effects of sleep disruption from the effects of physical perturbation required to disrupt sleep, a pair of rats were housed on a shared disk over two water pans, and their EEGs were recorded continuously. Whenever the experimental rat was about to fall asleep, the change in its EEG pattern would trigger the disk to spin such that both rats had to move in order to stay on the disk. (Rats do not sleep in water.) In this experimental paradigm, both rats received identical physical stimuli, but the control rat could take naps when the experimental rat's EEG pattern indicated wakefulness (**Figure 8–56**). When totally deprived of sleep, experimental (but not control) rats inevitably died within a few weeks. Thus, this experiment demonstrated that sleep is essential for life; but exactly why this is so was not clear, since sleep deprivation induced a host of physiological changes, all of which could contribute to its lethality. These included increased stress level, increased energy expenditure, weight loss, compromised thermal regulation, weakened immune system, and loss of gut integrity. It is difficult to pinpoint the primary cause of death and to separate causes and effects when so many parameters are changing at the same time.

Many interesting ideas have been proposed to explain why we sleep. One proposal is that animals may have evolved to perform at high levels during limited periods within the daily cycle, for instance, at times when food is most readily available; rest during other times might function to conserve energy. Indeed, many examples in nature can be found in which animals adapt their sleep patterns according to their ecological niche: as previously noted, fruit flies sleep during the night because they can best seek food in the daytime, while mice have adapted to nocturnal life and sleep during the day to avoid daytime predators. Another proposal is that sleep is necessary for restoration of key cellular components—such as machinery for protein synthesis, protein folding, and synthesis of lipids and membrane—that are used up during the waking period. Indeed, changes of gene expression consistent with this hypothesis have been identified: homologous genes from different species often exhibit similar shifts in expression pattern during sleep–wake cycles. However, neither of the above proposals explains why quiet rest without sleep is not sufficient for these processes, and REM sleep does not really save energy. A recent report suggested another interesting function for sleep: to clear metabolic waste products that accumulate in the awake brain. Live imaging of fluorescent tracer-injected mouse brains revealed that the interstitial space (extracellular space surrounding neurons and glia) increases by 60% during

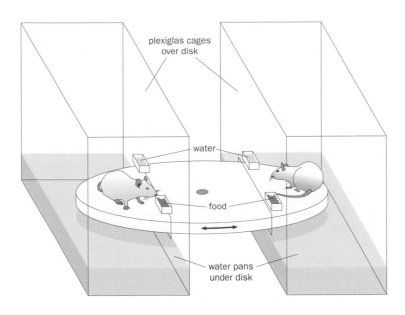

plexiglas cages
over disk

water

food

water pans
under disk

Figure 8–56 The disk-over-water apparatus to study the effect of sleep deprivation. An experimental and a control rat are housed in separate cages that share a disk. The disk spins whenever the EEG of the experimental rat indicates that it is about to fall asleep. When the disk spins, the rats cannot sleep because they must move to avoid falling into the water pans beneath the disk. However, the control rat can sleep during the time the experimental rat is alert. (Adapted from Rechtschaffen A, Gilliland MA, Bergmann BM, et al. [1983] *Science* 221:182.)

sleep compared with the awake period; this facilitates the exchange of interstitial fluid with cerebral spinal fluid and helps waste product clearance.

Perhaps the most intriguing proposal is that sleep facilitates memory, learning, and synaptic plasticity, which are topics that will be studied in depth in Chapter 10. Animal and human studies have shown that both our declarative memory (required to recall names and events) and procedural memory (required to perform specific motor tasks) are enhanced by sleep and even by short naps. Physiological recordings in rodents have identified the replay of hippocampal cell firing patterns resembling those observed during learning procedures that took place during the previous waking period, as if animals were rehearsing during sleep what they had just learned. In the fly model of sleep, levels of synaptic proteins become elevated as the waking period lengthens naturally or is prolonged experimentally; these protein levels are restored to the basal state after sleep. A common feature of studies investigating the function of sleep is that they are mostly correlative. It is difficult to perturb sleep specifically without affecting many other physiological functions, as the disk-over-water deprivation study exemplifies. There is little doubt that sleep has restorative effects required for proper brain function. Perhaps as we learn more about how the brain functions, we will gain further appreciation for why we need sleep.

SUMMARY

The nervous system employs three output systems: (1) skeletal muscle contractions that mediate all body movement; (2) smooth and cardiac muscle contractions that regulate the functions of internal organs; and (3) hormone excretion that regulates many physiological processes. The motor system utilizes the first form of output, whereas the autonomic nervous system, together with its control center in the hypothalamus, regulates the second and third forms of output.

The motor system is organized in a hierarchical manner. The powerful neuromuscular junction converts nearly every action potential from the motor neuron into muscle contraction—via a rise of $[Ca^{2+}]_i$ that triggers the sliding of actin and myosin fibers—such that motor control is equivalent to the control of motor neuron firing patterns. Each muscle is controlled by a pool of motor neurons; firing of motor neurons within a motor pool follows a size principle, which enables incremental strengthening of muscle contraction. Motor neurons integrate information from multiple sources, including input from local pre-motor neurons, descending commands from the brainstem motor control nuclei and motor cortex, and sensory feedback from proprioceptive neurons. In rhythmic motor programs such as locomotion, the rhythmic output pattern is produced by central pattern generators in the spinal cord in the absence of sensory feedback and is activated by brainstem motor control nuclei. The potential mechanisms by which central pattern generators operate are best understood in invertebrate systems, where the biophysical properties of constituent neurons and their connection patterns and strengths determine the rhythmic output patterns. Modern genetic and circuit analysis tools have begun to enable the dissection of complex circuits in the spinal cord and brainstem in motor controls.

Voluntary movement is controlled by the motor cortex along with the cerebellum and basal ganglia. Both the cerebellum and basal ganglia employ generic circuit designs. The cerebellum integrates information about motor commands from the motor cortex and about motor performance from the spinal cord, and adjusts motor system output. The basal ganglia receive cortical and thalamic inputs via striatal spiny projection neurons; activation of spiny projection neurons initiate and select motor programs through parallel direct and indirect pathways to control basal ganglia output. The motor cortex is grossly organized in a somatotopic manner, but the somatotopy breaks down at a fine scale. While each motor cortical neuron is broadly tuned to multiple motor tasks, population activity of motor cortical neurons can be predictive of movement parameters, such as the direction and trajectory of arm reaching. The population activity of motor cortical neurons has been used to control neural prosthetic devices with remarkable success.

The sympathetic and parasympathetic systems—the major output of the autonomic nervous system—regulate the function of internal organs, often with opposing effects. The sympathetic system promotes energy expenditure and mobilizes the body for rapid action, whereas the parasympathetic system promotes energy conservation during rest. Like the motor system, the autonomic output systems are regulated at multiple levels. The hypothalamus is a key control center, which functions to maintain homeostasis. The hypothalamic nuclei integrate information from the sensory and interoceptive systems, compare the body's actual state to biological set points, and regulate autonomic output, hormone secretion, and organismal behavior. In their control of the neuroendocrine system, hypothalamic neurons either secrete hormones directly from their axon terminals in the posterior pituitary, or secrete pre-hormones that induce or inhibit hormone secretion by endocrine cells in the anterior pituitary.

Eating is regulated by a number of hypothalamic nuclei, notably the POMC and AgRP neurons in the arcuate nucleus. POMC neurons inhibit eating by secreting α-MSH that activates the melanocortin-4 receptor on target neurons. AgRP neurons promote eating by antagonizing the action of POMC neurons and by inhibiting additional target neurons in the hypothalamus and brainstem. POMC and AgRP neurons are activated and inhibited, respectively, by leptin, an adipose tissue-derived feedback signal that inhibits eating. Whereas leptin and the pancreas-derived hormone insulin signal nutrient levels and act over long timescales to maintain energy balance, stomach-derived ghrelin and intestine-derived cholecystokinin (CCK) signal hunger and satiety, respectively, and act on short timescales to promote or inhibit eating.

Circadian rhythms are self-sustained oscillations of an organism's biochemistry, physiology, and behavior that have a near-24-hour period and can be entrained by light input. The central clockwork is highly conserved from insects to mammals and utilizes auto-inhibitory transcriptional feedback loops. In flies, entrainment is achieved by light-mediated degradation of circadian regulators. In mammals, light entrainment is mediated by visual input that regulates transcription of circadian regulators in the suprachiasmatic nucleus (SCN) of the hypothalamus, which acts as a master clock. While exhibiting pacemaker properties individually, SCN neurons form an interacting network to control peripheral clocks via its interactions with other hypothalamic nuclei that controls autonomic and neuroendocrine systems.

Sleep is homeostatically regulated, and it is timed by the circadian system. Sleep is universal in mammals, and sleeplike states are found in all vertebrates and in some invertebrates. In mammals, sleep can be divided into different stages according to electroencephalogram patterns, including rapid eye movement (REM) and non-REM sleep. The sleep–wake cycle is controlled by mutually inhibitory groups of neurons in the hypothalamus and brainstem, with the arousal pathway utilizing several parallel systems, including monoamine neurotransmitters and hypocretin neuropeptides. Sleep is vital to animals, although its exact functions still remain to be explored.

FURTHER READING

Reviews and books

Boyden ES, Katoh A & Raymond JL (2004) Cerebellum-dependent learning: the role of multiple plasticity mechanisms. *Annu Rev Neurosci* 27:581–609.

Gerfen CR & Surmeier DJ (2011) Modulation of striatal projection systems by dopamine. *Annu Rev Neurosci* 34:441–466.

Huxley HE (1965) The mechanism of muscular contraction. *Sci Am* 213:18–27.

Marder E & Bucher D (2007) Understanding circuit dynamics using the stomatogastric nervous system of lobsters and crabs. *Annu Rev Physiol* 69:291–316.

Mignot E (2008) Why we sleep: the temporal organization of recovery. *PLoS Biol* 6:e106.

Mohawk JA, Green CB & Takahashi JS (2012) Central and peripheral circadian clocks in mammals. *Annu Rev Neurosci* 35:445–462.

Shenoy KV, Sahani M & Churchland MM (2013) Cortical control of arm movements: a dynamical systems perspective. *Annu Rev Neurosci* 36:337–359.

Sternson SM (2013) Hypothalamic survival circuits: blueprints for purposive behaviors. *Neuron* 77:810–824.

Swanson LW (2012) Brain Architecture. Oxford University Press.

Motor system

Churchland MM, Cunningham JP, Kaufman MT et al. (2012) Neural population dynamics during reaching. *Nature* 487:51–56.

Cui G, Jun SB, Jin X et al. (2013) Concurrent activation of striatal direct and indirect pathways during action initiation. *Nature* 494:238–242.

Esposito MS, Capelli P & Arber S (2014) Brainstem nucleus MdV mediates skilled forelimb motor tasks. *Nature* 508:351–356.

Georgopoulos AP, Kettner RE & Schwartz AB (1988) Primate motor cortex and free arm movements to visual targets in three-dimensional space. II. Coding of the direction of movement by a neuronal population. *J Neurosci* 8:2928–2937.

Gilja V, Nuyujukian P, Chestek CA et al. (2012) A high-performance neural prosthesis enabled by control algorithm design. *Nat Neurosci* 15:1752–1757.

Graham Brown T (1911) The intrinsic factors in the act of progression in the mammal. *Proc R Soc Lond B* 84:308–319.

Graziano MS, Taylor CS & Moore T (2002) Complex movements evoked by microstimulation of precentral cortex. *Neuron* 34:841–851.

Henneman E, Somjen G & Carpenter DO (1965) Functional significance of cell size in spinal motoneurons. *J Neurophysiol* 28:560–580.

Hochberg LR, Bacher D, Jarosiewicz B et al. (2012) Reach and grasp by people with tetraplegia using a neurally controlled robotic arm. *Nature* 485:372–375.

Ito M, Sakurai M & Tongroach P (1982) Climbing fibre induced depression of both mossy fibre responsiveness and glutamate sensitivity of cerebellar Purkinje cells. *J Physiol* 324:113–134.

Kravitz AV, Freeze BS, Parker PR et al. (2010) Regulation of parkinsonian motor behaviours by optogenetic control of basal ganglia circuitry. *Nature* 466:622–626.

Kron SJ & Spudich JA (1986) Fluorescent actin filaments move on myosin fixed to a glass surface. *Proc Natl Acad Sci USA* 83:6272–6276.

McLean DL & Fetcho JR (2009) Spinal interneurons differentiate sequentially from those driving the fastest swimming movements in larval zebrafish to those driving the slowest ones. *J Neurosci* 29:13566–13577.

Miller JP & Selverston AI (1982) Mechanisms underlying pattern generation in lobster stomatogastric ganglion as determined by selective inactivation of identified neurons. II. Oscillatory properties of pyloric neurons. *J Neurophysiol* 48:1378–1391.

Prinz AA, Bucher D & Marder E (2004) Similar network activity from disparate circuit parameters. *Nat Neurosci* 7:1345–1352.

Stepien AE, Tripodi M & Arber S (2010) Monosynaptic rabies virus reveals premotor network organization and synaptic specificity of cholinergic partition cells. *Neuron* 68:456–472.

Talpalar AD, Bouvier J, Borgius L et al. (2013) Dual-mode operation of neuronal networks involved in left–right alternation. *Nature* 500:85–88

Velliste M, Perel S, Spalding MC et al. (2008) Cortical control of a prosthetic arm for self-feeding. *Nature* 453:1098–1101.

Eating

Atasoy D, Betley JN, Su HH et al. (2012) Deconstruction of a neural circuit for hunger. *Nature* 488:172–177.

Balthasar N, Coppari R, McMinn J et al. (2004) Leptin receptor signaling in POMC neurons is required for normal body weight homeostasis. *Neuron* 42:983–991.

Coleman DL (1973) Effects of parabiosis of obese with diabetes and normal mice. *Diabetologia* 9:294–298.

Coleman DL & Hummel KP (1969) Effects of parabiosis of normal with genetically diabetic mice. *Am J Physiol* 217:1298–1304.

Cowley MA, Smart JL, Rubinstein M et al. (2001) Leptin activates anorexigenic POMC neurons through a neural network in the arcuate nucleus. *Nature* 411:480–484.

Farooqi IS, Jebb SA, Langmack G et al. (1999) Effects of recombinant leptin therapy in a child with congenital leptin deficiency. *N Engl J Med* 341:879–884.

Halaas JL, Gajiwala KS, Maffei M et al. (1995) Weight-reducing effects of the plasma protein encoded by the obese gene. *Science* 269:543–546.

Hervey GR (1959) The effects of lesions in the hypothalamus in parabiotic rats. *J Physiol* 145:336–352.

Luquet S, Perez FA, Hnasko TS et al. (2005) NPY/AgRP neurons are essential for feeding in adult mice but can be ablated in neonates. *Science* 310:683–685.

Wu Q, Clark MS & Palmiter RD (2012) Deciphering a neuronal circuit that mediates appetite. *Nature* 483:594–597.

Zhang Y, Proenca R, Maffei M et al. (1994) Positional cloning of the mouse obese gene and its human homologue. *Nature* 372:425–432.

Circadian rhythms and sleep

Adamantidis AR, Zhang F, Aravanis AM et al. (2007) Neural substrates of awakening probed with optogenetic control of hypocretin neurons. *Nature* 450:420–424.

Dement W & Kleitman N (1957) Cyclic variations in EEG during sleep and their relation to eye movements, body motility, and dreaming. *Electroencephalogr Clin Neurophysiol* 9:673–690.

Hardin PE, Hall JC & Rosbash M (1990) Feedback of the *Drosophila* period gene product on circadian cycling of its messenger RNA levels. *Nature* 343:536–540.

Hunter-Ensor M, Ousley A & Sehgal A (1996) Regulation of the *Drosophila* protein timeless suggests a mechanism for resetting the circadian clock by light. *Cell* 84:677–685.

Konopka RJ & Benzer S (1971) Clock mutants of *Drosophila melanogaster*. *Proc Natl Acad Sci USA* 68:2112–2116.

Lin L, Faraco J, Li R et al. (1999) The sleep disorder canine narcolepsy is caused by a mutation in the hypocretin (orexin) receptor 2 gene. *Cell* 98:365–376.

Liu AC, Welsh DK, Ko CH et al. (2007) Intercellular coupling confers robustness against mutations in the SCN circadian clock network. *Cell* 129:605–616.

Peyron C, Tighe DK, van den Pol AN et al. (1998) Neurons containing hypocretin (orexin) project to multiple neuronal systems. *J Neurosci* 18:9996–10015.

Ralph MR, Foster RG, Davis FC et al. (1990) Transplanted suprachiasmatic nucleus determines circadian period. *Science* 247:975–978.

Rechtschaffen A, Gilliland MA, Bergmann BM et al. (1983) Physiological correlates of prolonged sleep deprivation in rats. *Science* 221:182–184.

Vitaterna MH, King DP, Chang AM et al. (1994) Mutagenesis and mapping of a mouse gene, Clock, essential for circadian behavior. *Science* 264:719–725.

Xie L, Kang H, Xu Q et al. (2013) Sleep drives metabolite clearance from the adult brain. *Science* 342:373–377.

CHAPTER 9
Sexual Behavior

When, as by a miracle, the lovely butterfly bursts from the chrysalis full-winged and perfect … it has, for the most part, nothing to learn, because its little life flows from its organization like a melody from a music box.

Douglas A. Spalding (1873)

Sex is nearly universal in the biological world. Even the bacterium *E. coli*, a unicellular prokaryote that reproduces asexually by rapid cell division, engages periodically in conjugation to exchange genetic material between individual cells and produce recombinant progeny. In plants and animals, sexual reproduction becomes increasingly prevalent as organisms become more complex, and diverse strategies are employed to ensure mating success. As an interesting example, bee orchids have evolved flowers that resemble female bees and scents that mimic virgin bee pheromone, such that they attract male bees to 'mate' with their flowers (**Figure 9–1**) and carry the orchid pollen from flower to flower. Thus, the bee orchid capitalizes on the bee's sexual behaviors to facilitate its own sexual reproduction.

In the preceding chapters, we have studied how animals perceive the world with their sensory systems, how they act through their motor system, and how the nervous system is wired during development. In the context of eating and sleeping, we have also started to touch upon a central question in neurobiology: how does the nervous system produce behavior? In this chapter, we use sexual behavior as a specific example to illustrate how sensory and motor systems as well as developmental processes are integrated to produce behaviors that are fundamental for the propagation of species.

Sexual behavior in model organisms offers important experimental advantages to researchers investigating the neural bases of behavior in general. First, sexual behavior is robust and often **stereotyped** (varying little from one individual to another); both of these traits facilitate quantitative behavioral analyses. Second, sexual behavior has a strong innate component, echoed by the epigraph from Spalding. It is therefore likely to be specified by genetic programs and can be subjected to powerful genetic analysis. Third, reproductive behaviors are **sexually dimorphic**, that is, they differ in females and males. Sexually dimorphic behavior originates from differences in the chromosomes that determine the sex (**sex chromosomes**) of an individual; such genetic differences in females and males offer an experimental entry point for dissecting the complex behaviors into simpler components.

In this chapter, we focus primarily on the sexual behavior of fruit flies and rodents, where the neural basis of sexual behavior has been best understood utilizing the experimental advantages discussed above. We will see that, in fruit flies, sexual dimorphisms in neural circuits and behavior are largely specified by the cell-autonomous actions of two transcription factors, whereas in mammals such dimorphisms require a combination of genetic and hormone controls both during development and in adults. By studying analogous behavior in animals that exhibit different levels of complexity, we can appreciate both the common principles and diverse strategies that ensure animals' reproductive success.

Figure 9–1 A sexual decoy. The flower of the bee orchid *Ophrus apifera* resembles a female bee. The bee orchid evolved this deceptive flower pattern to attract male bees as pollinators. (Courtesy of Perennou Nuridsany/Science Source.)

HOW DO GENES SPECIFY SEXUAL BEHAVIOR IN THE FLY?

We have already encountered many examples in which the fruit fly *Drosophila melanogaster* has been used in studies that address fundamental problems of neurobiology. The advanced genetic manipulations available in *Drosophila* make it an attractive model organism (see Section 13.2). Studies of sexual behavior in flies have been further facilitated by the finding that individual genes can have large influences on sexual behavior.

9.1 *Drosophila* courtship follows a stereotyped ritual that is instinctive

Fruit flies congregate at food sources, where they find potential mating partners, carry out courtship, and mate (on the ground, rather than in flight). A male fruit fly uses both visual and chemosensory cues to find his appropriate mate: a virgin female of the same species. What follows is a series of steps (**Figure 9–2**; see also **Movie 9–1**), with each step occurring in a stereotyped sequence. At the start, the male orients himself in front of the female to be noticed. He then positions himself beside or behind the target female, taps her abdomen, and sings a courtship song by extending and vibrating one of his wings. He follows this by licking her genitalia, and then attempts to copulate. If he succeeds in inducing the female to accept his advances, they mate; if the female rejects his overtures, they part ways. For *Drosophila melanogaster*, the entire ritual from orientation to attempted copulation usually lasts a few minutes.

The courtship ritual depends on several sensory modalities for communication between males and females. The orientation step utilizes visual and chemosensory (pheromone) cues. Tapping transmits olfactory, taste, and somatosensory information. (In flies, taste receptor neurons are present not only in the proboscis—the insect equivalent of a mouth—but also on the forelegs.) Singing engages audition in both the female (the target audience for the courtship song) and the male (who receives auditory feedback relating to his song). In fact, audition as a sense in insects is largely dedicated to sexual behavior. Licking involves taste for males and somatosensation for females. Somehow, the brains of female and male fruit flies integrate these sensory cues and orchestrate a complex, extended interaction with their mating partners—but how?

Before we delve into the neural mechanisms, it is important to note that the entire ritual is innate and genetically programmed. One can rear a male fly in isolation so that he has never seen or smelled a female. Within minutes of being

Figure 9–2 The courtship ritual of the fruit fly, *Drosophila melanogaster*. Fruit fly courtship follows a stereotyped sequence of steps that lasts for a few minutes. The male first finds the female and orients himself in front of her using visual and chemosensory cues. He then moves to her side and taps her with his foreleg, which contains taste receptor neurons. Next, the male sings courtship songs by vibrating one of his wings. He follows by licking her genitalia and attempting copulation. If the female is receptive, she reacts by slowing herself down, letting the male tap, listening to his song, and allowing him to lick. A successful courtship concludes with copulation. (Adapted from Greenspan RJ [1995] *Sci Am* 272:72–78. With permission from Macmillan Publishers Ltd.)

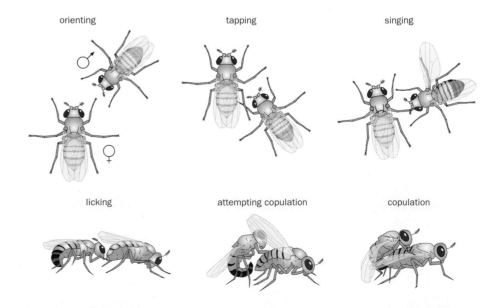

orienting   tapping   singing

licking   attempting copulation   copulation

introduced to a virgin female, the sight and smell of an appropriate partner will trigger mating behavior such that he can perform the entire ritual perfectly.

9.2 *Fruitless (Fru)* is essential for many aspects of sexual behavior

A first step in the genetic dissection of a complex biological process is to isolate single-gene mutations that affect specific aspects of the process of interest. Identifying such mutations makes it possible to establish causal relationships between the genes and the biological processes they control, as we have seen in the case of the gene *Period* and the control of circadian rhythms (see Section 8.19).

A gene called ***Fruitless (Fru)*** has been found to regulate all aspects of the male courtship ritual in *Drosophila*. Mutations that disrupt the production of a male-specific isoform of *Fru* (see below) affect each step of the courtship ritual in males, and hence render the male sterile. *Fru* mutants with distinct phenotypes have been identified. Some *Fru* mutant males fail to recognize the sex of the partner and court males and females indiscriminately. If a group of these mutant males are placed together, they often form courtship chains, in which each male is courting the male in front of him while being courted by the male behind him (**Figure 9–3**; Movie 9–1). Some *Fru* mutant males can produce courtship songs, but these songs are highly aberrant. Certain *Fru* mutant males are defective in the transfer of sperm and seminal fluid. Thus, *Fru* is required for many aspects of male courtship. By contrast, females are not affected by these mutations morphologically or behaviorally. *Fru* encodes a DNA-binding protein that controls the expression of other genes, consistent with its regulatory role. However, *Fru* was not recognized as a central regulator of sexual behavior until its link with a hierarchy of regulatory genes that determine the sex of fruit flies was established.

9.3 A sex-determination hierarchy specifies sex-specific splicing of *Fru* that produces male-specific Fru^M

In the fruit fly, sex is determined by the ratio of the copy numbers of X chromosomes and **autosomes** (non-sex chromosomes, A). Females (XX) have two copies of the X chromosome and two copies of each autosome and thus an X/A ratio of 1. Males (XY) have only one X chromosome and therefore an X/A ratio of ½. (The Y chromosome makes no contribution to sex determination in flies.) These different ratios translate into sex-specific gene expression through a regulatory hierarchy (**Figure 9–4A**). An X/A ratio of 1 in females leads to the production of a functional Sex-lethal (Sxl) protein; the corresponding gene is called ***Sex lethal*** because mutations cause only females to die. Sxl regulates the alternative splicing of its own transcript and of ***Tra (Transformer)*** mRNA; the *Tra* gene takes its name from the phenotype of mutant females, which look and act like males. Like *Sxl*, *Tra* and its partner *Tra2* encode splicing factors. Together, Tra and Tra2 proteins

Figure 9–3 Male *Fruitless (Fru)* mutants exhibit altered sexual behavior. Whereas wild-type males court only females, certain *Fru* mutant males court males and females indiscriminately. These *Fru* mutant males can form a courtship chain: each male is courting the male in front of him and being courted by the male behind him. (From Hall JC [1994] *Science* 264:1702–1714. With permission from AAAS.)

Figure 9–4 The sex determination hierarchy and its regulation of *Fruitless*. (A) In females, an X/A ratio of 1 leads to the expression of functional Sxl, which acts as a splicing factor and is required for producing functional Tra. Together with Tra2, Tra controls the alternative splicing of the *Dsx* and *Fru* genes, resulting in the expression of the Dsx^F protein. The male X/A ratio of ½ does not yield functional Sxl and consequently generates no Tra. This leads to male-specific splicing that produces Dsx^M and Fru^M. Dsx^F and Dsx^M specify the sexual female and male bodies, respectively. As we will learn in subsequent sections, Fru^M (with a contribution from Dsx) specifies sexual dimorphism of brain and behavior. **(B)** Top, the *Fruitless* transcript from the first promoter; exons are shown as blocks joined by introns (thin lines). Note the sex-specific splicing of the second exon. In males, the default splicing joins the blue portion of the second exon with the 3′ exons (shown here as a single gray exon); the resulting translation product is a functional, male-specific Fru^M protein (bottom). In females, the female-specific splicing caused by binding of Tra/Tra2 (indicated by the star) yields a transcript that includes the second exon's white segment, which has a stop codon and does not produce a functional protein. (Based on Baker BS, Taylor BJ & Hall JC [2001] *Cell* 105:13–24.)

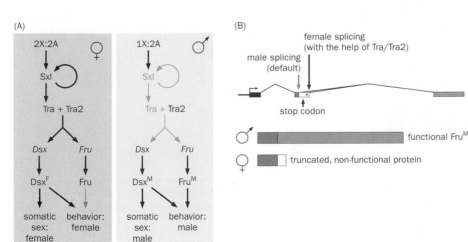

(A)

2X:2A ♀

Sxl

Tra + Tra2

Dsx   Fru

Dsx^F   Fru

somatic   behavior:
sex: female
female

1X:2A ♂

Sxl

Tra + Tra2

Dsx   Fru

Dsx^M   Fru^M

somatic   behavior:
sex: male
male

(B)

male splicing
(default)

female splicing
(with the help of Tra/Tra2)

stop codon

♂ ▬▬▬▬▬▬▬▬ functional Fru^M

♀ ▬▬□ truncated, non-functional protein

regulate the alternative splicing of **Doublesex** (**Dsx**) to produce a female-specific Dsx^F protein. In males, an X/A ratio of ½ does not support the production of functional Sxl protein, and consequently yields no functional Tra protein. A default male-specific Dsx^M protein is made in the absence of Tra/Tra2. Both Dsx^F and Dsx^M are transcription factors that regulate the expression of many downstream target genes to specify the differentiation of female and male bodies (and to some extent brains, as we will see later).

The *Fru* gene produces different Fru protein isoforms due to the use of multiple promoters and alternative splicing. The transcript from the first promoter exhibits a sex-specific splicing pattern controlled by Tra/Tra2. In females, the transcript generated by Tra/Tra2-directed splicing has a premature stop codon that prevents translation of a functional protein product. In males, which lack functional Tra/Tra2, the default splicing results in expression of male-specific **Fru^M** protein (Figure 9–4B).

9.4   Expression of Fru^M in females is sufficient to produce most aspects of male courtship behavior

As we have seen from its loss-of-function phenotypes, Fru^M is necessary for multiple aspects of the male courtship ritual. Because mutations in *Tra* and *Tra2* cause females to look and act like males, and because Fru^M production is regulated by *Tra* and *Tra2*, it was hypothesized that Fru^M may also be sufficient to specify male sexual behavior. This hypothesis was tested by deleting from the endogenous *Fru* gene the white segment of the second exon shown in Figure 9-4B. In these deletion mutants, which we refer to as *Fru^Δ*, both males and females utilize the male form of splicing and both produce functional Fru^M proteins (**Figure 9–5**A).

As expected, *Fru^Δ* males acted like normal males, since their splicing pattern was unaffected by the genetic modification. What happened to the *Fru^Δ* females, which were forced to express Fru^M protein by the genetic modification? Although they looked like females, they acted like males. *Fru^Δ* females vigorously courted other females and performed the male courtship ritual as normal males (Figure 9–5B, C; see also Movie 9–1), except that they did not copulate properly because their external anatomy was still female. This remarkable result

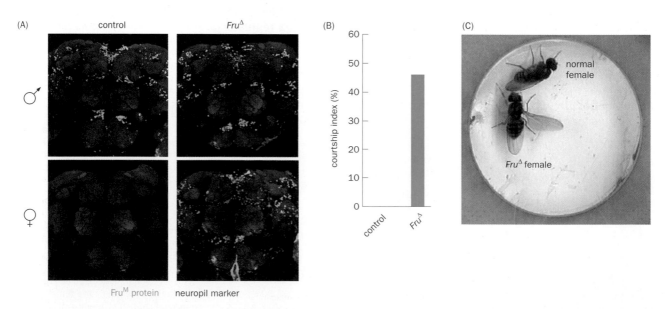

Figure 9–5 Expression of Fru^M in females confers male-typical sexual behavior to females. (A) In wild-type control flies, the Fru^M protein (green, visualized by antibody staining) is produced in males but not in females. By contrast, both male and female *Fru^Δ* flies produce Fru^M. Magenta, staining of a synaptic marker that highlights the neuropil organization of the *Drosophila* central brain. **(B)** Whereas control females do not court other females, *Fru^Δ* females robustly court wild-type females, as indicated by the high courtship index, a measure of the percentage of time spent on courtship when two flies are placed in a chamber. **(C)** A *Fru^Δ* female is courting a wild-type female by singing a courtship song. (Adapted from Demir E & Dickson BJ [2005] *Cell* 121:785–794. With permission from Elsevier Inc.)

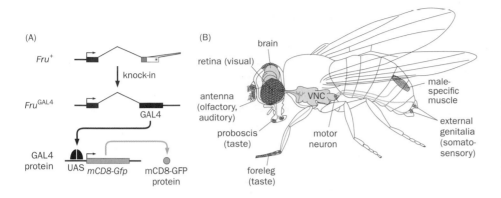

(A)

(B)

Figure 9–6 Strategies to investigate Fru^M expression and function. (A) *Fru*^GAL4, in which the yeast transcription factor GAL4 is knocked into the *Fru* locus after the first promoter, mimics the expression pattern of Fru^M. *Fru*^GAL4 can be used to drive expression of a variety of UAS-transgenes in cells that normally express Fru^M protein by introducing the *Fru*^GAL4 transgene and a UAS-transgene into the same fly through genetic crossing. Bottom: in flies that carry both *Fru*^GAL4 and *UAS-mCD8-Gfp* transgenes, GAL4 proteins bind to UAS to activate the expression of a membrane-tethered green fluorescence protein that allows visualization of axonal projections of Fru^M neurons. Other UAS-transgenes have been used to silence or activate Fru^M neurons, or to impair endogenous Fru^M expression. **(B)** Using *Fru*^GAL4 to drive expression of mCD8-GFP, it was found that *Fru*^GAL4 (and by inference Fru^M) is expressed in subsets of olfactory receptor neurons and auditory sensory neurons in the antenna, taste receptor neurons in the proboscis and foreleg, visual neurons in the retina, mechanosensory neurons in the external genitalia, and motor neurons in the ventral nerve cord (VNC) that innervate a male-specific muscle. Green lines highlight the axonal projections of these neurons. (B, adapted from Billeter JC, Rideout EJ, Dornan AJ et al. [2006] *Curr Biol* 16:R766–R776. With permission from Elsevier.)

demonstrated that changing the splicing pattern of a single gene from female to male is sufficient to confer many aspects of male-typical courtship behavior to females.

9.5 Activity of Fru^M neurons promotes male courtship behavior

How does the expression of Fru^M in females confer male-typical behavior? To answer this question, we need to step back and examine which cells in males normally express the Fru^M protein. Antibodies specific for the Fru^M isoform were used to label groups of neuronal nuclei in the male brain (see Figure 9–5A) and ventral nerve cord (the insect equivalent of the vertebrate spinal cord; see Figures 7–11 and 7–12). Fru^M is present in the nuclei of about 2000 neurons, constituting 2% of the ~100,000 total neurons in *Drosophila*.

Fru^M distribution can also be examined using an alternative binary transgene expression method (see Section 13.10 for details): knock-in of the yeast transcription factor GAL4 under the control of the first *Fru* promoter. This transgene, denoted as *Fru*^GAL4, can drive the expression of any transgene that contains a GAL4-binding upstream activation sequence (UAS), resulting in an expression pattern that mimics Fru^M (**Figure 9–6**A). Using *Fru*^GAL4 to drive the expression of a UAS-membrane-tethered green fluorescent protein that labels not only cell bodies but also axons and dendrites, it was discovered that Fru^M is also expressed in some sensory neurons that project to the CNS and some motor neurons that project to specific muscles. These include a subset of olfactory receptor neurons and most auditory neurons in the antenna, a subset of taste receptor neurons in the foreleg and proboscis, a subset of visual neurons, somatosensory neurons in the external genitalia, and the motor neurons that innervate a male-specific muscle (Figure 9–6B). Fru^M expression is therefore consistent with the likely function of these neurons in different aspects of courtship (see Section 9.1).

To assess the function of Fru^M neurons in courtship behavior, *Fru*^GAL4 was used to express an effector transgene that silences neuronal activity (a loss-of-function manipulation). A widely used neuronal silencer in flies is a temperature-sensitive mutant protein called Shibire^ts (Shi^ts). At low temperature (19°C), Shi^ts expression is largely innocuous. At high temperature (29°C), Shi^ts expression blocks synaptic vesicle recycling (see Section 3.9) and therefore shuts off synaptic transmission. Thus, Shi^ts can be used to conditionally silence neurons that express it by shifting the Shi^ts-expressing flies to the restrictive high temperature. When all Fru^M neurons were transiently silenced, courtship behavior was severely impaired (**Figure 9–7**), indicating that the activity of Fru^M neurons is acutely required in adult males for the execution of courtship behavior. Remarkably, many other aspects of fly behavior, including locomotion, flight, phototaxis, and olfaction and taste in general, were not affected under the same experimental conditions. Thus, Fru^M neurons appear to be largely dedicated to the performance of courtship in males.

Interestingly, artificial activation of Fru^M neurons (a gain-of-function manipulation) in isolated males can promote multiple steps of the courtship ritual—including tapping, singing, licking, and even attempted copulation—in the absence of a mating partner (**Figure 9–8**A)! This was achieved by expressing a

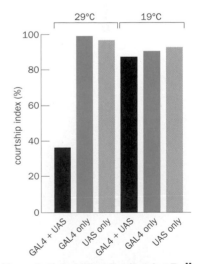

Figure 9–7 Conditionally silencing Fru^M neurons impairs courtship. Male flies that carry both *Fru*^GAL4 and *UAS-Shi*^ts transgenes (red, GAL4- + UAS-transgenes) exhibit reduced courtship index toward virgin females at the restrictive temperature (29°C) but are normal at the permissive temperature (19°C). Control males (GAL4- or UAS-transgene only) exhibit high courtship index at both temperatures. This indicates that blocking synaptic transmission from Fru^M neurons impairs courtship. (Adapted from Stockinger P, Kvitsiani D, Rotkopf S et al. [2005] *Cell* 121:795–807. With permission from Elsevier Inc.)

(A)

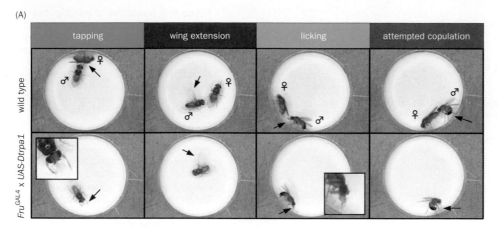

(B)

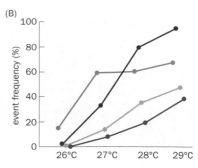

Figure 9–8 Artificial activation of Fru^M neurons promotes courtship behavior in isolated males. (A) Top, examples of four steps in the courtship ritual exhibited by wild-type males in the presence of a virgin female. Bottom, when Fru^M neurons are activated by dTRPA1-induced depolarization at high temperature, males can exhibit a similar courtship ritual without a partner. Arrows indicate tapping, unilateral wing extension, licking, and attempted copulation (with curled abdomen). Insets show high-magnification images in which a fly mimics tapping by extending one of its forelegs (left), or mimics licking by extending its proboscis (right). **(B)** The frequency of courtship behaviors of tapping, wing extension, licking, and attempted copulation (following the color code in A) increases as Fru^M neurons become increasingly depolarized at higher temperatures. (Adapted from Kohatsu S, Koganezawa M & Yamamoto D [2011] *Neuron* 69:498–508. With permission from Elsevier Inc.)

heat-activated cation channel, **dTRPA1**, in Fru^M neurons. (Some TRP channels in flies are used to sense temperatures as in mammals, see Section 6.31.) As the temperature rose, Fru^M neurons became increasingly depolarized and activated, and the frequency of isolated males exhibiting courtship behaviors increased (Figure 9–8B). Thus, both loss- and gain-of-function experiments indicate that Fru^M neurons play a central role in promoting male courtship behavior. The fact that males can elicit aspects of courtship behavior without a partner when Fru^M neurons are artificially activated provides a striking example of the fixed action patterns that were proposed to underlie innate behaviors (see Section 1.2).

9.6 Fru^M sensory neurons process mating-related sensory cues

How do Fru^M neurons control courtship behavior in males? In the next three sections, we discuss how researchers have dissected the complex courtship ritual into individual building blocks by identifying how subsets of Fru^M neurons function in sensory perception, central coordination, and motor control.

The 50 types of fly olfactory receptor neurons (ORNs) in *Drosophila* (see Section 6.13) include three types that express Fru^M and project to three antennal lobe glomeruli that are sexually dimorphic, having larger volumes in males than in females (**Figure 9–9**A). In males, conditional silencing of just these three ORN types (using an intersectional transgene expression strategy; see Section 13.12) significantly impairs courtship in the dark, indicating that the olfactory cues processed by these ORN classes play an important role in courtship. The functions of two ORN types that project to the DA1 and VL2a glomeruli have been identified. As was introduced in Section 6.15, the DA1 ORNs (which project axons to the DA1 glomerulus) are activated by 11-*cis*-vaccenyl acetate (cVA), a pheromone produced by males to inhibit male's courtship toward another male or a fertilized female. During mating, cVA is transferred into females together with seminal fluid, such that mated females signal their unavailability to potential suitors via cVA. Anatomical and physiological studies have outlined a putative cVA-processing pathway in males: at the DA1 glomerulus, DA1 ORNs synapse with second-order DA1 projection neurons (PNs). In the lateral horn, DA1 PNs synapse with third-order DC1 neurons, which in turn signal DN1 neurons that project into the ventral nerve cord and may serve to control behavioral output downstream from the cVA pheromone cues (Figure 9–9B). DA1 PNs, DC1, and DN1 neurons all express Fru^M, suggesting that Fru^M neurons tend to connect with other Fru^M neurons to process mating-related signals.

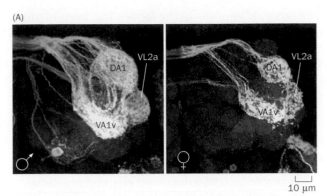

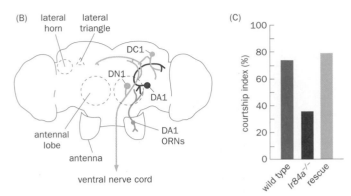

(A)

DA1
VL2a
VA1v
♂

VL2a
DA1
VA1v
♀

10 μm

(B) lateral
horn   lateral
triangle

DC1
DN1
DA1
DA1
ORNs

antennal
lobe

antenna

ventral nerve cord

(C)

courtship index (%)

wild type   Ir84a⁻/⁻   rescue

Figure 9–9 Fru^M olfactory receptor neurons process cues that inhibit male–mated female courtship and promote male–virgin female courtship. (A) Expression of *Fru^GAL4*-driven mCD8GFP in axon terminals of three types of ORNs (green); these ORNs project to three glomeruli (DA1, VA1v, VL2a) that are larger in males than females. Magenta, synaptic marker that labels all glomeruli. **(B)** A pathway comprising DA1 ORNs → DA1 PNs → DC1 neurons → DN1 neurons processes the cVA pheromone in the male brain, which inhibits male courtship of mated females. All neurons in this pathway express Fru^M. Their cell bodies (circles) and processes (lines) are depicted. DA1 ORNs synapse with DA1 PNs at the DA1 glomerulus in the antennal lobe. The DA1 → DC1, DC1 → DN1 connections occur at the neuropil

regions named the lateral horn (see Figure 6–27) and lateral triangle, respectively. **(C)** The Ir84a-expressing ORNs, which project to the VL2a glomerulus, promote male–virgin female courtship. Compared with wild type, mutant flies with no functional *Ir84* gene have a reduced courtship index. This is rescued by expressing an *Ir84a* transgene in VL2a ORNs. (A, adapted from Stockinger P, Kvitsiani D, Rotkopf S et al. [2005] *Cell* 121:795–807. With permission from Elsevier Inc.; B, adapted from Ruta V, Datta SR, Vasconcelos ML et al. [2010] *Nature* 468:686–690. With permission from Macmillan Publishers Ltd; C, adapted from Grosjean Y, Rytz R, Farine JP et al. [2011] *Nature* 478:236–240. With permission from Macmillan Publishers Ltd.)

The VL2a ORNs express an olfactory receptor (Ir84a) that is activated by aromatic odors such as phenylacetic acid that are enriched in food sources. Flies mutant for the Ir84a receptor have severely reduced male–virgin female courtship (Figure 9–9C), suggesting that food odors act as an aphrodisiac to promote mating. This is in line with the fact that fruit flies mate near their food sources. VL2a projection neurons arborize their axon terminals in the mating pheromone area rather than the fruit odor area in the lateral horn (see Section 6.16), suggesting that flies treat specific kinds of food odor as mating pheromones. In summary, analyses of Fru^M olfactory neurons have identified neural pathways that utilize olfactory cues to facilitate male-virgin female courtship and inhibit nonproductive male-male or male-mated female courtship.

The taste system in flies resembles the mammalian taste system we studied in Chapter 6, with specific taste receptor neurons that are activated by sweet or bitter tastants. An important difference is that a subset of fly taste receptor neurons appears to be dedicated to 'tasting' the hydrocarbon molecules on the bodies of other flies, which are used for sex discrimination and bidirectional control of courtship behavior. Some Fru^M taste receptor neurons on the foreleg are activated by male-enriched hydrocarbons and cVA, while others are activated by female-enriched hydrocarbons, as has been observed in Ca²⁺ imaging experiments (**Figure 9–10**). In principle, male flies can use the differential activity of these neurons to distinguish male and female partners during the tapping step of the courtship ritual. Genetic silencing and activation experiments have demonstrated that these neurons indeed facilitate male-female courtship and inhibit male-male courtship.

Figure 9–10 Fru^M taste neurons detect chemical cues from the body. Changes in the fluorescence intensity (ΔF/F) of a genetically encoded Ca²⁺ indicator (see Section 13.22) in response to different taste stimuli provide a means to measure the activity of two Fru^M taste receptor neurons (outlined) in a male foreleg. One neuron is activated by the female hydrocarbons HD and ND, whereas the other neuron is activated by male hydrocarbons 7T and cVA. Hexane is the solvent for the hydrocarbons and serves as a negative control. (From Thistle R, Cameron P, Ghorayshi A et al. [2012] *Cell* 149:1140–1151. With permission from Elsevier Inc.)

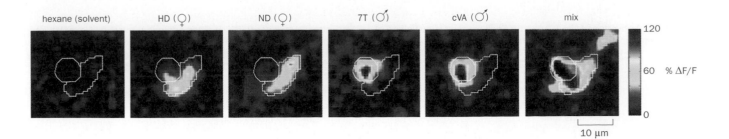

hexane (solvent)   HD (♀)   ND (♀)   7T (♂)   cVA (♂)   mix

120

60   % ΔF/F

0

10 μm

A distinct group of taste receptor neurons in the foreleg has been identified that function in species discrimination, inhibiting nonproductive interspecies courtship. Interestingly, while these neurons do not appear to express Fru^M, their axons contact Fru^M-expressing neurons in the CNS, suggesting that these sensory neurons form a circuit with Fru^M neurons to control species specificity in courtship.

9.7 Fru^M central neurons integrate sensory information and coordinate the behavioral sequence

The advanced genetic tools available in *Drosophila* have enabled researchers to begin dissecting the functions of different Fru^M neurons deep in the brain. For example, since artificial activation of all Fru^M neurons causes male flies to exhibit many aspects of courtship behavior in isolation, researchers have used mosaic expression techniques (see Section 13.16) to ask which subset of Fru^M neurons, when activated in males, is sufficient to elicit male-typical courtship behaviors such as licking and unilateral wing extension. This led to the identification of a P1 cluster of Fru^M neurons, which receives input from sensory systems and extends elaborate axonal arborizations in the brain (**Figure 9–11**A). Ca^{2+} imaging studies in tethered males indicated that P1 neurons are activated by chemosensory cues from the virgin female abdomen. Artificial activation of P1 neurons elicits tapping, unilateral wing extension, and licking behavior with high probability (Figure 9–11B). Thus, P1 neurons are part of an integrative center that transforms sensory input into behavioral output.

As noted previously, wild-type *Drosophila* males follow a stereotyped courtship sequence that lasts a few minutes. Even in the case of behavior elicited by artificial activation of dTRPA1-expressing Fru^M neurons, the different steps appear to follow a similar sequence (Figure 9–11B). How is this sequence coordinated, and what happens if it is disrupted? Insight into this question came from an experiment in which Fru^M expression was knocked down by **RNA interference** (**RNAi**; see Section 13.8 for details) in a small cluster of median bundle neurons in the central brain. Male Fru^M knockdown flies sped up the courtship dramatically. They began to court females as soon as they encountered them, skipped the orientation and tapping steps, and performed the other mating ritual behaviors—singing, licking, and attempting to copulate—simultaneously rather than

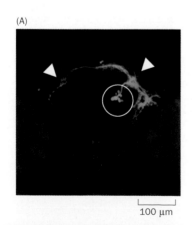
100 μm

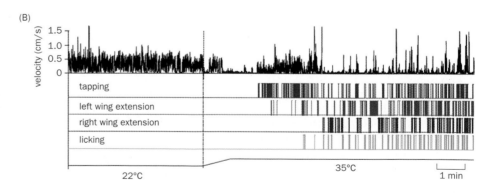

Figure 9–11 Activation of the P1 cluster Fru^M neurons is sufficient to elicit many aspects of courtship behavior. (A) A neuroblast clone that contains the P1 cluster of Fru^M neurons, showing the location of their cell bodies (circle) and their elaborate projections both ipsilaterally (on the right hemisphere) and contralaterally (on the left hemisphere), visualized by the MARCM method (see Sections 7.20 and 13.16). Arrowheads indicate the areas where P1 processes overlap with another group of Fru^M neurons that send descending projections to the ventral nerve cord. **(B)** An ethogram (quantitative plot of behavior over time) of a male fly with a P1 neuroblast clone expressing dTrpA1, which causes neuronal depolarization in response to heat. In response to a temperature step (and hence activation of the P1 neurons), the fly slows down (top) and begins to exhibit tapping, unilateral wing extension, and licking behaviors in isolation. (Adapted from Kohatsu S, Koganezawa M & Yamamoto D [2011] *Neuron* 69:498–508. With permission from Elsevier Inc.)

Table 9–1: Courtship differences between Fru^M knockdown in median bundle neurons and control *Drosophila* males

Courtship assays	Control males (*median bundle-GAL4 × UAS-Gfp*)	Fru^M knockdown males (*median bundle-GAL4 × UAS-Fru RNAi*)
Latency to initiation	94 ± 8 s	8 ± 1 s
Initiation to attempted copulation	111 ± 20 s	4 ± 0 s
Orienting	17/20	Not observed (0/20)
Tapping	15/20	Not observed (0/20)
Wing extension	20/20	20/20
Proboscis extension	19/20	20/20
Attempted copulation	17/20	20/20
Fertility	15/15	0/15

Rows 1 and 2 represent the time required to complete stages of courtship (mean ± standard error), while rows 3–8 present the number of flies that exhibited a specific behavior over the total number of flies examined. (Data from Manoli & Baker [2004] *Nature* 430:564.)

in sequence (**Table 9–1**). The entire ritual, from initiation to attempted copulation, was reduced to 4 seconds. This accelerated courtship came at a cost: males in which median bundle Fru^M was knocked down produced no progeny, likely as a result of failing to properly transfer sperm and seminal fluids. This experiment suggested that median bundle Fru^M neurons play an important role in coordinating the behavioral sequence, and this coordination is essential for the productivity of *Drosophila* mating.

Whereas *Drosophila* exhibits an elaborate mating ritual, other species of flies such as hoverfly and house fly employ a speedy mating strategy similar to that of *Drosophila* males in which Fru^M expression is knocked down in median bundle neurons. These different strategies likely reflect the different social environments of these species. *Drosophila* congregate near ripened fruits, so encounters with potential mates are plentiful. By contrast, insects that exhibit rapid aerial mating (**Figure 9–12**) must initiate and complete copulation quickly in order to have reproductive success. The fact that modulating Fru^M function in a small cluster of neurons can switch these mating modes suggests a common neural circuitry that has been modified through evolution to enable different mating strategies appropriate for particular lifestyles.

Researchers are only beginning to understand how sensory information is transformed into motor output and how behavioral sequences are coordinated during *Drosophila* courtship. For example, it is not known how P1 neurons integrate cues from sensory systems, or how median bundle neurons coordinate the different steps of courtship. One aspect that has started to emerge is the connectivity of P1 neurons with motor control, a topic to which we now turn.

9.8 Fru^M neurons in the ventral nerve cord regulate mating-related behavioral output

A hallmark of male *Drosophila* courtship is singing the courtship song (see Figure 9–2). Females do not extend a single wing, and males do so only when they are courting. Early studies of gynandromorphs—mosaic flies in which some tissues are male and others are female due to random loss of an X chromosome from a progenitor cell during early development—indicated that if certain parts of the thoracic ventral nerve cord are male, the mosaic flies can sing even if the rest of the CNS is female. Given that Fru^M is expressed in a subset of neurons in the ventral nerve cord, it was hypothesized that these neurons control the unilateral wing beat that produces the courtship song. This hypothesis was validated by an experiment utilizing light-induced uncaging of ATP to activate an ATP-gated cation channel expressed in Fru^M neurons (see Section 13.24). Activation of Fru^M

Figure 9–12 Midair mating of hoverflies. Fly species that mate in the air adopt a speedy mating strategy distinct from the elaborate mating ritual of *Drosophila*. (Courtesy of Flagstaffotos.)

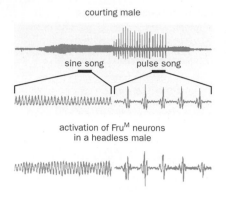

Figure 9–13 Fru^M neurons in the ventral nerve cord control the production of courtship song. Top, the courtship song of a normal courting fly (top) consists of alternating sequences of sine song and pulse song. Bracketed segments are magnified below. Bottom, ventral nerve cord Fru^M neurons in headless male flies that express an ATP-gated cation channel and injected with caged ATP can be activated by light, which induces ATP uncaging (see Figure 13–44 for details of this method). This produces a song that closely resembles the song of a courting male fly. (Adapted from Clyne JD & Miesenbock G [2008] *Cell* 133:354–363. With permission from Elsevier Inc.)

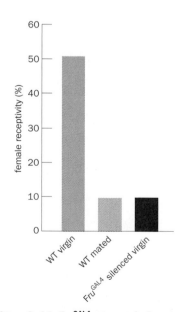

Figure 9–14 Fru^{GAL4} neurons in females promote courtship receptivity. Reversible silencing of *Fru*^{GAL4} neurons at high temperature (following the same procedure as Figure 9–7) causes experimental virgin females (column 3) to behave like wild-type (WT) mated females (column 2), with markedly reduced receptivity compared to WT virgins (column 1). (Adapted from Kvitsiani D & Dickson BJ [2006] *Curr Biol* 16:R355–R356. With permission from Elsevier Inc.)

neurons in the ventral nerve cord in headless flies was sufficient to produce a courtship song that resembles the song of a normal courting fly (**Figure 9–13**). The effectiveness of these songs can be tested in a behavioral assay. When a male's wings are removed, his copulation success is drastically reduced because he cannot attract his partner's attention with a courtship song. Playing back a recorded courtship song complements the mating defect of a wingless male, and the recorded song resulting from artificial activation of Fru^M neurons was as effective as the song of a wild-type courting fly.

Several groups of Fru^M neurons in the ventral nerve cord involved in courtship song production have been identified by intersectional transgene expression methods (see Section 13.12). Artificial activation of one group promotes wing extension but not beating, whereas activation of other groups produces songs with increased or decreased inter-pulse intervals, respectively. These neurons likely connect with each other and receive input from the P1 neurons via a specific type of descending neuron from the brain; activation of either P1 or these descending neurons produces complete songs. Further investigations may reveal whether ventral nerve cord Fru^M neurons constitute a courtship-song-producing central pattern generator (see Chapter 8), how these neurons are activated by descending control, and how coordinated activities of these neurons produce the courtship song.

9.9 Fru^M-equivalent neurons in females promote female receptivity to courtship

So far we have focused on the role of Fru^M and Fru^M neurons in male sexual behavior. How is female sexual behavior controlled? As noted previously, normal females do not express the Fru^M protein. Are the groups of neurons that express Fru^M in males entirely absent in females, or are these neurons present in females but unable to produce Fru^M protein as a result of sex-specific alternative splicing (see Figure 9–4)? *Fru*^{GAL4} was used to distinguish these possibilities, as it mimics transcription from the first *Fruitless* promoter but is not subjected to regulation by sex-specific alternative splicing (see Figure 9–6A). Researchers found that females did indeed express *Fru*^{GAL4} in specific populations of neurons that grossly resemble the Fru^M neurons of males; however, as we will learn in the next two sections, substantial sexual dimorphisms exist in both the number of *Fru*^{GAL4}-expressing neurons and their wiring patterns.

What do *Fru*^{GAL4} neurons do in females? Normally, virgin females are receptive to proper courtship stimuli from males, but mated females drastically reduce their receptivity to further mating by actively rejecting suitors. When *Fru*^{GAL4} neurons in virgin females were reversibly silenced at high temperature by the expression of *UAS-Shi^{ts}*, their receptivity to courtship was greatly diminished, becoming comparable to that of mated females (**Figure 9–14**). Thus, in females the activities of neurons that express *Fru*^{GAL4} promote courtship receptivity in virgin females.

The sensory input that leads to the behavioral switch between virgin and mated females has been identified. About six *Fru*^{GAL4} neurons that innervate the female reproductive tract express a G-protein-coupled receptor for a **sex peptide** that is transferred from the male together with the seminal fluid during mating. These sensory neurons project their axons to the ventral nerve cord, where they modulate central circuits that facilitate female receptivity (**Figure 9–15**). Sex peptide binding is inferred to inhibit the activity of these sensory neurons, and thereby influences subsequent mating behavior of females. This strategy, along with transferring cVA to mated females to inhibit mating by other males, has obvious evolutionary significance: successfully mated males can minimize sperm competition and maximize the likelihood of propagating their own genes.

9.10 Fru^M and Doublesex (Dsx) regulate sexually dimorphic programmed cell death

In previous sections, we have learned that Fru^M neurons regulate male courtship behavior by sensing and integrating mating-specific cues and regulating

pre-mating: SP sensors ON

post-mating: SP sensors OFF

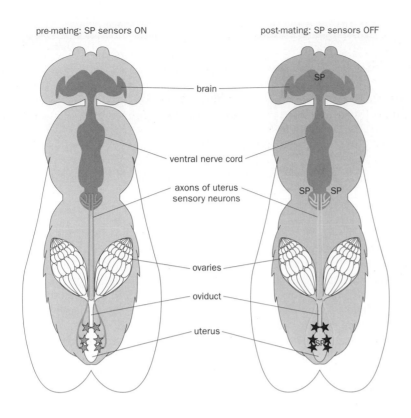

brain

ventral nerve cord

axons of uterus
sensory neurons

ovaries

oviduct

uterus

Figure 9–15 *Fru*^{GAL4} **sensory neurons in females sense the sex peptide and regulate courtship receptivity.** In the absence of sex peptide (SP) in the virgin female (left), *Fru*^{GAL4}-expressing sensory neurons from the uterus, which project to the ventral nerve cord, are active (green), and courtship receptivity is high. After mating, SP transferred from the male to the uterus of the mated female (right) inhibits the activity of these sensory neurons (red), and reduces female receptivity. SP may additionally influence the activities of neurons in the ventral nerve cord and brain directly by diffusing through the hemolymph (equivalent to the blood in vertebrates), as illustrated by the yellow gradient. (Adapted from Clyne JD & Miesenböck G [2009] *Neuron* 61:491–493. With permission from Elsevier Inc. See also Häsemeyer M, Yapici N, Heberlein U et al. [2009] *Neuron* 61:511–518 and Yang C, Rumpf S, Xiang Y et al. [2009] *Neuron* 61:519–526.)

mating-specific behavioral output. *Fru*^{GAL4} neurons in females (equivalent to Fru^M neurons in males) also promote courtship receptivity. How do males and females exhibit these dimorphic sexual behaviors?

Sexual dimorphism in *Drosophila* originates with the sex-specific splicing of *Fruitless* and *Doublesex* and is therefore controlled by the sex determination hierarchy (see Figure 9–4). Sex-specific splicing of *Fru* is a key regulator for sexual behavior, while sex-specific splicing of *Doublesex* (*Dsx*) is responsible for the sexual dimorphism of external morphology and also contributes to sexual behavior. When *Dsx*^{GAL4} (GAL4 knocked in at the *Dsx* locus) was used to visualize *Dsx*-expressing cells in the brain, it was found that Dsx is expressed in both male and female brains; indeed, many *Dsx*^{GAL4} neurons also express *Fru*^{GAL4}. How do *Fru* and *Dsx* cooperate to regulate sexually dimorphic behaviors? Given the principles of neural development we learned in Chapter 7, three possibilities come to mind: *Fru* and *Dsx* could differentiate the male and female nervous systems by generating (1) different numbers of neurons, (2) different wiring diagrams, or (3) different functional properties of neurons expressing one or both of these proteins. These possibilities are not mutually exclusive; while the third mechanism remains to be tested, the first two mechanisms have already been supported by ample evidence.

Overall, *Dsx*^{GAL4} is expressed in about 900 neurons in males but only about 700 neurons in females. However, in some areas of the nervous system there are more *Dsx*^{GAL4} neurons in females than males; one such example is the posterior ventral nerve cord, which is implicated in regulating female reproductive behaviors such as egg laying. When small subsets of *Fru*^{GAL4}-expressing neurons were examined in males and females, sexual dimorphism was found to be prevalent. In the examples we have studied, the P1 cluster in the central brain (see Figure 9–11) is present only in males. By contrast, the sensory neurons in the oviduct that bind the sex peptide (see Figure 9–15) are only found in females.

What mechanisms are used to produce sexually dimorphic neuronal numbers? As we learned in Chapter 7, neuronal number in the developing nervous system can be regulated by controlling the division of progenitor cells to produce new neurons and by programmed cell death after neurons are born. The latter was found to be the predominant mechanism for producing sexual dimorphism. For example, the P1 neurons in the central brain are normally present in the male

Figure 9–16 Programmed cell death accounts for sexual dimorphism of P1 neurons. The P1 cluster of Fru^M neurons (circled) is present in males **(A)** but absent in females **(B)**. When programmed cell death is inhibited in the neuroblast lineage that produces P1 in the left hemisphere, these neurons are retained in females **(C)**. (From Kimura K, Hachiya T, Koganezawa M et al. [2008] *Neuron* 59:759–769. With permission from Elsevier Inc.)

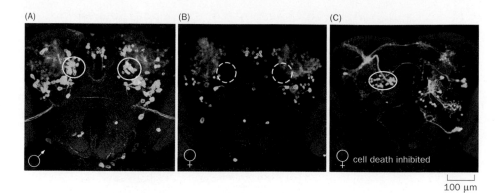

brain, where they express both Fru^M and Dsx^M, but are absent in the female brain (**Figure 9–16**A, B). When genes essential for programmed cell death were eliminated in the lineage that produces P1 neurons, these neurons persisted in the female brain as well (Figure 9–16C). Genetic analysis indicates that the expression of Dsx^F in developing P1 neurons of females activates a cell death program that eliminates these neurons after they are born.

In other cases of dimorphic neuronal numbers, programmed cell death is the default pathway in females, and is inhibited in males by the presence of Fru^M. For example, the male-specific muscle in the abdomen of adult *Drosophila*—named the muscle of Lawrence, or MOL (**Figure 9–17**A)—never forms in females or in *Fru* mutant males. Remarkably, Fru^M is necessary not in MOL, but in the motor neuron that innervates MOL (see Figure 9–6B). In the absence of Fru^M, the motor neuron undergoes programmed cell death; as a consequence, MOL is not induced to form. When Fru^M was expressed in a single MOL-innervating motor neuron in a *Fru* mutant male, MOL induction was restored (Figure 9–17B). Thus, Fru^M cell-autonomously instructs the survival of the motor neuron, which in turn induces the formation of a male-specific muscle.

In summary, sex-specific Fru and Dsx isoforms regulate the cell death program, causing sexually dimorphic neuronal populations in the male and female nervous systems. Indeed, differences in the number of *Dsx*^GAL4-expressing CNS neurons in males and females were largely eliminated when *Dsx*^GAL4 was used to drive a transgene that inhibits programmed cell death. Thus, control of programmed cell death is a predominant mechanism by which sexually dimorphic neuronal numbers are generated.

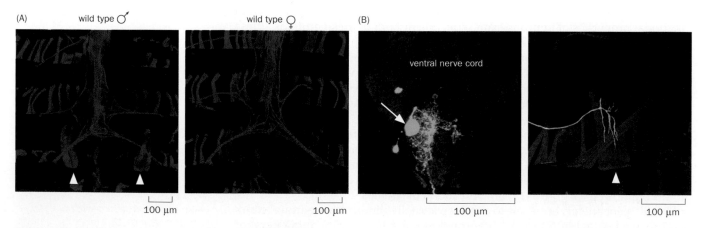

Figure 9–17 Fru^M in a motor neuron prevents its death and promotes the induction of a male-specific muscle. (A) The bilaterally symmetrical muscle of Lawrence (MOL) is present in the fifth abdominal segment in *Drosophila melanogaster* males (arrowheads in the left panel), but not in females (right panel) or *Fru* mutant males

(not shown). **(B)** Expression of Fru^M in a single motor neuron (arrow in the left panel) in a *Fru* mutant male resulted in the induction of MOL (arrowhead) innervated by the axon of that motor neuron (right panel). (From Nojima T, Kimura K, Koganezawa M et al. [2010] *Curr Biol* 20:836–840. With permission from Elsevier Inc.)

9.11 Dsx and Fru^M control sexually dimorphic neuronal wiring

In addition to differences in neuronal number, some Fru^M and Dsx neurons also exhibit sexually dimorphic wiring. For example, while eliminating programmed cell death permitted P1 neurons to survive in females (see Figure 9–16C), the projection patterns of these neurons differed from those of normal males. Additional expression of Fru^M in these surviving neurons instructed male-typical projection patterns. Thus, Dsx and Fru^M act together to regulate the differentiation of P1 neurons in males: the lack of Dsx^F is responsible for their survival, and the presence of Fru^M instructs their wiring patterns.

Fru^M and Dsx can also act together to regulate sexually dimorphic neuronal wiring. For example, taste receptor neurons on the foreleg exhibit sexually dimorphic wiring: their central axons cross the midline in males but not in females (**Figure 9–18**). Genetic analysis indicates that midline crossing is promoted by Fru^M and Dsx^M, but inhibited by Dsx^F. Fru^M likely exerts its action by inhibiting Robo, an axon guidance receptor that plays an evolutionarily conserved role in regulating midline crossing (see Section 7.5). This finding links the regulators of sexual behavior with an axon guidance molecule that controls the wiring of the nervous system.

A third example of sexually dimorphic neuronal wiring relates to the processing of the male-produced pheromone, cVA. As discussed in Section 9.6, cVA inhibits courtship of males with other males or mated females. In addition, cVA also promotes male–male aggression, an evolutionary conserved behavior related to mating; males exhibit aggression to defend territories and compete for female mating partners. Interestingly, cVA also promotes female courtship, as virgin females mutant for the cVA receptor exhibit reduced mating behavior. How does the same pheromone produce sexually dimorphic behaviors? Within the cVA processing pathway (see Figure 9–9B), the DA1 olfactory receptor neurons (ORNs) and projection neurons (PNs) in males and females produce indistinguishable physiological responses to cVA application. A recent study identified sexual dimorphism between connections of the DA1 PNs and 3rd order neurons in the lateral horn. A group of lateral horn neurons called 3A (which are the same as DC1 neurons in Figure 9–9B) exhibit sexually dimorphic dendrite projections such that only male 3A dendrites have substantial overlap with axons of DA1 PNs carrying the cVA signal (**Figure 9–19**A). Another group of lateral horn neurons, 3B, also exhibit sexually dimorphic dendrite projections such that only female 3B dendrites overlap substantially with DA1 PN axons (Figure 9–19B). Indeed, electrophysiological recordings indicated that cVA preferentially activates 3A neurons in males and 3B neurons in females. Both 3A and 3B neurons express Fru^M, and

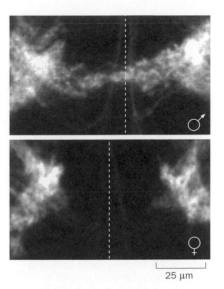

Figure 9–18 Sexually dimorphic wiring of taste receptor neurons. The central axonal projections of taste receptor neurons located in the foreleg cross the midline (dashed) in males but not females. Both Dsx and Fru^M contribute to this sexual dimorphism. (After Mellert DJ, Knapp JM, Manoli DS et al. [2010] *Development* 137:323–332. With permission from the Company of Biologists.)

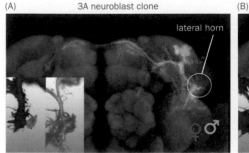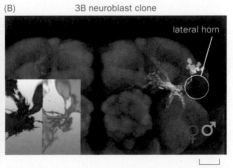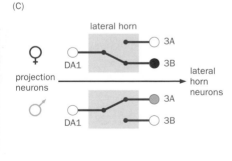

Figure 9–19 Sexually dimorphic switch in a pheromone-processing pathway. (A) A neuroblast clone that contains 3A neurons in a male (green) and in a female (magenta) are registered to a standard brain based on the neuropil staining (gray). Male and female 3A neurons exhibit substantial differences in their dendrite projections in the lateral horn. The inset shows that male but not female 3A dendrites overlap substantially with axons of DA1 projection neurons (yellow). **(B)** 3B neurons also exhibit sexual dimorphism, with female but not male dendrites overlapping substantially with axons of DA1 projection neurons (inset). **(C)** A circuit switch model. Along the olfactory pathway from projection neurons to lateral horn neurons, DA1 projection neurons connect with 3B lateral horn neurons in females (top) but with 3A lateral horn neurons in males (bottom). Male patterns of dendrite projection in both 3A and 3B neurons are controlled by Fru^M. (Adapted from Kohl J, Ostrovsky AD, Frechter S et al. [2013] *Cell* 155:1610–1623. With permission from Elsevier Inc.)

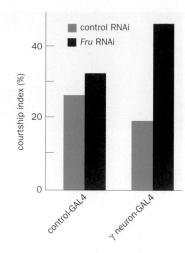

Figure 9–20 Fru^M regulates courtship conditioning. After being repeatedly rejected by mated females, males with RNA interference (RNAi) against *Gfp* or against *Fru* in control neurons reduce their courtship index. Flies in which Fru^M expression is knocked down in a subset of mushroom body neurons (the γ neurons) still exhibit a high courtship index, indicating a deficiency in courtship conditioning. (Adapted from Manoli DS, Foss M, Villella A et al. [2005] *Nature* 436:395–400. With permission from Macmillan Publishers Ltd.)

require Fru^M to produce their male forms of dendrite projection. Thus, Fru^M controls a circuit switch, promoting DA1 PN → 3A connections and inhibiting DA1 PN → 3B projections in males, resulting in sexually dimorphic wiring patterns (Figure 9–19C); this circuit switch may contribute to sexually dimorphic behavior elicited by the same pheromone.

9.12 Even innate behavior can be modified by experience

The epigraph by Spalding is only partially correct in its description of butterfly life, because learning influences even innate behaviors such as courtship. When a male fly tries to court a mated female, he is rejected repeatedly because mating changes the female's receptivity (see Section 9.9). A normal male learns from the experience of repeated rejections, and reduces his courtship attempt; this process is called **courtship conditioning**. The mushroom body neurons in the central brain, which receive olfactory input from projection neurons (see Figures 6–26 and 6–33), are essential for this experience-dependent modification of behavior. Remarkably, courtship conditioning is also regulated by Fru^M, which is expressed in a subset of mushroom body neurons. Courtship conditioning was blocked if Fru^M expression was knocked down using RNAi in this subset of mushroom body neurons: males did not learn from repeated failures and did not reduce courtship after being repeatedly rejected (**Figure 9–20**). Here is an example of the intricate interplay between nature and nurture: an innate behavior can nonetheless be modified by experience, and the genetic program that specifies innate behavior also controls its experience-dependent modification. The production of courtship song by songbirds provides another excellent example of the interplay between nature and nurture (**Box 9–1**).

In summary, studies of two key genes, *Fruitless* and *Doublesex*, have exemplified how regulatory genes specify complex behaviors by building into the nervous system circuit elements that enable sensory perception, motor action, sensory-motor integration, and experience-dependent modification. Molecular-genetic approaches have been used to assemble an impressive list of circuit elements. The next major challenge is to study how these circuit elements operate. What are the activities of neurons in these circuits during courtship? How do neurons that are implicated in different steps of courtship connect with each other? What are the principles that regulate information flow in these circuits to produce exquisitely coordinated behavior? By utilizing tools in electrophysiology, imaging, and circuit manipulation together with molecular-genetic and behavioral analysis (see Chapter 13), the answers to these questions will be within reach in the near future.

Two final questions: what can sexual behavior in flies teach us about sexual behavior in mammals, including our own species? Are there common principles in sexual behavior and their neural control that are shared across diverse organisms? Let's continue our journey to find out.

HOW ARE MAMMALIAN SEXUAL BEHAVIORS REGULATED?

In the second part of this chapter, we investigate the mechanisms that regulate mammalian sexual behavior. Just like fruit flies, the mammalian models studied to date—mostly rodents—engage in sexual and reproductive behavior that has a large innate component. Male rodents **mount** females, whereas female rodents exhibit **lordosis** behavior when sexually aroused, assuming a posture that facilitates sexual intercourse. In addition, males show aggressive behavior toward intruders (particularly sexually mature males) in order to defend their territory. Females exhibit maternal behavior after giving birth to pups, including nest building, pup retrieval, and nursing. How are these behaviors controlled by the nervous system? What is the origin of such sexually dimorphic behavior? As in the case of the fly, the genetic origins of sexual dimorphism in mammals are the sex chromosomes, which provide an experimental entry point to investigate the mechanisms of sexual behavior.

Box 9–1: Bird song: nature, nurture, and sexual dimorphism

Thousands of species of birds use song as a form of communication; often it is only the males that sing. A male songbird sings to females to signal his species and individual identity, his location, and his readiness to mate. His song also conveys to other males information with respect to territory and neighbor-versus-stranger status. Bird songs can be represented as notes and syllables in the time–frequency sound spectrograms (**Figure 9–21**A). Each species has its characteristic song(s). In the past decades, bird song research has made contributions to many areas of neurobiology.

The final song a mature bird sings is a product of extensive interplay between nature and nurture. During an early **sensory stage**, a young male bird hears and memorizes the song of a tutor, usually his father. This is followed by a **sensorimotor stage**, when the young bird starts to produce his own immature song. He uses auditory feedback to compare his own song with the tutor song template he has memorized. Through trial-and-error, his song comes to more closely resemble the tutor song, until it acquires its mature, crystallized form. In some species such as the white-crowned sparrow, the sensory and sensorimotor stages are completely separated in time, whereas in other species, such as the extensively studied zebra finches, these stages overlap (Figure 9–21B).

While many features of bird song are learned, there are interesting aspects that seem to be innate. If a bird has been raised in acoustic isolation during the sensory stage, he can only sing a rudimentary **innate song**, which is nevertheless species specific, reflecting the nature aspect of bird song acquisition. If a bird has been exposed only to the song of a different species during the sensory period, his song can take the other species' form, reflecting the influence of nurture. If a bird has been exposed to songs of his own species and others, he preferentially learns from and sings the song of his species; in other words, birds are predisposed to learn their species' song. Finally, auditory input is required not only during the sensory period, when the bird listens to the tutor song, but also during the sensorimotor stage, when the bird must compare what he sings to what he has memorized. If a bird is deafened after the sensory stage but before the sensorimotor stage, his song remains immature. Indeed, if a bird raised in acoustic isolation is additionally deafened during the sensorimotor stage, he sings a song that is different from the innate song of acoustically isolated birds that have not been deafened. Thus, trial-and-error learning during the sensorimotor stage is essential even for the innate song.

Lesion, anatomical, and physiological studies have identified neural circuits in the brain involved in song production, auditory feedback, and song learning (Figure 9–21C). Two pallial (dorsal forebrain) nuclei analogous to motor cortical areas in mammals, the **HVC** (high vocal center) and **RA** (robust nucleus of the arcopallium), are essential

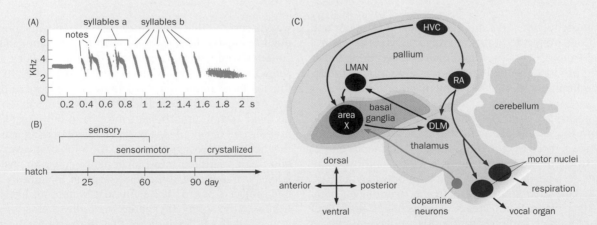

Figure 9–21 Bird song: composition, ontogeny, and neural circuits. (A) Time–frequency sound spectrogram of a white-crowned sparrow's song. A note is a continuous marking on the spectrogram; two or more notes may group together to form a syllable. **(B)** Developmental stages for zebra finches in days after hatching. During the sensory stage, a young bird listens to and memorizes a tutor song, usually from his father. During the sensorimotor stage, he sings an immature song and compares it with the tutor song he memorizes. The song takes its mature form at the crystallized stage. **(C)** A side view of a songbird brain depicting neural circuits for song production and learning. The song motor pathway (red), consisting of HVC (high vocal center), RA (robust nucleus of the

arcopallium), and the brainstem motor nuclei that regulate muscle contraction in the vocal and respiratory systems, is responsible for song production. The anterior forebrain pathway (black), consisting of LMAN (lateral magnocellular nucleus of the anterior nidopallium), area X, and DLM (medial dorsolateral thalamus), is essential for song learning. Dopamine neurons project to and modulate neurons in area X. (A, adapted from Konishi M [1985] *Annu Rev Neurosci* 8:125–170. With permission from Annual Reviews Inc.; B, adapted from Brainard MS & Doupe AJ [2002] *Nature* 417:351–358. With permission from Macmillan Publishers Ltd; C, adapted from Brainard MS & Doupe AJ [2013] *Ann Rev Neurosci* 36:489–517.)

(Continued)

Box 9–1: Bird song: nature, nurture, and sexual dimorphism

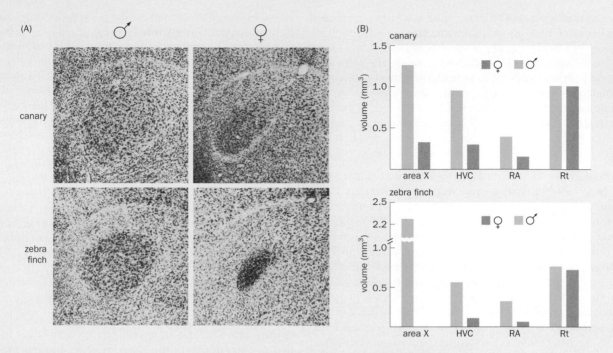

Figure 9–22 Sexual dimorphism in vocal control areas of songbirds. (A) The RA nucleus in the canary (top) and zebra finch (bottom) is larger in males (left) than in females (right). These sections are stained with cresyl violet, a basic dye that stains cell bodies. The RA nuclei are more darkly stained in the center of each section, surrounded by lightly stained bands. **(B)** Volumes of nuclei that are part of the song circuits (area X, HVC, and RA) are larger in males than in females for both canaries (top) and zebra finches (bottom). Area X is unrecognizable in zebra finches. The Rt nucleus (nucleus rotundus) is unrelated to vocal control and has a similar volume in males and females. (Adapted from Nottebohm F & Arnold AP [1976] *Science* 194:211–213. With permission from AAAS.)

for song production. HVC neurons project to the RA, and RA neurons project to brainstem motor nuclei that control muscle contraction in the vocal organ and breathing center for song production. While not essential for song production, forebrain nuclei **LMAN** (lateral magnocellular nucleus of the anterior nidopallium) and **area X** are instrumental for song learning. Area X is located in an anterior forebrain structure that is analogous to the basal ganglia in mammals. Area X receives input from the HVC and sends output to the RA via the DLM (a thalamic nucleus) and LMAN, to regulate the transmission of information between the HVC and RA. Like their counterparts in the mammalian basal ganglia (see Figure 8–22), area X receives modulatory input from midbrain dopamine neurons; this input contributes to trial-and-error-based learning. Auditory information is processed in specific forebrain areas analogous to the mammalian auditory cortex, which sends input to the song system through the HVC.

Scientists have used songbird models to investigate a wide range of fascinating topics, such as the neural mechanisms of song production and song learning. Indeed, bird song is one of the best models for human language. We end this box on a question related to the theme of this chapter: why is singing a male-typical trait in many bird species? In the 1970s, it

was discovered that brain nuclei associated with song production and learning exhibit robust sexual dimorphism in canaries, where males sing much more than females, and in zebra finches, where only males sing (**Figure 9–22**). In zebra finches, for example, the HVC, RA, and area X are rudimentary or unrecognizable in females. Mechanisms that underlie these sexual dimorphisms have been more extensively investigated in canaries. A particularly interesting feature is that new neurons are produced in the adult brain every year and being recruited into functional circuits; for instance, adult-born neurons in the HVC respond to auditory input and project axons to the RA. These new neurons likely contribute to the ability of canaries to modify their song every year. Further studies indicate that the male sex hormone plays an important role in producing sexual dimorphism; it promotes the survival of adult-born neurons in the HVC by up-regulating the expression of brain-derived neurotrophic factor (BDNF, see Section 7.15). Providing females with the male sex hormone or infusing female brains with BDNF both promote the survival of new HVC neurons in females. We will study the mechanisms of sex hormone action in mammalian sexual dimorphism and sexual behavior in great detail in the second part of this chapter.

9.13 The *Sry* gene on the Y chromosome determines male differentiation via testosterone production

Whereas sex in fruit flies is determined by the ratio of X chromosomes to autosomes (see Figure 9-4), in most mammals (including mice and humans), sex is determined by the presence or absence of the Y chromosome (**Figure 9–23**). People with a single X chromosome but no Y chromosome develop into females (Turner syndrome, occurring in 1 out of 2500 girls), whereas people with two X chromosomes and a Y chromosome develop into males (Klinefelter's syndrome, occurring in 1 out of 1000 boys).

A key event in mammalian sex differentiation occurs during mid-embryogenesis when a structure called the genital ridge differentiates into testes in males and ovaries in females. It was discovered in the 1950s that when the genital ridge was removed *in utero* prior to overt sexual differentiation, all embryos differentiated as females. This experiment suggested that female differentiation follows a default pathway, and the presence of testes instructs the male differentiation pathway. Thus, the Y chromosome must contain some genetic factor(s) that exert its sex determination effect by promoting testis development.

A single gene, **Sry** (*Sex determining region Y*), was identified in the 1990s as the illusive testis-determining factor (Figure 9–23). *Sry* is located on the Y chromosome in all mammals whose sex is determined by the presence or absence of the Y chromosome, and is expressed in the genital ridge during sexual differentiation. In humans, XY subjects with loss-of-function *Sry* mutations develop into females despite having the rest of the Y chromosome intact; this indicates that *Sry* is necessary for male differentiation. Moreover, XX mice carrying an *Sry* transgene on an autosome develop into males, with characteristic male reproductive system and male sexual behavior; this indicates that the presence of *Sry* in females is sufficient to convert them into males. *Sry* encodes a transcription factor that regulates gene expression. As we will see, however, the signals downstream of *Sry*-mediated control of sex determination—namely the sex hormones testosterone, estrogen, and progesterone—have more complex and interactive roles in their control of sexual dimorphisms and behaviors in mammals than do the cell-autonomous mechanisms that mediate sexual differentiation in flies.

A major function of the testes during embryonic development (Figure 9–23) is to produce **testosterone**, a steroid hormone that promotes the development of the male reproductive system (masculinization) and inhibits the development of the female reproductive system (de-feminization). In fact, much of the testes' effect on the development of the male reproductive system can be mimicked by applying exogenous testosterone during proper stages of embryonic development. So the key questions are: How does testosterone act to specify sexual differentiation? Does it also regulate sexual behavior? What is the relationship between sexual differentiation during development and male- and female-typical sexual behavior in adults? We will investigate these questions in the following sections.

9.14 Testosterone and estradiol are the major sex hormones

Testosterone (**Figure 9–24**A) was originally purified from adult male testes as an **androgen** (male sex hormone). Sexually inexperienced, **castrated males** (that is, males with the testes removed) do not exhibit male-typical sexual behaviors, such as mounting of females and aggression toward intruder males. Acute injection of testosterone into adult castrated males can restore these behaviors, indicating an activational effect of testosterone in male sexual behavior.

Testosterone is synthesized from cholesterol in the testes. Because of its hydrophobic nature, testosterone can freely diffuse across the plasma membrane to enter target cells, where it binds to the **androgen receptor**. This binding triggers nuclear translocation of the testosterone–androgen receptor complex, which acts as a sequence-specific DNA-binding transcription factor to activate or repress target gene expression (Figure 9–24B). A testosterone metabolite,

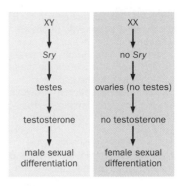

Figure 9–23 Sex determination and sexual dimorphism in mammals. In most mammals, the *Sry* gene on the Y chromosome specifies testis development. Testosterone produced by the testes regulates male differentiation. Female differentiation occurs as a default in the absence of the Y chromosome, the *Sry* gene, testes, and testosterone.

Figure 9–24 Sex hormones and their mechanisms of action. (A) Structure of testosterone, a male sex hormone (androgen). **(B)** As a steroid hormone, testosterone diffuses across the plasma membrane, binds to the androgen receptor, and causes nuclear translocation of the complex. The complex binds to the promoters of target genes to regulate their transcription. **(C)** Structure of estradiol, a female sex hormone (estrogen). **(D)** Estradiol has a mode of action analogous to testosterone and regulates the expression of estrogen-receptor-responsive genes. Intracellular estradiol can derive either from circulating estradiol, or from circulating testosterone that is converted to estradiol by the action of the aromatase enzyme. The androgen and estrogen receptors both act as homodimers, and there are two estrogen receptor variants; for simplicity the androgen and estrogen receptors are represented here by a single molecule. Not shown here, the steroid hormone progesterone acts in a similar manner.

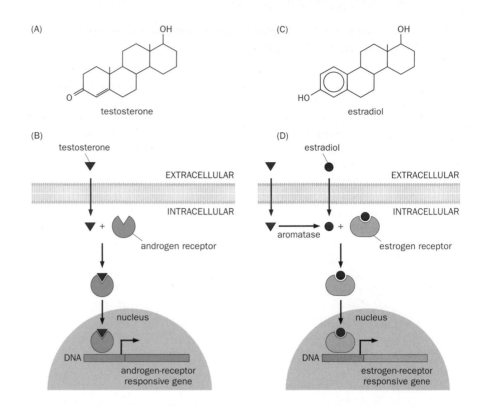

dihydrotestosterone (DHT), is a more potent activator of the androgen receptor and is responsible for external genital masculinization during development.

Estradiol (Figure 9–24C), the major **estrogen** (female sex hormone), is a steroid hormone made by the ovaries of sexually mature females. While estradiol is often regarded as the female counterpart of testosterone, we will see below that it also plays important roles during male development. Like the effect of testosterone on castrated males, application of estradiol to **ovariectomized females** (that is, females from which the ovaries have been removed) can restore female-typical sexual behavior, such as lordosis. (Co-application of **progesterone**, another steroid hormone that regulates female sexual behavior, augments the effect of estradiol.) Estradiol's mode of entry into cells—diffusion across the plasma membrane's lipid bilayer—likewise reflects its biochemical similarity to testosterone. However, once inside a cell, estradiol acts by binding its own receptors, the **estrogen receptors** α and β encoded by two different genes. The estrogen receptors are also DNA-binding transcription factors, but their sequence specificity differs from that of the androgen receptor, so that they regulate the expression of a different set of target genes (Figure 9–24D). In addition to transcriptional responses, estradiol is also known to exert more rapid effects on target tissues such as changes of intracellular Ca^{2+} concentration and protein phosphorylation. A third estrogen receptor, a G-protein-coupled receptor named GPER (for G-protein-coupled estrogen receptor) localized to the endoplasmic reticulum membrane, mediates some of these rapid effects.

Importantly, cells that express an enzyme called **aromatase** can convert testosterone (but not DHT) to estradiol (Figure 9–24D). In these cells, therefore, testosterone can exert its action by two means: (1) by binding to an androgen receptor directly, or (2) by binding to an estrogen receptor after being converted to estradiol. This explains the observation that estradiol application, just like testosterone application, stimulates males to fight with intruder males. Given these equivalent responses, the mechanism by which testosterone promotes aggression toward intruders most likely involves its conversion to estradiol and subsequent action through the estrogen receptors. However, estradiol application to females stimulates lordosis toward intruding males. The sexually dimorphic behavioral effects elicited by the application of the same exogenous hormone suggest inherent differences between male and female brains. What are these differences, and what is their origin?

9.15 Early exposure to testosterone causes females to exhibit male-typical sexual behavior

In 1959, an experiment was reported that tested the effects of early testosterone exposure on the sexual behavior of adults. Testosterone was injected into pregnant female guinea pigs. The genitalia of some female progeny were found to be partially or fully transformed to resemble those of males, in accord with the role of testosterone in regulating the development of the male reproductive system. In order to determine the long-lasting effect of prenatal exposure to testosterone, the gonads were removed from adult females that had been prenatally exposed to testosterone, as well as from control females that had not been exposed to testosterone. Sex hormones were then administered to these females to elicit sexual behavior. When given a combination of estradiol and progesterone to induce sexual arousal, control gonadectomized females exhibited robust lordosis; females that were exposed to testosterone prenatally had drastic impairment of lordosis, but displayed male-like mounting behavior. When given exogenous testosterone during testing, females that were exposed to testosterone prenatally, but not control gonadectomized females, displayed male-typical mounting behavior similar to that of castrated males receiving the same amount of exogenous testosterone. These changes in sexual behavior appeared to be permanent, and did not occur if testosterone was given to females after early development.

This experiment suggested that prenatal testosterone has an 'organizational' activity that biased the nervous system toward a male configuration, just like its effect on external genitalia. Females that were exposed to testosterone during embryonic development had their brains configured in a male-typical manner and therefore exhibited male sexual behavior when supplied as adults with testosterone or estrogen to 'activate' such behavior. This **organization–activation model** for the dual roles of sex hormones (**Figure 9–25**) has received much experimental support in the past 50 years, and has become a central principle in the field of endocrinology.

The organizational effect of testosterone has a sensitive window of action. The 1959 study indicated that the sensitive window of guinea pig is prenatal. Analogous experiments on rats and mice revealed that their sensitive period starts shortly before birth and spans the first 10 days after birth (Figure 9–25). This sensitive period coincides with peak testosterone production in male rodents. Due to the

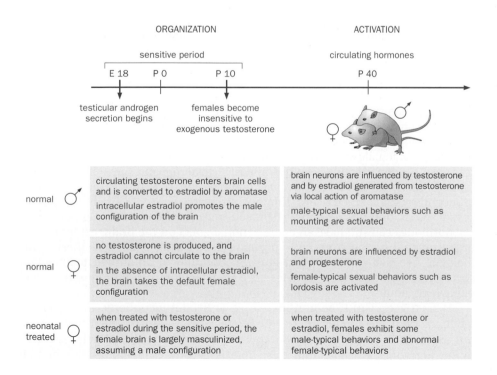

Figure 9–25 The organization–activation model for sex hormone action. Originally proposed by Phoenix et al. (*Endocrinology* 65:369, 1959) to explain the behavior of guinea pigs treated prenatally with testosterone, this model has been validated and expanded by many subsequent experiments. The timing of events shown here corresponds to that seen in rats and mice. E, embryonic; P, postnatal. Additional experimental evidence on the specific actions of sex hormones is discussed in Section 9.16. The organization period for certain brain areas may extend to puberty, as discussed in Section 9.17.

male

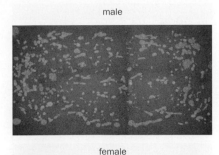

female

NE female

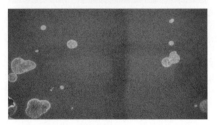

Figure 9–26 Territorial marking behavior in male, female, and neonatal estrogen-treated female mice. Male mice urinate throughout their cages (top panel), a typical territory-marking behavior. Female mice urinate in a corner (arrow in the middle panel). Female mice treated neonatally with estrogen (NE) partially adopt the male urination pattern (bottom). (From Wu MV, Manoli DS, Fraser EJ et al. [2009] *Cell* 139:61–72. With permission from Elsevier Inc.)

ease with which sex hormones can be applied in conjunction with gonadectomy in postnatal animals, rats and mice have become the model organisms of choice for studying hormonal regulation of sexual behavior.

9.16 Testosterone exerts its organizational effect mainly through the estrogen receptors in rodents

During embryonic development in rodents, the testosterone produced by the testes of male fetuses can circulate through the blood to reach the brain. By contrast, estradiol, which originates from the placenta, is bound to and sequestered by circulating α-fetoprotein, a protein secreted by fetal liver cells into the bloodstream. This means that estradiol cannot reach the brains of males or females via circulation. However, estradiol can be produced locally within the brain cells of males by the action of aromatase on testosterone (see Figure 9–24D). During the early postnatal period, while the testes continue to produce testosterone, estradiol production by female ovaries is negligible. Therefore, only male brains can be affected by estradiol during early development.

In principle, testosterone can exert its organizational effect in normal male brains (and in female brains that have early exposure to testosterone) through one or both of its mechanisms of action, namely activation of the androgen receptor, or aromatization to estradiol and subsequent activation of the estrogen receptors. Several lines of evidence suggest that the organizational effect is mainly through the second mechanism. First, female mice treated with estradiol during the first 10 days after birth exhibit male-typical behaviors such as territorial marking (**Figure 9–26**) and aggression toward intruders. Early estradiol treatment also reduces female receptivity to mating. Second, aromatase knockout male mice exhibit profound defects in male-typical behaviors. Although this experiment in itself does not distinguish whether the defects are caused by an organizational role of estradiol during development or an activational role of estradiol in adults, in conjunction with the first experiment it suggests an organizational role. Third, although androgen receptor is expressed in the adult brain, its brain expression during the sensitive period of organization appears negligible. Indeed, male mice with brain-specific knockout of the androgen receptor still exhibit qualitatively similar male-typical mating behavior, albeit at a reduced level.

9.17 Dialogues between the brain and gonads initiate sexual maturation at puberty and maintain sexual activity in adults

In mammals including rodents and humans, sexual maturation occurs at puberty. A specialized group of neurons in the preoptic area of the hypothalamus starts to release, in a pulsatile fashion, high levels of **gonadotropin-releasing hormone** (**GnRH**, also called LHRH for luteinizing-hormone-releasing hormone). GnRH travels via the portal vessels in the pituitary stalk to the anterior pituitary, where GnRH stimulates the release of **gonadotropins—luteinizing hormone** (**LH**) and **follicle-stimulating hormone** (**FSH**)—by pituitary endocrine cells. LH and FSH then circulate via the blood to reach the gonads, stimulating the maturation of the male testes or female ovaries (**Figure 9–27**; see also Table 8–1 and Section 8.14). In response, the gonads release sex hormones: testosterone in males and estradiol in females. Testosterone and estradiol act in the periphery to trigger the maturation of secondary sex characteristics. They also circulate back to the brain to influence sexual maturation during puberty. Indeed, recent evidence has suggested that sex hormones can still have an organizational effect on the brain, extending the organizational period from early postnatal (see Figure 9–25) to puberty. This dialogue between the brain and gonads continues in adulthood to maintain sexual activity, where sex hormones have an activation role in promoting sexual behavior in adults.

What triggers GnRH neurons to release GnRH during puberty, and how does feedback from the gonads control GnRH release in adult? (GnRH neurons do not express sex hormone receptors themselves so cannot be directly regulated by sex hormones.) Insights into these questions came from studies of a human

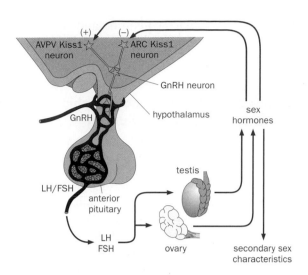

Figure 9–27 Communication between the brain and the gonads during puberty and in adults. Puberty is marked by activation of gonadotropin-releasing hormone (GnRH) neurons in the hypothalamus. Kisspeptins release from Kiss1 neurons in the arcuate nucleus (ARC) and anteroventral periventricular nucleus (AVPV) stimulate GnRH release from GnRH neurons, which express the Kiss1R. GnRH circulates to the anterior pituitary, where it stimulates local cells to produce luteinizing hormone (LH) and follicle-stimulating hormone (FSH). The bloodstream transports LH and FSH to the gonads (testes or ovaries), which respond by producing steroid sex hormones that regulate the differentiation of secondary sex characters during puberty. Sex hormones also circulate back to the brain to affect further sexual differentiation in puberty and to activate sexual behavior in adults. Specifically, negative feedback to Kiss1 neurons in ARC provides a homeostatic regulation of GnRH release, whereas positive feedback to Kiss1 neurons in AVPV in females facilitates a pre-ovulatory surge of GnRH release during the estrous cycle. (Adapted from Sisk CL & Foster DL [2004] *Nat Neurosci* 7:1040– 1047. With permission from Macmillan Publishers Ltd. See also Pinilla L, Aguilar E, Dieguez C et al. [2012] *Physiol Rev* 92:1235–1316.)

condition called **hypogonadotropic hypogonadism** (HH), in which patients display delayed, reduced, or absent puberty due to reduced gonadotropin levels. HH has diverse causes, including mutations in genes that produce GnRH, the GnRH receptor, or failure of GnRH neurons to migrate during development—remarkably, GnRH neurons are born in the epithelia of the vomeronasal organ in the accessory olfactory system (see Box 6–1) and migrate to preoptic areas of the hypothalamus during embryonic development. One cause of HH, discovered in 2003, was mutations in a gene encoding the G-protein-coupled receptor **Kiss1R** (previously named GPR54), a receptor for hypothalamic neuropeptides called **kisspeptins** (encoded by the *Kiss1* gene). Subsequently, mutations in the human *Kiss1* gene have also been found in some HH patients, and mouse knockouts for *Kiss1r* and *Kiss1* displayed hypogonadotropic hypogonadism phenotypes analogous to those seen in human patients. Importantly, a lack of luteinizing hormone release due to *Kiss1r* mutation in humans can be suppressed by supplying exogenous GnRH, suggesting that kisspeptins/Kiss1R act upstream of GnRH release. Indeed, Kiss1 neurons (neurons that produce kisspeptins) project axons to GnRH neurons and their axon terminals, and GnRH neurons express Kiss1R. Application of kisspeptins *in vivo* can stimulate GnRH release. Together, these data demonstrate that the kisspeptins/Kiss1R system serves as an important upstream activator of GnRH neurons during puberty (Figure 9–27). Additional factors that trigger the activation of GnRH and Kiss1 neurons during puberty include nutritional factors such as the leptin discussed in Chapter 8.

Because Kiss1 neurons express sex hormone receptors, they also provide a key step in the feedback regulation of GnRH release by gonads in adults. For example, testosterone and estradiol both down-regulate *Kiss1* expression in the arcuate nucleus, thereby providing a negative regulatory loop for homeostatic regulation (see Figure 8–34B) of GnRH release and sex hormone production. At the same time, *Kiss1* in the **anteroventral periventricular nucleus** (**AVPV**) of the hypothalamus is positively regulated by estradiol in rodents; this positive feedback loop has been proposed to produce the pre-ovulatory surge of gonadotropin release during estrous cycles (Figure 9–27).

In summary, extensive dialogs between the brain and gonads occur both during development and in adults. During early development, testosterone exerts a major organizational effect to configure the brain in a male-typical form. In rodents this is mainly achieved through testosterone being converted to estradiol and acting via estrogen receptors during a sensitive period of early development. In females, estrogen is excluded from the brain during early development and therefore the brain is configured in a female-typical form in the absence of any sex hormone activity. This organization period may extend to puberty, where a surge of sex hormones in both males and females due to GnRH release continue to configure the brain in a sex-typical manner. In adults, testosterone and estradiol act on the male or female brains to activate male- and female-typical reproductive behaviors, respectively. Kiss1-producing neurons, and likely additional hypothalamic neurons that express sex hormone receptors, provide pivotal feedbacks of sex hormones on the regulation of GnRH release, which in turn control sex hormone production (Figure 9–27).

9.18 Sex hormones specify sexually dimorphic neuronal numbers by regulating programmed cell death

Having introduced the organizational roles of sex hormones during development, we now explore the differences between male and female brains that might account for the differential responses of adult males and females to sex hormones. Just like the fruit flies we discussed earlier in the chapter, mammals exhibit sexual dimorphism in neuronal numbers and projection patterns.

Sexual dimorphism in neuronal number was first discovered in the **medial preoptic area** (**MPOA**) in the anterior hypothalamus, an area implicated by lesion studies as essential for male courtship behavior including mounting, intromission (insertion of the penis into the vagina), and ejaculation. Nuclear staining indicated that the MPOA was several-fold larger in male rats compared to female rats (**Figure 9–28**A). Castration immediately after birth caused a reduction in the size of the male MPOA, whereas treating females with testosterone at birth caused an increase in MPOA size (Figure 9–28B). Further studies indicated that this difference is caused by programmed cell death that occurs only in females. In the male brain, sex hormones act during the sensitive period around birth to inhibit cell death programs in MPOA neurons. Similar sexual dimorphism in the anterior hypothalamus has also been found in humans.

Since the discovery of sexual dimorphism in the MPOA, additional sexual dimorphisms have been found in other brain areas, including the **medial amygdala** and **bed nucleus of stria terminalis** (**BNST**). Both are basal nuclei in the forebrain that lesion studies have implicated in regulating male courtship behavior. As in the MPOA of the hypothalamus, males have more cells than females in both the medial amygdala and BNST, and sexual dimorphism is caused by the action of neonatal hormones preventing the programmed cell death of neurons in the male brain. Expression of an inhibitor of programmed cell death resulted in elevation of neuronal numbers in females equal to those of males.

Not all sexually dimorphic brain areas are bigger in males. The anteroventral periventricular nucleus (AVPV) in the hypothalamus is larger in females than males, consistent with its major role in regulating the female ovulatory cycle (see Section 9.17). Interestingly, this sexual dimorphism is also created during the hormone-sensitive period around birth. In this case, estradiol derived from the aromatization of testosterone acts to promote programmed cell death of neurons in males.

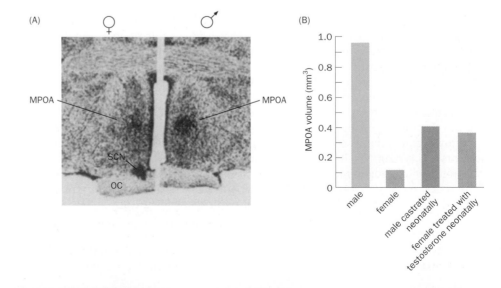

Figure 9–28 Sexual dimorphism in the medial preoptic area (MPOA) of the hypothalamus. (A) Nuclear staining of female (left) and male (right) rat brains revealed a striking size difference in a densely stained nucleus in the MPOA of the anterior hypothalamus (arrows). The absence of the suprachiasmatic nucleus (SCN) in the male brain is an artifact of the section plane. OC, optic chiasm. **(B)** Quantitative analyses show that the MPOA is eight times larger in males than females. The MPOA size is markedly reduced in males that were castrated as neonates, whereas females treated neonatally with testosterone have increased MPOA size. (Adapted from Gorski RA, Gordon JH, Shryne JE et al. [1978] *Brain Res* 148:333–346. With permission from Elsevier Inc.)

An interesting case of sexual dimorphism occurs in neurons of the spinal nucleus of the bulbocavernosus that innervate a muscle at the base of the penis (**Figure 9–29**). Both the bulbocavernosus muscle and its motor neurons die in neonatal females but not males. In this case dihydrotestosterone acts via the androgen receptor to prevent programmed cell death in the muscle, and the muscle provides trophic support required for the motor neuron to live (see Section 7.15). This cause-and-effect relationship is the opposite of what was observed for the fly muscle of Lawrence, in which the survival of a motor neuron in the male induces the differentiation of a male-specific muscle (see Figure 9–17).

As discussed in the previous section, some sexually dimorphic nuclei are continually influenced by sex hormones during puberty or even in adults. In songbird species for which singing is a male-typical trait, for instance, testosterone promotes the survival of adult-born neurons in male song nuclei, which contributes to the sexual dimorphic size of these nuclei (see Box 9–1). In most instances, however, the principal organizational role for sex hormones in causing sexually dimorphic neuronal numbers occurs during the sensitive period of early development through the regulation of programmed cell death.

9.19 Sex hormones also regulate sexually dimorphic neuronal connections

Recent molecular-genetic advances in mice have created new tools to visualize sexual dimorphism with better sensitivity than classic histology methods. One hypothesis is that genes regulating sexual behavior, such as the androgen receptor and aromatase, may have sexually dimorphic expression patterns. Using knock-in strategies in the mouse (see Section 13.7), markers can be expressed in the same pattern as these key regulatory genes (**Figure 9–30**A).

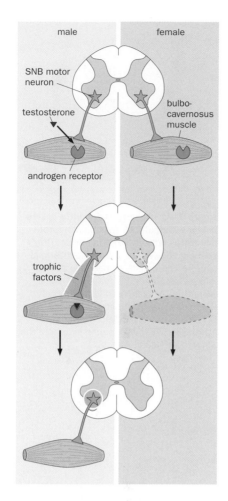

Figure 9–29 Sexual dimorphism of a penile muscle and its motor neurons. Sexual dimorphism in neurons of the spinal nucleus of the bulbocavernosus (SNB) is mediated by testosterone activation of the androgen receptor, which regulates gene expression to prevent programmed cell death of a penile muscle in males; trophic support provided by the muscle promotes the survival of its motor neurons. (Adapted from Morris JA, Jordan CL & Breedlove SM [2004] *Nat Neurosci* 7:1034–1039. With permission from Macmillan Publishers Ltd.)

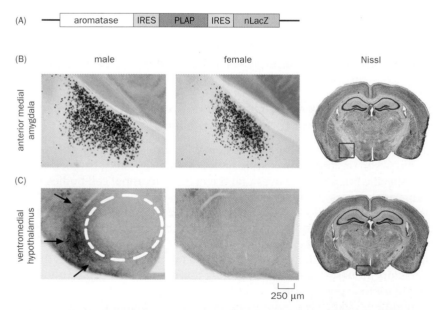

Figure 9–30 Visualizing aromatase-expressing neurons and their projections. (A) Schematic diagram of an engineered aromatase reporter gene introduced into the mouse. Aromatase-positive neurons express two additional marker genes from the IRES (internal ribosome entry site; see Figure 6–15). The placenta alkaline phosphatase (PLAP, an enzyme that labels plasma membrane) is used to visualize neuronal projections; the nuclear-targeted β-galactosidase (nβGal labels the nuclei. **(B)** In the anterior medial amygdala, males have more aromatase-expressing neurons than females, as revealed by βGal activity. **(C)** Near the ventromedial hypothalamic nucleus (circled), males have denser projections than females, as revealed by PLAP staining (arrows). The red boxes in the right panels of B and C indicate where the sections in the left panels were taken. (Adapted from Wu MV, Manoli DS, Fraser EJ et al. [2009] *Cell* 139:61–72. With permission from Elsevier Inc.)

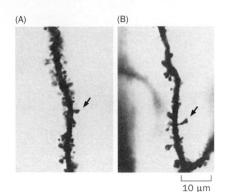

(A)   (B)

10 μm

Figure 9–31 Density of dendritic spines in hippocampal CA1 pyramidal neurons fluctuates with the estrous cycle. Golgi staining shows that in CA1 hippocampal neurons the density of dendritic spines (arrows) is about 30% greater in the proestrus **(A)** compared to the estrus **(B)** phase. (From Woolley CS, Gould E, Frankfurt M et al. [1990] *J Neurosci* 10:4035–4039. With permission from the Society for Neuroscience.)

Markers knocked in following the aromatase gene, for instance, revealed more aromatase-expressing neurons in the medial amygdala in males than in females (Figure 9–30B). Moreover, aromatase-expressing neurons have denser projections to several parts of the brain in males than in females, including the **ventromedial hypothalamic nucleus** (**VMH**; Figure 9–30C), a key center for regulating mating behavior (see Section 9.22). Females that undergo neonatal estrogen treatment displayed male-typical neuronal number and projection patterns, indicating that these dimorphisms are largely produced by the neonatal action of estrogen.

In addition to their organizational role during development, sex hormones can also regulate neuronal connections in adults. As we learned in Chapter 3, most of the excitatory synapses that provide input to cortical and hippocampal pyramidal neurons are formed on the dendritic spines. Thus, spine density can be used as a proxy for the density of excitatory synapses onto a pyramidal neuron. Interestingly, the density of dendritic spines in hippocampal CA1 pyramidal neurons of female rats fluctuates naturally according to the estrous cycle (**Figure 9–31**): spine densities are 30% greater in proestrus (which corresponds to higher estrogen levels in rats) than in estrus (which is associated with lower estrogen levels). Ovariectomization reduced spine density, which could be partially restored by estradiol application. Thus, the natural fluctuation of spine density appears to be regulated by estradiol. The function of this drastic change of synapse density is not known.

In summary, as in the fly brain discussed in the first part of the chapter, sexual dimorphisms occur extensively in the mammalian brain in the form of neuronal number and connections, and programmed cell death is a key mechanism used in both systems to regulate sexual dimorphisms in neuronal numbers. However, unlike the specification of sexual dimorphisms by transcription factors during development in flies, sexual dimorphisms in mammals are regulated by sex hormones, which can act at different stages of development and in adults. It is important to note that the causal roles of sexual dimorphisms in the mammalian brain in regulating sexual behavior have not been demonstrated in most cases. Nevertheless, the brain areas where sexual dimorphisms occur and their connection patterns provide clues as to how they might contribute to sexual dimorphic behavior.

9.20 Sexually dimorphic nuclei define neural pathways from olfactory systems to the hypothalamus

What are the connection patterns of the sexually dimorphic nuclei? Gross brain connections can be studied by classic neuroanatomical methods, such as injecting into specific brain regions dyes that travel from neuronal cell bodies to axons (**anterograde tracers**) or from axons and terminals back to cell bodies (**retrograde tracers**) (see Section 13.18). These studies have identified strong and bidirectional connections between the medial amygdala and BNST (**Figure 9–32**), both of which also project extensively to the hypothalamus including the MPOA. The medial amygdala and BNST are both direct targets of mitral cells in the accessory olfactory bulb (see Box 6–1), suggesting that sexually dimorphic nuclei play a role in pheromone perception.

In addition to receiving input from the accessory olfactory system, hypothalamic nuclei involved in regulating sexual behavior are also influenced by input from the main olfactory system. For example, studies utilizing trans-neuronal tracing methods (see Section 13.19) have revealed that the GnRH neurons in the hypothalamus, which play a central role in regulating sex hormone production (see Figure 9–27), also receive strong input from the main olfactory system (Figure 9–32). Altogether, these studies suggest that the main and accessory olfactory systems can both regulate sexually dimorphic behavior.

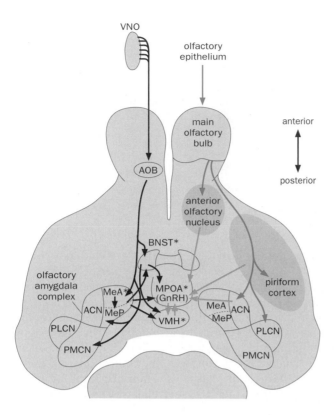

Figure 9–32 Neural pathways that link the main and accessory olfactory systems to the hypothalamus in a rodent brain. The accessory olfactory system (left) includes several sexually dimorphic nuclei (indicated by *) such as MeA (anterior medial amygdala) and BNST (bed nucleus of the stria terminalis); both project axons to the hypothalamic nuclei MPOA (medial preoptic area) and VMH (ventromedial hypothalamic nucleus). Information from the main olfactory system (right) projects to distinct targets that also send input to the hypothalamus, including to the GnRH (gonadotropin-releasing hormone) neurons in the MPOA. GnRH neurons also form reciprocal connections with VMH neurons. The red and blue pathways are derived from anterograde and retrograde tracing, respectively, whereas the green pathways to GnRH neurons were suggested by trans-neuronal tracing studies. The olfactory amygdala complex is divided into distinct nuclei, including the MeA and posterior medial amygdala (MeP), and the anterior, posterolateral, and posteromedial cortical amygdala (ACN, PLCN, PMCN). AOB, accessory olfactory bulb; VNO, vomeronasal organ. (Adapted from Yoon H, Enquist LW & Dulac C [2005] *Cell* 123:669–682. With permission from Elsevier Inc. See also Boehm U, Zou Z & Buck LB [2005] *Cell* 123:683–695.)

9.21 Whereas the main olfactory system is essential for mating, the accessory olfactory system discriminates sex partners in mice

Olfaction, and in particular the detection of pheromones, has long been implicated in regulating mating behavior. As we learned in Chapter 6, a cyclic nucleotide-gated channel expressed in the olfactory epithelia (CNGA2) is essential for olfactory transduction in the main olfactory system. Mice that lack CNGA2 are anosmic (see Figure 6–5). The accessory olfactory system utilizes a different transduction mechanism, culminating in the opening of the TRPC2 channel to depolarize sensory neurons in the vomeronasal organ (see Box 6–1). Thus, by studying *Cnga2* or *Trpc2* mutant mice, one can separately test the primary contributions of the main and accessory olfactory systems in sexual behavior.

Cnga2 mutant male mice exhibited virtually no mating behaviors, including sniffing to show interest, mounting, or intromission (**Figure 9–33**). They also did not exhibit aggressive behavior toward intruder males as normal male mice do. These experiments indicate that olfactory cues detected by the main olfactory system are essential to trigger mating and territorial defense responses. Indeed, physiological studies have identified mitral cells in the main olfactory bulb that are specifically tuned to components of female urine that are attractive to males.

By contrast, *Trpc2* knockout male mice exhibited normal mating behavior toward female partners. However, they also exhibited equally robust mating behavior toward intruder males (**Figure 9–34**A). *Trpc2* knockout male mice also produce ultrasonic vocalization toward other males, a behavior that is usually reserved for courtship and directed toward females. Thus, male mice without the accessory olfactory system function cannot seem to distinguish the sex of their partners. This behavioral deficit suggests the following model. For a male mouse, mating is a default behavior when it encounters another mouse. His accessory olfactory system enables him to recognize the partner mouse's sex and switch his behavior from mating to fighting if he detects that the other mouse is a male. In support of this model, sensory neurons and mitral cells of the accessory olfactory

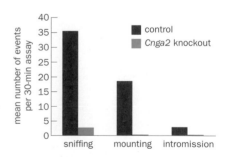

Figure 9–33 The main olfactory system is essential for mating. Signaling from the main olfactory system is essential for male mating behavior. Compared to control mice, *Cnga2* (encoding cyclic nucleotide-gated channels essential for olfactory transduction, see Section 6.1) knockout mice show little interest in sexually mature females (expressed as sniffing) and do not exhibit mounting or intromission. (Adapted from Mandiyan VS, Coats JK & Shah NM [2005] *Nat Neurosci* 12:1660–1662. With permission from Macmillan Publishers Ltd.)

Figure 9–34 The accessory olfactory system discriminates sex partners. **(A)** *Trpc2* knockout males mount females and intruder males with equal frequency given a simultaneous choice. The intruder males, which were castrated to reduce aggressive and sexual behavior, were swabbed with urine from intact males that normally would elicit aggressive rather than mating behavior from the resident male (see Figure 6–23). **(B)** *In vivo* single unit recording of an AOB mitral cell in a behaving CBA male mouse. The firing rate of an AOB mitral cell (*y* axis) was selectively elevated when the mouse investigated and formed physical contact (black horizontal bars on the *x* axis for time) with a BALBc male, but not a BALBc female, or B6 and CBA mice of either sex (all partner mice were anesthetized). BALBc, B6 (C57Bl6), and CBA are names of specific mouse strains. Other AOB mitral cells from this or other recorded mice also exhibited similar specificity to a combination of strain and sex. (A, adapted from Stowers L, Holy TE, Meister M et al. [2002] *Science* 295:1493–1500; B, adapted from Luo M, Fee MS & Katz LC [2003] *Science* 299:1196–1201.)

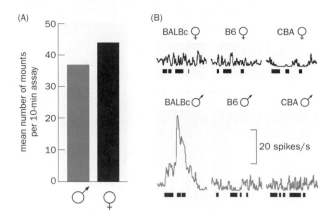

system have been identified that are specifically tuned to a combination of sex and strain signals from other mice (Figure 9–34B; see also Box 6–1).

What is the function of the accessory olfactory system in females? Analysis of *Trpc2* mutant female mice revealed a surprising phenotype. *Trpc2* mutant females produce ultrasonic vocalizations that resemble those of males, and they mount other females (see Movie 9–1). They are also impaired in female-typical behaviors, such as nest building and caring for pups. Thus, *Trpc2* mutant females act as if they had been partially sex-transformed. An implication from these observations is that females may have retained aspects of neural circuits for male-typical sexual behavior, and signals from the accessory olfactory system normally prevent the action of these male-typical circuits. This notion is supported by the observation that when given high doses of testosterone, wild-type females can exhibit mounting behavior toward sexually receptive females. Indeed, sex reversals in mating pattern, such as mounting by untreated females, have been observed in nature in over a dozen mammals including rats. The study of unisexual lizards provides an interesting perspective for the function of such sex reversal (see **Box 9–2**).

How do we reconcile these results with the organizational roles of sex hormones during development? A parsimonious model is that in both male and female brains, there exist primordial circuits that control male- and female-typical sexual behavior; we will see another example of this in the context of parental behavior in Section 9.23. These circuits may even share the same neurons (see Section 9.22). Sexual dimorphisms in the form of neuronal number and connection patterns, specified by the organizational function of steroid hormones during development, bias these circuits toward eliciting male- or female-typical sexual behaviors in response to hormone activation in adults. The function of the accessory olfactory system is to gate these gender-specific circuits based on the sex identity of self (in females) and partners (in males). Thus, the accessory olfactory system activates the circuit appropriate for the animal's own sex and inhibits the circuit for the opposite sex. The fact that most salient sexually dimorphic nuclei are targets of the accessory olfactory system supports this model.

9.22 The same neuronal population can control multiple behaviors in females and males

How do we causally link the function of specific neuronal populations in a brain area with specific behavior? As introduced in Section 1.14 and reinforced by studies of *Drosophila* sexual behavior discussed earlier in the chapter, loss-of-function and gain-of-function experiments can provide insights. Here, we use studies of the ventromedial hypothalamic nucleus (VMH) as an example to illustrate how researchers have investigated its function in rodent sexual behavior using a variety of methods.

VMH in rodents has been linked with the control of female lordosis, a posture for copulation that is triggered by tactile stimulation of the flank and perineal areas of females by mounting males. In rats, lordosis only occurs in the estrous cycle when females have a high level of circulating estradiol. By the 1970s, the following

Box 9–2: Courtship in unisexual lizards

The presence of primordial circuits that underlie sex-typical behaviors of both genders provides animals more flexibility in their behavioral repertoire to adapt to the changing environment. A fascinating case is illustrated in some whiptail lizard species, such as *Cnemidophorus uniparens*, which descend from bisexual ancestors but are unisexual in that all individuals are females. They reproduce by **parthenogenesis**, which does not involve exchange of genetic material and embryos develop from unfertilized eggs. Interestingly, these unisexual lizards still display courtship behavior. Indeed, their courtship ritual resembles closely the ritual displayed between male and female bisexual whiptail lizard, with one lizard acting 'male-like' by approaching and mounting a 'female' partner (**Figure 9–35**A). This courtship behavior has been shown to stimulate ovulation and thereby promote reproduction, despite no exchange of genetic material resulting.

Further studies have revealed interesting similarities and differences between unisexual lizards and most bisexual vertebrate species. There is no detectable level of testosterone in the unisexual lizards. However, their ovulation cycles persist, along with fluctuation of estrogen and progesterone levels. Indeed, the female-like or male-like behaviors coincide with stages with high- or low-estrogen levels during the ovulation cycle (Figure 9–35B). Evidence suggests that progesterone, which peaks after ovulation, activates male-like courtship behavior as does testosterone in most bisexual vertebrates. (Exogenously applied testosterone can also stimulate male-like behavior in unisexual lizards.) Hormone implants in different brain regions suggest that progesterone (and exogenous testosterone) acts in medial preoptic-equivalent area in the lizard brain to activate male-like behavior, whereas estrogen acts in ventromedial hypothalamus-equivalent area to activate female-like courtship behavior, much like bisexual mammals (see Sections 9.18–9.23). These studies provide interesting insight on how courtship behaviors and their hormone control adapt to changing conditions.

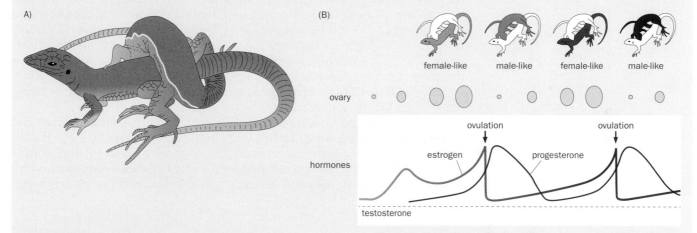

Figure 9–35 Courtship in unisexual lizard. (A) The pseudocopulation position exhibited by a pair of unisexual whiptail lizards, with the lizard on the top displaying a male-like mounting behavior and the lizard on the bottom assuming a female-like mating position; the ritual closely resembles copulation of bisexual whiptail lizards. **(B)** Individual lizards alternate in their display of female-like and male-like behavior coinciding with their ovulation cycles. The bottom panels show the size of ovary and hormone levels corresponding to the red lizard in the top panel. It displays female-like behavior when the estrogen level is high, and male-like behavior when the estrogen level is low but the progesterone level is high. There is no detectable testosterone in unisexual whiptail lizards. Separate experiments indicate that courtship stimulates ovulation. (Adapted from Crews D [1987] *Sci Am* 257:116–121. With permission from Macmillan Publishers Ltd.)

lines of evidence supported VMH's role in regulating lordosis: (1) Female rats with VMH lesion exhibit impaired copulation; (2) radioactively labeled estradiol is taken up strongly by VMH neurons; (3) estrogen implant at VMH can restore lordosis in ovariectomized rats. To further investigate a causal role of VMH in regulating lordosis, researchers used microelectrodes to stimulate VMH neurons in ovariectomized rats that were treated with estradiol. They found that electrical stimulation enhanced lordosis of female rats triggered by appropriate tactile stimuli (**Figure 9–36**A) or in response to mounting by male rats, indicating that VMH neurons facilitate lordosis. Interestingly, an increase in lordosis behavior occurred with an ~15 minute delay after electrical stimulation, and returned to baseline gradually after termination of stimulation. Complementing this gain-of-function

Figure 9–36 The ventromedial hypothalamic nucleus (VMH) facilitates lordosis in female rats. (A) Each data point on the graph represents lordosis behavior of an estradiol-treated, ovariectomized rat in response to five manual stimuli applied to the flank/perineal regions, on a scale from 0 (no lordosis) to 3 (maximal lordosis). Electrical stimulations of VMH neurons (10 Hz, 0.2 ms pulse at 50 μA; durations of the three experiments are shown at the bottom) increased the lordosis score within 15 min, and had a long-lasting effect after the stimuli were terminated. **(B)** Bilateral lesion of the VMH by passing large currents (1 mA, 15 s) through stimulating electrodes caused gradual reduction of the lordosis score within the next 48 hours. A slow recovery followed though not to the peak level, which may result from activation of compensatory pathway(s). Estradiol was injected daily to enable lordosis. (A, adapted from Pfaff DW & Sakuma Y [1979] *J Physiol* 288:189–202; B, adapted from Pfaff DW & Sakuma Y [1979] *J Physiol* 288:203–210.)

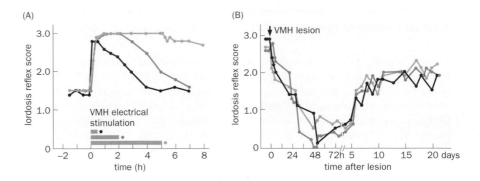

experiment, a loss-of-function condition was created by passing large currents through the electrodes to produce VMH lesions; this resulted in impaired lordosis, again with a time delay (Figure 9–36B). Together, these experiments suggest that rather than directly controlling the execution of lordosis, VMH neuronal activity promotes the activation of a downstream lordosis execution circuit, which integrates VMH signals over a relatively long period of time.

Subsequent experiments suggest that VMH contains heterogeneous cell populations, and activation of different subpopulations can have opposing effects on lordosis. In addition, electrical stimulation of male VMH and expression of immediate early gene studies suggest that this region also regulates sexual and aggressive behavior in males. Are these behaviors controlled by different VMH subpopulations? As we learned in the first part of the chapter, dissection of sexual behavior in fruit flies has been greatly aided by genetic manipulations of subsets of Fru^M neurons. Since sex hormone receptors regulate the development and activation of sexually dimorphic neural circuits in mammals, genetic manipulations of subsets of neurons expressing sex hormone receptors may provide insights into the function of specific nuclei in regulating mammalian sexual behavior, as exemplified by the following studies.

In a genetically modified mouse, the Cre recombinase gene was knocked into the *PR* locus (encoding the progesterone receptor, which augments the effect of estradiol on lordosis), such that neurons that normally express PR also express Cre (**Figure 9–37**A). Researchers could then inject Cre-dependent viruses into specific brain areas to mark and genetically manipulate only the PR-expressing

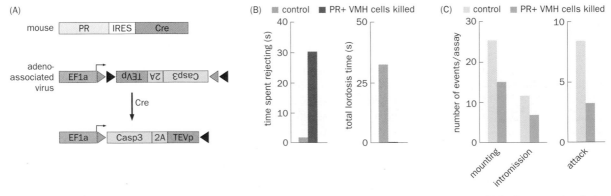

Figure 9–37 Progesterone receptor (PR)-expressing VMH neurons regulate female and male mating behavior as well as male aggression. (A) A molecular-genetic strategy to kill subpopulations of cells that express PR. *Cre* was knocked into the *PR* locus following the internal ribosome entry site (IRES), such that PR-expressing neurons also express the Cre recombinase. A transgene expressing a modified caspase-3 (Casp3), along with the TEV protease (TEVp) that activates it, was introduced by the adeno-associated virus (AAV) injected into the VMH, such that only VMH cells that expressed PR were selectively killed. EF1a is a promoter that allows ubiquitous expression. 2A is a self-cleaving peptide that allows expression of two proteins

(Casp3 and TEVp) from the same mRNA. Orange and red triangles are non-compatible *loxP* sites. Cre-induced recombination between orange–orange or red–red *loxP* sites inverts the intervening sequence, permitting caspase-3 expression. **(B)** Compared with control females (wild type injected with the AAV vector), females in which PR-expressing VMH neurons were killed rejected courting males and did not exhibit lordosis. **(C)** Compared with control males, males with ablated PR-expressing VMH neurons exhibited reduced mounting and intromission toward females, and reduced aggression toward intruder males. (Adapted from Yang CF, Chiang MC, Gray DC et al. [2013] *Cell* 153:896–909. With permission from Elsevier Inc.)

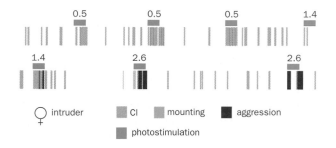

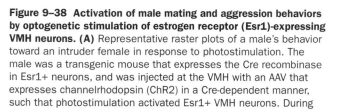

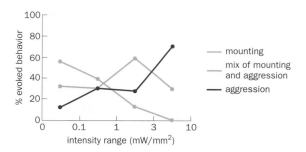

Figure 9–38 Activation of male mating and aggression behaviors by optogenetic stimulation of estrogen receptor (Esr1)-expressing VMH neurons. (A) Representative raster plots of a male's behavior toward an intruder female in response to photostimulation. The male was a transgenic mouse that expresses the Cre recombinase in Esr1+ neurons, and was injected at the VMH with an AAV that expresses channelrhodopsin (ChR2) in a Cre-dependent manner, such that photostimulation activated Esr1+ VMH neurons. During each 30-s photostimulation period (blue bars above, numbers indicating stimulation intensity in mW/mm²), the mouse increased the probability of switching from close investigation (CI) to mounting at low stimulation intensity, or to aggression at high stimulation intensity. **(B)** Quantification of percentage of evoked behavior (y axis) in response to a range of photostimulation intensities (x axis). (Adapted from Lee H, Kim DW, Remedios R et al. [2014] *Nature* 509:627–632. With permission from Macmillan Publishers Ltd.)

(PR+) neurons. One such manipulation was to kill these cells through activation of **caspase-3**, a key enzyme that promotes programmed cell death. When this technique was used to ablate PR+ neurons in VMH, females exhibited severely impaired courtship responses: they actively rejected courting males and did not engage in lordosis behavior (Figure 9–37B). Thus, PR+ VMH neurons promote lordosis. Some VMH neurons in males also express PR. In males, ablation of VMH PR+ neurons impaired male-typical behavior, including mounting and intromission during courtship with females, and aggression toward intruder males (Figure 9–37C). Thus, female- and male-typical sexual behaviors can be controlled by the same genetically and anatomically defined population of neurons in female and male brains, respectively.

How do PR+ VMH neurons in males regulate both mating with females and aggression toward intruder males? Another study explored this question by manipulating the activity of VMH neurons that express estrogen receptor 1 (Esr1) in male mice, which largely overlaps with PR+ VMH neurons. Viral transduction of Cre-dependent channelrhodopsin (ChR2) into VMH of mice that express Cre in Esr1+ neurons allowed photostimulation to selectively activate Esr1+ VMH neurons. Interestingly, photostimulation preferentially induced mating at low-stimulation intensity, but induced aggression at high-stimulation intensity (**Figure 9–38**), regardless of whether the partner is a male or a female. This suggests that Esr1+ VMH neurons act downstream of the accessory olfactory system, and photostimulation bypasses the requirement of the sex discrimination process. One interpretation for distinct behaviors elicited by different stimulation intensities is that the same population of Esr1+ VMH neurons controls both mating and aggression depending on their activity levels. An alternative interpretation is that a subpopulation of Esr1+ VMH with a low activation threshold controls mating, whereas another subpopulation with a high activation threshold controls aggression, and that activation of aggression suppresses the non-compatible mating behavior. These models can be distinguished by future studies that record activity of individual PR+/Esr1+ neurons during different behaviors. Tracing their input and output is also crucial to understand how these neurons are regulated by the accessory olfactory system, and how they regulate distinct behaviors in males and females.

9.23 Parental behavior is activated by mating and regulated by specific populations of hypothalamic neurons

In mammals, courtship represents just a subset of sexually dimorphic behaviors. For example, males defend their territory by exhibiting aggression toward intruders while females care for their pups. Most mammals are born immature and require extensive care from parents to survive and prosper. In rodents such as

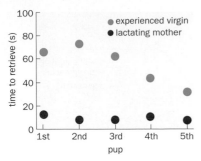

Figure 9–39 Pup retrieval behavior in females. Five pups were sequentially placed in a cage with an adult female mouse, and the average times it took for the female to retrieve the pup to the nest are plotted on the y axis. Lactating mothers are efficient in retrieving pups. Naive virgins ignore pups (not shown; see Movie 9–2), whereas experienced virgins that have cohabited with lactating mothers and pups retrieve pups at a reduced speed. (Adapted from Cohen L, Rothschild G & Mizrahi A [2011] *Neuron* 72:357–369. With permission from Elsevier Inc.)

Figure 9–40 Mating induces changes of infanticide versus parental behavior in male mice. Each singly housed mouse of the CF-1 strain was given two 1-day-old mice in their home cage at a time specified on the *x* axis. One of three behavioral responses was recorded in a 30-min period. Infanticide—one or both pups were dead; parental behavior—one or both pups were in a nest with the male hovering over them, and the pups were warm; ignore—neither pup was in the nest or wounded, and both pups were cold (newborn mice cannot thermoregulate). About 50% of sexually naive male mice commit infanticide (time 0, indicating before mating). An increase in infanticide occurs shortly after mating. This is followed by a switch to parental behavior between 12 and 50 days after mating. Infanticide increases again at 60 days after mating. (Data from vom Saal FS [1985] *Physiol Behav* 34:7–15.)

mice and rats, mothers usually do most parental caring by feeding and providing a warm and safe nest for their pups. Motherhood is associated with significant physiological and behavioral changes. For example, mothers efficiently retrieve displaced pups back into their nest whereas inexperienced virgin females do not (see **Movie 9–2**). After watching lactating mothers retrieving pups, virgin females also learn to do so, though with reduced speed (**Figure 9–39**). Significant changes have been reported in the auditory and olfactory systems of new mothers, which might make them more sharply tuned to alert calls and smells of displaced pups.

Male mice exhibit a qualitatively different behavior toward the young compared to females: a fraction of male mice (the size of the fraction is strain-dependent) attack and kill pups. Infanticide has been hypothesized to have an evolutionary advantage in facilitating the survival of infanticidal males' own progeny. Experiments have shown that infanticide speeds up mating of infanticidal males with mothers of the pups, as ovulation is inhibited in lactating mothers but is restored more quickly when the pups are gone. How do infanticidal males avoid killing their own pups? Experiments aimed at testing the influence of mating on male behavior toward pups revealed that, after mating, there was a transient period when males increased infanticidal behavior, possibly to remove competition with their own pups to come. Between 12 and 50 days after mating, however, a behavioral switch occurred such that males preferentially exhibited parental behavior. After this period, males resumed predominantly infanticidal behavior (**Figure 9–40**). The period of reduced infanticidal behavior coincides with gestation (19 days) and pre-weaning (3–4 weeks) of the males' own progeny, thus decreasing the chance that male mice kill their own pups.

Recent experiments identified a key role for the accessory olfactory system in regulating infanticide, as vomeronasal organ ablated or *Trpc2* mutant male mice exhibit drastically reduced infanticide and enhanced parental behavior. What neurons in the brain receive signals from the accessory olfactory system to control parental versus infanticidal behavior? Immediate early gene expression studies identified a group of neurons that express the neuropeptide **galanin** in the medial preoptic area (MPOA) of the hypothalamus to be preferentially activated after mice displayed parental behavior. Remarkably, ablating galanin+ MPOA neurons severely impaired parental behavior of fathers, whereas optogenetic stimulation of these neurons in virgin males inhibited their pup-attacking behavior (**Figure 9–41**A, B). Furthermore, ablating galanin+ MPOA neurons in mothers also impaired their maternal behavior, whereas ablating these neurons in virgin females caused them to attack pups (Figure 9–41C, D). Together, these experiments suggest that galanin+ MPOA neurons promote parental behavior in both males and females. They also inhibit pup-attacking behavior in virgin females. Ablating these neurons in fathers or mothers, while impairing their parental behavior, did not lead to pup attacking, suggesting that additional neurons are involved in inhibiting pup-attacking behavior during parenthood.

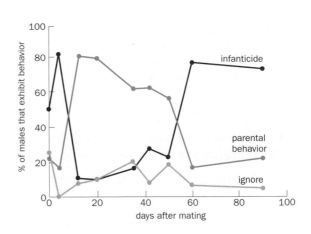

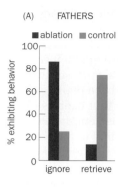

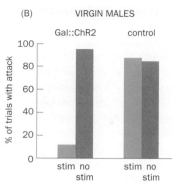

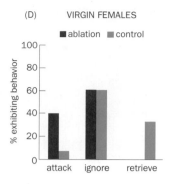

Figure 9–41 Galanin-expressing neurons in the medial preoptic area (MPOA) promote parental behavior in male and female mice. **(A)** Ablation of galanin+ MPOA neurons decreased pup retrieval behavior of fathers. **(B)** Photostimulation of galanin+ (Gal) MPOA neurons that express channelrhodopsin (ChR2) reduced pup-attacking behavior of virgin males. **(C)** Ablation of galanin+ MPOA neurons reduced pup-retrieval behavior in mothers. **(D)** Ablation of galanin+ MPOA neurons caused virgin females to attack pups. In all the experiments above, control mice expressed GFP in galanin+ MPOA neurons. Experimental animals expressed in galanin+ MPOA neurons either ChR2 for photostimulation (panel B) or diphtheria toxin (panels A, C, D) for killing. (Adapted from Wu Z, Autry AE, Bergan JF et al. [2014] *Nature* 509:325–330. With permission from Macmillan Publishers Ltd.)

9.24 Two neuropeptides, oxytocin and vasopressin, regulate pair bonding and parental behavior

Another fascinating behavior, which occurs in 3–5% of mammalian species, including humans, is pair bonding. Each member of a mating pair displays selective (although not necessarily exclusive) affiliation and copulation with his or her partner. Pair bonding is usually associated with bi-parental care of offspring. Although mice and rats, the most widely used mammalian models, do not exhibit pair bonding behavior, another rodent, the prairie vole, is probably the champion of monogamy. After losing a partner, a prairie vole in the wild might never mate with a different partner. Such pair-bonding behavior can be reproduced in the laboratory: when offered a choice, prairie voles that had mated spent significantly more time huddling with their partners than with new voles. Remarkably, a closely related species, the meadow vole, does not exhibit pair bonding (**Figure 9–42**).

Two related neuropeptides, **oxytocin** and **vasopressin**, have been implicated in regulating pair bonding and parental behavior. Both peptides are synthesized

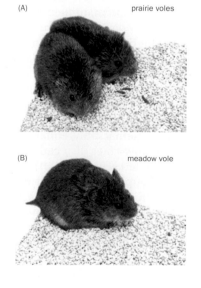

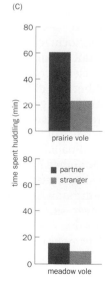

Figure 9–42 Pair bonding in prairie voles. (A) Prairie voles huddle with their sex partners after mating. **(B)** Meadow voles, a closely related species, do not exhibit pair bonding. **(C)** Quantification of pair bonding by comparing the relative times male voles spend with their previous mating partners (dark bars) versus with stranger females (light bars) when given a choice. Prairie voles preferred their previous mating partners, whereas meadow voles spent little time huddling with either female. (Adapted from Lim MM, Wang Z, Olazábal DE et al. [2004] *Nature* 429:754–757. With permission from Macmillan Publishers Ltd.)

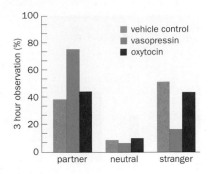

Figure 9–43 Infusion of vasopressin into the male brain is sufficient to establish pair bonding. Male prairie voles were infused with vehicle control (artificial cerebral spinal fluid), vasopressin, or oxytocin while placed together with ovariectomized female partners. Each male was then given a choice between their partner during infusion, a novel female (stranger), and an empty cage (neutral). Vasopressin infusion induced partner preference in subsequent choice assays. (Adapted from Winslow JT, Hastings N, Carter CS et al. [1993] *Nature* 365:545–548. With permission from Macmillan Publishers Ltd.)

by specialized groups of hypothalamic neurons. They are released from the posterior pituitary into the bloodstream (see Figure 8–35 and Table 8–1) and act as hormones to regulate a number of physiological processes such as labor and lactation (for oxytocin) or water retention (for vasopressin) through their G-protein-coupled receptors in the peripheral tissue. Oxytocin- and vasopressin-expressing neurons also project to multiple brain areas and release these peptides as neurotransmitters to modulate the activity of target neurons that express oxytocin and vasopressin receptors. Among their central functions as peptide neurotransmitters, oxytocin is known to promote mother–infant bonding in multiple mammalian species, whereas vasopressin is implicated in aggression toward intruders for territory defense. Indeed, vasopressin/oxytocin-like neuropeptides appear to have an ancestral role in regulating sexual behavior in invertebrates (**Box 9–3**).

The central roles of oxytocin and vasopressin in regulating pair bonding were demonstrated by perturbation experiments. Mating induces males to release vasopressin and females to release oxytocin. When oxytocin was infused into the brain of an ovariectomized (and therefore not sexually receptive) female prairie vole in the presence of a male, she would develop preference toward this male in subsequent partner choice tests without having mated. Likewise, when vasopressin was infused into the brain of a male prairie vole in the presence of an ovariectomized female, he would develop partner preference toward this female in the future without having mated (**Figure 9–43**). Conversely, when a male prairie vole was treated with a vasopressin antagonist immediately before exposure to a sexually mature female, he exhibited no preference toward this female even after having mated with her. Thus, in establishing a pair bond, application of oxytocin or vasopressin can replace the actual mating process.

For both males and females, mating is rewarding and causes dopamine release from ventral tegmental area dopamine neurons at their nucleus accumbens targets (see Section 8.9). Similar to the oxytocin/vasopressin studies just discussed, application of dopamine agonists facilitates, whereas application of dopamine antagonists inhibits, mating-induced pair bonding. A current model is that oxytocin and vasopressin are involved in the neural processing of the partner's social cues through the main and, particularly, the accessory olfactory systems that express a high level of oxytocin and vasopressin receptors; and the dopamine system provides the reinforcing value of the mating. Simultaneous activation of both systems could form an association between the neural encoding of partner's olfactory signature and dopamine-based reward. We will study reward-based learning in more detail in Section 10.24.

How is it that prairie voles exhibit pair bonding but the meadow voles do not? Unlike the prairie vole, injecting vasopressin into the brain of the male meadow vole did not increase his affiliation with a female present during the injection. Indeed, expression patterns of the oxytocin receptor and the V1a receptor (V1aR) for vasopressin are highly variable across different mammalian species, including prairie and meadow voles. For example, compared to the meadow vole, in the prairie vole the oxytocin receptor is more highly expressed in the nucleus accumbens whereas V1aR is expressed more highly in the **ventral pallidum** (a major output site for nucleus accumbens neurons) (**Figure 9–44**A).

To test whether the difference in V1aR expression has a functional consequence, virus-mediated gene transfer was used to increase V1aR expression in specific brain areas. Increasing V1aR expression in the ventral pallidum enhanced the partner preference of male prairie voles. Remarkably, enhancing V1aR expression in the ventral pallidum of male meadow voles also caused them to prefer females with whom they had previously mated (Figure 9–44B). Thus, differential expression of a single gene, a G-protein-coupled receptor for the neuropeptide vasopressin, contributes significantly to the species difference in pair-bonding behavior. Moreover, since expression was altered in adult meadow voles, this result also implies that the neural circuits that promote monogamous behavior are present in a polygamous species, and can be activated by neuropeptide signaling.

In summary, social interactions key to reproductive success, such as parental behavior and in some species pair bonding, use social cues from conspecifics to

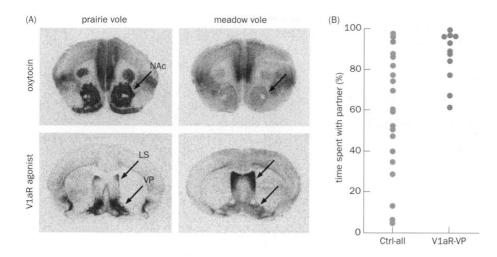

Figure 9–44 Increasing vasopressin receptor (V1aR) expression induces pair bonding in meadow voles. (A) Distribution of the oxytocin receptor or vasopressin 1a receptor (Va1R) in prairie and meadow voles as assayed by binding of radioactively labeled oxytocin (top panels) or V1aR antagonist (bottom panels) to brain sections. The oxytocin receptor is more highly expressed in the nucleus accumbens (NAc) of the prairie voles than meadow voles. V1aR is more highly expressed in the ventral pallidum (VP) in the prairie voles and in the lateral septum (LS) in the meadow voles. **(B)** Each dot represents a male meadow vole's degree of preference for a previous mating partner versus a naive female in a partner preference test. Control meadow voles do not exhibit a preference on average. Viral-mediated V1aR overexpression in the ventral pallidum (VP) causes a marked shift of meadow voles to prefer previous sex partners. (A, from Insel TR [2010] *Neuron* 65:768–779. With permission from Elsevier Inc.; B, adapted from Lim MM, Wang Z, Olazábal DE et al. [2004] *Nature* 429:754–757. With permission from Macmillan Publishers Ltd.)

regulate hypothalamic neuropeptide-expressing neurons, which in turn control behavioral output. We are only just beginning to understand the neural basis of parental and pair bonding behaviors. Many interesting questions remain to be answered. For example, what accounts for the different parental behaviors in males and females? How does mating control the behavioral switch and its timing in males? How do the accessory olfactory system, the galanin+ MPOA neurons, and oxytocin- and vasopressin-expressing neurons interact with each other? Studies in other species have implicated oxytocin signaling in social interactions in other contexts, such as trust and empathy in human. Thus, investigating neural mechanisms of sexual and parental behaviors may contribute to our general understanding of neural basis of social interactions.

Box 9–3: An ancestral function of oxytocin/vasopressin-like neuropeptide in sexual behavior

Oxytocin and vasopressin are ancient molecules. In early vertebrate evolution, a single gene encoding an oxytocin/vasopressin-like neuropeptide underwent a duplication event. Over time, the duplicated genes for oxytocin and vasopressin evolved separately to regulate female and male sexual and reproductive behavior, respectively. In addition to its roles in sexual behavior, vasopressin also regulates water retention in both sexes (see Table 8–1). Oxytocin has also been implicated in social behaviors beyond sexual and reproductive behavior. For instance, oxytocin knockout mice exhibit selective social memory deficit (failure to distinguish new versus familiar mice) with apparently normal non-social memory.

Oxytocin/vasopressin-like neuropeptides have also been identified in invertebrate species; one example is **nematocin**, derived from the nematode *C. elegans*. Although it is difficult to evaluate homology among neuropeptides because of their short amino acid sequences, the precursor proteins from which mature oxytocin, vasopressin, and nematocin are produced through proteolytic cleavages share conserved structures; these precursor proteins regulate the folding, sorting, and secretion of the mature peptides. Furthermore, the mature forms share structural features such as the disulfide bond formed between

vasopressin	C Y F Q N C P R G – NH$_2$
oxytocin	C Y I Q N C P L G – NH$_2$
nematocin	C F L N S C P Y R R Y – NH$_2$

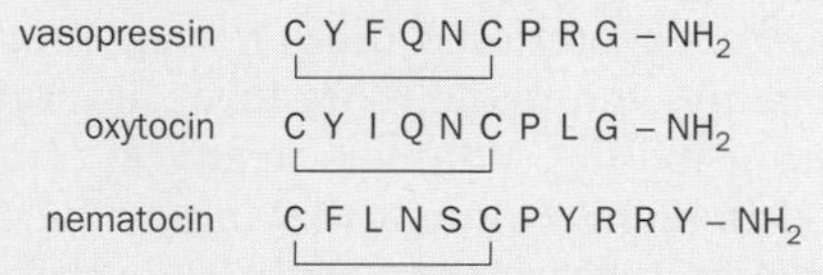

Figure 9–45 Comparison of mammalian vasopressin and oxytocin with nematocin from *C. elegans*. All three peptides have a conserved cysteine–cysteine disulfide bond between the first and sixth residues. (Adapted from Garrison JL, Macosko EZ, Bernstein S et al. [2012] *Science* 338:540–543.)

cysteines at the first and sixth residues (**Figure 9–45**). Lastly, *C. elegans* has two G-protein-coupled nematocin receptors, which are most closely homologous to mammalian oxytocin and vasopressin receptors. Interestingly, nematocin and its receptors also play important roles in *C. elegans* mating.

C. elegans has two sexes: males and hermaphrodites. Hermaphrodites produce both sperm and eggs and can self-fertilize in addition to mating with males. As in fruit flies, *C. elegans* mating follows a stereotyped sequence (**Figure 9–46**A). A wild-type male usually mates with the first hermaphrodite his tail encounters. Upon contact, the male turns about the hermaphrodite in a stereotyped manner, until it locates the vulva and transfers sperm. Mutant

(Continued)

Box 9–3: An ancestral function of oxytocin/vasopressin-like neuropeptide in sexual behavior

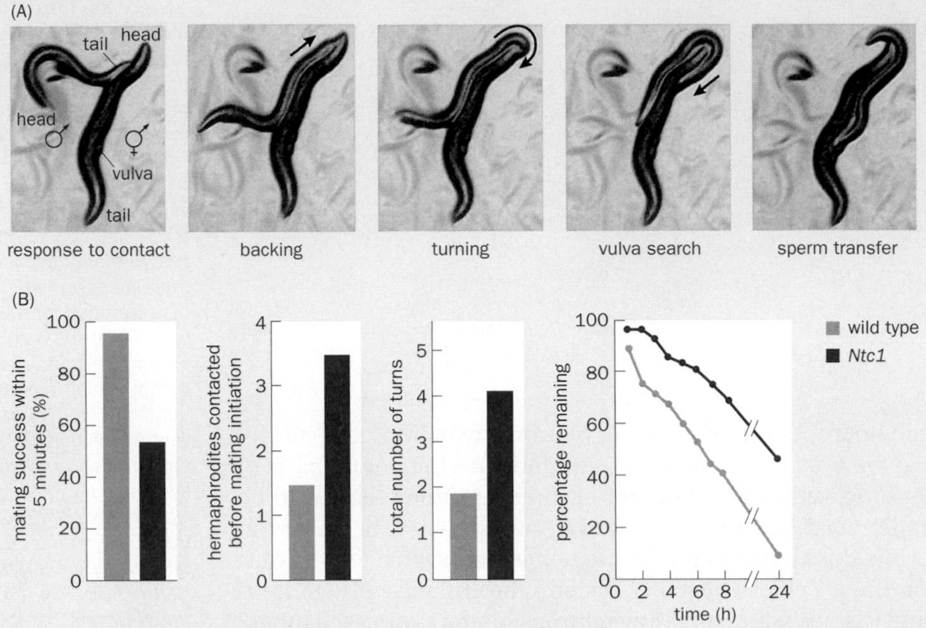

Figure 9–46 Nematocin is required for optimal execution of multiple steps of male mating. (A) Mating ritual of *C. elegans*. When the tail of a male encounters a hermaphrodite, he backs his body around the potential mate, initiates turning, and searches until his tail locates the vulva for sperm transfer. **(B)** Compared with wild-type controls (gray bar or trace), *Ntc1*-deficient males (red bar or trace) are less likely to mate successfully within 5 minutes of encountering a hermaphrodite (first panel). The mutant males also contacted more hermaphrodites before mating initiation (second panel), made more turns during mating (third panel), and a larger fraction failed to leave hermaphrodite-free food lawns to search for a mating partner (graph on the right). (Adapted from Garrison JL, Macosko EZ, Bernstein S et al. [2012] *Science* 338:540–543. With permission from AAAS.)

males without a functional *Nematocin* (*Ntc1*) gene exhibited abnormalities in all steps of mating (Figure 9–46B): they took longer to mate, encountered more hermaphrodites before initiating mating, and made more turns around hermaphrodites during mating. Whereas nearly all normal males leave a food-rich environment that lacks a hermaphrodite within 24 hours in search of a mate, only half of the *Ntc1* mutant males did so. Males lacking the nematocin receptors exhibited essentially the same phenotypes. Nematocin and its receptors are expressed in neurons that are responsible for different aspects of the mating process, and deletion of *Ntc1* in specific neurons yielded specific behavioral defects.

Thus, while not absolutely essential for mating, nematocin as a neuromodulator is required for the optimal execution of all mating steps.

Oxytocin/vasopressin-like neuropeptides can also induce mating-like behavior in medicinal leeches, which are non-self-fertilizing simultaneous hermaphrodites (that is, mating occurs between two hermaphrodites, each of which acts as the male and the female simultaneously). Thus, despite the diverse sex roles in different animals, oxytocin/vasopressin-like neuropeptides appear to have an ancestral and conserved function in regulating reproductive behavior.

SUMMARY

In fruit flies, sexual behavior is specified by sex-specific splicing of two transcription factors, Fruitless (Fru) and Doublesex (Dsx), with Fru playing a predominant role. Expression of the male-specific Fru^M in the female brain is sufficient to produce most male-typical courtship behaviors in females. Fru^M is expressed in about 2% of all neurons, including peripheral olfactory, taste, and somatosensory neurons that sense mating-related sensory cues, neurons in the brain that process and integrate these cues, and neurons in the ventral nerve cord that produce mating-related behaviors such as the courtship song. Fru^M neurons are largely dedicated to regulating sexual behavior, and tend to connect with each other. Fru^M-equivalent neurons in females promote female receptivity to courtship. During development, Fru^M and Dsx regulate sexually dimorphic execution of programmed cell death and neuronal wiring, which presumably underlie sexually dimorphic behaviors.

In rodents, sexual behavior is controlled predominantly by the dual actions of sex hormones, which play an organizational role during development and an activational role in adults. At neonatal stages, circulating testosterone acts mostly on estrogen receptors after being aromatized to estradiol to configure the brain in a male-typical manner by regulating sexually dimorphic programmed cell death and neuronal wiring. A surge of sex hormones during puberty may play additional organizational roles. Sex hormones activate sexual behavior in adults. Sexually dimorphic nuclei, including pheromone-processing nuclei in the accessory olfactory system and hypothalamic nuclei, regulate mating and other reproductive functions. The main and accessory olfactory systems are essential for initiating mating and discriminating sex partners, respectively.

Flies and rodents thus share substantial similarities in how sexual behaviors are regulated. In both cases, sexually dimorphic behaviors originate from differences in genetic sex, which result in sexually dimorphic expression of key transcription factors (for example, Fru^M and Dsx in flies and Sry in mammals) or sex hormones that act on transcription factors (for example, testosterone and estradiol acting on androgen and estrogen receptors, respectively). These transcription factors in turn generate sexual dimorphism in the numbers of specific neurons and/or their projection patterns. Recent genetic dissections of neural circuits underlying sexual behavior further suggest that in both flies and mice, the same genetically defined neuronal populations (for example, Fru^M or equivalent neurons in flies and estrogen/progesterone receptor-expressing ventromedial hypothalamic neurons in mice) regulate distinct behaviors in males and females. A notable difference between flies and mammals, apart from the molecular details of the key regulators themselves, is the employment of sex hormones and their two-stage action in rodents. This may reflect the larger size and more sophisticated organ systems of mammals and the prolonged period between early development and sexual maturity. Future research on the neural basis of sexual behavior in these model organisms will further reveal the common principles and diverse strategies that ensure their reproductive success.

Sexual behavior extends beyond male–female courtship in flies, mice, and rats, and encompasses considerably more variations than are seen in these model species. Among the examples we have discussed are self- and non-self-fertilizing hermaphrodites (nematodes and leeches, respectively), alternating sex roles (unisexual lizards), and pair bonding in prairie voles. The oxytocin/vasopressin family of neuropeptides optimizes male courtship behavior in nematodes and promotes pair bonding in prairie voles, suggesting that conserved molecules can regulate sexual behaviors in vastly different contexts. Likewise, sex hormones that control sexual differentiation and activation can also apply in different contexts, such as in regulating the survival of adult-born neurons that produce sexually dimorphic song nuclei in songbirds, and in the switching of sex roles in unisexual lizards. Studies of mating-related behavior in diverse organisms enrich our understanding of diversity of life and uncover themes and variations in how sexual behaviors are regulated.

Finally, as we transition to the next chapter, which examines memory and learning, it is important to note that while sexual behaviors are mostly innate, animals can nevertheless learn from individual experience and modify their behavior, as exemplified by courtship conditioning in flies. Indeed, song production in songbirds, which contributes to their reproductive success, has also been a paradigm for studying sensory and sensorimotor learning. While learning offers the flexibility that allows animals to adjust to changing environments, genetically programmed innate behaviors, selected by evolution, offer robustness and adaptation to an animal's niche. For example, the elaborate mating ritual of fruit flies and their apparent perception of a specific food odor as an aphrodisiac may have much to do with their frequent congregation near food sources. The reliance on chemical cues for mating in mice may be related to their nocturnal habit. Thus, genetically programmed innate behaviors and postnatally acquired learned behaviors interact dynamically throughout an animal's life, and future studies will further reveal how genes and neural circuits regulate, and are shaped by, this complex relationship.

FURTHER READING

Books and reviews

Baker BS, Taylor BJ & Hall JC (2001) Are complex behaviors specified by dedicated regulatory genes? Reasoning from *Drosophila*. *Cell* 105:13–24.

Dickson BJ (2008) Wired for sex: the neurobiology of *Drosophila* mating decisions. *Science* 322:904–909.

Morris JA, Jordan CL & Breedlove SM (2004) Sexual differentiation of the vertebrate nervous system. *Nat Neurosci* 7:1034–1039.

Yamamoto D & Koganezawa M (2013) Genes and circuits of courtship behaviour in *Drosophila* males. *Nat Rev Neurosci* 14:681–692.

Yang CF & Shah NM (2014) Representing sex in the brain, one module at a time. *Neuron* 82:261–278.

Young LJ & Wang Z (2004) The neurobiology of pair bonding. *Nat Neurosci* 7:1048–1054.

Sexual behavior in the fly

Clyne JD & Miesenböck G (2008) Sex-specific control and tuning of the pattern generator for courtship song in *Drosophila*. *Cell* 133:354–363.

Demir E & Dickson BJ (2005) *fruitless* splicing specifies male courtship behavior in *Drosophila*. *Cell* 121:785–794.

Fan P, Manoli DS, Ahmed OM et al. (2013) Genetic and neural mechanisms that inhibit *Drosophila* from mating with other species. *Cell* 154:89–102.

Grosjean Y, Rytz R, Farine JP et al. (2011) An olfactory receptor for food-derived odours promotes male courtship in *Drosophila*. *Nature* 478:236–240.

Hall JC (1979) Control of male reproductive behavior by the central nervous system of *Drosophila*: dissection of a courtship pathway by genetic mosaics. *Genetics* 92:437–457.

Kimura K, Hachiya T, Koganezawa M et al. (2008) Fruitless and doublesex coordinate to generate male-specific neurons that can initiate courtship. *Neuron* 59:759–769.

Kohatsu S, Koganezawa M & Yamamoto D (2011) Female contact activates male-specific interneurons that trigger stereotypic courtship behavior in *Drosophila*. *Neuron* 69:498–508.

Kohl J, Ostrovsky AD, Frechter S et al. (2013) A bidirectional circuit switch reroutes pheromone signals in male and female brains. *Cell* 155:1610–1623.

Manoli DS & Baker BS (2004) Median bundle neurons coordinate behaviours during *Drosophila* male courtship. *Nature* 430:564–569.

Manoli DS, Foss M, Villella A et al. (2005) Male-specific fruitless specifies the neural substrates of *Drosophila* courtship behaviour. *Nature* 436:395–400.

Nojima T, Kimura K, Koganezawa M et al. (2010) Neuronal synaptic outputs determine the sexual fate of postsynaptic targets. *Curr Biol* 20:836–840.

Ruta V, Datta SR, Vasconcelos ML et al. (2010) A dimorphic pheromone circuit in *Drosophila* from sensory input to descending output. *Nature* 468:686–690.

Stockinger P, Kvitsiani D, Rotkopf S et al. (2005) Neural circuitry that governs *Drosophila* male courtship behavior. *Cell* 121:795–807.

Thistle R, Cameron P, Ghorayshi A et al. (2012) Contact chemoreceptors mediate male–male repulsion and male–female attraction during *Drosophila* courtship. *Cell* 149:1140–1151.

von Philipsborn AC, Liu T, Yu JY et al. (2011) Neuronal control of *Drosophila* courtship song. *Neuron* 69:509–522.

Yang CH, Rumpf S, Xiang Y et al. (2009) Control of the postmating behavioral switch in *Drosophila* females by internal sensory neurons. *Neuron* 61:519–526.

Sexual behavior in mammals

Beach FA (1975) Hormone modification of sexually dimorphic behavior. *Psychoneuroendocrinology* 1:3–23.

Edwards DA & Burge KG (1971) Early androgen treatment and male and female sexual behavior in mice. *Hormone Behav* 2:49–58.

Gorski RA, Gordon JH, Shryne JE et al. (1978) Evidence for a morphological sex difference within the medial preoptic area of the rat brain. *Brain Res* 148:333–346.

Lee H, Kim DW, Remedios R et al. (2014) Scalable control of mounting and attack by Esr1+ neurons in the ventromedial hypothalamus. *Nature* 509:627–632.

Lim MM, Wang Z, Olazabal DE et al. (2004) Enhanced partner preference in a promiscuous species by manipulating the expression of a single gene. *Nature* 429:754–757.

Luo M, Fee MS & Katz LC (2003) Encoding pheromonal signals in the accessory olfactory bulb of behaving mice. *Science* 299:1196–1201.

Pfaff DW & Sakuma Y (1979) Facilitation of the lordosis reflex of female rats from the ventromedial nucleus of the hypothalamus. *J Physiol* 288:189–202.

Phoenix CH, Goy RW, Gerall AA et al. (1959) Organizing action of prenatally administered testosterone propionate on the tissues mediating mating behavior in the female guinea pig. *Endocrinology* 65:369–382.

Seminara SB, Messager S, Chatzidaki EE et al. (2003) The GPR54 gene as a regulator of puberty. *N Engl J Med* 349:1614–1627.

Stowers L, Holy TE, Meister M et al. (2002) Loss of sex discrimination and male-male aggression in mice deficient for TRP2. *Science* 295:1493–1500.

Winslow JT, Hastings N, Carter CS et al. (1993) A role for central vasopressin in pair bonding in monogamous prairie voles. *Nature* 365:545–548.

Woolley CS & McEwen BS (1992) Estradiol mediates fluctuation in hippocampal synapse density during the estrous cycle in the adult rat. *J Neurosci* 12:2549–2554.

Wu MV, Manoli DS, Fraser EJ et al. (2009) Estrogen masculinizes neural pathways and sex-specific behaviors. *Cell* 139:61–72.

Wu Z, Autry AE, Bergan JF et al. (2014) Galanin neurons in the medial preoptic area govern parental behaviour. *Nature* 509:325–330.

Yang CF, Chiang MC, Gray DC et al. (2013) Sexually dimorphic neurons in the ventromedial hypothalamus govern mating in both sexes and aggression in males. *Cell* 153:896–909.

Yoon H, Enquist LW & Dulac C (2005) Olfactory inputs to hypothalamic neurons controlling reproduction and fertility. *Cell* 123:669–682.

Nematodes, lizards, and songbirds

Brainard MS & Doupe AJ (2013) Translating birdsong: songbirds as a model for basic and applied medical research. *Annu Rev Neurosci* 36:489–517.

Crews D (1987) Courtship in unisexual lizards: a model for brain evolution. *Sci Am* 257:116–121.

Garrison JL, Macosko EZ, Bernstein S et al. (2012) Oxytocin/vasopressin-related peptides have an ancient role in reproductive behavior. *Science* 338:540–543.

Konishi M (1985) Birdsong: from behavior to neuron. *Annu Rev Neurosci* 8:125–170.

Rasika S, Alvarez-Buylla A & Nottebohm F (1999) BDNF mediates the effects of testosterone on the survival of new neurons in an adult brain. *Neuron* 22:53–62.

CHAPTER 10

Memory, Learning, and Synaptic Plasticity

学而时习之，不亦说乎？

Is it not a pleasure, to have learned something, and to practice it at regular intervals?

Confucius (~500 BC)

A hallmark of the nervous system is its ability to change depending on experiences. In the preceding chapters, we have learned how the nervous system processes sensory information and how it organizes motor output. However, the nervous system is much more than a giant sensorimotor circuit. In addition to acquiring sensory information from the environment and making appropriate responses, animals are constantly learning from their sensory experiences and from the consequences of their actions. These learning processes and events can cause lasting changes in the brain that make it possible to retain the learned information we call **memory**. Learning enables animals to adapt to their changing world much faster than by evolutionary mechanisms, and its importance to animals and humans cannot be overstated. Memory gives us much of our individuality, as we are profoundly shaped by what we can remember from our past experiences.

Memory and learning have fascinated human beings throughout our written history. The epigraph above, taken from the opening statement of the *Analects of Confucius*, reveals that the importance of practicing what has been learned was already recognized 2500 years ago. The French philosopher Rene Descartes described memory as an imprint made in the brain by external experience (**Figure 10–1**). Over a century ago, psychologists had already established important concepts, such as the distinct steps of the memory process including acquisition, storage, and retrieval. But our understanding of the neurobiological basis of memory and learning comes mostly from research conducted during the past few decades, fostered by our increasing knowledge about the workings of the brain at molecular, cellular, and systems levels.

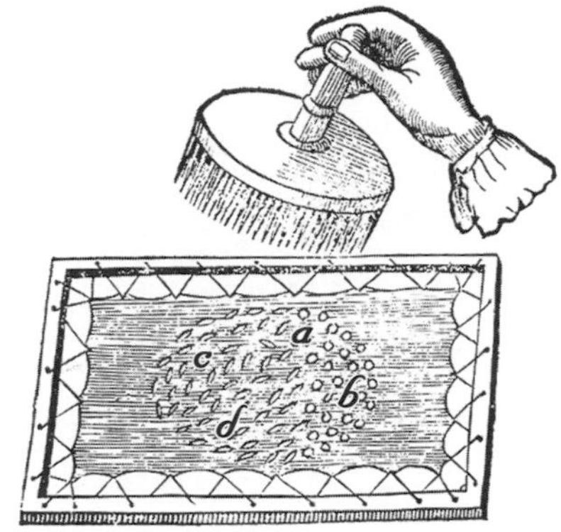

Figure 10–1 Memory as an imprint. According to Rene Descartes, memory could be considered as the imprints left on a linen cloth after needles had passed through it; some of the needle holes would stay open (as near points *a* and *b*), and for holes that close (as near points *c* and *d*), some traces would remain that make it easier to reopen them afterwards. (Adapted from Descartes R [1664] Treatise of Man.)

PRELUDE: WHAT IS MEMORY, AND HOW IS IT ACQUIRED BY LEARNING?

That different parts of the brain perform different functions seems an obvious concept today, but historically it took a long time for this concept to take root (see Section 1.10). Prior to the 1950s, the prevailing view was that memories for specific events and skills are distributed across large areas of the cerebral cortex. For example, in the 1920s, Karl Lashley carried out systematic lesions of the cerebral cortex of rats that had learned maze navigation to search for brain areas that, when removed, would affect the learned task. He did not identify a particular area that was necessary for memory; instead, task performance deteriorated progressively as increasingly larger areas were removed. From the 1950s onward, this concept of distributed memory changed, at least with regard to memory acquisition, as a result of studies in human patients, particularly the patient H.M.

10.1 Memory can be explicit or implicit, short-term, or long-term: Insights from amnesic patients

Henry Molaison (**Figure 10–2**), widely known as H.M. to protect his privacy until his death at the age of 82 in 2008, suffered from intractable seizures as a young

Figure 10–2 Henry Molaison (H.M.), a famous amnesic patient. Bilateral removal of the medial temporal lobes to alleviate his epilepsy resulted in profound defects in H.M.'s ability to form new memories of facts and events. (From Permanent Present Tense by Suzanne Corkin, copyright © 2013. Reprinted by permission of Basic Books, a member of The Perseus Books Group.)

man. In 1953, he underwent a bilateral surgical removal of the medial temporal lobes for the treatment of his seizure. While his seizures improved significantly, he emerged from the surgery with irreparable damage: he appeared to have lost his ability to form new memories. He did not recognize doctors who saw him frequently. Within half an hour of eating lunch, he could not remember a single item he had eaten; in fact, he could not remember having eaten lunch at all.

Extensive studies were performed on H.M. His personality and general intelligence, including perception, abstract thinking and reasoning abilities, were not affected by the surgery. In fact, his IQ improved slightly, from 104 pre-surgery to 112 post-surgery, likely because he was less affected by seizures after the surgery. However, he could not retain memory during intensive tasks such as trying to remember a three-digit number with repeated rehearsals; as soon as his attention shifted to a new task, he did not recall the old task or having ever been exposed to it. However, H.M. still had vivid memories of childhood and had largely intact memories of events until about 3 years prior to his surgery. He remembered the address of his old house (but not the address of the new house he moved to after the surgery).

Interestingly, not all forms of memory were impaired in H.M. In a mirror drawing task, subjects are asked to trace a line between the two borders of a double-outlined star (**Figure 10–3**A) while looking at their hands only in a mirror. Healthy people improve at this task with practice, so that the number of errors they make—defined by the number of times the traced line crosses one of the borders—decreases in later trials. H.M. could learn this task with a decreasing error rate just as normal subjects do. He showed steady improvement in this task across three days (Figure 10–3B), although each day he could not recall ever having performed the task before.

Studies of amnesic patients like H.M. have provided important insights into human memory. First, memory can be divided into two broad categories: explicit and implicit (**Figure 10–4**). **Explicit memory** (also called **declarative memory**) refers to memory that requires conscious recall, such as memories of names, facts, and events. When we use the term 'memory' in daily life, we are usually referring to explicit memory. **Implicit memory** (also called **non-declarative memory** or **procedural memory**) refers to memory in which previous experience aids in the performance of a task without conscious recall. The skill that H.M. acquired in the mirror drawing task and the ability to ride a bicycle involve implicit memory; so do habituation and sensitization, memory types that will be introduced later in this chapter. H.M. was selectively deficient for forming new explicit memories after his surgery.

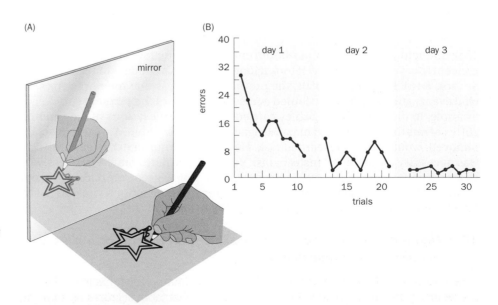

(A)

(B)

Figure 10–3 Memory of motor-skill learning displayed by H.M. (A) In this task, subjects are asked to view a double-outlined star in a mirror and draw a line in the space between its two borders. Subjects can only see their hands in the mirror. **(B)** With practice (number of trials, *x* axis), H.M. improved his performance in the mirror drawing task within and across days, as seen by the decreasing number of errors (occasions on which the traced line crosses a border, *y* axis). (B, adapted from Milner B, Squire LR & Kandel ER [1998] *Neuron* 20:445–468. With permission from Elsevier Inc.)

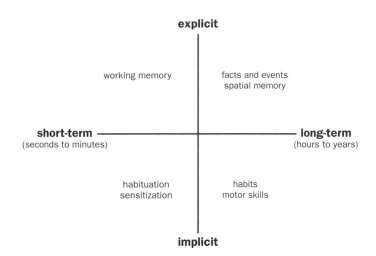

Figure 10–4 Different types of memory.
One major division of memory is explicit (for example, facts and events that require conscious recall) versus implicit (for example, habits and motor skills that do not require conscious recall). Another distinction among different types of memory is their duration: short-term memory lasts for seconds to minutes, while long-term memory can remain intact throughout the lifespan of a human or other animal.

Second, memory has different temporal phases, which are usually divided into short-term and long-term memory (Figure 10–4). **Working memory**, where facts are temporarily held (such as doing multi-step mental arithmetic, or remembering a telephone number before dialing before the era of smart phones), is a form of **short-term memory**. H.M. had intact working memory, which enabled him to hold normal conversations with others, but he could not convert facts and events into **long-term memory**. Implicit memory also has short- and long-term components. The exact temporal window can vary for different types of memories and in different organisms, but typically the memories we define as short-term are retained for seconds to minutes, whereas long-term memories can last for hours to years (Figure 10–4). As we will learn later in the chapter, there are mechanistic differences between short-term and long-term memory.

Third, distinct steps of the memory process and different types of memory require the function of specific parts of the brain. As we alluded to in the introduction, nineteenth century psychologists had divided memory into distinct steps. **Acquisition** is the initial formation of a memory as a consequence of experience and learning. **Retrieval** is the recall of a memory. **Storage** is the step in between acquisition and retrieval, where memory is held somewhere in the nervous system. More recently, a distinct step called **consolidation** has been proposed between acquisition and storage, during which newly acquired memory is solidified. Systematic comparisons of the lesions of H.M. and other amnesic patients have revealed that the region of the medial temporal lobe essential for the acquisition of new explicit memories is the **hippocampus**, located underneath the cortical surface of the temporal lobes (see Figure 1–8).

Importantly, H.M. still had largely intact explicit memory after surgery for the facts and events he had encountered prior to surgery. This suggests that the hippocampus is required for the acquisition of new explicit memories, but not for the long-term storage or retrieval of remote explicit memories. This also implies that the memories formed by utilizing the hippocampus are then stored elsewhere in the brain, such that they can be recalled even when hippocampal function is disrupted (as with H.M.). The fact that H.M. appeared to have intact working memory (which enabled him to hold conversations) and implicit memory (which enabled him to perform the mirror drawing task) implies that working memory and implicit memory also do not require the presence of the hippocampus. It is generally accepted that the prefrontal cortex plays a central role in working memory, whereas the cerebellum and the basal ganglia are instrumental for many types of motor learning (see Sections 8.8 and 8.9).

10.2 Hypothesis I: Memory is stored as strengths of synaptic connections in neural circuits

A key question that connects memory to the neurobiology we have studied in the preceding chapters is: What is the cellular basis of memory storage? Finding a satisfactory answer to this question would allow researchers to then study the mechanisms by which memory is acquired and retrieved. The leading hypothesis, which is strongly supported by the experimental evidence presented in this chapter, is that memory is stored as strengths of synaptic connections in neural circuits.

Let's first discuss this hypothesis from a theoretical perspective. Suppose that we have a synaptic connection matrix between five input neurons and five output neurons, which have the potential to form 25 synaptic connections. To simplify the discussion, we use a binary code for the connection matrix, where 1 designates a connection (purple dots in **Figure 10–5**, left) and 0 indicates the lack of a connection. Suppose further that the firing threshold of each output neuron obeys the following integration rule: if two or more of its connected presynaptic input neurons are firing simultaneously, it will fire its own action potential. The input–output function of this circuit, determined by the synaptic connection matrix, can in principle be used for event-triggered memory recall, where each input pattern can be considered an event and each output pattern can be considered a memory recall. Each input pattern is represented by a specific combination of firing patterns of the five input neurons at a given time. Three input patterns are shown as X_1, X_2, and X_3 (Figure 10-5, right), where 1 means that a presynaptic neuron is firing an action potential, and 0 means the presynaptic neuron is not firing an action potential. After passing through the connection matrix, each input pattern produces a corresponding output pattern, Y_1, Y_2, and Y_3, represented by the firing pattern of output neurons at a given time as determined by the integration rule. Through this synaptic connection matrix, each input pattern produces a defined output pattern; in other words, each event (X_1, X_2, X_3, and so forth), by interacting with this synaptic matrix, triggers the recall of specific memories (Y_1, Y_2, Y_3, and so forth) (**Movie 10–1**).

Instead of only five input and five output neurons like the above example, neural circuits in the mammalian brain usually comprise many more neurons. As

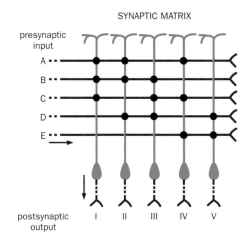

Figure 10–5 The synaptic weight matrix as a memory device. **Left**, a highly simplified model is used to illustrate how a synaptic matrix can store memory. In this synaptic matrix, axons of five presynaptic input neurons (A–E, red) form specific connections with dendrites of five postsynaptic output neurons (I–V, blue) that are represented by a binary code: each purple dot signifies a synaptic connection (value = 1); the absence of a purple dot indicates that no synaptic connection exists (value = 0). (In reality, rather than binary, synaptic connection strengths are continuous—from 0 or no connection to 1 or connection with maximal strength.) Blue cell bodies of the postsynaptic output neurons are shown below the matrix. Arrows indicate the direction of information flow. **Right**, this synaptic matrix can transform specific events, represented by the firing pattern of five input axons at any given time, to specific memory recalls represented by the firing pattern of output neurons. As examples, three specific input patterns, X_1, X_2, and X_3, are transformed to three corresponding output patterns, Y_1, Y_2, and Y_3. In these input and output patterns 1 and 0 represent an action potential or no action potential, respectively. The integration rule of each postsynaptic neuron is set such that it fires when two or more of its presynaptic partners are firing an action potential at a given time (in other words, when the matrix product is equal to or greater than 2). For example, for X_1, presynaptic neurons A, B, and D fire action potentials; neurons C and E do not. Neuron A synapses on output neurons I, II, and IV, neuron B synapses on neurons I, II, and III, and neuron D synapses on neurons II, III, and V. Thus, two presynaptic partner neurons fire on postsynaptic output neurons I, II, and III (matrix product ≥2), whereas only one presynaptic partner fires on neurons IV and V (matrix product <2). The resulting Y_1 is that neurons I, II, and III fire action potentials, while neurons IV and V do not. This 5 × 5 matrix has 2^{25} or ~30 million binary codes that can be used as a memory device to mediate event (X_N) triggered recall (Y_N).

the number of neurons increases, the number of possible synaptic connections goes up astronomically. Whereas the 5×5 matrix in Figure 10-5 has $2^{(5\times5)}$ or ~30 million possible binary codes, a 100×100 matrix has $2^{(100\times100)}$ or ~10^{3000} possible binary codes, more than there are atoms in the universe. At the same time, suppose that input patterns are represented by the simultaneous firing of 10 out of 100 input neurons; choosing 10 active input fibers out of 100 provides ~10^{13} different events. Even if the input fibers encode a different event each millisecond, the system can run for more than 300 years without repeating an event. Furthermore, we have simplified the synaptic connection matrix as consisting of 0–1 binary codes, but in reality the strength (or the weight) of synaptic connections can be any value between 0 (no connection) and 1 (maximal strength of connection). This greatly expands the coding capacity. In summary, these **synaptic weight matrices** can in principle store enormous amounts of information that can be used to transform specific input patterns (events) to specific output patterns (memory recalls). In Section 10.18, we will see a discrete example of how information in the synaptic weight matrix is read out by different downstream neurons to instruct distinct behavior.

As an example of synaptic weight matrices, let's examine the circuit organization of the mammalian hippocampus (**Figure 10-6**). The hippocampus receives input from the neocortex via the adjacent entorhinal cortex. Axons that project from neurons in the superficial layers of the entorhinal cortex, which constitute the **perforant path**, synapse onto the dendrites of **granule cells** in the **dentate gyrus**, the input part of the hippocampus. The axons of dentate gyrus granule cells, called **mossy fibers** because of their elaborate axon terminals, form synapses with the dendrites of CA3 pyramidal neurons, while the axons of CA3 pyramidal neurons form extensive recurrent connections via association fibers (that is, they synapse onto CA3 pyramidal neurons, including themselves). CA3 axons also form branches called **Schaffer collaterals**, which synapse onto the dendrites of CA1 pyramidal neurons. In addition to receiving trisynaptic input (perforant path → granule cells → CA3 → CA1), CA1 dendrites also receive direct input from the entorhinal cortex via the perforant path (Figure 10-6).

Thus, the hippocampus contains not just one but multiple synaptic matrices for information processing. These include the perforant path → granule cell synapses, the granule cell mossy fiber → CA3 synapses, the recurrent network among CA3 neurons, the CA3 Schaffer collateral → CA1 synapses, and the direct

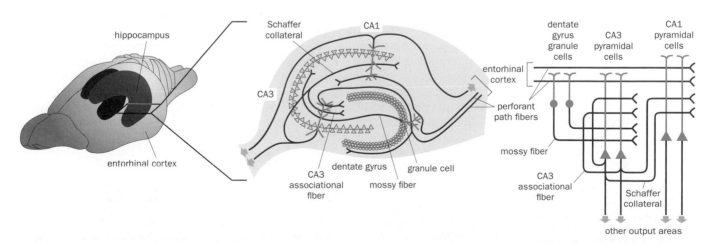

Figure 10–6 The hippocampal circuit. Left, location of the hippocampus and entorhinal cortex in the rat brain. A magnified section of the hippocampus (middle) and a circuit diagram (right) highlight the principal neurons (circles, granule cells; triangles, pyramidal neurons) and their major connections. Blue, dendrites and cell bodies; red, axons. Synapses can form where blue and red lines intersect. Perforant pathway axons from superficial layers of the entorhinal cortex can reach hippocampal CA1 pyramidal neurons directly via a monosynaptic connection, or indirectly via a trisynaptic connection in which the dentate gyrus granule cells and CA3 pyramidal neurons act as intermediates. CA3 pyramidal neurons also form extensive recurrent connections. Both CA3 and CA1 axons project to subcortical areas (middle panel, bottom left; right panel, bottom). In addition, CA1 axons project directly and via intermediate neurons (not shown) to deep layers of the entorhinal cortex (middle panel, top right).

perforant path → CA1 synapses. In the rat hippocampus, there are hundreds of thousands of CA1 and CA3 pyramidal neurons and over a million dentate gyrus granule cells. Each neuron is connected with thousands to tens of thousands of other neurons in these synaptic matrices, thus providing huge capacity for memory acquisition and storage.

10.3 Hypothesis II: Learning modifies the strengths of synaptic connections

If memory is stored as weights of synaptic matrices, then the essence of learning is to alter such weights based on experience. We have already studied one such mechanism—Hebb's rule—in Chapter 5. According to Hebb's rule, when the firing of a presynaptic neuron repeatedly participates in causing the postsynaptic neuron to fire, their synaptic connection becomes strengthened; conversely, when the firing of the presynaptic neuron repeatedly fails to elicit the firing of the postsynaptic neuron, their synaptic connection becomes weakened (see Figure 5–25). In principle, Hebb's rule can be used to modify the weights of synaptic connection matrices, including the formation of new synapses and the dismantling of existing ones. In a synaptic weight matrix (for example, see Figure 10–5), a change of synaptic weight at specific synapses means that the same input must produce different outputs before and after learning. The term **synaptic plasticity** is used to describe changes of the strengths of synaptic connections in response to experience and neuronal activity.

In summary, synaptic connections can be modified (that is, formed, dismantled, strengthened, or weakened), and neuroscientists hypothesize that these modifiable synaptic connections represent a major form of plasticity underlying memory and learning. We will devote the rest of this chapter to examining how well the experimental evidence supports this conceptual framework. In addition to synaptic plasticity, other plastic changes, such as the expression level and subcellular distribution of ion channels that underlie intrinsic properties of neurons (see Section 8.5), can also contribute to memory and learning. One specific example of an intrinsic property is the concentration of voltage-gated Na$^+$ channels at the axon initial segment, which determines the efficacy by which input (collective synaptic potentials) is transformed into output (action potentials) (see Sections 3.24–3.25).

Memory and learning have been studied on a variety of levels of organization, including genes and proteins, individual neurons and their synapses, the circuits comprising those neurons, and the animal behaviors effected by the activity of those circuits. Researchers can study memory and learning by taking two complementary approaches: a top-down approach that deconstructs complex phenomena to reveal the underlying mechanisms, or a bottom-up approach that starts with more basic, smaller-scale phenomena and explores how they relate to high-level events (**Figure 10–7**). A full understanding of the complexities of memory and learning requires investigations at all of these levels of organization. We begin at the level of neurons and synapses, focusing on the mechanisms that underlie synaptic plasticity.

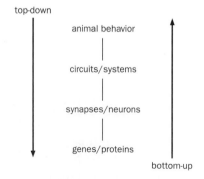

Figure 10–7 Memory and learning can be studied at multiple levels. When researchers start by observing a complex, high-level phenomenon and work to discover its underlying mechanisms, the approach is described as top-down or reductionist. By contrast, when researchers start by examining a low-level phenomenon and try to elucidate its relationship to more complex, high-level events, the approach is termed bottom-up or integrative.

HOW IS SYNAPTIC PLASTICITY ACHIEVED?

The ability of synapses to change their strengths according to experience is one of the most remarkable properties of the nervous system. Most mechanistic studies of synaptic plasticity in mammals have centered on the hippocampus; this focus has been prompted by human (see Section 10.1) and animal studies indicating that the hippocampus plays an essential role in memory acquisition, by the highly organized architecture of the synaptic input and output of hippocampal principal neurons (that is, excitatory projection neurons; see Figure 10–6), by the opportunity to investigate many synaptic connections *in vitro* using brain slices, and by the discovery of the plasticity phenomena to which we now turn.

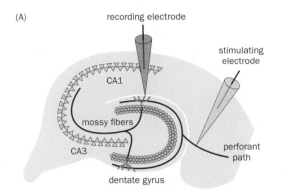

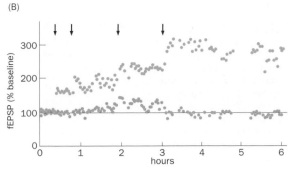

Figure 10–8 Long-term potentiation (LTP) induced by high-frequency stimulation. (A) Experimental setup. The stimulating electrode was placed at the perforant path, which consists of axons that innervate dentate gyrus granule cells. A second electrode was placed near the granule cell bodies to record the field excitatory postsynaptic potential (fEPSP), which represents the collective EPSPs from the population of granule cells near the recording electrode. Axons of dentate gyrus granule cells form the mossy fibers. **(B)** High-frequency stimulations (downward arrows, each representing 10 s of 15-Hz stimulation) caused an increase in the amplitude of fEPSPs produced afterward by single stimuli (green dots) compared to controls (yellow dots, no high-frequency stimulation). (Adapted from Bliss TVP & Lomo T [1973] *J Physiol* 232:331–356. With permission from the Physiological Society.)

10.4 Long-term potentiation (LTP) of synaptic efficacy can be induced by high-frequency stimulation

In the early 1970s, it was discovered that the connection strengths of hippocampal neurons could be altered in response to high-frequency stimulation (**Figure 10–8**). In these experiments, an extracellular recording electrode was implanted in the dentate gyrus of anesthetized rabbits to record the activity of granule cell populations near the electrode. A stimulating electrode was placed in the perforant path to provide synaptic input to the granule cells. A single stimulus applied to the stimulating electrode would depolarize the granule cell populations via the perforant path → granule cell synapses. This was recorded as a **field excitatory postsynaptic potential** (fEPSP; see Section 3.15 for EPSP and Section 13.20 for field potential), whose amplitude (or in later experiments, initial slope) is a measure of the strength of synaptic transmission between the stimulated axons of the perforant path and the granule cell population near the recording electrode. After brief trains of high-frequency stimulation were delivered through the stimulating electrode, each single stimulus thereafter produced an fEPSP with a two- to threefold greater magnitude than the baseline. This indicates that the strength of synaptic transmission (**synaptic efficacy** in short) between the perforant path axons and granule cells was enhanced as a result of the high-frequency stimulation. Importantly, this enhancement could last for many hours to several days (Figure 10–8). This phenomenon is thus called **long-term potentiation (LTP)**.

LTP in response to high-frequency stimulation has since been observed at all excitatory synapses in the hippocampus, including the mossy fiber → CA3 synapse, the CA3 → CA3 recurrent synapse, the CA3 Schaffer collateral → CA1 synapse (which we will refer to as the CA3 → CA1 synapse), and the perforant path → CA1 synapse (see Figure 10–6). LTP has also been found in many regions of the nervous system including the neocortex, striatum, amygdala, thalamus, cerebellum, and spinal cord. Importantly, LTP can be reproduced *in vitro* in **brain slices**, which largely preserve the local three-dimensional architecture of brain tissues *in vivo* while allowing easier experimental access for mechanistic studies. These studies have revealed that LTP at different synapses can exhibit different properties through distinct mechanisms. Below, we focus on LTP at the CA3 → CA1 synapse, which is one of the most studied synapses in the mammalian brain.

10.5 LTP at the hippocampal CA3 → CA1 synapse exhibits input specificity, cooperativity, and associativity

The reproduction of LTP in hippocampal slices has enabled many studies to probe its properties. In one experiment, two separate electrodes were placed on the

Figure 10–9 Input specificity, cooperativity, and associativity of long-term potentiation. In each experiment, two sets of presynaptic axons from CA3, *a1* and *a2*, form synapses with the same postsynaptic CA1 neuron *b*. **(A)** LTP exhibits input specificity. In the schematic shown here, only the *a1* → *b* synapses that have undergone high-frequency stimulation (represented by repeated vertical bars) exhibit LTP (* indicates potentiated synapses). **(B)** LTP exhibits cooperativity. Depolarization (blue) of postsynaptic cell *b* by current injection enables a weak stimulus (single shock) at axon *a1* to induce LTP. **(C)** LTP exhibits associativity. A weak stimulus at the *a2* → *b* synapses normally would not induce LTP at that synapse. However, when the timing of a weak *a2* stimulus coincides with high-frequency stimulation of *a1*, the *a2* → *b* synapse also becomes potentiated, because local depolarization at the *a1* → *b* synapses spreads to the *a2* → *b* synapses (blue represents the extent of depolarization spread). (See Bliss TVP & Collingridge GL [1993] *Nature* 361:31–39.)

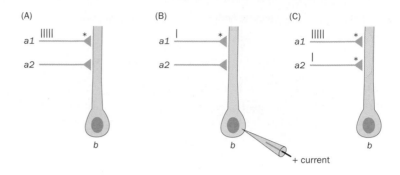

Schaffer collaterals to stimulate two sets of presynaptic axons (from two groups of CA3 neurons), *a1* and *a2*, which synapsed onto the dendrites of cell *b*, a CA1 postsynaptic neuron that was being recorded. LTP was induced by high-frequency stimulation of *a1* (**Figure 10–9**A). When the synaptic efficacy was measured afterwards, only the strength of the *a1* → *b* connection was potentiated, whereas the strength of the *a2* → *b* connection remained unchanged. Thus, LTP exhibits **input specificity**: it occurs at the synapses that have experienced high-frequency stimulation but does not occur at inactive synapses of the same postsynaptic neuron.

A second property of LTP was derived from experiments attempting to induce LTP by directly manipulating the postsynaptic neurons. When a weak axonal stimulation that was insufficient to induce LTP (such as a single stimulus, also called a shock) was paired with coincident injection of depolarizing currents into the postsynaptic cell from the recording electrode, LTP could be induced (Figure 10–9B). Thus, LTP is induced at a synapse when two events coincide: (1) the presynaptic cell fires and releases neurotransmitters and (2) the postsynaptic cell is in a depolarized state. This property is called **cooperativity** of LTP.

The cooperativity of LTP explains why high-frequency stimulation can induce LTP. Early in the train, action potentials from *a1* depolarize cell *b* at the *a1* → *b* synapses, such that the arrival of action potentials late in the train coincides with a depolarized state of the postsynaptic cell, hence potentiating the *a1* → *b* synapses. (Indeed, cooperativity was originally used to describe the phenomenon that high-frequency stimulation of one or few axons is insufficient to induce LTP, and 'cooperation' of many active axons is needed to induce LTP. The underlying mechanism is the same as defined above—to produce sufficient depolarization in the postsynaptic cell coincident with presynaptic axon firing.) Cooperativity can also explain a third property of LTP illustrated in the following experiment. While high-frequency stimulation was applied to *a1* to induce LTP at the *a1* → *b* synapses, *a2* was also stimulated at a level (for example, a single shock) that by itself would not reach the threshold of inducing LTP. The coincident stimulation was found to potentiate the *a2* → *b* synapses as well (Figure 10–9C). This is because high-frequency stimulation of *a1* causes depolarization in a region of cell *b* that includes the site of the *a2* → *b* synapses. If *a2* receives a weak stimulus (such as a single shock) during the time *b* is depolarized at the *a2* → *b* synapses, the synapses become potentiated. This potentiation of synapses that experience a weak stimulus by a coincident strong stimulus is called **associativity** of LTP.

These properties of LTP make it a suitable mechanism for adjusting the synaptic weight matrix that is hypothesized to underlie memory. Using Figure 10–5 as an example, input specificity allows the strengths of different synapses of a postsynaptic neuron with different input neurons to be altered independently, while cooperativity allows a given input to alter the strengths of synapses with a specific subset of co-active postsynaptic neurons. Together, these properties allow experience to adjust synaptic weights in the matrix on a synapse-by-synapse basis. Associativity makes it possible for coincident inputs to influence each other's synaptic strengths and is particularly well suited for associative learning, which we will discuss later in the chapter.

10.6 The NMDA receptor is a coincidence detector for LTP induction

The cooperativity of LTP is consistent with Hebb's rule (see also Section 10.3 and Figure 5–25). Indeed, this property made the CA3 → CA1 synapse the first known example of what is now called a **Hebbian synapse**, that is, a synapse whose strength can be enhanced by co-activating pre- and postsynaptic partners. Recall that we have already studied a molecule capable of implementing Hebb's rule: the **NMDA receptor**. The opening of the NMDA receptor channel requires simultaneous glutamate release from the presynaptic terminal and depolarization of the postsynaptic neuron to remove the blockade by Mg^{2+} (see Figure 3–24). This property accounts for the cooperativity and associativity of LTP. Indeed, ample evidence supports a key role for the NMDA receptor in the establishment of LTP (termed LTP induction) at the CA3 → CA1 synapse.

First, the NMDA receptor is highly expressed in developing and adult hippocampal neurons (**Figure 10–10**A). Second, pharmacological inhibition of the NMDA receptor by a specific NMDA receptor antagonist, **2-amino-5-phosphonovaleric acid** (**AP5**), blocked LTP induction in hippocampal slices without affecting baseline synaptic transmission. Third, when the gene encoding the required GluN1 subunit of the NMDA receptor was selectively knocked out in hippocampal CA1 neurons of mice (Figure 10–10B), LTP at the CA3 → CA1 synapse was abolished (Figure 10–10C), but basal synaptic transmission was unaffected. Because GluN1 was knocked out only in the postsynaptic CA1 neurons and remained functional in the presynaptic CA3 neurons, this experiment also demonstrated a postsynaptic requirement for the NMDA receptor in the induction of LTP at the CA3 → CA1 synapse.

10.7 Recruitment of AMPA receptors to the postsynaptic surface is the predominant mechanism of LTP expression

It is widely accepted that at most CNS synapses, LTP induction occurs through postsynaptic activation of the NMDA receptor. (A notable exception is the mossy fiber → CA3 synapse, where LTP induction is independent of the NMDA receptor and instead involves a largely presynaptic mechanism in which cAMP and protein kinase A act to regulate neurotransmitter release probability.) The means by which NMDA receptor activation leads to long-lasting increases in the synaptic efficacy, called LTP expression, has been the subject of intense debate. Two major

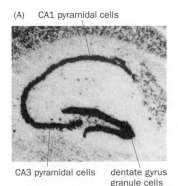

(A)   CA1 pyramidal cells

CA3 pyramidal cells   dentate gyrus granule cells

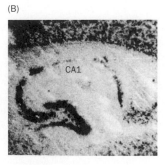

(B)

CA1

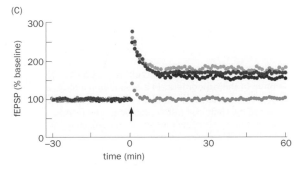

(C)

Figure 10–10 The NMDA receptor in the postsynaptic neuron is essential for LTP induction at the CA3 → CA1 synapse. (A) *In situ* hybridization shows that mRNA for the GluN1 subunit of the NMDA receptor is highly expressed in CA3 and CA1 pyramidal neurons as well as dentate gyrus granule cells in the hippocampus. GluN1 is also expressed in the cerebral cortex above CA1. **(B)** Conditional knockout of GluN1 using a transgene that expresses the Cre recombinase specifically in CA1 neurons (see Section 13.7) selectively disrupts GluN1 mRNA expression in the CA1 pyramidal neurons. **(C)** In

CA1-Cre-mediated GluN1 conditional knockout mice CA3 → CA1 LTP is blocked (blue trace) compared to normal LTP exhibited control mice that are wild type (yellow trace), that have the GluN1 conditional allele but lack the *CA1-Cre* transgene (red trace), or that have *CA1-Cre* transgene alone (brown trace). The upward arrow indicates high-frequency stimulation to induce LTP at *t* = 0. (Adapted from Tsien JZ, Huerta PT & Tonegawa S [1996] *Cell* 87:1327–1338. With permission from Elsevier Inc.)

types of mechanisms have been proposed: a presynaptic mechanism involving an increase in the probability that action potential arrival triggers neurotransmitter release (see Section 3.10), and a postsynaptic mechanism involving an increase in the sensitivity of the postsynaptic cell to the release of the same amount of neurotransmitter. These two mechanisms are not mutually exclusive.

At the CA3 → CA1 synapse, accumulating evidence suggests that the predominant mechanism of LTP expression is an increase in the number of AMPA-type glutamate receptors at the postsynaptic surface. As discussed in Chapter 3 (see Figure 3–24), the AMPA receptor is essential for basal synaptic transmission under conditions in which postsynaptic cells are insufficiently depolarized to activate the NMDA receptor. Following activation of the NMDA receptor during LTP induction, more AMPA receptors are inserted on the postsynaptic membrane. Subsequent glutamate release can thus trigger the opening of more AMPA receptors and hence stronger depolarization.

In fact, some glutamate synapses in the CNS, including a large fraction of the CA3 → CA1 synapses, initially contain only NMDA receptors on the postsynaptic surface. These synapses cannot be activated by presynaptic glutamate release alone and are therefore called **silent synapses**. However, coincident postsynaptic depolarization (presumably through AMPA receptors at other synapses) and presynaptic glutamate release activate the NMDA receptors at silent synapses and thereby cause the insertion of AMPA receptors into the postsynaptic membrane, transforming silent synapses into synapses that can be activated by presynaptic activity alone (**Figure 10–11**A–C). LTP expression involves both the activation of

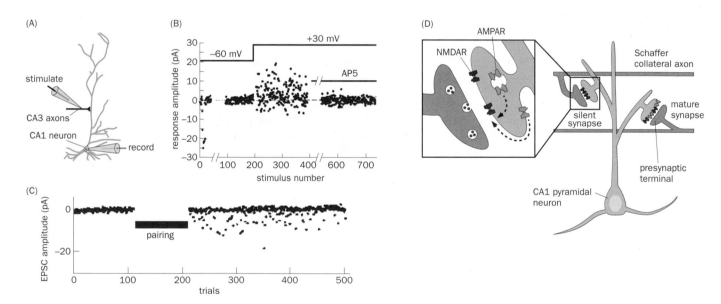

Figure 10–11 Silent synapses and their activation by LTP.
(A) Schematic of the experiment. In a hippocampal slice, a CA1 neuron's responses to stimulation of a set of CA3 axons were measured by whole-cell patch recording (see Box 13–2).
(B) Demonstration of silent synapses. At the beginning of this experiment, the CA1 cell was held at −60 mV, and after obtaining small excitatory postsynaptic currents (EPSCs) by stimulating CA3 axons, the stimulation strength was reduced (resulting in stimulating fewer axons) so the stimuli 100–200 did not produce any EPSCs. This means that no AMPA receptor was activated by the weak stimulus. However, when the cell was held at +30 mV, the same weak stimulus now evoked EPSCs that were blocked by AP5, indicating that the stimulated synapses contained NMDA receptors. In other words, the weak stimulus activated synapses that contained NMDA but not AMPA receptors. **(C)** Activating silent synapses. In this experiment, for the first 100 trials, CA1 neurons were held at −65 mV so that only AMPA currents could be induced by CA3 axon stimulation. Prior to pairing, EPSCs were not elicited, indicating that either the stimulated CA3 axons did not connect with the recorded CA1 neurons, or that they were connected via silent synapses. After repeated pairing of CA3

axon stimulation with depolarization of the postsynaptic CA1 neurons, a condition that induces LTP (see Figure 10–9B), a subset of CA3 stimulations elicited EPSCs, indicating that this subset was previously connected via silent synapses, which were activated (unsilenced) by the pairing of presynaptic stimulation and postsynaptic depolarization. Note that EPSCs were outward when the cell was clamped at +30 mV (B) and inward at −65 mV (C). This is because the reversal potentials for AMPA and NMDA receptors are near 0 mV (see Section 3.15).
(D) Schematic summary. Left, silent synapses have only NMDA receptors (NMDAR) at their postsynaptic surface. LTP causes a net insertion of AMPA receptors (AMPAR) at the postsynaptic surface via exocytosis of AMPA-receptor-containing vesicles, recruitment of AMPA receptors from extra-synaptic areas, or both (dashed arrows). Right, mature synapses contain both AMPA and NMDA receptors.
(B, adapted from Isaac JTR, Nicoll RA & Malenka RC [1995] *Neuron* 15:427–434. With permission from Elsevier Inc.; C, adapted from Liao D, Hessler NA & Mallnow R [1995] *Nature* 375:400–404. With permission from Macmillan Publishers Ltd; D, adapted from Kerchner GA & Nicoll RA [2008] *Nat Rev Neurosci* 9:813–825. With permission from Macmillan Publishers Ltd.)

silent synapses (Figure 10–11D) and an increased number of AMPA receptors in synapses that already have AMPA receptors.

In LTP and other forms of synaptic plasticity (discussed in following sections), AMPA receptor trafficking is subjected to many forms of regulation as a consequence of NMDA receptor activation. These include increasing the exocytosis of AMPA-receptor-containing vesicles leading to an increase in the number of cell-surface AMPA receptors, enhancing the binding of AMPA receptors to postsynaptic density scaffolding proteins to increase their residence time at the postsynaptic surface, facilitating lateral diffusion of AMPA receptors toward the synaptic surface, and altering the subunit compositions and phosphorylation status of AMPA receptors to increase their conductance. Exactly how these regulations are triggered by the activation of the NMDA receptor is the subject of intense research; we turn now to one mechanism that involves the activation of a specific protein kinase.

10.8 CaMKII auto-phosphorylation creates a molecular memory that links LTP induction and expression

As we learned in Chapter 3, a key property of the NMDA receptor, distinct from other glutamate receptors, is its high conductance for Ca^{2+} (see Figure 3–24). NMDA receptor activation causes an increase of $[Ca^{2+}]_i$ that activates a number of signaling pathways; for example, Ca^{2+}-activated adenylate cyclases increase the production of cAMP and activation of protein kinase A (see Figure 3–41). Another key signaling molecule is Ca^{2+}/calmodulin-dependent protein kinase II (**CaMKII**), which is activated by Ca^{2+}/calmodulin binding and is highly enriched in the postsynaptic density (see Figures 3–27 and 3–34). The holoenzyme of CaMKII consists of 12 subunits. Each subunit contains a catalytic domain plus an auto-inhibitory domain that binds to the catalytic domain and inhibits its function. Binding of Ca^{2+}/calmodulin to CaMKII transiently displaces the auto-inhibitory domain and thus activates the kinase. When $[Ca^{2+}]_i$ decreases, Ca^{2+}/calmodulin dissociates, deactivating CaMKII if no further modification occurs to CaMKII (**Figure 10–12**A, top).

The combination of the multi-subunit structure and auto-inhibitory domains that can be regulated by phosphorylation endows CaMKII with an interesting property. Active CaMKII can phosphorylate a threonine residue at amino acid 286 (T286) in the auto-inhibitory domain of a neighboring CaMKII subunit; T286 phosphorylation impairs the auto-inhibitory function, so that the activity of the phosphorylated subunits persists even after Ca^{2+}/calmodulin dissociates. Thus, if the initial Ca^{2+} signal is sufficiently strong to cause T286 phosphorylation at

Figure 10–12 Auto-phosphorylation of CaMKII and its requirement in LTP. **(A)** The CaMKII holoenzyme has 12 subunits; only six are shown here for simplicity. Top, binding of Ca^{2+}/calmodulin to a particular subunit transiently activates that subunit (* denotes an active subunit). When Ca^{2+}/calmodulin dissociates after $[Ca^{2+}]_i$ drops, the subunit becomes inactive. Bottom, if a sufficient number of CaMKII subunits become activated in response to a prolonged $[Ca^{2+}]_i$ elevation, specific threonine residues (T286) in multiple subunits are phosphorylated by neighboring subunits in the same complex. This cross-subunit phosphorylation maintains CaMKII in an activated state after $[Ca^{2+}]_i$ drops and Ca^{2+}/CaM complexes dissociate, until phosphatase activity overrides the auto-activation. **(B)** LTP in the CA3 → CA1 synapse can be induced by 10-Hz or 100-Hz high-frequency stimulation, or by two theta bursts (2TB) each consisting of four stimuli at 100 Hz with 200 ms separating the onset of each burst, which mimic endogenous firing of hippocampal neurons. In mutant mice in which T286 of CaMKII was replaced with an alanine residue (T286A), all these forms of LTP were disrupted. (A, adapted from Lisman J, Schulman H & Cline H [2002] *Nat Rev Neurosci* 3:175–190. With permission from Macmillan Publishers Ltd; B, adapted from Giese KP, Federov NB, Filipkowski RK et al. [1998] *Science* 279:870–873.)

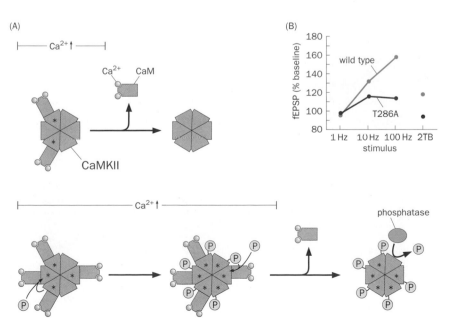

(A)

(B)

CaMKII

multiple subunits, subsequent CaMKII cross-phosphorylation can lead to sustained activity that outlasts Ca^{2+}/calmodulin binding. This process creates a 'memory' in the CaMKII molecule—a historical record of Ca^{2+} signaling—until phosphatases erase the memory through T286 dephosphorylation (Figure 10–12A, bottom). This molecular memory contributes to sustained changes in synaptic efficacy after transient NMDA receptor activation. Supporting this proposal, mice in which auto-phosphorylation of CaMKII at T286 was prevented by mutating the T286 residue to an alanine exhibited profound defects in LTP (Figure 10–12B).

Activation of CaMKII also appears to be sufficient for LTP induction. When a truncated, constitutively active form of CaMKII that lacks the auto-inhibitory domain was injected directly into CA1 pyramidal neurons, CA3 → CA1 synaptic transmission was potentiated. Furthermore, synapses potentiated by constitutively active CaMKII could no longer be induced to exhibit LTP by high-frequency stimulation, while synapses at which LTP had been induced by high-frequency stimulation could no longer be potentiated by constitutively active CaMKII (**Figure 10–13**). Thus, the two mechanisms of synaptic potentiation—high-frequency stimulation and CaMKII activation—occlude each other. These occlusion experiments provide strong evidence that CaMKII activation is an integral component of LTP induction and maintenance.

CaMKII activity contributes to the regulation of synaptic transmission strength through multiple mechanisms. For example, CaMKII-catalyzed phosphorylation of AMPA receptors increases their ion conductance and influences their trafficking (see Section 10.9 below). CaMKII also phosphorylates postsynaptic scaffolding proteins (see Section 3.16), which creates locking sites for AMPA receptors in the postsynaptic membrane. Another key output mediated by CaMKII and other signaling molecules, which is essential for long-lasting changes in synaptic efficacy, involves transcription factor activation and gene expression (see Figure 3–41). One process that these genes likely regulate is the structural alteration of synapses (see Section 10.13).

10.9 Long-term depression weakens synaptic efficacy

So far we have focused on LTP and its mechanisms of induction and expression. However, if synaptic connections could only be made stronger, the entire synaptic weight matrix (see Figure 10–5) would eventually become saturated, and there would be no room to encode new memories. In fact, many additional plasticity mechanisms co-exist with LTP so that the synaptic weight can be adjusted bidirectionally, as is discussed below and in the next section.

Figure 10–13 LTP induction occludes CaMKII-induced synaptic potentiation. Top, experimental design schematics; bottom, experimental data. The arrow that links the two schematics indicates that experiment B was a continuation of experiment A in the same preparation. **(A)** High-frequency stimulation was applied via the S1-stimulating electrode at the time indicated in the graph by the upward arrow. Only the S1 → CA1 neuron synapses were potentiated (brown trace) whereas the efficacy of the S2 → CA1 synapses remains unchanged (yellow trace), showing input specificity. An extracellular recording electrode was used to measure field excitatory postsynaptic potential (fEPSP) in response to S1 or S2 stimulation. **(B)** Subsequent to potentiation and extracellular recording in (A), a postsynaptic cell was patched for whole-cell recording, and constitutively active CaMKII enzyme was injected into the CA1 neuron through the patch electrode (at $t = 0$). Only the previously unpotentiated S2 synapses were potentiated, as indicated by gradually increased excitatory postsynaptic current (EPSC) in response to stimulation of S2 but not S1. Thus, CaMKII potentiation of the S1 synapses was occluded by prior LTP. (Adapted from Lledo P, Hjelmstad GO, Mukherji S et al. [1995] *Proc Natl Acad Sci USA* 92:11175–11179.)

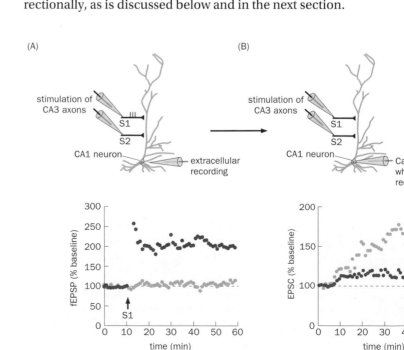

One counterbalancing mechanism is **long-term depression**, or **LTD**. Just like LTP, LTD has also been found in many CNS synapses (see Section 8.8 for an example of LTD at the parallel fiber → Purkinje cell synapse in the cerebellum). LTD can be induced at hippocampal CA3 → CA1 synapses by low-frequency stimulation of presynaptic axons; note that the same synapses exhibit LTP in response to high-frequency stimulation (**Figure 10–14**A). Like LTP induction, LTD induction is dependent on the NMDA receptor and Ca^{2+} influx. The increase of $[Ca^{2+}]_i$ resulting from low-frequency stimulation is lower than that resulting from high-frequency stimulation. This lower increase of $[Ca^{2+}]_i$ is thought to preferentially activate Ca^{2+}-dependent phosphatases, which do the opposite of what LTP-activated kinases do: the phosphatases reduce the number of AMPA receptors at the postsynaptic plasma membrane so that subsequent glutamate release from the presynaptic terminal induces a smaller depolarization.

LTD and LTP can affect the same synapse sequentially. Low-frequency stimulation can depress a synapse that has previously been potentiated by LTP; high-frequency stimulation can potentiate a synapse that has previously been depressed by LTD. Regulation of the phosphorylation status of the AMPA receptor GluA1 subunit at specific amino acid residues by CaMKII, protein kinase A (PKA), and protein kinase C (PKC) likely plays a role in LTP or LTD expression. One model proposes that in the context of LTP, GluA1 phosphorylation not only increases the channel conductance of AMPA receptors, but also stabilizes AMPA receptors newly added to the postsynaptic membrane, whereas GluA1 dephosphorylation triggers endocytosis of AMPA receptors from the postsynaptic membrane, leading to LTD (Figure 10–14B). Indeed, knock-in mice in which two such phosphorylation sites on GluA1 were replaced with alanines (so that neither could be phosphorylated) had significantly reduced LTP and LTD expression. These and other experiments support the notion that, at a given synapse, LTD and LTP represent a continuum of modifications of synaptic strength. The ability to control synaptic

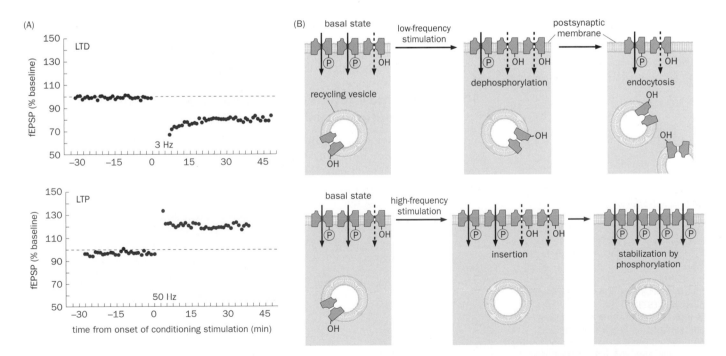

Figure 10–14 Long-term depression at the CA3 → CA1 synapse.
(A) Whereas high-frequency (50-Hz) stimulation induces LTP (bottom panel), low-frequency (3-Hz) stimulation of CA3 axons innervating a CA1 neuron causes long-term depression (LTD) of the efficacy of synaptic transmission (top panel). **(B)** In this model, AMPA receptors are in a dynamic equilibrium between cell surface and intracellular recycling vesicles. Low-frequency stimulation induces dephosphorylation of GluA1, which promotes endocytosis of AMPA receptors (top panel). High-frequency stimulation causes phosphorylation of GluR1, which stabilizes AMPA receptors at the postsynaptic membrane (bottom panel). In addition, phosphorylated GluA1 has higher AMPA channel conductance (solid arrow for larger ion flow) compared to non-phosphorylated GluA1 (dashed arrow). Together, low-frequency stimulation promotes LTD whereas high-frequency stimulation promotes LTP. (A, adapted from Dudek SM & Bear MF [1992] *Proc Natl Acad Sci USA* 89:4363–4367; B, adapted from Lee HK, Takamiya K, Han JS et al. [2003] *Cell* 112:631–643. With permission from Elsevier Inc.)

weights bidirectionally via LTP and LTD greatly increases the flexibility and storage capacity of synaptic memory matrices.

10.10 Spike-timing-dependent plasticity can adjust synaptic efficacy bidirectionally

Although high- and low-frequency stimulations are commonly used experimentally to induce synaptic plasticity, under physiological conditions, neurons are not usually activated at those precise frequencies. In reality, interconnected neurons can fire action potentials at many frequencies. Another plasticity mechanism that can influence synaptic strength is termed **spike-timing-dependent plasticity** (**STDP**). Originally discovered in the 1990s by researchers using patch clamp methods to study pairs of pyramidal neurons in rat cortical slices and in cultures of dissociated hippocampal neurons, STDP has since been found in many different preparations. In STDP, the precise timing of pre- and post-synaptic firing is critical in determining the sign of the synaptic strength change. For a typical synapse between two excitatory neurons, if the presynaptic neuron fires prior to the postsynaptic neuron within a narrow window (usually tens of milliseconds), and if these pairings are repeated, then subsequent synaptic efficacy increases. If repeated firing of the presynaptic neuron takes place within tens of milliseconds after the firing of the postsynaptic neuron, then the efficacy of subsequent synaptic transmission decreases (**Figure 10–15**). Thus, STDP incorporates features of both LTP and LTD. Indeed, it shares many similarities to LTP and LTD, such as dependence on NMDA receptor activation.

STDP is well suited for implementing Hebb's rule. If the presynaptic cell fires repeatedly before the postsynaptic cell, then it is likely that firing of the presynaptic cell contributes to the stimuli that cause the postsynaptic cell to fire; the synapses between the two cells should be strengthened. If the presynaptic cell fires repeatedly after the postsynaptic cell, then it is unlikely that the presynaptic cell contributes to causing the firing of the postsynaptic cell; synapses between the two cells should be weakened. In addition to serving a role in balancing potentiation and depression of synaptic strength in the synaptic weight matrix, the timing property of STDP can be used for other purposes, including activity-dependent wiring of the nervous system discussed in Chapters 5 and 7.

10.11 Dendritic integration in the postsynaptic neuron also contributes to synaptic plasticity

Not all forms of synaptic plasticity follow Hebb's rule as do LTP and STDP. In fact, synaptic plasticity can occur through dendritic integration without having to cause the firing of the postsynaptic neuron. We use a specific example involving hippocampal CA1 neurons to illustrate.

CA1 neurons receive direct perforant path input from the entorhinal cortex at their distal dendrites and Schaffer collateral input from CA3 neurons at more

Figure 10–15 Spike timing-dependent plasticity (STDP). If the presynaptic neuron repeatedly fires before the postsynaptic neuron, the synapse is potentiated (left). If the presynaptic neuron repeatedly fires after the postsynaptic neuron, the synapse is depressed (right). Data here were taken from retinotectal synapses in developing *Xenopus in vivo*, where the presynaptic neuron was a retinal ganglion cell and the postsynaptic neuron was a tectal neuron. (Adapted from Zhang IL, Tao HW, Holt CE et al. [1998] *Nature* 395:37–44. With permission from Macmillan Publishers Ltd.)

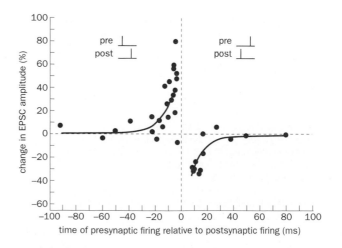

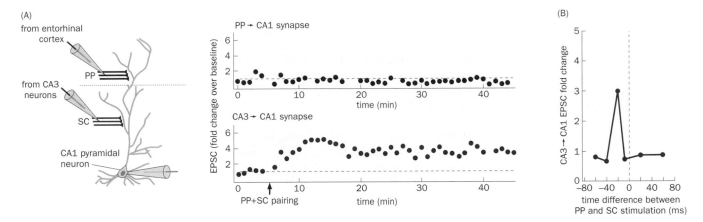

Figure 10–16 Input-timing-dependent plasticity (ITDP) in CA1 neurons. (A) Left, experimental setup. In a hippocampal slice, whole-cell patch clamp recording was performed on a CA1 neuron; stimulating electrodes were placed at the perforant path (PP) and the Schaffer collaterals (SC) that innervate the CA1 neuron's distal and proximal dendrites, respectively, which constitute different layers separated by the dotted line. Right, after paired sub-threshold stimulation of 1 Hz for 90 s, with PP stimulation preceding SC stimulation by 20 ms, average EPSC magnitude of the CA3 → CA1 synapse was enhanced whereas average EPSC magnitude of the PP → CA1 synapse remained unchanged. Thus, synaptic plasticity can be induced in the absence of postsynaptic cell firing. **(B)** Experiments with variable timing intervals: PP stimulation preceding SC stimulation by 20 ms was optimal for potentiating the CA3 → CA1 synapse. (Adapted from Dudman JT, Tsay D & Siegelbaum SA [2007] *Neuron* 56:866–879. With permission from Elsevier Inc.)

proximal dendrites (see Figure 10–6). An interesting means by which the perforant path → CA1 input contributes to CA1 neuronal activity is to influence the CA3 → CA1 synaptic efficacy. In a brain slice preparation in which whole-cell recording was performed on a CA1 pyramidal neuron, repeated pairing of perforant path and Schaffer collateral stimulations, with perforant path stimulation preceding the Schaffer collateral stimulation by ~20 ms, greatly potentiated the efficacy of CA3 → CA1 synapses (**Figure 10–16**A, bottom). The efficacy of the perforant path → CA1 synapses was unaffected (Figure 10–16A, top). Studies using varied time intervals indicated that the 20-ms difference was optimal for potentiating the CA3 → CA1 synapse (Figure 10–16B). This phenomenon has been termed input-timing-dependent plasticity (ITDP).

How do dendritic properties of CA1 neurons contribute to ITDP? Computational modeling suggests that 20 ms is the amount of time needed for perforant path → CA1 EPSCs from distal synapses to travel to the proximal dendrites, so that they can optimally summate with CA3 → CA1 EPSCs (see Section 3.24). This creates a prolonged depolarization at the proximal dendrite that is conducive to NMDA receptor activation and subsequent strengthening of the CA3 → CA1 synapse. What is the biological significance of the 20-ms difference in ITDP? As we saw in Figure 10-6, entorhinal cortical input can reach CA1 neurons through either the monosynaptic perforant path or the trisynaptic dentate gyrus → CA3 → CA1 loop. It takes about 20 ms longer for the entorhinal input to reach CA1 via the Schaffer collaterals than via the perforant path directly. Thus, the 20-ms difference coincides with a window during which individual CA1 neurons can assess the saliency of information processed by the trisynaptic loop by comparing it to direct input from entorhinal cortex. Thus, a combination of the properties of CA1 dendritic integration and the hippocampal circuit enables the perforant path input from the entorhinal cortex to selectively potentiate the efficacy of those CA3 → CA1 synapses that likely transmit the same entorhinal cortical input.

10.12 Postsynaptic cells can produce retrograde messengers to regulate neurotransmitter release by their presynaptic partners

Our discussions thus far have largely focused on postsynaptic mechanisms for modifying the efficacy of synaptic transmission, but synaptic plasticity can also engage presynaptic mechanisms. For example, synapses can be facilitated or depressed as a consequence of an increase or a decrease of the probability of neurotransmitter release in response to a train of action potentials (see Section 3.10). Longer-term changes of synaptic efficacy, such as LTP of the hippocampal mossy fiber → CA3 synapse, can also be induced by a presynaptic mechanism resulting in enhancement of neurotransmitter release probability. In other cases, however, modulation of presynaptic release probability is triggered by an initial change in the postsynaptic neuron. This implies that the postsynaptic neuron must send a retrograde messenger back to its presynaptic partner against the direction of the chemical synapse.

Endocannabinoids (endogenous cannabinoids) are among the best-studied retrograde messengers produced by postsynaptic neurons to regulate presynaptic neurotransmitter release probability. These lipophilic molecules, which include anandamide and 2-arachidonylglycerol, are ligands for a G-protein-coupled receptor, **CB1**, which is abundantly expressed in the brain and which was first identified as the receptor for cannabinoids from the marijuana plant (genus *Cannabis*). Upon depolarization, hippocampal CA1 pyramidal neurons rapidly produce endocannabinoids. In the 1990s, while some researchers discovered endocannabinoids and investigated their properties, others identified an interesting plasticity phenomenon called <u>d</u>epolarization-induced <u>s</u>uppression of <u>i</u>nhibition (DSI) in hippocampal CA1 pyramidal neurons. CA1 pyramidal neurons receive inhibitory input from GABAergic neurons in addition to receiving excitatory input from CA3 neurons and entorhinal cortex. During intracellular recording of CA1 neurons in hippocampal slices, it was found that depolarization elicited by intracellular current injection or high-frequency stimulation of incoming CA3 axons caused a transient suppression of inhibitory input to the CA1 neuron (**Figure 10–17**A).

Further experiments indicated that DSI required Ca²⁺ influx into the postsynaptic CA1 neuron yet did not affect the sensitivity of the CA1 neuron to exogenous GABA application. These data suggest that DSI is most likely mediated by

Figure 10–17 Depolarization-induced suppression of inhibition (DSI) and endocannabinoid signaling. (A) Following stimulation by a train of action potentials (indicated by the horizontal red bar), a hippocampal CA1 neuron exhibited DSI, as seen by a transient reduction of the frequency of spontaneous inhibitory postsynaptic potentials (IPSPs). Because the intracellular recording electrode was filled with KCl, diffusion of Cl⁻ from the electrode into the cell reversed the Cl⁻ gradient and caused IPSPs to be positive. **(B)** Schematic summary of endocannabinoid signaling in DSI. (1) CA1 neurons produce endocannabinoids in response to a rise of [Ca²⁺]ᵢ through voltage-gated Ca²⁺ channels or NMDA receptors (not shown) as a consequence of postsynaptic depolarization. (2) Endocannabinoids diffuse across the postsynaptic membrane and synaptic cleft, where they bind to the G-protein-coupled CB1 receptor enriched in the presynaptic terminals of GABAergic neurons. (3) Activation of CB1 releases Gβγ, which binds to and causes closure of presynaptic voltage-gated Ca²⁺ channels, resulting in inhibition of GABA release. (A, adapted from Pitler TA & Alger BE [1992] *J Neurosci* 12:4122–4132. With permission from the Society for Neuroscience; B, adapted from Wilson RI & Nicoll RA [2002] *Science* 296:678–682.)

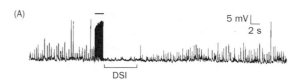

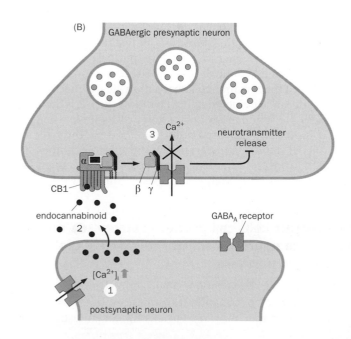

a reduction of GABA release from its presynaptic partners. Indeed, in the early 2000s, it was found that cannabinoid agonists could induce DSI in the absence of postsynaptic depolarization, whereas cannabinoid antagonists blocked DSI. Moreover, cannabinoid agonists and high-frequency stimulation of CA3 input occluded each other in causing DSI, and DSI was abolished in CB1 receptor knockout mice. These and other lines of evidence led to the model illustrated in Figure 10–17B. Depolarization of postsynaptic cells causes Ca^{2+} influx through voltage-gated Ca^{2+} channels (1), which triggers the synthesis of endocannabinoids from their precursors. These lipid-soluble endocannabinoids diffuse across the postsynaptic membrane and the synaptic cleft (2) to activate the CB1 receptor on the presynaptic membrane. CB1 activation triggers the release of G protein βγ subunits (3), which bind to and cause the closure of voltage-gated Ca^{2+} channels in the presynaptic terminal, thereby inhibiting neurotransmitter release. In principle, DSI should facilitate LTP at excitatory synapses. For example, depolarization of CA1 neurons due to excitatory input from CA3 would induce DSI, which would reduce inhibitory input onto the CA1 neurons, in turn facilitating depolarization and thus LTP induction.

In addition to CA1 pyramidal neurons, cerebellar Purkinje cells also exhibit DSI, as well as an analogous phenomenon called DSE (depolarization-induced suppression of excitation), depending on whether inhibitory or excitatory inputs are examined. Endocannabinoid signaling was also found to be responsible for cerebellar DSI and DSE. Given the wide range of brain tissues in which the CB1 receptor is expressed, it is likely that many synapses use this retrograde system to adjust presynaptic input based on the activity of the postsynaptic neurons. Unlike LTP and LTD, whose expression lasts many minutes to hours and days, DSI and DSE are transient (seconds, see Figure 10–17A) and only regulate short-term synaptic plasticity.

10.13 Long-lasting changes of connection strengths involve formation of new synapses

In addition to changing the probability of presynaptic release of neurotransmitters and the postsynaptic sensitivity to neurotransmitter release, which are two major mechanisms that account for synaptic plasticity we have discussed so far, long-lasting changes of synaptic efficacy can also be accomplished through structural changes to synapses. These include altering the size of existing synapses, forming new synapses, and dismantling old ones. These long-lasting changes typically depend on new gene expression (see **Box 10–1**). Structural changes in response to stimuli have been extensively documented in dendritic spines, where most excitatory synapses in the mammalian CNS are located, because of the relative ease of using fluorescence microscopy to image these structures in slice preparations and *in vivo* (see Section 13.22). For example, LTP induction was found to be accompanied by the growth of existing dendritic spines and the formation of new spines on CA1 pyramidal neurons in cultured hippocampal slices (**Figure 10–18**); this effect depended on the function of the NMDA receptor, suggesting that the structural changes are also mediated by signaling events initiated by Ca^{2+} entry.

LTP-associated structural changes have also been studied by serial electron microscopic reconstructions (see Section 13.19). High-frequency stimulation

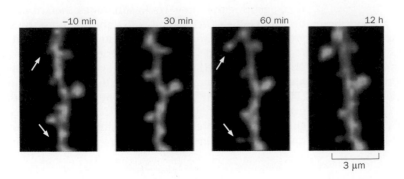

Figure 10–18 Growth of dendritic spines correlates with LTP. LTP is accompanied by the formation of two new spines (arrows) in CA1 pyramidal neurons from a cultured hippocampal slice that was imaged using two-photon microscopy. Time-lapse images were taken at −10, +30, +60 min, and +12 h relative to the onset of LTP induction (not shown). (From Engert F & Bonhoeffer T [1999] *Nature* 399:66–70. With permission from Macmillan Publishers Inc.)

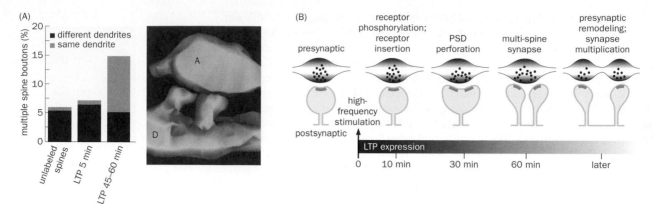

Figure 10–19 LTP correlates with formation of multiple-spine boutons. (A) Left, quantification of the fraction of axon terminals that contact more than one dendritic spine. Dendritic spines activated by LTP were labeled by a staining procedure that produces precipitates in EM micrographs of recently active spines to distinguish them from dendritic spines unrelated to LTP. A selective increase in the fraction of axon terminals that contact two dendritic spines from the same dendrite can be seen 45–60 min after LTP induction. Right, an example of serial EM reconstruction, showing two dendritic spines from the same dendrite, D, contacting the same presynaptic axon terminal, A. **(B)** A model of the temporal sequence of LTP expression. The initial enhancement of synaptic efficacy is caused by the phosphorylation of AMPA receptors and their insertion in the postsynaptic membrane. This is followed by a split of postsynaptic density (PSD), resulting in the formation of a multi-spine synapse. A further hypothetical split of the presynaptic terminal results in the duplication of synapses between the same two neurons. (A, adapted from Toni N, Buchs PA, Nikonenko I et al. [1999] *Nature* 402:421–425. With permission from Macmillan Publishers Ltd; B, adapted from Lüscher C, Nicoll RA, Malenka RC et al. [2000] *Nat Neurosci* 3:545–550. With permission from Macmillan Publishers Ltd.)

that induces LTP was found to cause a selective increase of axons that contact multiple dendritic spines from the same dendrites at a late (60-min) but not early (5-min) phase after the initial stimulation (**Figure 10–19**A). Thus, whereas early stages of LTP involve modulations of AMPA receptors at existing synapses, late-stage LTP can be manifested by structural modifications of synapses, namely the duplication of spines that are contacted by the same axons, possibly followed by a split of presynaptic axon terminals that results in the duplication of synapses (Figure 10–19B). Because these structural changes occur specifically between

Box 10–1: Synaptic tagging: maintaining input specificity in light of new gene expression

As discussed in Section 10.4, high-frequency stimulations (HFSs) can induce long-term potentiation (LTP) in hippocampus *in vivo* that lasts for many hours to days. Repeated HFSs of Schaffer collaterals can also induce LTP at the CA3 → CA1 synapses in hippocampal slices *in vitro* that lasts 8 hours or more. Further studies suggest that LTP in the *in vitro* model can be separated into two phases, an early-phase that decays within 3 hours and is protein synthesis-independent, followed by a late-phase (called **late LTP**) that requires new protein synthesis and new gene expression. This property echoes what we will learn in Section 10.16: short-term memory does not require new protein synthesis whereas long-term memory does.

A question arises as to how LTP maintains its input specificity (see Figure 10–9A) in light of new protein synthesis and new gene expression. For new protein synthesis, one solution could be the use of local protein synthesis from mRNA targeted to dendrites close to the postsynaptic compartments (see Section 2.2); indeed, activity-dependent local protein synthesis has been well documented. However, for new gene expression, activity-induced signals must go to the nucleus to trigger new transcription, and information

regarding which synapses initiated the signal is blind to the newly synthesized macromolecules (mRNAs and their protein products). To overcome this difficulty, a **synaptic tagging** hypothesis was proposed, which states that in parallel with enhancing synaptic efficacy, repetitive HFSs also produce a local synaptic tag that can selectively capture newly synthesized macromolecules distributed cell-wide, thereby conferring input specificity. The following experiments (**Figure 10–20**) provided strong support for the synaptic tagging hypothesis.

Two stimulating electrodes were placed at different depths of the CA1 dendritic field in a hippocampal slice preparation, ensuring that they would stimulate different populations of CA3 → CA1 synapses (S1 and S2) onto the same group of CA1 neurons, whose activity was monitored by a recording electrode. In the first experiment (Figure 10–20A) only S1 received HFSs, and only S1 synapses were potentiated, confirming input specificity. In the second experiment (Figure 10–20B), 35 min after HFSs at S1, protein synthesis inhibitors were applied to the slice (prior experiments had shown that this time lag would not inhibit late LTP formation at S1). 25 min later, HFSs were applied at S2 in the

Box 10–1: Synaptic tagging: maintaining input specificity in light of new gene expression

presence of protein synthesis inhibitors, which would normally block late LTP. However, S2 exhibited normal late LTP under this circumstance, thanks to the prior HFSs at S1. The simplest explanation is that HFSs at S2 produced a synaptic tag even in the presence of protein synthesis inhibitor, and the tag captured newly synthesized macromolecules due to HFSs at S1. In the third experiment, researchers tested how long the synaptic tag could last by first applying HFSs at S1 in the presence of a protein synthesis inhibitor, thus preventing it from inducing new gene expression but not inhibiting its ability to produce a synaptic tag. Then the protein synthesis inhibitor was washed away, and HFSs were applied to S2. If the two HFSs were separated by 3 hours, then S1 no longer exhibited late LTP (Figure 10–20C), suggesting that the synaptic tag is transient and lasts no more than 3 hours.

Although the molecular nature of the synaptic tag and the newly synthesized macromolecules they interact with are still incompletely understood (they may involve multiple molecular pathways in parallel), the concept of synaptic tag has been widely accepted. Similar phenomena have also been observed in the *Aplysia* model for learning and memory that we will discuss in later sections. While the Hebbian mechanisms of synaptic plasticity relies on the precise timing between activation of pre- and postsynaptic neurons (within tens of milliseconds of each other), the synaptic tagging model suggests that plasticity at one synapse may affect the plasticity of other synapses in the same neuron over a wider temporal window (for an hour or two). This may provide a cellular mechanism to explain why inconsequential events are remembered longer if they occur within a short window of well-remembered events.

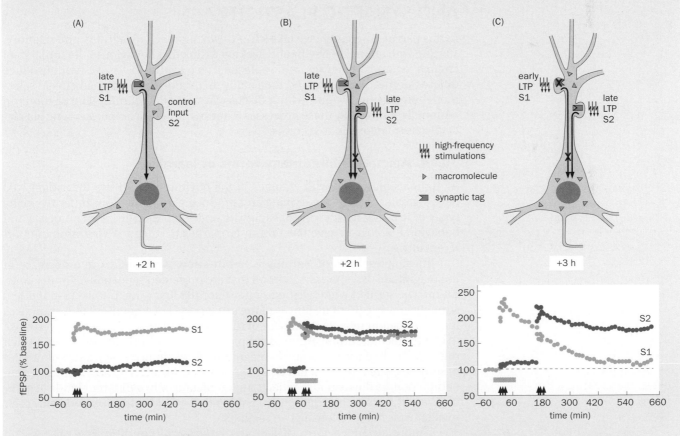

Figure 10–20 Experimental evidence for synaptic tagging hypothesis. The top schematics illustrate the experimental conditions and summarize the results at 2 or 3 hours after the first high-frequency stimulation (HFS) according to the synaptic tagging hypothesis. The bottom panels show the field EPSP changes over time. **(A)** HFSs were only applied to S1, and only S1 exhibited late LTP. This is because HFSs at S1, while inducing new gene expression (downward arrow to the nucleus), also produced a synaptic tag locally, which captured newly synthesized macromolecules necessary for late LTP. **(B)** 35 minutes after HFSs were applied to S1, protein synthesis inhibitors were added to the slice (duration represented by the horizontal bar in the bottom panel), during which time HFSs were applied to S2. Both S2 and S1 exhibited late LTP. This is because HFSs at S2, while incapable of inducing new gene expression (indicated by the cross on the downward arrow), nevertheless produced a synaptic tag, which captured newly synthesized macromolecules due to HFSs at S1. **(C)** HFSs were applied at S1 in the presence of protein synthesis inhibitors. Then HFSs were applied at S2 after protein synthesis inhibitors were washed away. When the two HFSs were 3 hours apart, late LTP at S1 was disrupted, presumably because the synaptic tag at S1 decayed (as indicated by the cross on the synaptic tag) by the time newly synthesized macromolecules due to HFSs at S2 arrived. (Adapted from Frey U & Morris RGM [1997] *Nature* 385:533–536. With permission from Macmillan Publishers Ltd.)

pre- and postsynaptic partners that have undergone LTP, this mechanism enhances the dynamic range of synaptic connections between a pair of neurons while at the same time maintaining the input specificity. This mechanism may be particularly important during development, when synapse formation and dendritic growth are influenced by experience, conveyed to the animal through patterned activity in sensory pathways (for example, see Box 5–3).

In summary, a wealth of mechanisms for synaptic plasticity, including changes in presynaptic neurotransmitter release probability and postsynaptic sensitivity to neurotransmitter release, as well as the structure and number of synapses, can be used to adjust the connection strengths between two neurons. These mechanisms allow experience and activity to adjust connection strengths both during development and in adulthood. Although we have focused largely on examples of mammalian hippocampal neurons and synapses, similar mechanisms likely occur throughout the nervous systems of both vertebrates and invertebrates. We next explore whether and how these plasticity mechanisms are linked to learning and memory.

WHAT IS THE RELATIONSHIP BETWEEN LEARNING AND SYNAPTIC PLASTICITY?

In this part of the chapter, we take a top-down approach to learning and memory, starting with animal behavior and seeking to link that behavior to the function of circuits, neurons, synapses, and molecules (see Figure 10–7). We first introduce different forms of learning and then study their underlying mechanisms in select model organisms. We end with a discussion of spatial learning and memory in mammals, noting how these processes relate to the hippocampal synaptic plasticity discussed in previous sections.

10.14 Animals exhibit many forms of learning

All animals must deal with changes in the environment. Those that adapt well have a greater chance of surviving and producing progeny. Consequently, many types of learning have evolved, each with specific properties. Psychologists and behavioral biologists have used these properties to categorize learning into different forms.

The simplest form of learning is **habituation**, which refers to a decrease in the magnitude of response to stimuli that are presented repeatedly. For instance, we may be startled when we hear a noise for the first time, but we respond less strongly to subsequent instances of the same noise—we 'get used' to it. Simple as it is, habituation reflects the ability of the nervous system to change its response to environmental stimuli. Another simple form of learning is **sensitization**, which refers to an increase of response magnitude to a stimulus after a different kind of stimulus, often noxious, has been applied. Sensitization is more complex than habituation, as the response reflects an interaction of two different kinds of stimuli. We will give specific examples of habituation and sensitization and study their mechanisms in the following two sections.

A more advanced form of learning is **classical conditioning** (also called Pavlovian conditioning), which refers to the ability of animals to produce a novel response to a previously neutral stimulus (the **conditioned stimulus**, or **CS**), after the CS has been repeatedly paired with a stimulus that always induces the response (the **unconditioned stimulus**, or **US**). A famous example is the experiment on salivation of dogs conducted by Ivan Pavlov, who discovered classical conditioning in the early twentieth century (**Figure 10–21**). Dogs always salivate in response to food in the mouth; this innate salivation constitutes the **unconditioned response**. After repeated pairing of food with a sound, which did not produce salivation before pairing, the sound alone induced salivation. In this example, food is the US, sound is the CS, and the process of pairing food and sound is called conditioning; the eventual salivation response to sound alone is called the **conditioned response**.

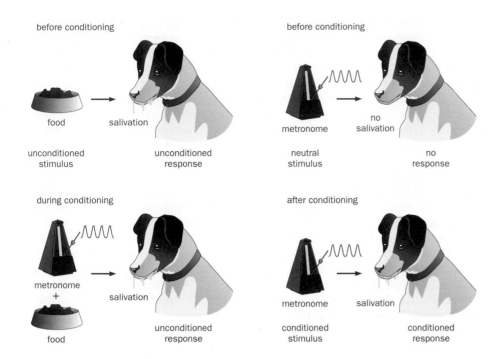

before conditioning

food → salivation

unconditioned stimulus   unconditioned response

before conditioning

metronome → no salivation

neutral stimulus   no response

during conditioning

metronome + food → salivation

unconditioned response

after conditioning

metronome → salivation

conditioned stimulus   conditioned response

Figure 10–21 Pavlov's experiment that established the concept of classical conditioning. Before conditioning, the dog salivates in response to food in the mouth (top left), but does not salivate when hearing a sound from a metronome (top right). During conditioning, which consists of repeated pairing of the sound and food (bottom left), the dog learns to associate the sound with food, such that after conditioning the dog salivates in response to the sound alone (bottom right). (See Pavlov IP [1926] Conditioned Reflexes. Dover Publications Inc.)

Whereas sensitization merely changes the magnitude of the response to a stimulus due to the presentation of a second kind of stimulus, classical conditioning establishes a novel and qualitatively different stimulus–response (for example, sound–salivation) relationship. Classical conditioning requires that an association form between the CS and US. In order for conditioning to be effective, the proper timing of the CS and the US is critical; the CS usually precedes the US. Therefore classical conditioning is a form of **associative learning**. It is observed across the animal kingdom, including humans.

Another major form of associative learning, distinct from classical conditioning, is **operant conditioning** (also called **instrumental conditioning**). In operant conditioning, a reinforcer is given only when the animal performs an appropriate behavior. For instance, a hungry rat in a cage can be trained to press a lever to obtain a food pellet. Initially the rat may not know the association between the lever pressing and the food pellet; after the reinforcer (food pellet) is given each time the rat presses the lever, the rat gradually associates the lever pressing (its own action) with the food reward (**Figure 10–22**). After operant conditioning, the rat selects one action over many other possible actions in order to receive the food pellet. A 'law of effect' was proposed in the early twentieth century to explain the association process: responses (behavior) that are followed by a reward will be repeated, whereas responses that are followed by a punishment will diminish.

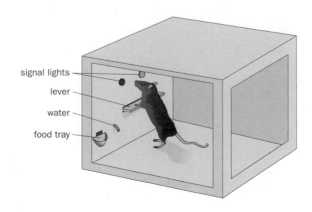

signal lights
lever
water
food tray

Figure 10–22 Basic design of an operant chamber. A hungry or thirsty rat placed in this chamber can learn through trial and error that pressing the lever results in the dispensing of either food or water (according to the particular experimental design); this reward reinforces the lever-pressing response. (See Skinner BF [1938] The Behavior of Organisms. B.F. Skinner Foundation.)

Timing is crucial in operant conditioning—as in classical conditioning—and the effect is greatest when the reinforcer is presented shortly after the behavior. Another property shared by classical and operant conditioning is **extinction**: in classical conditioning, when the CS is repeatedly *not* followed by the US, the conditioned response will diminish; in operant conditioning, when the behavior is repeatedly *not* followed by the reinforcer, the behavior will diminish. Operant conditioning is a prevalent learning mechanism in the animal kingdom and is widely used in the laboratory for training animals to perform tasks. Indeed, operant conditioning was used in many of the experiments discussed in this book, from motion perception to arm reaching (see Figures 4–52 and 8–27).

In our discussion so far, learning is viewed as the modification of behavior in response to experience, and the outcome of learning is measured by changes in behavior. There is a complementary view of learning. Psychologists use the term **cognitive learning** to refer to learning as an acquisition of new knowledge rather than simply modification of behavior. From this cognitive perspective, for instance, classical conditioning can be viewed as the animals having acquired the knowledge that the CS is followed by the US; the conditioned response is in fact a response to the predicted upcoming US rather than to the CS per se. While cognitive capabilities are usually thought to be specific to mammals with large cerebral cortices such as primates and particularly humans, the following example illustrates that even insects can master abstract concepts that qualify as cognitive learning.

Honeybees were trained to perform a task called delayed matching-to-sample, which is thought to utilize working memory (see Section 10.1). They first encountered a specific cue, such as a blue sign, after entering a Y-maze. After flying within the maze for a certain distance, they encountered the choice point, where the entrance into each arm of the Y-maze was marked by a blue sign or a yellow sign. If they chose to enter the arm marked by the same color as the color they encountered at the entrance of the maze, they would get a food reward (**Figure 10–23**A). After repeated training, bees not only can perform this task with a success rate well above chance (Figure 10–23B) but also can apply this skill to a completely new set of cues. For example, when the maze was outfitted with grid patterns that bees had not encountered previously, they could perform a pattern-matching task nearly as well as the original color-matching task (Figure 10–23C). Moreover, bees can apply the learned skill across different sensory modalities; for instance, training with a pair of odors improves the test results for matching a pair of colors. Lastly, bees can be trained to obtain a reward by entering the maze arm marked by a cue that differs from the one at the entrance—a task called delayed non-matching-to-sample—and can transfer the non-matching skill from colors to patterns. Thus, honeybees appear to be able to learn the abstract concepts of 'sameness' and 'difference' and use them to guide their behavior.

What are the neurobiological bases for these different forms of learning? Do they share common mechanisms? How are they related to the synaptic weight

Figure 10–23 Cognitive learning in honeybees. (A) Experimental setup. At the entrance to chamber B, bees first encounter a stimulus (for example, a blue sign). They then face a choice of two different C chambers, with the entrances marked with two different stimuli (for example, a blue sign and a yellow sign). With the sugar solution in one of the D chambers as reward, bees can be trained after repeated trials to choose either the C chamber marked with a stimulus that matches the entrance to the B chamber (delayed matching-to-sample, as shown here), or the C chamber marked with a stimulus that differs from the B chamber (delayed non-matching-to-sample, which would apply if the feeder were placed at D2). **(B)** Learning curves for bees that performed color- or pattern-matching tasks. Each block consisted of 10 consecutive training sessions. After six blocks, the percentage of correct choices for either task exceeded 70%, significantly above random chance (50%, dashed line). **(C)** After being trained for delayed-matching-to-sample in color, bees were tested for the pattern-matching task. Whether the entrance to B was marked with a vertical (left) or horizontal (right) grid pattern, bees preferentially chose the C chamber whose entrance was marked with the same pattern as the pattern at entrance to B. (Adapted from Giurfa M, Zhang S, Jenett A et al. [2001] *Nature* 410:930–933. With permission from Macmillan Publishers Ltd.)

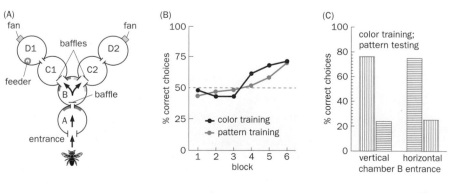

matrix hypothesis we introduced early in the chapter? We will now explore these questions, starting with simple forms of learning observed in a sea slug, *Aplysia*.

10.15 Habituation and sensitization in *Aplysia* are mediated by changes of synaptic strength

Aplysia has been used as a model for studying the cellular and molecular basis of learning and memory since the 1960s. *Aplysia* has only 20,000 neurons compared to about 10^8 neurons in the mouse. Many *Aplysia* neurons are large and individually identifiable such that electrophysiological recordings can easily be performed on multiple neurons in the same animal and with reproducible results across animals (as in the case of the crustacean stomatogastric ganglion discussed in Section 8.5). Importantly, *Aplysia* exhibits simple forms of learning and long-lasting memory that are similar to those found in more complex organisms.

The **gill-withdrawal reflex** has been used as a model behavior (**Figure 10–24**A). When a tactile stimulus is applied to the siphon, *Aplysia* reflexively withdraw their gill (and siphon) into the mantle shelf as a protective measure. This behavior shows habituation, as repeated siphon stimuli resulted in progressively smaller magnitudes of gill withdrawal (Figure 10–24B, left). However, if the habituated animal receives a noxious electric shock at the tail, the magnitude of gill withdrawal in response to the siphon stimulus applied shortly after the shock is drastically enhanced, indicating a sensitization of the gill-withdrawal reflex by the tail shock (Figure 10–24B, right).

The neural circuits underlying the gill-withdrawal reflex have been mapped (Figure 10–24C), thanks to the ease of electrophysiological recordings and manipulations. Siphon stimulation activates 24 sensory neurons; activating these neurons artificially was found to mimic siphon stimulation and induce the gill-withdrawal reflex. Six motor neurons control the muscle contraction that causes gill withdrawal. Activities of these motor neurons correlate with gill withdrawal, and direct electrical stimulation of these motor neurons is sufficient to cause gill withdrawal. These sensory and motor neurons form monosynaptic connections analogous to the sensorimotor circuit controlling our knee-jerk reflex (see Figure 1–19). A different group of sensory neurons transmits the tail-shock signal to a set of serotonin neurons, which in turn innervate the cell bodies of the

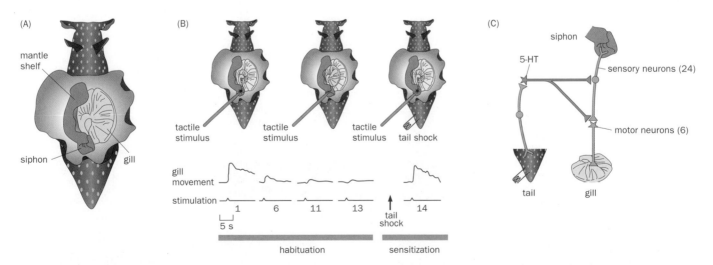

Figure 10–24 The gill-withdrawal reflex in *Aplysia* and the underlying neural circuits. (A) Schematic drawing of *Aplysia* highlighting the structures related to the gill-withdrawal reflex. **(B)** Top, schematic drawing of the gill-withdrawal reflex and its habituation (middle) and sensitization (right). Bottom, recording of the gill movement (top traces) shows progressive decrement in response to repetitive siphon stimulation (bottom traces). Numbers indicate repetitions. Shortly before the 14th stimulus, a tail shock

was applied, which caused an increase of response to stimulus 14. **(C)** Circuit diagram of the gill-withdrawal reflex. The 24 sensory neurons that innervate the siphon connect directly with the six motor neurons that innervate the gill muscle. Sensory neurons activated by tail shock connect with serotonin (5-HT) neurons, which in turn innervate the siphon sensory neurons and their presynaptic terminals onto the gill motor neurons. (Adapted from Kandel ER [2001] *Science* 294:1030–1038.)

24 sensory neurons that sense siphon stimulus and their presynaptic terminals on the motor neurons (Figure 10–24C). These connections would allow the tail shocks to modulate the activity of sensory neurons or neurotransmitter release from the sensory neurons to their motor neuron targets (see Figure 3–37).

Having mapped the neurons underlying the reflex circuit, researchers then asked the question: What is the nature of the circuit change responsible for behavioral habituation, that is, the reduction in the magnitude of gill withdrawal after repeated siphon stimulations? In principle, this could be caused by any of the following changes: (1) sensory neurons progressively reduce their response magnitude after repetitive stimuli, akin to sensory adaptation (see Section 4.7); (2) the efficacy of synaptic transmission between sensory and motor neurons is depressed; (3) the efficacy of synaptic transmission at the neuromuscular junctions is depressed; (4) the muscles become fatigued. A series of experiments using physiological recordings in conjunction with sensory stimulation and quantitative measure of behavioral responses were carried out to systematically examine these possibilities (**Figure 10–25**).

In Experiment 1, the gill-withdrawal responses to direct motor neuron stimulation were measured before and after behavioral habituation and were found to be the same. This ruled out the possibility that changes downstream of the motor neurons in the circuit, including a depression in synaptic efficacy at the neuromuscular junction or muscle fatigue, were responsible for habituation. To test for sensory adaptation at the peripheral sensory endings, a set of sensory stimuli were applied while motor neuron responses were recorded (Experiment 2). Responses became smaller as more stimuli were applied, correlating with behavioral habituation. As illustrated in Figure 10–25, during stimuli 10 through 18, a segment of the siphon nerve that connects the sensory nerve endings to the sensory neurons was bathed in a sodium-free solution to block action potential propagation. After the nerve block was relieved, the motor neuron response became larger, instead of becoming smaller as would be predicted if sensory adaptation at the periphery were responsible for habituation. This ruled out the possibility that habituation was due to an effect upstream of the sensory nerve. Collectively, these experiments suggested that changes at the sensorimotor synapses underlie behavioral habituation. Indeed, in studies carried out in an isolated ganglion, which facilitated stimulation and recording compared with intact *Aplysia*, motor neuron responses elicited by direct sensory neuron stimulation were found to undergo progressive depression after repeated trials

Figure 10–25 Neural mechanisms of habituation and sensitization of the *Aplysia* gill-withdrawal reflex. Top, diagram of information flow from siphon stimulation to gill withdrawal. Bottom, three experiments that investigate the neural mechanisms of behavioral habituation. **Experiment 1:** Gill responses (red traces) to direct motor neuron stimulation (spikes of motor neurons shown as blue traces) before and after habituation remained unchanged, arguing against the possibility that habituation affects processes downstream of the motor neuron. **Experiment 2:** Intracellular recording of a motor neuron (red traces) in response to a series of 20 siphon stimuli (blue line represents the duration of one stimulus). The first nine stimuli were applied under the normal condition, and the resulting motor neuron response was depressed (compare the top right with top left traces), correlating with habituated behavioral responses. During stimuli 10 through 18, action potentials from the siphon nerve were blocked so that the motor neuron did not respond (bottom left). After the siphon nerve was unblocked, the 20th stimulus gave a larger response than the 8th stimulus (compare the top right and bottom right traces), thus arguing against the possibility that habituation affects processes upstream of the siphon nerve. **Experiment 3:** In a reduced preparation consisting of an isolated ganglion that contains the sensory and motor neurons, motor neuron responses (red traces) were induced by intracellular stimulation of the sensory neuron that produced a single spike (blue traces). The top three pairs show three consecutive sensory neuron stimulations (mimicking behavioral habituation), which caused progressively reduced responses. In the bottom pair, the motor neuron response was facilitated due to stimulation of the axon bundle that includes axons of the serotonin neurons (mimicking behavioral sensitization) before the pairing. (Adapted from Kupfermann I, Castellucci V, Pinsker H et al. [1970] *Science* 167:1743–1745 and Castellucci V, Pinsker H, Kupfermann I et al. [1970] *Science* 167:1745–1748. With permission from AAAS.)

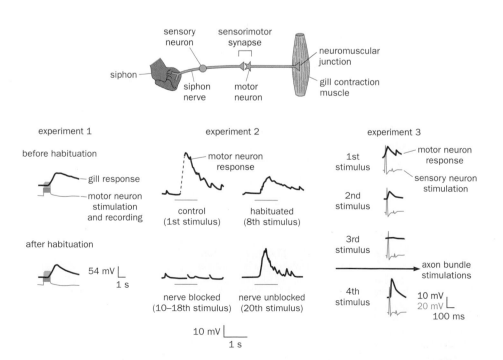

(Experiment 3), suggesting that depression of sensorimotor synaptic efficacy is the primary cause of behavioral habituation.

Analogous experiments were conducted to test the location of change during sensitization by tail shock. Remarkably, the same sensorimotor synapses that were depressed during habituation were potentiated during sensitization (Figure 10–25, Experiment 3). Together, these findings suggest that behavioral modifications, as measured by the magnitude of the gill-withdrawal reflex, are caused primarily by changes in the efficacy of synaptic transmission between sensory neurons and motor neurons—that is, habituation is caused by a depression and sensitization caused by a facilitation of the synaptic efficacy. These results provide compelling support for the hypothesis proposed in Section 10.3, namely that changes of synaptic strengths underlie learning.

10.16 Both short-term and long-term memory in *Aplysia* engage cAMP signaling

Studies of the *Aplysia* gill-withdrawal reflex have also provided important insights into the mechanisms of short-term and long-term memory. Behavioral studies in humans suggest that repeated training can strengthen memories, causing them to become long lasting (as noted in the epigraph of this chapter). Sensitization of the *Aplysia* gill-withdrawal reflex also exhibits these properties. Whereas one tail shock caused a transient increase in gill-withdrawal magnitude that returned to baseline within one hour, four shocks produced a memory (evidenced by a withdrawal response above baseline) that lasted at least a day. The memory produced by four trains of four shocks within a day was retained even after four days. Four trains of four shocks every day for four days produced a drastic increase in response magnitude that persisted for more than a week (**Figure 10–26A**).

In order to facilitate mechanistic studies, researchers established an *in vitro* co-culture system consisting of a siphon sensory neuron and a gill motor neuron (named L7, which can be identified in each animal based on its stereotyped size, shape, and location) that form synaptic connections in a dish. In this system, repeated stimulation of the sensory neuron caused progressive decreases of the magnitude of the postsynaptic potential (PSP) recorded from the motor neuron, mimicking behavioral habituation and consistent with the findings from studies in intact ganglion (for example, Figure 10–25, Experiment 3). Sensitization could also be recapitulated in the co-culture system by applying serotonin to the culture (Figure 10–26B). Whereas one pulse of serotonin application produced a short-term PSP facilitation that lasted for minutes, five repetitions of serotonin application separated by 15-minute intervals produced a long-term facilitation of the PSP that lasted for 24 hours (Figure 10–26C), comparable to the outcome of repeated tail shock (Figure 10–26A). These short-term and long-term facilitations of synaptic efficacy have been used as cellular models of short-term and long-term memory.

Figure 10–26 Long-term sensitization can be induced by repeated training or serotonin (5-HT) application, and is dependent on protein synthesis. **(A)** Duration of gill withdrawal above the baseline in response to three different tail-shock protocols as indicated. Increased training produced sensitization of the gill-withdrawal reflex that was longer lasting. **(B)** Behavioral habituation and sensitization can be recapitulated as changes of synaptic strength between a sensory and a motor neuron co-cultured *in vitro*. Here the relative magnitude of postsynaptic potential (PSP) of the L7 motor neuron in response to sensory neuron stimulation is plotted against the stimulus number. A progressive decline of the magnitude accompanied the application of successive stimuli. Application of 5-HT, which mimics tail shock, increased the PSP magnitude. Application of the protein synthesis inhibitor anisomycin (blue trace) had no effect on this short-term depression and facilitation compared to the control (brown trace). **(C)** A single 5-HT application (1 × 5-HT) did not produce long-term facilitation measured 24 hours later, whereas a sequence of five 5-HT applications did (5 × 5-HT). Application of the protein synthesis inhibitor anisomycin during the time of 5-HT application (5 × 5-HT + A) blocked the long-term facilitation. (A, adapted from Frost WN, Castelucci VF, Hawkins RD et al. [1985] *Proc Natl Acad Sci USA* 82:8266–8269. With permission from the authors; B & C, adapted from Montarolo PG, Goelet P, Castellucci VF et al. [1986] *Science* 234:1249–1254.)

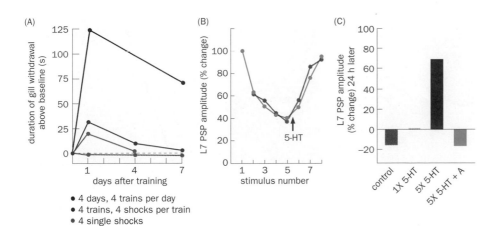

(A)

duration of gill withdrawal above baseline (s)

days after training

- 4 days, 4 trains per day
- 4 trains, 4 shocks per train
- 4 single shocks
- control

(B)

L7 PSP amplitude (% change)

5-HT

stimulus number

(C)

L7 PSP amplitude (% change) 24 h later

control   1X 5-HT   5X 5-HT   5X 5-HT + A

Are there mechanistic differences between short-term and long-term memory? Studies in different animals from mice to goldfish indicated that long-term (but not short-term) memory formation is inhibited by applying drugs that inhibit protein synthesis at the time of training, suggesting that long-term-memory formation selectively requires new protein synthesis. Likewise, in the co-culture system in *Aplysia*, applying protein synthesis inhibitors at the time of serotonin application blocked long-term (Figure 10–26C) but not short-term (Figure 10–26B) facilitation of synaptic transmission caused by serotonin application. Furthermore, applying a protein synthesis inhibitor before or after serotonin application did not affect long-term facilitation. These studies support the notion that protein synthesis is required during the acquisition step of long-term memory.

We now have a good understanding of the molecular mechanisms that mediate short- and long-term facilitation in this system. During short-term facilitation, serotonin acts on a G-protein-coupled receptor in the presynaptic terminal of the sensory neuron to elevate the intracellular cAMP concentration through the activation of an adenylate cyclase (see Sections 3.19 and 3.21). Indeed, intracellular injection of cAMP into the sensory neuron was sufficient to cause an enhancement of synaptic transmission between sensory and motor neurons. As was discussed in Chapter 3, cAMP is a second messenger that activates protein kinase A (PKA). One effect of PKA activation at the presynaptic terminal of the sensory neuron is the phosphorylation of a specific type of K^+ channel that is active during resting state, resulting in its closure. This raises the resting membrane potential and makes it easier for action potentials arriving from the cell body to cause the opening of voltage-gated Ca^{2+} channels at the presynaptic terminal of the sensory neuron, thus facilitating neurotransmitter release (**Figure 10–27**, bottom). Serotonin also activates other intracellular signaling pathways, notably protein kinase C (see Figure 3–34), which can phosphorylate other substrates such as voltage-gated K^+ channels, leading to spike broadening and increased neurotransmitter release per action potential. Thus, short-term facilitation alters synaptic strength by post-translational modification of ion channels, consistent with action that takes place on a timescale of seconds to minutes and does not require new protein synthesis.

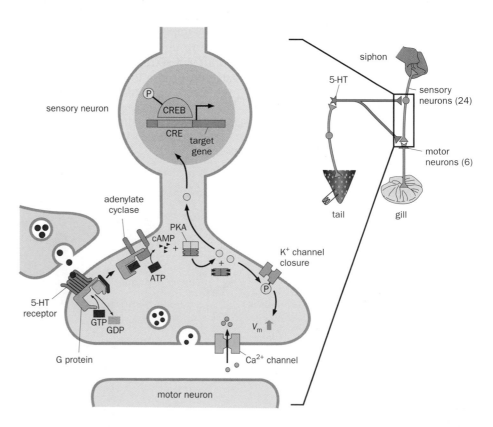

Figure 10–27 Short- and long-term facilitations in *Aplysia* both involve cAMP and PKA. During short-term facilitation, tail shock induces serotonin (5-HT) release at the presynaptic terminal of the sensory neuron, which activates a G-protein-coupled 5-HT receptor. One of the downstream mechanisms is the activation of adenylate cyclase, leading to cAMP production and PKA activation. PKA phosphorylates a specific type of presynaptic K^+ channel and causes its closure, which elevates the resting membrane potential and facilitates action potential-triggered neurotransmitter release. In conditions that produce long-term facilitation, the catalytic subunit of PKA enters the nucleus and phosphorylates nuclear substrates such as the transcription factor CREB and induces new gene expression. The circuit diagram of the gill-withdrawal reflex and sensitization is shown on the right; the box indicates where the scheme on the left is from. For simplicity, the 5-HT axon terminal at the cell body is skipped and the axon is shortened in the magnified diagram on the left. (Adapted from Kandel ER [2001] *Science* 294:1030–1038.)

Remarkably, cAMP and PKA are also key components for long-term facilitation (Figure 10–27, top). Here, a widely used signaling pathway involving the transcription factor CREB is engaged (see Figures 3–41 and 7–36B): PKA phosphorylation activates CREB, which binds to the CRE (cAMP response element) sequences near the promoters of target genes to activate their transcription. How does activation of PKA affect events both locally at the synapse and remotely in the nucleus? Whereas transient serotonin application causes transient PKA activation locally at the synapse, imaging experiments indicated that repeated or prolonged serotonin application induces translocation of the catalytic subunit of PKA to the nucleus, where it can phosphorylate nuclear substrates including CREB. Just as long-term potentiation in the mammalian hippocampus is accompanied by structural changes (see Section 10.13), long-term facilitation in *Aplysia* is also accompanied by growth of synaptic contacts between the sensory and motor neurons. Therefore, some of the molecules whose expression is regulated by CREB are likely responsible for regulating synaptic growth.

10.17 Olfactory conditioning in *Drosophila* requires cAMP signaling

Whereas *Aplysia* offers large cells for physiological studies of learning and memory, the fruit fly *Drosophila* provides an unbiased way to identify genes required for learning and memory by using genetic screening (see Section 13.6). In this procedure, flies with mutations in random genes (produced by treating flies with a chemical mutagen, for example) can be screened using a behavioral assay that tests learning and memory. Mutant flies that perform poorly can be isolated, and the corresponding gene can be mapped using molecular-genetic procedures.

Flies can be trained to associate odors with electrical shocks. In a widely used classical conditioning paradigm, flies are exposed to odorant *A* while being shocked. They are also exposed to odorant *B* without shock. In this case, odorant *A* is designated as the CS+ as it is a conditioned stimulus that is associated with the unconditioned stimulus (US), electric shocks, whereas odorant *B* is designated as the CS−. To test their odorant preference, flies are placed in a T-maze (**Figure 10–28**A; see also Movie 6–1), where they choose to enter one arm (exposed to odorant *A*) or the other (exposed to odorant *B*). Prior to the odorant-shock pairing, flies are as likely to choose odorant *A* as they are *B*. However, after the odorant-shock pairing, 95% of wild-type flies avoid the odorant associated with

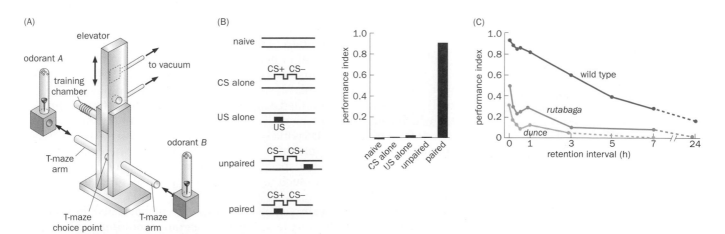

Figure 10–28 Olfactory conditioning in *Drosophila* and its disruption by mutations affecting cAMP metabolism. (A) Schematic of the olfactory conditioning procedure. About 100 flies in the training chamber were exposed to odorant A (CS+) paired with electric shock (US), and to odorant B without electric shock (CS−). These flies were then transferred via the sliding elevator to the bottom T-maze arms, where flies can freely choose a path to odorant A or odorant B. Performance index = [(number of flies in tube B − number of flies in tube A) / total number of flies] × 100. **(B)** Performance indices of flies under different training conditions. Flies learn the association only when US is paired with CS+. **(C)** Performance indices of wild-type and mutant flies. Performance indices represent learning when measured immediately after training (*t* = 0) and memory retention when measured as specific times thereafter. *rutabaga* and *dunce* mutant flies are defective in both learning and memory. (Adapted from Tully T & Quinn WG [1985] *J Comp Physiol* 157:263–277. With permission from Springer. See also Dudai Y, Jan Y, Byers D et al. [1976] *Proc Natl Acad Sci USA* 73:1684–1688 for the identification of the first learning mutant, *dunce*.)

shock (Figure 10–28B). Timing of the CS–US paring is crucial (Figure 10–28B), as would be predicted from a classical conditioning paradigm. In addition to learning, which is measured as the behavioral performance immediately after training, flies can also be tested for memory at specific times after training. One odorant–shock pairing (for 1 minute) produces a memory that lasts for several hours (Figure 10–28C). Repeated pairings with proper intervals (spaced training) can produce long-term memory that lasts for a week, similar to the *Aplysia* gill-withdrawal reflex following sensitization by tail shock (see Figure 10–26A).

Two of the first mutations identified through genetic screening, named *dunce* and *rutabaga*, affected both learning and memory. Performance of flies carrying either of these two mutations was drastically reduced compared with normal flies immediately after training, indicating a learning defect. In addition, they forgot quickly whatever they learned (Figure 10–28C). Separate tests showed that the abilities of these mutants to detect odorants and shocks were normal, indicating a specific defect in forming the odor–shock association. Molecular-genetic studies revealed that the *rutabaga* gene encodes an adenylate cyclase, an enzyme that catalyzes cAMP synthesis (see Figure 10–27), whereas the *dunce* gene encodes a phosphodiesterase, an enzyme that hydrolyzes cAMP (see Figure 6–4). Thus, proper regulation of cAMP metabolism is essential for learning and memory in a classical conditioning paradigm in *Drosophila*. Subsequent experiments found that perturbation of CREB, the transcription factor regulated by cAMP, affected long-term but not short-term memory of olfactory conditioning, again similar to sensitization of the *Aplysia* gill-withdrawal reflex (see Figure 10–27).

10.18 *Drosophila* mushroom body neurons are the site of CS–US convergence for olfactory conditioning

The identification of molecules required for *Drosophila* olfactory learning and memory also provided an entry point for cellular and circuit studies (see Figure 10–7). For example, it was found that both *dunce* and *rutabaga* genes have expression patterns that are highly enriched in mushroom body neurons, which are targets of olfactory projection neurons (see Figure 6–27). Indeed, expression of a wild-type *rutabaga* transgene in adult mushroom body neurons was sufficient to rescue the memory defects of *rutabaga* mutant flies, demonstrating that cAMP regulation in mushroom body neurons plays a crucial role in olfactory learning and memory.

A circuit model of olfactory learning has been proposed that is based on these studies and on the position of mushroom body neurons in the olfactory processing pathways (see Section 6.16). According to this model, odorants (the CS) are represented by ensembles of mushroom body neurons, whose connections with mushroom body output neurons are modified when the CS is paired with an unconditioned stimulus (the US) that is aversive (such as electric shocks) or appetitive (such as food). This plasticity is a cAMP-dependent process. Recent comprehensive mapping identified 21 types of mushroom body output neurons, most of which connect with one of 15 axonal compartments of mushroom body neurons. Information about US is likely carried by one or more of the twenty types of dopamine neurons, most of which also projects axons to one compartment. Behavioral studies suggest that specific types of mushroom body output neurons encode specific valence, such as aversive or appetitive, to guide behavior (**Figure 10–29**A).

As a specific example, we discuss below an experiment that tested the function of dopamine neurons in olfactory learning in an operant conditioning paradigm. In this paradigm, a single fly was allowed to walk freely in a chamber with two compartments, each of which contained a different odorant. During the training period, the fly received electric shocks whenever it entered the compartment containing one of the two odorants. Through its own actions, the fly learned to avoid the odorant associated with shock. To test the role of dopamine neurons in this learning paradigm, an ion channel that can be activated by light was selectively expressed in a subset of dopamine neurons. Researchers found that photoactivation of dopamine neurons could be used instead of electric shocks to train flies

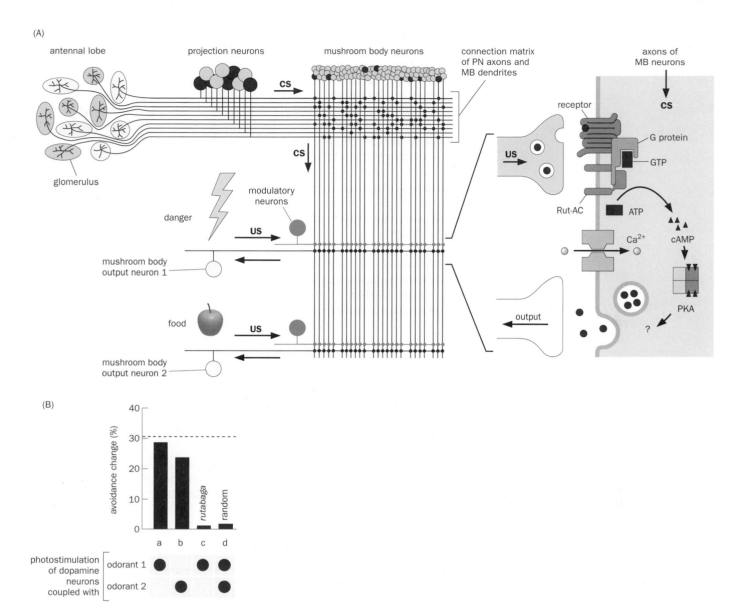

Figure 10–29 Neural circuit and mechanisms of *Drosophila* olfactory conditioning. (A) Left, a circuit model for olfactory conditioning. Odorants activate specific ensembles of glomeruli (ovals) in the antennal lobe, with active glomeruli shown in yellow (see Figure 6–27 for a schematic of the *Drosophila* olfactory system). Projection neurons (PNs) that innervate the active glomeruli become activated (red), which in turn activate a specific subset of mushroom body (MB) neurons. In this representative connection matrix between PN axons and MB neuron dendrites, each MB neuron is connected with three PN axons (dots); only when all three connected PNs are active would the MB neuron become active (red cell on top, with three red dots in the matrix). CS (odorant) information is represented by ensembles of active PNs and subsequently by ensembles of active MB neurons. Synapses between MB axons and dendrites of MB output neurons are modified by nearby input from modulatory neurons, such as dopamine neurons, that signal the presence of an aversive or appetitive US (bottom). Each dot represents a connection and each red or blue dot represents an active synapse. The co-activation of neurons representing the US and the CS modifies the synaptic efficacy between the MB neuron and the MB output neuron. Right, an enlarged diagram of the output synapses of MB neurons. Axons carrying the US information release modulatory neurotransmitters that activate G-protein-coupled receptors, resulting in the activation

of the Rutabaga adenylate cyclase (Rut-AC). This causes an increase in cAMP production and thereby activates PKA, leading to changes in synaptic efficacy through mechanisms yet to be explored. **(B)** Photostimulation of dopamine neurons expressing a light-activated ion channel (see Figure 13–44 for details) is used to train flies to avoid a specific odorant in an operant conditioning paradigm. Individual flies were trained to avoid one of the two odorants by giving an electric shock whenever the fly enters a compartment associated with the odorant. The horizontal line indicates the level of avoidance change after electric shock-based training. Photostimulation of specific dopamine neurons can substitute for shock and achieve similar effect. When photostimulation was repeatedly coupled with the fly entering the compartment containing odorant 1 (a) or odorant 2 (b), flies increased avoidance of the stimulus-paired odorant. This effect was abolished in the *rutabaga* mutant (c), or when photostimulation was random and not uniquely associated with one odorant compartment or the other (d). (A, adapted from Heisenberg M [2003] *Nat Rev Neurosci* 4:266–275. With permission from Macmillan Publishers Ltd. See also Aso Y, Hattori D, Yu Y et al. [2014] *Elife* 3:e04577 and Aso Y, Sitaraman D, Ichinose T et al. [2014] *Elife* 3:e04580; B, adapted from Claridge-Chang A, Roorda RD, Vrontou E et al. [2009] *Cell* 139:405–415. With permission from Elsevier Inc.)

to avoid a specific compartment (Figure 10–29B), consistent with the notion that dopamine neurons provide information about electric shocks. Both shock- and photostimulation-mediated training became ineffective in the *rutabaga* mutant, indicating that this operant conditioning paradigm also requires cAMP.

In summary, studies of olfactory conditioning in flies have produced a circuit and molecular model (Figure 10–29A) with remarkable similarities to sensitization of the gill-withdrawal reflex in *Aplysia* (see Figure 10–27). At the circuit level, information about olfactory conditioned stimuli enters the mushroom body neuron dendrites through excitatory input from olfactory projection neurons. Input from dopamine neurons, representing the US, likely modifies synapses that link mushroom body neurons to their downstream mushroom body output neurons. Indeed, the connections between the mushroom body neurons and the output neurons represent a specific example of the synaptic matrix discussed in Figure 10–5 (Movie 10–1). Here, input patterns represent specific odorants, and through the synaptic matrix produce at least two distinct output patterns, the activation of aversive or appetitive output neurons, leading to activation of two distinct behaviors. Before training, neutral odorants do not activate either of the output neurons. During learning, coincident activation of modulatory neurons modifies the connection strengths between mushroom body neurons and output neurons, such that after training, activation of specific mushroom body neuron ensembles alone (representing odorants) would activate either the aversive or appetitive output neurons depending on the training condition (see Movie 10–1).

At the molecular level, the US causes the activation of the G-protein-coupled dopamine receptor, which in turn activates the adenylate cyclase, leading to cAMP production and PKA activation in mushroom body neurons. Together, the *Aplysia* and *Drosophila* studies demonstrate an evolutionarily conserved role of cAMP in different forms of learning and memory. Indeed, cAMP and PKA also play important roles in synaptic plasticity (see Sections 10.7–10.9) as well as learning and memory in mammals (see Section 10.20), including the hippocampus-dependent learning that we now turn to. Many hippocampus-dependent learning paradigms and memory tasks take advantage of an important function of the hippocampus: spatial representation (**Box 10–2**).

Box 10–2: Place cells, grid cells, and representations of space

Navigation is essential for animals to find food and return home safely. Animals from ants and honeybees to mammals use two types of navigation strategies: a **landmark-based strategy**, where animals use external cues to determine their location, and a **path-integration strategy**, where animals use information based on the speed, duration, and direction of their own movement to calculate their current positions with respect to their starting position. Both strategies require that animals have an internal representation of space.

In mammals, the hippocampus and entorhinal cortex are central to spatial representation. A seminal discovery was made in the 1970s when researchers performed single-unit recordings of hippocampal neurons in freely moving rats navigating an arena or a maze. Individual cells were found to fire robustly when the rat was at a particular location in the maze, regardless of what behavior the animal was performing (for example, passing through from various directions, exploring, or just resting); different cells fired at different locations (**Figure 10–30**A). These cells are called **place cells**, and the physical location that elicits place-cell firing is known as the cell's **place field**.

We now know that virtually all hippocampal CA1 and CA3 pyramidal neurons are place cells. Their place fields are influenced by external landmarks. For example, after the place field is established in a circular arena, if external landmarks are rotated, the place fields also rotate, preserving their relative positions to the external landmarks. However, once place fields form, place cells fire at the same locations in the dark, and place fields in the same environment can be stable for over a month (see Movie 13–3). Since different place cells fire when the rat occupies different locations in the same arena, it is possible to reconstruct the path of a moving rat from simultaneous recordings of dozens of place cells using a multi-electrode array (Figure 10–30B); in other words, a few dozen place cells contain sufficient information to reconstruct the rat's path. At the same time, a single place cell can be active in different environments, with differing place fields in each. Thus, each environment is represented by a unique population of active place cells (a **cell assembly**), and each cell participates in multiple cell assemblies that represent multiple environments. These remarkable properties led to the proposal that hippocampal place cells collectively form cognitive maps that

Box 10–2: Place cells, grid cells, and representations of space

(A)

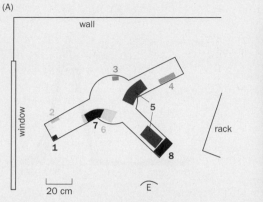

(B)

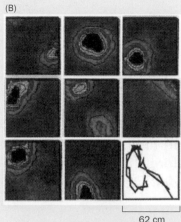

62 cm

Figure 10–30 Hippocampal place cells. (A) Map of a maze showing the place fields (numbered and illustrated in different colors) of eight place cells in the hippocampus of a freely moving rat. Each place field represents the regions within the maze in which a given place cell exhibited increased firing rate. E, location of the experimenter. **(B)** The activity of place cells can be used to construct a map of a rat's travels. A multi-electrode array was used to simultaneously record 80 hippocampal cells. The place fields of eight selected place cells are represented here as heat maps in eight squares: for each place field, the colors indicate the firing rate of the cell when the rat occupied a corresponding position in a 62 cm × 62 cm square arena (red, maximal firing rate; dark blue, no firing). Note that the place fields of different cells vary in size and are situated at different locations in the arena. Bottom right, the vector of firing rates of a neuronal population during a 30-second period was used to reconstruct the spatial trajectory of the rat. The calculated trajectory (red) closely matches the actual trajectory (black). (A, adapted from O'Keefe J [1976] *Exp Neurol* 51:78–109; B, From Wilson MA & McNaughton BL [1993] *Science* 261:1055–1058.)

animals can use to determine where they are in their environment and to aid their navigation using landmark-based and path-integration strategies. Unlike the topographic map discussed in the visual system (see Chapters 4 and 5), however, there is no obvious relationship between the positions of place cells in the hippocampus and the physical locations of their place fields.

How do hippocampal place cells acquire their firing properties? A partial answer to this question came from another remarkable discovery made in the mid-2000s. A large fraction of layer 2/3 neurons in the medial entorhinal cortex, which provide major input to the hippocampus (see Figure 10-6, right), were found to also have space-modulated firing patterns. The locations where these cells fire most are distributed across the environment in a periodic manner, forming grids that tile the entire space; each cell's peak firing rate occurs at the apices of the hexagonal grid unit (**Figure 10–31**A; see **Movie 10–2**). These cells are aptly named **grid cells**. Each grid cell has a characteristic grid size that remains constant in arenas of differing sizes and shapes (Figure 10–31A). Neighboring grid cells share similar grid sizes, but differ in the exact locations of the grid centers.

Grid cells and place cells share many similarities. The activities of simultaneously recorded grid cell populations, like those of place cells, can be used to reconstruct trajectories of movement (see Movie 10–2). As with place fields, grid patterns are influenced by external landmarks; when external landmarks are rotated in a circular arena, grid patterns rotate correspondingly. Grid patterns, like place fields, do not merely mirror sensory cues, since they are maintained when the animal moves in the dark. However, the properties of grid cells and place cells also differ in important ways. Grid cells tile space more efficiently: a few grid cells can cover a space that requires dozens or more place cells. After animals are introduced into a novel environment, grid cells retain their grid size but place fields may remap completely. Populations of grid cells maintain the positions of their grid centers relative to each other across different arenas, whereas place cells remap more randomly. These observations suggest that the grid cells provide a more fundamental metric of space for anchoring the place fields of hippocampal cells.

In addition to grid cells, the entorhinal cortex also contains **border cells**, which fire when the animal is at a specific edge of the arena (Figure 10–31B). Border cells provide information about the perimeters of the local environment, which can anchor grid patterns and place fields to geometric confines. **Head direction cells**, another intriguing cell type, fire when the animal's head is facing a specific direction independent of the animal's location in the arena (Figure 10–31C). Whereas grid cells and border cells have been found mostly in the entorhinal cortex, head direction cells are also present in brain areas that send input to the entorhinal cortex. Indeed, the entorhinal cortex receives diverse inputs representing visual, olfactory, and vestibular signals. In turn, intermingled populations of grid cells, border cells, and head direction cells in the entorhinal cortex all send direct projections via the perforant path (see Figure 10-6) to the hippocampus, which integrates these diverse streams of information to form place fields and send feedback signals

(Continued)

Box 10–2: Place cells, grid cells, and representations of space

(A)   arena 1   arena 2

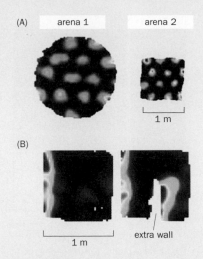

1 m

(B)

1 m   extra wall

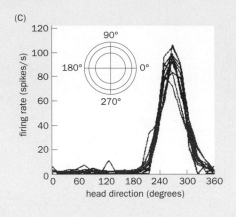

(C)

Figure 10–31 Grid cells, border cells, and head direction cells. **(A)** Firing patterns of two entorhinal cortex grid cells, one in a circular and another in a square arena. The color of each position within each arena reflects the firing rate of the grid cell when the rat occupied that position (red, maximal firing rate; dark blue, no firing). Periodic peaks in firing rate of each cell form hexagonal grids that tile each arena. **(B)** This entorhinal border cell fired selectively when the rat was located at the left border of a square arena (left). When an extra border was added, a new firing field along the new left border was created (right). **(C)** A head direction cell fired when the rat's head was facing a specific direction (peaking at ~270°, or when the rat's head was facing south) regardless of where the rat was located in the arena. Each of the 12 traces represents the firing rate of the cell when the rat was at one of the 12 divisions of the circular arena (inset at the top left). (A, from Hafting T, Fyhn M, Molden S et al. [2005] *Nature* 436:801–806. With permission from Macmillan Publishers Ltd; B, from Solstad T, Boccara CN, Kropff E et al. [2008] *Science* 322:1865–1868. With permission from AAAS; C, adapted from Taube JS, Muller RU & Ranck JB [1990] *J Neurosci* 10:420–435. With permission from the Society for Neuroscience.)

to the entorhinal cortex. Exactly how information is integrated and how the place code is read in order to guide navigation are still open questions.

The remarkable properties of place cells and grid cells in the hippocampal–entorhinal network, far removed from the sensory world, have provided a glimpse of how abstract information such as space is represented in the brain. What is the relationship between spatial representation and memory, another important function of the hippocampus? One hypothesis is that the hippocampal–entorhinal network is used in parallel for navigation and memory. Explicit memory often involves the binding of disparate details into a coherent event; this is conceptually similar to the process

by which hippocampal place cells extract spatial information from the activities of grid cells, border cells, and head direction cells. An alternative hypothesis is that the location of an experience is so essential to its explicit memory that the formation of a memory is intimately tied with the representation of space. Indeed, the use of 'memory palaces,' that is, the organization of events into imaginary spaces, is an ancient and effective mnemonic technique, and space-based tasks have been among the most effective ways to assay memory in mammals. As researchers learn more about the functions of the hippocampal–entorhinal network in memory and in spatial representation, the connections between these two systems will become clearer.

10.19 In rodents, spatial learning and memory depend on the hippocampus

Does synaptic plasticity underlie learning and memory in mammals as it does in *Aplysia*? Put in another way, do activity-dependent changes induced at given synapses during the formation of a specific memory serve as a basis for the information storage that underlies that specific memory? In the following sections we will explore these questions using the mammalian hippocampus as a model, because rich synaptic plasticity mechanisms have been discovered in the mammalian hippocampus (see Sections 10.4–10.13), and the human hippocampus is essential for forming explicit memory (see Section 10.1).

An essential step for linking memory and hippocampal synaptic plasticity is to establish hippocampus-dependent behavioral tasks that test memory in rodents, the animal model in which synaptic plasticity has been most intensely investigated. Given that the mammalian hippocampus contains spatial maps of the external world (see Box 10–2), a number of hippocampus-dependent behavioral

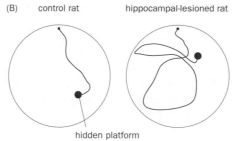

Figure 10–32 Spatial memory tested in the Morris water maze depends on the hippocampus. (A) Schematic of the Morris water maze. After training, a rat or a mouse can use distant spatial cues to help find the location of a hidden platform (dashed circle) in a large pool of milky water. **(B)** After training, a control rat swam directly to the hidden platform, whereas an experimental rat with hippocampal lesion found the platform after taking a circuitous route. (Adapted from Morris RGM, Garrud P, Rawlins JNP et al. [1982] *Nature* 297:681–683. With permission from Macmillan Publishers Ltd.)

assays have been established that require spatial recognition. One of the most widely used is the **Morris water maze** (**Figure 10–32**A), a navigation task in which rats (and mice) learn to locate a hidden platform in a pool of milky water to avoid having to swim. (Despite being able to swim, rats and mice prefer not to.) The rats cannot see, hear, smell, or touch the platform until they find it. Nevertheless, they can use distant cues in the room to learn the platform's spatial location, such that after training they can be placed at any position in the pool and will swim straight to the hidden platform (Figure 10–32B, left). Performance of this task is dependent on the hippocampus, as rats with hippocampal lesions no longer remember the location of the hidden platform even after extensive training (Figure 10–32B, right). When the platform was visible, both rats found it with equal ease. Of the forms of memory known in rodents, this spatial memory most closely resembles the explicit memory of humans.

10.20 Many manipulations that alter hippocampal LTP also alter spatial memory

The establishment of spatial memory tasks such as the Morris water maze enabled researchers to determine whether manipulations that affect synaptic plasticity in the hippocampus also affect spatial memory. One of the first such manipulations was to block the function of the NMDA receptor with a specific antagonist, AP5 (see Section 10.6). Infusion of AP5 into the hippocampus during the training session, at a concentration that blocked LTP *in vivo*, disrupted the subsequent recall of the platform position in the Morris water maze. When the hidden platform was removed after training, control rats focused their search preferentially in the quadrant where the hidden platform had been, whereas AP5-treated rats swam randomly (**Figure 10–33**A). Conditionally knocking out the essential NMDA

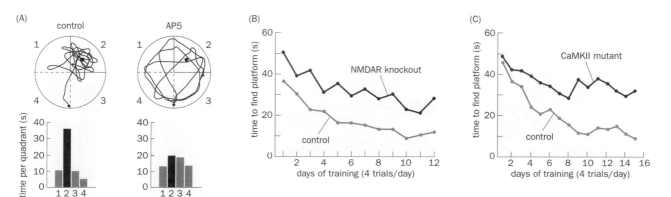

Figure 10–33 Manipulations that disrupt hippocampal LTP also interfere with performance in the Morris water maze. (A) Infusion of AP5, an antagonist of the NMDA receptor, disrupts the spatial memory of rats. Rats were trained with the hidden platform (circle) in quadrant 2. During the test, the platform was removed and the trajectory was recorded. Top, while the control rat focused its search near the phantom platform, the rat infused with AP5 during training swam randomly. Bottom, quantification of time spent in four quadrants. **(B)** Throughout the training regime, mice with CA1-specific knockout of the GluN1 subunit of the NMDA receptor (red trace) were slower to

find the hidden platform than *CA1-Cre* only control mice (blue trace). **(C)** Mice in which the CaMKII auto-phosphorylation site is mutated (red trace) also took longer to find the hidden platform compared with controls (blue trace). See Figures 10–10 and 10–12 for LTP defects of the same mice as in panels B and C. (A, adapted from Morris RGM, Anderson E, Lynch GS et al. [1986] *Nature* 319:774–776. With permission from Macmillan Publishers Ltd; B, adapted from Tsien JZ, Huerta PT & Tonegawa S [1996] *Cell* 87:1327–1338. With permission from Elsevier Inc.; C, adapted from Giese KP, Federov NB, Filipkowski RK et al. [1998] *Science* 279:870–873.)

receptor subunit GluN1 in CA1 pyramidal neurons, which blocked LTP at the CA3 → CA1 synapse (see Figure 10–10), also interfered with performance in the water maze assay (Figure 10–33B). These experiments demonstrated an essential function for the NMDA receptor in the hippocampus, and specifically in CA1 pyramidal neurons, in spatial learning and memory.

Many genetic manipulations in mice that disrupt hippocampal synaptic plasticity, notably LTP at the CA3 → CA1 synapse, also interfere with hippocampus-dependent spatial memory tasks. For example, mice that lack CaMKII or that have a point mutation in the CaMKII auto-phosphorylation site have impaired LTP at the CA3 → CA1 synapse (see Figure 10–12B) and perform poorly in the Morris water maze (Figure 10–33C). In concert with findings from *Aplysia* and *Drosophila*, the cAMP/PKA pathway is also essential for both hippocampal LTP and hippocampus-dependent memory. For example, in mice that carry double mutations for two Ca^{2+}-activated adenylate cyclase, which links Ca^{2+} entry to cAMP production (see Figure 3–41), CA3 → CA1 LTP was impaired (**Figure 10–34**A), as was a hippocampus-dependent memory task called passive avoidance. In this task, mice are placed in a chamber with two compartments, one of which is illuminated. Mice naturally prefer the dark, safer compartment. During training, entry into the dark chamber is paired with an electric shock. After training, mice are placed back in the illuminated compartment and the time it took for the mice to enter the dark compartment is a measure of their memory to avoid the shock-associated compartment. Adenylate cyclase double knockout mice performed poorly 30 minutes after training compared with control mice (Figure 10–34B).

Since the late 1990s, a number of genetic manipulations in mice have been reported to enhance memory performance compared to controls in a variety of memory tasks such as Morris water maze, passive avoidance, or fear conditioning that we will discuss in more detail later. For example, transgenic mice overexpressing the GluN2B subunit, which is normally preferentially expressed in developing neurons (see Box 5–3) and which has higher Ca^{2+} conductance than other GluN2 isoforms, exhibited superior performance in Morris water maze and several other memory tasks. Interestingly, GluN2B overexpressing mice and other genetically engineered mice with enhanced memory performance also exhibited enhanced hippocampal LTP. Together, these experiments have established a strong correlation between memory and hippocampal synaptic plasticity.

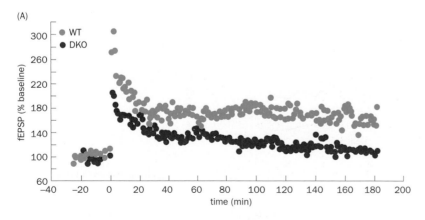

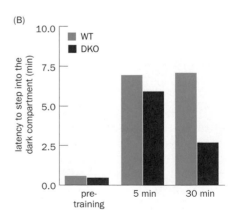

Figure 10–34 Interfering with the cAMP/PKA pathway affects hippocampal LTP and learning. (A) Compared with wild-type (WT) controls, mice in which two Ca^{2+}-dependent adenylate cyclase were doubly knocked out (DKO) exhibit reduced magnitude of LTP at the CA3 → CA1 synapse. **(B)** DKO mice also exhibit reduced memory in a passive avoidance task. In this assay, mice are placed in the illuminated compartment of a two-compartment chamber. Before training, mice quickly move into the dark compartment as a natural tendency to avoid predators. After training (pairing electric shocks with entrance to the dark compartment), DKO mice avoid the dark compartment similarly to controls 5 min after the training, but enter the dark compartment more quickly 30 min after the training, suggesting an impaired memory. (Adapted from Wong ST, Athos J, Figueroa XA et al. [1999] *Neuron* 23:787–798. With permission from Elsevier Inc.)

10.21 From correlation to causation: the synaptic weight matrix hypothesis revisited

While demonstrating strong correlations between hippocampal LTP and spatial memory in rodents, none of the genetic and pharmacological manipulations discussed in the previous section has proven that synaptic plasticity and memory are causally linked, that is, the synaptic changes cause the formation of memory. These manipulations could all affect synaptic plasticity and memory in parallel. To establish a causal link between modification of synaptic strength and learning, one would ideally perform experiments to specifically alter one and test the effect on the other.

One approach is to examine directly whether learning can induce hippocampal LTP. The key to this type of experiment is to identify which synapses in the hippocampus are related to a specific learning event. This difficult task was achieved in an experiment that combined passive avoidance training with use of a multi-electrode recording array in rats. These rats were implanted with a multi-electrode recording array at the CA1 dendritic fields and a stimulating electrode at the Schaffer collaterals (**Figure 10–35**A) such that synaptic transmission from Schaffer collaterals onto different populations of CA1 pyramidal neurons could be recorded before and after the training. Whereas none of the electrodes from control rats without training detected any potentiation, a small fraction of electrodes from trained rats detected potentiation after behavioral training (Figure 10–35B). Moreover, synapses that were potentiated by behavioral training became less likely to be potentiated further by subsequent high-frequency stimulation of the Schaffer collateral, a process known to induce LTP. Thus, learning can produce synaptic potentiation that partially occludes subsequent LTP at the same synapses.

Another approach to investigate the relationship between LTP and learning is to test whether saturation of LTP prevents further learning. Rats with unilateral hippocampal lesions (such that spatial memory must depend on the hippocampal tissues that remain) were implanted with a multi-electrode stimulating array at the perforant path in the unlesioned hippocampus. Repeated stimulation through this array could maximally induce and potentially saturate LTP at recording sites. Rats with nearly saturated LTP, measured *post hoc* by physiological recordings, were more impaired in the Morris water maze assay than were rats that still exhibited residual LTP. These experiments collectively provide stronger links between learning and changes in the strength of specific hippocampal synapses.

Let's revisit the hypotheses raised in Sections 10.2 and 10.3: memory is stored in the form of synaptic weight matrices in neural circuits, and learning is equivalent to altering the synaptic weight matrix as a result of experience. The strongest case for these hypotheses can be made in the context of simple forms of learning, such as the *Aplysia* gill-withdrawal reflex, where modifications of the strength of the sensory neuron–motor neuron synapse underlie behavioral habituation and sensitization (see Section 10.15). In the complex mammalian brain, the strongest evidence has come from studies of the hippocampus discussed in this and preceding sections. One way to strengthen the causal relationship between learning and alteration of synaptic weights would be to achieve the following: (1) identify the neurons and synapses in a circuit whose plasticity correlates with a learning experience; (2) determine the specific states of the synaptic weight matrix (for example, Figure 10–5) before the learning experience (state *A*) and after it (state *B*); (3) artificially change the synaptic weight matrix from state *A* to state *B* without learning; and (4) test whether the animal behaves as if the learning experience had occurred (that is, a mimicry experiment). This is a challenging task; the *in vivo* mimicry experiment has not been performed even in the *Aplysia* gill-withdrawal paradigm. Even though the complexity of the mammalian brain and the large number of neurons and synapses make this task even more challenging, researchers have employed modern circuit analysis tools to search for the potential physical substrates for memory (termed **memory traces** or **engrams**). **Box 10–3** provides an example of how such a search can be conducted.

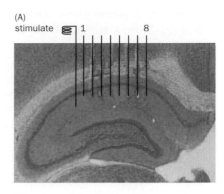

(A) stimulate 1 8

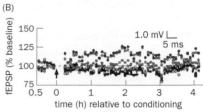

(B)

1.0 mV / 5 ms

Figure 10–35 Learning can induce LTP. **(A)** A multi-electrode array (electrodes 1–8) was placed in the CA1 area of a rat hippocampus to record the responses of CA1 neurons to stimulation of Schaffer collaterals. **(B)** After passive avoidance training, recordings from the two electrodes indicated by red and orange dots showed an enhancement of field excitatory postsynaptic potentials (fEPSPs); recordings from the remaining six electrodes (represented by dots of other colors) did not demonstrate fEPSP enhancement. Additional experiments (not shown) revealed that synapses potentiated by passive avoidance training were less responsive to subsequent potentiation by high-frequency stimulation using the stimulating electrode. (Adapted from Whitlock JR, Heynen AJ, Shuler MG et al. [2006] *Science* 313:1093–1097. With permission from AAAS.)

Box 10–3: How to find an engram

Searching for engrams has had a long history. As mentioned in the introduction to the Prelude of this chapter, Lashley's lesion study led him to conclude that the engram for maze running is widely distributed in a rat's cerebral cortex. Studies of human patients such as H.M. have led to the identification of hippocampus as the site essential for forming new explicit memory. Tools in modern neuroscience have the potential to reveal engrams at the level of neurons and synapses. Neural substrates that represent an engram for a memory task should become active during training, and their reactivation should mimic the recall of the memory. We use a specific example to illustrate how researchers have utilized a combination of transgenic mice, viral transduction, and optogenetic manipulation to search for an engram (**Figure 10–36**).

To identify the active population of neurons, the *Fos-tTA* transgenic mouse was used, in which the expression of a **tetracycline-repressible <u>t</u>ranscriptional <u>a</u>ctivator (tTA)** is controlled by the promoter of the immediate early gene *Fos*, such that tTA can be induced by neuronal activity (see Section 3.23 for this property of immediate early genes). tTA is a transcription factor that binds to DNA sequences called **tetracycline response elements (TRE)** to regulate gene expression; the activity of tTA is inhibited in the presence of a tetracycline analog, doxycycline (see Section 13.10 for more details of the tTA/TRE expression system). An adeno-associated virus (AAV) that enables the expression of channelrhodopsin (ChR2) under the control of TRE was used to transduce dentate gyrus granule cells, which provide input to CA3 pyramidal neurons (see Figure 10–6). These mice were tested for a hippocampus-dependent memory established by **contextual fear conditioning**. [In this paradigm, mice experience electric shocks during training in a specific environment (context *A*). Mice subsequently placed in the same environment exhibit a freezing response: they remain immobile, an adaptive response of rodents to avoid being

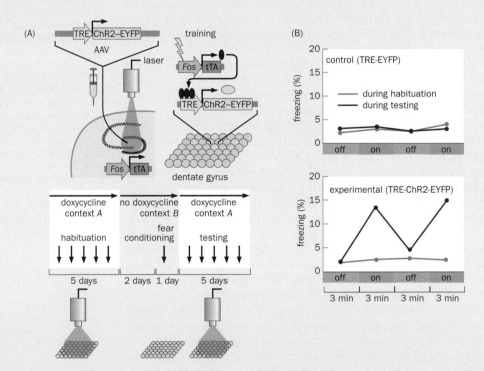

Figure 10–36 Optogenetic stimulation of a specific population of dentate gyrus granule cells activates fear memory.
(A) Experimental design. Adeno-associated virus (AAV) enables the expression of channelrhodopsin (ChR2) fused with an enhanced yellow fluorescent protein (EYFP) under the control of a tetracycline response element (TRE). AAV was injected into the dentate gyrus (indicated by the needle) of a transgenic mouse that express the tetracycline-repressible transcription activator (tTA, red ovals) from the *Fos* promoter so that its expression is induced by neuronal activity. The matrix of circles represents dentate gyrus granule cells. Mice were first exposed to and became habituated to context *A* and photostimulation in the presence of doxycycline; the dentate gyrus granule cells activated in context *A* did not express ChR2, because tTA activity is inhibited by doxycycline. Fear conditioning was induced in context *B* in the absence of doxycycline such that dentate gyrus granule cells activated by this experience expressed ChR2 (yellow circles); ChR2 expression persisted for several days even after the mice were treated with doxycycline again to prevent further tTA-induced gene expression. Mice were then reintroduced to context *A* to test whether optogenetic stimulation of ChR2-expressing cells could induce fear memory recall. **(B)** In control mice, in which tTA induced expression of EYFP, optogenetic activation (light-on period in green) did not induce fear memory, as assayed by the percentage of time spent freezing (top). In experimental mice, optogenetic stimulation induced freezing in a light-dependent manner (bottom) during the testing period but not during the earlier habituation period. (Adapted from Liu X, Ramirez S, Pang PT et al. [2012] *Nature* 484:381–385. With permission from Macmillan Publishers Ltd.)

Box 10–3: How to find an engram

seen by predators when facing danger in the wild. Mice placed in a different environment (context *B*, which differs from *A* in ceiling shape, flooring, and lighting) do not exhibit a freezing response.] The mice were first exposed to and became habituated in context *A* in the presence of doxycycline to prevent tTA/TRE-induced expression of ChR2. After doxycycline removal, mice were exposed to context *B*, during which they received electric shocks to induce contextual fear conditioning. This resulted in tTA and ChR2 expression in the population of dentate gyrus granule cells that were activated during fear conditioning in context *B*.

To test the effect of reactivation of neurons that were active during fear conditioning in context *B*, mice were given food containing doxycycline to prevent new tTA/TRE-induced ChR2 expression, and were introduced to context *A* with or without optogenetic stimulation (Figure 10–36A). Control mice did not freeze in context *A*. However, ChR2-expressing mice froze in context *A* in response to optogenetic stimulation, as if they were in context *B* (Figure 10–36B). Thus,

activation of a population of cells that were active during contextual fear conditioning was sufficient to induce fear recall in a different context, suggesting that this population of dentate gyrus granule cells contributes to the memory of context *B*.

This experiment did not show which synapses were modified and what additional properties in the circuits were changed to make mice fearful of context *B*. In principle, plasticity could occur anywhere in the neural pathway downstream of the granule cell population that leads to the motor behavior of freezing. In light of the hippocampal plasticity findings discussed in this chapter, it is likely that plasticity occurs in the downstream circuits within the hippocampus, such as at the dentate gyrus → CA3 synapse, the CA3 → CA3 recurrent synapse, the CA3 → CA1 synapse, or all of the above. Plasticity can also occur in the amygdala, whose function in fear conditioning will be discussed in Section 10.23.

WHERE DOES LEARNING OCCUR, AND WHERE IS MEMORY STORED IN THE BRAIN?

So far in this chapter, with the exception of the invertebrate systems, we have focused on the hippocampus as a model for studying mechanisms of synaptic plasticity and spatial (explicit) memory. However, synaptic plasticity occurs throughout the nervous system. For example, in Section 8.8 we discussed that the cerebellum plays an important role in motor skill learning, and long-term depression of parallel fiber–Purkinje cell synapses caused by co-stimulating parallel fibers and climbing fibers contributes to cerebellum-based motor learning. In the last four sections of this chapter, we will broaden our study of memory systems beyond the hippocampus with a few selected examples.

10.22 The neocortex contributes to long-term storage of explicit memory

Although the medial temporal lobe including the hippocampus is essential for the initial formation of explicit memory, it does not appear to be required for long-term memory storage and retrieval, as suggested by the ability of H.M. to recall memories of his childhood (see Section 10.1). Where is long-term explicit memory stored?

A widely accepted view is that the neocortex is involved in long-term explicit memory storage, and that specific types of memory engage specific cortical regions. This idea, first proposed in the late nineteenth century, states that remembering involves reactivating the sensory and motor components of the original event that led to the formation of the memory. Two types of human studies are consistent with this view. First, lesions of specific parts of the neocortex lead to loss of specific types of memory. For example, patients with damage to the color- or face-processing areas of the visual cortex not only lose their ability to perceive colors or recognize faces, but also exhibit retrograde memory deficits in specific domains. Patients with adult-onset prosopagnosia (inability to distinguish faces), for instance, not only exhibit defects in face perception, but also cannot remember faces that were familiar before the onset of the disorder.

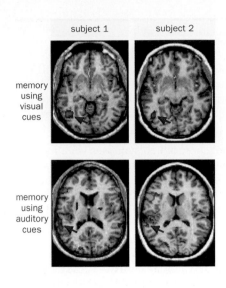

memory using visual cues

memory using auditory cues

subject 1   subject 2

Figure 10–37 Reactivation of specific sensory cortices during long-term memory recall. fMRI images of two subjects, each performing the task of vividly remembering an object when presented with a word that had been extensively paired with either a picture or a sound during prior training. For image-based recall, the high-order visual cortex is activated (arrows in the top panels). For sound-based recall, the high-order auditory cortex is activated (arrows in the bottom panels). These fMRI images in the top and bottom rows were taken at two different horizontal planes. Both subjects have a bias for using the left cortex. (From Wheeler ME, Petersen SE & Buckner RL [2000] *Proc Natl Acad Sci USA* 97:11125–11129. Copyright National Academy of Sciences.)

cortical modules

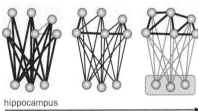

hippocampus

time

Figure 10–38 A model illustrating the interactions between the hippocampus and neocortex during long-term memory consolidation. During the initial memory formation (left), signals that pass through different cortical areas to the hippocampus establish links between the hippocampus and those cortical areas. Ongoing interactions between the hippocampus and the cortical areas after the initial memory formation gradually establish links among the different cortical areas (middle), until these intracortical links are sufficient to represent the remote memory, and the memory can be recalled independent of the hippocampus (right). Heavy and light lines represent strong and weak links, respectively. Links shown in gray at a given stage are not required for memory recall at that time. (Adapted from Frankland PW & Bontempi B [2005] *Nat Rev Neurosci* 6:119–130. With permission from Macmillan Publishers Inc.)

Second, functional imaging studies of healthy human subjects engaged in memory tasks revealed the reactivation of cortical areas relevant to specific memory tasks. In one study, for example, subjects were first extensively trained to associate words (for example, DOG) with either pictures (an image of a dog) or sounds (the bark of a dog). During subsequent testing, they were asked to vividly recall the items when given only the word as a cue, while their brains were being scanned via functional magnetic resonance imaging (fMRI). After the recall task/fMRI scan, subjects then indicated whether they vividly remembered an image or a sound; their answers to this question usually matched with their training. After pairing the word with an image during training, high-order visual cortical areas were selectively activated during the recall (**Figure 10–37**, top), whereas after sound-based training, the recall elicited the selective activation of high-order auditory cortical areas (Figure 10–37, bottom). These data suggest that the act of remembering indeed reactivates sensory-specific cortices.

How do the hippocampus and neocortex collaborate to form and store long-term memory? At present, we can only speculate. According to a prevalent hypothesis (**Figure 10–38**), signals that lead to the original hippocampus-dependent formation of explicit memory also activate primary and associative cortical areas. The hippocampus integrates distributed signals from multiple cortical areas during the initial memory formation. Over the course of long-term memory consolidation, the hippocampus 'trains' the establishment of new connections among cortical neurons, such that memories over the long term are no longer hippocampus-dependent. Exactly how this is achieved is not known.

Animal studies have strengthened the notion that neocortex plays a role in remote memory, and have shed light on the interactions between the hippocampus and neocortex. Let's consider the three sets of experiments on contextual fear memory (see Box 10–3), a hippocampus-dependent form of memory that resembles human explicit memory. In the first experiment, rats received electric shocks when placed in a specific environment; subsequently, their hippocampi were bilaterally lesioned 1, 7, 14, or 28 days after the training. Seven days after the surgery, rats were returned to the training environment to measure their fear memory. Whereas control rats exhibited fear memory under all conditions, rats with hippocampal lesions lacked the contextual fear response if lesioning was performed 1 day after training, but had less severe deficits in memory recall as the duration between training and lesioning lengthened (**Figure 10–39**A). This experiment suggests that fear memory becomes increasingly less dependent on the hippocampus as time passes after the initial training.

Where is long-term fear memory stored? In the second experiment, researchers used immediate early gene expression to identify brain regions that were activated during retrieval of remote fear memory. Several frontal cortical areas were shown to have elevated expression of the immediate early genes *Fos* and *Egr1* (see Section 3.23). Inactivation of these specific cortical areas by focal injection of **lidocaine** (an anesthetic that blocks action potential propagation by inhibiting voltage-gated Na$^+$ channels) identified the **anterior cingulate cortex (ACC)**, which is located near the midline of the frontal lobe, as a neocortical site

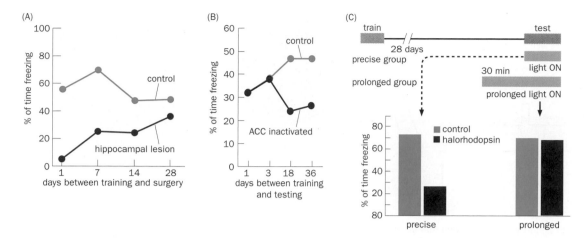

Figure 10–39 Interplay between the hippocampus and neocortex in contextual fear conditioning. (A) The *x* axis shows days elapsed between training and the surgery that bilaterally lesioned the hippocampi in experimental rats. All rats were tested for contextual fear conditioning (*y* axis) 7 days after the surgery. Compared with unlesioned controls that went through the same surgical procedure (blue trace), lesioned rats (red trace) did not exhibit fear memory (quantified by the percentage of time spent freezing) when lesioning was performed 1 day after training. The effect of lesioning became less pronounced as the period between training and lesioning lengthened. **(B)** Injection of lidocaine, an anesthetic that blocks action potentials, into the anterior cingulate cortex (ACC; red trace), reduced contextual fear memory compared with controls (blue trace) when drug administration and testing were performed 18 or 36 days, but not 1 or 3 days, after training. This finding suggests

that the ACC is required for recall of remote memory but not recent memory. **(C)** Compared with controls that expressed a fluorescent protein, optogenetic silencing of mouse hippocampal neurons expressing a modified halorhodopsin significantly reduced contextual fear memory if silencing (light ON) was precisely timed with testing (precise group). If silencing occurred 30 minutes prior to as well as during the testing (prolonged group), the reduction in fear memory disappeared. This suggests that a compensatory mechanism was used for fear memory if hippocampal neurons were silenced under the prolonged condition. (A, adapted from Kim JJ & Fanselow MS [1992] *Science* 256:675–677; B, adapted from Frankland PW, Bontempi B, Talton LE et al. [2004] *Science* 304:881–883; C, adapted from Goshen I, Brodsky M, Prakash R et al. [2011] *Cell* 147:678–689. With permission from Elsevier Inc.)

involved in remote memory. Inactivation of ACC during testing caused significant loss of fear memory when testing occurred 18 or 36 days, but not 1 or 3 days, after initial training (Figure 10–39B). Interestingly, human fMRI studies consistently find activation of the frontal cortex, including the ACC, during different kinds of memory recall.

In the third experiment, researchers expressed halorhodopsin in hippocampal CA1 neurons and then used optogenetic manipulation to reversibly silence these cells (see Section 13.25) during remote contextual fear memory recall. Surprisingly, as shown on the left in Figure 10–39C, acute silencing of hippocampal CA1 neurons during testing 28 days after training severely reduced fear memory, suggesting that the hippocampus was indeed required for remote memory recall. (This acute silencing did not cause detectable nonspecific effects on global brain activity and did not interfere with a hippocampus-independent memory.) This result seemingly contradicted prior studies (see Figure 10–39A); however, when the hippocampal neurons were silenced during the 30 minutes prior to testing as well as during the testing, remote memory was intact (Figure 10–39C, right), consistent with previous pharmacological or lesion experiments in which the hippocampus had been inactivated for a longer period or permanently. In related experiments, optogenetic silencing of the ACC confirmed its requirement in remote fear memory. These findings suggest that (1) retrieval of remote memory may normally involve dynamic interplay between the hippocampus and neocortex, and (2) remote memory can be retrieved by more than one mechanism, such that if the hippocampus is inactivated, readjustment can be made within 30 minutes to allow the cortical network to perform the task. Such redundancy and flexibility likely increase the robustness of memory systems.

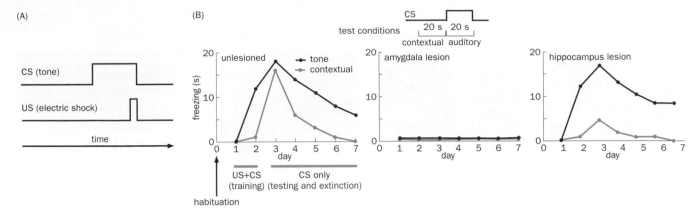

Figure 10–40 Both auditory and contextual fear conditioning require the amygdala. (A) Schematic of auditory fear conditioning, a form of classical conditioning in which the tone (~20 s) serves as a conditioned stimulus (CS) and the electric shock (~0.5–1 s) serves as an unconditioned stimulus (US) that co-terminates with the CS. **(B)** Left, learning curve as measured by average time spent freezing within a 20-s period (*y* axis) for control rats. On day 0, rats were introduced into the conditioning chamber without shock for habituation. On days 1 and 2, a 20-s tone was paired with a 0.5-s electric shock at the end of the tone, with two pairings per day. On days 3–7, only the tone was presented in the same conditioning chamber. As shown in the schematic above, contextual conditioning was measured as the freezing time during the 20 s immediately prior to the CS (tone) onset, whereas auditory conditioning was measured during the 20-s tone period. Middle, lesioning of the amygdala prior to training disrupted both contextual and auditory conditioning. Right, lesioning of the hippocampus disrupted contextual conditioning without affecting auditory conditioning. (B, adapted from Phillips RG & LeDoux JE [1992] *Behav Neurosci* 106:274–285. With permission from the American Psychological Association Inc.)

10.23 The amygdala plays a central role in fear conditioning

In the previous sections, we studied contextual fear conditioning as a hippocampus-dependent form of fear conditioning. There is also a different form of fear conditioning, called **cued fear conditioning**, in which an electric shock is applied at the end of a cue presentation during training. The most commonly used cue is a sound, in which case cued fear conditioning is called **auditory fear conditioning** (**Figure 10–40**A). Auditory fear conditioning is a form of classical conditioning (see Section 10.14); the shock and the tone serve as the US and CS, respectively. Lesion studies indicate that while contextual fear conditioning is dependent on the hippocampus, auditory fear conditioning is not. However, both forms of conditioning depend on the amygdala (Figure 10–40B). When compared side-by-side in control animals (Figure 10–40B, left panel), auditory conditioning was more rapidly acquired during training than contextual conditioning and was more resistant to extinction during the testing phase. After training, animals conditioned in one context to associate a tone with an electric shock also exhibit a robust fear response to the CS (tone) in a different context. These studies suggest that the amygdala is the location where the association of auditory CS and US occurs, whereas the hippocampus contributes specifically to the contextual aspect of the fear conditioning.

Extensive anatomical, physiological, and perturbation studies have delineated circuit diagrams that underlie fear conditioning (**Figure 10–41**). The amygdala complex consists of several major divisions: the lateral amygdala, the basal amygdala (collectively constituting the **basolateral amygdala**), and the **central amygdala**. (Note that this complex is adjacent to but distinct from the olfactory amygdala, which includes the cortical amygdala and medial amygdala and which receives direct input from mitral cells of the main and accessory olfactory systems; see Figure 9–32.) The central amygdala is the output site of the amygdala complex for fear and defensive responses. It sends descending projections to distinct sites in the hypothalamus and brainstem to regulate behavioral output (for example, freezing), autonomic nervous system response (for example, increased blood pressure), and neuroendocrine response (for example, stress hormone production). In auditory fear conditioning, information about the tone (CS) reaches the lateral amygdala via a direct pathway from the auditory thalamic nuclei, as well as

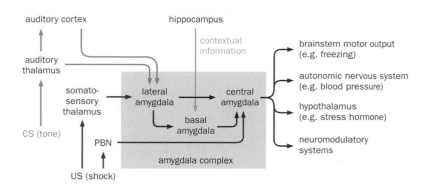

Figure 10–41 Circuit diagrams for fear conditioning. The tone signal (CS, blue) can reach the lateral amygdala directly from the auditory thalamus or indirectly from the auditory cortex. The shock signal (US, red) can reach the lateral amygdala via the somatosensory thalamus, or can reach the central amygdala directly through the pain pathway via the parabrachial nucleus (PBN). Contextual information from the hippocampus (green) enters through the basal amygdala. Within the amygdala complex, information flows from the lateral nucleus to the central nucleus either directly or via the basal amygdala. The central amygdala provides the output to brainstem and hypothalamus targets to regulate behavioral, autonomic, endocrine, and neuromodulatory systems. (See LeDoux JE [2000] *Annu Rev Neurosci* 23:155–184 and Pape HC & Pare D [2010] *Physiol Rev* 90:419–463.)

via an indirect pathway from high-order auditory cortex. Information can be sent by lateral amygdala neurons to the central amygdala either directly or indirectly via the basal amygdala. The foot shock (US) also reaches the amygdala by multiple pathways, including projections from the somatosensory thalamic nuclei to the lateral amygdala and from the pain pathway through the parabrachial nucleus (PBN) to the central amygdala (see also Figure 6–70B). In contextual fear conditioning, contextual input from the hippocampus enters the amygdala complex via the basal amygdala (Figure 10–41).

What is the neural basis of behavioral conditioning? The general framework is that the simultaneous presence of the CS and the US during training strengthens, through a Hebbian mechanism, the connection between neurons that represent the CS and neurons that produce fear response, such that the CS alone can elicit a fear response after the conditioning. The best evidence supporting this model has thus far come from studies of the synapses that connect auditory thalamic input neurons and excitatory projection neurons of the lateral amygdala, where strong correlations have been established between auditory fear conditioning and LTP of these synapses. For example, it has been shown that (1) auditory conditioning can enhance the response of lateral amygdala neurons to the shock-associated tone; (2) LTP can be induced by pairing presynaptic stimulation of thalamic axons with postsynaptic depolarization of lateral amygdala neurons (this postsynaptic depolarization can be achieved by the US during fear conditioning); and (3) LTP and fear conditioning share a common set of molecular mechanisms, including dependence on the postsynaptic NMDA receptor, CaMKII auto-phosphorylation, and AMPA receptor trafficking. This is very much analogous to the relationship between spatial learning and hippocampal LTP discussed earlier in this chapter. As in the hippocampus, plasticity can occur at multiple sites in the amygdala (**Box 10–4**).

Research on fear conditioning in rodent models has revealed the amygdala to be a center for emotional memory and for processing emotion-related signals, which has also been substantiated by human studies. For instance, fMRI studies have shown that the amygdala can be activated by stimuli that are emotionally negative (such as a fearful face) or emotionally positive (such as a pleasant picture). Similar to the rodent fear-conditioning model, the amygdala of human subjects is also activated by presentation of an image (for example, a blue square) that has previously been associated with mild electric shock to the wrist (**Figure 10–42**) but not by presentation of a comparable image (for example, a yellow circle) that has not been paired with a shock. In this fear-conditioning paradigm, patients with amygdala lesions do not exhibit physiological responses that normal subjects do, such as sweating, which can be measured by changes in skin conductance and which is due to activation of the sympathetic system as part of the fear response. Interestingly, the amygdala-lesioned patients remain aware of the explicit association of the CS (the blue square) and the US (mild electric shock), suggesting that amygdala-dependent fear conditioning utilizes a form of implicit memory distinct from the explicit memory that remains intact in these patients.

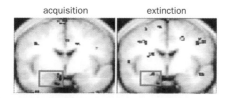

Figure 10–42 Acquisition and extinction of fear condition activate the amygdala in humans. Average fMRI images of 10 healthy human subjects during the acquisition (left) and extinction (right) phases of fear conditioning. The green box in each image highlights the right amygdala, showing activation (yellow to red colors) of the amygdala. During the acquisition phase, subjects were exposed to a blue square image paired with a mild electrical shock. During the extinction phase, previously trained subjects were exposed to a blue square without the shock. (From LaBar KS, Gatenby JC, Gore JC et al. [1998] *Neuron* 20:937–945. With permission from Elsevier Inc.)

Box 10–4: Microcircuits of the central amygdala

As is evident from the circuit diagram of fear conditioning (see Figure 10–41), CS and US can converge at multiple potential sites, where synaptic plasticity can contribute to fear conditioning. Indeed, each amygdala nucleus contains a heterogeneous population of neuronal types that have different properties and connections. For instance, the central amygdala can be divided into a lateral compartment (CEl), which receives input from the lateral and basal amygdala, and a medial compartment (CEm), which sends output to the brainstem to activate the freezing behavior. The lateral compartment further contains two separate populations of GABAergic neurons, CEl_{ON} and CEl_{OFF} cells, which mutually inhibit each other (**Figure 10–43**).

CEl_{ON} cells mostly restrict their projections within the CEl, and most projections from the CEl to the CEm are from CEl_{OFF} cells. After fear conditioning, presynaptic potentiation of excitatory input from the lateral and basal amygdala results in potentiation of the response of CEl_{ON} cells to tone stimuli. The responses of CEl_{OFF} cells to sound stimuli are depressed after fear conditioning due to presynaptic depression of excitatory input. Thus, the fear conditioning circuit in the amygdala contains a series of plastic synapses. Fear conditioning (1) enhances the excitatory input from the lateral and basal amygdala to the central amygdala, (2) enhances excitatory transmission to CEl_{ON} cells so that they are more activated by the tone, and (3) reduces excitatory transmission to CEl_{OFF} cells while increasing inhibition of CEl_{OFF} cells via activation of CEl_{ON} cells, so that the CEl_{OFF} cells become less activated by the tone. In combination, these changes result in a net disinhibition of the CEm output neurons (Figure 10–43), thus causing a fear response to the tone.

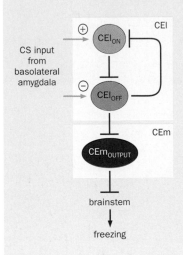

Figure 10–43 Microcircuits of the central amygdala. The central amygdala consists of a lateral compartment (CEl) and a medial compartment (CEm). Within the CEl, two separate GABAergic neuronal populations have been identified. CEl_{ON} neurons acquire a potentiated conditioned stimulus (CS) response after fear conditioning because of presynaptic potentiation (+) and less inhibition from CEl_{OFF} neurons; CEl_{OFF} neurons acquire a depressed CS response after fear conditioning because of presynaptic depression (–) and more inhibition from CEl_{ON} neurons. Thus, after fear conditioning, the CS signal from the basolateral amygdala preferentially activates CEl_{ON} neurons, which inhibit CEl_{OFF} neurons that normally inhibit CEm output neurons. This disinhibition ultimately activates CEm output neurons and the freezing response to the tone. Green, excitatory pathway; red, inhibitory pathway. (Adapted from Haubensak W, Kunwar PS, Cai H et al. [2010] *Nature* 468:270–276. With permission from Macmillan Publishers Inc.; See also Ciocchi S, Herry C, Grenier F et al. [2010] *Nature* 468:277–281 and Li H, Penzo MA, Taniguchi H et al. [2013] *Nat Neurosci* 16:332–339.)

10.24 Dopamine plays a key role in reward-based learning

In Section 10.14 we discussed the 'law of effect' in the context of operant conditioning: behaviors that are followed by a reward will be repeated, whereas behaviors that are followed by a punishment will be diminished. What is the neural basis for this effect? An interesting set of experiments to identify brain areas responsible for reward utilized electrical self-stimulation in an operant conditioning paradigm (**Figure 10–44**A). An electrode was implanted in a specific area of a rat's brain. Whenever the rat pressed a lever in an operant chamber, the electric circuit became connected and current from the electrode excited nearby neurons or axonal projections. When the electrode was placed in certain areas of the brain presumed to signal reward, the rat would keep pressing the lever in order to receive more electrical stimulations. When the stimulation was sufficiently strong, rats would keep pressing the lever at the expense of eating, drinking, or having sex; they also withstood substantial foot shock (an aversive stimulus) in order to receive more electrical stimulations (Figure 10–44B). Where are these presumed reward centers, the stimulation of which can override an animal's basic drives?

Systematic mapping revealed that the most effective self-stimulation sites coincide with midbrain dopamine neurons in the **ventral tegmental area** (**VTA**)

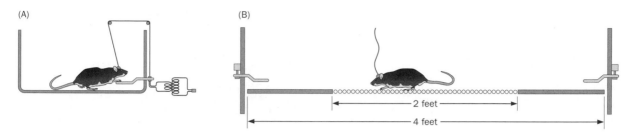

Figure 10–44 Electrical self-stimulation. (A) Design of the experiment. When the rat presses the lever, the electrode implanted in its brain connects to a source of current so that the neurons or axon bundles near the electrode tip become excited. If the excited neurons or axons signal reward, the rat will keep pressing the lever. **(B)** In this experiment, after the rat receives rewarding electric stimulation on one side of the arena, it must run across the center grid to the other side to receive more stimulation. After that rat had learned the task, electrical shock was applied in the central portion of the arena. If the electrode was implanted in the reward centers, rats withstood more foot shock when crossing the grid to receive electrical brain stimulation than they did when crossing the grid for food after 24 hours of food deprivation. (Adapted from Olds J [1958] *Science* 127:315–324.)

and **substantia nigra pars compacta** (**SNc**) and their projections to the striatum (see Figure 8–22), in particular to the ventral striatum, also called the **nucleus accumbens**. The involvement of dopamine neurons was further supported by additional experiments. For example, lesions of dopamine neurons or their fore-brain projections abolished the self-stimulation behavior, as did application of drugs that block dopamine synthesis. The addition of dopamine agonists in the nucleus accumbens, which bypassed the blockade of dopamine synthesis, restored reward behavior. Indeed, as we will learn in Chapter 11, most drugs of abuse act by enhancing the activity of midbrain dopamine neurons.

How does dopamine regulate reward and modify behavior? An important insight came from *in vivo* recording of dopamine neurons in alert monkeys performing behavioral tasks. Dopamine neurons normally fire in two different modes: in the **tonic** mode, dopamine neurons maintain a low and relatively constant basal firing rate; in the **phasic** mode, they fire in bursts in response to specific stimuli. In a specific example, a monkey was trained to touch a lever after a cue light was turned on. Following the cued lever touch, the monkey would receive a juice reward. Prior to and during the initial phase of training, dopamine neurons exhibited phasic firing in response to the juice reward. However, after extensive training, phasic firing was triggered by the cue (light on) that predicted the reward, but not by the actual reward delivery itself. After training, in trials in which the reward was omitted, the tonic firing rate was depressed at the time when reward was expected (**Figure 10–45**A). These data suggest that rather than signaling the reward per se, phasic firing of dopamine neurons signals a **reward prediction error**, that is, the difference between the actual reward and the predicted reward. Before training, reward came unexpectedly, resulting in a positive reward prediction error that triggered phasic firing. After the training, reward was predicted by the sensory cue, such that the sensory cue became the unexpected reward signal; when reward was actually delivered, it was fully predicted, and hence there was no reward prediction error and no phasic firing of dopamine neurons; when reward was omitted, a negative prediction error resulted in depression of the tonic firing.

Various learning theories have been proposed to account for these remarkable experimental findings. Let's first discuss an abstract model for reward-based learning (Figure 10–45B), and then place it in the context of a realistic dopamine circuit. In this abstract model, the connection between a signal neuron and a response neuron has an adjustable strength (ω). Through a negative feedback loop, the response magnitude (ωS) is compared to a reward signal (R). This difference, or reward prediction error carried by the dopamine neuron (blue in Figure 10–45B), is used to modify ω. Before training, ω is small such that the reward prediction error ($R - \omega S$) is large. Dopamine neurons fire and send a large signal to increase ω. As learning proceeds, ω increases until $R - \omega S = 0$; at that point, dopamine-neuron-mediated learning is accomplished and the dopamine

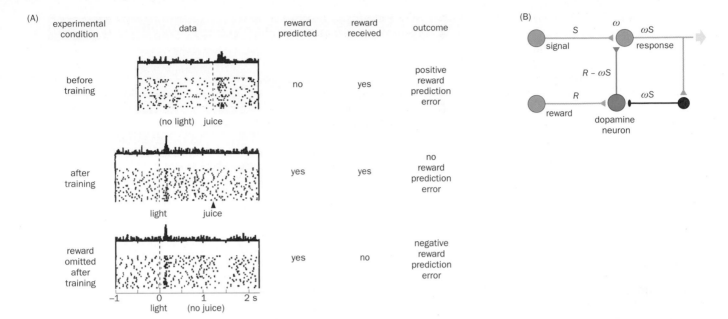

Figure 10–45 Dopamine neurons, reward prediction error, and reinforcement-based learning. (A) *In vivo* single-unit recordings of a midbrain dopamine neuron of a monkey trained to associate light onset with a reward (a drop of juice). Within each of the three blocks, each row is a separate trial and each vertical bar is an action potential. Above the individual trials is a histogram that combines the firing rates from all trials. Top, prior to training, phasic firing was triggered by reward (juice) delivery (vertical dashed line). Middle, after training, phasic firing was triggered by light onset (vertical dashed line) but not by reward delivery (arrowhead). Bottom, in trials when reward was omitted after training, tonic firing was depressed around the time reward was expected. Individual trials are aligned with reward delivery (top) or light onset (middle and bottom). Phasic firing of this dopamine neuron can be interpreted to signal the difference between the actual and expected reward, as summarized on the right. **(B)** An abstract circuit model for reward-based learning. A signal neuron produces a signal with the magnitude S, and connects with a response neuron through a synapse whose strength ω can be adjusted, resulting in a response whose magnitude is the product of ω and S. In addition to

sending information to downstream circuits (arrow), the response is also transmitted to a feedback inhibitory neuron (red), which in turn sends output to a dopamine neuron (blue). The dopamine neuron also receives an excitatory input that delivers a reward signal with a magnitude of R. Thus, the dopamine neuron positively adjusts ω with an output magnitude of $R - \omega S$ (for simplicity we assume that synaptic transmission is faithful and integration is linear). Before training, ω is small and $R - \omega S$ is large, resulting in a large magnitude of dopamine release to increase ω. As training proceeds, ω increases and $R - \omega S$ decreases; when $R - \omega S$ becomes zero, training is accomplished. Note that although an excitatory neuron (green) is used as the response neuron, in the midbrain dopamine circuit, the response neuron is GABAergic spiny projection neurons (SPNs). SPNs can either signal directly to dopamine neurons to deliver ωS, or through an intermediate GABAergic neuron to signal to the feedback inhibitory neuron. (A, adapted from Schultz W, Dayan P & Montague R [1997] *Science* 275:1593–1599. With permission from AAAS; B, adapted from Schultz W & Dickinson A [2000] *Ann Rev Neurosci* 23:473–500.)

neurons no longer exhibit phasic firing in response to the reward (as in the middle panel of Figure 10–45A).

As discussed in Section 8.9, a major projection region for midbrain dopamine neurons is the striatum, with VTA dopamine neurons projecting preferentially to the nucleus accumbens, and SNc dopamine neurons projecting to the rest of the striatum (see Figure 8–22). There, dopamine release regulates the strength of connections between the cortical and thalamic excitatory input and the spiny projection neurons (SPNs). Thus, the signal neuron in Figure 10–45B is equivalent to the cortical and thalamic projection neurons into the striatum. The response neuron is equivalent to the GABAergic SPNs. Some SPNs connect directly to dopamine neurons, thus could serve both as response neurons and feedback neurons (red in Figure 10–45B). Other candidates to carry feedback signals are the midbrain GABAergic neurons, which receive striatal input and inhibit dopamine neurons. (Because SPNs are GABAergic themselves, another GABAergic neuron is required to deliver the ωS signal with a positive sign to these midbrain GABAergic neurons that in turn synapse onto dopamine neurons.) The dopamine neurons additionally receive input from sensory systems signaling the reward, such as juice in the experiment discussed in Figure 10–45A, or pleasure derived from sex in the example of pair bonding discussed in Section 9.24. As supporting evidence for this model, dopamine has been shown to regulate various forms of plasticity at the cortical/thalamic → SPN synapse in *in vitro* slices. Whereas dopamine projection to the ventral striatum is associated with reward- and motivation-based learning, dopamine-guided synaptic plasticity in the dorsal striatum facilitates procedural learning and habit formation, likely through a similar circuit mechanism. Much remains to be learned about how striatal circuits are organized into subcircuits that carry out these distinct functions and whether striatal synaptic plasticity is causally linked with various forms of reinforcement-based learning.

Although midbrain dopamine neurons have been demonstrated to represent (that is, fire in response to) reward prediction errors in primates and rodents, recent studies have also identified heterogeneity among dopamine neurons, with some signaling aversive stimuli and others signaling salience of motivational stimuli; dopamine neurons in this latter group are activated by both strong appetitive and strong aversive signals and respond poorly to weak appetitive and weak aversive signals. This heterogeneity of dopamine neuron function may be accounted for by the heterogeneity of input to and output from different dopamine neurons. For example, according to a recent study, VTA dopamine neurons that project to the nucleus accumbens tend to signal the presence of appetitive stimuli, whereas those that project to the prefrontal cortex tend to signal the presence of aversive stimuli. Just as reward-based learning can increase the frequency of actions that lead to reward, aversion-based learning can reduce the frequency of actions that lead to punishment. Indeed, conceptually similar circuit designs can be applied to reinforcement-based learning that does not involve dopamine at all, such as the cerebellum-based motor learning (see Figure 8–21B).

10.25 Early experience can leave behind long-lasting memory traces to facilitate adult learning

We have seen that learning can occur and memory can be stored in neural circuits in many parts of the brain, including the hippocampus, cerebral cortex, amygdala, striatum, and cerebellum. Remarkably, memory can even be formed by artificially activating random populations of cortical neurons (**Box 10–5**). In the final section of this chapter, we further broaden the scope of learning and memory to developmental and structural plasticity by returning to the story of the barn owl introduced at the beginning of Chapter 1, integrating what we have learned about the organization and wiring of the brain in the intervening chapters.

Box 10–5: Memory can be formed by the activation of random populations of cortical neurons

Recent advances in genetically targeting specific neuronal populations *in vivo* for precise control of their activity has contributed much to our understanding of the neural basis of brain function and behavior. In particular, we have seen examples of the application of optogenetic approaches for dissecting memory circuits in model species ranging from flies (see Figure 10–29B) to mice (see Figure 10–36 and Figure 10–39C). Photostimulation of channelrhodopsin (ChR2)-expressing neurons has also been used to probe whether a random population of neurons can be associated with reward or punishment such that reactivation of those neurons changes the behavior of the animal. We discuss two examples of this approach below.

In the first example, a random population of piriform cortical neurons in mice was transduced with an adeno-associated virus to express ChR2, such that they fire action potentials in response to photostimulation. During training, the mouse was allowed to freely move in an arena, but whenever it moved to one side of the arena, foot shock was applied along with photostimulation. This elicited a robust flight response—mice ran quickly to the other side of the arena where no foot shock was applied. After training, photostimulation alone could elicit the flight response (**Figure 10–46**A). Thus, activation of a random population of piriform cortical neurons (~500) could serve as an effective CS with which the animal can be trained to associate a US (the shock) and subsequently to elicit a robust conditioned response. In separate experiments, photostimulation of ChR2-expressing piriform neurons was shown to also effectively serve as a CS for reward; indeed, activation of the same random population of ChR2-expressing neurons can be sequentially used as a CS for reward and subsequently as a CS for electrical shock.

(Continued)

Box 10–5: Memory can be formed by the activation of random populations of cortical neurons

Because the spatial representation of odor in the piriform cortex has no discernible order, these experiments have been interpreted as a support for the hypothesis that this brain area is a random network whose connectivity is sculpted by individual experience (see Sections 6.10 and 6.16). However, a second example shows that the ability of researchers to influence behavior by activating a random population of neurons is not restricted to the piriform cortex. In this experiment, a random set of layer 2/3 neurons in the mouse barrel cortex (see Box 5-3) was electroporated *in utero* to introduce a functional ChR2 gene, and thirsty mice were trained to associate photostimulation of ChR2-expressing neurons with a water reward at one of two choice ports (Figure 10–46B, left). In this task, the snout of the mouse must enter the center port in order for a drop of water to be delivered to the left or the right port. While the mouse's snout was at the center port, the mouse either received photostimulation, after which water would be delivered to the left port, or did not receive photostimulation, after which water would be delivered to the right port. After training, mice could report photostimulation reliably by choosing the correct port for

the water reward. The effectiveness of photostimulation depended on the number of ChR2-expressing neurons and the strength and the duration of photostimulation (Figure 10–46B, right). Single action potentials (elicited by 1-ms photostimulation) in about 300 ChR2-expressing layer 2/3 neurons of the barrel cortex served as a sufficient cue to bias the mouse to the reward port.

These examples highlight the nervous system's remarkable plasticity for learning: given sufficient strength of stimulation and sufficient training, association can be established between reward or punishment and the activity of random populations of neurons in different brain areas. It is likely that these photostimulations mimic the perception of a smell or a touch, so that the animals use the normal neural pathways that process olfactory or somatosensory information to associate the photostimulation with punishment or reward. These experiments also offer valuable estimates as to the number of cortical neurons that must be activated and the number of action potentials that must be fired in order for animals to associate neuronal activity with reward or punishment and consequently alter their behavior.

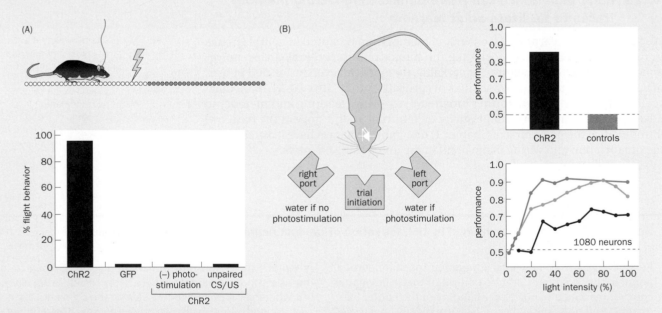

Figure 10–46 Forming a memory by activating a random population of cortical neurons. (A) Top, schematic of the experimental design. During training, when the mouse moved to the left side of the chamber, it received an electrical shock (yellow) while its piriform cortex, which contained channelrhodopsin (ChR2)-expressing neurons, was photostimulated. After two training sessions (each of which consisted of 10 pairings), mice readily exhibited flight behavior in response to photostimulation alone (bottom, left bar). In negative control groups, mice did not exhibit flight behavior when green fluorescent protein (GFP) instead of ChR2 was expressed, when photostimulation was omitted during training, or when photostimulation was not paired with shock. **(B)** Left, experimental design. The mouse was trained to place its snout at the center port to initiate the trial. Water was delivered to the left or the right port depending on whether the mouse received photostimulation or not during the trial. Right, After 4–7 sessions of

trials (200–800 trials per session), ChR2-expressing mice reliably reported photostimulation by making proper port choices compared to controls that did not express ChR2; the dotted line indicates port selection at chance levels. Performance on the y axis is the ratio of the number of corrected trials, which include turning left in photostimulation trials and turning right in no-photostimulation trials, over the total number of trials. Performance increased as the number of ChR2 neurons increased (shown at bottom is an example of 1080 ChR2-expressing neurons), with light intensity, and with the number of 1-ms pulses of light (red, 1 pulse; green, 2 pulses; blue, 5 pulses; separate experiments indicated that each pulse elicited at most one action potential in ChR2-expressing neuron at 100% light intensity). (A, adapted from Choi GB, Stettler DD, Kallman BR et al. [2011] *Cell* 146:1004–1015. With permission from Elsevier Inc.; B, adapted from Huber D, Petreanu L, Ghitani N et al. [2008] *Nature* 451:61–64. With permission from Macmillan Publishers Ltd.)

Recall that the owl's auditory map can adapt to match a visual map altered by wearing prisms and that this ability declines with age (see Section 1.3). Recall further that if an owl had an earlier experience of auditory map adjustment, its auditory map re-adapted to an altered visual map more easily in adulthood (see Figure 1–7). What is the neural basis for these phenomena? As we learned in Chapter 6, neurons in the nucleus laminaris of the owl's brainstem form a map that identifies sound locations on the horizontal plane based on interaural time differences (ITDs) (see Figure 6–55). This ITD map projects topographically to the central nucleus of the inferior colliculus (ICC). ICC axons project further to the external nucleus of the inferior colliculus (ICX). ICX neurons then project to the optic tectum, where integration of auditory and visual information occurs in a topographically aligned manner (**Figure 10–47**A, top). Anatomical tracing studies indicate that in juvenile prism-reared owls, ICC axonal projections to the ICX expand in the direction that matches the altered visual map in the optic tectum (Figure 10–47A, bottom). The expanded axons bear synaptic terminals and likely make functional connections with the postsynaptic neurons in the new topographic location, thus realigning the auditory map with that prism-altered visual map. Although the mechanisms underlying this axonal expansion have not been examined in detail, it is likely that the connections made by the expanded axons are stabilized by synchronous firing with postsynaptic neurons that process altered visual information, similar to the Hebbian-based synaptic strengthening in visual system wiring that we discussed in Chapter 5.

As discussed in Section 1.3, when the prisms were removed from the juvenile prism-reared owls, the auditory map was restored to normal so that it was realigned with the normal visual map. Indeed, ICC neurons still maintain their normal axonal projections in the ICX during the prism-rearing period (Figure 10-46A bottom); these normal projections, which become topographically mismatched during the prism-wearing period, receive preferential GABAergic inhibition such that they are preferentially silenced. The persistence of these normal connections during prism rearing may account for the rapid restoration of the normal auditory map after the prisms are removed. The ICC axons that expanded into the topographically abnormal area of the ICX due to juvenile prism rearing are also maintained into adulthood (Figure 10–47B), well after prism removal and complete

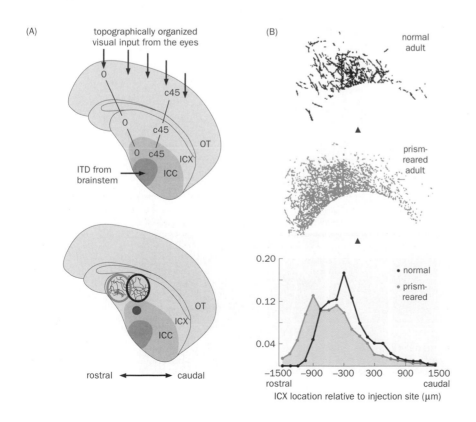

Figure 10–47 Adaptive axonal expansion in the inferior colliculus and auditory map adjustment in juvenile and adult owls. (A) Top, representation of auditory and visual information in the owl's brain. Brainstem inputs to the central (ICC) and external (ICX) nuclei of the inferior colliculus are topographically organized according to interaural time differences (ITDs): 0 represents ITD = 0, c45 represents the contralateral side leading by 45 μs. ICX neurons project to the optic tectum (OT), where they align with topographically organized visual input. Bottom, axons of ICC neurons from the area indicated by the dark dot normally project to a topographically appropriate area of the ICX (red circle), but expand rostrally (blue circle) to match with a visual map altered by prism experience in the juvenile owl. **(B)** Top and middle, anterograde tracing was used to examine ICC → ICX axonal projection in normal (top) and prism-reared (middle) adult owls. Arrowheads indicate the anterograde tracer injection site in the ICC. Bottom, normalized distribution of axonal projections in normal and prism-reared owls. The significant rostral shift in adult owls due to juvenile experience with wearing prisms (prism-reared) likely accounts for the rapid auditory map adjustment during a second prism experience in adulthood (see Figure 1–7B). (Adapted from Linkenhoker BA, con der Ohe CG & Knudsen EI [2005] *Nat Neurosci* 8:93–98. With permission from Macmillan Publishers Ltd. See also DeBello WM, Feldman DE & Knudsen EI [2001] *J Neurosci* 21:3161–3174.)

restoration of the normal auditory map as assayed by behavior. (It is unknown how the activity of these expanded axons is silenced after prism removal so that they do not interfere with behavior.) Thus, adaptive expansion of ICC axons as a consequence of juvenile prism rearing leaves behind an anatomical trace that likely facilitates the readjustment of the auditory map in response to a similar visual displacement event later in adulthood.

A conceptually similar experiment examined traces of structural change in mammalian visual cortical neurons in response to monocular deprivation. As we learned in Chapter 5, monocular deprivation within a critical developmental period has a profound effect on wiring of the visual cortex. In mice, for example, transient monocular deprivation for a few days during the critical period can modify the binocular area of the visual cortex, significantly shifting the relative representation of visual input from the two eyes in favor of the open eye, as assayed by intrinsic signal imaging of cortical responses to visual stimulation (see Figure 4–42). If the deprivation period is short, the normal balance of representation of the two eyes is restored after binocular vision is restored. Monocular deprivation in adult mice can also shift ocular dominance in response to a longer duration of monocular deprivation. Interestingly, ocular dominance shifts in response to monocular deprivation are more rapid in adult mice that previously experienced monocular deprivation than in those experiencing monocular deprivation for the first time, analogous to the finding in the owl.

To examine a structural basis for this monocular-deprivation-induced plasticity, repeated two-photon microscopic imaging was carried out through a window implanted in the binocular area of the mouse visual cortex (**Figure 10–48**A). The dendritic spines of pyramidal neurons were observed and quantified to determine spine gains and losses over time (Figure 10–48B). It was found that the first monocular deprivation resulted in significant addition of new spines, a proxy for new synapse formation (see Section 10.13); these spines are subsequently retained. The second deprivation, which caused a more rapid ocular dominance shift, did not change the spine density (Figure 10–48C). A likely interpretation of these data is that the anatomical traces left behind by the first monocular deprivation—the new spines—were reused for the ocular dominance shift during the second monocular deprivation, thus facilitating the neuron's more rapid adaptation. As with the owl experiments, questions remain as to whether and how the activities of the new spines are silenced during the intervening period between the first and second instances of monocular deprivation, such that the synaptic connections enabled by the new spines do not interfere with binocular vision during the intervening period.

Figure 10–48 Spine dynamics in the adult mouse visual cortex in response to monocular deprivation. (A) Experimental protocol. After window implantation, intrinsic signal imaging (see Figure 4–42B) was carried out to identify the binocular region. Repeated two-photon imaging of dendrites from a transgenic mouse that expresses GFP in a sparse population of neurons (see Section 13.16) in the binocular regions was then carried out every 4 days, covering two 8-day periods of monocular deprivation (MD1 and MD2). **(B)** Representative images of the same apical dendritic segment of a layer 5 pyramidal neuron. Blue and red arrowheads indicate spine loss and spine gain, respectively; these changes were inferred by comparing each image to the image acquired 4 days earlier. **(C)** A significant increase in spine gain is detected only during the MD1 (top) but not the MD2 (bottom), suggesting that new spines gained during MD1 may be used to adjust ocular dominance during MD2. (Adapted from Hofer SB, Mrsic-Flogel TD, Bonhoeffer T et al. [2009] *Nature* 457:313–317. With permission from Macmillan Publishers Ltd.)

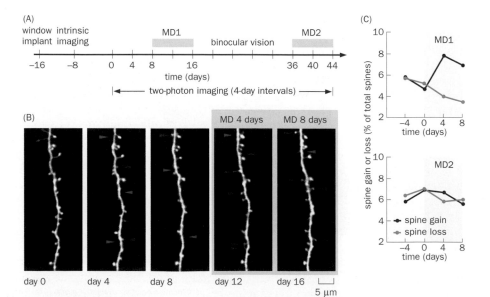

In summary, these experiments suggest that structural changes in neural circuits in response to specific experiences—whether changes in axonal arborization in the inferior colliculus or the formation of new dendritic spines in the visual cortex—can provide long-lasting memory traces to facilitate future learning. These structural changes may underlie a widely occurring phenomenon called **savings**, that is, less effort is required for an animal to re-learn something it has previously learned. Altogether, modern research discussed in this chapter has provided rich neurobiological bases for Descartes' needle-through-the-cloth analogy of memory (see Figure 10–1).

SUMMARY

In this chapter, we have studied memory and learning at multiple levels: molecules, synapses, neurons, circuits, systems, animal behaviors, and theories. From simple invertebrate systems to the complex mammalian brain, diverse experimental models have yielded data that support two central theses: (1) memory is primarily stored as strengths of synaptic connections in neural circuits, and (2) learning modifies synaptic weight matrices through a rich set of plasticity mechanisms.

A reductionist approach in *Aplysia*, using the gill-withdrawal reflex as a model behavior, suggested that depression and potentiation of the synaptic strength between the siphon sensory neurons and gill motor neurons mediate behavioral habituation and sensitization, respectively. Short-term sensitization of the gill-withdrawal reflex by tail shock is mediated by serotonin activation of cAMP/PKA and PKC signaling in the presynaptic terminal of the sensory neuron, modifying ion channels through phosphorylation that results in an elevated membrane potential and broadened spikes. Long-term sensitization involves prolonged activation of cAMP/PKA, causing phosphorylation of the CREB transcription factors, expression of new genes, and growth of new synapses between the sensory and motor neurons. Hence, in *Aplysia* as well as in many other animals, short-term memory does not require new protein synthesis whereas long-term memory requires new protein synthesis. Genetic analysis of *Drosophila* olfactory conditioning independently identified a central role for cAMP signaling in mushroom body neurons. Electric shock and food as the unconditioned stimuli modulate the strengths of synaptic connections between ensembles of mushroom body neurons representing conditioned stimuli (odorants) and output neurons through neuromodulators such as dopamine, whose receptors act through the cAMP cascade. cAMP/PKA also plays an important role in synaptic plasticity and memory in mice. Formation of new explicit memory in humans and spatial memory in rodents relies on the hippocampus, a medial temporal lobe structure that along with the nearby entorhinal cortex also plays a central role in spatial representation in mammals. A rich set of synaptic plasticity mechanisms has been identified in the hippocampus, and strong correlations have been established between hippocampal synaptic plasticity and spatial learning and memory.

Synapses onto the hippocampal CA1 pyramidal neurons in rats and mice have been used as a model to investigate general mechanisms of synaptic plasticity. Long-term potentiation (LTP) of the CA3 → CA1 synapse exhibits cooperativity that follows Hebb's rule: LTP is induced when presynaptic glutamate release coincides with postsynaptic depolarization. The NMDA receptor serves as a coincidence detector to execute Hebb's rule, and its function in CA1 neurons is required for both LTP induction and spatial memory. Ca^{2+} entry through the NMDA receptor activates protein kinases such as PKA and CaMKII. Auto-phosphorylation of the multi-subunit CaMKII can translate a transient Ca^{2+} signal into more persistent kinase activity. A central mechanism for LTP expression is an increase in AMPA receptor numbers at the postsynaptic membrane, which enhances response magnitude to presynaptic glutamate release. The CA3 → CA1 synaptic efficacy can also be regulated by long-term depression, which preferentially activates phosphatases to counteract the kinase activity. LTD, LTP, and spike-timing-dependent plasticity allow bidirectional adjustment of synaptic weights. Activity-dependent retrograde endocannabinoid signaling from CA1 neurons can regulate the release of neurotransmitters by their presynaptic GABAergic neurons. Finally, long-term changes in the strength of connections between pre- and postsynaptic neurons involve formation of new synapses as a result of long-lasting LTP.

The synaptic plasticity mechanisms in the hippocampus likely apply, with variations according to specific neuronal and circuit properties, to other synapses in the central nervous system where experience-dependent changes underlie many forms of learning and memory. For example, long-term storage of explicit memory may engage specific neocortical areas that process and relay information to the hippocampus during memory acquisition; these cortical circuits likely interact with the hippocampus during memory consolidation. The amygdala is a center for processing emotion-related memory. Auditory fear conditioning engages parallel pathways and plasticity in multiple synapses in the basolateral and central amygdala, whereas contextual fear conditioning engages additional synaptic plasticity in the hippocampus. The amygdala is also required for fear conditioning in humans as a form of implicit memory. Some midbrain dopamine neurons signal reward prediction errors; they exhibit phasic firing when the actual reward exceeds the predicted reward. This property can be used for reinforcement-based learning, in which the synapses between cortical/thalamic input neurons and striatal spiny projection neurons are modulated by dopamine. This reinforcement-based learning plays an important role in motivational behavior as well as motor skill learning and habit formation.

Learning has different forms including simple habituation and sensitization, associative learning such as classical conditioning and operant conditioning, reinforcement-based learning, cognitive learning, and structural plasticity in both developing and adult sensory systems in response to altered experience. Most forms of learning involve changes in the synaptic weight matrices of relevant neural circuits, whether by strengthening or weakening existing synapses, making new synapses, or dismantling old ones; additional forms of learning include changes in the intrinsic properties of neurons. These changes alter neural circuit function in information processing and ultimately cause behavioral changes that enable animals to better adapt to a changing environment.

FURTHER READING

Reviews and books

Bromberg-Martin ES, Matsumoto M & Hikosaka O (2010) Dopamine in motivational control: rewarding, aversive, and alerting. *Neuron* 68:815–834.

Dan Y & Poo MM (2006) Spike timing-dependent plasticity: from synapse to perception. *Physiol Rev* 86:1033–1048.

Heisenberg M (2003) Mushroom body memoir: from maps to models. *Nat Rev Neurosci* 4:266–275.

Huganir RL & Nicoll RA (2013) AMPARs and synaptic plasticity: The last 25 years. *Neuron* 80:704–717.

Janak PH & Tye KM (2015) From circuits to behavior in the amygdala. *Nature* 517:284–292

Kandel ER (2001) The molecular biology of memory storage: a dialogue between genes and synapses. *Science* 294:1030–1038.

Lisman J, Schulman H & Cline H (2002) The molecular basis of CaMKII function in synaptic and behavioural memory. *Nat Rev Neurosci* 3:175–190.

Martin SJ, Grimwood PD & Morris RG (2000) Synaptic plasticity and memory: an evaluation of the hypothesis. *Annu Rev Neurosci* 23:649–711.

Milner B, Squire LR & Kandel ER (1998) Cognitive neuroscience and the study of memory. *Neuron* 20:445–468.

Pavlov IP (1927) Conditioned Reflexes: An Investigation of the Physiological Activity of the Cerebral Cortex. Oxford University Press.

Hippocampus, synaptic plasticity, and spatial memory

Bliss TV & Lomo T (1973) Long-lasting potentiation of synaptic transmission in the dentate area of the anaesthetized rabbit following stimulation of the perforant path. *J Physiol* 232:331–356.

Dudek SM & Bear MF (1992) Homosynaptic long-term depression in area CA1 of hippocampus and effects of N-methyl-D-aspartate receptor blockade. *Proc Natl Acad Sci USA* 89:4363–4367.

Dudman JT, Tsay D & Siegelbaum SA (2007) A role for synaptic inputs at distal dendrites: instructive signals for hippocampal long-term plasticity. *Neuron* 56:866–879.

Engert F & Bonhoeffer T (1999) Dendritic spine changes associated with hippocampal long-term synaptic plasticity. *Nature* 399:66–70.

Frey U & Morris RGM (1997) Synaptic tagging and long-term potentiation. *Nature* 385:533.

Giese KP, Fedorov NB, Filipkowski RK et al. (1998) Autophosphorylation at Thr286 of the alpha calcium-calmodulin kinase II in LTP and learning. *Science* 279:870–873.

Hafting T, Fyhn M, Molden S et al. (2005) Microstructure of a spatial map in the entorhinal cortex. *Nature* 436:801–806.

Isaac JT, Nicoll RA & Malenka RC (1995) Evidence for silent synapses: implications for the expression of LTP. *Neuron* 15:427–434.

Liao D, Hessler NA & Malinow R (1995) Activation of postsynaptically silent synapses during pairing-induced LTP in CA1 region of hippocampal slice. *Nature* 375:400–404.

Liu X, Ramirez S, Pang PT et al. (2012) Optogenetic stimulation of a hippocampal engram activates fear memory recall. *Nature* 484:381–385.

Marr D (1971) Simple memory: a theory for archicortex. *Philos Trans R Soc Lond B Biol Sci* 262:23–81.

Morris RG, Anderson E, Lynch GS et al. (1986) Selective impairment of learning and blockade of long-term potentiation by an *N*-methyl-D-aspartate receptor antagonist, AP5. *Nature* 319:774–776.

Morris RG, Garrud P, Rawlins JN et al. (1982) Place navigation impaired in rats with hippocampal lesions. *Nature* 297:681–683.

O'Keefe J (1976) Place units in the hippocampus of the freely moving rat. *Exp Neurol* 51:78–109.

Tang YP, Shimizu E, Dube GR et al. (1999) Genetic enhancement of learning and memory in mice. *Nature* 401:63–69.

Toni N, Buchs PA, Nikonenko I et al. (1999) LTP promotes formation of multiple spine synapses between a single axon terminal and a dendrite. *Nature* 402:421–425.

Tsien JZ, Huerta PT & Tonegawa S (1996) The essential role of hippocampal CA1 NMDA receptor-dependent synaptic plasticity in spatial memory. *Cell* 87:1327–1338.

Whitlock JR, Heynen AJ, Shuler MG et al. (2006) Learning induces long-term potentiation in the hippocampus. *Science* 313:1093–1097.

Wilson MA & McNaughton BL (1993) Dynamics of the hippocampal ensemble code for space. *Science* 261:1055–1058.

Wilson RI & Nicoll RA (2001) Endogenous cannabinoids mediate retrograde signalling at hippocampal synapses. *Nature* 410:588–592.

Wong ST, Athos J, Figueroa XA et al. (1999) Calcium-stimulated adenylyl cyclase activity is critical for hippocampus-dependent long-term memory and late phase LTP. *Neuron* 23:787–798.

Learning and memory in diverse invertebrate and vertebrate systems

Aso Y, Sitaraman D, Ichinose T et al. (2014) Mushroom body output neurons encode valence and guide memory-based action selection in *Drosophila*. *Elife* 3:e04580

Bacskai BJ, Hochner B, Mahaut-Smith M et al. (1993) Spatially resolved dynamics of cAMP and protein kinase A subunits in *Aplysia* sensory neurons. *Science* 260:222–226.

Choi GB, Stettler DD, Kallman BR et al. (2011) Driving opposing behaviors with ensembles of piriform neurons. *Cell* 146:1004–1015.

Claridge-Chang A, Roorda RD, Vrontou E et al. (2009) Writing memories with light-addressable reinforcement circuitry. *Cell* 139:405–415.

Dudai Y, Jan YN, Byers D et al. (1976) *dunce*, a mutant of *Drosophila* deficient in learning. *Proc Natl Acad Sci USA* 73:1684–1688.

Flexner JB, Flexner LB & Stellar E (1963) Memory in mice as affected by intracerebral puromycin. *Science* 141:57–59.

Frankland PW, Bontempi B, Talton LE et al. (2004) The involvement of the anterior cingulate cortex in remote contextual fear memory. *Science* 304:881–883.

Giurfa M, Zhang S, Jenett A et al. (2001) The concepts of 'sameness' and 'difference' in an insect. *Nature* 410:930–933.

Goshen I, Brodsky M, Prakash R et al. (2011) Dynamics of retrieval strategies for remote memories. *Cell* 147:678–689.

Hofer SB, Mrsic-Flogel TD, Bonhoeffer T et al. (2009) Experience leaves a lasting structural trace in cortical circuits. *Nature* 457:313–317.

Huber D, Petreanu L, Ghitani N et al. (2008) Sparse optical microstimulation in barrel cortex drives learned behaviour in freely moving mice. *Nature* 451:61–64.

Kim JJ & Fanselow MS (1992) Modality-specific retrograde amnesia of fear. *Science* 256:675–677.

Kupfermann I, Castellucci V, Pinsker H et al. (1970) Neuronal correlates of habituation and dishabituation of the gill-withdrawal reflex in *Aplysia*. *Science* 167:1743–1745.

LaBar KS, Gatenby JC, Gore JC et al. (1998) Human amygdala activation during conditioned fear acquisition and extinction: a mixed-trial fMRI study. *Neuron* 20:937–945.

Lammel S, Lim BK, Ran C et al. (2012) Input-specific control of reward and aversion in the ventral tegmental area. *Nature* 491:212–217.

Linkenhoker BA, von der Ohe CG & Knudsen EI (2005) Anatomical traces of juvenile learning in the auditory system of adult barn owls. *Nat Neurosci* 8:93–98.

McGuire SE, Le PT, Osborn AJ et al. (2003) Spatiotemporal rescue of memory dysfunction in *Drosophila*. *Science* 302:1765–1768.

Montarolo PG, Goelet P, Castellucci VF et al. (1986) A critical period for macromolecular synthesis in long-term heterosynaptic facilitation in *Aplysia*. *Science* 234:1249–1254.

Olds J (1958) Self-stimulation of the brain; its use to study local effects of hunger, sex, and drugs. *Science* 127:315–324.

Phillips RG & LeDoux JE (1992) Differential contribution of amygdala and hippocampus to cued and contextual fear conditioning. *Behav Neurosci* 106:274–285.

Schultz W, Dayan P & Montague PR (1997) A neural substrate of prediction and reward. *Science* 275:1593–1599.

Tully T & Quinn WG (1985) Classical conditioning and retention in normal and mutant *Drosophila melanogaster*. *J Comp Physiol A* 157:263–277.

Wheeler ME, Petersen SE & Buckner RL (2000) Memory's echo: vivid remembering reactivates sensory-specific cortex. *Proc Natl Acad Sci USA* 97:11125–11129.

CHAPTER 11

Brain Disorders

And men ought to know that from nothing else but the brain come joys, delights, laughter and sports, and sorrows, griefs, despondency, and lamentations. And by this, in an especial manner, we acquire wisdom and knowledge, and see and hear, and know what are foul and what are fair, what are bad and what are good, what are sweet, and what unsavory... And by the same organ we become mad and delirious, and fears and terrors assail us... All these things we endure from the brain, when it is not healthy....

Hippocrates (~400 BC)

In this chapter, we examine how nervous system dysfunction causes neurological and psychiatric disorders. Brain disorders are greater causes of disability than any other class of diseases in our modern society. An important and obvious goal of studying brain disorders is to identify therapeutic strategies that will decrease disability and alleviate human suffering. In addition, research that focuses on specific diseases offers unique perspectives on normal brain development and function in the same way that studying genetic mutants can reveal the normal function of genes and the biological processes they control. Conversely, some of the most important progress made in understanding brain disorders has come from basic research seemingly unrelated to disease, as numerous disease examples introduced in previous chapters have shown. Thus, basic and disease-focused investigations mutually enhance each other to help us understand the function and dysfunction of the nervous system. While focusing on specific brain disorders, this chapter also seeks to integrate and extend the knowledge and principles presented in all previous chapters.

Rather than comprehensively addressing the vast array of brain disorders, we will focus primarily on select disorders, the principles of which can be applied broadly to other disorders not named here. Some disorders are selected because they have a large impact on human society; others are selected because their pathogenic mechanisms are better understood. We group these disorders as neurodegenerative, psychiatric, or neurodevelopmental. Although generally useful, these groupings also reflect our ignorance of the underlying disease mechanisms. For instance, as we will see, some of the classic psychiatric disorders were thought to arise in adulthood, but in fact have a developmental origin. We start with Alzheimer's disease, the most common neurodegenerative disorder.

ALZHEIMER'S DISEASE AND OTHER NEURODEGENERATIVE DISEASES

Alzheimer's disease (AD) is well known because of its prevalence: in the United States, it affects 1 in 20 people by age 65 and about half of the population above age 85. Thus, AD is a disease of aging. With the dramatic global increase in life-span over the past century, AD has become an escalating burden on society. As in all **neurodegenerative disorders**, the progression of AD causes an increasing number of neurons to become dysfunctional: synaptic connections are lost, dendrites and axons deteriorate, and neurons eventually die. As a result, the brain undergoes significant atrophy (**Figure 11–1**). Memory loss, an early and characteristic symptom of AD, is followed by loss of other cognitive and intellectual capabilities such as reasoning and language. Patients may also become depressed early in the disease when they are aware of their deterioration, and may exhibit personality changes and behavioral problems as the disease progresses. Gradually, patients lose their ability to cope with daily life and often require round-the-clock care for years before they succumb to death.

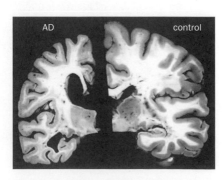

Figure 11–1 Brain atrophy in Alzheimer's disease (AD). Postmortem brain sections from an AD patient (left) and an age-matched, cognitively normal subject (right) show severe brain atrophy in the AD case. (Courtesy of Nigel Cairns, Washington University, Department of Neurology.)

Alzheimer's disease is one of the best-understood brain disorders in terms of its neuropathological underpinnings, thanks to biochemical and molecular-genetic studies since the 1980s. AD research has also offered valuable lessons that can be applied to other brain disorders. However, many key questions remain unanswered, and there is no effective treatment or prevention.

11.1 Alzheimer's disease is defined by brain deposition of numerous amyloid plaques and neurofibrillary tangles

In 1907, a German psychiatrist, Alois Alzheimer, reported a case of a patient who suffered from numerous psychiatric symptoms and severe memory loss, and who passed away four-and-a-half years after being admitted to an insane asylum. Using a newly invented silver-staining procedure to visualize postmortem brain specimens, Alzheimer described two major pathological features in the patient's cerebral cortex: numerous abnormal intracellular fibrils (now termed **neurofibrillary tangles**) in one-quarter to one-third of all neurons, and extracellular plaques (now termed **amyloid plaques**) distributed throughout the cerebral cortex. Today, although the clinical diagnosis based on history and examination of symptoms is usually correct, the combined presence in brain sections taken at autopsy of abundant neurofibrillary tangles and amyloid plaques remains the standard for a definitive pathological diagnosis of the disease that bears Alzheimer's name (**Figure 11–2**A). Both tangles and plaques are also found in the brains of non-symptomatic aged subjects, but their prevalence throughout the cerebral cortex, hippocampus, and amygdala increases drastically in AD brains.

What is the molecular nature of the neurofibrillary tangles and amyloid plaques? Could these pathological features offer clues to understanding this devastating disease? With these questions in mind, researchers biochemically characterized these structures and identified their molecular nature in the

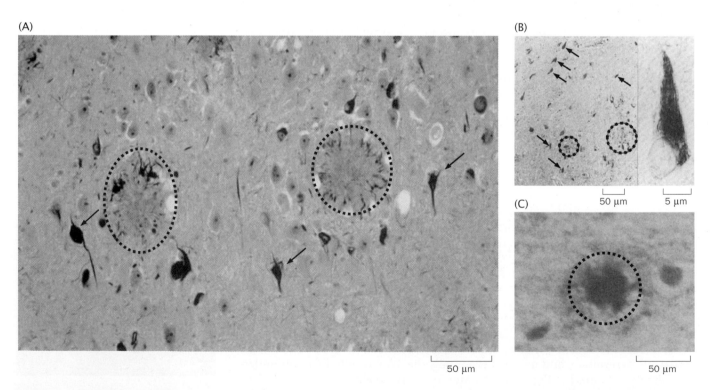

Figure 11–2 Neurofibrillary tangles and amyloid plaques in Alzheimer's disease. (A) Silver staining of a postmortem cortical section of an AD patient. Amyloid plaques (circles) and neurofibrillary tangles (arrows) are prevalent. **(B)** An antibody against the microtubule-associated protein tau strongly stains the neurofibrillary tangles (arrows; high magnification on the right), but does not stain the core of the amyloid plaques (circles). **(C)** An antibody against a peptide derived from amyloid β protein stains the core of an amyloid plaque intensely (circle). (A, from Selkoe DJ [1999] *Nature* 399:A23–A31. With permission from Macmillan Publishers Ltd.; B, from Grundke-Iqbal I, Iqbal K, Quinlan M et al. [1986] *J Biol Chem* 261:6084–6089. With permission from ASBMB; C, from Wong CW, Quaranta V & Glenner GG [1985] *Proc Natl Acad Sci USA* 82:8729–8732.)

mid-1980s. The neurofibrillary tangles consist of abnormal aggregates of hyper-phosphorylated microtubule-binding protein **tau** (Figure 11–2B). Amyloid plaques are composed mostly of a 39- to 43-amino-acid peptide named **amyloid β protein** (**Aβ**) for its strong tendency to form aggregates of β-pleated sheets (Figure 11–2C). While neurofibrillary tangles have also been found in several other neurodegenerative diseases, which are collectively called **tauopathies**, amyloid plaques are most characteristic of AD. Research focused on both Aβ and tau has provided important insights in our understanding of Alzheimer's disease. We start our story with Aβ.

11.2 Amyloid plaques mainly consist of aggregates of proteolytic fragments of the amyloid precursor protein (APP)

The peptide sequence of Aβ was used to isolate its gene from cDNA libraries, leading to the discovery that Aβ is part of a transmembrane protein called **amyloid precursor protein**, or **APP**. The predicted protein sequence indicated that APP has a large extracellular domain, a single transmembrane domain, and a small cytoplasmic domain (**Figure 11–3**). APP homologs have been found in the fly and worm, indicating that the proteins are evolutionarily conserved, although their exact cellular function is still the subject of investigation. The sequence of the Aβ peptide itself is not conserved in the fly or worm, or in two other APP paralogs in humans that share similar overall structure and sequence. These data suggest that the normal function of APP does not depend on Aβ; rather, Aβ's amino acid sequence renders it particularly prone to aggregation after it is excised from APP.

The location of Aβ within APP is peculiar: two-thirds of the peptide is at the C-terminal end of the extracellular domain and one-third is part of the predicted transmembrane domain (Figure 11–3). This implies that to produce Aβ, APP must be cleaved by two different proteases, one of which must cut APP in the middle of the transmembrane domain. The existence of proteases capable of cleaving within the membrane was unknown when APP was first discovered. Indeed, the study of APP processing has enriched our understanding of the cell biology of regulated proteolysis.

The first protease identified as being able to process APP, named **α-secretase**, cuts APP in the middle of the Aβ peptide and therefore prevents the production of the pathology-associated Aβ (Figure 11–3, left). This proteolysis is likely related to the physiological function of APP, as the fly homolog of APP also produces a secreted form of APP by proteolysis near the end of the extracellular domain. Subsequently, the proteases that cut APP at the N- and C-termini of Aβ, which produce the intact Aβ peptide, were identified and named **β-secretase** and

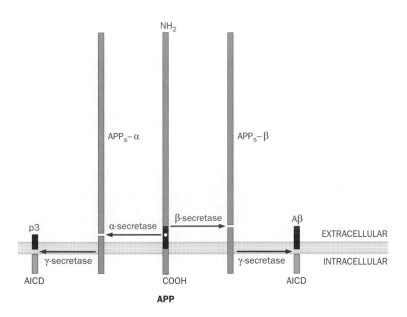

Figure 11–3 Proteolytic processing of amyloid precursor protein (APP) produces Aβ. APP is synthesized as a transmembrane protein (middle), with an N-terminal large extracellular domain, a single transmembrane domain, and a short C-terminal cytoplasmic domain. Aβ (red) spans the junction of the extracellular and transmembrane domains. APP is usually cleaved by α-secretase (left; dot indicates the α-secretase cleavage site) or β-secretase (right) to produce a secreted form (APP$_s$-α or APP$_s$-β, respectively). The remaining portion is further processed by the γ-secretase, to yield an intracellular fragment (AICD). Whereas the combined actions of α- and γ-secretase produce a protein of 3 kilodalton (p3), the combined actions of the β- and γ-secretases produce intact Aβ.

γ-**secretase**, respectively (Figure 11–3, right). The cleavage site of the γ-secretase is not fixed: it can produce Aβ of different lengths, ranging from 39 to 43 amino acids. The predominant forms have 40 or 42 amino acids and are called $A\beta_{40}$ and $A\beta_{42}$, respectively. γ-Secretase was subsequently found to be a general intramembrane protease complex important for many signaling events. Studies of APP and other γ-secretase substrates indicate that a major trigger for γ-secretase activation is an extracellular cleavage that produces a transmembrane protein with a short extracellular stub (such as those produced by α- or β-secretase). APP is normally expressed in many cell types, and is proteolytically processed by the α- or β-secretases followed by the γ-secretase. So what goes wrong in Alzheimer's disease?

11.3 Mutations in human APP and γ-secretase cause early-onset familial Alzheimer's disease

At this point you may ask: are APP and Aβ related to the cause of AD, or is Aβ plaque formation simply the consequence of a disease whose causes lie elsewhere? Genetic studies of early-onset **familial Alzheimer's disease (FAD)** have shed light on this important issue. Most AD cases are late onset (age 65 or older) and are **sporadic** (from the Latin word for scattered) because patients do not have an identifiable family history of the disease; however, as with many sporadic illnesses, genetic risk factors of smaller effect play an important role in AD, as we discuss below. Patients with early-onset FAD usually develop AD symptoms in their 40s or 50s, and have a clear inheritance pattern. Most types of FAD follow a Mendelian autosomal dominant inheritance pattern: an AD patient is heterozygous for the disease allele, and imparts the disease allele to 50% of his/her progeny (we will discuss genetics of human disease in more detail in Box 11–3). Those who inherit the disease allele invariably develop AD if they live long enough (**Figure 11–4**A). In these cases, genetic mapping has helped to pinpoint mutations in specific genes that cause AD.

The first FAD mutation was mapped onto the *App* gene itself (Figure 11–4A): a missense mutation that changed a valine to an isoleucine in the middle of the transmembrane domain near the C-terminus of $A\beta_{42}$ (Figure 11–4B). Subsequently, about 20 FAD mutations have been mapped onto the *App* gene. Interestingly, most mutations are clustered near the γ- or β-secretase cleavage sites, with some in the middle of the Aβ peptide (Figure 11–4B). Biochemical studies indicate that mutations near the γ-secretase site increase the ratio of $A\beta_{42}$ over

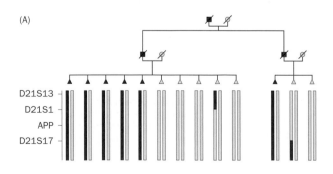

(A)

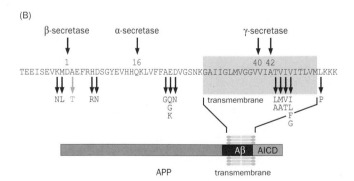

(B)

Figure 11–4 Mutations in the *App* gene cause familial Alzheimer's disease (FAD). (A) Pedigree of an early-onset FAD family (average onset 57 ± 5 years). Square, male; circle, female; triangle, either sex to preserve anonymity. Oblique lines indicate deceased individuals. Beneath the pedigree are maps of Chromosome 21 in which segments of the chromosomes were mapped according to the markers on the left. The linkage data suggest that chromosome segments in red were inherited from the disease-causing chromosome of the affected fathers (red squares). Inheriting from the father the red chromosome segment that includes mutant *APP* correlates perfectly with having AD (red triangles). **(B)** Summary of FAD mutations in the APP protein, most of which are located within or near Aβ. The cleavage sites for the three secretases are also indicated. Numbers indicate amino acid residues starting from the beginning of the Aβ peptide. The green arrow points to an AD-protective A → T (alanine to threonine) mutation that reduces Aβ production. (A, adapted from Goate A, Chartier-Harlin MC, Mullan M et al. [1991] *Nature* 349:704–706. With permission from Macmillan Publishers Ltd; B, adapted from Holtzman DM, John CM & Goate A [2011] *Sci Transl Med* 3:77sr1. See also Jonsson T, Atwal JK, Steinberg S et al. [2012] *Nature* 488:96–99.)

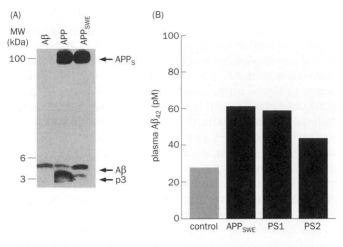

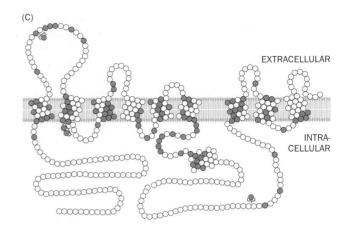

Figure 11–5 Mutations in APP and presenilin increase Aβ production.
(A) Left lane: synthetic Aβ as a control. Middle lane: transfecting cDNA expressing wild-type APP into cultured cells produced a major 3-kilodalton (kDa) band (p3) corresponding to the cleavage product of α- and γ-secretase, and a minor 4-kDa band corresponding to Aβ. Right lane: cells transfected with cDNA expressing APP$_{SWE}$ produced more Aβ than p3. MW, molecular weight in kilodalton. In these experiments, culture media containing radioactively labeled proteins from transfected cells were immunoprecipitated with an antibody against Aβ; immunoprecipitated proteins were run on a gel. **(B)** Compared with controls, the plasma Aβ$_{42}$ concentration (in picomoles per liter) was increased in AD patients with the APP$_{SWE}$ mutation or with pathogenic mutations in presenilin-1 (PS1) or presenilin-2 (PS2). **(C)** Structure of presenilin-1 and locations of FAD mutations. PS1 spans the membrane nine times; PS2, not shown here, has a similar structure. Mutations in PS1 that result in FAD were first reported by Sherrington et al. in 1995 (*Nature* 375:754); by 2010, more than 170 mutations that cause FAD had been identified in PS1 (red amino acid residues in the figure; red brackets denote insertions). (A, from Citron M, Oltersdorf T, Haass C et al. [1992] *Nature* 360:672–674. With permission from Macmillan Publishers Ltd.; B, adapted from Scheuner D, Eckman C, Jensen M et al. [1996] *Nat Med* 2:864–870. With permission from Macmillan Publishers Ltd.; C, adapted from De Strooper B & Annaert W [2010] *Ann Rev Cell Dev Biol* 26:235–260.)

Aβ$_{40}$. Although Aβ$_{40}$ is the dominant form produced by γ-secretase cleavage, Aβ$_{42}$ has a higher tendency to form aggregates. FAD mutations in the middle of Aβ may also increase the aggregation tendency. The FAD mutation near the β cleavage site (KM → NL amino acid changes; also called the Swedish mutation, or APP$_{SWE}$) leads to an increase of Aβ production (**Figure 11–5**A, B). Most of these mutations cause early-onset AD. Interestingly, an A → T amino acid change recently found near the β-secretase cleavage site (green arrow in Figure 11–4B), which reduces β-secretase cleavage and Aβ production *in vitro*, confers *protection* against late-onset AD and age-related cognitive decline. Thus, these genetic data strongly suggest that an increase in Aβ aggregation or production is causally linked to at least some forms of AD.

Another piece of evidence supporting the causal link between increased Aβ production and AD came from **Down syndrome**, which is caused by having an extra copy of Chromosome 21 in every cell. Down syndrome patients invariably have high levels of amyloid deposits in their 30s and 40s and exhibit Alzheimer's-like dementia in their 50s. The *App* gene is located on Chromosome 21, so Down syndrome patients have an extra copy of *App*. Indeed, people with smaller duplications of Chromosome 21 that cover the *App* gene also have early-onset AD symptoms, suggesting that increasing the *App* gene dose is sufficient to cause AD.

Genetic mapping studies have also identified two other loci, on human Chromosomes 14 and 1, respectively, that harbor autosomal dominant FAD mutations. The causal genes encode similar transmembrane proteins named **presenilin-1** (Figure 11–5C) and **presenilin-2**. Subsequent biochemical and genetic studies showed that the presenilins, together with three associated proteins, constitute the γ-secretase protein complex responsible for cleaving APP near the C-terminus of Aβ within the transmembrane domain. AD patients with presenilin mutations have increased Aβ$_{42}$ levels compared to control subjects (Figure 11–5B). Thus, mutations in APP and its processing enzymes point to increases in Aβ production or aggregation as a common cause that underlies AD pathogenesis, at least in early-onset FAD cases. This **Aβ hypothesis** does not exclude the possibility that other mechanisms can additionally contribute to

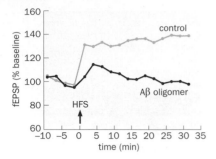

Figure 11–6 Oligomeric Aβ$_{42}$ disrupts hippocampal long-term potentiation (LTP). Application of 500-nM oligomeric Aβ$_{42}$ caused a marked reduction of LTP at the perforant path → dentate gyrus synapse in rat hippocampal slices, induced by high-frequency stimulation (HFS) at t = 0. Aβ$_{42}$ oligomers were applied to the experimental group at −45 min (not shown). See Figure 10–8 for a schematic of the LTP assay. (Adapted from Lambert MP, Barlow AK, Chromy BA et al. [1998] *Proc Natl Acad Sci USA* 95:6448–6453.)

Figure 11–7 Transgenic mice overexpressing human APP exhibit amyloid plaques and cognitive deficits. (A) Amyloid plaque stained with antibody against Aβ found in a 354-day-old transgenic mouse overexpressing APP with the Swedish mutation (APP$_{SWE}$). **(B)** As they age, mice transgenic for APP$_{SWE}$ (Tg$^+$) exhibit learning defects in the Morris water maze. Mice were trained to locate a hidden platform using spatial cues. When the hidden platform was removed, control mice (Tg$^-$) aged 9–10 months logged more target platform crossings (that is, spent more time near the platform's previous location, an indication of spatial memory) than did APP$_{SWE}$ transgenic mice of the same age. See Figure 10–32 for a schematic of this assay. (Adapted from Hsiao K, Chapman P, Nilsen S et al. [1996] *Science* 274:99–103.)

AD; for example, disruption of presenilin function might interfere with cellular processes unrelated to Aβ production that also contribute to AD pathogenesis.

How might excessive Aβ production and aggregation cause AD? While studies have suggested that amyloid deposits are toxic to neurons, the severity of AD symptoms does not always correlate with the density of amyloid plaques. This indicates that amyloid plaques may not be the only pathological factor. Indeed, as we will see in the next section, amyloid plaques and tau-enriched neurofibrillary tangles exhibit a synergistic relationship. Furthermore, diffuse oligomeric forms of Aβ have been found to be potent toxins to neurons. For example, an early study showed that long-term potentiation in a hippocampal slice was markedly impaired within 45 min of applying nanomolar concentrations of Aβ$_{42}$ oligomers synthesized *in vitro* (**Figure 11–6**), and significant neuronal death occurred within 24 hours of Aβ$_{42}$ oligomer application. Subsequent investigations revealed that oligomeric Aβ induces synaptic depression, spine loss, and abnormal synaptic plasticity, suggesting that non-aggregated Aβ oligomers may be potent AD-promoting agents. Future studies must further elucidate the relative contributions made by Aβ aggregates and different forms of oligomers to disease phenotypes *in vivo*.

11.4 Animal models offer crucial tools to investigate pathogenic mechanisms

Animal models are instrumental in human disease research. Appropriate animal models can validate causality between suspected pathogenic processes and disease outcomes. They can be used to trace disease progression, investigate disease mechanisms, and test the effects of therapeutic agents. The development of AD animal models offers excellent illustrations of the utility of this approach. The technical feasibility of performing precise genetic manipulations in mice (see Sections 13.7 and 13.10) has made this species the dominant mammalian model for AD and many other brain disorders. However, as will be discussed in more detail later, there are many differences between the mouse brain and human brain, such that even very useful mouse models do not tend to produce identical pathology or treatment responses in humans.

Mice normally do not exhibit AD pathology such as amyloid plaques and neurofibrillary tangles. This is probably because the mouse's lifespan of about 2 years is not long enough for abnormal protein aggregates to cause sufficient insult to the nervous system, a process that usually takes decades in humans. However, this protein pathology can be accelerated by overexpressing wild-type or FAD-mutation alleles of the human *App* gene in transgenic mice (**Figure 11–7**A). Moreover, some transgenic mice also develop age-dependent cognitive decline, such as deficits in spatial memory in the Morris water maze (Figure 11–7B). These experiments support the Aβ hypothesis: overproduction of human APP with FAD mutations is sufficient to produce pathology and cognitive defects consistent with AD.

Transgenic mice have been also used to investigate the relationship between APP and presenilin. Mice expressing a mutant form of presenilin by itself did not develop AD pathology. However, mice co-expressing transgenes with FAD mutations in human APP and human presenilin exhibited earlier onset of amyloid

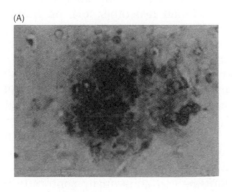

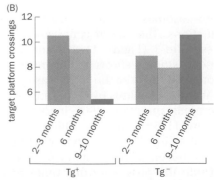

(A) tauP301L + APP_SWE   (B) tauP301L

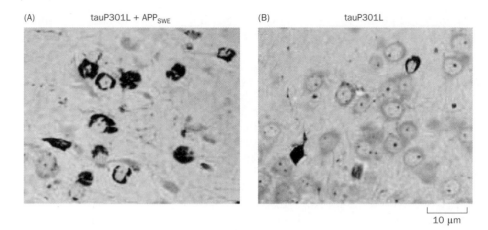

10 μm

Figure 11–8 Enhanced neurofibrillary tangle pathology in transgenic mice expressing mutant APP and tau. In humans, a leucine → proline mutation in tau (tauP301L) causes frontotemporal dementia with parkinsonism. Neurofibrillary tangles, as seen by intense silver staining (dark blue), are markedly increased in mice that were doubly transgenic for the Swedish allele of APP (APP_SWE) and tauP301L **(A)**, compared with transgenic mice that expressed tauP301L alone **(B)**. (From Lewis J, Dickson DW, Lin WL et al. [2001] *Science* 293:1487–1491. With permission from AAAS.)

plaque deposition and memory deficits compared with transgenic mice expressing mutant APP alone, indicating that mutations in presenilin and APP synergize to accelerate AD-like pathological changes.

Transgenic mice have also been used to investigate the relationship between amyloid plaques and neurofibrillary tangles, the two major pathological features of AD. As noted in Section 11.1, the major component of neurofibrillary tangles is the microtubule-associated protein tau, and abnormal accumulation of tau is a feature of several neurodegenerative tauopathies. Although mutations in human tau have not been found in AD patients, they have been found in patients with a tauopathy called frontotemporal dementia with parkinsonism (FTDP), which exhibits neurofibrillary tangle pathology. Transgenic mice that express human tau with a dominant FTDP mutation at high levels can recapitulate tau aggregation similar to the neurofibrillary tangles observed in human AD and FTDP patients, without amyloid plaques. Interestingly, mice in which tau overexpression was combined with overexpression of mutant APP (double transgenic) or with both mutant APP and mutant presenilin (triple transgenic) developed both plaques and tangles. Moreover, neurofibrillary tangles were much more prevalent and widespread in double- or triple-transgenic mice than in transgenic mice expressing mutant tau alone (**Figure 11–8**). Thus, while plaques and tangles can form by independent mechanisms, increased Aβ production facilitates neurofibrillary tangle formation. Indeed, removing one or both copies of the endogenous mouse gene encoding tau can alleviate behavioral and synaptic defects caused by APP overexpression. This suggests that some symptoms caused by the overproduction of APP or Aβ may be mediated by dysregulation of tau.

Despite the utility of mouse AD models in creating plaques and tangles and in investigating their relationships, they do not recapitulate the prominent neuronal death and brain atrophy of human AD (see Figure 11-1). This may result from the limited lifespan of mice or physiological differences between rodent and primate brains. Thus, development of primate models will be valuable to further explore AD pathogenesis and test therapeutic strategies.

11.5 An apolipoprotein E (ApoE) variant is a major risk factor for Alzheimer's disease

Mutations in genes encoding APP and presenilins, while very helpful in establishing a causal relationship between Aβ production and AD, account for less than 2% of AD cases, which are usually early onset. Almost all other AD cases are late onset and are sporadic. On the other hand, twin studies of large AD populations indicated a high degree of heritability (see Section 1.1), from 60 to 80%. This suggests that AD has a strong genetic component, although environmental factors are not negligible. What additional genes contribute to AD?

Thus far, genetic analyses have not revealed FAD genes with Mendelian inheritance patterns beyond those encoding APP and presenilins. This suggested that most late-onset AD cases are caused by combinations of multiple genetic factors or by genetic factors interacting with environmental factors. By far the most

(A)

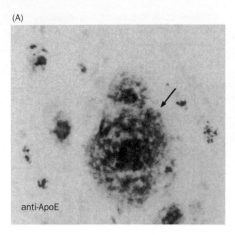

anti-ApoE

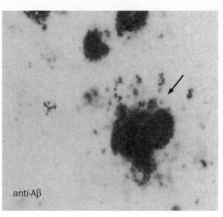

anti-Aβ

|— 50 μm —|

(B)

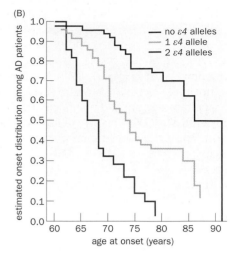

Figure 11–9 ApoE is found in amyloid plaques, and the ε4 allele is a major risk factor for AD. (A) Adjacent postmortem brain sections of an AD patient stained with antibodies against ApoE (left) and Aβ (right). ApoE is localized to the amyloid plaques (arrows). **(B)** As the copy number of *Apoe ε4* increases, the age of AD onset becomes younger. For instance, about 50% of AD patients lacking the ε4 allele had been diagnosed with AD by age 86. The median age of AD onset drops to 73 years for patients with one ε4 allele and 66 years for those with two ε4 alleles. (A, from Strittmatter et al. [1993] *Proc Natl Acad Sci USA* 90:1977–1981; B, adapted from Corder EH, Saunders AM, Strittmatter WJ et al. [1993] *Science* 261:921–923. With permission from AAAS.)

important genetic risk factor identified to date is an individual's allele composition for a gene called *Apoe* which encodes **apolipoprotein-E (ApoE)**. A component of high-density lipoproteins in the brain, ApoE is involved in lipid transport and metabolism. The most common allele of *Apoe* in the human population is *ε3*. A less common allele, *ε4*, differs from *ε3* by a single amino acid. In the early 1990s, ApoE was found to bind to Aβ and to be present in amyloid plaques (**Figure 11–9**A). Genetic studies that examined the relationship between *Apoe* allele composition and AD revealed that the *ε4* allele frequency increased from approximately 15% in the general population to 40% in AD patients. Compared to the common *ε3/ε3* allele combination, individuals with a single *ε4* allele have a more than threefold greater chance of developing AD, while individuals with two *ε4* alleles have a 12-fold greater chance of getting AD. Furthermore, among AD patients, as the copy number of *ε4* alleles increases, the age of disease onset decreases (Figure 11–9B).

Thus, *Apoe* is a **genetic susceptibility locus** for AD. Unlike the FAD mutations in APP or presenilins, which invariably cause AD if the carrier lives long enough, ApoE *ε4* does not definitively cause AD but increases the likelihood of developing the disease. However, given the relatively high frequency of the *ε4* allele in the human population, ApoE *ε4* contributes far more to the incidence of AD than do mutations in APP and presenilins. Interestingly, a less common allele, *Apoe ε2*, appears to be protective against AD, as the *ε2* allele frequency in the general population is 8%, but drops to 4% in the AD population.

The mechanism by which ApoE *ε4* increases the chance of developing AD is not well understood. One proposed function for ApoE is that by binding to Aβ, ApoE regulates its metabolism and clearance. Further work is required to clarify how the *ε4* and *ε3* alleles differentially affect Aβ metabolism, and whether ApoE affects Aβ-independent processes. Nevertheless, the discovery of ApoE's association with AD has provided a paradigm for studying complex genetic disorders, in which mutations in individual genes increase the susceptibility for a disease rather than causing the disease outright. We will see many examples of such susceptibility variants later in this chapter.

11.6 Microglia dysfunction contributes to late-onset Alzheimer's disease

Recent genome-wide association studies and whole-genome sequence analyses (see Box 11–3) have identified additional genes that confer risks to late-onset

Alzheimer's disease. While most of these genetic variants have less effect on AD risk levels than does ApoE *ε4*, subjects with one copy of a particular variant in the gene encoding TREM2 (t̲riggering r̲eceptor e̲xpressed on m̲yeloid cells 2) have an AD risk similar to those with one copy of *Apoe ε4*, although the TREM2 variant frequency in the general population is far lower (<1%) than that of ApoE *ε4*. TREM2 is normally expressed at high levels in immune cells, including brain microglia that play an important role in clearing damaged cells and debris (see Figure 1–9). Specifically, TREM2 is known to stimulate phagocytosis and suppress inflammation. As another example suggesting a link between microglia and AD, several studies have associated CD33, a cell-surface antigen in immune cells as well as microglia, with late-onset AD. CD33 inhibits microglia uptake of $A\beta_{42}$ in culture, and *Cd33* mutant mice exhibit reduced levels of Aβ plaques in a mouse model of AD. Together, these findings suggest that microglia dysfunction also contributes to AD pathogenesis, possibly via abnormal Aβ clearance.

11.7 How can we treat Alzheimer's disease?

An ultimate goal of studying human diseases is to find effective ways to cure or prevent them. By the time AD is typically diagnosed, neurodegeneration may be too extensive for the disease to be cured (**Figure 11–10**A), although it may still be possible to halt further progression and treat AD symptoms. The most likely path to reducing the impact of AD is early diagnosis and prevention. Delaying AD onset by even a few years could have a very significant impact on the quality of life for patients and the burden on their families and society.

While we are still far from developing a cure, past decades of AD research have provided valuable clues about potential treatments. For example, multiple lines of evidence suggest that key pathogenic events may include increased Aβ levels or enhanced Aβ aggregation. Thus, this pathway has been the primary focus of intervention. Strategies to reduce Aβ toxicity include developing drugs to inhibit β- or γ-secretase activity, increase clearance of Aβ, or neutralize Aβ activity (Figure 11–10B). It will be also important to determine the nature of potential Aβ-independent pathway(s) to identify more drug targets for AD intervention.

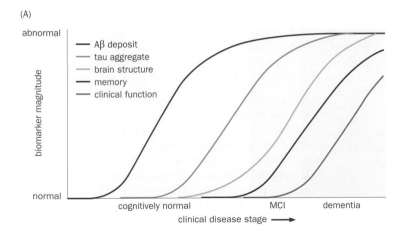

(A)

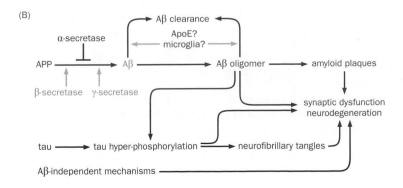

(B)

Figure 11–10 Temporal progression, pathogenic pathways, and potential sites of intervention for Alzheimer's disease. **(A)** A proposed scheme of temporal progression of AD, with *x* axis representing clinical stage, and *y* axis representing the magnitude of abnormalities of biomarkers such as Aβ deposit and tau aggregate. MCI, mild clinical impairment. **(B)** The top pathway summarizes the Aβ hypothesis. The roles of ApoE and microglia in Aβ clearance remain to be further elucidated. The middle pathway depicts the production of neurofibrillary tangles. The arrow between the top and middle pathways represents evidence that tau can mediate at least some of the Aβ toxicity in animal models (see Section 11.4). The bottom pathway emphasizes that AD pathogenesis may involve Aβ-independent mechanisms. Highlighted in blue are some of the therapeutic targets currently being pursued to reduce the Aβ level, including antibodies against Aβ and inhibitors or modifiers of β- and γ-secretase. Green and red, facilitative and inhibitory actions. (A, adapted from Jack CR, Knopman DS, Jagus WJ et al. [2013] *Lancet Neurol* 12:207–216. With permission from Elsevier Inc.)

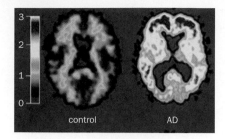

Figure 11–11 Positron emission tomography (PET) imaging can visualize Aβ deposits. Radioactively labeled benzothiazole, named Pittsburgh Compound-B (PIB), is used for PET imaging of AD brains because it enters the brain rapidly, binds selectively to aggregated Aβ deposits, and is cleared rapidly. Compared with a 67-year-old normal subject (left), a 79-year-old AD patient (right) shows elevated standardized values (color-coded on the left) for PIB uptake. (From Klunk WE, Engler H, Nordberg A et al. [2004] *Ann Neurol* 55:306–319. With permission from the American Neurological Association.)

Many factors need to be considered when developing drugs for brain disorders (see **Box 11–1**). For instance, while a drug should have its intended effects on its therapeutic target, it should have minimal side effects and toxicity at the therapeutic dose. As a specific example, several γ-secretase inhibitors have been developed that effectively interfere with Aβ production but failed in clinical testing due to severe side effects, possibly because γ-secretase has many substrates other than APP. (For example, one well-known γ-secretase substrate is the developmental signaling molecule Notch, which plays a key role in regulating cell fates as discussed in Section 7.3; Notch also has many important functions in adults.) It is necessary to identify γ-secretase inhibitors that specifically interfere with its activity toward APP cleavage, or to modify γ-secretase activity to bias the product toward shorter, less toxic Aβ.

Even if successful drugs are developed, early diagnosis will be crucial in order to halt AD pathogenesis at the earliest stage possible. An important step is to identify **biomarkers**, which are characteristics that are objectively measured and evaluated as indicators of normal biological processes, pathogenic processes, or pharmacologic responses to a therapeutic intervention. Important biomarkers for AD include **positron emission tomography** (**PET**) using radioactive compounds that bind fibrillary Aβ (**Figure 11–11**). (PET is a non-invasive three-dimensional imaging technique that traces the distribution of positron-emitting probes introduced into the body.) This would allow detection of amyloid deposits prior to the onset of cognitive and behavioral symptoms. Identifying biomarkers that signal AD progression, such as $A\beta_{42}$ and other metabolites in cerebrospinal fluid and plasma, can also contribute to early diagnosis. The goal is to have increasingly reliable diagnostic methods for treating preclinical AD before irreversible damage occurs (Figure 11–10A).

Box 11–1: Rational drug development to treat brain disorders

Many drugs currently used for treating brain disorders, including most drugs for psychiatric disorders, were discovered serendipitously, typically when clinicians noticed unintended but desirable effects during the treatment of another condition (see Sections 11.15–11.17). However, basic research in neuroscience and advances in the pharmaceutical and biotechnology industries during the past decades are changing this picture. Therapeutics can be developed through a rational process to make them more effective in treating the causes of a specific disease while limiting unwanted side effects. Identifying the mechanisms of disease pathogenesis is a prerequisite for rational therapeutic intervention. In this box, we focus on common steps of drug development (**Figure 11–12**), assuming that a key pathogenic process has already been identified.

The first step is to choose a specific molecular target for intervention. Some biological processes are more amenable than others to pharmacological intervention. For instance, cell-surface proteins are some of the preferred molecular targets because water-soluble chemicals and therapeutic antibodies can modulate their functions extracellularly. Once a target has been chosen, robust biological assays relevant to the disease process must be established to screen for candidate drugs, which usually belong to two large categories: small-molecule chemicals and large-molecule biologics such as peptides and antibodies. For small-molecule drugs, the first step is usually to establish *in vitro* assays that enable high-throughput screens of chemical libraries (which usually contain 10^5–10^6 synthetic, semi-synthetic, or

naturally occurring compounds). Once a promising compound is identified, many variants can be synthesized to increase biological activity and target accessibility *in vivo* and to reduce potential side effects. Large-molecule biologics, while not selected using high-throughput screening, nevertheless undergo a similar optimization process for affinity, activity, safety, and physiological processing.

While some of the optimization steps can be conducted *in vitro*, many steps must be performed in animal models *in vivo*. Two commonly used terms for body–drug interactions are **pharmacodynamics**, which characterizes what the drug does to the body, including the intended effects on target molecules and processes as well as unintended side effects, and **pharmacokinetics**, which characterizes what the body does with the drug, including its absorption, distribution, metabolism, and excretion. Disease proxies, such as the animal models discussed in Section 11.4, are very helpful to assay a drug's therapeutic effects and to establish proof-of-concept. Animal models are also essential to evaluate potential drug toxicity. Initial animal models are likely to be rodents, but toxicity is usually also evaluated in additional models with more human-like physiology, such as dogs or nonhuman primates.

If the target resides within the nervous system, an important step is to ensure that drugs, which usually access the body through circulation in the blood, can pass through the **blood–brain barrier** (**BBB**). Derived from endothelial cell tight junctions in the blood vessels of the brain, the BBB

Box 11–1: Rational drug development to treat brain disorders

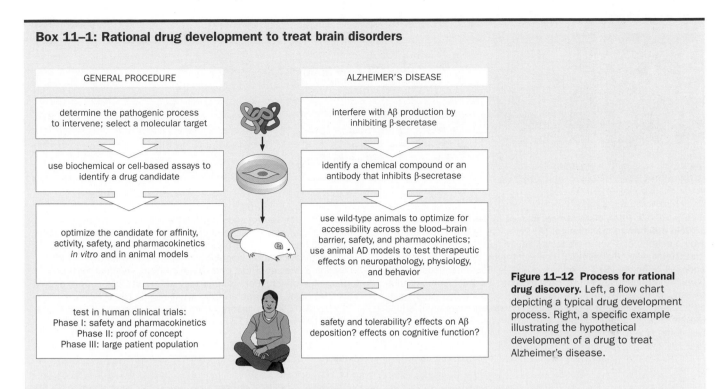

Figure 11–12 Process for rational drug discovery. Left, a flow chart depicting a typical drug development process. Right, a specific example illustrating the hypothetical development of a drug to treat Alzheimer's disease.

prevents the exchange of many substances between the blood and brain tissues. Small molecules can be chemically engineered to diffuse through the BBB or to enter cells directly, thereby reaching target molecules in any cellular compartments. Antibodies, on the other hand, usually access only cell-surface proteins and do not freely cross the BBB, although progress has been made recently to facilitate this process.

If a drug succeeds in preclinical tests in animals, including extensive safety studies, the next step is human clinical trials. Clinical trials are usually conducted in distinct phases, although some phases can be combined. Phase I studies are usually conducted with a small number of healthy volunteers and emphasize safety and drug metabolism. While Phase II studies continue to test the safety of a drug, they also gather preliminary data on the drug's effectiveness

in a relatively small number of patients, comparing the drug with a placebo control. Phase III studies collect more information about safety and effectiveness from a large patient population. In the United States, the Food and Drug Administration (FDA) oversees clinical trials and approves drugs for marketing. Drug development is a long process, averaging 10 years from initial target selection to approval for use in intended patients. This time window can be considerably longer for diseases that progress slowly, such as neurodegenerative diseases.

The process outlined above has been highly successful in identifying drugs that treat diseases such as cancer, immune disorders, and infectious diseases. Rationally designed drugs are also in the pipeline for treatment of neurodegeneration, including AD.

11.8 Prion diseases are caused by propagation of protein-induced protein conformational change

Just as abnormal Aβ aggregation is characteristic of Alzheimer's disease, research in the past decades has shown that **proteinopathy**—altered protein conformations, interactions, and homeostasis—appears to be a common feature of many neurodegenerative diseases. We discuss several examples in the next two sections. Among the most enigmatic causes of neurodegenerative disease are **prions** (pronounced *PREE-ons*, which stands for pro<u>teinaceous</u> in<u>fectious</u> particles). Three seemingly separate diseases share prions as the causative agent. The first is **scrapie**, which is known to infect sheep and goats after a prolonged incubation period following exposure to tissues from diseased animals; a variant of scrapie that affects cattle is colloquially known as 'mad cow disease.' The second is an infectious human disease called **kuru**, which occurred in Papua New Guinea tribes that observed ritual cannibalism. The third is a rare inherited human disease called **Creutzfeldt–Jakob disease** (**CJD**). All three diseases are associated

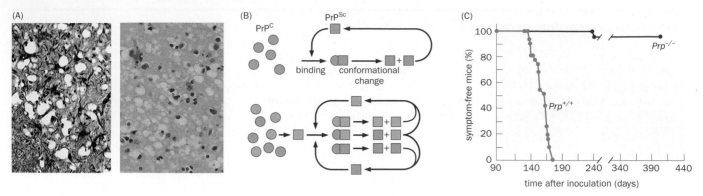

Figure 11–13 Prion diseases are caused by protein-induced protein conformational change. (A) Pathology seen in postmortem brain sections of a scrapie-affected sheep (left) and a human Creutzfeldt-Jakob disease (CJD) patient (right); note the numerous sponge-like holes. **(B)** The prion hypothesis. Top: infectious PrP^{Sc} (squares) can induce PrP^{C} (circles) to adopt the PrP^{Sc} conformation, thereby propagating PrP^{Sc}. Bottom, in genetic prion diseases such as CJD, mutant PrP^{C} proteins (orange circles) spontaneously adopt the PrP^{Sc} conformation (orange square) on occasion, thereby becoming able to convert both mutant and wild-type (green) PrP^{C} into PrP^{Sc}. **(C)** Control mice (*Prp$^{+/+}$*) that receive intracerebral inoculation of mouse-adapted prions invariably die within 6 months. However, their *Prp$^{-/-}$* littermates are resistant to prion infection; following intracerebral inoculation, most live beyond a year with no prion pathology. Thus, endogenous PrP is essential for the effect of prion infection. (A, courtesy of Robert Higgins (left) and the CDC (right); B, adapted from Prusiner SB [1991] *Science* 252:1515–1522; C, adapted from Büeler H, Aguzzi A, Sailer A et al. [1993] *Cell* 73:1339–1347. With permission from Elsevier Inc.)

with massive neurodegeneration and neuronal death that creates sponge-like holes in the brain. For this reason, they are called spongiform encephalopathies (**Figure 11–13**A).

The nature of the infectious scrapie agent was heavily debated for several decades. A breakthrough came in the 1980s when reliable animal models were developed as bioassays for biochemical purification of the scrapie agent. Animals that had been inoculated with infected brain tissues, or with fractions of infected tissues resulting from biochemical purification, exhibited spongiform encephalopathy. Moreover, brain extracts from these newly infected animals were highly infectious when inoculated into new animals. The infectivity of these extracts was not disrupted by treatments that destroyed nucleic acids, suggesting that, unlike known infectious agents such as viruses and bacteria, the scrapie agent does not have a nucleic acid genome. This led to the **prion hypothesis**: the infectious scrapie agent is proteinaceous in nature. This was considered heretical: how could a protein be infectious without an associated genome for replication?

The infectious agent was subsequently identified as PrP^{Sc}, a conformational variant of the cell-surface protein PrP. (The ^{Sc} superscript stands for scrapie.) It was later shown that a non-infectious conformation of PrP (termed PrP^{C} for cellular PrP) is normally produced by most cell types. The presence of PrP^{Sc} can cause PrP^{C} to adopt the PrP^{Sc} conformation (Figure 11–13B, top), which is a highly stable β-pleated sheet (recall that Aβ also adopts a β-pleated sheet conformation in AD). PrP^{Sc} was proposed to propagate from cell to cell and even through the digestive system when diseased tissues are ingested by healthy animals or humans, catalyzing the conversion of PrP^{C} into PrP^{Sc} along the way. The inherited Creutzfeldt–Jakob disease does not require infectious protein agents; instead, CJD results from mutations in the *Prp* gene that make PrP^{C} more prone to adopt the PrP^{Sc} conformation spontaneously (Figure 11–13B, bottom). Thus, the prion hypothesis unified the causes of scrapie, kuru, and CJD, which are collectively called **prion diseases** (**Movie 11–1**).

Strong support for the prion hypothesis came from *Prp* knockout mice, which were found to be resistant to PrP^{Sc} infection, indicating that PrP^{Sc} requires endogenous PrP^{C} to cause disease (Figure 11–13C). Indeed, the concept of protein-induced protein conformational change and its propagation has subsequently been applied to other neurodegenerative diseases and to underlie widespread phenomena in the normal physiology of organisms ranging from yeast to humans.

11.9 Aggregation of misfolded proteins is associated with many neurodegenerative diseases

Huntington's disease (**HD**) is a dominantly inherited disease that usually strikes patients during midlife; it is named after George Huntington who first described the inheritance pattern in 1872. The earliest symptoms are often depression or mood swings, followed by abnormal movements, since the striatum is the most vulnerable brain area in HD. Patients later develop cognitive deficits. Death occurs 10–20 years after symptom onset. Genetically speaking, HD is one of the simplest neurological diseases, as it is caused by alterations in a single gene encoding a widely expressed protein named **huntingtin**. The cause of the disease is an expansion of a CAG trinucleotide repeat in the middle of the gene's coding sequence, resulting in an extended poly-glutamine (polyQ) repeat near the N-terminus of the huntingtin protein. (The nucleotide triplet cytidine–adenosine–guanosine codes for amino acid glutamine, abbreviated 'Q.') Healthy individuals have 6 to 34 glutamine repeats, whereas HD patients have 36 to 121 repeats (**Figure 11–14**). Greater numbers of polyQ repeats correlate with earlier onset of HD symptoms.

Since the discovery of polyQ repeats in HD, expanded polyQ repeats in eight other proteins have been shown to cause neurodegenerative diseases, all of which are dominantly inherited, including six forms of **spinocerebellar ataxia** that affect motor functions (**Table 11–1**). In most cases, the polyQ repeat number that distinguishes healthy and diseased conditions is around 35. Because of this common feature, it had been hypothesized that the expanded polyQ repeats themselves could cause disease. Indeed, *in vivo* transgenic overexpression of expanded polyQ repeats alone is sufficient to cause degeneration of mouse and even *Drosophila* neurons. Proteins with long polyQ repeats form aggregates called inclusion bodies, which may be present in the nucleus, cytoplasm, or axons, depending on the specific protein affected. The host proteins in which these polyQ repeats reside also play essential roles for pathogenesis *in vivo*. Inclusion bodies of abnormally aggregated host protein isoforms that contain expanded polyQ repeats can recruit additional proteins that normally interact with the host proteins. As a result, the normal function of these interacting proteins is disrupted, accounting for the dominant nature of the expanded-polyQ disorders. The participation of specific host proteins in the pathogenesis of polyQ disorders also explains why different polyQ diseases have different symptoms, as each disrupts a different set of interacting proteins. In the case of HD, disruption of transcription, axonal transport, and mitochondrial function may all contribute to the eventual dysfunction and degeneration of neurons in the striatum, giving rise to characteristic uncontrolled movements (Huntington's chorea).

Amyotrophic lateral sclerosis (**ALS**, also known as Lou Gehrig's disease) is a rapidly progressing motor neuron disease that usually kills patients within

normal (CAG)$_{21}$ HD (CAG)$_{48}$

Figure 11–14 Huntington's disease is caused by expanded poly-glutamine (polyQ) repeats. DNA sequences of a healthy subject and a Huntington's disease (HD) patient, showing the expanded CAG repeats that result in mutant huntingtin protein having longer polyQ repeats. (From MacDonald ME, Ambrose CM, Duyao MP et al. [1993] *Cell* 72:971–983. With permission from Elsevier Inc.)

Table 11–1: Diseases caused by poly-glutamine repeat

Disease[1]	Gene product	Normal repeat length	Expanded repeat length
HD	huntingtin	6–34	36–121
SCA1	ataxin1	6–44	39–82
SCA2	ataxin2	15–24	32–200
SCA3	ataxin3	13–36	61–84
SCA6	voltage-gated Ca^{2+} channel subunit	4–19	10–33
SCA7	ataxin7	4–35	37–306
SCA17	TATA-binding protein	25–42	47–63
SBMA	androgen receptor	9–36	38–62
DRPLA	atrophin	7–34	49–88

[1] Abbreviations: HD, Huntington's disease; SCA, spinocerebellar ataxia; SBMA, spinobulbar muscular atrophy; DRPLA, dentatorubral-pallidoluysian atrophy.

(After Orr & Zoghbi [2007] *Annu Rev Neurosci* 30:575.)

(A)

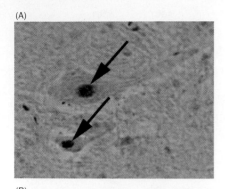

(B)

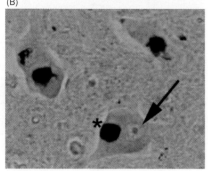

Figure 11–15 Abnormal TDP-43 cytoplasmic inclusions in amyotrophic lateral sclerosis (ALS). Spinal cord sections from an unaffected individual **(A)** and an ALS patient **(B)** immunostained for TPD-43 (brown). TDP-43 is normally localized to the nuclei of motor neurons (arrows in A), but in some motor neurons of ALS patients, TDP-43 is absent from the nucleus (arrow in B) and accumulates in cytoplasmic inclusion bodies (* in B). (Adapted from Figley MD & Gitler AD [2013] *Rare Diseases* 1: e24420. With permission from Landes Bioscience.)

a few years after symptoms emerge. Like AD, only a small fraction (~10%) of ALS cases are caused by dominantly inherited mutations in a handful of genes, whereas 90% are sporadic. Mutations that cause dominant familial ALS affect a range of different proteins, including SOD1 (superoxide dismutase 1, an enzyme), TDP-43 (TAR DNA-binding protein of 43 kilodalton, a DNA/RNA-binding protein), FUS (fused in sarcoma, another nucleic-acid-binding protein). The most common familial ALS is caused by an expansion of hexanucleotide repeats in the intron of a previously unstudied open reading frame, *C9orf72*. Each of the mutant proteins associated with ALS exhibits abnormal aggregation within affected motor neurons. In the case of TDP-43, for instance, normal TDP-43 is enriched in the nucleus, where it binds to DNA and RNA and regulates RNA processing; in ALS conditions, both mutant and wild-type TDP-43 accumulate in cytoplasmic inclusion bodies (**Figure 11–15**). This disrupts the normal function of TDP-43 in the nucleus, and at the same time recruits RNAs and proteins that interact with TDP-43 into the inclusion bodies. This may account for why TDP-43 mutations are dominant.

TDP-43-containing aggregates in cytoplasmic inclusions have been found in motor neurons not only from familial ALS with TDP-43 mutations, but also from the other familial ALS types such as those caused by the expansion of hexanucleotide repeats in *C9orf72*, as well as from the majority of sporadic ALS cases. *De novo* mutations in TDP-43 have also been found in some sporadic ALS cases. Thus, TDP-43 aggregation may represent a common pathological event in ALS that can have diverse causes.

11.10 Parkinson's disease results from death of substantia nigra dopamine neurons

Parkinson's disease (**PD**), first described by James Parkinson in 1817, is the second most common neurodegenerative disease after AD. PD primarily affects movement control. Characteristic PD symptoms include shaking, rigidity, slowness, and difficulty with walking. The primary cause for these symptoms is the death of dopamine neurons in a midbrain structure called the **substantia nigra** (Latin: black substance), which is named after the high levels of melanin pigment present in these dopamine neurons in healthy human subjects (**Figure 11–16**A). Indeed, PD studies have benefited from and contributed to our understanding of dopamine regulation of the striatal circuits that control movement.

As we learned in Section 8.9, striatal GABAergic spiny projection neurons (SPNs) control movement through two parallel pathways: a direct pathway that sends inhibitory signals to the globus palladus internal segment (GPi) and the substantia nigra pas reticulata (SNr), and an indirect pathway that inhibits the globus palladus external segment (GPe) and thereby relieves inhibition of the GPi/SNr (see Figure 8–22). Dopamine neurons in the substantia nigra pas compacta (SNc) project to the striatum and regulate the direct and indirect pathways in opposite directions: they facilitate the direct pathway through the dopamine D_1 receptor and inhibit the indirect pathway through the dopamine D_2 receptor. Thus, loss of SNc dopamine neurons in PD is predicted to cause hypo-activation of the direct pathway and hyper-activation of the indirect pathway. Both lead to excessive activation of basal ganglia output neurons in the GPi/SNr, which in turn causes excessive inhibition of their target neurons in the thalamus and brainstem and thereby inhibits movement (Figure 11–16B). This useful framework has prompted effective treatment such as deep brain stimulation that will be discussed in Section 11.13.

11.11 α-Synuclein aggregation and spread are prominent features of Parkinson's pathology

What causes dopamine neuron death in PD? In many ways PD parallels what we have learned about AD and ALS. Most PD cases are late-adult onset without a clear inheritance pattern (that is, they are sporadic), but a small fraction of PD cases are familial. In 1997, the first identified familial PD gene, inherited in an autosomal

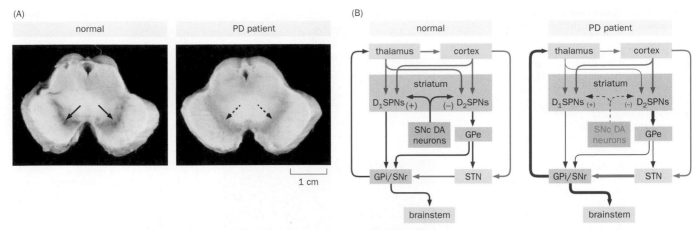

Figure 11–16 Parkinson's disease is caused by loss of midbrain dopamine neurons, leading to dysregulation of basal ganglia circuits. (A) Coronal sections of postmortem midbrains from a normal subject (left) and a PD patient (right). Arrows point to the substantia nigra, enriched in pigmented dopamine neurons that appear dark in normal brains and are selectively lost in the PD brains (dotted arrows). **(B)** A comparison of simplified basal ganglia circuit models for normal and PD brains. Left, dopamine neurons in the substantia nigra pars compacta (SNc DA neurons) positively regulate synaptic transmission between cortical/thalamic input and spiny projection neurons expressing the D_1 dopamine receptor (D_1SPNs), which project directly to the globus pallidus internal segment and the substantia nigra pars reticulata (GPi/SNr). At the same time, SNc DA neurons negatively regulate synaptic transmission between cortical/thalamic input and SPNs expressing the D_2 receptor (D_2SPNs), which

project indirectly to GPi/SNr via the globus pallidus external segment (GPe) and the subthalamic nucleus (STN). Right, loss of SNc DA neurons in PD patients reduces the activity of D_1SPNs and enhances the activity of D_2SPNs, both of which contribute to hyperactivation of GABAergic output neurons in the GPi/SNr and excessive inhibition of the target neurons of GPi/SNr. Red, inhibitory projection; green, excitatory projection; blue, DA projection. Thicker or thinner arrows in the PD diagram, respectively, indicate an increase or a decrease of pathway strengths in PD compared with normal. As will be discussed in Section 11.13, lesion or deep brain stimulation of STN can alleviate PD symptoms by reducing the hyperactivation of GPi/SNr. (A, courtesy of the Duke University School of Medicine; B, see Bergman H, Wichmann T & DeLong MR [1990] *Science* 249:1436–1438 and Limousin P, Krack P, Pollak P et al. [1998] *N Engl J Med* 339:1105–1111.)

dominant fashion, was found to encode a presynaptic protein called **α-synuclein**. Soon afterwards, α-synuclein was found to be the major component of **Lewy bodies**, intracellular inclusions that have been a defining feature of PD pathology since F. H. Lewy first described them in 1912 (although not all forms of PD exhibit Lewy body pathology). Familial PD mutations in α-synuclein (single amino acid changes) promote aggregation of α-synuclein *in vitro*. Indeed, increasing the copy number of the wild-type α-synuclein gene in humans is sufficient to cause PD with Lewy body pathology. Thus, PD can be caused by overproduction of wild-type α-synuclein or by production of mutant α-synuclein prone to aggregation.

Postmortem analyses of brains from patients who died at different stages of PD suggest that the distribution of Lewy body pathology follows a stereotyped spatiotemporal sequence: neurons with α-synuclein pathology are found mostly in the brainstem of early-stage PD patients, but are more widespread in the forebrains of patients with more advanced PD. In fact, substantia nigra dopamine neurons are not the first to exhibit α-synuclein pathology in this sequence, although they might be the most vulnerable. This suggests that α-synuclein aggregates may spread from neuron to neuron. This hypothesis was spurred by findings from a PD patient whose brain received a transplant of fetal dopamine neurons, a treatment strategy that we will discuss in Section 11.13 below. When the patient died 14 years after surgery, the transplanted fetal neurons contained α-synuclein-enriched Lewy body-like structures. Additional support for this hypothesis came from more recent studies in mice: focal injection into wild-type mice of α-synuclein fibrils that were preformed *in vitro* caused the spread of α-synuclein pathology across the brain, including to SNc dopamine neurons (**Figure 11–17**). Remarkably, injecting α-synuclein fibrils into α-synuclein knockout mice did not cause spread of α-synuclein pathology, suggesting that recruitment of the endogenous α-synuclein is essential for the pathology, as in the case of prion diseases (see Figure 11–13C). These studies, which together suggest that PD may involve cell-to-cell spread of pathogenic proteins, raise important and as-yet-unresolved questions regarding the mechanisms by which such spread might occur for cytoplasmic proteins such as α-synuclein.

Figure 11–17 Cell-to-cell spread of α-synuclein pathology. (A) Serial coronal brain maps of a wild-type mouse 180 days after injection of preformed α-synuclein fibrils (PFF) into the dorsal striatum revealed the development of α-synuclein pathology across the brain. Cells with α-synuclein aggregates are shown in red; the injection site is marked by a light red circle in the 2nd map from left. **(B)** A high-magnification image of the substantia nigra, showing two dopamine neurons with aggregated α-synuclein in Lewy body-like inclusions (arrows). Dopamine neurons are visualized in green by immunostaining against tyrosine hydroxylase, a marker for dopamine neurons. α-synuclein is visualized in red using an antibody that preferentially stains aggregated α-synuclein. **(C)** Substantia nigra pars compacta (SNc) cells that exhibited α-synuclein pathology were located primarily on the ispilateral side of the injection. Injecting phosphate-buffered saline (PBS) into wild-type (wt) or injecting PFF into α-synuclein knockout ($Snca^{-/-}$) mice did not cause α-synuclein pathology. (Adapted from Luk KC, Kehm V, Carroll J et al. [2012] *Science* 338:949–953. With permission from AAAS.)

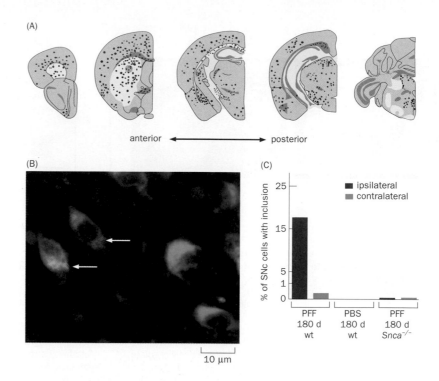

11.12 Mitochondrial dysfunction is central to the pathogenesis of Parkinson's disease

Prior to molecular-genetic studies of familial PD in the 1990s, Parkinson's disease was thought to be environmentally induced. A striking example came in the early 1980s, when MPTP (1-methyl-4-phenyl-1,2,3,6-tetrahydropyridine), a contaminant in the chemical synthesis of the opioid-like drug MPPP (1-methyl-4-phenyl-propionoxypiperidine), caused PD-like symptoms in a number of drug users. Subsequent animal and biochemical studies confirmed MPTP toxicity and established the mechanisms: MPTP freely crosses the BBB and is converted to MPP⁺ (1-methyl-4-phenylpyridinium) by **monoamine oxidase**, which normally oxidizes monoamine neurotransmitters leading to their degradation. MPP⁺ is selectively accumulated via the plasma membrane dopamine transporter in dopamine neurons, where it potently inhibits mitochondrial complex I function, thus selectively killing the dopamine neurons (**Figure 11–18**). Biochemical assays of postmortem tissue from sporadic PD patients have also indicated complex I deficiency, thereby implicating mitochondrial defects in dopamine neurons as a cause for PD.

While the majority of PD cases are sporadic, human genetic studies since the 1990s have revealed mutations in multiple genes that are associated with familial PD. In addition to the dominant α-synuclein mutations discussed above, familial PD can also be caused by autosomal recessive mutations, that is, disease occurs only if mutations in both alleles lead to loss of gene function (see Box 11–3 for more details). PD-linked recessive mutations have been identified in two evolutionarily conserved genes: *Pink1*, which encodes a mitochondrion-associated kinase, and *Parkin*, which encodes an enzyme in the ubiquitin-proteasome system for protein degradation. The relationship between *Pink1* and *Parkin* was first revealed by studying their homologs in the fruit fly *Drosophila*. *Pink1* mutant flies could not fly, died young, and exhibited degeneration of muscles and dopamine neurons with abnormal mitochondrial morphology and function (**Figure 11–19**). *Parkin* mutant flies exhibited similar defects. Notably, defects in *Pink1* mutants were rescued by overexpression of Parkin, but defects in *Parkin* mutants were unaffected by overexpression of Pink1. These data suggest that Pink1 and Parkin act in a common pathway to regulate mitochondrial function, with Parkin acting downstream of Pink1. This relationship has subsequently been confirmed in mammalian systems. Further studies in *Drosophila* and mammals suggest that

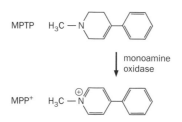

Figure 11–18 PD-like symptoms caused by a chemical toxin. After passing through the blood–brain barrier, MPTP (1-methyl-4-phenyl-1,2,3,6-tetrahydropyridine) is converted to MPP⁺ (1-methyl-4-phenylpyridinium) by monoamine oxidase. MPP⁺ selectively accumulates in dopamine neurons through the action of the plasma membrane dopamine transporter, inhibits mitochondrial function, and kills the dopamine neurons.

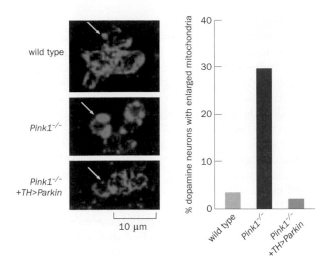

Figure 11–19 Pink1 and Parkin act in a common pathway in mitochondrial function. Clusters of *Drosophila* dopamine neurons visualized by expressing a mitochondrion-targeted GFP transgene under the control of the tyrosine hydroxylase (TH) promoter. In each panel, an arrow points to an individual mitochondrion. In the *Pink1* mutant (*Pink1$^{-/-}$*), mitochondria were abnormally enlarged compared with wild type. This mitochondrial defect in *Pink1* mutant dopamine neurons was rescued by overexpression of Parkin, driven by the TH promoter (*TH>Parkin*). Results are quantified at right. These data suggest that Pink1 acts upstream of Parkin in the same pathway. (Adapted from Park J, Lee SB, Lee S et al. [2006] *Nature* 441:1157–1161. With permission from Macmillan Publishers Ltd. See also Clark IE, Dodson MW, Jiang C et al. [2006] *Nature* 441:1162–1166 and Yang Y, Gehrke S, Imai Y et al. [2006] *Proc Natl Acad Sci USA* 103:10793–10798.)

the Pink1/Parkin pathway regulates multiple aspects of mitochondrial dynamics, including mitochondrial fission, fusion, and movement along microtubules, as well as removal of damaged mitochondria. Studies of Pink1 and Parkin reinforced a central role for mitochondrial dysfunction in Parkinson's disease pathogenesis.

In addition to α-synuclein, Parkin, and Pink1, mutations in several other genes have been identified in familial PD, together accounting for 10–15% of PD cases. The relationships among these different familial PD genes are the focus of intensive investigation. Given our discussions above implicating α-synuclein and mitochondrial dysfunction in both familial and sporadic PD, some of the outstanding questions include: Does α-synuclein aggregation cause mitochondrial dysfunction? Does mitochondrial dysfunction cause α-synuclein aggregation? Do these two processes act synergistically? How might they eventually lead to the death of dopamine neurons?

11.13 Treating Parkinson's disease: L-dopa, deep brain stimulation, and cell-replacement therapy

Despite the molecular complexity of Parkinson's disease, the selective loss of substantia nigra dopamine neurons as a common outcome suggested a possible treatment strategy: boosting dopamine levels. Starting in the 1960s, the successful use of **L-dopa** to treat PD symptoms has inspired generations of researchers and clinicians to find better treatments for PD and other devastating brain disorders.

Dopamine is synthesized from the amino acid L-tyrosine in two enzymatic steps (**Figure 11–20**). The first step, catalyzed by **tyrosine hydroxylase**, converts L-tyrosine to L-dopa, which is then acted upon by aromatic L-amino acid decarboxylase to yield dopamine. Tyrosine hydroxylase is the rate-limiting enzyme in the biosynthesis of **catecholamines**, a class of molecules that includes the neurotransmitters dopamine, norepinephrine, and epinephrine (Figure 11–20; see also Section 3.11; epinephrine also acts as a hormone, see Section 3.19). First known as an intermediate precursor for norepinephrine and epinephrine, dopamine was not recognized as a neurotransmitter until the late 1950s. Shortly thereafter, it was discovered that dopamine levels were markedly reduced in the striatum of postmortem PD brains, suggesting a selective defect of the dopamine system in PD.

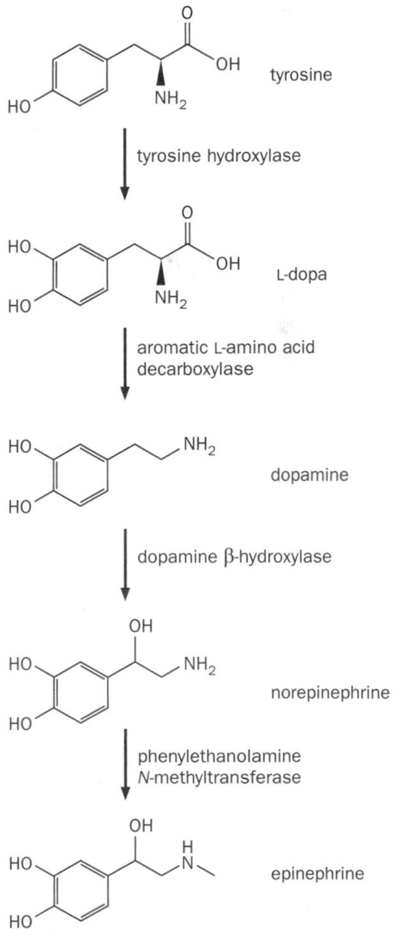

Figure 11–20 Biosynthetic pathway of catecholamines. Tyrosine hydroxylase is the rate-limiting enzyme in the pathway. Whereas dopamine does not cross the blood–brain barrier, L-dopa does. Injection of L-dopa has been used to boost the synthesis of dopamine by the remaining dopamine neurons as a therapy for dopamine deficiency in Parkinson's disease.

In Parkinson's disease, loss of dopamine neurons causes a decline in dopamine release. Thus, one strategy to increase dopamine release by the remaining dopamine neurons is to bypass the rate-limiting step of dopamine synthesis. It was found that dopamine cannot effectively cross the blood–brain barrier, but its immediate precursor L-dopa can. Through trial and error to optimize the therapeutic dose and reduce side effects, a treatment protocol was developed in the 1960s that is still widely used today: L-dopa administration drastically improves movement control in most early-stage PD patients.

Unfortunately, L-dopa is only effective at ameliorating PD symptoms for a few years. The eventual decline in L-dopa efficacy is likely because it must be converted to dopamine by the remaining dopamine neurons, whose progressive death is not halted by the treatment. An alternative treatment strategy is **deep brain stimulation** (**DBS**), which is designed to compensate for the alteration of circuit dynamics in PD due to the loss of dopamine modulation and excessive output of GPi/SNr (see Figure 11–16B). For this treatment, electrodes are surgically implanted to stimulate neurons and axons in specific nuclei. The exact mechanisms by which deep brain stimulation affects the basal ganglia circuitry in the PD brain are likely to be complex. Excitatory and inhibitory neurons as well as axons-in-passage are all stimulated simultaneously, and firing of target neurons can be affected in opposite directions depending on stimulation frequency and distance to the electrode. Clinically, deep brain stimulation of neurons and axon fibers in the STN or GPi can alleviate movement-related symptoms in late-stage PD patients when L-dopa treatment becomes less effective. In the case of STN stimulation, DBS likely causes inhibition of STN output, thus counteracting the increased excitatory input and decreased inhibitory input to GPi/SNr in PD patients (see Figure 11–16B); indeed, earlier studies had shown that STN lesion or high-frequency stimulation could alleviate PD-like symptoms in monkeys treated with MPTP, providing an important basis for human clinical trials. The success of DBS in treating PD has inspired its clinical trials for treating a number of psychiatric disorders such as depression and obsessive-compulsive disorders, even though there is far less knowledge about the underlying circuitry in these psychiatric disorders compared with PD.

A radically different strategy for PD treatment is to replace dying dopamine neurons with new dopamine neurons. This **cell-replacement therapy** has several requirements. First, a reliable source of dopamine neurons must be identified. Additionally, transplanted dopamine neurons must survive in the host, have access to their targets, and release appropriate levels of dopamine. Studies in animal PD models have shown that fetal tissues derived from midbrain areas that contain dopamine neurons can survive after being grafted into the host striatum. (To bypass the requirement for correct axonal projections to targets, dopamine neurons are usually transplanted directly to the striatum.) Remarkably, these grafted cells can release dopamine and improve motor control. Subsequently, small clinical trials in humans have reported improvement of clinical symptoms and long-term increase in dopamine release after transplantation (**Figure 11–21**).

There are several limitations to using human fetal tissue in cell-replacement therapy. Large numbers of dopamine neurons are required for effective therapy,

Figure 11–21 Embryonic dopamine neuron transplantation as a treatment of Parkinson's disease. (A) PD patients fluctuate between 'off' periods when motor function is severely impaired, and 'on' periods when motor function is relatively normal. This figure shows the self-report of a PD patient describing the fraction of awake time spent in off periods (red trace; scale on left axis) and the number of off periods per day (blue trace; scale on right axis) in the months before and after receiving transplanted dopamine neurons derived from fetal tissue. (Time of transplantation is indicated by the dashed line.) The duration and frequency of off periods are markedly reduced after transplantation. **(B)** Positron emission tomography scan of radioactive fluorodopa uptake into dopamine neuron terminals in the striatum of a normal subject (left) and a PD patient before (middle) and 12 months after (right) transplantation of fetal dopamine neurons. Before surgery, radioactivity (shown in red) in the brain of the PD patient was restricted to the caudate (medial striatum); after the bilateral transplantation, radioactivity was extended to the putamen (lateral striatum), more similar to the distribution in the normal control. (A, adapted from Lindvall O, Brundin P, Widner H et al. [1990] *Science* 247:574–577; B, from Freed CR, Greene PE, Breeze RE et al. [2001] *N Engl J Med* 344:710–719. With permission from the Massachusetts Medical Society.)

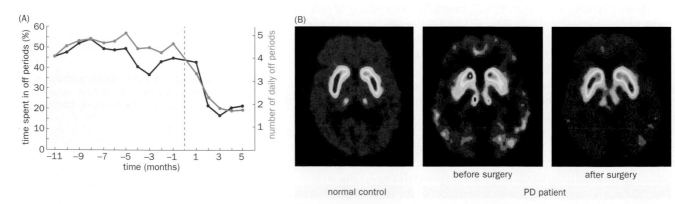

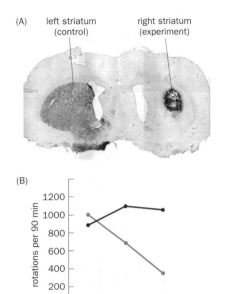

(A) left striatum (control)   right striatum (experiment)

(B)

Figure 11–22 Transplantation of dopamine neurons derived from induced pluripotent stem (iPS) cells can improve motor function in animal models.
(A) Coronal section of a rat brain at the level of the striatum. The right hemisphere lacks endogenous dopamine neurons as a result of chemical ablation, and serves as a host for transplantation of iPS-derived dopamine neurons. The left hemisphere serves as a control. Tyrosine hydroxylase staining (dark signal) shows that transplanted iPS-derived dopamine neurons survive and produce tyrosine hydroxylase 4 weeks after transplantation. **(B)** Application of a drug that enhances dopamine action induces rotation in animals in which dopamine neurons were ablated in one hemisphere (red trace); this is caused by asymmetric activation of striatal circuits in two hemispheres. This defect was ameliorated after transplanting iPS-derived dopamine neurons (blue trace). (Adapted from Wernig M, Zhao JP, Pruszak J et al. [2008] *Proc Natl Acad Sci USA* 105:5856–5861. Copyright National Academy of Sciences, USA.)

and these neurons need to be free of other cell types. Contamination by additional cell types is a major side effect of fetal transplantation that has prevented further clinical development. Moreover, optimal survival of transplanted tissues from different individuals requires patients to take immunosuppressant drugs, which can compromise their immune system and increase susceptibility to infection.

Recent progress in stem cell research has provided an exciting potential means by which large numbers of dopamine neurons might be produced *in vitro*, in some cases by using somatic cells derived from the patients themselves and thereby avoiding the problem of tissue rejection. For example, embryonic stem cells and induced pluripotent stem cells derived from fibroblasts (see **Box 11–2**) can differentiate into tyrosine hydroxylase-positive dopamine neurons *in vitro*. These induced dopamine neurons can survive after being grafted into host striatum that lacks endogenous dopamine neuron projections and can improve motor function of the recipient animal (**Figure 11–22**). Although much work is needed to assess the safety, reliability, and robustness of this approach, cell-replacement therapy may offer a promising avenue for treating Parkinson's disease in the future.

Box 11–2: Producing neurons from embryonic stem cells, induced pluripotent cells, and fibroblasts

As we learned in Section 7.1, different neuronal types are produced from progenitors that are located in defined parts of the neuroepithelia and are specified by signals that pattern the nervous system. The ectodermal progenitors from which neural progenitors originate derive from **pluripotent cells**, which can give rise to all cell types in the embryo. These pluripotent cells, called **embryonic stem (ES) cells**, can be cultured *in vitro*. By overexpressing specific genes and culturing ES cells with specific growth factors under conditions mimicking development *in vivo*, researchers can coax ES cells to differentiate in culture into specific types of neurons, including dopamine neurons that express tyrosine hydroxylase and release dopamine (**Figure 11–23**, left). These *in vitro* differentiated neurons can in principle be used for cell-replacement therapies. The drawbacks associated with ES cell-based therapy include ethical difficulties

related to the use of human embryos and potential tissue rejection after transplantation into patients.

A major breakthrough came in 2006, when researchers reported the ability to convert embryonic and adult fibroblasts into **induced pluripotent stem (iPS) cells** by forced expression of four transcription factors involved in maintaining the pluripotency of ES cells (Figure 11–23, middle top). These iPS cells resemble ES cells in many ways—they can be induced to differentiate into many different cell types *in vitro*, and they can support germ line transmission after transplantation into blastocysts just like ES cells (see Section 13.7). Fibroblast-derived iPS cells can also be induced to re-differentiate into dopamine neurons (Figure 11–23, middle bottom) and, when transplanted into the striatum, can ameliorate Parkinson-like symptoms in rodent models

(Continued)

Box 11–2: Producing neurons from embryonic stem cells, induced pluripotent cells, and fibroblasts

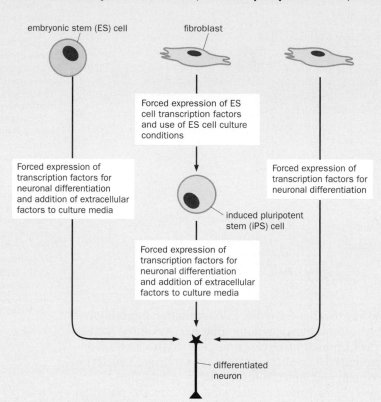

Figure 11–23 Multiple ways of producing differentiated neurons *in vitro*. Left, embryonic stem (ES) cells that are forced to express specific transgenes can be induced to become specific neuronal types by providing required extracellular factors and defined cell-culture conditions. For example, dopamine neurons can be produced from ES cells that first overexpress a transcription factor essential for dopamine neuron differentiation and subsequently undergo a multi-stage culture protocol with specific growth factors (see Kim JH, Auerbach JM, Rodriguez-Gomez JA et al. [2002] *Nature* 418:50–56). Middle, fibroblasts can be de-differentiated into iPS cells by forced expression of a cocktail of transcription factors normally expressed in ES cells (see Takahashi K & Yamanaka S [2006]

Cell 126:663–676). The resulting iPS cells can then be induced to re-differentiate into dopamine neurons following a protocol similar to the ES cell → dopamine neuron path shown on the left (see Wernig M, Zhao JP, Pruszak J et al. [2008] *Proc Natl Acad Sci USA* 105:5856–5861). Right, fibroblasts can be converted directly to neurons by transfecting a cocktail of neuronal differentiation factors (see Vierbuchen T, Ostermeier A, Pang ZP et al. [2010] *Nature* 463:1035–1041), including dopamine neurons (see Caiazzo M, Dell'Anno MT, Dvoretskova E et al. [2011] *Nature* 476:224–227 and Pfisterer U, Kirkeby A, Torper O et al. [2011] *Proc Natl Acad Sci USA* 108:10343–10348).

(see Figure 11–22). In principle, if patients carry familial PD mutations, such mutations can also be corrected at the iPS-cell stage by replacing the mutant gene with a wild-type copy through homologous recombination (see Section 13.7) before expansion and re-differentiation.

More recently, researchers described methods to transdifferentiate fibroblasts directly into neurons by over-expressing a set of transcription factors known to be important for neuronal differentiation, bypassing the de-differentiation and re-differentiation procedures with iPS cells as an intermediate (Figure 11–23, right). Transcription factor cocktails have been identified that can induce fibroblasts from healthy subjects and PD patients to differentiate into specific types of neurons, including dopamine neurons. These methods pave the way toward the eventual goal of using patient-derived cells to produce neurons for cell-replacement therapy, which would avoid the problem of transplant rejection. Compared to these direct induction procedures, iPS-cell-based strategies benefit from easy proliferation of iPS cells *in vitro* and can therefore generate

large cell populations for cell-replacement therapy. On the other hand, some iPS cells may have the potential to produce tumors after transplantation due to their pluripotency and proliferation capacity. Researchers are optimizing these procedures for production efficiency, effectiveness, and safety.

Neurons produced from fibroblasts *in vitro*, whether through direct induction or via iPS cells, have wide applications in disease research far beyond cell-replacement therapies. For example, neurons derived from patients with specific brain disorders can be examined, in culture or *in vivo* after transplantation into animal models, for potential defects in morphological development, electrophysiological characteristics, synaptic transmission, and synaptic plasticity. Once specific phenotypes have been identified, researchers can use *in vitro* assays, including high-throughput drug screens, to identify strategies for correcting the phenotypes. Drugs that improve the phenotypes in culture can serve as prime candidates for testing in animal models and for clinical trials (see Box 11–1).

11.14 The various neurodegenerative diseases have common themes and exhibit unique properties

Despite being associated with distinct proteins and disease symptoms, a broad suite of neurodegenerative diseases—including AD, prion diseases, polyQ diseases, ALS, and most forms of PD—share in common the abnormal aggregation of misfolded proteins or cleaved fragments. Familial mutations tend to facilitate such aggregation. These protein aggregates, or their intermediates, are either toxic by themselves or alter the localization or function of their normal interacting partners, thus disrupting protein homeostasis and causing toxic gain-of-function phenotypes. In some cases, loss-of-function of the misfolded protein may further exacerbate the gain-of-function effects. Much is to be learned about how misfolding occurs, what structure(s) define the toxic species, and which downstream effects are specific to each disease.

In the case of the prion, PrP in the pathogenic conformation (PrP^{Sc}) serves as a seed to convert normal PrP^C into additional pathogenic PrP^{Sc}, causing the disease to spread and become infectious. Although no other neurodegenerative disorders are known to be infectious, the concept of seeding-induced conformational changes that result in misfolded protein aggregates and cell-to-cell spread of misfolded proteins may apply to other diseases such as PD (Section 11.11) and contribute to their progression. Much remains to be investigated about the mechanisms of cell-to-cell spread and the factors that promote or inhibit such spread.

One property that distinguishes different diseases is the neuronal types that degenerate in each disease. AD causes widespread degeneration encompassing many types of neurons in the cerebral cortex, hippocampus, and amygdala. HD primarily causes striatal neuron degeneration. ALS preferentially affects motor neurons. PD results from degeneration of dopamine neurons in the substantia nigra, at least initially. Many of the causal genes that are mutated in familial forms of the diseases, such as APP and presenilins in AD, PrP in prion diseases, huntingtin in HD, SOD1 and TDP-43 in ALS, and α-synuclein in PD, are ubiquitously expressed. It remains largely a mystery how mutations in these widely expressed genes primarily damage specific neuronal cell types, thus causing specific diseases. One contributing factor could be that each disease is caused by aggregated proteins interacting with and interrupting the function of a unique set of partners that are preferentially required in specific neuronal types.

Sadly, the development of effective therapies for neurodegenerative diseases has thus far been limited to early-stage PD. However, approaches being pioneered in PD research and intense ongoing research discussed in previous sections may ultimately yield successful treatments for a broader range of neurodegenerative diseases.

PSYCHIATRIC DISORDERS

Disorders of the nervous system have traditionally been divided into neurological and psychiatric. Neurological disorders are usually associated with structural, biochemical, or physiological symptoms, as in the neurodegenerative diseases we studied. By contrast, psychiatric disorders have historically included those that affect the mind—how we perceive, feel, think, and act—without established physical basis. As the brain and the mind are inseparable and as we gain more understanding about both, the distinctions between neurology and psychiatry become increasingly blurred and somewhat arbitrary. However, traditionally defined psychiatric and neurological disorders have historically been studied using different approaches. For example, studies of neurodegenerative diseases, traditionally considered neurological disorders, start with pathology. Scientists try to understand the mechanisms underlying pathological changes, in the hope of designing treatments to interfere with the pathogenic process. By contrast, most therapeutic drugs for psychiatric disorders have been discovered through fortunate chance. By studying how such drugs act, researchers attempt to uncover the mechanisms that may underlie these disorders. Below, we illustrate this path of discovery for four classes of psychiatric disorders: schizophrenia, mood disorders, anxiety disorders, and addiction.

11.15 Schizophrenia can be partially alleviated by drugs that interfere with dopamine function

Schizophrenia is among the most costly psychiatric disorders, with a lifetime prevalence of 1% in the general population. The onset is usually during adolescence or early adulthood, and patients are typically affected for the rest of their lives. The most typical symptoms include hallucinations and delusions, often leading to paranoia. These psychotic disturbances (**psychosis**) are the symptoms most commonly associated with the disorder and are characterized as positive symptoms, referring to their presence in patients but not in healthy people. Schizophrenia is also associated with a set of negative symptoms including social withdrawal and lack of motivation, as well as cognitive impairment in memory, attention, and executive functions. The negative symptoms and cognitive impairments are usually more disabling to patients' quality of life because there is currently no treatment.

Before the 1950s, there was no treatment for schizophrenia other than confining patients to mental asylums, often for decades. Then came the fortuitous discovery of the first antipsychotic drugs: chlorpromazine and reserpine. **Chlorpromazine** was chemically synthesized as a potential anesthetic, and **reserpine** is the active ingredient purified from the snakeroot plant for treating hypertension. Both drugs were found to alleviate positive symptoms of schizophrenia patients, albeit with similar side effects, namely motor control deficits similar to those of Parkinson's disease.

Subsequent studies showed that reserpine acts by interfering with the metabolism of all three monoamine neurotransmitters: dopamine, norepinephrine, and serotonin (**Figure 11–24**A; see also Section 3.11 and Figure 3–16). After release into the synaptic cleft, these monoamine neurotransmitters are taken back into the presynaptic terminal by specific **plasma membrane monoamine transporters** (**PMATs**, see Section 3.8 and Section 11.16) of neurotransmitters. A **vesicular monoamine transporter** (**VMAT**) shared by all three monoamines then transports these neurotransmitters from the cytosol into synaptic vesicles for the future rounds of synaptic transmission. Reserpine acts as an inhibitor of VMAT, thus blocking monoamine neurotransmitter recycling. Monoamine transmitters retained in the cytosol after VMAT blockade are inactivated by the

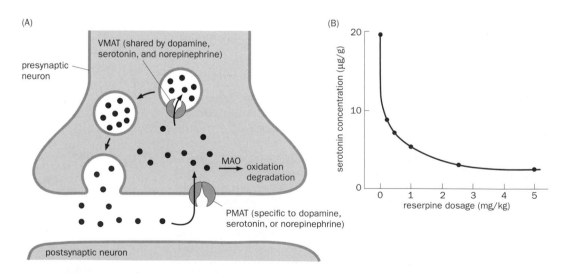

Figure 11–24 Metabolism of monoamine neurotransmitters in the presynaptic terminal and the effect of reserpine. (A) Schematic drawing of monoamine neurotransmitter metabolism in the presynaptic terminal. After release to a synaptic cleft, each monoamine neurotransmitter is taken up by its specific plasma membrane monoamine transporter (PMAT) to the presynaptic cytosol. There, neurotransmitter molecules are either taken up by the vesicular monoamine transporter (VMAT) into synaptic vesicles for reuse, or oxidized by the monoamine oxidase (MAO) for degradation. **(B)** Reserpine's effect was first discovered based on its ability to deplete serotonin levels. In this experiment, the level of serotonin released from the intestine (where serotonin is used as a major neurotransmitter in the enteric nervous system) was found to decrease progressively after rabbits received increasing amount of reserpine. Reserpine exerts this effect by inhibiting VMAT. (Adapted from Pletscher A, Shore PA & Brodie BB [1955] *Science* 122:374–375.)

monoamine oxidase enzyme, which removes the amine group and thereby allows its products to be further metabolized (Figure 11–24A). By inhibiting VMAT, reserpine effectively depletes the levels of dopamine, norepinephrine, and serotonin (Figure 11–24B). Indeed, depletion of norepinephrine in the sympathetic neurons that promote cardiac muscle contraction (see Section 8.12) underlies reserpine's effect on hypertension. Its ability to deplete dopamine also explains why reserpine treatment causes Parkinson-like symptoms.

Chlorpromazine acts differently. In the 1970s, competitive binding assays were established to test drug effectiveness: a given drug would compete against a particular neurotransmitter for binding to brain membrane extracts presumed to contain high concentrations of neurotransmitter receptors (**Figure 11–25**A). When chlorpromazine and other antipsychotic drugs were tested in such competitive binding assays against different neurotransmitter receptors, it was found that a drug's affinity for dopamine receptors (but not for serotonin or adrenergic receptors) correlated well with its effectiveness as an antipsychotic (Figure 11–25B). Subsequent work suggested that the blockade of a specific dopamine receptor subtype, the D_2 dopamine receptor, correlated best with antipsychotic effects. Unfortunately, blocking the D_2 dopamine receptor also leads to Parkinson-like symptoms. New generations of antipsychotic drugs with milder side effects on movement control have been developed, although all of these have significant metabolic side effects.

The fact that reserpine and chlorpromazine both reduce dopamine function through distinct mechanisms suggested that dopamine system abnormalities contribute to the positive symptoms of schizophrenia. This hypothesis was further supported by cases of drug-induced psychosis. Two commonly abused drugs, cocaine and amphetamine, are known to induce delusional paranoia similar to the positive symptoms of schizophrenia. These **psychostimulants** produce transient euphoria and suppress fatigue, but are potently addictive because they modulate the brain's reward system. As we will learn in Section 11.18, both amphetamine and cocaine increase dopamine levels at their target sites. As in schizophrenia, the positive symptoms induced by these psychostimulants can be effectively treated with antipsychotic drugs that reduce dopamine receptor function. Additional support for the involvement of dopamine in schizophrenia came

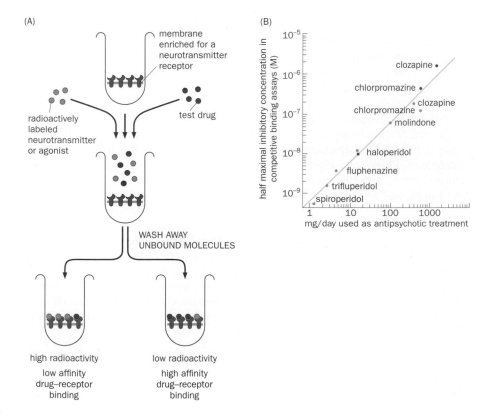

Figure 11–25 Competitive binding assay used to test drug action.
(A) Illustration of the competitive binding assay to determine whether a drug binds specifically to receptors for a given neurotransmitter or agonist. A fixed amount of radioactively labeled neurotransmitter (or known agonist for the receptor) and variable amounts of drugs are used for competitive binding (a 1:1 ratio of the drug and agonist is illustrated for simplicity). Retention of a large fraction of the radioactivity by the receptors indicates low affinity of the drug to the receptor (left), whereas retention of a small fraction of radioactivity indicates that the drug outcompetes the radiolabeled neurotransmitter or agonist and binds the receptor with greater affinity (right). **(B)** The efficacy of various antipsychotic drugs, as measured by the amount required for effective treatment, correlates well with their ability to compete with radioactively labeled dopamine (blue data points) or haloperidol (itself an antipsychotic drug; red data points) for binding to dopamine receptors from brain extract. The diagonal line represents a 1:1 relation between the blocking molarity and the clinical dose. (Adapted from Seeman P, Chau-Wong M, Tedesco J et al. [1975] *Proc Natl Acad Sci USA* 72:4376–4380.)

Figure 11–26 Self-portrait by Vincent van Gogh. A brilliant artist, van Gogh suffered from illnesses that may have included bipolar disorder—he was prolific during normal and possibly manic phases, but committed suicide at the age of 37, probably during a depressive episode.

iproniazid

imipramine

fluoxetine (Prozac)

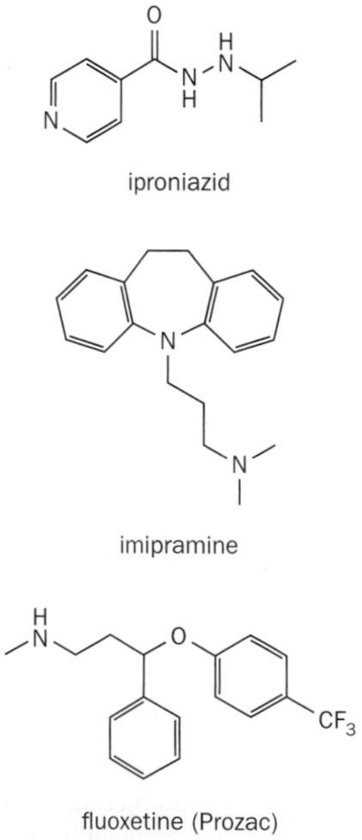

Figure 11–27 Structures of representative antidepressants. Resembling the structure of monoamines (see Figure 3–16), iproniazid inhibits monoamine oxidase, whereas imipramine inhibits the plasma membrane transporters for serotonin and norepinephrine, and fluoxetine selectively inhibits the plasma membrane transporter for serotonin.

from PET imaging studies that correlated acute psychotic states with increased dopamine levels.

Current antipsychotic drugs are not effective for treating about a third of schizophrenia patients, suggesting a considerable degree of heterogeneity in the disorder. For those that respond, antipsychotic drugs are only effective in reducing positive symptoms, and have no effect on negative symptoms or cognitive impairment. This is likely because schizophrenia alters more than just the dopamine system. Indeed, drugs that act as NMDA receptor antagonists, such as phencyclidine (PCP) and ketamine, also induce psychosis that resembles schizophrenia, with associated negative symptoms, leading to the proposal that a reduction of NMDA receptor function also contributes to schizophrenia.

Although informative, studies investigating drug action have not revealed the root causes of schizophrenia. Recent structural magnetic resonance imaging studies indicate that schizophrenia is associated with significant thinning of the cerebral cortex. The most affected cortical areas include the **prefrontal cortex**, an executive control center that integrates multisensory information, processes working memory, and performs complex executive functions such as goal selection and decision making. Cortical thinning has been suspected to be due to excessive synaptic pruning that is normally associated with cortical development (see Section 7.14), suggesting a neurodevelopmental origin of schizophrenia. Indeed, prior to first psychosis, schizophrenia patients usually already exhibited considerable social, mood, and cognitive impairment collectively known as schizophrenia prodrome. As will be discussed in Section 11.19, schizophrenia has a strong genetic contribution. Identifying genetic factors and studying their mechanisms of action may shed more light on the causes of this calamitous mental disorder and thereby suggest more effective therapeutic strategies.

11.16 Mood disorders have been treated by manipulating monoamine neurotransmitter metabolism

We all experience moments of happiness and sadness. However, a sizable fraction of the general population suffers from mood disorders that at times appear to take over their lives. Mood disorders fall into two major categories: bipolar disorder and major depression. Individuals with **bipolar disorder** swing between manic and depressive phases. During the manic phase, patients feel grandiose and tireless; this is then interrupted by the depressive phase of feeling sad, empty, and worthless. The artist Vincent van Gogh (**Figure 11–26**) is one of many historical figures who are suspected to have suffered from bipolar disorder, which has a lifetime prevalence of 1%. **Major depression** has only the depressive phase (and is thus also termed unipolar), and is more common than bipolar disorder, with a lifetime prevalence of more than 5%. Both of these mood disorders can be life threatening, as they account for a large fraction of suicides.

Like the first antipsychotics, the first antidepressant was discovered serendipitously in the 1950s. **Iproniazid** (**Figure 11–27**) was originally introduced to treat tuberculosis. Physicians reported that iproniazid-treated patients were happier despite the drug's ineffectiveness as a tuberculosis treatment. This and other clues from animal studies led to trials of iproniazid for depression, with the finding that the drug significantly improved patients' depressive states. Further studies showed that iproniazid acts by inhibiting monoamine oxidase (see Figure 11–24A), hence increasing the concentration of monoamines in presynaptic terminals and synaptic clefts. These findings suggest that elevation of one or more of the monoamine neurotransmitters—serotonin, norepinephrine, or dopamine—may have a therapeutic effect on depression. However, because monoamine oxidase inhibitors indiscriminately increase monoamine levels, they have many side effects. Over the years, they have been replaced by tricyclic antidepressants such as **imipramine**, named for their characteristic three-ring molecular structures (Figure 11–27).

Imipramine was synthesized as a variant of chlorpromazine with the hope of identifying a more effective antipsychotic. Although imipramine had no effect on treating psychosis, it had a pronounced effect on depression. Further studies led to the discovery that imipramine and other tricyclic antidepressants inhibit

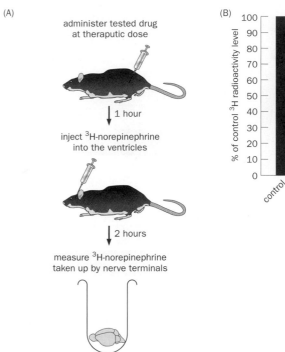

(A) administer tested drug at theraputic dose

1 hour

inject ³H-norepinephrine into the ventricles

2 hours

measure ³H-norepinephrine taken up by nerve terminals

Figure 11–28 Antidepressants inhibit norepinephrine uptake by the brain. (A) Assay procedure. Because norepinephrine does not cross the blood–brain barrier, it was injected into ventricles to access neurons. ³H-norepinephrine molecules retained in the brain 2 hours after injection indicated nerve terminal uptake. Those that were not taken up by nerve terminals were presumably to be metabolized. **(B)** Compared to controls, application of three clinically effective antidepressants markedly reduced brain uptake of radioactively labeled norepinephrine. In the same assay, clinically ineffective antidepressants did not inhibit norepinephrine uptake (not shown). (Adapted from Glowinski J & Axelrod J [1964] *Nature* 204:1318–1319. With permission from Macmillan Publishers Ltd.)

the plasma membrane monoamine transporters (PMATs; see Figure 11-24A) that allow neurotransmitter reuptake from presynaptic terminals. Originally discovered by studying the action of norepinephrine on the targets of sympathetic nerves, this pump-like reuptake turns out to be a general mechanism by which the action of neurotransmitters (particularly monoamines) is terminated. The effect of various drug treatments on reuptake can be determined using a quantitative assay based on how much experimentally administered radioactively labeled norepinephrine brain tissues retain (**Figure 11-28**A). Antidepressants such as imipramine reduced the norepinephrine level retained in brain tissues compared with controls and with other drugs that lack antidepressant effects (Figure 11-28B), indicating that imipramine inhibits norepinephrine reuptake.

Each monoamine neurotransmitter has its own PMAT encoded by a distinct gene. Inhibiting its reuptake system prolongs a neurotransmitter's actions. Imipramine affects PMATs for norepinephrine as well as for serotonin. Drugs developed subsequently, such as **fluoxetine** (brand name Prozac; Figure 11-27) block serotonin reuptake selectively. These **SSRIs** (**selective serotonin reuptake inhibitors**) are the most widely used antidepressants today. Thus, enhancing the actions of monoamine neurotransmitters, notably serotonin, can have significant effects in relieving depressive states. As discussed in Box 8-1, serotonin (and norepinephrine) neurons are clustered in the brainstem nuclei, but their axons project throughout the central nervous system from the forebrain to the spinal cord, enabling these neurons to modulate many excitatory and inhibitory target neurons. The primary target neurons and circuits relevant for mood regulation remain to be elucidated.

11.17 Modulating GABAergic inhibition can alleviate symptoms of anxiety disorders

Anxiety disorders, the most prevalent class of psychiatric disorders, include generalized anxiety, various kinds of phobias and panic disorders caused by irrational fear, and obsessive-compulsive disorder (OCD). Generalized anxiety disorder alone has a lifetime prevalence of more than 5%; patients with this disorder exhibit persistent worries about impending misfortunes, often with physical symptoms such as fatigue, muscle tension, and sleep disturbance.

Barbiturates and their derivatives were the earliest classes of drugs to treat anxiety disorders. Barbiturates are also potent sedatives, and a more serious

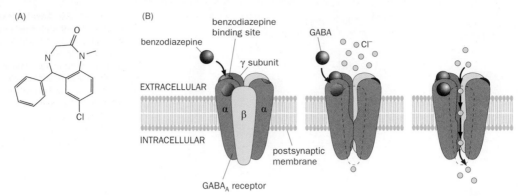

Figure 11–29 Benzodiazepines act as allosteric agonists on the GABA_A receptor. (A) Structure of diazepam (Valium), a benzodiazepine. **(B)** Schematic of benzodiazepine action as an allosteric agonist for the GABA_A receptor. Here, the GABA_A receptor pentamer consists of two α subunits, two β subunits, and one γ subunit, the most commonly occurring subunit composition. Binding of a benzodiazepine to the GABA_A receptor at the interface of α and γ subunits does not open the channel (left), but enhances the receptor's affinity for GABA. The GABA_A receptor channel is opened by binding of two GABA molecules at the interface of the α and β subunits (middle and right). To visualize the channel opening, the β subunit at the front is removed in the middle and right panels.

concern is that overdose is lethal. Once again, fortuitous discoveries since the 1950s produced a new class of molecules called **benzodiazepines** (**Figure 11–29**A), which are effective in relieving anxiety symptoms but which have a reduced sedating effect. Importantly, benzodiazepine overdose induces lengthy sleep, but the lethal dose is much higher than that of barbiturates. Benzodiazepines have therefore largely replaced the use of barbiturates today.

Interestingly, barbiturates and benzodiazepines act on the same types of molecule to exert their anxiolytic (anxiety-reducing) effects: both classes of drugs bind to the ionotropic GABA_A receptors and enhance GABA transmission (see Section 3.17). Barbiturates, benzodiazepines, and GABA have distinct binding sites on the GABA_A receptors. At high concentrations, barbiturates can activate the GABA_A receptor independent of GABA, causing chloride influx through the GABA_A channels and hyperpolarization of target neurons. In contrast, benzodiazepines enhance the receptor's affinity for GABA (as well as for GABA agonists such as barbiturates or alcohol) but do not activate the GABA_A receptor by itself. Thus, benzodiazepines act as **allosteric agonists** by enhancing the action of endogenous GABA (Figure 11–29B). This accounts for the relative safety of benzodiazepines: their maximal effect is limited by the amount of endogenous GABA.

Although benzodiazepines have fewer side effects than barbiturates, they still induce sedation. Is it possible to isolate the anxiolytic and sedative effects? Molecular-genetic studies of benzodiazepine action have offered some clues. The benzodiazepine-binding site is located at the interface of the α and γ subunits of the pentameric GABA_A receptor (Figure 11–29B; see also Figure 3–21). In humans and mice, the six separate genes that encode the GABA_A receptor subunits α1 through α6 are differentially expressed in the brain. Subunits α1, α2, α3, and α5 possess a specific histidine residue (located at position 101 of α1 and α2) that renders them sensitive to benzodiazepines, whereas α4 and α6 have an arginine at the same position, which disrupts benzodiazepine binding. In other words, benzodiazepines only affect brain regions that express high levels of α1-, α2-, α3- or α5-containing GABA_A receptors but not those that are enriched only for α4- or α6-containing GABA_A receptors. Replacing the conserved histidine (H) with arginine (R) can cause a benzodiazepine-sensitive subunit to become insensitive without affecting GABA_A receptor function. This property offered a means to determine the contributions of individual α subunits to the effects of benzodiazepines *in vivo*.

Using the genetic knock-in approach (see Section 13.7), researchers produced mice with a specific H101 to R101 substitution in the α1 subunit. These mice (designated as H101R) were no longer sedated by benzodiazepines but retained benzodiazepines' anxiolytic effects. In contrast, benzodiazepines could still sedate mice with the H101R substitution in the α2 subunit but did not show anxiolytic effects (**Figure 11–30**). These experiments thus suggested that in normal mice,

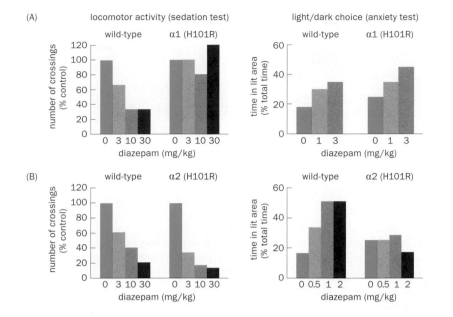

(A) locomotor activity (sedation test)   light/dark choice (anxiety test)

(B)

Figure 11–30 Selective functions of GABA_A receptor subunits α1 and α2 in promoting sedation and relieving anxiety. (A) Left, when the α1-subunit of the GABA_A receptor was rendered incapable of binding benzodiazepines by changing a key histidine residue to arginine (H101R), diazepam no longer induced sedation as assayed by locomotor activity, whereas locomotor activity of wild-type animals was reduced by diazepam in a dose-dependent manner. Right, benzodiazepine still had anxiolytic effects on mutant mice, as assayed by the time mice spent in the lit area of an open field. (B) When the α2 subunit of the GABA_A receptor was rendered incapable of binding to benzodiazepine by the H101R mutation, diazepam still induced sedation in these genetically modified mice as it did in wild type (left), while the anxiolytic effects of diazepam were abolished (right). In both panels, locomotor activity was measured by the number of line crossings mice made when they moved in an open field. Anxiety was measured by the amount of time mice spent in areas that were lit versus dark (anxious mice excessively avoid lit areas). Additional anxiety test using the elevated plus-maze gave similar results (not shown). See Section 13.29 for details about these behavioral assays. (A, adapted from Rudolph U, Crestani F, Benke D et al. [1999] *Nature* 401:796–800. With permission from Macmillan Publishers Ltd; B, adapted from Löw K, Crestani F, Keist R et al. [2000] *Science* 290:131–134.)

benzodiazepines promote sedation through α1-containing GABA_A receptors and relieve anxiety through α2-containing GABA_A receptors. These differential effects are likely caused by differential expression of α subunits in brain regions that control the functions of different brain circuits. Indeed, one of the highest α2-expressing areas is the amygdala, a center for regulating emotion and fear responses (see Section 10.23). Recent studies using optogenetic manipulations have begun to identify specific projections in the amygdala that mediate anxiolytic effects. In principle, drugs that specifically elevate the function of GABA_A receptors containing α2 (but not α1) should be more effective and specific anxiolytics than the benzodiazepines currently in use.

GABA is the major inhibitory neurotransmitter in the brain and mediates many diverse physiological functions. In addition to treating anxiety, drugs that affect the GABAergic system have been used to treat epilepsy (see Box 11–4), pain, and sleep problems. Studies of benzodiazepine action illustrate how investigating drug action can help us tease apart GABA's diverse functions, which in turn will help inform the design of drugs that are better targeted to treat specific disorders.

A limitation with benzodiazepines for treating anxiety disorders is that long-term use can lead to addiction (see Section 11.18). Increasing doses are required over time to achieve similar effects, and a halt of treatment results in withdrawal symptoms. SSRIs used to treat depression (see Section 11.16) also have anxiolytic effects but are not addictive, and are being used to treat certain anxiety disorders. It is not known how altering serotonin levels alleviates anxiety.

11.18 Addictive drugs hijack the brain's reward system by enhancing the action of VTA dopamine neurons

Addictive substances, such as alcohol from fermentation, opium from poppy plants, and cocaine from coca leaves, have been with human society for thousands of years. Successful chemical syntheses of active components and new methods for efficient delivery have both contributed to the recent rise in the prevalence of drug abuse. The result is a significant problem for humanity, spanning many nations and cultures. **Drug addiction** is defined as compulsive drug use despite long-term negative consequences. It is also associated with loss of self-control and propensity to relapse. What is the neurobiological basis of addiction?

Remarkably, almost all drugs of abuse have one common effect: they increase dopamine concentration at the output targets of **ventral tegmental area** (**VTA**) dopamine neurons, including at the VTA itself. Two major output targets of VTA dopamine neurons are the **nucleus accumbens** (ventral striatum),

Figure 11–31 Drugs of abuse increase dopamine concentration in the target areas of VTA dopamine neurons. A simplified circuit diagram illustrating connections between the ventral tegmental area (VTA), nucleus accumbens, prefrontal cortex, and other connected areas. Blue, dopamine neurons; red, GABAergic neurons; green, glutamatergic neurons. Also summarized are the action sites of most common drugs of abuse, all of which enhance dopamine concentrations in target areas. +, enhancement; −, suppression. For example, cocaine blocks dopamine reuptake from presynaptic terminals of dopamine neurons. Benzodiazepines enhance dopamine neuron firing by disinhibiting VTA GABAergic neurons. Nicotine causes increased release of glutamate at the terminals of excitatory input to dopamine neurons, and directly depolarizes dopamine neurons. (Based on Lüscher C & Malenka RC [2011] *Neuron* 69:650–663 and Sulzer D [2011] *Neuron* 69:628–649.)

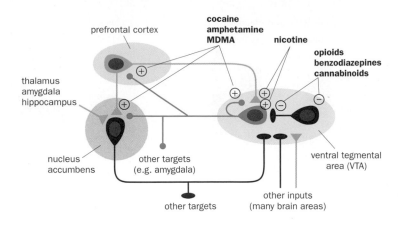

an area best known for processing reward information (see Section 10.24), and the prefrontal cortex responsible for executive functions such as goal selection and decision making (**Figure 11–31**). VTA dopamine neurons also receive inputs from many parts of the brain, including glutamatergic excitatory input from the prefrontal cortex and GABAergic inhibitory input from local and nucleus accumbens neurons.

Different drugs of abuse enhance dopamine action through distinct mechanisms. For example, nicotine enhances excitatory input onto VTA dopamine neurons by presynaptic excitation—it activates nicotinic ACh receptors (see Section 3.13) located on the glutamatergic presynaptic terminals, causing increased release of glutamate and hence greater excitation of dopamine neurons. Nicotine can also excite dopamine neurons directly through nicotinic ACh receptors on the dopamine neurons themselves. In contrast, opioids, benzodiazepines, and cannabinoids act by hyperpolarizing and thereby inhibiting local GABAergic neurons in the VTA, causing disinhibition of dopamine neurons. Ethanol is known to boost dopamine concentration at the VTA and nucleus accumbens, but the exact mechanisms and site(s) of action remain unclear. The psychostimulant drugs cocaine and amphetamine act by enhancing dopamine's effects at presynaptic terminals of dopamine neurons. Cocaine blocks the **plasma membrane dopamine transporter** (**DAT**) for dopamine reuptake, thus increasing the dopamine concentration in the synaptic cleft post-release. Amphetamines (including MDMA, or 3,4-methylenedioxy-*N*-methylamphetamine, commonly known as ecstasy) have more complex effects: (1) they reverse the normally unidirectional transport mediated by the DAT (that is, from synaptic cleft to presynaptic cytosol), causing presynaptic vesicle-independent release of dopamine into the synaptic cleft; (2) they enhance dopamine biosynthesis; and (3) they inhibit dopamine degradation. All these effects enhance dopamine action (Figure 11–31).

How does enhancement of the dopamine action of VTA neurons lead to addiction? As we learned in Section 10.24, projections from VTA dopamine neurons to the nucleus accumbens play a critical role in reward-based learning. Specifically, studies in primates and rodents have shown that many VTA dopamine neurons encode reward prediction errors. This error signal is hypothesized to direct synaptic plasticity in target neurons in the nucleus accumbens and prefrontal cortex for reinforcement-based learning. If VTA dopamine neurons signal a reward, the action or behavior that immediately preceded the reward is reinforced through dopamine modulation of downstream circuits (see Figure 10–44). Drugs of abuse bypass natural signals that activate these dopamine neurons, thus dissociating the reward system from its natural stimuli. Specifically, by increasing dopamine concentration at dopamine neurons' presynaptic terminals, drug consumption mimics dopamine neuron activation; this reinforces the preceding actions, include drug consumption itself. Thus, addictive drugs hijack the brain's reward system and exploit mechanisms that otherwise regulate learning and motivational behaviors.

What are the cellular and molecular mechanisms underlying the long-lasting behavioral changes in drug addiction? As discussed in Chapter 10, learning is

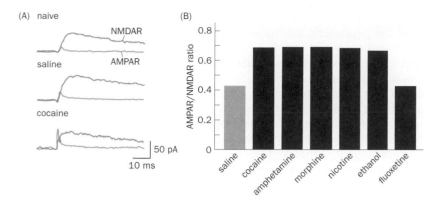

Figure 11–32 Exposure to drugs of abuse causes long-lasting enhancement of excitatory input to VTA dopamine neurons. (A) Dopamine neurons in VTA slices were recorded by whole-cell patch clamp 24 hours after a single *in vivo* exposure to cocaine to measure the magnitude of excitatory postsynaptic current conducted by the AMPA and NMDA receptors (AMPAR and NMDAR). The total AMPAR- and NMDAR-mediated excitatory postsynaptic current (total current) in response to stimulating input axons was measured at +40 mV (therefore the Mg^{2+} block of the NMDAR was relieved, and current was outward; see Section 3.15); then the NMDAR antagonist AP5 was added so only the AMPAR current in response to the same input stimulation was measured. The NMDAR current was calculated by subtracting AMPAR current from the total current. Compared to naive and saline injection controls, cocaine exposure caused an increase in the AMPAR current and an enhanced AMPAR/NMDAR current ratio, indicative of a potentiated synapse (see Section 10.7). **(B)** The AMPAR/NMDAR current ratio is similarly enhanced following *in vivo* exposure to five addictive substances, but not to exposure of non-addictive drugs such as fluoxetine. (A, adapted from Ungless MA, Whistler JL, Malenka RC et al. [2001] *Nature* 411:583–587. With permission from Macmillan Publishers Ltd; B, adapted from Saal D, Dong Y, Bonci A et al. [2003] *Neuron* 37:577–582. With permission from Elsevier Inc.)

mediated by synaptic plasticity in relevant neural circuits. Drugs of abuse likely act to alter synaptic weights in circuits involving VTA dopamine neurons and their targets. For example, a single *in vivo* exposure to cocaine induced marked enhancement of excitatory input onto VTA dopamine neurons, as measured by an increased ratio of AMPA receptor (AMPAR)-mediated current to NMDA receptor (NMDAR)-mediated current in subsequent whole-cell patch clamp recording in VTA slices *in vitro* (**Figure 11–32**A). This effect is similar to long-term potentiation in hippocampal synapses (see Section 10.7). Indeed, the cocaine-induced increase of AMPAR/NMDAR ratio was NMDAR-dependent and occluded subsequent LTP induced *in vitro*. Other drugs of abuse, including morphine, nicotine, and ethanol, cause a similar increase in the AMPAR/NMDAR ratio of VTA dopamine neurons (Figure 11–32B). Addictive drugs can likewise affect excitatory synapses onto spiny projection neurons in the nucleus accumbens, which receive input from diverse brain areas (see Figure 11–31). The VTA–nucleus accumbens–prefrontal cortex circuits are complex and heterogeneous: for instance, while some VTA dopamine neurons signal reward prediction errors, other dopamine neurons signal aversion, yet other dopamine neurons signal salience of stimuli (see Section 10.24). Identifying the specific synaptic connections and subcircuits that are modulated by drugs of abuse will be crucial in establishing causal relationships between synaptic changes and addictive behaviors.

11.19 Human genetic studies suggest that many genes contribute to psychiatric disorders

As we see from the above sections, studies on the action of various therapeutic and abuse-related drugs have enriched our understanding of normal brain function. A major limitation of relying on drug action to reveal the pathophysiology of psychiatric disorders, however, is that our understanding is limited to drug targets, which may only be related to the symptoms of the disorders rather than to their root cause. Twin studies in schizophrenia, mood, and anxiety disorders all point to significant genetic contributions (for example, **Figure 11–33**), with a heritability (the proportion of phenotypic differences contributed by genetic differences; see Section 1.1) of up to 80% for schizophrenia and bipolar disorders. Heritability for major depression and general anxiety disorders is lower (30–40%). Family studies also suggest that different psychiatric disorders may share common genetic factors. For example, family members of a schizophrenic patient have an increased chance of developing not only schizophrenia but also bipolar disorder. Likewise, major depression and generalized anxiety disorders often run together in families. These studies implicate strong genetic contributions to these psychiatric disorders. At the same time, factors other than inheritance also contribute significantly—these could be environmental factors, epigenetic influences, or *de novo* mutations that occur in the parental germ line or during the early embryonic development of patients (**Box 11–3**). Because environmental factors are multifaceted and more difficult to trace, hope has been placed on identifying genes that contribute to psychiatric disorders as a means to gain new insight into their origins and find new targets for drug development.

Figure 11–33 The Genain quadruplets. Born in 1930, each of these identical quadruplet sisters was diagnosed with schizophrenia of variable severity by age 24. Several other family members also suffered from mental illnesses, suggesting a strong genetic component in these disorders. (From Rosenthal, D. [1963] *The Genain Quadruplets: A Case Study and Theoretical Analysis of Heredity and Environment in Schizophrenia*. Basic Books, New York.)

So far, a simple Mendelian inheritance pattern has not been identified in any of the psychiatric disorders we discussed thus far. (This contrasts with Huntington's disease and certain familial forms of Parkinson's or Alzheimer's diseases, which are caused by dominant or recessive mutations in single genes.) This suggests that each psychiatric disorder is caused by multiple genetic factors and that each factor in itself only increases susceptibility, similar to the case of ApoE *ε4* in Alzheimer's disease (see Section 11.5). The recent genomic revolution (see Section 13.14) has generated new tools for identifying genetic variations that contribute to complex diseases, including psychiatric disorders (see Box 11–3). These studies have suggested that schizophrenia and bipolar disorder may result from modest contributions of multiple genetic variants, with total variants estimated to be in the hundreds. These variations can take the form of point mutations or differences in gene copy numbers; some are inherited, whereas others are produced *de novo*. The identities of some of the candidate susceptibility genes (**Table 11–2**) also suggest that abnormal neuronal signaling and neural development are major contributors to psychiatric disorders, and that each of these genes may increase susceptibility to multiple disorders.

For example, the *Drd2* locus, which encodes the <u>d</u>opamine <u>r</u>eceptor <u>D</u>$_2$ widely expressed in the brain including the spiny projection neurons that constitute the indirect pathway from the striatum (see Figure 8–22), was recently identified in a large-scale genome-wide association study (GWAS; see Box 11–3 for more details) as a risk factor for schizophrenia. As discussed in Section 11.15, the D$_2$-type dopamine receptor is a major target for all effective antipsychotic drugs used today; thus, the GWAS finding provides a satisfying link between recent genomic approaches and decades of drug-based investigations. GWAS of schizophrenia also identified genetic loci that encode other neuronal signaling molecules, including multiple glutamate receptors and voltage-gated Ca^{2+} channels that play important roles in synapse-to-nucleus signaling (see Section 3.23). Genes important for neural development have also been associated with schizophrenia and bipolar disorders. These include genetic loci that encode Satb2, a transcription factor that regulates cerebral cortex neuronal fate and axonal projections (see Section 7.4), and Neurexin-1 and Teneurin-4, transmembrane proteins that are enriched in synapses and play important roles in synapse development, organization, and wiring specificity in mice and flies (see Sections 7.11 and 7.23). A strong

Table 11–2: Selected candidate genes associated with psychiatric disorders[1]

Protein encoded by candidate gene	Identified on the basis of[2]	Associated with	Also associated with	Physiological functions
Drd2[3]	Genome-wide association study (GWAS)	schizophrenia		dopamine receptor D$_2$ that couples dopamine binding to G protein signaling
Ca$_v$1.2[3]	GWAS	schizophrenia	autism spectrum disorders	voltage-gated Ca^{2+} channel with large conductance, used in neuronal and synapse-to-nucleus signaling, and cardiac muscle contraction
Satb2[3]	GWAS	schizophrenia		transcription factor that regulates neuronal fate and axon targeting in cerebral cortex
Neurexin-1[4]	copy number variation	schizophrenia	autism spectrum disorders	cell-surface protein for synapse development and trans-synaptic signaling
Teneurin-4[5]	GWAS	bipolar disorder	schizophrenia	cell-surface protein for wiring specificity and synapse development
Laminin-α2[6]	whole-exome sequencing	schizophrenia		extracellular matrix protein for cell adhesion, axon growth, and synapse development

[1] This list represents selected findings from a large body of human genetic studies of psychiatric disorders.
[2] See Box 11–3 for definitions.
[3] Data from Schizophrenia Working Group of the Psychiatric Genomics Consortium (2014) *Nature* 511:421.
[4] Data from Rujescu et al. (2009) *Hum Mol Genet* 18:988.
[5] Data from Psychiatric GWAS Consortium Bipolar Disorder Working Group (2011) *Nat Genet* 43:977.
[6] Data from Xu et al. (2012) *Nat Genet* 44:1365.

link between neural development and psychiatric disorders was further suggested by the fact that some psychiatric disorder susceptibility genes are also associated with neurodevelopmental disorders such as autism spectrum disorders, which we will study later in the chapter.

Rapid advances in human genetics will undoubtedly uncover many more susceptibility genes for psychiatric disorders. Progressing from these genetic variations to mechanistic understanding and rational drug design for treatment, however, poses significant challenges. As discussed earlier, animal models are instrumental for studying disease mechanisms and testing therapeutic strategies. However, many symptoms of neuropsychiatric disorders, such as hallucination, delusion, or depression, are difficult to model in animals; the small effects of each susceptibility gene further complicate efforts to create effective animal models. Researchers are developing behavioral paradigms (see Section 13.29) and physiological assays in animals that mimic specific aspects of psychiatric disorders. In addition, investigating the physiological and developmental functions of susceptibility genes may ultimately bring new insights into why their disruption contributes to psychiatric disorders and how different susceptibility genes may interact with each other and with environmental factors in affected patients.

Box 11–3: How to collect and interpret human genetics data for brain disorders

Brain disorders caused by single-gene mutations that follow Mendelian inheritance patterns are the simplest to study from a genetics perspective (**Figure 11–34**). We've already discussed **autosomal dominant** mutations (the phenotypes of which can result from toxic gain-of-function effects

from the mutant allele, or loss-of-function effects due to insufficient amount of normal gene products produced from the wild-type allele) and **autosomal recessive** mutations (the phenotypes of which result from loss-of-function effects due to disruption of both alleles) that give rise to familial forms of neurodegenerative diseases. Single-gene mutations can also be **sex-linked**, that is, the mutant gene is located on the X chromosome. (The human Y chromosome carries few genes.) Sex-linked mutations affect males more severely than females, because mutations on a male's single X chromosome exert their effects in every cell. In females, by contrast, one of the two X chromosomes is randomly inactivated in each cell since early development (**random X-inactivation**), so that sex-linked mutations are expressed in about half of a female's cells. Red–green colorblindness (see Section 4.13) is a good example of a sex-linked trait. Genes that cause Mendelian disorders are usually mapped by pedigree analyses and by molecular markers that are distributed across the genome (for example, see Figure 11–4A).

Most brain disorders that are defined by symptoms or pathology do not follow simple Mendelian inheritance patterns and likely have heterogeneous causes. These disorders can in principle be caused by (1) inheritance of multiple genetic variants that interact with each other, (2) *de novo* **mutations** that occur in the parental germ line and—with the exception of X-chromosome mutations inherited by a female—affect all of the patient's cells, (3) *de novo* **somatic mutations** that occur in progenitor cells and therefore affect a subset of the patient's cells derived from the progenitors, (4) environmental factors, and (5) any of the above factors acting in combination. Of these, only factor 1 (and a small fraction of factor 2; see below) contributes to heritability. Therefore, if a genetic disorder such as schizophrenia or bipolar disorder has a high heritability but no clear Mendelian inheritance pattern, then multiple inherited

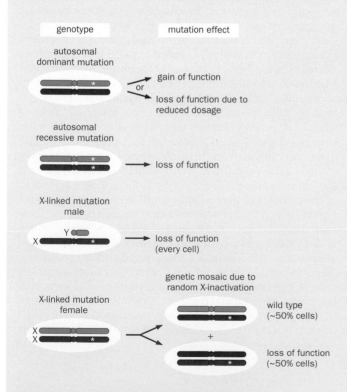

Figure 11–34 Three types of Mendelian inheritance.
Left, genotypes are represented by pairs of homologous chromosomes, with one chromosome inherited from the father (blue) and one from the mother (red), in each cell (yellow oval). * designates a mutation. Right, summary of mutation effects. Black chromosomes indicate inactivated X chromosomes.

(Continued)

Box 11–3: How to collect and interpret human genetics data for brain disorders

mutations, interacting either with each other or with additional factors, must contribute to the disorder.

A conceptually simple way of identifying genes that contribute to a given disorder is to perform **a genome-wide association study** (**GWAS**), taking advantage of **single nucleotide polymorphisms** (**SNPs**) that are present throughout the human genome. Any individual has about 3.5 million SNPs compared with the reference human genome. If a SNP is close to or within a gene whose mutations contribute to a disease, then it should be tightly linked with the disease-contributing mutation in the general population. DNA samples collected from many patients (usually thousands or more) can be compared with those from a similar number of healthy controls (ideally healthy relatives, or populations with the same ethnicity and geographic distribution) to identify the SNPs that are most strongly linked with the disease. The strength of the association can be quantified by parameters such as the **odds ratio**, which is defined as the probability of having the disease among people with the SNP divided by the probability of having the disease among people without the SNP. Given that most brain disorders have multiple genetic causes, the odds ratio is a complex function of both the heterogeneity of the patient population and the penetrance of the linked mutation that contributes to the disease. For schizophrenia and bipolar disorder, identified disease-associated SNPs have odds ratios between 1.10 and 1.25. By comparison, GWAS studies identified *Apoe ε4* as having an odds ratio of about 3.5 for Alzheimer's disease.

While SNPs have historically been detected by DNA microarray analyses (see Section 13.13), recent advances in sequencing technology have made it possible to sequence whole exomes (that is, the roughly 1% of genomic DNA sequences that corresponds to exons) or whole genomes of patients and control subjects; these methods offer powerful ways to identify disease-causing DNA variants. Whole-exome sequencing and whole-genome sequencing have revealed that *de novo* mutations contribute significantly to many brain disorders. *De novo* mutations usually occur spontaneously in the germ line of parents, with a bias toward the paternal germ line because spermatogenesis involves many more cell divisions than oogenesis, and hence presents more opportunities for DNA replication errors. By definition, *de novo* mutations do not affect the phenotypes of parents and do not contribute to heritability (except in the rare case where the mutations occur early in parental germ line development and affect the sperm or eggs inherited by more than one progeny, contributing to sibling similarities). Thus, whole-exome or whole-genome sequencing of patients with a specific disease and their healthy parents should reveal *de novo* mutations that contribute to the disease. A complication is that *de novo* mutations occur even in healthy individuals, with an incidence of about one gene-disrupting *de novo* mutation per individual; in fact, each of us has about 100 inherited gene-disruption mutations in our genome. As a result, identifying which *de novo* mutations contribute to a given disease involves complex statistical analysis, taking into consideration the sequence conservation and possible physiological functions of affected proteins. In general, if *de novo* mutations affect the same gene in more than one patient with the same disease (as was the case for laminin α2 in Table 11–2), then the probability that these mutations contribute to the disease increases.

Among *de novo* mutations, **copy number variations** (**CNVs**) make a major contribution to brain disorders. CNVs are deletions or duplications of chromosome segments that vary in length from 500 base pairs to several megabases (Mb) and may contain coding sequences that range from a small fraction of a single gene to many genes. CNVs can also be inherited if carriers bear progeny. Frequently occurring CNVs are associated with repeat elements in the genome that cause errors in DNA recombination during the meiotic cell cycles that produce sperm and eggs. Healthy humans usually carry an average of about 1000 polymorphic CNVs; having one or three copies of most genes has no significant impact on health. However, some genes are dosage-sensitive, such that losing a copy or gaining an extra copy can contribute to or cause specific disorders. For instance, a spontaneous deletion of a 3-Mb segment on Chromosome 17 that is flanked by genomic repeats affects 1 in 15,000–25,000 people and causes a neurodevelopmental disorder called **Smith-Magenis syndrome**, characterized by mild-to-moderate intellectual disability, delayed speech, sleep disturbances, and impulse control and other behavioral problems. Despite the fact that the common deletion contains more than 30 genes, losing one copy of a single gene called *Rai1* (retinoic acid induced 1) within the common deletion interval is sufficient to cause most of the symptoms. Remarkably, duplication of this genomic region (which occurs with the same frequency as the common deletion) results in **Potocki-Lupski syndrome**, which is likely caused by an increased dose of *Rai1* and is associated with mild intellectual disability and autistic symptoms. Thus, the gene dosage of *Rai1*, which encodes a nuclear protein that regulates gene expression, is critical for proper brain development and function. As another example, deletions of one copy of part of the gene encoding Neurexin-1 markedly increase the odds of developing schizophrenia and autism (see Table 11–2).

NEURODEVELOPMENTAL DISORDERS

Whereas neurodegenerative and psychiatric disorders usually have an adult or adolescent onset, the symptoms of neurodevelopmental disorders first appear in infancy or early childhood. Depending on the types of symptoms, neurodevelopmental disorders are categorized as intellectual disabilities (ID, previously referred to as mental retardation), autism spectrum disorders (ASD), communication disorders, attention deficit/hyperactivity disorders, learning disorders, or motor disorders. Despite this classification, recent work suggests that different neurodevelopmental disorders and some psychiatric disorders share similar underlying genetic causes. Below, we start with a general discussion of ID and ASD, two developmental disorders that are significant both in their frequency of occurrence and in their profound effects on patients and their caregivers. We then focus in greater detail on two specific syndromes that include symptoms of both ID and ASD: Rett syndrome and fragile-X syndrome. Approaches pioneered by research on these two syndromes will likely apply to studies of other neurodevelopmental disorders.

11.20 Intellectual disabilities and autism spectrum disorders are caused by mutations in many genes

Intellectual disability is characterized by deficits in general mental abilities such as reasoning, problem-solving, planning, abstract thinking, judgment, and learning from experience. ID patients usually have an intelligence quotient (IQ) of 70 or less, which is two standard deviations below the age-matched population mean (see Figure 1–2). ID is estimated to affect 1–3% of the general population.

Genetic factors, including chromosomal abnormalities and monogenic causes, account for a large fraction of ID cases, especially for those with IQs below 50. ID can also be one feature of **syndromic disorders** characterized by defined constellations of behavioral, cognitive, and physical symptoms. For example, Down syndrome is caused by having an extra copy of Chromosome 21 and is the most common genetic form of ID (affecting 1 in 500–1000 births). ID can also be caused by genetic mutations in the absence of recognizable syndromes or global structural abnormalities of the brain; these are called non-syndromic ID (NS-ID). Because the primary symptom of NS-ID is intellectual impairment, the corresponding genes may function more specifically in processes related to learning and intellectual capabilities, and are thus of considerable interest to scientists who seek to understand the biological bases of cognitive functions.

Genetic mapping studies in the past two decades have identified several dozen genes that when mutated cause NS-ID, 80% of which reside on the X chromosome. Because males have only one copy, mutations on the X chromosome affect all cells (see Figure 11–34) and are therefore technically easier to identify than autosomal recessive mutations. However, mutations in X-chromosome genes are estimated to account for only a small fraction of NS-ID cases; thus genetic causes for NS-ID may involve hundreds of genes distributed throughout the genome. The ID-associated genes identified to date encode proteins that include transcriptional regulators and cell-adhesion and signaling molecules important for brain wiring, as well as molecules known to regulate synapse development and function. Below we discuss one specific example.

A prominent class of proteins involved in ID are involved in Rho GTPase signaling. These proteins transduce extracellular signals to regulate cytoskeletal changes that underlie axon growth and guidance, dendrite morphogenesis, and synapse development (see Box 5–2). Rho GTPase signaling pathway members associated with NS-ID or syndromic ID include guanine nucleotide exchange factors (GEFs) that activate GTPases, GTPase activating proteins (GAPs) that deactivate GTPases, and protein kinases downstream of GTPases (**Figure 11–35**A). One of the first X-linked NS-ID genes to be identified encodes a protein called oligophrenin, which acts as a GAP for Rho GTPases. Oligophrenin is widely expressed in the nervous system and is distributed in axons, dendrites, and dendritic spines (Figure 11–35B). RNAi knockdown in cultured rat hippocampal neurons resulted in decreased spine length (Figure 11–35C) and impaired synaptic transmission and synaptic plasticity. Oligophrenin knockout mice exhibited a variety of cognitive defects, including impaired spatial learning in the Morris water maze assay (Figure 11–35D; see also Figure 10–32). Thus, in the case of oligophrenin, the cognitive deficits observed in human patients may be caused in part by impaired synapse structure and function.

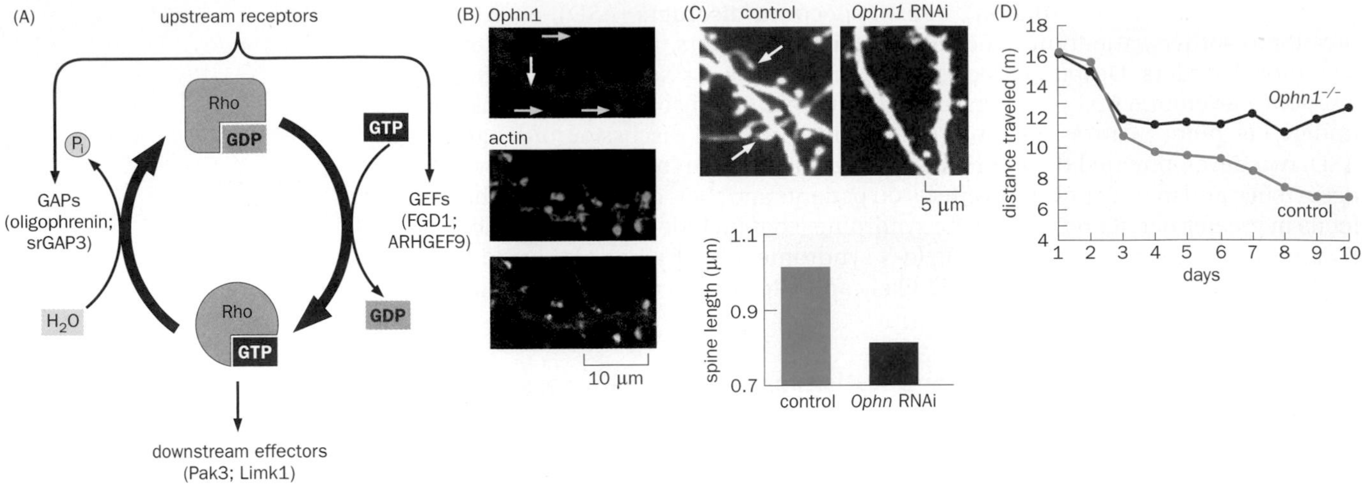

Figure 11–35 Defects in Rho GTPase signaling can cause intellectual disabilities. (A) Schematic of Rho GTPase signaling pathway. Mutations in two GTPase activating proteins (GAPs), two guanine nucleotide exchange factors (GEFs), and two downstream kinases are associated with intellectual disabilities. The protein names corresponding to these mutations are in parentheses. **(B)** Oligophrenin1 (Ophn1), a RhoGAP, is highly concentrated in axons (cyan arrow), dendrites (white arrow), and dendritic spines (yellow arrows) of cultured rat hippocampal neurons. These cultures are doubly stained with antibodies against Ophn1 in red, and actin in green that highlights the dendritic spines enriched for F-actin.

(C) Compared with wild-type dendritic spines (arrows), dendritic spines in neurons treated with RNAi against *Ophn1* show reduced length, as quantified below. **(D)** Compared with controls, *Ophn1* mutant mice travel longer distances to reach the hidden platform during daily sessions in the Morris water maze. (A, based on Pavlowsky A, Chelly J & Billuart P [2012] *Mol Psychiatry* 17:682–693; B & C, adapted from Govek EE, Newey SE, Akerman CJ et al. [2004] *Nat Neurosci* 7:364–372. With permission from Macmillan Publishers Ltd; D, adapted from Khelfaoui M, Denis C, van Galen E et al. [2007] *J Neurosci* 27:9439–9450.)

Autism spectrum disorders (**ASD**) cover a wide range of symptoms that in total affect >1% of children. At their core, ASD is characterized by deficits in communication and reciprocal social interactions. ASD patients often show an absence or reduction in sharing of interests and emotions and have difficulty adapting their behavior to different environments. They also exhibit restricted and repetitive patterns of activities and excessive adherence to routines. About 70% of ASD patients also have ID, but others have normal intelligence, and some exhibit exceptional ability in mathematical calculation, memory, art, or music.

Genetic factors are a predominant cause for ASD. For example, compared with the general population, the relative risk of a child being diagnosed with ASD is >25-fold greater if a sibling is affected. Recent genome-wide association studies, CNV analyses, and whole-exome and whole-genome sequencing studies (see Box 11–3) have identified many independent genetic lesions associated with ASD. Similar to ID, genes associated with ASD encode proteins that regulate synapse development and synaptic function, transcription, and chromatin structures (see also Section 11.26). Most ASD-associated genes are risk factors and little is known yet about how they affect neural development. On the other hand, studies of specific syndromes whose symptoms overlap with ID and ASD can shed light on the underlying neurobiological mechanisms. These syndromes are usually caused by mutations in single genes with complete penetrance, and animal models often recapitulate significant aspects of the symptoms, making it possible to study pathogenic processes and reveal the underlying mechanisms. Below, we use studies of Rett syndrome and fragile-X syndrome to illustrate these points.

11.21 Rett syndrome is caused by defects in MeCP2, a regulator of global gene expression

First described by Andreas Rett in the 1960s in severely disabled girls who exhibited a common set of symptoms such as incessant hand wringing, **Rett syndrome** is a neurodevelopmental disorder that affects 1 in 10,000–15,000 girls during early childhood. Rett patients usually develop normally for the first 6–18 months, often achieving milestones such as walking and first words at a normal age. Their

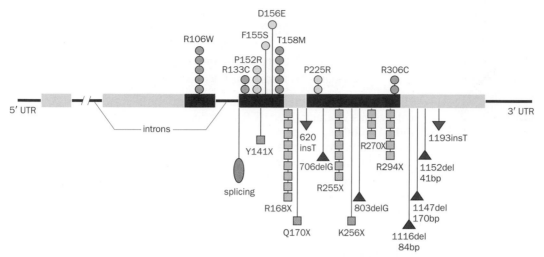

Figure 11–36 Rett syndrome is caused by mutations in the *Mecp2* gene, which encodes methyl-CpG-binding protein 2. Distribution of mutations in the coding region of the *Mecp2* gene (boxed region is the coding region, interrupted by two introns) identified in Rett patients. Above the gene structure are missense mutations, each of which changes the identity of a single amino acid. Below are a splicing mutation (oval), nonsense mutations (squares) that prematurely terminate translation, insertions (downward triangles; the inserted nucleotide is followed by 'ins'), and deletions (upward triangles). Mutations observed in more than one patient are represented by multiple symbols at the same position; most of these occur at mutational hot spots. The methyl-binding domain and transcriptional repression domain of MeCP2 are colored blue and red, respectively. 5′ UTR and 3′ UTR, 5′ and 3′ untranslated regions. (Adapted from Amir RE, Van Den Veyver IB, Schultz R et al. [2000] *Ann Neurol* 47:670–679. With permission from John Wiley & Sons.)

development then slows, arrests, and regresses. Patients exhibit social withdrawal, loss of language, and other autistic features. The onset of mental deficits is also accompanied by motor symptoms such as hand wringing. The condition subsequently stabilizes and patients usually live to adulthood, although with severe and persistent disability.

In 1999, genetic mapping and candidate gene sequencing revealed that Rett syndrome is caused by mutations in an X-linked gene encoding a protein called **methyl-CpG-binding protein 2** (**MeCP2**; **Figure 11–36**). Loss-of-function *Mecp2* mutations usually lead to prenatal or infant lethality in boys, who have only one X chromosome. Girls with a loss-of-function *Mecp2* mutation are genetic mosaics for MeCP2 function and develop Rett syndrome: of the two X chromosomes per cell, one is randomly inactivated (see Figure 11–34), so about half of their cells have defective MeCP2, and the severity of the disorder is influenced by the pattern of random X-chromosome inactivation. Rett syndrome is almost always caused by *de novo* mutations in *Mecp2*, as patients are so severely disabled that they rarely have children. Since the discovery that *Mecp2* mutations underlie Rett syndrome, specific missense mutations in *Mecp2* (that presumably have weaker effects than a complete loss-of-function) have been associated with sporadic ASD and schizophrenia. The proper level of MeCP2 expression is important, as duplication of *Mecp2* also causes severe neurodevelopmental defects.

MeCP2 is a nuclear protein that binds to methylated DNA at CpG sites (CpG refers to a cytidine followed by a guanosine in the DNA sequence). DNA methylation, a major form of epigenetic regulation of gene expression, is usually associated with target gene repression; for instance, methylation is the primary contributor to random X-chromosome inactivation. In the mammalian genome, CpGs are usually methylated except where they are present in large clusters (CpG islands), which are often associated with active transcription. Because methylation states affect gene expression, MeCP2 provides a link between chromatin structure and gene expression. MeCP2 is most abundantly expressed in the brain. In the mouse, MeCP2 expression increases greatly during the first 5 weeks of life as the final stages of neural development take place. Biochemical analysis indicated that MeCP2 binds CpG tracks along the entire genome, suggesting that MeCP2 acts as a general regulator of chromatin structure, which in turn affects global gene expression. Indeed, the number of MeCP2 molecules per neuronal

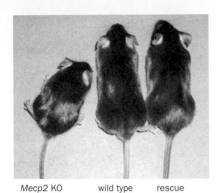

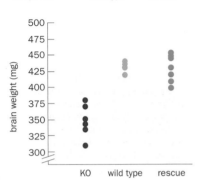

Figure 11–37 MeCP2 acts primarily in post-mitotic neurons. Compared with wild-type male mice at 8 weeks, *Mecp2* knockout (KO) males have reduced size (top) and reduced brain weight (bottom). Both phenotypes are rescued by transgenic expression of MeCP2 in post-mitotic neurons of KO mice. (Adapted from Luikenhuis S, Giacometti E, Beard CF et al. [2004] *Proc Natl Acad Sci USA* 101:6033–6039. Copyright the National Academy of Sciences, USA.)

nucleus is similar to that of histones, the principal proteins that complex with DNA to form chromatin.

11.22 MeCP2 acts predominantly in post-mitotic neurons to regulate their maturation and function

How does loss of a global regulator of chromatin structure cause the neurological deficits characteristic of Rett patients? As discussed earlier in this chapter, animal models can provide important insights into human disease. MeCP2 is present in all vertebrates and is highly conserved in mammals. Indeed, *Mecp2* knockout mice mimic many aspects of Rett syndrome. As in humans, *Mecp2* in the mouse is located on the X chromosome. *Mecp2* mutant male mice grow normally for the first several weeks. Between 3 and 8 weeks of age, mutant male mice start to exhibit motor coordination defects, impaired growth, and reduced brain weight compared to normal mice (**Figure 11–37**). These symptoms become progressively worse, and most mutant males die by age 12 weeks. Female mice heterozygous for the *Mecp2* mutation, a genetic condition equivalent to that of girls with Rett syndrome, initially develop normally. They start to exhibit symptoms such as mild motor defects and inertia when several months old. Importantly, conditional knockout of *Mecp2* only in neurons and glia resulted in phenotypes essentially identical to those of *Mecp2* knockout mice. Conversely, restoring MeCP2 function only in post-mitotic neurons rescued many neurological phenotypes (Figure 11–37) and prevented the death of *Mecp2* mutant males. These experiments indicate that MeCP2 acts predominantly in post-mitotic neurons to exert its function. Furthermore, overexpression of MeCP2 in post-mitotic neurons in wild-type mice also caused severe motor dysfunction, echoing the symptoms caused by human *Mecp2* duplication.

Detailed analyses of the *Mecp2*-deficient mouse model (mostly in males) have revealed a host of defects, including reductions in the size of the brain, neurons, and dendritic spines and alterations in dendritic morphology, synaptic transmission, and synaptic plasticity. Conditional knockout (see Section 13.7) of *Mecp2* in specific neuronal populations indicated that MeCP2 plays important roles in excitatory, inhibitory, modulatory, and peptidergic neurons, as well as in glia, each of which contributes to a subset of phenotypes found in mice lacking *Mecp2* in all cells. Deletion of *Mecp2* in GABAergic neurons caused the most severe phenotypes, including many Rett syndrome features such as repetitive and compulsive behaviors (**Figure 11–38**A), motor dysfunction, learning deficits, and premature death. Some of the defects observed in *Mecp2* mutant GABAergic neurons may be caused by reduced expression of glutamic acid decarboxylases (Gad1 and Gad2), the enzyme for GABA synthesis (Figure 11–38B). Reduction of GABAergic inhibition could also contribute to the seizures (see Box 11–4) often associated with Rett syndrome.

(A)

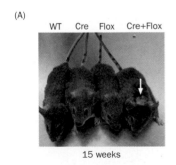

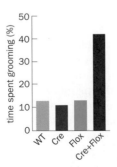

(B)

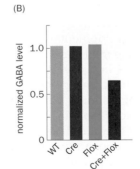

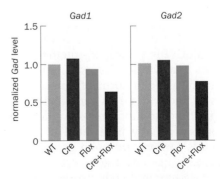

Figure 11–38 MeCP2 regulates functions of GABAergic neurons. **(A)** When *Mecp2* was knocked out only in GABAergic neurons using a vesicular-inhibitory-amino-acid-transporter-Cre transgene, the conditional knockout animals (Cre + Flox) frequently exhibited fur loss (arrow) due to excessive time spent self-grooming (quantified at right); controls include wild type (WT), Cre only (Cre), conditional knockout allele without Cre (Flox). **(B)** Compared with controls, conditional knockout animals had reduced levels of the neurotransmitter GABA (left) and reduced mRNA levels for *Gad1* and *Gad2*, encoding two enzymes for GABA synthesis (right). (Adapted from Chao HT, Chen H, Samaco RC et al. [2010] *Nature* 468:263–269. With permission from Macmillan Publishers Ltd.)

11.23 Restoring MeCP2 expression in adulthood reverses symptoms in a mouse model of Rett syndrome

A key question in neurodevelopmental disorders is whether the symptoms are reversible. A given neurodevelopmental disorder may be caused by early and irreversible defects in nervous system development, with more challenging consequences for therapeutic intervention. Alternatively, it may reflect that the developing and adult nervous systems require an ongoing supply of the disrupted gene's product for maturation and function, whose symptoms might be reversible. The mouse model for Rett syndrome provided an opportunity to ask whether MeCP2 is continuously required in the adult nervous system and to determine whether defects caused by *Mecp2* deficiency can be alleviated by restoring MeCP2 expression in adults.

To answer the first question, a temporally controlled knockout scheme was employed using a drug-inducible variant of the Cre recombinase, CreER (see Section 13.7), to remove *Mecp2* gene upon drug application only in adults. Mice that lost *Mecp2* as adults developed symptoms resembling those observed in mice born with *Mecp2* knockout, including abnormal motor coordination and premature death (**Figure 11–39**). This experiment indicated that MeCP2 is continuously required in adults.

To determine whether late expression of MeCP2 can rescue developmental *Mecp2* deficiency, one can in principle use a *Mecp2* transgene under the control a drug-inducible promoter to test whether late expression of MeCP2 can remedy the knockout phenotypes. However, overexpression of MeCP2 also causes significant neurological defects. To restore MeCP2 expression at physiological level, a conditional transcriptional stop cassette was knocked in between the promoter and the coding sequence of *Mecp2* such that MeCP2 is normally not expressed from this modified allele. Upon drug-induced activation of CreER, the stop cassette can be excised by the recombinase (see Section 13.10), and transcription is reactivated from the endogenous promoter such that MeCP2 is expressed at physiological levels (**Figure 11–40**A). Remarkably, Rett-like symptoms as well as premature death were rescued after reactivation of MeCP2 in young adult males carrying the modified allele (Figure 11–40B). In female mice that were heterozygous for the modified allele and therefore similar to girls with Rett syndrome in genotype, defects in hippocampal long-term potentiation were also reverted after MeCP2 reactivation (Figure 11–40C). These results do not suggest a specific therapy to treat Rett patients; indeed we are still far from any therapy. However, these findings offer hope for effective intervention even after symptom onset.

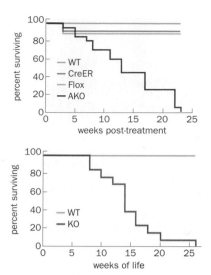

Figure 11–39 MeCP2 is required in adulthood. Top, mice in which the *Mecp2* gene is knocked out in adulthood (AKO, red) die prematurely compared with controls (WT, wild type; CreER, CreER only; Flox, conditional allele only). Bottom, survival curves for mice born with *Mecp2* knockout mutation (KO, brown) and wild-type controls (green). Note the similarity in AKO and KO survival curves from the time point at which *Mecp2* gene product is lost. (Adapted from McGraw CM, Samaco RC, & Zaghbi HY [2011] *Science* 333:186.)

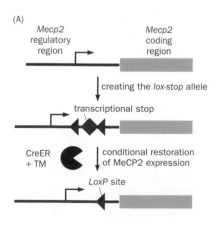

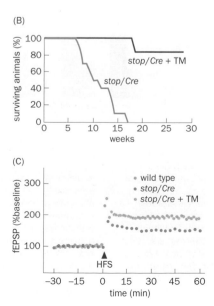

Figure 11–40 Reversing Rett symptoms by restoring MeCP2 expression in young adult mice. (A) Strategy to restore MeCP2 expression conditionally. Top, the endogenous *Mecp2* locus. Middle, a transcriptional stop flanked by two *loxP* sites is inserted between the transcriptional start and the coding region, such that no MeCP2 protein is made from this *lox-stop* allele. Bottom, upon tamoxifen (TM)-inducible CreER excision of the transcriptional stop, MeCP2 is expressed at the endogenous level (see Sections 13.7 and 13.10 for more details of the techniques). **(B)** Male *stop/Cre* mice survived after TM injection in adulthood (red curve), but died prematurely without TM injection (blue curve). **(C)** Restoration of MeCP2 expression in adult females (orange) rescued the long-term potentiation defects exhibited by *Mecp2* heterozygous females (blue) to wild-type level (green), as seen by the magnitude of field excitatory postsynaptic potential (fEPSP) increase in response to high-frequency stimulation (HFS). (Adapted from Guy J, Gan J, Selfridge J et al. [2007] *Science* 315:1143–1147.)

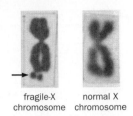

Figure 11–41 Fragile-X chromosome.
X chromosomes from a fraction of cells of fragile-X syndrome (FXS) patients exhibit a gap near the tip of the long arm (arrow on the left), which is not seen in normal X chromosomes (right). (Adapted from Lubs HA [1969] *Am J Hum Genet* 21:231–244. With permission from American Society of Human Genetics.)

11.24 Fragile-X syndrome is caused by loss of an RNA-binding protein that regulates translation

Fragile-X syndrome (**FXS**) is a leading cause of inherited intellectual disability (ID), affecting about 1 in 5000 boys. FXS patients exhibit reduced IQ and a significant developmental delay in speech and motor skills. Many patients exhibit autistic features, such as a tendency to avoid eye contact and repetitive, stereotyped behaviors; indeed, about 20–30% of FXS patients are diagnosed with autism spectrum disorders (ASD) based on behavioral criteria. The name 'fragile-X' came from the unusual chromosomal gap on the X chromosome of patients (**Figure 11–41**). The defective gene that causes FXS was molecularly identified in 1991 and named *Fmr1* (fragile-X mental retardation 1). A polymorphic CGG trinucleotide repeat was found in the 5′ untranslated region of the *Fmr1* gene. Healthy individuals contain 6 to 54 CGG repeats, but the number expands to >200 in FXS patients. CGG repeats between 55 and 200 are called premutations; these repeats are at high risk of expansion. FXS is inherited either from mothers who carry premutations, or from mothers who carry the full mutation but are not severely affected themselves. Boys with the syndrome are most severely affected, while the severity of the syndrome in girls varies widely depending on X-inactivation patterns. Expanded CGG repeats cause extensive methylation near the *Fmr1* promoter and silencing of *Fmr1* gene expression. Therefore, FXS is caused by loss of a single protein, **FMRP**, encoded by the *Fmr1* gene.

FMRP is an evolutionarily conserved RNA-binding protein. A point mutation affecting a conserved residue in the RNA-binding domain of FMRP causes FXS with symptoms similar to those in patients with silenced *Fmr1* expression, demonstrating the importance of RNA binding for normal FMRP function. FMRP is highly expressed during embryonic development and is continuously expressed in all neurons throughout life. Within neurons, FMRP is localized to the cytoplasm, axons, dendrites, and postsynaptic compartments, and thus can regulate local protein translation (see Section 2.2). FMRP is enriched in polyribosomes, where it binds to mRNA molecules and represses translation *in vitro* and *in vivo* (**Figure 11–42**A, B). Normally most FMRP is phosphorylated at a conserved serine residue (S499 in the mouse). S499-phosphorylated FMRP represses translation; dephosphorylation relieves this repression. Upon specific signals such as synaptic activity, FMRP is dephosphorylated, which transiently reduces its translational

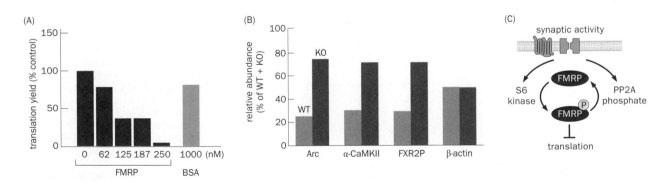

Figure 11–42 Regulation of protein synthesis in dendrites and synapses by fragile-X mental retardation protein (FMRP). **(A)** In an *in vitro* translation assay, purified FMRP repressed translation of total brain mRNA in a dose-dependent manner, whereas a control protein, bovine serum albumin (BSA), did not. **(B)** Synapse-enriched protein extracts from *Fmr1* knockout mice contained higher amounts of the proteins Arc (a postsynaptic signaling protein), α-CaMKII (α subunit of the Ca²⁺/calmodulin-dependent kinase), and FXR2P (fragile-X mental retardation syndrome-related protein 2, encoded by a *Fmr1* paralog) than do similar extracts from wild-type mice. By contrast, the level of β-actin protein was unaffected. This suggested that FMRP normally represses translation of select mRNAs such as those that produce Arc, α-CaMKII, and FXR2P proteins. **(C)** Only phosphorylated FMRP represses translation. S6 kinase and phosphatase 2A (PP2A) respectively phosphorylate and dephosphorylate FMRP. Through its effects on PP2A and S6 kinase, synaptic activity can regulate FMRP phosphorylation, and hence local translation. Type I metabotropic glutamate receptors activate PP2A more rapidly than they activate S6 kinase, thus causing transient FMRP dephosphorylation and activation. (A, adapted from Li Z, Zhang Y, Ku L et al. [2001] *Nucleic Acid Res* 29:2276–2283. With permission from Oxford University Press; B, adapted from Zalfa F, Giorgi M, Primerano B et al. [2003] *Cell* 112:317–327. With permission from Elsevier Inc.; C, based on Santoro MR, Bray SM & Warren ST [2012] *Annu Rev Patho Mec Dis* 7:219–245.)

repression activity, thereby allowing rapid local translation (Figure 11–42C). This activity-dependent regulation of local translation contributes to synaptic plasticity and learning.

Biochemical studies that used cross-linking of bound RNA with FMRP followed by immuno-purification of FMRP have identified many specific target mRNAs associated with FMRP. These include microtubule-associated proteins involved in axonal and dendritic transport, presynaptic proteins that regulate synaptic vesicle release, postsynaptic scaffolding proteins, and components of the NMDA and metabotropic glutamate receptor (mGluR) signaling pathways for neurotransmitter reception. Interestingly, FMRP also binds to the mRNAs for many ASD-associated proteins, providing a molecular link between FXS and ASD.

11.25 Reducing mGluR signaling ameliorates fragile-X symptoms in animal models

As with Rett syndrome, *Fmr1* knockout mice recapitulated certain symptoms and pathologies of human FXS. For example, postmortem analysis revealed that dendritic spines in FXS patients tend to be longer, more tortuous, and more numerous compared to those of control subjects. Similar phenotypes have been found in *Fmr1* knockout mice, which also exhibit deficits in a variety of learning and synaptic plasticity assays.

Of particular interest in *Fmr1* research is long-term depression (LTD) in the hippocampus mediated by the type I metabotropic glutamate receptors (mGluRs). The type I mGluRs, mGluR1 and mGluR5, are known to regulate local protein translation at the postsynaptic density, leading to AMPA-type glutamate receptor endocytosis and hence long-term depression (see also Section 10.9). In *Fmr1* knockout mice, mGluR-dependent LTD was enhanced (**Figure 11–43**A). Given that FMRP represses translation, one interpretation of this result is that FMRP-mediated translational repression normally counterbalances mGluR-induced translation. A critical function of FMRP may be to repress translation of proteins that are normally synthesized following mGluR activation in the context of synaptic plasticity and learning (Figure 11–43B).

This 'mGluR hypothesis' has received support in animal model studies. For example, a prediction of this hypothesis is that a reduction of mGluR function should ameliorate FXS symptoms. Indeed, removing one copy of the gene encoding mGluR5, which is the major form of type I mGluR expressed in the hippocampus, ameliorated several phenotypes in *Fmr1* knockout mice, such as increased dendritic spine density (**Figure 11–44**A). Treatment of *Fmr1* knockout mice with an mGluR5 antagonist likewise improved certain behavioral phenotypes. Since *Fmr1* is a highly conserved gene, its homolog in *Drosophila* (*Dfmr1*) has also been used to explore its physiological function. *Dfmr1* mutant flies exhibit a variety of phenotypes ranging from defects in synaptic transmission to abnormalities in courtship and courtship conditioning. *Dfmr1* males do not mate as efficiently as controls, and they do not learn to reduce mating after being rejected by mated females (see Section 9.12). Remarkably, mGluR5 antagonist treatment ameliorated both phenotypes in *Dfmr1* mutants (Figure 11–44B). Thus, drugs that reduce mGluR signaling may represent a viable means of alleviating FXS symptoms.

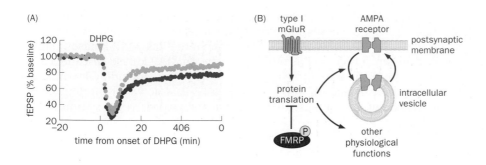

Figure 11–43 FMRP action in the context of metabotropic glutamate receptor (mGluR) signaling.
(A) Application of dihydroxyphenylglycine (DHPG), an agonist of type I mGluR, induces long-term depression (LTD) of the CA3 → CA1 synapse in hippocampal slices. The magnitude of LTD is enhanced in *Fmr1* knockout mice (red) compared to wild type (green). fEPSP, field excitatory postsynaptic potential. **(B)** A model of the interaction between mGluR signaling and FMRP. Type I mGluR-induced LTD involves a net removal of AMPA-type glutamate receptors from the postsynaptic surface and requires protein translation. Phosphorylated FMRP normally represses the translation of the same set of proteins. In the absence of FMRP, these normally repressed proteins are at elevated levels prior to mGluR activation and thus enhance mGluR-induced LTD. (A, adapted from Huber KM, Gallagher SM, Warren ST et al. [2002] *Proc Natl Acad Sci USA* 99:7746–7750; B, adapted from Bear MF, Huber KM, & Warren ST [2004] *Trends Neurosci* 27:370–377. With permission from Elsevier Inc.)

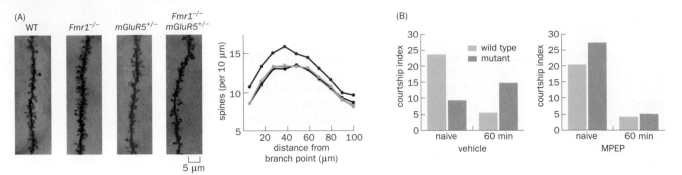

Figure 11–44 Reducing mGluR function ameliorates defects due to loss of FMRP in animal models. (A) The left panels show representative dendrites visualized by Golgi staining. The right panel shows the quantification of spine density. *Fmr1* knockout mice (*Fmr1*−/−; red) had more dendritic spines compared to wild type (WT; orange). Loss of one copy of the mGluR5 gene (*mGluR5*+/−; green) in a WT background did not affect spine density, but loss of one copy of the mGluR5 gene in an *Fmr1* knockout background (purple) rescued the supernumerary spine defect of *Fmr1* knockout. **(B)** Left, loss of *Drosophila Dfmr1* caused a reduction of courtship in naive males (compare the first two columns) and a defect in courtship conditioning (see Section 9.12). Control males reduce courtship after being paired with mated females, and remember this experience 60 minutes later by displaying a reduced courtship index (compare the first and third columns). *Dfmr1* mutant flies do not reduce courtship after being paired with mated females (compare the second and fourth columns). Right, both phenotypes are rescued by treatment with 2-methyl-6-(phenylethynyl)pyridine (MPEP), an mGluR5 antagonist. (A, adapted from Dölen G, Osterweil E, Shankaranarayana Rao BS et al. [2007] *Neuron* 56:955–962. With permission from Elsevier Inc.; B, adapted from McBride SMJ, Choi CH, Wang Y et al. [2005] *Neuron* 45:753–764. With permission from Elsevier Inc.)

However, clinical trials using mGluR antagonist to treat FXS patients have not been successful so far, highlighting the difficulty of translating findings in animal models to therapies in humans (see also Section 11.4).

11.26 Synaptic dysfunction is a common cellular mechanism that underlies neurodevelopmental and psychiatric disorders

Studies of Rett and fragile-X syndromes have reinforced the theme we introduced in Section 11.20: disruption of synaptic development and function may be a common cellular mechanism for many neurodevelopmental disorders. Further support for such a common mechanism comes from the identification of other syndromic, non-syndromic, or sporadic ASD and ID genes that affect different aspects of synaptic signaling (**Figure 11–45**). For example, as we discussed in Chapters 3 and 7, neurexins and neuroligins form trans-synaptic complexes that regulate synapse assembly and organization. Independent genetic studies have identified mutations in human genes that encode several neurexin and neuroligin isoforms as being associated with ASD. In addition, disruption of Shank3, a postsynaptic scaffolding protein (see Section 3.16), has been associated with syndromic as well as sporadic ASD. Expanding on FMRP's role in regulating synaptic protein translation, a key regulator of protein translation in response to extracellular signals is **mTOR** (**m**ammalian **t**arget **o**f **r**apamycin), which in turn is negatively regulated by a complex consisting of **Tsc1** and **Tsc2** (**t**uberous **s**clerosis 1 and 2). A large fraction of patients with **tuberous sclerosis**, which is characterized by nonmalignant tumors in the brain and other organs and which results from mutations in *Tsc1* or *Tsc2*, exhibit ASD symptoms, reinforcing the notion that abnormal protein translation may contribute to ASD.

Many factors that are involved in postsynaptic events, including synapse-to-nucleus signaling (see Section 3.23), have also been implicated in neurodevelopmental disorders. For example, a gain-of-function mutation in a voltage-gated Ca²⁺ channel, Ca$_V$1.2, causes **Timothy syndrome**, with cardiac arrhythmia and autistic symptoms. Disruption of components of the Ras/MAP kinase pathway, including the small GTPase Ras or its downstream MAP kinase cascade, causes Noonan syndrome with multiple developmental defects including learning disability. Disruption of Neurofibromin 1 (NF1), a GTPase activating protein (that is, negative regulator) of Ras, causes neurofibromatosis type I, whose symptoms also include learning disability. The Ras-MAP kinase pathway regulates protein translation and is also a key signaling pathway from the synapse to the nucleus (see Figure 3–41). One nuclear effector for synapse-to-nucleus signaling is MeCP2,

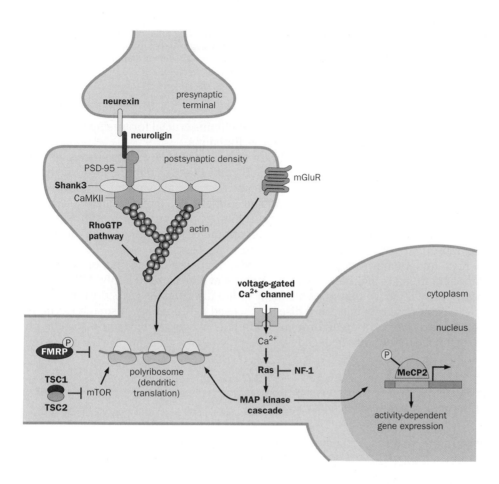

Figure 11–45 Defects in many proteins involved in synaptic signaling contribute to neurodevelopmental disorders. Mutations in genes that encode proteins named here in bold have been implicated either in various syndromes with increased risks of intellectual disability or autism spectrum disorder (ID/ASD), or in non-syndromic or sporadic ID/ASD. For example, mutations in synaptic adhesion proteins neurexin and neuroligin, a voltage-gated Ca²⁺ channel, and a postsynaptic scaffolding protein Shank3 are associated with ASD; mutations in several components of the Rho GTPase signaling pathways cause ID; mutations in translation regulator TSC1 and TSC2 yield ASD symptoms; mutations in small GTPase Ras, its negative regulator NF-1, and effector kinases are associated with learning disabilities; and mutations in the nuclear effector of the MAP kinase cascade, MeCP2, cause Rett syndrome with features of both ASD and ID. See also Figures 3–27 and 3–41.

encoded by the causal gene for Rett syndrome. MeCP2 is phosphorylated at multiple sites in response to neuronal activity, and phosphorylation regulates its gene repressive activity.

Defects in synaptic signaling may also be a major cause for psychiatric disorders such as schizophrenia and bipolar disorders. Indeed, mutations in several genes, including those that encode Ca$_v$1.2, Neurexin-1, and MeCP2, have been associated with both schizophrenia and ASD (see Table 11–2). As new genes associated with brain disorders are being discovered at a rapid pace thanks to recent advances in human genetics (see Box 11–3), the overlap between genes associated with psychiatric and neurodevelopmental disorders will likely increase, as will the link between these disorders and synaptic signaling.

11.27 Studies of brain disorders and basic neurobiology research advance each other

If synaptic dysfunction is a common cellular mechanism for neurodevelopmental and psychiatric disorders, why do disruptions of different genes (and sometimes different mutations of the same gene) give rise to different disorders? Do these mutations affect all synapses equally, such that the final symptoms of these disorders are caused by overall suboptimal synaptic function? Or are synapses in specific brain areas with special circuit functions differentially affected in different disorders? We are still far from having satisfactory answers to these questions, and the answers may differ for different disorders. For example, as we learned in Section 11.22, MeCP2 disruption in inhibitory neurons appears to cause the mutation's strongest effects in a mouse model. An imbalance between excitation and inhibition has also been suggested to underlie epilepsy (see **Box 11–4**), ASD, and schizophrenia. As researchers use more sophisticated tools to dissect the contributions of different subpopulations of neurons, we will surely find better answers to the questions raised above. At the same time, studies that focus on specific diseases may shed light on how the normal brain functions—revealing,

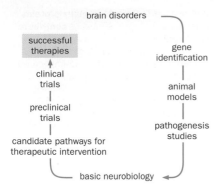

for instance, the specific brain regions and circuits crucial for intelligence, social interactions, and other complex cognitive functions.

Studies of genetically defined brain disorders have also introduced general strategies for treating these disorders (**Figure 11–46**). Identification of defective genes underlying brain disorders leads to the establishment of appropriate animal models. This enables mechanistic studies of the pathogenic process that enrich our understanding of basic neurobiology and at the same time suggest candidate pathways for therapeutic intervention. Development and clinical trials of appropriate drugs may eventually lead to successful therapies. For disorders whose underlying causes are largely unidentified, are multigenic, or are largely nongenetic, parts of this discovery-to-treatment path can still apply. Although we don't have effective therapies for most of the disorders described in this chapter, new advances in basic and disease-focused neurobiology research are being made each day, and breakthrough treatments for disabling brain disorders are anticipated in the coming decades.

Box 11–4: Epilepsy is a disorder of neuronal network excitability

We have encountered seizures and epilepsy many times in this book. With the proper framework of studying brain disorders established, we are now ready to discuss the symptoms, causes, and treatment strategies of seizures and epilepsy. A **seizure** is an episode involving abnormal synchronous firing of large groups of neurons; about 1 in 20 people has at least one seizure in the lifetime. **Epilepsy** is a chronic condition characterized by recurrent seizures, which affects about 1% of the human population. When cortical neurons are engaged in abnormal synchronous firing, the activities can often be detected on **electroencephalograms** (**EEGs**); EEGs record electrical potential differences between surface electrodes placed on specific locations of the scalp, which report the collective electrical activities of many nearby cortical neurons underneath the surface electrodes (**Figure 11–47**).

Seizures are typically categorized into either focal (partial) or generalized. **Focal seizures** are defined by clinical symptoms or EEG changes that indicate an initial activation of neurons in a relatively small, discrete region of the brain. Depending on the brain region, the symptoms can be a temporary loss of sensation, an odd sensory experience, a temporary loss of movement control, or confusion. **Generalized seizures** affect multiple, bilateral regions of the brain. In primary generalized seizures, the entire cortex seems to be activated at the same time, whereas a secondary generalized seizure results when a focal seizure spreads to larger areas of the brain. A generalized **absence seizure** (formerly called petit mal) is characterized by a brief lapse of consciousness (about 10 seconds or less) and a cessation of motor activities without loss of posture. A generalized **tonic-clonic seizure** (previously called grand mal) is associated with loss of consciousness and a predictable sequence of motor activity:

patients first stiffen and extend all extremities (tonic phase), then undergo full-body spasms during which muscles alternately flex and relax (clonic phase).

As with many brain disorders discussed in this chapter, epilepsy has diverse causes, including head injury, infection, strokes, brain cancers, and brain surgery. Epilepsy can also result from inherited or *de novo* mutations in several dozen identified genes, some of which appear monogenic whereas others confer risks. Other brain disorders, notably neurodevelopmental disorders, such as Rett syndrome, can have epilepsy as a symptom. Despite the diverse causes, epilepsy shares a common phenotype: an abnormal balance between the actions of excitatory and inhibitory neurons (the **E-I balance**) results in hyperactivation of excitatory neurons that spread the abnormal excitation across the network. This notion is best illustrated by examining epilepsies caused by defective ion channels (**channelopathies; Table 11–3**).

Given the key roles of voltage- and ligand-gated ion channels in regulating neuronal excitability that we have learned in Chapters 2 and 3, it is not surprising that mutations disrupting ion channels can cause abnormal neuronal firing. For instance, voltage-gated K^+ channels are key for repolarization of neurons after excitation; a reduction of their function can cause abnormal excitation of mutant neurons. Likewise, reduction of $GABA_A$ receptor function can cause epilepsy because neurons do not receive proper inhibitory signals. Let's examine a specific example in more detail: loss of one copy of the gene encoding a voltage-gated Na^+ channel, $Na_v1.1$. This human condition, called Dravet syndrome or severe myoclonic epilepsy of infancy (Table 11–3), has been recapitulated in the mouse model—mice heterozygous for a knockout allele of $Na_v1.1$ exhibit spontaneous seizures and sporadic death. Voltage-gated Na^+ channels are generally

Box 11–4: Epilepsy is a disorder of neuronal network excitability (Continued)

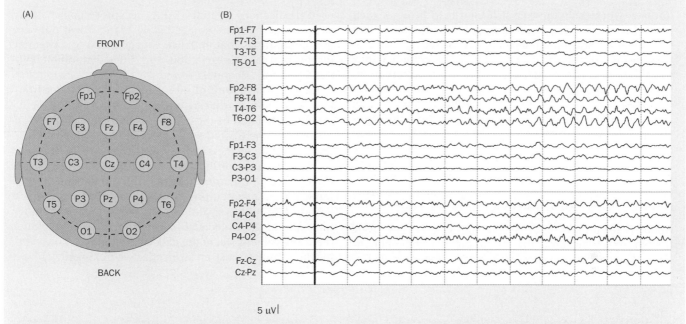

5 μV|

Figure 11–47 Detecting seizure onset by electroencephalograms. **(A)** Schematic of surface electrode placement on the scalp. The electrode positions are named according to cortical regions (F, frontal; T, temporal; P, parietal; O, occipital; C: central, that is, closer to the vertex of the head). **(B)** EEG record of a patient suffering from focal epilepsy presumed to originate from the right temporal lobe. Each row is the electrical potential difference between two designated electrodes according to panel A. The solid vertical line indicates the onset of seizure. Before the onset,

EEGs are small in amplitude and mostly asynchronous, reflecting brain activity during normal awaking period. After seizure onset, EEG records between multiple pairs of electrodes on the right hemisphere show a large-amplitude synchronous pattern. Note that the exact time of onset or the location of epileptic activity cannot be determined with scalp recordings; for this, intracranial electrodes need to be implanted in the presumed area of seizure activity. (Courtesy of Dr. Josef Parvizi, Stanford University Medical Center.)

responsible for producing action potentials, and therefore their reduction should inhibit rather than promote excitability. However, whereas most excitatory neurons express multiple genes encoding voltage-gated Na$^+$ channels, Na$_v$1.1 is highly expressed in GABAergic inhibitory neurons. Thus, a reduction of Na$_v$1.1 activity preferentially reduces the excitability of inhibitory neurons and thereby increases the network excitability. This example illustrates why mutations in a given ion channel cause specific types of epilepsy; this is likely because the neuronal types that uniquely express a particular ion channel (and therefore can least compensate for the loss) differ for each gene implicated in epilepsy.

Furthermore, the phenotypic severity of the same mutation often differs in individual patients or in mice with different genetic backgrounds, highlighting that even monogenic mutations may be subjected to complex interactions with other factors.

Channelopathies can reveal some of the mechanisms that produce seizures, but these mutations account for only a small fraction of epilepsy cases. Although the root causes in other cases are less clear, perturbation of excitation-inhibition balance may also be the culprit. For example, the post-injury neuronal process sprouting and synapse

Table 11–3: Representative examples of ion channel mutations that cause epilepsy

Affected protein	Disorder
α_4 subunit of nicotinic ACh receptor	autosomal dominant nocturnal frontal lobe epilepsy[1]
K$_v$7.2 or K$_v$7.3 (voltage-gated K$^+$ channels)	benign familial neonatal seizures[1]
α_1 subunit of GABA$_A$ receptor	juvenile myoclonic epilepsy[2]
Ca$_v$2.1 (voltage-gated Ca^{2+} channel)	absence epilepsy and episodic ataxia[2]
Na$_v$1.1 (voltage-gated Na$^+$ channel)	severe myoclonic epilepsy of infancy[2]

This is only a partial list of known mutations in ion channels that cause focal (denoted by [1]) or generalized (denoted by [2]) seizures. All mutations above are autosomal dominant, and epilepsy in most cases is a result of loss-of-function effect (reducing the dose) of the gene product. Data from Lerche et al. (2013) *J Physiol* 591:753.

(Continued)

Box 11–4: Epilepsy is a disorder of neuronal network excitability

formation that results from physical injuries to the brain, such as strokes and surgical removal of brain tissues, may differ for excitatory and inhibitory neurons, thereby perturbing the delicate E-I balance. Another important factor that may contribute to recurrent seizures is the act of seizure itself: many neurons firing in synchrony can cause significant changes in the involved circuits according to the plasticity rules discussed in Chapter 10. These activity-dependent changes may in turn decrease the threshold for future seizures. Indeed, excess excitation of glutamatergic neurons, with abnormally high glutamate release and NMDA receptor activation, can result in excessive elevation of intracellular Ca^{2+} concentration of their postsynaptic target neurons, which can trigger **excitotoxicity** and neuronal death.

About two-thirds of epilepsy patients can be effectively treated by medication, thanks to the common phenotype of excessive network excitability. The most widely used medications include $GABA_A$ receptor agonists such as benzodiazepines (see Section 11.17) to boost network inhibition, drugs that enhance voltage-gated Na^+ channel inactivation to curb excitation, and drugs that inhibit voltage-gated Ca^{2+} channels to reduce synaptic transmission efficacy. About one-third of epilepsy patients suffer from intractable seizures that are not responsive to current medications. A fraction of these patients that suffer from focal seizures can be treated with brain surgery. The identification of seizure focus is the key, which is usually achieved by intracranial recording and stimulation during surgery. (Indeed, we have learned a great deal about the functions of individual human neurons as a result of this procedure; see Section 1.10.) If the seizure focus regulates non-vital functions and ideally is located in non-dominant hemisphere, then surgical removal of the affected brain tissues or severing their connections can be an effective treatment.

SUMMARY

Defined by a specific set of symptoms, each brain disorder has a unique pattern of genetic (and sometimes environmental) contributions. Huntington's disease (HD), Rett syndrome, and fragile-X syndrome are each caused by disruption of single genes. Mutations may follow Mendelian inheritance as in HD, or may be produced *de novo* as in Rett syndrome. More complex disorders such as Alzheimer's disease (AD), Parkinson's disease (PD), amyotrophic lateral sclerosis (ALS), intellectual disability (ID), and epilepsy are heterogeneous in their origin. Only a fraction of these disorders are caused by mutations in specific genes that follow Mendelian inheritance; most cases are sporadic and have incompletely defined causes that include genetic risk factors, *de novo* mutations, and environmental factors. Even more complex disorders, including all of the psychiatric disorders we discussed and the non-syndromic autism spectrum disorders (ASD), are mostly sporadic in the sense that genetic causes with full penetrance have not been identified. Whereas schizophrenia, bipolar disorder, and ASD have strong genetic contributions, environmental factors may play a more significant role in drug addiction, depression, and anxiety disorders. Future studies must carefully define the contributions of and interactions between genetic and environmental factors in the context of specific brain disorders.

A common pathological feature of neurodegenerative diseases is alteration in protein conformation, interactions, and homeostasis. AD is characterized by extracellular Aβ deposition and intracellular tau aggregation. Most PD cases involve aggregated α-synuclein. Multiple ALS-causing mutations result in the aggregation of distinct mutant proteins, with TDP-43 aggregation occurring in most sporadic cases. HD and spinocerebellar ataxia are caused by toxic gain-of-function effects associated with aggregation of polyglutamine repeats in distinct proteins. Prion diseases are caused by propagation of pathogenic PrP^Sc, which converts nonpathogenic PrP^C to PrP^Sc aggregates. The ultimate symptoms for different neurodegenerative diseases reflect the distinct neuronal types affected. AD and prion diseases affect a broad range of neuronal types, whereas PD symptoms are primarily caused by death of substantia nigra dopamine neurons, and ALS preferentially affects motor neurons.

The causal roles of mutant genes in monogenic diseases are well established, and efforts to understand diseases with more complex genetic contributions can benefit from investigating the subset caused by Mendelian mutations. For example, genetic alterations that increase APP expression, Aβ production, or propensity for Aβ oligomerization are sufficient to cause AD in humans and AD-like pathology in mouse models, suggesting that Aβ and its oligomers play a causal role in AD pathological process. Disease-causing PD mutations and drug-induced PD symptoms both implicate the importance of mitochondrial function in maintaining dopamine neuron health. While most neurodegenerative diseases do not have effective treatments, PD symptoms can be alleviated at least temporarily by L-dopa injection and deep brain stimulation, and may potentially benefit from cell-replacement therapy.

Our current understanding of psychiatric disorders has benefited from studying the actions of serendipitously discovered drugs that have therapeutic effects. For example, most antipsychotic drugs that reduce the positive symptoms of schizophrenic patients act as antagonists of the dopamine D_2 receptor. The most effective antidepressants block the action of serotonin reuptake into the presynaptic terminals. Enhancing GABAergic inhibition mediated by specific $GABA_A$ receptors is effective in reducing anxiety. Studies on the cellular effects of addictive drugs and reward-based learning have suggested that addictive drugs act by hijacking the dopamine-based reward system. As these neurotransmitter systems have broad actions in diverse brain areas, investigating specific neural circuits that mediate these drug actions and that are abnormal in psychiatric disorders are likely key for generating better treatments in the future.

Animal models that recapitulate certain disease symptoms can be used to investigate disease mechanisms and potential therapeutic strategies. This approach has elucidated the causes of syndromic neurodevelopmental disorders such as Rett syndrome and fragile-X syndrome. Rett syndrome is caused by disruption of MeCP2, a global regulator of gene expression that is particularly important in post-mitotic neurons. MeCP2 is required both during postnatal development and in adults, and reactivation of MeCP2 in adult mice can ameliorate defects caused by developmental disruption of MeCP2. Fragile-X syndrome is caused by disruption of FMRP, an RNA-binding protein involved in translational regulation. FMRP's substrates include many ASD-associated genes. Recent human genetic studies have identified increasing numbers of genes associated with psychiatric and neurodevelopmental disorders. These studies suggest that despite their diverse symptoms, many disorders share synaptic dysfunction as a common cellular mechanism and potential target for further research and treatment efforts.

FURTHER READING

Reviews

Dawson TM, Ko HS & Dawson VL (2010) Genetic animal models of Parkinson's disease. *Neuron* 66:646–661.

Holtzman DM, Morris JC & Goate AM (2011) Alzheimer's disease: the challenge of the second century. *Sci Transl Med* 3:77sr71.

Ling SC, Polymenidou M & Cleveland DW (2013) Converging mechanisms in ALS and FTD: disrupted RNA and protein homeostasis. *Neuron* 79:416–438.

Lüscher C & Malenka RC (2011) Drug-evoked synaptic plasticity in addiction: from molecular changes to circuit remodeling. *Neuron* 69:650–663.

McCarroll SA, Feng G & Hyman SE (2014) Genome-scale neurogenetics: methodology and meaning. *Nat Neurosci* 17:756.

Nestler E, Hyman SE, Holtzman DM et al. (2015) Molecular Pharmacology: A Foundation for Clinical Neuroscience, 3rd ed. McGraw-Hill.

Online Mendelian Inheritance in Man. http://www.omim.org.

Prusiner SB (1991) Molecular biology of prion diseases. *Science* 252:1515–1522.

Santoro MR, Bray SM & Warren ST (2012) Molecular mechanisms of fragile X syndrome: a twenty-year perspective. *Annu Rev Pathol* 7:219–245.

Snyder S (1996) Drugs and the Brain. Scientific American Books, Inc.

Zoghbi HY & Bear MF (2012) Synaptic dysfunction in neurodevelopmental disorders associated with autism and intellectual disabilities. *Cold Spring Harb Perspect Biol* 4:a009886.

Neurodegenerative disorders

Bergman H, Wichmann T & DeLong MR (1990) Reversal of experimental parkinsonism by lesions of the subthalamic nucleus. *Science* 249:1436–1438.

Braak H, Del Tredici K, Rub U et al. (2003) Staging of brain pathology related to sporadic Parkinson's disease. *Neurobiol Aging* 24:197–211.

Büeler H, Aguzzi A, Sailer A et al. (1993) Mice devoid of PrP are resistant to scrapie. *Cell* 73:1339–1347.

Goate A, Chartier-Harlin MC, Mullan M et al. (1991) Segregation of a missense mutation in the amyloid precursor protein gene with familial Alzheimer's disease. *Nature* 349:704–706.

Hsiao K, Chapman P, Nilsen S et al. (1996) Correlative memory deficits, Aβ elevation, and amyloid plaques in transgenic mice. *Science* 274:99–102.

Jonsson T, Atwal JK, Steinberg S et al. (2012) A mutation in APP protects against Alzheimer's disease and age-related cognitive decline. *Nature* 488:96–99.

Kang J, Lemaire HG, Unterbeck A et al. (1987) The precursor of Alzheimer's disease amyloid A4 protein resembles a cell-surface receptor. *Nature* 325:733–736.

Kim JH, Auerbach JM, Rodriguez-Gomez JA et al. (2002) Dopamine neurons derived from embryonic stem cells function in an animal model of Parkinson's disease. *Nature* 418:50–56.

Klunk WE, Engler H, Nordberg A et al. (2004) Imaging brain amyloid in Alzheimer's disease with Pittsburgh Compound-B. *Ann Neurol* 55:306–319.

Langston JW, Ballard P, Tetrud JW et al. (1983) Chronic Parkinsonism in humans due to a product of meperidine-analog synthesis. *Science* 219:979–980.

Lewis J, Dickson DW, Lin WL et al. (2001) Enhanced neurofibrillary degeneration in transgenic mice expressing mutant tau and APP. *Science* 293:1487–1491.

Limousin P, Krack, P, Pollak P et al. (1998) Electrical stimulation of the subthalamic nucleus in advanced Parkinson's disease. *N Engl J Med* 339:1105–1111.

Lindvall O, Brundin P, Widner H et al. (1990) Grafts of fetal dopamine neurons survive and improve motor function in Parkinson's disease. *Science* 247:574–577.

Luk KC, Kehm V, Carroll J et al. (2012) Pathological alpha-synuclein transmission initiates Parkinson-like neurodegeneration in nontransgenic mice. *Science* 338:949–953.

Neumann M, Sampathu DM, Kwong LK et al. (2006) Ubiquitinated TDP-43 in frontotemporal lobar degeneration and amyotrophic lateral sclerosis. *Science* 314:130–133.

Park J, Lee SB, Lee S et al. (2006) Mitochondrial dysfunction in Drosophila PINK1 mutants is complemented by parkin. *Nature* 441:1157–1161.

Roberson ED, Scearce-Levie K, Palop JJ et al. (2007) Reducing endogenous tau ameliorates amyloid beta-induced deficits in an Alzheimer's disease mouse model. *Science* 316:750–754.

Strittmatter WJ, Saunders AM, Schmechel D et al. (1993) Apolipoprotein E: high-avidity binding to beta-amyloid and increased frequency of type 4 allele in late-onset familial Alzheimer disease. *Proc Natl Acad Sci USA* 90:1977–1981.

The Huntington's Disease Collaborative Research Group (1993) A novel gene containing a trinucleotide repeat that is expanded and unstable on Huntington's disease chromosomes. *Cell* 72:971–983.

Wernig M, Zhao JP, Pruszak J et al. (2008) Neurons derived from reprogrammed fibroblasts functionally integrate into the fetal brain and improve symptoms of rats with Parkinson's disease. *Proc Natl Acad Sci USA* 105:5856–5861.

Wong CW, Quaranta V & Glenner GG (1985) Neuritic plaques and cerebrovascular amyloid in Alzheimer disease are antigenically related. *Proc Natl Acad Sci USA* 82:8729–8732.

Psychiatric disorders

Glowinski J & Axelrod J (1964) Inhibition of uptake of tritiated-noradrenaline in the intact rat brain by imipramine and structurally related compounds. *Nature* 204:1318–1319.

Pletscher A, Shore PA & Brodie BB (1955) Serotonin release as a possible mechanism of reserpine action. *Science* 122:374–375.

Rudolph U, Crestani F, Benke D et al. (1999) Benzodiazepine actions mediated by specific gamma-aminobutyric acid(A) receptor subtypes. *Nature* 401:796–800.

Saal D, Dong Y, Bonci A et al. (2003) Drugs of abuse and stress trigger a common synaptic adaptation in dopamine neurons. *Neuron* 37:577–582.

Schizophrenia Working Group of the Psychiatric Genomics Consortium (2014) Biological insights from 108 schizophrenia-associated genetic loci. *Nature* 511:421–427.

Seeman P, Chau-Wong M, Tedesco J et al. (1975) Brain receptors for antipsychotic drugs and dopamine: direct binding assays. *Proc Natl Acad Sci USA* 72:4376–4380.

Neurodevelopmental disorders

Amir RE, Van den Veyver IB, Wan M et al. (1999) Rett syndrome is caused by mutations in X-linked MECP2, encoding methyl-CpG-binding protein 2. *Nat Genet* 23:185–188.

Chao HT, Chen H, Samaco RC et al. (2010) Dysfunction in GABA signalling mediates autism-like stereotypies and Rett syndrome phenotypes. *Nature* 468:263–269.

Darnell JC, Van Driesche SJ, Zhang C et al. (2011) FMRP stalls ribosomal translocation on mRNAs linked to synaptic function and autism. *Cell* 146:247–261.

De Rubeis S, He X, Goldberg AP et al. (2014) Synaptic, transcriptional and chromatin genes disrupted in autism. *Nature* 515:209–215.

Dölen G, Osterweil E, Shankaranarayana Rao BS et al. (2007) Correction of fragile X syndrome in mice. *Neuron* 56:955–962.

Guy J, Gan J, Selfridge J et al. (2007) Reversal of neurological defects in a mouse model of Rett syndrome. *Science* 315:1143–1147.

Huber KM, Gallagher SM, Warren ST et al. (2002) Altered synaptic plasticity in a mouse model of fragile X mental retardation. *Proc Natl Acad Sci USA* 99:7746–7750.

Khelfaoui M, Denis C, van Galen E et al. (2007) Loss of X-linked mental retardation gene oligophrenin1 in mice impairs spatial memory and leads to ventricular enlargement and dendritic spine immaturity. *J Neurosci* 27:9439–9450.

Lubs HA (1969) A marker X chromosome. *Am J Hum Genet* 21:231–244.

Luikenhuis S, Giacometti E, Beard CF et al. (2004) Expression of MeCP2 in postmitotic neurons rescues Rett syndrome in mice. *Proc Natl Acad Sci USA* 101:6033–6038.

Skene PJ, Illingworth RS, Webb S et al. (2010) Neuronal MeCP2 is expressed at near histone-octamer levels and globally alters the chromatin state. *Mol Cell* 37:457–468.

Zalfa F, Giorgi M, Primerano B et al. (2003) The fragile X syndrome protein FMRP associates with BC1 RNA and regulates the translation of specific mRNAs at synapses. *Cell* 112:317–327.

CHAPTER 12
Evolution of the Nervous System

In the preceding chapters we have mostly asked questions that deal with function: How does it work? How do neurons communicate with each other? How do we see and smell? How is the wiring of the brain established during development and altered by experience? In the study of any biological system, there is a second type of question that deals with evolution: How did it arise?

Evolution has produced wonders. Consider the human eye. As we learned in Chapters 4 and 5, our rods are exquisitely sensitive, able to detect a single photon in near darkness, yet also able to discern contrast over a 10^4 range of ambient light level. Cones extend this range by an additional factor of 10^7, enabling us to see over a 10^{11}-fold range of ambient light level, and give us high-acuity and trichromatic color vision. The processing of visual signals by dozens of retinal cell types enables information about contrast, color, and motion to be extracted and transmitted to the brain. All these features are products of a developmental process that generates not only the exquisite structure of the eye but also the precise network of connections that link retinal neurons with each other and with their brain targets. How did these properties arise during evolution? What external forces shaped their emergence? What mechanisms underlie their changes over time?

The eye exemplifies the complex and multifaceted nature of evolutionary investigations. While we can definitively answer many 'how does it work' questions through carefully designed experiments, we cannot experiment on the distant past to address 'how did it arise' questions. However, studies of evolution can be aided by analyzing the results of numerous past and present-day 'experiments' carried out in nature, which have generated the wonderful diversity of life. Furthermore, as we will learn, all biological evolution must occur at the level of DNA in order for the traits to be passed to future generations; as a result, evolutionary history has been documented in DNA sequences that can be extracted from extant (and recently extinct) life forms. By combining analysis of DNA sequences along with anatomical, physiological, and behavioral traits in different life forms, and incisive simulation experiments, we can deduce answers to an increasing number of 'how did it arise' questions with an increasing level of confidence. Answering these questions deepens our understanding of the biological world by revealing clues about the unitary nature of life and the relationships of living organisms. It offers a rationale for our use of model organisms, such as the fly and mouse, to understand the human brain. It enhances our understanding of how the nervous system works by providing a historical perspective. It also fulfills our desire to know where we come from.

In this chapter, we will focus on selected topics relating to the evolution of neuronal communication, sensory systems, and neural development. Our treatment of these topics is not intended to be comprehensive, but instead to encompass examples for which there are sufficient data to construct a plausible evolutionary history, and from which we can learn about general principles of evolution. Before addressing these specific topics regarding nervous system evolution, we will first introduce several key concepts and approaches in evolutionary analysis that form the foundation for studying the specific problems presented later in the chapter.

GENERAL CONCEPTS AND APPROACHES IN EVOLUTIONARY ANALYSIS

The publication of *On the Origin of Species by Means of Natural Selection* by Charles Darwin in 1859 was a landmark in human history. Darwin's theory of evolution offered a mechanism by which life forms could change over time and provided the first satisfactory explanation of biological diversity without resorting to supernatural causes. Its profound influence on biology cannot be overestimated, as reflected by the evolutionary geneticist Theodosius Dobzhansky's remark that "Nothing in biology makes sense except in the light of evolution." In essence, Darwin proposed that (1) living organisms are not static but evolving; (2) evolution is a gradual and continuous process; (3) all living organisms descended from a common ancestor; and (4) **natural selection** is the primary force that directs the evolutionary process and its outcomes (**Figure 12–1**).

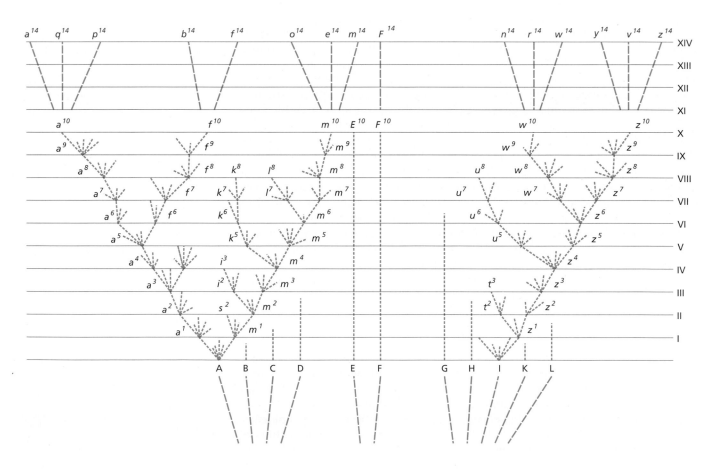

Figure 12–1 Darwin's depiction of natural selection. In this sole diagram in the *Origin of Species*, letters A to L represent the species of a hypothetical genus at the start. The horizontal distance represents the degree of their resemblance (for example, D shares more similarities with C than it does with E). The vertical distance between an adjacent pair of horizontal lines labeled by Roman numerals represents the span of 1000 generations (a number chosen arbitrarily by Darwin). The dashed lines that extend upward from species represent the species' progeny, and the fan-shaped dashed lines that radiate from many species indicate distinct variations among the progeny. For example, after 1000 generations, species A, which generates more varieties than other species, has produced two well-marked varieties, a^1 and m^1, both of which continue to produce many varieties, perhaps as an inherited trait from species A. Superscripts represent varieties at specific thousand-generations (for example, a^5 represents a variety at the 5000th generations). Most progeny lines become extinct (dashed lines that terminate) in the natural selection process; only a small subset continues to produce progeny and new varieties. Species that produce more varieties (such as A and I) are more likely to have surviving progeny lines, but this is not absolute (for example, species F). At the 14,000 generations, 15 varieties (or new species as they are sufficiently further apart from each other) have survived; the traits of these survivors differ more widely than in the founding species, as represented by the increased horizontal spread. (Adapted from Darwin C [1859] On the Origin of Species by Means of Natural Selection. John Murray).

Evolution relies on two interrelated processes: variation and selection. **Variation**, the continuous generation of differences in traits that can be inherited, is the first requirement for evolution. Genetic mechanisms were unknown to Darwin; Gregor Mendel's laws of inheritance, originally published in 1866, were not discovered by the scientific community until the beginning of the twentieth century. Subsequent advances in our knowledge of genes, mutations, molecular biology, and genomics in the twentieth and early twenty-first centuries have given us a rich understanding of how variations in DNA underlie the variations in phenotype that Darwin described. The second requirement for evolution is the selection of variations through an individual's struggle for existence and for transmitting their genetic material to the next generation. In any population, an individual whose set of genetic variations provides a better chance of reproductive success has improved odds of passing its versions of those genes to the next generation.

To reach his conclusions, Darwin compared the morphological traits of different organisms shaped by human selective breeding or by natural selection to infer their evolutionary function. Comparative studies of morphological, physiological, and behavioral traits are still widely used today to reveal similarities and differences among living organisms. The revolution in molecular biology added nucleotide and protein sequences to the list of traits that can be effectively and quantitatively compared across different organisms; these advances have provided a deeper level of understanding of the evolutionary process.

12.1 Phylogenetic trees relate all living organisms in a historical context

From *E. coli* to elephants, all organisms share the common building blocks of life. We string together the same nucleotides to form a genetic blueprint. We assemble the same amino acids into proteins that serve as the major executors of cellular functions. We employ a nearly universal genetic code for translating nucleotide sequences into protein sequences. Coupled with the biochemical similarities shared by all cells in processes ranging from energy metabolism to macromolecule synthesis, these findings provide evidence beyond doubt that all living organisms descend from a common ancestor. By comparing sequence similarities and the divergences in ancient molecules, such as ribosomal RNAs and enzymes involved in basic metabolism, **phylogenetic trees** that unify all living organisms have been constructed. Assuming a constant mutation rate, a technique called **molecular clock** uses the rates of sequence changes, calibrated against fossil records, to estimate the times at which species diverged, that is, the times at which branching occurred on the phylogenetic tree.

Life on Earth started about four billion years ago as single-celled **prokaryotes**. Prokaryotes today belong to two large branches: the eubacteria and archaea. The first **eukaryotes** (cells with a nuclear membrane that separates the nucleus from the cytoplasm) originated from prokaryotes about 2.5 billion years ago and along with eubacteria and archaea form three domains of life. A precondition in the evolution of the nervous system was the emergence of multicellular organisms, which occurred more than a billion years ago in multiple branches of the eukaryotes, including the one that gave rise to animals (**Figure 12–2**). With multicellularity came the differentiation of cells with specialized functions, such as sensor cells for detecting external stimuli, effector cells for producing movement, and connecting cells that link the sensor and effector cells. The first nervous system that featured such interconnected cells arose prior to the divergence of **cnidarians** (animals that are radially symmetrical, such as hydra, jellyfish, and corals) and **bilaterians** (animals that are bilaterally symmetrical with three germ layers, including all vertebrates and most invertebrate species present today), or even earlier (see **Box 12–1**).

Within the bilaterian lineage, the increased number of peripheral nerve nets and need for overall coordination gave rise to the centralization of neurons, forming the prototypes of the central nervous system and the brain. This centralization likely occurred prior to the divergence of **protostomes** ('mouth first,' referring

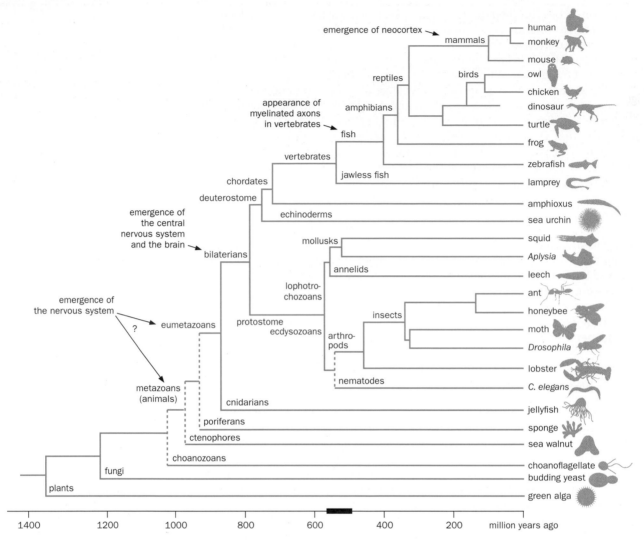

Figure 12–2 Phylogenetic tree that includes the animal kingdom.
Times of divergence were estimated based on molecular clock data
at www.timetree.org as of April 2014. Dates (see scale at bottom) are
mean estimates as different molecules and analysis methods yield
considerable variations in dating. These dates should be considered
rough approximations, especially those that pre-date the Cambrian
period highlighted in red on the timescale; prior to the Cambrian,
fossil records for calibrating the dates are scarce. Dashed vertical
lines indicate that these placements represent only the phylogenetic
relationships and not the timing of events, because of the uncertainty
of the dates. Some representative species that are mentioned in this
chapter or elsewhere in the book are shown at the right. Red arrows
point to several milestones of nervous system evolution that will be
discussed in the chapter. The question mark indicates the uncertainty
as to when the nervous system first emerged (see Box 12–1).

to the mouth appearing first during the stage of embryonic development that
connects the future digestive system to the outside world) and **deuterostomes**
('mouth second,' with the anus appearing first). During the **Cambrian** period
(542–488 million years ago), major phyla within the animal kingdom diversified,
as evidenced by an abundance of corresponding fossils. However, molecular
clock analyses suggest that the roots of bilaterian diversification originated long
before in the pre-Cambrian sea. One of the branches, the **chordates** (animals
with a notochord), gave rise to the modern vertebrate animals we know today—
fish, amphibians, reptiles, birds, and mammals—with mammals and birds
evolving independently from reptiles. Other branches led to many phyla of invertebrates that have been used as model organisms for neurobiological research
(Figure 12–2).

The construction of accurate phylogenetic trees not only unifies the animal
kingdom, but also provides an important basis for evolutionary analysis of the
gain and loss of specific traits, as we discuss in the next section.

Box 12–1: When did the nervous system first emerge?

The nervous system has traditionally been thought to first emerge in **eumetazoans**, a taxon that includes the most recent common ancestor of cnidarians and bilaterians (see Figure 12-2). This is because the nervous system is absent from poriferans (sponges), a more distantly related species of eumetazoans. However, recent phylogenetic analyses of ctenophores (comb jellies), culminating in the complete genome sequencing of *Mnemiopsis leidyi* and *Pleurobrachia bachei* (commonly known as sea walnut and sea gooseberry, respectively), have provided compelling evidence that these comb jellies, despite their morphological similarities with jellyfish, are more distantly related to jellyfish than are sponges. Yet, unlike sponges, comb jellies possess neurons, synapses, and nerve nets. These findings suggest that the nervous system might have first emerged well before poriferans

and eumetazoans diverged, and had subsequently been lost in poriferans (**Figure 12–3**A). Alternatively, the nervous systems in eumetazoans and ctenophores may have evolved independently (Figure 12–3B), using common building blocks that were present before any of the nervous systems appeared (see Sections 12.6–12.9).

While resolving these alternative hypotheses will require further data and analyses, these new findings illustrate that one of the following two principles must be at work. First, loss of traits (including the entire nervous system) can be just as significant as gain of traits in the course of evolution. Second, similar solutions to a common problem can be independently arrived at in different branches in the tree of life, a concept called convergent evolution. We will encounter examples of both repeatedly in the chapter.

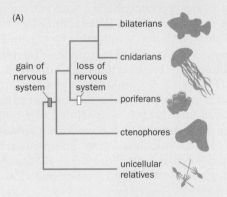

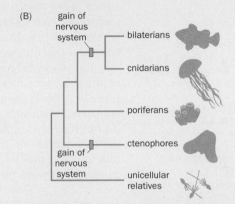

Figure 12–3 Two different views on the first emergence of the nervous system. The nervous system was previously thought to have emerged in eumetazoans, the most recent common ancestors of bilaterians and cnidarians (see Figure 12–2). The sponges (poriferans) were believed to have branched off earlier and been more distantly related to the other multicellular animals. However, recent genomic analysis indicates that the cnidarians and bilaterians are more closely related to sponges, which lack

a nervous system, than they are to comb jellies (ctenophores), which have a nervous system. Thus, either **(A)** the nervous system emerged prior to the divergence of ctenophores and the rest of the animals, and was subsequently lost in poriferans; or **(B)** nervous systems independently emerged in ctenophores and in eumetazoans. (See Ryan JF, Pang K, Schnitzler CE et al. [2013] *Science* 342:1336 and Moroz LL, Kocot KM, Citarella MR et al. (2014) *Nature* 510:109–114.)

12.2 Cladistic analysis distinguishes processes of evolutionary change

As exemplified in Box 12-1, evolutionary processes may include both the gain and loss of traits. How do we distinguish between these two types of events? Let's consider the following hypothetical example. Suppose that after comparing three animals, *a*, *b*, and *c*, we found that trait *T* is present in *a* and *b* but is absent in *c*. *T* might be a new trait acquired by animals *a* and *b*. Alternatively, *T* might have already existed in the last common ancestor of *a*, *b*, and *c*, and was lost from *c* during subsequent evolution. How do we distinguish between these alternatives?

The phylogenetic relationship between animals *a*, *b*, and *c* can provide useful clues (**Figure 12–4**). If we know that the phylogenetic tree follows Figure 12-4A, then it is more likely that *T* was lost by *c* rather than acquired by *a* and *b*, as the former scenario requires only one event—the loss of *T* after *b* and *c* diverged—whereas the latter scenario requires two separate events—the independent acquisition of *T* by the *a* and *b* clades. (A **clade** is a branch in the tree of life, consisting of an ancestor plus all of its descendants.) If we know that the phylogenetic tree follows Figure 12-4B, then both scenarios are equally probable, since each can be accounted for by a single event—the gain of *T* in the common ancestor of *a*

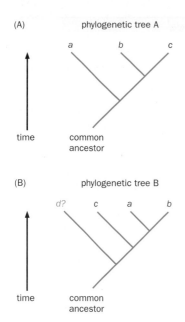

(A) phylogenetic tree A

(B) phylogenetic tree B

Figure 12–4 Illustration of cladistic analysis. Animals with trait T (a and b) are in red; animals that lack trait T (c) are in black. The likelihood that T is gained in a and b, or lost in c, can be inferred if we know the phylogenetic relationship between the three animals. **(A)** In this phylogenetic tree, T is more likely to have been lost in c (requiring a single event) rather than gained in a and b (requiring two events). **(B)** In this phylogenetic tree, the outcome is explained equally well by either of the two evolutionary processes: the loss of T in c, or the gain of T in the common ancestor of a and b. Analysis of whether T is present or absent in the outgroup d can help determine which evolutionary process is more likely to have given rise to T distribution among animals in case B.

and b after it diverged from c, or the loss of T in the c clade, assuming here for simplicity that a loss and a gain are equally likely events. To further distinguish between these possibilities, we can test whether T is present or not in a more distantly related species d, which is called an **outgroup**. The absence of T in d would favor a gain in the a/b clade, whereas the presence of T in d would favor a loss in the c clade.

This type of evolutionary comparison in the context of phylogenetic relationships is called **cladistic analysis**. The method we just used to deduce whether T was gained by a/b or lost by c is known as **maximum parsimony**, a means of generating phylogenetic predictions by giving preference to the smallest number of evolutionary changes needed to explain the data. As the example of Figure 12–4 illustrates, conclusions about evolutionary changes become more robust when more animals are included in the cladistic analysis. In fact, the phylogenetic trees, assumed to be known in the discussion in Figure 12–4 above, are usually constructed by methods such as maximum parsimony by comparing many traits (for example, protein and nucleic acid sequences) across different organisms.

Let's apply cladistic analysis to a real-world example: the evolution of gyri (ridges) and sulci (furrows) in the mammalian neocortex. Because the human neocortex is full of gyri and sulci (**gyrencephalic**), whereas the neocortex of smaller mammals such as the mouse is smooth (**lissencephalic**), one might think that gyri and sulci evolved along with the complexity of the primate brain. However, it turns out that all three major branches of mammals—the monotremes, marsupials, and placental mammals—have examples of both gyrencephalic and lissencephalic neocortices (**Figure 12–5**). Thus, gyri and sulci most likely arose in all three branches independently. It is less likely that they were independently lost by subsets of mammals in all three branches, as reptiles (the outgroup of mammals) do not have gyri, sulci, or neocortex. One possible explanation for the independent development of gyri and sulci in all three mammalian branches is that the mechanisms that underlie gyri and sulci formation are ancestral, and the degree to which they are utilized might be what is under selection. Indeed, the degree

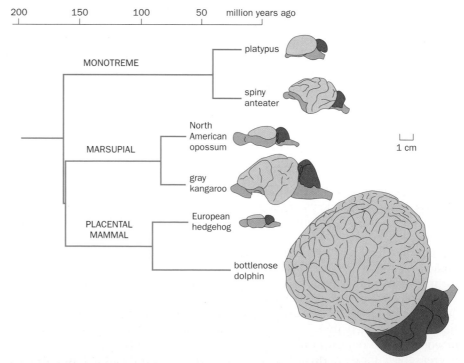

Figure 12–5 Gyri and sulci in the neocortex likely arose independently in three branches of mammals. Examples of animals with neocortices (tan) that are either smooth (lissencephalic) or folded (gyrencephalic); both traits are present in all three mammalian branches. All brains are drawn to the same scale, and large brains tend to be folded. The estimated times of divergence are according to www.timetree.org.

of gyri and sulci correlates with the brain size. As we will discuss in Section 12.21, gyri and sulci can simply be by-products of neuronal number expansion.

12.3 Gene duplication, diversification, loss, and shuffling provide rich substrates for natural selection

What is the basis of evolutionary changes? Evolutionary changes must occur at the level of DNA in order to be passed to future generations. Despite numerous efforts, there is little evidence for directed changes in DNA (that is, changes in favor of a particular trait; for an exception in prokaryotes, see Box 13–1). Instead, **mutations**—changes in DNA that insert, delete, or alter the identity of one or more base pairs—occur blindly. Natural selection tends to increase the frequency of those changes that are **adaptive**, that is, changes that render an individual and its progeny more likely to survive and reproduce in a particular environment. (Note that the term adaptation in the context of evolution has an entirely different meaning than the adaptation of sensory systems we discussed in Chapters 4 and 6.) In addition to natural selection, **allele** frequency of a gene (an allele is a specific version of a gene) can also change due to **genetic drift**, in which chance events that result in death or failure to reproduce can lead to the loss of an allele from a small population and an increased frequency of the remaining allele(s); genetic drift is caused by the fact that the alleles in the progeny are a random sample of those in the parents, reflecting the stochastic aspect of evolution. Whether via natural selection, genetic drift, or a combination, when the frequency of the allele representing the change becomes 1, the change is **fixed**.

Population geneticists use the concept of **fitness** to quantify the ability of individuals to pass their alleles to their progeny. The fitness of a specific allele can be defined as the ratio of the allele frequency in a population after one generation of selection over the allele frequency before the selection. A fitness value greater than 1 indicates that the specified allele is, on average, beneficial for the carrier's survival and reproduction in a given environment. (Fitness can also be similarly defined with respect to a specified phenotype.) Note that fitness is contingent upon environment. For example, sickle-cell anemia is a human genetic disorder caused by homozygosity of a point mutation in the β-globin gene that affects the morphology and function of red blood cells. Despite its deleterious effect when homozygous, the mutant allele can occur at relatively high frequencies in populations that live in areas where malaria infection is prevalent, because heterozygous carriers of the mutant allele are more resistant to malaria infection than non-carriers. Thus, in different environments, the mutant β-globin allele confers different fitness.

What are the mechanisms of evolutionary changes at the level of DNA? A major step in evolutionary innovation is gene duplication. If a gene is essential for survival, the effects of blind mutations are more likely to be detrimental than beneficial (that is, they decrease rather than increase fitness). If a gene is duplicated, however, then the extra copy can undergo changes more freely without compromising the survival of the animal. Even though most mutations are neutral or detrimental, beneficial changes can occasionally arise. Through trial and error across many generations, the duplicated copy of the gene may evolve to carry out the original gene's function more effectively, or it may acquire one or more entirely new functions. The evolution of cone opsin genes that brought trichromatic color vision to the Old World monkeys and apes is one such example (see Section 4.12 and later in this chapter). The most common means of producing gene duplication is through errors in DNA recombination. Unequal crossing over between nonhomologous DNA sequences that share similarity can produce an extra copy of a chromosomal segment in one daughter cell and a deletion of the same chromosomal segment in another daughter cell (**Figure 12–6**A).

When the function that a gene serves can be better served by other genes, or when the function becomes obsolete to the organism (for example, as a result of environmental changes), there is no longer selection pressure to maintain the gene's integrity. As mutations accumulate, the gene may become a nonfunctional pseudogene. We have seen examples of this in our discussion of genes that encode

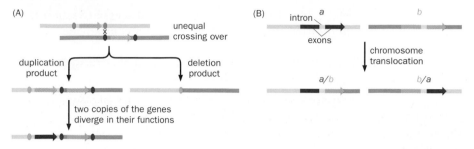

Figure 12–6 Example mechanisms for gene gain, loss, or shuffle. (A) Top, simplified schematic for unequal crossing over. During the meiotic cell cycle, error in the pairing of homologous chromosomes (gray bars) at the DNA repeat sites (light and dark blue ovals, which share sequence similarity) produces unequal crossing over (×), which gives rise to one germ cell with a duplication of the ancestral gene (yellow) and another with a gene deletion. Bottom, as each new gene acquires different mutations, the two genes diverge (one red, the other green), adding to the genome's diversity. **(B)** Exon shuffling. A translocation between two non-homologous chromosomes (two different shades of gray) is illustrated. If the breakpoint for each chromosome is located in the intron (cyan) of a gene, exons can be shuffled between the two genes (*a* and *b*). This process separates functional domains that were unified in the old proteins and pairs them with different functional domains to create new proteins (encoded by *a/b* and *b/a*).

odorant receptors (see Section 6.4). Eventually, genes that are no longer useful can be deleted by errors in DNA recombination (for example, Figure 12–6A), since events leading to their deletion are no longer selected against.

Another prevalent mechanism for creating new genes is through **DNA shuffling**. Part or all of the protein-coding sequence of one gene can be fused with that of another gene. The resulting fused gene may become able to perform some aspects of the functions previously carried out by two separate genes, or it may acquire novel properties. Conversely, a gene encoding two different domains of a protein can be split into two separate genes, each of which encodes one domain. Such shuffling can be achieved by chromosomal duplications or translocations that place part or all of a gene into a new genomic context.

Genes in eukaryotes contain many introns that separate the protein-coding exons. Translocations between the introns of two genes cause **exon shuffling** (Figure 12–6B), a special case of DNA shuffling. The emergence of repeating modules in ion channels is an example of how genes encoding individual modules can be duplicated and shuffled (see Section 12.6). DNA shuffling can also cause proteins to be expressed under the control of new regulatory elements, thus altering gene expression patterns.

12.4 Altering patterns of gene expression is an important mechanism for evolutionary change

In addition to producing new proteins or changing protein-coding sequences, recent studies have highlighted the importance of altering gene expression patterns in the evolutionary process. Let's examine some lessons we have learned from comparing the whole-genome sequences of humans and other model organisms. First, the number of protein-coding genes does not increase considerably as the nervous system becomes more complex. For example, whereas the sea anemone (a cnidarian), nematode *C. elegans*, and fruit fly *Drosophila melanogaster* have about 18,000, 19,000, and 15,000 protein-coding genes, respectively, the human genome contains only 21,000. (By comparison, the rice genome contains about 50,000 protein-coding genes.) Second, whole-genome comparison reveals that at least 6% of the genome has been highly conserved between mice and humans since their divergence about 100 million years ago (see Figure 12–2), whereas only 1.5% of the human genome consists of protein-coding sequences. Among the known functions of conserved non-protein-coding sequences are tissue-specific *cis*-**regulatory elements** (DNA elements that regulate the expression of genes on the same chromosome, such as transcriptional enhancers, repressors, and insulators) and DNA sequences that encode microRNAs (see Section 13.8) and other non-protein-coding RNAs. Note that all of these non-protein-coding

Figure 12–7 Evolution of three-spined sticklebacks. Marine sticklebacks have colonized numerous freshwater lakes and streams since the last Ice Age. Comparison of marine and freshwater sticklebacks provides a rich repertoire of phenotypic differences from which to trace evolutionary changes that have taken place over 10,000 generations as freshwater sticklebacks have adapted to their new environment. Compared to their marine counterparts (top of each pair), different freshwater varieties demonstrate loss of the pelvic hind fin (left pair; arrow indicates the pelvic hind fin), significant reductions in bony armor (middle pair), or lighter skin color (right pair). Different species of sticklebacks can be crossed to produce hybrids, allowing researchers to examine evolutionary changes using molecular-genetic and genomic methods. Each of the changes in the three examples above has been traced to variations in the *cis* elements that regulate transcription of a developmental control gene. (From Kingsley DM [2009] *Sci Am* 300:52–59. With permission from Macmillan Publishers Ltd.)

sequences and products participate in the regulation of gene expression. Third, the genome of our closest living relative, the chimpanzee, differs from our own by only 1% in nucleotide sequence. Most proteins in the human and chimpanzee genomes are identical or differ by a single amino acid substitution. Of the 40 million genetic differences between chimps and humans—comprising roughly 35 million single-nucleotide substitutions and 5 million deletions/insertions—most represent genetic drift, but some of the better-studied differences that are subject to **positive selection** include changes in noncoding RNA and *cis*-regulatory elements for transcription. (Positive selection refers to the process by which an allele that is beneficial to an organism becomes more prevalent in the population.)

Even though we are far from a complete understanding of the function of the non-protein-coding portion of the genome, these examples point to the importance of changes in gene expression patterns during evolution. Indeed, in examples ranging from the wing spot patterns of butterflies to the skeletal structures of three-spined sticklebacks (**Figure 12–7**), researchers have repeatedly identified changes in the *cis*-elements that regulate the expression patterns of developmental control genes as the underlying cause of phenotypic variations.

As discussed in Box 11–3, recent human genome sequencing studies revealed that, in comparison to the reference genome, each human contains approximately 3.5 million single-nucleotide polymorphisms, 100 gene disruption mutations, and 1000 polymorphic copy number variations. An estimated 100 new mutations are introduced with each conception. These numbers highlight the abundance of genetic variations present in our own species. Similar abundant genetic variations are likely present in other species as well. These genetic variations in copy numbers, protein-coding sequences, and gene regulatory elements provide rich substrates for natural selection.

12.5 Natural selection can act on multiple levels in the developing and adult nervous systems to enhance fitness

A major question in evolutionary biology is how inherited variations, which occur at the level of DNA and genes, contribute to natural selection. Although there has been considerable debate historically, scientists now agree that the object of selection in most cases is the individual organism. An individual that is better at finding food, avoiding predators, securing mates, and caring for offspring has a better chance of passing its version of genes (that is, its alleles) to future generations. In other words, whereas variations occur at the level of genotypes, natural selection acts on an individual's collective phenotypes. (There are notable exceptions to the individual organism being the object of selection. In an ant colony, for instance, the individual worker ants are not the objects of selection as they do not produce offspring. Nevertheless, their behavior contributes to the well-being of the queen, who produces all the progeny. Thus, the object of selection in an ant colony is likely the set of genes that contributes to the collective behavior of the entire colony.)

Figure 12–8 Natural selection can exert its effect on multiple levels in the development and function of the nervous system. Since genes can act on the level of protein, cell, circuit, behavior, and developmental process to affect nervous system function, natural selection can exert its primary effects on these levels to determine the fitness of individuals. Individuals with greater fitness are more likely to pass their versions of genes to the next generation (red feedback arrow) for a new cycle of selection. While there is a left-to-right hierarchy from protein to cell to circuit to behavior, with developmental processes largely in parallel, the reciprocal arrows show that changes at a higher level can also affect properties at a lower level.

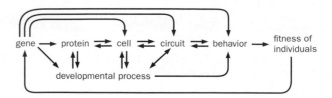

We have defined fitness with respect to individual alleles or phenotypes in Section 12.3. Fitness can also be defined with respect to the individual organism (or genome, which collectively has allelic variations in many different genes). In this context, fitness is described as the number of second-generation descendants (that is, the 'grandchildren') that the type of individual with a particular genome can expect to have. Note that this definition of fitness focuses on a type of individual to guard against stochastic events that might affect the offspring number of any particular individual. Also, because it is based on the expected number of grandchildren, this definition of fitness removes influences other than the individual's genome (for example, maternal effects, in which an individual's phenotype is affected by its mother's phenotype or genotype rather than by the individual's genome) and takes into account the well-being and fertility of the individual's offspring.

Genes can influence an individual's fitness at multiple levels of nervous system organization—molecules, cells, circuits, behaviors, and developmental processes. Likewise, natural selection, while acting on individuals as discussed earlier, can exert its primary effect on each of these levels (**Figure 12–8**): on proteins such as a sensory receptor that improves an individual's ability to detect predators; on cells such as a photoreceptor neuron that transduces light signals into electrical activity more efficiently; on circuits such as a wiring pattern that offers greater flexibility for color discrimination; on behaviors such as a daily rhythm of activity that is better adapted to the circadian cycle of the environment; and on developmental processes such as a pattern of cell division that produces more neurons for a larger brain. The combination of alleles that collectively improves the fitness of an individual will have a better chance of being passed to the next generation during natural selection (the red feedback arrow in Figure 12–8). Numerous cycles of selection occurring simultaneously in many branches of the phylogenetic tree (see Figures 12–1 and 12–2) during the past billion years have produced myriad nervous systems in animals that occupy diverse niches in the web of life.

Having introduced general concepts and approaches in evolutionary analysis, we will now investigate how neuronal communication, sensory systems, and the structure and development of the nervous system have arisen.

EVOLUTION OF NEURONAL COMMUNICATION

The emergence of functionally specialized cells in metazoans enabled the development of different cell types, such as sensory neurons to detect environmental signals, motor neurons to control muscle contraction, and muscles to execute motion. Specialized sensory and motor neurons enable better detection of food and predators and allow more sophisticated behaviors, which facilitate exploration of additional environmental niches and provide selective advantages. Simpler nervous systems in ctenophores and cnidarians are not centralized; instead, neurons are scattered around the body in the form of nerve nets. The central nervous system and brain emerged in bilaterians, likely in response to the increasing need to integrate sensory input and coordinate movement that accompanied the greater complexity of bilaterians' body plan and movement. Accompanying these developments were changes in the speed and sophistication of neuronal communication (see Movie 2–1). As a way to approach the question of how different features of the nervous system arose, we examine in the following sections the origin of the key components that allow signals to propagate within and between neurons.

12.6 Ion channels appeared sequentially to mediate electrical signaling

As we learned in Chapter 2, rapid neuronal communication relies on electrical signaling. Action potentials, a primary means of propagating signals from the soma to axon terminals, are based on the sequential opening of voltage-gated Na^+ and K^+ channels. Since both protostomes (for example, a squid) and deuterostomes (for example, a mouse) have nearly identical mechanisms for producing action potentials, Na^+- and K^+-based action potentials likely existed in early bilaterians prior to the protostome/deuterostome split (see Figure 12–2). Tracing the origin of voltage-gated ion channels suggested potential steps by which the building blocks that produce the action potential might have evolved.

Among the voltage-sensitive ion channels, K^+ channels are the most ancient since they are prevalent in prokaryotes (**Figure 12–9**A). Many bacterial species have K^+ channels with two transmembrane domains (2TM). Some bacteria also have K^+ channels with six transmembrane domains (6TM) that resemble the voltage-gated K^+ channels of animals, with S1–S4 added to the 2TM pore. Extensive **horizontal gene transfer** (that is, gene transfer from one organism to another through mechanisms other than reproduction, such as via viral transduction) during the early history of life makes it difficult to determine the origin of K^+ channels, but they likely existed well before the emergence of eukaryotes. K^+ channels function in prokaryotes to maintain ion gradients across the cell membrane and to create a negative resting membrane potential, a feature that appears to be shared by prokaryotic and eukaryotic cells.

Ca^{2+} channels were the next to appear, likely via the duplication and divergence of a gene for a 6TM K^+ channel, followed by two rounds of duplication and fusion (see Figure 12–6). The resulting 24TM channel comprises four tandem repeats of an ancestral 6TM structure. Single-cell yeast and green algae have

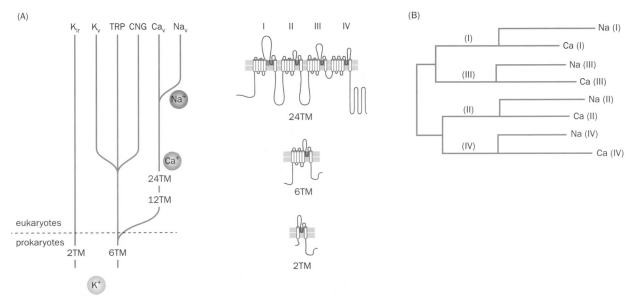

Figure 12–9 Evolutionary path of ion channels inferred from sequence comparison. (A) Of the three voltage-gated cation channels, K^+ channels are the most ancient. The two major types of K^+ channels are designated 2TM or 6TM, according to the number of their transmembrane (TM) domains (see schematic on the right, taken from Figure 2–34), and are found in both prokaryotes and eukaryotes. Voltage-gated Ca^{2+} channels diverged from 6TM K^+ channels through two rounds of gene duplication and fusion to yield a 24TM channel, from which voltage-gated Na^+ channels later derived. The four repeats of the voltage-gated Na^+ and Ca^{2+} channels are designated by Roman numerals, as shown in the schematic. Sequence comparisons suggest that TRP (transient receptor potential) and CNG (cyclic nucleotide-gated) channels also derived from 6TM K^+ channels. K_{ir}, inward-rectifier K^+ channel; K_v, Ca_v, and Na_v, voltage-gated K^+, Ca^{2+}, and Na^+ channels.

(B) Each of the four repeats of the voltage-gated Na^+ channel is most similar to the corresponding repeat of the voltage-gated Ca^{2+} channel. Furthermore, repeat I is more similar to repeat III than it is to repeats II and IV. These data suggest that the gene for the 24TM Ca^{2+} channel originated from an ancestral 6TM channel gene that underwent two sequential rounds of duplication, while the gene for the 24TM Na^+ channel arose from duplication of an ancestral 24TM Ca^{2+} channel gene. The tree was built based on the most parsimonious amino acid sequence alignment of the transmembrane segments of the rat $Na_v1.1$ and $Ca_v1.1$ channels. The length of each branch reflects the number of amino acid substitutions. (A, adapted from Hille [2001] Ion Channels of Excitable Membranes. With permission from Sinauer; B, adapted from Strong M, Chandy KG & Gutman GA [1993] *Mol Biol Evol* 10:221–242.)

24TM Ca^{2+} channels, suggesting that 24TM Ca^{2+} channels likely appeared in early eukaryotes before the split of the plant, fungal, and animal lineages (Figure 12–9A). The appearance of Ca^{2+} channels enables the control of intracellular Ca^{2+} concentration, which is important for many eukaryotic intracellular signaling events. Indeed, Ca^{2+}-binding proteins such as calmodulin are prevalent in all eukaryotic branches, suggesting that Ca^{2+} has been used as a signaling molecule since ancient times. In some animals, voltage-gated Ca^{2+} channels work together with voltage-gated K^+ channels to produce action potentials. However, this requires fluctuation of intracellular Ca^{2+} concentration, the magnitude of which is constrained by the use of Ca^{2+} as an important signaling molecule.

Voltage-gated Na^+ channels appeared on the evolutionary scene most recently, likely from the duplication and diversification of a gene encoding the 24TM Ca^{2+} channel. Comparison of amino acid sequences revealed that each of the Na^+ channel's repeat segments is more similar to the corresponding repeat of the Ca^{2+} channel than to the three other repeats in its own structure (Figure 12–9B). This suggests that the two duplications of a 6TM channel that gave rise to the 24TM channel occurred prior to the divergence of Na^+ channels from Ca^{2+} channels. The emergence of voltage-gated Na^+ channels freed Ca^{2+} from a necessary role in conducting action potentials. Ca^{2+} could be dedicated to regulate many other processes, such as biochemical reactions through Ca^{2+}-dependent kinases and phosphatases, and synaptic transmission. Moreover, since cells can tolerate greater fluctuations of Na^+ than Ca^{2+}, action potentials could be produced with larger membrane potential changes, thus enhancing the signal-to-noise ratio for long-distance propagation of electrical signals. Recent genome analysis identified genes encoding homologs of voltage-gated Na^+ channels in single-celled choanoflagellates, suggesting that the duplication event that gave rise to voltage-gated Na^+ channels occurred before the nervous system emerged. Thus, nerve cells likely co-opted a preexisting gene for a more specialized usage, rather than relying on the emergence of a new gene.

Interestingly, *C. elegans* has voltage-gated Ca^{2+} channels but no Na^+ channels. Given that the voltage-gated Na^+ channels are highly conserved between squid, insects, and mammals, it is almost certain that the last common bilaterian ancestor (see Figure 12–2) already had a voltage-gated Na^+ channel, which was subsequently lost in the clade leading to *C. elegans*. This may be because voltage-gated Ca^{2+} channels are sufficient for propagating electric signals in small animals such as *C. elegans*. The cost of maintaining the voltage-gated Na^+ channels in the genome may have outweighed the benefit.

Tracing the evolutionary history of ion channels has illustrated several principles that we will encounter again and again in the rest of the chapter. (1) Gene duplication followed by diversification is a prevalent mechanism for the evolution of new functions. (2) New functions can be built on more ancient molecules. (3) Evolution involves both gene loss and gene gain, which together shape the genomes of present-day animals.

12.7 Myelination evolved independently in vertebrates and large invertebrates

Larger animals have a higher demand for rapid propagation of action potentials. Two different solutions have evolved (see Section 2.13). The first is an increase in the diameter of the axon, as the conduction speed of action potentials is proportional to the square root of the axon diameter. We have seen a striking example in the squid giant axon, whose rapid conduction enables the squid to escape from danger swiftly. Likewise, *Drosophila* and zebrafish possess the giant fiber and Mauthner fiber, respectively, both of which are large diameter axons that connect the brain with motor neurons in the ventral nerve cord or spinal cord used for escape reflexes.

The second and more efficient solution is glial wrapping of axons, which increases the membrane resistance, reduces the membrane capacitance, and enables saltatory propagation to speed up action potential conduction and to conserve energy. Axon myelination in all vertebrates shares many common

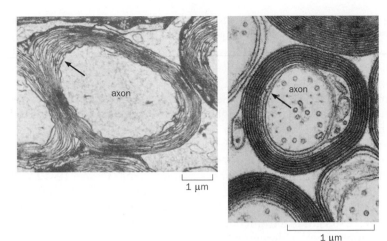

Figure 12–10 Convergent evolution of axon myelination in invertebrates and vertebrates. Electron micrograph of a cross section of a myelinated axon in the nerve cord of a sea prawn (*Palaemonetes vulgaris*) (left) and in the dog spinal cord (right). Arrows point to the myelin sheaths. (Left, from Heuser JE & Doggenweiler CF [1966] *J Cell Biol* 30:381–403. With permission from Rockefeller University Press. Right, courtesy of Cedric Raine.)

properties and likely evolved only once in jawed fish, as jawless fish such as lamprey (see Figure 12–2) do not have myelinated axons. Indeed, the appearance of myelination must have been a crucial event in the evolution of nervous systems in diverse forms of large vertebrates. Glial wrapping of axons likely evolved independently in several invertebrate clades including annelids and arthropods, because the properties and protein compositions of their glial membranes are distinct from each other and from those of vertebrates. Despite independent origins, the morphology of myelinated axons in some invertebrates closely resembles myelinated axons in vertebrates (**Figure 12–10**). Thus, myelination provides a striking example of **convergent evolution**, that is, independent evolution of similar features in animals from different clades of the phylogenetic tree.

12.8 Synapses likely originated from cell junctions in early metazoans

How did the synapse—the core of interneuronal communication—arise? Tracing the evolutionary origins of proteins found in chemical synapses by comparing the genomes of simple eukaryotes has offered interesting insights (**Figure 12–11**).

Figure 12–11 Origin of synaptic proteins. **(A)** Phylogeny of animals based on alignment of 229 conserved genes from the sequenced genomes of 18 species. The bold line represents the metazoan branch. Both genera and common taxon names are given. **(B)** Origin of individual synaptic proteins; the color code matches the phylogeny shown in panel A. See also Figure 3–10 and Figure 3–27 for the organization of presynaptic terminal and postsynaptic specialization, respectively. For proteins we have not introduced in Chapter 3: VGLUT is a vesicular transporter for glutamate; Cask is a presynaptic scaffolding protein; Homer, Shank, GKAP, GRASP are all postsynaptic scaffolding proteins; Cortactin is an actin-binding protein. α-, β- and δ-catenins are intracellular partners of cadherin. (Adapted from Srivastava M, Simakov O, Chapman J et al. [2010] *Nature* 466:720–726. With permission from Macmillan Publishers Ltd.)

(A)

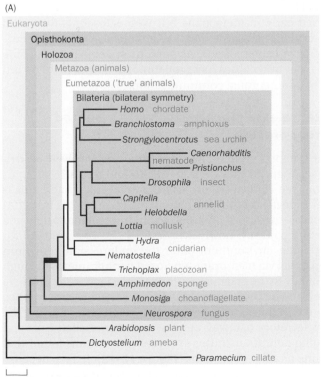

(B)

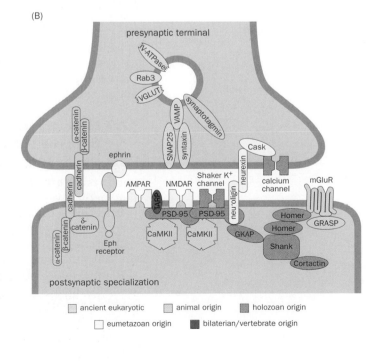

0.1 changes per site

(A)

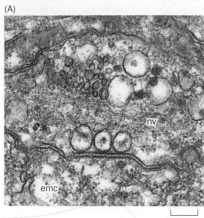

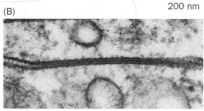

200 nm

(B)

100 nm

Figure 12–12 A chemical synapse and a gap junction in a cnidarian.
(A) Electron micrograph of a chemical synapse between a nerve cell (nv) and an epitheliomuscular cell (emc, a cell type that has properties of both epithelium and muscle) in *Hydra*. The three vesicles docked at the presynaptic membrane have larger diameter (~150 nm) than synaptic vesicles of bilaterians (~40 nm, see Figure 3–3). **(B)** Electron micrograph of a gap junction between epitheliomuscular cells in *Hydra*. (From Chapman JA, Kirkness EF, Simakov O et al. [2010] *Nature* 464:592–596. With permission from Macmillan Publishers Ltd.)

The chemical synapse is a cell–cell junction between presynaptic and post-synaptic compartments. Proteins that hold synaptic compartments together likely evolved from those that establish cell junctions in ancestral multicellular metazoans. For example, members of the cadherin family of homophilic cell adhesion molecules are abundant in the synapses of animals ranging from insects to mammals (see Figure 3–10). Because they are present in all animals, the origin of cadherins pre-dates the emergence of the nervous system. Heterophilic adhesion partners such as neurexins and neuroligins, which are prominent synapse-organizing molecules in both vertebrates and insects, first emerged in eumetazoans, a group that includes the last common ancestor of bilaterians and cnidarians. Indeed, chemical synapses in cnidarians (**Figure 12–12**A) share morphological features with the synapses we have studied extensively in bilaterians. Ephrin and Eph receptors, which play important roles in synapse formation in addition to their axon guidance functions discussed in Section 5.4, first functioned as a pair in eumetazoans although Eph receptors appeared earlier, presumably as partners for other molecule(s). The addition of heterophilic trans-synaptic interaction molecules may have contributed to the asymmetry and directionality of chemical synapses, which are distinct from the symmetrical cell junctions that form between epithelial cells. Once the trans-synaptic communication molecules in the presynaptic and postsynaptic membranes were no longer identical, it became possible for their cytoplasmic domains to interact with distinct sets of scaffolding proteins present in the pre- and postsynaptic compartments. Some of these scaffolding proteins are ancient, others co-evolved with synapses, and still others were added later as regulation of synaptic transmission became more elaborate (Figure 12–11).

What about the electrical synapses that allow direct coupling of neurons through gap junctions (see Figure 1–14; Box 3–5)? The protein classes that mediate most gap junctions in vertebrates, the connexins (see Box 3–5), appear to be a chordate innovation. But gap junctions are also prevalent in invertebrates, where they are mediated by a distinct class of proteins called innexins (invertebrate connexins; their vertebrate orthologs do not appear to function as gap junctions). Innexins were found in the genomes of cnidarians but not in sponges; the earliest function of these proteins likely involved the formation of gap junctions between epithelial cells (Figure 12–12B). Gap junctions are also prevalent in ctenophores, likely mediated by innexin family of proteins, suggesting that proteins that mediate gap junctions appeared early in metazoans but were subsequently lost in sponges. Taken together, evidence suggests that, like chemical synapses, electrical synapses also evolved from cell junctions. As nervous systems have become more complex, these two types of synapses have collaborated as independent but complementary systems of interneuronal communication.

12.9 Neurotransmitter release mechanisms were co-opted from the secretory process

Among the most ancient synaptic proteins are those that are involved in synaptic vesicle exocytosis (see Figure 12–11). The v-SNARE and t-SNARE proteins that mediate vesicle fusion (see Figure 3–8) are present in all eukaryotes. This is because neurotransmitter exocytosis utilizes a general mechanism that is fundamental to all eukaryotic cells—the fusion of vesicles with intracellular or plasma membranes. In the secretory pathway, newly synthesized proteins destined for the cell surface or extracellular environment enter the endoplasmic reticulum (ER) upon translation and travel via membrane-enclosed compartments through the Golgi apparatus and finally to the plasma membrane (see Figure 2–2). Vesicle budding and fusion are key steps in this secretory pathway. A striking convergence in scientific inquiry occurred in the early 1990s when cell biologists studying the machinery of the general secretory pathway in yeast and mammalian cells and neurobiologists investigating neurotransmitter exocytosis in neurons found that they were studying the same families of proteins (**Figure 12–13**). In fact, because of the abundance of synaptic vesicles in the brain, neurotransmitter exocytosis

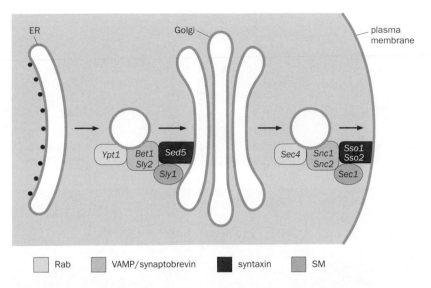

| Rab | VAMP/synaptobrevin | syntaxin | SM |

Figure 12–13 Conservation of secretion in yeast and neurons. This schematic summarizes genes identified by yeast geneticists as being required for two different steps of membrane fusion in the secretory pathway: from the endoplasmic reticulum (ER) to the Golgi, and from the Golgi to the plasma membrane. Similar to neurotransmitter release, each step is regulated by a Rab GTPase and requires a VAMP/synaptobrevin-like v-SNARE, a syntaxin-like t-SNARE, and an SM protein (see Figures 3–8 and 3–10). The names of specific genes in yeast are in italics; their colors correspond to the protein classes they encode shown at the bottom. (Adapted from Bennett MK & Scheller RH [1993] *Proc Natl Acad Sci USA* 90:2559–2563.)

has become a favorite model system for cell biologists to study the general mechanisms of vesicle fusion.

Across the synaptic cleft, many proteins enriched in the postsynaptic density also pre-date the emergence of the nervous system (see Figure 12–11), suggesting that they had ancestral functions prior to being recruited for neuronal communication. For example, postsynaptic scaffolding proteins such as PSD-95, Homer, and Shank first emerged in single-celled choanoflagellates. G-protein-coupled metabotropic glutamate receptors first arose in sponges. Ionotropic glutamate receptors such as AMPA and NMDA receptors first appeared in the eumetazoans, although an ancestral glutamate receptor is found in all animals. Ionotropic glutamate receptors have undergone expansion in ctenophores as glutamate is used as a major neurotransmitter in nervous systems that are most distant from ours (see Figure 11–2). The ancestral functions of these postsynaptic proteins may have been to sense chemicals in the environment, as will be discussed in the next part of the chapter.

In summary, proteins with specialized functions in neuronal communication were gradually recruited or added during evolution as the nervous system became more complex. Indeed, the assembly of the first nervous system relied mostly on the recruitment of pre-existing proteins rather than on multiple new proteins that arose all at once, echoing the epigraph by Francis Jacob: "evolution does not produce novelties from scratch."

EVOLUTION OF SENSORY SYSTEMS

As a primary means of detecting food, mates, and predators, sensory systems are under strong selection pressure throughout an individual organism's struggle for existence and reproductive success. Indeed, sensory systems provide dramatic examples of how interactions between certain animals and their environment over time can build upon the five common senses discussed in Chapters 4 and 6 to form distinct, extraordinary, niche-specific senses. Living in swamps and marshes, star-nosed moles rarely see and are functionally blind, but their snouts are surrounded by 11 pairs of mechanosensory appendages (rays) that allow them to capture small prey with incredible speed (**Figure 12–14**A; **Movie 12–1**). Rattlesnakes possess extremely sensitive pit organs for detecting infrared radiation (heat) that signals the presence of their warm-blooded prey nearby (Figure 12–14B). The TRPA1 channel that the pit organs employ is the most sensitive heat-activated channel identified to date. (TRPA1 channels mainly serve as irritant receptors in nociceptive neurons in mice; see Section 6.31. The fruit fly TRPA1 channel is also a heat sensor; see Figure 13–43.) As nocturnal animals, bats possess special organs to produce ultrasonic sounds and detect the resulting echoes to navigate within their environment and identify their prey (Figure 12–14C; see

Figure 12–14 Extraordinary senses.
(A) Star-nosed moles extend 11 pairs of appendages (the arrow points to one appendage) that surround the nose and use them to detect and capture small prey in swamps. **(B)** Rattlesnakes express a highly heat-sensitive TRPA1 channel in nerve terminals innervating the pit organ (red arrow), which is located between the nostril (black arrow) and the eye and is used to sense infrared radiation from warm-blooded prey. **(C)** Bats use echoes of the ultrasonic sounds they emit to navigate and hunt. **(D)** Monarch butterflies fly thousands of miles each fall from the northern United States and Canada to Mexico. To guide their migration, they use a circadian-rhythm-adjusted sun compass and possibly an additional mechanism for detecting the magnetic field. (A, from Catania KC [2012] *Curr Opin Neurobiol* 22:251–258. With permission from Elsevier Inc.; B, from Gracheva EO, Ingolia NT, Kelly YM et al. [2010] *Nature* 464:1006–1011. With permission from Macmillan Publishers Ltd; C, courtesy of Brock Fenton; D, from Reppert SM, Gegear RJ & Merlin C [2010] *Trends Neurosci* 33:399–406. With permission from Elsevier Inc.)

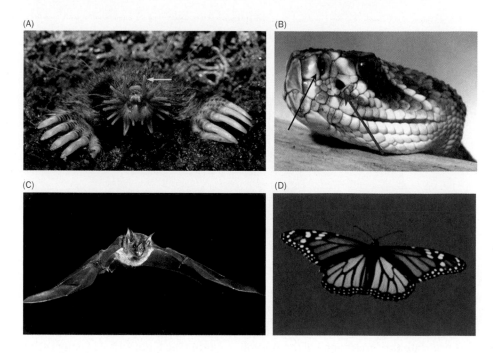

Section 6.28). Finally, monarch butterflies integrate diverse information—the direction of the sun as sensed by the eye, the circadian cycle with a clock entrained by light detected with the antennae, and probably the magnetic field (magneto-ception)—to help achieve their remarkable migration across the North American continent (Figure 12–14D).

How did sensory systems evolve? As noted above, the development of sophisticated sensory systems that have better sensitivity, speed, and resolution should aid animals in survival and procreation. However, such systems also cost more resources to build and maintain. These conflicting forces, together with historical constraints, have shaped the evolution of sensory systems. Below we use chemical senses (**Box 12–2**) and vision as examples to illustrate how sensory receptors, sensory neurons, and sensory circuits came about.

Box 12–2: Chemotaxis: from bacteria to animals

Chemical senses are probably the most ancient senses and are well developed in bacteria. Like all living beings, bacteria require nutrients. When a capillary tube containing attractants (such as amino acids) is inserted into a suspension of *E. coli*, the bacteria quickly accumulate near the mouth of the capillary in a process called **chemotaxis** (moving toward or away from a chemical source). How do bacteria achieve this?

Tracing the trajectory of individual bacteria provided important insights. In a solution of uniform concentration (an isotropic solution), bacteria exhibit two kinds of motion: runs of smooth swimming, and tumbles that result in re-orientation (**Figure 12–15**A). During a run, bacteria travel a fairly straight path, but they change direction in a random fashion during a tumble, resembling a 'random walk.' In a gradient of attractant, bacteria behave differently depending on the direction in which they are moving. When swimming away from the attractant, a bacterium behaves in a manner similar to that observed in an isotropic solution, exhibiting frequent tumbles. However, when a bacterium

swims toward an attractant, its runs are interrupted less frequently by tumbles (Figure 12–15B). This **biased random walk** strategy effectively moves bacteria toward the attractant source. Conversely, as bacteria travel down a gradient of a repellent, they tumble less frequently than in isotropic solution, thus tend to swim away from the repellent source.

The molecular mechanisms that enable bacteria to sense chemicals and modify their motion have been worked out in detail (**Figure 12–16**). Attractants and repellents are sensed by specific transmembrane receptors, which are associated with an adaptor protein, CheW, and a histidine kinase, CheA. CheA phosphorylates itself and then CheY, a response regulator. Phosphorylated CheY diffuses to the flagellar motors and changes the direction of their rotation to promote tumbles. Attractant binding to receptors inhibits CheA autophosphorylation, CheY activity, and tumbles. In contrast, repellent binding to receptors promotes CheA autophosphorylation, CheY activity, and tumbles. The CheZ phosphatase ensures a rapid turnover of phosphorylated CheY, which is essential for re-adjusting bacterial behavior.

Box 12–2: Chemotaxis: from bacteria to animals

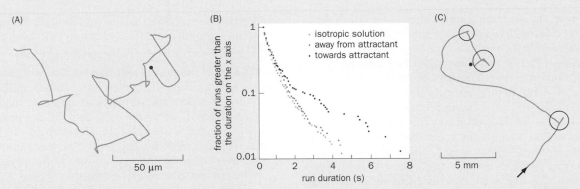

(A)

(B)

(C)

50 µm

fraction of runs greater than the duration on the x axis

- isotropic solution
- away from attractant
- towards attractant

run duration (s)

5 mm

Figure 12–15 Biased random walk in bacterial and C. elegans chemotaxis. (A) The trajectory of a wild-type *E. coli* cell in an isotropic solution was recorded for 29.5 s; linear runs are interrupted by tumbles that change the cell's direction of movement. **(B)** Fractions of runs greater than a given duration (*x* axis) are quantified for runs in isotropic solution (yellow), down an aspartate (an attractant) gradient (blue), or up an aspartate gradient (red). On average, runs are longer when bacteria move up an attractant gradient. **(C)** Track (spanning 170 s) of a *C. elegans* moving toward

a NaCl spot (dot) on a plate. Arrow indicates the start of the track. Circles indicate pirouettes, and the rest of the track is defined as runs. Note that runs can curve toward the NaCl source. (A & B, adapted from Berg HC & Brown DA [1972] *Nature* 239:500–504. With permission from Macmillan Publishers Ltd; C, adapted from Iino Y & Yoshida K [2009] *J Neurosci* 29:5370–5380. With permission from the Society for Neuroscience. See also Pierce-Shimomura JT, Mores TM & Lockery SR [1999] *J Neurosci* 19:9557–9569.)

E. coli can sense attractants such as aspartate at minute concentrations of 10 nM, which is equivalent to 10 molecules dissolved in a volume equaling that of a bacterium. At the same time, bacteria can differentiate between a broad range of concentrations—from 10 nM to 1 mM, or 10^5 fold—so

as to swim toward an attractant source. Signal amplification and sensory adaptation, two concepts introduced in Chapter 4, are key to solving these problems. Abundant chemotactic receptors and all components of the signaling pathway are clustered at one end of the cell (Figure 12–16A)

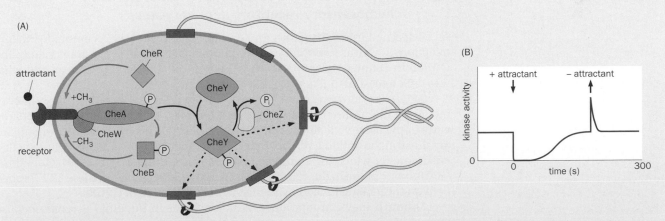

(A)

attractant

CheR

CheY

$+CH_3$

CheA

P

CheW

$-CH_3$

P

CheB

CheY

P

CheY

P

P$_i$

CheZ

receptor

(B)

+ attractant   − attractant

kinase activity

time (s)

300

Figure 12–16 Mechanisms of bacterial chemotaxis. (A) A simplified schematic of the attractive chemotaxis pathway. The receptor, the CheW adaptor, the CheA histidine kinase, the CheR methyltransferase, and the CheB methylesterase are all clustered near the pole of the bacterial cell. Autophosphorylation of the CheA histidine kinase causes the transfer of phosphate (P) to the response effector CheY, which diffuses across the cell to change the direction of the flagellar motors and thereby promotes tumbles. The CheZ phosphatase returns CheY to the dephosphorylated state. Attractant binding to the receptor inhibits CheA activity, CheY phosphorylation, and tumbles. Adaptation is mediated by two enzymes, CheR that adds a methyl group to the receptor ($+CH_3$) and reduces its sensitivity to the attractant, and CheB that removes the methyl group ($-CH_3$) and enhances its sensitivity. CheA phosphorylates CheB and enhances its activity. Thus, high attractant concentration inhibits the activity of CheA and CheB,

which promotes receptor methylation and decreases its sensitivity. Low attractant concentration elevates the activity of CheA and CheB, which promotes receptor demethylation and thereby increases its sensitivity. **(B)** Activity of the CheA kinase upon attractant addition and withdrawal. Addition of attractant causes a rapid decrease of CheA activity. The kinase activity slowly adapts due to CheR-mediated methylation of the receptor, which decreases the inhibition of CheA by attractant-bound receptor, so that more receptors must be activated to keep CheA in an inhibited state. Withdrawal of attractant causes rapid increase of CheA activity. This increase is ultimately reversed by CheA phosphorylation of the methylesterase CheB, which activates CheB and accelerates its removal of methyl groups from the receptor. Demethylation increases the receptor's sensitivity so that attractant binding more effectively inhibits CheA. (Adapted from Sourjik V [2004] *Trends Microbiol* 12:569–576. With permission from Elsevier Inc.)

(Continued)

Box 12–2: Chemotaxis: from bacteria to animals

to facilitate signal amplification. Regulation of receptor methylation at multiple sites by the methyltransferase CheR (which adds methyl groups, causing the receptor to become less sensitive) and the methylesterase CheB (which removes methyl groups, thereby restoring receptor sensitivity) allows receptors to adapt to different concentrations of an attractant (Figure 12–16B). Furthermore, CheB is a substrate of CheA, and its demethylation activity is up-regulated by phosphorylation and serves as a feedback mechanism. Attractant binding to the receptor inhibits CheA, leading to a decrease of CheB activity and a decrease of receptor sensitivity, such that a higher concentration of attractant is required to keep inhibiting CheA. Sensory adaptation allows bacteria to swim toward progressively higher concentrations of attractants, because only when bacteria are exposed to ever-increasing levels of attractant over time can tumbles be effectively inhibited.

Thus, although bacteria are separated from animals by more than two billion years of evolution and use different molecules and molecular circuits, they face the same problems of signal detection, transduction, and response we encountered in our discussions of animal sensory systems in Chapters 4 and 6. Bacterial chemotaxis solves these

problems by relying on strategies similar to those employed in the early steps of animal sensory systems: signal amplification, sensory adaptation, and feedback control.

Our discussions above indicate that bacteria assess the spatial gradient by comparing the concentration at the present time versus an earlier time (a temporal strategy). This is because bacteria are too small to compare the concentration between the two ends of the cell at any given time (a spatial strategy). With a small body size, the nematode *C. elegans* also employs a biased-random-walk strategy similar to bacteria for chemotaxis, consisting of runs and turns termed pirouettes (Figure 12–15C). However, *C. elegans* can also curve their run trajectory toward an attractant, suggesting that they can detect the direction of chemical signals during their run. This is likely achieved by swinging their heads to sample concentration differences across space. Insects place their olfactory organs at the distal ends of their two antennae to sample concentration difference across a larger distance, so that they can in principle compare odorant concentrations received from the two antennae at the same time to assist their chemotactic behavior. Evidence suggests that some mammals (including humans) can use inter-nostril comparisons to help localize an odor source.

12.10 G-protein-coupled receptors (GPCRs) are ancient chemosensory receptors in eukaryotes

As we learned from Chapters 4 and 6, mammalian sensory receptors for vision, olfaction, as well as sweet, bitter, and umami tastes are all seven-transmembrane G-protein-coupled receptors (GPCRs). When did GPCRs first appear and what was their ancestral function?

GPCRs are not found in prokaryotes, but are present in all eukaryotic branches including the protists, plants, fungi, and animals. Thus, GPCRs likely appeared early in the eukaryotic lineage. The best-studied GPCRs in unicellular organisms are those in the budding yeast *S. cerevisiae*, a model genetic organism and the first eukaryote for which the complete genome was sequenced (in 1996). Only three GPCRs were found in the budding yeast genome, and studies of these GPCRs shed light on the general function and mechanisms of action of GPCRs.

Although they can reproduce asexually through budding, *S. cerevisiae* also engage in sexual reproduction through the production and reception of peptide pheromones. Haploid budding yeast has two mating types, a and α. a-cells produce the a-factor and the receptor for the α-factor, while α-cells produce the α-factor and the receptor for the a-factor. Two cells of opposite mating type can release mating factors that activate each other's receptors. Receptor activation leads each cell to extend toward its partner a cytoplasmic projection—called a shmoo because of its resemblance to the Al Capp cartoon character—and triggers the two cells to fuse (**Figure 12–17**A). Genetic screens identified genes that constitute a mating pathway (Figure 12–17B); many of these genes were named *Ste* for their sterile phenotypes. Molecular-genetic analysis revealed that the two most upstream genes in the mating pathway encode receptors for the a-factor and the α-factor, respectively; these receptors constitute two of the three GPCRs in the yeast genome. Signals are then transduced through trimeric G proteins (see Section 3.18) and amplified by the MAP kinase cascade (see Box 3–4) to induce transcription and cell cycle arrest. In addition, G proteins also activate a separate

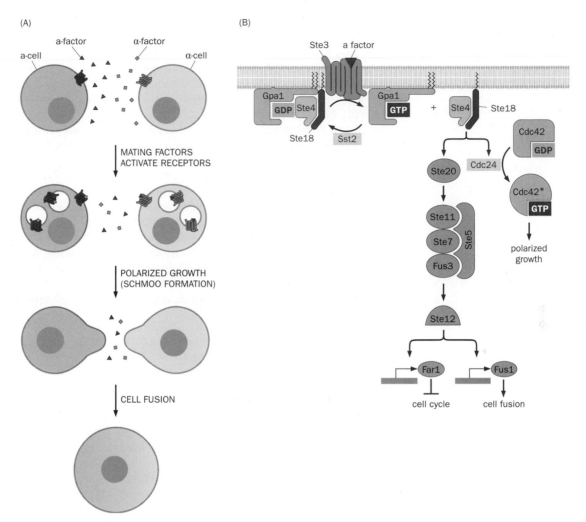

Figure 12–17 G-protein-coupled receptor (GPCR) pathways in yeast mating. **(A)** Schematic of the mating process of budding yeast. a-cells secrete a-factor, a peptide pheromone, and express a GPCR as a receptor for a different peptide pheromone, α-factor. α-cells secrete α-factor and express a GPCR as a receptor for the a-factor. The release of mating factors by cells of opposite mating type triggers receptor activation, internalization of the partner cell's mating factor along with its receptor, and activation of intracellular pathways that lead to the polarized growth of a- and α-cells toward each other and subsequent cell fusion. **(B)** Signal transduction pathways in mating yeast. Activation of Ste2 (α-factor receptor, not shown in the schematic) or Ste3 (a-factor receptor, shown here) triggers the dissociation of the βγ subunits (Ste4/Ste18) from the α subunit (Gpa1) of the trimeric G protein (see the similarity with Figure 3–31C). βγ subunits activate Ste20, a kinase that triggers the activation of the MAP kinase cascade consisting of Ste11, Ste7, and Fus3 (see Box 3–4). Ste5 is a scaffolding protein that assists the

formation of the MAP kinase complex. The kinase cascade leads to the phosphorylation and activation of the Ste12 transcription factor, which turns on many genes including Far1 (which causes cell cycle arrest) and Fus1 (which is essential for fusion). In addition, the βγ subunits also locally activate Cdc24, a guanine nucleotide exchange factor for the Rho-family small GTPase Cdc42. Cdc42 regulates polarized growth through its action on the actin cytoskeleton, thereby promoting shmoo formation at the site of the highest concentration of the partner cell's mating factor. The G protein cycle is terminated by GTP hydrolysis of Gpa1-GTP assisted by the GTPase activation protein Sst2, causing re-association of βγ subunits with Gpa1-GDP. In addition to the trimeric G protein cascade, many downstream pathways similar to those shown here participate in neuronal signaling, including the MAP kinase cascade in synapse-to-nucleus signaling (see Section 3.23) and small GTPase Cdc42 signaling in growth cone guidance (see Box 5–2). (Based on Herskowitz I [1995] *Cell* 80:187–197 and Akrowitz RA [2009] *Cold Spring Harb Perspect Biol* 1:a001958.)

pathway to direct the polarized growth of shmoos via the Rho-family small GTPase Cdc42 (Figure 12–17B).

Unlike motile bacteria, which compare attractant concentrations at different times to regulate run versus tumble (see Box 12–2), yeast cells are mostly stationary. They must detect the pheromone gradient spatially in order to direct their polarized growth toward a mating partner, which is the source of the highest concentration of peptide hormone. Yeast cells employ several strategies to ensure reproductive success: mating factors are lipid modified to diffuse slowly over short distances; recipient cells secrete proteases to degrade the partner's mating factor so as to sharpen the gradient; and mating factors are internalized once they bind to receptors (see Figure 12–17A) to further limit their range of diffusion.

The third GPCR in the budding yeast is a receptor for sugars. It senses extracellular glucose or sucrose and in response activates a Gα protein distinct from the one used in the mating pathway. The downstream pathway is a type commonly associated with GPCR signaling, with the Gα protein activating an adenylyl cyclase that produces cAMP and thereby activates PKA (see Figure 3–33). GPCRs in other fungi and protists that have been studied all participate in sensing sugars, amino acids, other nutrients, or pheromones.

Given the striking conservation between the molecular components and signaling mechanisms of GPCRs employed in present-day single-cell and multicellular organisms, it is likely that their single-cell common ancestors already employed the signaling pathway, and the ancestral function of GPCRs in unicellular organisms is chemoreception of nutrients and pheromones. The sensory systems of multicellular organisms essentially inherited these functions. When animals left the predominantly aqueous environment where early unicellular organisms resided, the GPCRs diversified in function to detect volatile chemicals. GPCRs also expanded for use in intercellular communications, as all metabotropic neurotransmitter receptors are GPCRs.

12.11 Chemosensory receptors in animals are predominantly GPCRs

Following the precedent set in unicellular organisms, most multicellular organisms continue to dedicate the majority of GPCRs to sensory systems that detect chemicals. In *C. elegans*, hundreds of predicted GPCRs are expressed in ciliated chemosensory neurons that allow these nematodes to detect soluble and volatile chemicals in the environment (see FIgure 6–24). Some mammalian species have evolved more than 1000 odorant receptors alone (see Figure 6–10). Mammals that possess a functional accessory olfactory system (for example, rodents) also employ up to a few hundred additional GPCRs for detecting chemicals produced by conspecifics, predators, and prey (see Box 6-1).

Chemoreceptors in the sensory neurons of multicellular organisms must convert the binding of a chemical ligand into an electrical signal in order to transmit information to the rest of the nervous system. From *C. elegans* to mammals, modulation of two types of channels—cyclic nucleotide-gated (CNG) channels and transient receptor potential (TRP) channels (see Box 2-4)—are among the most commonly employed effectors of GPCR activation in chemosensory neurons (**Figure 12–18**). Both CNG and TRP channels derived from K+ channels in early eukaryotes (see Figure 12–9). Opening of these channels causes cation influx and depolarization of sensory neurons, which triggers neurotransmitter release from sensory cells and thereby transmits information to second-order neurons. Signals can also be sent by closing these channels, causing hyperpolarization of the sensory neurons and decreased neurotransmitter release, as appears to be the case for *C. elegans* olfaction (see Section 6.12).

12.12 Two distinct families of ligand-gated ion channels cooperate to sense odors in insects

While GPCRs perform most of the chemical sensing in the animal kingdom, insect olfaction provides an exception that is instructive about evolutionary processes. As we learned in Chapter 6, both insect and mammalian olfactory systems use glomeruli to organize axonal input from olfactory receptor neurons (ORNs)

Figure 12–18 Chemosensory GPCRs activate transduction pathways that change the open probability of CNG or TRP channels. Most chemosensory GPCRs activate signal transduction pathways that eventually lead to the opening of CNG or TRP channels and depolarization of the sensory neurons. Both CNG and TRP channels are tetramers of 6TM channels derived from 6TM K+ channels (see Figure 12–9A). See Figure 6–4 for a specific example of this generic pathway in mammalian olfaction.

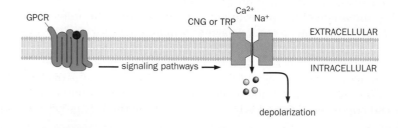

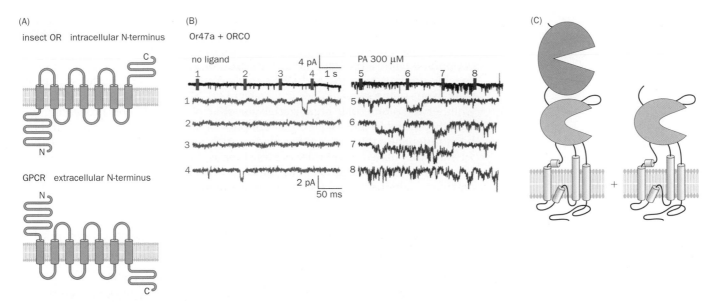

(A) insect OR   intracellular N-terminus

GPCR   extracellular N-terminus

(B) Or47a + ORCO

no ligand   4 pA 1 s

PA 300 µM

2 pA
50 ms

(C)

+

Figure 12–19 Insect odorant receptors appear to be ligand-gated cation channels. (A) Insect ORs have an inverted membrane topology compared to GPCRs. **(B)** Co-expression in *Xenopus* oocytes of *Drosophila* Or47a, a specific OR activated by pentyl acetate (PA), and ORCO, a co-receptor, led to inward currents in an excised membrane patch (see Box 13–2) in response to PA application (red) compared to control (blue). Traces in the numbered regions are expanded below. **(C)** A subset of *Drosophila* olfactory receptor neurons uses heteromultimers of ionotropic receptors as odorant receptors. The common subunit at left shares the same structure as the subunit of the ionotropic glutamate receptor (see Figure 3–26); the specific subunit at right also shares similarity with the ionotropic glutamate receptor subunit but lacks the N-terminal domain. (A, adapted from Benton R, Sachse S, Michnick SW et al. [2006] *PLoS Biol* 4:e20; B, adapted from Sato K, Pellegrino M, Nakagawa T et al. [2008] *Nature* 452:1002–1006. With permission from Macmillan Publishers Ltd; C, based on Abuin L, Bargeton B, Ulbrich MH et al. [2011] *Neuron* 69:44–60.)

(see Figure 6–27). Like the GPCR odorant receptors identified in mammals and *C. elegans*, the first olfactory receptors (ORs) identified in *Drosophila* were predicted to contain seven transmembrane domains. Thus, it was assumed that insect ORs were also GPCRs. However, further studies indicated that, unlike GPCRs, which have an extracellular N-terminus and an intracellular C-terminus, insect ORs have the opposite configuration (**Figure 12–19**A). In addition to expressing a unique OR, each ORN also expresses a co-receptor (ORCO) that has the same 7TM topology as other ORs and is highly conserved among insects. Together, OR and ORCO can act as a ligand-gated ion channel (Figure 12–19B).

A second family of *Drosophila* odorant receptors belongs to an entirely new class of proteins named ionotropic receptors (IRs), which resemble ionotropic glutamate receptors. IRs function as heterotetramers, composed of a common subunit that retains the architecture of an ionotropic glutamate receptor subunit (see Figure 3–26) and a specific subunit that is also similar to an ionotropic glutamate receptor subunit but does not have the N-terminal domain (Figure 12–19C). Whereas the OR family is specific to the insect lineage, IRs are present in a broad range of invertebrates, including nematodes and mollusks. IRs in *Aplysia* and *C. elegans* are also expressed in chemosensory neurons. Thus, IRs may have an ancestral chemosensory function in all protostomes. It is unclear why insects use ionotropic receptors instead of GPCRs as olfactory receptors. One possibility is that ligand-gated ion channels transmit signals faster than GPCRs. As discussed in Chapter 6, ion channels are used in mammals for sour, salt, and trigeminal chemosensation (for example, TRPV1 for hot spice).

These findings provided interesting insights into the evolution of sensory systems. First, the extant olfactory receptor repertoire in *Drosophila* appears to have been acquired in a piecemeal fashion during evolution, with 20% of ORNs utilizing the more ancestral IR family and the remaining 80% using the more recently evolved OR family. Despite employing distinct receptors, these ORNs cooperate to provide complementary coverage of the chemical world, and their axons project to complementary glomerular targets within the same antennal lobe (see Figure 6–27). Second, the identification of chemoreceptors that resemble ionotropic glutamate receptors suggests a close link between chemosensation and detection of neurotransmitter release in postsynaptic neurons—after all, both of these processes transform ligand binding into an electrical signal. Third, despite using completely different odorant receptors, the olfactory systems of insects and vertebrates share strikingly similar organization, with most ORNs expressing one specific odorant receptor, and ORNs that express the same receptor projecting their axons to the same glomeruli (see Figure 6–27). It remains to be determined whether the glomerular organization for olfactory information processing was already present in the last common ancestor of insects and vertebrates (and has been independently lost in other clades, such as nematodes and mollusks), or whether this shared glomerular organization is the product of convergent evolution to solve a similar problem in insects and mammals. The finding that insect and mammalian olfactory systems employ different classes of receptors favors the convergent evolution hypothesis.

12.13 Retinal- and opsin-based light-sensing apparatus evolved independently at least twice

Light sensing, like chemical sensing, is an ancient faculty. Light enables phototrophic archaea, bacteria, algae, and plants to generate chemical energy, directs the motion of a wide range of organisms, and regulates physiology in a circadian fashion in nearly all forms of life. In each of these processes, light reacts with a chromophore, the structure of which is altered after it absorbs a photon, causing an associated protein to change its conformation.

The process in microorganisms that most closely resembles animal vision is **phototaxis**, the ability to move toward or away from a light source. Light sensing in bacterial phototaxis is mediated by **sensory rhodopsins**, which activate membrane-anchored histidine kinases to transduce signals that regulate the flagellar motor in a manner similar to chemotaxis (see Box 12–2). Remarkably, bacterial sensory rhodopsins consist of an opsin protein with seven transmembrane helices and a retinal as the chromophore just like animal rhodopsins. Do animal vision and bacterial phototaxis share a common evolutionary origin?

Several lines of evidence argue against this hypothesis. First, although they both possess seven transmembrane helices, the opsins in the prokaryotic sensory rhodopsins (type I rhodopsins) and animal rhodopsins (type II rhodopsins) share no sequence similarities and differ in the arrangement of their transmembrane helices (**Figure 12–20**A). Second, although both use retinal as the chromophore,

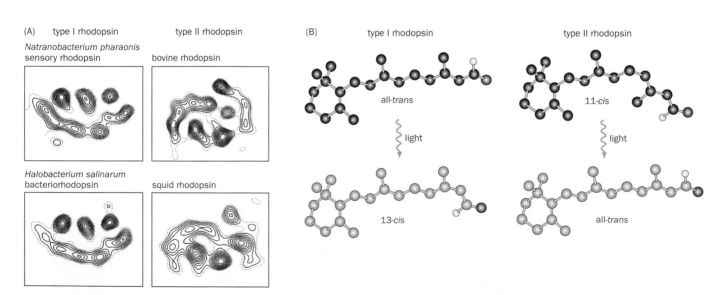

Figure 12–20 Differences between type I and type II rhodopsins.
(A) Electron density maps of the transmembrane helices of two type I rhodopsins from archaea (left) differ from those of type II rhodopsins from animals (right). **(B)** In type I rhodopsins, light absorption triggers isomerization of all-*trans* retinal to 13-*cis* retinal (left), whereas in type II rhodopsins, light absorption triggers isomerization of 11-*cis* retinal to all-*trans* retinal (right). The black circle is a from a lysine residue of the opsin (see Figure 4–6). Adapted from Spudich JL, Yang CS, Jung KH et al. [2000] *Annu Rev Cell Dev Biol* 16:365–392. With permission from Annual Reviews.)

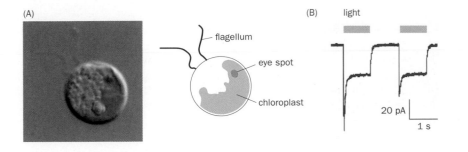

(A)

(B)

Figure 12–21 Light sensing in
***Chlamydomonas.* (A)** In addition to
harvesting light energy for photosynthesis
in the chloroplast, this single-cell green
alga uses light-sensing pigments in the
eyespot to direct flagellar movement
for phototaxis. In the latter process,
light is detected by two type I sensory
rhodopsins, named channelrhodopsin-1
and channelrhodopsin-2 (ChR1 and
ChR2). **(B)** Excised patch clamp recording
of a piece of plasma membrane from a
Xenopus oocyte that expresses ChR2.
In the presence of exogenous all-*trans*
retinal, light induces an inward current
from the membrane patch that contains
ChR2. (A, courtesy of Moritz Meyer;
B, adapted from Nagel G, Szellas T,
Huhn W et al. [2003] *Proc Natl Acad Sci
USA* 100:13940–13945. Copyright The
National Academy of Sciences, USA.)

photon absorption triggers different retinal isomerization reactions, converting all-*trans* retinal to 13-*cis* retinal in type I rhodopsins, whereas 11-*cis* retinal is converted to all-*trans* retinal in type II rhodopsins (Figure 12–20B). Third, only type II rhodopsins are G-protein-coupled receptors.

Phylogenetic analysis indicates that prokaryotic sensory rhodopsins derived from more ancient type I rhodopsins that harvest light energy. **Bacteriorhodopsin**, a light-driven proton pump, and **halorhodopsin**, a light-driven chloride pump, are used by some prokaryotes to convert solar energy to chemical energy in the form of ionic gradients across the membrane (see Section 2.4). Type I rhodopsins are also present in many eukaryotic microbes, including algae, fungi, and amoeba. For example, the single-cell green alga *Chlamydomonas reinhardtii* has two type I rhodopsins that are enriched in the eyespot (**Figure 12–21**A) and play a role in phototaxis. When expressed in *Xenopus* oocytes or mammalian cells these proteins function as light-activated cation channels (Figure 12–21B), and therefore were named **channelrhodopsins**. Remarkably, because mammalian neurons contain endogenous retinal, expression of *Chlamydomonas* channelrhodopsin-2 (ChR2) in mammalian neurons can cause light-induced depolarization and hence neuronal activation. Halorhodopsin, on the other hand, has been adopted for inactivating neurons because it pumps Cl⁻ into mammalian neurons in response to light, resulting in hyperpolarization. These light-activated channels and pumps have become powerful tools for manipulating the activity of genetically defined neurons, as we have encountered numerous times in preceding chapters (see also Section 13.25 for more details).

Type II rhodopsins are present in cnidarians and bilaterians but not in sponges, and thus likely first emerged in eumetazoans. The overall structure of type II opsin protein is much closer to other eukaryotic GPCRs than to type I opsins. Thus, nature has coupled opsin with retinal to form a light-sensing apparatus at least twice independently: once in the ancient prokaryotes, generating a light sensor that later spread to eukaryotic microbes, and again during early metazoan evolution, modifying chemosensory GPCRs to produce what would become a critical component in all visual systems of animals.

12.14 Photoreceptor neurons evolved in two parallel paths

We have focused on the evolution of sensory receptors in our discussion so far. Emergence of a light-sensing molecule is just one step in the evolution of animal vision. Rhodopsins must be placed in appropriate sensory neurons (photoreceptors), and light activation of rhodopsin must be linked with signal transduction and amplification pathways to convert light sensing to electrical signals. Two distinct types of photoreceptors are widely used in animals. Vertebrate rods and cones, which we studied in detail in Chapter 4, belong to the **ciliary type**, so named because the outer segments, into which the light-sensing opsins are packed, derive from the **primary cilium**, a short, single, non-motile cilium that projects from the surface of many animal cell types and is used in other contexts as a signaling center. The eyes of most invertebrates, including the compound eyes of flies (see Figure 5–35) and the sophisticated eyes of cephalopods, use photoreceptors of the **rhabdomeric type**, in which the apical surface folds into

Figure 12–22 Ciliary versus rhabdomeric photoreceptors. Ciliary photoreceptors and rhabdomeric photoreceptors, which are used respectively in vertebrate rods and cones and in the eyes of many invertebrate phyla, differ in many features. Ciliary photoreceptors pack opsins in the primary cilia; they utilize c-subclass opsins and a phototransduction pathway involving a transducin-type Gα subunit that activates phosphodiesterase (PDE) to hydrolyze and degrade cGMP, which causes the closure of cyclic nucleotide-gated (CNG) cation channels and hyperpolarizes the membrane potential. Rhabdomeric photoreceptors pack rhodopsins in microvilli; they utilize r-class opsins and a phototransduction pathway involving a G_q-type Gα subunit that activates phospholipase C (PLC) to produce diacylglycerol (DAG), leading to the opening of transient receptor potential (TRP) channels that depolarizes the membrane potential. (Adapted from Fernald RD [2006] *Science* 313:1914–1918.)

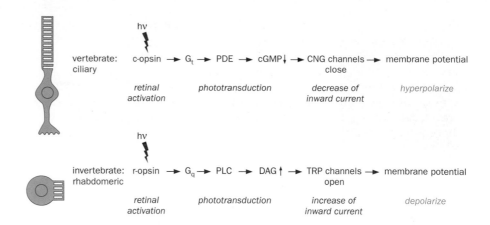

microvilli to house opsins. In addition to these morphological distinctions, the two photoreceptor types also differ in many other aspects (**Figure 12–22**).

Although both photoreceptors use G-protein-coupled type II rhodopsins, their opsins belong to two distinct subfamilies based on sequence similarities. Rhabdomeric photoreceptors contain r-opsins, while ciliary photoreceptors employ c-opsins. The phototransduction pathways also differ significantly. Ciliary photoreceptors use the heterotrimeric G protein transducin (G_t) to couple photon absorption to the activation of phosphodiesterase (PDE), which reduces cGMP levels, causing the closure of cGMP-gated channels (Figure 12–22, top; see also Figure 4–10). Thus, photon absorption leads to hyperpolarization of the photoreceptor. The phototransduction pathway in *Drosophila*, which likely applies to other rhabdomeric photoreceptor neurons, uses a different G protein, G_q, that couples photon absorption to the activation of a phospholipase C (PLC) to produce diacylglycerol (DAG) (see Figure 3–34); this triggers the opening of a TRP cation channel through as-yet-unidentified mechanisms (the founding member of TRP channels was discovered in this context). Thus, photon absorption leads to depolarization of the photoreceptor (Figure 12–22, bottom). How did these two types of photoreceptors arise?

Studies in recent years have found that ciliary and rhabdomeric photoreceptors have coexisted in both vertebrate and invertebrate clades (**Figure 12–23**). In fact, opsins in cnidarians such as jellyfish resemble more closely the c-opsins used in vertebrate photoreceptors, and phylogenetic analysis suggests that c-opsin and r-opsin diverged very early, pre-dating the cnidarian–bilaterian split. In both the

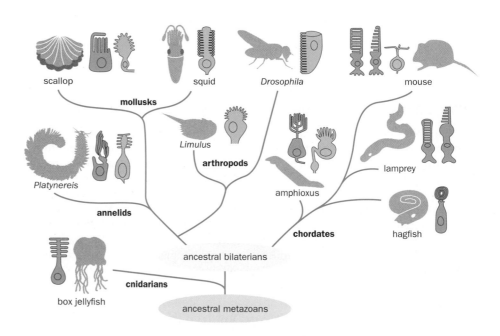

Figure 12–23 Parallel evolution of ciliary and rhabdomeric photoreceptors. Ciliary (green) and rhabdomeric (blue) photoreceptors coexist in many phyla of animals, indicating an early origin for both types of photoreceptors. (Adapted from Fain GL, Hardie R & Laughlin SB [2010] *Curr Biol* 20:R114–R124. With permission from Elsevier Inc.)

invertebrate and vertebrate branches of the bilaterian lineage, remnants of both types of photoreceptors have been found.

Although rhabdomeric photoreceptors dominate most invertebrate phyla, ciliary photoreceptors coexist in certain animals. In scallops, hyperpolarizing and depolarizing photoreceptors have been found side-by-side in the same retina. Like most invertebrates, the marine annelid *Platynereis* uses r-opsin in the eye. However, c-opsin has been found in *Platynereis* brain neurons associated with rhabdomeric structures, likely for sensing light intensity for circadian rhythm regulation.

In the chordate clade, amphioxus (a basal chordate; see Figure 12–2) also possesses both rhabdomeric and ciliary photoreceptors. In vertebrates, including humans, remnant rhabdomeric photoreceptors coexist with the dominant ciliary photoreceptors. As discussed in Box 4–2, the intrinsically photosensitive retinal ganglion cells (ipRGCs), whose functions include circadian rhythm entrainment and pupil constriction, can directly sense light independent of rods and cones. ipRGCs uses melanopsin, an r-opsin, for light detection, and its transduction requires a phospholipase C and TRP channels just like the rhabdomeric photoreceptors in *Drosophila*. These data suggest that mammalian ipRGCs and invertebrate rhabdomeric photoreceptors derived from a common ancestor.

While it remains unknown why vertebrate vision is predominantly based on ciliary photoreceptors and invertebrate vision on rhabdomeric photoreceptors, the parallel evolution of these two photoreceptor types illustrates that multiple solutions exist for a common problem, with individual organisms using these solutions in different ways based on their evolutionary history. Indeed, evolutionary analyses of eye morphology suggest that each of the photoreceptor types has been independently integrated into many different types of eyes (**Box 12–3**). The collaboration of different light-sensing strategies in the same animal also illustrates that natural selection does not always produce a 'winner takes all' scenario.

Box 12–3: Darwin and the evolution of the eye

The eye has occupied a special place in the history of the evolutionary theory. In a chapter entitled 'Difficulties on Theory' from the first edition of *Origin of Species*, Darwin considered the eye as an "organ of extreme perfection." He stated that "To suppose that the eye, with all its inimitable contrivances for adjusting the focus to different distances, for admitting different amounts of light, and for the correction of spherical and chromatic aberration, could have been formed by natural selection, seems, I freely confess, absurd in the highest possible degree." Indeed, attackers of the evolutionary theory have often used the eye as an example of the improbability of natural selection. But Darwin went on, "Yet reason tells me, that if numerous gradations from a perfect and complex eye to one very imperfect and simple, each grade being useful to its possessor, can be shown to exist; if further, the eye does vary ever so slightly, and the variations be inherited, which is certainly the case; and if any variation or modification in the organ be ever useful to an animal under changing conditions of life, then the difficulty of believing that a perfect and complex eye could be formed by natural selection, though insuperable by our imagination, can hardly be considered real."

In the 150 years since *Origin of Species* was published, studies on the structure, function, and development of the eye have validated Darwin's intuition, revealing the details of eye evolution at multiple levels of organization, from molecules to circuits. In research based primarily on morphological criteria, a careful comparative study of photoreceptors and eyes in different organisms concluded that the eye as a photo-sensing apparatus has evolved independently 40 to 65 times across animal phyla. Twenty phylogenetic lines include animals that have a regular series of 'ever more perfect' eyes among still-living relatives. The 'ever more perfect' eye has also been a subject of a simulation experiment, with the objective of finding the number of generations required for the evolution of an eye's optical geometry. The study supposed that the driving force is for higher spatial resolution (visual acuity), and that changes take the form of sequential 1% steps of modification between two successive stages. For example, doubling the length of a structure would take about 70 1% steps of modification (since $1.01^{70} \approx 2$). An estimated 1829 steps were needed to convert a simple patch of photosensitive epithelium into an eye with a lens focusing on a curved layer of photosensitive cells as in the vertebrate retina (**Figure 12–24**). A conservative estimate using inheritance of quantitative traits gave about 360,000 generations, or less than half a million years if the generation time is about 1 year. Considering that both r-opsin-based and

(Continued)

Box 12–3: Darwin and the evolution of the eye

stage

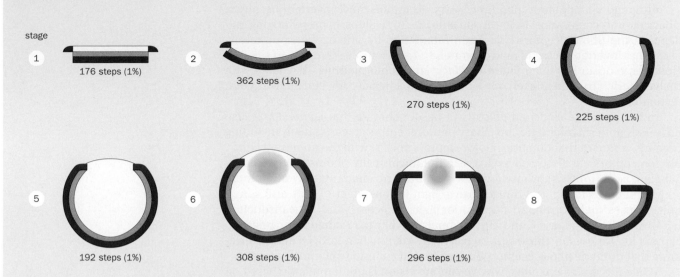

Figure 12–24 Simulating the evolution of optical geometry.
The evolution from a patch of light-sensitive cells to an eye with a focused lens is divided arbitrarily into eight stages. Changes between successive stages are divided into 1% steps of modification. In stage 1, the initial structure (referred to as 'retina') consists of a layer of light-sensitive cells (orange) sandwiched between a transparent protective layer (yellow) and a layer of dark pigment (red). In stages 2 and 3, the retina invaginates to form a sphere. In stages 4 and 5, the retina continues to grow without changing its radius, causing the retinal pit to deepen and an aperture to form. In stages 6 through 8, a graded-index lens appears with gradual shortening of the focal length; the focal length equals the distance to the retina by stage 8, producing a sharply focused system. The spatial resolution increases along each of the modification steps. (Adapted from Nilsson DE & Pelger S [1994] *Proc R Soc Lond B* 256:53–58.)

c-opsin-based photoreceptors most likely existed before the cnidarian–bilaterian divergence around 1 billion years ago, a half-million-year period is a very short time in which to perfect optical geometry.

Hermann von Helmholtz, a nineteenth century physicist and authority on vision, recognized that the human eye is not perfect after all. After citing numerous defects such as spherical aberration and blind spots created by the optic nerve head, Helmholtz stated that "Now it is not too much to say that if an optician wanted to sell me an instrument which had all these defects, I should think myself quite justified in blaming his carelessness in the strongest terms, and giving him back his instrument." However, Helmholtz went on, appreciating Darwin's evolutionary theory: "We have now seen that the eye in itself is not by any means so complete an optical instrument as it first appears: its extraordinary value depends on the way in which we use it: its perfection is practical, not absolute, consisting not in the avoidance of every error, but the fact that all its defects do not prevent its rendering the most important and varied services."

In essence, natural selection does not produce perfect products; it produces 'good enough' products that help animals to survive and reproduce. This is because natural selection is constrained by evolutionary history and by the competing energy costs and fitness benefits of maintaining complex structures. If one were to chart a fitness landscape with peaks and troughs, natural selection can only bring organisms to a local peak of fitness from any given starting point; it does not have the foresight to overcome further troughs to reach distant and perhaps higher peaks.

12.15 Diversification of cell types is a crucial step in the evolution of the retinal circuit

The fitness benefit of vision is that, in environmental niches exposed to light, the extraction of useful features from light signals enables animals to identify behaviorally relevant objects such as food, predators, and mates. This feature extraction starts in retinal neural circuits downstream from the photoreceptors. As we learned in Chapter 4, the mammalian retina has more than 60 well-characterized neuronal types. Photoreceptors (rods and cones) provide the input; retinal ganglion cells send the output to the brain; bipolar cells connect photoreceptors with retinal ganglion cells; horizontal and amacrine cells participate in transforming light intensity signals received from the input cells into contrast, motion, and color signals transmitted by the output cells. How did such exquisite circuits that integrate so many types of neurons come about?

We are far from being able to answer this question satisfactorily. A key step must be the creation of different cell types. It is generally believed that as animals evolve sophisticated functions, cells become more and more specialized such that each increasingly distinct cell type performs a subset of the functions once performed by an ancestral cell that was multifunctional. Viewed from this perspective, unicellular microbes capable of phototaxis may be considered the most versatile living cells. It is, at the same time, a 'sensory neuron' that detects light, an 'interneuron' that integrates sensory information, a 'motor neuron' that converts the information into instructions for motion, and a 'muscle cell' that produces motion (flagellar movement). Multifunctionality comes at a cost: each function is constrained by other functions the cell must perform. One way to improve existing functions or add new ones is to duplicate and diversify the cells, similar to the results obtained through the duplication and diversification of genes (see Figure 12–6A). As cellular diversity increases, each cell can perform a specialized function with greater sophistication, free from the constraint of having to perform other functions.

The origin of the bipolar cell provides an illuminating example in the evolution of vertebrate retinal cell types and circuits. Bipolar cells share many similarities with photoreceptors in their shape, developmental sequence, synaptic organization, and gene expression patterns, suggesting that bipolar cells might be phylogenetic 'sister cells' to photoreceptors. Indeed, in hagfish, a basal vertebrate (see Figure 12–23), bipolar cells are absent and photoreceptors directly connect with retinal ganglion cells. It is likely that bipolar cells originated from photoreceptors in early vertebrates through division of labor. The functions of light sensing and signal propagation, both of which were once carried out by photoreceptors, were gradually distributed across two different cell types, one of which became specialized in light sensing and lost its connection to the retinal ganglion cells, while the other became specialized in communicating signals from the light-sensing cells to the retinal ganglion cells and lost the capacity to sense light (**Figure 12–25**).

Diversification and division of labor may also account for the production of different subtypes of retinal cells, including retinal ganglion cells, amacrine cells, and photoreceptors. For example, rods likely evolved from photoreceptors that resemble modern-day cones during the early evolution of chordates (Figure 12–25). A more recent example of photoreceptor diversification is the elaboration of cones that endowed some primates including humans with trichromatic color vision, which we will study in the next two sections.

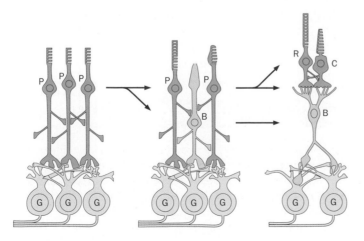

Figure 12–25 A likely scenario for the diversification of retinal cell types. In this hypothetical model, bipolar cells (B) derived from early multifunctional photoreceptors (P) and gradually became specialized in connecting photoreceptors to retinal ganglion cells (G) while losing their light-sensing capability, whereas photoreceptor cells gradually lost the connection with ganglion cells and became more specialized in light sensing. Likewise, specialized cones (C) and rods (R) probably derived from an earlier photoreceptor that resembled modern cones. (Adapted from Arendt D [2008] *Nat Rev Genet* 9:868–882. With permission from Macmillan Publishers Ltd.)

12.16 Trichromatic color vision in primates originated from variations and duplications of a cone opsin gene

As we learned in Chapter 4, color vision is enabled by the presence of multiple types of cone cells, each expressing an opsin with a distinct spectral sensitivity. The ancestral vertebrates already appeared to have four cone opsin genes, likely a result of two genome duplication events in stem vertebrates after they diverged from amphioxus. This is then followed by duplication of one of the cone opsin genes that produced the gene for the rod opsin in jawed fish after they diverged from jawless fish such as lampreys (see Figure 12–2). Many extant vertebrates, including jawed fish, reptiles, and birds, have retained four cone opsins in addition to the rod opsin, and therefore are tetrachromatic. However, during early mammalian evolution, genes encoding two cone opsins were lost, likely because early mammals were nocturnal and there was a lack of selection pressure for the multitude of cone opsins. Thus, most mammals are dichromatic, having a short-wavelength cone opsin and a longer-wavelength cone opsin encoded by the *S* and *L'* genes, respectively (**Figure 12–26**).

Humans are trichromatic, possessing three types of cones: S (blue), M (green), and L (red) (Figure 12–26; see Section 4.11). The genes encoding M-opsin and L-opsin derived from a recent duplication of the ancestral *L'* gene about 35 million years ago in the ancestor that gave rise to the catarrhines (that is, the Old World monkeys and the apes, including humans). Indeed, the *M* and *L* genes in humans and other catarrhines are located next to each other on the X chromosome, suggesting that they derived from an unequal crossing over of the ancestral *L'* gene (see Figure 12–6A). This duplication was accompanied by diversification of the *M* and *L* genes (see below for more details). As few as three amino acid changes in the transmembrane helices of the cone opsin can account for the spectral shift of the M- and L-cones, which have maximal absorbance around 530 nm and 563 nm, respectively (see Figure 4–19).

As the acquisition of trichromacy in primates was a more recent event than most evolutionary changes we have discussed thus far, more details of this evolutionary change have been discerned. Studies on the New World monkeys, which are indigenous to the Americas, provided particularly interesting insights. Most

Figure 12–26 Phylogenetic relationships in the vertebrate opsin gene family. Circles indicate gene duplication events. The leftmost three that gave rise to four cone opsin genes likely corresponded to the two rounds of genome duplication in the ancestral vertebrate (>500 million years ago). The duplication that gave rise to the rod opsin gene occurred in jawed fish after their divergence from jawless fish. The bottom duplication of the *L'* cone opsin gene occurred more recently (about 35 million years ago). The *Rh2* and *S2* cone opsin genes were lost in the mammalian clade. (Adapted from Bowmaker JK [2008] *Vis Res* 48:2022–2041. With permission from Elsevier Inc.)

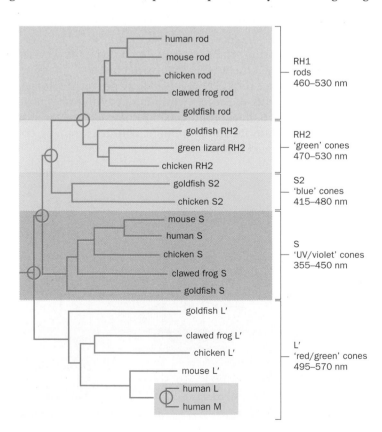

New World monkeys, like most mammals, have only the *S* and *L*′ genes. However, multiple polymorphic alleles have been found in the *L*′ gene encoding opsins that confer maximal absorbance ranging from 535 nm to 563 nm. Because the *L*′ gene is located on the X chromosome, males only have one copy and are dichromatic. Females homozygous for a particular *L*′ allele are dichromatic as well. However, females heterozygous for two polymorphic *L*′ alleles express each allele in complementary patches of cones due to random X-inactivation (see Box 11–3), which renders them trichromatic (**Figure 12–27A**).

Interestingly, one New World monkey species—the howler monkey—has also duplicated the ancestral *L*′ gene independently of the Old World monkeys and apes, such that howler monkeys are uniformly trichromatic. Sequence comparisons indicate that the *M* and *L* alleles in howler monkeys are very similar to polymorphic *L*′ alleles in other New World monkeys that do not have the duplication event. These data suggest that, at least in the case of howler monkeys, the diversification of the L′ opsins actually preceded the duplication event (Figure 12–27B). Indeed, the three key amino acids that account for the spectral shift of M- and L-cones are identical in howler monkeys and in Old World monkeys and apes. This suggests that different alleles in the ancestral *L*′ gene might have already existed prior to the divergence of New World monkeys from Old World monkeys and apes more than 35 million years ago, and that a process similar to the acquisition of trichromacy in howler monkeys (Figure 12–27B) might account for the acquisition of trichromacy in Old World monkeys and apes as well.

In order to make use of spectrally distinct opsins for color vision, two additional properties are necessary. First, the M- and L-opsins must be expressed in distinct cone cells. This problem was solved in trichromatic female New World monkeys by random X-inactivation, such that each cone expresses only one allele. In uniformly trichromatic primates, the adjacent *M* and *L* genes share the same **locus control region** (LCR), a *cis*-regulatory DNA element that confers cone-specific opsin expression (Figure 12–27B). In a given cell, the LCR can only be paired with one of the two opsin genes, and this pairing appears to be random, thus giving rise to random distribution of M- and L-cones (see Figure 4–20).

Second, the retinal circuitry should be able to extract the difference between signals detected by the M- and L-cones. This is accomplished by the midget bipolar cell, which transmits signals from the cones in the fovea with an extremely small receptive field, contacting only a single cone. As a result, the bipolar cell and the retinal ganglion cell downstream of the cone can compare a center dominated by a single cone—and therefore a single color—with a surround that samples mostly a random mix of M- and L-cone signals (see Figure 4–33).

Figure 12–27 Origin of trichromacy in primates. (A) All New World monkeys carry two autosomal copies of the short-wavelength opsin gene (S). The S gene is genetically specified to be expressed only in S-cones and not in L′-cones. In some species of New World monkeys, the L′ gene acquired polymorphisms that endow cones with different spectral sensitivities: M (green) and L (red). Males carry one copy of the L′ gene on each X chromosome so they are dichromatic (top). Females that are homozygous for the M or L alleles are also dichromatic (middle). However, females that are heterozygous for the M and L alleles (bottom) are trichromatic because random X-inactivation yields a retinal mosaic of cones that express either L or M. Gray segments represent the chromosomal DNA surrounding the opsin genes. **(B)** A likely sequence of events that gave rise to trichromacy in primates. Mutations in the L′ gene first produced polymorphism in spectral sensitivities, as is seen in many New World monkeys (first step). In howler monkey, a recent duplication caused each X chromosome to harbor both the L and M alleles, such that males and females are uniformly trichromatic (second step). A similar event might have occurred in an ancestor of Old World monkeys and apes. The same LCR (locus control region) activates expression of either the M or L opsin gene (but not both) in a given cone cell.

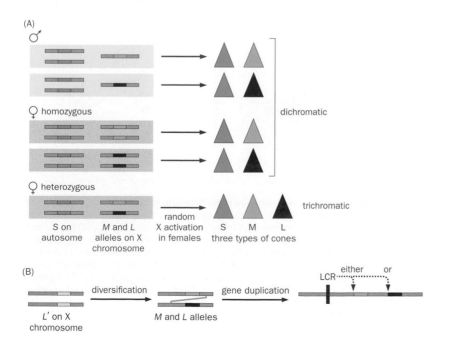

(A)

♂

♀ homozygous

dichromatic

♀ heterozygous

random X activation in females

trichromatic

S on autosome | M and L alleles on X chromosome | S   M   L three types of cones

(B)

L′ on X chromosome — diversification → M and L alleles — gene duplication → LCR either or

Figure 12–28 Trichromatic vision helps primates find fruits. For trichromats, ripe red fruits stand out clearly against the forest canopy (top). It is more difficult for dichromats with only one long-wavelength cone to find fruits in the same scene (bottom). (From Regan BC, Julliot C, Simmen B et al. [2001] *Phil Trans R Soc London B* 356:229–283. With permission from the Royal Society.)

Trichromacy in primates appears to confer a strong evolutionary advantage. Evidence for this includes the maintenance of multiple polymorphic *L'* alleles in New World monkeys and the duplication of the *L'* gene in catarrhines and, independently, in howler monkeys, both of which have since become fixed in the genome. A predominant theory is that the distinction of green and red allows primates—many of which are frugivores—to identify red, orange, and yellow fruits among green leaves and use colors to judge their ripeness (**Figure 12–28**). In fact, the evolution of trichromacy is closely linked with the evolution of colored fruits. Frugivores use color vision to feed on fruits, while plants bearing colored fruits that can be seen, eaten, and transported by animals with color vision enjoy the advantages of seed dispersal. As an added dividend, if the seeds are small enough to be ingested, they come to rest in a nutrient-rich environment.

12.17 Introducing an extra cone opsin in dichromatic animals enables superior spectral discrimination

While most evolutionary studies compare traits in different species to infer the likely processes that led to these differences, elegant experiments that simulate the evolutionary process can provide strong support to the inferred processes and bring new insights. We end our discussion of sensory system evolution with two such experiments.

Mice are dichromatic, having one S-opsin and one longer-wavelength opsin that we shall call M (for medium wavelength, which confers maximal absorbance near 510 nm, equivalent to the L' opsin in Figure 12–26). Using the knock-in technique (see Section 13.7), a human gene encoding the L-opsin (with maximal absorbance of 556 nm) replaced one of the endogenous *M* alleles of the mouse, which is also located on the X chromosome. Female mice heterozygous for the *L* and *M* alleles thus have three opsin genes—the endogenous autosomal *S*, the endogenous *M* on one X chromosome, and the knocked-in *L* on the other X chromosome—which, as a result of random X-inactivation, produce three types of cones, a situation similar to that of heterozygous female New World monkeys (see Figure 12–27A). Spectral sensitivity measurements indicated that female mice that carry both the *L* and the *M* alleles have an expanded spectrum compared to mice that express only the *M* or the *L* allele. Do these 'trichromatic' mice possess superior color vision compared to normal, dichromatic mice?

A spectral discrimination task was designed to ask mice to report their detection of spectral differences. In a three-alternative forced-choice task, mice were trained to indicate which color was different from the two other colors in order to obtain a reward (**Figure 12–29**A). Colors for two ports were set at 600 nm, and

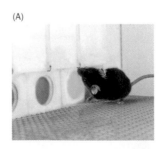

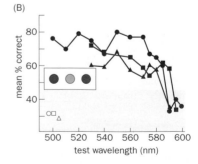

Figure 12–29 Superior spectral discrimination of trichromatic mice. (A) In a color discrimination task, thirsty mice were trained to indicate which port displayed a color that was different from two other ports in order to receive a water reward. The light wavelength was set at 600 nm for two ports and was variable for the third port. **(B)** In the color discrimination task (symbolized by the three color circles) described in panel A, a control dichromatic mouse and two trichromatic mice with imbalanced M:L ratios (open symbols) performed the 500 nm versus 600 nm discrimination test at the chance level (~33%; the dark yellow area indicates chance performance at the 95% confidence level). Three trichromatic mice with balanced M:L ratios (filled symbols) discriminated wavelengths in the 500–580 nm range versus 600 nm significantly above chance. (A, from Jacobs GH & Nathans J [2009] *Sci Am* 300:56–63. With permission from Macmillan Publishers Ltd; B, adapted from Jacobs GH, Williams GA, Cahill H et al. [2007] *Science* 315:1723–1725. With permission from AAAS.)

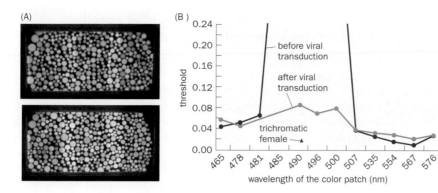

(A)

(B)

Figure 12–30 The effects of viral expression of a third cone opsin on spectral discrimination by adult dichromatic squirrel monkeys. (A) Color discrimination task. Monkeys were trained to touch a color patch that appeared randomly against a gray background in one of three compartments separated by two light bands in order to receive a juice reward. **(B)** Performance of a male squirrel monkey before (black line) and after (blue line) viral transduction of a human L-opsin gene. Before viral transduction, the dichromatic monkey could not discriminate color patches with wavelengths near 490 nm, as indicated by very high threshold values (derived from psychometric measurements of the monkey's performance with varying degrees of color saturation). After expressing the human L-opsin, the male squirrel monkey was able to discriminate color patches of all wavelengths tested. Red triangle indicates the threshold for a trichromatic female control. (Adapted from Mancuso K, Hauswirth WW, Li Q et al. [2009] *Nature* 461:784–787. With permission from Macmillan Publishers Ltd.)

the remaining port varied within a range of 500–600 nm. Control dichromatic mice completely failed the task, making correct choices at chance level (33%) for discrimination of 500 nm versus 600 nm. However, heterozygous female mice with a balanced M:L ratio performed significantly above chance level even for discrimination of 580 nm versus 600 nm (Figure 12–29B), indicating that these genetically engineered trichromatic mice indeed have superior color discrimination capabilities.

In a conceptually similar experiment, viral transduction was used to express the human L-opsin gene in the retina of the genetically dichromatic male adult squirrel monkey, a New World monkey. These monkeys were trained in a three-alternative forced-choice task; in order to receive a juice reward, they were required to touch a patch of colored dots surrounded by gray dots of varying size and intensity (**Figure 12–30**A). Even after extensive training, the dichromatic monkeys still could not identify a color patch with a wavelength near 490 nm, a so-called 'spectral neutral point.' However, monkeys that were virally transduced to express human L-opsin could discriminate the color patches from the gray background at all wavelengths tested (Figure 12–30B). Thus, expression of L-opsin in adult retina was sufficient to confer new color vision behavior to dichromatic squirrel monkeys. This experiment also provided a proof-of-principle for gene therapy to correct color blindness in humans.

Whereas trichromatic primates in principle have had millions of years to evolve their color perception circuits to match their polymorphic cone genes, in the mouse simulation experiment, a new color-processing channel was introduced within a single generation. The fact that these mice can immediately take advantage of the new opsin for spectral discrimination and to inform behavior speaks strongly to the flexibility of the nervous system to adapt to abrupt changes in sensory input. The squirrel monkey experiment further demonstrates that a new spectral channel introduced in adulthood can be utilized almost immediately. In both cases, spectral information must be processed properly in the retina, likely via the mechanisms analogous to what we discussed in Section 12.16. Moreover, the new spectral information must be successfully utilized by circuits in the visual cortex to enhance behavioral performance. Investigating how the visual cortices of trichromatic mice process spectral information may offer interesting insights into how color is processed beyond the retina.

EVOLUTION OF NERVOUS SYSTEM STRUCTURE AND DEVELOPMENT

We now turn to the final part of the chapter, which deals with the evolution of nervous system structures. An important angle from which to approach this question is through the study of development. After all, evolution of new structures must occur through changes in developmental processes. In recent decades, advances in developmental biology have revealed a remarkable degree of conservation in developmental mechanisms across the animal kingdom. Such is the case for the establishment of the body plan, our first topic of discussion.

12.18 All bilaterians share a common body plan specified by conserved developmental regulators

Bilaterians, which include all animals with a central nervous system, have two major branches: the protostomes that include most invertebrate phyla and the deuterostomes that include all vertebrates (see Figure 12–2). Among the substantial differences between protostome and deuterostome body plans is the location of their central nervous system: whereas the spinal cord is located on the dorsal side in vertebrates, the nerve cords of protostomes are usually located on the ventral side.

In 1822, Étienne Geoffroy Saint-Hilaire first suggested that vertebrate and invertebrate body plans share similarities; for example, an upside-down lobster looks like a vertebrate (**Figure 12–31**A). Although he lost debates against the establishment in those pre-Darwin days, elucidation of the molecular mechanisms of development in the 1990s validated Geoffroy Saint-Hilaire's hypothesis. In the fruit fly embryo, the ventral neuroectoderm that gives rise to the ventral nerve cord expresses the secreted protein Sog (short gastrulation), which antagonizes a dorsally expressed secreted protein Dpp (Decapentaplegic) that promotes the dorsal fate (Figure 12–31B). In the frog embryo, the dorsal structure that includes the neuroectoderm is first determined during gastrulation (see Section 7.1) by the dorsally expressed Chordin protein, which antagonizes Bmp4 (bone morphogenetic protein 4) that promotes the ventral fate (Figure 12–31C). Remarkably, Dpp and Bmp4 are homologous proteins, as are Sog and Chordin. Moreover, expression of frog Chordin in fly embryos promotes ventral fate, whereas expression of fly Sog in frog embryos promotes dorsal fate. These data demonstrate a functional equivalence of Sog/Chordin and Dpp/Bmp4 in specifying the dorsoventral axis in the frog and fly embryos, even though the embryos have inverted polarities. Furthermore, following the global dorsoventral axis patterning, homologous transcription factors in the fly and frog are expressed at different distances from the ventral or dorsal midline, respectively, and are used for patterning the dorsoventral axis within the fly ventral nerve cord and the vertebrate spinal cord, respectively (Figure 12–31B and C, right). These data suggest that the central nervous systems of vertebrates and invertebrates have a common origin within a common body plan. The inversion of the dorsoventral axis in vertebrates, from the more basal state of the invertebrates, may have occurred during early chordate evolution.

Figure 12–31 Vertebrates have an inverted dorsoventral axis compared to invertebrates. (A) Étienne Geoffroy Saint-Hilaire's drawing of an upside-down lobster illustrates that after this inversion, the body plan of an invertebrate resembles that of an vertebrate, with the central nervous system (cns) above the digestive system including the stomach (S), liver (li), and intestine (in), which are further above the heart (he) and blood vessels (bl). Muscles (mu) flank the CNS. **(B, C)** Cross sections through embryos of a fly (B) and a frog (C) show that both are patterned along the dorsoventral axis by homologous secreted proteins Dpp/Bmp4 and Sog/Chordin, with Sog promoting CNS on the ventral side of the fly embryo and Chordin promoting CNS on the dorsal side of the frog embryo by antagonizing the activity of Dpp/Bmp4 (red inhibitory signs). Moreover, at a later stage the neuroectoderm in each embryo is patterned by the expression of three transcription factors that are homologous between fly and frog: Msh/Msx, Ind/Gsh2, and Vnd/Nkx2. These transcription factors define the dorsoventral axis within the ventral nerve cord and spinal cord, respectively. Note that in the fly, neural progenitors delaminate from the epidermis to form the nervous system, whereas in the frog, neural progenitors invaginate from the epidermis (see Figure 7–2). Thus, the three pairs of transcription factors follow the same (rather than the opposite) dorsoventral orientation. (A, from De Robertis EM & Sasai Y [1996] *Nature* 380:37–40. With permission from Macmillan Publishers Ltd.)

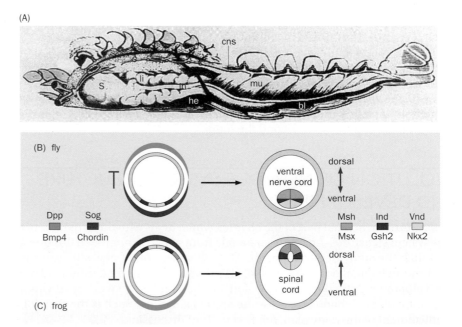

That all bilaterians share a common body plan is further demonstrated by the role of *Hox* gene clusters in patterning the anteroposterior body axis. *Hox* genes were originally discovered in *Drosophila* as mutants that exhibit **homeotic transformation**, that is, transformation of one body part to another. In the *Antennapedia* mutants, a pair of antennae is transformed into a pair of legs; in the *Ultrabithorax* mutants, the thoracic segment that gives rise to the wing is duplicated at the expense of another thoracic segment. Molecular-genetic analysis indicated that the corresponding genes are part of a gene cluster; all members of the cluster encode transcription factors that share a common DNA-binding domain called the **homeobox** (hence the name ***Hox* gene** cluster). Each *Hox* gene is expressed in specific body segments along the anteroposterior axis, and there is a co-linear relationship between the expression domain in the body and the location of the gene on the chromosome (**Figure 12–32**, top). Remarkably, the *Hox* gene clusters of vertebrates follow the same co-linear relationship as in the fly. However, vertebrates have four *Hox* clusters that likely resulted from the two rounds of genome duplication during early chordate evolution (Figure 12–32, bottom) that also gave rise to four cone opsin genes as noted earlier (see Figure 12–26).

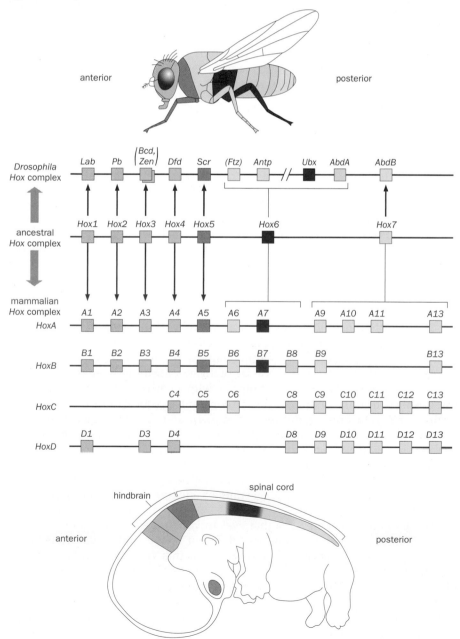

Figure 12–32 *Hox* gene clusters pattern the anteroposterior body axis. Each square represents a *Hox* gene. The top row shows the *Hox* gene cluster in *Drosophila*, including *Antennapedia* (*Antp*) and *Ultrabithorax* (*Ubx*). Other abbreviations: *Lab*, *Labial*; *Pb*, *Proboscipedia*; *Bcd*, *Bicoid*; *Zen*, *Zenknullt*; *Dfd*, *Deformed*; *Scr*, *Sex combs reduced*; *Ftz*, *Fushi tarazu*; *Abd*, *Abdominal*. The bottom four rows show the four *Hox* gene clusters in mammals. Colors represent the relationship between individual *Hox* genes and the segment identities they control in their respective organisms (top and bottom). Although *Hox* gene clusters regulate many aspects of vertebrate patterning, only the CNS segments are shown for simplicity. In both cases, there is a co-linear relationship between the location of a particular *Hox* gene on the chromosome and its expression domain along the anteroposterior axis in the body. Compared with the ancestral *Hox* gene cluster (second row), there is considerable addition and deletion of individual members in both fly and mammalian *Hox* clusters. Some genes (in parenthesis) have been co-opted to play functions other than regulating body segments along the anteroposterior axis. There is also a split between *Antp* and *Ubx* on the fly chromosome (indicated by //). (Adapted from Alberts B, Johnson A, Lewis J et al. [2015] Molecular Biology of the Cell, 6th ed. Garland Science.)

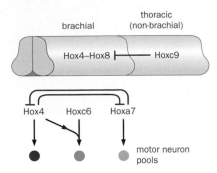

Figure 12–33 Hox proteins regulate spinal segment and motor neuron identities. Expression of Hoxc9 in the thoracic segments posterior to the brachial segments represses the expression of Hox4 through Hox8 and confines their expression to the brachial and anterior segments. Within the brachial segments, mutually repressive interactions between Hox4 and Hoxa7, and cooperation between Hoxc6 and Hox4, help specify the identity of three different motor neuron pools. (Adapted from Philippidou P & Dasen JS [2013] *Neuron* 80:12–34. With permission from Elsevier Inc.)

Although *Hox* gene expression patterns are established by distinct mechanisms during development in flies and vertebrates, once such patterns are established, *Hox* genes auto-activate their own expression and repress the expression of other *Hox* genes so as to maintain and confine their expression within specific body segments. Thus, a loss-of-function mutation in one *Hox* gene can cause misexpression of another *Hox* gene. *Hox* genes also regulate the expression of many downstream genes to produce the identity of individual body segments. This is why mutations in these genes give striking homeotic transformation phenotypes.

In addition to specifying the identity of body segments along the anteroposterior axis, *Hox* genes are also used at later developmental stages to specify neuronal identity (cell fate) in both flies and vertebrates. For example, after specifying the spinal cord segments that innervate the arms (brachial segments), the mutual repression and cooperation of *Hox* genes within the brachial segments specify the identity of motor pools that produce the motor neurons innervating different muscles in the arms (**Figure 12–33**). It is unclear, however, whether these later developmental functions of *Hox* genes in the fly and vertebrate CNS are inherited from an ancestral function in their common ancestor, or are a result of the independent adoption of additional uses of *Hox* genes whose ancestral functions are to specify body segments along the anteroposterior axis. Regardless, these findings reinforce the notion we highlighted in Chapter 7: the same molecules can be used multiple times for distinct developmental processes.

Besides the *Hox* gene cluster, homeobox DNA-binding domains are also present in many other transcription factors; these transcription factors often play evolutionarily conserved roles in regulating development, as illustrated in the next section.

12.19 Eye development is controlled by evolutionarily conserved transcription factors

Developmental genetic analysis in *Drosophila* has identified a number of transcription factors that initiate the formation of its compound eyes. One such factor is **Eyeless**, mutation of which leads to a complete absence of the compound eye. Remarkably, misexpression of Eyeless in progenitors that give rise to other tissues can cause eye formation in ectopic places, such as the antenna and underneath the wing (**Figure 12–34**), similar to the homeotic transformations discussed above. Several additional genes that exhibit loss- and gain-of-function phenotypes similar to the *Eyeless* gene have been identified in subsequent studies. *Eyeless* and these other genes together form a network that acts to specify the formation of the fly eye.

The Eyeless protein belongs to the Pax family of transcription factors, each member of which contains two DNA-binding domains: a paired box and a homeobox. Eyeless is most similar to mammalian **Pax6**. Interestingly, deletion of the *Pax6* gene in mice results in the absence of the eye. Losing a single copy of *Pax6* in humans causes a condition called aniridia, with a partial or complete absence of iris. Losing two copies of *Pax6* in humans prevents eye formation and causes stillbirth. Strikingly, expressing mouse Pax6 in *Eyeless* mutant flies rescues the development of normal eyes, and misexpression of mouse Pax6 in flies can produce ectopic eyes similar to those that result from misexpression of *Drosophila* Eyeless. These data indicate that *Eyeless* and *Pax6* are evolutionarily conserved genes that regulate eye formation. Indeed, even some jellyfish species have *Pax* genes that are expressed in the eyespots, suggesting an ancient association of Pax and eye development.

We learned in Section 12.14 that eyes in vertebrates and most invertebrates have different morphologies and use different opsins and signal transduction pathways. These findings have led to proposals that vertebrate and invertebrate eyes have multiple origins (see Box 12–3). However, research regarding the role of *Pax* genes in eye development seems to suggest that all eyes have a common origin. A parsimonious model that reconciles these different views is that the Pax transcription factors regulated the development of photosensitive cells early in animal evolution, before the cnidarian–bilaterian split, and have retained this

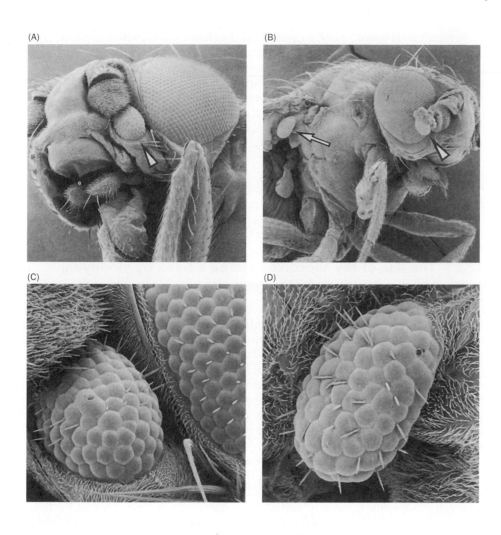

(A) (B)

(C) (D)

Figure 12–34 Misexpression of *Eyeless* leads to ectopic eyes. Scanning electron micrographs reveal that misexpression of *Eyeless* in progenitors for the antenna caused an eye to form instead of an antenna (arrowheads in A and B; A is magnified in C, with the ectopic eye transformed by the antenna to the left and normal eye to the right). Misexpression of *Eyeless* in progenitors for the wing caused eye formation under the wing (arrow in B, magnified in D). Misexpression of mouse *Pax6* in flies produced similar ectopic eyes (not shown). (From Halder G, Callaerts P & Gehring WJ [1995] *Science* 267:1788–1792. With permission from AAAS.)

function across different animal phyla. These photosensitive cells evolved independently in multiple taxa, giving rise to rhabdomeric photoreceptors (present in invertebrates plus a subset of vertebrate retinal ganglion cells) and ciliary photoreceptors (present in vertebrates plus photoreceptors and brain neurons in some invertebrates). At the same time, eyes evolved independently with different morphologies that converged on the basis of shared optical principles.

Both the loss-of-function and, particularly, the misexpression phenotypes of *Eyeless/Pax6* and *Hox* genes illustrate how altering the expression pattern of key developmental regulators can easily give rise to large structural changes. Because these transcription factors sit at the top of a regulatory hierarchy that controls the expression of many downstream target genes, altering their expression patterns may represent a key mechanism for evolutionary changes (see Section 12.4).

12.20 The mammalian neocortex underwent rapid expansion recently

In the remaining sections, we discuss the structure that endows us with complex cognitive functions: the mammalian **neocortex**. The neocortex can be traced back to a homologous but much simpler structure in the dorsal telencephalon of reptiles called the **dorsal cortex**. In the past 200 million years of mammalian evolution, neocortex has expanded both vertically—to six layers, from the three thin layers in reptilian dorsal cortex—and horizontally, encompassing dozens of areas responsible for the sensory, motor, and associational functions we have studied in the preceding chapters. Fossil records suggest that much of the expansion might have occurred during the past 65 million years, following the extinction of dinosaurs, when mammals started to evolve from mostly small and nocturnal animals to the great diversity seen today. For example, reconstruction of the skull of an

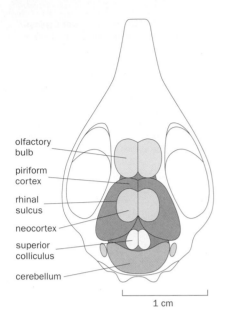

olfactory
bulb

piriform
cortex

rhinal
sulcus

neocortex

superior
colliculus

cerebellum

1 cm

Figure 12–35 Brain organization of an early mammal reconstructed from the fossil record. In this dorsal view of a mammalian skull dating from approximately 85 million years ago, the neocortex is seen at the top separated by the rhinal sulcus from the piriform cortex. Compared to most extant mammals, the structures of the fossil's brain that are related to the processing of olfactory information, such as the olfactory bulb and piriform cortex, occupy much larger areas than the neocortex. (Adapted from Kaas JH [2007] Evolution of Nervous Systems. Elsevier.)

early mammal (**Figure 12–35**) suggests that the rhinal sulcus, which separates the dorsally located neocortex from the ventrally located piriform cortex, had a much more dorsal position than is seen in extant mammals, indicating that the neocortex occupied a much smaller fraction of the forebrain in this early mammal. (The piriform cortex, which processes olfactory information as discussed in Chapter 6, has a three-layered structure and therefore is not part of the neocortex.)

What accounts for the rapid expansion of mammalian neocortex in a relatively short time during animal evolution (**Figure 12–36**)? One contributing factor may be the modular nature of cortical circuit organization (see Box 4–3), which enables the expansion of cortical areas when mechanisms that produce increased numbers of neurons are in place. A second contributing factor may be the flexibility and plasticity of the neocortex in accommodating different input. In the following sections, we discuss possible mechanisms by which the neocortex could expand in size and in the number of specialized areas it contains.

12.21 The size of the neocortex can be altered by modifying the mechanisms of neurogenesis

Humans have the largest neocortex expansion among all mammals—the human neocortex occupies about 75% of the total brain volume. By comparison, the neocortex of the mouse lemur, one of the smallest primates, accounts for about 40% of its total brain volume. As we learned in Section 7.2, excitatory neurons in the neocortex can be produced directly from radial glia in the ventricular zone or via intermediate progenitors that are products of the radial glia. At an earlier stage, radial glia are produced by neuroepithelial progenitors through symmetrical expansion (see Figure 7–4). In principle, an increase in the number of cell divisions (and a concomitant expansion of the volume of the skull) can produce a larger brain with more cortical neurons.

Several mechanisms have been proposed for increasing the number of cell divisions during cortical neurogenesis. The first mechanism is to increase the pool of radial glia by lengthening the period during which neuroepithelial progenitors undergo exponential symmetric cell division; each extra division doubles the total neuronal number. The second mechanism is to increase the number of cell divisions that yield intermediate progenitors, such that each radial glial cell produces more post-mitotic neurons. While both mechanisms have received experimental support, recent imaging studies of human cortical neurogenesis *in vitro* identified a third mechanism. In addition to radial glia in the ventricular zone, the developing human neocortex has an expanded subventricular zone, which contains an additional type of radial glia cell called an **outer radial glia** (**oRG**). These oRGs can divide asymmetrically to produce oRGs and intermediate progenitors. The processes of oRGs that extend to the pia surface can serve as additional

Figure 12–36 Images of select mammalian brains. All brains are shown from a slanted top view, with rostral toward the left. The 4-cm scale applies to the three primate brains; the 2-cm scale applies to the rest. The phylogenetic relationships of the taxa are drawn in blue. Images are from http://brainmuseum. org. These samples are from University of Wisconsin and Michigan State Comparative Mammalian Brain Collections, as well as National Museum of Health and Medicine, funded by the National Science Foundation and the National Institutes of Health.

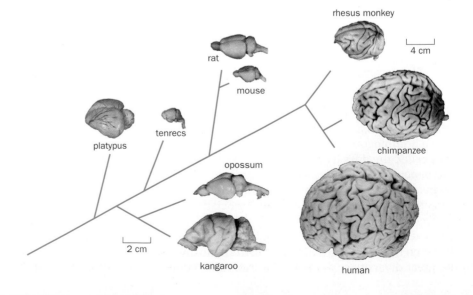

rhesus monkey

4 cm

rat

mouse

tenrecs

platypus

chimpanzee

opossum

2 cm

kangaroo

human

guideposts for radial migration of neurons to the cortical plate, thus expanding the cortical surface for a given unit area of ventricular surface (**Figure 12–37**). oRGs likely derived from ventricular radial glia, reinforcing the importance of cell type diversification in evolutionary processes discussed in Section 12.15.

Brains with large neocortices and an expanded number of neurons develop gyrencephalic cortices (see Figure 12–36). Compared to the smooth lissencephalic cortices, gyrencephalic cortices have an increased surface area for a given brain volume and therefore can accommodate more cortical processing units. Another advantage of gyrencephalic cortices is the reduced axon length for cortical–cortical connections. How are gyri and sulci produced? Several experiments in mice have suggested that simply increasing the number of cortical neurons is sufficient to produce a cortex with gyrencephalic features. In one example, a stable form of β-catenin was overexpressed in neuroepithelial progenitors in transgenic mice. (In addition to playing a role in cell adhesion along with cadherins [see Figure 12–8], β-catenin is a signaling molecule that can regulate cell proliferation.) This caused an expansion of the progenitor pool, and subsequent massive overgrowth of the cerebral cortex. Interestingly, this overgrowth did not cause a thickening of the cortical plate, but rather caused undulations of the neocortex that resembled the gyri and sulci of larger brains (**Figure 12–38**). In separate experiments, genes essential for programmed cell death were deleted in mice, resulting in an increased number of progenitors and neurons. Cerebral cortices of these mutant mice were also enlarged, with gyri- and sulci-like undulations. Cortical development was not examined in detail in these mice because the genetic manipulations were lethal *in utero* due to pleiotropic effects, inadequate cranial volume (the skull did not enlarge to accommodate the increased numbers of neurons), or both. Nevertheless, these experiments suggested the relative ease with which gyri and sulci might evolve; this may account for their independent appearance multiple times during mammalian evolution (see Figure 12–5). In order to connect these hints from mouse experiments to the actual process of human brain evolution, however, one needs to determine which differences

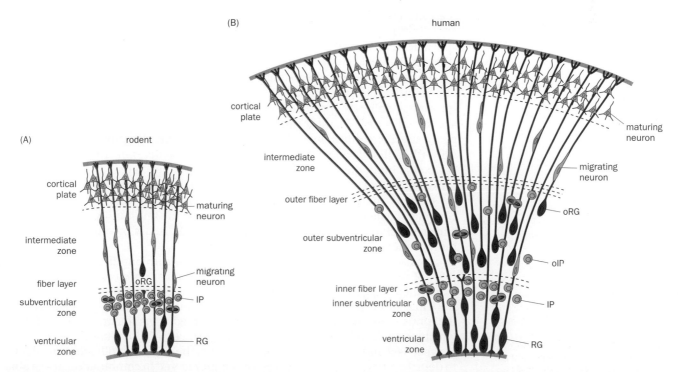

Figure 12–37 Schematic of cortical neurogenesis in rodent and human. In both species, cortical neurogenesis initiates at the ventricular zone, where radial glia (RG) produce neurons or intermediate precursors (IP). IP divisions in the subventricular zone produce more neurons (see also Figure 7–4). A salient difference in humans compared with rodents is the expansion of the outer subventricular zone, which not only contains outer intermediate precursors (oIP) but also outer radial glia (oRG) that can produce oIPs and neurons, thus greatly expanding the proliferation capacity. oRGs have subsequently been found in rodent brains, but are much smaller in number compared to ventricular RGs. (Adapted from Lui JH, Hansen DV & Kriegstein AR [2011] *Cell* 146:18–36. With permission from Elsevier Inc.)

Figure 12–38 Enlarged mouse brain with gyri- and sulci-like folds due to an increase in neural progenitor proliferation. **(A)** Cross section of a control littermate and **(B)** a transgenic mouse expressing a stable form of β-catenin in neuroepithelial progenitor cells, both at embryonic day 15 stained with cresyl violet (which stains all cell bodies). The ventricle is denoted by a star in the control. Cells around the ventricle form a dense layer lining the ventricle, the ventricular zone. The cortical plate between the ventricular zone and brain surface (arrow) eventually develops into the cerebral cortex. In transgenic mice, cortices have gyri- and sulci-like folds. (From Chenn A & Walsh CA [2002] *Science* 297:365–369. With permission from Macmillan Publishers Ltd.)

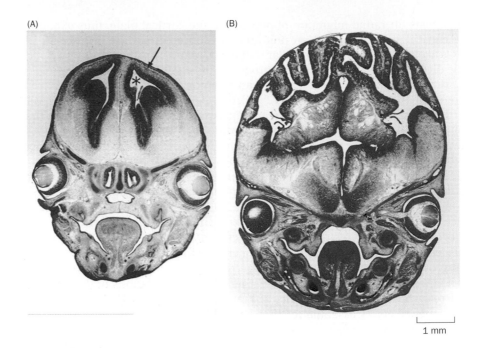

1 mm

in the human and mouse genomes ultimately account for the variations of the developmental mechanisms employed, a challenging task for the future.

12.22 Cortical area specialization can be shaped by input patterns

Mammals with a larger neocortex not only have more neurons, but also develop more specialized cortical areas and elaborate connections. For example, the prefrontal cortex in primates is especially enlarged compared to that of rodents, contributing to their more sophisticated executive functions. Humans have specialized areas such as Broca's and Wernicke's areas for language processing (**Box 12–4**; see also Figure 1–23). How have these specializations evolved?

There has been an interesting debate about the developmental mechanisms of neocortical area specialization. In one extreme view, radial glia at the ventricular zone are already specified based on their locations with respect to the body axes (see Figure 7–3); these progenitors then give rise to specified neurons occupying specific cortical areas. In another extreme view, the neocortex is a blank slate when first developed; area specializations are determined by the subcortical axonal input they receive. The truth is likely somewhere in between, with the gross patterning of motor, sensory, and associational cortices being determined by secreted morphogens and intrinsic transcription factors, as we learned in Chapter 7, whereas input patterns can influence the fine-tuning of cortical areas with respect to their functions. The examples we discuss below provide support for the latter proposal.

As discussed in Chapter 5, visual experience can profoundly affect the connectivity of the primary visual cortex (V1), such as ocular dominance. The following experiment suggested neocortical plasticity at a larger scale after surgery-induced rewiring. This experiment involved ablating the auditory axonal projections to the thalamus in newly born ferrets such that the medial geniculate nucleus (MGN), which normally relays auditory input to the primary auditory cortex (A1), no longer received auditory input. At the same time, the superior colliculi were ablated to deprive retinal ganglion cell axons of a major target. In combination, these ablations created a situation in which MGNs were connected with retinal ganglion cell axons, enabling A1 to receive visual input (**Figure 12–39**A). Remarkably, A1 neurons seemed to not only respond to visual stimuli, but also exhibited orientation selectivity that resembled normal V1 neurons. Moreover, the orientation-selective neurons in rewired A1 showed an arrangement reminiscent of pinwheel maps

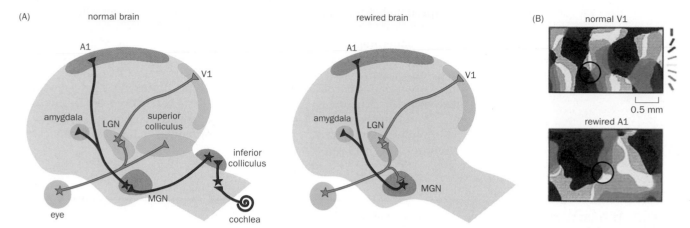

Figure 12–39 Rewiring the auditory cortex to process visual information. (A) Left, schematic of the axonal projections of the visual (blue) and auditory (red) systems in a normal ferret. The lateral geniculate nucleus (LGN) relays retinal input to primary visual cortex (V1), whereas the nearby medial geniculate nucleus (MGN) relays auditory input from inferior colliculus to the primary auditory cortex (A1). Right, after bilateral ablation of superior colliculi and severing of inferior collicular input to the MGN early during development, some retinal input was channeled into the MGN and sent to the auditory cortex. **(B)** A1 neurons in the rewired brain can not only respond to visual stimuli, but also in some cases even exhibit orientation selectivity and a functional organization similar to V1 neurons. Although different in detail, the orientation maps in V1 and rewired A1 display similarities like pinwheel structures. The bars to the right of the top panel assigns a color to each orientation; circles highlight the centers of two pinwheels. (A, adapted from Sur M & Ruberstein JLR [2005] *Science* 310:805–810; B, from Sharma J, Angelucci A & Sur M [2000] *Nature* 404:841–847. With permission from Macmillan Publishers Ltd.)

observed in V1 (Figure 12–39B; see also Figure 4–42). This experiment suggested that certain properties of V1 are not entirely predetermined but can be influenced by the nature of the thalamic input. In principle, then, evolution can act on variations in axonal pathways to shape the properties of the neocortex.

The rewiring example above, while striking, reflects a highly artificial case. Many examples in nature do support the notion that functional specialization in the neocortex is shaped by the input it receives. For example, bats that use echolocation to identify prey have expanded areas in the auditory cortex specifically tuned to the ultrasonic frequencies they emit (see Section 6.28), but non-echolocating bats do not. To analyze the rich array of tactile information, rodents that use whiskers on their snout to explore their environment develop barrels in layer 4 of their somatosensory cortex, with each barrel corresponding to a single whisker (see Box 5–3). As nearly blind predators, star-nosed moles rely on special mechanosensory rays surrounding the snout to catch small prey (**Figure 12–40**A, see also Figure 12–14A); a large fraction of the mole's somatosensory cortex is devoted to representing individual rays (Figure 12–40B). Rays 1 through 10 are used to search for prey, and the 11th pair functions as the 'fovea' equivalent, allowing for rapid fine discrimination of potential food objects (see Movie 12–1). Interestingly, cortical representation of the 11th pair of rays is enlarged compared to other appendages (Figure 12–40B), just as primate V1 is disproportionally dedicated to the representation of the retinal fovea (see Figure 4–38).

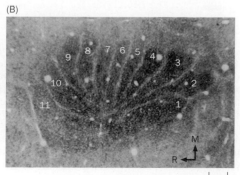

Figure 12–40 Brain representation of mechanosensory rays in the star-nosed mole. (A) Scanning electron micrograph of 11 pairs of rays surrounding the nostrils. **(B)** Horizontal section showing cytochrome oxidase staining of the somatosensory cortex. Each of the 11 bands of dark staining represents a sensory ray on the contralateral side. The 11th band occupies a disproportionally larger cortical area, reflecting that the 11th pair of rays is used for fine discrimination of objects prior to eating. R, rostral; M, medial. (From Catania KC & Kaas JH [1995] *J Comp Neurol* 351:549–567. With permission from John Wiley & Sons, Inc.)

The detailed mechanisms by which these cortical specializations arise are interesting subjects for future exploration. Natural selection can act on intrinsic cortical patterning mechanisms, axonal input, or both. The visual-to-auditory rewiring experiment, along with the superior color discrimination in trichromatic mice (see Section 12.17), suggest that the neocortex can modify its functional organization according to changes in input patterns in a single generation, likely as a consequence of its modular circuit design (see Box 4–3). This flexibility enables the neocortex to accommodate many kinds of input patterns and provides a powerful mechanism by which environmental changes can impact brain structure and organization.

Box 12–4: Transcription factor FoxP2 and the evolution of language

What genes did natural selection act on during the evolution of the human brain? Although we are far from any satisfactory answers, recent convergence of human genetic studies of brain disorders and comparative analysis of genomic data across species has begun to shed light on this question. Here we discuss **FoxP2** and the evolution of language as an example.

Language appears to be uniquely human. Whereas humans seem to be born with an innate ability to use language, our closest living relative, the chimpanzees, communicate using vocalizations that are more rudimentary than the babbling of human babies. In 2001, molecular-genetic studies of patients with speech difficulties revealed that carrying one copy of a mutant *Foxp2* gene was the cause of verbal dyslexia that affected members of a three-generation family. Loss of one copy of *Foxp2* was later identified independently in different groups of speech-impaired patients. During a task of covert (silent) generation of verbs in response to hearing the corresponding nouns, fMRI activity in unaffected individuals was restricted to the left hemisphere of Broca's area and language-related areas of the striatum. However, fMRI activity of verbally dyslexic patients with one mutant copy of *Foxp2* was scattered in both hemispheres concomitant with under-activation of relevant areas (**Figure 12–41**A). Thus, the proper level of *Foxp2* expression appears to be essential for processing speech and language.

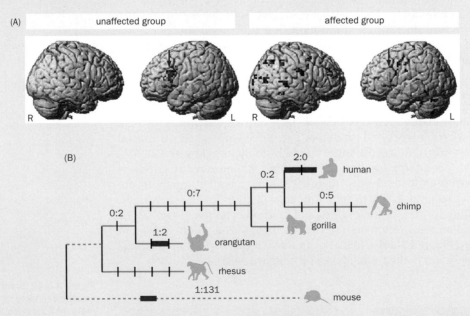

Figure 12–41 FoxP2 and the evolution of language. (A) False-colored group average in fMRI activation pattern of family members that do not carry (left pair of images) or carry (right pair of images) one mutant allele of *Foxp2*, recorded during the performance of a covert verbal task. Arrows point to Broca's area of the left (L) hemisphere, where fMRI signals concentrate in the unaffected individuals. In affected individuals, fMRI signals are scattered in both hemispheres but are diminished in Broca's area. **(B)** Synonymous and non-synonymous substitutions in *Foxp2*. Each blue bar (separated by vertical tick marks) represents a synonymous substitution (a silent single-nucleotide change that does not affect the amino acid sequence). Each red box represents a non-synonymous substitution (a single-nucleotide change that alters the amino acid sequence). The ratio of synonymous to non-synonymous substitutions is a measure of positive selection. A larger ratio reflects a stronger positive selection. (A, adapted from Liegeois F, Baldeweg T, Connelly A et al. [2003] *Nat Neurosci* 6:1230–1237. With permission from Macmillan Publishers Ltd; B, adapted from Enard W, Przeworski M, Fisher SE et al. [2002] *Nature* 418:869–872. With permission from Macmillan Publishers Ltd.)

Box 12–4: Transcription factor FoxP2 and the evolution of language

Foxp2 encodes a transcription factor that is highly conserved in vertebrates and is widely expressed in many brain areas, including the cerebral cortex, basal ganglia, and cerebellum. Interestingly, RNAi knockdown of FoxP2 levels in Area X of zebra finch (a nucleus in the basal ganglia involved in song learning; see Box 9-1) causes defective imitation of the tutor song by young birds, suggesting a role of FoxP2 in regulating vocalization in songbirds. However, mice homozygous for *Foxp2* knockout exhibit pleiotropic defects and early death, consistent with its broad expression pattern. Thus, FoxP2 functions more widely than as a specific regulator of vocalization.

The coding sequences of human and mouse FoxP2 differ by only three amino acids, two of which were recently acquired in the clade leading to humans after its split from the chimpanzee about 6.5 million years ago. Analysis of the ratio of **non-synonymous substitutions** (nucleotide changes that result in amino acid changes) to **synonymous substitutions** (nucleotide changes that do not result in amino acid changes, generally used as a measure of genetic drift) suggests that the human-specific amino acids have undergone positive selection (Figure 12–41B). To test the possible functions of these changes, knock-in mice were created in which the two human-specific amino acids replaced the endogenous amino acids in the mouse. These knock-in mice had a normal life span and mostly normal behavior and physiology, but exhibited significant differences in striatal neuronal morphology and ultrasonic vocalization patterns. Thus, humanized FoxP2 largely retains the functions of the mouse FoxP2 but also possesses unique properties.

Recent sequencing of the genomes of Neanderthals and Denisovans, two extinct species that were the closest relatives of our own sublineage of *Homo sapiens*, revealed that the human-specific amino acids were present in both species. Thus, these changes must have been introduced prior to the divergence of modern humans from Neanderthals and Denisovans 400,000–800,000 years ago. Yet DNA sequences around exon 7 of human *Foxp2*, which contains the two human-specific amino acids, appear to have been affected within the past 50,000–200,000 years by a **selective sweep**, in which nucleotide variations are reduced or eliminated as a result of strong positive selection of a nearby chromosome locus. This prompted searches for additional changes in the *Foxp2* gene that occurred more recently during human evolution. A single nucleotide in intron 8, which is part of a binding site for a transcription factor POU3F2 and is conserved among all vertebrates including Neanderthals, was found to be substituted in nearly all modern-day humans. *In vitro* studies showed that the ancestral intron 8 has a transcriptional enhancer activity in the presence of POU3F2, and the human variant has significantly reduced enhancer activity. It will be interesting to test whether this substitution alters the level or pattern of *Foxp2* expression *in vivo*, and how this contributes to language evolution.

In summary, substantial evidence suggests that the transcription factor FoxP2 may have contributed to the evolution of human language. Rather than innovating a new protein for a new function, natural selection appears to have acted on a highly conserved protein with pleiotropic functions by changing specific amino acids and regulatory sequences to tweak FoxP2 into serving a newly evolved function in humans. If FoxP2 indeed plays a crucial role in language evolution, as seems likely, it must act with many other players, including its upstream regulators and downstream targets, the study of which may bring new insights. While illustrating how problems as complex as language evolution can now be studied using a multidisciplinary approach, the FoxP2 story also highlights the challenges of reconstructing evolutionary history and establishing causality in such analyses.

SUMMARY

Evolution relies on two interrelated processes. The first process is the production of heritable variations at the level of DNA. Mutations in coding and regulatory regions of genes can alter protein function and expression pattern, respectively. Gene duplications create new templates that mutations can alter to generate functions distinct from those of the ancestral gene. Gene loss can remove an existing function. DNA shuffling can produce novel protein functions by fusing together domains from separate genes, or novel expression patterns by juxtaposing protein-coding sequences to new regulatory sequences. These events occur at sufficiently high frequencies to produce populations with large genetic variations and thus provide a rich substrate for evolutionary changes.

The second process is the selection of genetic variations that improve the fitness of individuals, that is, the ability to pass their version of genes to future generations. The function of the nervous system is under strong selection pressure. Animals that are better at finding food and mates, avoiding predators, and caring for their offspring are more likely to pass their variations to future generations. Numerous reiterations of the above two processes in all branches of the phylogenetic tree of animals over the last billion years have created a rich diversity of nervous systems.

Mutations occur randomly, beneficial mutations are more infrequent than neutral or detrimental ones, and selection is relative to the competing populations (that is, better than average is often good enough) and is influenced by environmental conditions at the time. These properties constrain what can happen in the evolution of nervous systems. We summarize examples discussed in this chapter in nine general lessons below.

1. Evolutionary changes are mostly gradual and sequential. Take the propagation of electrical signals as an example. The combined action of voltage-gated K^+ and Ca^{2+} channels can produce action potentials, allowing electrical signals to be rapidly transmitted across the cell. The use of voltage-gated Na^+ channels increased the speed of signal propagation, and freed Ca^{2+} for regulating other cellular events. The emergence of the myelin sheath further improved the speed and reliability of signal propagation over long distances.

2. New functions emerge more often from modifying the coding and regulatory sequences of existing genes than from acquiring new genes. For example, primitive synapses were built mostly with proteins that existed in animals without a nervous system, and initial neurotransmitter release machinery may have co-opted existing machineries for secretion.

3. Useful mechanisms that arose early in the phylogenetic tree can be preserved throughout subsequent evolutionary history, forming the basis of evolutionary conservation. For example, molecules used for establishing dorsoventral and anteroposterior body axes emerged in early metazoans and have been conserved in all bilaterians, surviving a dorsoventral body-axis inversion in early chordates.

4. Useful mechanisms can also be extended to solve new problems. For example, the emergence of G-protein-coupled receptors (GPCRs) in early eukaryotes created an effective way of sensing the external environment and modifying intracellular signaling. While the budding yeast has three GPCRs for detecting mating pheromones and nutrients, most animals have many hundreds of GPCRs for sensing not only environmental chemicals (conservation as in point 3) but also light, as well as for detecting neurotransmitters, neuropeptides, and hormones in intercellular communications.

5. Animals in different lineages can independently evolve similar solutions to a common problem. Such convergent evolution can occur at multiple levels, including molecules, cells, circuits, and strategies. For instance, retinal-based light sensing has been adopted independently by prokaryotes and multicellular eukaryotes. Vertebrates and invertebrates independently acquired myelination. The glomerular organization for convergent axonal projections of sensory neurons expressing the same odorant receptor appears to have emerged independently in insects and vertebrates. A biased-random-walk strategy is utilized for chemotaxis in both bacteria and nematodes; bacterial chemotaxis further employs signal amplification and adaptation similar to those used by sensory systems in animals.

6. Multiple solutions can evolve to solve the same problem and can coexist to play complementary functions. Electrical and chemical synapses have coexisted since the beginning of the nervous system as two complementary means of interneuronal communication. The prototypes of rhabdomeric and ciliary photoreceptors were likely present before the cnidarian–bilaterian split and have coexisted in both vertebrate and invertebrate lineages. Two distinct families of odorant receptors cooperate in the same *Drosophila* olfactory system.

7. Evolutionary changes are constrained by history. Rhabdomeric and ciliary photoreceptors dominate vision in invertebrates and vertebrates, respectively, probably because their ancestors adopted one type versus the other by chance. Insects expanded families of ionotropic receptors to detect odorants because their ancestors did not use GPCRs. Primates reacquired trichromacy from dichromatic ancestors because early mammals had lost two of the four cone opsin genes present in other vertebrates.

8. Diversification of cell types is an important step for evolving complex nervous systems. Diversification of photoreceptors and bipolar cells in the vertebrate retina allowed the dedication of photoreceptors for sensing light and bipolar cells for processing signals. Diversification of rods and cones enabled each type to specialize with different sensitivity, speed, and dynamic range, thus expanding the collective capacity for detecting visual signals. Diversification of radial glia facilitated neurogenesis and cortical expansion.

9. Flexibility of neural circuits is instrumental to the evolution of complex nervous systems. The retinal circuits serving the primate fovea are capable of extracting new color information as soon as a new cone with different spectral sensitivity appears. This may have contributed to the convergent evolution of trichromacy in Old World monkeys and apes and in one New World monkey species. The modular nature of the neocortical circuits and their ability to be patterned by input pathways may account for their rapid expansion.

While most of these principles apply to the evolution of all biological systems, the evolution of the nervous system has provided striking examples of natural selection in action. Studying 'how did the brain arise' deepens our understanding of 'how does the brain work' by providing a historical perspective and by considering brain function in the context of an interconnected web of life. We are at an exciting time to explore these rich and complex relations.

FURTHER READING

General concepts and approaches

Darwin C (1859) On the Origin of Species by Means of Natural Selection. John Murray.

Jacob F (1977) Evolution and tinkering. *Science* 196:1161–1166.

Kingsley DM (2009) From atoms to traits. *Sci Am* 300(1):52–59.

Mayr E (1997) The objects of selection. *Proc Natl Acad Sci USA* 94:2091–2094.

Woese CR, Kandler O & Wheelis ML (1990) Towards a natural system of organisms: proposal for the domains Archaea, Bacteria, and Eucarya. *Proc Natl Acad Sci USA* 87:4576–4579.

Origin of the nervous system and evolution of neuronal communication

Bennett MK & Scheller RH (1993) The molecular machinery for secretion is conserved from yeast to neurons. *Proc Natl Acad Sci USA* 90:2559–2563.

Chapman JA, Kirkness EF, Simakov O et al. (2010) The dynamic genome of *Hydra*. *Nature* 464:592–596.

Hartline DK & Colman DR (2007) Rapid conduction and the evolution of giant axons and myelinated fibers. *Curr Biol* 17:R29–35.

Moroz LL, Kocot KM, Citarella MR et al. (2014) The ctenophore genome and the evolutionary origin of neural systems. *Nature* 510:109–114.

Novick P, Field C & Schekman R (1980) Identification of 23 complementation groups required for post-translational events in the yeast secretory pathway. *Cell* 21:205–215.

Ryan JF, Pang K, Schnitzler CE et al. (2013) The genome of the ctenophore *Mnemiopsis leidyi* and its implications for cell type evolution. *Science* 342:1242592.

Sollner T, Whiteheart SW, Brunner M et al. (1993) SNAP receptors implicated in vesicle targeting and fusion. *Nature* 362:318–324.

Srivastava M, Simakov O, Chapman J et al. (2010) The *Amphimedon queenslandica* genome and the evolution of animal complexity. *Nature* 466:720–726.

Strong M, Chandy KG & Gutman GA (1993) Molecular evolution of voltage-sensitive ion channel genes: on the origins of electrical excitability. *Mol Biol Evol* 10:221–242.

Evolution of sensory systems

Arendt D (2008) The evolution of cell types in animals: emerging principles from molecular studies. *Nat Rev Genet* 9:868–882.

Arkowitz RA (2009) Chemical gradients and chemotropism in yeast. *Cold Spring Harb Perspect Biol* 1:a001958.

Benton R, Vannice KS, Gomez-Diaz C et al. (2009) Variant ionotropic glutamate receptors as chemosensory receptors in *Drosophila*. *Cell* 136:149–162.

Berg HC & Brown DA (1972) Chemotaxis in *Escherichia coli* analysed by three-dimensional tracking. *Nature* 239:500–504.

Catania KC (2012) Tactile sensing in specialized predators – from behavior to the brain. *Curr Opin Neurobiol* 22:251–258.

Collin SP, Knight MA, Davies WL et al. (2003) Ancient colour vision: multiple opsin genes in the ancestral vertebrates. *Curr Biol* 13:R864–865.

Fernald RD (2006) Casting a genetic light on the evolution of eyes. *Science* 313:1914–1918.

Gorman AL & McReynolds JS (1969) Hyperpolarizing and depolarizing receptor potentials in the scallop eye. *Science* 165:309–310.

Jacobs GH & Nathans J (2009) The evolution of primate color vision. *Sci Am* 300:56–63.

Jacobs GH, Williams GA, Cahill H et al. (2007) Emergence of novel color vision in mice engineered to express a human cone photopigment. *Science* 315:1723–1725.

Julius D & Nathans J (2012) Signaling by sensory receptors. *Cold Spring Harb Perspect Biol* 4:a005991.

Mancuso K, Hauswirth WW, Li Q et al. (2009) Gene therapy for red–green colour blindness in adult primates. *Nature* 461:784–787.

Montell C & Rubin GM (1989) Molecular characterization of the *Drosophila trp* locus: a putative integral membrane protein required for phototransduction. *Neuron* 2:1313–1323.

Nagel G, Szellas T, Huhn W et al. (2003) Channelrhodopsin-2, a directly light-gated cation-selective membrane channel. *Proc Natl Acad Sci USA* 100:13940–13945.

Nilsson DE & Pelger S (1994) A pessimistic estimate of the time required for an eye to evolve. *Proc Biol Sci* 256:53–58.

Pierce-Shimomura JT, Mores TM & Lockery SR (1999) The fundamental role of pirouettes in *Caenorhabditis elegans* chemotaxis. *J Neurosci* 19:9557–9569.

Porter J, Craven B, Khan RM et al. (2007) Mechanisms of scent-tracking in humans. *Nat Neurosci* 10:27–29.

Ramdya P & Benton R (2010) Evolving olfactory systems on the fly. *Trends Genet* 26:307–316.

Regan BC, Julliot C, Simmen B et al. (2001) Fruits, foliage and the evolution of primate colour vision. *Philos Trans R Soc Lond B Biol Sci* 356:229–283.

Reppert SM, Gegear RJ & Merlin C (2010) Navigational mechanisms of migrating monarch butterflies. *Trends Neurosci* 33:399–406.

Salvini-Plawen LV & Mayr E (1977) On the evolution of photoreceptors and eyes. *Evol Biol* 10:207–263.

Spudich JL, Yang CS, Jung KH et al. (2000) Retinylidene proteins: structures and functions from archaea to humans. *Annu Rev Cell Dev Biol* 16:365–392.

Yau KW & Hardie RC (2009) Phototransduction motifs and variations. *Cell* 139:246–264.

Evolution of nervous system structure and development

Carroll SB (2005) Endless Forms Most Beautiful. Norton.

Catania KC & Kaas JH (1995) Organization of the somatosensory cortex of the star-nosed mole. *J Comp Neurol* 351:549–567.

Chenn A & Walsh CA (2002) Regulation of cerebral cortical size by control of cell cycle exit in neural precursors. *Science* 297:365–369.

De Robertis EM & Sasai Y (1996) A common plan for dorsoventral patterning in Bilateria. *Nature* 380:37–40.

Enard W, Przeworski M, Fisher SE et al. (2002) Molecular evolution of FOXP2, a gene involved in speech and language. *Nature* 418:869–872.

Halder G, Callaerts P & Gehring WJ (1995) Induction of ectopic eyes by targeted expression of the eyeless gene in *Drosophila*. *Science* 267:1788–1792.

Hansen DV, Lui JH, Parker PR et al. (2010) Neurogenic radial glia in the outer subventricular zone of human neocortex. *Nature* 464:554–561.

Holley SA, Jackson PD, Sasai Y et al. (1995) A conserved system for dorsal-ventral patterning in insects and vertebrates involving sog and chordin. *Nature* 376:249–253.

Maricic T, Gunther V, Georgiev O et al. (2013) A recent evolutionary change affects a regulatory element in the human FOXP2 gene. *Mol Biol Evol* 30:844–852.

Philippidou P & Dasen JS (2013) Hox genes: choreographers in neural development, architects of circuit organization. *Neuron* 80:12–34.

Scott MP (2000) Development: the natural history of genes. *Cell* 100:27–40.

Sharma J, Angelucci A & Sur M (2000) Induction of visual orientation modules in auditory cortex. *Nature* 404:841–847.

CHAPTER 13

Ways of Exploring

Progress in science depends on new techniques, new discoveries, and new ideas, probably in that order.

Sydney Brenner, 1980

We have repeatedly seen throughout this book how new techniques have led to the discovery of fundamental principles in neurobiology. In this final chapter, we discuss in greater detail some of the key techniques that have advanced our understanding of the nervous system. Studying how these techniques work will enable you to better understand the experiments discussed in the book and to apply these techniques to explore new terrain in neurobiology. I hope that this chapter will also inspire some of you to invent new ways of exploring that will, in turn, bring new discoveries, new ideas, and new principles.

ANIMAL MODELS IN NEUROBIOLOGY RESEARCH

A major goal of neurobiology is to understand how the human brain works. Because the human brain is so complex, and because our ability to perform well-controlled experiments in humans is limited for ethical reasons, most neurobiologists use animal models to conduct their research. As we have seen in previous chapters, many of the principles identified in animal models are generally applicable to all nervous systems, including the human brain. At the same time, the variations observed in different animal models (see Figure 12-2) can be equally informative, as they reveal how the evolution of different nervous systems enabled animals to better adapt to their environmental niches.

What do scientists look for in animal models? According to one scientist, William Quinn, an ideal animal for neurobiology research "should have no more than three genes, a generation time of twelve hours, be able to play the cello or at least recite classical Greek, and learn these tasks with a nervous system containing only 10 large, differently colored, and therefore easily recognizable neurons." Of course such an 'ideal' animal does not exist, but this statement reflects the qualities that neurobiologists look for in an animal model: a simple genome and short generation time to facilitate gene manipulations and genetic studies; complex brain functions and behaviors to extrapolate findings more easily to humans; and large, easily identifiable neurons, the activities of which can be recorded and manipulated individually or together to study the principles of information processing within the neural circuits they constitute.

Before discussing specific techniques, we first have a brief overview of the commonly used animal models, upon which subsequent discussions of specific techniques are based.

13.1 Some invertebrates provide large, identifiable neurons for electrophysiological investigations

Recording the electrical signals from individual neurons and manipulating their activities are essential for investigating the mechanisms by which the nervous system functions (see Sections 13.20–13.25 for more details). The larger the neuron, the more easily researchers can record its activity by placing an electrode

(A)

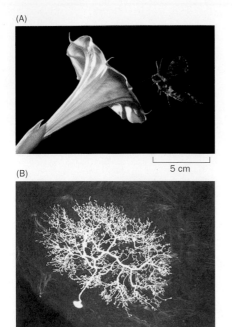

5 cm

(B)

200 µm

Figure 13–1 Invertebrate animals with large neurons aid neurophysiological investigations. The hawkmoth *Manduca sexta* has been used as a model for studying olfactory and pheromone signaling because it has a superb sense of smell and large olfactory system neurons, from which it is easy to obtain physiological recordings. **(A)** A nectar-feeding *Manduca*. **(B)** A local interneuron of *Manduca* arborizes its processes in the antennal lobe, the first olfactory processing center in the insect brain. The neuron has been filled with a fluorescent dye by an intracellular recording electrode. Compare the scale here with that of a similar neuron in *Drosophila* (see Figure 13–23C). (A, courtesy of John G. Hildebrand and Charles Hedgcock, R.B.P; B, from Reisenman CE, Dacks AM & Hildebrand JG [2011] *J Comp Physiol A* 197:653–665. With permission from Springer.)

inside it. A good example is the giant axon of the squid *Loligo*, which was used to discover the ionic basis of the action potential (see Sections 2.9 and 2.10). *Loligo* also offered giant synapses for intracellular recordings from the presynaptic terminals, which validated the role of Ca^{2+} entry in the control of neurotransmitter release (see Section 3.4).

In addition to offering simple preparations for elucidating fundamental principles of neuronal communication, invertebrates have been used to investigate the mechanisms by which neural circuits process and store information. These studies take advantage of the relatively small number of neurons (compared to vertebrate nervous systems), their large size, and their stereotyped arrangement; these properties enable the electrophysiological recording and manipulation of neurons that are individually identifiable, facilitating the comparison of experimental results across different members of the same species. For example, studying the *Aplysia* gill-withdrawal reflex enabled the discovery that changes of synaptic connection strengths underlie behavioral habituation and sensitization (see Section 10.15), and the stomatogastric ganglion of lobsters and crabs has been used to elucidate the mechanisms of central pattern generation that underlie rhythmic movement (see Section 8.5). Many other invertebrate animals such as snails, leeches, locusts, cockroaches, and moths have been used to probe the neural basis of sensation and motor control (for example, **Figure 13–1**).

13.2 *Drosophila* and *C. elegans* allow sophisticated genetic manipulations

The two invertebrate animals that we have encountered most often in this book are the fruit fly *Drosophila melanogaster* and the nematode *Caenorhabditis elegans*. These species are popular among neurobiologists because researchers can employ efficient genetic tools to manipulate their genes and identifiable neuronal populations with precision (see Sections 13.6–13.12). In contrast to the invertebrate models discussed above, these animals do not offer researchers the benefit of large neuronal size; in fact, *C. elegans* and *D. melanogaster* have the smallest neurons of any animal models commonly used for neurobiology research. Among animals with nervous systems of similar complexity, neuronal size usually correlates with the size of the animal, which is inversely correlated with its generation time. Model organisms for genetic research have been selected for short generation times—about 10 days for *Drosophila* and 3 days for *C. elegans*; hence their small bodies and neurons.

Drosophila has served as a genetic model organism for more than a century. Research first conducted in *Drosophila* laid the foundation for many fundamental concepts in genetics, such as the nature of genes, mutations, chromosomes, and the basis of linkage mapping. *Drosophila* has roughly 10^5 neurons, considerably fewer than the mouse (~10^8 neurons) or human (~10^{11} neurons) brain, but a number sufficient to mediate sophisticated neural computation and behavior. Studies in *Drosophila* can also be compared with studies in other insects such as moths, honeybees, ants, locusts, and mosquitoes that act as pollinators or pests in agriculture or as vectors of human diseases; after all, insects constitute the most diverse order in the animal kingdom.

Sydney Brenner, the author of this chapter's epigraph, introduced *C. elegans* in the 1960s for the purpose of studying the nervous system and behavior of a simple organism. *C. elegans* has since been used in many other fields of biological research, contributing to fundamental discoveries such as mechanisms of programmed cell death and RNA interference. Not only is *C. elegans* well suited to genetic manipulation, its transparent body is also advantageous for developmental and imaging studies. *C. elegans* is the only organism for which the entire **connectome**—that is, the complete set of synaptic connections linking its 302 neurons—has been deciphered using serial electron microscopy (**Figure 13–2**); this invaluable advance has guided developmental and neural circuits research (for example, see Figure 6–26).

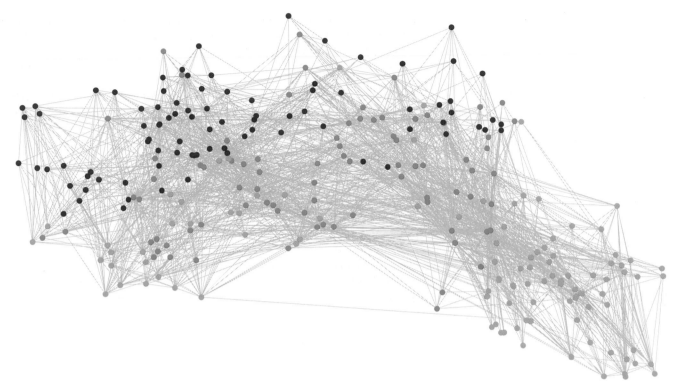

Figure 13–2 Wiring diagram of *C. elegans*. Depicted here are the 279 *C. elegans* somatic neurons (dots) and their synaptic connections (gray lines, 6393 in total), as reconstructed from serial electron microscopic sections. (Of a total of 302 neurons, 20 that constitute the pharyngeal nervous system and 3 that do not make synapses with other neurons are not shown here. Electrical synapses and neuromuscular synapses are also not shown.) Red, sensory neurons; blue, interneurons; green, motor neurons. The vertical axis represents signal flow (from top to bottom), whereas the horizontal axis represents the connectivity closeness of neurons in the combined chemical and electrical synapse network. (Courtesy of Dmitri Chklovskii. See also White JG, Southgate E, Thomson JN et al. [1986] *Phil Trans R Soc Lond B* 314:1–340; Varshney LR, Chen BL, Paniagua E et al. [2011] *PLoS Comp Biol* 7:e1001066.)

13.3 Diverse vertebrate animals offer technical ease or special faculties

Cold-blooded vertebrate animals including fish, amphibians, and reptiles are useful for producing robust explant preparations in which to study many neurobiological problems. Unlike mammalian tissues, these *in vitro* preparations often do not require constant temperature and oxygenation to maintain tissue integrity. Being vertebrates, their nervous systems share organizational similarities with the human nervous system that are not found in invertebrate models. Studies in amphibian models have contributed to many fundamental discoveries in neurobiology, such as wiring specificity in the retinotectal system (see Sections 5.1 and 5.2) and mechanisms of synaptic transmission (see Sections 3.1 and 3.2). Zebrafish (*Danio rerio*) has in recent years been a popular vertebrate model organism because its body is transparent in the larval stage, which facilitates developmental and imaging studies (**Figure 13–3**), while its relatively short generation time is well suited for genetic studies.

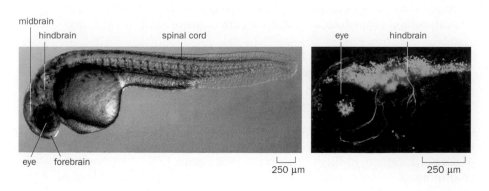

Figure 13–3 The transparency of zebrafish larvae facilitates developmental and imaging analysis. Left, differential interference microscopic image of a living zebrafish embryo at 36 hours post-fertilization. Major nervous system structures are indicated. Right, GFP (green) expression in neurons of a living zebrafish at 3 days post-fertilization. (Left, from Schier AF & Talbot WS [2005] *Annu Rev Genet* 39:561–613. With permission from Annual Reviews; Right, courtesy of Thomas Glenn & William Talbot.)

Whereas some animal models have been selected for technical ease, others have been chosen for their special faculties. One principle of neuroethology is to select a model animal in which the behavior of interest is robustly displayed. For instance, barn owls have been used to study audition because of their superb ability to locate sounds (see Sections 1.3, 6.26, and 10.25). Songbirds have been used to study vocalization and learning because they have advanced vocal communication systems and a sophisticated song-learning process (see Box 9-1). Uncovering the neural mechanisms that underlie a particular property in a species well suited to its study can benefit researchers that investigate the same property in other animals.

13.4 Mice, rats, and nonhuman primates are important models for mammalian neurobiology research

Among mammalian species, rats and mice have been the predominant animal models for many branches of biology, including neurobiology. A major advantage of mice is that they are the only mammals for which the production of transgenic and gene knockout animals is routine; this facilitates genetic manipulation and allows genetically identified neuronal populations to be recorded and manipulated with precision (see Sections 13.6–13.12). Rats have been used longer than mice as models in neurobiology research. Many behavioral paradigms such as operant conditioning (see Figure 10–22) were first developed in rats. Genetic tools first developed in mice are now being expanded to rats, whereas many physiological and behavioral paradigms originated in rats are being adapted for mice.

In addition to studying intact animals, reduced preparations from mice and rats have been widely used in neurobiology research. For instance, neurons can be dissociated and cultured *in vitro* for studying a wide range of topics such as the development of neuronal polarity (see Figure 7–16) and molecular mechanisms of synapse formation (see Figure 7–25) and synaptic transmission (see Figure 3–9). Acute or cultured brain slices have been widely used to study neuronal connectivity (see Figures 3–49 and 4–46), electrical signaling (see Figure 3–44), synaptic transmission (see Figure 3–23), and synaptic plasticity (see Figures 10–11 and 10–18). These *in vitro* preparations offer ease of experimental manipulations, such as performing patch clamp recordings of multiple neurons while controlling the extracellular environment.

Compared with mice and rats, nonhuman primates such as rhesus monkeys have brain structures (see Figure 12–36), gene expression patterns, and physiology that are more similar to those of humans; likewise, their cognitive abilities are superior to those of rodents. Many sophisticated psychophysical and cognitive tests, such as decision-making tasks (see Figure 4–54), were first developed in primate models. The visual system of trichromatic Old World monkeys and apes is very similar to our own (see Figure 4–19). Nonhuman primates are also valuable models for human disease and for drug testing, because, compared with other animals, their physiology is more similar to that of humans.

When working with animals, researchers are obligated to follow certain ethical practices; these include replacing animals with non-animal systems whenever possible, using the smallest number of animals necessary to obtain the desired information, and using all available methods to minimize pain and distress to animals being used for research. These practices apply particularly to vertebrate animals, whose proper use is regulated by governments and research institutions.

13.5 Human studies are facilitated by a long history of medicine and experimental psychology and by the recent genomic revolution

A long history of medicine, which provides many examples of human neurobiology and neuropathology, has contributed uniquely to our understanding of the nervous system. Lesions due to injury in patients provided clues to the existence of language centers in the human brain (see Figure 1–23). Electrophysiological recordings of epilepsy patients elucidated the topographic organization of the sensory and motor cortices (see Figure 1–25). Studies of amnesic patients such as H.M. revealed different memory systems and their brain localization (see Section 10.1). Likewise, experimental psychology using healthy humans as subjects has contributed substantially to our understanding of perception (see Figure 4–3), cognition, and behavior. Functional brain imaging studies have greatly improved our understanding of normal human brain organization (see Figures 1–24, 10–37, and 10–42), and can be used to monitor disease progression (see Figure 11–11) and therapeutic effect (see Figure 11–21). Genetic variations in humans have helped researchers to identify key genes that are essential for basic neurobiological processes (such as bitter taste; see Section 6.19). Mutations that cause brain disorders are making important contributions to our knowledge of how the normal nervous system develops and functions (see Chapter 11).

With the sequencing of the human genome now complete and the cost of sequencing individual genomes becoming significantly more affordable, we can anticipate a wealth of data correlating genetic variations with many kinds of phenotypes, from brain disorders to personality traits. These data provide fascinating entry points to many new areas of neurobiological investigation.

GENETIC AND MOLECULAR TECHNIQUES

Genes, the basic functional units in the genome, encode the RNAs and proteins that execute all cellular functions. Many biological processes can be viewed as the consequences of a series of actions by individual genes. Thus, by manipulating individual genes, one can dissect complex biological processes into discrete steps. This genetic approach has made fundamental contributions to all branches of biology, including the study of the nervous system. While the gene-centric approach has been more widely used in molecular and cellular neuroscience, the genetic framework has been extended to cell-type-based approach that is becoming instrumental in investigating problems in circuit and systems neuroscience and in animal behavior.

The most fundamental genetic manipulation is to disrupt the function of an individual gene—that is, to create a **loss-of-function mutation** in a gene of interest without affecting any other genes in the genome. Researchers have taken two general approaches to link a gene with its function inferred from loss-of-function phenotypes: forward genetics, which traces an observed phenotype to a gene, and reverse genetics, which follows a gene to its associated phenotype (**Figure 13–4**).

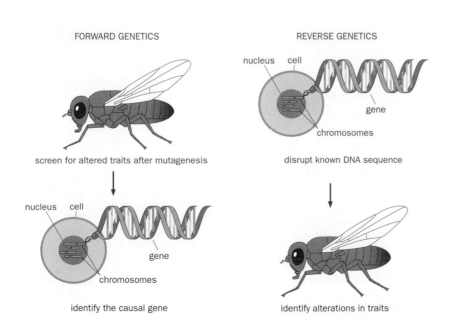

FORWARD GENETICS

screen for altered traits after mutagenesis

identify the causal gene

REVERSE GENETICS

disrupt known DNA sequence

identify alterations in traits

Figure 13–4 Forward and reverse genetics. In forward genetics, researchers start by observing an altered trait (phenotype) to identify the gene responsible for causing the phenotype of interest. In reverse genetics, researchers start with a gene of interest and disrupt the gene function to examine the phenotypic consequences.

13.6 Forward genetic screens use random mutagenesis to identify genes that control complex biological processes

An experimental approach that dominated much of twentieth century genetics, the **forward genetic screen** has provided key insights into many complex biological processes, from cell division and protein secretion to development of multicellular organisms. Forward genetic screens employ a strategy of **random mutagenesis** to identify the genes involved in a biological process of interest. Suppose that a series of unknown genes play essential roles in the process of interest. Researchers can use chemical mutagens, radiation, or transposon insertion (insertion of a transposable DNA element in a gene that disrupts its function) to mutagenize a population of animals, such that each treated animal carries a different set of random mutations in a small number of genes or in a single gene. Researchers can then screen for mutations that disrupt the biological process of interest based on the phenotypes exhibited by the offspring of the mutagenized animals (**Figure 13–5**).

The mutated gene that causes the phenotype can be traced using a variety of molecular-genetic methods depending on the nature of the mutagen. Mutations caused by transposon insertions can readily be mapped by identifying the DNA sequences that neighbor the insertion sites. Mutations induced by chemicals or radiation can be mapped by molecular-genetic procedures such as **positional cloning**. In this strategy, a large number of meiotic recombinant chromosomes are produced, and the linkage between the mutant phenotype and genetic or molecular markers with known positions in the genome is used to identify where the mutated gene resides; the closer the mutation is to a particular marker, the less frequently the mutation and marker are separated by recombination events. The causal gene can be validated by identifying the disruptive mutation in the candidate gene and by rescue of the mutant phenotype using a wild-type transgene (we will discuss transgenes in Section 13.10). With the development of high-throughput genome sequencing (see Section 13.14), researchers can also compare the whole-genome sequences of mutants and wild-type controls to identify the causal genes.

Forward genetic screens are particularly powerful in tackling problems for which the cellular and molecular pathways are poorly understood. Researchers

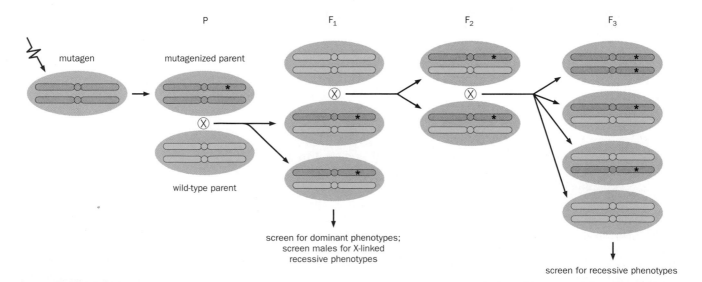

Figure 13–5 A simplified scheme for a forward genetic screen to identify single-gene mutations that cause specific phenotypes. After mutagen treatment, the mutagenized individual (P) is crossed with wild type; the mutation is indicated as * on the chromosome. Individual progeny from the next generation (F$_1$, for first filial generation) can be screened directly for mutations that exhibit dominant phenotypes (bottom), or can be crossed with wild type to produce a larger population of progeny (F$_2$) heterozygous for the mutation, from which homozygous mutant progeny (F$_3$) can be bred and screened for recessive phenotypes (top). The simplest recessive mutations to screen for are X-linked alleles in males; because they have only one X chromosome, males can be screened for X-linked recessive traits in F$_1$ instead of F$_3$. Note that for simplicity only a subset of progeny from each cross that is relevant for the progression of mutagenized chromosome is drawn out in the F$_1$ and F$_2$ generations.

rely on mutant phenotypes to identify genes involved in a particular biological process without any bias or knowledge as to what kinds of genes are expected. The identifications of the *Drosophila Period* gene and mouse *Clock* gene provide striking examples of how forward genetic screens have led to our current understanding of the molecular mechanisms that control circadian rhythms (see Figure 8–45). The same procedures used to identify mutant phenotypes and their causal genes resulting from mutagenesis can also be applied to mutations that arise spontaneously, as in the case of shaking flies (see Section 2.15), obese mice (see Section 8.16), narcoleptic dogs (see Section 8.23), and inherited human disorders (see Chapter 11).

13.7 Reverse genetics disrupts pre-designated genes to assess their functions

We now discuss **reverse genetics**, a term that refers to strategies for disrupting a pre-designated gene (see Figure 13–4, right). Many molecular components of the nervous system were identified by means other than mutant phenotypes arising from forward genetic screens or spontaneous mutations. For example, the Na⁺ channel, synaptotagmin, rhodopsin, and ephrin were identified by biochemical purification of proteins enriched in the electric organ, presynaptic terminals, bovine retina, and developing tectum, respectively (see Sections 2.15, 3.6, 4.3, and 5.4). The TRP channels for sensing temperature were identified from expression cloning (see Section 6.31). Most ion channels and neurotransmitter receptors were first identified based on their sequence homology with known proteins of similar function. With genome sequences completed for most model organisms, researchers can search databases to identify candidate genes that might perform certain functions based on expression patterns and predicted protein sequences. A key approach to test the function of a candidate gene in a suspected biological process is to create loss-of-function mutations and examine the phenotypes of the resulting mutant animals.

The most widely used method for deleting a specific gene of interest is by **homologous recombination**, in which a piece of endogenous DNA essential for the function of a gene is replaced by a piece of *in vitro* engineered DNA, the ends of which have sequences identical (hence the term 'homologous') to the endogenous DNA. Homologous recombination is an intrinsic property essential for meiosis in germ-line cells; it also occurs in other cell types, including embryonic stem (ES) cells. Among multicellular animals, the homologous recombination-based gene disruption procedure known as gene **knockout** was first developed in mice and has since become routine in that species (**Figure 13–6**). The first step is to create an *in vitro* engineered DNA construct that carries a drug-resistance gene flanked on both sides by pieces of DNA (homology arms) derived from the endogenous gene of interest. This construct is then introduced into ES cells, where recombination at both homology arms causes the replacement of an essential part of the gene of interest with the drug-resistance gene (Figure 13–6A). ES cell clones that carry the knockout allele are identified based on their drug resistance, expanded, and injected into blastocyst-stage host embryos. These embryos are subsequently implanted into surrogate mothers, where they develop to produce chimeric pups in which a fraction of germ-line cells derives from the *in vitro* engineered ES cells. (A chimera contains some cells with the genotypes of injected ES cells, and other cells with the genotype of the host embryo.) These chimeras are bred with wild-type animals to generate offspring in which all cells carry the knockout allele (Figure 13–6B), and subsequent breeding of the offspring can yield mice homozygous for the knockout allele.

Since the basic knockout procedure was established in the 1980s, many variations and extensions have been added to make this technique more versatile. For instance, instead of disrupting a gene, single nucleotide changes can be made to test *in vivo* the contribution of specific amino acid residues to protein function (for example, see Figure 3–9B). It is also possible to insert any *in vitro* engineered construct into a predetermined genomic locus. Both procedures are referred to as **knock-in**. Among its many uses, a knock-in mouse can express a marker gene

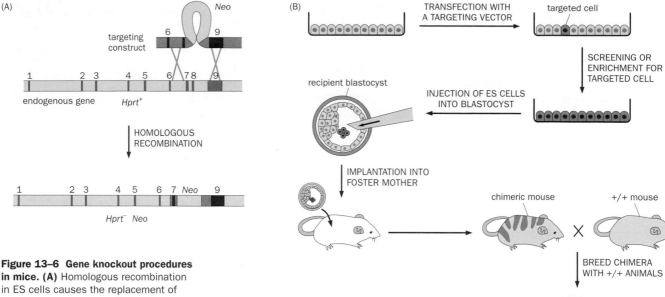

Figure 13–6 Gene knockout procedures in mice. (A) Homologous recombination in ES cells causes the replacement of an essential part of a target gene with a marker gene that encodes drug resistance (in this case *Neo*, for resistance to neomycin). Here, *Neo* replaced the sequences that correspond to exon 8 of the *Hprt* gene, which encodes an enzyme for nucleotide biogenesis. Homologous recombination at two crosses results in deletion of the DNA segment corresponding to exon 8. The resulting recombinant chromosome is deficient for *Hprt* and confers neomycin resistance. Light gray and light red: homologous introns of the *Hprt* gene; dark gray and dark red, homologous exons of the *Hprt* gene. **(B)** Modified ES cells can be used to create knockout mice by following the steps in this flow chart. Modified ES cells and their derivatives are red, whereas cells in the host blastocyst are blue. Note that the final product in the scheme is heterozygous for the modified ES cell genome, but appears red if the marker for the modified ES cell (for example, a coat color) is dominant. (Adapted from Capecchi MR [1989] *Science* 244:1288–1292.)

in the spatiotemporal pattern defined by an endogenous gene's promoter (see Figure 6–15); we will discuss many applications of this technology below.

One of the most important extensions of the knockout technique is the production of **conditional knockout** mice. This technique was first developed using the bacteriophage Cre/*loxP* system. **Cre recombinase** is a bacteriophage-derived enzyme that catalyzes recombination between two sequence-specific DNA elements called ***loxP*** sites. When two *loxP* sites are in the same orientation, a recombination event will delete the intervening sequence. (When the *loxP* sites are in opposite orientations, a recombination event will invert the intervening sequence.) In conditional knockout, two *loxP* sites are inserted in the same orientation by homologous recombination into introns that flank essential exons of a gene of interest. An allele in which essential exons are located between two *loxP* sites is termed a floxed allele (short for "flanked by *loxP*"). In the absence of Cre-mediated recombination, these *loxP*-containing introns are spliced out of RNA transcripts and do not affect gene expression. The gene can be knocked out (that is, the floxed exons can be excised) only in cells where Cre has been active (**Figure 13–7**). Researchers have generated hundreds of transgenic 'Cre lines' with different spatiotemporal patterns of Cre expression, such that gene deletion can be achieved in specific cell types and occurs only after the Cre transgene is first expressed. In addition to the Cre/*loxP* system, other site-specific recombinase systems can be used. For instance, the yeast **FLP recombinase** mediates recombination between two ***FRT*** (FLP recognition target) sites through a mechanism analogous to Cre-mediated recombination between two *loxP* sites.

An important extension of conditional knockout technology is to engineer Cre so that its activity can be temporally controlled, such as by addition of a drug. One way to achieve such temporal control is to regulate translocation of Cre into the nucleus where recombination takes place. As discussed in Section 9.14, the estrogen receptor normally remains in the cytoplasm but translocates to the nucleus in the presence of estradiol (see Figure 9–24). **CreER** is a fusion of the Cre recombinase and the portion of the estrogen receptor responsible for cytoplasmic retention. Similar to the endogenous estrogen receptor, CreER remains in the cytoplasm when not bound to its ligand, but translocates to the nucleus in the presence of the estrogen analog **tamoxifen**. (The estrogen-binding site of CreER is modified so that it binds tamoxifen but not endogenous estrogen.) Thus, the CreER–tamoxifen system allows temporal control of recombination of a floxed allele, and hence control of the precise time at which an endogenous gene is deleted.

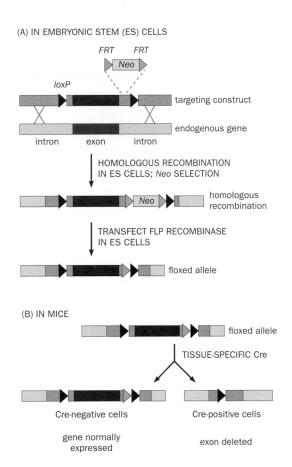

(A) IN EMBRYONIC STEM (ES) CELLS

Figure 13–7 Conditional knockout in mice. (A) An example of floxed allele production in ES cells. In the targeting construct, a pair of *loxP* sites is inserted into two introns flanking an essential exon of interest. In addition, the *Neo* gene, flanked by a pair of *FRT* sites, is inserted into one of the introns. Recombination at crosses between homologous sequences produces the desired recombinant after neomycin selection. Subsequent transient expression of the FLP recombinase induces recombination between two *FRT* sites, thus removing the neomycin resistance gene to produce the floxed allele. Because the two *loxP* sites and one *FRT* site are all inserted in the intron, the floxed allele does not affect the expression of the target gene of interest. **(B)** In mice that contain the floxed allele and a transgene expressing Cre recombinase, cells that do not express Cre are unaffected, but the essential exon is removed by Cre/*loxP* mediated recombination in cells in which Cre has been active, thus creating conditional knockout of the gene of interest.

In addition to mice, homologous recombination techniques have also been used successfully for gene deletions in *Drosophila* and rats. The rate-limiting step is to screen for rare recombination event; this has been achieved in mice and rats by developing ES cell culture so that such screens can be performed *in vitro*. In *Drosophila,* the homologous recombination procedure has been sufficiently streamlined so that it is possible to screen recombination events directly *in vivo*. For most model organisms, however, techniques of gene disruption using homologous recombination have not been established. The recent development of genome engineering tools has the potential to enable genetic manipulations, such as the production of knockout and knock-in animals, to be performed in species other than the traditional genetic model organisms (**Box 13–1**).

Box 13–1: Genome engineering by the CRISPR–Cas9 system

Genome engineering refers to the general process of altering the genome at a predetermined locus, whether by deleting a piece of endogenous DNA, inserting a piece of foreign DNA, or creating a specific base-pair change. The knockout and knock-in procedures discussed in Section 13.7 are genome-engineering procedures that employ homology arms to guide alterations using the homologous recombination system intrinsic to germ-line or embryonic stem cells. An alternative strategy is to induce, at a genomic locus of interest, double-strand DNA breaks that activate endogenous DNA repair systems, and in so doing introduce sequence

alterations. In genome engineering, double-strand breaks are typically induced by DNA-sequence-specific targeting of exogenous nucleases, such as zinc finger nucleases (ZFNs), transcription activator-like effector nucleases (TALENs), or the most recently developed and likely most versatile system for this approach: the CRISPR–Cas9 system.

Discovered in the 2000s, **CRISPR** (clustered regularly interspaced short palindromic repeat) is an adaptive immune system present in many bacteria and archaea. CRISPR is a genomic locus that contains repetitive DNA elements

(Continued)

Box 13–1: Genome engineering by the CRISPR–Cas9 system

derived from the genomes of invading pathogens such as viruses or plasmids. These DNA repeats are then made into small RNA molecules that guide nucleases to degrade the genomes of the invading pathogens through sequence-specific base pairing. Thus, bacteria previously exposed to a pathogen can rapidly defend against future infection by the same pathogen. (Since the modification occurs at the level of genomic DNA, this anti-pathogen trait is inherited by progeny; this constitutes a rare case where directed changes in DNA sequence can contribute to natural selection; see Section 12.3).

Although there are several variants of the CRISPR system, the type II system present in bacteria such as *Streptococcus pyogenes* utilizes a single protein called **Cas9** (<u>C</u>RISPR-<u>as</u>sociated 9) with two nuclease domains that cut both DNA strands to produce a double-strand break (**Figure 13–8**). Cas9 is brought to a specific site on the target DNA through the action of an engineered **guide RNA**, which contains sequences that base pair with the DNA target (part of the guide RNA is normally transcribed from the CRISPR locus). Double-strand breaks created by the CRISPR–Cas9 system can be repaired by the **nonhomologous end joining** system in the absence of any homologous DNA sequence to serve as a template. Such repairs usually introduce small deletions or duplications at the break point; if the breaks occur in the coding sequence, repairs have a two-thirds chance of creating a frame-shift mutation that disrupts the protein-coding sequence after the breakpoint. Double-strand

breaks can also be repaired by homologous recombination, which utilizes a donor DNA that shares sequence identity on both sides of the break as a template; homologous recombination-based repair can produce any arbitrary changes to the DNA sequence, from single-base-pair changes to insertions of a *loxP* site or a transgene, at a predetermined site in the genome (Figure 13–8).

The CRISPR–Cas9 system has been shown to target double-strand DNA breaks and repair events to specific DNA sequences in human cell lines, including induced pluripotent cells (see Box 11–2), as well as to the germ lines of *C. elegans, Drosophila,* zebrafish, mouse, and monkey *in vivo.* The efficiency is remarkably high such that multiple guide RNAs can be injected into the same early mouse embryo to create mutations of both copies of multiple genes simultaneously, without requiring ES cell culture, transfection, screening, and injection into blastocysts (see Figure 13–6). The CRISPR–Cas9 system has also been used to create large deletions (between sequences targeted by two guide RNAs) and to insert transgenes such as a fluorescent protein. One limitation is the potential for off-target effects due to the presence of sequences elsewhere in the genome that are similar to the intended target, although techniques have been developed to minimize this effect. Given the rapid development and improvement of the CRISPR–Cas9 system, and the promise it already exhibits, CRISPR–Cas9 is likely to become a major genome-engineering tool.

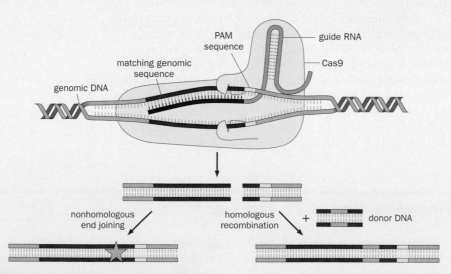

Figure 13–8 The CRISPR–Cas9 system for genome engineering. Any eukaryotic DNA that contains a PAM sequence (<u>p</u>rotospacer-<u>a</u>ssociated <u>m</u>otif, which is usually two or three nucleotides and thus occurs frequently) can be a target for the CRISPR–Cas9 system illustrated here. (CRISPR stands for <u>c</u>lustered <u>r</u>egularly <u>i</u>nterspaced <u>s</u>hort <u>p</u>alindromic <u>r</u>epeat, Cas for <u>C</u>RISPR-<u>a</u>ssociated.) A guide RNA that contains sequences complementary to a piece of DNA from the target gene of interest brings the Cas9 enzyme to the target site on the chromosome through DNA–RNA base pairing (purple and red). The two nuclease domains of Cas9 create a double-strand break in the genomic DNA. This double-strand break can be repaired by the nonhomologous end joining system, through which small deletions or insertions may be created at the repair site (indicated by the star). The double-strand break can also be repaired by the homologous recombination system using a donor DNA as a template, through which specific modifications such as the insertion of a transgene (green) can result. (Adapted from Charpentier E & Doudna JA [2013] *Nature* 495:50–51. With permission from Macmillan Publishers Ltd; see also Ran FA, Hsu PD, Wright J et al. [2013] *Nat Protocol* 8:2281–2308.)

13.8 RNA interference (RNAi)-mediated knockdown can also be used to assess gene function

A genetic technique widely used in recent years to examine loss-of-function phenotypes is **RNA interference** (**RNAi**). Stemming from a discovery originally made in *C. elegans* in the late 1990s, RNAi technology takes advantage of a naturally occurring process that efficiently degrades double-stranded RNAs (dsRNAs). RNAi serves as a cellular defense system against invaders such as viruses that produce dsRNAs at least transiently in their life cycle. It utilizes a cascade of RNA-processing enzymes and machinery that is highly conserved in eukaryotic cells for the production and function of endogenous **microRNAs**; these are short, non-coding RNAs (21–26 nucleotides) used to regulate gene expression by triggering the degradation and inhibiting the translation of mRNAs with complementary sequences. As such, RNAi has been exploited by scientists to reduce expression of any gene of interest following the introduction of exogenous dsRNAs.

For experimental gene silencing, dsRNAs can be produced by the base pairing of sense and antisense transcripts with sequences corresponding to a target gene of interest; the dsRNA or genes encoding its components can be delivered to the tissue of interest by microinjection or viral transduction (**Figure 13–9**A). Alternatively, RNAi-mediated gene silencing can be performed by expressing a transgene that encodes the homologous region to the target gene of interest in an inverted repeat (that is, a sequence followed by its reverse complement). Because the two halves of the repeat can base pair with each other, the transgene's RNA product folds into a hairpin, forming a dsRNA substrate for further processing (Figure 13–9B). Both approaches make use of the cell's microRNA-production machinery, which cleaves the dsRNA to produce **siRNA**— double-stranded short interfering RNA, with a length similar to microRNA (21–26 nucleotides). The siRNA directs a protein complex to degrade the target mRNA through base pairing.

Since inhibition of gene expression by RNAi tends to be incomplete, the procedure is referred to as causing a knockdown rather than a knockout of the target gene of interest. RNAi, like CRISPR–Cas9 (see Box 13–1), has the potential for off-target effects based on unintended targeting of similar sequences, although RNAi targets homologous RNA instead of DNA. Proper controls are necessary, such as the use of multiple and non-overlapping target sequences or a rescue of RNAi phenotypes by expression of an RNAi-resistant transgene (that is, a transgene that does not contain sequences complementary to those of the dsRNA). The advantage of RNAi over gene knockout is its increased speed and potential for high-throughput screening; this enabled RNAi to be employed not only in reverse genetics but also in genetic screening. Candidate genes that are identified via RNAi screens are often validated subsequently by gene knockout.

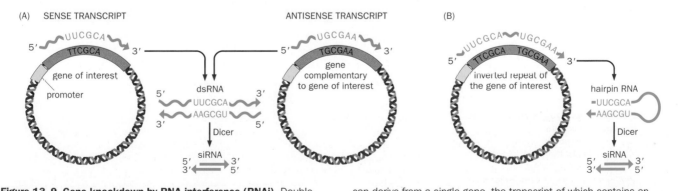

Figure 13–9 Gene knockdown by RNA interference (RNAi). Double-stranded small interfering RNAs (siRNAs) cause degradation and translation inhibition of target mRNAs bearing the same sequence, and therefore can be used to knockdown the expression of an endogenous gene. **(A)** siRNAs can derive from two genes encoding sense and antisense RNA transcripts from the same DNA sequence. **(B)** siRNAs can derive from a single gene, the transcript of which contains an inverted repeat (a sequence followed by its reverse complement); the transcript folds back onto itself and the complementary sequences base pair to produce a hairpin. In both cases, double-stranded RNA molecules are cleaved and processed by enzymes in the microRNA-processing pathway (such as Dicer) to produce siRNA.

13.9 Genetic mosaic analysis can pinpoint which cell is critical for mediating gene action

In multicellular organisms, determining which cell requires the function of a gene of interest can provide valuable information about the gene's mechanism of action. The general procedures discussed above for generating loss-of-function mutations, whether produced by random mutagenesis or engineered by gene targeting, result in the breeding of homozygous mutants. Because the gene of interest is disrupted in all cells, these methods do not help researchers identify the specific cell types in which the gene acts to contribute to developmental, cellular, and circuit functions of interest or animal behavior. Conditional knockout using Cre expressed under tissue-specific promoters can narrow this search by revealing the tissue(s) or defined cell populations in which a gene of interest is required. An alternative method is to create genetic mosaics using **mitotic recombination**. In this procedure, DNA recombination occurs between two homologous parental chromosomes in a somatic cell, such that one of the daughter cells can be homozygous for part of one parental chromosome. If an animal is heterozygous for a recessive mutation in a gene of interest and is thus phenotypically normal, one daughter cell (and all its descendants) can be made homozygous for the mutation and thus become phenotypically mutant (**Figure 13–10**); this creates a **genetic mosaic** animal—that is, an animal that contains cells of more than one genotype.

If cells of distinct genotypes can be differentially labeled, then phenotypic analysis of such mosaic animals can provide information about whether the gene of interest is **cell autonomous** (that is, acts only within the cell that produces the gene product) or **nonautonomous** (that is, acts on cells that do not produce the product) to regulate a given biological process. We have seen examples of genes that act either cell autonomously or nonautonomously in our studies of cell fate determination (see Figure 5–36), wiring specificity (see Figures 5–29, 5–38, 7–41), and mating behavior (see Section 9.8). The rate of mitotic recombination is very low naturally, but can be markedly enhanced by X-ray irradiation or by introducing into the genome a recombinase and its recognition sites, such as the Cre/*loxP* or FLP/*FRT* systems discussed above. When two recombinase recognition sites are at an identical location on two homologous chromosomes in the same orientation, a recombination event between the two sites produces recombination of the two chromosomes (see Figure 13–23 for an illustration). Site-directed recombinases also enable spatiotemporal control of mitotic recombination events via the control of recombinase expression. In addition to determining cell autonomy

Figure 13–10 Mitotic recombination can create genetic mosaics. A pair of homologous chromosomes is shown. The parent cell is heterozygous for a recessive mutation (*) in a gene of interest. If DNA recombination occurs between the homologous chromosomes (red cross) following DNA replication, chromosomal segregation in the subsequent cell division can create daughter cells that are homozygous for either the mutant (left) or wild-type (right) alleles.

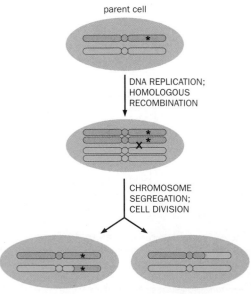

parent cell

DNA REPLICATION;
HOMOLOGOUS
RECOMBINATION

CHROMOSOME
SEGREGATION;
CELL DIVISION

two daughter cells with different genotypes

of gene function, genetic mosaics can be used to label single cells, trace cell lineage, and access specific neuronal populations for genetic manipulation (see Section 13.16).

13.10 Transgene expression can be controlled in both space and time in transgenic animals

The ability to introduce an *in vitro* engineered piece of DNA into an organism—that is, to create a **transgenic organism**, has revolutionized biology. A protein-coding **transgene** usually consists of enhancer/promoter elements that direct the spatiotemporal expression pattern, a 5′-untranslated region downstream from the transcription start site, a coding sequence that dictates the production of a specific protein, and a 3′-untranslated region that include the poly-adenylation (poly-A) signal to regulate mRNA stability and nuclear export (**Figure 13–11**).

Transgenic animals serve two broad purposes in neurobiology research. The first is to examine the function of an endogenous gene *in vivo*. For instance, expression of a transgene that encodes a wild-type protein can be used to rescue a loss-of-function mutant phenotype and thereby confirm a causal relationship between the disruption of a gene and a given phenotype. Expression in defined spatiotemporal patterns of a transgenic hairpin construct that produces RNAi effects against a specific mRNA (see Figure 13–9B) can be used to assess gene knockdown phenotypes. The gene of interest can also be misexpressed at different levels or with different spatiotemporal patterns to test **gain-of-function** effects. Finally, transgenes with specific modifications can be used *in vivo* in both loss- and gain-of-function contexts to assess the structure–function relationships of the gene of interest [that is, which specific domain(s) or amino acid(s) are required for a given function].

The second broad application of transgenes is to express molecular tools, such as cell markers to visualize neuronal morphology and projection patterns, Ca^{2+} or voltage indicators to record neuronal activity, and effectors such as light- or chemical-activated channels to silence or activate neuronal activity (see Sections 13.16, 13.18, and 13.21–13.25).

For both applications, a crucial component is to control the transgene's spatiotemporal expression pattern, which is usually regulated by enhancer sequences surrounding the coding sequence of a gene. Various mechanisms can be employed to mimic the expression pattern of an endogenous gene or to create an artificial expression pattern. One simple method to drive transgene expression in a particular pattern is to use the DNA sequences 5′ to the coding sequence of an endogenous gene whose expression pattern is being mimicked (**Figure 13–12**A). Since regulatory elements can sometimes be distributed far away from the coding sequence, using a **bacterial artificial chromosome** (**BAC**)—a cloning vector that

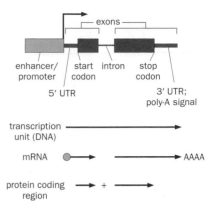

Figure 13–11 Anatomy of a transgene. A protein-coding transgene typically includes an enhancer/promoter upstream from the transcription unit that directs the spatiotemporal pattern of transcription. The transcription unit shown here comprises a 5′ untranslated region (5′ UTR); the protein-coding region separated by an intron (or introns) that is spliced out of the transcript during post-transcriptional processing; and a 3′ UTR containing a poly-A signal. The protein-coding region constitutes the region of mRNA between the start codon (translation initiation site) and the stop codon (translation termination site). The extents of the transcription unit, mRNA after splicing, and coding region are shown at the bottom. The green dot represents the 5′ cap (see also Figure 2–2 and Section 2.1). Sometimes introns, 5′ and 3′ UTRs, and sequences 3′ to the transcription unit may contain additional regulatory elements for gene expression.

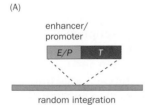 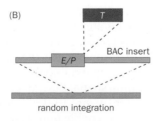 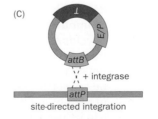 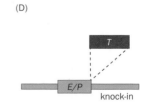

Figure 13–12 Methods of regulating the patterns of transgene expression. In each panel, the genomic DNA from an endogenous gene whose expression pattern is to be mimicked is shown in green. **(A)** The transcription unit of interest (*T*, brown) is placed under the control of an enhancer/promoter (*E/P*) element, and the resulting transgene is randomly integrated into a host chromosome (gray). **(B)** *T* is inserted into a large piece of genomic DNA downstream from the enhancer/promoter. The large size of the genomic DNA segment, which is inserted into a bacterial artificial chromosome (BAC) cloning vector capable of integrating several hundred kilobase pairs of DNA,

increases the probability that the segment contains distant regulatory elements. The modified BAC insert is then randomly integrated into the host chromosome. **(C)** Transcription unit *T*, under the control of an enhancer/promoter, is integrated at a predetermined locus in the host genome by expressing a bacteriophage integrase that catalyzes an irreversible recombination between the *attP* and *attB* sites. **(D)** *T* is knocked in at the endogenous locus of the gene whose expression pattern is to be mimicked (see Section 13.7). (Adapted from Luo L, Callaway EM & Svoboda K [2008] *Neuron* 57:634–660. With permission from Elsevier Inc.)

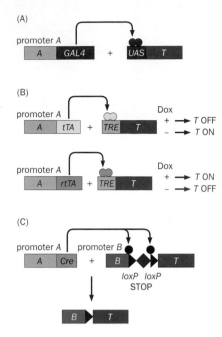

Figure 13–13 Binary expression of transgenes. Circles indicate protein products from the first (driver) transgene that act on the second (responder) transgene to regulate its expression. **(A)** Transcription unit *T* is expressed indirectly under the control of promoter *A* through the GAL4/UAS system using two transgenes. **(B)** Here, *T* is expressed indirectly under the control of promoter *A* through the tTA/TRE or rtTA/TRE systems; exogenous application of doxycycline (Dox), a tetracycline analog, provides an additional mechanism for temporal control. tTA is only active in the absence of Dox, whereas rtTA is only active in the presence of Dox. **(C)** Binary expression can be achieved by using Cre recombinase to excise the transcription and translation stop signals (diamond) between two *loxP* sites (triangles), enabling the expression of *T*. Often a ubiquitous (widely active) promoter is chosen as promoter *B*. (Adapted from Luo L, Callaway EM & Svoboda K [2008] *Neuron* 57:634–660. With permission from Elsevier Inc.)

can accommodate hundreds of kilobases of DNA—to drive transgene expression often improves the likelihood of reproducing the expression pattern of an endogenous gene (Figure 13–12B).

Transgenic animals are usually created by injecting DNA into early embryos (for example, the pronuclei in single-cell embryos of mammals); the injected transgenes then integrate randomly into host chromosomes. Expression of randomly integrated transgenes (usually multiple copies in tandem repeats) can be subject to the influence of endogenous regulatory sequences near the integration sites, causing transgene expression patterns to be variable or unpredictable. Site-specific integration of transgenes, in which an integrase catalyzes the insertion of a single copy of a transgene into a predetermined genomic locus via DNA recombination, offers greater consistency of expression. To facilitate this, one part of the integrase recognition sequence is knocked into the host chromosome at a predetermined locus, and another part of the integrase recognition sequence is inserted into a vector that carries the transgene. Integrase-mediated recombination between the two recognition sequences causes the insertion of the transgene into a predetermined locus (Figure 13–12C). (The integrase itself is either co-injected as an mRNA, or is expressed from a separate transgene.) The most faithful mimicry of endogenous gene expression pattern is achieved by knock-in to insert the transgene at the genomic locus of the endogenous gene (Figure 13–12D). Generating knock-ins is more laborious and currently can be performed only in a limited number of organisms, such as flies and mice. However, the genome-engineering tools discussed in Box 13–1 hold promise for the development of faster procedures that can be used in any organism for which transgenesis is possible.

The regulatory and protein-coding components of a gene of interest can also be expressed separately as two transgenes, a strategy called **binary expression**. For instance, the regulatory elements can be used to drive the yeast transcription factor **GAL4**, and the coding sequence of a gene of interest can be driven by a **UAS** (upstream activating sequence of GAL4). When the two transgenes are present in the same animal, the gene of interest will be expressed in the same cells that express GAL4 (**Figure 13–13A**). This GAL4/UAS binary expression system is widely used in *Drosophila* (for example, see Figure 9–6). A binary expression system often used in mice consists of the transcription factor **tTA** (tetracycline-regulated trans-activator) and its binding sequence **TRE** (tetracycline response element). A gene under the control of a TRE is activated only in cells that express tTA, which is driven by the regulatory elements of interest (Figure 13–13B). In addition, this system can be regulated using a drug: tTA activates TRE only in the *absence* of tetracycline. (A variant of tTA, reverse tTA or **rtTA**, activates TRE only in the *presence* of tetracycline). Since tetracycline and its analog **doxycycline** (Dox) are small molecules that readily diffuse across cells and the blood–brain barrier, drug treatment provides temporal control of transgene expression. Another binary expression system widely used in mice is the Cre/*loxP* system described previously: a gene of interest can be placed after a transcriptional/translational stop sequence flanked by *loxP* sites (called a *loxP-stop-loxP* sequence) following a ubiquitous promoter (a promoter that is strongly active in a wide range of cells, tissues, and developmental stages). In cells that lack Cre activity, transcription and/or translation of the transgene is disrupted by the *stop* sequence such that the transgene is not expressed. Only in Cre-active cells will the transcription and translation stops be excised by recombination so that the transgene can be expressed under control of the ubiquitous promoter (Figure 13–13C). As discussed in Section 13.7, CreER can be used instead of Cre to enable temporal control of transgene expression in this system.

The binary systems illustrated in Figure 13–13 have increased flexibility and versatility compared to the single transgene expression systems shown in Figure 13–12. For instance, simple genetic crosses can be employed to combine a given **responder transgene** (under the control of UAS, TRE, or promoter-*loxP-stop-loxP*) with different **driver transgenes** that express GAL4, tTA, or Cre to produce animals in which the responder transgene is expressed in different spatiotemporal patterns. Likewise, the same driver transgene can be combined with many responder transgenes in different animals. As a specific example, UAS-hairpin RNA transgenes (see Figure 13–9B) have been produced for almost all of the approximately 15,000 protein-coding genes in the *Drosophila* genome, such that researchers can use specific GAL4 drivers to knockdown genes one at a time in cell types of interest. This can facilitate unbiased genetic screening to identify genes that are necessary for any given biological process.

13.11 Transgene expression can also be achieved by viral transduction and other transient methods

Section 13.10 discussed methods of transgene expression that rely on the integration of transgenes into the germ line to produce transgenic animals. In this way, the same expression pattern can be reproduced in different animals across multiple generations. Other methods have been used to express transgenes transiently in somatic cells. Transient methods have the drawback that expression levels and patterns may differ from animal to animal. However, they are simpler and faster, especially in animals that have long generation times or for which germ-line transgenesis techniques have not been established.

One transient method is to directly inject DNA or mRNA encoding a gene of interest, usually into large cells such as those in early embryos. Incorporation of DNA into the host cell genome confers all progeny derived from the cell the potential to express the transgene, whereas mRNA injection is usually limited to studying early development (as mRNAs are diluted with cell division and degraded with time). Another transient method is **electroporation** where DNA containing the transgene is introduced into cells in a specific brain region of a host animal by placing a micropipette containing the DNA near the cells of interest and applying electrical currents to facilitate the transfer of negatively charged DNA molecules into the cells.

A widely used transient method of transgene expression in neurobiology, especially in mammals, is via **viral transduction**. Here, transgenes can be expressed from viral vectors that are used to produce high-titer viruses, which are usually delivered to specific region of interest by **stereotactic injection** (that is, the use of a three-dimensional coordinate system to inject substances such as viruses into a small target region). The most commonly used viruses in neurobiology include **adeno-associated virus** (**AAV**), **lentivirus**, or **herpes simplex virus** (**HSV**), each of which can be used for specific purposes based on its characteristic properties (**Table 13–1**). These viral vectors have been engineered to minimize the deleterious effects of transduction and to allow spatiotemporal control of transgene expression using strategies similar to those described above, such as Cre-dependent expression (for example, see Figure 9–37A). Viral vectors are the predominant mechanisms employed to alter gene expression in humans for the purpose of **gene therapy**, that is, the use of DNA as a therapeutic agent to treat disease.

Table 13–1: Properties of commonly used viral vectors for gene expression in the nervous system

Property	Adeno-associated virus (AAV)	Lentivirus	Herpes simplex virus (HSV)
Genetic material	single-strand DNA	RNA	double-strand DNA
Capacity	~ 5 kilobases	~ 8 kilobases	~150 kilobases
Speed of expression	weeks	weeks	days
Duration of expression	years	years	weeks to months
Tropism (cell types susceptible to viral transduction)	from broad to highly preferential depending on the serotype	usually pseudotyped[1] with coat proteins from other viruses for broad tropism	broad tropism for neurons

[1] To pseudotype a virus, the gene encoding the endogenous coat protein is deleted, and viruses are assembled in cell lines that co-express genes encoding a coat protein from a different virus.

Data from Luo L, Callaway EM & Svoboda K [2008] *Neuron* 57:634–660.

13.12 Accessing specific neuronal types facilitates functional circuit dissection

In complex nervous systems, a cell type rather than an individual cell is often the unit of neural circuit organization. The ability to monitor and manipulate neuronal activities of specific cell types is critical for neural circuit analysis. Thus, establishing genetic access to specific cell types for recording, silencing, or activation has become a fundamental experimental approach (see Sections 13.21–13.26). The most common strategy employed to gain genetic access to a specific cell type is to identify genes that are expressed in that cell type and then use regulatory elements from those genes to drive responder transgenes in the cell type of interest. For example, the promoters of odorant receptors allow genetic access to specific types of olfactory receptor neurons (see Figures 6-15 and 6-28); likewise, the regulatory elements of *fruitless* permit genetic access to many types of neurons that express endogenous *fruitless* (see Figure 9-6).

As the expression pattern of *fruitless* illustrates, a given gene may be expressed in many cell types (see Sections 9.6–9.9). Likewise, most neuronal cell types have no corresponding endogenous genes that are expressed exclusively within those cells and nowhere else. Thus, additional methods have been employed to identify regulatory elements or binary system drivers that are expressed in specific subpopulations of neurons. One approach is to use only a fraction of an endogenous enhancer's elements to drive transgene expression, based on the assumption that distinct, separable regulatory elements control the endogenous gene expression in different cell types. A second approach is to use **intersectional methods**: if promoter A drives gene expression in cell types X and Y, and promoter B drives gene expression in cell types Y and Z, then one can create an AND logic gate (if A and B, then C) for expression only in Y (**Figure 13–14**A), or a NOT logic gate (if A and not B, then C) for expression only in X (Figure 13–14B). Indeed, the dissection of the *fruitless* circuit for mating behavior has extensively utilized these intersectional approaches (for example, Figure 9–9A). Other methods include the use of the timing of neuronal birth and cell lineage to access specific neuronal populations, with the assumption that specific populations of neurons may be born within a specific developmental window and/or may arise from a common ancestor (see Figure 7–40 and Section 13.16). Yet other methods use activity of neurons to gain genetic access to specific populations by utilizing, for example, properties of immediate early genes (see Figure 10–36).

13.13 Gene expression patterns can be determined by multiple powerful techniques

We have repeatedly referred to recapitulating patterns of endogenous gene expression in previous sections. How is a gene expression pattern revealed in the

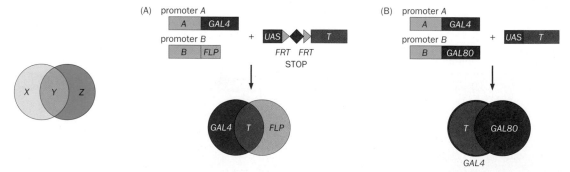

Figure 13–14 Refining transgene expression by intersectional methods. In both examples, promoter A drives gene expression in cell populations X and Y, while promoter B drives gene expression in populations Y and Z. **(A)** An AND logic gate strategy in which target gene T is expressed only in population Y, utilizing a combination of the GAL4/UAS binary expression system and the FLP/*FRT* recombination system. T can be expressed only in cells in which GAL4 is available to drive UAS expression and FLP is available to remove the stop signal. **(B)** An NOT logic gate strategy in which target gene T is expressed only in population X. GAL80 is an inhibitor of GAL4; in cells expressing both GAL80 and GAL4, transcription from UAS is repressed. (Adapted from Luo L, Callaway EM & Svoboda K [2008] *Neuron* 57:634–660. With permission from Elsevier Inc.)

first place? Many techniques have been developed for this purpose, depending on whether researchers want to characterize the expression pattern of mRNA or protein, the level of quantitative precision and spatial resolution required, and whether the expression pattern of one gene or many genes is to be characterized.

The first technique involves isolating the mRNAs or proteins from specific tissues, using gel electrophoresis to separate each tissue's mixture of mRNAs or proteins according to physical properties such as molecular weight or net electrical charge, and transferring the separated contents of the gel to a nylon or nitrocellulose membrane for probing. A labeled gene-specific nucleic acid probe can be hybridized to membrane-bound mRNAs to produce a **northern blot** (for example, see Figure 6–9B), whereas probing of membrane-bound proteins to generate a **western blot** is based on antigen recognition by specific labeled antibody probes. (The compass-point names of these and related methods were inspired by **Southern blotting**, a similar technique developed by Edwin Southern, in which DNA immobilized on a membrane hybridizes with sequence-specific DNA probes.) Northern and western blots provide information about the molecular weights and relative abundance of specific mRNAs or proteins, respectively, within a sample, but their spatial resolution is limited to tissue type.

To characterize mRNA distribution in intact tissue, fixed tissue sections can be hybridized with gene-specific probes, a procedure called *in situ* **hybridization** that we have encountered many times (for example, see Figure 5–7). Recently, *in situ* hybridization has been applied systematically to map the expression patterns in the adult mouse brain for all of the estimated 20,000 genes in the mouse genome (**Figure 13–15**), creating a valuable gene expression database. Protein distribution can likewise be determined by a technique called **immunostaining** that has also been used in numerous experiments described in this book. In this approach, fixed tissues are incubated with **primary antibodies** that bind to specific proteins, followed by **secondary antibodies** that selectively recognize primary antibodies made by specific animal species. The secondary antibodies can be tagged with an enzyme to produce a color substrate (for example, see Figure 7–12A), or with fluorescent molecules that allow simultaneous visualization of multiple proteins (for example, see Figure 2–26B). Although these techniques provide cellular and subcellular resolutions (especially in the case of immunostaining; see also Section 13.17), *in situ* hybridization and immunostaining are not as easily quantified as northern or western blots.

The complete sequencing of whole genomes brought new methods that allow quantitative determination of the expression levels of all genes in specific tissues or cell types. With **DNA microarray** technology, oligonucleotides or gene-specific probes are individually immobilized onto specific spots on a solid substrate; these spots can be arranged densely such that probes corresponding to expressed sequences from the entire genome can be packed onto a single chip (which can contain more than a million probes per cm^2). mRNA mixture from a specific tissue can be labeled and hybridized to the chip, and expression levels of all genes can be read out as signal intensities from individual spots approximately proportional to the level of RNA label. Genes with similar expression profiles can then be grouped together for expression pattern analysis (for example, **Figure 13–16**). A more recent alternative to DNA microarrays is **RNA-seq**, in which mRNA molecules from a given tissue are simply sequenced one by one in a massively parallel fashion using next generation sequencing methods (see Section 13.14). Each

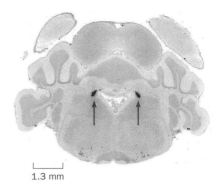

1.3 mm

Figure 13–15 Determining gene expression pattern by *in situ* hybridization. The expression pattern of dopamine β-hydroxylase, an enzyme that converts dopamine to norepinephrine, in a coronal section of the mouse brain determined by hybridizing the brain section with a probe that specifically recognizes mRNAs that produce dopamine β-hydroxylase. Arrows point to the bilateral locus coeruleus, where most norepinephrine neurons in the brain reside (see Box 8–1). Expression patterns of all mouse genes in the brain have been determined systematically by *in situ* hybridization. (From Lein ES, Hawrylycz MJ, Ao N et al. [2007] *Nature* 445:168–176. With permission from Macmillan Publishers Ltd; see also http://mouse.brain-map.org.)

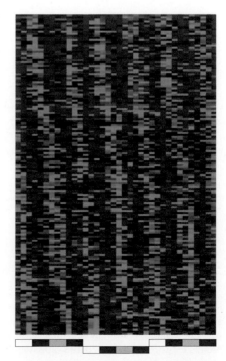

Figure 13–16 Determining gene expression patterns of many genes by microarray analysis. Each row represents a single *Drosophila* gene. Columns represent mRNA levels for given genes during specific phases of the circadian cycle indicated at the bottom (white, day; black, night; gray, subjective day, that is, day according to circadian rhythm under constant darkness condition). *Drosophila* heads were collected every four hours in three two-day time blocks to extract mRNA, from which probes were made for hybridization to the microarray. Green or red represent high or low expression, respectively. These selected genes all exhibit a circadian cycle in their expression, peaking at different phases of the cycle. (From Claridge-Chang A, Wijnen H, Naef F et al. [2001] *Neuron* 32:657–671. With permission from Elsevier Inc.)

mRNA molecule produces a single read of nucleotide sequences that are sufficiently long to identify the gene from which the mRNA is transcribed. The entire set of reads provides qualitative information about which genes are expressed in a given tissue sample and quantitative data regarding the numbers of specific mRNA molecules present in the sample.

Gene expression profiling methods such as microarray and RNA-seq are more informative and powerful when they can be applied at cell-type resolution (that is, to highly purified populations of specific cell types). Many methods have been developed for purifying cell types of interest. For instance, as an extension of traditional physical dissection, laser-capture microdissection allows histologically or fluorescently labeled cells from fixed tissue sections to be cut out with a laser beam for mRNA extraction. Other methods of purifying specific cells include sorting dissociated cells based on fluorescence or cell-surface markers, purifying mRNA specifically from cells that express transgenes encoding a tagged poly-A binding protein or ribosomal subunit, or using a micropipette to select genetically labeled fluorescent cells. The combination of high-fidelity amplification by polymerase chain reaction and high-sensitivity sequencing technology has enabled reliable gene expression profiling from very small populations of cells, even down to a single cell.

13.14 Genome sequencing reveals connections across species and identifies genetic variations that contribute to diseases

Along with recombinant DNA technology, DNA sequencing has transformed modern biology. Since the first development in the 1970s, DNA sequencing technology has seen rapid growth, thanks in large part to the Human Genome Project initiated in the late 1980s. In the subsequent two decades, and in particular with the introduction of many kinds of massively parallel sequencing platforms in the mid-2000s, the cost of sequencing has fallen dramatically while the speed has increased by many orders of magnitude (**Figure 13–17**).

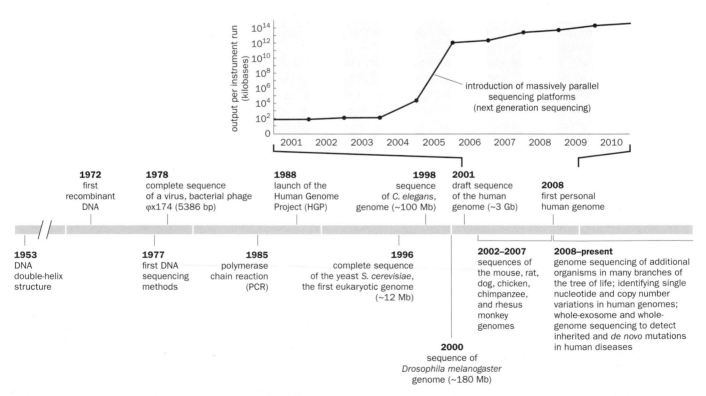

Figure 13–17 Timeline of genome sequencing and related advances. Below the timeline are selected milestones. The graph above illustrates the exponential growth of sequencing technology in the 10 years since the draft human sequences were first published. (The top graph is adapted from Mardis ER [2011] *Nature* 470:198–203. With permission from Macmillan Publishers Ltd.)

In parallel with determining the human genome sequence, the first drafts of which were completed in 2001, whole-genome sequences have been determined for many organisms in all branches of the tree of life. The impact of these data and knowledge on research has been enormous. For example, when researchers identified a gene of interest in a model organism in the 1980s or 1990s, it often took months to years to determine how many similar genes might exist within the same organism or in other organisms and to compare how similar the gene was to its homologs (for example, see Section 4.12). These questions can now be answered definitively in minutes by searching genome sequence databases. Indeed, as we learned in Chapter 12, comparative genomics in different species provides insights into how individual genes arise during evolution and how different organisms are related to each other in the tree of life.

Likewise, comparing the genomes of different individuals within the same species, such as humans, should reveal the genetic contributions to individuality. While its contribution to understanding variation in human traits is still being explored, comparative human genomics has already greatly expanded our understanding of the genetic bases of diseases, including many brain disorders that are inherited or caused by *de novo* mutations (see Box 11–3). It has also launched a new era of personalized medicine, in which treatment strategies are customized based on genetic etiology rather than symptoms; specific treatments may be more successful with patients that share genetic etiologies rather than just similar symptoms.

ANATOMICAL TECHNIQUES

In order to comprehend how the nervous system operates, it is necessary to understand its structure at different levels. In the following sections, we will examine the major anatomical techniques that have advanced our knowledge of nervous system structures. We begin with general histological methods that have provided overviews of nervous system organization. We then review techniques for visualizing individual neurons, the building blocks of the nervous system. We probe further into the fine structures of individual neurons. Lastly, we study methods that determine how neurons connect with each other to construct the wiring diagram of the nervous system.

13.15 Histological analyses reveal the gross organization of the nervous system

The anatomical organization of the nervous system is typically examined in **histological sections**; frozen or chemically fixed tissues are sliced into sections using microtomes, with the thickness of the slices ranging from several to several hundred micrometers, so that the sectioned tissues can be examined under a light microscope. As we introduced in Chapter 1, three commonly used sections are coronal, sagittal, and horizontal, which are perpendicular to the anterior–posterior (rostral–caudal), medial–lateral, and dorsal–ventral axes of the body, respectively (see Figure 1–8C).

Histological sections are typically stained to create contrast and highlight specific structures for microscopic examination. Starting in the nineteenth century, long before molecular techniques such as *in situ* hybridization and immunostaining became available (see Section 13.13), histologists invented staining methods to label cell bodies, axon fibers, or myelin sheaths; this early work revealed the overall organization of the gray and white matters in the CNS, as well as subdivisions within the gray matter. One of the most widely used staining methods for cell bodies is the **Nissl stain**, which utilizes basic (that is, proton-accepting, positively charged) dyes such as cresyl violet that bind to RNA molecules (which are negatively charged) and thereby highlight the rough endoplasmic reticulum in cytoplasm enriched for ribosomal RNAs. When applied to brain sections, Nissl stain provides a comprehensive overview of the density, size, and distribution of neurons and glia, and is commonly used to construct brain atlases and as a

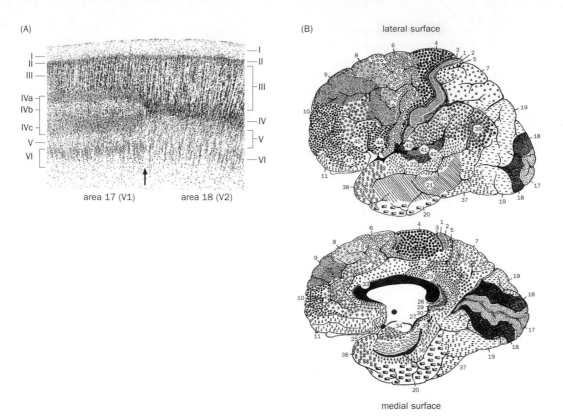

(A)

I
II
III
IVa
IVb
IVc
V
VI

I
II
III
IV
V
VI

area 17 (V1)   area 18 (V2)

(B)

lateral surface

medial surface

Figure 13–18 Nissl stain and cortical area divisions. (A) Nissl stain of a section across the border of the primary (area 17, or V1) and secondary (area 18, or V2) human visual cortices. Layer IV of V1 is characterized by the presence of three sublayers (IVa, IVb, IVc), whereas V2 has a single layer IV. Arrow indicates the border of V1 and V2. **(B)** The type of histological staining shown in panel A has been used to visualize cytoarchitectonic differences that define distinct areas within the cerebral cortex; for example, the human cerebral cortex is divided here into 50 areas, each represented by a specific symbol. (Adapted from Brodmann K [1909] *Vergleichende Localisationslehre der Grosshirnrinde in ihren Prinzipien dargestellt auf Grund des Zellenbaues*. Barth: Leipzig.)

counterstain for *in situ* hybridization, immunostaining, and additional anatomical methods, some of which are discussed below. Other staining methods selectively label cell nuclei using dyes that preferentially bind to DNA. The Nissl and nuclear staining methods enabled many discoveries regarding nervous system organization, such as the layering of the lateral geniculate nucleus and neocortex (see Figures 4–37 and 4–45A), the whisker-barrel pattern in the rodent somatosensory cortex (see Figure 5–27), and the sexual dimorphism of certain mammalian brain nuclei (see Figure 9–28). Indeed, using **cytoarchitectonics**, an approach based on differences in the laminar (layered) organization of cortical neurons as well as the density and thickness of each layer (**Figure 13–18**A), early twentieth century histologists divided the cerebral cortex into distinct areas (Figure 13–18B); these divisions are still used today.

Histological sections allow ready penetration of staining reagents throughout the sectioned tissues and high-resolution microscopic examination and imaging of neural structures within the sections. However, deciphering large-scale anatomical organization, such as axonal projections from one brain region to another (see Section 13.18), requires the reconstruction of three-dimensional volumes from individual two-dimensional sections. By contrast, **whole-mount** preparations enable investigators to examine the nervous system as a whole, whether in a dissected specimen or an intact organism. High-resolution fluorescence imaging of whole-mount tissues can be obtained by using **confocal fluorescence microscopy** (which is often shortened to confocal microscopy; confocal means 'having a common focus'). In confocal microscopy, a laser beam is focused on a spot with a volume on the order of one cubic micrometer and a pinhole is employed near the detector to ensure that fluorescence is collected only from this spot (**Figure 13–19**A). By scanning the laser across a plane to image many focal spots, confocal microscopy can produce thin optical sections of thick tissues, with

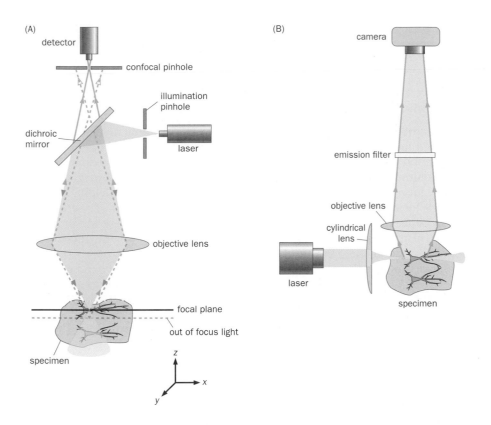

Figure 13–19 Confocal and light-sheet fluorescence microscopy. These schemes are simplified to highlight the unique features of each system. **(A)** In confocal microscopy, a small pinhole is placed in front of a detector such that only fluorescence emission from the focal plane (solid green lines and arrows), but not emission from out-of-focus planes (for example, dashed green lines and open arrows), is collected by the detector. The dichroic mirror reflects the short-wavelength excitation light (blue) but transmits the long-wavelength emission light (green). In a typical imaging experiment, the laser beam scans across different x–y positions on the focal plane such that the detector can reconstruct a two-dimensional image for that focal plane. Then the focal plane is adjusted along the z axis to produce two-dimensional images of other focal planes, eventually producing a three-dimensional confocal stack representing the three-dimensional volume of the specimen. For details, see Conchello JA & Lichtman JW (2005) *Nat Methods* 2:920–931. **(B)** In light-sheet microscopy, the laser illuminates the specimen from the side and, through the cylindrical lens, produces a thin sheet of excitation light at the focal plane of the objective lens. Only tissues in the focal plane are exposed to the excitation light, and all fluorescence emission in the focal plane is collected simultaneously by a detector (camera) to form a two-dimensional image without scanning. By systematically moving the focal plane and excitation beam along the z axis, a three-dimensional stack can be produced to represent the three-dimensional volume of the specimen. For details, see Keller PJ & Dodt HU (2012) *Curr Opin Neurobiol* 22:138–143.

the thickness of the optical section ranging from a fraction of a micrometer to several micrometers. Three-dimensional structures can be reconstructed from a series of such optical sections obtained at consecutive x–y planes along the z axis (the axis perpendicular to the imaging plane). Many images shown in this book are the products of confocal microscopy of whole-mount brains (for example, see Figure 6–33). In addition to imaging of whole-mount preparations, confocal microscopy is also widely used to obtain thin optical sections from thicker physically sectioned tissues.

In the past, the use of whole-mount preparations has typically been restricted to tissues that are less than a few hundred micrometers thick, limited by penetration of staining reagents and the opacity of tissues due to light scattering. Common whole-mount preparations have included intact *C. elegans*, dissected brain and ventral nerve cord of *Drosophila*, or nervous systems from organisms at early stages of development, such as a zebrafish larva or a mouse embryo. More recently, the development of a number of tissue-clearing methods has enabled high-resolution fluorescence imaging of larger pieces of intact tissues, up to several millimeters in each dimension (for example, **Figure 13–20; Movie 13–1**). **Light-sheet fluorescence microscopy**, which illuminates only the focal plane with a thin sheet of laser beam from the side (Figure 13–19B), offers more rapid imaging (because no scanning is required) and less photobleaching of fluorescent probes (because out-of-focus planes are not illuminated) compared with confocal microscopy. Because of these properties, light-sheet microscopy is particularly

Figure 13–20 CLARITY-based tissue clearing for fluorescence imaging. In this method, intact tissue is fixed in the presence of hydrogel monomers that covalently link DNA, RNA, and proteins into a mesh during subsequent polymerization. Lipids, which are the major source of opacity for fluorescence imaging, are not covalently linked, and are removed during the subsequent clearing process by passive diffusion or electrophoresis in the presence of detergent. The resulting tissue is nearly transparent for fluorescence imaging across several millimeters in each dimension. Shown here is a piece of tissue from a *Thy1-Gfp* transgenic mouse (see Figure 13–22 for details) imaged with confocal microscopy, showing labeled neurons and their axonal fibers across the neocortex, hippocampus, and thalamus. (From Chung K, Wallace J, Kim SY et al. [2013] *Nature* 497:332–337. With permission from Macmillan Publishers Ltd.)

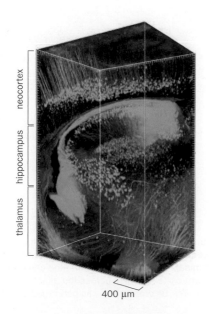

neocortex

hippocampus

thalamus

400 μm

advantageous for fluorescence imaging of large blocks of tissues and live tissues. As discussed below, these whole-mount imaging methods can facilitate the study of many aspects of the nervous system's anatomical organization.

13.16 Visualizing individual neurons opens new vistas in understanding the nervous system

As we learned in Chapter 1, development of the Golgi staining method for visualizing the morphologies of individual neurons was a milestone in neurobiology. This method was used by Ramón y Cajal and his contemporaries to establish individual neurons as the building blocks of the nervous system and provided important insights into how information flows within and between individual neurons. Methods for visualizing individual neurons are still important in modern neurobiology, since they allow researchers to correlate the morphology and projection patterns of individual neurons with their cell identity, development, physiological properties, and function in neural circuits.

Because of its simplicity, the Golgi staining method is still used today to characterize the neuronal compositions in a given brain region and to analyze the effects of gene mutations on neuronal morphology (see Figure 11-44A). However, Golgi staining has a number of limitations. It does not reliably stain long-distance axonal projections and fine terminal processes; it cannot be used to visualize neurons in live brains; and it cannot specifically stain an intended cell type because of its random nature. A number of methods that overcome these limitations have been developed.

Individual neurons can be labeled by injecting dyes during intracellular recording experiments (see Section 13.22). Injection of small molecules such as Lucifer yellow (a fluorescent dye) or biocytin allows neurons to be visualized either in live brains as neurons are being recorded or by *post hoc* staining. Dye-filling methods allow a cell's electrophysiological properties to be correlated with its morphology. They also permit tracing of long-distance projections of individual neurons, since no other neurons in the brain are labeled to interfere with such tracing (for example, **Figure 13–21**).

The use of **green fluorescent protein** (**GFP**) from jellyfish as a gene expression marker in living cells has revolutionized many fields of biology. Following the initial introduction of GFP as a marker, fluorescent proteins of different colors

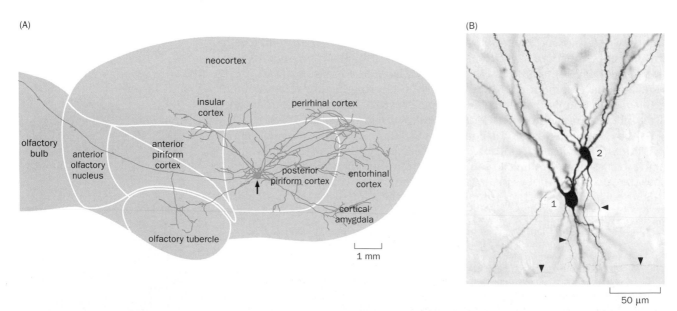

(A)

(B)

Figure 13–21 Using intracellular dye fill to trace the axonal projection of a single neuron. (A) The axonal arborization of a single pyramidal neuron (arrow points to the cell body) in the rat posterior piriform cortex ramifies across the olfactory cortex and other adjacent areas. The reconstruction was performed after intracellular injection of a tracer *in vivo* followed by several days of recovery before histology. **(B)** A photomicrograph of two numbered cells that have been injected with tracer and stained afterwards. Arrowheads, axons. (Adapted from Johnson DMG, Illig KR, Behan M et al. [2000] *J Neurosci* 20:6974–6982. With permission from the Society for Neuroscience.)

Figure 13–22 Labeling of individual neurons in *Thy1-Gfp* mice. Green fluorescent protein (GFP) expression is driven by the promoter of the *Thy1* gene, which is preferentially expressed in excitatory neurons. Due to random integration of transgenes (see Figure 13–12A), this particular transgenic line labels a very small subset of isolated hippocampal pyramidal neurons, allowing visualization of their dendritic trees and spines (inset). (From Feng G, Mellor RH, Bernstein M et al. [2000] *Neuron* 28:41–51. With permission from Elsevier Inc.)

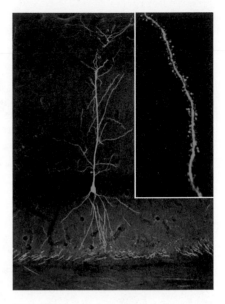

were discovered and engineered. In conjunction with a variety of genetic methods, these fluorescent proteins enable visualization of individually labeled neurons in live brains. The simplest genetic strategy for visualizing single neurons is one that uses a promoter to drive fluorescent protein expression in sparsely distributed neurons such that only one neuron is visualized in a given region. While such specific promoters are extremely rare in the central nervous system, integration of transgenes into random locations in the genome, for reasons that are still unclear, can result in the fluorescence reporter being expressed in isolated neurons (**Figure 13–22**). Along with the two-photon microscopy (to be introduced in Section 13.22), this sparse labeling has allowed individual neurons to be visualized in live brains over long periods of time to trace the dynamics and stability of dendritic branches and spines (see Figure 10–48). Individual neurons can also be labeled by sparse activation of a recombinase in animals that express a marker gene in a recombination-dependent manner (see Figure 13-13C). In addition to being able to label live neurons, these genetic methods can specify the type of neurons being labeled through the use of specific regulatory elements to drive expression of the marker gene (see Section 13.10).

Individual neurons can also be labeled by genetic mosaic methods based on mitotic recombination between the homologous chromosomes of somatic cells (see Figure 13-10). The **MARCM** method (for mosaic analysis with a repressible cell marker; **Figure 13–23**) has been widely used in *Drosophila* to label individual neurons (see Figure 1-15E) or groups of neurons that share the same lineage (see Figure 7-40). In addition to allowing individual neurons to be labeled, MARCM enables simultaneous deletion of endogenous genes or expression of transgenes in the labeled neurons. An analogous mitotic recombination-based method called MADM (for mosaic analysis with double markers) can achieve similar purposes in mice. Labeling can be restricted to specific cell types by using tissue-specific promoters to drive the recombinase, and can be produced at a desired frequency by controlling the expression level or activity of the recombinase. Because of their versatility, these genetic mosaic methods have been used not only to label individual neurons of a specific type, but also to examine the relationship between lineage and wiring properties (see Figure 7-40) and to determine cell-autonomous gene functions in neuronal morphogenesis and wiring specificity (see Figures 5–29, 5–38, and 7–41).

13.17 Fine structure studies can identify key facets of molecular organization within neurons

Most observations about nervous system structure have been made using light microscopy, which can resolve structures as small as 200 nm, a value that is defined by the diffraction limit of visible light. This resolution is sufficient for visualizing neuronal cell bodies (which have diameters ranging from 3 μm for small neurons in *Drosophila* and *C. elegans* to 10–20 μm for typical vertebrate neurons to roughly 1 mm for giant neurons in *Aplysia*) as well as most dendritic and axonal processes (which are hundreds of nanometers to a few micrometers in diameter). However, resolving fine subcellular structures of neurons, including individual synapses in densely labeled tissues, requires higher resolution techniques.

Electron microscopy (EM) has served neurobiology well since its first use in the 1950s. EM was instrumental in the discovery of the synaptic cleft, the observation of synaptic vesicle fusion with presynaptic membranes, and the finding that glial membranes wrap axons (see Figures 3–3, 3–4, and 2–26). EM has also been used to study many aspects of neuronal cell biology, including cytoskeletal organization, intracellular trafficking (see Figure 2-6), membrane compartment

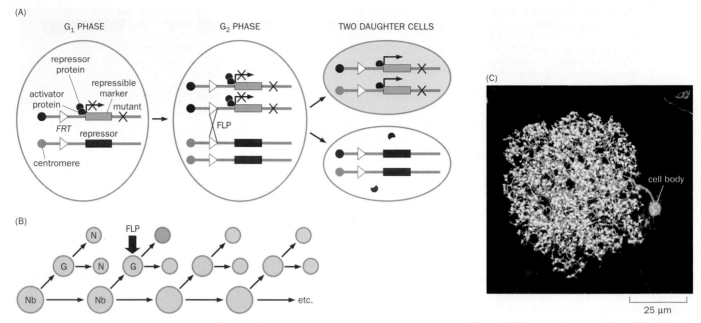

Figure 13–23 Single neuron labeling and genetic manipulation by the MARCM method. (A) In MARCM (mosaic analysis with a repressible cell marker), a cell marker gene (green bar) is under the control of a repressible promoter—it is activated by the activator protein (blue blob) only in the absence of a repressor protein (red blob). The schematic shows a pair of homologous chromosomes in a cell at three cell-cycle stages: before DNA replication (G_1 phase), after DNA replication (G_2 phase), and after cell division that produces two daughter cells. The transgene expressing the repressor protein (red bar) is located on the chromosome *in trans* to a mutation of interest (×). FLP/*FRT*-mediated chromosomal recombination (cross in the middle panel) followed by cell division results in the loss of the repressor transgene in one of the two daughter cells, thus allowing the marker to be expressed only in that cell, which is also homozygous mutant for a gene of interest. In the original version, the marker is expressed by the GAL4/UAS binary expression system and the repressor is GAL80 (see Figures 13–13 and 13–14). Other repressible binary expression systems have since been developed. **(B)** A widely occurring cell division pattern that produces neurons in the insect central nervous system. The neuroblast (Nb) undergoes asymmetric division to produce another neuroblast and a ganglion mother cell (G), which divides once more to produce two post-mitotic neurons (N). If FLP/*FRT*-mediated recombination is induced in G, then a single neuron is labeled. **(C)** An example of a MARCM-labeled single *Drosophila* olfactory local interneuron that ramifies throughout the entire antennal lobe. The cell is labeled in green with a membrane-targeted GFP to visualize the neuronal morphology, and in red with an epitope-tagged synaptotagmin that labels its presynaptic terminals, which appear yellow because the red synaptotagmin is expressed within the confine of a green cell. Blue is neuropil staining. (A & B, adapted from Lee T & Luo L [1999] *Neuron* 22:451–461. With permission from Elsevier Inc.; C from Chou YH, Spletter ML, Yaksi E et al. [2010] *Nat Neurosci* 13:439–449. With permission from Macmillan Publishers Ltd.)

organization, and pre- and postsynaptic terminal structures. Most EM images in the book have been taken using **transmission electron microscopy** in which high-voltage electron beams transmit through ultra-thin sections of biological specimen (typically under 100 nm) to create images. **Scanning electron microscopy** produces images by scanning the surface of a biological specimen, collecting information regarding the interaction of the electron beam with the surface areas (for example, see Figure 12–34).

As discussed in Section 13.13, light microscopy can be combined with antibody staining to study the tissue-level and subcellular distributions of individual proteins. Likewise, EM can be combined with antibody staining in a procedure called **immuno-EM** to visualize individual proteins at an ultra-structural level. This method can provide important clues about the actions of individual molecules (for example, **Figure 13–24**). High-resolution EM has also been used to analyze the atomic structures of proteins such as the nicotinic acetylcholine receptor (see Figure 3–20).

While EM has become the gold standard for ultra-structural analysis, the recent development of a number of **super-resolution fluorescence microscopy** techniques has provided impressive views of the fine structures of neurons at a resolution beyond the diffraction limit of conventional light microscopy. In the variants called **STORM** (**st**ochastic **o**ptical **r**econstruction **m**icroscopy) and **PALM** (**p**hoto**a**ctivated **l**ocalization **m**icroscopy), spatial precision is achieved by photoactivating a random small subset of photo-switchable fluorophores at any one time, such that the position of each fluorophore can be localized to a

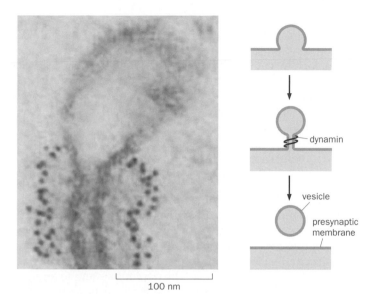

100 nm

Figure 13–24 Immuno-EM localization of dynamin. Dynamin molecules coat tubular membrane invaginations, visualized here using gold particles in an *in vitro* preparation of nerve terminal membrane. (In this procedure, primary anti-dynamin antibody binds dynamin, and a gold particle-conjugated secondary antibody binds to the primary antibody.) This suggests a function of dynamin in the fission of synaptic vesicles from the plasma membrane during endocytosis, as shown in the schematic at right (red, dynamin localization). This role for dynamin is consistent with the protein's loss-of-function phenotypes (see Figure 3–14). (Adapted from Takei K, McPherson PS, Schmid S et al. [1995] *Nature* 374:186–190. With permission from Macmillan Publishers Ltd.)

precision much finer than the resolution limits set by diffraction. The fluorophores are subsequently deactivated (bleached), and a different subset of fluorophores is photoactivated to allow a second round of localization. Repeated rounds of localization enable the reconstruction of the entire imaging field, but with a resolution that can be <20 nm in the $x–y$ plane and <50 nm in the z axis in brain sections, far superior than the resolution of confocal microscopy (**Figure 13–25**A). As a result, for example, STORM can be used to determine the distances of different synaptic molecules from the synaptic cleft and the orientations of certain synaptic proteins with respect to the synaptic cleft (Figure 13–25B, C). Likewise, application of **STED** (**stimulated emission depletion microscopy**),which achieves super resolution by depleting fluorescence from surrounding regions while leaving a central focal spot of the sample active to emit fluorescence, enabled the spatial distribution of synaptic molecules at the fly neuromuscular junction to be reconstructed (see Figure 3–11). Compared with immuno-EM for localizing specific molecules with fine resolution, super-resolution fluorescence microscopy is relatively easy to perform, and is especially well suited for imaging multiple proteins that are differentially labeled in the same sample. Super-resolution microscopy can also be used for live imaging to study protein dynamics whereas EM is limited to fixed tissues.

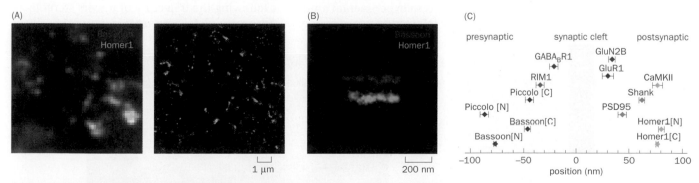

Figure 13–25 Application of super-resolution fluorescence microscopy to map synaptic protein organization. (A) Double labeling of the presynaptic scaffold protein Bassoon (red) and the postsynaptic scaffold protein Homer1 (green) in a section of mouse olfactory bulb's glomerular layer, imaged using confocal fluorescence microscopy (left) and stochastic optical reconstruction microscopy (STORM), a super-resolution microscopic technique (right). Whereas red and green signals exhibit considerably overlap and appear fuzzy at this magnification in the confocal image, they appear distinct in the STORM image.
(B) High-magnification view of STORM imaging, which resolves clearly

the distributions of Bassoon and Homer1 across the synaptic cleft.
(C) Estimate of the distribution of different synaptic proteins relative to the synaptic cleft, based on the distribution of antibodies against these proteins relative to each other as visualized by STORM. Colored dots are average positions of individual proteins with respect to the synaptic cleft, and vertical bars are standard deviations of the means. To determine the orientation of the molecules with respect to the synaptic cleft, different antibodies were used against N-termini [N] and C-termini [C]. (Adapted from Dani A, Huang B, Bergan J et al. [2010] *Neuron* 68:843. With permission from Elsevier Inc.)

Conventional light microscopy, super-resolution fluorescence microscopy, and EM can be used to systematically study the locations of individual molecules with respect to each other and to the subcellular compartments of neurons. It is conceivable that such investigations might someday provide a realistic model of single neurons at the level of molecular complexes, from the soma to the axonal and dendritic terminals. These molecular neuroanatomical techniques can also probe how protein complexes change with neuronal activity and reveal what goes wrong in brain disorders.

13.18 Mapping neuronal projections allows the tracking of information flow across different brain regions

While fine structural analyses elucidate how molecular complexes are involved in the functioning of individual neurons, deciphering a nervous system's wiring diagram—that is, the representation of how individual neurons connect with each other to form a complex nervous system—poses a great challenge in neurobiology today. All neurons can be classified into one of two categories: **projection neurons** send information from one region of the nervous system to another, whereas the axonal projections of **local neurons** (also referred to as **interneurons**; see Section 1.9) are confined within a given region. Wiring diagrams are being studied at different scales with varying resolutions. Below we discuss methods that are used to map the long-distance connections of projection neurons. In the next section, we study how projection neurons and local neurons form synaptic connections within a local region.

At a global scale, **diffusion tensor imaging** (**DTI**) is a magnetic resonance imaging technique that allows noninvasive imaging of fiber bundles that connect different structures. The basic idea is that water diffusion in the white matter occurs primarily along the path of axons, whereas water diffusion in the gray matter occurs almost equally in all directions. By acquiring a series of images, each sensitive to diffusion in a specific direction, DTI can determine the motion of water at any given volume in the white matter, which is then used to estimate the trajectories of the axonal fibers passing through that volume. Information obtained across the white matter can then be used to reconstruct the flow lines that approximate the trajectories of axon bundles (**Figure 13–26**). The resolution of DTI is on the scale of a few millimeters, so it can only depict major pathways in the white matter in large brains such as the human brain; the method is less effective in resolving the trajectories of axons within the gray matter. Still, DTI has already been used to map gross brain connectivity in healthy subjects and abnormalities in patients with neurological disorders.

A widely used method to determine connections between different brain regions in experimental animals is following the trajectory of tracers stereotactically injected at a specific region. **Anterograde tracers** are taken up primarily by

Figure 13–26 Diffusion tensor imaging. In this sagittal view of the human brain, axon bundles that run mostly along the medial–lateral axis are colored in red, axon bundles that run mostly along the anterior–posterior axis are colored in green, and axon bundles that run through the brainstem are colored in blue. (Courtesy of the Laboratory of Neuro Imaging and Martinos Center for Biomedical Imaging, Consortium of the Human Connectome Project, www.humanconnectomeproject.org)

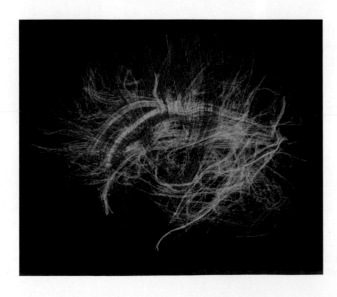

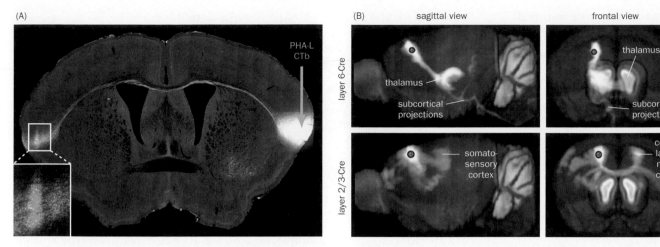

Figure 13–27 Examples of methods that trace long-distance neuronal projections. (A) A mix of phytohemagglutinin (PHA-L, green), an anterograde tracer, and cholera toxin subunit b (CTb, magenta), a retrograde tracer, was injected into the right insular cortex (arrow) of a mouse brain. This coronal section is stained in blue with a fluorescent Nissl stain that labels all cell bodies. PHA-L-labeled axons project to the left (contralateral) insular cortex, where CTb-labeled cell bodies (magenta dots in inset magnified rectangle) represent retrogradely labeled insular cortical cells in the left hemisphere that project to the injection site. Green or magenta areas elsewhere represent projection sites from or to the right insular cortical neurons, respectively. **(B)** Adeno-associated viruses that express Cre-dependent GFP (see Figure 13–13C) were injected into the motor cortices of mice expressing Cre recombinase in layer 6 or layer 2/3 neurons. The projection patterns of layer 6 (top) and layer 2/3 (bottom) cortical neurons are shown in sagittal views at left and frontal (coronal) views at right. Circles represent the injection sites. Cre-expressing layer 6 neurons project primarily to the ipsilateral thalamus (with some contralateral thalamic projections) and subcortical regions. Cre-expressing layer 2/3 neurons project mostly within the neocortex. (A, courtesy of Hongwei Dong. See also http://www.mouseconnectome. org/ and Zingg B, Hintiryan H, Gou L et al. [2014] *Cell* 156:1096–1111; B, courtesy of Honghui Zeng. See also http://connectivity.brain-map. org/ and Oh SW, Harris JA, Ng L et al. [2014] *Nature* 508:207–214.)

neuronal cell bodies and dendrites, and travel down the axons to label their projection sites. Classic anterograde tracers include radioactively labeled amino acids (which can also be released at the axon terminal and taken up by postsynaptic neurons; see Figure 4–43) and phytohemagglutinin (PHA-L), a lectin from the red kidney bean *Phaseolus vulgaris* (**Figure 13–27**A). By contrast, **retrograde tracers** are mostly taken up by axon terminals and transported back to the cell bodies. Classic retrograde tracers include horseradish peroxidase and cholera toxin subunit b (CTb; Figure 13–27A). Whether a tracer travels in an anterograde or retrograde fashion is mostly determined empirically. Retrograde tracers likely bind to receptors that are selectively enriched in axon terminals, are taken up by the axon terminals via endocytosis, and utilize the endogenous retrograde axonal transport system to reach cell bodies (see Section 2.3). Anterograde tracers are likely taken up either selectively by receptors in the cell body or nonselectively, with the much greater cell body volume favoring greater uptake by cell bodies than by axon terminals. Anterograde and retrograde tracing are usually performed in conjunction with *in situ* hybridization or more often immunostaining, as well as Nissl stain; the combination of location, axonal projection, and gene/protein expression are often used in conjunction to define neuronal types. Much of our knowledge about the connections between different regions in the mammalian brain has been obtained through experiments utilizing these tracers.

A limitation of classic tracers is that all cells at the injection site take up the tracer, and therefore the projection patterns revealed by these methods represent contributions from a combination of different cell types. Ample data indicate that different cell types within the same region can have distinct projection patterns. For instance, at least 20 different types of retinal ganglion cells are intermingled in the retina with projections to distinct brain targets (for example, only ipRGCs project to the suprachiasmatic nucleus; see Box 4–2); in addition, those types that project to the same target region can have distinct connection patterns within that target region. Tracing the connections of neurons not only from a particular location but also of a particular cell type can thus provide higher resolution in mapping projection patterns. In the mouse, cell-type-specific anterograde tracing can be implemented by injecting an AAV (see Table 13–1) that expresses a marker in a Cre-dependent manner into a transgenic line that expresses Cre recombinase

Figure 13–28 Visualizing many single neurons in the same brain using image registration. Seven neurons, one from each of seven *Drosophila* brains, were individually labeled by the MARCM method using seven different GAL4 lines (representing six different neurotransmitter types); the neurons' projection patterns were determined in whole mount using confocal microscopy. The seven brains were then individually registered to a standard brain using a presynaptic marker counterstain (not shown), and the transformation used to register the counterstain was applied to individually labeled neurons to yield the image. Arrows indicate the seven cell bodies. (From Chiang AS, Lin CY, Chuang CC et al. [2011] *Curr Biol* 21:1–11. With permission from Elsevier.)

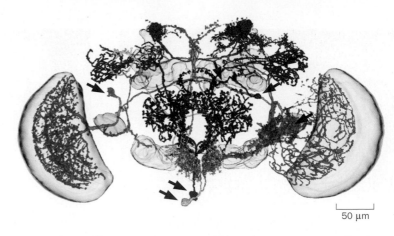

50 µm

in a specific cell type (for example, Figure 13–27B). Since many cell-type-specific Cre lines are available, axonal projections from different cell types within a given region can be studied independently.

Projection patterns of individual neurons can be further deciphered by a number of strategies. For example, a genetic strategy called 'brainbow' has been developed in mice to label individual neurons with different colors using stochastic Cre-dependent expression of different levels of three fluorescent proteins; the labeling of many neurons in different colors can allow multiple neurons to be visualized and traced within the same brain (see Figure 1–12C). An alternative strategy is to generate many singly labeled neurons, each from a different brain, and then use image registration methods to align those images in a common standard brain, such that the projection patterns of individual neurons can be compared (for example, **Figure 13–28**; see also Section 6.16). While its success is dependent on the accuracy of the brain image registration, this latter strategy has the advantage that individual neurons can be labeled using markers from transgenes with different regulatory elements representing different cell types. The relationships between different cell types, including potential pre- and postsynaptic partners, can thus be examined. However, the overlap of the axons of neuron *A* with the dendrites of neuron *B* is necessary but not sufficient for them to be synaptic partners. To reconstruct wiring diagrams with synaptic resolution, it is necessary to use higher resolution anatomical methods discussed in the next section and the physiological methods discussed in Section 13.26.

13.19 Mapping synaptic connections reveals neural circuitry

The most reliable anatomical method to determine whether or not two neurons form synaptic connections is EM, because synaptic connections can be directly visualized in EM. In principle, reconstructions of serial EM sections can establish a neuronal wiring diagram at the synaptic level (also called a **connectome**); indeed, this has been completed for the entire nervous system of *C. elegans* (see Figure 13–2). Serial EM reconstruction of a connectome involves making thin sections (50 nm or less) of the tissue of interest, imaging the sections, segmenting individual images into cell bodies and axonal and dendritic profiles that belong to individual neurons, aligning consecutive images to create a three-dimensional volume, and reconstructing individual segments across image stacks to create three-dimensional volumes of individual neurons (**Figure 13–29; Movie 13–2**). Synaptic connections between different profiles in EM sections can then be assigned to specific neurons. For larger nervous systems this procedure is not only labor-intensive but also demands a high level of technical precision at each step. Even a very small error rate in a single section may propagate with the tracing of axonal and dendritic profiles across many thousands of sections and compromise the quality of the reconstructed connectome.

Intensive efforts are currently focused on tackling these challenges. For example, in a method called serial block-face scanning electron microscopy, a microtome is integrated with a scanning electron microscope; the cut face of a tissue block is imaged, and then each imaged section is sliced away to allow the next

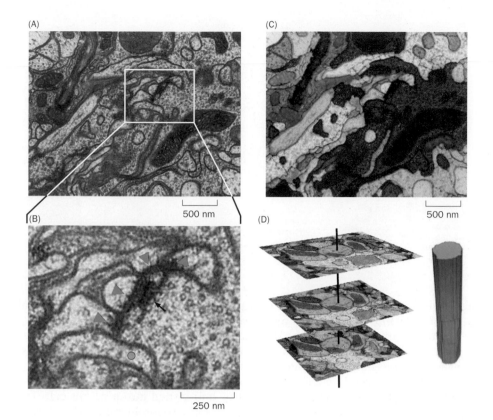

Figure 13–29 Constructing a wiring diagram using serial electron microscopy.
(A) A representative EM micrograph of the *Drosophila* medulla in the optic lobe (see Figure 5–35 for a light microscopic image and a schematic). **(B)** High-magnification view of the box in panel A, showing a presynaptic terminal (red arrow) in contact with four postsynaptic profiles (blue arrowheads). Another profile (green dot) is in contact with the presynaptic terminal but lacks a postsynaptic density in this and adjacent sections, so it was not scored as a synaptic partner. **(C)** The micrograph in panel A is segmented into profiles of neurites (axons and dendrites) by different colors. **(D)** Neurites are reconstructed by linking profiles in thousands of consecutive sections (of which three are shown here) to construct the three-dimensional object shown on the right (see Movie 13–2). (From Takemura S, Bharioke A, Lu Z et al. [2013] *Nature* 500:175-181. With permission from Macmillan Publishers Ltd.)

section in the tissue block to be imaged. Because successive images are acquired from the same tissue block, images are automatically aligned. This and other methods have enabled reconstruction of synaptic connections from many hundreds of cells in the retina and primary visual cortex of the mouse, as well as from the fly optic medulla (Figure 13–29). These studies have provided new insights into the mechanisms by which retinal neurons acquire their direction-selective responses (see Figure 4–32), for example.

While serial EM reconstruction provides the ultimate anatomical resolution for constructing synaptic connections within a local region, it is not yet feasible to reconstruct connections between brain regions in large nervous systems such as those of mammals. In 2014, the largest reconstruction volume was about 10^6 μm³, a volume equivalent to a cube in which the length of each side is roughly equal to the diameter of a human hair. The mouse or human brain contains roughly 10^5 or 10^9 of that volume, respectively. Other genetic and viral tracing methods have been invented to identify synaptic connections between neurons separated by large distances. A powerful strategy to determine both local and long-distance connections is **trans-synaptic tracing**. An ideal trans-synaptic tracer should have the following properties: if the tracer is expressed from a given neuron (a **starter cell**), a retrograde tracer should be received only by the complete set of neurons that are presynaptic to the starter cell, whereas an anterograde tracer should be received only by the complete set of neurons that are postsynaptic to it. Neurons, axons, or dendrites that are near the starter cell but do not form synaptic connections with the starter cell should not receive the tracer.

Trans-synaptic tracing strategies are still being developed and refined. The most efficient tracers thus far derive from neurotropic viruses such as **rabies virus** and herpes simplex virus, which spread within the nervous systems of their hosts naturally by crossing synapses. Usually, once a virus infects a neuron, it spreads not only to that neuron's direct synaptic partners, but also to the synaptic partners of subsequently infected neurons, making it difficult to distinguish whether two neurons are connected directly or through intermediate neurons. A strategy has been developed to prevent the retrograde trans-synaptic rabies virus from crossing more than one synapse, thereby restricting infection to those cells that are directly presynaptic to the starter cell (**Figure 13–30**). To achieve this, the viral

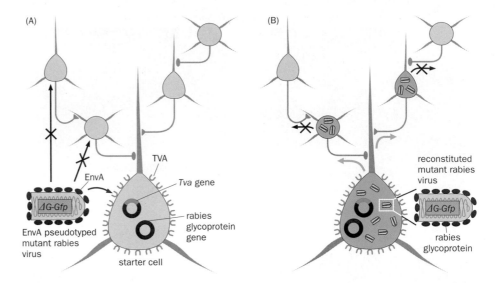

gene encoding the glycoprotein (a coat protein of the virus), which is essential for viral entry into a host cell, is replaced with a gene encoding GFP in the viral genome. This mutant rabies virus is pseudotyped (see Table 13–1) with the envelope protein (EnvA) from an avian virus that cannot transduce mammalian cells. However, if transgenes encoding TVA (the receptor for EnvA) and rabies glycoprotein are expressed in a specific neuron (Figure 13–30A), that neuron can be transduced by the pseudotyped rabies virus and turned into a starter cell. The rabies glycoprotein transgene complements the glycoprotein gene deleted from the mutant viral genome to produce functional rabies viruses, which spread to the starter cell's presynaptic partners. However, since the presynaptic partners do not express rabies glycoproteins, mutant rabies viruses cannot spread further (Figure 13–30B). This monosynaptic strategy for tracing the presynaptic partners of specific starter cell types has been applied to map synaptic connections in many parts of the mammalian nervous system (see Figures 6–21 and 8–10).

In summary, many methods have been developed to decipher the wiring diagrams of nervous systems at different scales with various resolutions. These methods range from noninvasive human brain imaging to the complete reconstruction of synaptic connections in *C. elegans*. However, many technical challenges must be overcome in order to comprehensively map the synaptic connections of larger neural systems. Mapping of electrical synapses poses an additional challenge as they are not as readily identifiable as chemical synapses, even in electron micrographs. Furthermore, a wiring diagram based on anatomical connections alone is just a first step toward understanding a neural circuit. To decipher how a neural circuit operates, investigators must also assess whether synapses are excitatory or inhibitory, how strong they are, and how they are influenced by the actions of modulatory neurotransmitters and neuropeptides (for example, see Section 6.12). Understanding neural circuit function requires research tools that can measure and manipulate the activities of neurons in the wiring diagram in the context of animal behavior; we will discuss these methods in the next part of the chapter.

RECORDING AND MANIPULATING NEURONAL ACTIVITY

Signals in the nervous system spread predominantly by membrane potential changes. Thus, the ability to record membrane potential as a means of measuring neuronal activity is critical to our understanding of nervous system function. In this part of the chapter, we discuss the principal methods for recording neuronal activity.

While observations and measurements are the foundations of discovery, well-designed perturbation experiments are necessary to elucidate the underlying mechanisms. We will also discuss the loss- and gain-of-function approaches investigators have employed to silence and excite neurons of interest and thereby

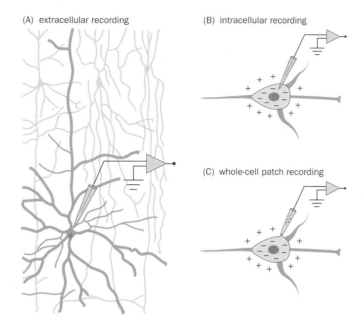

Figure 13–31 Three principal methods of electrophysiological recordings. **(A)** In extracellular recording, the tip of an electrode is sufficiently close to the neuronal cell body such that when the neuron fires action potentials, the electrode can detect the extracellular voltage changes. **(B)** In intracellular recording, a sharp electrode penetrates into the cell to measure intracellular membrane potentials directly. **(C)** In whole-cell patch recording, the patch electrode forms a tight seal with the plasma membrane to measure intracellular membrane potentials (see Figure 13–37B for more details).

identify the roles that specific neurons or populations of neurons play in the function of neural circuits and the behavior of animals.

13.20 Extracellular recordings can detect the firing of individual neurons

Three principal electrophysiological methods have been employed to record neuronal activity. In **extracellular recording**, an electrode, often made of metal wire that is insulated except at the tip, is placed at close range outside a neuronal cell body to record voltage changes when the neuron fires an action potential (**Figure 13–31**A). **Intracellular recording** typically utilizes a sharp electrode, usually made of glass with a very fine, open, solution-filled tip, which penetrates into the cell to directly record the intracellular membrane potential (Figure 13–31B). **Whole-cell patch recording** is a special form of intracellular recording in which the glass electrode forms a tight seal with the plasma membrane of the recorded cell (Figure 13–31C). We discuss extracellular recording in this section, and intracellular and whole-cell patch recording in the following section.

Although the simplest form of recording in electrophysiology, extracellular recording has proven to be a powerful method. When a cell fires an action potential, the ionic flow creates voltage changes not only inside the cell but also in its immediate surroundings. When an electrode tip is close to a neuronal soma (Figure 13–31A), the signal picked up by the electrode predominantly reflects the spiking activity of this single nearby neuron (**Figure 13–32**A). Action potentials with a specific amplitude and waveform as detected in extracellular recording define a 'unit;' thus extracellular recording aimed at detecting the firing paterns of individual neurons is also called **single-unit recording**. Sometimes an extracellular electrode can detect and distinguish action potentials from more than one neuron because of the distinct amplitudes or waveforms of action potentials from neurons at varied distances to the electrode tip. (This effect is amplified by the use of a **tetrode**, an extracellular electrode that contains four independent wires enabling four independent recordings of spiking activities of neurons nearby the electrode. Firing patterns of up to 20 neurons may be resolved by tetrode recording.) Single-unit extracellular recording of neuronal activity has been the key method for making many conceptual advances in neurobiology. As we learned in Chapter 4, for example, extracellular recording at successive stages of the visual processing pathway revealed how the brain's representation of a visual scene is transformed as the signals travel from the retina to the visual cortex. Extracellular recording is still the predominant method used for recording neuronal activities *in vivo* today.

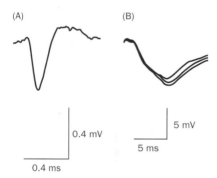

(A)   (B)

0.4 mV

5 mV

5 ms

0.4 ms

Figure 13–32 Examples of single-unit and field potential recordings. Extracellular electrophysiological recordings have a significant history in neurobiology. Shown here are **(A)** a single 'spike' of a cat retinal ganglion cell recorded by an extracellular electrode in response to light stimulation, and **(B)** three superimposed traces of local field excitatory postsynaptic synaptic potentials (fEPSPs) in the dendritic layer of the dentate gyrus in response to electrical stimulation of axons in the perforant path. Panel A is from the original recording that described the receptive fields of retinal ganglion cells (see Figure 4–24). Panel B is from the original recordings that led to the discovery of long-term potentiation (see Figure 10–8). Note that in both examples, the voltages recorded from extracellular electrodes become more negative as neurons are activated. This is because positive ions flow into the cell during the rising phase of the action potential (panel A) or when postsynaptic cells are depolarized (panel B). Note also that the time course of the spike in panel A is much faster than the field EPSP in panel B. (A, adapted from Kuffler SW [1953] *J Neurophysiol* 16:37–68; B, adapted from Bliss TVP & Lømo T [1973] *J Physiol* 232:331–356.)

Figure 13–33 Multi-electrode array. Multi-electrode arrays similar to this 10 × 10 silicon-based prototype have been widely used for recordings of cortical neurons in neural prosthetics (see Section 8.11). (From Campbell PK, Jones KE, Huber RJ et al. [1991] *IEEE Trans Biomed Eng* 38:758–768. With permission from IEEE.)

Extracellular recording can also be used to measure **local field potentials**, which are local potential variations measured relative to a distal ground. Local field potentials can be measured with an electrode for single-unit extracellular recording, but high-frequency signals such as those produced by action potentials are removed by signal-processing methods such as a low-pass filter. The remaining low-frequency signals reflect collective dendritic and synaptic activities of many neurons near the electrode tip. We have seen the application of field potential recording to studies of hippocampal long-term potentiation *in vivo* (Figure 13–32B; see also Figure 10–8) or in freshly prepared brain slices (acute slices) *in vitro* (see Figure 10–10).

In order to explore how groups of neurons act together to encode and process information, various kinds of **multi-electrode arrays** have been developed that record many neurons at the same time. These consist of many independent electrodes horizontally arranged in grids (for example, **Figure 13–33**) or vertically along different depths (see Figure 4–47A). Multi-electrode arrays can record tens to hundreds of neurons simultaneously *in vitro*, such as in brain slices or explants; for example, recording simultaneously from many retinal ganglion cells allowed the retinal wave to be characterized (see Figure 5–21A). Multi-electrode arrays can also be implanted *in vivo* to record the activities of many neurons in awake, behaving animals (see Figures 8–26 and 8–29) or humans (see Figure 8-30).

Electroencephalography (**EEG**) is essentially a noninvasive field potential recording method that usually involves multi-electrode arrays. In a typical EEG setting, the electrodes are attached to the surface of the scalp. Because the distance is greater between EEG electrodes and the recorded neurons compared with conventional extracellular recording, EEG can only detect the synchronized activities of tens of thousands of neurons or more. EEG enabled the discovery of brain waves of various frequencies and is an extremely useful tool for assessing brain states such as different forms of sleep (see Figure 8–51A) or epileptic conditions (see Figure 11–47).

Extracellular recordings are typically performed blind to the type of neuron that is being recorded. They have a bias toward detecting the dominant neuron types and the most active neurons, because extracellular recording procedures often involve the researcher shifting electrode positions until a spike is detected. Spike waveforms and firing properties have been used to classify cell types from extracellular recordings, but different cell types may have overlapping properties in most regions of the brain, making such classification ambiguous. Recently, this limitation has been addressed by a phototagging method, where single-unit recording was performed in a tissue in which neurons of specific types can be stimulated using optogenetics (see Section 13.25). The cell types of the recorded units can be classified based on whether or not they respond to optical stimulation. For example, the ventral tegmental area (VTA) has a mixture of dopamine and GABA neurons. When mice that express Cre only in GABA neurons are injected in the VTA with viral vectors encoding a channelrhodopsin-2 (ChR2) transgene that requires Cre for expression, GABA neurons can be selectively stimulated using light. Likewise, mice that express Cre only in dopamine neurons can be used to selectively activate dopamine neurons using light. To distinguish between dopamine and GABA neurons, recordings can be collected from these two types of transgenic mice in which either dopamine or GABA cells can be optically activated. Researchers were able to determine whether a recorded neuron in a given Cre mouse was GABAergic or dopaminergic based on whether or not it could be depolarized by photostimulation (**Figure 13–34**). Using this approach, it

Figure 13–34 The use of phototagging to identify cell types in extracellular recordings. Extracellular recording of ventral tegmental area neurons was performed in a transgenic mouse that expressed Cre in dopamine neurons and that was infected with an adeno-associated virus expressing a Cre-dependent channelrhodopsin (ChR2). The top trace shows that each photostimulation (cyan bar) resulted in a spike of the recorded neuron; two spikes are temporally magnified below. These results are consistent with recording of a dopamine neuron that expresses ChR2 channels and produces action potentials in response to depolarizations caused by light-induced opening of those channels. (Adapted from Cohen JY, Haesler S, Vong L et al. [2012] *Nature* 482:85–88. With permission from Macmillan Publishers Ltd.)

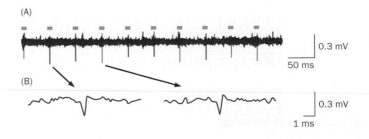

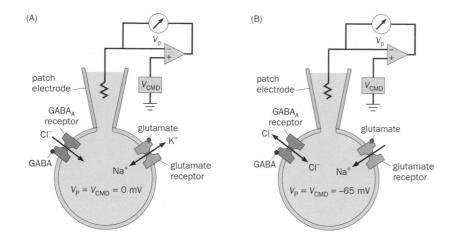

Figure 13–35 Using voltage clamp to dissect inhibitory versus excitatory input. **(A)** When the voltage is clamped at 0 mV in whole-cell recording, at the reversal potential of glutamate receptors (see Section 3.15), even if glutamate receptor channels are open upon glutamate binding, cation influx and efflux balance out and do not contribute to the net current across the cell membrane. The measured current is mostly contributed by Cl⁻ flow through the GABA_A receptor in response to GABA, thereby reflecting inhibitory input. **(B)** When the voltage is clamped at −65 mV, around the reversal potential of the GABA_A and glycine receptors (that is, the equilibrium potential of Cl⁻; see Section 3.17), the measured current reflects mostly excitatory input from the glutamate receptor channel. V_{CMD}, commanding voltage; V_P, voltage of the patch electrode.

was found that VTA dopamine and GABA neurons exhibit distinct properties with regard to reward presentations (see Section 10.24). A caveat of the phototagging method is that spikes can also be driven indirectly in non-ChR2-expressing cells; extremely short latency between light onset and spike (~1 ms), or certainty that there is no local excitation, are required for confidence in cell identification.

13.21 Intracellular and whole-cell patch recordings can measure synaptic input in addition to firing patterns

Compared to extracellular recordings, intracellular recordings with a sharp electrode (see Figure 13–31B) offer much higher sensitivity and signal-to-noise ratio for detecting electrical signals. Intracellular recordings can detect not only firing patterns but also sub-threshold membrane potential changes resulting from excitatory or inhibitory input received by the recorded neuron. Many discoveries about neuronal communication we discussed in Chapters 2 and 3, including the ionic basis of the action potential and the mechanisms of synaptic transmission, were made using intracellular recording methods.

Whole-cell patch recording (or simply whole-cell recording; see Figure 13–31C) is one of several variations of the patch clamp method (**Box 13–2**). Here, the interior of the electrode forms a continuous compartment with the cytoplasm, which allows the electrode to measure the membrane potential with a high sensitivity comparable to intracellular recording with a sharp electrode. The patch electrode affords greater access to the cell—the tip diameter is usually 1 μm or so for a patch electrode, an order of magnitude greater than the diameter of an intracellular sharp electrode. As a result, whole-cell patch electrodes not only can record the activity of a neuron but also can pass currents that change a neuron's membrane potential, effectively achieving voltage clamp (see Figure 2–21). In this manner, inhibitory input to a neuron can be recorded in isolation when the neuron is voltage clamped at the reversal potential of the glutamate receptors, such that excitatory input does not make a contribution (**Figure 13–35A**). Likewise, excitatory input to a neuron can be recorded without interference from inhibitory input when the neuron is voltage clamped at the reversal potential of Cl⁻ (Figure 13–35B). These techniques are often combined with pharmacological blockers for specific receptors to dissect excitatory and inhibitory inputs to specific neurons in slices or explant preparations (for example, see Figure 4–31). Dissection of inhibitory and excitatory input using the strategies outlined in Figure 13–35 can also be performed *in vivo*.

Because the solution that fills the patch electrode is continuous with the cytoplasm of the cell, macromolecules such as dyes or fluorescent markers can be included in the patch electrode solution to label the recorded cell (**Figure 13–36**); signaling molecules can also be included to alter the properties of the recorded cell (see Figure 10–13B). The filling of recorded cells during whole-cell recording (and during intracellular recording with a sharp electrode) can be used to

Figure 13–36 Dye fill during whole-cell patch recording from the dendrite and the cell body. In this example, two patch electrodes are used for whole-cell recording at the cell body (bottom) and at the apical dendrite (top) of a cortical pyramidal cell in a rat brain slice. Dual patching of the dendrite and cell body of the same neuron is confirmed by the mixing of two different fluorescent dyes, one from each electrode. This dual patch approach can be used to study synaptic potential propagation from the dendrite to the cell body, and action potential back-propagation from the cell body to the dendrite. (From Stuart GJ & Sakmann B [1994] *Nature* 367:69–72. With permission from Macmillan Publishers Ltd.)

Box 13–2: Patch clamp recordings can serve many purposes

Patch clamp recording requires the formation of a very tight seal (with a resistance in the range of gigaohms, or 10^9 ohms) between a glass patch electrode, also called a patch pipette, and the plasma membrane of a target neuron (**Figure 13–37**A). Due to this high resistance, the very small currents that pass through individual ion channels in the membrane underneath the patch electrode can be recorded. Indeed, the measurement of single-channel conductances in a defined extracellular environment (the internal solution of the patch electrode) in this **cell-attached recording** mode (Figure 13–37B) was the first application of the patch clamp technique (see Figure 2–30). The patch clamp technique can also be used in a number of additional modes. For example, the membrane patch underneath the electrode can be excised (Figure 13–37C). The **excised patch** can be placed in a defined solution so that both the extracellular and intracellular environments of the ion channels in the patch can be controlled. Excised patch recording is widely used to study the biophysical and biochemical properties of ion channels (for example, see Figure 4–9). In the whole-cell recording mode, a gentle suction applied to a cell-attached patch ruptures the membrane underneath the patch electrode, such that the interior of the patch electrode and the cytoplasm of the recorded cell form a single compartment (Figure 13–37D).

Because the intracellular solution and cytoplasm are continuous in the whole-cell recording mode, the cellular contents may be diluted with solution from the patch electrode during prolonged recording. A procedure called **perforated patch** can be used to minimize this concern; in this method, the membrane between the patch electrode and the cell is not fully ruptured as in whole-cell mode, but the internal solution of the patch electrode contains chemicals that make small holes in the underlying neuronal membrane, such that the patch electrode can record the current and voltage of the cell with minimal exchange of macromolecules. In another variation called **loose-patch recording**, a patch electrode is placed against the cell membrane without forming a gigaohm seal. Recording in loose-patch mode thus does not affect the cellular contents. However, this approach does not provide sufficient sensitivity to record sub-threshold activity; it can only record action potentials. Loose-patch recording is commonly used for *in vivo* recordings. Compared with conventional extracellular recording, it does not have a bias toward active neurons, as the criterion it uses to assess whether the electrode is approaching a cell is a sudden increase in resistance, rather than the recording of action potentials. Furthermore, DNA can be added to the internal solution of the patch electrode and introduced specifically into the recorded cell by electroporation after the recording. One application of this single-cell electroporation technique is to introduce DNA encoding a fluorescent protein so that the same cell can be located and recorded at a later time, for instance after a specific experience.

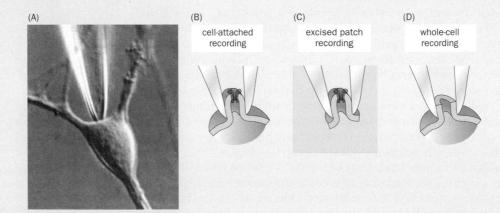

Figure 13–37 Multi-functionality of patch clamp recording. **(A)** Photomicrograph of a patch electrode in contact with the plasma membrane of a neuron in culture. **(B–D)** Three patch clamp modes are schematized. In a cell-attached patch (B), the patch electrode forms a tight seal with a neuron's plasma membrane, which allows measurement of ion flow through a single channel in the patch of membrane underneath the electrode. In an excised patch (C), the piece of membrane underneath the electrode is excised from the cell and can be placed in defined medium to study the properties of the channel. In whole-cell recording (D), the membrane patch under the electrode is ruptured, such that the interior of the electrode and the recorded cell becomes a single compartment. (Adapted from Neher E & Sakmann B [1992] *Sci Am* 266:44–51. With permission from Macmillan Publishers Ltd.)

determine the locations, morphologies, and projection patterns of recorded neurons (see Figures 13–1 and 13–21). Such information is highly valuable in determining how the structures and functions of neurons are correlated (for example, see Figure 4–45).

Another important advantage of intracellular recording is that it allows a genetically defined population of cells to be targeted for recording. A given brain region most often contains a mixture of different neuronal types with varying densities. As discussed in the previous section, blind recording cannot distinguish unequivocally between different cell types, and it may not allow rare cell types to be recorded at all. The job of an electrophysiologist is made easier if the type of neuron targeted for recording is pre-labeled with fluorescent protein using genetic strategies. For example, recording specific pairs of genetically labeled pre- and postsynaptic partner neurons in the *Drosophila* olfactory system enabled investigators to examine how the representation of olfactory information is transformed as signals travel between the presynaptic olfactory receptor neurons and the postsynaptic projection neurons (see Section 6.14). With the advance of genetic technology to access specific neuronal types (see Section 13.12), these targeted electrophysiology experiments have become increasingly powerful in revealing how neural circuits process information.

13.22 Optical imaging can measure the activity of many neurons simultaneously

To appreciate a symphony, it's not enough to hear one instrument at a time; the listener must be able to hear all of the orchestra's musical instruments simultaneously. Likewise, a deep understanding of how neural circuits encode and process information requires that researchers capture the simultaneous activity of many (ideally all) neurons in the circuit. Even with multi-electrode arrays, investigators can only record at most hundreds of neurons at a time. In addition, the spacing of recorded neurons is constrained by the spacing of the electrodes. The only method currently available that can in principle record the activity of all neurons within a region at cellular resolution is **optical imaging**, which uses changes of fluorescence or other optical properties as indicators of neuronal activity.

Since neurons communicate by membrane potential changes, the ideal indicator would be one that reports voltage changes directly. Indeed, many variants of **voltage-sensitive dyes**, which change fluorescence intensity or other optical properties in response to membrane potential changes, have been developed. However, the voltage-sensitive dyes invented thus far have been limited by low signal-to-noise ratios and high levels of phototoxicity to be widely applicable for measuring neuronal activity with cellular resolution *in vivo*. Other optical sensors have been developed to enable imaging of neurotransmitter release and receptor conformation changes as proxies for neuronal activity. By far the most widely used sensors of neuronal activity are Ca^{2+} **indicators**, which translate changes of intracellular Ca^{2+} concentration ($[Ca^{2+}]_i$) into changes in fluorescence signal. A rise in $[Ca^{2+}]_i$ usually accompanies neuronal activation due to the activation of postsynaptic neurotransmitter receptors that are permeable to Ca^{2+} and the opening of voltage-gated Ca^{2+} channels in response to depolarization in both cell bodies and presynaptic terminals.

Some of the Ca^{2+} indicators used to record neuronal activity are made from synthetic chemicals, while others are protein based. Chemical indicators typically link a Ca^{2+}-chelating moiety with a fluorophore. As a specific example, binding of Ca^{2+} to the chemical indicator **fura-2** shifts the wavelength of maximal fluorescence excitation about 30 nm shorter (**Figure 13–38**A). Thus, the ratio of fluorescence intensity measured at the excitation wavelengths of 350 nm and 380 nm can be used as a sensitive measure of $[Ca^{2+}]_i$. Protein-based Ca^{2+} indicators are also called **genetically encoded Ca^{2+} indicators** because they can be expressed as a transgene in specific cell types. For example, the cameleon indicator reports $[Ca^{2+}]_i$ utilizing **fluorescence resonance energy transfer (FRET)**, a mechanism of energy transfer between two fluorophores. The efficiency of FRET is inversely proportional to the sixth power of distance and therefore can be used to determine

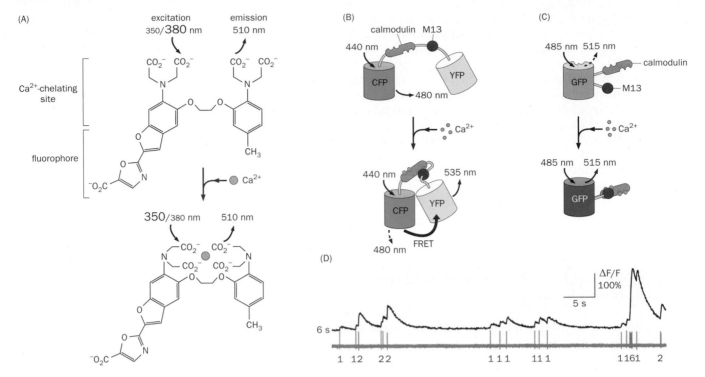

Figure 13–38 Chemical and genetically encoded Ca²⁺ indicators.
(A) Fura-2 is a chemical Ca²⁺ indicator consisting of a fluorophore fused with a Ca²⁺-chelating site from the Ca²⁺ buffer EGTA. When $[Ca^{2+}]_i$ is low, excitation at 380 nm produces stronger fluorescence emission than excitation at 350 nm (symbolized by the size of the numbers); when $[Ca^{2+}]_i$ is high, the converse is true. Thus ratiometric imaging at 350/380 excitation wavelengths serves as a sensitive reporter of $[Ca^{2+}]_i$. **(B)** Design principle of cameleon, the first genetically encoded Ca²⁺ indicator. When $[Ca^{2+}]_i$ increases, the binding of calmodulin (CaM) to the CaM-binding peptide M13 brings two fluorescent proteins closer, thus increasing fluorescence resonance energy transfer (FRET) as measured by the ratio of 535 nm emission (reflecting the contribution of FRET) over 480 nm emission (no FRET). **(C)** Design principle of GCaMP. A permutated GFP is restored to its native three-dimensional structure with an associated increase in fluorescence after Ca²⁺-triggered binding of M13 and CaM, which are

at the ends of GCaMP. The fluorescence intensity of GCaMP can thus be a readout of $[Ca^{2+}]_i$. **(D)** Single action potentials reliably induce fluorescence changes of GCaMP6 in mouse visual cortical neurons *in vivo*, as measured by simultaneous loose-patch recording to identify action potentials (bottom) and GCaMP6 fluorescence intensity change (top). ΔF/F is the ratio of fluorescence intensity change (ΔF) over basal fluorescence intensity (F). Note that when action potentials occur in rapid succession (indicated by numbers below), individual action potentials cannot be resolved as individual peaks by imaging. (A–C, adapted from Grienberger NL & Konnerth A [2011] *Neuron* 73:862–885. With permission from Elsevier Inc.; D, adapted from Chen TW, Wardill TJ, Sun Y et al. [2013] *Nature* 499:295–300. With permission from Macmillan Publishers Ltd. See also Grynkiewicz G, Poenie M & Tsien RY [1985] *J Biol Chem* 260:3440–3450; Miyawaki A, Llopis J, Heim R et al. [1997] *Nature* 388:882–887; Nakai J, Ohkura M & Imoto K [2001] *Nat Biotech* 19:137–141.)

the distance between two fluorophores. Cameleon is an *in vitro* engineered protein in which calmodulin and its target M13 peptide link two fluorescent proteins, cyan and yellow fluorescent proteins (CFP and YFP). Excitation of CFP can cause emission from CFP, or from YFP via FRET if YFP is sufficiently close. Ca²⁺-triggered binding of calmodulin and the M13 peptide decreases the distance between CFP and YFP, and hence increases the emission from YFP (Figure 13–38B). Genetically encoded Ca²⁺ indicators can also report $[Ca^{2+}]_i$ by increasing fluorescence intensity. For instance, in an engineered protein called **GCaMP**, the positions of two halves of the green fluorescent protein GFP are switched, and the swapped halves are linked with calmodulin and M13. Ca²⁺-triggered binding of calmodulin to M13 restores the original GFP conformation and hence increases the fluorescence intensity (Figure 13–38C).

Chemical indicators have offered superior sensitivities compared to genetically encoded indicators in the past decades and have been widely used in experiments described in this book (see Figures 4–42C and 5–21B). However, protein engineering has produced a new generation of genetically encoded Ca²⁺ indicators, such as GCaMP6, that can detect single action potentials *in vivo* more reliably than chemical indicators (Figure 13–38D), achieving an important milestone in optical imaging. Moreover, selective expression of genetically encoded Ca²⁺ indicators can be specified by cell type and can enable repeated imaging of the same neuron over the course of months, which is not possible with chemical indicators. These developments will make many new discoveries possible in the coming years.

In addition to activity indicators, optical imaging relies on suitable microscopes for visualizing fluorescence changes in live tissues with high spatial resolution, strong signals, and minimal photodamage to the imaged tissues. Conventional fluorescence microscopes have poor resolution along the z axis and are best used with relatively thin tissues such as retinal explants (for example, Figure 5–21B). As described earlier, laser-scanning confocal microscopes have great resolution along the z axis because fluorescence emission is collected via a small pinhole to block out-of-focus fluorescence from other planes (see Figure 13–19A). However, tissues above or below the imaging plane are nevertheless exposed to the laser excitation during scanning; the resulting photodamage includes heating of the tissues and photobleaching of the fluorescent indicators. In addition, light scattering in brain tissues limits the depth of effective imaging by confocal microscopy to about 100 μm from the surface.

The most widely used method for *in vivo* optical imaging is **laser-scanning two-photon microscopy**, which relies on the absorption of two long-wavelength photons simultaneously in order to excite a fluorophore (**Figure 13–39**A). Only at the focal plane is the density of photons high enough to cause substantial fluorescence emission, such that photodamage is limited to the focal plane (Figure 13–39B). Furthermore, the longer excitation wavelength allows light to penetrate more deeply into the light-scattering brain tissue. Lastly, because fluorescence excitation is limited to the focal plane, any emission photons collected by the objectives, including those resulting from scattering, can be used without the need of a pinhole (Figure 13–39C; compared with Figure 13–19A). Thus, two-photon microscopes collect emitted photons more efficiently than confocal microscopes. Many optical imaging studies discussed in early chapters utilized two-photon microscopy, including Ca^{2+} imaging of neuronal activity (see Figure 4–42C) and structural imaging of neuronal morphology (see Figure 10–48). But even two-photon microscopes can effectively image only those tissues that are within about 500 μm from the surface. This restriction prevents imaging of most of the nervous system of even small mammals such as mice. Additional imaging methods, such as fluorescence endoscopy, are being developed to allow imaging of deeper tissues (see **Box 13–3** for a comparison of different methods).

In addition to fluorescence-based imaging, other imaging techniques such as **intrinsic signal imaging** (see Figure 4–42B and Figure 6–16) and **functional magnetic resonance imaging** (**fMRI**) (see Figures 1–24, 4–51, and 10–37) also report neuronal activity. These techniques use blood flow near the excited neurons as an indicator of neuronal activity, as increased neuronal activity is associated with changes of blood flow and oxygenation. These imaging methods have poorer spatial and temporal resolution compared to fluorescence imaging of individual neurons. For instance, the current fMRI resolution is about 2 mm in the linear dimension; 8 mm³ of tissue contains hundreds of thousands of neurons. Nevertheless, an important advantage of fMRI is its noninvasiveness, making it a favored tool for imaging neuronal activities in the human brain.

Figure 13–39 Laser-scanning two-photon microscopy. The schematic of a two-photon microscope is similar to that of a confocal microscope (see Figure 13–19A) except that the pinhole next to the detector is omitted. **(A)** Simultaneous absorption of two long-wavelength (lower energy, red) photons brings a fluorescence indicator molecule to the excited state, from which it subsequently relaxes while emitting fluorescence (green). **(B)** Schematic of a microscope objective on top of a brain tissue sample (cyan) focusing on a portion of a dendrite. The near-simultaneous absorption of two infrared photons (the left and middle photon paths) excites fluorophores at the focal plane; the excitation light outside the focal plane due to scattering (the photon path on the right) is insufficiently intense to allow efficient two-photon excitation, so excitation is effectively restricted to the imaging spot at the focal plane (highlighted in yellow). **(C)** All emitted photons (green paths) collected by the objective, including those resulting from scattering (indicated by *), contribute to the fluorescence signal from the imaging spot. (Adapted from Svoboda K & Yasuda R [2006] *Neuron* 50:823–839. With permission from Elsevier Inc. See also Denk W, Strickler JH & Webb WW [1990] *Science* 248:73–76.)

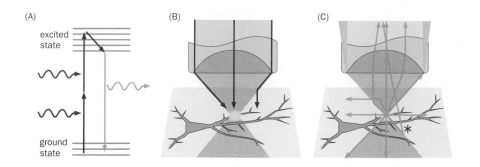

Box 13–3: From *in vitro* preparations to awake, behaving animals: a comparison of recording methods

What method should researchers choose for recording neuronal activity? The answer depends on the biological question the experiment intends to address and the preparation used to address the question. In this box, we compare the pros and cons of the recording methods discussed in Sections 13.20–13.22 in the context of different experimental preparations: from neuronal cell cultures and tissue explants *in vitro* to anesthetized animals and awake, behaving animals *in vivo* (**Table 13–2**).

Electrode-based recording methods directly measure membrane potentials, and hence have superb sensitivity and temporal resolution. They are widely used in reduced preparations such as cultured neurons or *in vitro* explants such as brain slices, providing insight into the properties of individual neurons and helping researchers investigate synaptic transmission and plasticity as assayed from single neurons or a group of neurons in aggregate (measured by local field potentials). The reduced preparations also offer easier access to target neurons, so intracellular recording methods are preferred since they have superb sensitivity for detecting sub-threshold membrane potential changes and support an array of approaches through which researchers can manipulate the recorded neurons by injecting currents or molecules. Intracellular recordings (particularly whole-cell recording) have been the predominant method for studies using slice preparations of the mammalian brain. Optical imaging methods can be employed in reduced preparations if recording the activity of many individual neurons is desirable (for instance to observe retinal waves) or when subcellular resolution of neuronal activity is needed.

Recording from *in vivo* preparations poses more challenges regardless of method because intact systems have greater inherent complexity, the neurons to be recorded are harder to access, and the animals must be kept alive (and, in some cases, awake and behaving) during the recording process. However, obtaining recordings from living animals, including from awake, behaving animals, is essential for addressing many questions in neurobiology related to perception, cognition, and the neural basis of behavior (for example, see Section 4.29). Recording with extracellular electrodes has historically been the predominant method *in vivo*, particularly in awake, behaving animals, due to its overall superior ability to maintain stable recordings (that is, to record from the same cells over long periods of time), to penetrate deep into tissues, and to record many neurons simultaneously (Table 13–2). For example, the discovery of place cells in the hippocampus and grid cells in the entorhinal cortex relied on extracellular recordings in freely moving rodents (see Box 10–2). Intracellular recordings, whether with a sharp electrode or whole-cell patch electrode, are difficult to maintain when animals are awake and moving, because physical movement can shift the electrode. Likewise, optical imaging requires stability and is sensitive to movement. However, intracellular recording offers superior sensitivity, and optical imaging can record many neurons at once and over long periods of time using genetically encoded indicators. Thus, researchers have developed methods to improve stability for *in vivo* recording.

One way to provide the mechanical stability essential for both intracellular recording and optical imaging is to use

Table 13–2: Comparison of electrophysiological and optical imaging methods for recording neuronal activity

Property	Electrophysiology		Optical imaging with Ca^{2+} indicators[1]
	Extracellular recording	**Intracellular recording**	
Sensitivity to electrical signal	spikes	spikes and sub-threshold activity	generally less sensitive[2]
Spatial resolution	cellular to network	cellular to subcellular[3]	cellular and subcellular
Temporal resolution	<1 millisecond	<1 millisecond	10s to 100s of milliseconds for a single imaging plane
Number of neurons recorded simultaneously	up to hundreds	at most a few	thousands or more
Stability during movement	good	poor	poor
Depth of recording	any depth	easier superficially	limited[4]
Duration of recording	days to weeks	10s of minutes	hours with chemical indicators; months with protein indicators
Cell-type-specific recording	poor	good	excellent with protein indicators
Biases	active neurons; dominant cell types	large cells	cells that take up or express the indicators well

[1] Most other indicators share similar properties.

[2] GCaMP6 can detect single action potentials, but cannot resolve high-frequency spikes (see Figure 13–38D).

[3] Whole-cell patch recording can be applied to large dendrites (see Figure 13–35).

[4] <10 μm with conventional fluorescence microscopy for cellular resolution; ~100 μm with confocal microscopy; ~500 μm with two-photon microscopy.

Box 13–3: From *in vitro* preparations to awake, behaving animals: a comparison of recording methods

head-fixed animals, that is, to restrain head movement with respect to the recording equipment such as a microscope or a micromanipulator that holds an electrode. (The head-fixed preparation was originally developed for extracellular recording of neurons in the visual cortex of awake, behaving monkeys; if the monkey cannot move its head and is trained to fixate, then a stimulus on the screen always falls on the same spot on the retina; see Figure 4–52.) For instance, a head-fixed mouse can be trained to associate an odor with a water reward and to move its tongue to fetch the reward while its motor cortical neurons are being optically imaged during the learning process (**Figure 13–40**A). A more sophisticated preparation involves virtual reality feedback. For example, a head-fixed mouse navigates on a spherical treadmill made of a ball floating on air; the mouse's movement is used to adjust the visual scene as if the animal was navigating in a real environment (Figure 13–40B; **Movie 13–3**). The mouse can be trained to run on linear tracks or make turning choices in virtual reality while its neurons are being recorded with a whole-cell patch electrode or a two-photon microscope. An alternative to the head-fixed preparation is the use of a miniature fluorescence microscope. Miniature microscopes have been designed that can be attached to the head of a freely moving mouse, making it possible to record neuronal activity while the animal explores its environment

(Figure 13–40C and **Movie 13–4**). Further development and refinement of these preparations will expand researchers' ability to record neuronal activity under different experimental settings.

Even optical imaging, which allows simultaneous recording of thousands of neurons, can only provide information about the activities of a very small fraction of the nervous system. A method that can in principle provide information about the activity of neurons throughout the brain at cellular resolution is the expression of immediate early genes (see Section 3.23) after the animal has experienced sensory stimuli or behavioral episodes. Important limitations of this method include the requirement that it be done *post mortem* in fixed brain tissue, the slow temporal resolution (transcription operates on a timescale of minutes or more, whereas neuronal activity operates on a timescale of milliseconds), and its indirectness—it is unclear what kind of activity patterns trigger immediate early gene expression. Nevertheless, with proper design, immediate early gene expression has been used successfully for many applications, from identifying sensory receptors for specific stimuli (see Figure 6–22B) to allowing genetic access to memory traces (see Figure 10–36).

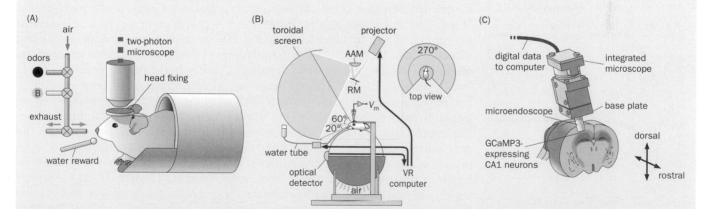

Figure 13–40 Three examples of recording neuronal activity in behaving mice. (A) In this head-fixed preparation, a metal plate is surgically attached to the head of a mouse. During a two-photon imaging experiment, the head plate is mounted onto the microscope to prevent movement of the mouse's head relative to the objective, thus stabilizing the imaging field. A head-fixed thirsty mouse can be trained to extend its tongue only when odor A but not odor B is presented in order to receive a water reward. The motor cortical area that controls tongue extension can be imaged during the learning process. **(B)** In this virtual reality (VR) preparation, a head-fixed mouse is placed on a ball floating on air while being presented with visual stimuli through a projector. The motion of the mouse causes the ball to move; the movement vector is measured by an optical detector and fed into a VR computer to control the projector such that the mouse's movement changes the scene in the projector as if it were moving in real world (see Movie 13–3). Mice could

be trained to run along linear tracks in this setting while their hippocampal place cells were subjected to whole-cell recordings of membrane potentials (V_m) or two-photon imaging (not shown). AAM, angular amplification mirror; RM, reflecting mirror. **(C)** A miniature fluorescence microscope that weights 1.9 g can be attached to the head of a freely moving mouse. With an attached microendoscope, this miniature microscope can image over a period of more than a month the place fields of hundreds of CA1 pyramidal cells expressing the genetically encoded Ca^{2+} indicator GCaMP3 (see also Movie 13–4). (A, adapted from Komiyama T, Sato TR, O'Connor DH et al. [2010] *Nature* 464:1182–1186. With permission from Macmillan Publishers Ltd.; B, adapted from Harvey CD, Collman F, Dombeck DA et al. [2009] *Nature* 461:941–946. With permission from Macmillan Publishers Ltd; C, adapted from Ziv Y, Burns LD, Cocker ED et al. [2013] *Nat Neurosci* 16:264–266. With permission from Macmillan Publishers Ltd.)

13.23 Neuronal inactivation can be used to reveal which neurons are essential for circuit function and behavior

Inactivation of neurons can determine their necessity to the normal function of the nervous system. The crudest method of inactivating neurons is a lesion, which removes or destroys a chunk of nervous tissue, either by accident or by design. For instance, lesions in the aphasia patients of Broca and Wernicke (see Section 1.10) and in H.M. (see Section 10.1) helped pinpoint brain regions important for speech and episodic memory in humans. Lesions in animal models, which can be produced by passing large currents through an electrode or by injecting toxic chemicals at stereotactic positions, offer opportunities to systematically examine the requirement of specific regions for brain function and behavior (for example, see Figure 10–40). Whereas lesions cause permanent damage, researchers can also transiently inactivate specific brain regions by injecting pharmacological agents such as GABA$_A$ receptor agonists such as muscimol (see Box 3-2) to enhance inhibition, glutamate receptor antagonists to reduce excitation, or Na$^+$ channel inhibitors to block action potentials (for example, see Figure 10–39B). Experiments using lesions and drugs can assess the functions of specific brain regions but cannot differentiate the roles of specific neuronal types within a region.

In some invertebrate animals, many neurons are individually identifiable and make decisive contributions to circuit function and animal behavior. For example, individual neurons can be identified in *C. elegans* and subsequently ablated using a high-intensity laser; this approach, which takes advantage of *C. elegans'* transparency and the small number of neurons in its nervous system, has been used to identify neurons involved in many specific functions, such as the detection of volatile chemicals (see Section 6.11). In animals with large neurons, injecting hyperpolarizing currents with an intracellular or patch electrode can transiently inactivate a neuron to assess its contribution to circuit function or animal behavior (see Section 8.5).

In animals with more complex nervous systems, such as vertebrates, specific neural functions are often carried out by populations of neurons with similar anatomical and physiological properties. It is therefore more difficult to assess neuronal function by inactivating individual neurons. Genetic approaches can overcome this limitation. Since populations of neurons with similar functions often share similar gene expression patterns, researchers can use genetic methods (see Sections 13.10–13.12) to express in a neuronal population of interest an effector transgene capable of silencing most or all of these neurons at once. Based on our knowledge of neuronal communication (see Chapters 2 and 3), many effective approaches to inactivating neurons have been developed. For example, the gene encoding tetanus toxin (see Box 3-2), which cleaves synaptobrevin, a SNARE protein essential for neurotransmitter release (see Figure 3–8), has been used for blocking synaptic transmission from target neurons (**Figure 13–41**). Another widely used method of reducing neuronal action potential firing is to overexpress Kir2.1, an inward rectifier K$^+$ channel (see Box 2–4); because the K$^+$ equilibrium potential is always more hyperpolarized than the resting potential, increasing K$^+$ conductance causes hyperpolarization of target neurons and makes it more difficult for them to reach firing threshold.

Killing or long-term silencing of neurons may induce compensatory changes in neural circuits. Therefore, a particularly informative method of inactivating neurons to assess their normal function is to silence them acutely and reversibly. Although effectors such as tetanus toxin and Kir2.1 can be temporally regulated at the transcriptional level when they are expressed as transgenes (see Section 13.10), such regulation is usually slow (hours to days). Expression of a temperature-sensitive mutant of Shibire protein (Shi^{ts}), which reversibly blocks synaptic vesicle recycling only at high temperatures (see Figure 3–14), has been a powerful tool in fruit flies to inactivate target neurons transiently (within minutes of a temperature shift; for example, see Figure 9–7). However, the Shi^{ts} strategy cannot be used in mammals, which have a constant body temperature.

Chemicals can be used to selectively silence neurons that express corresponding receptors; this **chemogenetic** approach has been increasingly used in

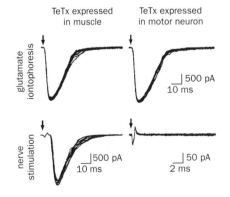

Figure 13–41 Transgenic expression of tetanus toxin (TeTx) blocks synaptic transmission. In each panel, 10 current traces from voltage clamp recordings of *Drosophila* muscles are superimposed. When TeTx was expressed in muscles (left) as a control, both glutamate iontophoresis (see Figure 3–1) and nerve stimulation induced robust postsynaptic currents. When TeTx was expressed in motor neurons, muscles still responded to glutamate iontophoresis, but no longer produced postsynaptic currents in response to nerve stimulation, indicating a blockage of glutamate release from motor axon terminals. Downward arrows: onset of iontophoresis (top panels) or nerve stimulation (bottom panels). (Adapted from Sweeney ST, Broadie K, Keane J et al. [1995] *Neuron* 14:341–351. With permission from Elsevier Inc.)

mammals. We discuss two examples here, both of which utilize chemicals that can cross the blood–brain barrier and have rapid onset and metabolism. The first example, named DREADD (for $\underline{d}$esigner $\underline{r}$eceptors $\underline{e}$xclusively $\underline{a}$ctivated by a $\underline{d}$esigner $\underline{d}$rug), uses a mutant metabotropic acetylcholine (ACh) receptor called hM_4D, which binds a chemical, CNO ($\underline{c}$lozapine-$\underline{N}$-$\underline{o}$xide), but not endogenous ACh. CNO does not have endogenous targets, but binding of CNO and hM_4D leads to hyperpolarization of neurons through opening of an inward rectifier K^+ channel, which results in effective silencing of hM_4D-expressing neurons (**Figure 13–42A**). The second approach uses a mutant ligand-binding domain

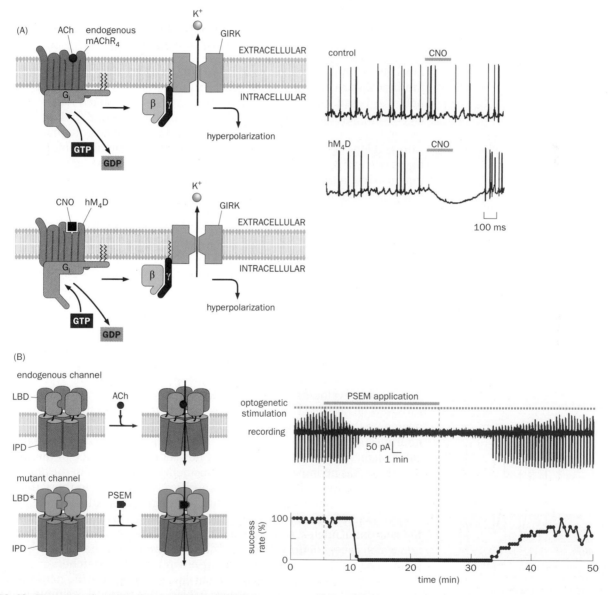

Figure 13–42 Chemogenetic approaches to silence neuronal activity. **(A)** Top left, binding of acetylcholine (ACh) to the endogenous metabotropic ACh receptor ($mAChR_4$) is coupled to a G_i protein that activates an inward rectifier K^+ channel (GIRK), causing hyperpolarization of the neuron. Bottom left, a mutant $mAhR_4$ (hM_4D) no longer binds ACh but binds $\underline{c}$lozapine-$\underline{N}$-$\underline{o}$xide (CNO) with high affinity; CNO binding to hM_4D triggers neuronal hyperpolarization. Right, voltage traces showing that CNO application (horizontal bar) induced hyperpolarization and inhibited spontaneous firing of an hM_4D-expressing cultured hippocampal neuron (lower trace) but had no effect on a normal neuron (upper trace). **(B)** Left, whereas the normal ligand-binding domain (LBD) of the nicotinic ACh receptor (nAChR) binds to ACh but not $\underline{p}$harmacologically $\underline{s}$elective $\underline{e}$ffector $\underline{m}$olecule (PSEM), the mutant LBD (LBD*) binds to PSEM but not ACh. When fused to ion pore domains

(IPD) of different channels, LBD* can confer PSEM regulation of different ion conductances. Top right, cell-attached recording of a hypothalamus AgRP neuron (see Section 8.17) expressing channelrhodopsin (ChR2) and an LBD* fused to the IPD of a glycine receptor in a slice preparation, showing that PSEM application inhibits optogenetically induced neuronal firing. Blue, photostimulation period (each rectangle represents bursts of 10 light pulses within 1 s every 30 s); yellow bar, PSEM application period. Bottom right, quantification of the success rate of action potentials induced by optogenetic stimulation. (A, adapted from Armbruster BN, Li X, Pausch MH et al. [2007] *Proc Natl Acad Sci USA* 104:5163–5168. Copyright The National Academy of Sciences, USA; B, adapted from Magnus CJ, Lee PH, Atasoy D et al. [2011] *Science* 333:1292–1296.)

(LBD) of the ionotropic ACh receptor that binds to a chemical called PSEM (pharmacologically <u>s</u>elective <u>e</u>ffector <u>m</u>olecule) but not endogenous ACh. When this mutant LBD was fused to the ion pore domain of a glycine receptor, the hybrid receptor became a PSEM-gated Cl⁻ channel, which can cause effective neuronal silencing in response to PSEM application (Figure 13–42B).

While chemogenetic methods such as the PSEM-based approach can induce reversible neuronal silencing within 20 min of drug application (Figure 13–42B), optogenetics offers a method of manipulating neuronal activity with a higher temporal precision that matches the timescale of electrical signaling (see Section 13.25).

13.24 Neuronal activation can establish sufficiency of neuronal activity in circuit function and behavior

Electrical stimulation can be used to mimic the activation of neurons. Indeed, the first discovery that the nervous system communicates using electrical signals was based on the observation that muscles contracted when nerves were electrically stimulated (see Section 1.8). Many fundamental findings in neuronal communication, such as the phenomena and mechanisms of synaptic transmission and long-term potentiation, involved the experimental stimulation of nerve fibers (for example, Figures 3–1 and 10–8). Experiments using electrical stimulation also led to the discovery of the sensory and motor homunculi in the human brain (see Section 1.11), suggested brain regions related to reward processing (see Section 10.24), and helped establish causal relationships between neuronal activity and visual perception (see Section 4.29). Electrical stimulation in humans can be used to treat brain disorders such as Parkinson disease (see Section 11.13).

In most *in vivo* preparations, electrical stimulation has been delivered using an extracellular electrode. In principle, the strength and frequency of electrical stimulation can be controlled to match the endogenous firing patterns of neurons. However, in the complex milieu of the central nervous system, where different types of neurons and passing axons are intermingled, it is difficult to control what types of neurons are being activated by a stimulating electrode. Indeed, electrical stimulation usually activates a mixture of excitatory neurons, inhibitory neurons, and projecting axons that pass near the electrode tip, making it difficult to assign the effects of stimulation to the activities of specific types of neurons. The recent development of genetically encoded effectors that can activate neurons and be targeted to specific types (see Sections 13.10 to 13.12) has begun to overcome this limitation. These effectors depolarize and therefore activate target neurons in response to an experimentally applied trigger, such as heat, a chemical, or light. Ideally, the effector does not act in the absence of the trigger, and the trigger has no impact on the systems being investigated in the absence of the effector. This way, activation can be specifically restricted to those neurons that express the effector molecule at the time of trigger application.

For example, an effective way to activate neurons in the fruit fly is to express in the neurons of interest a transgene that encodes a temperature-gated TrpA1 channel, which causes neuronal depolarization in response to heat (**Figure 13–43**). By simply changing the temperature of the environment, researchers can investigate the behavioral consequences of selectively activating specific neurons in freely moving flies (for example, see Figure 9–11B). This method is simple and noninvasive, but it can only be employed if the behavior being investigated is insensitive to temperature, and it cannot be used in animals such as mammals that have a constant body temperature. The temporal resolution of neuronal activation is limited by how fast temperature changes can occur (usually over the course of seconds to minutes).

Neurons can also be activated by chemicals. Indeed, both of the chemogenetic approaches discussed in Section 13.23 can also be adapted to activate neurons rather than silence them. For example, by creating CNO-binding mutations in a metabotropic AChR isoform that couples to G_s instead of G_i, CNO application results in an increase of intracellular cAMP, which can depolarize certain neurons

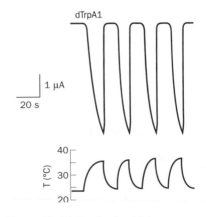

Figure 13–43 Turning heat into depolarization. Expression of a *Drosophila* TrpA1 channel in *Xenopus* oocytes results in heat-induced depolarization. In response to heat pulses (bottom), inward (depolarizing) currents are induced in *Xenopus* oocytes voltage clamped at –60 mV (top). (Adapted from Hamada FN, Rosenzweig M, Kang K et al. [2008] *Nature* 454:217–220. With permission from Macmillan Publishers Ltd.)

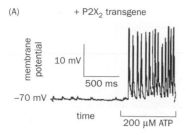
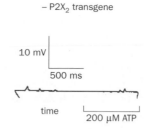
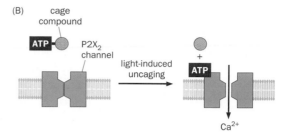

Figure 13–44 Neuronal activation by supplying ATP to neurons that express an ATP-gated channel. (A) Intracellular recording of *Drosophila* larva muscle in response to ATP application in a dissected neuromuscular preparation. Left, when motor neurons express the $P2X_2$ transgene, ATP application depolarizes the motor neuron, evoking neurotransmitter release and endplate potentials in muscles. Right, in the absence of the $P2X_2$ transgene, ATP application does not induce endplate potentials in muscles. The smaller depolarization events that occur under both conditions are miniature endplate potentials reflecting spontaneous transmitter release (see Section 3.2). **(B)** Schematic of neuronal activation in behaving flies. Flies were injected with caged ATP, which does not bind to $P2X_2$ channels. Light triggers the uncaging of ATP, which binds to and activates the $P2X_2$ channel, causing Ca^{2+} influx and activation of $P2X_2$-expressing neurons. (Adapted from Lima SQ & Miesenbock G [2005] *Cell* 121:141–152. With permission from Elsevier Inc.)

through cyclic nucleotide-gated channels (see Figure 6–4). By fusing a mutant LBD to the ion pore domain of a cation channel, PSEM application can likewise cause neuronal activation.

Neurons can also be activated by combinations of chemicals and light. For example, a mammalian ATP-gated $P2X_2$ channel has been used in *Drosophila* for neuronal activation. *Drosophila* does not have its own ATP-gated channels, nor does it have sufficient extracellular ATP to activate a mammalian $P2X_2$ channel produced from a transgene. Thus, application of exogenous ATP can be a trigger to activate *Drosophila* neurons that express transgenic $P2X_2$ (**Figure 13–44A**). The temporal resolution of this approach is limited by the slow time course of ATP application and clearance. To achieve fast activation of neurons in behaving animals, caged ATP (ATP modified chemically so it cannot activate its receptor) can be injected into the central nervous systems of transgenic flies that express the $P2X_2$ channel. A flash of light triggers uncaging of the ATP (removal of chemical modification and thereby release of free ATP) and activation of the $P2X_2$-expressing neurons (Figure 13–44B). This strategy has been used to effectively induce various behaviors, from mimicking aversive stimuli (see Figure 10–29B) to singing courtship songs (see Figure 9–13), depending on the type of neurons in which $P2X_2$ was expressed. Thus, these experiments have helped to establish causal relationships between the activity of specific types of neurons and the behaviors they control.

13.25 Optogenetics allows control of the activity of genetically targeted neurons with millisecond precision

Broadly speaking, **optogenetics** is an approach for altering neuronal activity by using light to activate an effector that is genetically targeted to specific neurons. Thus, it includes the strategy we just discussed—using light to uncage ATP and activate neurons that express an ATP-gated channel. In most cases, optogenetics refers to the use of microbial opsins as the effectors because of their simplicity, effectiveness, and general applicability. Since the term was first introduced in 2006, optogenetics has had a huge impact in many areas of neuroscience, especially on how researchers evaluate the function of neural circuits and their roles in behavior.

The most remarkable and widely used optogenetic effector for neuronal activation is **channelrhodopsin-2 (ChR2)**, a protein first found in the green alga *Chlamydomonas reinhardtii*. As we discussed in Section 12.13, ChR2 is a seven-transmembrane type I rhodopsin; instead of coupling to G proteins, it is itself a cation channel that opens in response to blue-light stimulation (**Figure 13–45A**; see also Figure 12–21B). Although ChR2 requires all-*trans* retinal as a cofactor, mammalian neurons have sufficient endogenous all-*trans* retinal to support

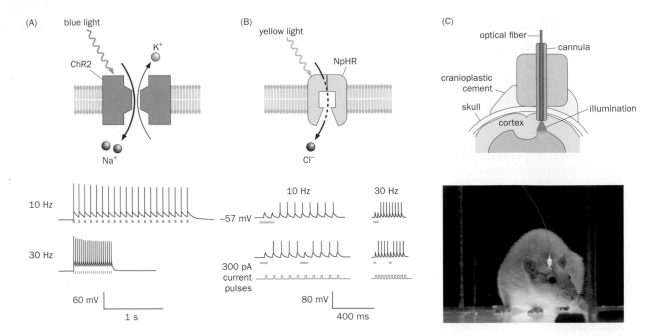

Figure 13–45 Optogenetics for precise temporal control of neuronal activity. (A) Top, channelrhopsin-2 (ChR2) from a green alga is a cation channel that is gated by blue light and causes neuronal depolarization from the resting potential because it allows more Na^+ influx than K^+ efflux. Bottom, voltage traces show that in culture the firing of a hippocampal neuron that expresses ChR2 can be precisely controlled by 10-ms blue light pulses (blue dashes at the bottom) up to 30 Hz. **(B)** Top, halorhodopsin (NpHR) from an archaean species is a Cl^- pump that is activated by yellow light. Bottom, voltage traces show that yellow light pulses (yellow dashes at the bottom) can cancel with millisecond precision the action potentials produced by depolarizing current pulses in cultured neurons. **(C)** Optogenetics *in vivo*. A cannula is surgically introduced into the brain at a region of interest. Viral vectors that carry opsin transgenes can be introduced through the cannula at the desired depth, allowing ChR2 or NpHR to be expressed in the neurons of interest prior to the experiment. During the experiment, an optical fiber delivering blue or yellow light is introduced into the cannula, so that light-induced behavior can be observed in a freely moving rodent (shown at the bottom, see Movie 13–5). (A, adapted from Boyden ES, Zhang F, Bamberg E et al. [2005] *Nat Neurosci* 8:1263–1268. With permission from Macmillan Publishers Ltd; B, adapted from Zhang F, Wang LP, Brauner M et al. [2007] *Nature* 446:633–639. With permission from Macmillan Publishers Ltd; C, schematic adapted from Zhang F, Aravanis AM, Adamantidis A et al. [2007] *Nat Rev Neurosci* 8:577–581. With permission from Macmillan Publishers Ltd. Image courtesy of Karl Deisseroth.)

ChR2 function; thus, expression of ChR2 alone is sufficient to allow robust depolarization and firing of mammalian neurons in response to light. Indeed, pulses of blue-light stimulation can precisely direct the firing of ChR2-expressing neurons at rates as high as 30 Hz (Figure 13–45A, bottom), achieving control of target neuron activity with millisecond precision. In nervous systems that lack sufficient all-*trans* retinal (such as *Drosophila* and *C. elegans*), supplemental retinal provided through food is often sufficient for ChR2 function. Many ChR2 variants have been developed by *in vitro* mutagenesis with enhanced expression, photocurrent, and faster or slower off time constant (how long the current decays after light is off) for different applications.

Optogenetic approaches can also be used to reversibly inactivate neurons with high temporal precision. For this purpose, an archaean type I rhodopsin called **halorhodopsin**, which is a yellow-light-activated inward Cl^- pump (Figure 13–45B), can be expressed in neurons. Yellow light stimulation hyperpolarizes neurons that express halorhodopsin, making it more difficult for these neurons to fire action potentials in response to depolarizing signals (see Figure 4–47). Indeed, in cultured neurons expressing halorhodopsin, short pulses of yellow light were sufficient to block with high temporal precision action potentials induced by depolarizing current pulses (Figure 13–45B, bottom). **Archaerhodopsin** (Arch), a light-driven outward proton pump from an archaean species, has also been widely used as an optogenetic effector for light-induced neuronal silencing.

In order to achieve optogenetic manipulation of neurons in behaving animals, fiber optics-based systems have been developed to activate or silence neurons with laser pulses in specific brain regions that have previously been transduced by viruses that allow the expression of optogenetic effectors (Figure 13–45C);

optogenetic effectors can also be expressed by transgenic animals. The optical fibers are usually ~200 μm in diameter, thin enough to minimize damage but still able to effectively excite ChR2-expressing neurons in approximately one cubic millimeter at the fiber's tip. Thus, the behaviors of freely moving animals can be assayed while specific populations of neurons in specific brain regions are activated or silenced (see **Movie 13–5**).

Because optogenetic methods manipulate neuronal activity with the same timescale as fast neuronal communication (with spikes and in the millisecond range), they are well suited for probing neuronal signaling and computation. Still, a limitation of the optogenetic approach is that for neuronal activation to achieve its biological effect, a large fraction of the neurons of interest must be accessible to light delivered by optical fiber; for neuronal inactivation, access to the full population of relevant neurons may be required. Fiber optic implantation is also associated with physical damage. By comparison, the chemogenetic approaches discussed in Section 13.23, while not as temporally precise, can assess the entire population of neurons expressing the relevant receptor. They are minimally invasive if the chemicals can cross the blood–brain barrier and do not have side effects in the absence of the effector. Thus, these methods complement each other.

In summary, classic methods to activate or silence neurons, including electrical stimulation, lesion, and pharmacology, have now been supplemented by a variety of genetically encoded effectors to activate or inactivate specific neuronal types with light, heat, or chemicals in a number of model organisms. Coupled with the development of genetic tools that provide access to an increasing number of cell types, these approaches are making a significant impact on our understanding of how neural circuits operate and how they control behavior.

13.26 Synaptic connections can be mapped by physiological and optogenetic methods

Having studied methods for recording and manipulating neuronal activity, we now return to the subject of mapping neuronal connections discussed in Section 13.19 to study how physiological methods can be employed to address this important problem.

A widely used method for determining whether two neurons are directly connected is to place an electrode in each of the neurons and record the two neurons simultaneously in response to manipulating the activity of one of them (**Figure 13–46**). If injecting a depolarizing current into neuron *A* to produce an action potential causes depolarization of neuron *B* within the timeframe of a monosynaptic connection (usually a few milliseconds), we can conclude that neuron *A* forms excitatory synapses directly onto neuron *B* (Figure 13–46A). If firing of neuron *A* causes hyperpolarization of neuron *B* within a few milliseconds, then neuron *A* forms inhibitory synapses onto neuron *B* (Figure 13–46B). Likewise, stimulating neuron *B* and recording from neuron *A* can test whether these neurons are connected by reciprocal *B* → *A* chemical synapses. If hyperpolarization of neuron *A* causes hyperpolarization of neuron *B*, and vice versa, then we can conclude that these two neurons are connected by an electrical synapse (Figure 13–46C), as chemical synapses do not transmit hyperpolarizing signals (see Box 3–5). Paired recordings have been performed extensively in invertebrates with large neurons using intracellular electrodes (see Figure 8–13) and in mammalian brain slices using whole-cell patch recording (see Figure 4–46). Such tests provide definitive evidence that two neurons are functionally connected, can determine the type (for example, excitatory, inhibitory, electrical) and strength of the connection, but are labor intensive for mapping large-scale neuronal connections and are not applicable to map long-range connectivity.

A higher throughput mapping method developed in mammalian brain slices utilizes laser uncaging of neurotransmitters. For example, a brain slice can be placed in medium that contains caged glutamate (glutamate modified chemically so it cannot activate its receptor). A focal laser stimulation leads to uncaging (removal of the chemical modification) and thus local release of glutamate, which causes neurons near the laser stimulation site to fire action potentials. If

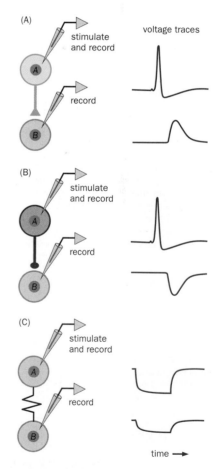

Figure 13–46 Mapping neuronal connections with paired recording. **(A)** Depolarizing an excitatory neuron *A* causes it to fire an action potential that elicits a depolarizing membrane potential in its postsynaptic partner neuron *B*. **(B)** Depolarizing an inhibitory neuron *A* causes it to fire an action potential that elicits a hyperpolarizing membrane potential in its postsynaptic partner neuron *B*. **(C)** Hyperpolarizing neuron *A* causes hyperpolarization of neuron *B* via an electrical synapse.

Figure 13–47 Mapping neuronal connections with optical and electrophysiological methods. Arrows indicate the direction of signal flow. For simplicity, axons of neurons that do not form a synapse with the recorded neurons are not drawn. **(A)** Photo uncaging. The brain slice is incubated with caged glutamate. When a scanning laser (blue) reaches a neuron that is presynaptic to the recorded neuron, local release of glutamate causes this neuron to fire action potentials, which produce excitatory postsynaptic potentials in the recorded postsynaptic neuron (the *y* axes of synaptic potential and action potential are not at the same scale). A two-dimensional input map can be produced after the laser systematically scans through the entire slice. In the illustration, neurons 1 and 2 are presynaptic partners of the recorded neuron, but neuron 3 is not. **(B)** ChR2-assisted circuit mapping. A defined subpopulation of neurons (green) express channelrhodopsin (ChR2), causing them to fire action potentials in response to blue light stimulation. In this scheme, stimulating neuron 1 does not activate the recorded neuron because it does not express ChR2. Stimulating neuron 2 activates the recorded neuron because it expresses ChR2 and synapses onto the recorded neurons. Note that this method can also be used to map connections between ChR2-expressing presynaptic neurons whose cell bodies lie outside the slice (bottom right) and the recorded neuron, because stimulating their axons (3) and terminals (not shown) is often sufficient to elicit synaptic response in postsynaptic neurons. Although drawn in the same schematic, the axon-stimulation experiment is performed separately, to ensure that the only source of ChR2-expressing neurons is defined and located outside the slice. (Adapted from Luo L, Callaway EM & Svoboda K [2008] *Neuron* 57:634–660. With permission from Elsevier Inc. The original methods were described in Callaway EM & Katz LC [1993] *Proc Natl Acad Sci USA* 90:7661–7665 and Petreanu L, Huber D, Sobczyk A et al. [2007] *Nat Neurosci* 10:663–668.)

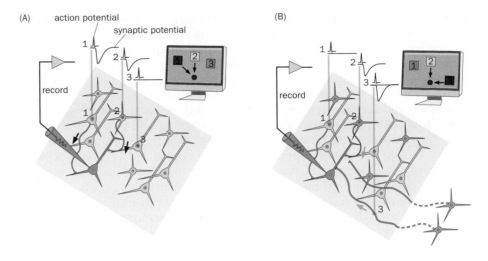

one or a few neurons near the laser stimulation site form monosynaptic excitatory connections with a postsynaptic neuron of interest that is being recorded using intracellular or whole-cell recording, then laser uncaging will produce excitatory postsynaptic potentials in the recorded neuron. After the laser scans through a defined area of the brain slice, investigators can create a two-dimensional map of all excitatory neurons in the area that connect with the target neuron that is being electrically recorded (**Figure 13–47A**).

A more widely used approach to this type of experiment replaces laser uncaging of a neurotransmitter with photoactivation of ChR2-expressing neurons (Figure 13–47B), a method called CRACM (**ChR2**-**a**ssisted **c**ircuit **m**apping). The advantage of CRACM is that the type of presynaptic neurons can be genetically defined by the promoter that drives ChR2 expression. Moreover, unlike paired recordings and laser uncaging discussed above, which are limited to mapping connections within the brain slice, CRACM can also map long-range connections. If the only source of ChR2 expression is from a defined neuronal population outside the brain slice, photoactivation of ChR2 molecules in their axons and terminals is often sufficient to cause neurotransmitter release at synaptic terminals, which can be detected by recording of postsynaptic neurons located within the slice.

None of these physiological mapping methods can be easily applied *in vivo*, at least in the complex mammalian brain. Thus, these methods complement anatomical methods such as serial EM reconstruction and trans-synaptic tracing (see Section 13.19) in mapping synaptic connections.

BEHAVIORAL ANALYSES

A major goal of neurobiology is to understand how behaviors arise from the molecular and cellular properties of neurons, the wiring specificity of neural circuits established during development and modified by experience, and the spatiotemporal patterns of neuronal activity at the times when behaviors occur. Insightful and quantitative analysis of animal behavior is instrumental for studying many neurobiological problems, from sensory perception and motor control to emotion and cognition. Unlike in human studies, where verbal reports are revealing, animal studies depend on observation and measurement of behavior in order to infer what animals sense, feel, learn, and understand.

In neurobiology research, behavioral analyses serve three broad purposes. First, behavioral analyses aim to explain the behavior itself (for example, see Chapter 9): what is the function of the behavior for animal survival and reproduction, what external factors influence it, what are its constituent motor actions, and what are the underlying neural bases? Second, behavioral analyses are used as quantitative assays for the functions of brain regions, circuits, and neurons in specific neurobiological processes being investigated, such as sensory perception (see Figure 4–52) or learning and memory (see Figure 10–32). Third, behavioral analyses are used to test the effects of manipulating specific genes (see Figure 10–33) or to assess animal models of human brain disorders (see Figure 11–7). Given the inseparable links between genes, neurons, circuits, and behaviors (see Figure 10–7), these purposes have considerable overlap. In the following sections, we first highlight two general approaches in behavioral analysis that are applicable to all of these purposes and then discuss behavioral assays commonly used to assess the functions of genes, neurons, and circuits and to model human brain disorders.

13.27 Studying animal behavior in natural environments can reveal behavioral repertoires and their adaptive value

From an evolutionary perspective, behaviors are products of natural selection that allow animals to interact with their environments in ways that improve their probability for survival and reproduction. Thus, an influential approach to animal behaviors is to study them in the natural environment. This neuroethological approach can reveal an animal's behavioral repertoire (what behaviors an animal is capable of exhibiting), the relationships between different behaviors (for example, whether one behavior precedes or follows another in a sequence, or whether two behaviors are mutually exclusive in their occurrence), and the adaptive values of specific behaviors.

The principal methods of neuroethology include observation and measurement in carefully designed field studies. We use the study of honeybee dancing as an example to illustrate. Honeybees are social insects that can perform sophisticated behavioral tasks (see Figure 10–23). They are also expert nectar collectors and pollinators. Once forager bees find a good source of nectar, sometimes kilometers away from their hive, they communicate with their fellow bees (forager recruits) to direct these hive-mates to the same place. How do bees achieve this? Researchers have set up observation hives with glass windows so that they can observe the behaviors of foragers in the hive environment. Once foragers locate a good source of nectar, they return to the hive and perform dances to convey information to forager recruits. When the source of the nectar is more than 50 m from the hive, foragers typically perform a tail-wagging dance following a trajectory that resembles the Arabic numeral 8 (**Figure 13–48**A). By placing scented feeding bowls at different distances and directions from the hives and measuring the dances of bees and the subsequent foraging of forager recruits, researchers reached the following set of conclusions. The richness of the nectar is indicated by the vigor of the dance. The distance to the nectar source is signaled by the duration of tail wagging, which takes place as the forager bee dances in a straight path between the 8s two circular halves (Figure 13–48B). The orientation of the figure-8 dance in the hive signals the direction of the outgoing flight with respect to the sun (Figure 13–48C). Finally, the scent that foragers carry informs forager recruits about the kind of nectar they should look for once they arrive at the vicinity of the nectar source. Experiments using foraging behavior as readout validated these conclusions (Figure 13–48D). The adaptive value for efficiently locating nectar is obvious for the bee colonies, and efficient nectar foraging is also beneficial to the plants that produce the nectar and are pollinated by the bees.

The spirit of the neuroethological approach can be extended to the laboratory, where animal behaviors can be observed, recorded, and quantitatively measured in settings that resemble the natural environment but offer greater technical ease compared with field studies. For instance, the complete recording of *Drosophila* mating behaviors in a laboratory setting (see Figure 9–2 and Movie 9–1) enabled these behaviors to be dissected into discrete components. Quantitative plots of different behaviors exhibited over time can be used to compare individuals receiving different experimental treatments and thereby to study the neural mechanisms underlying these behaviors (see Figure 9–11B). The development of high-speed video recording and automatic video analysis have further enhanced the sensitivity and throughput of behavioral observations and measurements.

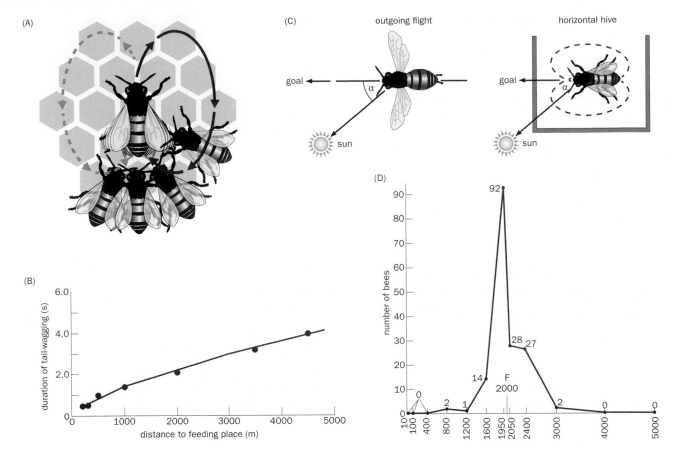

Figure 13–48 Behavioral analysis of honeybee foraging.
(A) Illustration of the figure-8 dance of a forager bee in an observation hive. The forager alternates between the right half (solid red arrows and trajectory) and the left half (dashed gray arrows and trajectory) of the figure 8. The four forager recruits take in the information by moving with the forager and maintaining close contact with her, particularly during the straight part of the dance, when the forager exhibits tail wagging. **(B)** The relationship between the duration of tail wagging, measured from film recordings, and the foraging distance signaled by the dancing bee. **(C)** The angle (α) of the outgoing flight with respect to the sun (left) is signaled by the angle at which the figure-8 dance orients with respect to the sun in this horizontal hive (right). **(D)** In this field experiment, foragers were fed at a scented plate (F) 2000 m away from the hive. Afterwards, similarly scented plates without food were placed at many distances (numbers on the x axis) and the visits of forager recruits were quantified; the y axis values adjacent to the data points reveal that the forager recruits preferred distances around 2000 m. (Adapted from von Frisch K [1974] *Science* 185:663–668.)

13.28 Studying behaviors in highly controlled conditions facilitates investigation of their neural basis

Behavior is influenced by many factors: external stimuli, internal drives and brain states, and the individual animal's genetic make-up and life experience. Thus, another influential approach, which at face value might seem to be the polar opposite of neuroethology, is to study behavior under as much experimental control as possible so that known factors that might influence behavior can be varied one at a time in order to study their contributions to the behavior. To achieve this, researchers carry out behavioral studies in inbred animal strains to decrease genetic variability—all individuals within an inbred strain are essentially genetically identical—and use animals of the same sex and age reared under similar conditions to decrease variability in experience. Behaviors can be performed in a fixed apparatus (for example, see Figure 10–22) to reduce the variability of external factors, and standardized conditions can be implemented to control for internal factors, such as the circadian cycle of the animal and the time when the animal last ate or drank prior to a behavioral experiment.

This approach has yielded great insights. The discovery of classical and operant conditioning and the identification of factors that affect those learning processes were made using this approach (see Section 10.14). The operant conditioning paradigm has been particularly influential in studying internal factors such as drives that affect a behavior. Behavioral paradigms are often designed to take advantage of internal drives. For instance, in two commonly employed paradigms, thirsty animals are motivated by a potential water reward to choose to perform an action (such as pressing a button) or to do nothing (go/no-go task), or to choose one of the two actions (two-alternative forced choice task, such as a saccade toward one of the two alterative targets). These tasks can be used to study sensory perception, decision-making, motor execution, and memory (see Figures 4–53, 4–54, 6–73, 8–27, and 10–46). The combination of behavioral analysis with methods for recording and manipulating the activity of relevant neurons and circuits (see Box 13-3) has enabled researchers to establish causal links between behavior, neuronal activity, and circuit function.

The two general approaches outlined above are complementary, and many behavioral paradigms incorporate the merits of both. Ideally, the choice of which behaviors to study under highly controlled experimental conditions and the design of the assays used to study these behaviors should be based on an understanding of the behavioral repertoire of animals in their natural environment. For instance, the Morris water maze (see Figure 10–32) takes advantage of the aptitude of rodents for using external landmarks to navigate and their preference not to swim. The **closed-loop** design of behavioral experiments, in which the behavior of an animal changes the environmental stimuli that induce the behavior in the first place (for example, see Figure 10–44), can allow animals to explore naturalistic virtual environments in highly controlled settings. For instance, an influential closed-loop preparation was developed while studying visual control of the flight behavior of houseflies. In a cylinder that delivers panoramic visual stimuli, a test fly is suspended from a torque compensator, which measures the rotation of the fly as it attempts to orient toward visual stimuli, and sends a signal to control the rotation of the panorama cylinder (**Figure 13–49**). In this way, the fly's intended movement alters the visual scene as if the fly were actually moving through the scene. This design allows researchers to study in an immobilized fly some of the flight behaviors that might be observed in a more naturalistic setting, such as tracking of objects, and at the same time makes possible the precise delivery of visual stimuli and the quantitative measurement of the fly's behavioral output. This setup also enables researchers to investigate the underlying neural bases of behavior by performing electrophysiological recordings of brain neurons.

Conceptually similar approaches to the closed-loop design for studying the flight behavior of a housefly have also been developed for other animals. For example, navigation of head-fixed mice through complex virtual reality environments has facilitated investigation of the neuronal activity associated with spatial navigation using optical imaging and whole-cell recording methods that cannot easily be applied to freely moving mice (see Figure 13–40B). Indeed, feedback signals can be obtained from recordings of motor neurons rather than the motion itself, as in the use of paralyzed zebrafish larvae. In this preparation, a zebrafish larva was visually stimulated by the backward motion of a grating on a computer screen to simulate the flow of water; this virtual water flow induced a compensatory forward swim in the fish. Extracellular recordings of the motor neurons were used as a measure of this fictive motion, and the recorded signals were fed back to control the apparent speed of the grating, forming a closed loop (**Figure 13–50**A). When researchers adjusted the gain of the feedback signals, fish adjusted their swimming speed correspondingly to match the virtual change of the water flow speed (Figure 13–50B). This preparation enabled researchers to image neurons from the entire brains of paralyzed zebrafish larvae using a genetically encoded Ca^{2+} indicator, and to identify neurons in the cerebellum and the inferior olive (see Section 8.8) whose activities correlate with the adjustment of speed in response to changes in gain.

In summary, by combining the insights derived from neuroethological approaches with the merits of experimental controls, researchers can develop sophisticated behavioral paradigms that enable quantitative analysis of behavior (or fictive behavior) while recording and manipulating neuronal activity. Thanks to the techniques described in previous sections, these approaches are becoming increasingly powerful for dissecting the neural bases of complex behaviors.

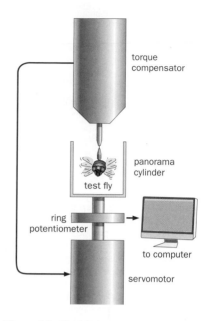

Figure 13–49 A simplified scheme of a closed-loop design to study the flight behavior of flies. A housefly is suspended from a torque compensator. Signals from the torque compensator provide a quantitative measure of the fly's intended behavioral output. These signals are sent to a servomotor, which controls the rotation of the panorama cylinder through a ring potentiometer. Thus, the fly's intended behavioral output controls its visual environment. (Adapted from Reichardt W & Poggio T [1976] *Q Rev Biophy* 3:311–375.)

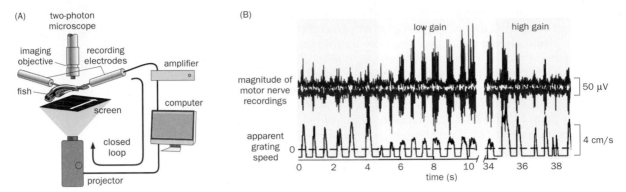

Figure 13–50 Closed-loop control of fictive swimming by a zebrafish larva. (A) Experimental setup. A paralyzed zebrafish larva receives visual stimuli from a computer screen, on which the image of a grating moves to mimic the backward flow of water (white arrow). Forward fictive swims are measured by extracellular recordings of the motor neurons. These signals are fed back in real time to control the apparent grating speed. In this set up, neuronal activities in the brain can be measured using a two-photon microscope and genetically encoded Ca^{2+} indicator expressed in all neurons. **(B)** Example recordings of a larva's response to experimentally altered feedback gain. The blue and red traces represent simultaneous recordings of motor nerves from two sides of the larva. For ease of viewing the red trace is flipped vertically relative to the blue trace. When the gain is low (pale yellow background), the fish increases its fictive swimming speed as illustrated by an increase in motor nerve output; when the gain is high (bright yellow background), the fish decreases its fictive swimming speed as illustrated by a decrease in motor nerve output. The bottom trace indicates the apparent grating speed. (Adapted from Ahrens MB, Li JM, Orger MB et al. [2012] *Nature* 485:471–477. With permission from Macmillan Publishers Ltd.)

13.29 Behavioral assays can be used to evaluate the functions of genes and neurons and to model human brain disorders

In the final section, we highlight behavioral assays that are commonly used to assess phenotypes associated with the disruption of specific genes or the activation or inactivation of specific neuronal populations. These assays are also frequently used to study animal models of human brain disorders and the effects of pharmacological intervention. We focus on assays designed for rats and mice, as they are the most widely used mammalian models (see Section 13.4).

Behavioral assays can be used to assess general sensorimotor functions. The simplest assay is to record an animal's behavior in its home cage with a video camera over several days followed by manual or automatic analysis of videos. This recording allows scientists to assess general motor activity, circadian rhythms, and eating, drinking, sleeping, and nest-building patterns in the laboratory housing environment with minimal handling. Another assay is to place the animal in an open field, which is essentially a box with walls but no top, and videotape, analyze, and quantify its trajectory within a given period of time. Motor coordination can also be tested by recording the footprints of an animal with painted feet (see Figure 8–19), or by measuring the length of time an animal can remain on a rotating rod (**Figure 13–51**), the speed of which can be held constant or can increase over time so that the task becomes progressively more difficult. These assays evaluate the basic functions of the nervous system, including those of the cerebral cortex, cerebellum, basal ganglia, and spinal cord. More specific assays can be used to examine specific sensory functions; for example, the hot plate assay tests temperature and pain perception by measuring the time it takes for an animal to flick a tail or lick a hind paw placed on a plate as the temperature of the plate rises.

Behavioral assays have also been designed to assess animals' cognition, such as learning and memory. The Morris water maze and the contextual and auditory fear conditioning assays (see Figures 10–32 and 10–40) are widely used to test the functions of the hippocampus and amygdala in learning and memory. Another often-used assay is the radial arm maze, which tests spatial and working memory. The maze consists of a number of arms (usually eight), with food pellets located at the ends of the arms (**Figure 13–52**A). The maze is placed in a room in which numerous visual cues are scattered outside the maze. After becoming habituated to the apparatus and the room, food-restricted rats placed at the center of the maze efficiently visit all of the arms to collect the food, with minimal revisiting of arms from which food had already been collected. Behavioral experiments

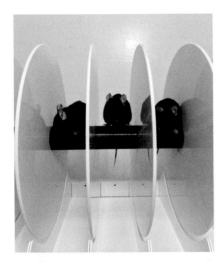

Figure 13–51 The rotarod assay for testing motor coordination. At the beginning of the experiment, mice are placed on a stationary rod; the separate compartments shown in the photograph allow simultaneous testing of multiple animals. The rod is connected to a motor with an adjustable speed. Typical experiments include running the motor at a constant speed (for example, 10 revolutions per minute, or rpm) or a speed that increases (for example, from 5 to 30 rpm), and measuring the period of time each mouse remains on the rod. In addition to testing motor coordination, the rotarod can also test motor skill learning (see Figure 8-18D). (Courtesy of Mehrdad Shamloo.)

(A)

(B)

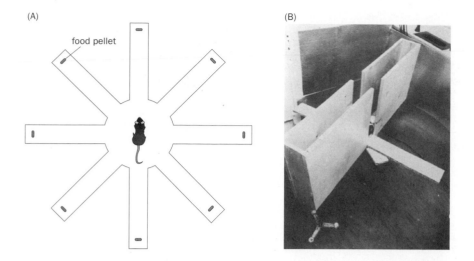

food pellet

Figure 13–52 Mazes used to test memory and anxiety. (A) Top view diagram of a radial arm maze for testing spatial and working memory. Food pellets are located at the ends of the arms. A food-restricted rat placed in the center of the maze will visit each arm to consume the pellets with minimal repetition, utilizing extra-maze cues (not shown) for navigation. **(B)** Photograph of an elevated plus-maze with a rat located at the intersection, between the opened and closed arms. The fractions of time spent in the closed arms and open arms can be used to quantify anxiety in rats or mice and can be modulated by anxiolytic and anxiogenic drugs. (A, adapted from Olton DS & Samuelson RJ [1976] *J Exp Psychol Animal Behav Proc* 2:97–116. With permission from the American Psychological Association; B, from Pellow S & File SE [1986] *Pharmacol Biochem Behav* 24:525–529. With permission from Elsevier Inc.)

indicate that rats rely mostly on extra-maze cues rather than intra-maze cues such as odor trails they leave behind or sequential strategies such as visiting adjacent arms one-by-one. By recording the trajectories of animals visiting different arms in the maze, scientists can assess the functions of the hippocampus, which has an essential role in spatial memory, and the prefrontal cortex, which has been implicated in working memory.

Perhaps the most challenging behavioral assays are those aimed at modeling human psychiatric disorders, because it is difficult to assess whether a mouse or a rat is experiencing a condition that resembles schizophrenia, depression, anxiety, or autism. However, scientists have devised behavioral paradigms to measure phenotypes that could reflect some aspects of these mental states. For instance, one measure of anxiety-like states is based on the open field test mentioned above. When mice are placed in an open field, their natural tendency is to stay near its periphery, presumably because mice are more susceptible to predation when they are exposed than when they are hidden. Normal mice venture into the center occasionally. Mice that visit the center with reduced frequency can be considered anxious. Another commonly used anxiety assay is the elevated plus-maze (Figure 13–52B). Here, a mouse (or a rat) is placed onto a four-arm maze that is elevated from the floor, with two arms covered on the sides with high walls and the other two arms exposed. Although mice prefer the closed arms, normal mice also spend a fraction of time exploring the open arms. Putatively anxious mice spend significantly less time in the open arms. Administering anxiolytic drugs such as benzodiazepines (see Section 11.17) can increase the tendency of animals to enter the center of an open field or the open arms of an elevated plus-maze, whereas anxiogenic drugs do the opposite, suggesting that these behavioral assays test phenomena that may be related to anxiety in humans.

Assays have also been designed to test social behaviors; these assays are often used in studies of mouse models of human autism-spectrum disorders. For example, a three-chamber social interaction assay for mice (**Figure 13–53**) resembles the partner preference assay for prairie voles discussed in Section 9.24. Here, a mouse is first allowed to habituate in a three-chamber apparatus where it can move freely between chambers. Then, a small wire cage is placed in each of the two side chambers, one containing a live mouse and the other empty. Most strains of mice spend more time in the chamber containing the conspecific than in the chamber with the empty cage. The relative time spent in each of the two chambers can be a measure of sociability; a reduction of time in the side chamber with the mouse corresponds to reduced sociability. An additional test of social novelty is to place one mouse in each of the side chambers, one familiar to the mouse being tested and the other unfamiliar. Mice usually spend more time investigating the unfamiliar mouse, so the relative time spent in the two chambers can be a measure of preference for social novelty (Figure 13–53). To validate that the observed behavior is social in nature, control experiments with non-social objects or odors

Figure 13–53 Three-chamber social interaction assays. (A) Schematic drawings of testing procedures. Prior to social interaction, a test mouse becomes habituated to an empty three-chamber environment in which it can move freely between the chambers (top). Next, additional mice can be added to investigate social interactions. The number of visits and total time spent in the chamber containing a live mouse compared to the chamber with an empty cage can be used to assess sociability (middle). The number of visits and total time spent in the chamber containing a familiar mouse (orange) compared to the chamber with an unfamiliar mouse (brown) can be used to assess social novelty (bottom). **(B)** Photograph of a social novelty assay being conducted in a three-chamber social interaction apparatus. (From Moy SS, Nadler JJ, Perez A et al. [2004] *Genes Brain Behav* 3:287–302. With permission from John Wiley & Sons.)

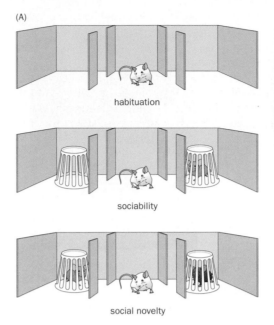

habituation

sociability

social novelty

must be conducted in an identical experimental setup; these control experiments can exclude possible confounding variables, such as the failure to recognize or discriminate between different odors or objects.

Behavior is sensitive to a large number of factors, so it cannot be overemphasized that in performing each of the assays described above, care must be taken to control the experimental conditions (discussed in Section 13.28) and thereby minimize unintended influences on behavior by factors not being tested. Proper control groups should always accompany the experimental group. Multiple behavioral assays are often used in combination to identify the effects of specific perturbations on the nervous systems, as defects in more complex assays can have multiple interpretations. For instance, if an animal exhibits a defect in the auditory fear conditioning assay, the animal might have a defect in hearing, in sensing electrical shocks, or in associating these two events (that is, learning). Control experiments on hearing and shock sensitivity are necessary in order to determine whether or not the defect is related to learning. The interpretation of behavioral results also depends on the nature of experimental perturbation. For instance, if the experimental manipulation is to knockout a gene of interest or to apply a systemic pharmacological agent, then in principle the entire nervous system could be affected and could contribute to the observed behavioral phenotypes. If the experimental manipulation is a conditional knockout of a gene in specific neuronal types or the activation and silencing of specific brain regions or neuronal populations, then the behavioral phenotypes reflect alterations of the manipulated brain regions or neuronal populations.

SUMMARY AND PERSPECTIVES

Neurobiology research has been carried out in diverse animal models. These animal models have been chosen for the technical ease with which they can be studied (for example, their neurons recorded, genes manipulated, or explants studied *in vitro*), for the special faculties they exhibit, for the sophistication of their behavior, or for their resemblance to humans. Indeed, recent revolutions in genomics and non-invasive imaging, combined with a long history of medicine and experimental psychology, have made humans an increasingly attractive organism to study neurobiology. Whereas many molecular and cellular processes in the operation of nervous systems are well conserved across animals, we do not yet know the extent to which information processing at the circuitry and systems levels follows general principles that apply across nervous systems of varying complexity. Regardless, studies in diverse animal models enrich our understanding of the diversity of life and the evolution of the nervous system.

All neurobiological processes are ultimately the direct or indirect consequences of gene actions. The two most widely used molecular-genetic manipulations in neurobiological research are the disruption of endogenous genes and the expression of transgenes. Gene disruption can be performed in the context of a forward genetic screen or by utilizing reverse genetic methods such as homologous-recombination-mediated knockout or RNAi-mediated knockdown. Many methods have been employed for expressing transgenes with sophisticated spatiotemporal controls. Both gene disruption and transgene expression can reveal how genes function in specific neurobiological processes. Transgene expression is also instrumental in providing genetic access to specific neuronal populations for the purposes of investigating their anatomical organization, their physiological properties, and the functional consequences of manipulating their activity. Future challenges include not only expanding access to specific cell types with increasing precision in genetic models such as flies and mice, but also broadening genetic tools to other animal models.

Classic anatomical methods such as cell staining, axon tracing, and single-cell labeling have provided foundations for our present understanding of how the nervous system is organized. A deeper appreciation of this organization requires building further connections between molecules, neurons, and the brain. One frontier is the fine-structural analysis of how individual molecules form complexes in different parts of the neuron and the dynamics of these molecular complexes; this will deepen our understanding of the mechanisms by which individual neurons function. Another frontier is the construction of connection diagrams for complex nervous systems, ultimately to the resolution of individual synapses; this will serve as a blueprint to decipher the information processing principles of neural circuits.

Extracellular, intracellular, and patch recordings of electrical activity have contributed fundamentally to our understanding of how information flows within individual neurons, across synapses, in small circuits, and over large networks. These electrophysiological methods have more recently been supplemented by optical imaging, which enables the simultaneous recording of many neurons of specific cell types and over longer periods using genetically encoded Ca^{2+} indicators. Future challenges include combining the sensitivity and temporal resolution of electrophysiological recording with the breadth, cell-type specificity, and duration of optical imaging, to expand the methods available to record neuronal activity in behaving animals, and to develop a conceptual framework to convert rich data into an understanding of the principles of neural circuit operation and the neural basis of behavior.

A crucial approach that links neuronal activity, circuit function, and behavior is the manipulation of neuronal activity with precise spatiotemporal control. Classic lesion, pharmacology, and electrical stimulation methods to inactivate and activate brain regions have been supplemented in recent years with sophisticated control of neuronal activity by light, heat, and chemicals. New tools developed in recent years, in particular the ability to control neuronal activity with light, have made it possible to activate and inactivate genetically defined neuronal population at spatiotemporal scales that begin to match those of neuronal signaling.

Ultimately, the combination of methods we have studied in this chapter, such as deleting or misexpressing specific genes in defined cell types, measuring, activating, and silencing the activity of specific neuronal populations with high spatiotemporal precision, and quantitative analysis of animal behavior, will help establish causal links between genes, neurons, circuits, and animal behaviors. These links will deepen our understanding of the nervous system in health and disease. With the rapid pace of tool developments and their wide applications, neurobiology research has never seen a more exciting time.

FURTHER READING

Reviews

Capecchi MR (1989) Altering the genome by homologous recombination. *Science* 244:1288–1292.

Fenno L, Yizhar O & Deisseroth K (2011) The development and application of optogenetics. *Annu Rev Neurosci* 34:389–412.

Jorgenson LA, Newsome WT, Anderson DJ et al. (2015) The BRAIN Initiative: developing technology to catalyse neuroscience discovery. *Phil Trans B* 370:20140614. See also http://www.braininitiative.nih.gov/2025/BRAIN2025.pdf

Luo L, Callaway EM & Svoboda K (2008) Genetic dissection of neural circuits. *Neuron* 57:634–660.

Neher E & Sakmann B (1992) The patch clamp technique. *Sci Am* 266:44–51.

Scanziani M & Häusser M (2009) Electrophysiology in the age of light. *Nature* 461:930–939.

von Frisch K (1974) Decoding the language of the bee. *Science* 185:663–668.

Molecular, genetic, and anatomical methods

Brand AH & Perrimon N (1993) Targeted gene expression as a means of altering cell fates and generating dominant phenotypes. *Development* 118:401–415.

Dani A, Huang B, Bergan J et al. (2010) Superresolution imaging of chemical synapses in the brain. *Neuron* 68:843–856.

Denk W & Horstmann H (2004) Serial block-face scanning electron microscopy to reconstruct three-dimensional tissue nanostructure. *PLoS Biol* 2:e329.

Feil R, Wagner J, Metzger D et al. (1997) Regulation of Cre recombinase activity by mutated estrogen receptor ligand-binding domains. *Biochem Biophys Res Commun* 237:752–757.

Feng G, Mellor RH, Bernstein M et al. (2000) Imaging neuronal subsets in transgenic mice expressing multiple spectral variants of GFP. *Neuron* 28:41–51.

Fire A, Xu S, Montgomery MK et al. (1998) Potent and specific genetic interference by double-stranded RNA in *Caenorhabditis elegans*. *Nature* 391:806–811.

Fodor SP, Read JL, Pirrung MC et al. (1991) Light-directed, spatially addressable parallel chemical synthesis. *Science* 251:767–773.

Golic KG & Lindquist S (1989) The FLP recombinase of yeast catalyzes site-specific recombination in the *Drosophila* genome. *Cell* 59:499–509.

Gong S, Zheng C, Doughty ML et al. (2003) A gene expression atlas of the central nervous system based on bacterial artificial chromosomes. *Nature* 425:917–925.

Gordon JW, Scangos GA, Plotkin DJ et al. (1980) Genetic transformation of mouse embryos by microinjection of purified DNA. *Proc Natl Acad Sci USA* 77:7380–7384.

Groth AC, Fish M, Nusse R et al. (2004) Construction of transgenic *Drosophila* by using the site-specific integrase from phage phiC31. *Genetics* 166:1775–1782.

Gu H, Marth JD, Orban PC et al. (1994) Deletion of a DNA polymerase beta gene segment in T cells using cell type-specific gene targeting. *Science* 265:103–106.

King DP, Zhao Y, Sangoram AM et al. (1997) Positional cloning of the mouse circadian clock gene. *Cell* 89:641–653.

Lander ES, Linton LM, Birren B et al. (2001) Initial sequencing and analysis of the human genome. *Nature* 409:860–921.

Lee T & Luo L (1999) Mosaic analysis with a repressible cell marker for studies of gene function in neuronal morphogenesis. *Neuron* 22:451–461.

Lein ES, Hawrylycz MJ, Ao N et al. (2007) Genome-wide atlas of gene expression in the adult mouse brain. *Nature* 445:168–176.

Rubin GM & Spradling AC (1982) Genetic transformation of *Drosophila* with transposable element vectors. *Science* 218:348–353.

Sanger F, Nicklen S & Coulson AR (1977) DNA sequencing with chain-terminating inhibitors. *Proc Natl Acad Sci USA* 74:5463–5467.

Takemura SY, Bharioke A, Lu Z et al. (2013) A visual motion detection circuit suggested by *Drosophila* connectomics. *Nature* 500:175–181.

Wang H, Yang H, Shivalila CS et al. (2013) One-step generation of mice carrying mutations in multiple genes by CRISPR/Cas-mediated genome engineering. *Cell* 153:910–918.

White JG, Southgate E, Thomson JN et al. (1986) The structure of the nervous system of the nematode *Caenorhabditis elegans*. *Philos Trans R Soc Lond B Biol Sci* 314:1–340.

Wickersham IR, Lyon DC, Barnard RJ et al. (2007) Monosynaptic restriction of transsynaptic tracing from single, genetically targeted neurons. *Neuron* 53:639–647.

Recording neuronal activity, manipulating neuronal activity, and behavioral analyses

Ahrens MB, Li JM, Orger MB et al. (2012) Brain-wide neuronal dynamics during motor adaptation in zebrafish. *Nature* 485:471–477.

Aravanis AM, Wang LP, Zhang F et al. (2007) An optical neural interface: in vivo control of rodent motor cortex with integrated fiberoptic and optogenetic technology. *J Neural Eng* 4:S143–156.

Armbruster BN, Li X, Pausch MH et al. (2007) Evolving the lock to fit the key to create a family of G protein-coupled receptors potently activated by an inert ligand. *Proc Natl Acad Sci USA* 104:5163–5168.

Boyden ES, Zhang F, Bamberg E et al. (2005) Millisecond-timescale, genetically targeted optical control of neural activity. *Nat Neurosci* 8:1263–1268.

Callaway EM & Katz LC (1993) Photostimulation using caged glutamate reveals functional circuitry in living brain slices. *Proc Natl Acad Sci USA* 90:7661–7665.

Chen TW, Wardill TJ, Sun Y et al. (2013) Ultrasensitive fluorescent proteins for imaging neuronal activity. *Nature* 499:295–300.

Chow BY, Han X, Dobry AS et al. (2010) High-performance genetically targetable optical neural silencing by light-driven proton pumps. *Nature* 463:98–102.

Cohen JY, Haesler S, Vong L et al. (2012) Neuron-type-specific signals for reward and punishment in the ventral tegmental area. *Nature* 482:85–88.

Denk W, Strickler JH & Webb WW (1990) Two-photon laser scanning fluorescence microscopy. *Science* 248:73–76.

Grynkiewicz G, Poenie M & Tsien RY (1985) A new generation of Ca^{2+} indicators with greatly improved fluorescence properties. *J Biol Chem* 260:3440–3450.

Hamada FN, Rosenzweig M, Kang K et al. (2008) An internal thermal sensor controlling temperature preference in *Drosophila*. *Nature* 454:217–220.

Hamill OP, Marty A, Neher E et al. (1981) Improved patch-clamp techniques for high-resolution current recording from cells and cell-free membrane patches. *Pflugers Arch* 391:85–100.

Harvey CD, Collman F, Dombeck DA et al. (2009) Intracellular dynamics of hippocampal place cells during virtual navigation. *Nature* 461:941–946.

Johns DC, Marx R, Mains RE et al. (1999) Inducible genetic suppression of neuronal excitability. *J Neurosci* 19:1691–1697.

Kitamoto T (2001) Conditional modification of behavior in *Drosophila* by targeted expression of a temperature-sensitive shibire allele in defined neurons. *J Neurobiol* 47:81–92.

Komiyama T, Sato TR, O'Connor DH et al. (2010) Learning-related fine-scale specificity imaged in motor cortex circuits of behaving mice. *Nature* 464:1182–1186.

Lima SQ & Miesenbock G (2005) Remote control of behavior through genetically targeted photostimulation of neurons. *Cell* 121:141–152.

Magnus CJ, Lee PH, Atasoy D et al. (2011) Chemical and genetic engineering of selective ion channel-ligand interactions. *Science* 333:1292–1296.

Maimon G, Straw AD & Dickinson MH (2010) Active flight increases the gain of visual motion processing in *Drosophila*. *Nat Neurosci* 13:393–399.

Miyawaki A, Llopis J, Heim R et al. (1997) Fluorescent indicators for Ca^{2+} based on green fluorescent proteins and calmodulin. *Nature* 388:882–887.

Nakai J, Ohkura M & Imoto K (2001) A high signal-to-noise $Ca^{(2+)}$ probe composed of a single green fluorescent protein. *Nat Biotechnol* 19:137–141.

Petreanu L, Huber D, Sobczyk A et al. (2007) Channelrhodopsin-2-assisted circuit mapping of long-range callosal projections. *Nat Neurosci* 10:663–668.

Sweeney ST, Broadie K, Keane J et al. (1995) Targeted expression of tetanus toxin light chain in *Drosophila* specifically eliminates synaptic transmission and causes behavioral defects. *Neuron* 14:341–351.

Zhang F, Wang LP, Brauner M et al. (2007) Multimodal fast optical interrogation of neural circuitry. *Nature* 446:633–639.

Ziv Y, Burns LD, Cocker ED et al. (2013) Long-term dynamics of CA1 hippocampal place codes. *Nat Neurosci* 16:264–266.

GLOSSARY

absence seizure A seizure characterized by a brief lapse of consciousness (about 10 seconds or less) and a cessation of motor activities without loss of posture.

accessory olfactory bulb A brain region adjacent to the olfactory bulb, it is the axonal projection target of sensory neurons from the vomeronasal organ. (Figure 6–22)

accessory olfactory system (vomeronasal system)
An anatomically and biochemically distinct system from the main olfactory system, it detects and analyzes nonvolatile chemicals and peptides such as pheromones and cues from predators. (Figure 6–22)

acetylcholine (ACh) Neurotransmitter released by vertebrate motor neurons at the neuromuscular junction. It is also used in the CNS as an excitatory or modulatory neurotransmitter, and in the autonomic nervous system. In some invertebrates such as *Drosophila*, it is the major excitatory neurotransmitter in the CNS. (Figure 3–1; Table 3–2)

acetylcholine receptor (AChR) Receptor for the neurotransmitter acetylcholine. The nicotinic AChRs (nAChRs) are non-selective cation channels; they are the postsynaptic receptor at the vertebrate neuromuscular junction and function as excitatory receptors at some CNS synapses. The metabotropic AChRs (muscarinic AChRs or mAChRs) are G-protein-coupled receptors that play a modulatory role. (Figure 3–20 for nAChR)

acetylcholinesterase An enzyme enriched in the cholinergic synaptic cleft that degrades acetylcholine.

acquisition (of memory) The initial formation of a memory as a consequence of experience and learning.

action potential An elementary unit of nerve impulses that axons use to convey information across long distances. It is all-or-none, regenerative, and propagates unidirectionally in the axon. It is also called a spike. (Figure 2–18; Figure 2–19)

active electrical property A membrane property that is due to voltage-dependent changes in ion conductance. It can reduce or eliminate the attenuation of electrical signals across a distance that occurs due to passive electrical properties.

active transport Movement of a solute across a membrane against its electrochemical gradient via a transporter that uses external energy, such as ATP hydrolysis, light, or movement of another solute down its electrochemical gradient. (Figure 2–8)

active zone An electron-dense region of the presynaptic terminal that contains clusters of synaptic vesicles docked at the presynaptic membrane, ready for release. (Figure 3–3; Figure 3–10)

activity-dependent transcription The process by which neuronal activity regulates gene expression.

adaptation (in evolution) Genetic or phenotypic changes that render an individual and its progeny more likely to survive and reproduce in a particular environment.

adaptation (in sensory systems) The adjustment of the system's sensitivity according to the background level of sensory input.

adeno-associated virus (AAV) A DNA virus widely used to deliver transgenes into post-mitotic neurons. It has a capacity to include about 5 kb of foreign DNA. (Table 13–1)

adenylate cyclase A membrane-associated enzyme that synthesizes cyclic AMP (cAMP) from ATP. (Figure 3–33)

advanced sleep phase syndrome A disorder characterized by very early morning waking and an early evening sleep onset.

afferent An axon that projects from peripheral tissue to the CNS. It can also be generalized to describe an input axon to a particular neural center within the CNS.

agonist A molecule that mimics the action of an endogenous molecule such as a neurotransmitter.

agrin A protein secreted by motor neurons that induces aggregation of acetylcholine receptors in the muscle. (Figure 7–24)

AgRP neuron A neuron in the hypothalamic arcuate nucleus that releases the orexigenic peptides agouti-related protein (AgRP) and neuropeptide Y. (Figure 8–42)

AII amacrine cell A type of amacrine cell that links rod bipolars to the pathways that process cone signals. (Figure 4–34)

AKAP (A kinase anchoring protein) An anchoring protein associated with protein kinase A.

all-or-none Having the property of being binary in occurrence. It applies to action potentials, which have the same amplitude and waveform regardless of the strength of the inducing stimulus as long as the stimulus is above threshold.

allele A specific version of a gene.

allelic exclusion A phenomenon in which mRNAs of a gene are transcribed exclusively from one chromosome of a homologous pair. *See also* **allele**.

allodynia A phenomenon whereby gentle touch or innocuous temperature causes pain when applied to inflamed or injured tissue.

allosteric agonist A molecule that facilitates binding of an endogenous ligand to its receptor. An allosteric agonist binds to a site on a receptor that is different from the site that binds the endogenous ligand.

Alzheimer's disease (AD) A neurodegenerative disorder prevalent in the aging population. It is defined by the combined presence of abundant amyloid plaques and neurofibrillary tangles in postmortem brains, with symptoms including gradual loss of memory, impaired cognitive and intellectual capabilities, and reduced ability to cope with daily life. (Figure 11–2)

amacrine cell An inhibitory neuron whose actions influence the signals that are transmitted from the bipolar cells to the retinal ganglion cells. (Figure 4–28)

AMPA receptor A glutamate-gated ion channel that conducts mostly Na^+ and K^+ and can be selectively activated by the drug AMPA (2-amino-3-hydroxy-5-methylisoxazol-4-propanoic acid). It is a heterotetramer containing two or more kinds of subunits (GluA1, GluA2, GluA3, and GluA4) encoded by four genes. (Figure 3–24; Figure 3–26)

amygdala An almond-shaped structure underneath the temporal lobe best known for its role in processing emotion-related information. (Figure 1–8; Figure 10–41)

amyloid plaque An extracellular deposit consisting primarily of aggregates of amyloid β protein. (Figure 11–2)

amyloid precursor protein (APP) A single-pass transmembrane protein from which the amyloid β protein is derived by proteolytic processing. (Figure 11–3)

amyloid β protein (Aβ) A major component of the amyloid plaques in Alzheimer's disease, it is a 39–43-amino-acid peptide with a strong tendency to form aggregates rich in β-pleated sheets. (Figure 11–4)

amyotrophic lateral sclerosis (ALS) A rapidly progressing motor neuron disease that is usually terminal within a few years after symptoms emerge. It is also known as Lou Gehrig's disease.

analog signaling Signaling that uses continuous values to represent information.

androgen A male sex hormone, such as testosterone and its derivatives.

androgen receptor A cytosolic protein that upon binding of an androgen (such as testosterone) translocates to the nucleus, where it acts as a transcription factor. (Figure 9–24)

anions Negatively charged ions such as Cl^-.

ankyrinG An intracellular scaffolding protein that is highly concentrated in the axon initial segment and nodes of Ranvier.

anosmic Unable to perceive odors.

antagonist A molecule that counters the action of an endogenous molecule. For example, by binding to nAChR in competition with ACh and inhibiting nAChR function, curare acts as an antagonist of ACh.

antagonistic muscles Muscles that perform opposite actions, such as an extensor and a flexor that control the same joint. (Figure 8–8)

antennal lobe The first olfactory processing center in the insect brain. (Figure 6–27)

anterior cingulate cortex (ACC) A neocortical area located near the midline of the frontal lobe. It has extensive connections with the hippocampus and is implicated in long-term memory storage.

anterior pituitary *See* **pituitary**.

anterograde From the cell body to the axon terminal.

anterograde tracer A molecule used to trace axonal connections. They are taken up primarily by neuronal cell bodies and dendrites and travel down the axons to label their projection sites. (Figure 13–27)

anterolateral column pathway An axonal pathway from the spinal cord to the brainstem, it consists of axons from lamina I dorsal horn projection neurons on the contralateral side of the spinal cord. It mainly relays pain, itch, and temperature signals to the brain. (Figure 6–70)

anteroventral periventricular nucleus (AVPV) A hypothalamic nucleus in the preoptic area that plays a pivotal role in regulating the female ovulatory cycle. (Figure 9–27)

antidromic spike An action potential that propagates from the axon terminal to the cell body in artificial situations in which experimenters electrically stimulate the axon or its terminal.

antiporter A coupled transporter that moves two or more solutes in opposite directions. Also called an exchanger. (Figure 2–10)

anxiety disorders A group of psychiatric disorders that includes generalized anxiety disorders (characterized by persistent worries about impending misfortunes), phobias and panic disorders (characterized by irrational fears), and obsessive-compulsive disorder.

AP5 (2-amino-5-phosphonovaleric acid) A widely used selective NMDA receptor antagonist.

apolipoprotein E (ApoE) A high-density lipoprotein in the brain involved in lipid transport and metabolism. A specific polymorphic isoform (ε4) is a major risk factor for Alzheimer's disease. (Figure 11–9)

Arc A cytoskeletal protein present at the postsynaptic density that regulates trafficking of glutamate receptors. It is a product of the immediate early gene *Arc*.

archaerhodopsin A light-activated outward proton pump in archaea, it can be used to silence neuronal activity in a heterologous system by light. *See also* **optogenetics**. (Figure 13–45)

arcuate nucleus A ventromedial hypothalamic nucleus that regulates food intake and energy expenditure.

area X A basal ganglia structure in the songbird that is essential for song learning. (Figure 9–21)

aromatase An intracellular enzyme that converts testosterone to estradiol. (Figure 9–24)

ascending arousal system A neural system consisting of parallel projections from the brainstem and hypothalamus to the forebrain that are essential for maintaining wakefulness. It includes cholinergic projections from the tegmental nuclei, norepinephrine projections from the locus coeruleus, serotonin projections from the raphe nuclei, histamine projections from the tuberomammillary nucleus, and hypocretin projections from the lateral hypothalamus. (Figure 8–52)

association cortex Cortical areas that integrate information from multiple sensory areas and link sensory systems to motor output.

associative learning A type of learning involving the formation of an association between two events, such as the formation of an association between an unconditioned stimulus and a conditioned stimulus in classical conditioning or the formation of an association between a behavior and a reinforcer in operant conditioning.

associativity (of LTP) A property of long-term potentiation (LTP) whereby activation of a synapse that alone would be too weak to produce LTP can nonetheless lead to LTP if it coincides with the strong, LTP-inducing activation of a different synapse onto the same postsynaptic cell. (Figure 10–9)

astrocyte A glial cell present in the gray matter. It plays many roles including synaptic development and function. (Figure 1–9)

asymmetric cell division A cell division in which the two daughter cells are of different types from birth.

ataxia An abnormality in coordinated muscle contraction and movement.

attention The cognitive function in which a subset of sensory information is subjected to further processing at the expense of other information.

attractant A molecular cue that guides axons toward its source. (Figure 5–9)

auditory cortex The part of the cerebral cortex that first receives auditory sensory information. It is located in the temporal lobe. (Figure 1–23)

auditory fear conditioning A classical conditioning procedure in which aversive, fear-inducing stimuli, such as electric shocks, are paired with sound stimuli during training; animals will subsequently exhibit fear responses, such as freezing, in response to sound stimuli alone. It depends on the amygdala but not the hippocampus. Sound can be replaced with other sensory cues such as odor, and the learning procedure is generally called cued fear conditioning.

auditory nerve A bundle of axons from spiral ganglion neurons that transmits auditory information to the brainstem. It also contains efferents from the brainstem that synapse primarily onto outer hair cells. (Figure 6–49)

autism spectrum disorders (ASDs) Neurodevelopmental disorders characterized by deficits in communication and reciprocal social interactions. Patients also exhibit restricted interests and repetitive behaviors.

autocrine Of or related to a form of signaling in which a recipient cell receives a signal produced by itself.

autonomic nervous system The collected parts of the nervous system that regulate the function of internal organs, including the contraction of smooth and cardiac muscles and the activities of glands.

autosomal dominant Of a mutation, having a Mendelian inheritance pattern in which mutation of only one allele of a gene located on an autosome is sufficient to produce a phenotype. It can result from a toxic gain-of-function effect of the mutant allele or a loss-of-function effect due to an insufficient amount of the normal gene product being produced by the remaining wild-type allele. (Figure 11–34)

autosomal recessive Of a mutation, having a Mendelian inheritance pattern in which mutation of both alleles of a gene located on an autosome is required to produce a phenotype. It usually results from a loss-of-function effect of the mutation. (Figure 11–34)

autosome A non-sex chromosome.

axon A long, thin process of a neuron, it often extends far beyond the soma and propagates and transmits signals to other neurons or muscle at its presynaptic terminals. (Figure 1–9)

axon guidance molecules Extracellular cues and cell surface receptors that guide axons along their path towards the appropriate targets. (Figure 5–9)

axon myelination The process by which glial cells wrap their cytoplasmic extensions around axons to increase conduction velocity. (Figure 2–27)

Aβ fiber A heavily myelinated somatosensory axon. (Figure 6–63)

Aβ hypothesis The idea that an increase of amyloid β (Aβ) protein production or accumulation is a common cause of Alzheimer's disease.

Aδ fiber A lightly myelinated somatosensory axon. (Figure 6–63)

bacterial artificial chromosome (BAC) A cloning vector (circular DNA molecule that can be grown in bacteria) that can accommodate hundreds of kilobases of foreign DNA. (Figure 13–12)

bacteriorhodopsin A light-driven proton pump in archaea. (Figure 12–20)

ball-and-chain A model of voltage-gated channel inactivation in which a cytoplasmic portion of the channel protein ('ball'), connected to the rest of the channel by a polypeptide chain, blocks the channel pore after the ion channel opens. (Figure 2–32)

barrel A discrete anatomical unit in layer 4 of the rodent primary somatosensory cortex that represents a whisker. The cortical region containing barrels for all whiskers is called barrel cortex. The corresponding discrete units in the brainstem and thalamus are called barrelettes and barreloids, respectively. (Figure 5–27)

barrel cortex *See* **barrel**.

basal ganglia A collection of nuclei underneath the cerebral cortex, it includes the striatum, globus pallidus, subthalamic nucleus, and substantia nigra and is essential for motor initiation and control, habit formation, and reward-based learning. (Figure 1–8; Figure 8–22)

basilar membrane An elastic membrane at the base of hair cells in the cochlea. (Figure 6–50)

basket cell A type of GABAergic neuron, it wraps its axon terminals around the cell bodies of pyramidal cells in the cerebral cortex or Purkinje cells in the cerebellar cortex. (Figure 1–15; Figure 3–46)

basolateral amygdala A brain region consisting of two subdivisions, the lateral amygdala and basal amygdala. It receives input from the thalamus, cortex, and hippocampus and sends output to the central amygdala and other brain regions. It is involved in regulating emotion-related behavior. (Figure 10–41)

battery An electrical element that maintains a constant voltage, or electrical potential difference, across its two terminals and that can thus serve as an energy source. (Figure 2–13)

BDNF (brain-derived neurotrophic factor) *See* **neurotrophins** and **Trk receptors**.

bed nucleus of stria terminalis (BNST) A sexually dimorphic brain region that receives direct input from accessory olfactory bulb mitral cells. Its diverse functions include regulation of male courtship behavior. (Figure 9–32)

benzodiazepines A class of drugs that act as allosteric agonists of $GABA_A$ receptors. They are widely used to treat anxiety, pain, epilepsy, and sleep problems. (Figure 11–29)

biased random walk In chemotaxis, a strategy employed by bacteria to move towards an attractant source (or away from a repellent source). When swimming away from an attractant, bacteria exhibit frequent tumbles (reorientation); when swimming towards an attractant, they tumble less frequently. It is also employed by *C. elegans* for chemotaxis. (Figure 12–15)

bilaterians Animals that are bilaterally symmetrical and that have three germ layers. They include all vertebrates and most invertebrate species alive today. (Figure 12–2)

binary expression Expression of a transgene using a strategy in which the regulatory elements (which determine where the transgene is expressed) and the coding sequence are separated into two transgenes. (Figure 13–13)

binocular vision A form of vision involving integration of inputs from the two eyes that carry information about the same visual field location. It is important for depth perception.

binomial distribution A discrete probability distribution that describes the frequency (f) that k events occur in n independent trials, given the probability that an event occurs in each trial is p. $f(k; n, p) = [n!/k!(n-k)!]\, p^k (1-p)^{n-k}$. (Box 3–1)

biomarker A biological characteristic that is objectively measured and evaluated as an indicator of a normal biological process, a pathogenic process, or a response to a therapeutic intervention. (Figure 11–10)

bipolar Having two processes leaving the cell body.

bipolar cell (in retina) An excitatory neuron that transmits information from the photoreceptors to the retinal ganglion cells and amacrine cells. (Figure 4–25; Figure 4–28)

bipolar disorder A mood disorder in which patients alternate between manic phases (characterized by feelings of grandiosity and tirelessness) and depressive phases (characterized by feelings of sadness, emptiness, and worthlessness).

bitter A taste modality that functions primarily to warn the animal of potential toxic chemicals. It is usually aversive.

blastula The product of cleavage, it is an early-stage embryo consisting of a hollow ball of thousands of cells. (Figure 7–2)

blood–brain barrier (BBB) Derived from endothelial cell tight junctions in the blood vessels of the brain, it prevents the exchange of many substances between the blood and brain tissues.

blue-ON bipolar cell An ON bipolar cell that selectively connects with S-cones. It is activated by short-wavelength light and inhibited by longer-wavelength light. (Figure 4–33)

bone morphogenetic proteins (BMPs) A family of secreted proteins that act as morphogens to pattern embryonic tissues, such as the tissues along the anterior–posterior axis of the telencephalon and the dorsal–ventral axis of the spinal cord.

border cell A cell in the entorhinal cortex that fires when an animal is at a specific edge of an arena.

Boss (*Bride of sevenless*) Originally identified from a mutation in *Drosophila* that lacks photoreceptor R7, it is a gene that acts cell-nonautonomously in R8 to specify R7 fate. It encodes a transmembrane ligand for the Sevenless receptor tyrosine kinase. (Figure 5–36)

botulinum toxins A family of proteases produced by *Clostridium botulinum*. Different isoforms cleave synaptobrevin, syntaxin, or SNAP-25 at distinct sites.

bradykinin A peptide released during inflammation, it binds to specific G-protein-coupled receptors on the peripheral terminals of nociceptive neurons. (Figure 6–71)

brain The rostral part of the central nervous system located in the head. It is the command center for nervous system functions. (Figure 1–8)

brain slice A fresh section of brain tissue (usually about a few hundred micrometers thick) that largely preserves the three-dimensional architecture for physiological studies of neuronal and local circuit properties *in vitro*.

brainstem A structure that comprises the midbrain, pons, and medulla. (Figure 1–8)

Broca's area An area in the left frontal lobe involved in language production. Patients with lesions in this area have difficulty speaking. (Figure 1–23)

α-bungarotoxin A snake toxin from the venom of *Bungarus* that is a competitive inhibitor of the nicotinic acetylcholine receptor.

C fiber An unmyelinated somatosensory axon. (Figure 6–63)

Ca^{2+} indicator A molecule whose optical properties are dependent on intracellular Ca^{2+} concentration. Ca^{2+} indicators are used as optical sensors of neuronal activity. (Figure 13–38)

cable properties *See* **passive electrical properties**.

cadherin A Ca^{2+}-dependent homophilic cell-adhesion protein.

callosal projection neuron (CPN) A cortical neuron that extends its axon across the corpus callosum to the contralateral cortex. (Figure 7–10)

calmodulin (CaM) A Ca^{2+}-binding protein that transduces Ca^{2+} signals to many effectors. (Figure 3–34)

CaM kinase II (CaMKII) A Ca^{2+}/calmodulin-dependent serine/threonine kinase that is highly enriched in the postsynaptic densities of excitatory synapses and that regulates synaptic plasticity, such as long-term potentiation. (Figure 3–34; Figure 10–12)

Cambrian A geological period between 542 and 488 million years ago when major phyla within the animal kingdom diversified, as evidenced by an abundance of corresponding fossils. (Figure 12–2)

cAMP-dependent protein kinase A serine/threonine kinase composed of two regulatory and two catalytic subunits. Binding of cAMP to the regulatory subunits leads to dissociation of the catalytic subunits, which can then phosphorylate their substrates. It is also called A kinase, protein kinase A, or PKA. (Figure 3–33)

capacitance (*C*) The ability of a capacitor to store charge; defined as $C = Q/V$, where Q is the electric charge stored when the voltage across the capacitor is V.

capacitor An electrical element consisting of two parallel conductors separated by a layer of insulator. It is a charge-storing device. (Figure 2–13)

capping The process by which a modified guanosine nucleotide is added to the 5′ end of the RNA. (Figure 2–2)

Capricious A *Drosophila* transmembrane protein that contains extracellular leucine-rich repeats and that instructs wiring specificity of axons and dendrites. (Figure 5–39; Figure 7–41)

cardiac muscle Muscle that controls heartbeat.

Cas9 A key protein in the type II CRISPR system, it is an RNA-guided endonuclease containing two separate nuclease domains that generate a double-strand break in DNA complementary to a bound RNA. It is used by bacteria for adaptive immunity and used experimentally for genome engineering. *See also* **CRISPR** and **guide RNA**. (Figure 13–8)

caspase-3 A key protease that triggers apoptosis, a form of programmed cell death.

castrated male A male from which the testes have been removed.

catecholamines A class of chemicals that includes the neurotransmitters dopamine, norepinephrine, and epinephrine. (Figure 11–20)

cations Positively charged ions such as K^+ and Na^+.

CB1 A G-protein-coupled receptor originally identified as the receptor for cannabinoids from the marijuana plant. It serves as a receptor for endocannabinoids under physiological conditions.

CCK (cholecystokinin) A neuropeptide produced in the small intestine in response to a rise in fatty acid concentration. It acts as a satiety signal to inhibit eating. (Figure 8–43)

cDNA library A collection of cloned cDNAs, or complementary DNAs, synthesized from mRNA templates derived from a specific tissue.

cell adhesion molecule Cell-surface proteins that bind to their partners in opposing cells or extracellular matrix to facilitate cell–cell or cell–matrix adhesion.

cell assembly A group of neurons whose firing patterns collectively encode information, such as locations of an animal in an environment.

cell-attached patch recording (cell-attached recording) A variant of the patch clamp recording method in which the patch pipette forms a high resistance (gigaohm) seal with the plasma membrane of an intact cell, allowing measurement of ion flow through a small number of channels or a single channel in the patch of membrane underneath the electrode. (Figure 2–30; Figure 13–37)

cell autonomous Of a gene, acting in the cell that produces the gene product.

cell fate The outcome of the developmental decision as to what type of cell it is.

cell lineage The developmental history of a cell, including the identities of all progenitors from which a cell was derived.

cell nonautonomous Of a gene, acting in a cell that does not produce the gene product.

cell theory The idea that all living organisms are composed of cells as basic units.

cell-replacement therapy A treatment strategy in which cells differentiated *in vitro* are transplanted into the body to replace dying cells, such as dopamine neurons in Parkinson's disease.

cell-surface receptor A membrane protein that binds to extracellular ligands and subsequently sends a signal into the recipient cell. (Figure 3-38)

center–surround (receptive field) A property of a visual system neuron's receptive field in which light in the receptive field center and light just outside of the receptive field center are antagonistic. (Figure 4-24). The concept has also been extended beyond the visual system.

central amygdala The output nucleus of the amygdala complex, it receives input from the basolateral amygdala and sends GABAergic output to brainstem nuclei, the autonomic nervous system, the hypothalamus, and neuromodulatory systems to regulate emotion-related behavior. (Figure 10-41; Figure 10-43)

central dogma The principle that genetic information flows from DNA to RNA to protein.

central pattern generator (CPG) A CNS circuit that is capable of producing rhythmic output for coordinated contraction of different muscles without sensory feedback. (Figure 8-12; Figure 8-13)

cerebellum A structure located dorsal to the pons and medulla, it plays an important role in motor coordination, motor learning, and cognitive functions. (Figure 1-8; Figure 8-20)

cerebral cortex The outer layer of the neural tissue in the rostral part of the mammalian brain. It is associated with higher functions including sensory perception, cognition, and control of voluntary movement. (Figure 1-8; Figure 1-23)

CGRP (calcitonin gene-related peptide) A peptide that promotes inflammation when released by the peripheral terminals of sensory neurons. (Figure 6-71)

chandelier cell A type of GABAergic neuron in the cerebral cortex that forms synapses onto the initial axon segments of cortical pyramidal cells. (Figure 3-46)

channel A transmembrane protein or protein complex that forms an aqueous pore that allows specific solutes to pass through directly when it is open. (Figure 2-8)

channelopathies Diseases caused by mutations in ion channels.

channelrhodopsin A member of a class of light-activated cation channels in single-cell green algae used for chemotaxis. (Figure 12-21). *See also* **channelrhodopsin-2 (ChR2)**.

channelrhodopsin-2 (ChR2) A light-activated cation channel from a single-celled green alga, it is widely used to activate neurons in heterologous systems by light. *See also* **optogenetics**. (Figure 12-21; Figure 13-45)

characteristic frequency The sound frequency to which a given cell in the auditory system is most sensitive.

Charcot–Marie–Tooth (CMT) disease A PNS demyelinating disease characterized by progressive deficits in sensation or movement that preferentially affects neurons with longer axons. Genetic alterations in ~30 genes have been identified as causes of CMT disease.

chemical gradient Concentration difference of the solute on the two sides of the membrane, which contributes to the direction and magnitude of solute movement across the membrane. If the solute is not charged, the chemical gradient alone determines the movement direction: from higher concentration to lower concentration. (Figure 2-9)

chemical synapse A specialized junction between two neurons or between a neuron and a muscle where communication between cells is mediated by the release of neurotransmitters. It comprises a presynaptic terminal and a postsynaptic specialization separated by a synaptic cleft. (Figure 1-14; Figure 3-3)

chemoaffinity hypothesis Proposed by Roger Sperry, it states that growing axons use cell-surface proteins to find their path and connect with appropriate synaptic partners.

chemogenetics An approach that uses chemicals to activate or silence neurons that express receptors specifically engineered to be sensitive to those chemicals.

chemotaxis Movement toward or away from a chemical source.

chlorpromazine A first-generation antipsychotic drug, it is an antagonist of the D_2 dopamine receptor.

chordates Animals with a notochord. (Figure 12-2)

chromatic aberration The phenomenon in which a lens refracts different wavelengths of light differently and thus cannot focus all wavelengths with equal sharpness.

chromophore The light-absorbing portion of a molecule.

ciliary type A type of photoreceptor in which opsins are packed into the primary cilium-derived outer segment. (Figure 12-22)

circadian pacemaker neuron A neuron whose activity in isolation oscillates in a circadian fashion (i.e. with a circa 24 hour period).

circadian rhythms Self-sustained oscillations in an organism's behavior, physiology, and biochemistry, with a period close to 24 hours.

***cis*-regulatory elements** DNA elements, such as transcriptional enhancers, repressors, and insulators, that regulate the expression of genes on the same chromosome.

Cl⁻ channels An ion channel family that allows selective passage of Cl^-.

clade A branch in the tree of life, consisting of an ancestor species plus all of its descendant species.

cladistic analysis The study of emergence and change of traits of organisms in the context of their phylogenetic relationships.

classical conditioning A form of learning in which repeated pairing of a conditioned stimulus (CS) with an unconditioned stimulus (US) causes a subject to exhibit a novel conditioned response (CR) to the CS. Prior to learning, the CS does not produce the CR, and after learning the CR resembles the unconditioned response (UR), which is elicited without conditioning by the US. It is also called Pavlovian conditioning.

cleavage A series of rapid cell divisions in early embryogenesis that convert a single large zygote cell into thousands of smaller cells. (Figure 7-2)

climbing fiber An axon that climbs dendritic trees of individual Purkinje cells. It originates from a neuron in the inferior olive. (Figure 8-20)

Clock A gene identified from a forward genetic screen in mice for mutations that produce circadian rhythm phenotypes. It encodes a transcriptional activator, CLOCK, that positively regulates the expression of genes whose products feedback to negatively regulate CLOCK function. Its fly homolog serves a similar function. (Figure 8-45)

clonal analysis A method of analyzing the relationships of cells by birth. It involves labeling a progenitor in a way that all its progeny are also labeled.

closed-loop Of a system, having input to the system that is modified by the output of the system. In the context of a behavioral paradigm, having environmental stimuli that induce a behavior in an animal to also change in response to the animal's behavior.

cnidarians Animals that are radially symmetrical, such as hydra, jellyfish, and corals. (Figure 12-2)

CNS (central nervous system) The brain and spinal cord in vertebrates; the brain and nerve cord in some invertebrates.

cochlea A coiled structure in the inner ear that contains fluid-filled chambers and the organ of Corti. (Figure 6–45)

cochlear nuclei Brainstem nuclei where the auditory nerve terminates, consisting of the dorsal and ventral cochlear nuclei. (Figure 6–54)

coding space A theoretical space used to describe the activity of a neuronal population. The firing rate of each neuron in the population constitutes one dimension/axis in this space, and the activity state of the entire population is represented as a point in this space. (Figure 6–30)

cognitive learning A theory of learning with the emphasis of learning as an acquisition of new knowledge rather than simply a modification of behavior.

coincidence detector (in auditory system) A cell that is maximally activated by simultaneous auditory signals from the left ear and the right ear.

coincidence detector (in synaptic transmission) A receptor that opens only in response to concurrent neurotransmitter binding *and* postsynaptic depolarization, such as the NMDA receptor.

collateral An axon branch.

color-opponent RGC A retinal ganglion cell that differentiates signals from cones with distinct spectral sensitivities. The blue-yellow opponent RGC (in all mammals) differentiates short- and longer-wavelength light signals; the green–red opponent RGC (in trichromatic primates) differentiates two long-wavelength light signals. (Figure 4–33)

Comm (Commissureless) A *Drosophila* protein that acts in the secretory pathway to down-regulate cell-surface expression of Robo. (Figure 7–12)

commissural neuron A neuron that projects its axon to the contralateral side of the body. In the vertebrate spinal cord, midline crossing of commissural neurons has been used as a model to study axon guidance.

compact myelin Closely packed layers of glial plasma membranes wrapped around axons.

complex cell A functionally defined primarry visual cortex neuronal type present in all layers except layer 4. It has no mutually antagonistic ON and OFF regions and is excited by light bars on a dark background or dark bars on an illuminated background. The stimulus bars must be in a specific orientation but can fall on any part of the receptive field. (Figure 4–40)

conditional knockout The process of disrupting a gene in a specific spatiotemporal pattern or an animal in which a gene has been disrupted in a specific spatiotemporal pattern. The most common strategy for generating conditional knockouts in mice uses Cre/*loxP*-based recombination. It usually involves inserting a pair of *loxP* elements in introns that flank (an) essential exon(s) of a gene of interest. The gene of interest is only disrupted in cells in which Cre is active or in cells derived from a progenitor in which Cre was active. (Figure 13–7)

conditioned response (CR) *See* **classical conditioning**.

conditioned stimulus (CS) *See* **classical conditioning**.

conductance (g) The degree to which an object or substance passes electricity, it is the inverse of resistance: $g = 1/R$.

conductor An object or substance that passes electric current.

cone A cone-shaped photoreceptor in the vertebrate retina, it contributes to high acuity, motion, and color vision. (Figure 4–2)

confocal fluorescence microscopy (confocal microscopy) A fluorescence microscopy technique in which a detector pinhole is used to collect fluorescence emissions originating only from a focal spot restricted in all three dimensions. By scanning the laser across a plane to record fluorescence emissions from many focal spots, it can produce a thin optical section of whole-mount tissue or a thick tissue section. (Figure 13–19)

connectome A representation of the complete set of synaptic connections among a group of neurons of interest. (Figure 7–28; Figure 13–2)

connexin A protein component of gap junctions in vertebrates. (Figure 3–48)

ω-conotoxin A small peptide from marine snails that specifically blocks presynaptic voltage-gated Ca^{2+} channels and thus inhibits neurotransmitter release.

consolidation (of memory) A step in the process of memory formation that occurs between acquisition and storage, during which a newly acquired memory is solidified.

contextual fear conditioning A learning procedure in which a rodent is subjected to aversive, fear-inducing stimuli, such as electric shocks, in a specific environment (i.e. context). When placed in the same context subsequently, the animal will exhibit a fear response, such as freezing. It depends on both the hippocampus and amygdala.

continuous map A type of neural map in which input field neurons that neighbor each other connect with target field neurons that neighbor each other, as exemplified by the relationship between the retina and the tectum. (Figure 7–33)

contralateral Of the other side of the midline. For example, a contralateral axonal projection is an axon that crosses the midline and terminates on the side of the nervous system opposite the soma.

convergent evolution The independent evolution of similar features in animals from different clades of the phylogenetic tree.

cooperativity (of LTP) A property of long-term potentiation (LTP) whereby LTP can be induced at a synapse if the presynaptic cell releases neurotransmitter while the postsynaptic cell is in a depolarized state, even if transmitter release from the presynaptic cell alone (in the absence of postsynaptic depolarization) is insufficient to induce LTP. (Figure 10–9)

copy number variations (CNVs) Deletions or duplications of chromosome segments that can vary in length from 500 base pairs to several megabases and may contain coding sequences that range from a small fraction of a single gene to many genes.

coronal section A section plane that is perpendicular to the rostral–caudal axis; also called frontal or transverse sections.

corpus callosum A structure composed of axon bundles that link the two cerebral hemispheres. (Figure 7–10)

cortical amygdala Part of the olfactory amygdala complex that receives direct mitral cell input. (Figure 6–19)

corticothalamic projection neuron (CTPN) A cortical neuron, found in layer 6, that projects its axon to the thalamus. (Figure 7–10)

cotransporter A transporter that uses the movement of one solute down its electrochemical gradient to drive the transport of another solute up its concentration gradient.

courtship conditioning The process by which a normal *Drosophila* male learns to reduce his attempts at courtship following repeated rejections by mated females.

CRE (cAMP-response element) *See* **CREB**.

Cre recombinase A bacteriophage-derived enzyme that catalyzes recombination between two sequence-specific DNA elements called *loxP* sites. (Figure 13–7; Figure 13–13)

CREB (cAMP-response element binding protein) A transcription factor that binds the cAMP response element (CRE), a DNA *cis*-regulatory element in the promoter regions of target genes. It is a substrate for several kinases, including cAMP-dependent protein kinase. (Figure 3–41)

CreER A fusion of the Cre recombinase with the portion of the estrogen receptor responsible for ligand-dependent nuclear trafficking. CreER enters the nucleus only in the presence of tamoxifen, an estrogen analog, and therefore catalyzes recombination in a tamoxifen-dependent manner.

Creutzfeldt-Jakob disease (CJD) *See* **prion diseases**.

CRISPR A clustered regularly interspaced short palindromic repeat, it is a genomic locus in some bacteria and archaea that contains repeated DNA elements derived from the genomes of invading pathogens. It is used by bacteria for adaptive immunity, and components of the CRISPR system are used experimentally for genome engineering. *See also* **Cas9** and **guide RNA**. (Figure 13–8)

critical period A sensitive period during development when experience plays an important role in shaping the wiring properties of the brain.

cryptochrome A protein that acts as a negative regulator of circadian gene expression in mice but as a light sensor for entrainment of circadian rhythms in flies. (Figure 8–46)

curare A plant toxin that is a competitive inhibitor of the nicotinic acetylcholine receptor.

cyclic AMP (cAMP) An intracellular second messenger synthesized from ATP by adenylate cyclase. (Figure 3–33)

cyclic GMP (cGMP) A cyclic nucleotide derived from GTP, one of its functions is to activate the cyclic nucleotide-gated cation channel in vertebrate photoreceptors in the absence of light. (Figure 4–10)

cyclic nucleotide-gated (CNG) channels Non-selective cation channels whose gating is regulated by the concentration of a specific intracellular cyclic nucleotide. (Figure 2–34)

cytoarchitectonics An approach to describe tissue organization using differences in cell density and distribution. (Figure 13–18)

DCC/Unc40 Homologous proteins in vertebrates (DCC, for deleted in colon cancer) and *C. elegans* (Unc40) that act as receptors for netrin/Unc6 and mediate attraction in the absence of Unc5. The *Drosophila* homolog is Frazzled. (Figure 5–10)

de novo mutation A mutation produced in the parental germ line that is present in all of the offspring's cells.

deep brain stimulation (DBS) A treatment strategy used for a number of neurological and psychiatric conditions in which electrodes are surgically implanted to stimulate neurons and axons in specific brain nuclei.

deep cerebellar nuclei The output nuclei of the cerebellum, they receive input from Purkinje cell axons as well as from the collaterals of the mossy and climbing fibers. (Figure 8–20)

delay line A thin axon fiber that carries auditory signals to target neurons at different locations along the axon with different time delays. (Figure 6–55)

Delta A transmembrane ligand that activates Notch. (Figure 7–7)

demyelinating disease A disease in which damage to the myelin sheath decreases the axonal membrane resistance between nodes of Ranvier, leading to disruption of ion channel organization in the nodal region and reduction in action potential conduction speed.

dendrites Thick, bushy processes of a neuron that receive and integrate synaptic inputs from other neurons. (Figure 1–9)

dendritic spine A small protrusion on a dendrite of certain neurons that receives synaptic input from a partner neuron. The thin spine neck creates chemical and electrical compartments for each spine such that it can be modulated independently from neighboring spines. (Figure 1–9; Figure 3–45)

dendritic tiling A phenomenon in which the dendrites of certain neuronal types collectively cover the entire field exactly once so they can sample the field without redundancy. For example, certain types of retinal neurons collectively cover the retina exactly once. Certain types of somatosensory neurons cover the body surface exactly once. (Figure 4–29)

dendrodendritic synapse A synapse between dendritic processes of two neurons. The reciprocal synapses between the olfactory bulb granule cell dendrites and mitral cell secondary dendrites were the first discovered examples. (Figure 6–18)

dense-core vesicle An intracellular vesicle containing neuropeptides, they are larger and more electron-dense than synaptic vesicles, which contain small molecule neurotransmitters.

dentate gyrus The input part of the hippocampus, consisting of granule cells and their dendrites, which receive input from the entorhinal cortex. (Figure 10–6)

depolarization A change in the electrical potential inside the cell toward a less negative value.

depressing synapse A synapse at which successive presynaptic action potentials trigger progressively smaller postsynaptic responses. (Figure 3–15)

deuterostomes Animals in which the anus appears before the mouth during development. They include all vertebrates. *See also* **protostomes**. (Figure 12–2)

developmental axon degeneration The process by which axons are fragmented into pieces that are subsequently engulfed by surrounding glia during normal development.

diacylglycerol (DAG) A lipid second messenger that binds to and activates protein kinase C (PKC). (Figure 3–34)

diffusion tensor imaging (DTI) A magnetic resonance imaging technique that allows noninvasive imaging of axon bundles in the white matter based on the direction of water diffusion in a given volume. (Figure 13–26)

digital signaling Signaling that uses discrete values (0s and 1s) to represent information.

direct pathway (in basal ganglia) An axonal projection from a subset of spiny projection neurons that link the striatum directly to the basal ganglia output nuclei, GPi and SNr. (Figure 8–22)

direction-selective retinal ganglion cell (DSGC) A retinal ganglion cell whose firing pattern is influenced by the direction of motion of a stimulus. (Figure 4–30)

discrete map A type of neural map in which input or target neurons or their processes are spatially organized into discrete units (such as glomeruli or layers) representing different qualities (such as cell types). (Figure 7–33)

disinhibition The reduction of the inhibitory output of an inhibitory neuron. (Figure 1–21)

dizygotic twins Non-identical (fraternal) twins who share only 50% of their genes, because they originated from two independent eggs fertilized by two independent sperm.

DNA (deoxyribonucleic acid) Long double-stranded chains of nucleotides. The nucleotides consist of the sugar deoxyribose, a phosphate group, and one of four nitrogenous bases: adenine (A), cytosine (C), guanine (G), or thymidine (T).

DNA microarray A solid substrate containing up to millions of immobilized spots of different oligonucleotides or gene-specific probes. Labeled nucleic acid samples, such as mRNAs extracted from a specific tissue or genomic DNA from an individual, can be hybridized to a DNA microarray to quantify the abundance of different species of nucleic acid molecules in samples. It can be used to determine gene expression patterns or profiles of single-nucleotide and copy number polymorphisms. (Figure 13–16)

DNA shuffling A process by which part or all of the protein-coding sequence of one gene is fused with that of another gene, usually following chromosomal duplication or translocation. The specific type of DNA shuffling that occurs when translocational breakpoints are within introns of two genes is called exon shuffling. (Figure 12–6)

L-dopa The intermediate metabolite between tyrosine and dopamine in the catecholamine biosynthetic pathway. (Figure 11–20)

dopamine A monoamine neuromodulator derived from the amino acid tyrosine. (Figure 11–20; Table 3–2)

Doppler effect A phenomenon whereby the sound frequency detected by an observer increases if the sound-emitting object moves toward the observer and decreases if the sound-emitting object moves away from the observer.

dorsal column pathway An axonal pathway from the spinal cord to the brainstem, it consists of ascending branches of proprioceptive neurons and Aβ-LTMRs, as well as axons of some dorsal horn projection neurons. (Figure 6–70)

dorsal cortex The evolutionary precursor to the mammalian neocortex in reptiles. It consists of three thin layers (as opposed to the six-layered structure of the mammalian neocortex).

dorsal horn The dorsal part of the spinal gray matter devoted to processing somatosensory information. (Figure 6–70)

dorsal horn projection neuron A neuron located in the dorsal horn of the spinal cord that projects its axon to the brainstem to relay touch signals. (Figure 6–70)

dorsal root The place where somatosensory axons enter the spinal cord. (Figure 8–6)

dorsal root ganglia (DRG) Clusters of somatosensory neurons located along an axis parallel to the spinal cord used for sensation of the body (as opposed to the face). (Figure 6–63)

dorsal stream A visual processing pathway from primary visual cortex to the parietal cortex. It is responsible for analyzing motion and depth; the 'where' stream. (Figure 4–48)

dorsal–ventral Of a body axis, from back to belly.

Doublesex (Dsx) A *Drosophila* gene that encodes sex-specific transcription factors produced by sex-specific alternative splicing. The Dsx isoform determines sex-specific somatic structures and also regulates sexual behavior. (Figure 9–4)

Down syndrome A syndrome caused by the presence of an extra copy of Chromosome 21. It is the most common form of intellectual disability with an established genetic etiology.

doxycycline (Dox) A tetracycline analog that readily diffuses across cell membranes and the blood–brain barrier. It is widely used for temporal regulation of gene expression through the tTA/rtTA/TRE system. (Figure 13–13)

driver transgene In binary expression, it is the transgene that expresses a transcription factor or a recombinase under the control of a tissue-specific or temporally regulated promoter. (Figure 13–13)

driving force The force that pushes an ion into or out of a cell, it equals the difference between the membrane potential of the cell and the equilibrium potential of the ion.

drug addiction Compulsive drug use that persists despite long-term negative consequences. It is often associated with loss of self-control and propensity to relapse.

Dscam (Down syndrome cell adhesion molecule) Encoded by a gene on human chromosome 21, which is trisomic in Down syndrome, it is an evolutionary conserved protein that, in insects, exhibits extraordinary molecular diversity due to alternative splicing. (Figure 7–20)

dTRPA1 A *Drosophila* TRP channel that is activated by high temperature. (Figure 13–43)

dye-coupling The diffusion of a small-molecule dye from one cell to another through gap junctions. It is used as a criterion to identify the presence of gap junctions between two cells.

dynamic range In sensory systems, the ratio between the largest and smallest values of a given dimension of sensory stimuli that can be detected and distinguished.

dynamical state A point in a coding space, representing the status of a dynamical system at a given time. *See also* **coding space**. (Figure 8–28)

dynamical system A physical system whose future state is a function of its current state, its input, and possibly some noise. It can be represented as time-dependent change of neural states in a coding space. *See also* **coding space**. (Figure 8–28)

dynein A minus-end-directed, microtubule-based motor protein. (Figure 2–6)

eardrum A membrane at the intersection of the mammalian outer ear and middle ear whose vibrations are transmitted by the bones in the middle ear to the cochlea in the inner ear. (Figure 6–45)

echolocation The ability of certain species to use echoes of their own ultrasonic sound pulses to locate objects.

ectoderm The outer germ layer that gives rise to the skin and nervous system. (Figure 7–2)

efferent An axon that projects from the CNS to peripheral targets. It can also be generalized to describe an output axon from a particular neural center within the CNS.

efficacy of synaptic transmission (synaptic efficacy) The strength of a synaptic connection, it is ususally measured by the mean magnitude of the postsynaptic response to a defined presynaptic stimulus.

E-I balance The relative strength of synaptic excitation versus synaptic inhibition.

electrical circuit Connected electrical elements that contain at least one closed current path.

electrical gradient Electrical potential difference between the two sides of the membrane, which contributes to the direction and magnitude of movement of a charged solute across the membrane. It promotes the movement of a charged solute toward the side with the opposite charge. (Figure 2–9)

electrical synapse A cell–cell junction enriched in gap junction channels. It transmits (usually bidirectionally) both depolarizing and hyperpolarizing signals between the two cells. *See also* **gap junction**. (Figure 1–14)

electrochemical gradient A combination of chemical and electrical gradients, which determines the direction and magnitude of movement of a charged solute across the membrane. (Figure 2–9)

electroencephalography (EEG) A method for recording the electrical potential differences between surface electrodes placed on specific locations of the scalp. It reports the collective electrical activities of many cortical neurons underneath the surface electrodes. (Figure 8–51; Figure 11–47)

electromotility A property of the cochlear outer hair cells whereby hyperpolarization causes the cells to lengthen, and depolarization causes them to shorten, along their long axis. (Figure 6–52)

electron microscopy A microscopic technique that uses beams of electrons to create an image of a specimen. It has much higher resolution than light microscopy and can resolve structures that are separated by a nanometer or less. *See also* **transmission electron microscopy** and **scanning electron microscopy**.

electroporation A procedure in which DNA containing a transgene is introduced into cells by applying electrical current to facilitate the transfer of negatively charged DNA molecules into the cells. In animals, this can be achieved by placing a micropipette containing the DNA near the cells of interest and applying electrical current.

embryonic stem (ES) cells Pluripotent cells derived from early embryos that can be propagated indefinitely *in vitro* and that can give rise to all cell types of an embryo *in vivo*. (Figure 11–23)

end-plate current The current that crosses a muscle cell membrane in response to release of acetylcholine from a presynaptic motor neuron.

end-plate potential (EPP) Depolarization produced in a postsynaptic muscle cell by acetylcholine released from a presynaptic motor neuron in response to an action potential. (Figure 3–1)

endocannabinoids Endogenous cannabinoids, which are lipophilic molecules such as anadamide and 2-arachidonylglycerol. They can be produced in response to a rise of intracellular Ca^{2+} concentration in certain postsynaptic neurons and diffuse across the synapse to affect presynaptic neurotransmitter release by binding to the CB1 G-protein-coupled receptor.

endocrine Of or related to a form of signaling in which a recipient cell receives a signal produced by a remote source and delivered through systemic circulation.

endocrine system A system consisting of glands that release hormones into the bloodstream, so that those hormones can circulate throughout the body.

endocytosis The process by which cells retrieve, via budding of intracellular vesicles from the plasma membrane, fluid and proteins from the extracellular space and transmembrane proteins from the cell's plasma membrane. (Figure 2–2)

endoderm The inner germ layer that gives rise to a variety of tissues such as the liver, the inner linings of the gut, and the respiratory tract. (Figure 7–2)

endoplasmic reticulum (ER) A network of membrane-enclosed compartments in eukaryotic cells where secreted and transmembrane proteins are made and into which secreted and transmembrane proteins are translocated. It also serves as a store for intracellular Ca^{2+}. (Figure 2–2)

endosome A membrane-enclosed organelle produced by endocytosis. It carries newly internalized extracellular materials and transmembrane proteins. (Figure 2–2)

engram Physical substrate for memory, it is also called memory trace.

enteric nervous system A division of the autonomic nervous system that is associated with the gastrointestinal tract and that regulates digestion rather independently of the rest of the autonomic nervous system.

entorhinal cortex The part of the temporal cortex overlying the hippocampus. It provides major input to and receives output from the hippocampus. It plays a major role in representing spatial information. (Figure 10–6)

entrainment The process by which a stimulus, such as light, resets the phase of the circadian clock.

Eph receptors Receptor tyrosine kinases that bind ephrins with their extracellular domains. Two Eph receptor subtypes, the EphA and EphB receptors, typically bind ephrin-As and ephrin-Bs, respectively, but this specificity is not absolute. They can also serve as ligands during reverse signaling. (Figure 5–7; Figure 5–12)

ephrins Cell-surface proteins that usually act as ligands for Eph receptors to mediate repulsion during axon guidance. The ephrin family consists of two subfamilies: ephrin-As are attached to the extracellular face of the plasma membrane by GPI, and ephrin-Bs are transmembrane proteins. They can also serve as receptors during reverse signaling. (Figure 5–7; Figure 5–12)

epigenetic modifications Molecular modifications to DNA and chromatin, such as DNA methylation and various forms of post-translational modification of histones. They do not modify the DNA sequence but can alter gene expression.

epilepsy A medical condition characterized by recurrent seizures. *See also* **seizure**.

epinephrine A hormone produced primarily by chromaffin cells in the adrenal gland that mediates systemic responses to extreme conditions, such as the systemic response associated with fright, fight, and flight. It also acts as a modulatory neurotransmitter in a small group of neurons in the brainstem. (Figure 11–20)

epithelial Na^+ channel (ENaC) A member of a class of Na^+ channels involved in Na^+ reabsorption by epithelial cells, it is also essential in mammals for the taste of low concentrations of salts. Its invertebrate homologs participate in mechanotransduction. (Figure 6–41)

EPSC (excitatory postsynaptic current An inward current produced by binding of an excitatory neurotransmitter to its receptor. (Figure 3–23)

EPSP (excitatory postsynaptic potential) A transient depolarization of a postsynaptic cell associated with an excitatory postsynaptic current (EPSC). (Figure 3–23)

equilibrium potential The membrane potential at which there is no net flow of an ion across the membrane, because the electrical and chemical forces are equal in magnitude but opposite in direction.

estradiol A steroid hormone produced by the ovaries of sexually mature females. It can also be produced outside the ovaries by the actions of aromatase on testosterone. (Figure 9–24)

estrogen A female sex hormone, such as estradiol.

estrogen receptor A cytosolic protein that upon binding of an estrogen (such as estradiol) translocates to the nucleus, where it acts as a transcription factor. (Figure 9–24)

eukaryote Organism made of cell(s) with a nuclear membrane that separates the genetic material from the rest of the cellular components.

eumetazoan A taxon that includes cnidarians, bilaterians, and the most recent common ancestor of cnidarians and bilaterians. (Figure 12–2)

exchanger *See* **antiporter**.

excised patch   A patch clamp configuration in which the membrane patch underneath the electrode is excised from the cell and placed in a defined medium. It is often used to study the biophysical and biochemical properties of the ion channel(s) in the membrane patch. (Figure 13–37)

excitability   A property of a neuron that defines how readily it fires action potentials.

excitable cell   A cell that produces action potentials, such as a neuron or a muscle cell. It can also refer to any cell that uses electrical signaling to receive, integrate, propagate, and transmit information.

excitation–contraction coupling   A process by which action potentials in muscle cells lead to muscle contraction. It involves actin/myosin-mediated contraction triggered by a rise of intracellular Ca^{2+} concentration. (Figure 8–5)

excitatory neuron   A neuron that, when activated, depolarizes its postsynaptic target cells and makes them more likely to fire action potentials.

excitatory neurotransmitter   A neurotransmitter that depolarizes postsynaptic target cells and makes them more likely to fire action potentials.

excitotoxicity   Toxicity to neurons caused by excessive stimulation by excitatory neurotransmitters such as glutamate, which results in a large or persistent increase in intracellular Ca^{2+} concentration.

exocrine system   A system consisting of glands that excrete fluids, such as sweat or tears, locally through specific ducts.

exocytosis   The process by which intracellular vesicles fuse with the plasma membrane to release secreted proteins into the extracellular space, and to deliver lipids and transmembrane proteins to the plasma membrane. (Figure 2–2)

exon   The part of an RNA molecule that is retained in mRNA after splicing. (Figure 2–2)

exon shuffling   See **DNA shuffling**.

explicit memory   A form of memory that requires conscious recall, such as memory for names, facts, and events. It is also called declarative memory. (Figure 10–4)

expression cloning   A strategy for cloning a gene by transfecting cells with pools of cDNAs and using a functional assay to identify the pool that contains the cDNA of interest. The assay is reiterated with progressively divided pools of cDNAs until a single cDNA is identified. (Figure 6–68)

extensor   A muscle whose contraction increases the angle of a joint. (Figure 8–8)

extinction   In classical conditioning, a decrease in the conditioned response caused by repeated exposure to the conditioned stimulus without the unconditioned stimulus. In operant conditioning, a decrease in a reinforced action or an increase in a punished action when the action is repeatedly not reinforced or punished, respectively.

extracellular recording   A technique for recording voltage changes, such as action potentials from a single neuron or synaptic activity from a population of neurons. It utilizes an electrode, often made of metal wire that is insulated except at the tip, which is placed at close range to a neuronal cell body or a synapse-rich region. (Figure 13–31)

exuberant connection   Excess connection made during development that is not retained in adulthood.

Eyeless   A *Drosophila* transcription factor belonging to the Pax family. It contains a homeobox and a paired box and is required for eye development. Its ectopic expression in other structures, such as the antenna or the wing precursors, can induce ectopic eye formation. *See also* **Pax6**.

facilitating synapse   A synapse at which successive presynaptic action potentials trigger progressively larger postsynaptic responses. (Figure 3–15)

familial Alzheimer's disease (FAD)   A small subset of Alzheimer's disease cases that follows a Mendelian (autosomal dominant) inheritance pattern.

fast axonal transport   Intracellular transport at a speed of 50–400 mm per day; cargos subject to fast axonal transport include organelles, as well as transmembrane and secreted proteins. (Figure 2–4)

fear conditioning   *See* **contextual fear conditioning** and **auditory fear conditioning**.

feedback inhibition   A circuit motif in which an excitatory neuron both provides output to and receives input from an inhibitory neuron. (Figure 1–21)

feedforward excitation   A circuit motif in which serially connected excitatory neurons propagate information across multiple regions of the brain. (Figure 1–21)

feedforward inhibition   A circuit motif in which a postsynaptic neuron receives both direct excitatory input from a presynaptic neuron and disynaptic inhibitory input from the same excitatory neuron via an inhibitory interneuron. (Figure 1–21)

fertilization   The fusion of sperm and egg to create a genetically new organism. (Figure 7–2)

Fezf2   A transcription factor that specifies subcerebral projection neuron identity. (Figure 7–10)

fibroblast growth factor (FGF)   A member of a family of secreted growth factors that act as morphogens to pattern early embryos during development.

field excitatory postsynaptic potential (fEPSP)   Excitatory postsynaptic potentials recorded from a population of neurons near the tip of an extracellular electrode. fEPSPs evoked by stimulation of axonal inputs to a population are often used as a measure of the strength of synaptic transmission between the stimulated inputs and neurons near the recording electrode. (Figure 10–8)

filamentous actin (F-actin)   A major cytoskeletal element composed of two parallel helical strands of actin polymers. They are also called microfilaments. (Figure 2–5; Figure 8–3)

filopodia   A thin, protruding process of the growth cone made of bundled F-actin. (Figure 5–15)

fissure   A deep invagination of the cortical surface that separates areas of the cerebral cortex.

fitness   With respect to an allele (or phenotype), fitness is the ratio of the frequency of the allele (or phenotype) in a population after one generation of selection to the frequency of the allele (or phenotype) in the same population before the selection. With respect to an individual, fitness is the number of second-generation descendants the type of individual with a particular genome is expected to have.

fixed (of an allele)   The state of an allele when every member of the population is homozygous for the allele.

fixed action pattern   An instinctive sequence of behaviors, it is largely invariant and runs to completion once triggered.

flavor   A synthesis of taste and olfaction.

flexor   A muscle whose contraction decreases the angle of a joint. (Figure 8–8)

floor plate   A structure at the ventral midline of the spinal cord. (Figure 5–10; Figure 7–8)

FLP recombinase A yeast-derived enzyme that catalyzes recombination between two sequence-specific DNA elements called *FRT* (FLP recognition target) sites. (Figure 13–7; Figure 13–23)

fluorescence resonance energy transfer (FRET) A phenomenon in which energy is transferred between two fluorophores with an efficiency inversely proportional to the sixth power of the distance between them. It can be used experimentally to determine the distance between two fluorophores. It is also known as Förster resonance energy transfer (FRET).

fluoxetine A widely used antidepressant that acts as a selective serotonin reuptake inhibitor. Its brand name is Prozac. (Figure 11–27). *See also* **SSRI**.

Fmr1 *See* **fragile-X syndrome**.

FMRP *See* **fragile-X syndrome**.

focal seizures Seizures that affect a relatively small, discrete region of the brain.

follicle-stimulating hormone (FSH) *See* **gonadotropins**.

forebrain The rostral-most division of the three divisions of the embryonic brain. It gives rise to the cerebral cortex, basal ganglia, hippocampus, amygdala, thalamus, and hypothalamus. (Figure 1–8; Figure 7–3)

forward genetic screen A procedure to identify genes that are necessary for a biological process. It usually involves (1) inducing mutations in a population of experimental animals (such as through radiation, transposon insertion, or treatment with a chemical mutagen) so that each animal carries a different set of random mutations in a small number of genes or a single gene, and (2) identifying mutations that disrupt the biological process of interest based on the phenotypes exhibited by the offspring of the mutagenized animals. (Figure 13–4)

forward signaling *See* **reverse signaling**.

Fos An immediate early gene that encodes a transcription factor. Its expression is commonly used as an indicator of recently activated neurons.

fovea The central part of the primate retina that has a high density of cones. (Figure 4–14)

fragile-X syndrome (FXS) A leading cause of inherited intellectual disability, it is caused by expanded trinucleotide repeats in the 5′ untranslated region of the *Fmr1* gene, which encodes an RNA binding protein called the fragile X mental retardation protein (FMRP).

Frazzled *See* **DCC/Unc40**.

frequency tuning The property whereby a cell in the auditory system is best activated by sounds of a particular frequency. It is usually represented as a V-shaped curve on a frequency–intensity plot.

frontal eye field (FEF) A neocortical area that receives extensive feedforward connections from both the dorsal and ventral streams and sends feedback projections to many visual cortical areas. (Figure 4–48)

frontal lobe One of the four cerebral cortex lobes, it is located at the front of the brain rostral to the central sulcus. (Figure 1–23)

FRT *See* **FLP recombinase**.

Fruitless (Fru) A *Drosophila* gene that regulates all aspects of male courtship rituals. The splicing of one of its transcripts is regulated by a hierarchy of sex-determining splicing factors: females express a non-functional splice isoform of the protein while males express a functional form of the protein (Fru^M) that acts as a transcription factor. (Figure 9–4)

functional architecture The physical arrangement of neurons in a brain region based on their functional properties.

functional magnetic resonance imaging (fMRI) A non-invasive functional brain imaging technique, it monitors signals originating from changes in blood flow that are closely related to local neuronal activity. It is also called BOLD (blood-oxygen-level dependent) fMRI.

fundamental frequency The frequency of the lowest frequency component of a periodic waveform.

fura-2 A small molecule Ca^{2+} indicator whose optimal excitation wavelength shifts from 380 nm to 350 nm when Ca^{2+} is bound. The ratio of fluorescence intensity measured at excitation wavelengths of 350 nm and 380 nm can be used as a sensitive measure of Ca^{2+} concentration.

fusiform face area A specific area of human temporal cortex that is preferentially activated by images of human faces.

GABA A glutamate derivative that is the predominant inhibitory neurotransmitter in vertebrates and invertebrates. (Figure 3–16; Table 3–2)

GABA$_A$ receptor An ionotropic receptor that is gated by GABA and mediates fast inhibition. (Figure 3–21; Figure 11–29)

GABA$_B$ receptor A metabotropic receptor that is activated by GABA and that mediates slow inhibition.

gain control Modulation of the slope of a system's input–out function, it is often used to restrict output to a limited dynamic range.

gain-of-function experiments Experiment in which a specific component is added to the system. They are often used to test whether the added component is sufficient for the system to function in a specific context.

GAL4 A yeast transcription factor that binds to a DNA element called a UAS (upstream activation sequence) in the promoter regions of genes to activate the transcription of those genes. (Figure 13–13)

galanin A neuropeptide with diverse functions, including the promotion of parental behavior.

ganglion A cluster of neurons located in the peripheral nervous system.

ganglionic eminences Developing ventral telencephalon structures that include the medial, caudal, and lateral ganglionic eminences (MGE, CGE, LGE). They are the birthplaces of cortical GABAergic neurons (MGE and CGE), GABAergic interneurons in the basal ganglia and amygdala (MGE and CGE), and olfactory bulb interneurons and most GABAergic projection neurons in the striatum (LGE). (Figure 7–5)

gap junction The morphological correlate of the electrical synapse, which usually contains hundreds of closely clustered channels that bring the plasma membrane of two neighboring cells together and allow passage of ions and small molecules between the two cells. (Figure 3–38)

gastrin-releasing peptide receptor (GRPR) A G-protein-coupled receptor that is activated by gastrin-releasing peptide and is involved in processing itch signals.

gastrula The product of gastrulation, it is an embryo with a three-layered structure consisting of ectoderm, mesoderm, and endoderm. (Figure 7–2)

gastrulation The process by which an embryo is transformed from a ball of cells into a structure with three distinct layers: ectoderm, mesoderm, and endoderm. (Figure 7–2)

GCaMP A GFP-based genetically encoded Ca^{2+} indicator whose fluorescence increases in response to a rise of Ca^{2+} concentration.

GCAP (guanylate cyclase activating protein) A calcium-binding protein which in its calcium-free form binds to and activates guanylate cyclase.

gene A segment of DNA that carries the instructions for how and when to make specific RNAs and proteins. (Figure 2–2)

gene expression profiling Determining the genes expressed in a sample on a whole-genome scale using methods such as microarray and RNA-seq.

gene therapy The use of DNA and/or genome modification to treat disease.

generalized seizures Seizures that affect multiple, bilateral regions of the brain.

genetic drift The process in which chance events that result in death of an organism or failure of an organism to reproduce can lead to the loss of an allele from a small population and an increase in the prevalence of the remaining allele(s).

genetic mosaic animal An animal that contains cells of more than one genotype. (Figure 5–36; Figure 13–10)

genetic susceptibility locus A genomic locus with variant(s) that increase the probability of carriers developing a trait (such as a disease).

genetically encoded Ca^{2+} indicators Proteins whose fluorescence properties change before and after binding to Ca^{2+}. *See also* **Ca^{2+} indicator**. (Figure 13–38)

genome engineering The general process of altering the genome at a predetermined locus, such as deleting a piece of endogenous DNA, inserting a piece of foreign DNA, or creating a specific base-pair change.

genome-wide association study (GWAS) A strategy for identifying genes associated with a specific trait by comparing DNA samples collected from many people with or without the trait. The DNA samples are used to identify single nucleotide polymorphisms throughout the entire genome that are most strongly linked with the trait.

ghrelin A neuropeptide produced by stomach-associated glands in response to a reduced glucose level. It acts as a hunger signal to stimulate eating. (Figure 8–43)

G_i (inhibitory G protein) A $G\alpha$ variant that binds to adenylate cyclase and inhibits its activity.

gill-withdrawal reflex A reflex in the sea slug *Aplysia* in which the gill is withdrawn into the mantle shelf when a tactile stimulus is applied to its siphon. It has been used as a model system to investigate the mechanisms that underlie simple forms of learning and memory.

glia Nonneuronal cells of the nervous system, they play essential roles for the development and function of neurons.

glomerulus A discrete, ball-like structure in the vertebrate olfactory bulb or insect antennal lobe where ORN axons form synapses with the dendrites of their postsynaptic target neurons. (Figure 6–3; Figure 6–17)

GluN1 *See* **NMDA receptor**.

GluN2 *See* **NMDA receptor**.

glutamate An amino acid that is the predominant excitatory neurotransmitter in vertebrates. (Figure 3–16; Table 3–2)

glutamic acid decarboxylase (GAD) An enzyme that converts glutamate into GABA.

glycine An amino acid that is an inhibitory neurotransmitter released by a subset of brainstem and spinal cord neurons in vertebrates. (Figure 3–16; Table 3–2)

glycine receptor An ionotropic receptor that is gated by glycine and mediates fast inhibition. (Figure 3–21)

Goldman-Hodgkin-Katz (GHK) equation An equation that relates the membrane potential at equilibrium to the membrane permeabilities and concentrations of multiple ions on the two sides of a membrane. A variant of the GHK equation relates the membrane potential at equilibrium to the equilibrium potential and conductance of each ion.

Golgi outpost Fragments of the Golgi apparatus that are located in neuronal dendrites. (Figure 7–18)

Golgi staining A histological staining method, it uses solutions of silver nitrate and potassium dichromate, which react to form a black precipitate (microcrystals of silver chromate). This precipitate accumulates stochastically in a small fraction of the nerve cells so that these cells, and most or all of their elaborate extensions, can be visualized against unstained tissue.

gonadotropin-releasing hormone (GnRH) A pre-hormone released by hypothalamic neurons (called GnRH neurons) that stimulates the release of gonadotropins by anterior pituitary endocrine cells. (Figure 9–27)

gonadotropins A family of hormones that includes luteinizing hormone (LH) and follicle-stimulating hormone (FSH). Released by anterior pituitary endocrine cells, these hormones stimulate the maturation of male testes and female ovaries during puberty. In adults, they stimulate the testes to release testosterone and the ovaries to release estradiol. (Figure 9–27)

GPCR (G-protein-coupled receptor) A member of a receptor family with 7 transmembrane domains that, upon ligand binding, activate trimeric G proteins, which in turn activate intracellular signaling cascades.

GPe (globus pallidus external segment) An intermediate nucleus in the basal ganglia indirect pathway. It contains GABAergic neurons that project to the GPi, SNr, and STN. (Figure 8–22)

GPi (globus pallidus internal segment) One of the two major output nuclei of the basal ganglia. It contains GABAergic neurons that project to the thalamus. (Figure 8–22)

GPI (glycosylphosphatidylinositol) A lipid anchor that can covalently attach to an extracellular protein to anchor it to the plasma membrane. GPI-anchored protein can be released from the membrane by phosphatidylinositol-specific phospholipase C (PI-PLC), which cleaves the bond between the GPI group and the protein.

G_q A $G\alpha$ variant that activates phospholipase C, in turn leading to activation of the inositol-phospholipid signaling pathway. (Figure 3–34)

graded potentials (local potentials) Membrane potentials that can change in continuous values, as opposed to the all-or-none property of the action potential. (Figure 2–18)

granule cells Neurons that are granular in appearance because they are densely packed, including three prominent types. The cerebellar granule cells are the most numerous type of neuron in the brain; their cell bodies and dendrites reside in the granular layer of the cerebellar cortex where they receive mossy fiber input; their axons ascend into the molecular layer, where each bifurcates to become a parallel fiber to send glutamatergic output to Purkinje cells. (Figure 8–20). The granule cells in the hippocampus are the major cellular constituents of the dentate gyrus; they receive input from the entorhinal cortex via the perforant path and send glutamatergic output to CA3 pyramidal neurons. (Figure 1–12; Figure 10–6). The olfactory bulb granule cells constitute a large subtype of olfactory bulb interneurons that receive input from the secondary dendrites of mitral cells and send GABAergic output back to mitral cells. (Figure 6–17).

gray matter The parts of the CNS that are enriched with neuronal cell bodies, dendrites, axon terminals, and synapses and that appear gray.

green fluorescent protein (GFP) A jellyfish protein that emits green fluorescence when excited by blue light. It is widely used as a marker for gene expression and for live imaging.

grid cell A cell in the entorhinal cortex whose activity depends on an animal's location in an arena, with peak firing rate occurring at the apices of an imaginary hexagonal grid superimposed on the arena floor. (Figure 10–31)

growth cone A dynamic structure at the tip of a developing neuronal process, it enables the extension of the process and guides its direction.

G$_s$ (stimulatory G protein) A Gα variant that binds to adenylate cyclase and stimulates its activity. (Figure 3–33)

GTPase An enzyme that hydrolyzes GTP, converting it to GDP.

GTPase activating protein (GAP) A protein that switches GTPases off by accelerating the GTPases' endogenous activity, which converts GTP to GDP. (Figure 3–32)

guanine nucleotide exchange factor (GEF) A protein that switches GTPases on by catalyzing the exchange of GDP for GTP. (Figure 3–32)

guanylate cyclase An enzyme that produces cGMP from GTP.

guide RNA In the CRISPR/Cas9 system, an RNA molecule that brings Cas9 to a target DNA sequence, where Cas9 generates a double-strand break. The guide RNA must contain sequences that base-pair with the target DNA. *See also* **CRISPR** and **Cas9**. (Figure 13–8)

gustatory nerve A bundle of axons that originate from the basal ends of the taste receptor cells. The nerve projects to the nucleus of the solitary tract in the brainstem, and thus relays taste information from the tongue to the brain. (Figure 6–35)

gyrencephalic Of cortex, having gyri and sulci. (Figure 12–5)

Gα, Gβ, Gγ *See* **trimeric GTP-binding protein**.

habituation A decrease in the magnitude of responses to stimuli that are presented repeatedly.

hair cell The primary sensory cell for audition, it converts mechanical stimuli—movement of stereocilia at its apical end—into electrical signals. (Figure 6–47; Figure 6–50)

halorhodopsin A light-activated inward chloride pump in archaea, it can be used to silence neuronal activity in heterologous system by light. *See also* **optogenetics**. (Figure 13–45)

harmonics Sounds with frequencies that are integer multiples of the fundamental frequency.

HCN (hyperpolarization-activated cyclic nucleotide-gated) channels Non-selective cation channels that are activated by hyperpolarization and whose gating is influenced additionally by the concentration of a specific intracellular cyclic nucleotide. (Figure 2–34)

head direction cell A cell that fires when an animal's head is facing a specific direction in space, regardless of the animal's location in the environment.

Hebb's rule A postulate by Donald Hebb that describes how learning can be transformed into a lasting memory, it states: "When an axon of cell *A* is near enough to excite a cell *B* and repeatedly or persistently takes part in firing it, some growth process or metabolic change takes place in one or both cells such that *A*'s efficiency, as one of the cells firing *B*, is increased."

Hebbian synapse A synapse whose strength can be enhanced by co-activation of pre- and postsynaptic partners.

hedonic value The degree to which something is pleasant or unpleasant, which usually correlates with the degree to which something is potentially beneficial or harmful to an animal.

hemispheres The two sides of the brain.

heritability A measure of the contribution of genetic differences to trait differences within a population. It can be measured in twin studies as 2 × (the correlation of the trait between pairs of monozygotic twins – the correlation of the trait between pairs of dizygotic twins).

herpes simplex virus (HSV) A DNA virus used to deliver transgenes into post-mitotic neurons. It has a capacity to include ~150 kb of foreign DNA. (Table 13–1)

heterophilic binding Binding of two different proteins, usually two different membrane proteins expressed from adjacent cells across the cell junction.

hindbrain The caudal-most division of the three divisions of the embryonic brain. It gives rise to the pons, medulla, and cerebellum. (Figure 1–8; Figure 7–3)

hippocampus A structure underneath the cortical surface of the temporal lobe. It has been most studied for its role in the acquisition of explicit memory and representation of space. (Figure 1–8; Figure 10–6)

histamine A monoamine neuromodulator derived from the amino acid histidine. (Figure 3–16; Table 3–2)

histological sections Slices of frozen or chemically fixed tissue produced by microtomes, with thicknesses ranging from several to several hundred micrometers. They can be stained using a number of different methods and examined under a light microscope.

homeodomain Originally discovered in proteins whose disruption causes transformation of one body part into another, it is a DNA-binding domain shared by all Hox proteins and many other transcription factors. It is also called a homeobox.

homeostasis The maintenance of a steady state of a physiological parameter—such as blood pressure, body temperature, or nutritional level—by feedback physiological and behavioral responses. (Figure 8–34)

homeotic transformation Transformation of one body part to another, such as the transformation of a pair of antennae to a pair of legs in *Drosophila antennapedia* mutants.

homologous recombination Exchange of nucleotide sequences between two identical or highly similar DNA molecules. It occurs naturally in certain cells due to its role in specific biological processes, such as in germ-line cells during meiotic crossing over. It is also used experimentally for genome engineering, such as the generation of knockout and knock-in alleles. (Figure 13–6; Figure 13–8)

homophilic binding Binding of two identical proteins, usually two membrane proteins expressed from adjacent cells across the cell junction.

horizontal cell An inhibitory neuron in the vertebrate retina whose actions influence the signals that are transmitted from the photoreceptors to the bipolar cells. (Figure 4–26)

horizontal gene transfer Gene transfer from one organism to another through mechanisms other than reproduction, such as via viral transduction.

horizontal sections A section plane that is perpendicular to the dorsal–ventral axis.

Hox **gene** A member of a family of evolutionarily conserved genes that are arranged in genomes in clusters and that encode homeobox-containing transcription factors. *Hox* genes define the anterior–posterior body axes of most invertebrates and all vertebrates and also regulate neuronal fate at later developmental stages. (Figure 12–32)

5-HT *See* **serotonin**.

HTMR (high-threshold mechanoreceptor) A mechanosensory neuron that senses pain caused by strong mechanical stimuli. (Figure 6–64)

huntingtin *See* **Huntington's disease.**

Huntington's disease (HD) A dominantly inherited disease that usually strikes patients during midlife. It is characterized initially by depression or mood swings and subsequently by abnormal movements due to degeneration of striatal neurons. It is caused by expanded poly-glutamine repeats in the huntingtin protein. (Figure 11–14)

HVC (high vocal center) A dorsal forebrain nucleus in the songbird essential for song production. (Figure 9–21)

hyperpolarization A change in the electrical potential inside the cell toward a more negative value.

hypocretin (orexin) A neuropeptide expressed by specific lateral hypothalamus neurons, it is important for regulating sleep and eating.

hypogonadotropic hypogonadism A disorder characterized by delayed, reduced, or absent puberty due to reduced gonadotropin levels.

hypothalamus A collection of nuclei ventral to the thalamus, it controls many bodily functions including eating, digesting, metabolic rate, drinking, salt intake, reproduction, body temperature, emergency response, and circadian rhythms. It executes many of these functions by regulating the autonomic nervous system and neuroendocrine system. (Figure 1–8; Figure 8–34)

identified neuron A neuron that can be recognized across individuals of the same species due to its stereotyped location, size, and shape.

Ig CAM (immunoglobulin cell adhesion molecule) A cell adhesion molecule that contains immunoglobulin domains on its extracellular side.

imipramine A tricyclic antidepressant that inhibits the plasma membrane monoamine transporters. (Figure 11–27)

immediate early genes (IEGs) A class of genes whose transcription is rapidly induced by external stimuli without requiring new protein synthesis.

immuno-EM A combination of immunostaining and electron microscopy used to visualize the distribution of individual proteins at an ultrastructural level. (Figure 13–24)

immunostaining A staining method that uses antibodies to visualize the distributions of proteins in fixed tissues. The most common form uses sequential application of two antibodies: a primary antibody that binds the protein of interest and a fluorescence- or enzyme-conjugated secondary antibody that binds to the primary antibody. Protein distribution can be visualized by fluorescence or a color substrate produced by the enzyme conjugated to the secondary antibody.

implicit memory A form of memory in which previous experience aids in the performance of a task without conscious recall. It is also called non-declarative memory or procedural memory. (Figure 10–4)

***in situ* hybridization** A method for determining mRNA distribution in tissues by hybridizing labeled gene-specific nucleic acid probes to fixed histological sections or whole-mount tissues.

***in vitro* mutagenesis** A molecular biology technique used to alter the sequence of a gene in a test tube.

inactivation (of ion channels) A decrease of ion conductance through a channel after an initial increase. The ion channel when inactivated is in a distinct state from when it is closed.

indirect pathway (in basal ganglia) An axonal projection from a subset of spiny projection neurons that terminate in the GPe and STN. (Figure 8–22)

induced pluripotent stem (iPS) cells Pluripotent cells produced experimentally from differentiated cells by a variety of means, such as forced expression of key transcription factors involved in maintaining the pluripotency of embryonic stem cells. (Figure 11–23)

induction A mechanism for determining cell fate in which a cell is born with the same potential to develop into different cell types as its sibling or cousins, and its fate is acquired by receiving external signals (i.e. the cell's fate is 'induced' by external cues).

inferior colliculus A midbrain nucleus that integrates auditory signals from brainstem nuclei. It sends auditory output to the thalamus and to the nearby superior colliculus/ tectum. (Figure 6–54)

inferior olive A nucleus in the medulla containing neurons whose axonal projections to the cerebellum form climbing fibers. (Figure 8–20)

inhibitory neuron A neuron that, when activated, hyperpolarizes its postsynaptic target cells and makes them less likely to fire action potentials.

inhibitory neurotransmitter A neurotransmitter that hyperpolarizes postsynaptic target cells and makes them less likely to fire action potentials.

initial segment of the axon The segment of the axon closest to the neuronal cell body, it is usually the site of action potential initiation.

innate A trait or behavior that is genetically programmed and that is thus with an organism from birth rather than acquired by experience.

innate song The song a songbird would sing if raised in acoustic isolation during the sensory stage of song learning.

innexins A protein component of gap junctions in invertebrates.

inositol 1,4,5-triphosphate (IP_3) A second messenger that binds to the IP_3 receptor on the endoplasmic reticulum (ER) membrane to trigger the release of ER-stored Ca^{2+} into the cytosol. (Figure 3–34)

input specificity (of LTP) A property of long-term potentiation (LTP) whereby LTP occurs only at synapses that have experienced an LTP-inducing stimulus and not at unstimulated synapses on the same postsynaptic neuron. (Figure 10–9)

insular cortex A part of the cerebral cortex that represents taste, pain, and interoception. (Figure 6–35; Figure 8–33)

insulator An object or substance that does not allow electric current to pass. It is equivalent to a resistor with infinite resistance.

insulin A peptide hormone produced by the pancreas in response to a rise in blood glucose level after meals. It regulates carbohydrate metabolism throughout the body, and also regulates food intake through its actions on target neurons in the brain. (Figure 8–43)

intellectual disability A condition characterized by deficits in general mental abilities such as reasoning, problem-solving, planning, abstract thinking, judgment, and learning.

interaural level difference (ILD) The level difference of a sound that is received in the left ear and the right ear, used for sound localization.

interaural time difference (ITD) The difference in the arrival time of a sound at the left ear and the right ear, used for sound localization.

intermediate progenitor A progenitor cell produced by division of a radial glial cell. It divides further to give rise to post-mitotic neurons. (Figure 7-4)

interneuron A neuron with the axon confined to the specific CNS region that houses the neuron's cell body, it is also called local neuron in this context. It may also refer to any neuron that is not a motor or a sensory neuron.

interoception The sense of the state of internal organs.

intersectional methods (in genetics) Strategies that use two orthogonal binary expression systems to refine patterns of transgene expression. (Figure 13-14)

interstitial branching Extending a collateral from the side of a growing process. (Figure 7-19)

intracellular recording A procedure for measuring the membrane potential of a cell using an electrode inserted into or continuous with the cytoplasm. (Figure 13-31)

intracellular vesicle A small, membrane-enclosed organelle in the cytoplasm of a eukaryotic cell. (Figure 2-2)

intrinsic properties The electrophysiological properties of a neuron determined by the composition, concentration, subcellular distribution, and biophysical properties of ion channels it expresses.

intrinsic signal imaging A method for measuring neuronal activity based on changes in the optical properties of tissue surrounding active neurons, primarily as a result of changes in blood oxygenation in those regions. (Figure 4-42)

intrinsically photosensitive retinal ganglion cell (ipRGC) A type of RGC that expresses melanopsin and that can be directly depolarized by light. (Figure 4-36)

intron The part of an RNA molecule that is removed during splicing. (Figure 2-2)

inward-rectifier K⁺ channels A subfamily of K⁺ channels that preferentially pass inward currents over outward currents: i.e. these channels pass current at membrane potentials more hyperpolarized than E_K but allow minimal outward currents at membrane potentials more positive than E_K. (Figure 2-34)

ion channel A channel that allows the passage of one or more specific species of ion.

ionotropic receptor A neurotransmitter receptor that functions as a neurotransmitter-gated ion channel to allow rapid (within a few milliseconds) membrane potential changes in response to neurotransmitter binding. (Figure 3-21)

iontophoresis A technique by which ions or charged chemicals are locally applied from a micropipette via a current pulse.

IP₃ receptor An IP₃-gated Ca²⁺ channel on the ER membrane. (Figure 3-34)

iproniazid The first antidepressant discovered serendipitously in the 1950s. It is an inhibitor of monoamine oxidase. (Figure 11-27)

IPSC (inhibitory postsynaptic current) An outward current produced by binding of an inhibitory neurotransmitter to its receptor. The fast component is usually mediated by Cl⁻ influx through the GABA_A receptor or glycine receptor.

ipsilateral Of the same side of the midline. For example, an ipsilateral axonal projection is an axon that does not cross the midline and therefore terminates on the same side of the nervous system as the soma.

IPSP (inhibitory postsynaptic potential) A transient hyperpolarization of a postsynaptic cell associated with an inhibitory postsynaptic current (IPSC).

I–V curve A graphical representation of the relationship between the current that passes through a piece of ion-channel-containing membrane (I) and the voltage across the membrane (V). (Figure 3-17)

K⁺ channels Ion channels that allow selective passage of K⁺, they constitute the most diverse channel family. (Figure 2-34)

K⁺–Cl⁻ cotransporter A transporter that couples K⁺ and Cl⁻ export to help maintain the Cl⁻ gradient across the membrane. (Figure 2-12)

kainate receptor A glutamate-gated ion channel that conducts Na⁺ and K⁺ and that can be selectively activated by the drug kainate (kainic acid).

kinesins A family of microtubule-based motor proteins that are mostly plus-end-directed. (Figure 2-6; Figure 2-7)

Kiss1R A G-protein-coupled receptor for kisspeptins, it is also called GPR54.

kisspeptins A family of neuropeptides encoded by the *Kiss1* gene that play an important role in activating GnRH neurons. (Figure 9-27)

knee-jerk reflex The involuntary forward movement of the lower leg due to the contraction of the quadriceps femoris muscle (an extensor) and relaxation of the hamstring muscle (a flexor). A tap of the knee (patellar ligament) stretches the muscle spindle in the quadriceps muscle and activates the proprioceptive sensory neurons. Sensory neuron activation initiates the reflex through monosynaptic excitation of motor neurons that excite the quadriceps femoris muscle and disynaptic inhibition of motor neurons that excite the hamstring muscle. (Figure 1-19)

knock-in A variation of the knockout procedure in which an *in vitro* engineered gene—either a transgene or a variant of an endogenous gene—is inserted into a specific chromosomal locus; the procedure can produce changes to endogenous genes as small as a single base pair.

knockout A genetic engineering procedure that inactivates a specific gene. In the mouse, it is usually achieved by homologous recombination in embryonic stem cells to create a mutation in the target gene. The resulting mutant mouse is called a knockout mouse for that particular gene. (Figure 13-6)

kuru *See* **prion diseases**.

lamellipodia A veil-like meshwork of the growth cone made of branched F-actin. (Figure 5-15)

lamina (in insect visual system) The first neuropil layer underneath the retina in the insect compound eye. (Figure 5-35)

landmark-based strategy A navigational strategy in which animals use external cues to determine their locations.

laser-scanning two-photon imaging *See* **two-photon microscopy**.

late LTP A long-lasting phase of long-term potentiation (LTP), usually lasting longer than 3 hours and requiring new protein synthesis and likely new gene expression.

lateral geniculate nucleus (LGN) A thalamic nucleus that receives visual input from retinal ganglion cell axons and sends output to the primary visual cortex. (Figure 4-35; Figure 4-37)

lateral horn A second-order olfactory center for odor-mediated innate behavior in the insect brain. It and the mushroom body are the two major output sites for projection neuron axons. (Figure 6-27)

lateral inhibition (in cell fate determination) The process by which neighboring cells are prevented from adopting identical fates through cell–cell interactions, such as those mediated by Notch/Delta. (Figure 7–7)

lateral inhibition (in information processing) A circuit motif in which an inhibitory neuron receives excitatory input from one or several parallel streams of excitatory neurons, and sends inhibitory output to many or all of the postsynaptic targets of these excitatory neurons. It is widely used in sensory systems. (Figure 1–21)

lateral intraparietal area (LIP) A cortical area in the primate parietal lobe implicated in making the decision to move eyes in a particular direction. (Figure 4–48)

length constant (space constant, λ) A key parameter that defines the passive electrical properties of electrical signaling. It is equal to the distance along a neuronal process over which the amplitude of a membrane potential change decays to $1/e$ or about 37% of its original value.

lentivirus A retrovirus that can infect post-mitotic neurons. It has a capacity to include ~8 kb foreign DNA. (Table 13–1)

leptin A hormone secreted by the fat tissues that negatively regulates food intake through its actions on specific neurons in the brain. (Figure 8–38)

Lewy bodies Intracellular inclusions that are a defining pathological feature of most forms of Parkinson's disease.

lidocaine An anesthetic that blocks action potential propagation by inhibiting voltage-gated Na$^+$ channels.

ligand A molecule that binds to its receptor.

ligand-gated ion channel A transmembrane protein complex that directly conducts ions in response to the binding of a neurotransmitter or other ligand.

light microscopy The most widely used microscopic technique in biology. It uses beams of visible light (photons) to create an image of a specimen and, with the exception of some methods for super-resolution fluorescence microscopy, can only resolve structures greater than 200 nm apart.

light-sheet fluorescence microscopy A fluorescence microscopy technique in which only the focal plane (i.e. a single plane in the z-dimension) is illuminated with a thin sheet of a laser beam from the side. All fluorescence emissions in the focal plane are collected simultaneously by a detector. (Figure 13–19)

lissencephalic Of cortex, being smooth. (Figure 12–5)

LKB1 A protein kinase essential for determining axon fate during the establishment of neuronal polarity.

LMAN (lateral magnocellular nucleus of the anterior nidopallium) A forebrain nucleus in the songbird that is essential for song learning but not for song production. (Figure 9–21)

lobula complex Neuropil underneath the medulla in the insect compound eye. (Figure 5–35)

local field potential Electrical potential at an extracellular recording site relative to a distal ground. Usually filtered to remove high-frequency signals, it reflects collective dendritic and synaptic activities of many neurons near the electrode. (Figure 10–8)

local interneuron (LN) (in insect olfactory system) A neuron whose processes are restricted to the antennal lobe. (Figure 6–27)

local neuron *See* **interneuron**.

local protein synthesis Translation of mRNA into protein in a neuron's cytoplasmic extensions (usually dendrites) rather than in the cell body.

locus coeruleus A brainstem nucleus consisting of norepinephrine neurons that project widely across the brain. (Figure 8–54)

long-range cue (in axon guidance) A secreted protein that can act at a distance from its cell of origin. (Figure 5–9)

long-term depression (LTD) A long-lasting decrease of synaptic efficacy that can be induced experimentally by specific stimulus conditions.

long-term memory Memory that lasts hours to years. (Figure 10–4)

long-term potentiation (LTP) A long-lasting enhancement of synaptic efficacy. It can be induced experimentally under a variety of conditions, such as high-frequency stimulation of input axons. (Figure 10–8)

long-term synaptic plasticity A change in the efficacy of synaptic transmission that lasts hours to the lifetime of the animal.

loose-patch recording A technique in which a patch electrode is placed against the cell membrane without forming a gigaohm seal. It can only be used to record spiking activity (not sub-threshold activity), but, unlike whole-cell recording, does not affect the intracellular content of the recorded cell.

lordosis A posture that female rodents assume when sexually aroused. It facilitates sexual intercourse.

loss-of-function experiment An experiment in which a specific component is disrupted, often used to determine if the missing component is necessary for the system to function.

loss-of-function mutation A mutation that disrupts the function of a gene.

lower envelope principle The idea that the limits of psychophysical performance are determined by the sensitivities of the most sensitive individual neurons. (Figure 6–72)

loxP *See* **Cre combinase**.

LRP4 (low-density lipoprotein receptor-related protein-4) Along with MuSK, it is an agrin receptor in muscle. It also signals back to motor axons to trigger presynaptic differentiation in a MuSK-independent manner. (Figure 7–24)

LTMRs (low-threshold mechanoreceptors) Touch sensitive somatosensory neurons that innervate hair follicles, specialized epithelial cells, and encapsulated corpuscles in the skin. They respond to vibration, indentation, pressure, and stretch of the skin, as well as to the movement or deflection of hairs. (Figure 6–63)

luteinizing hormone (LH) *See* **gonadotropins**.

lysosome A membrane-enclosed organelle that contains enzymes for protein degradation. (Figure 2–2)

M pathway A visual processing pathway from the retina to the visual cortex that originates from retinal ganglion cells with large receptive fields and engages lateral geniculate nucleus cells in the magnocellular layers. It carries information about luminance and has excellent contrast and temporal sensitivity. (Figure 4–48)

macular degeneration A disease that causes photoreceptors in the fovea to die, impairing high-acuity vision.

major depression A mood disorder characterized by persistent feelings of sadness, emptiness, and worthlessness.

major urinary protein (MUP) A highly stable protein found in the urine, which is used by some species to mark an individual's territory for a long duration.

MARCM (mosaic analysis with a repressible cell marker) A genetic mosaic method in *Drosophila* used to label individual neurons or groups of neurons that share the same lineage and at the same time to delete an endogenous gene or express a transgene specifically in these labeled neurons. (Figure 13–23)

Martinotti cell A type of GABAergic neuron in the cerebral cortex that forms synapses onto the distal dendrites of cortical pyramidal cells. (Figure 3–46)

massively parallel processing An information processing method, it utilizes a large number of units to perform a set of coordinated computations in parallel. It is a key feature of the nervous system.

maximum parsimony A means of generating phylogenetic predictions by selecting the interpretation of the experimental data that posits the fewest number of evolutionary changes among all the potential interpretations.

MC4R A G-protein-coupled receptor that is activated by α-MSH. (Figure 8–42)

mechanosensory neurons Somatosensory neurons that are activated by mechanical force and are responsible for proprioception, touch, and a subset of pain sensations.

mechanotransduction The process in sensory cells by which mechanical stimuli are converted to electrical signals.

mechanotransduction channel An ion channel that is gated by mechanical force.

MeCP2 (methyl-CpG-binding protein 2) A nuclear protein that binds to DNA at methylated CpG sites (i.e. adjacent cytosine and guanine nucleotides). It is highly expressed in developing and adult neurons. (Figure 11–36) *See also* **Rett syndrome.**

medial amygdala Part of the olfactory amygdala complex that receives direct input from accessory olfactory bulb mitral cells. It is sexually dimorphic and regulates male courtship behavior. (Figure 9–32)

medial geniculate nucleus A thalamic nucleus that processes and relays auditory signals to the auditory cortex. (Figure 6–54)

medial–lateral Of a body axis, from midline to side.

medial preoptic area (MPOA) A sexually dimorphic nucleus in the anterior hypothalamus that regulates male courtship behavior. (Figure 9–28; Figure 9–32)

medulla The caudal-most part of the brainstem between the pons and the spinal cord.

medulla (in insect visual system) A neuropil that lies beneath the lamina in the insect compound eye. (Figure 5–35)

Meissner corpuscle A specialized structure closely associated with the peripheral ending of the rapidly adapting type I (RAI) LTMR. (Figure 6–64)

melanopsin An opsin expressed by vertebrate intrinsically photosensitive retinal ganglion cells (ipRGCs). It is a member of the c-opsin subfamily, whose members are most widely used in invertebrate visual systems.

membrane potential The electrical potential difference between the inside of the cell and the extracellular environment.

memory The process in which information is encoded, stored, and retrieved. It can also be defined as the lasting changes in the brain that retain the learned information.

Merkel cell A specialized epithelial cell at the junction of the dermis and epidermis. It is closely associated with the peripheral ending of the slowly adapting type I (SAI) LTMR. (Figure 6–64)

mesencephalic locomotor region (MLR) A midbrain region where electrical stimulation evokes locomotor activity.

mesoderm The middle germ layer that gives rise to the skeletal system, connective tissues, muscle, and the circulatory system. (Figure 7–2)

messenger RNA (mRNA) A mature RNA molecule that has undergone 5′ capping, 3′ polyadenylation, and splicing to remove introns and that is exported to the cytoplasm to direct protein synthesis. (Figure 2–2)

metabotropic receptor A neurotransmitter receptor that regulates ion channel conductance indirectly through intracellular signaling cascades, modulating membrane potential over a timescale of tens of milliseconds to seconds. (Figure 3–22)

microglia A glial cell that functions as the resident immune cell of the nervous system. It engulfs damaged cells and debris. (Figure 1–9)

microneurography A neurophysiological technique used to record neuronal activity in the peripheral nerves of awake human subjects.

microRNA A short, noncoding RNA (21–26 nucleotides in length) widely used in eukaryotic organisms to regulate gene expression. It triggers the degradation and inhibits the translation of mRNAs with complementary sequences. *See also* **RNA interference**.

microstimulation Delivery of small currents through an extracellular electrode with the goal of activating a limited number of nearby neurons.

microtubule A major cytoskeletal element composed of hollow cylinders of 13 parallel protofilaments made of α- and β-tubulin. (Figure 2–5)

midbrain The rostral-most part of the brainstem, it includes the tectum (superior and inferior colliculus in mammals) dorsally and the tegmentum ventrally. It is also the middle part of the three divisions of the embryonic brain caudal to the forebrain and rostral to the hindbrain. It is also called the mesencephalon. (Figure 1–8; Figure 7–3)

middle temporal visual area (MT) A high-order visual cortical area in the dorsal stream specialized for analyzing motion signals. (Figure 4–48)

midget ganglion cell A retinal ganglion cell with a small receptive field used for high-acuity vision and green–red color vision. (Figure 4–33)

miniature end-plate potential (mEPP) Small depolarization of the muscle cell in response to spontaneous neurotransmitter release from the motor neuron. (Figure 3–2)

mitogen-activated protein (MAP) kinase cascade A kinase cascade that acts downstream of the small GTPase Ras and other signaling molecules. The cascade consists of three serine/threonine kinases represented by Raf, Mek, and Erk (also called mitogen-activated kinase). Ras-GTP activates Raf, which phosphorylates and activates Mek, which in turn phosphorylates and activates Erk. (Figure 3–39)

mitotic recombination Exchange of a portion of homologous maternal and paternal chromosomes during mitotic cell division. It can create daughter cells homozygous for alleles on portions of the paternal or maternal chromosomes. (Figure 13–10; Figure 13–23)

mitral cell A second-order neuron in the vertebrate olfactory bulb, it receives input from ORNs and sends output to the olfactory cortex. It differs from a tufted cell, also a second-order neuron in the vertebrate olfactory bulb, in its cell body location in the olfactory bulb and axon termination pattern in the olfactory cortex. (Figure 6–17)

modulatory neurons Neurons that release modulatory neurotransmitters. They can act on both excitatory and inhibitory neurons to up- or down-regulate their excitability or synaptic transmission.

modulatory neurotransmitter (neuromodulator)
A neurotransmitter that can bidirectionally change the membrane potential, excitability, or neurotransmitter release of its postsynaptic target neurons.

molecular clock A technique that utilizes the rates of sequence changes, calibrated against fossil records, to estimate the times at which two species diverged.

monoamine neurotransmitter A neurotransmitter, such as serotonin, dopamine, norepinephrine, and histamine, derived from an aromatic amino acid.

monoamine oxidase An enzyme that oxidizes dopamine, norepinephrine, and serotonin, leading to their degradation. (Figure 11–24)

monozygotic (identical) twins Twins produced from the same fertilized egg or zygote, they share 100% of their genomes.

morphine The active ingredient of opiates.

morphogen A diffusible signaling protein that can cause cells located at different distances from the source to adopt different fates.

Morris water maze A navigation task in which rats and mice learn to locate a hidden platform in a pool of milky water using distant cues in the room.

Mosaic analysis A method for analyzing the cell types in which the function of a gene is important by creating genetic mosaic animals containing both wild-type and mutant cells that are usually differentially marked.

mossy fiber An axon that has elaborate terminal arborizations. The two most prominent types are found in the cerebellum and hippocampus. The cerebellar mossy fiber is an axon that terminates in the granular layer of the cerebellar cortex, where it synapses onto granule cells. It originates from a neuron residing in the pons, medulla, or spinal cord. (Figure 8–20). The hippocampal mossy fiber is an axon of a dentate gyrus granule cell, which synapses onto CA3 pyramidal neuron dendrites. (Figure 10–6)

motor homunculus A map in the primary motor cortex that corresponds to movement of specific body parts. Nearby areas in the motor cortical areas represent movement control of nearby body parts. (Figure 1–25)

motor neuron A type of neuron that extends dendrites within the CNS (the spinal cord or brainstem in vertebrates) and projects its axon out of the CNS to innervate a muscle. (Figure 1–15; Figure 8–9)

motor pool A cluster of motor neurons that innervate the same muscle. (Figure 8–6)

motor protein A protein that converts energy from ATP hydrolysis to movement along the cytoskeletal polymers.

motor system The collected parts of the nervous system that control the contraction of skeletal muscles and thereby enable movement and maintain body posture.

motor unit A motor neuron and the set of muscle fibers it innervates. (Figure 8–6)

motor unit size The number of muscle fibers a motor neuron innervates.

mount A posture that male rodents assume when sexually aroused. It facilitates sexual intercourse.

MrgprA3 A G-protein-coupled receptor that is activated by the pruritogen chloroquine.

α-MSH (α-melanocyte-stimulating hormone) A neuropeptide released by POMC neurons in the arcuate nucleus that reduces food intake.

mTOR (mammalian target of rapamycin) A key protein in intracellular signaling pathways that plays an important role in regulating protein translation. (Figure 11–45)

Müller glia A glial cell in the retina where the conversion of all-*trans* retinal to 11-*cis* retinal occurs to assist the recovery process in cones.

multi-electrode array A device used to record the spiking activities of many individual neurons. The electrodes can be arrayed either horizontally or vertically. (Figure 4–47; Figure 13–33)

multiple sclerosis (MS) A common adult-onset CNS demyelinating disease, it is characterized by inflammatory plaques in the white matter caused by immune cell attack of myelin. The cause is still mostly unknown.

multipolar Having more than two processes leaving the cell body.

muscarinic AChR *See* **acetylcholine receptor (AChR)**.

muscimol A mushroom-derived toxin that is a potent activator of the GABA$_A$ receptor.

muscle fiber A muscle cell.

muscle spindle A special apparatus in muscle cells that sense muscles stretches. It has embedded endings of peripheral branches of the proprioceptive somatosensory neurons. (Figure 1–19; Figure 6–63)

mushroom body A second-order olfactory center for odor-mediated learning and memory in the insect brain, it and the lateral horn are the two major output sites for projection neuron axons. (Figure 6–27; Figure 10–29)

MuSK A muscle-specific receptor tyrosine kinase, it acts together with LRP4 as an agrin receptor to promote acetylcholine receptor clustering. (Figure 7–24)

mutation A change in DNA, including the insertion, deletion, or alteration of one or more base pairs.

myelin sheath Cytoplasmic extensions of oligodendrocytes and Schwann cells, they wrap around the axons with multi-layered glial plasma membrane to increase resistance and decrease capacitance for action potential propagation. *See also* **axon myelination**. (Figure 2–26; Figure 2–27)

myofibril A thread-like longitudinal structure in muscle cells composed of repeating sarcomeres and responsible for muscle contraction. (Figure 8–3)

myosin An F-actin-based motor protein. (Figure 8–3)

Na$^+$–K$^+$ ATPase A pump that uses energy derived from ATP hydrolysis to pump Na$^+$ out of a cell and K$^+$ into a cell against their respective electrochemical gradients. It helps maintain the Na$^+$ and K$^+$ concentration differences across the membrane. (Figure 2–12)

β2 nAChR A subunit of nicotinic acetylcholine receptors that, among other functions, is essential for cholinergic retinal wave propagation. *See also* **acetylcholine receptor (AChR)**.

narcolepsy A disorder characterized by difficulty staying awake during the day, especially following moments of happiness or excitement. It is caused either by a deficiency of the neuropeptide hypocretin or by dysfunction of hypocretin-expressing neurons.

nasal (in retinal map) In the direction of the nose.

natural selection A key mechanism of evolution, it is the process by which genetic variations that confer individuals a better chance of reproductive success become more common in a population over time. (Figure 12–1)

nature In the context of *nature versus nurture*, 'nature' is the contribution of genetic inheritance to brain function and behavior.

nematocin *C. elegans* ortholog of vertebrate oxytocin and vasopressin. (Figure 9–45)

neocortex The largest part of the mammalian cerebral cortex, it typically contains six layers and is evolutionarily the newest part of the cerebral cortex.

Nernst equation An equation that relates the equilibrium potential of an ion to the concentrations of the ion on the two sides of a membrane.

nerve A discrete bundle of axons in the peripheral nervous system.

nerve growth factor (NGF) A prototypical neurotrophin, it is a target-derived secreted protein that supports the survival and axon growth of sensory and sympathetic neurons. (Figure 7–31; Figure 7–32)

nerve impulse Historical name for transient changes in membrane potential that propagate along axons, it is the same as action potential.

netrin/Unc6 Homologous secreted proteins originally identified by biochemical purification (netrins in vertebrates) and genetic screen (Unc6 in *C. elegans*). They are widely used axon guidance cues first discovered in the context of midline guidance in both *C. elegans* and vertebrates. (Figure 5–10)

neural circuit An ensemble of interconnected neurons that act together to perform specific functions.

neural crest cells A special group of cells at the junction of the dorsal neural tube and the overlying epidermal cells. They migrate away from the neural tube to produce diverse cell types, including cells of the peripheral nervous system. (Figure 7–2)

neural plasticity Changes of the nervous system in response to experience and learning.

neural plate The layer of ectodermal cells overlaying the notochord that invaginates and gives rise to the neural tube during neurulation. (Figure 7–2)

neural progenitor A dividing cell that gives rise to neurons and glia. In vertebrates, it is usually located near the ventricle in the developing vertebrate CNS. (Figure 7–4)

neural prosthetic device A device that can substitute a sensory or motor function that has been disrupted due to an injury or a disease. For example, population activity of neurons in the motor cortex can be used to control an external device such as a robotic arm or a computer cursor to help patients who suffer from paralysis or motor neuron diseases. (Figure 8–29)

neural tube A hollow tube surrounded by layers of neuroectodermal cells, it is the embryonic precursor to the vertebrate CNS. (Figure 7–2)

neuraxis Axis of the CNS. The rostral–caudal neuraxis follows the curvature of the embryonic neural tube; the dorsal–ventral neuraxis is perpendicular to the rostral–caudal neuraxis. (Figure 1–8)

neurexin A protein on the presynaptic membrane that mediates synaptic adhesion. A major binding partner is neuroligin. (Figure 7–25)

neuroblast A neuronal progenitor.

neurodegenerative disorders Disorders characterized by progressive neuronal dysfunction, including loss of synapses, atrophy of dendrites and axons, and death of neurons.

neuroendocrine system The collected parts of the nervous system that control the secretion of hormones to regulate an animal's physiology and behavior in response to sensory stimuli and brain states.

neuroethology A branch of science that emphasizes the study of animal behavior in the natural environment.

neurofascin An immunoglobulin superfamily molecule that, among other functions, serves as a cue in Purkinje cells to instruct targeting of basket cell axons and presynaptic terminals to the correct subcellular domain.

neurofibrillary tangle An intracellular fibril consisting of an abnormal accumulation of hyper-phosphorylated tau, a microtubule-binding protein. (Figure 11–2)

neurofilament An intermediate filament (a cytoskeletal polymer with a diameter between F-actin and microtubules) in vertebrate neurons. It is concentrated in axons and provides stability to axons.

neurogenic inflammation Inflammation triggered by release of neuropeptides such as substance P and calcitonin gene-related peptide from the peripheral terminals of sensory neurons.

neuroligin A protein on the postsynaptic membrane that mediates synaptic adhesion. A major binding partner is neurexin. (Figure 7–25)

neuromuscular junction The synapse between a motor neuron's presynaptic terminals and a skeletal muscle cell. (Figure 3–1; Figure 7–28)

neuron (nerve cell) An electrically excitable cell that receives, integrates, propagates, and transmits information as the working unit of the nervous system.

neuron doctrine The principle that individual neurons are the working units of the nervous system.

neuronal polarity The distinction between axons and dendrites.

neuronal process Cytoplasmic extension of a neuron.

neuropeptide A polypeptide a few to a few dozen amino acids in length that acts as a neurotransmitter.

neuropil A structure composed mostly of synapses.

neuropilin-1 (Nrp1) A co-receptor for semaphorin.

neurotransmitter reuptake The process by which neurotransmitters in the synaptic cleft are transported either into nearby glial cells or back into the presynaptic cytosol and into synaptic vesicles. (Figure 3–12)

neurotransmitters Molecules that are stored in synaptic vesicles (or dense-core vesicles in the case of neuropeptides) in the presynaptic terminals, are released into the synaptic cleft triggered by presynaptic depolarization, and activate ionotropic or metabotropic receptors on a postsynaptic target cell. (Figure 3–16; Table 3–2)

neurotrophic hypothesis The idea that the survival of developing neurons depends on neurotrophins produced by the neurons' postsynaptic targets.

neurotrophin-3 (NT3) *See* **neurotrophins**.

neurotrophin-4 (NT4) *See* **neurotrophins**.

neurotrophins A family of secreted signaling proteins that regulate the survival, morphology, and physiology of target neurons through binding to specific receptors on those neurons. Mammalian neurotrophins include nerve growth factor (NGF), brain-derived neurotrophic factor (BDNF), neurotrophin-3 (NT3), and neurotrophin-4 (NT4). (Figure 3–39; Figure 7–32). *See also* **Trk receptors, p75NTR**.

neurulation The developmental process in vertebrate embryos leading to formation of the neural tube, which gives rise to the nervous system. (Figure 7–2)

nicotinic AChR *See* **acetylcholine receptor (AChR)**.

Nissl stain A stain that labels RNA and thus highlights the rough endoplasmic reticulum in cytoplasm. Nissl stains are basic (that is, proton-accepting, positively charged) dyes such as cresyl violet that bind to RNA molecules (which are negatively charged). (Figure 13–18)

NMDA receptor A glutamate-gated ion channel that conducts Na^+, K^+, and Ca^{2+} and can be activated by the drug NMDA (N-methyl-D-aspartate). Its opening requires both binding of glutamate and postsynaptic depolarization. It is a heterotetramer of two GluN1 subunits encoded by a single gene, and two GluN2 subunits, of which there are four variants (GluN2A, GluN2B, GluN2C, GluN2D) each encoded by separate genes. (Figure 3–24; Figure 3–25)

nociception The sense of pain.

nociceptive neuron A somatosensory neuron that senses pain.

nodes of Ranvier Periodic gaps in the myelination of an axon, usually 200 μm to 2 mm apart, where the axon surface is exposed to the extracellular ionic environment. They contain high concentrations of voltage-gated Na^+ and K^+ channels that regenerate action potentials. (Figure 2–26)

nonhomologous end joining An endogenous DNA repair system, it re-joins the two ends of a DNA molecule with a double-strand break. It often creates a small deletion or duplication at the breakpoint as a result of the repair process. (Figure 13–8)

non-spiking neuron A neuron that uses graded potentials rather than action potentials to transmit information.

non-synonymous substitutions Nucleotide changes in DNA that result in corresponding amino acid changes in the protein encoded by that DNA.

norepinephrine A monoamine neuromodulator derived from dopamine. (Figure 11–20; Table 3–2)

northern blot A method for determining the amount of a specific RNA in an RNA mixture. RNAs are separated by gel electrophoresis and are then transferred to a membrane; labeled nucleic acid probes are then hybridized to the membrane to visualize specific RNA molecules that hybridize to the probe. It can be used to determine RNA expression patterns. (Figure 6–9)

Notch A transmembrane receptor widely involved in diversifying cell fate during development. Binding of a ligand to Notch triggers proteolytic cleavage of Notch in the transmembrane domain to release the Notch intracellular domain, which can then enter the nucleus to regulate gene expression. (Figure 7–7)

notochord A midline mesodermal structure in vertebrate embryos ventral to the spinal cord that produces secreted cues for patterning the spinal cord. (Figure 7–2; Figure 7–8)

NREM sleep Non-rapid eye movement (NREM) sleep, or sleep stages other than REM sleep. (Figure 8–51)

nucleus accumbens The major part of the ventral striatum, it receives input preferentially from the prefrontal cortex, thalamus, hippocampus, and amygdala. (Figure 11–31)

nucleus laminaris (NL) A brainstem nucleus in the barn owl that analyzes interaural time differences. It is analogous to the medial superior olivary nucleus in mammals.

nucleus of the solitary tract (NTS) A nucleus in the brainstem that receives input from the taste system as well as sensory information from internal organs. (Figure 6–35)

null direction The direction of stimulus motion that elicits the lowest firing rate of a direction-sensitive visual system neuron.

Numb A *Drosophila* protein that is segregated asymmetrically to daughter cells during sensory organ precursor and neuroblast divisions. It is essential for conferring different fates to the two daughter cells of an asymmetric division. (Figure 7–6)

nurture In the context of *nature versus nurture*, 'nurture' is the contribution of environmental factors to brain function and behavior.

occipital lobe One of the four cerebral cortex lobes; it is located at the back of the brain. (Figure 1–23)

octopamine A neurotransmitter in some invertebrate nervous systems that is chemically similar to norepinephrine in vertebrates.

ocular dominance Preference for receiving and/or representing visual input from one eye over the other eye. In the primary visual cortex of some mammals, such as cats and monkeys, cells in the same vertical columns share the same ocular dominance, thus producing ocular dominance columns. (Figure 4–43)

ocular dominance column *See* **ocular dominance**.

odds ratio In genetics, a measure of the effect of a genetic variant on the likelihood of having a particular trait, such as a disease. It is calculated by dividing the probability of having the trait among people with the genetic variant by the probability of having the trait among people without the genetic variant.

odorant A molecule that elicits olfactory perception, it is usually volatile.

odorant receptor A receptor on the surface of olfactory cilia that binds odorants. (Figure 6–9)

OFF bipolar A bipolar cell that expresses ionotropic glutamate receptors and is depolarized by glutamate release from photoreceptors. Its membrane potential changes follow the sign of photoreceptors such that it is hyperpolarized by light. (Figure 4–25)

Ohm's law An equation that relates current (I) to voltage (V) and resistance (R); $I = V/R$.

olfactory bulb The first olfactory processing center in the vertebrate brain. (Figure 6–3; Figure 6–17)

olfactory cilium A dendritic branch of an olfactory receptor neuron enriched for odorant receptors. (Figure 6–3)

olfactory cortex Brain regions that receive direct input from mitral/tufted cells, including the anterior olfactory nucleus, piriform cortex, olfactory tubercle, cortical amygdala, and entorhinal cortex. (Figure 6–19)

olfactory epithelium The epithelial layer in the nose that houses the olfactory receptor neurons. (Figure 6–3)

olfactory processing channel A discrete information processing unit in the olfactory system consisting of olfactory receptor neurons (ORNs) that express a given odorant receptor, the glomerular target of those ORNs, and second-order neurons that send dendrites to the same glomerulus.

olfactory receptor neuron (ORN) The primary sensory neuron in the olfactory system, it converts odorant binding to odorant receptor proteins into an electrical signal that is relayed to the brain via its axon. (Figure 6–3)

oligodendrocyte A glial cell in the CNS that wraps axons with its cytoplasmic extension to form myelin sheath. (Figure 1–9)

ommatidium A repeating unit of the arthropod compound eye. In *Drosophila*, each ommatidium contains eight photoreceptors. (Figure 5–35)

ON bipolar A bipolar cell that expresses metabotropic glutamate receptors and is inhibited by glutamate release from photoreceptors. Its membrane potential changes are opposite in sign to those of the photoreceptors such that it is depolarized by light. (Figure 4–25)

open probability The proportion of time that an individual ion channel is open and able to conduct current.

operant conditioning A form of learning in which a subject associates performance of a specific action (e.g. pressing a lever) with a particular outcome, such as delivery of a reinforcer (e.g. food) or a punishment (e.g. an electrical shock).

opioid receptors A subfamily of G-protein-coupled receptors that serve as receptors for opioids, including morphine and endogenous opioid neuropeptides. They are widely distributed across the nervous system.

opioids Molecules that have effects similar to opiates such as morphine. They include opiates from opium poppy and endogenous neuropeptides such as encephalin, endorphin, and dynorphin.

opsin A member of a family of G-protein-coupled receptors expressed in photoreceptors of multicellular organisms, it is associated with retinal and converts photon absorption to the activation of a trimeric GTP-binding protein. In microbes, it is a member of light-induced channels or pumps, which are not G-protein-coupled receptors.

optic chiasm The midline structure where a fraction of retinal ganglion cell axons cross to the side of the brain contralateral to the eye of origin. (Figure 4–35)

optic lobe The part of the insect brain that consists of the retina, lamina, medulla, and lobula complex and that is used to analyze visual signals. (Figure 5–35)

optic nerve The bundle of retinal ganglion cell axons, it sends visual information from the eye to the brain. (Figure 4–35)

optic tract The bundles of retinal ganglion cell axons distal to the optic chiasm. (Figure 4–35)

optical imaging An approach that uses changes of fluorescence or other optical properties as indicators of neuronal activity.

optogenetics The set of methods used to manipulate neuronal activity by using light to activate a genetically encoded effector, most commonly microbial opsins (e.g. channelrhodopsin-2, archaerhodopsin, halorhodopsin). (Figure 13–45)

organ of Corti An organ in the cochlea that consists of hair cells, the surrounding support cells, and the basilar membrane. (Figure 6–45)

organization–activation model A central principle in endocrinology, it proposes that sex hormones have two different types of effects: 'organizational' effects during development, which configure the brain in a sex-typical manner, and 'activational' effects in adults, which stimulate male- or female-typical sexual behaviors. (Figure 9–25)

otolith organ A sensory organ in the vestibular system that senses linear acceleration and stationary head tilts. (Figure 6–59)

outer radial glia (oRG) A type of radial glia whose cell bodies are located in the subventricular zone. They serve along with ventricular zone radial glia as neural progenitors. They are greatly expanded in number in human neocortex compared with mouse neocortex and likely contribute to increased neuronal production in mammals with large neocortices. *See also* **radial glia**. (Figure 12–37)

outer segment A cytoplasmic extension of a rod or a cone, it contains a highly specialized photon detection apparatus made of tightly stacked membrane disks enriched in opsins. (Figure 4–2)

outgroup A group of organisms that is closely related to but falls outside of a set of organisms of interest. It is used as a reference group in determining the phylogenetic relationships among a set of organisms.

ovariectomized female A female from which the ovaries have been removed.

oxytocin A hormone secreted by hypothalamic neurons in the posterior pituitary and a neuropeptide released by some CNS neurons. It regulates maternal and social behavior.

P pathway A visual processing pathway from the retina to the visual cortex that originates from retinal ganglion cells with small receptive fields and engages lateral geniculate nucleus cells in the parvocellular layers. It carries information about high-acuity and color vision. (Figure 4–48)

p75NTR A 75 kilodalton neurotrophin receptor that has a low affinity for all neurotrophins and is also a receptor for all proneurotrophins. (Figure 7–32)

pacemaker A cell that can produce rhythmic output in the absence of input.

Pacinian corpuscle An encapsulated structure closely associated with the ending of the rapidly adapting type II (RAII) LTMR. (Figure 6–64)

PALM *See* **super-resolution fluorescence microscopy**.

parabiosis The joining of the circulatory systems of two animals so that they have limited exchange of substances in systemic circulation.

parabrachial nucleus A brainstem nucleus that transmits ascending signals from the visceral sensory system and pain somatosensory system to the thalamus, amygdala, hypothalamus, and brainstem autonomic centers. (Figure 6–70; Figure 8–33)

paracrine Of or related to a form of signaling in which a recipient cell receives a signal produced by nearby cells.

parallel fiber The portion of the axon of a cerebellar granule cell that runs in parallel to the pial surface and crosses Purkinje cell dendrites at a right angle. (Figure 8–20)

parasympathetic nervous system A branch of the autonomic nervous system that facilitates energy conservation. Activation of the parasympathetic nervous system slows down the heart rate, decreases blood flow, constricts airways in the lung, and stimulates salivation and digestion. (Figure 8–31; Figure 8–32)

paraventricular hypothalamic nucleus (PVH) A hypothalamic nucleus involved in multiple physiological functions, including the release of oxytocin and vasopressin into the bloodstream through axonal projections in the posterior pituitary and the descending control of autonomic nervous system functions. (Figure 8–43)

parietal lobe One of the four cerebral cortex lobes, it is located behind the frontal lobe and above the occipital lobe. (Figure 1–23)

Parkinson's disease (PD) A common neurodegenerative disease caused by death of substantia nigra dopamine neurons. It primarily affects movement control, with symptoms that include shaking, rigidity, slowness, and difficulty walking. (Figure 11–16)

parthenogenesis A reproductive process in which embryos develop from unfertilized eggs and which therefore does not involve exchange of genetic materials.

passive electrical properties Membrane properties in the absence of voltage-dependent conductance. Two salient examples are: (1) a sharp change of electrical signal (e.g. a current pulse) becomes more spread temporally as the signal travels along a neuronal process because of membrane capacitance; (2) the magnitude of electrical signal becomes attenuated across a distance because of membrane conductance. They are also called cable properties. (Figure 2–16)

passive transport Movement of a solute across a membrane down its electrochemical gradient via a channel or a transporter. (Figure 2–8)

patch clamp recording An electrophysiological recording technique that utilizes a glass electrode (patch pipette) to form a high-resistance seal with the membrane. It has several variants, including cell-attached patch, excised patch, and whole-cell recording. (Figure 13–37)

patch pipette *See* **patch clamp recording.**

path-integration strategy A navigational strategy in which animals use the speed, duration, and direction of their own movement to calculate their current position with respect to their starting position.

Pax6 A member of the Pax family of transcription factors, it contains a homeobox and a paired box. It regulates the patterning of the cerebral cortex and spinal cord and is required for eye development in mammals. Its *Drosophila* homolog is Eyeless. *See also* **Eyeless.**

PDZ domain Acronym for a domain shared by PSD-95, Discs large (a *Drosophila* protein implicated in cell proliferation and associated with postsynaptic density), and ZO-1 (an epithelial tight junction protein). It is a protein–protein interaction domain that binds to a specific protein sequence motif that is present at the C-terminal end of many transmembrane receptors.

percept A specifically perceived object or the brain representation of the object.

perforant path Axons of neurons in the superficial layer of the entorhinal cortex that project to the hippocampus. (Figure 10-6)

peri-stimulus time histogram (PSTH) A graph that plots firing rates of neurons as a function of time after stimulus onset.

periaqueductal gray (PAG) A midbrain gray matter structure surrounding the cerebral aqueduct. It serves many functions including the descending control of pain and the execution of defensive behavior such as freezing. (Figure 6–70)

periglomerular cell A member of diverse types of interneurons that receive direct input from olfactory receptor neuron (ORN) axons or from apical dendrites of mitral cells and that send (mostly inhibitory) output to targets within the same glomerulus, or in nearby glomeruli. (Figure 6–17)

Period A fruit fly gene discovered based on mutations that speed up, slow down, or disrupt circadian rhythms. It encodes a protein that participates in the negative regulation of its own transcription, and its mammalian homologs serve a similar function. (Figure 8–45)

permeability The ability of a membrane to conduct specific ions, determined principally by the number of open channels capable of conducting those ions.

perturbation experiment An experiment in which key parameters in a biological system are altered, usually under the experimenter's control, in order to study the consequences.

pharmacodynamics The effects of a drug in the body, including the intended effects on target molecules and processes as well as the unintended side effects.

pharmacokinetics The effects of the body's biological processes on a drug, including the drug's absorption, distribution, metabolism, and excretion.

phase locking A property whereby the spikes of auditory neurons occur at a specific phase of each cycle of a sound wave. (Figure 6–51)

phasic Of a neuronal firing pattern, bursts of action potentials in response to specific stimuli.

pheromone A substance produced by an individual to elicit a specific reaction from other individuals of the same species.

phosphodiesterase (PDE) An enzyme that hydrolyzes cyclic AMP (cAMP) to AMP, or cGMP to GMP.

phospholipase C (PLC) A membrane-associated enzyme that is activated by G_q and cleaves inositol-phospholipids to produce inositol 1,4,5-triphosphate (IP_3) and diacylglycerol (DAG). (Figure 3–34)

photoreceptor A cell that converts light into electrical signals. (Figure 4–2; Figure 12–23)

phototaxis Movement toward or away from a light source.

phototransduction The biochemical reactions triggered by photon absorption. (Figure 4–10)

phrenology A discipline created by Franz Joseph Gall with the goal of mapping the functions of brain areas by studying the shape and size of bumps and ridges on the skull, which were thought to be correlated with an individual's talents and character traits. (Figure 1–22)

phylogenetic tree A branching diagram showing the relationships among different organisms. It is constructed based on the similarities and differences of different organisms' traits, such as nucleotide and protein sequences. (Figure 12–2)

picrotoxin A plant toxin that is a potent blocker of the $GABA_A$ receptor.

Piezo A mechanotransduction channel with over 30 transmembrane segments for each subunit. (Figure 6–66)

pigment cell A cell in the pigment epithelium layer of the retina adjacent to the outer segments of photoreceptors that reflects light and converts all-*trans* retinal back to 11-*cis* retinal to assist the recovery process in rods.

piriform cortex The largest olfactory cortical region, it is a three-layered cortex separated from more dorsally located neocortex by the rhinal sulcus. (Figure 6–19)

pituitary The endocrine center of the brain, it is located ventral to the hypothalamus. The posterior pituitary contains axon terminals of hypothalamic neurons that directly release hormones into the bloodstream. The anterior pituitary contains endocrine cells that release hormones into the bloodstream in response to prehormones originated from hypothalamic neurons and transmitted by specialized portal vessels. (Figure 8–35)

place cell A hippocampal cell that fires maximally when the animal is at a particular place in an environment.

place field The physical location in an environment that elicits maximal firing of a particular place cell.

placebo effect In the context of pain perception, the phenomenon whereby the perception of pain can be reduced in some patients by the mistaken belief that they have received a treatment thought to reduce pain.

plasma membrane dopamine transporter (DAT) *See* **plasma membrane monoamine transporters.**

plasma membrane monoamine transporters (PMATs) A family of proteins on the presynaptic membrane that transport serotonin [serotonin transporter (SERT)], dopamine [dopamine transporter (DAT)], or norepinephrine [norepinephrine transporter (NET)] from synaptic cleft into the presynaptic cytosol. (Figure 11–24). *See also* **plasma membrane neurotransmitter transporter.**

plasma membrane neurotransmitter transporter A transmembrane protein on the presynaptic or glial plasma membrane that transports neurotransmitters from the extracellular space into the cell using energy from the co-transport of Na^+ down its electrochemical gradient. (Figure 3–12)

plexin A member of a class of proteins that serve as receptors for the axon guidance cues, semaphorins.

pluripotent cell A cell that has the potential to develop into all cell types of an embryo.

PNS (peripheral nervous system) Neural tissue and cells outside the central nervous system (CNS), including the nerves that connect the CNS with the body and internal organs as well as isolated ganglia outside of the CNS.

Poisson distribution A discrete probability distribution in which the frequency (*f*) that *k* events occur can be determined by a single parameter λ (the mean frequency of occurrence, which equals the product of *n* and *p* in the binomial distribution). $f(k; \lambda) = (\lambda^k / k!) e^{-\lambda}$. It is an approximation of the binomial distribution when *n* is large and *p* is small. (Box 3–1)

polyadenylation The process by which a long sequence of adenosine nucleotides is added to the 3′ end of the mRNA. (Figure 2–2)

polymerase chain reaction (PCR) A highly sensitive DNA amplification technique that uses a pair of oligonucleotide primers to amplify the DNA segment between the sequences corresponding to the primers through cycles of DNA replication.

polymodal neuron In the somatosensory system, a neuron that responds to stimuli of more than one sensory modality.

polymorphism In the context of genetics, a DNA sequence variation among individuals of the same species.

POMC neuron A neuron in the arcuate nucleus that expresses pro-opiomelanocortin (POMC), a precursor protein for multiple peptides including the anorexigenic peptide α-melanocyte-stimulating hormone (α-MSH). (Figure 8–42)

pons The middle part of the brainstem caudal to the midbrain and rostral to the medulla. (Figure 1–8)

population vector (in movement control) The sum of the preferred direction vectors of a population of neurons weighted by the firing rate of each neuron. The preferred direction of a neuron is a vector in a three-dimensional space pointing in the direction towards which movement elicits the highest firing rate of the neuron. (Figure 8–27)

positional cloning A molecular genetic technique that uses molecular and genetic markers on specific chromosomes to identify a gene that causes a particular phenotype or disease.

positive selection The process by which an allele that is beneficial to an organism becomes more prevalent in a population.

positron emission tomography (PET) A non-invasive three-dimensional imaging technique for measuring the distribution of positron-emitting probes introduced into the body.

posterior pituitary *See* **pituitary**.

postganglionic neuron A neuron whose cell body is located in a sympathetic or parasympathetic ganglion in the peripheral nervous system and whose axon innervates effectors such as smooth muscle, cardiac muscle, and glands. (Figure 8–32)

postsynaptic specialization A structure on a postsynaptic target cell that is adjacent to a presynaptic terminal, it is enriched for neurotransmitter receptors as well as signaling and scaffolding molecules. It is also called postsynaptic density because it is electron dense in electron microscopic images.

Potocki-Lupski syndrome A neurodevelopmental disorder characterized by mild intellectual disability and autistic symptoms. It is caused by duplication of a chromosome segment (that includes *Rai1* and many other genes) reciprocal to the common deletion that causes Smith-Magenis syndrome.

power stroke The process by which myosin and actin filaments move relative to each other. It involves the conversion of chemical energy from ATP hydrolysis into mechanical force by the myosin motor. (Figure 8–4)

precedence effect The ability of a first-arriving sound to suppress the perception of later-arriving sounds.

preferred direction The direction of stimulus motion that elicits the highest firing rate of a direction-sensitive visual system neuron.

prefrontal cortex A neocortical area anterior to the motor cortex, it is an executive control center that integrates multisensory information, mediates working memory, and performs complex executive functions such as goal selection and decision making.

preganglionic neuron A neuron whose cell body is located within the CNS and whose axon synapses onto the postganglionic neurons in the sympathetic or parasympathetic ganglion. (Figure 8–32)

premotor cortex Areas of motor cortex anterior to the primary motor cortex. Its neurons send axons primarily to primary motor cortex.

premotor neuron A spinal cord or brainstem neuron that is presynaptic to motor neurons and thereby participates directly in controlling the firing of the motor neurons. (Figure 8–10)

presenilin One of two members of a family (consisting of presenilin-1 and presenilin-2) of multi-pass transmembrane proteins that function as subunits of the γ-secretase complex. They were originally identified based on mutations that cause familial Alzheimer's disease. (Figure 11–5)

prestin A protein that mediates electromotility in the cochlear outer hair cells.

presynaptic facilitation The process by which neurotransmitter release from cell *A* onto the presynaptic terminal of cell *B* leads to an increase in neurotransmitter release from cell *B*.

presynaptic inhibition The process by which neurotransmitter release from cell *A* onto the presynaptic terminal of cell *B* leads to a decrease in neurotransmitter release from cell *B*.

presynaptic terminal A structure at the end (or along the trunk) of an axon that is specialized for releasing neurotransmitters onto target cells. (Figure 1–9)

pretectum A brainstem structure that receives retinal ganglion cell axon input and regulates pupil, lens, and eye movement reflexes. (Figure 4–35)

primary antibody An antibody that selectively recognizes a specific protein.

primary auditory cortex (A1) The part of the cerebral cortex that first receives auditory sensory information. (Figure 6–54)

primary cilium A short, single, non-motile cilium that projects from the surface of many animal cell types and is used often as a signaling center.

primary motor cortex (M1) The part of the cerebral cortex that sends descending axons directly to motor neurons to control muscle contraction. (Figure 1–25)

primary somatosensory cortex The part of the cerebral cortex that first receives somatosensory information from the body. (Figure 1–25)

primary visual cortex (V1) The visual cortical area that receives direct input from the lateral geniculate nucleus. (Figure 4–38; Figure 4–45)

principal component analysis (PCA) A statistical method used to reduce the dimensionality of a dataset. The axes of the reduced dataset are called principal components and their orientations in the non-reduced space are selected to maximize the spread of the data along each principal component—data are most spread along the axis of the first principal component, followed by the axis of the second principal component, and so forth.

prion diseases   Diseases characterized by the propagation across the brain of prion protein (PrP) that adopts a specific conformation (PrP^{Sc}), which aggregates and causes massive neurodegeneration and neuronal death. PrP^{Sc} spreads by interacting with PrP in its innocuous cellular conformation (PrP^C) to induce a conformational change and convert it into PrP^{Sc}. Prion diseases include scrapie in sheep and goats, mad cow disease in cows, kuru (a human disease that occurred in certain tribes that observed ritual cannibalism), and Creutzfeldt-Jakob disease (CJD, a human disease in which mutations in the *Prp* gene make PrP^C more prone to adopt the PrP^{Sc} conformation spontaneously). (Figure 11–13)

prion hypothesis   The idea that the infectious agent in scrapie is solely proteinaceous in nature.

programmed cell death (apoptosis)   A form of cell death in which a cell kills itself by initiating a cell-death program.

projection neuron   A neuron with an axon that projects outside the CNS region that houses the neuron's cell body. In the insect olfactory system, it is a second-order neuron (PN) that receives input from olfactory receptor neuron (ORN) axons and sends output to higher olfactory centers, analogous to a vertebrate mitral/tufted cell. (Figure 6–27)

prokaryote   A single-cell organism without a nucleus. Prokaryotes are members of one of two domains of life: eubacteria or archaea.

proprioception   The sense of body position and movement.

proprioceptive neurons   Somatosensory neurons that have peripheral endings embedded in the muscle spindles, tendons, and joints for sensation of muscle stretch and tension. (Figure 6–63)

prostaglandin   A lipid released during inflammation, it binds to specific G-protein-coupled receptors on the peripheral terminals of nociceptive neurons. (Figure 6–71)

protein   A specific sequence of amino acids linked by peptide bonds to form a chain.

protein kinase A (PKA)   *See* **cAMP-dependent protein kinase**.

protein kinase C (PKC)   A serine/threonine kinase with diverse substrates that is activated by binding of both diacylglycerol and Ca^{2+}. (Figure 3–34)

protein phosphatase   An enzyme that removes phosphates from phosphorylated proteins, thus counteracting the actions of kinases.

proteinopathy   A disease caused by altered protein conformations, interactions, and homeostasis.

protocadherin   A member of a class of cell adhesion molecules in vertebrates whose structures and biochemical properties resemble those of cadherins.

protostomes   Animals in which the mouth appears before the anus during development. They include most invertebrate phyla. *See also* **deuterostomes**. (Figure 12–2)

pruriception   The sense of itch.

pruritogen   A chemical that causes the sensation of itch.

PSD-95 (postsynaptic <u>d</u>ensity protein of <u>95</u> kilodalton) A postsynaptic scaffolding protein highly enriched at the glutamatergic synapse. (Figure 3–27; Figure 7–25)

pseudogene   A gene that has been rendered nonfunctional by stop codons in the coding sequences or by other disrupting mutations. Such disrupting mutation(s) is prevalent in a given species.

psychometric function   The quantitative relationship between a parameter of a physical stimulus and the response or perception of a subject.

psychophysical study   An experimental approach that characterizes the relationship between physical stimuli and the sensations or behaviors they elicit.

psychosis   The mental state characterized by hallucinations and/or delusions.

psychostimulant   A drug that transiently produces euphoria and suppresses fatigue.

pump   A transporter that uses external energy, such as ATP hydrolysis or light, to actively move a solute across a membrane against its electrochemical gradient. (Figure 2–10)

Purkinje cell   GABAergic neuron of the cerebellar cortex with highly branched planar dendritic trees. It receives excitatory input from parallel fibers (axons of cerebellar granule cells) and climbing fibers from inferior olive neurons, and sends output to the deep cerebellar nuclei. (Figure 1–11; Figure 8–20)

pyramidal neuron   A type of glutamatergic neuron that has a pyramid-shaped cell body with an apical dendrite and several basal dendrites that further branch. It is abundant in mammalian cerebral cortex and hippocampus. (Figure 1–15)

quantal hypothesis of neurotransmitter release   The idea that neurotransmitters are released in discrete packages of relatively uniform size.

quantal yield   The number of synaptic vesicle exocytosis events in response to a single action potential.

R-C circuit   A circuit that contains both resistors and capacitors. (Figure 2–14)

RA (<u>r</u>obust nucleus of the <u>a</u>rcopallium) A dorsal forebrain nucleus in the songbird essential for song production. It functions downstream of the HVC. (Figure 9–21)

Rab   A member of a family of small monomeric GTPases involved in intracellular vesicle trafficking.

rabies virus   A neurotropic RNA virus that spreads within the nervous system of its host naturally by crossing synapses. It has been modified for the use of retrograde trans-synaptic tracing. (Figure 13–30)

radial glia   Progenitor cell in the ventricular zone that extends two radial processes—one to the ventricle and the other to the pial surface of the developing cortex. These radial processes serve as substrates for neuronal migration. (Figure 7–4)

random mutagenesis   *See* **forward genetic screen**.

random X-inactivation   A process in which one of the two X chromosomes in female mammals is randomly inactivated in each cell during early development.

raphe nuclei   Brainstem nuclei enriched for serotonin neurons that project widely across the brain. (Figure 8–54)

Ras   A member of a family of small monomeric GTPases involved in signaling pathways necessary for cell growth and differentiation.

readily releasable pool   A small subset of synaptic vesicles that are docked at the active zone, primed by an ATP-dependent process to achieve a high-energy configuration that includes pre-assembled SNARE complexes.

receptive field In the visual system, the area of the visual field that influences the activity of a given neuron. In the somatosensory system, the area of the body where stimuli can influence the firing of a neuron.

receptor A protein that binds and responds to a specific signaling molecule.

receptor potential A type of graded potential induced at the peripheral endings of sensory neurons by sensory stimuli.

receptor tyrosine kinase (RTK) A transmembrane protein with an N-terminal extracellular ligand-binding domain and a C-terminal intracellular tyrosine kinase domain. Upon ligand binding, receptor tyrosine kinases add phosphates to tyrosine residues of target proteins.

recording electrode An electrode used to measure membrane potential changes.

recovery (photoreceptor) The process by which light-activated photoreceptor cells return to the dark state. (Figure 4–11)

recurrent (cross) inhibition A circuit motif in which two parallel excitatory pathways mutually inhibit each other via inhibitory interneurons. (Figure 1–21)

refractory period A time window after an action potential during which another action potential cannot be initiated. (Figure 2–25)

regeneration (axon) Re-extension of axon after injury, including formation of synaptic connections with their original partners.

regenerative Having the property of propagating without attenuation in amplitude, it applies to action potentials. (Figure 2–25)

regulator of G protein signaling (RGS) A protein that acts as a GTPase activating protein for a trimeric GTP-binding protein.

release probability The probability that an active zone will release one or more synaptic vesicles following an action potential.

releasers The essential features of a stimulus that activate a fixed action pattern.

REM sleep A stage of sleep that is characterized by rapid eye movement. (Figure 8–51)

repellent A molecular cue that guides axons away from its source. (Figure 5–9)

reserpine A first-generation antipsychotic drug, it is an inhibitor of monoamine oxidase.

reserve pool A large subset of vesicles in the axon terminal available to replenish the readily releasable pool.

resistance (R) The degree to which an object or substance opposes the passage of electrical current, it is the inverse of conductance: $R = 1/g$.

resistor An electrical element through which passage of current is limited. Current flow through a resistor produces a voltage difference across its two terminals. (Figure 2–13)

responder transgene In binary expression, it is the transgene containing the coding sequence for the protein or RNA of interest, along with binding or recombinase sites for the transcription factor or recombinase, respectively, encoded by the driver transgene. (Figure 13–13)

resting potential The membrane potential of a neuron at rest (i.e. in the absence of action potentials or synaptic input), which is typically between –50 and –80 millivolts relative to the extracellular fluid. (Figure 2–11)

reticular theory The idea that the processes of nerve cells fuse and form a giant net that constitutes the working unit of the nervous system. It has been mostly disproven (with the exception of electrical synapses, that allow limited exchange of ions and small molecules between partner neurons).

retina A layered structure at the back of the vertebrate eye made of five major neuronal types (photoreceptors, horizontal cells, bipolar cells, amacrine cells, and retinal ganglion cells) and support cells. Collectively, these cells convert light into electrical signals, extract biologically relevant signals from the outputs of photoreceptors, and transmit these signals to the brain. (Figure 4–2)

retinal A chromophore covalently linked with an opsin, it changes its configuration after photon absorption. (Figure 4–6; Figure 12–20)

retinal ganglion cell (RGC) The output cell of the retina that transmits information from the eye to the brain. (Figure 4–2; Figure 4–28)

retinal wave The spread of spontaneous excitation of retinal neurons, including retinal ganglion cells and amacrine cells, across the developing retina. (Figure 5–21)

retinotopy The topographical arrangement of cells in the visual pathway according to the position of the retinal ganglion cells that transmit signals to them.

retrieval (of memory) The recall of a memory.

retrograde From the axon terminal to the cell body.

retrograde flow The flow of F-actin from the leading edge of the growth cone to its center powered by the myosin motor. It contributes to dynamic changes of growth cone shape. (Figure 5–15)

retrograde tracer A molecule used to trace axonal connections. It is taken up primarily by axon terminals and transported back to the cell bodies. (Figure 13–27)

retrograde trans-synaptic tracing *See* **trans-synaptic tracing**.

Rett syndrome A neurodevelopmental disorder in girls caused by disruption of an X-linked gene encoding methyl-CpG binding protein 2 (MeCP2). Patients usually develop normally for the first 6–18 months. Their development then slows, arrests, and regresses, with severe deficits that include social withdrawal, loss of language, and motor symptoms. *See also* **MeCP2**.

reversal potential (E_{rev}) The membrane potential at which the current flow through an ion channel changes direction.

reverse genetics The strategy or process of disrupting a pre-designated gene to identify its loss-of-function phenotypes. (Figure 13–4)

reverse signaling The process by which a protein that 'normally' functions as a ligand functions as a receptor and a protein that 'normally' functions as a receptor functions as a ligand. The originally discovered ('normal') signaling mode is referred to as forward signaling. (Figure 5–12)

reward prediction error A theoretical value representing the difference between a received reward and the predicted reward. It is represented by a population of midbrain dopamine neurons.

rhabdomeric type A type of photoreceptor in which the apical surface folds into microvilli that house opsins. (Figure 12–22)

Rho A member of a family of small monomeric GTPases involved in actin cytoskeleton regulation.

rhodopsin A photosensitive molecule in the rod consisting of opsin covalently attached to retinal, a chromophore derived from vitamin A. (Figure 4–6)

RNA editing   A post-transcriptional modification that alters a nucleotide sequence of an RNA transcript after it is synthesized.

RNA splicing   The process by which introns are removed from RNA molecules. In the case of alternative splicing a subset of exons is removed as well. (Figure 2–2)

RNA-seq   A technique in which RNA molecules from a given tissue are sequenced one by one in a massively parallel fashion using next generation sequencing methods. It is used to obtain information about which genes are expressed and at what level in a genome-wide scale.

RNAi (RNA interference)   A genetic technique for knocking down the expression of a gene of interest by producing a double-stranded RNA with a sequence corresponding to that of the gene of interest. (Figure 13–9)

Robo (Roundabout)   A receptor for Slit. (Figure 7–13)

rod   A rod-shaped photoreceptor in the vertebrate retina, it is a very sensitive photon detector specialized for night vision. (Figure 4–2)

rostral–caudal (anterior–posterior)   Of a body axis, from head to tail.

rtTA   *See* **tTA**.

Ruffini ending   An encapsulated structure closely associated with the ending of the slowly adapting type II (SAII) LTMR. (Figure 6–64)

ryanodine receptor   A Ca^{2+} channel on the ER membrane that is activated by an increase in intracellular Ca^{2+} concentration and that thus amplifies cytosolic Ca^{2+} signals. It is also activated by the plant-derived agonist ryanodine. (Figure 3–41)

saccade   A rapid movement of the eyes between fixation points.

sagittal section   A section plane that is perpendicular to the medial–lateral axis.

saltatory conduction   The process by which an action potential in a myelinated axon 'jumps' from one node of Ranvier to the next. (Figure 2–26)

salty   A taste modality that functions primarily to reveal the salt content of food. It is usually appetitive at a low concentration and aversive at a high concentration.

sarcomere   The contractile element of a myofibril composed of overlapping F-actin (thin filaments) and myosin (thick filaments). (Figure 8–3)

sarcoplasmic reticulum   A special endoplasmic reticulum derivative that extends throughout muscle cells. Ca^{2+} released from the sarcoplasmic reticulum mediates the excitation–contraction coupling. (Figure 8–5)

Satb2   A transcription factor that specifies callosal projection neuron identity. (Figure 7–10)

savings   A phenomenon whereby less effort is required for an animal to re-learn something it has previously learned and then forgotten.

scanning electron microscopy (SEM)   A form of electron microscopy that produces images by scanning the surface of a biological specimen, collecting information regarding the interaction of the electron beam with the surface areas.

Schaffer collateral   An axonal branch of a hippocampal CA3 pyramidal neuron that synapses onto CA1 pyramidal neurons. (Figure 10–6)

schizophrenia   A psychiatric disorder characterized by a set of positive symptoms (those not present in control people, such as hallucinations and delusions), negative symptoms (those, such as social withdrawal and lack of motivation, that reflect an absence of some characteristic that is normally present), and cognitive impairment (such as deficiencies in memory, attention, and executive functions).

Schwann cell   A glial cell in the PNS that wraps axons with its cytoplasmic extension to form myelin sheath. (Figure 2–27)

sciatic nerve   A nerve consisting of sensory and motor axons that innervates the leg.

scrapie   *See* **prion diseases**.

secondary antibody   An antibody that recognizes selectively primary antibodies made by specific animal species. It is usually conjugated to a fluorophore or to an enzyme that produces a color substrate.

secondary dendrite   A mitral cell dendrite that extends laterally, it is used to form reciprocal synapses with granule cells and other olfactory bulb interneurons to spread information across different olfactory processing channels. It is distinct from the primary (apical) dendrite of mitral cells that extends to the glomerulus. (Figure 6–17)

α-secretase   An extracellular protease that cleaves amyloid precursor protein (APP) in the middle of the amyloid β (Aβ) peptide and prevents the production of pathology-associated Aβ. (Figure 11–3)

β-secretase   An extracellular protease that cleaves amyloid precursor protein (APP) at the N-terminus of amyloid β (Aβ) to produce, along with γ-secretase, the intact Aβ peptide. (Figure 11–3)

γ-secretase   An intra-membrane protease that cleaves α- or β-secretase-processed amyloid precursor protein (APP) at the C-terminus of amyloid β (Aβ). (Figure 11–3)

secreted protein   A protein that is destined for export from the cell. (Figure 2–2)

seizure   An episode involving abnormal synchronous firing of large groups of neurons. (Figure 11–47)

selective sweep   The reduction or elimination of nucleotide variations as a result of strong positive selection of a nearby chromosome locus.

selectivity filter   The part of an ion channel pore that is responsible for discriminating between the different ionic species so that only some species can pass through the channel. (Figure 2–33)

self-avoidance   The process in which different axonal or dendritic branches from the same neuron are repelled by each other to avoid overlap of processes from a single cell.

Sema1A, Sema2A, Sema2B (Semaphorins-1A, -2A, -2B)   Axon guidance molecules of the semaphorin family in invertebrates; Sema1A is a transmembrane isoform, whereas Sema2A and Sema2B are secreted isoforms.

Sema3A (Semaphorin-3A)   A secreted axon guidance molecule of the semaphorin family in vertebrates.

semaphorins   Evolutionarily conserved and widely used axon guidance cues, they consist of secreted and transmembrane variants and mostly act as repellents. Some transmembrane variants can also act as receptors. (Figure 5–9)

semicircular canal   A sensory organ in the vestibular system that senses angular acceleration in a specific plane. (Figure 6–59)

sensitization   An increase in the magnitude of a response to a stimulus after a different kind of stimulus, often noxious, has been applied.

sensorimotor stage   The period of song learning in birds when a young bird starts to produce his own immature song, which he compares with the tutor song template he has memorized. He then adjusts his own song until it closely matches the tutor's song. (Figure 9–21)

sensory homunculus A map in the primary somatosensory cortex, it corresponds to sensation of specific body parts. Nearby somatosensory cortical areas represent sensation from nearby body surface. (Figure 1–25)

sensory neuron A neuron that directly responds to external stimuli, such as light, sound, chemical, thermal, or mechanical stimuli.

sensory organ precursor (SOP) In *Drosophila*, a progenitor cell whose asymmetric divisions give rise to different cells (a socket cell, hair cell, sheath cell, and sensory neuron) in the external sensory organ. (Figure 7–6)

sensory rhodopsin Type I rhodopsin used in prokaryotes for phototaxis. (Figure 12–20)

sensory stage The period of song learning in birds when a young bird hears and memorizes the song of a tutor. (Figure 9–21)

serial electron microscopic (EM) reconstruction A method in which consecutive electron micrographs of thin sections are aligned to produce a three-dimensional volume. (Figure 13–29)

serial processing An information processing method in which processing units are arranged in sequential steps.

serine/threonine kinase An enzyme that adds a phosphate onto specific serine or threonine residues of target proteins.

serotonin A monoamine neurotransmitter derived from the amino acid tryptophan that primarily acts as neuromodulator. It is also called 5-HT for 5-hydroxytryptamine. (Figure 3–16; Table 3–2)

Sevenless Originally identified from a mutation in *Drosophila* that lacks photoreceptor R7, it is a gene that acts cell-autonomously in R7 to specify the R7 fate. It encodes a receptor tyrosine kinase. (Figure 5–36)

sex chromosome The chromosome whose presence or number determines the sex of an organism.

Sex lethal (Sxl) A *Drosophila* gene that encodes a splicing factor which acts at the top of the sex-determination hierarchy. (Figure 9–4)

sex-linked Of a mutation, having a Mendelian inheritance pattern characteristic of genes located on a sex chromosome. (Figure 11–34)

sex peptide In *Drosophila*, a peptide transferred with sperm from males to females during mating. It reduces female receptivity to courtship.

sexually dimorphic Of a trait, differing between females and males.

SH2 (Src homology 2) domain A domain present in many signaling proteins, it binds phosphorylated tyrosines in the context of specific amino acid sequences.

Shaker Identified as a mutation in *Drosophila* that causes defects in a fast and transient K⁺ current in muscles and neurons. Its corresponding gene encodes a voltage-gated K⁺ channel.

short-range cue (in axon guidance) A cell-surface protein that can exert its guidance effects only when axons contact the cell that produces it. (Figure 5–9)

short-term memory Memory that lasts seconds to minutes. (Figure 10–4)

short-term synaptic plasticity A change in the efficacy of synaptic transmission that lasts milliseconds to minutes.

sign In sensory physiology, the direction in which a neuron's activity or membrane potential is changed by a stimulus (for example, the sign is positive if a neuron is depolarized by a

stimulus, and the sign is negative if a neuron is hyperpolarized by a stimulus).

signal transduction The process by which an extracellular signal is relayed via intracellular pathways to varied effectors to produce specific biological effects.

silent synapse A glutamatergic synapse that contains NMDA but not AMPA receptors on the postsynaptic membrane. It can be activated by presynaptic glutamate release that coincides with postsynaptic depolarization but not by presynaptic glutamate release alone.

simple cell A functionally defined neuronal type enriched in layer 4 of the primary visual cortex. It is best excited by a bar of light in a specific orientation, and it has separate ON and OFF regions that, when stimulated together, cancel each other's effect. (Figure 4–39)

single channel conductance (γ) Conductance of a single ion channel when open.

single nucleotide polymorphism (SNP) A single nucleotide of DNA in the genome that varies between members of a species.

single-unit recording Extracellular recording of the firing pattern of an individual neuron. *See also* **extracellular recording**. (Figure 13–31)

siRNA (short interfering RNA) Double-stranded RNA with a length similar to microRNA (21–26 nucleotides). It directs a protein complex to degrade the target mRNA through base pairing. *See also* **RNA interference**.

size principle The idea that within a motor pool, motor neurons that have smaller motor unit sizes (with smaller axon diameters and cell bodies) fire before neurons with larger motor unit sizes during muscle contraction. (Figure 8–7)

Slit A secreted protein best studied as a repulsive ligand involved in midline axon guidance in many species, from insects to vertebrates. (Figure 7–13)

slow axonal transport Intracellular transport at a speed of 0.2–8 mm per day. Cargos subject to slow axonal transport mostly include cytosolic proteins and cytoskeletal components. (Figure 2–4)

SM protein A protein related to yeast Sec1 and mammalian Munc18, it binds SNAREs and is essential for vesicle fusion.

small bistratified RGC A blue–yellow color opponent retinal ganglion cell. *See also* **color-opponent RGC**. (Figure 4–33)

Smith-Magenis syndrome A neurodevelopmental disorder characterized by mild-to-moderate intellectual disability, delayed speech, sleep disturbances, impaired impulse control, and other behavioral problems. It is caused by mutations that disrupt the function of one copy of a single gene called *Rai1* (retinoic acid induced 1) or by loss of one copy of a chromosome segment that includes *Rai1*.

smooth muscle Muscle that controls movement of tissue within the digestive, respiratory, vascular, excretory, and reproductive systems.

SNAP-25 A t-SNARE attached to the plasma membrane via lipid modification. (Figure 3–8)

SNAREs (soluble NSF-attachment protein receptors) Proteins on intracellular vesicles and target membranes that form a complex and mediate membrane fusion. (Figure 3–8)

SNc (substantia nigra pars compacta) A midbrain nucleus containing dopamine neurons that project mainly to the dorsal striatum. (Figure 8–22)

SNr (substantia nigra pars reticulata) One of the two major output nuclei of the basal ganglia, it contains GABAergic neurons that project to the thalamus, superior colliculus, and brainstem motor control nuclei. (Figure 8–22)

solute   A water-soluble molecule such as an inorganic ion, nutrient, metabolite, or neurotransmitter.

soma   Cell body of a neuron or any cell.

somatic mutation   A mutation that occurs in a progenitor cell and that thus affects only the cells derived from that progenitor.

somatosensory system   The collected parts of the nervous system that provide bodily sensation.

somatostatin   A neuropeptide whose transcription is regulated by a signaling cascade involving cAMP, PKA, and CREB. It is a marker for a subset of cortical GABAergic neurons.

Sonic Hedgehog (Shh)   A morphogen that determines cell fate by regulating the expression of specific transcription factors in many developmental contexts. For instance, floor-plate-derived Shh is responsible for determining the different fates of neuronal progenitors located at different positions along the dorsal–ventral axis of the ventral spinal cord. It is also used as a midline attractant for commissural axons. (Figure 7–8)

sour   A taste modality that functions primarily to warn the animal of potentially spoiled food. It is usually aversive.

Southern blotting   A method for determining the amount of a specific DNA in a DNA mixture. DNA molecules are separated by gel electrophoresis and are then transferred to a membrane; labeled nucleic acid probes are then hybridized to the membrane to visualize specific DNA molecules that hybridize to the probe.

spatial integration (in dendrites)   The summation of postsynaptic potentials produced by synchronous activation of synapses located at different spatial locations on the postsynaptic neuron. (Figure 3–43)

spectral sensitivity   The relationship between a response (e.g. of a photosensitive cell or molecule) and the wavelength of the stimulus light.

spike   *See* **action potential**.

spike-timing-dependent plasticity (STDP)   A change of synaptic efficacy induced when pre- and postsynaptic neurons repeatedly fire within a restricted time window: synaptic efficacy is potentiated if the presynaptic neuron fires prior to the postsynaptic neuron, and synaptic efficacy is depressed if the presynaptic neuron fires after the postsynaptic neuron.

spinal cord   The caudal part of the vertebrate CNS enclosed by the vertebral column. (Figure 1–8)

spinocerebellar ataxia   One of a collection of neurodegenerative diseases, which share motor defects such as ataxia and are caused by poly-glutamine expansion in a number of proteins. (Table 11–1)

spinocervical tract pathway   An axonal pathway from the dorsal spinal cord to the lateral cervical nucleus that relays a subset of touch signals, particularly from hairy skin.

spiny projection neuron   The most numerous type of neuron in the striatum, it is a GABAergic neuron that projects either directly or indirectly to the output nuclei of the basal ganglia. It is also called medium spiny neuron. (Figure 8–22)

spiral ganglion neuron   A bipolar neuron whose peripheral axon receives auditory information from a hair cell in the cochlea and whose central axon transmits information to the brainstem as part of the auditory nerve. (Figure 6–49)

spontaneous neuronal activity   Firing of neurons in the absence of environmental stimuli.

sporadic   Of a human disease, occurring in a patient without an identifiable family history of the disease.

Sry (Sex determining region Y)   A gene located on the Y chromosome in mammals, it encodes a transcription factor that determines testes differentiation and other male-specific characteristics.

SSRI (selective serotonin reuptake inhibitor)   An inhibitor of the plasma membrane serotonin transporter. It prolongs the action of serotonin in the synaptic cleft.

starburst amacrine cell (SAC)   A class of GABAergic inhibitory neurons in the retina that also release acetylcholine. It is a crucial cell type that shapes the responses of direction-selective retinal ganglion cells. It also participates in generating retinal waves essential for activity-dependent wiring of the visual system. (Figure 4–31)

starter cell   *See* **trans-synaptic tracing**.

STED   *See* **super-resolution fluorescence microscopy**.

stereocilium   A rigid bundled F-actin-based cylinder located on the apical surface of a hair cell. Stereocilia on the same hair cell are arranged in rows of increasing height like a staircase. (Figure 6–47; Figure 6–50)

stereotactic injection   The use of a three-dimensional coordinate system to inject substances such as viruses into a small target region of tissue in an animal.

stereotyped axon pruning   The pruning of exuberant axons with an invariable outcome.

stereotypy   A trait or behavior that is largely invariant in different individual organisms.

stimulating electrode   An electrode used to pass current into a neuron, usually with the goal of changing the membrane potential of a neuron or its processes.

stomatogastric ganglion (STG)   A crustacean ganglion that controls stomach contraction. It has been used as a model system to study central pattern generators and rhythmic activity in neuronal circuits. (Figure 8–13)

storage (of memory)   A step in between acquisition and retrieval, in which a memory is encoded as a persistent representation somewhere in the nervous system.

STORM   *See* **super-resolution fluorescence microscopy**.

striatum   The part of the basal ganglia that receives convergent input from the cerebral cortex and thalamus. Also called caudate–putamen because in some species, the striatum has two separate regions called caudate and putamen, respectively. (Figure 8–22)

subcerebral projection neuron (SCPN)   A cortical neuron, found in layer 5, that projects its axon to subcortical targets, such as the pons, superior colliculus, and spinal cord. (Figure 7–10)

substance P   A neuropeptide that promotes inflammation when released by the peripheral terminals of sensory neurons. (Figure 6–71)

substantia nigra   A midbrain structure named after the high levels of melanin pigments present in the dopamine neurons of healthy human subjects. (Figure 11–16). *See also* **SNc** and **SNr**.

subthalamic nucleus (STN)   An intermediate nucleus in the basal ganglia indirect pathway, it contains glutamatergic neurons that project to the GPi and SNr. These neurons receive GABAergic input from the GPe and glutamatergic input from the cerebral cortex. (Figure 8–22)

sub-threshold stimulus   A stimulus that is insufficient to cause a neuron to generate an action potential. (Figure 2–18)

super-resolution fluorescence microscopy   A set of fluorescence microscopy techniques capable of imaging specimens at resolutions below the diffraction limit of light. For example, STED (stimulated emission depletion microscopy) achieves super resolution by exciting fluorophores in a region of tissue smaller than the diffraction limit through depletion of fluorescence in an annulus surrounding a central focal spot. STORM (stochastic optical reconstruction microscopy) and PALM (photoactivated localization microscopy) achieve super resolution by photoactivating a random small subset of photo-switchable fluorophores at any one time, such that the position of each fluorophore can be localized to a precision much finer than the resolution limits set by diffraction; repeated rounds of imaging and deactivation enable the reconstruction of the entire imaging field. (Figure 13–25)

superior colliculus A multi-layered midbrain structure in mammals that receives retinal ganglion cell axonal input, as well as input from other sensory systems. It regulates head orientation and eye movement and is analogous to the tectum in non-mammalian vertebrates. (Figure 4–35)

superior olivary nuclei Brainstem nuclei in mammals where auditory signals from the left ear and right ear first converge. The medial superior olivary nucleus (MSO) analyzes interaural time differences, whereas the lateral superior olivary nucleus (LSO) analyzes interaural sound level differences. (Figure 6–57)

suprachiasmatic nucleus (SCN) A hypothalamic nucleus that is the master regulator of circadian rhythms and light entrainment in mammals. (Figure 8–34; Figure 8–49)

supra-threshold stimulus A stimulus that can cause a neuron to generate an action potential. (Figure 2–18)

sweet A taste modality that functions primarily to detect the sugar content of food. It is usually appetitive.

sympathetic nervous system A branch of the autonomic nervous system that facilitates energy expenditure, such as in the case of an emergency response. Activation of the sympathetic nervous system speeds up the heart rate, increases blood flow, relaxes airways in the lungs, inhibits salivation and digestion, and stimulates the production of the hormone epinephrine (adrenaline) from the adrenal glands. (Figure 8–31; Figure 8–32)

symporter A coupled transporter that moves two or more solutes in the same direction. (Figure 2–10)

synapse A site at which information is transferred from one neuron to another neuron or a muscle cell. It consists of a presynaptic terminal and a postsynaptic specialization separated by a synaptic cleft.

synapse elimination The process by which extra synapses are removed during development. It is best described at the vertebrate neuromuscular junction where the innervation of muscle cells by multiple motor neurons is refined during early postnatal development so that each muscle cell is innervated by a single motor neuron in adults. (Figure 7–27)

synaptic cleft A 20–100 nm gap that separates the presynaptic terminal of a neuron from its target cell. (Figure 1–14; Figure 3–3)

synaptic efficacy *See* **efficacy of synaptic transmission**.

synaptic failure An event in which an action potential in a presynaptic neuron does not produce a postsynaptic response.

synaptic plasticity The ability to change the efficacy of synaptic transmission, usually in response to experience and neuronal activity.

synaptic potential A type of graded potential produced at postsynaptic sites in response to neurotransmitter release by presynaptic partners.

synaptic tagging The hypothesis that induction of LTP at a synapse causes the production of a 'tag' at the synapse and that newly synthesized macromolecules necessary for stabilization of LTP are selectively captured by the tag. The hypothesis explains how the input specificity of LTP is maintained despite the cell-wide distribution of newly synthesized macromolecules required for LTP.

synaptic transmission The process of neurotransmitter release from the presynaptic neuron and neurotransmitter reception by the postsynaptic neuron.

synaptic vesicle A small, membrane-enclosed organelle (typically about 40 nm in diameter) enriched at the presynaptic terminal. They are filled with neurotransmitters and, upon stimulation, fuse with the plasma membrane to release neurotransmitters into the synaptic cleft. (Figure 3–4; Figure 3–7)

synaptic weight matrix A network of synapses between ensembles of input neurons and output neurons, where the strength (weight) of each synapse can vary between 0 (no connection) and 1 (maximal strength connection). (Figure 10–5)

synaptobrevin A transmembrane SNARE on the synaptic vesicle (i.e. a v-SNARE), it is also named VAMP. (Figure 3–8)

synaptotagmin A Ca^{2+}-binding transmembrane protein on the synaptic vesicle that serves as a Ca^{2+} sensor to trigger neurotransmitter release.

syndromic disorder A disorder characterized by a defined constellation of behavioral, cognitive, and physical symptoms.

synonymous substitutions Nucleotide changes in DNA that do not result in amino acid changes in the protein encoded by that DNA. They are used in calculations of genetic drift.

syntaxin A transmembrane SNARE on the target plasma membrane (i.e. a t-SNARE). (Figure 3–8)

α-synuclein A protein normally enriched in the presynaptic terminal. It is a major component of Lewy bodies, a defining pathological feature of most forms of Parkinson's disease.

T1R1 A G-protein-coupled receptor and a subunit (along with T1R3) of the mammalian umami taste receptor. (Figure 6–41)

T1R2 A G-protein-coupled receptor and a subunit (along with T1R3) of the mammalian sweet taste receptor. (Figure 6–41)

T1R3 A G-protein-coupled receptor and a shared subunit of the mammalian umami and sweet taste receptors. (Figure 6–41)

T2Rs A family of G-protein-coupled receptors that are the mammalian bitter taste receptors. (Figure 6–41)

tamoxifen *See* **CreER**.

tastant A nonvolatile and hydrophilic molecule in saliva that elicits taste perception.

taste bud A cluster of tens of taste receptor cells, with their apical endings facing the surface of the tongue. (Figure 6–35)

taste pore Collected apical endings of taste receptor cells in a taste bud. (Figure 6–35)

taste receptor cell A sensory neuron on the surface of the tongue and oral cavity, it converts tastant binding to taste receptor proteins to an electrical signal that is transmitted to the peripheral terminals of the gustatory nerve. (Figure 6–35)

tau A microtubule binding protein that is highly enriched in axons.

tauopathies Neurodegenerative diseases characterized by the presence of neurofibrillary tangles, which consist of aggregates of hyperphosphorylated tau.

Tbr1 A transcription factor that specifies corticothalamic projection neuron identity. (Figure 7–10)

tectorial membrane A membrane on the apical side of hair cells apposed to the stereocilia. (Figure 6–50)

tectum The major target of retinal ganglion cells in the brains of amphibians and lower vertebrates. It is a midbrain structure analogous to the mammalian superior colliculus. (Figure 5–5)

telencephalon The anterior part of the forebrain, including the olfactory bulb, cerebral cortex, hippocampus, and basal ganglia. (Figure 7-3)

temporal (in retinal map) In the direction of the temple.

temporal integration (in dendrites) The summation of postsynaptic potentials produced by activation of synapses within a finite time window. (Figure 3-43)

temporal lobe One of the four cerebral cortex lobes, it is located at the side of the brain. (Figure 1-23)

testosterone A steroid hormone that promotes the development of the male reproductive system (masculinization) and inhibits the development of the female reproductive system (de-feminization). In adults, it stimulates sexual behaviors. (Figure 9-24)

tetanus toxin A protease produced by *Clostridium tetani* that cleaves synaptobrevin at a specific site, thereby inhibiting neurotransmitter release.

tetraethylammonium (TEA) A chemical that selectively blocks voltage-gated K$^+$ channels.

tetrode An extracellular electrode containing four wires that enable four independent recordings of spiking activities of neurons nearby the electrode tip. The firing patterns of up to ~20 neurons can be resolved based on their different action potential amplitudes and waveforms.

tetrodotoxin (TTX) A toxin that potently blocks voltage-gated Na$^+$ channels across animal species and is widely used experimentally to silence neuronal firing. It is produced by symbiotic bacteria in puffer fish, rough-skinned newt, and some octopi. (Figure 2-29)

thalamocortical axons (TCAs) Axons of thalamic neurons that project to the cortex.

thalamus A structure situated between the cerebral cortex and the midbrain, it relays sensory and motor signals to the cerebral cortex through its extensive bidirectional connections with cortex. (Figure 1-8)

theory of dynamic polarization The idea that every neuron has (1) a receptive component, the cell body and dendrites; (2) a transmission component, the axon; and (3) an effector component, the axon terminals. According to this theory, originally proposed by Ramón y Cajal, neuronal signals flow from dendrites and cell bodies to the axon.

thermosensation The sense of temperature.

thermosensory neuron A somatosensory neuron that senses temperature.

threshold (of action potential) The membrane potential above which an action potential is generated. (Figure 2-18)

thrombospondin (TSP) A member of a family of secreted proteins with diverse functions. It can be produced by astrocytes to stimulate synapse formation.

time constant (τ) The product of resistance and capacitance in an *R-C* circuit, it is a measure of the rate at which both a capacitor charges or discharges and the voltage across a resistor changes in response to changes in current. In neurons, τ corresponds to the time required for the membrane potential change to reach 63% $(1 - 1/e)$ of its maximal value in response to a sudden change of current flow.

Timeless A fruit fly gene discovered based on mutations that affect circadian rhythms. It encodes a protein that participates in the negative regulation of its own transcription. (Figure 8-46)

Timothy syndrome Characterized by cardiac arrhythmia and autistic symptoms, it is caused by mutation in the gene encoding a voltage-gated Ca^{2+} channel, Ca$_V$1.2. (Figure 11-45)

tip link The connection between adjacent stereocilia, it consists of cadherin-23 on the taller stereocilium and protocadherin-15 on the shorter stereocilium. (Figure 6-47)

tonic Of a neuronal firing pattern, regularly timed and repetitive.

tonic-clonic seizure A seizure associated with loss of consciousness and a predictable sequence of motor activity: patients first stiffen and extend all extremities (tonic phase) and then undergo full-body spasms during which muscles alternately flex and relax (clonic phase).

tonotopic map The ordered arrangement of cells in the auditory system in physical space according to their frequency tuning. The cochlea and multiple brain regions contain tonotopic maps. (Figure 6-49)

topographic map An ordered representation in the brain of features of either the external world or the animal's interaction with the world. For examples, *see* **retinotopy**, **sensory homunculus**, and **motor homunculus**.

touch sensory neurons *See* **LTMRs**.

transcription The process by which RNA polymerase uses DNA as a template to synthesize RNAs. (Figure 2-2)

transcription factor A DNA-binding protein that regulates transcription of target genes.

transcription unit The part of the gene that serves as a template for RNA synthesis. (Figure 2-2)

transcytosis The process by which transmembrane or extracellular proteins are first retrieved by endocytosis in one cellular compartment and then delivered for exocytosis at another cellular compartment.

transducin A trimeric GTP-binding protein that links light-activated rhodopsin (or cone opsin) to phosphodiesterase activation in vertebrate photoreceptors. (Figure 4-8)

Transformer (Tra) A *Drosophila* gene that encodes a splicing factor which acts downstream of *Sex lethal* (*Sxl*) but upstream of *Doublesex* (*Dsx*) and *Fruitless* (*Fru*). (Figure 9-4)

transgene An *in vitro* engineered gene that is introduced into somatic cells or the germ line of an organism. (Figure 13-11)

transgenic organism An organism that contains a transgene, usually in the germ line.

translation The process by which an mRNA is decoded by ribosomes for protein synthesis. (Figure 2-2)

transmembrane protein A protein that is destined to span the lipid bilayer of a membrane. (Figure 2-2)

transmission electron microscopy (TEM) A form of electron microscopy in which high voltage electron beams transmitted through ultra-thin (typically under 100 nm) sections of biological specimens are used to create images.

transporter A transmembrane protein or protein complex that has two separate gates that open and close sequentially to allow solutes to move from one side of the membrane to the other. (Figure 2-8)

trans-synaptic tracing A method that labels the synaptic partners of a given neuron or a population of neurons of interest (starter cell or cells). A retrograde trans-synaptic tracer labels presynaptic partners of starter cells, whereas an anterograde trans-synaptic tracer labels postsynaptic partners of starter cells. (Figure 13-30)

transverse tubules (T tubules) An invagination of the plasma membrane that extends into the muscle cell interior, bringing the plasma membrane close to the sarcoplasmic reticulum, such that depolarization effectively triggers Ca^{2+} release from the sarcoplasmic reticulum throughout the entire large muscle cell. (Figure 8-5)

TRE (tetracycline response elements) The DNA sequences to which tTA or rtTA bind. (Figure 13-13). *See also* **tTA.**

trichromat Organisms that have three different cones for color vision—the S-cone, M-cone, and L-cone.

trigeminal ganglia Clusters of somatosensory neurons near the brainstem involved in sensation of the face.

trimeric GTP-binding protein (G protein) A GTP-binding protein complex composed of a Gα, a Gβ, and a Gγ subunit with an intrinsic GTPase activity in Gα. It has many variants, which couple different GPCRs to diverse signaling pathways. *See also* **G$_s$, G$_i$,** and **G$_q$.**

Trk receptors A family of neurotrophin receptors that are receptor tyrosine kinases. It includes TrkA, TrkB, and TrkC. (Figure 3-39; Figure 7-32)

TRP channels Non-selective cation channels that share sequence similarities with the *Drosophila* transient receptor potential (TRP) protein. (Figure 2-34)

TRPM8 A non-selective cation channel that is activated by menthol and by temperatures < 26°C. (Figure 6-68)

TRPV1 A non-selective cation channel that is activated by capsaicin and by temperatures > 43°C. (Figure 6-68)

Tsc1, Tsc2 *See* **tuberous sclerosis.**

t-SNARE A SNARE located on the target membrane, such as syntaxin.

tTA (tetracycline-repressible transcriptional activator) A bacterial transcription factor widely used in heterologous systems, including transgenic mice, to control expression of a transgene. It drives expression of target genes whose promoters contain a tetracycline response element (TRE), but its activity is repressed by tetracycline or its analog doxycycline. A variant called rtTA (reverse tTA) activates TRE-driven transgenes in the presence but not the absence of doxycycline. (Figure 13-13)

tuberomammillary nucleus A hypothalamic nucleus rich in histamine neurons. (Figure 8-52)

tuberous sclerosis Characterized by non-malignant tumors in the brain and other organs as well as by symptoms of autism spectrum disorders, it is caused by mutations in genes encoding *Tsc1* or *Tsc2*, the products of which are negative regulators of mTOR-mediated translational control. (Figure 11-45)

tufted cell *See* **mitral cell.**

two-photon microscopy A microscopy technique that relies on simultaneous absorption of two long-wavelength photons in order to excite a fluorophore. Compared with confocal microscopy, it produces less photo-damage because only at the focal plane is the density of photons high enough to cause substantial fluorescence emission. Like confocal microscopy, it relies on laser scanning of imaging spots across a plane to produce an optical section. (Figure 13-39)

type III neuregulin-1 (Nrg1-III) An axonal cell-surface protein, the expression level of which determines the degree of axon myelination by Schwann cells.

tyrosine hydroxylase An enzyme that converts L-tyrosine to L-dopa, it is the rate-limiting enzyme in the catecholamine biosynthetic pathway. (Figure 11-20)

UAS *See* **GAL4.**

ubiquitin-proteasome system A protein degradation system present in all eukaryotes.

umami A taste modality that functions primarily to detect the amino acid content of food. It is usually appetitive.

Unc5 A co-receptor for netrin/Unc6 that acts together with DCC/Unc40 to mediate repulsion.

Unc6 *See* **netrin/Unc6.**

Unc40 *See* **DCC/Unc40.**

unconditioned response (UR) *See* **classical conditioning.**

unconditioned stimulus (US) *See* **classical conditioning.**

unipolar Having one process leaving the cell body that gives rise to both dendritic and axonal branches. (Figure 1-15)

V1 *See* **primary visual cortex.**

vagus nerve A cranial nerve in the parasympathetic nervous system that connects the brainstem with internal organs. (Figure 8-32)

variations Differences in genes or inheritable traits.

vasopressin A hormone secreted by hypothalamic neurons in the posterior pituitary and a neuropeptide released by certain CNS neurons. It regulates water balance and social behavior.

V-ATPase A proton pump on the synaptic vesicle that pumps protons (H$^+$) into the vesicle against their electrochemical gradient using energy derived from ATP hydrolysis. (Figure 3-12)

ventral horn The ventral part of the spinal gray matter where motor neurons reside. (Figure 8-6)

ventral nerve cord An invertebrate CNS structure posterior to the brain. It is analogous to the vertebrate spinal cord. (Figure 7-11; Figure 7-12)

ventral pallidum A basal ganglia region that is a major target of GABAergic projection neurons from the nucleus accumbens. (Figure 9-44)

ventral root The place where motor axons exit the spinal cord. (Figure 8-6)

ventral stream A visual processing pathway from the primary visual cortex to the temporal cortex. It is responsible for analyzing form and color; the 'what' stream. (Figure 4-48)

ventral tegmental area (VTA) A midbrain nucleus containing dopamine neurons that project mainly to the ventral striatum (nucleus accumbens) and prefrontal cortex. (Figure 8-22; Figure 11-31)

ventricle A cavity derived from the lumen of the neural tube. It is filled with cerebrospinal fluid. (Figure 7-5)

ventricular zone A layer of cells adjacent to the ventricles. (Figure 7-4)

ventromedial hypothalamic nucleus (VMH) A hypothalamic nucleus whose best characterized roles include regulating female lordosis and male mounting and aggression. (Figure 9-32)

vesicular monoamine transporter (VMAT) A transmembrane protein on synaptic vesicles that transports dopamine, norepinephrine, and serotonin from the presynaptic cytosol into synaptic vesicles. (Figure 11-24). *See also* **vesicular neurotransmitter transporter.**

vesicular neurotransmitter transporter A transmembrane protein on the synaptic vesicle that transports neurotransmitters from the presynaptic cytosol into the vesicle using energy from the transport of protons down their electrochemical gradient. (Figure 3-12)

vestibular ganglion neuron A bipolar neuron whose peripheral axon receives vestibular information from a cell in an otolith organ or a semicircular canal and whose central axon transmits information to the brainstem as part of the vestibular nerve.

vestibular nerve A collection of axons from vestibular ganglion neurons that transmits vestibular information to the brainstem. (Figure 6-59)

vestibular nuclei Brainstem nuclei where the vestibular nerve terminates. They also receive input from other sensory systems such as the somatosensory systems. (Figure 6–60)

vestibular system The collected parts of the nervous system that sense the movement and orientation of the head and use this information to regulate a variety of functions including balance, spatial orientation, coordination of head and eye movements, and perception of self-motion.

vestibulo-ocular reflex (VOR) A reflexive eye movement that stabilizes images on the retina during head movement by moving the eyes in the direction opposite to the head movement. (Figure 6–61)

viral transduction The process by which a virus infects a host cell, introducing its genome. It is widely used for transgene expression in somatic cells.

visceral motor neurons Pre- and postganglionic neurons in the autonomic nervous system.

visceral motor system *See* **autonomic nervous system**.

visceral sensory neuron A sensory neuron whose peripheral branch innervates an internal organ and whose central branch extends to the spinal cord or brainstem. (Figure 8–33)

visual cortex The part of the cerebral cortex that is dedicated to analyzing visual information.

visual field The portion of external world that can be seen at a given time.

voltage clamp An experimental technique used to measure the ion currents through the membrane while holding (i.e. 'clamping') the membrane potential at a set level. (Figure 2–21)

voltage-gated Ca^{2+} channel An ion channel that allows selective passage of Ca^{2+} and whose conductance is regulated by the membrane potential. (Figure 2–34)

voltage-gated ion channel An ion channel whose conductance changes as a function of the membrane potential. At a single channel level, a channel is either open or closed; membrane potential change alters its open probability. (Figure 2–30)

voltage-sensitive dye A molecule whose optical properties change in response to membrane potential changes.

volume transmission The secretion of neurotransmitters (usually neuromodulators) into the extracellular space outside the confines of morphologically defined synapses, where they can affect multiple nearby target cells.

vomeronasal organ (VNO) A special structure located at the front of the nose that houses sensory neurons of the accessory olfactory system. (Figure 6–22)

vomeronasal system *See* **accessory olfactory system**.

v-SNARE A SNARE located on a vesicle, such as synaptobrevin.

VOR gain The ratio of rotation of the eyes to the rotation of the head in the vestibulo-ocular reflex.

Wallerian degeneration The process by which distal axons are eliminated after they are severed from the somata.

Weber's Law In sensory perception, the property that the just-noticeable difference between two sensory stimuli is proportional to the magnitude of the stimulus.

Wernicke's area An area in the left temporal lobe involved in language comprehension. Patients with lesions in this area have difficulty understanding language. (Figure 1–23)

western blot A method for determining the amount of a specific protein in a protein mixture. Proteins are separated by gel electrophoresis and are then transferred to a membrane; labeled antibodies are then used to visualize specific proteins bound by the antibody. It can be used to determine protein expression patterns.

white matter The parts of the CNS that are enriched with oligodendrocytes and myelinated axons and that appear white because of the high lipid content of the myelin.

whole-cell patch recording (whole-cell recording) A form of intracellular recording in which a glass electrode forms a high-resistance seal with the plasma membrane of the recorded cell. After formation of the seal, the membrane underneath the patch electrode is ruptured, such that the interior of the patch electrode and the cytoplasm form a single compartment. *See also* **patch clamp recording**. (Figure 13–37)

whole-mount A tissue specimen that has not been sectioned.

Wnts A family of secreted proteins that act as morphogens to pattern embryonic tissues, such as the tissues along the anterior–posterior axis of vertebrates and *C. elegans*. They can also serve as cues for axon guidance and for directing where synapses form along an axon.

working memory Short-term explicit memory, such as temporary retention of facts. (Figure 10–4)

zygote A fertilized egg. (Figure 7–2)

INDEX

Note: Page numbers and ranges suffixed B, F, or T indicate that material relevant to the topic appears only in a Box, Figure or Table on that page. Where a text treatment on the same page is already indexed, non-text material is not so distinguished.

When acronyms or their expansions are used consistently, the preferred form in the text becomes the sole index entry; where both appear, acronyms are usually preferred.